Handbook of Environmental Engineering

This new edition provides a practical view of pollution and its impact on the natural environment. Driven by the hope of a sustainable future, it stresses the importance of environmental law and resource sustainability and offers a wealth of information based on real-world observations and expert experience. It presents a basic overview of environmental pollution, emphasizes key terms, and addresses specific concepts in advanced algebra, fundamental engineering, and statistics. In addition, it considers socioeconomic, political, and cultural influences and provides an understanding of how to effectively treat and prevent air pollution, implement industrial hygiene principles, and manage solid waste, water, and wastewater operations.

The *Handbook of Environmental Engineering* is written in a down-to-earth style for a wide audience, as it appeals to technical readers, consultants, policymakers, as well as a wide range of general readers.

Features:

- Updated throughout, with a new chapter on modern trends in environmental engineering, the book further emphasizes climate change effects on water/wastewater infrastructure
- Examines the physical, chemical, and biological processes fundamental to understanding the environment fate and engineered treatment of environmental contaminants
- Presents technologies to prevent pollution at the source as well as treatment and disposal methods for remediation
- Identifies multiple environmental pollutants and explains the effects of each
- Includes the latest environmental regulatory requirements.

Handbook of Environmental Engineering

Second Edition

Frank R. Spellman

Taylor & Francis Group
Boca Raton London New York

CRC Press is an imprint of the
Taylor & Francis Group, an **informa** business

Designed cover image: © Shutterstock

Second edition published 2023
by CRC Press
6000 Broken Sound Parkway NW, Suite 300, Boca Raton, FL 33487-2742

and by CRC Press
4 Park Square, Milton Park, Abingdon, Oxon, OX14 4RN

CRC Press is an imprint of Taylor & Francis Group, LLC

© 2023 Taylor & Francis Group, LLC

First edition published by CRC Press 2016

Reasonable efforts have been made to publish reliable data and information, but the author and publisher cannot assume responsibility for the validity of all materials or the consequences of their use. The authors and publishers have attempted to trace the copyright holders of all material reproduced in this publication and apologize to copyright holders if permission to publish in this form has not been obtained. If any copyright material has not been acknowledged please write and let us know so we may rectify in any future reprint.

Except as permitted under U.S. Copyright Law, no part of this book may be reprinted, reproduced, transmitted, or utilized in any form by any electronic, mechanical, or other means, now known or hereafter invented, including photocopying, microfilming, and recording, or in any information storage or retrieval system, without written permission from the publishers.

For permission to photocopy or use material electronically from this work, access www.copyright.com or contact the Copyright Clearance Center, Inc. (CCC), 222 Rosewood Drive, Danvers, MA 01923, 978-750-8400. For works that are not available on CCC please contact mpkbookspermissions@tandf.co.uk

Trademark notice: Product or corporate names may be trademarks or registered trademarks and are used only for identification and explanation without intent to infringe.

Library of Congress Cataloging-in-Publication Data
Names: Spellman, Frank R., author.
Title: Handbook of environmental engineering / Frank R. Spellman.
Description: 2nd edition. | Boca Raton: CRC Press, 2023. | Includes bibliographical references and index.
Identifiers: LCCN 2022040610 (print) | LCCN 2022040611 (ebook) | ISBN 9781032288079 (hardback) | ISBN 9781032288086 (paperback) | ISBN 9781003298601 (ebook)
Subjects: LCSH: Pollution prevention–Handbooks, manuals, etc. | Environmental engineering–Handbooks, manuals, etc. | Environmental management–Handbooks, manuals, etc.
Classification: LCC TD176.4 .S644 2023 (print) | LCC TD176.4 (ebook) | DDC 628–dc23/eng/20221012
LC record available at https://lccn.loc.gov/2022040610
LC ebook record available at https://lccn.loc.gov/2022040611

ISBN: 978-1-032-28807-9 (hbk)
ISBN: 978-1-032-28808-6 (pbk)
ISBN: 978-1-003-29860-1 (ebk)

DOI: 10.1201/9781003298601

Typeset in Times
by codeMantra

Contents

A Note to the Reader: Will it be Milk Toast or Hardtack?................................xxvii
Preface...xxxi
About the Author xxxiii

Chapter 1 Introduction 1
    Environment 2
    Pollution Defined 4
        Key Terms 4
        Pollution: Effects Easy to See, Feel, Taste or Smell 15
        Unexpected Pollution 16
        Pollution and Environmental Science/Health/Engineering 16
        Environmental Pollution and Technology: The Connection 18
    Environmental Administrative Laws and Regulations 22
        Administrative Procedure Act (APA) 23
        The Regulatory Process 23
            Creating a Law 23
            Putting the Law to Work 23
            Creating a Regulation 24
        NEPA: Environmental Impact Statements (EIS) 24
    A Word about Machine and Deep Learning 25
    The Bottom Line 26
    Notes 27
    References 27

Chapter 2 Units, Standards of Measurement, and Conversions 29
    Units of Measurement: The Basics 29
        Conversion Factors 30
            Weight, Concentration, and Flow 31
        Conversions 33
        Typical Conversion Examples 34
            Temperature Conversions 39
        MKS System (SI Units) 41
    Energy Units and Math Operations 42
        Environmental Engineering: Perspective on Energy 42
        The 411 on Energy 43
            Types of Energy 43
            Measuring Energy 45
        Clean Energy Parameters, Calculations and References 46
            Electricity Reduction (Kilowatt-Hours) 46
            Gallons of Gasoline Consumed 47
            Passenger Vehicles per Year 47
            Miles Driven by the Average Passenger Vehicle per Year 48
            Therms of Natural Gas 49
            Barrels of Oil Consumed 50
            Tanker Trucks Filled with Gasoline 50

Number of Incandescent Bulbs Switched to Compact Fluorescent
Bulbs (CFLs) 50
Home Electricity Use 54
Home Energy Use 54
Number of Tree Seedlings Grown for 10 Years 56
Acres of U.S. Forests Storing Carbon for One Year 56
Acres of U.S. Forest Preserved from Conversion to Croplands 58
Propane Cylinders Used for Home Barbecues 60
Railcars of Coal Burned 61
Pounds of Coal Burned 61
Tons of Waste Recycled Instead of Landfilled 62
Number of Garbage Trucks of Waste Recycled Instead of Landfilled 62
Coal-Fired Power Plant Emissions for One Year 63
Number of Wind Turbines Installed 63
Notes 64
References 64

Chapter 3 Math Operations Review 65
Introduction 65
Basic Math Terminology and Definitions 65
Sequence of Operations 66
Sequence of Operations—Rules 66
Sequence of Operations—Examples 67
Percent 68
Significant Digits 72
Powers and Exponents 74
Averages (Arithmetic Mean) 76
Ratio 78
Dimensional Analysis 81
Threshold Odor Number (TON) 86
Geometrical Measurements 87
Geometrical Calculations 88
Perimeter and Circumference 88
Area 91
Volume 94
Force, Pressure, and Head Calculations 98
Force and Pressure 98
Head 98
Static Head 99
Friction Head 99
Velocity Head 100
Pressure/Head 100
Head/Pressure 100
Review of Advanced Algebra Key Terms/Concepts 104
Key Definitions Include 104
Quadratic Equations 107
The Quadrtic Equation and Environmental Practice 107
Quadratic Equations: Theory and Application 107
Derivation of the Quadratic Equation Formula 108

		Using the Quadratic Equation	110

Trigonometric Ratios 110
    Trigonometric Functions and the Environmental Practitioner 110
        Trigonometric Ratios or Functions 111
References 112

Chapter 4   Environmental Modeling and Algorithms 113

Introduction 113
Basic Steps for Developing an Effective Model 114
What Are Models Used For? 114
Media Material Content 115
    Material Content: Liquid Phases 117
Phase Equilibrium and Steady State 120
Math Operations and Laws of Equilibrium 120
    Solving Equilibrium Problems 120
    Laws of Equilibrium 122
        Ideal Gas Law 122
        Dalton's Law 122
        Raoult's Law 124
        Henry's Law 124
Chemical Transport Systems 124
Algorithms: What Are They? 125
    Expressing Algorithms 126
    General Algorithm Applications 127
        The Problem 128
    Environmental Practice Algorithm Applications 128
Dispersion Models 129
Screening Tools 130
References 130

Chapter 5   Statistics Review 131

Statistical Concepts 131
    Probability and Statistics 132
    Measure of Central Tendency 133
Symbols, Subscripts, Basic Statistical Terms and Calculations 133
    The Mean 134
    Median 135
    Mode 135
    Range 135
Distribution 137
    Normal Distribution 137
Standard Deviation 139
Coefficient of Variation 141
Standard Error of the Mean 141
Covariance 142
Simple Correlation Coefficient 143
Variance of a Linear Function 144
Sampling—Measurement Variables 146
    Simple Random Sampling 146

Standard Errors	146
Confidence Limits	147
Sample Size	148
Stratified Random Sampling	149
Sample Size	151
Sampling—Discrete Variables	152
Random Sampling	152
Sample Size	153
Cluster Sampling for Attributes	154
Transformations	155
Chi-Square Tests	155
Test of Independence	155
Test of a Hypothesized Count	156
Bartlett's Test of Homogeneity of Variance	157
Comparing Two Groups by the t Test	158
The t Test for Unpaired Plots	158
Sample Size	160
The t Test for Paired Plots	161
Number of Replicates	162
Comparison of Two or More Groups by Analysis of Variance	162
Complete Randomization	162
Multiple Comparisons	165
F Test with Single Degree of Freedom	165
Scheffe's Test	166
Unequal Replication	167
Randomized Block Design	168
Latin Square Design	171
Factorial Experiments	173
Sums of Squares	175
The Split Plot Design	179
Calculations	180
Date-Burning SS	182
Missing Plots	185
Regression	186
Simple Linear Regression	186
How Well Does the Regression Line Fit the Data?	189
Coefficient of Determination	190
Confidence Intervals	190
Multiple Regression	192
Tests of Significance	196
Coefficient of Multiple Determination	197
The c-Multipliers	198
Curvilinear Regressions and Interactions	198
Curves	198
Interactions	199
Group Regressions	200
Testing for the Common Regressions	201
Analysis of Covariance in a Randomized Block Design	203
Adjusted Means	206
Tests among Adjusted Means	206
Note	207
References	207

Contents

Chapter 6 Fundamental Engineering Concepts 209
    Deep and Gloomy Woods 209
    Introduction 209
    Resolution of Forces 210
        Slings 212
        Inclined Plane 215
    Properties of Materials and Principles of Mechanics 217
        Properties of Materials 217
            Friction 221
            Pressure 222
            Specific Gravity 222
            Force, Mass, and Acceleration 222
            Centrifugal and Centripetal Forces 223
            Stress and Strain 223
        Principles of Mechanics 223
            Statics 223
            Welds 223
            Dynamics 224
            Hydraulics and Pneumatics—Fluid Mechanics 224
            Soil Mechanics 224
    References 231

Chapter 7 Air Pollution 233
    The 411 on Air 233
        Definition of Key Terms 235
            Definitions 235
    Components of Air: Characteristics and Properties 237
        Atmospheric Nitrogen 238
            Nitrogen: Physical Properties 238
            Nitrogen: Uses 238
            Nitrogen Oxides 238
        Atmospheric Oxygen 240
            Oxygen: Physical Properties 240
            Oxygen: Uses 240
            Ozone: Just another Form of Oxygen 240
        Atmospheric Carbon Dioxide 241
            Carbon Dioxide: Physical Properties 241
            Carbon Dioxide: Uses 242
        Atmospheric Argon 242
            Argon: Physical Properties 242
            Argon: Uses 243
        Atmospheric Neon 243
            Neon: Physical Properties 243
            Neon: Uses 243
        Atmospheric Helium 243
            Helium: Physical Properties 243
        Atmospheric Krypton 243
            Krypton: Physical Properties 244
            Krypton: Uses 244

- Atmospheric Xenon 244
  - Xenon: Physical Properties 244
  - Xenon: Uses 244
- Atmospheric Hydrogen 245
  - Hydrogen: Physical Properties 245
- Atmospheric Water 246
- Atmospheric Particulate Matter 247
- Air for Combustion 247
- Air for Power 248
- Gas Physics 248
  - Material Balance 248
  - Compressibility 249
  - Gas Laws 249
    - Boyle's Law 249
    - Charles's Law 250
    - Ideal Gas Law 251
  - Flow Rate 252
  - Gas Conversions 252
    - Major Constituents 252
    - Both Major and Minor Constituents 253
  - Gas Velocity 253
    - Gas Stream Treatment (Residence) Time 254
  - Gas Density 254
  - Heat Capacity and Enthalpy 254
  - Heat and Energy in the Atmosphere 255
    - Adiabatic Lapse Rate 255
  - Viscosity 256
  - Flow Characteristics 256
- Particle Physics 257
  - Characteristics of Particles 257
    - Surface Area and Volume 257
    - Aerodynamic Diameter 259
    - Particle Size Categories 260
    - Regulated Particulate Matter Categories 260
    - Size Distribution 261
    - Particle Formation 262
- Collection Mechanisms 264
- Atmospheric Chemistry and Related Phenomena 266
  - Photochemical Reaction—Smog Production 267
- Air Quality 268
  - Earth's Heat Balance 268
  - Insolation (EPA, 2005) 269
    - Solar Constant 269
    - Transparency 270
    - Daylight Duration 270
    - Angle of Sun's Rays 270
  - Heat Distribution 270
    - Differential Heating 270
    - Transport of Heat 271
    - Global Distribution of Heat 271
- Air Quality Management 272

Contents

- Clean Air Act (CAA) 272
  - Title 1: Attainment and Maintenance of NAAQS 273
  - Title 2: Mobile Sources 274
  - Title 3: Air Toxics 275
  - Title 4: Acid Deposition 275
  - Title 5: Permits 276
  - Title 6: Ozone and Global Climate Protection 276
  - Title 7: Enforcement 277
- Clean Air Act Amendments (EPA, 2005) 277
  - State Implementation Plans (SIPs) 277
  - New Source Review 278
  - Air Quality Monitoring 279
  - Visibility 280
  - Pollutant Dispersion 280
  - Vapor Plume Induced ICING 280
- Air Pollution Mechanics 281
  - Atomopheric Disperson, Transportation and Deposition (EPA, 2005) 281
    - Weather 282
    - Turbulence 283
    - Air Parcels 283
    - Buoyancy Factors 283
    - Lapse Rates 284
    - Mixing 285
    - Topography 285
    - Inversions 285
    - Plume Behavior 286
    - Plume Rise Equation 286
    - Transport 287
  - Dispersion Models 287
- Major Air Pollutants 289
  - Sulfur Dioxide (SO_2) 290
  - Nitrogen Oxides (NO_x) 290
  - Carbon Monoxide (CO) 291
  - Volatile Organic Compounds (VOCs—Hydrocarbons) 291
  - Ozone and Photochemical SMOG 292
  - Carbon Dioxide 293
  - Particulate Matter 293
  - Lead 293
- Air Pollution Control Technology 294
  - Air Pollution Control: Choices 295
- Air Pollution Control Equipment and Systems 295
  - Removal of Dry Particulate Matter 296
    - Gravity Settlers 297
    - Cyclone Collectors 297
    - Electrostatic Precipitators (ESPs) 298
    - Wet (Venturi) Scrubbers 298
    - Baghouse (Fabric) Filters 298
  - Removal of Gaseous Pollutants: Stationary Sources 299
    - Absorption 300
    - Adsorption 301
    - Condensation 301

Combustion (Incineration) 302
Indoor Air Quality 303
    Legionnaires' Disease 304
    Sick Building Syndrome 305
Indoor Air Pollution 306
    Common Indoor Air Pollutants in the Home 307
        Radon 307
        Environmental Tobacco Smoke 308
        Biological Contaminants 308
        Combustion by-Products 308
        Household Products 309
        Pesticides 309
        Asbestos in the Home 309
    Building Factors Affecting Indoor Air Quality 309
    Types of Workplace Air Pollutants 310
    Sources of Workplace Air Pollutants 310
    Indoor Air Contaminant Transport 311
Indoor Air Dispersion Parameters 314
    Parameters 314
Common Airflow Pathways 315
Major IAQ Contaminants 316
    Asbestos Exposure (OSHA, 2002) 316
        Permissible Exposure Limits 317
        Exposure Monitoring 317
        Competent Person 317
        Regulated Areas 317
        Methods of Compliance 317
        Respirators 317
        Labels 319
        Protective Clothing 319
        Training 319
        Recordkeeping 319
        Hygiene Facilities and Practices 319
        Medical Exams 319
    Silica Exposure 320
        Guidelines for Control of Occupational Exposure to Silica 320
    Formaldehyde (HCHO) Exposure 320
    Lead Exposure 321
        Health Effects of Lead 321
        Lead Standard Definitions 322
        Worker Lead Protection Program 323
    Mold Control 323
        Mold Prevention 325
        Mold Remediation 325
        Mold Cleanup Methods 325
References 326

Chapter 8 Water Pollution 329

Introduction 329
The 411 on Water 329

Surface Water 331
   Lakes—Rivers and Streams—Estuaries—Wetlands 331
   Surface Water Pollutants 332
      Biochemical Oxygen Demand (BOD) 332
      Nutrients 333
      pH 333
      Suspended Solids 333
      Oil and Grease 333
      Pathogenic Organisms 333
      Toxic Pollutants 334
      Nontoxic Pollutants 334
Emerging Contaminants 334
   Endocrine Disruptors 336
      Biological Effects 337
      Reproductive Effects 340
      Neurological Effects 340
      Immunological Effects 342
   PPCPs 343
      Pharmaceuticals in the Environment 346
      Wastewater Treatment and PPCPs 353
      Landfills 355
      Drinking Water 355
      Domestic Animals 356
   Drug Classes and Environmental Occurrences 356
      Hormones/Mimics 356
      Antibiotics 357
      Blood Lipid Regulators 358
      Nonopiod Analgestics/Nonsteroidal Anti-Inflammatory Drugs 359
      Beta-Blockers/β_2-Sympathomimetics 359
      Antidepressants/Obsessive-Compulsive Regulators 359
      Antiepileptics 361
      Antineoplastics 361
      Impotence Drugs 362
      Tranquilizers 362
      Retinoids 362
      Diagnostic Contrast Media 362
   Personal Care Products in the Environment 363
      Fragrances (Musks) 363
      Disinfectants/Antiseptics 366
      Preservatives 366
      Sunscreen Agents 366
      Nutraceuticals/Herbal Remedies 367
   Illicit Drugs in Wastewater 368
      Screening for Big Bucks 368
Surface Water Pollution: The Impact 369
Groundwater 369
   Groundwater Uses and Sources 370
      Aquifers 370
      Groundwater Flow 371
      Groundwater Pollution 372
   Wetlands 372

Water Treatment	373
Purpose of Water Treatment	373
Stages of Water Treatment	374
Pretreatment	375
Hardness Treatment	382
Corrosion	384
Coagulation	387
Flocculation	390
Sedimentation	390
Filtration	390
Disinfection	398
Arsenic Removal from Drinking Water	430
Arsenic Exposure	430
Arsenic Removal Technologies	431
Prescriptive Processes	431
Adsorptive Processes	434
Membrane Processes	434
Alternative Technologies	435
Wastewater Treatment	437
The Wastewater Treatment Model	438
Wastewater Terminology and Definitions	438
Measuring Plant Performance	441
Plant Performance and Efficiency	442
Percent Volatile Matter Reduction in Sludge	442
Hydraulic Detention Time	443
Hydraulic Detention Time in Days	443
Hydraulic Detention Time in Hours	443
Detention Time in Minutes	444
Wastewater Sources and Characteristics	444
Wastewater Sources	445
Classification of Wastewater	445
Wastewater Characteristics	445
Wastewater Collection Systems	448
Gravity Collection System	448
Force Main Collection System	448
Vacuum System	448
Pumping Stations	449
Preliminary Wastewater Treatment	450
Screening	450
Shredding	453
Grit Removal	453
Preaeration	457
Chemical Addition	457
Equalization	458
Primary Wastewater Treatment (Sedimentation)	458
Types of Sedimentation Tanks	459
Sedimentation Calculations	460
Effluent from Settling Tanks	463
Secondary Treatment	463
Treatment Ponds	464
Trickling Filters	476

Contents

- Rotating Biological Contactors (RBCs) 484
- Activated Sludge 487
  - Activated Sludge Terminology 488
  - Activated Sludge Process 491
  - Factors Affecting the Operation of the Activated Sludge Process 492
  - Microorganism Growth Curve 492
  - Activated Sludge Formation 493
  - Activated Sludge Modifications 494
  - Activated Sludge Process Control Parameters 500
- Disinfection of Wastewater 501
  - Chlorine Disinfection 501
  - Wastewater Chlorination Process Description 503
- Ultraviolet Irradiation 507
  - Advantages and Disadvantages 508
- Ozonation 509
  - Advantages and Disadvantages 509
  - Applicability 510
  - Operation and Maintenance 510
- Bromine Chloride 511
- No Disinfection 511
- Advanced Wastewater Treatment 511
  - Chemical Treatment 512
  - Microscreening 512
  - Filtration 513
  - Membrane Bioreactors 513
    - Advantages 513
    - Disadvantages 514
  - Biological Nitrification 514
  - Biological Denitrification 514
  - Carbon Adsorption 515
  - Land Application 515
  - Biological Nutrient Removal (BNR) 516
  - Enhanced Biological Nutrient Removal (EBNR) 518
- Wastewater Solids (Sludge/Biosolids) Handling 518
  - Background Information on Sludge 519
  - Sources of Sludge 519
  - Sludge Characteristics 520
    - Sludge Pathogens and Vector Attraction 521
  - Sludge Pumping Calculations 523
    - Estimating Daily Sludge Production 523
    - Sludge Pumping Time 523
    - Gallons of Sludge Pumped per Day 524
    - Pounds Sludge Pumped per Day 524
    - Pounds Solids Pumped per Day 524
    - Pounds Volatile Matter (VM) Pumped per Day 524
    - Sludge Production in Pounds/Million Gallons 525
    - Sludge Production in Wet Tons/Year 525
  - Sludge Thickening 526
    - Gravity Thickening 526
    - Flotation Thickening 527
    - Solids Concentrators 527

Process Calculations (Gravity/Dissolved Air Flotation) ... 528
Sludge Stabilization ... 530
Aerobic Digestion ... 530
Anaerobic Digestion ... 533
Composting ... 537
Sludge Dewatering ... 539
Process Control Calculations ... 544
Land Application of Biosolids ... 548
Process Control Calculations ... 548
Notes ... 550
References ... 551

Chapter 9 Soil Quality ... 563
Introduction ... 564
Soil: What Is It? ... 566
Soil Basics ... 567
Soil Properties ... 567
Soil Formation ... 568
Soil Fertility ... 569
Soil Pollution ... 570
Gaseous and Airborne Pollutants ... 571
Infiltration of Contaminated Surface Water ... 571
Land Disposal of Solid and Liquid Waste Materials ... 571
Stockpiles, Tailings, and Spoils ... 572
Dumps ... 572
Salt Spreading on Roads ... 572
Animal Feedlots ... 572
Fertilizers and Pesticides ... 573
Accidental Spills ... 573
Composting of Leaves and Other Wastes ... 573
Industrial Practices and Soil Contamination ... 574
Contamination from Oil Field Sites ... 574
Contamination from Chemical Sites ... 574
Contamination from Geothermal Sites ... 575
Contamination from Manufactured Gas Plants ... 575
Contamination from Mining Sites ... 575
Contamination from Environmental Terrorism ... 576
USTs: The Problem ... 576
Corrosion Problems ... 577
Faulty Construction ... 577
Faulty Installation ... 577
Piping Failures ... 577
Spills and Overfills ... 578
Compatibility of Contents and UST ... 578
Risk Assessment ... 578
Exposure Pathways ... 579
Remediation of UST Contaminated Soils ... 579
In Situ Technologies ... 580
In Situ Volatilization (ISV) ... 580
In Situ Biodegradation ... 581

Contents

	In Situ Leaching and Chemical Reaction	582
	In Situ Vitrification	583
	In Situ Passive Remediation	583
	In Situ Isolation/Containment	583
Non-*In Situ* Technologies	584
	Land Treatment	584
	Thermal Treatment	584
	Asphalt Incorporation and Other Methods	584
	Solidification/Stabilization	586
	Chemical Extraction	586
	Excavation	587
Note	587
References	587

Chapter 10 Solid and Hazardous Waste ... 589

Introduction ... 590
Solid Waste Regulatory History (United States) ... 591
Solid Waste Characteristics ... 592
Sources of Municipal Solid Wastes (MSW) ... 593
  Residential Sources of MSW ... 593
  Commercial Sources of MSW ... 594
  Institutional Sources of MSW ... 594
  Construction and Demolition Sources of MSW ... 594
  Municipal Services Sources of MSW ... 594
  Treatment Plant Site Sources of MSW ... 594
What Is a Hazardous Substance? ... 595
  Hazardous Materials ... 595
  Hazardous Substances ... 596
  Extremely Hazardous Substances ... 596
  Toxic Chemicals ... 596
  Hazardous Wastes ... 596
  Hazardous Chemicals ... 597
Again, What Is a Hazardous Substance? ... 597
What Is a Hazardous Waste? ... 597
  EPA Lists of Hazardous Wastes ... 598
Where Do Hazardous Wastes Come from? ... 599
Why Are We Concerned about Hazardous Wastes? ... 599
Hazardous Waste Legislation ... 600
  Resource Conservation and Recovery Act ... 600
  CERCLA ... 601
Waste Control Technology ... 601
  Waste Minimization ... 602
    Substitution of Inputs ... 602
    Process Modifications ... 603
    Good Operating Practices ... 603
    Recycling ... 603
  Steps to Recycling Materials ... 603
    Collection and Processing ... 603
    Manufacturing ... 604
    Purchasing New Products Made from Recycled Materials ... 604

Recycling Hazardous Wastes ... 605
Treatment Technologies ... 605
    Biological Treatment ... 605
    Thermal Processes ... 606
    Activated Carbon Sorption ... 607
    Electrolytic Recovery Techniques ... 607
    Air Stripping ... 607
    Stabilization and Solidification ... 607
    Filtation and Separation ... 608
Ultimate Disposal ... 608
    Deep-Well Injection ... 608
    Surface Impoundments ... 609
    Waste Piles ... 609
    Landfilling ... 610
References ... 611

Chapter 11 Industrial Hygiene ... 613

What Is Industrial Hygiene? ... 613
Industrial Hygiene Terminology ... 613
    Industrial Hygiene Terms and Concepts ... 613
History of Industrial Hygiene (OSHA, 1998) ... 628
OSHA/NIOSH and Industrial Hygiene ... 629
Industrial Hygiene: Workplace Stressors ... 630
Industrial Hygiene: Areas of Concern ... 633
    Industrial Toxicology ... 633
        What Is Toxicology? ... 634
    Air Contaminants ... 635
    Routes of Entry ... 635
Industrial Health Hazards ... 636
    Environmental Controls ... 637
        Engineering Controls ... 638
        Work Practice Controls ... 638
        Administrative Controls ... 638
        Personal Protective Equipment (PPE) ... 639
Hazard Communication ... 640
    HAZCOM and the Environmental Professional ... 640
    Modification of the Hazard Communication Standard (HCS) ... 641
Occupational Environmental Limits (OELS) ... 646
    OELs ... 646
Air Monitoring/Sampling ... 648
    Air Sample Volume ... 648
    Limit of Detection (LOD) ... 648
    Limit of Quantification (LOQ) ... 649
        Precision, Accuracy, and Bias ... 649
    Calibration Requirements ... 649
    Types of Air Sampling ... 649
    Analyltical Methods for Gases and Vapors ... 649
    Air Monitoring Versus Air Sampling ... 651
    Air Sampling for Airborne Particulates ... 652
    Dusts ... 652

Contents

- Duration of Exposure 653
- Particle Size 653
  - Stoke's Law 653
  - Airborne Dust Concentration 654
- Particulate Collection 654
- Analysis of Particulates 654
- Health and Environmental Impacts of Particulates 654
- Control of Particulates 655
- Air Sampling for Gases and Vapors 655
  - Types of Air Samples 655
  - Methods of Sampling 656
  - Air Sampling Collection Processes 656
  - Calibration of Air Sampling Equipment 656
  - Direct Reading Instruments for Air Sampling 657
    - Types of Direct Reading Instruments 657
    - Calibration of Direct-Reading Instruments 659
  - Air Monitoring: Confined Space Entry 659
- Noise & Vibration 660
  - OSHA Noise Control Requirements 660
  - Noise and Hearing Loss Terminology 661
  - Occupational Noise Exposure 665
  - Determining Workplace Noise Levels 665
  - Engineering Control for Industrial Noise 666
  - Audiometric Testing (OSHA, 1998) 667
  - Noise Units, Relationships & Equations 668
  - Industrial Vibration Control 671
- Radiation 672
  - Radiation Safety Program Acronyms and Definitions 672
    - Abbreviations Typically Used in Radiation Safety Programs 672
    - Typical Radiation Program Definitions 673
  - Ionizing Radiation 674
    - Effective Half-Life 675
  - Alpha Radiation 675
    - Alpha Radiation Detectors 676
  - Beta Radiation 676
    - Beta Detection Instrumentation 676
    - Shielding for Beta Radiation 676
  - Gamma Radiation and X-Rays 677
    - Gamma Detection Instrumentation 677
    - Shielding for Gamma and X-Rays 677
  - Radioactive Decay Equations 677
  - Radiation Dose 679
  - Non-Ionizing Radiation 679
    - Optical Density (OD) 681
  - OSHA's Radiation Safety Requirements 681
  - Radiation Exposure Controls 683
  - Radiation Exposure Training 683
- Thermal Stress 684
  - Thermal Comfort 684
  - The Heat Index 685
  - The Body's Response to Heat 686

    Heat Disorders and Health Effects 687
    Cold Hazards 688
    Wind-Chill Factor 689
    Cold Stress Prevention 689
  Ventilation 690
    Concepts of Ventilation 690
      Local Exhaust Ventilation 693
      General and Dilution Ventilation 693
  Personal Protective Equipment (PPE) 694
    OSHA's PPE Standard 694
      OSHA's PPE Requirements 695
      Hazard Assessment 696
      PPE Training Requirement 696
    Head Protection 697
    Hand Protection 697
    Eye and Face Protection 697
    Foot Protection 697
    Full Body Protection: Chemical Protective Clothing 698
    Description of Protective Clothing 698
      Protective Clothing Applications 698
      The Clothing Ensemble 699
      Levels of Protection 699
      Clothing Selection Factors 701
      Classification of Protective Clothing 701
      Material Chemical Resistance 702
      Decontamination Procedures 702
      Types of Contamination 703
      Decontamination Methods 703
    Inspection, Storage, and Maintenance of Protective Clothing 703
      Inspection 703
      Storage 704
      Maintenance 704
  Respiratory Protection 704
  Engineering Design and Controls for Safety 705
    Codes and Standards 707
    Plant Layout 708
    Illumination 708
    High Hazard Work Areas 709
    Personal and Sanitation Facilities 709
  Note 710
  References 710

Chapter 12 Green Engineering 711

  Introduction 711
  Wind Energy 712
  Environmental Impact of Wind Turbines 713
    The Good, Bad, and Ugly of Wind Energy 713
    Air Quality (Including Global Climate Change and Carbon Footprint) 713
    Cultural Resources 714
    Ecological Resources 714

Contents

- Water Resources (Surface Water and Groundwater) 715
  - Water Use 715
  - Water Quality 715
  - Flow Alteration 715
- Land Use 715
- Soils and Geologic Resources (Including Seismicity/Geo Hazards) 716
- Paleontological Resources 716
- Transportation 716
- Visual Resources 716
- Socioeconomics 717
- Environmental Justice 717
- Hazardous Materials and Waste Management 717
- Wind Energy Operations Impacts 718
  - Air Quality (Including Global Climate Change and Carbon Footprint) 718
  - Cultural Resources 718
  - Ecological Resources 718
  - Water Resources (Surface Water and Groundwater) 718
  - Land Use 719
  - Soils and Geologic Resources (Including Seismicity/Geo Hazards) 719
  - Paleontological Resources 719
  - Transportation 719
  - Visual Resources 719
  - Socioeconomics 720
  - Environmental Justice 720
  - Hazardous Materials and Waste Management 720
  - Impact on Wildlife 720
- Energy from the Sun 723
  - Environmental Impacts of Solar Energy 723
    - Land Use/Siting Impact 723
    - Water Use Impact (Water Footprint) 723
    - Hazardous Waste Impact 724
    - Ecological Impact 725
- Hydropower 726
  - Ecological Impact of Hydropower 728
    - Physical Barrier to Fish Migration 729
    - Flow Alteration, Flow Fluctuation, Regulated and Unregulated Rivers and Salmonids 729
  - Biological Impact of Flow Fluctuations 730
    - Increases in Flow 730
    - Stranding 730
    - Increased Predation 732
    - Aquatic Invertebrates 733
    - Redd Dewatering 733
    - Spawning Interference 734
    - Hydraulic Response to Flow Fluctuations 734
    - Types of Hydropower Activity that Fluctuate Flows 734
- Biomass/Bioenergy 735
  - Impact of Biomass Construction, Production and Operation 739
    - Air Quality 740
    - Cultural Resources 740
    - Ecological Resources 740

      Water Resources 741
      Land Resources 741
      Soils and Geologic Resources 742
      Paleontological Resources 743
      Transportation 744
      Visual Resources 744
      Socioeconomics 744
      Environmental Justice 744
    Biomass Feedstock Production Impact 745
      Air Quality 745
      Cultural Resources 745
      Ecological Resources 746
      Water Resources 746
      Land Resources 746
      Soils and Geologic Resources 746
      Paleontological Resources 746
      Transportation 747
      Visual Resources 747
      Socioeconomics 747
      Environmental Justice 747
    Biomass Energy Operations Impact 747
      Air Quality 747
      Cultural Resources 748
      Ecological Resources 748
      Water Resources 748
      Land Use 749
      Soils and Geologic Resources 749
      Paleontological Resources 749
      Transportation 750
      Visual Resources 750
      Socioeconomics 750
      Environmental Justice 750
Geothermal Energy 750
    Environmental Impact of Geothermal Power Development 751
    Geothermal Energy Exploration and Drilling Impact 753
      Air Quality 753
      Cultural Resources 753
      Ecological Resources 753
      Water Resources 754
      Land Use 754
      Soils and Geologic Resources 754
      Paleontological Resources 755
      Transportation 755
      Visual Resources 755
      Socioeconomics 755
      Environmental Justice 755
    Geothermal Energy Construction Impact 755
      Air Quality 756
      Cultural Resources 756
      Ecological Resources 756
      Water Resources 756

Land Use 757
Soils and Geologic Resources 757
Paleontological Resources 757
Transportation 757
Visual Resources 758
Socioeconomics 758
Environmental Justice 758
Geothermal Energy Operation and Maintenance Impact 758
Air Quality 759
Cultural Resources 759
Ecological Resources 759
Water Resources 759
Land Use 759
Soils and Geologic Resources 760
Paleontological Resources 760
Transportation 760
Visual Resources 760
Socioeconomics 760
Environmental Justice 760
Marine and Hydrokinetic Energy 761
Ocean Tides, Currents, and Waves 761
Tides 761
Currents 762
Waves 762
Wave Energy 762
Wave Energy Conversion Technology 762
Point Absorber 764
Terminator 764
Attenuator 764
Tidal Energy 764
Tidal Energy Technologies 764
Ocean Thermal Energy Conversion 765
Potential Environmental (General) Impacts 765
Alteration of Current and Waves 766
Alteration of Substrates and Sediment Transport and Deposition 768
Impact of Habitat Alternation on Benthic Organisms 769
Displacement of Benthic Organisms by Installation of the Project 769
Alteration of Habitats for Benthic Organisms during Operation 769
Impact of Noise 770
Impact of Electromagnetic Fields (EMF) 776
Effect of Electromagnetic Fields on Aquatic Organisms 778
Toxic Effect of Chemicals 782
Toxicity of Paints, Anti-Fouling Coatings, and Other Chemicals 783
Interference with Animal Movement and Migration 783
Collision and Strike 786
Effect of Rotor Blade Strike on Aquatic Animals 786
Impact of Ocean Thermal Energy Conversion (OTEC) 788
Potential Environmental (Specific) Impacts 789
Hydrokinetic Energy Site Evaluation Impact 790
Air Quality (Including Carbon Footprint and Global Climate Change) 790
Cultural Resources 790

- Ecological Resources 790
- Water Resources (Surface Water and Groundwater) 790
- Land Use 790
- Soils and Geologic Resources 791
- Paleontological Resources 791
- Transportation 791
- Visual Resources 791
- Socioeconomics 791
- Environmental Justice 791
- Acoustics (Noise) 791
- Hazardous Materials and Waste Management 791
- Hydrokinetic Energy Facility Construction Impact 791
  - Air Quality (Including Global Climate Change and Carbon Footprint) 792
  - Cultural Resources 792
  - Ecological Resources 793
  - Water Resources (Surface Water and Groundwater) 794
  - Land Use 794
  - Soils and Geologic Resources 794
  - Visual Resources 794
  - Paleontological Resources 795
  - Transportation 795
  - Socioeconomics 795
  - Environmental Justice 795
  - Acoustics (Noise) 796
  - Hazardous Materials and Waste Management 796
- Hydrokinetic Energy Facility Operations & Maintenance Impact 796
  - Air Quality (Including Global Climate Change and Carbon Footprint) 796
  - Cultural Resources 797
  - Ecological Resources 797
  - Water Resources (Surface Water and Groundwater) 798
  - Land Use 798
  - Soils and Geologic Resources 798
  - Paleontological Resources 798
  - Transportation 799
  - Visual Resources 799
  - Socioeconomics 799
  - Environmental Justice 799
  - Acoustics (Noise) 799
  - Hazardous Materials and Waste Management 799
- Fuel Cells 800
  - Hydrogen Fuel Cell 800
    - Hydrogen Storage 801
    - How a Hydrogen Fuel Cell Works 802
  - Environmental Impact of Fuel Cells 802
- Carbon Capture and Sequestration 802
  - The 411 on Carbon Capture and Sequestration 803
    - Terrestrial Carbon Sequestration 803
  - Geologic Carbon Sequestration 806
  - Potential Impact of Terrestrial Sequestration 807
  - Potential Impact of Geologic Sequestration 808
  - Geologic Sequestration Exploration Impact 808

Contents

- Air Quality 808
- Cultural Resources 808
- Ecological Resources 808
- Water Resources (Surface Water and Groundwater) 809
- Land Use 809
- Soils and Geologic Resources 809
- Paleontological Resources 809
- Transportation 809
- Visual Resources 809
- Socioeconomics 810
- Environmental Justice 810
- Acoustics (Noise) 810
- Hazardous Materials and Waste Management 810
- Geologic Sequestration Drilling/Construction Impact 810
  - Air Quality 810
  - Cultural Resources 810
  - Ecological Resources 811
  - Water Resources (Surface Water and Groundwater) 811
  - Land Use 812
  - Soils and Geologic Resources 812
  - Paleontological Resources 813
  - Transportation 813
  - Visual Resources 813
  - Socioeconomics 813
  - Environmental Justice 814
- Acoustics (Noise) 814
- Hazardous Materials and Waste Management 814
- Geologic Sequestration Operations Impact 814
  - Air Quality 815
  - Cultural Resources 815
  - Ecological Resources 815
  - Water Resources (Surface Water and Groundwater) 815
  - Land Use 816
  - Soils and Geologic Resources 816
  - Paleontological Resources 816
  - Transportation 817
  - Visual Resources 817
  - Socioeconomics 817
  - Environmental Justice 817
  - Acoustics (Noise) 817
  - Hazardous Materials and Waste Management 817
- The Bottom Line 818
- Notes 818
- References 819

Index 831

A Note to the Reader: Will it be Milk Toast or Hardtack?

As I stated in the first edition of this book, I remember the moment well. It was a few minutes before I was to enter my classroom for the first time that semester to present my opening lecture on *An Introduction to Environmental Engineering* to a room full of college students (a large mix of juniors, seniors, and graduate students). I was standing at my office window, looking out onto the campus grounds. I just stood there, bringing on one of those thinking-too-hard headaches. Because I had developed the tendency to think on two wavelengths at the same time (i.e., thinking about general thoughts and subliminally about what I was viewing at that moment—I call this compound thinking), a super headache manifested itself as always.

Anyway, back to that day and my compound thinking. On a general wavelength, I was thinking about my pending lecture presentation. More specifically, more to the point…I was thinking about how to grab the students' attention; that is, how to hook them. I knew that to actually deliver the message, I had to hook the students' attention—an absolute must. I knew I needed to hook them; otherwise, I would lose them to one of those annoying Twitter tweets, or songs, or daydreams (boredom was not an option and totally impossible in my presence). So, I had the choice to take the milk toast presentation method: yes, I could point out that the environment is in trouble; yes, that they should be careful about that glass of water they might drink; yes, they should think about the air they are breathing; yes, they should think about the ozone hole; yes, they should think about Arctic ice melt; and, yes, they should think about their current practices as a contributor in the world's foremost "throw-away society." And, no, it is not hopeless. No, we are not doomed. No, we are not headed to the so-called Sixth Extinction. The fact is, I knew I could present all this. The truth is, I did not believe (and still don't) that environmental professionals should focus on reactionary actions only to mitigate any of it. This type of presentation is a milk toast presentation. A milk toast presentation is just that: milk toast with no body or substance, a timid, unassertive, spineless (jellyfish) approach lacking in boldness and vigor. One thing is certain; if you set out to save the world from human-derived destruction, you best leave the milk toast approach at home. Why would anyone want to fix the problem after it occurs?

At that moment, I remember rubbing that throbbing forehead of mine as I contributed to that other compound ingredient of my headache—that is, the visual aspect. I contributed to it, sort of, in a cursory way… subliminally, you understand…well, sort of…by staring through my office window at the campus surroundings. I looked at what should have been a glorious early Fall campus panoramic. Instead, I visually took in a disgusting trash-laden panorama of the campus and then directly at a large tree with its full-leafed branches loaded with candy wrappers, remnants of this or that…hallmarks of the throw-away society; just a bunch of those fluttering banners of debris, those candy wrappers and lost plastic bags, scraps of newspaper, the rings of recently, and not so recently, gulped six packs, and to my right, on a singular branch, hung a signature of passion or lust, and exclamation point: a used condom…all hanging around and blowing in the wind. Then my glance shifted to the flower bed and the beauty of fully bloomed flowers, interrupted here and there with plastic, along with paper wraps of previous uses…and, of course, those trail marker beer cans and bottles here, there, and everywhere. All of this bordering and bisecting one of the numerous campus parking lots, the debris actually land-marking those campus pathways often taken. It was funny that the cars and trucks parked there did not register in my brain…only the discarded paper, plastic, beer cans/bottles, cigarette butts, and other smoking debris registered on the lid and margin of my aching brain cells.

I remember then, at that moment, that my thought process had shifted gears to my preferred hardtack approach (i.e., I would shift to scaring the Be Jesus out of those bright-eyed students). More specifically, I had decided that I would present the obvious or that which was not so obvious. Simply, I would point out the need for environmental studies because of a litany of environmental issues, harmful aspects of human activity on the biophysical environment, and, as such, how they relate to the anthropogenic effects on the natural environment, which were combining to kill us all off, eventually. I planned to loosely divide causes, effects, and mitigation, pointing out that effects are interconnected and that they can cause new effects. To accomplish all this, I initially decided to present, in one form or another, a semester of presentations that would include the following causes and effects of environmental problems:

Causes

- Human overpopulation
- Hydrological issues
- Intensive farming
- Land use
- Nanotechnology
- Nuclear issues
- Biodiversity issues
- Ecophagy

Effects

- Climate change (I think this is a natural phenomenon not human-caused but is exasperated by humans)
- Environmental degradation
- Environmental health
- Environmental issues with energy
- Environmental issues with war
- Overpopulation
- Genetic engineering
- Pollution—air, water, and soil
- Resource depletion
- Toxicants
- Waste

Of course, I had also planned to point out (and did point out) to the environmental students that all of the above causes and effects were about to be their focus in life, their mantra, their burden, their passion (hopefully). As prospective environmental professionals, they were in this class (and those to follow) to learn about the causes and effects of environmental issues but also to learn how to mitigate them, not only through practicing various mitigation measures and complying with environmental laws but, more importantly, by performing practices to prevent the causes and thus prevent environmental damage in the first place. Simply put, at least in my view, it was important to me to point out the environmental issues and imprint them on those young minds while expressing the importance of *engineering out* a potential problem before it manifests itself. Prevention is always better than mitigation (mitigation being actions that include recycling, treatment, and/or disposal). Proactive is better than reactive. Building one's own box to think outside of makes better sense than thinking outside of someone else's creation.

So, on that day, I was faced with that same never-ending dilemma: How was I to harness the students' attention and hook them to the cause of preserving our environment? How was I to make the lasting point that our environment was (is) in trouble? Intuitively, I knew scaring the Be Jesus out of them was the easy approach to making them tune into my lectures, but I also knew that words only have so much weight—whether they are scary or not. I rubbed my head and, indirectly, the headache, and then I had my Eureka moment. Right there in front of me…you know the old saying, "A picture is worth a thousand words," and it often provides the cement needed to imprint to memory the sights beheld. All that was visible outside my window was all I needed to show them. And I did. I ushered them into my office and single-filed them to the window to glance out of it and onto the campus grounds, and then sent them back to their desks in the classroom. Of course, I had directed them to write a 150-word description of what they had observed through the window (I always make my students write, write, and write some more). Knowing environmental engineering without knowing how to write is like knowing dentistry without knowing how to pull teeth. They were instructed to email their papers to me. I have found that anytime I put a student on his or her computer, I have my foot in the door to their attention span and their learning mechanism. By the way, 77 out of 79 of the digital responses I received related descriptions of exactly what I wanted them to observe outside that window. And what they saw is what this handbook is: a ready reference and guidebook of information and instruction on how to engineer out a problem before it occurs.

In this note to the reader, a final word about this handbook. This is not your granddad's handbook. This handbook presents the facts, as it must, but is also opinionated with statements based on the author's observations, experiences, mistakes, misjudgments, and a few triumphs along the way. You may not agree with the opinions stated within. But if this is the case, you will be thinking on your own. And this is what I call the hardtack approach; that is, it makes you think! Is there any other way?

By the way, if my students do not learn a thing from me other than the following statement, then MISSION ACCOMPLISHED!

Probably the first known written admonition regarding the need for accident prevention is contained in Hammurabi's Code, about 1750 B.C. It states: "If a builder constructs a house for a person and does not make it firm and the house collapses and causes the death of the owner, the builder shall be put to death."

Preface

Hailed on its first publication as a masterly account combining physics, chemistry, biology, and mathematics with applications to environmental problems (real-world problems), this book is also unique in the sense that it also includes industrial hygiene—a discipline that is not only important but has become an ingredient in the mix that makes up what environmental engineering is all about. Additionally, the environmental engineer is becoming more involved in the practice of green engineering, which is gaining prominence with the increasing trend toward renewable energy sources and the conservation of energy. Also, in this current trend toward environmental professionals becoming more involved with green engineering, there is a sometimes-important issue. The issue Is the personal safety of green energy personnel. In this new edition, green energy and green energy safety concerns are included in the industrial hygiene section of this book.

Note that this second edition continues the trend, model and/or paradigm presenting environmental engineering as pure science and not as a "feel good science" to fit one particular view or the other. The truth be known, for a long time, environmental engineers suffered from the lack of a well-defined identity. The field is so broad with numerous "ingredients" that almost require a recipe to conduct practice, mitigation procedures, and to teach. But today, now, this is not the case. Environmental engineers work to sustain human existence by balancing human needs and their impacts on the environment with the natural state of the environment as per nature's grand design, its grand scheme. It goes without over-emphasizing the obvious and the truth that sustaining human existence and natural processes in a state of harmony is no easy undertaking. In the face of global pollution, diminishing natural resources, increased population growth (especially in disadvantaged countries), geopolitical warfare, terrorism, global climate change (cyclical and/or human-caused), natural disasters, improper land use, generation, and disposal of waste and other environmental problems, it is basic to the argument or thesis of *Handbook of Environmental Engineering* that we live in a world that is undergoing rapid ecological transformation. Because of these rapid changes, the role of environmental engineering has become increasingly prominent. Moreover, advances in technology (e.g., nanotechnology, artificial intelligence, machine learning and deep learning and emerging technology) have created a broad array of modern environmental issues. To mitigate these issues, we must capitalize on environmental protection and remediation opportunities presented by technology.

This is where the trained environmental engineer comes into play. Simply put, the environmental engineer must be a Jack or Jill of many fields, specialties, and disciplines and have some knowledge in many branches of learning. This is the case, of course, because the field is broad and the science required to mitigate the problems faced by environmental engineers requires knowledge in many engineering fields, including chemical, civil, sanitary, and mechanical engineering. The well-trained environmental engineer must have a very large toolbox; that is, he or she must be a generalist—they must know a little bit about a whole bunch. Based on personal observation I can state without reservation that increased demand for undergraduate training in environmental engineering has led to growth in the number of undergraduate programs offered, and the number of practicing environmental engineers.

Common to all environmental specialists is recognition of the importance of obtaining a strong quantitative background in environmental engineering, science, and management principles that govern environmental processes. Of course, all the training in the world will never suffice by itself, unless we blend it with that uncommon trait known as common sense tempered with loads of practical on-the-job experience.

At this time, assorted commentators cite technological advances as the culprit humankind has employed to negatively impact nature's way and the environment in deleterious ways. The fact is environmental engineers respond to the needs of society with these same technological innovations; that is, we use technology to solve technologically driven problems.

And this is the point the author is making. Actually, this is the compound part of the answer to the question: How do environmental engineers employ and put to use technology to absolutely correct or mitigate environmental problems? One way in which environmental engineers do this is by using their tools; the *Handbook of Environmental Engineering* is one of these tools and is also highly accessible and user-friendly. Do not be fooled; however, this text is a blend of technical rigor mixed with broad accessibility and is presented in the author's traditional conversational style—I talk to the reader. The author's ultimate goal is to make certain there is absolutely no failure to communicate.

Frank R. Spellman
Norfolk, VA

About the Author

Frank R. Spellman, PhD, is a retired Assistant Professor of Environmental Health at Old Dominion University, Norfolk, Virginia, and the author of more than 157 books covering topics ranging from concentrated animal feeding operations (CAFOs) to all areas of environmental science and occupational health. Many of his texts are readily available online, and several have been adopted for classroom use at major universities throughout the United States, Canada, China, Europe, and Russia; two have been translated into Spanish for South American markets. Dr. Spellman has been cited in more than 850 publications. He serves as a Professional Expert Witness for three law groups and as an Incident/Accident Investigator for the U.S. Department of Justice and a northern Virginia law firm. In addition, he consults on homeland security vulnerability assessments for critical infrastructures including water/wastewater facilities nationwide and conducts pre-Occupational Safety and Health Administration (OSHA)/Environmental Protection Agency EPA audits throughout the country. Dr. Spellman receives frequent requests to co-author with well-recognized experts in several scientific fields; for example, he is a contributing author of the prestigious text *The Engineering Handbook*, 2nd ed. (CRC Press). Dr. Spellman lectures on sewage treatment, water treatment, and homeland security and lectures and safety topics throughout the country and teaches water/wastewater operator short courses at Virginia Tech (Blacksburg, Virginia). He holds a BA in Public Administration, a BS in Business Management, an MBA, an MS, and PhD in Environmental Engineering.

1 Introduction

One of the penalties of an [environmental] education is that one lives alone in a world of wounds. Much of the damage inflicted on the land is quite invisible to laymen. An [environmentalist] must either harden his[/her] shell and make believe the consequences of science are none of his[/her] business or he[/she] must be the doctor who sees the marks of death in a community that believes itself well and does not want to be told otherwise.

The government tells us we need flood control and comes to straighten the creek in our pasture. The engineer on the job tells us the creek is now about to carry off more flood water, but in the process, we lost our old willows where the cows switched flies in the noon shade, and where the owl hooted on a winter night. We lost the little marshy spot where our fringed gentians bloomed.

> Some engineers are beginning to have a feeling in their bones that the meanderings of a creek not only improve the landscape but are a necessary part of hydrologic functioning. The [environmental professional] clearly sees that for similar reasons we can get along with less channel improvement on Round River.
>
> **A. Leopold (1970, p. 165)**

> The best advice I give to any rookie environmental engineer is that which already has been rendered by wiser and a lot more far-seeing people than me. The advice: "primuum non nocere."
>
> **Frank R. Spellman (1998)**

With regard to training for environmental engineers, students as well as practitioners in the field need to obtain an up-to-date level of knowledge and understanding in the following interrelated fields:

- knowledge and understanding of physical, chemical, and biological processes fundamental to understanding the environment fate and engineered treatment of environmental contaminants
- knowledge and understanding of the sources and nature of waste materials that contribute to air, oil, and water pollution and relevant management and control technologies
- knowledge and understanding of science, impacts mitigation, adaption, and policy relevant to climate change and global environmental sustainability, energy planning, alternative energy technologies (e.g., hydraulic fracking), sustainable development, and next generation processes
- knowledge and understanding of the transport and transformation of contaminants through environmental pathways
- knowledge and understanding of the pollution prevention and technologies and designs associated with the treatment and disposal of waste materials
- knowledge and understanding of the connection between the engineering and scientific aspects of environmental problems and decision-making processes
- knowledge and understanding of not only green energy but also about the personal/personnel hazards involved when incorporating green technology

Understanding all of these areas is achieved through a quantitative educational-training program and years of on-the-job experience built around the common theme of engineering and science in support of environmental decision-making and management.

The question is: How do environmental engineers employ and put to use technology to correct or mitigate environmental problems? The simple and compound answers? The simple answer first: Environmental engineers use their tools. Tools? Yes; absolutely. Environmental engineers have a toolbox full of tools.

What are these tools?

Good question. The short list of these tools include: knowledge of fluid mechanics, ecology, principles of toxicology, risk assessment, management principles, hydrogeology, modeling contaminant migration through multimedia systems, aquatic chemistry, environmental microbiology, applied statistical analyses, open channel hydraulics, field methods in habitat analysis and wetland delineation, principles of estuarine environment, hydrology, resources modeling, environmental sampling, sediment transport and river mechanics, geomorphic and ecological foundations of steams restoration, atmospheric chemistry, environmental chemicals, economic foundations for public decision making, business law for engineers, environmental impact assessment, geographic information systems (GIS), global environmental sustainability, water resources management, sustainable development, green engineering, energy planning, renewable energy, smart growth strategies for sustainable urban development and revitalization, environmental safety and health, and many other specialties, such as artificial intelligence, machine learning, and deep thinking, all of which are new to the toolbox but essential in the ever changing, advancing technology.

ENVIRONMENT

When we say the "environment," we are talking about a key definition in environmental engineering. So, what do we mean exactly? Think about it. Environment can mean many different things to many different people. From the individual point of view, the environment can be defined as his or her environment. From the environmental professional's point of view, the environment he or she works with needs a more specific definition. To the environmental professional the word environment may take on global dimensions (i.e., the atmosphere, hydrosphere, lithosphere, and biosphere), may refer to a very localized area in which a specific problem must be addressed, or may, in the case of contained environments, refer to a small volume of liquid, gaseous, or solid materials within a treatment plant unit process (Peavy et al., 1985). In the current digital era, for example, some may view "environment" as the office environment, creative environment, learning environment, corporate environment, virtual environment, aquatic environment, tropical environment, social environment, conservation environment, or in this digital age maybe we are referring to the desktop environment, integrated development environment, or runtime environment, machine learning environment, artificial intelligent environment, and so on. Obviously, again, when we use the term "environment" we need to be more specific. In this text, we are specific, actually more specific by defining the *environment* as the natural environment, which includes all living (all life-forms) and nonliving things that influence life (organisms). That is, our environment is that ten-mile-thick layer on the 200-million-square-mile surface of this planet. Remember, without that ten-mile-thick layer (the environment), which contains air, soil, and water (the environment), Earth would be a sterile hunk of orbiting rock. Without air, water, and soil, there is nothing we can—or could—relate to (Spellman, 1998).

SIDEBAR 1.1: ENVIRONMENTAL EQUILIBRIUM AND THE FLY IN THE OINTMENT

We had not walked any part of the Appalachian Trail with its nearly 50 mountains spanning 14 states and 8 national forests for more than several years. Though we had never walked its entire 2,160-mile length, from Springer Mountain in Georgia to Katahdin, Maine at once, over the course of several years a long time ago, we had in piecemeal fashion covered most of it, and hiked many of the several hundred trails that parallel and join it as well. But we had moved out of easy reach of the Trail, and for years had only our memories of it.

Introduction

For us, the lure of sojourning the Appalachian Trail had always been more than just an excuse to get away from it all—whatever "it" happened to be at the time. The draw, the magnetism, the pull, the appeal of the Trail was more—much more to us than that, though we have always found its magic difficult to define. Maybe it is a combination of elements—recollections, pleasant memories, ephemeral surprises found and never forgotten. Memories waking from the miles-deep sleep of earned exhaustion to the awareness of peace… inhaling deep draughts of cool, clean mountain air, breathing through nostrils tickled with the pungency of pure, sweet pine… eardrums soothed by the light tattoo of fresh rain pattering against taut nylon… watching darkness lift, then suddenly replaced with cloud-filtered daylight, spellbound by the sudden, ordinary miracle of a new morning… anticipating our expected adventure and realizing the pure, unadulterated treasure of pristine wilderness we momentarily owned, with minds not weighed down by the everyday mundanities of existence. That is what we took away from our Trail experiences, years ago, what we remembered about living on the trail, on our untroubled sojourn through one of the last pure wilderness areas left in the United States. Those memories were magnets. They drew us inexorably to the Trail—back again and again.

But, of course, the Trail had another drawing card—the Natural World and all its glory. The Trail defined that for us. The flora that surrounds you on the Trail literally encapsulates you as it does in any dense forest, and brings you fully into its own world, shutting out all the other worlds of your life. For a brief span of time, along the Trail, the office was gone, cities, traffic, the buzz, and grind of work melted away into forest. But this forest was different, and its floral inhabitants created the difference. Not only the thickets of rhododendrons and azaleas (in memory, always in full bloom), the other forest growths drew us there: the magnificent trees—that wild assortment of incomparable beauty that stood as if for forever—that was the Trail.

This was how it had been no more than 25 years ago, but now things were different; things had changed for the worse; there was fly in the ointment: pollution—human-caused. To say that we were and are shocked at what we found recently along the Trail—along most of its length—is true, and we can only describe it as wounding heartache, as achingly sad to us as the discovery of the physical debilitation of a long-beloved friend. Even though still lined (and in some places densely packed) with Fraser fir, red spruce, sugar maples, shagbark hickory, northern red oak, quaking aspens, tulip poplars, white basswood, yellow buckeyes, black gums, old growth beech, mountain laurel, and those incomparable dogwoods whose creamy-white bracts light up the woods in early spring—the world along the Trail was different. Let us paint you a picture of the differences. Walking various segments of the trail and its arteries in North Carolina, Virginia, and Maryland we observed:

1. Standing dead Fraser fir and red spruce
2. Stands of pollution-killed trees where fallen gray tree trunks crisscrossed each other in a horrible game of giant jackstraws
3. Standing dead red spruce silhouetted by polluted fog
4. Understories of brambles looking up at dead sugar maples
5. Foliage areas bleached by ozone
6. Trees of all varieties starved to death; the needed soil nutrients leached away by decades of acid deposition, and the trees weakened until they were no longer capable of withstanding the assaults of even ordinary disease and bad weather.
7. Logged wasteland areas
8. Branch dieback on northern red oak
9. Premature leaf-drop on quaking aspens
10. Thinning crowns on sugar maples
11. Tipped-over tulip poplars with rotted roots

12. Chemically green ponds in areas where active strip mining occurs
13. An orange waterfall next to an abandoned mine
14. An overview, where 25 years earlier we viewed the surrounding landscape for 50 miles, now veiled in thick, stagnant, polluted fog with visibility reduced to 2 or 3 miles

When asked to describe pollution—the fly in the ointment of what should be a pristine environment—most people have little trouble doing so, having witnessed some form of it firsthand. They usually come up with an answer that is a description of its obvious effects. But the pollution is complicated and although it can be easily described it cannot be easily defined, because what pollution is and is not a judgment call. In nature, however, even the most minute elements are intimately connected with every other element, and so too are pollution's effects.

In this book, we describe pollution more fully and explain the difficulty involved with defining it, beginning a process that will allow you to create your own definition of pollution—though each reader's definition will vary.

POLLUTION DEFINED

When we need a definition for any environmental term, the first place we look is in pertinent USEPA publications. With the term pollution, however, we did not find the EPA definition particularly helpful or complete.

USEPA (1989) defines pollution as "Generally... the presence of matter or energy whose nature, location or quantity produced undesired environmental effects... impurities producing an undesirable change in an ecosystem." Under the Clean Air Act (CWA), for example, the term is defined as "the human-made or human-induced alteration of the physical, biological, and radioactive integrity of water" (Spellman, 1998). Though their definition is not inaccurate, it leaves out too much to suit our needs. USEPA does, however, provide an adequate definition of the term *pollutant* as "Any substance introduced into the environment that adversely affects the usefulness of a resource." Pollution is often classed as point source or nonpoint source pollution. However, USEPA's definition of *pollution* seems general so as to be useless, perhaps because it fails to add material on what such a broadly inclusive term may cover. Definitions from other sources presented similar problems. One of the problems with defining pollution is that is has many manifestations (see Figure 1.1).

The manifestations of pollution shown in Figure 1.1 attempts to illustrate what pollution is but also works to confound the difficulty. The main problem with the manifestations of pollution is that they are too general. Beyond its many manifestations, why is pollution so difficult to define? The element of personal judgment mentioned earlier contributes to the difficulty. Anyone who seriously studies pollution quickly realizes that there are five major categories of pollution each with their own accompanying subsets (types); these are shown in Table 1.1 (The type of pollution listed is defined and discussed later).

The categories and types of pollution listed in Table 1.1 can also be typed or classified as to whether they are *biodegradable* (subject to decay by microorganisms) or *nonbiodegradable* (cannot be decomposed by microorganisms). Moreover, nonbiodegradable pollutants can also be classified as *primary pollutants* (emitted directly into the environment) or *secondary pollutants* (result of some action of a primary pollutant).

Key Terms

To understand the basic concepts of environmental pollution, you'll need to learn the core vocabulary. Here are some of the key terms that were used in this chapter and are discussed in greater detail throughout this text. Remember what Voltaire said: "If you wish to converse with me, please define your terms."

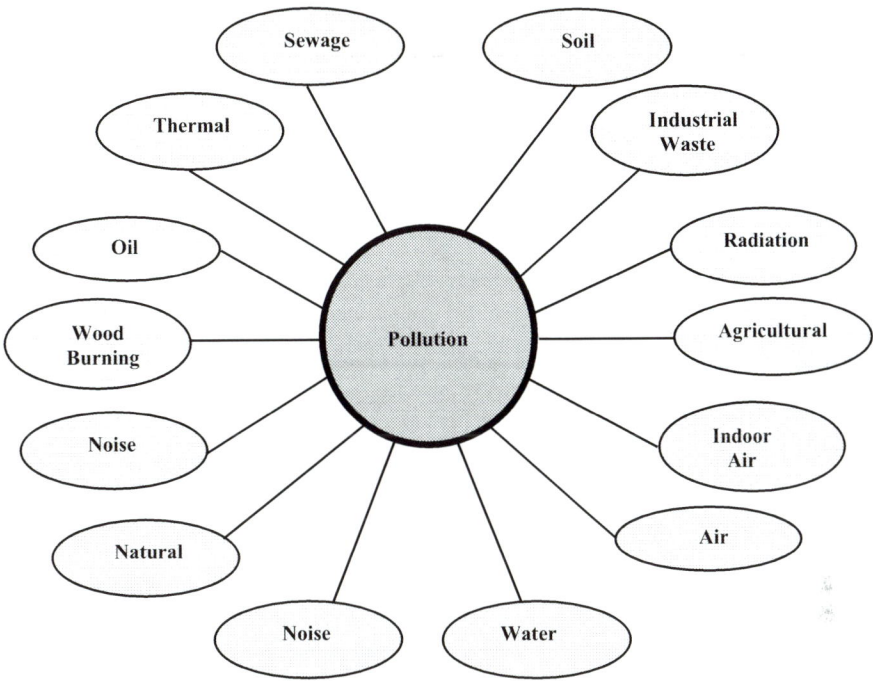

FIGURE 1.1 Manifestations of pollution.

Scientists (and environmental engineers) gather information and draw conclusions about the workings of the environment by applying the *scientific method*, a way of gathering and evaluating information. It involves observation, speculation (hypothesis formation), and reasoning.

The science of pollution may be divided among the study of air pollution (atmosphere), water pollution (hydrosphere), soil pollution (geosphere), and life (biosphere). Again, the emphasis in this text is on the first three, air, water, and soil, because without any of these, life as we know it is impossible.

The *atmosphere* is the envelope of thin air around the earth. The role of the atmosphere is multifaceted: (1) it serves as a reservoir of gases; (2) it moderates the earth's temperature; (3) it absorbs energy and damaging ultraviolet (UV) radiation from the sun; (4) it transports energy away from equatorial regions; and (5) it serves as a pathway for vapor-phase movement of water in the hydrologic cycle. *Air*, the mixture of gases that constitutes the earth's atmosphere, is by volume at sea level, 78.0% nitrogen, 21.0% oxygen, 0.93% argon, and 0.03% carbon dioxide, together with very small amounts of numerous other constituents.

The *hydrosphere* is the water component of the earth, encompassing the oceans, seas, rivers, streams, swamps, lakes, groundwater, and atmospheric water vapor. *Water* (H_2O) is a liquid that when pure is without color, taste, or odor. It covers 70% of the earth's surface and occurs as standing (oceans, lakes) and running (rivers, streams) water, rain, and vapor. It supports all forms of Earth's life.

The *geosphere* consists of the solid earth, including *soil*—the *lithosphere*, the topmost layer of decomposed rock and organic matter that usually contains air, moisture, and nutrients, and can therefore support life.

The *biosphere* is the region of the earth and its atmosphere in which life exists, an envelope extending from up to 6,000 m above to 10,000 m below sea level. Living organisms and the aspects of the environment pertaining directly to them are called *biotic* (biota), and other portions, nonliving part, of the physical environment are *abiotic*.

A series of biological, chemical, and geological processes by which materials cycle through ecosystems are called *biogeochemical cycles*. We are concerned with two types, the *gaseous* and

TABLE 1.1
Categories and Types of Pollution

Pollution Categories	Type of Pollution
Air pollution	Acid rain
	Chlorofluorocarbon
	Global warming
	Global dimming
	Global distillation
	Particulate
	Smog
	Ozone depletion
Water pollution	Eutrophication
	Hypoxia
	Marine pollution
	Marine debris
	Ocean acidification
	Oil spill
	Ship pollution
	Surface runoff
	Thermal pollution
	Wastewater
	Waterborne diseases
	Water quality
	Water stagnation
Soil contamination	Bioremediation
	Electrical resistance heating
	Herbicide
	Pesticide
	Soil Guideline Values (SGVs)
Radioactive contamination	Actinides in the environment
	Environmental radioactivity
	Fission product
	Nuclear fallout
	Plutonium in the environment
	Radiation poisoning
	Radium in the environment
	Uranium in the environment
Others	Invasive species
	Light pollution
	Noise pollution
	Radio spectrum pollution
	Visual pollution

the *sedimentary*. Gaseous cycles include the carbon and nitrogen cycles. The main *sink* (the main receiving area for material: for example, plants are sinks for carbon dioxide) of nutrients in the gaseous cycle is the atmosphere and the ocean. The sedimentary cycles include sulfur and phosphorous cycles. The main sink for sedimentary cycles is soil and rocks of the earth's crust.

Formerly known as natural science, and now more commonly known as ecology, *ecology* is critical to the study of environmental science, as the study of the structure, function, and behavior of the natural systems that comprise the biosphere. The terms "ecology" and "interrelationship"

Introduction

are interchangeable; they mean the same thing. In fact, ecology is the scientific study of the inter-relationships among organisms and between organisms, and all aspects, living and non-living, of their environment.

Ecology is normally approached from two viewpoints: (1) the environment and the demands it places on the organisms in it or (2) organisms and how they adapt to their environmental conditions. An *ecosystem*, a cyclic mechanism, describes the interdependence of species in the living world (the biome or community) with one another and with their non-living (abiotic) environment. An ecosystem has physical, chemical, and biological components, as well as energy sources and pathways.

An ecosystem can be analyzed from a functional viewpoint in terms of several factors. The factors important in this discussion include: *biogeochemical cycles, energy,* and *food chains.* Each ecosystem is bound together by biogeochemical cycles through which living organisms use energy from the sun to obtain or "fix" non-living inorganic elements such as carbon, oxygen, and hydrogen from the environment, and transform them into vital food, which is then used and recycled. The environment in which a particular organism lives is a *habitat.* The role of an organism in a habitat is its *niche*.

Obviously, in a standard handbook on a major profession such as environmental engineering, we need to define the "niche" they (environmental engineers) play in their "habitat" (i.e., the role environmental engineers' play in their profession). Environmental engineers use the principles of engineering, biology, and chemistry to develop solutions to environmental problems. They are involved in efforts to improve recycling, waste disposal, public health, and water, soil, and air pollution control. More specifically, as is made clear in Figure 1.2, is an integration of science and engineering principles to improve the natural environment. Further, as shown in Figure 1.2,

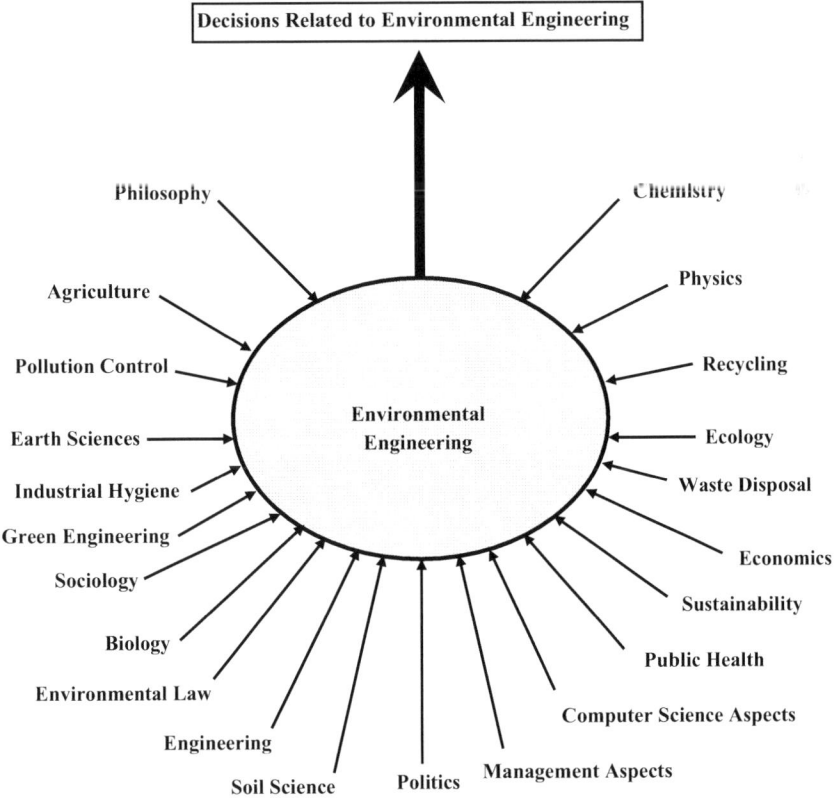

FIGURE 1.2 Components of environmental engineering/components of decision making.

environmental engineers provide healthy water, air, and land for human habitation and for other organisms. Although this text takes the proactive approach (engineer out the problem of pollution event before it occurs) the reality is environmental engineers also work to clean up pollution sites (the reactive mode, for sure). Notice that in Figure 1.2, the components of environmental engineering, I have included environmental law and sustainability. These two specialties have become not just buzzwords in today's environmental world, but also absolutely necessary skill sets in the practitioner of environmental engineers' toolbox to be successful. We could say (and I do) that to be well-rounded in practical knowledge and in practice, period, the environmental engineering practitioner must know environmental law and understand resource sustainability.

Environmental engineers are also concerned with finding plausible solutions in the field of public health, such as vector-borne (arthropod-borne) diseases, and implementing laws which promote adequate sanitation in urban, rural, and recreational areas.

To demonstrate, explain, and/or illustrate what environmental engineers do is to provide an example. Example 1.1 is just one hypothetical example of environmental professionals at work in the real world.

Example 1.1: Salmon and the Rachel River

The Rachel River, a hypothetical river system in the northwestern United States, courses its way through an area that includes a Native American Reservation. The river system outfalls to the Pacific Ocean and the headwaters begin deep and high within the Cascade Mountain Range of Washington State. For untold centuries, this river system provided a natural spawning area for salmon. The salmon fry thrived in the river, and eventually grew the characteristic dark blotches on their bodies and transformed from fry to parr. When the time came to make their way to the sea, their bodies now larger and covered with silver pigment, the salmon, now called smolt, inexorably migrated to the ocean, where they thrived until time to return to the river and spawn (about 4 years later). In spawning season, the salmon instinctively homed their way toward the odor generated by the Rachel River (their homing signal), and up the river to their home waters, as their life cycle instincts demand.

Before non-Native Americans (settlers) arrived in this pristine wilderness region, nature, humans, and salmon lived in harmony and provided for each other. Nature gave the salmon the perfect habitat; the salmon provided Native Americans with sustenance. Native Americans provided to both their natural world and the salmon the respect they deserved.

After the settlers came to the Rachel River Valley, changes began. The salmon still ran the river, humans still fed on the salmon, but circumstances quickly altered. The settlers wanted more land, and Native Americans were forced to give way; they were destroyed, or diseased and died, or forcibly moved to other places, to reservations, where the settlers did all they could to erase Native American beliefs and cultural inheritance.

The salmon still ran the river.

After the settlers drove out the Native Americans, the salmon continued to run, for a while. But more non-Native Americans poured into the area. As the area became more crowded, the salmon still ran, but now their home, their habitat, the Rachel River, started to show the effects of modern civilization's influence. The "civilized" practice and philosophy was "If I don't want it any more, it's trash. Throw it away" and the river provided a seemingly endless dump—out of the way, out of sight, out of mind. And they threw their trash away, all the mountains of trash they could manufacture, into the river.

The salmon still ran the river.

More time passed. More people moved in, and the more people, the bigger their demands. In its natural course, sometimes the river flooded, creating problems for the settler populations. Besides, everyone wanted power to maintain modern lifestyles—and hydropower poured down the Rachel River to the ocean constantly. So, they built flood control systems and a dam to convert hydropower to hydroelectric power. (Funny. The Native Americans didn't have a problem with flood control. When the river rose, they broke camp and moved to higher ground). Hydroelectric power? If you don't build your life around things, you don't need electricity to make them work.

Introduction

With the sun, the moon and the stars and their healthy, vital land at hand, who would want hydroelectric power? (Not the Native Americans).

The salmon still ran.

Building dams and flood control systems takes time, but humans, though impatient, have a way of conquering and using time (and anything else that gets in the way) to accomplish their tasks, goals, objectives—and construction projects. As the years passed, the construction moved on to completion, and finally ended. The salmon still ran—but in reduced numbers and size. Soon local inhabitants couldn't catch the quantity and quality of salmon they had in the past. When the inconvenience finally struck home, they began to ask, "Where are the salmon?"

But no one seemed to know. Obviously, the time had come to call in the scientists, the experts. So, the inhabitant's governing officials formed a committee and funded a study and hired some scientists to tell them what was wrong. "The scientists will know the answer. They'll know what to do," they said, and that was partly true. Notice they didn't try to ask the Native Americans. They also would have known what to do. The salmon had already told them.

The scientists came and studied the situation, conducted testing, tested their tests, and decided that the salmon population needed to increase. They determined increased salmon population could be achieved by building a fish hatchery, which would take the eggs from spawning salmon, raise the eggs to fingerling-sized fish, release them into specially built basins, and later, release them to restock the river.

A lot of science goes into the operation of a fish hatchery. It can't operate successfully on its own (though Mother Nature never has a serious problem with it when left alone) but must be run by trained scientists and technicians following a proven protocol based on biological studies of salmon life cycles.

When the time was right, the salmon were released into the river—meanwhile, other scientists and engineers realized that some mechanism had to be installed in the dam to allow the salmon to swim downstream to the ocean, and the reverse, as well. In salmon lives (since they are an anadromous species—they spend their adult lives at sea but return to freshwater to spawn), what goes down must go up (upstream). Those salmon would eventually need some way of getting back up past the dam and into home water, their spawning grounds. So, the scientists and engineers devised, designed, built and installed fish ladders in the dam, so that the salmon could climb the ladders, scale the dam and return to their native waters to spawn and die.

In a few seasons, the salmon again ran strong in the Rachel River. The scientists had temporarily—and at a high financial expenditure—solved the problem. Nothing in life or in Nature is static or permanent. All things change. They shift from static to dynamic, in natural cycles that defy human intervention, relatively quickly, without notice—like a dormant volcano, or the Pacific Rim tectonic plates. In a few years, local Rachel River residents noticed an alarming trend. Studies over a 5-year period showed that no matter how many salmon were released into the river, fewer and fewer returned to spawn each season.

So, they called in the scientists again. "Don't worry. The scientists will know. They'll tell us what to do."

The scientists came in, analyzed the problem, and came up with five conclusions:

1. The Rachel River is extremely polluted both from point and nonpoint sources.
2. The Rachel River Dam has radically reduced the number of returning salmon to the spawning grounds.
3. Foreign fishing fleets off the Pacific Coast are depleting the salmon.
4. Native Americans were removing salmon downstream before they even get close to the fish ladder at Rachel River Dam.
5. A large percentage of water is withdrawn each year from the river for cooling machinery in local factories. Large rivers with rapid flow rates usually can dissipate heat rapidly and suffer little ecological damage unless their flow rates are sharply reduced by seasonal fluctuations. This was not the case, of course, with the Rachel River. The large input of heated water from Rachel River area factories back into the slow-moving Rachel River created an adverse effect called *thermal pollution*. Thermal pollution and salmon do not mix. In the first place, increased water temperatures lowers the dissolved oxygen (DO) content by decreasing the solubility of oxygen in the river water, and warmer river water

also causes aquatic organisms to increase their respiration rates and consume oxygen faster, increasing their susceptibility to disease, parasites, and toxic chemicals. Although salmon can survive in heated water—to a point—many other fish (the salmon's food supply) cannot. Heated discharge water from the factories also disrupts the spawning process and kills the young fry.

The scientists prepared their written findings and presented them to city officials, who read them and were (at first) pleased. "Ah!" they said. "Now we know why we have fewer salmon!"

But their short-lived pleasure faded. They did indeed have the causal factors defined—but what was the solution? The scientists looked at each other and shrugged. "That's not my job," they said. "Call in the environmental engineers."

The salmon still ran, but not up the Rachel River to its headwaters.

Within days, the city officials hired an environmental engineering firm to study the salmon depletion problem. The environmentalists came up with the same causal conclusions as the scientists (which they also related to the city officials), but they also related the political, economic, and philosophical implications of the situation to the city powers. The environmental engineers explained that most of the pollution constantly pouring into the Rachel River would soon be eliminated when the city's new wastewater treatment plant came online, and that specific *point source pollution* would be eliminated. They explained that the state agricultural department and their environmental staff were working with farmers along the lower river course to modify their farming practices and pesticide treatment regimens to help control the most destructive types of *nonpoint source pollution*. The environmental engineers explained that the Rachel River dam's present fish ladder was incorrectly configured but could be modified with minor retrofitting.

They explained that the over-fishing by the foreign fishing fleets off the Pacific Coast was a problem that the Federal government was working to resolve with the governments involved. The environmental engineers explained that the State of Washington and the Federal Government were addressing the problem with the Native Americans fishing the down-river locations before the salmon ever reached the dam. Both governmental entities were negotiating with the local tribes on this problem, and the local tribes had pending litigation against the state and the federal government on who actually owned fishing rights to the Rachel River and the salmon.

For the final problem, thermal pollution from the factories making the Rachel River unfavorable for spawning, decreasing salmon food supply, and/or killing off the young salmon fry, the environmental engineers explained that to correct this problem, the outfalls from the factories would have to be changed—relocated. The environmental engineers also recommended the construction of a channel basin whereby the ready-to-release salmon fry could be released in a favorable environment, at ambient stream temperatures, and would have a controlled one-way route to safe downstream locations where they could thrive until time to migrate to the sea.

After many debates and many newspaper editorials, the city officials put the matter to a vote—and voted to fund the projects needed to solve the salmon problem in the Rachel River. It is important to note that some short-term projects are already showing positive signs of change, long-term projects are underway, and the Rachel River is on its way to recovery.

In short, environmental engineers are professionals who study to find "the" answer to a problem through scientific analysis and study. Their interest is in pure science. The environmental engineers can arrive at the same causal conclusions as the scientists but are also able to factor in socio-economic, political, and cultural influences as well.

But wait! It still isn't over. Concerns over the disruption of the wild salmon gene pool by hatchery trout are drawing attention from environmentalists, conservationists, and wildlife biologists. Hatchery- or farm-raised stock of any kind is susceptible to problems caused by, among other things, a lack of free genetic mixing, spread of disease, infection and parasites, and reinforcement of negative characteristics—when escaped hatchery salmon breed with wild salmon, the genetic strain is changed, diseases can be transmitted... many problems arise.

Let's fast-forward and continue our listing and definition of key terms. The following key terms, many listed in Table 1.1, are defined below.[1]

Introduction

- **Acid rain**: is any form of precipitation made more acidic from falling though air pollutants (primarily sulfur dioxide) and dissolving them.
- **Actinides in the environment**: refers to the sources, environmental behavior, and effects of radioactive actinides in the environment.
- **Air quality index**: is a standardized indicator of the air quality in a given location.
- **Atmospheric dispersion modeling**: is the mathematical simulation of how air pollutants disperse in the ambient atmosphere.
- **Bioremediation**: is any process that uses microorganisms, fungi, green plants, or their enzymes to return the natural environment altered by contaminants to its original condition.
- **Chlorofluorocarbons (CFCs)**: are synthetic chemicals that are odorless, nontoxic, non-flammable, and chemically inert.
- **Cleaner products**: Cleaner products or clean products refers to consumer and industrial products that are less polluting and less harmful to the environment and less toxic and less harmful to human health.
- **Electrical resistance heating remediation**: is an in situ environmental remediation method that uses the flow of alternating current electricity to heat soil and groundwater and evaporate contaminants.
- **Emerging pollution (contaminants); PPCPs**: are any synthetic or naturally occurring chemical or any microorganism that is not commonly monitored in the environment but has the potential to enter the environment and cause known or suspected adviser ecological and/or human health effects. PPCP—Pharmaceuticals and Personal Care Products comprise a very broad, diverse collection of thousands of chemical substances, including prescription and over-the-counter therapeutic drugs, fragrances, cosmetics, sun-screen agents, diagnostic agents, nutra-pharmaceuticals, biopharmaceuticals, and many others.
- **Environmental radioactivity**: is the study of radioactive material in the human environment.
- **Environmentally safe products, environmentally preferable products, or green products**: The terms environmentally safe products, environmentally preferable products, or green products refer to products that are less toxic and less harmful to human health and the environment when their polluting effects during their entire life cycle are considered.
- **Fission product**: are the atomic fragments left after large nucleus fissions.
- **Global dimming**: is the gradual reduction in the amount of global direct irradiance at the Earth's surface.
- **Global distillation (or grasshopper effect)**: is the geochemical process by which certain chemicals, most notably persistent organic pollutants (POPs), are transported from warmer to colder regions of the Earth.
- **Global warming**: is the long-term average temperature of the earth.
- **Herbicide**: is used to kill unwanted plants.
- **Indoor air quality**: is a term referring to the air quality, within and around buildings and structures, especially as it relates to the health and comfort of building occupants.
- **Life cycle analysis**: life cycle analysis is a study of the pollution generation characteristics and the opportunities for pollution prevention associated with the entire life cycle of a product or process. Any change in the product or process has implications for upstream stages (extraction and processing of raw materials, production, and distribution of process inputs) and for downstream stages (including the components of a product, its use, and its ultimate disposal.
- **Eutrophication**: is a natural process in which lakes receive inputs of plant nutrients as a result of natural erosion and runoff from the surrounding land basin.
- **Hypoxia**: is a phenomenon that occurs in aquatic environments as dissolved oxygen (DO) becomes reduced in concentration to appoint detrimental to aquatic orgasms living in the system.

- **Invasive species**: are non-indigenous species (e.g., plants or animals) that adversely affect the habitats they invade economically, environmental, or ecologically.
- **Light pollution**: is excessive or obtrusive artificial light (photo-pollution or luminous pollution).
- **Marine pollution**: occurs when harmful effects, or potentially harmful effects, can result from the entry into the ocean of chemicals, particles, industrial, agricultural, and residential waste, or the spread of invasive organisms.
- **Marine debris**: is human-created waste that has deliberately or accidentally become afloat in a waterway, lake, ocean, or sea.
- **Noise pollution**: is unwanted sound that disrupts the activity or balance of human or animal life.
- **Nuclear fallout**: is the residual radiation hazard form a nuclear explosion, so named because it "falls out" of the atmosphere into which it is spread during the explosion.
- **Ocean acidification**: is the on-going decrease in the pH of the Earth's oceans, caused by their uptake of anthropogenic carbon dioxide form the atmosphere (Caldeira and Wickett, 2003).
- **Oil spill**: is the release of a liquid petroleum hydrocarbon into the environment due to human activity and is a form of pollution.
- **Ozone depletion**: while ozone concentrations vary naturally with sunspots, the seasons, and latitude, these processes are well understood and predictable. Scientists have established records spanning several decades that detail normal ozone levels during these natural cycles. Each natural reduction in ozone levels has been followed by a recovery. Recently, however, convincing scientific evidence has shown that the ozone shield is being depleted well beyond changes due to natural processes.
- **Particulate**: normally refers to fine dust and fume particles; travels easily through air.
- **Pesticide**: is a substance or mixture of substances used to kill pests.
- **Plutonium in the environment**: is an article (part) of the actinides series in the environment.
- **Pollution/pollutants**: pollution and pollutants refer to all nonproduct output, irrespective of any recycling or treatment that may prevent or mitigate releases to the environment (includes all media).
- **Pollution prevention**: pollution prevention refers to activities to reduce or eliminate pollution or waste at its source or to reduce its toxicity. It involves the use of processes, practices, or products that reduce or eliminate the generation of pollutants and waste or that protect natural resources through conservation or more efficient utilization. Pollution prevention does not include recycling, energy recovery, treatment, and disposal. Some practices commonly described as in-process recycling may qualify as pollution prevention.
- **Radiation poisoning**: is a form of damage to organ tissue due to excessive exposure to ionizing radiation.
- **Radio spectrum pollution**: is the straying of waves in the radio and electromagnetic spectrums outside their allocations that cause problems for some activities.
- **Radium and radon**: radium is highly radioactive and its decay product, radon gas, is also radioactive.
- **Resource protection**: In the context of pollution prevention, resource protection refers to protecting natural resources by avoiding excessive levels of waste and residues, minimizing the depletion of resources, and assuring that the environment's capacity to absorb pollution is not exceeded.
- **Smog**: term used to describe visible air pollution; a dense, discolored haze containing large quantities of soot, ash, and gaseous pollutants such as sulfur dioxide and carbon dioxide.
- **Soil Guideline Values (SGVs)**: are a series of measurements and values used to measure contamination of the soil.

- **Source reduction**: Source reduction is defined in the *Pollution Prevention Act of 1990* as "any practice which (1) reduces the amount of any hazardous substance, pollutant, or contaminant entering any waste stream or otherwise released into the environment (including fugitive emissions) prior to recycling, treatment, and disposal; and (2) reduces the hazards to public health and the environment associated with the release of such substances, pollutants, or contaminants. The term includes equipment or technology modifications, process or procedure modifications, reformulations or design of products, substitution of raw materials, and improvements in housekeeping, maintenance, training, or inventory control." Source reduction does not entail any form of waste management (e.g., recycling and treatment). The act excludes from the definition of source reduction any practice which alters the physical, chemical, or biological characteristics or the volume of a hazardous substance, pollutant, or contaminant through a process or activity which itself is not integral to and necessary for the production of a product or the providing of a service.
- **Surface runoff**: is the water flow which occurs when soil is infiltrated to full capacity and excess water, form rain, snowmelt, or other sources flows over the land.
- **Thermal pollution**: is increase in water temperature with harmful ecological effects on aquatic ecosystems.
- **Toxic chemical use substitution**: toxic chemical use substitution or material substitution describes replacing toxic chemicals with less harmful chemicals even though relative toxicities may not be fully known. Examples include substituting a toxic solvent in an industrial process with a less toxic chemical and reformulating a product to decrease the use of toxic raw materials or the generation of toxic by-products. The term also refers to efforts to reduce or eliminate the commercial use of chemicals associated with health or environment risks, including the substitution of less hazardous chemicals for comparable uses and the elimination of a particular process or product from the market without direct substitution.
- **Toxics use reduction**: toxics use reduction refers to the activities grouped under source reduction where the intent is to reduce, avoid, or eliminate the use of toxics in processes and products so that the overall risks to the health of workers, consumers, and the environment are reduced without shifting risks between workers, consumers, or parts of the environment.
- **Uranium**: is a naturally occurring element found in low levels within all rock, soil, and water.
- **Visual pollution**: is the unattractive or unnatural (human-made) visual elements of a vista, a landscape, or any other thing that a person might not want to look at.
- **Waste**: In theory, waste applies to the non-product output of processes and discarded products, irrespective of the environmental medium affected. In practice, since the passage of the RCRA, most uses of waste refer exclusively to the hazardous and solid wastes regulated under RCRA and do not include air emissions or water discharges regulated by the Clean Air Act or the Clean Water Act.
- **Waste minimization**: waste minimization initially included both treating waste to minimize its volume or toxicity and preventing the generation of waste at the source. The distinction between treatment and prevention became important because some advocates of deceased waste generations believed that an emphasis on waste minimization would deflect resources away from prevention towards treatment. In the current RCRA biennial report, waste minimization refers to source reduction and recycling activities and now excludes treatment and energy recovery.
- **Waste reduction**: this term is used by the Congressional Office of Technology Assessment synonymously with source reduction. However, many groups use the term to refer to waste minimization. Therefore, determining the use of waste reduction is important when it is encountered.

- **Wastewater**: is the liquid wastestream primarily produced by the five major sources: human and animal waste, household wastes, industrial waste, stormwater runoff, and groundwater infiltration.
- **Waterborne diseases**: are caused by pathogenic microorganisms which are directly transmitted when contaminated drinking water is consumed.
- **Water quality**: is the physical, chemical, and biological characteristics of water.
- **Water stagnation**: is water at rest; it allows for the growth of pathogenic microorganisms to take place.

The preceding list of key terms and definitions along with Figures 1.1 and 1.2, Table 1.1 provide some help in defining pollution and point out that environmental engineers make decisions related to preventing, controlling, and/or mitigating pollution and its effects. But we are still trying to nail down a definitive meaning of pollution. Accordingly, to clear the fog, maybe it will help to look at a few more definitions for the term pollution.

According to E. A. Keller (1988), pollution is "a substance that is in the wrong place in the environment, in the wrong concentrations, or at the wrong time, such that it is damaging to living organisms or disrupts the normal functioning of the environment" (p. 496). Again, this definition seems incomplete, though it makes the important point that often pollutants are or were useful—in the right place, in the right concentrations, at the right time.

Let's take a look at some of the definitions for pollution that have been used over the years.

1. Pollution is the impairment of the quality of some portion of the environment by the addition of harmful impurities.
2. Pollution is something people produce in large enough quantities that it interferes with our health or well-being.
3. Pollution is any change in the physical, chemical, or biological characteristics of the air, water, or soil that can affect the health, survival, or activities of human beings or other forms of life in an undesirable way. Pollution does not have to produce physical harm; pollutants such as noise and heat may cause injury, but more often cause psychological distress, and aesthetic pollution such as foul odors and unpleasant sights affects the senses.

Pollution that initially affects one medium frequently migrates into the other media; air pollution falls to earth, contaminating soil and water; soil pollutants migrate into groundwater; acid precipitation, carried by air, falls to earth as rain or snow, altering the delicate ecological balance in surface waters.

In our quest for the definitive definition, the source of last resort was consulted: the common dictionary. According to one dictionary, pollution is a synonym for contamination. A contaminant is a pollutant—a substance present in greater than natural concentrations as a result of human activity and having a net detrimental effect upon its environment or upon something of value in the environment. Every pollutant originates from a source. A receptor is anything that is affected by a pollutant. A sink is a long-time repository of a pollutant. What is actually gained from the dictionary definition is that since pollution is a synonym for contamination, contaminants are things that contaminate the three environmental mediums (air, water, and/or soil) in some manner. The bottom line is that we have come full circle; the impact and the exactness of what we stated in the beginning of this text: "Pollution is a judgment call."

Why is pollution a judgment call? It is a judgment call because people's opinions differ in what they consider to be a pollutant based on their assessment of benefits and risks to their health and their economic well-being. For example, visible and invisible chemicals spewed into the air or water by an industrial facility might be harmful to people and other forms of life living nearby. However, if the facility is required to install expensive pollution controls, forcing the industrial facility to shut down or move away, workers who would lose their jobs and merchants who would lose their livelihoods might feel that the risks from polluted air and water are minor weighed against the benefits of profitable employment. The same level of pollution can also affect two people quite differently. Some forms of air pollution, for example, might cause slight irritation to a healthy person but cause life-threatening problems to someone with chronic obstructive pulmonary disease (COPD) like emphysema. Differing priorities lead to differing perceptions of pollution (concern at the level of pesticides in foodstuffs generating the need for wholesale banning of insecticides is unlikely to help the starving). No one wants to hear that cleaning up the environment is going to have a negative impact on them. The fact is public perception lags behind reality because reality is sometimes unbearable.

POLLUTION: EFFECTS EASY TO SEE, FEEL, TASTE OR SMELL

Although pollution is difficult to define, its adverse effects are often relatively easy to see.[2] For example, some rivers are visibly polluted, have an unpleasant odor, or apparent biotic population problems (fish kill, for example). The infamous Cuyahoga River in Ohio became so polluted it twice caught on fire from oil floating on its surface. Air pollution from automobiles and unregulated industrial facilities is obvious. In industrial cities, soot often drifts onto buildings and clothing and into homes. Air pollution episodes can increase hospital admissions and kill people sensitive to the toxins. Fish and birds are killed by unregulated pesticide use. Trash is discarded in open dumps and burned, releasing impurities into the air. Fumes generated in city traffic plague commuters daily. Ozone levels irritate the eyes and lungs. Sulfate hazards obscure the view. And it is important to point out that alternative energy sources championed by environmentalists and others, such as wind turbines and solar farms, are also pollutants, contaminants, or environmental hazards (depending on how you define them). You may have questions about this statement. How can it be true that wind and solar cause harm to the environment? Well, if we could ask the bald eagle who makes the mistake of flying into rotating turbine blades or flies too close to solar collectors/reflectors that cook it to a crisp, then the answer would be apparent. The answer is apparent to the bald eagle and other unlucky avian creatures that come into contact with these green energy sources.

Even if you are not in a position to see pollution, you are still made aware of it through the news or social media. How about the 1984 Bhopal Incident, the 1986 Chernobyl nuclear plant disaster, the 1991 pesticide spill into the Sacramento River, the Exxon Valdez, or the 1994 oil spill in Russia's Far North? Most of us do remember some of them, even though most of us did not directly witness any of these tragedies. Events, whether man-made (Bhopal) or natural disasters (Mt. St. Helens erupting), sometimes impact us directly but if not directly, they still get our attention. Worldwide, we see constant reminders of the less dramatic, more insidious, continued, and increasing pollution of our environment. We see or hear reports of dead fish in stream beds, or litter in national parks or college campuses; or decaying buildings and bridges, leaking landfills, and dying lakes and forests. On the local scale, air quality alerts may have already begun in your community.

Some people experience pollution more directly, firsthand—what we call "in your face," "in your nose," "in your mouth," "in your skin" type pollution. Consider the train and truck accidents that result in the release of toxic pollutants that force us to evacuate our homes. We become ill after drinking contaminated water or breathing contaminated air or eating contaminated (salmonella-laced) peanut butter products. We can no longer swim at favorite swimming holes because of sewage contamination. We restrict fish, shellfish, and meat consumption because of the presence of harmful chemicals, cancer-causing substances, and hormone residues. We are exposed to nuclear contaminants released into the air and water from uranium-processing plants.

Unexpected Pollution

On 9/11, if you were not present in New York City or the Pentagon or in that Pennsylvania farm field and not up close and personal with any of these events, then you might not be aware of the catastrophic unleashing of various contaminants into the environment because of the plane crashes. Or maybe you did not have access to television coverage where it clearly showed the massive cloud of dust, smoke and other ground-level debris engulfing New York City. Maybe you have not had a chance to speak with any of the emergency response personnel who climbed through the contaminated wreckage looking for survivors; responders who were exposed to chemicals and various hazardous materials, many of which we still are not certain of their exact nature. Days later when rescue turned to recovery, you may not have noticed personnel garbed in moon-suits (level A hazmat response suits) with instruments attempting to sample and monitor the area for harmful contaminants. If you had not witnessed or known about any of the reactions after the 9/11 event, then it might be reasonable to assume that you might not be aware that these were indeed pollution-emitting events.

In addition to terrorism, vandalism, ecophagy (grey goo scenario), and other deliberate acts, we pollute our environment with apparent abandon. Many of us who teach various environmental science and health subjects to undergraduate and graduate students often hear students complain that the human race must have a death wish. Students quickly adopt this view based on their research and intern work with various environmental-based service entities. During their exposure to all facets of pollution—air, water, and soil contamination, they come to understand that everything we do on Earth contributes to pollution of some sort or another to one or all of the three environmental mediums.

Science and technology notwithstanding, we damage the environment through the use, misuse, and abuse of technology. Frequently we use technological advances before we fully understand their long-term effects on the environment. We weigh the advantages a technological advance can give us against environmental effects and discount the importance of the environment, through greed, hubris, lack of knowledge, and/or through absolute stupidity. We often only examine short-term plans without fully developing how problems may be handled years later. We assume that when the situation becomes critical, technology will be there to fix it. Recently, common verbiage from the so-called experts professes that every problem has a solution and what is needed today is innovation, innovation, and more innovation. No, what is needed first is discovery, then invention, and then finally innovation. We simply believe that scientists and engineers will eventually figure out how to prevent or mitigate pollution of Mother Earth; thus, we ignore the immediate consequences of our technological abuse and the contaminants emanating from such.

Consider this: while technological advances have provided us with nuclear power, the light bulb and its energy source, plastics, the internal combustion engine, air conditioning and refrigeration, iPads, tablets, laptops, digital cameras, flat screen TVs (and scores of other advances that make our modern lives pleasant and comfortable) these advances have affected the earth's environment in ways we did not expect, in ways we eventually come to deplore, and in ways we may not be able to live with. In this Handbook, the argument is made that this same science and technology that created or exacerbated pollution events, can, in turn, be used to prevent or mitigate the misuse of science and technology.

Pollution and Environmental Science/Health/Engineering

In order to prevent and/or mitigate pollution events, highly trained interdisciplinary-trained environmental engineers are needed to monitor air, water, and soil quality. Generally, professionals responsible for environmental pollution monitoring, prevention or control are thoroughly trained in several aspects of environmental science and/or environmental health.

To precisely define *Environmental Science*, as an interdisciplinary study of how the earth works, how we are affecting the earth's life-support systems (environment), and how to deal with the environmental problems we face, we must first break down the phrase and look at both terms separately. The *environment* includes all living and nonliving (such as air, soil, and water) things that influence organisms. *Science* is the observation, identification, description, experimental investigation, and theoretical explanation of natural phenomena. When we combine the two, we are left with a complex interdisciplinary study that must be defined both narrowly and broadly—and then combined—to allow us an accurate definition.

The narrow definition of *Environmental Science* is the study of the human impact on the physical and biological environment of an organism. In this sense, environmental scientists are interested in determining the effects of pesticides on croplands, learning how acid rain affects vegetation, or determining the impact of introducing an exotic species of game fish into a pond or lake, and so on.

Beginning in the early 1960s, environmental science evolved out of the studies of natural science, biology, ecology, conservation, and geography. Increasing awareness of the interdependence that exists among all the disparate elements that make up our environment led to the field of study that contains aspects of all of those elements. While well-trained and experienced environmental scientists are generalists, and while they may have concentrated their study on a particular specialty, solidly trained environmental scientists have one thing in common: they are well grounded in biological and physical ideas combined with ideas from the social sciences—sociology, economics, and political science, into a new, interdisciplinary field: Environmental Science. We can say that a fully trained environmental engineer is well-grounded in environmental science.

Environmental health practitioners, like environmental scientists and environmental engineers, are trained in the major aspects of environmental science; however, they are also concerned with all aspects of the natural and human-built environment that may affect human health. Unlike the relatively new environmental science and engineering professions, the environmental health profession has its modern-day roots in the sanitary and public health movement of the United Kingdom in the 1880s. Environmental health practitioners address human-health-related aspects of both the natural and the human-made environment. Environmental health concerns are shown in Figure 1.3. Notice that the environmental health concerns shown in Figure 1.3 have much in common with the concerns of environmental scientists and environmental engineers.

In their broadest sense, environmental science, environment health, and environmental engineering encompass the social and cultural aspects of the environment. As a mixture of several traditional sciences, political awareness, and societal values, environmental science/health/engineering demands an examination of more than the concrete physical aspects of the world around us—and

FIGURE 1.3 Environmental health concerns.

many of those political, societal, and cultural aspects are far more slippery (the so-called "feel good" aspects) than what we can prove as scientific fact.

In short, we can accurately say that environmental science, environmental health, and environmental engineering are pure sciences because they include the study of all the mechanisms of environmental processes: The study of the air, water, and soil. But they are also an applied science because they examine problems with the goal of contributing to their solution: The study of the effects of human endeavors and technology thereon. Obviously, to solve environmental problems and understand the issues, environmental scientists, environmental health, and environmental engineering practitioners need a broad base of information from which to draw.

Again, the environment in which we live has been irreversibly affected by advancements in technology—and has been affected for as long as humans have wielded tools to alter their circumstances. As a result of rapid industrialization, over-population, and other human activities like deforestation for agriculture (and the practice of agriculture itself), Earth has become loaded with diverse pollutants that were released as by-products. We will continue to alter our environment to suit ourselves as long as we remain a viable species, but to do so wisely, we need to closely examine what we do and how we do it.

ENVIRONMENTAL POLLUTION AND TECHNOLOGY: THE CONNECTION

As long as capitalism drives most modern economies, people will desire material things—precipitating a high level of consumption. For better or for worse, the human desire to lead the "good life" (which Americans may interpret as a life enriched by material possessions) is a fact of life. Arguing against someone who wants to purchase a new, modern home with all the amenities, who wants to purchase the latest, greatest automobile, is difficult. Arguing against the person wanting to make a better life for his or her children by making sure they have all they need and want to succeed in their chosen pursuit is even harder. How do you argue against such goals with someone who earns his or her own way and spends his or her hard-earned money at will? Look at the tradeoffs, though. The tradeoff often affects the environment. That new house purchased with hard-earned money may sit in a field of radon-rich soil or on formerly undeveloped land. That new SUV may get only 8 miles to the gallon. The boat they use on weekends gets even worse mileage and exudes waste into the local lake, river, or stream. Their weekend retreat on the five wooded acres is part of the watershed of the local community and disturbs breeding and migration habitat for several species.

The environmental tradeoffs never enter the average person's mind. Most people don't commonly think about it. In fact, most of us don't think much about the environment until we damage it, until it becomes unsightly until it is so fouled that it offends us. People can put up with a lot of environmental abuse, especially with our surroundings—until the surroundings no longer please them. We treat our resources the same way. How often do we think about the air we breathe, the water we drink, and the soil our agribusiness conglomerates plant our vegetables in? Certainly not often enough; not as much as we should.

The typical attitude toward natural resources is often deliberate ignorance. Only when someone must wait in line for hours to fill the vehicle gas tank does gasoline become a concern. Only when he can see—and smell—the air we breathe, and cough when we inhale do air becomes a visible resource. Water, the universal solvent, causes no concern (and very little thought) until shortages occur, or until it is so foul that nothing can live in it or drink it. Only when we lack water, or the quality is poor do we think of water as a resource to "worry" about. Is soil a resource or is it "dirt?" Unless you farm, or plant a garden, soil is only "dirt." Whether you pay any heed to the soil/dirt debate depends on what you use soil for—and on how hungry you are.

Resource utilization and environmental degradation are tied together. While people depend on resources and must use them, this use impacts the environment. A **resource** is usually defined as anything obtained from the physical environment that is of use to man—the raw materials that support life on earth. Some resources, such as edible growing plants, water (in many places), and fresh

Introduction

air, are directly available to man. But most resources, like coal, iron, oil, groundwater, game animals and fish are not. They become resources only when man uses science and technology to find them, extract them, process them, and convert them, at a reasonable cost, into usable and acceptable forms. Natural gas, found deep below the earth's surface, was not a resource until the technology for drilling a well, fracking and installing pipes to bring it to the surface became available. For centuries, man stumbled across stinky, messy pools of petroleum and had no idea of its potential uses or benefits. When its potential was realized, man exploited petroleum by learning how to extract it and convert (refine) it into heating oil, gasoline, sulfur extract, road tar and other products.

Earth's natural resources and processes sustain other species, and we are known as *Earth's Natural Capital*. This includes air, water, soil, forests, grasslands, wildlife, minerals, and natural cycles. Societies are the primary engines of resource use, converting materials and energy into wealth, delivering goods and services, and creating waste or pollution. This provision of necessities and luxuries is often conducted in ways that systematically degrade the Earth's natural capital—the ecosystems that support all life.

Excluding *perpetual resources* (solar energy, tides, wind, and flowing water) two different classes (types) of resources are available to us: renewable and nonrenewable (see Figure 1.4). *Renewable resources* (fresh air, fresh water, fertile soil, plants, and animals via genetic diversity) can be depleted in the short run if used or contaminated too rapidly, but normally will be replaced through natural processes in the long run. Water is a good example. Water is a renewable resource. The amount of water on earth is constant, although its distribution is not. Water that is not available for safe use (i.e., contaminated) is of no value to humankind. Moreover, even though impurities are left behind when water evaporates, the water cycle does not ensure that clean water is always available. Thus, water must be managed; we must sample, monitor, and test it to ensure its safety for consumption.

Because renewable resources are relatively plentiful, we often ignore, overlook, destroy, contaminate, and/or mismanage them.

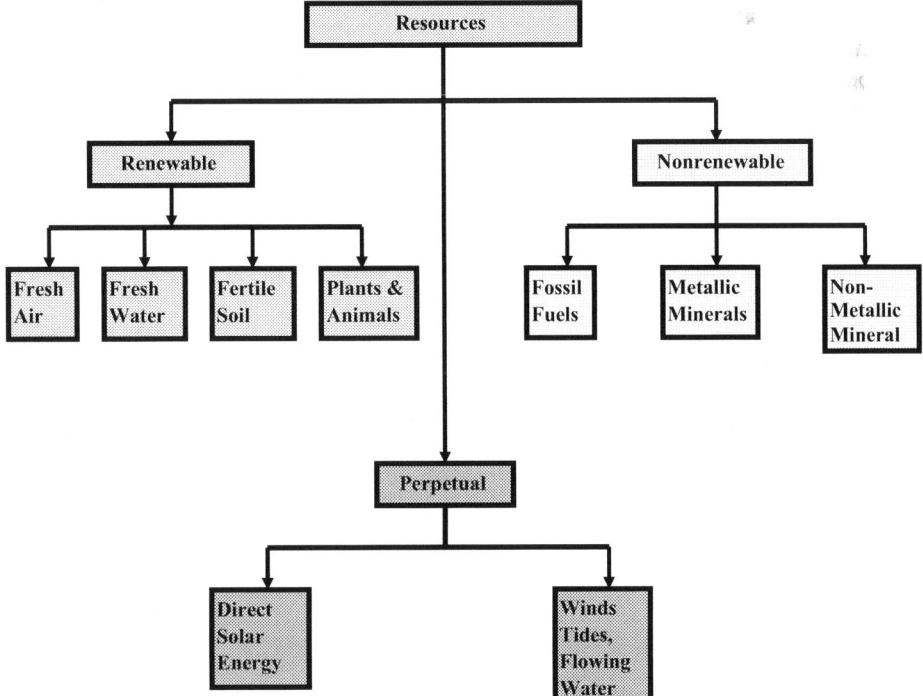

FIGURE 1.4 Major types of resources.

Mismanage? Yes. Classifying anything as "renewable" is a double-edged sword. Renewable resources are renewable only to a point. Timber or grass used for grazing must be managed for *maximum sustainable yield* (the highest rate at which a renewable resource can be used without impairing or damaging its ability to be fully renewed). If timber or grass yield exceeds this rate, the system gives ever-diminishing returns. Recovery is complicated by the time-factor, which is life cycle dependent. Grass can renew itself in a season or two. Timber takes decades. Any length of time is problematic when people get impatient.

Remember, one of the contributing factors to the plight of the Rachel River salmon was over-fishing. When a fishery is pushed past its limit, if the catch is maintained by collecting greater and greater numbers of younger salmon, no increase is possible. If the same practices are used on a wild species, extinction can result. We have no more passenger pigeons, heath hens, Carolina parakeets, dodos, solitaires, or great auks—and many other species are at risk, right now, including the Attwater Prairie Chicken.

Exceeding the maximum sustainable yield is only the tip of the iceberg—other environmental, social, and economic problems may develop. Let's look at *overgrazing* (depleting) grass on livestock lands. The initial problem occurs when the grass and other grazing cover is depleted. But secondary problems kick in fast. Without grass, the soil erodes quickly. In very little time, so much soil is gone that the land is no longer capable of growing grass—or anything else. Productive land converted to non-productive deserts (*desertification*) is a process of *environmental degradation*—and it impacts social and economic factors. Those who depend on the grasslands must move on, and moving on costs time, energy, and money—and puts more land at risk. Should the same level of poor stewardship of land resources continue on more acreage?

Environmental degradation is not limited to salmon and grass. Let's look at a few other examples. Along with over-fishing and overgrazing, land can also be over-cultivated. Intense over-cultivation reduces soil nutrients and increases erosion to the point where agricultural productivity is reduced, leading to over-fertilization, which eventually damages water supply. If irrigation of agricultural lands proceeds without proper drainage, the excessive accumulation of water or salts in the soil decreases productivity. Environmental degradation takes place when trees are removed from large areas without adequate re-planting. The results: the destruction of wildlife habitat, increased soil erosion, and flooding. Habitat fragmentation is another problem, related to habitat destruction. When the habitat is fragmented, species that require distance from human activity are affected. Take, for example, both Greater and Lesser Prairie Chickens in Kansas.

Radiotelemetry research demonstrates that prairie chickens are extremely sensitive to human activity. The birds seldom use sand sage within a quarter mile of an inhabited house; a house built on a one-acre site actually eliminates 160 acres of habitat for the birds.

Human habitation isn't the only factor, however. Natural gas compression facilities—and southwest Kansas has a bunch—are noisy, clanging affairs, usually a couple of acres in size. The birds won't use habitat within half a mile of these areas—another 640 acres down the tubes. Lesser chickens seldom venture within a mile of a coal-fired power plant, even though the sand sage habitat surrounding it may be the best on the range. A 30-acre power plant chews up 2,500 additional acres of chicken habitat. In addition, lesser chickens rarely nest or raise broods in habitat blocks <2,000 to 4,000 acres in size, nor do the birds frequent habitats along well-traveled roads. Do the math, and pretty soon you see the magnitude of the problems associated with preserving or supplying large, open blocks of the right habitat for chickens (Taylor, 2005).

Land is often environmentally degraded when a metropolitan area expands. In high-growth areas, productive land is covered with concrete, asphalt, buildings, or water or silt to such an extent that agricultural productivity declines and wildlife habitat is lost.

Nonrenewable resources such as copper, coal, tin, and oil, among many others (see Figure 1.4) have built up or evolved during various geological time-spans. They can't be replaced at will—only over a similar time-span. In this age of advanced technology, we often hear that (for example) when high-grade tin ore runs out (when 80% of its total estimated supply has been removed and used) low-grade tin ore (the other 20%) will become economically workable. This erroneous view neglects the facts of energy resource depletion and increasing pollution with lower-grade burdens. In short, to find, to extract, and to process the remaining 20% generally costs more than the result is worth. Even with unlimited supplies of energy (impossible according to the *Laws of Thermodynamics*, what if we could extract that last 20%? When it is gone, nothing is going to bring it back except time measured in centuries and millennia, paired with the elements that produce the resource.

Advances in technology have allowed us to make great strides in creating "the good life." These same technological advances have increased environmental degradation. But not all the news is bad. Technological advances have also let us (via recycling and reuse) to conserve finite resources—aluminum, copper, iron, plastics, and glass, for example. *Recycling* involves collecting household waste items (aluminum beverage cans, for example) and reprocessing usable portions. *Reuse* involves using a resource over and over in the same form (refillable beverage bottles, water).

We discussed the so-called "good life" earlier: modern homes, luxury cars and boats, the second home in the woods. With the continuing depletion of natural resources, prices must be forced upwards until economically, attaining "good life," or even gaining a foothold toward it, becomes difficult or impossible—and maintaining it becomes precarious. Ruthless exploitation of natural resources and the environment—overfishing a diminishing species (look at countless marine species populations, for example), intense exploitation of energy and mineral resources, cultivation of marginal land without proper conservation practices, degradation of habitat by unbalanced populations or introduced species, and the problems posed by further technological advances—will result in environmental degradation that will turn the "good life" into something we don't want to even think about. Our prevailing attitude of fly now, and pay later is, along with cowboy science, not pertinent here.

So—what's the answer? Are we looking for the bluebird? What are we to do? What should we do? Can we do anything? Should we even care or think about it?

Well, as is pointed out in Example 1.2, in order to preserve our natural resources and prevent environmental pollution, there are those, like Garrett Hardin, for example, who would have us privatize the "commons" so to speak,

Example 1.2: Tragedy of the Commons Revisited

Garrett Hardin's influential article, "The Tragedy of the Commons," first published in the journal *Science* in 1968, describes a dilemma in which multiple individuals acting independently in their own self-interest (sounds so American, does it not?) can ultimately, through over-exploitation, destroy a shared limited resource even when it is clear that it is not in anyone's long term interest for this to happen.

With regard to environmental pollution, Hardin points out that the tragedy of the commons, in a reverse way, is not taking something from the commons, but of putting something into it (e.g., sewage, or chemical, radioactive, and heat wastes in water; noxious and dangerous contaminants into the atmosphere; and various forms of visual pollution in line of sight). We can't readily fence the air, water, and soil we depend on; thus, the tragedy of the commons must be prevented by coercive laws or taxing devices that make it cheaper for the polluter to treat his pollutants than discharge them untreated. Hardin blames the pollution problem as a consequence of population.

To avoid the pollution problems Hardin alludes to in his "commons," we must seek means or ways to "prevent fouling our own nest" (Hardin, 1968). In light of this view, some would have us all "return to nature." Those people suggest returning to Thoreau's Walden Pond on a large scale, to give up the "good life," to which we have become accustomed. They think that giving up the cars, boats, fancy homes, the bulldozers that make construction and farming easier, the pesticides that protect our crops, the medicines that improve our health and save our lives—the myriad material improvements that make our lives comfortable and productive—will solve the problem. Is this approach the answer—or even realistic? To a small (vocal and impractical) minority, it is—although, for those who realize how urban Walden Pond was, the idea is amusing. To the rest of us? Get real! This is a pipe dream, founded in romance, not logic. It cannot, should not, and will not happen. We can't abandon ship—we must prevent the need for abandoning our society from ever happening. Properly managed, technological development is a boon to civilization and will continue to be. Technological development isn't the problem—improper use of technology is. Using technology in ways that degrade our environment by introducing undesirable changes in our ecosystems is absolutely untenable—we must prevent this from occurring. But, at the same time, we must continue to make advances in technology, we must find further uses for technology, and we must learn to use technology for the benefit of mankind and the environment. Technology and the environment must work hand in hand, not stand opposed. We must also foster respect for, and care for, what we have left.

Just how bad are the problems of technology's influence on environment?

Major advances in technology have provided us with enormous transformation and pollution of the environment. While transformation is generally glaringly obvious (damming a river system, for example), as mentioned, "polluting" or "pollution" is not always as clear. Remember that to *pollute* means to impair the purity of some substance or environment. *Air pollution* and *water pollution* refer to alteration of the normal compositions of air and water (their environmental quality) by the addition of foreign matter (gasoline, sewage).

Technological practices that have contributed to environmental transformation and pollution include:

- Extraction, production, and processing of raw natural resources, such as minerals, with accompanying environmental disruption.
- Manufacturing enormous quantities of industrial products that consume huge amounts of natural resources and produce large quantities of hazardous waste and water/air pollutants.
- Agricultural practices result in intensive cultivation of land, irrigation of arid lands, drainage of wetlands, and application of chemicals.
- Energy production and use are accompanied by disruption and contamination of soil by strip mining, emission of air pollutants, and pollution of water by the release of contaminants from petroleum production, and the effects of acid rain.
- Transportation practices (particularly reliance on the airplane) that cause scarring of land surfaces from airport construction, emission of air pollutants, and greatly increased demands for fuel (energy) resources.
- Transportation practices (particularly reliance on automobiles) that cause loss of land by road and storage construction, emission of air pollutants and increased demand for fuel (energy) resources.

Throughout this handbook, we discuss the important aspects of the impact of technology on the environment.

ENVIRONMENTAL ADMINISTRATIVE LAWS AND REGULATIONS

We have spent some time discussing pollution and briefly touched on its ramifications and impacts. So, the obvious question becomes: How do we protect the environment from pollution and polluters? Actually, there are many ways to prevent or to mitigate pollution. We discuss many of them later in this handbook. But for now, it is important that the government is not a potted plant and sitting on its laurels and ignoring pollution. The Environmental Protection Agency (EPA) works 24/7 to either prevent pollution or to mitigate it via the regulatory process. Environmental laws, regulations, standards, and other regulatory tools are used to maintain or clean up our environment. In this section we summarize the Administrative Procedure Act (APA) and outline in a "broadbrush" fashion some of the procedures under which laws are developed and applied; we discuss the

Pollution Control Laws (Clean Air Act, Clean Water Act, Resource Conservation and Recovery Act, and Toxic Substances Control Act.

ADMINISTRATIVE PROCEDURE ACT (APA)

The Administrative Procedure Act (APA) 5 USC § et seq. (1946) governs the process by which federal agencies develop and issue regulations; that is, it sets forth various stands for all agency actions. It includes requirements for publishing notices of purposes and final rulemaking in the Federal Register and provides opportunities for the public to comment on notices of proposed rulemaking. The APA requires most rules to have a 30-day delayed effective date. In addition to setting forth rulemaking procedures, the APA addresses other agency actions such as issuance of policy statements, licenses, and permits. It also provides standards for judicial review if the person has been adversely affected or aggrieved by an agency review.

THE REGULATORY PROCESS

Earlier we stated that the EPA works to protect the environment. To accomplish this, the EPA uses a variety of tools and approaches, like partnerships, education programs, and grants. One of the most significant tools is writing regulations. Regulations are mandatory requirements that can apply to individuals, businesses, state or local governments, non-profit organizations, or others. Although it is Congress that passes the laws that govern the United States, Congress has also authorized EPA and other federal agencies to help put those laws into effect by creating and enforcing regulations. Below, you'll find a basic description of how laws and regulations are developed, what they are, and where to find them, with an emphasis on environmental laws and regulations.

Creating a Law

Step 1: Congress writes a bill: a member of Congress proposes a bill. A bill is a document that, if approved, will become law.

Step 2: The president approves or vetoes the bill: if both houses of congress approved a bill, it goes to the president who has the option to either approve it to veto it. If approved, the new law is called an act or statue. Some of the better-known laws related to the environment are the Clean Air Act, the Clean Water Act, and the Safe Drinking Water Act.

Step 3: The act is codified in the *United States Code*: once an act is passed, the House of Representative standardizes the text of the law and publishes it in the United States Code (U.S.C). The U.S.C. is the codification by subject matter of the general and permanent laws of the United States. Since 1926, the U.S.C. has been published every 6 years. In between editions, annual cumulative supplements are published in order to present the most current information.

Putting the Law to Work

Once a law is official, here's how it is put into practice: Laws often do not include all the details needed to explain how an individual, business, state or local government, or others might follow the law. The *United States Code* would not tell you, for example, what the speed limit is in front of your house. In order to make the laws work on a day-to-day level, Congress authorizes certain government agencies—including EPA—to create regulations.

Regulations set specific requirements about what is legal and what isn't. Regulations explain the technical, and operational. And legal details necessary to implement laws. For example, a regulation issued by EPA to implement the Clean Air Act might explain what levels of a pollutant—such as sulfur dioxide—adequately protect human health and the environment. It would tell industries how much sulfur dioxide they can legally emit into the air, and what the penalty will be if they emit too much. Once the regulation is in effect, EPA then works to help Americans comply with the law and to enforce it.

Creating a Regulation

When developing regulations, the first thing we do is ask if regulation is needed at all. Every regulation is developed under slightly different circumstances, but this is the general process:

> **Step 1: EPA proposes a regulation**: the Agency researches the issues and, if necessary, proposes a regulation, also known as a Notice of Proposed Rulemaking (NPRM). The proposal is listed in the *Federal Register (FR)* so that members of the public can consider it and send their comments to EPA. The proposed rule and supporting documents are also filled EPA's official docket on Regulations.gov.
> **Step 2: EPA considers your comments and issues a final rule**: Generally once EPA considers the comments received when the proposed regulation was issued, it revises the regulation accordingly and issues a final rule. This final rule is also published in the FR and in EPA's official docket on Regulations.gov.
> **Step 3: The regulation is codified in the *code of federal regulations***: once a regulation is completed and has been printed in the FR as a final rule, it is codified when it is added to the Code of Federal Regulations (CFR). The CFR is the official record of all regulations created by the federal government. It is divided into 50 volumes, called titles, each of which focuses on a particular area. Almost all environmental regulations appear in Title 40. The FCR is revised yearly, with one fourth of the volumes updated every 3 months. Title 40 is revised every July 1.

NEPA: Environmental Impact Statements (EIS)

The purpose of the National Environmental Policy Act (NEPA) of 1969 encourages harmony between humans and the environment; promotes efforts to prevent or eliminate environmental damage; and to enrich humans' understanding of important ecological systems and natural resources. NEPA requires that affected entities consider the potential environmental consequences of its decision before deciding to proceed and provide opportunities for public involvement, which includes participating in scoping, reviewing the draft and final Environmental Impact Statement (EIS), and attending public hearings. An EIS evaluates the environmental actions that an agency plans to undertake with respect to a comprehensive program or set of actions. The purpose of the EIS is to objectively analyze and evaluate the potential significant impacts on environmental resources from research activities pertinent to whatever action is planned. The EIS will include descriptions of

- Proposed Action
- Purpose and need for the Proposed Action
- Alternatives
- Affected environment
- Environmental Consequences of the Proposed Action and alternatives
- Required mitigations or recommended best management practices (or BMPs)

The advantages of the EIS document include:

- Full disclosure of the potential effects related to all research that may be authorized
- Incorporation of comprehensive analyses to evaluate cumulative effects effectively
- Formulation of comprehensive mitigation efforts and suggested BMP
- Reduction of the need to re-address environmental consequences, mitigation measures, and BMPs at the permit-specific level.

Introduction

The following environmental factors are normally considered during the EIS process:

- Wildlife
    - Protected Species
        - Threatened and Endangered Species
        - Marine Mammals
        - Migratory Birds
    - Non-protected Species
- Special Biological Resource Areas
    - National Marine Sanctuaries
    - Essential Fish Habitat
    - Designated Critical Habitat
- Coastal Zone Management
- Water Resources
- Human Safety
- Socioeconomics
- Noise
- Air Quality
- Cultural Resources
- Cumulative Impacts

The EIS step-by-step process is shown in Figure 1.5.

A WORD ABOUT MACHINE AND DEEP LEARNING

To say that technology is a dynamic (ever-changing and advancing) discipline and that it is difficult to keep up with the daily, maybe hourly advances we are making in the discipline is to make a gross understatement; but it is also a truism. The point is that if the current environmental engineering professor, environmental engineering professional, or environmental engineer or engineering student is to stay tuned in to, current with what is 'happening' in the real world of technological advance then he or she must practice like the doctor who performs while medical advancements are occurring all the time; like the lawyer who must stay tuned into the legal advances, reverses, case law, and the whims and wants of state legislatures – you know the ones who love to initiate and pass new laws, regulations, rules, and so forth especially if they are named for or after the originator.

An advancing technical area that is presently making its way into the field of environmental engineering (and several other disciplines) is artificial intelligence (AI) with its subset of machine learning and its branch deep learning. Deep learning deals with the development and application of modern neural networks. Neural network refers to or relates to a network of electronic components designed to mimic the operation of the human brain. One example of a neural network is the feed-forward neutral network which is an artificial neural network wherein connections between the nodes do not form a cycle (Zell et al., 1994). In this feed-forward neutral network information always moves in direction; it never goes backwards.

In addition, with the need for environmental engineers and programs in environmental learning to get on board in their dynamic field(s), so to speak, with AI, machine, deep learning and feed-forward neutral network information as it applies to engineering and the environment current personnel within the environmental engineering field must also become familiar with Python programming, algorithms, TensorFlow/PyTorch, linear algebra, optimization/probability theory, GPU hardware, Convolution Neural Networks, Recurrent Neural Networks, Generative Adversarial networks/Variation Autoencoder, Physics-informed learning, Reinforcement Learning/Active Learning/Transfer learning, and other machine learning techniques along with terms such as *Perceptron*.

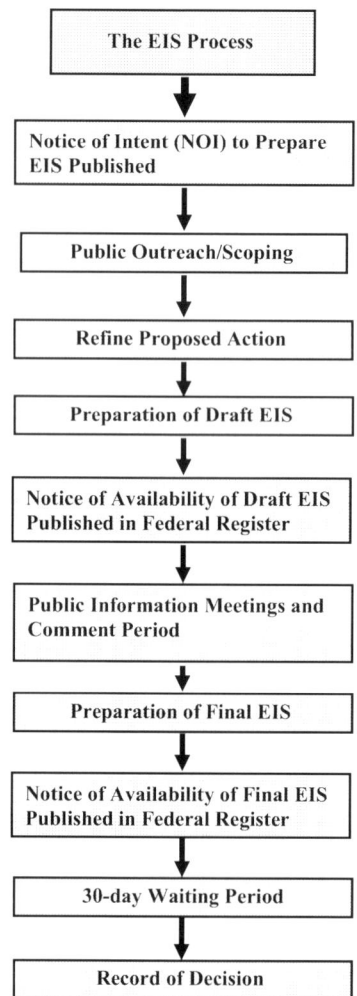

FIGURE 1.5 The EIS process.

THE BOTTOM LINE

When you throw a stone into a pool of quiet water, the ensuing ripples move out in concentric circles from the point of impact. Eventually, those ripples, much dissipated, reach the edge of the pond, where they break, disturbing the shore environment. When we alter our environment, similar repercussions affect the world around us—and some of these actions can—or will be—felt across the world. We use technology to alter our environment to suit our needs. That same technology can be put into effect so that our environment is protected from unrecoverable losses. Environmental engineers must maintain an acute sense of awareness concerning the global repercussions of problems we create for the environment—to extend the boundaries of the problem beyond our own backyard.

Historically, as long as human population numbers remained small, human pollutants could be adequately processed by the environment. But as population numbers increased, human waste began to overwhelm natural systems. As societies become technologically sophisticated, humans begin producing large numbers and volumes of new substances each year. The environment's ability to absorb and process these substances (to self-purify) has natural limitations. Our environment can only handle so much. Complicating the situation even further, consumption of resources per person in the developed world increases daily—sometimes dramatically. A prevailing attitude of "consumerism in a throw-away society" leads to resource abuse, resulting in ever-increasing amounts of pollution released to the environment. Consequently, pollution has become an increasing source of environmental problems.

Human progress has led inexorably to the soiling of air, water, and soil. Is this trend likely to continue? Maybe, maybe not. We do not know the answer to this vital question. Does anyone know the answer? We hope so. However, some things we do know. We know that the total de-industrialization of the developed world is improbable, and barring some major catastrophic event, unlikely—out of the question. A more practical view of pollution is required, one which allows sustainable development of healthy and vigorous industrial activity to provide the goods and services required by modern civilization.

We also know that we must do all this with a sense of balance.

What kind of balance? Another good question.

The balance we are referring to is actually the tradeoff we are willing to accept for those things that we do. More specifically, if we are to build a new factory that produces widgets that during the production phase will contribute an unreasonable level of pollutants to our air, water, and soil—we have to ask ourselves a question: Is it worth it? Is the tradeoff between creating jobs and enhancing the material livelihood of several individuals worth the potential damage to the environment?

We need to find a middle ground. We must balance our desires against possible or potential results. Thus, what we really have is a balancing act… a delicate balance—one we must work hard to maintain.

So, what is the bottom line of pollution and its effects on us? That is what this handbook is all about—providing facts, data and explaining and illustrating the bottom line. Throughout we consistently point to two over-riding and connected themes in this text. First, when it comes to pollution and its potential harmful effects on our lives, one thing is certain: We do not know the extent of the problem— we do not know what we do not know (scary, when you think about it). Second, the tragedy that underlies politics with the "pollution thing" (remember, in the real world everything revolves around politics), can be summed up by the words of philosopher Georg Hegel, "The nature of tragedy is not the conflict between right and wrong but between right and right."

For now, read this handbook with a mind open to the facts. Further, when you hear someone state that what we need to solve all of Earth's problems (social and otherwise) is innovation, innovation, and innovation, you can accept this view or you can settle for what this handbook preaches: What it is really is discovery, invention and then innovation to the nth degree. Make your own choices, your own decisions, and your own judgment because the real bottom line is that when it comes to defining pollution you are required to make a judgment—pollution is a judgment call.

NOTES

1. *Source*: U.S. Environmental Protection Agency (1992). *Pollution Prevention 1991: Research Program*, EPA/600/R-92/189 (September). Washington, D.C.: Office of Research and Development.
2. Pollution information provided here and in the following sections is based on Spellman, F.R. & Stoudt, M. (2013). *Environmental Science: Principles and Practice*. Lanham, MD: Government Institutes Press.

REFERENCES

APA (1946). Administrative Procedure Act. 5 USC et seq. Pub.L. 79-404, 60 Stat. 237. Washington, DC: United States Congress.

Calderia, K., & Wickett, M.E. (2003). Anthropogenic Carbon and Ocean pH. Nature, **425**, 365. https://doi.org/10.1038/425365a.

Hardin, G. (1968). The tragedy of the commons. *Science*, **162**: 1243–1248.

Keller, E.A. (1988). *Environmental Geology*, Columbus, OH: Merrill.
Leopold, A. (1970). *A Sand County Almanac*. New York: Ballentine Books.
Peavy, H.S., Rowe, D.R., & Tchobanoglous, G. (1985). *Environmental Engineering*. New York: McGraw-Hill, Inc.
Spellman, F.R. (1998). *The Science of Environmental Pollution*. Lancaster, PA: Technomic Publishing.
Taylor, J.D. (2005). *Gunning the Eastern Uplands*. Lancaster, PA: Bonasa Press.
USEPA (1989). Definition Environment. Accessed December 12, 2021. Posted at https://archive,epa.gov/osa/hsrb/web/pdf,
Zell, A., Mache, N., Huebner, R., Mammier, G., Vogt, M., Schmolz, L.M., & Herrmann, K.U. (1994). SNNS. In: *Neural Network Simulation Environments*, Skrzypek, J. (Ed.). New York: Springer, pp. 165–186.

2 Units, Standards of Measurement, and Conversions

An engineering handbook that does not include units and conversions is like a book without a table of contents or index. For example, converting units from grams to slugs, centistokes to square feet per second, or pounds per million gallons to milligrams per liter can be accomplished automatically, right away, if you have a mind containing a library of facts and figures and conversions and data and all kinds of stuff. However, if you are normal, there are times when even the most adept, confident, competent, and brilliant engineer or engineer want-to-be must refer to some reference to find some facts, even if these be their own name and address. I always hand out a printed copy of units and conversion in my college environmental classes. Most students welcome the handout; however, I have had a few who shunned it—at first. They stated they knew how to convert units and could do it from memory. My reply? "Well, now that means you know how convert abamperes to statampere, right?" Oh my! My, my, my! The look on their faces was worth hundreds, thousands... no, it was totally priceless: They wore that mask of total confusion and unknowing. I loved it... still do. Students who do not know everything are my kind of students; they are teachable. Thus, because not many of us are human computers with digital memories, I have included units, standards of measurement and conversions in this handbook. Moreover, because this is the age of energy consumption, renewable energy production, and hydraulic fracturing to capture more energy, I have included energy conversion calculations within. By the way, can you tell me (or can anyone) the difference, in gallons, between a barrel of U.S liquid and a barrel of petroleum? Don't know? Don't fear, if you use this chapter, you will know the difference.

Frank R. Spellman (2002)

UNITS OF MEASUREMENT: THE BASICS

Basic knowledge of units of measurement and how to use and convert them is essential. Environmental engineers should be familiar both with the U.S. Customary System (USCS) or English System and the International System of Units (SI). Some of the important units are summarized in Table 2.1; these are a few of the basic SI and USCS units of measurement that will be encountered in practice.

In the study of environmental engineering math operations (and in actual practice), it is quite common to encounter both extremely large quantities and extremely small ones. The concentrations of some toxic substances may be measured in parts per million or billion or trillion (ppm or ppb or ppt), for example.

- **Key point**: PPM may be roughly described as an amount contained in a full shot glass in the bottom of a standard-sized swimming pool—the full shot glass is 1 ppm relative to the rest of the water contained in the pool.

To describe quantities that may take on such large or small values, it is useful to have a system of prefixes that accompanies the units. Some of the more important prefixes are present in Table 2.2.

TABLE 2.1
Commonly Used Units

Quantity	SI Units	USCS Units
Length	Meter	Feet (ft)
Mass	Kilogram	Pound (lb)
Temperature	Celsius	Fahrenheit (F)
Area	Square meter	Square foot (ft^2)
Volume	Cubic meter	Cubic foot (ft^3)
Energy	Kilojoule	British thermal unit (Btu)
Power	Watt	Btu/hr
Velocity	Meter/second	mile/hour (mi/hr)

TABLE 2.2
Common Prefixes

Quantity	Prefix	Symbol
10^{-12}	pico	p
10^{-9}	nano	n
10^{-6}	micro	μ
10^{-3}	milli	M
10^{-2}	centi	C
10^{-1}	deci	D
10	deca	da
10^2	hecto	h
10^3	kilo	k
10^3	mega	M

CONVERSION FACTORS

Sometimes we have to convert between different units. Suppose that a 60-inch piece of wood is attached to a 6-foot piece of wood. How long are they together? Obviously, we cannot find the answer to this question by adding 60 to 6. The reason is that the two figures are given in different units. Before we can add the two numbers, we have to convert one of them to the units of the other. Then, when we have two numbers in the same units, we can add them. In order to perform this conversion, we need a *conversion factor*. That is, in this case, we have to know how many inches make up a foot; that is, 12 inches equals 1 foot. Knowing this, we can perform the calculation in two steps as follows:

a. 60 inches is really 60/12 = 5 feet
b. 5 feet + 6 feet = 11 feet

From the example above it can be seen that a conversion factor changes known quantities in one unit of measure to an equivalent quantity in another unit of measure.

In making the conversion from on unit to another, we must know two things:

1. The exact number that relates the two units
2. Whether to multiply or divide by that number

Units, Standards of Measurement & Conversions

When making conversions, confusion over whether to multiply or divide is common; on the other hand, the number that relates the two units is usually known and, thus, is not a problem. Understanding the proper methodology—the "mechanics"—to use for various operations requires practice and common sense.

Along with using the proper "mechanics" (and practice and common sense) in making conversions, probably the easiest and fastest method of converting units is to use a conversion table.

The simplest of the conversion requires that the measurement be multiplied or divided by a constant value. For instance, in wastewater treatment, if the depth of biosolids on a drying bed is 0.85 feet, multiplying by 12 inches per foot converts the measured depth to inches (10.2 inches). Likewise, if the depth of the solids blanket in the secondary clarifier is measured as 16 inches, dividing by 12 inches per foot converts the depth measurement to feet (1.33 feet).

Table 2.3 lists many of the conversion factors used in environmental engineering. Note that Table 2.3 is designed with a unit of measure in the left and right columns and a constant (conversion factor) in the center column (Table 2.4, a more extensive conversion table is presented later).

- **Key point**: To convert in the opposite direction (i.e., inches to feet), divide by the factor rather than multiply.

Weight, Concentration, and Flow

Using Table 2.3 to convert from one-unit expression to another and vice versa is good practice. However, in making conversions to solve process computations in water/wastewater treatment, we must be familiar with conversion calculations based upon a relationship between weight, flow or volume, and concentration. The basic relationship is

$$\text{Weight} = \text{Concentration} \times \text{flow or volume} \times \text{factor} \qquad (2.1)$$

Table 2.4 summarizes weight, volume, and concentration calculations. With practice, many of these calculations become second nature to operators; the calculations are important relationships and are used often in water/wastewater treatment process control calculations, so on-the-job practice is possible.

The following conversion factors are used extensively in water/wastewater operations and are commonly needed to solve problems on licensure examinations; the environmental engineer should keep them accessible for all plant operators.

- 7.48 gallons per ft^3
- 3.785 liters per gallon
- 454 grams per pound
- 1,000 mL per liter
- 1,000 mg per gram
- 1 ft^3/sec (cfs) = 0.6465 MGD

Key point: Density (also called specific weight) is mass per unit volume, and may be registered as lb/cu ft, lb/gal, grams/mL, grams/cu meter. If we take a fixed-volume container, fill it with a fluid, and weigh it, we can determine the density of the fluid (after subtracting the weight of the container).

- 8.34 pounds per gallon (water)—(density = 8.34 lb/gal)
- one milliliter of water weighs 1 gram—(density = 1 gram/mL)
- 62.4 pounds per ft^3 (water)—(density = 8.34 lb/gal)
- 8.34 lb/gal = mg/L (converts dosage in mg/L into lb/day/MGD)
    Example: 1 mg/L × 10 MGD × 8.3 = 83.4 lb/day

TABLE 2.3
Conversion Table

To Convert	Multiply By	To Get
Feet	12	Inches
Yards	3	Feet
Yards	36	Inches
Inches	2.54	Centimeters
Meters	3.3	Feet
Meters	100	Centimeters
Meters	1,000	Millimeters
Square yards	9	Square feet
Square feet	144	Square inches
Acres	43,560	Square feet
Cubic yards	27	Cubic feet
Cubic feet	1,728	Cubic inches
Cubic feet (water)	7.48	Gallons
Cubic feet (water)	62.4	Pounds
Acre-feet	43,560	Cubic feet
Gallons (water)	8.34	Pounds
Gallons (water)	3.785	Liters
Gallons (water)	3,785	Milliliters
Gallons (water)	3,785	Cubic centimeters
Gallons (water)	3,785	Grams
Liters	1,000	Milliliters
Days	24	Hours
Days	1,440	Minutes
Days	86,400	Seconds
Million gallons/day	1,000,000	Gallons/day
Million gallons/day	1.55	Cubic feet/second
Million gallons/day	3.069	Acre-feet/day
Million gallons/day	36.8	Acre-inches/day
Million gallons/day	3,785	Cubic meters/day
Gallons/minute	1,440	Gallons/day
Gallons/minute	63.08	Liters/minute
Pounds	454	Grams
Grams	1,000	Milligrams
Pressure, psi	2.31	Head, ft (water)
Horsepower	33,000	Foot-pounds/minute
Horsepower	0.746	Kilowatts

- 1 psi = 2.31 feet of water (head)
- 1 foot head = 0.433 psi
- °F = 9/5(°C + 32)
- °C = 5/9(°F − 32)
- average water usage: 100 gallons/capita/day (gpcd)
- persons per single-family residence: 3.7

Units, Standards of Measurement & Conversions

TABLE 2.4
Weight, Volume, and Concentration Calculations

To Calculate	Formula
Pounds	Concentration, mg/L × tank volume MG × 8.34 lb/MG/mg/L
Pounds/day	Concentration, mg/L × flow, MGD × 8.34 lb/MG/mg/L
Million gallons/day	$\dfrac{\text{Quantity, lb/day}}{(\text{Concentration, mg/L} \times 8.34 \text{ lb/mg/L/MG})}$
Milligrams/liter	$\dfrac{\text{Quantity, lb}}{(\text{Tank volume, MG} \times 8.34 \text{ lb/mg/L/MG})}$
Kilograms/liter	Concentration, mg/L × volume, MG × 3.785 lb/MG/mg/L
Kilograms/day	Concentration, mg/L × flow, MGD × 3.785 lb/MG/mg/L
Pounds/dry ton	Concentration, mg/kg × 0.002 lb/d.t./mg/kg

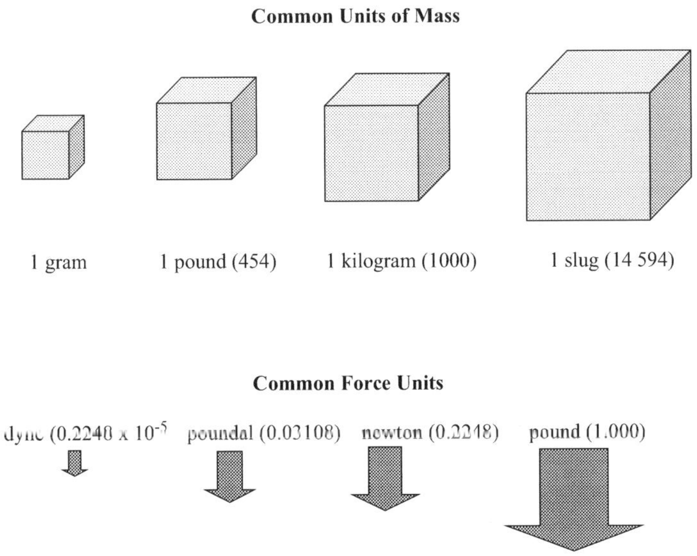

Common Units of Mass

1 gram   1 pound (454)   1 kilogram (1000)   1 slug (14 594)

Common Force Units

dyne (0.2248 × 10⁻⁵)   poundal (0.03108)   newton (0.2248)   pound (1.000)

FIGURE 2.1 Common units of mass and force. (Adapted from Lindeburg, M.R. (2009). *Environmental Engineering Reference Manual*, 2nd ed. Belmont, CA: Power to Pass.)

Before illustrating a few sample conversion problems, it is important to point out that some common units do not represent the same quantities. This is important because USA-type engineers not totally familiar with SI or *mks systems* sometimes are confused by the various unit comparisons. To illustrate this point, for example, as shown in Figure 2.1, the pound and slug and the newton and pound do not represent the same quantity of matter and force units, respectively.

CONVERSIONS

Use Tables 2.3 and 2.4 to make the conversions indicated in the following example problems. Other conversions are presented in appropriate sections of the text.

Typical Conversion Examples

Example 2.1

Convert cubic feet to gallons.

$$\text{Gallons} = \text{Cubic feet}, \text{ft}^3 \times \text{gal/ft}^3$$

Sample Problem:

How many gallons of biosolids can be pumped to a digester that has 3,600 cubic feet of volume available?

$$\text{Gallons} = 3{,}600\,\text{ft}^3 \times 7.48\,\text{gal/ft}^3 = 26{,}928\,\text{gal}$$

Example 2.2

Convert gallons to cubic feet.

$$\text{Cubic feet} = \frac{\text{gal}}{7.48\,\text{gal/ft}^3}$$

Sample Problem:

How many cubic feet of biosolids are removed when 18,200 gallons are withdrawn?

$$\text{Cubic feet} = \frac{18{,}200\,\text{gal}}{7.48\,\text{gal/ft}^3} = 2{,}433\,\text{ft}^3$$

Example 2.3

Convert gallons to pounds.

$$\text{Pounds}, \text{lb} = \text{gal} \times 8.34\,\text{lb/gal}$$

Sample Problem:
If 1,650 gallons of solids are removed from the primary settling tank, how many pounds of solids are removed?

$$\text{Pounds} = 1{,}650\text{-gal} \times 8.34/\text{gal} = 13{,}761\,\text{lb}$$

Example 2.4

Convert pounds to gallons.

$$\text{Gallons} = \frac{\text{lb}}{8.34\,\text{lb/gal}}$$

Sample Problem:
How many gallons of water are required to fill a tank that holds 7,540 pounds of water?

$$\text{Gallons} = \frac{7{,}540\,\text{lb}}{8.34\,\text{lb/gal}} \,904\,\text{gal}$$

Units, Standards of Measurement & Conversions

Example 2.5

Convert milligrams/liter to pounds.

- **Key point**: Concentrations in milligrams per liter or parts per million determined by laboratory testing are typically converted to quantities of pounds, kilograms, pounds per day, or kilograms per day.

$$\text{Pounds} = \text{Concentration, mg/L} \times \text{volume, MG} \times 8.34 \text{ lb/mg/L/MG}$$

Sample Problem:
The solids concentration in the aeration tank is 2,580 mg/L. The aeration tank volume is 0.95 MG. How many pounds of solids are in the tank?

$$\text{Pounds} = 2,580 \text{ mg/L} \times 0.95 \text{ MG} \times 8.34 \text{ lb/mg/L/MG} = 20,441.3 \text{ lb}$$

Example 2.6

Convert milligrams per liter to pounds per day.

$$\text{Pounds/day} = \text{Concentration, mg/L} \times \text{flow, MGD} \times 8.34 \text{ lb/mg/L/MG}$$

Sample Problem:
How many pounds of solids are discharged per day when the plant effluent flow rate is 4.75 MGD and the effluent solids concentration is 26 mg/L?

$$\text{Pounds/day} = 26 \text{ mg/L} \times 4.75 \text{ MGD} \times 8.34 \text{ lb/mg/L/MG} = 1,030 \text{ lb/day}$$

Example 2.7

Convert milligrams per liter to kilograms per day.

$$\text{kg/day} = \text{Concentration, mg/L} \times \text{volume, MG} \times 3.785 \text{ kg/mg/L/MG}$$

Sample Problem:
The effluent contains 26 mg/L of BOD_5. How many kilograms per day of BOD_5 are discharged when the effluent flow rate is 9.5 MGD?

$$\text{kg/day} = 26 \text{ mg/L} \times 9.5 \text{ MG} \times 3.785 \text{ kg/mg/L/MG} = 934 \text{ kg/day}$$

Example 2.8

Convert pounds to milligrams per liter.

$$\text{Concentration, mg/L} = \frac{\text{Quantity, lb}}{\text{Volume, MG} \times 8.34 \text{ lb/mg/L/MG}}$$

Sample Problem:
The aeration tank contains 89,990 pounds of solids. The volume of the aeration tank is 4.45 MG. What is the concentration of solids in the aeration tank in mg/L?

$$\text{Concentration, mg/L} = \frac{89,990 \text{ lb}}{4.45 \text{ MG} \times 8.34 \text{ lb/mg/L/MG}} = 2,425 \text{ mg/L}$$

Example 2.9

Convert pounds per day to milligrams per liter.

$$\text{Concentration, mg/L} = \frac{\text{Quantity, lb/day}}{\text{Volume, MGD} \times 8.34 \text{ lb/mg/L/MG}}$$

Sample Problem:
The disinfection process uses 4,820 pounds per day of chlorine to disinfect a flow of 25.2 MGD. What is the concentration of chlorine applied to the effluent?

$$\text{Concentration, mg/L} = \frac{4,820}{25.2 \text{ MGD} \times 8.34 \text{ lb/mg/L/MG}} = 22.9 \text{ mg/L}$$

Example 2.10

Convert pounds to flow in million gallons per day.

$$\text{Flow} = \frac{\text{Quantity, lb/day}}{\text{Concentration, mg/L} \times 8.34 \text{ lb/mg/L/MG}}$$

Sample Problem:
9,640 pounds of solids must be removed from the activated biosolids process per day. The waste-activated biosolids concentration is 7,699 mg/L. How many million gallons per day of waste-activated biosolids must be removed?

$$\text{Flow} = \frac{9,640 \text{ lb}}{7,699 \text{ mg/L} \times 8.34 \text{ lb/MG/mg/L}} = 0.15 \text{ MGD}$$

Example 2.11

Convert million gallons per day (MGD) to gallons per minute (gpm).

$$\text{Flow} = \frac{\text{Flow, MGD} \times 1,000,000 \text{ gal/MG}}{1,440 \text{ min/day}}$$

Sample Problem:
The current flow rate is 5.55 MGD. What is the flow rate in gallons per minute?

$$\text{Flow} = \frac{5.55 \text{ MGD} \times 1,000,000 \text{ gal/MG}}{1,440 \text{ min/day}} \; 3,854 \text{ gpm}$$

Example 2.12

Convert million gallons per day (MGD) to gallons per day (gpd).

$$\text{Flow} = \text{Flow, MGD} \times 1,000,000 \text{ gal/MG}$$

Sample Problem:
The influent meter reads 28.8 MGD. What is the current flow rate in gallons per day?

$$\text{Flow} = 28.8 \text{ MGD} \times 1,000,000 \text{ gal/MG} = 28,800,000 \text{ gpd}$$

Units, Standards of Measurement & Conversions

Example 2.13

Convert million gallons per day (MGD) to cubic feet per second (cfs).

$$\text{Flow, cfs} = \text{Flow, MGD} \times 1.55 \text{ cfs/MGD}$$

Sample Problem:
The flow rate entering the grit channel is 2.89 MGD. What is the flow rate in cubic feet per second?

$$\text{Flow} = 2.89 \text{ MGD} \times 1.55 \text{ cfs/MGD} = 4.48 \text{ cfs}$$

Example 2.14

Convert gallons per minute (gpm) to million gallons per day (MGD).

$$\text{Flow, MGD} = \frac{\text{Flow, gpm} \times 1{,}440 \text{ min/day}}{1{,}000{,}000 \text{ gal/MG}}$$

Sample Problem:
The flow meter indicates that the current flow rate is 1,469 gpm. What is the flow rate in MGD?

$$\text{Flow, MGD} = \frac{1{,}469 \text{ gpm} \times 1{,}440 \text{ min/day}}{1{,}000{,}000 \text{ gal/MG}} = 2.12 \text{ MGD (rounded)}$$

Example 2.15

Convert gallons per day (gpd) to million gallons per day (MGD).

$$\text{Flow, MGD} = \frac{\text{Flow, gal/day}}{1{,}000{,}000 \text{ gal/MG}}$$

Sample Problem:
The totalizing flow meter indicates that 33,444,950 gallons of wastewater have entered the plant in the past 24 hrs. What is the flow rate in MGD?

$$\text{Flow} = \frac{33{,}444{,}950 \text{ gal/day}}{1{,}000{,}000 \text{ gal/MG}} = 33.44 \text{ MGD}$$

Example 2.16

Convert flow in cubic feet per second (cfs) to million gallons per day (MGD).

$$\text{Flow, MGD} = \frac{\text{Flow, cfs}}{1.55 \text{ cfs/MG}}$$

Sample Problem:
The flow in a channel is determined to be 3.89 cubic feet per second (cfs). What is the flow rate in million gallons per day (MGD)?

$$\text{Flow, MGD} = \frac{3.89 \text{ cfs}}{1.55 \text{ cfs/MG}} = 2.5 \text{ MGD}$$

Example 2.17

Problem:

The water in a tank weighs 675 pounds. How many gallons does it hold?

Solution:

Water weighs 8.34 lbs/gal. Therefore:

$$\frac{675 \text{ lb}}{8.34 \text{ lb/gal}} = 80.9 \text{ gallons}$$

Example 2.18

Problem:

A wooden piling with a diameter of 16 inches and a length of 16 feet weighs 50-lb/cu ft. If it is inserted vertically into a body of water, what vertical force is required to hold it below the water surface?

Solution:

If this piling had the same weight as water, it would rest just barely submerged. Find the difference between its weight and that of the same volume of water. That is the weight needed to keep it down.

$$62.4 \text{ lb/cu ft (water)} - 50.0 \text{ lb/cu ft (piling)} = 12.4 \text{ lb/cu ft difference}$$

$$\text{Volume of piling} = 0.785 \times 1.33^2 \times 16 \text{ ft} = 22.21 \text{ cu ft}$$

$$12.4 \text{ lb/cu ft} \times 22.21 \text{ cu ft} = 275.4 \text{ lb (needed to hold it below water surface)}$$

Example 2.19

Problem:

A liquid chemical with a specific gravity (SG) of 1.22 is pumped at a rate of 40 gpm. How many pounds per day are being delivered by the pump?

Solution:

Solve for pounds pumped per minute; change to lb/day.

$$8.34 \text{ lb/gal water} \times 1.22 \text{ SG liquid chemical} = 10.2 \text{ lb/gal liquid}$$

$$40 \text{ gal/min} \times 10.2 \text{ lb/gal} = 408 \text{ lb/min}$$

$$408 \text{ lb/min} \times 1{,}440 \text{ min/day} = 587{,}520 \text{ lb/day.}$$

Example 2.20

Problem:

A cinder block weighs 70 pounds in the air. When immersed in water, it weighs 40 pounds. What is the volume and specific gravity of the cinder block?

Solution:

The cinder block displaces 30 pounds of water; solve for cu ft of water displaced (equivalent to the volume of the cinder block).

Units, Standards of Measurement & Conversions

$$\frac{30\text{-lb water displaced}}{62.4 \text{ lb/cu ft}} = 0.48 \text{ cu ft water displaced}$$

Cinder block volume = 0.48 cu ft; this weighs 70 lb.

$$\frac{70 \text{ lb}}{0.48 \text{ cu ft}} = 145.8 \text{ lb/cu ft density of cinder block}$$

$$\text{Specific gravity} = \frac{\text{Density of cinder block}}{\text{Density of water}}$$

$$= \frac{145.8 \text{ lb/cu ft}}{62.4 \text{ lb/cu ft}}$$

$$= 2.34$$

Temperature Conversions

Most water/wastewater operators are familiar with the formulas used for Fahrenheit and Celsius temperature conversions:

$$°C = 5/9(°F - 32)$$

$$°F = 9/5(°C) + 32$$

The difficulty arises when one tries to recall these formulas from memory. Probably the easiest way to recall these important formulae is to remember three basic steps for both Fahrenheit and Celsius conversions:

1. Add 40°
2. Multiply by the appropriate fraction (5/9 or 9/5)
3. Subtract 40°

Obviously, the only variable in this method is the choice of 5/9 or 9/5 in the multiplication step. To make the proper choice, you must be familiar with the two scales. The freezing point of water is 32° on the Fahrenheit scale and 0° on the Celsius scale. The boiling point of water is 212° on the Fahrenheit scale and 100° on the Celsius scale.

What does all this mean?

- **Key point**: Note, for example, that at the same temperature, higher numbers are associated with the Fahrenheit scale and lower numbers with the Celsius scale. This important relationship helps you decide whether to multiply by 5/9 or 9/5. Let's look at a few conversion problems to see how the three-step process works.

Example 2.21

Suppose that we wish to convert 240°F to Celsius. Using the three-step process, we proceed as follows:

1. **Step 1**: add 40°

$$240° + 40° = 280°$$

2. **Step 2**: 280° must be multiplied by either 5/9 or 9/5. Because the conversion is to the Celsius scale, we will be moving to a number *smaller* than 280. Through reason and observation, obviously, if 280 were multiplied by 9/5, the result would be almost the same as multiplying by 2, which would double 280 rather than make it smaller. If we multiply by 5/9, the result will be about the same as multiplying by ½, which would cut 280 in half. Because in this problem we wish to move to a smaller number, we should multiply by 5/9:

$$(5/9)(380°) = 156.0°C$$

3. **Step 3**: Now subtract 40°

$$156.0°C - 40.0°C = 116.0°C$$

Therefore, 240°F = 116.0°C

Example 2.22

Problem:

Convert 22°C to Fahrenheit.

1. **Step 1**: add 40°

$$22° + 40° = 62°$$

Because we are converting from Celsius to Fahrenheit, we are moving from a smaller to a larger number, and 9/5 should be used in the multiplications:

2. **Step 2**:

$$(9/5)(62°) = 112°$$

3. **Step 3**: Subtract 40°

$$112° - 40° = 72°$$

Thus, 22°C = 72°F

Obviously, knowing how to make these temperature conversion calculations is useful. However, in practical *in situ* or non-*in situ* operations, you may wish to use a temperature conversion table.

DID YOU KNOW?

U.S. environmental engineers find that identifying certain quantities used in the U.S. can be confusing. For example, at the beginning of this chapter, the question was asked if you knew how many gallons are contained in a barrel of U.S. liquid as compared to a barrel of petroleum. Along with answering this question and to clear the air on other quantities, several are listed below:

Units, Standards of Measurement & Conversions

- atmosphere standard atmosphere
- barrel 31.5 gallons (U.S., liquid)
- barrel 42 gallons (petroleum)
- Btu tradition (thermochemical) value
- Calories thermochemical
- Gallons U.S., liquid
- Horse power U.S. or mechanical
- ounce savoirdupois
- pints U.S., liquid
- pound savoirdupois
- quarts U.S., liquid
- tons short tons of 2,000 pounds

MKS System (SI Units)

The *mks system* (so named because it uses the meter, kilogram, and second as base units) is comprised of *SI units*. All other units are derived from the base units, which are listed in Table 2.5. This is a fully consistent system; there is only one recognized unit for each variable (physical quantity).

DID YOU KNOW?

An abampere, in electricity, is a centimeter-gram-second unit of electromagnetic current, equivalent to 10 Å. A Statampere is the electric unit of current equal to the current produced by an electromotive force of one statvolt acting through a resistance of one statohm. An abampere is multiplied by 2.99793×10^{10} to obtain a statampere.

TABLE 2.5
SI Base Units

Quantity	Name	Symbol
Length	Meter	m
Mass	Kilogram	kg
Time	Second	s
Electric current	Ampere	A
Temperature	Kelvin	K
Amount of substance	Mole	mol
Luminous intensity	Candela	cd

ENERGY UNITS AND MATH OPERATIONS[1]

Eventually, growth in the globe's population and material economy will confront humanity. When this occurs (and it will), we will either adjust and survive or we will simply join the ranks of the dinosaurs, Dodo birds, Passenger Pigeons, Golden Toads, or those other several species presently experiencing the Sixth Extinction.

Frank R. Spellman (2012)

During a recent Rabbit and Grasshopper conversation:

Grasshopper stated to his friend, Rabbit: "To fix human energy, population, unemployment, and many of their economic problems, what they need right now, my friend, is innovation, innovation, innovation…"

After deliberate and well-practiced thumping of his foot, Rabbit replied to his friend, Grasshopper: "No my long-legged friend… to fix all of humankind's economic problems what they need first is discovery, discovery, discovery and then invention, invention, invention… followed up by innovation, innovation, innovation… hmmmm, human leadership, brain power, common sense, and accountability would also help."

Grasshopper: hmmmmmmmmmm, buzzzzzzzz, etc., "Well, they ain't too smart… them humans… All they need to do us ask us. We know how to economize… and how to do the rest."

Rabbit replied: "Right on, Grasshopper!"

–A friend of all grasshoppers and rabbits (2014).

ENVIRONMENTAL ENGINEERING: PERSPECTIVE ON ENERGY

The motive force, the capacity to do the work behind the operation of just about anything and everything is energy. Energy is essential for most activities of modern society. Whether we use energy in the form of wood, fossil fuels and electricity, the goal is to make life comfortable and convenient—that is, to maintain the so-called "good life." We use electricity for our lights and fans, air-conditioner, water heater and room heaters, oven, microwave, washing machine, driers, cell phones, I-Phones, computers, and toasters. We use fossil fuel to run buses, trucks, trains, airplanes, and ships, and thus transportation uses a large percentage of all the energy used.

Most people understand the importance of energy in their lives. But few understand the interface between energy and its usage and its mining or its discovery, invention, and/or innovative uses and the role of the environmental engineer in all of this (heck, many environmental engineers do not understand the interface). Water use and the safety and health of energy workers are environmental engineers' primary concern. Water requirements of energy production, especially as used in hydraulic fracking operations, are, at present, the main concern. Specifically, the environmental engineer is concerned with the fate of water used to frack for gas and oil. What is to be done with the fracked contaminated wastewater? Environmental engineers are those tasked with answering this question because environmental engineers must consider the water use of our energy options using life-cycle analysis and coming up with new ways to recycle water quickly and reduce the overall water requirements through new technologies. Thus, because environmental engineers must know something about energy and energy production, the following section is provided.

The 411 on Energy[2]

Defining energy can be accomplished by providing a technical definition or by a characterization in layman's terms. Because the purpose of this book is to reach technical readers as well as a wide range of general readers, definitions provided herein and hereafter are best described as technical-nontechnical-based. Consider the definition of energy, for example; it can be defined in a number of ways. In the broad sense, energy means the capacity of something, a person, an animal, or a physical system (machine) to do work and produce change. In layman's terms, energy is the amount of force or power when applied can move one object from one position to another. It can also be used to describe someone doing energetic things such as running, talking, and acting in a lively and vigorous way. It is used in science to describe how much potential a physical system has to change. It also is used in economics to describe the part of the market where energy itself is harnessed and sold to consumers. For our purposes in this text, we simply define energy in technical-nontechnical-based terms as something that can do work or the capacity of a system to do work.

There are two basic forms of energy: kinetic and potential energy. *Kinetic energy* is energy at work or in motion; that is, moving energy—a car in motion or a rotating shaft has kinetic energy. In billiards, a player gives the cue ball kinetic energy when she strikes the ball with the cue. As the ball rolls, it exerts kinetic energy. When the ball comes into contact with another ball, it transmits its kinetic energy, allowing the next ball to be accelerated. *Potential energy* is stored energy, like the energy stored in a coiled or stretched spring or an object stationed above a table. A roller coaster has the greatest potential energy when it is stopped at the top of a long drop. Another example of potential energy is when a can of carbonated soda remains unopened. The can is pressurized with gas that is not in motion but that has potential energy. Once the can is opened, the gas is released, and the potential energy is converted to kinetic energy.

According to the Conservation Law of Energy, energy cannot be made or destroyed, but can be made to change forms. Moreover, when energy changes from one form to another, the amount of energy stays the same. Let's consider an example of the Conservation Law of Energy: The energy of something is measured to start with; the energy changes from potential (stored) energy to kinetic (moving) and back again; at the end, the energy is measured again. The energy measured at start is the same as that measured at the end; it will always be the same. One caveat to this explanation is that we now know that matter can be made into energy through processes modified or amplified to become the Law of Conservation of Matter and Energy.

Types of Energy

There are many types of energy. For example:

- Kinetic (motion) energy
- Water energy
- Potential (at rest) energy
- Elastic energy
- Nuclear energy
- Chemical energy
- Sound energy
- Internal energy
- Heat/Thermal energy
- Light (radiant) energy
- Electric energy

Energy sources can also be categorized as renewable or nonrenewable. EIA (2009) points out that when we use electricity in our home, the electrical power was probably generated by burning coal, by a nuclear reaction, or by a hydroelectric plant at a dam. Therefore, coal, nuclear and hydropower are called energy sources. When we fill up a gas tank, the source might be petroleum or ethanol made by growing and processing corn.

As mentioned, energy sources are divided into two groups—*renewable* (an energy source that can be easily replenished) and *nonrenewable* (an energy source that we are using up and cannot recreate; petroleum, for example, was formed millions of years ago from the remains of ancient sea plants and animals). In the United States, most of our energy comes from nonrenewable energy sources. Coal, petroleum, natural gas, propane, and uranium are nonrenewable energy sources. They are used to make electricity, to heat our homes, to move our cars, and to manufacture all kinds of products. Renewable and nonrenewable energy sources can be used to produce secondary energy sources including electricity and hydrogen. Renewable energy sources include:

- Solar
- Hydro
- Wind
- Geothermal
- Ocean Thermal Energy Conversion
- Tidal Energy
- Hydrogen Burning
- Biomass Burning

Renewable energy (an energy source that can be easily replenished) is the focus of this text. Unfortunately (depending on your point of view), nonrenewable energy sources on Earth are available in limited quantity and may vanish within the next 100 years. Moreover, keep in mind that nonrenewable sources *are not* environmentally friendly and can have serious effects on our health. Notwithstanding the environmental and health impacts of using nonrenewable energy sources, it is important to point out both sides of the argument; i.e., the benefits derived, and non-benefits obtained for using these sources. For example, nonrenewable energy sources are beneficial in that:

Nonrenewable Energy Benefits

- Nonrenewable sources are easy to use.
- A small amount of nuclear energy will produce a large amount of power.
- Nonrenewable energy sources have little competition.
- Nonrenewable energy sources are relatively cheap when converting from one type of energy to another.

Non-Benefits

- Nonrenewable sources will expire someday.
- The speed at which such resources are being used can bring about serious environmental changes.
- Nonrenewable sources release toxic gases in the air when burnt and can further exacerbate on-going, cyclical climate change.
- Because nonrenewable sources are becoming scarcer, prices of these sources will begin to soar.

The benefits of using renewable energy sources include:

- Wind, sun, ocean, and geothermal energy are available in abundant quantities and free to use.
- Renewable sources have low carbon emissions; therefore, they are considered environmentally friendly.
- Renewable energy helps stimulate the economy and creates job opportunities.
- Renewable energy sources enable the country to become energy independent, not having to rely on foreign (often hostile) sources.

Non-benefits of renewable energy sources include:

- Initial set-up costs of renewable energy sources are quite high.
- Solar energy is limited to day-time use and not during the night or rainy season.
- Geothermal energy can bring toxic chemicals beneath the earth's surface onto the top and can create environmental damage.
- Hydroelectric dams are expensive to build and can affect natural flow and wildlife.
- Wind energy production requires high winds and therefore has to be sited properly. Also, they are tall structures that can affect the bird population.

The use of energy in the United States is shared by four major sectors of the economy. Each end-use sector consumes electricity produced by the electric power sector (EIA, 2012):

- **Commercial**: 18%–includes buildings such as offices, malls, stores, schools, hospitals, hotels, warehouses, restaurants, places of worship, and more.
- **Industrial**: 32%–includes facilities and equipment used for manufacturing, agriculture, mining, and construction.
- **Residential**: 21%–consists of homes and apartments.
- **Transportation**: 28%–comprises vehicles that transport people or goods, such as cars, trucks, buses, motorcycles, trains, subways, aircraft, boats, barges, and even hot air balloons.

Primary energy consumption in the United States was almost three times greater in 2012 than in 1949. In all but 18 of the years between 1949 and 2012, primary energy consumption increased over the previous year.

The year 2009 provided a sharp contrast to the historical trend, in part due to the economic recession. Real gross domestic product (GDP) fell 2% compared to 2008, and energy consumption declined by nearly 5%, the largest single year decline since 1949. Decreases occurred in all four major end-use sectors: residential—3%, commercial—3%, industrial—9%, and transportation—3% (EIA, 2012).

Measuring Energy

Energy can be measured. That is, the amount of energy a thing has can be given a number. As in other kinds of measurements, there are measurement units. The units of measurement for measuring energy are used to make the numbers understandable and meaningful.

The SI unit for both energy and work is the joule (J) (as in jewels, like diamonds and emeralds). It is named after James Joule who discovered heat is a type of energy. One joule is equal to one newton-meter. In terms of SI base units, 1 J is equal to 1 kg m^2/s^2. The energy unit of measurement for electricity is the kilowatt-hour (kWh). One kWh is equivalent to 3,600,000 J (3,600 kJ or 3.6 MJ).

TABLE 2.6
Btu Conversion Factors

Energy Source	Physical Units and Btu (Weighted Averages, EIA 2008)
Electricity	1 kilowatt-hour = 3,412 Btu
Natural gas	1 cubic foot = 1,028 Btu
	1 cubic foot = 0.01 Therms
Motor gasoline	1 gallon = 124,000 Btu
Diesel fuel	1 gallon = 139,000 Btu
Heating oil	1 gallon = 139,000 Btu
Propane	1 gallon = 91,333 Btu
Wood	1 cord = 20,000,000 Btu

A common way to measure energy is expressed in Btu. This stands for British thermal unit (see Table 2.6). Btu is the amount of heat energy it takes to raise the temperature of one pound of water by 1°F, at sea level. mBtu stands for one million Btus, which can also be expressed as one decatherm (10 Therms). mBtu is occasionally used as a standard unit of measurement for natural gas and provides a convenient basis for comparing the energy content of various grades of natural gas and other fuels. One cubic foot of natural gas produces ~1,000 Btu, so 1,000 cu. ft. of gas is comparable to 1 mBtu. mBtu is occasionally expressed as mmBtu, which is intended to represent a thousand-thousand Btus.

$$1,000 \text{ joules} = 1 \text{ Btu}$$

$$1,000 \text{ joules} = 1 \text{ kilojoule} = 1 \text{ Btu}$$

$$1 \text{ therm} = 100,000 \text{ Btu}$$

CLEAN ENERGY PARAMETERS, CALCULATIONS AND REFERENCES

Harris (2006) points out that energy is an input fundamental to economic systems. Our current economic practice depends overwhelmingly on nonrenewable fossil fuels (90% of our energy supply) including oil, coal, and natural gas. As environmental professionals, we are concerned not only with the cost of energy but also with the cost to the environment, in general and as whole, as a result of using nonrenewable energy supplies. In the following sections, calculations related to the conversion of greenhouse gas emission numbers into different types of equivalent units and other pertinent calculations/conversions are discussed below.

Note: With regard to global warming potentials (GWPS, some of the equivalencies in the calculator are reported as CO_2 equivalents (CO_2E). These are calculated using GWPs from the Intergovernmental Panel on Climate Change's Fourth Assessment report.

Electricity Reduction (Kilowatt-Hours)

The Greenhouse Gas Equivalencies Calculator uses the Emissions and Generation resource Integrated Database (eGRID) U.S. annual non-baseload carbon dioxide output emission rate to convert reductions of kilowatt-hours into avoided units of carbon dioxide emissions. Most users of the Equivalencies Calculator who seek equivalencies for electricity-related emissions want to know equivalencies for emissions reduction from energy-efficient or renewable energy programs. These programs are not generally assumed to affect baseload emissions (the emissions from power plants that run all the time), but rather non-baseload generation (power plants that are brought online as necessary to meet demand). For that reason, the Equivalencies Calculator uses a non-baseload emission rate.

EMISSIONS FACTOR

6.89551×10^{-4} metric tons CO_2/kWh

Source:

USEPA (2014). eGRID, U.S. annual non-baseload CO_2 output emission rate, year 2010 data. Washington, DC: U.S. Environmental Protection Agency.

Notes:

- This calculation does not include any greenhouse gases other than CO^2.
- This calculation does not include line losses.
- Individual subregion non-baseload emissions rates are also available on the eGRID WEB site.
- To estimate indirect greenhouse gas emissions from electricity use, please use Power Profiler or use eGRID subregion annual output emission rates as a default emission factor.

Gallons of Gasoline Consumed

To obtain the number of grams of carbon dioxide emitted per gallon of gasoline combusted, the heat content of the fuel per gallon is multiplied by the kg CO_2 per heat content of the fuel. In the preamble to the joint EPA/Department of Transportation rulemaking on May 7, 2010 that established the initial National Program fuel economy standards for model years 2012–2016, the agencies stated that they had agreed to use a common conversion factor of 8,887 g of CO_2 emissions per gallon of gasoline consumed (Federal Register, 2010). This value assumes that all the carbon in the gasoline is concerted to CO_2 (IPCC, 2006).

CALCULATION

8,887 grams of CO_2/gallon of gasoline = 8.887×10^{-3} metric tons CO_2/gallon of gasoline.

Sources:

Federal Register (2010). Light-Duty Vehicle Greenhouse Gas Emissions Standards and Corporate Average Fuel Economic Standards; Final Rule, page 25,330 (PDF) (407 pp 5.7MB).

IPPC (2006). *2006 IPCC Guidelines for National Greenhouse Gas Inventories*. Intergovernmental Panel on Climate Change, Geneva, Switzerland.

Passenger Vehicles per Year

Passenger vehicles are defined as 2-axle 4-tire vehicles, including passenger cars, vans, pickup trucks, and sport/utility vehicles. In 2011, the weighted averaged combined fuel economy of cars and light trucks combined was 21.4 miles per gallon (FHWA, 2013). The average vehicle miles traveled in 2011 was 11,318 miles per year. In 2011, the ratio of carbon dioxide emissions to total greenhouse gas emissions (including carbon dioxide, methane, and nitrous oxide, all expressed as carbon dioxide equivalents) for passenger vehicles was 0.988 (EPA, 2013a,b). The amount of carbon dioxide emitted per gallon of motor gasoline burned is 8.89×10^{-3} metric tons, as calculated in the "Gallons of gasoline consumed" section above.

To determine annual greenhouse gas emissions per passenger vehicle, the following methodology was used: vehicle miles traveled (VMT) was divided by average gas mileage to determine gallons of gasoline consumed per vehicle per year. Gallons of gasoline consumed was multiplied by carbon dioxide divided per gallon of gasoline to determine carbon dioxide emitted per vehicle per year. Carbon dioxide emissions were then divided by the ratio of carbon dioxide emissions to total vehicle greenhouse gas emission to account for vehicle methane and nitrous oxide emissions.

CALCULATION

$$8.89 \times 10^{-3} \text{ metric tons CO}_2/\text{gallon gasoline} \times 11{,}318 \text{ VMT}_{\text{car/truck average}}$$

$$\times 1/21.4 \text{ miles per gallon}_{\text{car/truck average}} \times CO_2, CH_4, \text{ and } N_2O/0.988\,CO_2$$

$$= \mathbf{4.75 \text{ metric tons CO}_2\text{E/vehicle/year}}$$

Note: Due to rounding, performing the calculations given in the equations may not return the exact results shown.

Sources:

EPA (2013a). Inventory of U.S. Greenhouse Gas Emissions and Sinks: 1990–2011. Chapter 3 (energy), Tables 3.12–3.14. U.S. Environmental Protection Agency, Washington, DC. U.S. EPA #430-R-13-001 (PDF) (505 pp, 12.3 MB).

EPA (2013b). Inventory of U.S. Greenhouse Gas Emissions and Sinks: 1990–2011. Annex 6 (Additional Information), Table A-275. U.S. Environmental Protection Agency, Washington, DC. U.S. EPA #430-R-13-001 (PDF) (23 pp, 672kb).

FHWA (2013). Highway Statistics 2011. Office of Highway Policy Information, Federal Highway Administration, Table VM-1.

Miles Driven by the Average Passenger Vehicle per Year

Passenger vehicles are defined as 2-axle 4-tire vehicles, including passenger cars, vans, pickup trucks, and sport/utility vehicles. In 2011, the weighted average combined fuel economy of cars and light trucks combined was 21.4 miles per gallon (FHWA, 2013). In 2011, the ratio of carbon dioxide emission to total greenhouse gas emissions (including carbon dioxide, methane, and nitrous oxide, all expressed as carbon dioxide equivalents) for passenger vehicles was 0.988 (EPA, 2013a,b). The amount of carbon dioxide emitted per gallon of motor gasoline burned is 8.89×10^{-3} metric tons, as calculated in the Gallons of Gasoline Consumed section above.

To determine annual greenhouse gas emission per mile, the following methodology was used: carbon dioxide emissions per gallon of gasoline were divided by the average fuel economy of vehicles to determine carbon dioxide emitted per mile traveled by a typical passenger vehicle per year. Carbon dioxide emissions were then divided by the ratio of carbon dioxide emission to total vehicle greenhouse gas emissions to account for vehicle methane and nitrous oxide emissions.

CALCULATION

8.89×10^{-3} metric tons CO_2/gallon gasoline $\times 1/21.4$ miles per gallon$_{\text{car/truck average}}$
$\times 1 CO_2, CH_4,$ and $N_2O/0.988\ CO_2 = \mathbf{4.20 \times 10^{-4}}$ **metric tons CO_2 E/mile**

Sources:

EPA (2013a). Inventory of U.S. Greenhouse Gas Emissions and Sinks: 1990–2011. Chapter 3 (energy), Tables 3.12–3.14. U.S. Environmental Protection Agency, Washington, DC. U.S. EPA #430-R-13-001 (PDF) (505 pp, 12.3 MB).

EPA (2013b). Inventory of U.S. Greenhouse Gas Emissions and Sinks: 1990–2011. Annex 6 (Additional Information), Table A-275. U.S. Environmental Protection Agency, Washington, DC. U.S. EPA #430-R-13-001 (PDF) (23 pp, 672kb).

FHWA (2013). Highway Statistics 2011. Office of Highway Policy Information, Federal Highway Administration, Table VM-1.

Therms of Natural Gas

Carbon dioxide emissions per therm are determined by multiplying heat content times the carbon coefficient times the fraction oxidized times the ratio of the molecular weight ratio of carbon dioxide to carbon (44/12).

The average heat content of natural gas is 0.1 mmBtu per therm (EPA, 2013). The average carbon coefficient of natural gas is 14.46 kg carbon per mmBtu (EPA, 2013). The fraction oxidized to CO_2 is 100% (IPCC, 2006).

Note: When using this equivalency, keep in mind that it represents the CO_2 equivalency for natural gas burned as a fuel, not natural gas released to the atmosphere. Direct methane emissions released to the atmosphere (without burning) are about 21 times more powerful than CO_2 in terms of their warming effect on the atmosphere.

CALCULATION

$0.1\ \text{mmBtu}/1\ \text{therm} \times 14.46\ \text{kg C/mmBtu} \times 44\ \text{kg CO}_2/12\ \text{kg C}$
$\times 1$ metric ton/1,000 kg = **0.005302 metric tons CO_2/therm**

Sources:

EPA (2013). Inventory of U.S. Greenhouse Gas Emissions and Sinks: 1990–2011. Annex 2 (Methodology for estimating CO2 emission from fossil fuel combustion), Table A-36, U.S. Environmental Protection Agency, Washington, DC. U.S. EPA #430-$-13-001 (PDF) (429 pp, 10.6 MB.

IPCC (2006). 2006 IPCC Guidelines for National Greenhouse Gas Inventories. Intergovernmental Panel on Climate Change, Geneva, Switzerland.

Barrels of Oil Consumed

Carbon dioxide emissions per barrel of crude oil are determined by multiplying heat content times the carbon coefficient times the fraction oxidized times the ratio of the molecular weight of carbon dioxide to that of carbon (44/12).

The average heat content of crude oil is 5.80 mmBtu per barrel (EPA, 2013). The average carbon coefficient of crude oil is 20.31 kg carbon per mmBtu (EPA, 2013). The fraction oxidized is 100% (IPCC, 2006).

CALCULATION

$$5.80 \text{ mmBtu/barrel} \times 20.31 \text{ kg C/mmBtu} \times 44 \text{ kg CO}_2/12 \text{ kg C}$$
$$\times 1 \text{ metric ton}/1{,}000 \text{ kg} = \mathbf{0.43 \text{ metric tons CO}_2\text{/barrel}}$$

Sources:

EPA (2013). Inventory of U.S. Greenhouse Gas Emissions and Sinks: 1990–2011. Annex 2 (Methodology for estimating CO2 emission from fossil fuel combustion), P. A-68, Table A-45 and Table A-38, U.S. Environmental Protection Agency, Washington, DC. U.S. EPA #430-$-13-001 (PDF) (429 pp, 10.6 MB).

IPCC (2006). 2006 IPCC Guidelines for National Greenhouse Gas Inventories. Intergovernmental Panel on Climate Change, Geneva, Switzerland.

Tanker Trucks Filled with Gasoline

The amount of carbon dioxide emitted per gallon of motor gasoline burned is 8.89×10^{-3} metric tons, as calculated in the gallons of gasoline consumed section above. A barrel equals 42 gallons. A typical gasoline tanker truck contains 8,500 gallons.

CALCULATION

$$8.89 \times 10^{-3} \text{ metric tons CO}_2\text{/gallon} \times 8{,}500 \text{ gallons/tanker truck}$$
$$= \mathbf{75.54 \text{ metric tons CO}_2\text{/tanker truck}}$$

Sources:

Federal Register (2010). Light-Duty Vehicle Greenhouse Gas Emission Standards and Corporate Average Fuel Economy Standards; Final Rule, page 25,330 (PDF) (407 pp, 5.7 MB).

IPCC (2006). 2006 IPCC Guidelines for National Greenhouse Gas Inventories. Intergovernmental Panel on Climate Change, Geneva, Switzerland.

Number of Incandescent Bulbs Switched to Compact Fluorescent Bulbs (CFLs)

A 13-watt compact fluorescent light (CFL) bulb produces the same light output as a 60-watt incandescent light bulb. Annual energy consumed by a light bulb is calculated by multiplying the power (60 watts) by the average daily use (3 hours/day) by the number of days per year (365). Assuming an average daily use of 3 hours per day, an incandescent bulb consumes 65.7

kWh per year, and a compact fluorescent light bulb consumes 14.2 kWh per year (EPA, 2013). Annual energy savings from replacing an incandescent light bulb with an equivalent compact fluorescent bulb are calculated by subtracting the annual energy consumption of the compact fluorescent light bulb (14.2 kWh) from the annual energy consumption of the incandescent bulb (65.7 kWh).

Carbon dioxide emissions reduced per light bulb switched from an incandescent bulb to a compact fluorescent bulb are calculated by multiplying annual energy savings by the national average non-baseload carbon dioxide output rate for delivered electricity. The national average non-baseload carbon dioxide output rate for generated electricity in 2010 was 1,519.6 lbs CO_2 per megawatt-hour (EPA, 2014), which translates to about 1,637.5 lbs CO_2 per megawatt-hour for delivered electricity (assuming transmission and distribution losses at 7.2%) (EIA, 2013a,b; EPA, 2014).

CALCULATION

47 watts × 3 hours/day × 365 days/year × 1 kWh/1,000 Wh = **51.5 kWh/year/bulb replaced**

51.5 kWh/bulb/year × 1,637.5 pounds CO_2/MWh delivered electricity

× 1 MWh/1,000 kWh × 1 metric ton/2,204.6 lbs = $\mathbf{3.82 \times 10^{-2}}$ **metric tons CO_2/bulb replaced**

Sources:

EIA (2013a). 2013 Annual Energy Outlook. Table A4 (PDF) (2 pp, 234k). U.S. Energy Information Administration, Washington, DC.

EIA (2013b). 2013 Annual Energy Outlook. Table A8 (Total generation, use, and imports used for 7.5% T&D loss factor)(PDF) (2 pp, 230K). U.S. Energy Information Administration, Washington, DC.

EPA (2013). Savings Calculator for ENERGY STAR Qualified Light Bulbs. U.S. Environmental Protection Agency, Washington, DC.

EPA (2014). eGRID Year 2010 Data. U.S. Environmental Protection Agency, Washington, DC.

Example 2.23: The 411 on Fluorescent Lamps and Ballasts[3]

The Energy Policy Act of 1992, Executive Order 13123, and the Federal Acquisition Regulation, Part 23, Section 704 (48 CFR 23.704) institute guidelines for Federal agencies to purchase energy-efficient products. Lighting accounts for 20%–25% of the United States' electricity consumption. Retrofitting with automatic controls and energy-efficient fluorescent lamps and ballasts yields paybacks within 2–5 years. However, the best reason for retrofitting on old lighting system—increasing the productivity of workers—is often overlooked.

FLUORESCENT LIGHTING NOMENCLATURE

The pattern for interpreting fluorescent lamp names is FWWCCTDD where:

F: Fluorescent lamp
WW: Nominal power in watts (4, 5,8,12, 15, 20, 33, and so forth)

CC: The color. W=white, CW=cool white, WW=warm white, and so forth.
T: Tubular bulb.
DD: Diameter of the tube in eighths of an inch. A T8 bulb has a diameter of 1 inch, a T12 bulb has a diameter of 1.5 inches, and so forth.

For example, an F40T12 lamp is a 40-wattt fluorescent lamp with a 1.5-inch tubular bulb.

BACKGROUND ON COSTS

With electricity costing 8 cents per kilowatt hour, a typical 40-watt T12 fluorescent lamp in use $64 worth of electricity over its life. The purchase price of the bulb ($2) accounts for just 3% of the life-cycle costs of owning and operating the lighting systems. Energy accounts for 86% of the cost (i.e., operating cost breakdown for F40T12 fluorescent lamps accounts for energy at 86%, maintenance at 11%, and the lamp itself 3%). These calculations readily justify the cost of more expensive lamps that produce better quality light, save energy, and increase productivity.

The effect of lighting on human performance and productivity is complex. Direct effects of poor lighting include the inability to resolve detail, fatigue, and headaches. Lighting may indirectly affect someone's mood or hormonal balance. [Note: readers are referred to the Hawthorne Effect a phenomenon whereby workers improve or modify an aspect of their behavior in response to the fact of change in their lighting environment].

A small change in human performance dwarfs all costs associated with lighting. The typical annual costs of 1 square foot of office space are:

- **Heating and cooling**: $2
- **Lighting**: $0.50
- **Floor space**: $100
- **Employee salary and benefits**: $400

Cutting lighting consumption in half saves about 25 cents per square foot each year. A 1% increase in human productivity would save $4 per square foot each year. Costs will vary from facility to facility, but the relative magnitudes of these costs aren't likely to change. The focus needs to be on providing quality lighting to meet occupants' needs. However, it is possible to improve lighting quality while reducing energy costs thanks to improvements in lighting technology.

THE BEST FLUORESCENT LAMP AND BALLAST

A light's "warmness" is determined by its color temperature, expressed in degrees kelvin (degrees in kelvin; kelvin is never capitalized, unless used inappropriately or improperly. The kelvin scale is an absolute, thermodynamic temperature scale using as its null point absolute zero, the temperature at which all normal thermal motion ceases in the classical description of thermodynamics. The kelvin is defined as the fraction of 1/273.16 of the thermodynamic temperature of the triple point of water (exactly 0.01°C or 32.0018°F). In other, it is defined such that the triple point of water is exactly 273.16 K). The higher the correlated color temperature, the cooler the light. Office should use intermediate or neutral light. This light creates a friendly, yet businesslike environment. Neutral light sources have a correlated color temperature of 3,500 K. The color rendition index measures the quality of light. The higher the color rendition index, the better people see for a given amount of light. Currently available 4-foot fluorescent lamps have indexes of 70–98. Lamps with different correlated color temperatures and color rendition indexes should not be used in the same space. Specify the correlated color temperature and color rendition index when lamps are purchased.

The best lighting system for each operating dollar is realized with T8 fluorescent lamps that have a color rendition index of 80 or higher. Compared to standard T12 fluorescent lamps, T8 lamps have better balance between the surface area containing the phosphors that fluoresce and the arc stream that excites them. This means that T8 lamps produce more light for a given amount of energy. In Europe, T5 lamps are popular. The T5 lamps are more efficient than T8 lamps but cost more than twice as much. The availability of T5 lamps and fixtures is limited in the United States. T8 lamps are currently preferred.

A quick comparison of light output shows how important it is to specify the ballast factor and whether the ballast is electronic of magnetic. Electronic ballasts last twice as long as magnetic ballasts use less energy, have a lower life-cycle cost, and operated the lamp at much higher frequencies. Operating fluorescent lamps at higher frequencies improves their efficiency and eliminated the characteristics 60-cycle buzz and strobe-lighting effect associated with fluorescent lights. The 60-cycle strobe lighting effect may case eye fatigue and headaches. Electronic ballasts are especially desirable in shops with rotating equipment. The 60-cycle strobe-lighting effect produced by magnetic ballasts can cause rotating equipment to appear stationary. All new buildings and retrofits should use electronic ballasts.

FLUORESCENT LAMP AND BALLAST LIFE

Most fluorescent lamps have a rated life of 12,000–20,000 hours. The rated life is the time it takes for half of the bulbs to fail when they are cycled on for 3 hours and off for 20 minutes. Cycling fluorescent lamps off and on will reduce lamp life. On the other hand, turning a lamp off when it is not needed will reduce its operating hours and increase its useful life. Electricity—not lamps—accounts for the largest percentage of the operating cost of a lighting system. It is economical to turn off fluorescent lights if they are not being used.

According to the Certified Ballast Manufacturers Association, the average magnetic ballast lasts about 75,000 hours, or 12–15 years with normal uses. The optimum economic life of a fluorescent lighting system with magnetic ballasts is usually about 15 years. At this point, dirt on reflectors and lenses has significantly reduced light output. Other factors may make it desirable to retrofit a lighting system before the end of the 12- to 15-year life cycle. Those factors include increased productivity, utility rebates, and high energy costs.

ECONOMIC ANALYSIS

When considering the benefits of retrofitting, more lamps per existing fixture yield more energy savings per fixture, and a better payback. Higher than average energy or demand of the initial installation costs or a utility rebate will also produce a faster payback. Ballast factor can be used to adjust light levels. A high ballast factor increases lumens (a measure of light output), allowing fewer lamps to provide the same amount of light. For example, when electronic ballasts with a high ballast factor are used, two-lamp fixtures will produce as much light as three-lamp fixtures. This reduces the cost of the fixtures and improves the payback. An economic analysis of retrofitting three-lamp fixtures and magnetic ballasts with two-lamp fixtures with a high-ballast-factor electronic ballast yields a payback of slightly more than 2 years.

With regard to fluorescent lamp retrofit payback, a simple payback (SPB) is the time, in years, it will take for your savings (in present value) to equal the cost of the initial installation (in present value). The following calculations do not account for interest rates.

Example 2.24

Problem:

Compute your SPB using the following formula:

$$\frac{\text{Cost of installed equipment} - \text{Deferred maintenance} - \text{Rebates})}{(\text{Total energy dollar savings per year})} = \text{SPB}$$

Costs to replace a T12 lamp magnetic ballast system with a T8 lamp electronic ballast system:
New fixtures (includes fixture, two T8 lamps, and electronic ballast) cost $30 per fixture.
Installation costs $10 per fixture.
Deferred cost of cleaning existing fixtures is $5 per fixture.
The power company offers a one-time $8 per fixture rebated when replacing magnetic-ballasted T12 lamps with electronic-ballasted T8 lamps.

Solution:

$$\text{Total project cost for 100 fixtures} = (\$30 + \$10 - \$5 - \$8)(100\,\text{fixtures}) = \$2{,}700$$

$$\text{Total energy dollar savings per year} = \text{Lighting energy savings} + \text{cooling savings} - \text{heating costs}$$

$$= \$1{,}459 + \$120 - \$262$$

$$= \$1{,}317\,\text{per year}$$

$$\text{SPB} = \$2{,}700 / (\$1{,}317\,\text{per year}) = 2.05\,\text{years}$$

It is obvious that retrofitting an existing lighting system that uses F40T12 lamps and magnetic ballasts with F32T8 lamps and electronic ballasts can provide a very attractive payback.

Home Electricity Use

According to the Energy Information Administration, in 2012, 113.93 million homes in the United States consumed 1,375 billion kilowatt-hours of electricity (EIA, 2013a). On average, each home consumed 12,069 kWh of delivered electricity (EIA, 2013a). The national average carbon dioxide output rate for electricity generated in 2010 was 1,232.4 lbs CO_2 per megawatt-hour (EPA, 2014), which translates to about 1,328.0 lbs CO_2 per megawatt-hour for delivered electricity, assuming transmission and distribution losses of 7.2% (EIA, 2013b).

Annual home electricity consumption was multiplied by the carbon dioxide emission rate (per unit of electricity delivered) to determine annual carbon dioxide emissions per home.

CALCULATION

12,069 kWh per home × 1,232.4 lbs CO_2 per megawatt-hour generated

× $1/(1-0.072)$ MWh delivered/MWh generated × 1 MWh/1,000 kWh

× 1 metric ton/2,204.6 lb = 7.270 metric tons CO_2/home.

Sources:

EIA (2013a). 2013 Annual Energy Outlook. Table A4 (PDF) (2 pp, 234k). U.S. Energy Information Administration, Washington, DC.

EIA (2013b). 2013 Annual Energy Outlook. Table A8 (Total generation, use, and imports used for 7.5% T&D loss factor)(PDF) (2 pp, 230K). U.S. Energy Information Administration, Washington, DC.

EPA (2014). eGRID year 2010 data. U.S. Environmental Protection Agency, Washington, DC.

Home Energy Use

In 2012, there were 113.93 million homes in the United States (EIA, 2013a). On average, each home consumed 12,069 kWh of delivered electricity. Nationwide household consumption of natural gas, liquefied petroleum gas, and fuel oil totaled 4.26, 0.51, and 0.51 quadrillion Btu, respectively, in 2012 (EIA, 2013a). Averaged across households in the United States, this amounts to 52,372 cubic feet of natural gas, 70 barrels of liquefied petroleum gas, and 47 barrels of fuel oil per home.

The national average carbon dioxide output rate for generated electricity in 2010 was 1,232 lbs CO_2 per megawatt-hour (EPA, 2014), which translates to about 1,328.0 lbs CO_2 per megawatt-hour for delivered electricity (assuming transmission and distribution losses at 7.2%) (EIA, 2013a,b; EPA, 2014).

The average carbon dioxide coefficient of natural gas is 0.0544 kg CO_2 per cubic foot (EPA, 2013c). The fraction oxidized to CO_2 is 100% (IPCC, 2006).

The average carbon dioxide coefficient of distillate fuel oil is 429.61 kg CO_2 per 42-gallon barrel (EPA, 2013b). The fraction oxidized to CO_2 is 100% (IPCC, 2006).

The average carbon dioxide coefficient of liquefied petroleum gases is 219.3 kg CO_2 per 42-gallon Barrel (EPA, 2011b). The fraction oxidized is 100% (IPCC, 2006).

Total single-family home electricity, natural gas, distillate fuel oil, and liquefied petroleum gas consumption figures were converted from their various units to metric tons of CO_2 and added together to obtain total CO_2 emissions per home.

CALCULATION

1. **Electricity**: 12,069 kWh per home × 1,232 lbs CO_2 per megawatt-hour generated × (1/(1−0.072)) MWh generated/MWh generated/MWh delivered × 1 MWh/1,000 kWh × 1 metric ton/2,204.6 lb = 7.270 metric tons CO_2/home.
2. **Natural gas**: 52,372 cubic feet per home × 0.0544 kg CO_2/cubic foot × 1/1,000 kg/metric ton = 2.85 metric tons CO_2/home.
3. **Liquid petroleum gas**: 70.4 gallons per home × 1/42 barrels/gallon × 219.3 kg CO_2/barrel × 1/1,000 kg/metric ton = 0.37 metric tons CO_2/home.
4. **Fuel oil**: 47 gallons per home × 1/42 barrels/gallon × 429.61 kg CO_2/barrel × 1/1,000 kg/metric ton = 0.48 metric tons CO_2/home.

Total CO_2 emissions for energy use per home: 7.270 metric tons CO_2 for electricity + 2.85 metric tons CO_2 for natural gas + 0.37 metric tons CO_2 for liquid petroleum gas + 0.48 metric tons CO_2 for fuel oil = 10.97 metric tons CO_2 per home per year.

Sources:

EIA (2013a). *2014 Annual Energy Outlook Early Release*. Table A4 (PDF) (2 pp, 234K about PDF).

EIA (2013b). 2014 *Annual Energy Outlook Early Release*, Table A8 (Total generation, use, and imports used for 7.2% T&D loss factor) (PDF) (2 pp, 230K About PDF).

EPA (2011). *Inventory of U.S Greenhouse Gas Emissions and Sinks: Fast Facts 1990–2009*. Conversion Factors to Energy Units (Heat Equivalents) Heat Contents and Carbon Content Coefficients of Various Fuel Types. U.S. Environmental Protection Agency, Washington, DC. USEPA #430-F-11–007 (PDF) (2 pp, 430K, About PDF).

EPA (2013a). *Inventory of U.S. Greenhouse Gas Emissions and Sinks: 1990–2011.* Annex 2 (Methodology for estimating CO_2 emissions from fossil fuel combustion), Tables A-36 and A-279. U.S. Environmental Protection Agency, Washington, DC. U.S. EPA #430-R-13-001 (PDF) (429 pp, 10.6MB, About PDF).

EPA (2013b). *Inventory of U.S. Greenhouse Gas Emissions and Sinks: 1990–2011*. Annex 2 (Methodology for estimating CO_2 emissions from fossil fuel combustions), Table A-5, U.S. Environmental Protection Agency, Washington, DC. U.S. EPA #430-R-13-001 (PDF) (429 pp, 10.6 MB, about PDF).

EPA (2014). *eGRID 2010 Data*. U.S. Environmental Protection Agency, Washington, DC.

IPCC (2006). *2006 IPCC Guidelines for National Greenhouse Gas Inventories*. Intergovernmental Panel on Climate Change, Geneva, Switzerland.

Number of Tree Seedlings Grown for 10 Years

A medium-growth coniferous tree, planted in an urban setting and allowed to grow for 10 years, sequesters 23.2 lbs of carbon. This estimate is based on the following assumptions:

- The medium-growth coniferous trees are raised in a nursery for 1 year until they become 1 inch in diameter at 4.5 feet above the ground (the size of tree purchased in a 15-gallon container).
- The nursery-grown trees are then planted in a suburban/urban setting; the trees are not densely planted.
- The calculation considers "survival factors" developed by U.S. Department of Energy (1998). For example, after 5 years (1 year in the nursery and four in the urban setting), the probability of survival is 68%; after 10 years, the probability declines to 59%. For each year, the sequestration rate (in lbs per tree) is multiplied by the survival factor to yield a probability-weighted sequestration rate. These values are summed for the 10-year period, beginning from the time of planting, to derive the estimate of 23.2 lbs of carbon per tree.

Note that the following caveats to these assumptions:

- While most trees take 1 year in a nursery to reach the seedling stage, trees grown under different conditions and trees of certain species may take longer: up to 6 years.
- Average survival rates in urban areas are based on broad assumptions, and the rates will vary significantly depending on site conditions.
- Carbon sequestration is dependent on the growth rate, which varies by location and other conditions.
- This method estimates only direct sequestration of carbon and does not include the energy savings that result from buildings being shaded by urban tree cover.

To convert to units of metric tons CO_2 per tree, multiply by the ratio of the molecular weight of carbon dioxide to that of carbon (44/12) and the ratio of metric tons per pound (1/2,204.6).

CALCULATION

23.2 lb C/tree × (44 units of CO_2 ÷ 12 units C) × 1 metric ton ÷ 2,204.6 lb = **0.039 metric ton CO_2 per urban tree planted**.

Source:

US DOE (1998). Method of Calculating Carbon Sequestration by Trees in Urban and Suburban Settings, Voluntary Reporting of Greenhouse Gases, U.S. Department of Energy, Energy Information administration (16 pp, 111 K, about PDF).

Acres of U.S. Forests Storing Carbon for One Year

Growing forests accumulate and store carbon. Through the process of photosynthesis, trees remove CO_2 from the atmosphere and store it as cellulose, lignin, and other compounds. The rate of accumulation is equal to growth minus removals (i.e., harvest for the production of paper and wood) minus decomposition. In most U.S. forests, growth exceeds removals and decomposition, so the amount of carbon stored nationally is increasing overall.

Calculation for U.S. Forests

The *Inventory of U.S. Greenhouse Gas Emissions and Sinks: 1990–2010* (EPA, 2012) provides data on the net greenhouse gas flux resulting from the use and changes in forest land areas. Note that the term "flux" is used here to encompass both emission of greenhouse gases to the atmosphere, and removal of C from the atmosphere. Removal of C from the atmosphere is also referred to as "carbon sequestration." Forest land in the United States includes land that is at least 10% stocked with trees of any size. Timberland is defined as unreserved productive forest land producing or capable of producing crops of industrial wood. Productivity is at a minimum rate of 20 cubic feet of industrial wood per acre per year. Net changes in carbon attributed to harvest wood products are not included in the calculation. The remaining portion of forest land is classified as "reserved forest land," which is forest withdrawn from timber use by statute or regulation, of "other forest land," which includes forests that are incapable of growing timber at a rate of <20 or more cubic feet per acre per year (Smith et al., 2010).

CALCULATION

Annual net change in carbon stocks per area in yearn

$$= \frac{(\text{Carbon stocks})_{(t+1)} - \text{Carbon stocks}_t}{\text{Area of land remaining in the same land-use category}}$$

Step 1: Determine the carbon stock change between years by subtracting carbon stocks in year t from carbon stocks in year $(t+1)$. (This includes carbon stocks in the above-ground biomass, below-ground biomass, dead wood, litter, and soil organic carbon pools.)

Step 2: Determine the annual net change in carbon stocks (i.e., sequestration) per area by dividing the carbon stock change in U.S. forests from step 1 by the total area of U.S. forests remaining in forests in year $n+1$ (i.e., the area of land that did not change land-use categories between the time periods).

Note that applying these calculations to data developed by the USDA Forest Service of the *Inventory of U.S. Greenhouse Gas Emissions and Sinks: 1990–2010* yields a result of 150 metric tons of carbon per hectare (or 61 metric tons of carbon per acre) for the carbon stock density of U.S. forests in 2010, with an annual net change in carbon stock per acre in 2010 of 0.82 metric tons of carbon sequestered per hectare per year (or 0.33 metric tons of carbon sequestered per acre per year). These values include carbon in the five forest pools: above-ground biomass, below-ground biomass, deadwood, litter, and soil organic carbon, and are based on state-level Forest Inventory and Analysis (FIA) data. Forest carbon stocks and carbon stock change are based on the stock difference methodology and algorithms described by Smith, Heath, and Nichols (2010).

Conversion Factor for Carbon Sequestered Annually by 1 Acre of Average U.S. Forest

CALCULATION

-0.33 metric ton C/acre/year $\left(44 \text{ units } CO_2 \div 12 \text{ units C}\right)$

$= -1.22$ metric ton CO_2 sequestered annually by one acre of average U.S. forest.

Note: Negative values indicate carbon sequestration.

Note that this is an estimate for "average" U.S. forests in 2010; i.e., for U.S. forests as a whole in 2010. Significant geographical variations underlie the national estimates, and the values calculated here might not be representative of individual regions or states. To estimate carbon sequestered for addition acres in one year, simply multiply the number of acres by 1.22 mt CO_2 acre/year. From 2000 to 2010, the average annual sequestration per area was 0.73 metric tons C hectare/year (or 0.30 metric tons C acre/year) in the United States, with a minimum value of 0.36 metric tons C hectare/year (or 0.15 metric tons C acre/year) in 2000, and a maximum value of 0.83 metric tons C hectare/year (or 0.34 metric tons C acre/year) in 2006.

Sources:

> EPA (2012). *Inventory of U.S. Greenhouse Gas Emissions and Sinks: 1990–2010*. U.S. Environmental Protection Agency, Washington, DC. USEPA #430-R-12-001 (PDF) (481 pp, 16,6MB).
> IPCC (2006). *Guidelines for National Greenhouse Gas Inventories, Volume 4, Agriculture, Forestry and other Land Use*. Task Force on National Greenhouse Gas Inventories. Geneva: Intergovernmental Panel on Climate Change.
> Smith, J., Heath, L., & Nichols, M. (2010). *U.S. Forest Carb on Calculation Tool User's Guide: Forestland Carbon Stocks and Net Annual Stock Change*. General Technical Report NRS-13 revised, U.S. Department of Agriculture Forest Service, Northern Research Station.

Acres of U.S. Forest Preserved from Conversion to Croplands

Based on data developed by the USDA Forest Service for the *Inventory of U.S. Greenhouse Gas Emissions and Sinks: 1990–2010*, the carbon stock density of U.S. forests in 2010 was 150 metric tons of carbon per hectare (or 61 metric tons of carbon per acre) (EPA, 2012). This estimate is composed of the five carbon pools: aboveground biomass (52 metric tons C/hectare), belowground biomass (10 metric tons C/hectare), dead wood (9 meters tons C/hectare), litter (17 metric tons C/hectare), and soil organic carbon (62 metric tons C/hectare).

The *Inventory of U.S. Greenhouse Gas Emissions and Sinks: 1990–2010* estimates soil carbon stock changes using U.S.-specific equations and data from the USDA Natural Resource Inventory and the century biogeochemical model (EPA, 2012). When calculating carbon stock exchanges in biomass due to conversion from forestland to cropland, the IPCC guidelines indicate that the average carbon stock change is equal to the carbon stock change due to removal of biomass from the outgoing land use (i.e., forestland) plus the carbon stocks from 1 year of growth in the incoming land use (i.e., cropland), or the carbon in biomass immediately after the conversion minus the carbon in biomass prior to the conversion plus the carbon stocks from 1 year of growth in the incoming land use (i.e., cropland) (IPCC, 2006). The carbon stock in annual cropland biomass after 1 year is 5 metric tons C per hectare, and the carbon content of dry aboveground biomass is 45% (IPCC, 2006). Therefore, the carbon stock in cropland after 1 year of growth is estimated to be 2.25 metric tons C per hectare (or 0.91 metric tons C per acre).

The averaged reference soil carbon stock (for high-activity clay, low-activity clay, and sandy soils for all climate regions in the United States) is 40.83 metric tons C/hectare (EPA, 2012). Carbon stock change in soils is time-dependent, with a default time period for transition between equilibrium soil organic carbon values of 20 years for mineral soils in cropland systems (IPCC, 2006). Consequently, it assumed that the change in equilibrium mineral soil organic carbon will be annualized over 20 years to represent the annual flux. The IPCC (2006) guidelines indicate that there are insufficient data to provide a default approach or parameters to estimate carbon stock change from dead organic matter pools or below-ground carbon stocks in perennial cropland (IPCC, 2006).

Calculation for Converting U.S. Forests to U.S. Cropland

Annual Change in Biomass Carbon Stocks on Land Converted to Other Land-Use Category

$$\Delta CB = \Delta C_G + C_{Conversion} - \Delta C_L$$

Where:
- ΔCB = annual change in carbon stocks in biomass on land converted to another land-use category
- ΔC_G = annual increase in carbon stocks in biomass due to growth on land converted to another land-use category (i.e., 2.25 metric tons C/hectare)
- $C_{Conversion}$ = initial change in carbon stocks in biomass on land converted to another land-use category. The sum of the carbon stocks in aboveground, belowground, deadwood, and litter biomass (−88.47 metric tons C/hectare). Immediately after conversion from forestland to cropland, biomass is assumed to be zero, as the land is cleared of all vegetation before planting crops).
- ΔC_L = annual decrease in biomass stocks due to losses from harvesting, fuel wood gathering, and disturbances on land converted to other land-use category (assumed to be zero).

Therefore: $\Delta CB = \Delta C_G + C_{Conversion} - \Delta C_L = -86.22$ metric tons C/hectare/year of biomass carbon stocks are lost when forestland is converted to cropland.

Annual Change in Organic Stocks in Mineral Soils

$$\Delta C_{Mineral} = (SOC_O - SOC_{(O-T)}) \div D$$

Where:

- $\Delta C_{Mineral}$ = annual change in carbon stocks in mineral soils
- $SOC_{(O-T)}$ = soil organic carbon stock at beginning of inventory time period (i.e., 62 mt C/hectare)
- D = Time dependence of stock change factors which is the default time period for transition between equilibrium SOC values (i.e., 20 years for cropland systems)

Therefore: $\Delta C_{Mineral} = (SOC_O - SOC_{(O-T)}) \div D = (40.83 - 62) \div 20 = -1.06$ metric tons C/hectare/year of soil organic C lost.

Source:

IPPC (2006). *2006 IPCC Guidelines for National Greenhouse Gas Inventories.* Intergovernmental Panel on Climate Change, Geneva, Switzerland.

Consequently, the change in carbon density from converting forestland to cropland would be −86.22 metric tons of C/hectare/year of biomass plus −1.06 metric tons c/hectare/year of soil organic C, equaling a total loss of 87.28 metric tons C/hectare/year (or −35.32 metric tons C/acre/year).

To convert to carbon dioxide, multiply by the ratio of the molecular weight of carbon dioxide to that of carbon (44/12), to yield a value of −320.01 metric tons CO_2 hectare/year (or −129.51 metric tons CO_2 acre/year).

> Conversion Factor for Carbon Sequestered Annually by 1 acre of Forest Preserved from Conversion to Cropland
>
> −35.32 metric tons C/acre/year (44 units CO_2 ÷ 12 units C) = −129.51 metric tons CO_2/acre/year
>
> Note: Negative values indicate CO_2 that is not emitted.

To estimate CO_2 not emitted when an acre of forest is preserved from conversion to cropland, simply multiply the number of acres of forest not converted by −129.51 mt CO_2e/acre/year. Note that this calculation method assumes that all of the forest biomass is oxidized during clearing (i.e., none of the burned biomass remains as charcoal or ash). Also note that this estimate only includes mineral soil carbon stocks, as most forests in the contiguous United States are growing on mineral soils. In the case of mineral soil forests, soil carbon stocks could be replenished or even increased, depending on the starting stocks, how the agricultural lands are managed, and the period over which lands are managed.

Sources:

> EPA (2012). *Inventory of U.S. Greenhouse gas Emissions and Sinks: 1990–2010*. U.S. Environmental Protection Agency, Washington, DC. USEPA #430-R-12-001 (PDF) (481 pp, 16,6MB).
> IPCC (2006). *Guidelines for National Greenhouse Gas Inventories, Volume 4, Agriculture, Forestry and Other Land Use*. Task Force on National Greenhouse Gas Inventories. Geneva: Intergovernmental Panel on Climate Change.

Propane Cylinders Used for Home Barbecues

Propane is 81.7% carbon (EPA, 2013). The fraction oxidized is 100% (IPCC, 2006). Carbon dioxide emissions per pound of propane were determined by multiplying the weight of propane in a cylinder times the carbon content percentage times the fraction oxidized times the ratio of the molecular weight of carbon dioxide to the of carbon (44/12). Propane cylinders vary with respect to size; for the purpose of this equivalency calculation, a typical cylinder for home use was assumed to contain 18 pounds of propane.

> **CALCULATION**
>
> 18 pounds propane/1 cylinder × 0.817 pounds C/pound propane × 0.4536 kilograms/pound
>
> × 44 kg CO_2/12 kg C × 1 metric ton/1,000 kg = **0.024 metric tons CO_2 / cylinder**

Sources:

EPA (2013). Inventory of U.S. Greenhouse Gas Emissions and Sinks: 1990–2011, Annex 2 (Methodology for estimating CO2 emissions from fossil fuel combustion), Table A-48, P. A-82, U.S. Environmental Protection Agency, Washington, DC. U.S. EPA #430-R-13-001 (PDF).

IPCC (2006). 2006 IPCC Guidelines for national Greenhouse Gas Inventories. Intergovernmental Panel on Climate Change, Geneva, Switzerland.

Railcars of Coal Burned

The average heat content of coal consume din the U.S. in 1013 was 21.48 mmBtu per metric ton (EPA, 2014). The average carbon coefficient of coal combusted for electricity generation in 2012 was 26.05 kilograms of carbon per mmBtu (EPA, 2013). The fraction oxidized is 100 percent (IPCC, 2006).

Carbon Dioxide emissions per ton of coal were determined by multiplying heat content times the carbon coefficient times the fraction oxidized times the ratio of the molecular weight of carbon dioxide to that of carbon (44/12). The amount of coal in an average railcar was assumed to be 100.19 short tons, or 90.89 metric tons (Hancock, 2001).

CALCULATION

21.48 mmBtu/metric ton coal × 26.05 kg/C/mmBtu × 44 kg CO_2/12 kg C

× 90.89 metric tons coal/railcar × 1 metric ton/1,000 kg = **186.50 metric tons CO_2 / railcar**

Sources:

EIA (2014). *Monthly Energy Review*. February 2014. Approximate Heat Content of Coal and Coal Coke, Table A-5. (1 pp, 73.2kb, About PDF).

EPA (2013). *Inventory of U.S. Greenhouse Gas Emissions and Sinks: 1990–2011, Annex 2 (Methodology for estimating CO2 emissions from fossil fuel combustion)*, Table A-48, P. A-82, U.S. Environmental Protection Agency, Washington, DC. U.S. EPA #430-R-13-001 (PDF).

Hancock (2001). Hancock, K and Sreekanth, A. *Conversion of Weight of Freight to Number of Railcars*. Transportation Research Board, Paper 01–2056.

IPCC (2006). *2006 IPCC Guidelines for national Greenhouse Gas Inventories*. Intergovernmental Panel on Climate Change, Geneva, Switzerland.

Pounds of Coal Burned

The average heat content of coal consumed in the U.S. in 2013 was 21.48 mmBtu per metric ton (EIA, 2014). The average carbon coefficient of coal combusted for electricity generation in 2012 was 26.05 kilograms car kg C/mmBtu × 44 kg CO_2 bon per mmBtu (EPA, 2013). The fraction oxidized is 100% (IPCC, 2006).

Carbon dioxide emissions per pound of coal were determined by multiplying heat content times the carbon coefficient times that fraction oxidized times the ratio of the molecular weight of carbon dioxide to that of carbon (44/12).

CALCULATION

21.48 mmBtu/metric ton coal × 26.05 kg C/mmBtu × 44 kg CO_2/12 kg C

× 1 metric ton coal/2,204.6 pound of coal × 1 metric ton/1,000 kg

= **9.31 × 10^{-4} metric tons CO_2 / pound of coal**

Sources:

EIA (2014). *Monthly Energy Review.* February 2014. Approximate Heat Content of Coal and Coal Coke, Table A-5. (1 pp, 73.2kb, About PDF).

EPA (2013). *Inventory of U.S. Greenhouse Gas Emissions and Sinks: 1990–2011, Annex 2 (Methodology for estimating CO2 emissions from fossil fuel combustion),* Table A-48, P. A-82, U.S. Environmental Protection Agency, Washington, DC. U.S. EPA #430-R-13-001 (PDF).

IPCC (2006). *2006 IPCC Guidelines for national Greenhouse Gas Inventories.* Intergovernmental Panel on Climate Change, Geneva, Switzerland.

Tons of Waste Recycled Instead of Landfilled

To develop the conversion factor for recycling rather than landfilling waste, emission factors from EPA's WAste Reduction Model (WARM) were used (EPA, 2012). These emission factors were developed following a life-cycle assessment methodology using estimation techniques developed for national inventories of greenhouse gas emissions. According to WARM, the net emission reduction from recycling mixed recyclables (e.g., paper, metals, plastics), compared with a baseline in which the material is landfilled, is 0.73 metric tons of carbon dioxide equivalent by multiplying by 44/12, the molecular weight ratio of carbon dioxide to carbon.

CALCULATION

0.76 metric tons of carbon equivalent/ton × 44 kg CO_2/12 kg C

= 2.79 metric tons CO_2 equivalent/ton of waste recycled instead of landfilled

Source:

EPA (2013). *Waste Reduction Model (WARM).* U.S. Environmental Protection Agency. Washington, DC.

Number of Garbage Trucks of Waste Recycled Instead of Landfilled

The carbon dioxide equivalent emission avoided from recycling instead of landfilling 1 ton of waste is 2.67 metric tons CO_2 equivalent per ton, as calculated in the "Tons of waste recycled instead of landfilled" section above.

Carbon dioxide emissions reduced per garbage truck full of waste were determined by multiplying emissions avoided from recycling instead of landfilling 1 ton of waste by the amount of waste in an average garbage truck. The amount of waste in an average garbage truck was assumed to be 7 tons (EPA, 2002).

Units, Standards of Measurement & Conversions

CALCULATION

2.79 metric tons CO_2 equivalent/ton of waste recycled instead of landfilled × 7 tons/garbage truck

= 19.51 metric tons CO_2E / garbage truck of waste recycled instead of landfilled

Sources:

EPA (2002). *Waste Transfer Stations: A Manual for Decision-Making.* Washington, DC: U.S. Environmental Protection Agency.

EPA (2013). *Waste Reduction Model (WARM).* Washington, DC: U.S. Environmental Protection Agency.

Coal-Fired Power Plant Emissions for One Year

In 2010, a total of 454 power plants used coal to generate at least 95% of their electricity (EPA, 2014). These plants emitted 1,729,127,770.8 metric tons of CO_2 in 2010. Carbon dioxide emissions per power plant were calculated by dividing the total emissions from power plants whose primary source of fuel was coal by the number of power plants.

CALCULATION

1,729,127,770.8 metric tons of CO_2 × 1/454 power plants

= 3,808,651 metric tons CO_2 / power plant

Source:

EPA (2014). *eGRID 2010 Data.* Washington, DC: U.S. Environmental Protection Agency.

Number of Wind Turbines Installed

In 2012, the average nameplate capacity of wind turbines installed in the U.S. was 1.94 MW (DOE, 2013). The average wind capacity factor in the U.S. in 2012 was 31% (DOE, 2013). Electricity generation from an average wind turbine was determined by multiplying the average nameplate capacity of a wind turbine in the U.S. (1.94 MW) by the average U.S. wind capacity factor (0.31) and by the number of hours per year. It was assumed that the electricity generated from an installed wind turbine would replace marginal sources of grid electricity. The U.S. annual non-baseload CO_2 output, emission rate to convert reductions of kilowatt-hours into avoided units of carbon dioxide emissions is 6.89551×10^{-4}. Carbon dioxide emissions avoided per wind turbine installed were determined by multiplying the average electricity generated per wind turbine in a year by the national average non-baseload grid electricity CO_2 output rate (EPA, 2012).

CALCULATION

$1.94_{MW\,average\,capacity} \times 0.31 \times 8{,}760$ hours/year $\times 1{,}000$ kWh/MWh $\times 6.89551$

$\times 10^{-4}$ metric tons CO_2/kWh reduced = **3,633 metric tons CO_2/wind turbine installed**

Sources:

DOE (2013). 2012 *Wind Technologies Market Report* (PDF) (92 pp, 3.4 MB). U.S. Department of Energy, Energy Efficiency and Renewable Energy Division, Washington, DC.

EPA (2012). *Inventory of U.S. Greenhouse gas Emissions and Sinks: 1990–2010.* U.S. Environmental Protection Agency, Washington, DC. USEPA #430-R-12-001 (PDF) (481 pp, 16,6MB).

NOTES

1. Much of the information in this section is from F.R. Spellman (2015) *Economics for the Environmental Professional.* Boca Raton, FL: CRC Press.
2. Much of the information in this section is from F.R. Spellman (2014). *The Environmental Impacts of Renewable Energy.* Boca Raton, FL: CRC Press.
3. From USDA (2001). *Fluorescent Lamp Retrofits: Savings or Fantasy?* Accessed 06/05/14 @ http://fs.fed.us/t-d/pubs/htmlpubs/htm01712310/index.htm.

REFERENCES

Harris, J.M. (2006). *Environmental and Natural Resource Economics*, 2nd ed. Boston, MA: Houghton Mifflin Company.

Spellman, F.R. (2002). *The Science of Environmental Pollution.* Lancaster, PA: Technomic Publishing Company.

Spellman, F.R. (2012). *The Science of Environmental Pollution*, 2nd ed. Lancaster, PA: Technomic Publishing Company.

3 Math Operations Review

Not everything that counts can be counted, and not everything that can be counted counts.

Albert Einstein

INTRODUCTION

Most calculations required by environmental engineers (as with many others) start with the basics, such as addition, subtraction, multiplication, division, sequence of operations, and others. Although many of the operations are fundamental tools within each environmental practitioner's toolbox, these tools must be used on a consistent basis to remain sharp in their use. Environmental practitioners should master basic math definitions and the formation of problems: daily operations require the calculation of percentage, average, simple ratio, geometric dimensions, threshold odor number, force, pressure, and head, and the use of dimensional analysis and advanced math operations. With regard to advanced math operations, it is true that an in-depth knowledge of algebra, linear algebra, vectors, trigonometry, analytic geometry, differential calculus, integral calculus, and differential equations is required in certain environmental engineering design and analysis operations; however, we leave discussion of these higher operations to the math textbooks; only the basics are covered in this chapter and for those proficient in math operations skipping this chapter may be an option of choice.

BASIC MATH TERMINOLOGY AND DEFINITIONS

The following basic definitions will aid in understanding the material in this chapter.

Integer, or an *integral number*, is a whole number. Thus, 1, 2, 3, 4, 5, 6, 7, 8, 9, 10, 11, and 12 are the first 12 positive integers.
Factor, or *divisor*, of a whole number is any other whole number that exactly divides it. Thus, 2 and 5 are factors of 10.
Prime number in math is a number that has no factors except itself and 1. Examples of prime numbers are 1, 3, 5, 7, and 11.
Composite number is a number that has factors other than itself and 1. Examples of composite numbers are 4, 6, 8, 9 and 12.
Common factor, or *common divisor*, of two or more numbers is a factor that will exactly divide each of them. If this factor is the largest factor possible, it is called the *greatest Common divisor*. Thus, 3 is a common divisor of 9 and 27, but 9 is the greatest common divisor of 9 and 27.
Multiple of a given number is a number that is exactly divisible by the given number. If a number is exactly divisible by two or more other numbers, it is a common multiple of them. The least (smallest) such number is called the *lowest common multiple*. Thus, 36 and 72 are common multiples of 12, 9, and 4; however, 36 is the lowest common multiple.
Even number is a number exactly divisible by 2. Thus, 2, 4, 6, 8, 10, and 12 are even integers.
Odd number is an integer that is not exactly divisible by 2. Thus, 1, 3, 5, 7, 9, and 11 are odd integers.
Product is the result of multiplying two or more numbers together. Thus, 25 is the product of $5 \times \%$. Also, 4 and 5 are factors of 20.

Quotient is the result of dividing one number by another. For example, 5 is the quotient of 20 divided by 4.

Dividend is a number to be divided; a *divisor* is a number that divides. For example, in $100 \div 20 = 5$, 100 is the dividend, 20 is the divisor, and 5 is the quotient.

Area is the area of an object, measured in square units.

Base is a term used to identify the bottom leg of a triangle, measured in linear units.

Circumference is the distance around an object, measured in linear units. When determine for other than circles, it may be called the *perimeter* of the figure, object, or landscape.

Cubic units are measurements used to express volume, cubic feet, cubic meters, etc.

Depth is the vertical distance from the bottom of the tank to the top. This is normally measured in terms of liquid depth and given in terms of sidewall depth (SWD), measured in linear units.

Diameter is the distance from one edge of a circle to the opposite edge passing through the center, measured in linear units.

Height the vertical distance from the base or bottom of a unit to the top or surface.

Linear units are measurements used to express distances: feet, inches, meters, yards, etc.

Pi, (π) is a number in the calculations involving circles, spheres, or cones: $\pi = 3.14$.

Radius is the distance from the center of a circle to the edge, measured in linear units.

Sphere is a container shaped like a ball.

Square units are measurements used to express area, square feet, square meters, acres, etc.

Volume is the capacity of the unit, how much it will hold, measured in cubic units (cubic feet, cubic meters) or in liquid volume units (gallons, liters, and million gallons).

Width is the distance from one side of the tank to the other, measured in linear units.

Key words:

- **of**: means to multiply
- **and**: means to add
- **per**: means to divide
- **less than**: means to subtract

SEQUENCE OF OPERATIONS

Mathematical operations such as addition, subtraction, multiplication, and division are usually performed in a certain order or sequence. Typically, multiplication and division operations are done prior to addition and subtraction operations In addition, mathematical operations are also generally performed from left to right using this hierarchy. The use of parentheses is also common to set apart operations that should be performed in a particular sequence.

Consider the expression $2 + 3 \times 4$; you might answer: 20 or if you know the rules, you will answer (correctly) 14. The preceding expression may be rendered $2 + (3 \times 4)$, but the brackets are unnecessary if you know the rules, multiplication has precedence without the parentheses.

Note: It is assumed that the reader has a fundamental knowledge of basic arithmetic and math operations. Thus, the purpose of the following section is to provide a brief review of the mathematical concepts and applications frequently employed by environmental practitioners.

SEQUENCE OF OPERATIONS—RULES

Rule 1: In a series of additions, the terms may be placed in any order and grouped in any way. Thus, $4 + 3 = 7$ and $3 + 4 = 7$; $(4 + 3) + (6 + 4) = 17$, $(6 + 3) + (4 + 4) = 17$, and $[6 + (3 + 4) + 4 = 17$.

Math Operations Review

Rule 2: In a series of subtractions, changing the order or the grouping of the terms may change the result. Thus, $100-30=70$, but $30-100=-70$; $(100-30)-10=60$, but $100-(30-10)=80$.

Rule 3: When no grouping is given, the subtractions are performed in the order written, from left to right. Thus, $100-30-15-4=51$; or by steps, $100-30=70$, $70-15=55$, $55-4=51$.

Rule 4: In a series of multiplications, the factors may be placed in any order and in any grouping. Thus, $[(2\times3)\times5]\times6=180$ and $5\times[2\times(6\times3)]=180$.

Rule 5: In a series of divisions, changing the order or the grouping may change the result. Thus, $100\div10=10$, but $10\div100=0.1$; $(100\div10)\div2=5$, but $100\div(10+2)=20$. Again, if no grouping is indicated, the divisions are performed in the order written, from left to right. Thus, $100\div10\div2$ is understood to mean $(100\div10)\div2$.

Rule 6: In a series of mixed mathematical operations, the convention is as follows: whenever no grouping is given, multiplications and divisions are to be performed in the order written, then additions and subtractions in the order written.

SEQUENCE OF OPERATIONS—EXAMPLES

In a series of additions, the terms may be placed in any order and grouped in any way.
Examples:

$$3+6=10 \text{ and } 6+4=10$$
$$(4+5)+(3+7)=19, (3+5)+(4+7)=19, \text{ and } [7+(5+4)]+3=19$$

In a series of subtractions, changing the order or the grouping of the terms may change the result.
Examples:

$$100-20=80, \text{ but } 20-100=-80$$
$$(100-30)-20=50, \text{ but } 100-(30-20)=90$$

When no grouping is given, the subtractions are performed in the order written—from left to right.
Example:

$$100-30-20-3=47$$
or by steps, $100-30=70$, $70-20=50$, $50-3=47$

In a series of multiplications, the factors may be placed in any order and in any grouping.
Example:

$$[(3\times3)\times5]\times6 = 270 \text{ and } 5\times[3\times(6\times3)] = 270$$

In a series of divisions, changing the order or the grouping may change the result.
Examples:

$$100\div10 = 10, \text{ but } 10\div100 = 0.1$$
$$(100\div10)\div2 = 5, \text{ but } 100\div(10\div2) = 20$$

If no grouping is indicated, the divisions are performed in the order written—from left to right.

Example:

$$100 \div 5 \div 2 \text{ is understood to mean} (100 \div 5) \div 2$$

In a series of mixed mathematical operations, the rule of thumb is, whenever no grouping is given, multiplications and divisions are to be performed in the order written, then additions and subtractions in the order written.

Consider the following classic example of sequence of operations (Staple, 2012):

Problem:

$$\text{Simplify } 4 - 3\left[4 - 2(6 - 3)\right] \div 2.$$

Solution:

$$4 - 3\left[4 - 2(6 - 3)\right] \div 2$$
$$= 4 - 3\left[4 - 2(3)\right] \div 2$$
$$= 4 - 3\left[4 - 6\right] \div 2$$
$$= 4 - 3\left[-2\right] \div 2$$
$$= 4 + 6 \div 2$$
$$= 4 + 3$$
$$= 7$$

PERCENT

The words "percent" mean "by the hundred." Percentage is often designated by the symbol %. Thus, 15% means 15 percent or 15/100 or 0.15. These equivalents may be written in the reverse order: 0.15 = 15/100 = 15%. In environmental engineering (e.g., water/wastewater treatment), percent is frequently used to express plant performance and for the control of biosolids treatment processes. When working with percent, the following key points are important:

1. Percent is another way of expressing a part of a whole.
2. As mentioned, the percent means "by the hundred," so a percentage is the number out of 100. To determine percent, divide the quantity we wish to express as a percent by the total quantity then multiply by 100.

$$\text{Percent}(\%) = \frac{\text{Part}}{\text{Whole}} \tag{3.1}$$

For example, 22 percent (or 22%) means 22 out of 100, or 22/100. Dividing 22 by 100, results in the decimal 0.22:

$$22\% = \frac{22}{100} = 0.22$$

Math Operations Review

3. When using percentage in calculations (such as when used to calculate hypochlorite dosages and when the percent available chlorine must be considered), the percentage must be converted to an equivalent decimal number; this is accomplished by dividing the percentage by 100.

   For example, calcium hypochlorite (HTH) contains 65% available chlorine.
   What is the decimal equivalent of 65%? Since 65% means 65 per hundred divides 65 by 100: 65/100, which is 0.65.

4. Decimals and fractions can be converted to percentages. The fraction is first converted to a decimal and then the decimal is multiplied by 100 to get the percentage.

   For example, if a 50-foot-high water tank has 26 ft of water in it, how full is the tank in terms of the percentage of its capacity?

$$\frac{26 \text{ ft}}{50 \text{ ft}} = 0.52 \, (\text{decimal equivalent})$$

$$0.52 \times 100 = 52$$

The tank is 52% full.

Example 3.1

Problem:

The plant operator removes 6,500 gal of biosolids from the settling tank. The biosolids contain 325 gal of solids. What is the percent solids in the biosolids?

Solution:

$$\text{Percent} = \frac{325 \text{ gal}}{6,500 \text{ gal}} \times 100$$

$$= 5\%$$

Example 3.2

Problem:

Convert 65% to decimal percent.

Solution:

$$\text{Decimal percent} = \frac{\text{Percent}}{100}$$

$$= \frac{65}{100}$$

$$= 0.65$$

Example 3.3

Problem:

Biosolids contain 5.8% solids. What is the concentration of solids in decimal percent?

Solution:

$$\text{Decimal percent} = \frac{5.8\%}{100} = 0.058$$

Key point: Unless otherwise noted, all calculations in the text using percent values require the percent to be converted to a decimal before use.

  Key point: To determine what quantity a percent equals, first convert the percent to a decimal and then multiply by the total quantity.

$$\text{Quantity} = \text{Total} \times \text{Decimal percent} \tag{3.2}$$

Example 3.4

Problem:

Biosolids drawn from the settling tank is 5% solids. If 2,800 gallons of biosolids are withdrawn, how many gallons of solids are removed?

Solution:

$$\text{Gallons} = \frac{5\%}{100} \times 2{,}800\,\text{gal} = 140\,\text{gal}$$

Example 3.5

Problem:

Convert 0.55 to percent.

Solution:

$$0.55 = \frac{55}{100} = 0.55 = 55\%$$

In converting 0.55 to 55%, we simply moved the decimal point two places to the right.

Example 3.6

Problem:

Convert 7/22 to a decimal percent to a percent.

Solution:

$$\frac{7}{22} = 0.318 = 0.318 \times 100 = 31.8\%$$

Example 3.7

Problem:

What is the percentage of 3 ppm?

Math Operations Review

Key point: Because 1 L of water weighs 1 kg (1,000 g=1,000,000 mg), milligrams per liter is parts per million (ppm).

Solution:

Since 3 parts per million (ppm) = 3 mg/L,

$$3 \text{ mg/L} = \frac{3 \text{ mg}}{1 \text{ L} \times 1{,}000{,}000 \text{ mg/L}} \times 100\%$$

$$= \frac{3}{10{,}000}\%$$

$$= 0.0003\%$$

Example 3.8

Problem:

How many mg/L is a 1.4%?

Solution:

$$1.4 = \frac{1.4}{100}, \text{ since the weight of 1 L water to } 10^6$$

$$\frac{1.4}{100} \times 1{,}000{,}000 \text{ mg/L}$$

$$= 14{,}000 \text{ mg/L}$$

Example 3.9

Problem:

Calculate pounds per million gallons for 1 ppm (1 mg/L) of water.

Solution:

Because 1 gal of water = 8.34 lb

$$1 \text{ ppm} = \frac{1 \text{ gal}}{10^6 \text{ gal}}$$

$$= \frac{1 \text{ gal} \times 8.34 \text{ lb/gal}}{\text{mil gal}}$$

$$= 8.34 \text{ lb/mil gal}$$

Example 3.10

Problem:

How many pounds of activated carbon (AC) are needed with 42 lb of sand to make the mixture 26% AC?

Solution:

Let x be the weight of AC

$$\frac{x}{42+x} = 0.26$$

$$x = 0.26(42+x)$$

$$x = 10.92 + 0.26x$$

$$(1-0.26)x = 10.92$$

$$x = \frac{10.92}{0.74} = 14.76 \text{ lb}$$

Example 3.11

Problem:

A pipe is laid at a rise of 140 mm in 22 m. What is the grade?

Solution:

$$\text{Grade} = \frac{140 \text{ mm}}{22 \text{ m}} \times 100(\%)$$

$$= \frac{140 \text{ mm}}{22 \times 1{,}000 \text{ mm}} \times 100\%$$

$$= 0.64\%$$

Example 3.12

Problem:

A motor is rated as 40 horsepower (hp). However, the output horsepower of the motor is only 26.5 hp. What is the efficiency of the motor?

Solution:

$$\text{Efficiency} = \frac{\text{hp output}}{\text{hp input}} \times 100\%$$

$$= \frac{26.5 \text{ hp}}{40 \text{ hp}} \times 100\%$$

$$= 66\%$$

SIGNIFICANT DIGITS

When rounding numbers, the following key points are important:

1. Numbers are rounded to reduce the number of digits to the right of the decimal point. This is done for convenience, not for accuracy.
2. **Rule**: a number is rounded off by dropping one or more numbers from the right and adding zeroes if necessary to place the decimal point. If the last figure dropped is 5 or more, increase the last retained figure by 1. If the last digit dropped is <5, do not increase the last retained figure. If the digit 5 is dropped, round off the preceding digit to the nearest *even* number.

Math Operations Review

Example 3.13

Problem:

Round off the following to one decimal: 34.73; 34.77; 34.75; 34.45; 34.35.

Solution:

$$34.73 = 34.7$$
$$34.77 = 34.8$$
$$34.75 = 34.8$$
$$34.45 = 34.4$$
$$34.35 = 34.4$$

Example 3.14

Problem:

Round off 10,546 to 4, 3, 2, and 1 significant figures.

Solution:

$$10,546 = 10,550 \text{ to 4 significant figures}$$
$$10,546 = 10,500 \text{ to 3 significant figures}$$
$$10,546 = 11,000 \text{ to 2 significant figures}$$
$$10,547 = 10,000 \text{ to 1 significant figures}$$

In determining significant figures, the following key points are important:

1. The concept of significant figures is related to rounding.
2. It can be used to determine where to round off.
    Key point: No answer can be more accurate than the least accurate piece of data used to calculate the answer.
3. **Rule**: significant figures are those numbers that are known to be reliable. The position of the decimal point does not determine the number of significant figures.

Example 3.15

Problem:

How many significant figures are in a measurement of 1.35 in?

Solution:

Three significant figures: 1, 3, and 5.

Example 3.16

Problem:

How many significant figures are in a measurement of 0.000135?

Solution:

Again, three significant figures: 1, 3, and 5. The three zeros are used only to place the decimal point.

Example 3.17

Problem:

How many significant figures are in a measurement of 103,500?

Solution:

Four significant figures: 1, 0, 3, and 5. The remaining two zeros are used to place the decimal point.

Example 3.18

Problem:

How many significant figures are in 27,000.0?

Solution:

There are six significant figures: 2, 7, 0, 0, 0, 0. In this case, the 0 means that the measurement is precise to 1/10 unit. The zeros indicate measured values and are not used solely to place the decimal point.

POWERS AND EXPONENTS

In working with powers and exponents, the following key points are important:

1. *Powers* are used to identify *area*, as in square feet, and volume as in *cubic feet*.
2. Powers can also be used to indicate that a number should be squared, cubed, etc. This later designation is the number of times a number must be multiplied times itself. For example, when several numbers are multiplied together, as $4 \times 5 \times 6 = 120$, the numbers, 4, 5, and 6 are the *factors*; 120 is the *product*.
3. If all the factors are alike, as $4 \times 4 \times 4 \times 4 = 256$, the product is called a *power*. Thus, 256 is a power of 4, and 4 is the *base* of the power. A *power* is a *product* obtained by using a base a certain number of times as a factor.
4. Instead of writing $4 \times 4 \times 4 \times 4$, it is more convenient to use an *exponent* to indicate that factor 4 is used as a factor four times. This exponent, a small number placed above and to the right of the base number, indicates known many times the base is to be used as a factor. Using this system of notation, the multiplication $4 \times 4 \times 4 \times 4$ is written as 4^4. The ^{4} is the *exponent*, showing that 4 is to be used as a factor 4 times.
5. These same considerations apply to letters (*a, b, x, y,* etc.) as well. For example:

$$z^2 = (z)(z) \quad \text{or} \quad z^4 = (z)(z)(z)(z)$$

Key point: When a number or letter does not have an exponent, it is considered to have an exponent of one.

Math Operations Review

The Powers of 1:

$1^0 = 1$

$1^1 = 1$

$1^2 = 1$

$1^3 = 1$

$1^4 = 1$

The Powers of 10:

$10^0 = 1$

$10^1 = 10$

$10^2 = 100$

$10^3 = 1,000$

$10^4 = 10,000$

Example 3.19

Problem:

How is the term 2^3 written in expanded form?

Solution:

The power (exponent) of 3 means that the base number (2) is multiplied by itself three times.

$$2^3 = (2)(2)(2)$$

Example 3.20

Problem:

How is the term $(3/8)^2$ written in expanded form?

Key point: When parentheses are used, the exponent refers to the entire term within the parentheses. Thus, in this example, $(3/8)^2$ means

Solution:

$$(3/8)^2 = (3/8)(3/8)$$

Key point: When a negative exponent is used with a number or term, a number can be reexpressed using a positive exponent.

$$6^{-3} = 1/6^3$$

Another example is

$$11^{-5} = 1/11^5$$

Example 3.21

Problem:

How is the term 8^{-3} written in expanded form?

$$8^{-3} = \frac{1}{8^3} = \frac{1}{(8)(8)(8)}$$

Key point: Any number or letter such as 3^0 or X^0 does not equal 3×1 or $X1$, but simply 1.

AVERAGES (ARITHMETIC MEAN)

Whether we speak of harmonic mean, geometric mean, or arithmetic mean, each is designed to find "the center," or the "middle" of the set of numbers. They capture the intuitive notion of a "central tendency" that may be present in the data. In statistical analysis, an "average of data" is a number that indicates the middle of the distribution of data values.

An *average* is a way of representing several different measurements as a single number. Although averages can be useful by telling "about" how much or how many, they can also be misleading, as we demonstrate below. You find two kinds of averages in environmental engineering calculations: the *arithmetic* mean (or simply *mean*) and the *median*.

Definition: The mean (what we usually refer to as an average) is the total of values of a set of observations divided by the number of observations. We simply add up all of the individual measurements and divide them by the total number of measurements we took.

Example 3.22

Problem:
When working with averages, the **mean** *(again, what we usually refer to as an average)* is the total of values of a set of observations divided by the number of observations. We simply add up all of the individual measurements and divide them by the total number of measurements we took. For example, the operator of a waterworks or wastewater treatment plant takes a chlorine residual measurement every day, and part of his/her operating log is shown in Table 3.1. Find the mean.

Solution:

Add up the seven chlorine residual readings: 0.9, 1.0, 0.9, 1.3, 1.1, 1.4, 1.2 = 7.8. Next, divide by the number of measurements, in this case, seven: 7.8 ÷ 7 = 1.11. The mean chlorine residual for the week was 1.11 mg/L.

TABLE 3.1
Daily Chlorine Residual Results

Day	Chlorine Residual (mg/L)
Monday	0.9
Tuesday	1.0
Wednesday	0.9
Thursday	1.3
Friday	1.1
Saturday	1.4
Sunday	1.2

Math Operations Review

Example 3.23

Problem:

A water system has four wells with the following capacities: 115, 100, 125, and 90 gpm. What is the mean?

Solution:

$$\frac{115\,\text{gpm} + 100\,\text{gpm} + 125\,\text{gpm} + 90\,\text{gpm}}{4} = \frac{430}{4} = 107.5\,\text{gpm}$$

Example 3.24

Problem:

A water system has four storage tanks. Three of them have a capacity of 100,000 gallons each, while the fourth has a capacity of 1 million gallons. What is the mean capacity of the storage tanks?

Solution:

The mean capacity of the storage tanks is

$$\frac{100,000 + 100,000 + 100,000 + 1,000,000}{4} = 325,000\,\text{gal}$$

Notice that no tank in this example has a capacity anywhere close to the mean.

Example 3.25

Problem:

Effluent BOD test results for the treatment plant during August are shown below. What is the average effluent BOD for August?

Test 1	22 mg/L
Test 2	33 mg/L
Test 3	21 mg/L
Test 4	13 mg/L

Solution:

$$\text{Average} = \frac{22\,\text{mg/L} + 33\,\text{mg/L} + 21\,\text{mg/L} + 13\,\text{mg/L}}{4} = 22.3\,\text{mg/L}$$

Example 3.26

Problem:

For the primary influent flow, the following composite-sampled solids concentrations were recorded for the week. What is the average SS?

Monday	310 mg/L SS
Tuesday	322 mg/L SS
Wednesday	305 mg/L SS
Thursday	326 mg/L SS
Friday	313 mg/L SS
Saturday	310 mg/L SS
Sunday	320 mg/L SS
Total	2,206 mg/L SS

Solution:

$$\text{Average SS} = \frac{\text{Sum of all measurements}}{\text{Number of measurements used}}$$

$$= \frac{2,206 \text{ mg/L SS}}{7}$$

$$= 315.1 \text{ mg/L SS}$$

RATIO

A **ratio** is an established relationship between two numbers; it is simply one number divided by another number. For example, if someone says, I'll give you four to one the Redskins over the Cowboys in the Super Bowl, what does that person mean? Four to one, or 4:1, is a ratio. If someone gives you 4 to 1, it's his/her $4 to your $1.

As another more pertinent example, an average of 3 cu ft of screenings are removed from each million gallons of wastewater treated, the ratio of screenings removed (cu ft) to treated wastewater (MG) is 3:1. Ratios are normally written using a colon (such as 2:1) or written as a faction (such as 2/1).

When working with ratio, the following key points are important to remember.

1. One place where fractions are used in calculations is when **ratios** are used, such as calculating.
2. A ratio is usually stated in form *A* is to *B* as *C* is to *D*, and we can write it as two fractions that are equal to each other:

$$\frac{A}{B} = \frac{C}{D}$$

3. Cross-multiplying solves ratio problems; that is, we multiply the left numerator (*A*) by the right denominator (*D*) and say that is equal to the left denominator (*B*) times the right numerator (*C*):

$$A \times D = B \times C$$

$$AD = BC$$

4. If one of the four items is unknown, dividing the two known items that are multiplied together by the known item that is multiplied by the unknown solves the ratio. For example, If 2 pounds of alum are needed to treat 500 gallons of water, how many pounds of alum will we need to treat 10,000 gallons? We can state this as a ratio: 2 pounds of alum is to 500 gallons of water as "pounds of alum is to 10,000 gallons."

Math Operations Review

This is set up in this manner:

$$\frac{1 \text{ lb alum}}{500\text{-gal water}} = \frac{x \text{ lb alum}}{10{,}000\text{-gal water}}$$

Cross-multiplying:

$$(500)(x) = (1) \times (10{,}000)$$

Transposing:

$$x = \frac{1 \times 10{,}000}{500}$$

$$= 20 \text{ lb alum}$$

For calculating proportion, for example, 5 gallons of fuel costs $5.40. How much does 15 gallons cost?

$$\frac{5\text{-gal}}{\$5.40} = \frac{15 \text{ gal}}{\$y}$$

$$5 \times y = 15 \times 5.40 = 81$$

$$y = \frac{81}{5} = \$16.20$$

Example 3.27

Problem:

If a pump will fill a tank in 20 hours at 4 gpm (gallons per minute), how long will it take a 10-gpm pump to fill the same tank?

Solution:

First, analyze the problem. Here, the unknown is some number of hours. But should the answer be larger or smaller than 20 hours? If a 4-gpm pump can fill the tank in 20 hours, a larger pump (10-gpm) should be able to complete the filling in <20 hours. Therefore, the answer should be <20 hours.

Now set up the proportion:

$$\frac{x \text{ hours}}{20 \text{ hours}} = \frac{4 \text{ gpm}}{10 \text{ gpm}}$$

$$x = \frac{(4)(20)}{10}$$

$$x = 8 \text{ hours}$$

Example 3.28

Problem:

Solve for x in the proportion problem given below.

Solution:

$$\frac{36}{180} = \frac{x}{4{,}450}$$

$$x = 890$$

Example 3.29

Problem:

Solve for the unknown value x in the problem given below.

Solution:

$$\frac{3.4}{2} = \frac{6}{x}$$

$$(3.4)(x) = (2)(6)$$

$$x = \frac{(2)(6)}{3.4}$$

$$= 3.53$$

Example 3.30

Problem:

1 lb of chlorine is dissolved in 65 gallons of water. To maintain the same concentration, how many pounds of chlorine would have to be dissolved in 150 gallons of water?

Solution:

$$\frac{1 \text{ lb}}{65\text{-gal}} = \frac{x \text{ lb}}{150 \text{ gal}}$$

$$(65)(x) = (1)(150)$$

$$x = \frac{(1)(150)}{65}$$

$$= 2.3 \text{ lb}$$

Example 3.31

Problem:

It takes 5 workers 50 hours to complete a job. At the same rate, how many hours would it take eight workers to complete the job?

Solution:

$$\frac{5 \text{ workers}}{8 \text{ workers}} = \frac{x \text{ hours}}{50 \text{ hours}}$$

$$x = \frac{(5)(50)}{8}$$

$$= 31.3 \text{ hours}$$

Example 3.32

Problem:

If 1.6 L of activated sludge (biosolids) with volatile suspended solids (VSS) of 1,900 mg/L is mixed with 7.2 L of raw domestic wastewater with BOD of 250 g/L, what is the F/M (food/microorganisms) ratio?

Solution:

$$\frac{F}{M} = \frac{\text{Amount of BOD}}{\text{Amount of VSS}}$$

$$= \frac{250\,\text{mg/L} \times 7.2\,\text{L}}{1,900\,\text{mg/L} \times 1.6\,\text{L}}$$

$$= \frac{0.59}{1}$$

$$= 0.59$$

DIMENSIONAL ANALYSIS

Dimensional analysis is a problem-solving method that uses the fact that one without changing its value can multiply any number or expression. It is a useful technique used to check if a problem is set up correctly. In using dimensional analysis to check a math setup, we work with the dimensions (units of measure) only—not with numbers.

An example of dimensional analysis that is common to everyday life is the unit pricing found in many hardware stores. A shopper can purchase a 1-pound box of nails for 98 cents in 1 store, whereas a warehouse store sells a 5-pound bag of the same nails for 3 dollars and 50 cents. The shopper will analyze this problem almost without thinking about it. The solution calls for reducing the problem to the price per pound. The pound is selected without much thought because it is the unit common to both stores. A shopper will pay 70 cents a pound for nails in the warehouse store or 98 cents in the local hardware store. Implicit in the solution to this problem is knowing the unit price, which is expressed in dollars per pound ($/lb).

To use the dimensional analysis method, we must know how to perform three basic operations:

Note: Unit factors may be made from any two terms that describe the same or equivalent "amounts" of what we are interested in. For example, we know that 1 in = 2.54 cm.

Basic operation:

1. To complete a division of units, always ensure that all units are written in the same format; it is best to express a horizontal fraction (such as gal/ft²) as a vertical fraction.
   Horizontal to vertical:

$$\text{gal/cu ft} \quad \text{to} \quad \frac{\text{gal}}{\text{cu ft}}$$

$$\text{psi} \quad \text{to} \quad \frac{\text{lb}}{\text{sq in.}}$$

The same procedures are applied in the following examples.

$$\text{ft}^3/\text{min} \quad \text{becomes} \quad \frac{\text{ft}^3}{\text{min}}$$

s/min   becomes $\dfrac{s}{sq\ in.}$

2. We must know how to divide by a fraction. For example,

$$\dfrac{\dfrac{lb}{d}}{\dfrac{min}{d}} \quad becomes \quad \dfrac{lb}{d} \times \dfrac{d}{min}$$

In the above, notice that the terms in the denominator were inverted before the fractions were multiplied. This is a standard rule that must be followed when dividing fractions. Another example is

$$\dfrac{mm^2}{\dfrac{mm^2}{m^2}} \quad becomes \quad mm^2 \times \dfrac{m^2}{mm^2}$$

3. We must know how to cancel or divide terms in the numerator and denominator of a fraction. After fractions have been rewritten in the vertical form and division by the fraction has been re-expressed as multiplication, as shown above, then the terms can be canceled (or divided) out.

Key point: For every term that is canceled in the numerator of a fraction, a similar term must be canceled in the denominator and vice versa, as shown below:

$$\dfrac{kg}{\cancel{d}} \times \dfrac{\cancel{d}}{min} = \dfrac{kg}{min}$$

$$\cancel{mm^2} \times \dfrac{m^2}{\cancel{mm^2}} = m^2$$

$$\dfrac{\cancel{gal}}{min} \times \dfrac{ft^3}{\cancel{gal}} = \dfrac{ft^3}{min}$$

Question: How are units that include exponents calculated?

When written with exponents, such as ft^3, a unit can be left as 1 or put in expanded form, (ft)(ft)(ft), depending on other units in the calculation. The point is that it is important to ensure that square and cubic terms are expressed uniformly, as sq ft, cu ft, or as ft^2. For dimensional analysis, the latter system is preferred.

For example, let's say that we wish to convert 1,400 ft^3 volume to gallons, and we will use 7.48 gal/ft^3 in the conversions. The question becomes do we multiply or divide by 7.48?

In the above instance, it is possible to use dimensional analysis to answer this question; that is, are we to multiply or divide by 7.48? To determine if the math setup is correct, only the dimensions are used.

First, try dividing the dimensions:

$$\dfrac{ft^3}{gal/ft^3} = \dfrac{ft^3}{\dfrac{gal}{ft^3}}$$

Math Operations Review

Then, the numerator and denominator are multiplied to get

$$= \frac{ft^6}{gal}$$

Thus, by dimensional analysis, we determine that if we divide the two dimensions (ft³ and gal/ft³), the units of the answer are ft⁶/gal, not gal. It is clear that division is not the right way to go in making this conversion.

What would have happened if we had multiplied the dimensions instead of dividing?

$$(ft^3)(gal/ft^3) = (ft^3)\left(\frac{gal}{ft^3}\right)$$

Then, multiply the numerator and denominator to obtain

$$= \frac{(ft^3)(gal)}{ft^3}$$

and cancel common terms to obtain

$$= \frac{(ft^3)(gal)}{ft^3}$$

Obviously, by multiplying the two dimensions (ft³ and gal/ft³), the answer will be in gallons, which is what we want. Thus, because the math setup is correct, we would then multiply the numbers to obtain the number of gallons.

$$(1,400 \text{ ft}^3)(7.48 \text{ gal/ft}^3) = 10,472 \text{ gal}$$

Now let's try another problem with exponents. We wish to obtain an answer in square feet. If we are given the two terms—70 ft³/s and 4.5 ft/s—is the following math setup correct?

$$(70 \text{ ft}^3/s)(4.5 \text{ ft/s})$$

First, only the dimensions are used to determine if the math setup is correct. By multiplying the two dimensions, we get

$$(ft^3/s)(ft/s) = \left(\frac{ft^3}{s}\right)\left(\frac{ft}{s}\right)$$

Then, multiply the terms in the numerators and denominators of the fraction:

$$= \frac{(ft^3)(ft)}{(s)(s)}$$

$$= \frac{ft^4}{s^2}$$

The math setup is incorrect because the dimensions of the answer are not square feet. Therefore, if we multiply the numbers, as shown above, the answer will be wrong.

Let's try the division of the two dimensions instead.

$$ft^3/s = \dfrac{\dfrac{ft^3}{s}}{\dfrac{ft}{s}}$$

Invert the denominator and multiply to get

$$= \left(\dfrac{ft^3}{(s)}\right)\left(\dfrac{s}{(ft)}\right)$$

$$= \dfrac{(ft)(ft)(ft)(s)}{(s)(ft)}$$

$$= \dfrac{(ft)(ft)(ft)(s)}{(s)(ft)}$$

$$= ft^2$$

Because the dimensions of the answer are square feet, this math setup is correct. Therefore, by dividing the numbers as was done with units, the answer will also be correct.

$$\dfrac{70 \ ft^3/s}{4.5 \ ft/s} = 15.56 \ ft^2$$

Example 3.33

Problem:

We are given two terms 5 m/s and 7 m² –and the answer to be obtained is in cubic meters per second (m³/s). Is multiplying the two terms the correct math setup?

Solution:

$$(m/s)(m^2) = \dfrac{m^2}{s} \times m^2$$

Multiply the numerators and denominator of the fraction:

$$= \dfrac{(m)(m^2)}{s}$$

$$= \dfrac{m^2}{s}$$

Because the dimensions of the answer are cubic meters per second (m³/s), the math setup is correct. Therefore, multiply the numbers to get the correct answer.

$$5(m/s)(7 \ m^2) = 35 \ m^3/s$$

Math Operations Review

Example 3.34

Problem:

Solve the following problem:
   Given: The flow rate in a water line is 2.3 ft³/s. What is the flow rate expressed as gallons per minute?
   Set up the math problem and then use dimensional analysis to check the math setup:

$$(2.3 \text{ ft}^3/\text{s})(7.48 \text{ gal/ft}^3)(60 \text{ s/min})$$

Then, use dimensional analysis is used to check the math setup:

$$(\text{ft}^3/\text{s})(\text{gal/ft}^3)(\text{s/min}) = \left(\frac{\text{ft}^3}{\text{s}}\right)\left(\frac{\text{gal}}{\text{ft}^3}\right)\left(\frac{\text{s}}{\text{min}}\right)$$

$$= \frac{\text{ft}^3}{\text{s}} \frac{\text{gal}}{\text{ft}^3} \frac{\text{s}}{\text{min}}$$

$$= \frac{\text{gal}}{\text{min}}$$

The math setup is correct as shown above. Therefore, this problem can be multiplied to get the answer in the correct units.

$$(2.3 \text{ ft}^3/\text{s})(7.48 \text{ gal/ft}^3)(60 \text{ s/min}) = 1032.24 \text{ gal/min}$$

Example 3.35

Problem:

During an 8-hour period, a water treatment plant treated 3.2-mil gal of water. What is the plant total volume treated per day, assuming the same treatment rate?

Solution:

$$\frac{3.2 \text{ mil gal}}{8 \text{ hours}} \times \frac{24 \text{ hours}}{\text{day}}$$

$$= \frac{3.2 \times 24}{8} \text{ MGD}$$

$$= 9.6 \text{ MGD}$$

Example 3.36

Problem:

A 1 MGD equals how many cubic feet per second (cfs)?

Solution:

$$1 \text{ MGD} = \frac{10^6}{1 \text{ day}}$$

$$\frac{10^6\text{-gal} \times 0.1337 \text{ ft}^3/\text{gal}}{1 \text{ day} \times 86,400 \text{ s/day}}$$

$$= \frac{133,700 \text{ ft}^3}{86,400 \text{ s}}$$

$$= 1.547 \text{ cfs}$$

Example 3.37

Problem:

A 10-gal empty tank weighs 4.6 lb. What is the total weight of the tank filled with 6 gal of water?

Solution:

$$\text{Weight of water} = 6\text{-gal} \times 8.34 \text{ lb/gal}$$

$$= 50.04 \text{ lb}$$

$$\text{Total weight} = 50.04 + 4.6 \text{ lbs}$$

$$= 54.6 \text{ lb}$$

Example 3.38

Problem:

The depth of biosolids applied to the biosolids drying bed 10 in. What is the depth in centimeters (2.54 cm = 1 in)?

Solution:

$$10 \text{ in} = 10 \times 2.54 \text{ cm}$$

$$= 25.4 \text{ cm}$$

THRESHOLD ODOR NUMBER (TON)

The environmental practitioner responsible for water supplies soon discovers that taste and odor are the most common customer complaint. Odor is typically measured and expressed in terms of a *threshold odor number* (TON), the ratio by which the sample has to be diluted for the odor to become virtually unnoticeable. In 1989, the USEPA issued a "Secondary Maximum Contaminant Level" (SMCL) of 3 TON for odor.

Note: Secondary Standards are parameters not related to health.

When a dilution is used, a number can be devised in clarifying odor.

$$\text{TON}(\text{threshold odor number}) = \frac{V_r + V_p}{V_r} \tag{3.3}$$

where:

V_T = volume tested
V_P = volume of dilution with odor-free distilled water

For $V_p = 0$,   TON = 1 (lowest value possible)

For $V_p = V_T$,   TON = 2

For $V_p = 2V_T$ TON = 3, etc.

Example 3.39

Problem:

The first detectable odor is observed when a 50-mL sample is diluted to 200 mL with odor-free water. What is the TON of the water sample?

Solution:

$$\text{TON} = \frac{200}{V_T} = \frac{200 \text{ mL}}{50 \text{ mL}}$$

$$= 4$$

GEOMETRICAL MEASUREMENTS

Environmental engineers involved in fisheries, water/wastewater treatment plants, and other operations dealing with tanks, basins, and ponds operations must know the area and volume of all tanks, basins, and ponds they deal with. For example, in water and wastewater treatment plant operations, the plant configuration usually consists of a series of tanks and channels. Proper design and operational control require the environmental practitioner and plant operator to perform several process control calculations. Many of these calculations include parameters such as the circumference or perimeter, area, or the volume of the tank or channel as part of the information necessary to determine the result. Many process calculations require computation of surface areas. Moreover, in fisheries operations, exact measurement of area and volume is essential in order to calculate stoking grates and chemicals applications. Stocking fish into a pond of uncertain area can result in poor production, more disease and possibly death. Chemical treatments can be ineffective if volume/area is underestimated and potentially lethal if it is overestimated (Masser and Jensen, 1991). To aid in performing these calculations, the following definitions and relevant equations used to calculate areas and volumes for several geometric shapes are provided.

- **Area**: the area of an object, measured in square units.
- **Base**: the term used to identify the bottom leg of a triangle, measured in linear units.
- **Circumference**: the distance around an object, measured in linear units. When determined for other than circles, it may be called the perimeter of the figure, object, or landscape.
- **Cubic units**: measurements used to express volume, cubic feet, cubic meters, etc.
- **Depth**: the vertical distance from the bottom of the tank to the top. Normally measured in terms of liquid depth and given in terms of side wall depth (SWD), measured in linear units.
- **Diameter**: the distance from one edge of a circle to the opposite edge passing through the center, measured in linear units.
- **Height**: the vertical distance from one end of an object to the other, measured in linear units.

Relevant Geometric Equations

Circumference of a circle	$C = \pi d = 2\pi r$
Perimeter of a square with side a	$P = 4a$
Perimeter of a rectangle with sides a and b	$P = 2a + 2b$
Perimeter of a triangle with sides a, b, and c	$P = a + b + c$
Area A of a circle with radius r ($d = 2r$)	$A = \pi d^2/4 = \pi r^2$
Area of duct in square feet when d is in inches	$A = 0.005454 d^2$
Area of A of a triangle with base b and height h	$A = 0.5 bh$
Area of A of a square with sides a	$A = a^2$
Area of A of a rectangle with sides a and b	$A = ab$
Area A of an ellipse with major axis a and minor axis b	$A = \pi ab$
Area A of a trapezoid with parallel sides a and b and height h	$A = 0.5(a+b)h$
Area A of a duct in square feet when d is in inches	$A = \pi d^2/576$
	$= 0.005454 d^2$
Volume V of a sphere with a radius r ($d = 2r$)	$V = 1.33 \pi r^3$
	$= 0.1667 \pi d^3$
Volume V of a cube with sides a	$V = a^3$
Volume V of a rectangular solid (sides a and b and height c)	$V = abc$
Volume V of a cylinder with a radius r and height h	$V = \pi r^2 h$
	$= \pi d^2 h/4$
Volume V of a pyramid	$V = 0.33$

- **Length**: the distance from one end of an object to the other, measured in linear units.
- **Linear units**: measurements used to express distances: feet, inches, meters, yards, etc.
- **Pi, π**: a number in the calculations involving circles, spheres, or cones ($\pi = 3.14$).
- **Radius**: the distance from the center of a circle to the edge, measured in linear units.
- **Sphere**: a container shaped like a ball.
- **Square units**: measurements used to express area, square feet, square meters, acres, etc.
- **Volume**: the capacity of the unit, how much it will hold, measured in cubic units (cubic feet, cubic meters) or in liquid volume units (gallons, liters, million gallons).
- **Width**: the distance from one from one side of the tank to the other, measured in linear units.

GEOMETRICAL CALCULATIONS

Perimeter and Circumference

On occasion, determining the distance around grounds or landscapes is required. In order to measure the distance around property, buildings, ponds, and basin-like structures, it is necessary to determine either perimeter or circumference. The *perimeter* is the distance around an object; a border or outer boundary. *Circumference* is the distance around a circle or circular object, such as a clarifier. Distance is linear measurement, which defines the distance (or length) along a line. Standard units of measurement like inches, feet, yards, and miles and metric units like centimeters, meters, and kilometers are used.

The perimeter of a rectangle (a four-sided figure with four right angles) is obtained by adding the lengths of the four sides (see Figure 3.1).

$$\text{Perimeter} = L_1 + L_2 + L_3 + L_4 \tag{3.4}$$

Math Operations Review

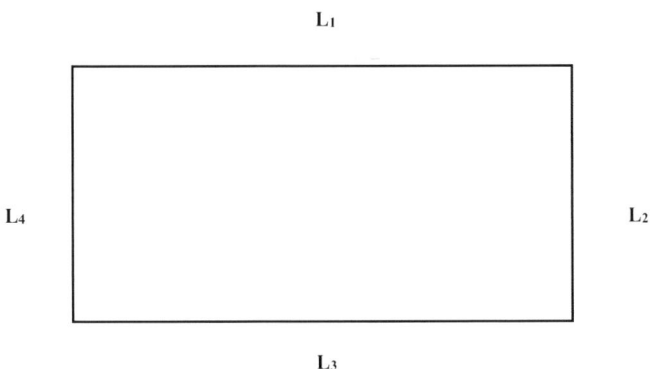

FIGURE 3.1 Perimeter.

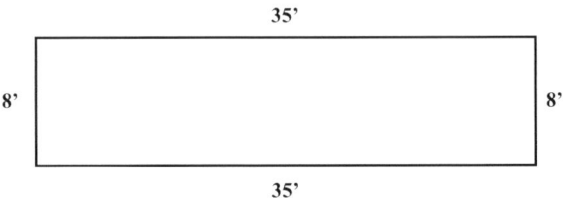

FIGURE 3.2 See Example 3.40.

Example 3.40

Problem:

Find the perimeter of the rectangle shown in Figure 3.2

Solution:

$$P = 35' + 8' + 35' + 8'$$
$$= 86'$$

Example 3.41

Problem:

What is the perimeter of a rectangular field if its length is 100 ft, and its width is 50 ft?

Solution:

$$\text{Perimeter} = 2 \times \text{length} + 2 \times \text{width}$$
$$= 2 \times 100 \text{ ft} + 2 \times 50 \text{ ft}$$
$$= 200 \text{ ft} + 100 \text{ ft}$$
$$= 300 \text{ ft}$$

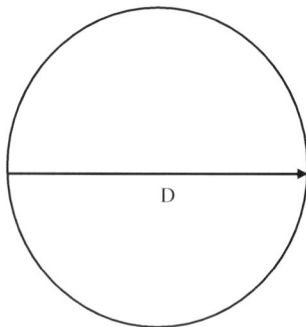

FIGURE 3.3 Diameter of circle.

Example 3.42

Problem:

What is the perimeter of a square whose side of 8 in?

Solution:

$$\text{Perimeter} = 2 \times \text{length} + 2 \times \text{width}$$
$$= 2 \times 8 \text{ in} + 2 \times 8 \text{ in}$$
$$= 16 \text{ in} + 16 \text{ in}$$
$$= 32 \text{ in}$$

The circumference is the length of the outer border of a circle. The circumference is found by multiplying pi (π) times the diameter (D) (diameter is a straight line passing through the center of a circle—the distance across the circle; see Figure 3.3).

$$C = \pi D \tag{3.5}$$

where
   C = circumference
   π = Greek letter pi = 3.1416
   D = diameter

Use this calculation if, for example, the circumference of a circular tank must be determined.

Example 3.43

Problem:

Find the circumference of a circle that has a diameter of 25 ft (π = 3.14)

Solution:

$$C = \pi \times 25'$$
$$= 3.14 \times 25'$$
$$= 78.5'$$

Example 3.44

Problem:

A circular chemical holding tank has a diameter of 18 m. What is the circumference of this tank?

Solution:

$$C = \pi \times 18\,m$$
$$= (3.14) \times (18\,m)$$
$$= 56.52\,m$$

Example 3.45

Problem:

An influent pipe inlet opening has a diameter of 6′. What is the circumference of the inlet opening in inches?

Solution:

$$C = \pi \times 6'$$
$$= 3.14 \times 6'$$
$$= 18.84\,ft$$

Area

For area measurements in fisheries and water/wastewater operations, four basic shapes are particularly important: circles, rectangles, triangles, and irregular shapes. Area is the amount of surface an object contains or the amount of material it takes to cover the surface. The area on top of a chemical tank is called the *surface area*. The area of the end of ventilation dust is called the *cross-sectional area* (the area at right angles to the length of ducting). Area is usually expressed in square units, such as square inches (in^2) or square feet (ft^2). Land may also be expressed in terms of square miles (sections) or acres (43,560 ft^2) or in the metric system as *hectares*. In fisheries operations, pond stocking rates, limiting rates and other important management decisions are based on surface area (Masser and Jensen, 1991).

If contractor's measurements or the country field office of the U.S Department of Agricultural Soil Conservation service does not have records on basin, lake or pond measurements, surveying basins, tanks, lagoons, and ponds using a transit is the most accurate way to determine area. Less accurate but acceptable methods of measuring basin or pond area ware chaining and pacing. Inaccuracies in these methods come from mismeasurements and measurement over uneven or sloping terrain. Measurements made on flat, or level areas are the most accurate.

Chaining uses a surveyor's chain or tape of known length. Stakes are placed at each end of the tape. The stakes are used to set or located the starting point for each progressive measurement and to maintain an exact count on the number of times the tape was moved. Sight down the stakes to keep the measurement in a straight line. The number of times the tape was moved multiplied by the length of the tape equals total distance.

Pacing uses the average distance of a person's pace or stride. To determine your pace length, measure a 100-foot distance and pace it, counting the number of strides. Pace in a comfortable and natural manner. Repeat the procedure several times and get an average distance for your stride. It is

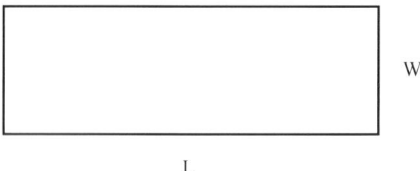

FIGURE 3.4 Rectangle.

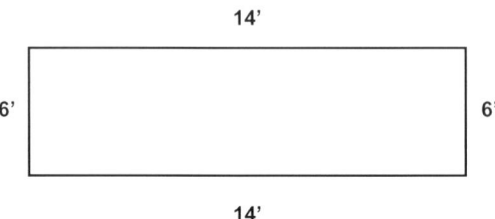

FIGURE 3.5 For Example 3.46.

good practice to always pace a distance more than once and average the number of paces (Masser and Jennings, 1991). The formula for calculating distances from pacing is:

$$\text{Distance}(\text{ft}) = \text{Total number of paces} \times \text{length of average pace}$$

A **rectangle** is a two-dimensional box. The area of a rectangle is found by multiplying the length (L) times width (W); see Figure 3.4.

$$\text{Area} = L \times W \tag{3.6}$$

Example 3.46

Problem:

Find the area of the rectangle shown in Figure 3.5.

Solution:

$$\text{Area} = L \times W$$
$$= 14' \times 6'$$
$$= 84 \text{ ft}^2$$

To find the **area of a circle**, we need to introduce one new term, the *radius*, which is represented by r. In Figure 3.6; we have a circle with a radius of 6 ft. The radius is any straight line that radiates from the center of the circle to some point on the circumference. By definition, all radii (plural of radius) of the same circle are equal. The surface area of a circle is determined by multiplying π times the radius squared.

$$\text{Area of circle} = \pi r^2 \tag{3.7}$$

Math Operations Review

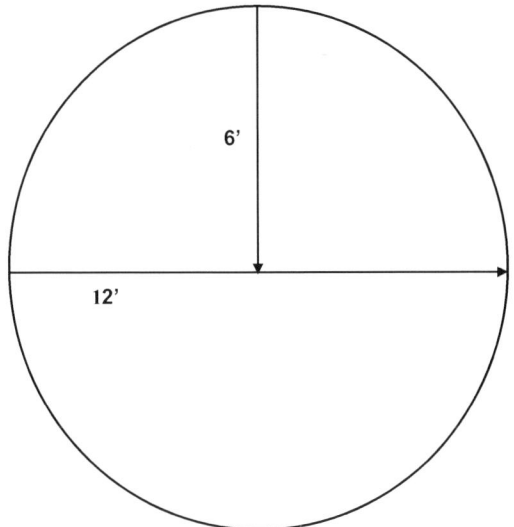

FIGURE 3.6 For Example 3.47.

where
 A = area

$$\pi = \text{pi } (3.14)$$

r = radius of circle—radius is one-half the diameter.

Example 3.47

Problem:

What is the area of the circle shown in Figure 3.6?

$$\text{Area of circle} = \pi r^2$$
$$= \pi 6^2$$
$$= 3.14 \times 36$$
$$= 113 \text{ ft}^2$$

If we were assigned to paint a water storage tank, we must know the surface area of the walls of the tank—we need to know how much paint is required. That is, we need to know the area of a circular or cylindrical tank. To determine the tank's surface area, we need to visualize the cylindrical walls as a rectangle wrapped around a circular base. The area of a rectangle is found by multiplying the length by the width; in this case, the width of the rectangle is the height of the wall, and the length of the rectangle is the distance around the circle, the circumference.

Thus, the area of the side walls of the circular tank is found by multiplying the circumference of the base ($C = \pi \times D$) times the height of the wall (H):

$$A = \pi \times D \times H$$
$$= \pi \times 20 \text{ ft} \times 25 \text{ ft}$$
$$= 3.14 \times 20 \text{ ft} \times 25 \text{ ft}$$
$$= 1{,}570 \text{ ft}^2$$

(3.8)

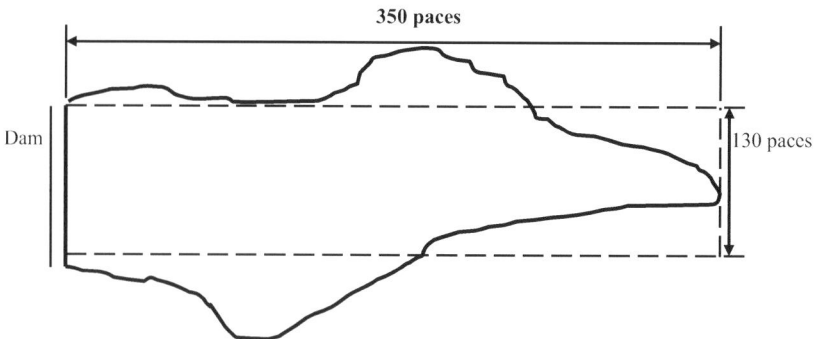

FIGURE 3.7 Pond area calculation.

To determine the amount of paint needed, remember to add the surface area of the top of the tank, which is 314 ft². Thus, the amount of paint needed must cover 1,570 ft² + 314 ft² = 1,884 or 1,885 ft². If the tank floor should be painted, add another 314 ft².

Many ponds are watershed ponds that have been built by damming valleys. These ponds are irregular in shape. If no good records exist on the pond, then a reasonable estimate can be made by changing or pacing off the pond margins and using the following procedures to calculate area.

1. Draw the general shape of the pond on graph paper.
2. Draw a rectangle on the pond shape that would approximate the area of the pond if some water were eliminated and placed onto an equal amount of land. This will give you a rectangle on which to base the calculation of area (see Figure 3.7).
3. Mark the corners of the rectangle (from the drawing) on the ground around the pond and chain or pace its length and width. For example, a length of 375 paces and a width of 130 paces and a pace length of 2.68 (for example) would be equal to 1,005 ft (375 paces × 2.68 ft/pace (pace length) by 348.4 ft.
4. Multiply the length times width to get the approximate pond area. For example, 1,005 ft × 348.4 ft = 350,142 ft² or 8.04 acres (350,142 ÷ 43,500).

Volume

The amount of space occupied by or contained in an object, volume (see Figure 3.8), is expressed in cubic units, such as cubic inches (in³), cubic feet (ft³), acre feet (1 acre foot = 43,560 ft³), etc. The volume of a rectangular object is obtained by multiplying the length times the width times the depth or height.

$$V = L \times W \times H \tag{3.9}$$

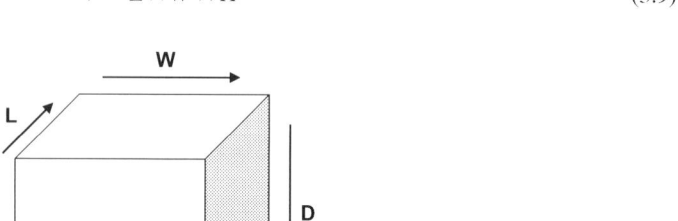

FIGURE 3.8 Volume.

Math Operations Review

where
L=length
W=width
D or H=depth or height

Example 3.48

Problem:

A unit rectangular process basin has a length of 15′, width of 7′, and depth of 9′, what is the volume of the basin?

Solution:

$$V = L \times W \times D$$
$$= 15 \text{ ft} \times 7 \text{ ft} \times 9 \text{ ft}$$
$$= 945 \text{ ft}^3$$

For environmental practitioners involved with fisheries and water/wastewater operators, representative surface areas are most often rectangles, triangles, circles, or a combination of these. Practical volume formulas used in fisheries and water/wastewater calculations are given in Table 3.2.

In determining the volume of round pipe and round surface areas, the following examples are helpful.

Example 3.49

Problem:

Find the volume of a 3-in round pipe that is 300-ft long.

Solution:

1. **Step 1**: Change the diameter of the duct from inches to feet by dividing by 12.

$$D = 3 \div 12 = 0.25 \text{ ft}$$

2. **Step 2**: Find the radius by dividing the diameter by 2.

$$R = 0.25 \text{ ft} \div 2 = 0.125$$

TABLE 3.2
Volume Formulas

Sphere Volume	$= (\pi/6) \text{ (diameter)}^3$
Cone volume	$= 1/3$ (volume of a cylinder)
Rectangular tank volume	$=$ (area of rectangle) (D or H)
	$=$ (LW) (D or H)
Cylinder volume	$=$ (area of cylinder) (D or H)
	$= \pi^2$ (D or H)
Triangle volume	$=$ (area of triangle) (D or H)
	$=$ (bh/2) (D or H)

3. **Step 3**: Find the volume.

$$V = L \times \pi r^2$$

$$= 300 \text{ ft} \times 3.14 \times 0.0156$$

$$= 14.72 \text{ ft}^2$$

Example 3.50

Problem:

Find the volume of a smokestack that is 24 in diameter (entire length) and 96 in tall. Find the radius of the stack. The radius is one-half the diameter.

$$24 \text{ in} \div 2 = 12 \text{ in}$$

Find the volume.

Solution:

$$V = H \times \pi r^2$$

$$= 96 \text{ in} \times \pi (12 \text{ in})^2$$

$$= 96 \text{ in} \times \pi (144 \text{ in}^2)$$

$$= 43{,}407 \text{ ft}^3$$

To determine the volume of a cone and sphere, we use the following equations and examples.

$$\text{Volume of cone} = \frac{\pi}{12} \times \text{diameter} \times \text{diameter} \times \text{height} \tag{3.10}$$

$$\frac{\pi}{12} = \frac{3.14}{12} = 0.262$$

Key point: The diameter used in the formula is the diameter of the base of the cone.

Example 3.51

Problem:

The bottom section of a circular settling tank has the shape of a cone. How many cubic feet of water are contained in this section of the tank if the tank has a diameter of 120 ft and the cone portion of the unit has a depth of 6 ft?

Solution:

$$\text{Volume, ft}^3 = 0.262 \times 120 \text{ ft} \times 120 \text{ ft} \times 6 \text{ ft} = 22{,}637 \text{ ft}^3$$

$$\text{Volume of sphere} = \frac{3.14}{6} \times \text{diameter} \times \text{diameter} \times \text{diameter} \tag{3.11}$$

$$\frac{\pi}{6} = \frac{3.14}{6} = 0.524$$

Math Operations Review

Example 3.52

Problem:

What is the volume of cubic feet of a gas storage container that is spherical and has a diameter of 60 ft?

Solution:

$$\text{Volume, ft}^3 = 0.524 \times 60 \text{ ft} \times 60 \text{ ft} \times 60 \text{ ft} = 113{,}184 \text{ ft}^3$$

Circular process and various water/chemical storage tanks are commonly found in water/wastewater treatment. A circular tank consists of a circular floor surface with a cylinder rising above it (see Figure 3.9). The volume of a circular tank is calculated by multiplying the surface area times the height of the tank walls.

Example 3.53

Problem:

If a tank is 20 ft in diameter and 25 ft deep, how many gallons of water will it hold?
   Hint: In this type of problem, calculate the surface area first, multiply by the height, and then convert it to gallons.

Solution:

$$r = D \div 2 = 20 \text{ ft} \div 2 = 10 \text{ fts}$$

$$A = \pi \times r^2$$

$$A = \pi \times 10 \text{ ft} \times 10 \text{ ft}$$

$$A = 314 \text{ ft}^2$$

$$V = A \times H$$

$$V = 314 \text{ ft}^2 \times 25 \text{ ft}$$

$$V = 7{,}850 \text{ ft}^3 \times 7.5 \text{ gal/ft}^3 = 58{,}875 \text{ gal}$$

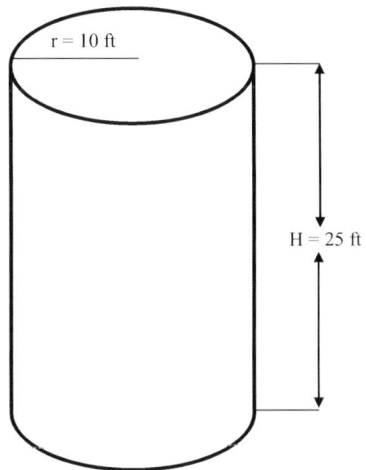

FIGURE 3.9 Circular or cylindrical water tank.

FORCE, PRESSURE, AND HEAD CALCULATIONS

Before we review calculations involving force, pressure, and head, we must first define these terms.

- **Force**: the push exerted by water on any confined surface. Force can be expressed in pounds, tons, grams, or kilograms.
- **Pressure**: the force per unit area. The most common way of expressing pressure is in pounds per square inch (psi).
- **Head**: the vertical distance or height of water above a reference point. Head is usually expressed in feet. In the case of water, head and pressure are related.

FORCE AND PRESSURE

Figure 3.10 helps to illustrate these terms. A cubical container measuring one foot on each side can hold one cubic foot of water. A basic fact of science states that one cubic foot of water weighs 62.4 pounds and contains 7.48 gallons. The force acting on the bottom of the container would be 62.4 pounds per square foot. The area of the bottom in square inches is:

$$1 \text{ ft}^2 = 12 \text{ in} \times 12 \text{ in} = 144 \text{ in}^2$$

Therefore, the pressure in pounds per square inch (psi) is:

$$\frac{62.4 \text{ lb/ft}^2}{1 \text{ ft}^2} = \frac{62.4 \text{ lb/ft}^2}{144 \text{ in}^2/\text{ft}^2} = 0.433 \text{ lb/in}^2 \text{ (psi)}$$

If we use the bottom of the container as our reference point, the head would be one foot. From this we can see that one foot of head is equal to 0.433 psi—an important parameter to remember. Figure 3.11 illustrates some other important relationships between pressure and head.

Important point: Force acts in a particular direction. Water in a tank exerts force down on the bottom and out of the sides. Pressure, however, acts in all directions. A marble at a water depth of one foot would have 0.433 psi of pressure acting inward on all sides.

Using the preceding information, we can develop Equations (3.12) and (3.13) for calculating pressure and head.

$$\text{Pressure (psi)} = 0.433 \times \text{head (ft)} \qquad (3.12)$$

$$\text{Head (ft)} = 2.31 \times \text{Pressure (psi)} \qquad (3.13)$$

HEAD

As mentioned, head is the vertical distance the water must be lifted from the supply tank or unit process to the discharge. The total head includes the vertical distance the liquid must be lifted (static

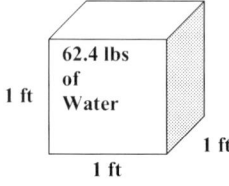

FIGURE 3.10 One cubic foot of water weights 62.4 lb.

Math Operations Review

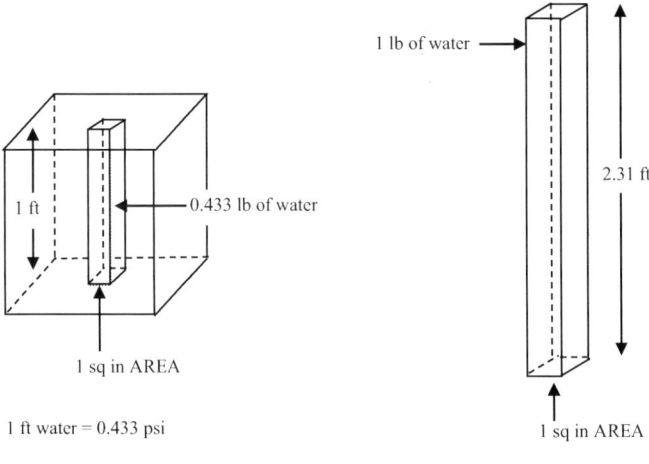

FIGURE 3.11 Shows the relationship between pressure and head.

head), the loss to friction (friction head), and the energy required to maintain the desired velocity (velocity head).

$$\text{Total head} = \text{Static head} + \text{friction head} + \text{velocity heads} \qquad (3.14)$$

Static Head
Static head is the actual vertical distance the liquid must be lifted.

$$\text{Static head} = \text{Discharge elevation} - \text{supply elevation} \qquad (3.15)$$

Example 3.54

Problem:

The supply tank is located at elevation 108 ft. The discharge point is at elevation of 205 ft. What is the static head in feet?

Solution:

$$\text{Static head, ft} = 205\,\text{ft} - 108\,\text{ft} = 97\,\text{ft}$$

Friction Head
Friction head is the equivalent distance of the energy that must be supplied to overcome friction. Engineering references include tables showing the equivalent vertical distance for various sizes and types of pipes, fittings, and valves. The total friction head is the sum of the equivalent vertical distances for each component.

$$\text{Friction head, ft} = \text{Energy losses due to friction} \qquad (3.16)$$

Velocity Head

Velocity head is the equivalent distance of the energy consumed in achieving and maintaining the desired velocity in the system.

$$\text{Velocity head, ft} = \text{Energy losses to maintain velocity} \quad (3.17)$$

Total Dynamic Head (Total System Head)

$$\text{Total head} = \text{Static head} + \text{friction head} + \text{velocity head} \quad (3.18)$$

Pressure/Head

The pressure exerted by water/wastewater is directly proportional to its depth or head in the pipe, tank, or channel. If the pressure is known, the equivalent head can be calculated.

$$\text{Head, ft} = \text{Pressure, psi} \times 2.31\,\text{ft/psi} \quad (3.19)$$

Example 3.55

Problem:

The pressure gauge on the discharge line from the influent pump reads 75.3 psi. What is the equivalent head in feet?

Solution:

$$\text{Head, ft} = 75.3 \times 2.31\,\text{ft/psi} = 173.9\,\text{ft}$$

Head/Pressure

If the head is known, the equivalent pressure can be calculated by:

$$\text{Pressure, psi} = \frac{\text{Head, ft}}{2.31\,\text{ft/psi}} \quad (3.20)$$

Example 3.56

Problem:

The tank is 15 ft deep. What is the pressure in psi at the bottom of the tank when it is filled with wastewater?

Solution:

$$\text{Pressure, psi} = \frac{15\,\text{ft}}{2.31\,\text{ft/psi}}$$

$$= 6.49\,\text{psi}$$

Before we look at a few example problems dealing with force, pressure, and head, it is important to review the key points related to force, pressure, and head.

Math Operations Review

1. By definition, water weighs 62.4 pounds per cubic foot.
2. The surface of any one side of the cube contains 144 square inches (12 in × 12 in = 144 in^2). Therefore, the cube contains 144 columns of water 1-foot tall and 1-inch square.
3. The weight of each of these pieces can be determined by dividing the weight of the water in the cube by the number of square inches.

$$\text{Weight} = \frac{62.4 \text{ lb}}{144 \text{ in}^2} = 0.433 \text{ lbs/in}^2 \text{ or } 0.433 \text{ psi}$$

4. Because this is the weight of one column of water one foot tall, the true expression would be 0.433 pounds per square inch per foot of head or 0.433 psi/ft.

Key point: 1 foot of head = 0.433 psi.

In addition to remembering the important parameter, 1 foot of head = 0.433 psi, it is important to understand the relationship between pressure and feet of head—in other words, how many feet of head 1 psi represents. This is determined by dividing 1 by 0.433.

$$\text{Feet of head} = \frac{1 \text{ ft}}{0.433 \text{ psi}} = 2.31 \text{ ft/psi}$$

If a pressure gauge were reading 12 psi, the height of the water necessary to represent this pressure would be 12 psi × 2.31 ft/psi = 27.7 ft.

Key point: Both the above conversions are commonly used in water/wastewater treatment calculations. However, the most accurate conversion is 1 ft = 0.433 psi. This is the conversion we use throughout this text.

Example 3.57

Problem:

Convert 40 psi to feet head.

Solution:

$$\frac{40 \text{ psi}}{1} \times \frac{\text{ft}}{0.433 \text{ psi}} = 92.4 \text{ ft}$$

Example 3.58

Problem:

Convert 40 ft to psi.

Solution:

$$40 \frac{\text{ft}}{1} \times \frac{0.433 \text{ psi}}{1 \text{ ft}} = 17.32 \text{ psi}$$

As the above examples demonstrate, when attempting to convert psi to feet, we divide by 0.433, and when attempting to convert feet to psi, we multiply by 0.433. The above process can be most helpful in clearing up the confusion on whether to multiply or divide. There is another way, however—one that may be more beneficial and easier for many operators to use. Notice that the relationship between psi and feet is almost two to one. It takes slightly more than two feet to make one psi. Therefore, when looking at a problem where the data is in pressure the result should be in feet, and the answer will be at least twice as large as the starting number. For instance, if the pressure were 25 psi, we intuitively know that the head is over 50 ft. Therefore, we must divide by 0.433 to obtain the correct answer.

Example 3.59

Problem:

Convert a pressure of 45 psi to feet of head.

Solution:

$$45\frac{\text{psi}}{1} \times \frac{1\text{ ft}}{0.433\text{ psi}} = 104\text{ ft}$$

Example 3.60

Problem:

Convert 15 psi to feet.

Solution:

$$15\frac{\text{psi}}{1} \times \frac{1\text{ ft}}{0.433\text{ psi}} = 34.6\text{ ft}$$

Example 3.61

Problem:

Between the top of a reservoir and the watering point, the elevation is 125 ft. What will the static pressure be at the watering point?

Solution:

$$125\frac{\text{psi}}{1} \times \frac{1\text{ ft}}{0.433\text{ psi}} = 288.7\text{ ft}$$

Example 3.62

Problem:

Find the pressure (psi) in a 12-foot-deep tank at a point 5 ft below the water surface.

Solution:

$$\text{Pressure}(\text{psi}) = 0.433 \times 5\text{ ft}$$
$$= 2.17\text{ psi}$$

Example 3.63

Problem:

A pressure gauge at the bottom of a tank reads 12.2 psi. How deep is the water in the tank?

Solution:

$$\text{Head (ft)} = 2.31 \times 12.2 \text{ psi}$$
$$= 28.2 \text{ ft}$$

Example 3.64

Problem:

What is the pressure (static pressure) 4 miles beneath the ocean surface?

Solution:

Change miles to ft, then to psi.

$$5,280 \text{ ft/mile} \times 4 = 21,120 \text{ ft}$$

$$\frac{21,120 \text{ ft}}{2.31 \text{ ft/psi}} = 9.143 \text{ psi}$$

Example 3.65

Problem:

A 150-ft diameter cylindrical tank contains 2.0-MG water. What is the water depth? At what pressure would a gauge at the bottom read in psi?

Solution:

- **Step 1**: Change MG to cu ft.

$$\frac{2,000,000 \text{ gal}}{7.48} = 267,380 \text{ cu ft}$$

- **Step 2**: Using volume, solve for depth.

$$\text{Volume} = 0.785 \times D^2 \times \text{depth}$$
$$267,380 \text{ cu ft} = 0.785 \times (150)2 \times \text{depth}$$
$$\text{Depth} = 15.1 \text{ ft}$$

Example 3.66

Problem:

The pressure in a pipe is 70 psi. What is the pressure in feet of water? What is the pressure in psf?

Solution:

- **Step 1**: Convert pressure to feet of water.

$$70 \text{ psi} \times 2.31 \text{ ft/psi} = 161.7 \text{ ft of water}$$

- **Step 2**: Convert psi to psf.

$$70 \text{ psi} \times 144 \text{ sq in/sq ft} = 10,080 \text{ psf}$$

Example 3.67

Problem:
The pressure in a pipeline is 6,476 psf. What is the head on the pipe?

$$\text{Head on pipe} = \text{ft of pressure}$$
$$\text{Pressure} = \text{Weight} \times \text{height}$$
$$6,476 \text{ psf} = 62.4 \text{ lbs/cu ft} \times \text{height}$$
$$\text{Height} = 104 \text{ ft}$$

REVIEW OF ADVANCED ALGEBRA KEY TERMS/CONCEPTS

Advanced algebraic operations (linear, linear differential, and ordinary differential equations) have in recent years become an essential part of the mathematical background required by environmental engineers, among others. While it is not the authors' intent to provide complete coverage of the topics (environmental practitioners are normally well grounded in these critical foundational areas), it is important to review the key terms and concepts germane to the topics.

KEY DEFINITIONS INCLUDE

Algebraic multiplicity of an eigenvalue—the algebraic multiplicity of an eigenvalue c of a matrix A is the number of times the factor $(t - c)$ occurs in the characteristic polynomial of A.

Basis for a subspace—a basis for a subspace W is a set of vectors $\{\mathbf{v}_1, \ldots, \mathbf{v}_k\}$ in which W such that:

1. $\{\mathbf{v}_1, \ldots, \mathbf{v}_k\}$ is linearly independent; and
2. $\{\mathbf{v}_1, \ldots, \mathbf{v}_k\}$ spans W.

Characteristic polynomial of a matrix: the characteristic polynomial of a n-by-n matrix A is the polynomial in t given by the formula $\det(A - tI)$.

Column space of a matrix: is the subspace spanned by the columns of the matrix considered as a set of vectors (also see row space).

Consistent linear system: a system of linear equations is consistent if it has at least one solution.

Defective matrix: a matrix A is defective if A has an eigenvalue whose geometric multiplicity is less than its algebraic multiplicity.

Diagonalizable matrix: a matrix is diagonalizable if it is similar to a diagonal matrix.
Dimension of a subspace: the dimension of a subspace W is the number of vectors in any basis of W. (If W is the subspace $\{0\}$, we say that its dimension is 0.)
Echelon form of a matrix: a matrix is in row echelon form if:

1. all rows that consist entirely of zeros are grouped together at the bottom of the matrix; and
2. the first (counting left to right) nonzero entry in each nonzero row appears in a column to the right of the first nonzero entry in the preceding row (if there is a preceding row).

Eigenspace of a matrix: the eigenspace associated with the eigenvalue c of a matrix A is the null space of $A - cI$.
Eigenvalue of a matrix: an eigenvalue of a matrix A is a scalar c such that $A\mathbf{x}=c\mathbf{x}$ holds for some nonzero vector $\mathbf{x}$.
Eigenvector of a matrix: an eigenvector of a square matrix A is a nonzero vector $\mathbf{x}$ such that $A\mathbf{x}=c\mathbf{x}$ holds for some scalar c.
Elementary matrix: is a matrix that is obtained by performing an elementary row operation on an identity matrix.
Equivalent linear systems: two systems of linear equations in n unknowns are equivalent if they have the same set of solutions.
Geometric multiplicity of an eigenvalue: the geometric multiplicity of an eigenvalue c of a matrix A is the dimension of the eigenspace of c.
Homogeneous linear system: a system of linear equations $A\mathbf{x}=\mathbf{b}$ is homogeneous if $\mathbf{b}=\mathbf{0}$.
Inconsistent linear system: a system of linear equations is inconsistent if it has no solution.
Inverse of a matrix: matrix B is an inverse for matrix A if $AB=BA=I$.
Invertible matrix: a matrix is invertible if it has no inverse.
Least squares solution of a linear system: a least-squares solution to a system of linear equations $A\mathbf{x}=\mathbf{b}$ is a vector $\mathbf{x}$ that minimizes the length of the vector $A\mathbf{x} - \mathbf{b}$.
Linear combination of vectors: a vector $\mathbf{v}$ is a linear combination of the vectors $\mathbf{v}_1, ..., \mathbf{v}_k$ if there exist scalars $a_1, ..., a_k$ such that $\mathbf{v}=a_1\mathbf{v}_1+\cdots+a_k\mathbf{v}_k$.
Linear dependence relation for a set of vectors: a linear dependence relation for the set of vectors $\{\mathbf{v}_1, ..., \mathbf{v}_k\}$ is an equation of the form $a_1\mathbf{v}_1+\cdots+a_k\mathbf{v}_k=\mathbf{0}$, where the scalars $a_1, ..., a_k$ are zero.
Linearly dependent set of vectors: the set of vectors $\{\mathbf{v}_1, ..., \mathbf{v}_k\}$ is linearly dependent if the equation $a_1\mathbf{v}_1+\cdots+a_k\mathbf{v}_k=\mathbf{0}$ has a solution where not all the scalars $a_1, ..., a_k$ are zero (i.e., if $\{\mathbf{v}_1, ..., \mathbf{v}_k\}$ satisfies a linear dependence relation).
Linearly independent set of vectors: the set of vectors $\{\mathbf{v}_1, ..., \mathbf{v}_k\}$ is linearly independent if the only solution to the equation $a_1\mathbf{v}_1 +\cdots+a_k\mathbf{v}_k=\mathbf{0}$ is the solution where all the scalars $a_1, ..., a_k$ are zero. (i.e., if $\{\mathbf{v}_1, ..., \mathbf{v}_k\}$ does not satisfy any linear dependence relation).
Linear transformation: a linear transformation from V to W is a function T from V to W such that:

1. $T(\mathbf{u}+\mathbf{v})=T(\mathbf{u})+T(\mathbf{v})$ for all vectors $\mathbf{u}$ and $\mathbf{v}$ in V
2. $T(a\mathbf{v})=aT(\mathbf{v})$ for all vectors $\mathbf{v}$ in V and all scalars a

Nonsingular matrix: a square matrix A is nonsingular if the only solution to the equation $A\mathbf{x}=\mathbf{0}$ is $\mathbf{x}=\mathbf{0}$.
Null space of a matrix: the null space of a m by n matrix A is the set of all vectors $\mathbf{x}$ in R^n such that $A\mathbf{x}=\mathbf{0}$.
Null space of a linear transformation: the null space of a linear transformation T is the set of vectors $\mathbf{v}$ in its domain such that $T(\mathbf{v})=\mathbf{0}$.

Nullity of a matrix: the nullity of a matrix is the dimension of its null space.
Nullity of a linear transformation: the nullity of a linear transformation is the dimension of its null space.
Orthogonal complement of a subspace: the orthogonal complement of a subspace S of R^n is the set of all vectors **v** in R^n such that **v** is orthogonal to every vector in S.
Orthogonal set of vectors: a set of vectors in R^n is orthogonal if the dot product of any two of them is 0.
Orthogonal matrix: a matrix A is orthogonal if A is invertible and its inverse equals its transpose; i.e., $A^{-1}=A^T$.
Orthogonal linear transformation: a linear transformation T from V to W is orthogonal if $T(\mathbf{v})$ has the same length as **v** for all vectors **v** in V.
Orthonormal set of vectors: a set of vectors in R^n is orthonormal if it is an orthogonal set and each vector has length 1.
Range of a linear transformation: the range of a linear transformation T is the set of all vectors $T(\mathbf{v})$, where **v** is any vector in its domain.
Rank of a matrix: the rank of a matrix A is the number of nonzero rows in the reduced row echelon form of A; i.e., the dimension of the row space of A.
Rank of a linear transformation: the rank of a linear transformation (and hence of any matrix regarded as a linear transformation) is the dimension of its range. Note: A theorem tells us that the two definitions of rank of a matrix are equivalent.
Reduced row echelon form of a matrix: a matrix is in reduced row echelon form if:

1. the matrix is in row echelon form
2. the first nonzero entry in each nonzero row is the number 1
3. the first nonzero entry in each nonzero row is the only nonzero entry in its column

Row equivalent matrices: two matrices are row equivalent if one can be obtained from the other by a sequence of elementary row operations.
Row operations: the elementary row operations performed on a matrix are:
- interchange two rows
- multiply a row by a nonzero scalar
- add a constant multiple of one row to another

Row space of a matrix: the row space of a matrix is the subspace spanned by the rows of the matrix considered as a set of vectors.
Similar matrices: matrices A and B are similar if there is a square invertible matrix S such that $S^{-1}AS=B$.
Singular matrix: a square matrix A is singular if the equation $A\mathbf{x}=\mathbf{0}$ has a nonzero solution for **x**.
Span of a set of vectors: the span of the set of vectors $\{\mathbf{v}_1, ..., \mathbf{v}_k\}$ is the subspace V consisting of all linear combinations of $\mathbf{v}_1, ..., \mathbf{v}_k$. One also says that the subspace V is spanned by the set of vectors $\{\mathbf{v}_1, ..., \mathbf{v}_k\}$ and that this set of vectors spans V.
Subspace: a subset W of R^n is a subspace of R^n if:

1. the zero vector is in W
2. **x**+**y** is in W whenever **x** and **y** are in W
3. $a\mathbf{x}$ is in W whenever **x** is in W and a is any scalar.

Symmetric matrix: a matrix A is symmetric if it equals its transpose; i.e., $A=A^T$.

QUADRATIC EQUATIONS

$$ax^2 + bx + c = 0$$

$$X = \frac{-b \pm \sqrt{b^2 - 4ac}}{2a}$$

When studying a discipline that does not include mathematics, one thing is certain, the discipline under study has nothing or little to do with environmental practice.

THE QUADRTIC EQUATION AND ENVIRONMENTAL PRACTICE

The logical question might be: Why is the quadratic equation important in environmental practice? The logical answer: The quadratic equation is used in environmental practice to find solutions to problems primarily dealing with length and time determinations. Stated differently: The quadratic equation is a tool—an important tool that belongs in every environmental practitioner's toolbox.

To the student of mathematics, this explanation might seem somewhat strange. Math students know, for example, that there will be two solutions to a quadratic equation. In environmental disciplines such as environmental engineering, many times only one solution is meaningful. For example, if we are dealing with a length, a negative solution to the equation may be mathematically possible but it is not the solution we would use. Negative time obviously would also pose the same problem.

So, what is the point? The point is that we often need to find "a" solution to certain mathematical problems. In environmental problems involving the determination of length and time using quadratic equation, we will end up with two answers. In some instances, a positive answer and a negative answer may result. One of these answers is usable; thus, we would use it. Real engineering is about modeling situations that occur naturally and using the model to understand what is happening, or maybe to predict what will happen in future. The quadratic equation is often used in modeling because it is a beautifully simple curve (Borne, 2008).

Key terms
a = coefficient of the x^2
b = coefficient of the x term
c = number in quadratic equation (not a coefficient of any x term)

Simple equations: equations in which the unknown appears only in the first degree.
Pure quadratic equation: an equation in which the unknown appears only in the second degree.
Affected quadratic equation: an equation containing the first and second degree of an unknown.

QUADRATIC EQUATIONS: THEORY AND APPLICATION

The equation $6x = 12$ is a form of equation most of us are familiar with. Such an equation, an equation in which the unknown appears only in the first degree, is called *simple equation* or *linear* equation.

Those experienced in mathematics know that not all equations reduce to this form. For instance, when an equation has been reduced, the result may be an equation in which the square of the unknown equals some number as in $x^2=5$. Such an equation, an equation in which the unknown appears only in the second degree, is called a *pure quadratic* equation.

In some cases, when an equation is simplified and reduced, the resultant form contains the square and the first power of the unknown, which equal some number; $x^2-5x=24$ is such an equation. An equation containing the first and second degree of an unknown s called an *affected quadratic* equation.

Quadratic equations, and certain other forms, can be solved with the aid of factoring. The procedure for solving a quadratic equation by factoring is as follows:

1. Collect all terms on the left and simplify to the form $ax^2+bx+c=0$.
2. Factor the quadratic expression.
3. Set each factor equal to zero.
4. Solve the resulting linear equations.
5. Check the solution in the original equation.

Example 3.68

Problem:

Solve

$$x^2 - x - 12 = 0$$

$$(x-4)(x+3) = 0 \quad \text{(Factor)}$$

$$x - 4 = 0 \quad x + 3 = 0 \quad \text{(Set each factor to zero)}$$

$$x = 4 \quad x = -3 \quad \text{(solve)}$$

The roots are $x=4$ and $x=-3$. We can check them in the original equation by substitution. Therefore, we have

$$(4)^2 - (4) - 12 = 0 \quad (-3)^2 - (-3) - 12 = 0$$

$$0 = 0 \quad\quad\quad 0 = 0$$

Many times, factoring is either too time-consuming or not possible. The formula shown below is called the *quadratic formula*. It expresses the quadratic equation in terms of its coefficients. The quadratic formula allows us to quickly solve for X with no factoring.

$$X = \frac{-b \pm \sqrt{b^2 - 4ac}}{2a} \tag{3.21}$$

To use the quadratic equation, just substitute the appropriate coefficients into the equation and solve.

Derivation of the Quadratic Equation Formula

The equation $ax^2 + bx + c = 0$, where a, b, and c are any numbers, positive or negative, representing any quadratic equation in one unknown. When this general equation is solved, the solution can be used to determine the unknown value in any quadratic equation. The solution follows.

Math Operations Review

Example 3.69

Problem:

Solve $ax^2+bc+c=0$ for x

Solution:

$$ax^2 + bx + x = 0$$

Subtract c from both members:

$$ax^2 + bx = -c$$

Divide both members by a:

$$x^2 = \frac{b}{a}x = -\frac{c}{a}$$

Complete the square:

$$X^2 + \frac{b}{a}X + \frac{b^2}{4a^2} - \frac{b^2}{4a^2} = -\frac{c}{a}$$

$$\left(X + \frac{b}{2a}\right)^2 - \frac{b^2}{4a^2} - \frac{c}{a}$$

Place the right member over the lowest common denominator:

$$\left(X + \frac{b}{2a}\right)^2 = \frac{b^2 - 4ac}{4a^2}$$

Extract the square root of both members:

$$X + \frac{b}{2a} = \pm\sqrt{\frac{b^2 - 4ac}{4a^2}}$$

$$X + \frac{b}{2a} = \pm\frac{\sqrt{b^2 - 4ac}}{2a}$$

Subtract $b/2a$ from both members:

$$X + \frac{b}{2a} = \pm\frac{\sqrt{b^2 - 4ac}}{2a}$$

Thus the quadratic formula:

$$X = \frac{-b \pm \sqrt{b^2 - 4ac}}{2a}$$

Using the Quadratic Equation

Example 3.70: Use the Quadratic Formula to Solve the Following Problem

Problem:

After conducting a study and deriving an equation representing time, we arrive at the following equation (Note: set the equation equal to zero and all like terms combined).

$$x^2 = 5x + 6 = 0$$

Solution:

$$X = \frac{-b \pm \sqrt{b^2 - 4ac}}{2a}$$

From our equation, $a=1$ (the coefficient of x^2), $b=-5$ (the coefficient of x), and $c=6$ (the constant or third term.
   Substituting these coefficients in the quadratic formula:

$$X = \frac{(-5) \pm \sqrt{(-5)^2 - 4(1)(6)}}{2(1)}$$

$$= \frac{5 \pm \sqrt{25 - 24}}{2}$$

$$= \frac{5 \pm 1}{2}$$

$$= 3, 2$$

Note: The roots may not always be rational (integers), but the procedure is the same.

TRIGONOMETRIC RATIOS

$$\sin A = a/c \quad \cos A = b/c \quad \tan A = ab$$

TRIGONOMETRIC FUNCTIONS AND THE ENVIRONMENTAL PRACTITIONER

Typically, environmental practitioners are called upon to make calculations involving the use of various trigonometric functions. Consider slings, for example, they are commonly used between cranes, derricks, and/or hoists and the load, so that the load may be lifted and moved to a desired location. For the environmental professional responsible for safety and health, knowledge of the properties and limitations of the slings; the type and condition of the material being lifted; weight and shape of the object being lifted; angle of the lifting sling to the load being lifted; and the environment in which the lift is to be made, are all important considerations to be evaluated before the safe transfer of material can take place.
   Later, we put many of the following principles to work in determining sling load and working load on a ramp (inclined plane); that is, the resolution of force-type problems. For now, we discuss the basic trigonometric functions used to make these calculations.

TABLE 3.3
Definition of Trigonometric Ratios

$$\text{sine of angle } A = \frac{\text{Measure of leg opposite angle } A}{\text{Measure of hypotenuse}} \qquad \sin A = \frac{a}{c}$$

$$\text{cosine of angle } A = \frac{\text{Measure of leg adjacent angle } A}{\text{Measure of hypotenuse}} \qquad \cos A = \frac{b}{c}$$

$$\text{tangent of angle } A = \frac{\text{Measure of leg opposite angle } A}{\text{Measure of leg adjacent angle } A} \qquad \tan A = \frac{a}{b}$$

Trigonometric Ratios or Functions

In trigonometry, all computations are based upon certain ratios (i.e., trigonometric functions). The trigonometric ratios or functions are sine, cosine, tangent, cotangent, secant, and cosecant. It is important to understand the definition of the ratios given in Table 3.3 and defined in terms of the lines shown in Figure 3.12.

Important point: In a right triangle, the side opposite the right angle is the longest side. This side is called the *hypotenuse*. The other two sides are the *legs*.

Example 3.71

Problem:

Find the sine, cosine, and tangent of angle Y in Figure 3.13.

Solution:

$$\sin Y = \frac{\text{Opposite leg}}{\text{Hypotenuse}} \qquad \cos Y = \frac{\text{Adjacent leg}}{\text{Hypotenuse}} \qquad \tan Y = \frac{\text{Opposite leg}}{\text{Adjacent leg}}$$

$$= \frac{9}{15} \text{ or } 0.6 \qquad = \frac{12}{15} \text{ or } 0.8 \qquad = \frac{9}{12} \text{ or } 0.75$$

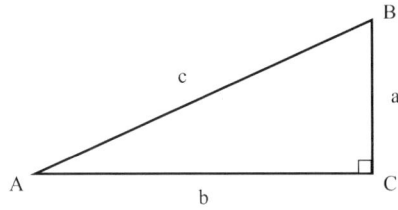

FIGURE 3.12 Right triangle.

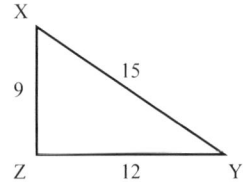

FIGURE 3.13 For Example 3.71.

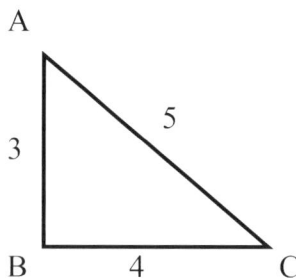

FIGURE 3.14 For Example 3.73.

Example 3.72

Problem:

Using Figure 3.13, find the measure of the angle X to the nearest degree.

Solution:

$$\sin X = \frac{\text{Opposite leg}}{\text{Hypotenuse}}$$

$$= \frac{12}{15} \text{ or } 0.8$$

Use a scientific calculator to find the angle measure with a sine of 0.8.
Enter: 0.8 [2nd] or [INV]
Result: 53.13010235
So, measure angle $X = 53°$.

Example 3.73

Problem:

For the triangle shown in Figure 3.14, find sin C, cos C, and tan C.

Solution:

$$\sin C = 2/5; \quad \cos C = 4/5; \quad \tan C = 3/4.$$

REFERENCES

Masser, M.P. & Jensen, J.W. (1991). Calculating Area and Volume of Ponds and Tanks. USDA Grant 89-38500-4516 to Southern regional Aquaculture Center, Washington, DC.

Staple, E. (2012). The Order of Operations: More Examples. Purplemath. Available from http.www.pruplemath.com/modules/orderops2.htm.

4 Environmental Modeling and Algorithms

Who has measured the waters in the hollow of the hand, or with the breadth of his hand marked off the heavens? Who has held the dust of the earth in a basket, or weighed the mountains on the scales and the hills in a balance?

Isaiah 40:12

Algorithm: The word comes from the Persian author Abu Ja'far Mohammed ibn Musa al-Khomwarizmi who authored a book with arithmetic rules dating from about 825 A.D.

INTRODUCTION

There is a growing interest in the field of environmental monitoring and quantitative assessment of environmental problems. For some years now, the results of environmental models and assessment analyses have been influencing environmental regulation and policies. These results are widely cited by politicians in forecasting the consequences of greenhouse gas emissions like carbon dioxide (CO_2) and in advocating dramatic reductions of energy consumption at local, national, and international levels. For this reason, and because environmental modeling is often based on extreme conceptual and numerical intricacy and uncertain validity, environmental modeling has become one of the most controversial topics of applied mathematics.

Having said this, environmental modeling continues to be widely used in environmental practice with its growth only limited by the imagination of the modelers. Environmental practice problem-solving techniques incorporating the use of modeling are widely used in watershed modeling, surface water information, flood hazard mapping, climate modeling, groundwater modeling and others. However, it is important to keep in mind that modelers often provide models where the user produces a product whereby the result is used to describe what it is based on and the reasons for thinking so. In this chapter, we do not provide a complete treatment of environmental modeling. (For the reader who desires such a treatment, we highly recommend N. Nirmalakhandan's (2020) *Modeling Tools for Environmental Engineers and Scientists,* 2020 and NIST's (U.S. Commerce Department's Technology Administration) *Engineering Statistics,* 2012. Much of the work presented in this chapter is modeled after these works.) We present an overview of quantitative operations implicit in environmental modeling processes.

Environmental modeling has become an important tool within the environmental engineer's well-equipped toolbox. Using an analogy, we can say that if a typical skilled handyperson's toolbox contains a socket set ratchet and several different sized wrench attachments, then the well-equipped environmental practitioner's toolbox includes a number of environmental models (socket set ratchets) with a varying set of algorithms (socket wrench attachments). Although a complete treatment or discussion of algorithms is beyond the scope of this book, we do provide basic underlying explanations of what algorithms are and examples of their applications in cyber space (in the real world?). For those interested in a more complete discussion of algorithms, there are many excellent texts on the general topic. We list several of these resources in the recommended reading section at the chapter's end.

BASIC STEPS FOR DEVELOPING AN EFFECTIVE MODEL

The basic steps used for model-building are the same across all modeling methods. The details vary somewhat from method to method, but an understanding of the common steps, combined with the typical underlying assumptions needed for the analysis, provides a framework in which the results from almost any method can be interpreted and understood.

The basic steps of the model-building process are:

1. model selection
2. model fitting
3. model validation

These three basic steps are used iteratively until an appropriate model for the data has been developed. In the model selection step, plots of the data, process knowledge and assumptions about the process are used to determine the form of the model to be fit to the data. Then, using the selected model and possibly information about the data, an appropriate model-fitting method is used to estimate the unknown parameters in the model. When the parameter estimates have been made, the model is then carefully assessed to see if the underlying assumptions of the analysis appear plausible. If the assumptions seem valid, the model can be used to answer the scientific or engineering questions that prompted the modeling effort. If the model validation identifies problems with the current model, however, then the modeling process is repeated using information from the model validation step to select and/or fit an improved model.

The three basic steps of process modeling described in the paragraph above assume that the data have already been collected and that the same data set can be used to fit all of the candidate models. Although this is often the case in mode-building situations, one variation on the basic model-building sequence comes up when additional data are needed to fit a newly hypothesized model based on a model fit to the initial data. In this case, two additional steps, experimental design, and data collection, can be added to the basic sequence between model selection and mode-fitting.

WHAT ARE MODELS USED FOR?

Models are used for four main purposes:

1. estimation
2. prediction
3. calibration
4. optimization

A brief explanation of the different uses of models is discussed below (NIST, 2012).

- **Estimation**: The goal of estimation is to determine the value of the regression function (i.e., the average value of the response variable, for a particular combination of the values of the predictor variables. Regression function values can be estimated for any combination of predictor variable values, including values for which no data have been measured or observed. Function values estimated for points within the observed space of predictor variable values are sometimes called interpolations. Estimation of regression function values for points outside the observed space of predictor variable values, called extrapolations, are sometimes necessary, but require caution.
- **Prediction**: The goal of prediction is to determine either
  1. the value of a new observation of the response variable or
  2. the values of a specified proportion of all future observations of the response variable for a particular combination of the values of the predictor variables. Predictions can be

made for any combination of predictor variable values, including values for which no data have been measured or observed. As in the case of estimation, predictions made outside the observed space of predictor variable values are sometimes necessary, but require caution.
- **Calibration**: The goal of calibration is to quantitatively relate measurements made using one measurement system to those of another measurement system. This is done so that measurements can be compared in common units or to tie results from a relative measurement method to absolute units.
- **Optimization**: Optimization is performed to determine the values of process inputs that should be used to obtain the desired process output. Typical optimization goals might be to maximize the yield of a process, to minimize the processing time required to fabricate a product, or to hit a target product specification with minimum variation in order to maintain specified tolerances.

MEDIA MATERIAL CONTENT

Media material content is a measure of the material contained in a bulk medium, quantified by the ratio of the amount of material present to the amount of the medium. The terms mass, moles, or volume can be used to quantify the amounts. Thus, the ratio can be expressed in several forms such as mass or moles of material per volume of medium resulting in mass or molar concentration; moles of material per mole of medium, resulting in mole fraction; and volume of material per volume of medium (resulting in volume fraction).

When dealing with mixtures of materials and media, the use of different forms of measures in the ratio to quantify material content may become confusing. In regard to mixtures, the ratio can be expressed in concentration units. The *concentration* of a chemical (liquid, gaseous, or solid) substance expresses the amount of substance present in a mixture. There are many different ways to express concentration.

Chemists use the term *solute* to describe the substance of interest and the term *solvent* to describe the material in which the solute is dissolved. For example, in a can of soft drink (a solution of sugar in carbonated water), there are ~12 tablespoons of sugar (the solute) dissolved in the carbonated water (the solvent). In general, the component that is present in the greatest amount is termed the solvent.

Some of the more common concentration units are:

1. **Mass per unit volume**: Some concentrations are expressed in milligrams per milliliter (mg/mL) or milligrams per cubic centimeter (mg/cm^3). Note that 1 mL=1 cm^3 is sometimes denoted as a "cc". Mass per unit volume is handy when discussing how soluble a material is in water or a particular solvent. For example, "the solubility of substance X is 4 grams per liter".
2. **Percent by mass**: Also called weight percent or percent by weight, this is simply the mass of the solute divided by the total mass of the solution and multiplied by 100%:

$$\text{Percent by mass} = \frac{\text{Mass of component}}{\text{Mass of solution}}(100\%) \tag{4.1}$$

The mass of the solution is equal to the mass of the solute plus the mass of the solvent. For example, a solution consisting of 30 g of sodium chloride and 70 g of water would be 30% sodium chloride by mass: [(30 g Nick)/(30 g NaCl+70 g water)]×100%=30%. To avoid confusion about whether a solution is percent by weight or percent by volume, the symbol "w/w" (for weight to weight) is often used after the concentration such as "10% potassium iodide solution in water (w/w)".

3. **Percent by volume**: Also called volume percent or percent by volume, this is typically only used for mixtures of liquids. Percent by volume is simply the volume of the solute divided by the sum of the volumes of the other components multiplied by 100%.

    If we mix 30 mL of ethanol and 70 mL of water, the percent ethanol by volume will be 30% BUT the total volume of the solution will NOT be 100 mL (although it will be close). That's because ethanol and water molecules interact differently with each other than they do with themselves. To avoid confusion about whether we have a percent by weight or percent by volume solution, we could label this as "30% ethanol in water (v/v)" where v/v stands for "volume to volume."

4. **Molarity**: Molarity is the number of moles of solute dissolved in one liter of solution. For example, if we have 90 grams of glucose (molar mass = 180 grams per mole) this is (90 g)/(180 g/mol) = 0.50 moles of glucose. If we place this in a flask and add water until the total volume = 1 L, we will have a 0.5 molar solution. Molarity is usually denoted with a capital M, i.e., a 0.50 M solution. Recognize that molarity is moles of solute per liter of solution, not per liter of solvent. Also, recognize that molarity changes slightly with temperature because the volume of a solution changes with temperature.

5. **Molality (m, used for calculations of colligative properties)**: Molality is the number of moles of solute dissolved in one kilogram of solvent. Notice the two key differences between molarity and molality. Molality uses mass rather than volume and uses solvent instead of solution.

$$\text{Molality} = \frac{\text{Moles of solute}}{\text{Kilograms of solution}} \qquad (4.2)$$

Unlike molarity, molality is independent of temperature because mass does not change with temperature. If we were to place 90 grams of glucose (0.50 moles) in a flask and then add one kilogram of water, we would have a 0.50 molar solution. Molality is usually denoted with a small m, i.e., a 0.50 m solution.

6. **Parts per million (PPM)**: Parts per million works like percent by mass but is more convenient when there is only a small amount of solute present. PPM is defined as the mass of the component in solution divided by the total mass of the solution multiplied by 10^6 (one million):

$$\text{Parts per million} = \frac{\text{Mass of component}}{\text{Mass of solution}} (1,000,000) \qquad (4.3)$$

A solution with a concentration of 1 ppm has t gram of substance for every million grams of solution. Because the density of water is 1 g per mL and we are adding such a tiny amount of solute, the density of a solution at such a low concentration is ~1 g per mL. Therefore, in general, one ppm implies one mg of solute per liter of solution.

Finally, recognize that 1% = 10,000 ppm. Therefore, something that has a concentration of 300 ppm could also be said to have a concentration of (300 ppm)/(10,000 ppm/percent) = 0.03% percent by mass.

7. **Parts per billion (PPB)**: This works like above, but we multiply by one billion (10^9); caution: the word billion has different meanings in different countries). A solution with 1 ppb of solute has 1 microgram (10^{-6}) of material per liter.

8. **Parts per trillion (PPT)**: This works like parts per million and parts per billion except that we multiply by one trillion (10^{12}). There are few, if any, solutes which are harmful at concentrations as low as 1 ppt.

Environmental Modeling and Algorithms

The following notation and examples can help in formalizing these different forms; subscripts for components are $i = 1, 2, 3, \ldots N$; and subscripts for phases are g=gas, a=air, l=liquid, w=water, s=solids and soil.

MATERIAL CONTENT: LIQUID PHASES

Mass concentration, molar concentration, or mole fraction can be used to quantify material content in liquid phases.

$$\text{Mass conc. of component } i \text{ in water} = p_{i,w} \frac{\text{Mass of material}, i}{\text{Volume of water}} \tag{4.4}$$

$$\text{Molar conc. of component } i \text{ in water} = C_{i,w} = \frac{\text{Mass of material}, i}{\text{Volume of water}} \tag{4.5}$$

Because moles of material=mass/molecular weight, MW, mass concentrations, $p_{i,w}$, are related by the following:

$$C_{i,w} = \frac{p_{i,w}}{MW_i} \tag{4.6}$$

For molarity—M, $[X]$ is molar concentration of "X."

Mole fraction, X, of a single chemical in water can be expressed as follows:

$$\text{Mole fraction, } X = \frac{\text{Moles of component/chemical}}{\text{Total moles of sol.}(\text{moles of chemical} + \text{moles of water})} \tag{4.7}$$

For dilute solutions, the moles of chemical in the denominator of the above can be ignored in comparison to the moles of water, n_w, can be approximated by:

$$X = \frac{\text{Moles of chemical}}{\text{Moles of water}} \tag{4.8}$$

If X is <0.02, an aqueous solution can be considered dilute. On mass basis, similar expressions can be formulated to yield mass fractions. Mass fractions can also be expressed as a percentage or as other ratios such as parts per million (ppm) or parts per billion (ppb).

The mole fraction of a component in a solution is simply the number of moles of that component divided by the total moles of all the components. We use the mole fraction because the sum of the individual fractions should equal 1. This constraint can reduce the number of variables when modeling mixtures of chemicals. Mole fractions are strictly additive. The sum of the mole fractions of all components is equal to one. Mole fraction, X_i, of component i in an N-component mixture is defined as follows:

$$X_i = \frac{\text{Moles of } i}{\left(\sum_{1}^{N} n_i\right) + n_w} \tag{4.9}$$

$$\text{The sum of all the mole fractions} = \left(\sum_{1}^{N} X_w\right) \tag{4.10}$$

For dilute solutions of multiple chemicals (as is the case of single chemical systems), mole fraction X_i of component i in an N-component mixture can be approximated by the following:

$$X_i = \frac{\text{Moles of } i}{n_w} \quad (4.11)$$

Note that the preceding ratio is known as an intensive property because it is independent of the system and the mass of the sample. An *intensive* property is any property that can exist at a point in space. Temperature, pressure, and density are good examples. On the other hand, an *extensive* property is any property that depends on the size (or extent) of the system under consideration. Volume is an example. If we double the length of all edges of a solid cube, the volume increases by a factor of eight. Mass is another. The same cube will undergo an eight-fold increase mass when the length of the edges is doubled.

Note: The material content in solid and gas phases is different from those in liquid phases. For example, the material content in solid phases is often quantified by a ratio of masses and is expressed as ppm or ppb. The material content in gas phases is often quantified by a ratio of moles or volumes and is expressed as ppm or ppb. It is preferable to report gas phase concentrations at standard temperature and pressure (STP—0°C and 769 mm Hg or 273 K and 1 atm).

Example 4.1

Problem:

A certain chemical has a molecular weight of 80. Derive the conversion factors to quantify the following:

(1) 1 ppm (volume/volume) of the chemical in the air in molar and mass concentration form.
(2) 1 ppm (mass ratio) of the chemical in water in mass and molar concentration form.
(3) 1 ppm (mass ratio) of the chemical in soil in mass ratio form.

Solution:

(1) **Gas phase**: The volume ratio of 1 ppm can be converted to the mole or mass concentration form using the assumption of Ideal Gas, with a molar volume of 22.4 L/g mole at STP conditions (273 K and 1.0 atm.).

$$1 \text{ ppm}_v \equiv \frac{1 \text{ m}^3 \text{ of chemical}}{1{,}000{,}000 \text{ m}^3 \text{ of air}}$$

$$\equiv \frac{1 \text{ m}^3 \text{ of chemical}}{1{,}000{,}000 \text{ m}^3 \text{ of air}} \left(\frac{\text{mol}}{22.4 \text{ L}} \right) \left(\frac{1{,}000 \text{ L}}{\text{m}^3} \right) = 4.46 \times 10^{-5} \frac{\text{mol}}{\text{m}^3}$$

$$\equiv 4.46 \times 10^{-5} \frac{\text{mol}}{\text{m}^3} \left(\frac{80 \text{ g}}{\text{gmol}} \right) \equiv 0.0035 \frac{\text{g}}{\text{m}^3} \equiv 3.5 \frac{\text{mg}}{\text{m}^3} \equiv 3.5 \frac{\mu\text{g}}{\text{L}}$$

The general relationship is 1 ppm=(MW/22.4) mg/m³.

(2) **Water phase**: The mass ratio of 1 ppm can be converted to mole or mass concentration form using the density of water, which is 1 g/cc at 4°C and 1 atm.

Environmental Modeling and Algorithms

$$1 \text{ ppm} = \frac{1 \text{ g of chemical}}{1,000,000 \text{ g of water}}$$

$$\equiv \frac{1 \text{ g of chemical}}{1,000,000 \text{ g of water}} \left(1 \frac{g}{cm^3}\right)\left(\frac{100^3 \text{ cm}^3}{m^3}\right) \equiv 1 \frac{g}{m^3} \equiv 1 \frac{mg}{L}$$

$$\equiv 1 \frac{g}{m^3}\left(\frac{mol}{80 \text{ g}}\right) \equiv 0.0125 \frac{mol}{m^3}$$

(3) **Soil phase**: The conversion is direct

$$1 \text{ ppm} = \frac{1 \text{ g of chemical}}{1,000,000 \text{ g of soil}}$$

$$= \frac{1 \text{ g of chemical}}{1,000,000 \text{ g of soil}}\left(\frac{1,000 \text{ g}}{kg}\right)\left(\frac{1,000 \text{ mg}}{g}\right) = 1 \frac{mg}{kg}$$

Example 4.2

Problem:

Analysis of a water sample from a pond gave the following results: volume of sample = 2 L, concentration of suspended solids in the sample = 15 mg/L, concentration of dissolved chemical = 0.01 moles/L, and concentration of the chemical adsorbed onto the suspended solids = 400 µg/g solids. If the molecular weight of the chemical is 125, determine the total mass of the chemical in the sample.

Solution:

Dissolved concentration = Molar concentration × MW

$$= 0.001 \frac{Mol}{L}\left(\frac{125 \text{ g}}{gmol}\right)$$

$$= 0.125 \frac{g}{L}$$

Dissolved mass in sample = Dissolved concentration × volume

$$= \left(0.125 \frac{g}{L}\right) \times (2 \text{ L})$$

$$= 0.25 \text{ g}$$

Mass of solids in sample = Concentration of solids × volume

$$= \left(25 \frac{mg}{L}\right) \times (2 \text{ L}) = 50 \text{ mg}$$

$$= 0.05 \text{ g}$$

Adsorbed mass in sample = Adsorbed concentration × Mass of solids

$$= \left(400 \frac{\mu g}{g}\right) \times (0.05 \text{ g})\left(\frac{g}{10^6 \mu g}\right)$$

$$= 0.00020 \text{ g}$$

Thus, total mass of chemical in the sample = 0.25 g + 0.00020 g = 0.25020 g

PHASE EQUILIBRIUM AND STEADY STATE

The concept of phase equilibrium (balance of forces) is an important one in environmental modeling. In the case of mechanical equilibrium, consider the following example. A cup sitting on a tabletop remains at rest because the downward force exerted by the earth's gravity action on the cup's mass (this is what is meant by the "weight" of the cup) is exactly balanced by the repulsive force between atoms that prevents two objects from simultaneously occupying the same space, acting in this case between the table surface the cup. If you pick up the cup and raise it above the tabletop, the additional upward force exerted by your arm destroys the state of equilibrium as the cup moves upward. If one wishes to hold the cup at rest above the table, one adjust the upward force to exactly balance the weight of the cup, thus restoring equilibrium.

For more pertinent examples (chemical equilibrium, for example) consider the following. Chemical equilibrium is a dynamic system in which chemical changes are taking place in such a way that there is no overall change in the composition of the system. In addition to partial ionization, equilibrium situations include simple reactions such as when the air in contact with a liquid is saturated with the liquid's vapor, meaning that the rate of evaporation is equal to the rate of condensation. When a solution is saturated with a solute, this means that the rate of dissolving is just equal to the rate of precipitation from the solution. In each of these cases, both processes continue. The equality of rate creates the illusion of static conditions. The point is that no reaction actually goes to completion.

Equilibrium is best described by the Principle of Le Chatelier, which sums up the effects of changes in any of the factors influencing the position of equilibrium. It states that a system in equilibrium, when subjected to stress resulting from a change in temperature, pressure, or concentration, and causing the equilibrium to be upset, will adjust its position of equilibrium to relieve the stress and reestablish equilibrium.

What is the difference between steady state and equilibrium? Steady state implies no changes with passage of time. Likewise, equilibrium can also imply no change of state with passage of time. In many situations this is the case—the system not only is at steady state, but also is at equilibrium. However, this is not always the case. In some cases where the flow rates are steady, but the phase contents, for example, are not being maintained at the "equilibrium values," the system is at steady state but not at equilibrium.

MATH OPERATIONS AND LAWS OF EQUILIBRIUM

Earlier we observed that no chemical reaction goes to completion; there are qualitative consequences of this insight that go beyond the purpose of this text. However, in this text, we are interested in the basic quantitative aspects of equilibria. The chemist usually starts with the chemistry of the reaction and fully utilizes chemical intuition before resorting to mathematical techniques. That is, science should always precede mathematics in the study of physical phenomena. Note, however, that most chemical problems do not have to be an exact, closed-form solution and the direct application of mathematics to a problem can lead to an impasse.

Several basic math operations and fundamental laws from physical chemistry and thermodynamics serve as the tools, blueprints, and foundational structures of mathematical models. They can be used and applied to environmental systems under certain conditions, serving to solve various problems. Many laws serve as important links between the state of the system, chemical properties, and their behavior. As such, some of the basic math operations used to solve basic equilibrium problems and laws essential for modeling the fate and transport of chemicals in natural and engineered environmental systems are reviewed in the following sections.

Solving Equilibrium Problems

In the following math operations, we provide examples of the various forms of combustion of hydrogen to yield water to demonstrate the solution of equilibrium problems. The example reactions are

Environmental Modeling and Algorithms

represented by the following equation. *Note*: Consider the reaction at 1000.0 K where all constituents are in the gas phase and the equilibrium constant is 1.15×10^{10} atm^{-1}.

$$2H^2(g) + O^2(g) = 2H^2O(g)$$

and the equilibrium constant expression:

$$K = [H_2O]^2 / [H_2]^2 [O_2] \tag{4.12}$$

where concentrations are given as partial pressures in atm. Note that K is very large and consequently the concentration of water is large and/or the concentration of at least one of the reactants is very small.

Example 4.3

Problem:

Consider a system at 1000.0 K in which 4.00 atm of oxygen is mixed with 0.500 atm of hydrogen and no water is initially present.

Note that oxygen is in excess, and hydrogen is the limiting reagent. Because the equilibrium constant is very large, virtually all the hydrogen is converted to water yielding [H$_2$O] = 0.500 atm and [O$_2$] = 4.000 − 0.5(0.500) = 3.750 atm. The final concentration of hydrogen, a small number, is an unknown, the only unknown.

Solution:

Using the equilibrium constant expression, we obtain

$$1.15 \times 1{,}010 = (0.500)^2 / [H_2]^2 (3.750)$$

from which we determine that [H$_2$] = 2.41×10^{-6} atm. Because this is a small number, our initial approximation is satisfactory.

Example 4.4

Problem:

Again, consider a system at 1000.0 K, where 0.250 atm of oxygen is mixed with 0.500 atm of hydrogen and 2,000 atm of water.

Solution:

Again, the equilibrium constant is very large and the concentration of at least reactants must be reduced to a very small value.

$$[H_2O] = 2.000 + 0.500 = 2.500 \text{ atm.}$$

In this case, oxygen and hydrogen are present in a 1:2 ratio, the same ratio given by the stoichiometric coefficients. Neither reactant is in excess and the equilibrium concentrations of both will be very small values. We have two unknowns, but they are related by stoichiometry. Because neither product is in excess and one molecule of oxygen is consumed for two of hydrogen, the ratio [H$_2$]/[O$_2$] = 2/1 is preserved during the entire reaction and [H$_2$] = 2[O$_2$].

$$1.15 \times 1{,}010 = 2.5002 / (2[O_2])^2 [O_2]$$

$$[O_2] = 5.14 \times 10^{-4} \text{ atm} \quad \text{and} \quad [H_2] = 2[O_2] = 1.03 \times 10^{-3} \text{ atm}$$

LAWS OF EQUILIBRIUM

Some of the laws essential for modeling the fate and transport of chemicals in natural and engineered environmental system include:

- Ideal Gas Law
- Dalton's Law
- Raoults' Law
- Henry's Law

Ideal Gas Law

An ideal gas is defined as one in which all collisions between atoms or molecules are perfectly elastic and in which there are no intermolecular attractive forces. One can visualize it as collections of perfectly hard spheres, which collide but which otherwise, do not interact with each other. In such a gas, all the internal energy is in the form of kinetic energy and any change in the internal energy is accompanied by a change in temperature. An ideal gas can be characterized by three state variables: absolute pressure (P), volume (V), and absolute temperature (T). The relationship between them may be deduced from kinetic theory and called the Ideal Gas Law:

$$PV = nRT = NkT \tag{4.13}$$

where

n = number of moles
R = universal gas constant = 8.3145 J/mol K or 0.821 L-atm/de-mol
N = number of molecules
K = Boltzmann constant = 1.38066×10^{-23} J/K
$K = R/N_A$ = where
N_A = Avogadro's number = 6.0221×10^{23}

Note: At Standard Temperature and Pressure (STP) the volume of 1 mole of ideal gas is 22.4 L, a volume called the *molar volume of a gas*.

Example 4.5

Problem:

Calculate the volume of 0.333 moles of gas at 300 K under a pressure of 0.950 atm.

Solution:

$$V = \frac{nRT}{P} = \frac{0.333 \text{ mol} \times 0.0821 \text{ L atm/K mol} \times 300 \text{ K}}{0.959 \text{ atm}} = 8.63 \text{ L}$$

Most gases in environmental systems can be assumed to obey this law. The ideal gas law can be viewed as arising from the kinetic pressure of gas molecules colliding with the walls of a container in accordance with Newton's laws. But there is also a statistical element in the determination of the average kinetic energy of those molecules. The temperature is taken to be proportional to this average kinetic energy; this invokes the idea of kinetic temperature.

Dalton's Law

Dalton's law states that the pressure of a mixture of gases is equal to the sum of the pressures of all of the constituent gases alone. Mathematically, this can be represented as:

$$P_{\text{Total}} = P_1 + P_2 + \cdots + P_n \tag{4.14}$$

where

P_{Total} = Total Pressure
$P_1 \ldots$ = Partial Pressure

and

$$\text{Partial } P = \frac{n_i RT}{V} \quad (4.15)$$

where n_j is the number of moles of component j in the mixture.

Note: Although Dalton's Law explains that the total pressure is equal to the sum of all of the pressures of the parts, this only is absolutely true for ideal gases, but the error is small for real gases.

Example 4.6

Problem:

The atmospheric pressure in a lab is 102.4 kPa. The temperature of water sample is 25°C, with pressure as 23.76 torr. If we use a 250 mL beaker to collect hydrogen from the water sample, what are the pressure of the hydrogen, and the moles of hydrogen using the ideal gas law?

Solution:

Step 1: Make the following conversions—A torr is 1 mm of mercury at standard temperature. In kilopascals, that would be 3.17 (1 mm mercury = 7.5 kPa).

$$\text{Convert } 250\,\text{mL} \text{ to } 0.250\,\text{L and } 25°C \text{ to } 298\,\text{L}.$$

Step 2: Use Dalton's law to find the hydrogen pressure.

$$P_{Total} - P_{Water} + P_{Hydrogen}$$

$$102.4\,\text{kPa} = 3.17\,\text{kPa} + P_{Hydrogen}$$

$$P_{Hydrogen} = 99.23\,\text{kPa or } 99.2\,\text{kPa}$$

Step 3: Recall that the Ideal Gas Law is:

$$PV = nRT$$

where P is pressure, V is volume, n is moles, R is the Ideal Gas Constant (0.821 L-atm/mol-K or 8.31 L-kPa/mol-K), and T is temperature.

Therefore,

$$99.2\,\text{kPa} \times 0.250\,\text{L} = n \times 8.31\,\text{L-kPa/mol} \times 298\,\text{K}$$

Rearranged:

$$n = 99.2\,\text{kPa} \times 0.250\,\text{L}/8.31\,\text{L-kPa/mol-K}/298\,\text{K}$$

$$n = 0.0100\,\text{mol or } 1.00 \times 10^{-2}\,\text{mol Hydrogen}$$

Raoult's Law

Raoult's law states that the vapor pressure of mixed liquids is dependent on the vapor pressures of the individual liquids and the molar fraction of each present. Accordingly, for concentrated solutions where the components do not interact, the resulting vapor pressure (P) of component "a" in equilibrium with other solutions can be expressed as

$$P = x_a P_a \tag{4.16}$$

where
  P = resulting vapor pressure
  x = mole fraction of component "a" in solution
  P_a = vapor pressure of pure "a" at the same temperature and pressure as the solution

Henry's Law

Henry's law states that the mass of a gas that dissolves in a definite volume of liquid is directly proportional to the pressure of the gas provided the gas does not react with the solvent. A formula for Henry's Law is:

$$p = Hx \tag{4.17}$$

where x is the solubility of a gas in the solution phase, H is Henry's constant and p is the partial pressure of a gas above the solution.

Hemond and Fechner-Levy (2000) point out that Henry's law constant, H, is a partition coefficient usually defined as the ratio of a chemical's concentration in air to its concentration in water at equilibrium. Henry's law constants generally increase with increased temperature, primarily due to the significant temperature dependency of chemical vapor pressures; solubility is much less affected by the changes in temperature normally found in the environment.

H can be expressed either in a dimensionless for or with units. Table 4.1 lists Henry's law constants for some common environmental chemicals.

CHEMICAL TRANSPORT SYSTEMS

In environmental modeling, environmental practitioners have a fundamental understanding of the phenomena involved with the transport of certain chemicals through the various components of the environment. The primary transport mechanism at the macroscopic level (referred to as *dispersive* transport) is by molecular diffusion driven by concentration gradients. Mixing and bulk movement (referred to as *advective* transport) of the medium are primary transport mechanisms at the macroscopic level.

Advective and dispersive transports are fluid-element driven; that is, for example, advection is the movement of dissolved solute with flowing groundwater. The amount of contaminant being transported is a function of its concentration in the groundwater and the quantity of groundwater flowing, and advection will transport contaminants at different rates in each stratum. Diffusive transport, on the other hand, is the process by which a contaminant in water will move from an area of greater concentration toward an area where it is less concentrated. Diffusion will occur as long as a concentration gradient exists, even if the fluid is not moving, and as a result, a contaminant may spread away from the place where it is introduced into a porous medium.

In today's computer age, environmental engineers have the advantage of choosing from a wide plethora of mathematical models available. These models enable environmental engineers and students with minimal computer programming skills to develop computer-based mathematical models for natural and engineered environmental systems. Commercially available syntax-free authoring software can be adapted to create customized, high-level models of environmental phenomena in groundwater, air, soil, aquatic, and atmospheric systems.

TABLE 4.1
Henry's Law Constants (H)

Chemical	Henry's Law Constant (atm×m³/mol)	Henry's Law Constant (Dimensionless)
Aroclor 1254	2.7×10^{-3}	1.2×10^{-1}
Aroclor 1260	7.1×10^{-3}	3.0×10^{-1}
Atrazine	3×10^{-9}	1×10^{-7}
Benzene	5.5×10^{-3}	2.4×10^{-1}
Benz[a]anthracene	5.75×10^{-6}	2.4×10^{-4}
Carbon tetrachloride	2.3×10^{-2}	9.7×10^{-1}
Chlorobenzene	3.7×10^{-3}	1.65×10^{-1}
Chloroform	4.8×10^{-3}	2.0×10^{-1}
Cyclohexane	0.18	7.3
1,1-Dichloroethane	6×10^{-3}	2.4×10^{-1}
1,2-Dichloroethane	10^{-3}	4.1×10^{-2}
cis-1,2-Dichloroethene	3.4×10^{-3}	0.25
trans-1,2-Dichlorethene	6.7×10^{-3}	0.23
Ethane	4.9×10^{-1}	20
Ethanol	6.3×10^{-6}	
Ethylbenzene	8.7×10^{-3}	3.7×10^{-1}
Lindane	4.8×10^{-7}	2.2×10^{-5}
Methane	0.66	27
Methylene chloride	3×10^{-3}	1.3×10^{-1}
n-Octane	2.95	121
Pentachlorophenol	3.4×10^{-6}	1.5×10^{-4}
n-Pentane	1.23	50.3
Perchloroethane	8.3×10^{-3}	3.4×10^{-1}
Phenanthrene	3.5×10^{-5}	1.5×10^{-3}
Toluene	6.6×10^{-3}	2.8×10^{-1}
1,1,1-Trichloroethane (TCA)	1.8×10^{-2}	7.7×10^{-1}
Trichloroethene (TCE)	1×10^{-2}	4.2×10^{-1}
o-Xylene	5.1×10^{-3}	2.2×10^{-1}
Vinyl chloride	2.4	99

Source: Adapted from Lyman et al. (1990).

ALGORITHMS: WHAT ARE THEY?

An *algorithm* is a specific mathematical calculation procedure; a computable set of steps to achieve the desired result. More specifically, according to Cormen et al. (2002), "an algorithm is any well-defined computational procedure that takes some value, or set of values, as input and produces some value, or set of values, as output." In other words, an algorithm is a recipe for an automated solution to a problem. A computer model may contain several algorithms. According to Knuth, the word "algorithm" is derived from the name "al-Khowarizmi," a ninth-century Persian mathematician.

Algorithms should not be confused with computations. While an algorithm is a systematic method for solving problems, and computer science is the study of algorithms (although the algorithm was developed and used long before any device resembling a modern computer was available), the act of executing an algorithm—that is, manipulating data in a systematic manner—is called *computation*.

For example, the following algorithm (attributed to Euclid, circa 300 B.C., and thus known for millennia) for finding the greatest common divisor of two given whole numbers may be stated as follows:

- Set a and b to the values A and B, respectively.
- Repeat the following sequence of operations until b has value 0:
  - let r take the value of a mod b;
  - let a take the value of b;
  - let b take the value of r.
- The greatest common divisor of A and B is final value of a.

Note: The operation a mod b gives the 'remainder' obtained by dividing a by b.

The problem—that of finding the greatest common divisor of two numbers, is specified by stating what is to be computed; the problem statement itself does not require that any particular algorithm be used to compute the value. Such method-independent specifications can be used to define the meaning of algorithms: the meaning of an algorithm is the value that it computes.

Several methods can be used to compute the required value; Euclid's method is just one. The chosen method assumes a set of standard operations (such as basic operations on the whole number and a means to repeat an operation) and combines these operations to form an operation that computes the required value. Also, it is not at all obvious to the vast majority of people that the proposed algorithm does actually compute the required value. That is one reason why a study of algorithms is important—to develop methods that can be used to establish what a proposed algorithm achieves.

EXPRESSING ALGORITHMS

Again, although an in-depth discussion is beyond the scope of this text, we point out that the analysis of algorithms often requires us to draw upon a body of mathematical operations. Some of these operations are as simple as high-school algebra, but others may be less familiar to the average environmental engineer. Along with learning how to manipulate asymptotic notations and solving recurrences, several other concepts and methods must be learned to analyze algorithms.

For example, methods for evaluating bounding summations (e.g., when an algorithm contains an iterative control construct such as a *while* or *for* loop, its running time can be expressed as the sum of the times spent on each execution of the body of the loop), which occur frequently in the analysis of algorithms are important. Many of the formulas commonly used in analyzing algorithms can be found in any calculus text. In addition, to analyze many algorithms, we must be familiar with the basic definitions and notations for sets, relations, functions, graphs, and trees. A basic understanding of elementary principles of counting (permutations, combinations, and the like) is important as well. Most algorithms used in environmental engineering require no probability for their analysis; however, familiarity with these operations can be useful.

Because mathematical and scientific analysis (and many environmental engineering functions) is so heavily based upon numbers, computation has tended to be associated with numbers. However, this need not be the case: algorithms can be expressed using any formal manipulation system, that is, any system which defines a set of entities and a set of unambiguous rules for manipulating those entities. For example, the system called the *SKI calculus* consists of three combinators (entities), called, coincidentally S, K, and I. The computation rules for calculus are:

- $Sfgx = fx(gx)$
- $Kxy = x$
- $Ix = x$

where f, g, x, y are strings of the three entities. The SKI calculus is computationally complete; that is, any computation that can be performed using any formal system can be performed using the SKI calculus (equivalently, all algorithms can be expressed using the SKI calculus). Not all systems of computation are equally as powerful though—some problems that can be solved using one system can't be solved using another. Further, it is known that there are problems that cannot be solved using any formal computation system.

GENERAL ALGORITHM APPLICATIONS

Practical applications of algorithms are ubiquitous. All computer programs are expressions of algorithms, where the instructions are expressed in the computer language being used to develop the program. Computer programs are described as expressions of algorithms as an algorithm is a general technique for achieving some purpose which can be expressed in a number of different ways. Algorithms exist for many purposes and are expressed in many different ways. Examples of algorithms include recipes in cookery books, servicing instructions in a computer's hardware manual, knitting patterns, digital instructions to a welding robot as to where each weld should be made, or cyber-speak to any system working cyberspace.

Algorithms can be used in sorting operations, for example to reorder a list into some defined sequence. It is possible to express the same algorithms as instructions to a human who had a similar requirement to reorder some list, for example, to sort a list of tax record reports into a sequence determined by the date of birth on the record. These instructions could employ the *insertion sort algorithm*, or the *bubble sort algorithm* or one of many other available algorithms. Thus, an algorithm as a general technique for expressing the process of completing a defined task is independent of the precise manner in which it is expressed.

Sorting is by no means the only application for which algorithms have been developed. Practical applications of algorithms include the following examples:

- **Internet routing**: single-source shortest paths
- **Search engine**: string matching
- **Public-key cryptography and digital signatures**: number-theoretic algorithms
- **Allocate scarce resources in the most beneficial way**: linear programming
- Algorithms are at the core of most technologies used in contemporary computers
  - The hardware design use algorithms
  - The design of any GUI relies on algorithms
  - Routing in networks relies heavily on algorithms
  - Compilers, interpreters, or assemblers make extensive use of algorithms

There are a few classic algorithms that are commonly used to illustrate their function, purpose, and applicability. One of these classics is known as the **Byzantine Generals**. Briefly, this algorithm is about the problem of reaching a consensus among distributed units if some of them give misleading answers. To be memorable, the problem is couched in terms of generals deciding on a common plan of attack. Some traitorous generals may lie about whether they will support a particular plan and what other generals told them. Exchanging only messages, what decision-making algorithm should generals use to reach a consensus? What percentage of liars can the algorithm tolerate and still correctly determine a consensus (Black, 2012)?

In the following, we use another classic algorithm to illustrate (because of its general usefulness, and because it is easy to explain to just about anyone) how an algorithm can be applied to real world situations is known as the **Traveling Salesman** problem. The Traveling Salesman problem is the most notorious NP-complete problem (that is, NP-complete means that no polynomial-time algorithm has yet been discovered for an NP-complete problem, nor has anyone yet been able to prove that no polynomial-time algorithm can exist for any one of them).

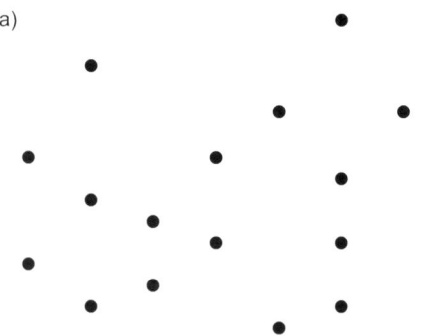

FIGURE 4.1a Input.

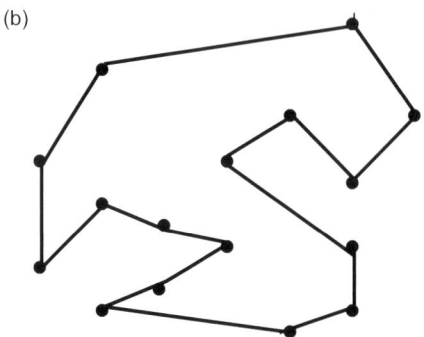

FIGURE 4.1b Output.

The Problem

Imagine a traveling salesman that has to visit each of a given set of cities by car. Using Figure 4.1a below, find the cycle of minimum cost (Figure 4.1b) visiting all of the vertices exactly once.

We have pointed out some of the functions that algorithms can perform. The question arises: "Can every problem be solved algorithmically?" The simple and complex answer is NO. For example, there are problems for which no generalized algorithmic solution can possibly exist (unsolvable). Also, there are problems for which no efficient solution is known: *NP-Complete Problems;* that is, it is unknown if efficient algorithms exist for NP-complete problems; if an efficient algorithm exists for any one of them, then efficient algorithms exist for all of them (e.g., Traveling-Salesman Problem). Finally, there are problems that we don't know how to solve algorithmically.

From the discussion above, it should be apparent that computer science is NOT about word processing and spreadsheets; it can include *applications* and not just *software applications.*

ENVIRONMENTAL PRACTICE ALGORITHM APPLICATIONS

Although algorithms can be used in transportation applications (traveling-salesman problem), many of their most important applications are applied to environmental engineering functions. For example, consider a robot arm assigned to weld all the metal parts on an automobile in an assembly line. The shortest tour that visits each weld point exactly once defines the most efficient path for the robot. A similar application arises in minimizing the amount of time taken by a design engineer or draftsperson to draw a given structure.

Algorithms have found widespread application in all branches of environmental practice. In environmental engineering, for example, USEPA uses computer models using various algorithms to monitor chemical spills and ultimate fate data. In the following, we provide selected model summary descriptions of applications used in Dispersion Modeling. Specifically, we discuss how USEPA (and others) employ preferred/recommended models (i.e., refined models that are recommended for a specific type of regulatory application) in monitoring air quality; that is, ambient pollutant concentrations and their temporal and spatial distribution. Further information on this important topic can be accessed at USEPA's Technology Transfer Network Support Center for Regulatory Air Models at www.epa.gov/scram001/tt22.htm website.

DISPERSION MODELS

The following algorithm-based models described in the following are currently listed in Appendix A of the *Guidelines on Air Quality Models* (published as Appendix W of 40 CFR Part 51):

Dispersion Models

- **BLP (Buoyant Line and Point Source Model)**: a Gaussian plume dispersion model designed to handle unique modeling problems associated with aluminum reduction plants, and other industrial sources where plume rise and downwash effects from stationary line sources are important.
- **CALINE3**: a steady-state Gaussian dispersion model designed to determine air pollution concentrations at receptor locations downwind of "at-grade," fill," "bridge," and "cut section" highways located in relatively uncomplicated terrain.
- **CALPUFF**: a multi-layer, multi-species non-steady-state puff dispersion model that simulates the effects of time- and space-varying meteorological conditions on pollution transport, transformation, and removal. CALPUFF can be applied on scales of tens to hundreds of kilometers. It includes algorithms for subgrid-scale effects (such as terrain impingement), as well as longer-range effects (such as pollutant removal due to wet scavenging and dry deposition, chemical transformation, and visibility effects of particulate matter concentrations).
- **CTDMPLUS**: (Complex Terrain Dispersion Model Plus Algorithms for Unstable Situations) a refined point source Gaussian air quality model for use in all stability conditions for complex terrain (i.e., terrain exceeding the height of the stack being modeled as contrasted with *simple terrain* which is defined as an area where terrain features are all lower in elevation than the top of the stack of the source). The model contains, in its entirety, the technology of CTDM for stable and neutral conditions.
- **ISC3 (Industrial Source Complex Model)**: a steady-state Gaussian plume model which can be used to assess pollutant concentrations from a wide variety of sources associated with an industrial complex. This model can account for the following: settling and dry deposition of particles; downwash; point, area, line, and volume sources; plume rise as a function of downwind distance; separation of point sources; and limited terrain adjustment. ISC3 operates in both long-term and short-term modes.
- **OCD (Offshore and Coastal Dispersion Model)**: a straight-line Gaussian model developed to determine the impact of offshore emissions from point, area, or line sources on the air quality of coastal regions. OCD incorporates overwater plume transport and dispersion as well as changes that occur as the plume crosses the shoreline. Hourly meteorological data are needed from both offshore and onshore locations.

SCREENING TOOLS

Note: *Screening tools* are relatively simple analysis techniques to determine if a given source is likely to pose a threat to air quality. Concentration estimates from screening techniques precede a refined modeling analysis and are conservative.

- **CAL3QHC/CAL3QHCR (CALINE3 with queuing and hot spot calculations)**: a CALINE3-based CO model with a traffic model to calculate delays and queues that occur at signalized intersections; CAL3QHCR requires local meteorological data.
- **COMPLEX 1**: a multiple point source screening technique with terrain adjustment that incorporates the plume impaction algorithm of the VALLEY model.
- **CTSCREEN (Complex Terrain Screening model)**: a Gaussian plume dispersion model designed as a screening technique for regulatory application to plume impaction assessments in complex terrain. CTSCREEN is a screening version of the CTDMPLUS model.
- **LONGZ**: a steady-state Gaussian plume formulation for both urban and rural areas in flat or complex terrain to calculate long-term (seasonal and/or annual) ground-level ambient air concentrations attributable to emissions from up to 14,000 arbitrarily placed sources (stack, buildings, and area sources).
- **SCREEN3**: a single source Gaussian plume model which provides maximum ground-level concentrations for point, area, flare, and volume sources, as well as concentrations in the cavity zone, and concentrations due to inversion break-up and shoreline fumigation. SCREEN3 is a screening version of the ISC3 model.
- **SHORTZ**: a steady-state bivariate Gaussian plume formulation for both urban and rural areas in flat or complex terrain to calculate ground-level ambient air concentrations. It can calculate 1-, 2-, 3-hour, etc. average concentrations due to emissions from stacks, buildings, and area sources for up to 300 arbitrarily placed sources.
- **VALLEY**: a steady-state, complex terrain, univariate Gaussian plume dispersion algorithm designed for estimating either 24-hour of annual concentrations resulting from emissions from up to 50 (total) point and area sources.
- **VISCREEN**: calculates the potential impact of a plume of specified emissions for specific transport and dispersion conditions.

REFERENCES

Black, P.E. (2012). Byzantine generals, in *Dictionary of Algorithms and Data Structures*. US. National Institute of Standards and Technology. accessed 12/12/21. https://www.nist.gov/publications/dictionary-algorithms-data-structure.

Cormen, T.H., Leiserson, C. E., Rivest, R.L., & Stein, C. (2002). *Introduction to Algorithms, 2nd ed.* New Delhi: Prentice-Hall of India.

Fechner-Levy (2000). *Henry Law Constant*. http://iupac.org/henry-law-constants. accessed 12/12/21.

Lyman, W.J., Reehl, W.R., & Rosenblatt, D.H. (1990). *Handbook of Chemical Property Estimation Methods*, 2nd ed. Washington, DC: American Chemical Society.

Nirmalakhandan, N. (2002). *Modeling Tools for Environmental Engineers and Scientists*. Boca Raton, FL: CRC Press.

NIST (2012). *Engineering Statistics*. Washington, DC: United States Department of Commerce.

5 Statistics Review

There are three kinds of lies: lies, damned lies, and statistics.

Benjamin Disraeli

To the uninitiated it may often appear that the statistician's primary function is to prevent or at least impede the progress of research. And even those who suspect that statistical methods may be more boon than bane are at times frustrated in their efforts to make use of the statistician's wares.

Frank Freese

No aphorism is more frequently repeated in connection with field trials, than that we must ask Nature few questions, or, ideally, one question, at atime. The writer is convinced that this view is wholly mistaken.

Ronald Fisher

STATISTICAL CONCEPTS

Despite the protestation of Disraeli and the wisdom of Freese, the environmental practice includes the study of and use of statistical analysis of the results.[1] The principal concept of statistics is that of variation. In conducting typical environmental health functions (biostatistics—a wide range in the application of statistics to an even wider range of topics in biology) such as toxicological or biological sampling protocols for air contamination, and other environmental functions applied to agriculture, forestry, fisheries, and other specialized areas variation is commonly found. This chapter provides the environmental engineer with a survey of the basic statistical and data analysis techniques that can be used to address many of the problems that he or she will encounter on a daily basis. It covers the data-analysis process, from research design to data collection, analysis, reaching of conclusions, and, most importantly, presentation of findings.

Finally, it is important to point out that statistics can be used to justify the implementation of a program, identify areas that need to be addressed, or justify the impact that various environmental health and safety programs have on losses and accidents. A set of occupational health and safety data (or other data) is only useful if it is analyzed properly. Better decisions can be made when the nature of the data is properly characterized. For example, the importance of using statistical data in selling the environmental health and safety plan or safety program or some other environmental operations—the innovation and the winning over of those who control the purse strings—cannot be overemphasized.

With regard to Freese's opening statement, much of the difficulty is due to not understanding the basic objectives of statistical methods. We can boil these objectives down to two:

1. The estimation of population parameters (values that characterize a particular population).
2. The testing of hypotheses about these parameters.

A common example of the first is the estimation of coefficients a and b in the linear relationship. $Y = a + bX$, between the variables Y and X. To accomplish this objective, one must first define the population involved and specify the parameters to be estimated. This is primarily the research worker's job. The statistician helps devise efficient methods of collecting the data and calculating the desired estimates.

Unless the whole population is examined, an estimate of a parameter is likely to differ to some degree from the population value. The unique contribution of statistics to research is that it provides ways of evaluating how far off the estimate may be. This is ordinarily done by computing confidence limits, which have a known probability of including the true value of the parameter. Thus, the mean diameter of the trees in a pine plantation may be estimated from a sample as 9.2 in, with 95% confidence limits of 8.8 and 9.6 in. These limits (if properly obtained) tell us that, unless a one-in-twenty chance has occurred in sampling, the true mean diameter is somewhere between 8.8 and 9.6 in.

The second basic objective in statistics is to test some hypotheses about the population parameters. A common example is a test of the hypothesis that the regression coefficient in the linear model.

$$Y = a + bX$$

has some specified value (say zero). Another example is a test of the hypothesis that the difference between the means of two populations is zero.

Again, it is the research worker who should formulate meaningful hypotheses to be tested, not the statistician. This task can be tricky. The beginner would do well to work with the statistician to be sure that the hypothesis is put in a form that can be tested. Once the hypothesis is set, it is up to the statistician to work out ways of testing it and to devise efficient procedures for obtaining the data (Freese, 1969).

PROBABILITY AND STATISTICS

Those who work with probabilities are commonly thought to have an advantage when it comes to knowing, for example, the likelihood of tossing coins heads up six times in a row, or the chances of a crapshooter making several consecutive winning throws ("passes"), and other such useful bits of information. It is fairly well known that statisticians work with probabilities; thus, they are often associated with having the upper hand, so to speak, on predicting outcomes in games of chance. However, statisticians also know that this assumed edge they have in games of chance is often dependent on other factors.

The fundamental role of probability in statistical activities is often not appreciated. In putting confidence limits on an estimated parameter, the part played by probability is fairly obvious. Less apparent to the neophyte is the operation of probability in the testing of hypotheses. Some of them say with derision, "You can prove anything with statistics"—remember what Disraeli said about statistics. Anyway, the truth is, you can prove nothing; you can at most compute the probability of something happening and let the researcher draw his own conclusions.

Let's return to our game of chance to illustrate this point. In the game of craps, the probability of the shooter winning (making a pass) is ~0.493—assuming, of course, a perfectly balanced set of dice and an honest shooter. Suppose now that you run up against a shooter who picks up the dice and immediately makes seven passes in a row! It can be shown that if the probability of making a single pass is really 0.493, then the probability of seven or more consecutive passes is about 0.007 (or 1 in 141). This is where statistics ends; you draw your own conclusions about the shooter. If you conclude that the shooter is pulling a fast one, then in statistical terms you are rejecting the hypothesis that the probability of the shooter making a single pass is 0.493.

In practice, most statistical tests are of this nature. A hypothesis is formulated, and an experiment is conducted, or a sample is selected to test it. The next step is to compute the probability of the experimental or sample results occurring by chance if the hypothesis is true. If this probability is less than some preselected value (perhaps 0.05 or 0.01), the hypothesis is rejected. Note that nothing has been proved—we haven't even proved that the hypothesis is false. We merely inferred this because of the low probability associated with the experiment or sample results.

Our inferences may be incorrect if we are given inaccurate probabilities. Reliable computation of these probabilities requires knowledge of how the variable we are dealing with is distributed (i.e., what the probability is of the chance of occurrence of different values of the variable). Accordingly, if we know that the number of beetles caught in light traps follows what is called the Poisson distribution, we can compute the probability of catching X or more beetles. But, if we assume that his variable follows the Poisson when it actually follows the negative binomial distribution, our computed probabilities may be in error.

Even with reliable probabilities, statistical tests can lead to the wrong conclusions. We will sometimes reject a hypothesis that is true. If we always test at the 0.05 level, we will make this mistake on the average of 1 time in 20. We accept this degree of risk when we select the 0.05 level of testing. If we're willing to take a bigger risk, we can test at the 0.10 or the 0.25 level. If we're not willing to take this much risk, we can test at the 0.01 or 0.001 level.

Researchers can make more than one kind of error. In addition to rejecting a hypothesis that is true (a Type 1 error), he can make the mistake of not rejecting a hypothesis that is false (a Type II error). In crap shooting, it is a mistake to accuse an honest shooter of cheating (Type I error—rejecting true hypothesis), but it is also a mistake to trust a dishonest shooter (Type II error—failure to reject a false hypothesis).

The difficulty is that for a given set of data, reducing the risk of one kind of error increases the risk of the other kind. If we set 15 straight passes as the critical limit for a crap shooter, then we greatly reduce the risk of making a false accusation (probability about 0.00025). But in so doing, we have dangerously increased the probability of making a Type II error—failure to detect a phony. A critical step in designing experiments is the attainment of an acceptable level of probability of each type of error. This is usually accomplished by specifying the level of testing (i.e., probability of an error of the first kind) and then making the experiment large enough to attain an acceptable level of probability for errors of the second kind.

It is beyond the scope of this book to go into basic probability computations, distribution theory, or the calculation of Type II errors. But anyone who uses statistical methods should be fully aware that he or she is dealing primarily with probabilities (not necessarily lies or damnable lies) and not with immutable absolutes. Remember, 1-in-20 chances do actually occur—about 1 time out of 20.

Measure of Central Tendency

As mentioned earlier, when talking statistics, it is usually because we are estimating something with incomplete knowledge. Maybe we can only afford to test 1% of the items we are interested in, and we want to say something about what the properties of the entire lot are; possibly because we destroy the sample by testing it. In that case, 100% sampling is not feasible if someone is supposed to get the items after we are done with them.

The questions we are usually trying to answer are "What is the central tendency of the item of interest?" and "How much dispersion about this central tendency can we expect?"

Simply, the average or averages that can be compared are measures of *central tendency* or central location of the data.

SYMBOLS, SUBSCRIPTS, BASIC STATISTICAL TERMS AND CALCULATIONS

In statistics, *symbols* such as X, Y, and Z are used to represent different sets of data. Hence, if we have data for five companies, we might let

X = company income
Y = company materials expenditures
Z = company savings

Subscripts are used to represent individual observations within these sets of data. Thus, X_i represents the income of the *i*th company, where *i* takes on values 1, 2, 3, 4, and 5. In this notation, X_1, X_2, X_3, X_4, and X_5 stand for the incomes of the first company, the second company, and so on. The data are arranged in some order, such as by size of income, the order in which the data were gathered, or any other way suitable to the purposes or convenience of the investigator.

The subscript *i* is a variable used to index the individual data observations. Therefore, X_i, Y_i, and Z_i represent the income, materials expenditures, and savings of the *i*th company. For example, X_2 represents the income of the second company, Y_2 materials expenditures of the second (same) company, and Z_5 the savings of the fifth company.

Suppose that we have data for two different samples, the net worth of 100 companies and the test scores of 30 students. To refer to individual observations in these samples, we can let X_i denote the net worth of the *i*th company, where *i* assumes values from 1 to 100. (This later idea is indicated by the notation $i = 1, 2, 3, ..., 100$.) We can also let Y_j denote the test score of the *j*th student, where $j = 1, 2, 3, ..., 20$. The different subscript letters make it clear that different samples are involved. Letters such as *X, Y,* and *Z* generally represent the different variables or types of measurements involved, whereas subscripts such as *i, j, k,* and *l* designated individual observations (Hamburg, 1987).

Next, we turn our attention to the method of expressing summations of sets of data. Suppose we want to add a set of four observations, denoted X_1, X_2, X_3, and X_4. A convenient way of designating this addition is

$$\sum_{i=1}^{4} X_i = X_1 + X_2 + X_3 + X_4$$

where the symbol Σ (Greek capital "sigma") means "the sum of." Thus, the symbol $\sum_{i=1}^{4} X_i$ is read "the sum of the X_i's, *i* going from 1 to 4." For example, if $X_1=5$, $X_2=1$, $X_3=8$, and $X_4=6$,

$$\sum_{i=1}^{4} X_i = 5 + 1 + 8 + 6 = 20$$

In general, if there are *n* observations, we write

$$\sum_{i=i}^{n} X_i = X_1 + X_2 + X_3 + \cdots + X_n$$

Basic statistical terms include mean or average, median, mode, and range. The following is an explanation of each of these terms.

THE MEAN

Mean is one of the most familiar and commonly estimated population parameters. It is the total of the values of a set of observations divided by the number of observations. Given a random sample, the population mean is estimated by

$$\overline{X} = \frac{\sum_{i=1}^{n} x}{n}$$

where X_i = The observed value of the *i*th unit in the sample.

Statistics Review

n = The number of units in the sample.

$\sum_{i=1}^{n} x_i$ means to sum up all n of the X-values in the sample.

If there are N units in the population, the total of the X-values over all units in the population would be estimated by

$$\hat{T} = N\bar{X}$$

The circumflex ($\hat{T}$) over the T is frequently used to indicate an estimated value as opposed to the true but unknown population value.

It should be noted that this estimate of the mean is used for a simple random sample. It may not be appropriate if the units included in the sample are not selected entirely at random.

MEDIAN

The median is the value of the central item when the data are arrayed in size.

MODE

The mode is the observation that occurs with the greatest frequency and thus is the most "fashionable" value.

RANGE

The range is the difference between the values of the highest and lowest terms.

Example 5.1

Problem.

Given the following laboratory results for the measurement of dissolved oxygen (DO) in water, find the mean, median, mode, and range.
   Data:
   6.5 mg/L, 6.4 mg/L, 7.0 mg/L, 6.9 mg/L, and 7.0 mg/L

Solution:

To find the mean:

$$\bar{X} = \frac{\sum_{i=1}^{n} x_i}{n}$$

$$\text{Mean}(\bar{X}) = \frac{(6.5\,\text{mg/L} + 6.4\,\text{mg/L} + 7.0\,\text{mg/L} + 6.0\,\text{mg/L} + 7.0\,\text{mg/L})}{5}$$

$$= 6.58\,\text{mg/L}$$

Mode = 7.0 mg/L (number that appears most often)
   Arrange in order: 6.4 mg/L, 6.5 mg/L, 6.9 mg/L, 7.0 mg/L, 7.0 mg/L
   Median = 6.9 mg/L (central value)
   Range = 7.0 mg/L – 6.4 mg/L = 0.6 mg/L

The importance of using statistically valid sampling methods cannot be overemphasized. Several different methodologies are available. A careful review of these methods (with an emphasis on designing appropriate sampling procedures) should be made before computing analytic results. Using appropriate sampling procedures along with careful sampling techniques will provide basic data that is accurate.

The need for statistics in environmental practice is driven by the discipline itself. As mentioned, environmental studies often deal with entities that are variable. If there were no variation in collected data, there would be no need for statistical methods.

Over a given time interval there will always be some variation in sampling analyses. Usually, the average and the range yield the most useful information. For example, in evaluating the indoor air quality (IAQ) in a factory, a monthly summary of air flow measurements, operational data, and laboratory tests for the factory would be used. Another example is seen in evaluating a work center or organization monthly on-the-job report of accidents and illnesses, a monthly summary of reported injuries, lost-time incidents and work-caused illnesses for the entity would be used.

In the preceding section, we used the term "sample" and the scenario sampling to illustrate the use and definition of mean, mode, median, and range. Though these terms are part of the common terminology used in statistics, the term sample in statistics has its own unique meaning. There is an exact delineation between both the term sample and the term population. In statistics, we most often obtain data from a sample and use the results from the *sample* to describe an entire population. The *population* of a sample signifies that one has measured a characteristic for everyone or everything that belongs in a particular group. For example, if one wishes to measure of that characteristic of the population defined as environmental professionals, one will have to obtain a measure of that characteristic for every environmental professional possible. Measuring a population is difficult, if not impossible.

We use the term *subject* or *case* to refer to a member of a population or sample. There are statistical methods for determining how many cases must be selected in order to have a credible study. *Data* is another important term: it consists of the measurements taken for the purposes of statistical analysis. Data can be classified as either qualitative or quantitative. *Qualitative* data deal with characteristic of the individual or subject (e.g., gender of a person or the color or a car). *Quantitative* data describe a characteristic in terms of a number (e.g., the age of a horse or the number of lost-time injuries an organization had over the previous year).

Along with common terminology, the field of statistics also generally uses some common symbols. Statistical notation uses Greek letters and algebraic symbols to covey meaning about the procedures that on should follow in order to complete a particular study or test. Greek letters are used as statistical notation for a population, while English letters are used for statistical notation for a sample. Table 5.1 summarizes some of the more common statistical symbols, terms and procedures used in statistical operations.

TABLE 5.1
Commonly Used Statistical Symbols and Procedures

Term or Procedure	Symbol	
	Population Symbol	Sample Notation
Mean	$\bar{\mu}$	$\bar{x}$
Standard deviation	σ	s
Variance	σ^2	s^2
Number of cases	N	N
Raw umber or value	X	X
Correlation coefficient	R	r
Procedure	**Symbol**	
Sum of	Σ	
Absolute value of x	$|x|$	
Factorial of n	$n!$	

DISTRIBUTION

If an environmental professional wishes to conduct a research study, and the data is collected and a group of raw data is obtained, to make sense out of the data, the data must be organized into a meaningful format.

The formatting begins by putting the data into some logical order, then grouping the data. Before the data can be compared to other data it must be organized. Organized data are referred to *distributions*.

As mentioned, when confronted with masses of ungrouped data (i.e., listings of individual figures), it is difficult to generalize about the information the masses contain. However, if a frequency distribution of the figures is formed, many features become readily discernible. A frequency distribution records the number of cases that fall in each class of the data.

Example 5.2

An environmental health and safety professional gathers data on the medical costs of 24 on-the-job injury claims for a given year. The raw data collected was as shown in Table 5.2. To develop a frequency distribution, the investigator takes the values of the claims and places them in order. Then the investigator counts the frequency of occurrences for each value as shown in Table 5.3.

In order to develop a frequency distribution, groupings are formed using the values in the table above, ensuring that each group has an equal range. The safety engineer grouped the data into ranges of 1,000. The lowest range and highest range are determined by the data. Because it was decided to group by 1,000, values will fall in the ranges of $0 to $4,999, and the distribution will end with this. The frequency distribution for this data appears in Table 5.4.

NORMAL DISTRIBUTION

When large amounts of data are collected on certain characteristics, the data and subsequent frequency can follow a distribution that is bell-shaped in nature—the *normal distribution*. The normal distributions are a very important class of statistical distributions. As stated, all normal distributions are symmetric and have bell-shaped curves with a single peak (see Figure 5.1).

To speak specifically of any normal distribution, two quantities have to be specified: the mean μ (pronounced mu), where the peak of the density occurs, and the standard deviation σ (sigma). Different values of μ and σ yield different normal density curves and hence different normal distributions.

Although there are many normal curves, they all share an important property that allows us to treat them in a uniform fashion. All normal density curves satisfy the following property, which is often referred to as the *Empirical Rule*.

TABLE 5.2
Value and Frequency of Claims

$60	$1,500	$85	$120
$110	$150	$110	$340
$2,000	$3,000	$550	$560
$4,500	$85	$2,300	$200
$120	$880	$1,200	$150
$650	$220	$150	$4,600

TABLE 5.3
Frequency Distribution

Value	Frequency
$60	1
$85	2
$110	2
$120	2
$150	3
$200	1
$220	1
$340	1
$550	1
$560	1
$650	1
$880	1
$1,200	1
$1,500	1
$2,000	1
$2,300	1
$3,000	1
$4,500	1
$4,600	1
Total	24

TABLE 5.4
Frequency Distribution

Range	Frequency
$0–999	17
$1,000–1,999	2
$2,000–2,999	2
$3,000–3,999	1
$4,000–4,999	2
Total	24

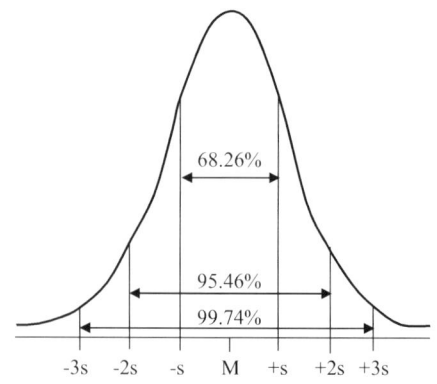

FIGURE 5.1 Normal distribution curve showing the frequency of a measurement.

68% of the observations fall within one **standard deviation** of the **mean**; that is, between $\mu-\sigma$ and $\mu+\sigma$.

95% of the observations fall within two **standard deviations** of the **mean**, that is, between $\mu-2\sigma$ and $\mu+2\sigma$.

98% of the observations all within three **standard deviations** of the **mean**; that is, between $\mu-3\sigma$ and $\mu+3\sigma$.

Thus, for a normal distribution, almost all values lie within three standard deviations of the mean (see Figure 5.1). It is important to stress that the rule applies to all normal distributions. Also, remember that it applies *only* to normal distributions.

Note: Before applying the empirical rule it is a good idea to identify the data being described, and the value of the mean and standard deviation. A sketch of a graph summarizing the information provided by the empirical rule should also be made.

Example 5.3

Problem:

The scores for all high school seniors taking the math section of the Scholastic Aptitude Test (SAT) in a particular year had a mean of 490 and a standard deviation of 100. The distribution of SAT scores is bell-shaped.

1. What percentage of seniors scored between 390 and 590 on this SAT test?
2. One student scored 795 on this test. How did this student do compared to the rest of the scores?
3. A rather exclusive university only admits students who were among the highest 16% of the scores on this test. What score would a student need on this test to be qualified for admittance to this university?

The data being described are the math SAT scores for all seniors taking the test one year. Because this is describing a population, we denote the mean and standard deviation as $\mu=490$ and $\sigma=100$, respectively. A bell-shaped curve summarizing the percentages given by the empirical rule is shown in Figure 5.2.

1. From Figure 5.2, about 68% of seniors scored between 390 and 590 on this SAT test.
2. Because about 99.7% of the scores are between 190 and 790, a score of 795 is excellent. This is one of the highest scores on this test.
3. Because about 68% of the scores are between 390 and 590, this leaves 32% of the scores outside the interval. Because a bell-shaped curve is symmetric, one-half of the scores, or 16%, are on each end of the distribution.

STANDARD DEVIATION

The standard deviation, s or σ (sigma), is often used as an indicator of precision. The *standard deviation* is a measure of the variation (the spread in a set of observations) in the results; that is, it gives us some idea whether most of the individuals in a population are close to the mean or spread out. In order to gain better understanding and perspective of the benefits derived from using statistical methods in safety engineering, it is appropriate to consider some of the basic theory of statistics. In any set of data, the true value (mean) will lie in the middle of all the measurements taken. This is true, providing the sample size is large and only random error is present in the analysis. In addition, the measurements will show a normal distribution as shown in Figure 5.1.

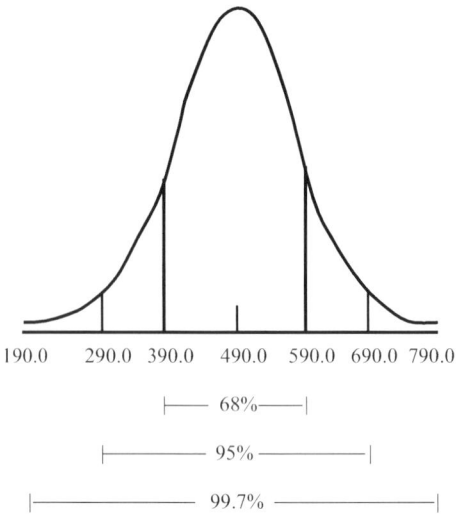

FIGURE 5.2 Sample Scholastic Aptitude Test (SAT) math percentages given by the empirical rule.

Figure 5.1 shows that 68.26% of the results fall between $M+s$ and $M-s$, 95.46% of the results lie between $M+2s$ and $M-2s$, and 99.74% of the results lie between $M+3s$ and $M-3s$. Therefore, if precise, then 68.26% of all the measurements should fall between the true value estimated by the mean, plus the standard deviation and the true value minus the standard deviation. The following equation is used to calculate the sample standard deviation:

$$s = \sqrt{\frac{\sum (X - \bar{X})^2}{n-1}}$$

where:
s = standard deviation
n = number of samples
X = the measurements from X to X_n
$\bar{X}$ = the mean
Σ = means to sum the values from X to X_n

Example 5.4

Problem:

Calculate the standard deviation, σ, of the following dissolved oxygen values (Table 5.5):

$$9.5, 10.5, 10.1, 9.9, 10.6, 9.5, 11.5, 9.5, 10.0, 9.4$$

$$\bar{X} = 10.0$$

$$\sigma = \sqrt{\frac{1}{(10-1)}(3.99)}$$

$$= \sqrt{\frac{3.99}{9}} = 0.67$$

TABLE 5.5
Dissolved Oxygen Values

X	$X - \bar{X}$	$(X - \bar{X})^2$
9.5	−0.5	0.25
10.5	0.5	0.25
10.1	0.1	0.01
9.9	−0.1	0.01
10.6	0.6	0.36
9.5	−0.5	0.25
11.5	1.5	2.25
9.5	−0.5	0.25
10.0	0	0
9.4	−0.6	0.36
		3.99

COEFFICIENT OF VARIATION

In nature, populations with large means often show more variation than populations with small means. The coefficient of variation (C) facilitates comparison of variability abut different sized means. It is the ratio of the standard deviation to the mean. A standard deviation of 2 for a mean coefficient of variation would be 0.20 or 20% in each case. If we have a standard deviation of 1.414 and a mean of 9.0, the coefficient of variation would be estimated by

$$C = \frac{s}{\bar{X}} = \frac{1.414}{9.0} = 0.157, \text{ or } 15.7\%$$

STANDARD ERROR OF THE MEAN

As mentioned, there is usually variation among the individual units of a population. Again, the standard deviation is a measure of this variation. Because the individual units vary, variation may also exist among the means (or any other estimates) computed from samples of these units. Take, for example, a population with a true mean of 10. If we were to select four units at random, they might have a sample mean of 8. Another sample of four units from the same population might have a mean of 11, another 10.5, and so forth. Clearly, it would be desirable to know the variation likely to be encountered among the means of samples from this population. A measure of the variation among sample means is the standard error of the mean. It can be thought of as a standard deviation among sample means; it is a measure of the variation among sample means, just as the standard deviation is a measure of the variation among individuals. The standard error of the mean may be used to compute confidence limits for a population mean.

The computation of the standard error of the mean (often symbolized by $s_{\bar{x}}$) depends on the manner in which the sample was selected. For simple random sampling without replacement (i.e., a given unit cannot appear in the sample more than once) from a population having a total of N units the formula for the estimated standard error of the mean is

$$s_x = \sqrt{\frac{s^2}{n}\left(1 - \frac{n}{N}\right)}$$

In a forestry example, if we had $n=10$ and found that $s=1.414$ or $s^2=2$ in the population that contains 1,000 trees, the estimated mean diameter ($X=9.0$) would have a standard error of

$$s_{\bar{x}} = \sqrt{\frac{2}{10}\left(1 - \frac{10}{1,000}\right)} = \sqrt{0.198}$$

$$= 0.445$$

Note: The term $(1 - n/N)$ is called the *finite population correction* or fpc. The fpc is the finite population correct and is used when the sampling fraction (the number of elements or respondents sampled relative to the population) becomes large. The fpc is used in the calculation of the standard error of the estimate. If the value of the fpc is close to 1, it will have little impact and can be safely ignored.

COVARIANCE

Very often, each unit of a population will have more than a single characteristic. In forest practice, for example, trees may be characterized by their height, diameter, and form class (i.e., the amount of taper). The covariance is a measure of the association between the magnitudes of two characteristics. If there is little or no association, the covariance will be close to zero. If the large values of one characteristic tend to be associated with the small values of another characteristic, the covariance will be negative. If the large values of one characteristic tend to be associated with the large values of another characteristic, the covariance will be positive. The population covariance of X and Y is often symbolized by σ_{xy}; the sample estimate by s_{xy}.

Let's return to a forestry practice example. Suppose that the diameter (inches) and age (years) have been obtained for a number of randomly selected trees. If we symbolize diameter by Y and the age by X, the sample covariance of diameter and age is given by

$$s_{xy} = \frac{\sum XY - \frac{\left(\sum X \sum Y\right)}{n}}{(n-1)}$$

This is equivalent to the formula

$$s_{xy} = \frac{\sum (X - \bar{X}) \cdot (Y - \bar{Y})}{(n-1)}$$

If $n=12$ and the Y and X values were as follows:

													Sums
Y	4	9	7	7	5	10	9	6	8	6	4	11	86
X	20	40	30	45	25	45	30	40	20	35	25	40	395

Then

$$s_{xy} = \frac{(4)(20) + (9)(40) + \cdots + (11)(40) - \frac{(86)(395)}{12}}{12 - 1}$$

$$= \frac{2,960 - 2,830.83}{11} = 11.74$$

The positive covariance is consistent with the well-known and economically unfortunate fact that larger diameters tend to be associated with older ages.

SIMPLE CORRELATION COEFFICIENT

The magnitude of the covariance, like that of the standard deviation, is often related to the size of the variables themselves. Units with large X and Y values tend to have larger covariances than units with small X and Y values. Also, the magnitude of the covariance depends on the scale of measurement; in the previous example, had diameter been expressed in millimeters instead of inches, the covariance would have been 298.196 instead of 11.74.

The simple correlation coefficient, a measure of the degree of linear association between two variables, is free of the effects of scale of measurement. It can vary from -1 and $+1$. A correlation of 0 indicates that there is no linear association (there may be a very strong nonlinear association, however). A correlation of $+1$ or -1 would suggest a perfect linear association. As for the covariance, a positive correlation implies that the large values of X are associated with the large values of Y. If the large values of X are associated with the small values of Y, the correlation is negative.

The population correlation coefficient is commonly symbolized by ρ (rho), and the sample-based estimate r. The population correlation coefficient is defined to be

$$\rho = \frac{\text{Covariance of } X \text{ and } Y}{\sqrt{(\text{Variance of } X)(\text{Variance of } Y)}}$$

For a simple random sample, the sample correlation coefficient is computed as follows:

$$r = \frac{s_{xy}}{s_x \cdot s_y} = \frac{\sum xy}{\sqrt{\left(\sum x^2\right)\left(\sum y^2\right)}}$$

where s_{xy} = Sample covariance of X and Y
s_x = Sample standard deviation of X
s_y = Sample standard deviation of Y
Σ_{xy} = Corrected sum of XY products

$$= \sum XY - \frac{\left(\sum X\right)\left(\sum Y\right)}{n}$$

Σx^2 = Corrected sum of squares for X

$$= \sum X^2 - \frac{\left(\sum X\right)^2}{n}$$

Σy^2 = Corrected sum of squares for Y

$$= \sum Y^2 - \frac{\left(\sum Y\right)^2}{n}$$

For the values used to illustrate the covariance we have:

$$\sum xy = (4)(20)+(9)(40)+\cdots+(11)(40)-\frac{(86)(395)}{12} = 129.1667$$

$$\sum y^2 = 4^2+9^2+\cdots+11^2-\frac{86^2}{12} = 57.6667$$

$$\sum x^2 = 20^2+40^2+\cdots+40^2-\frac{395^2}{12} = 922.9167$$

So,

$$R = \frac{129.1667}{\sqrt{(57.6667)(922.9167)}} = \frac{129.1667}{230.6980} = 56$$

DID YOU KNOW?

The computed value of a statistic such as the correlation efficiency depends on which particular units were selected for the sample. Such estimates will vary from sample to sample. More important, they will usually vary from the population value which we try to estimate.

VARIANCE OF A LINEAR FUNCTION

Routinely we combine variables or population estimates in a linear function. For example, if the mean timber volume per acre has been estimated as $\bar{X}$, then the total volume on M acres with be $M\bar{X}$; the estimate of total volume is a linear function of the estimated-mean volume. If the estimate of cubic volume per acre in sawtimber is $\bar{X}_1$ and of pulpwood above the sawtimber top is $\bar{X}_2$, then the estimate of total cubic foot volume per acre is $\bar{X}_1 + \bar{X}_2$. If on a given tract the mean volume per half-acre is $\bar{X}_1$ for spruce and the mean volume per quarter-acre is $\bar{X}_2$ for yellow birch, then the estimated total volume per acre of spruce and birch would be $2\bar{X}_1 + 4\bar{X}_2$.

In general terms, a liner function of three variables (say X_1, X_2, and X_3) can be written as

$$L = a_1 X_1 + a_2 X_2 + a_3 X_3$$

where a_1, a_2, and a_3 are constants.

If the variances are $s_1^2, s_2^2,$ and s_3^2 (for X_1, X_2 and X_3 respectively) and the covariances are $S_{1,2}$, $S_{1,3}$ and $S_{2,3}$, then the variance of L is given by

$$s_L^2 = a_1^2 s_1^2 + a_2^2 s_2^2 + a_3^2 s_3^2 + 2(a_2 a_2 s_{1,2} + a_1 a_3 s_{1,3} + a_2 a_3 s_{2,3})$$

The standard deviation (or standard error) of L is simply the square root of this.
The extension of the rule to cover any number of variables should be fairly obvious.

Example 5.5

Problem:

The sample mean volume per forested acre for a 10,000-acre tract is $\bar{X} = 5{,}680$ board feet with a standard error of $s_{\bar{x}} = 632$ (So $s_{\bar{x}}^2 = 399{,}424$). The estimated total volume is

$$L = 10{,}000(\bar{X}) = 56{,}800{,}000 \text{ board feet.}$$

The variance of this estimate would be

$$s_L^2 = (10{,}000)^2 (s_{\bar{x}}^2) = 39{,}942{,}400{,}000{,}000$$

Since the standard error of an estimate is the square root of its variance, the standard error of the estimated total is

$$s_L = \sqrt{s_L^2} = 6{,}320{,}000$$

Example 5.6

Problem:

In 1995 a random sample of 40 one-quarter-acre circular plots was used to estimate the cubic foot volume of a stand of pine. Plot centers were monumented for possible relocation at a later time. The mean volume per plot was $\bar{X}_1 = 225 \text{ ft}^3$. The plot variance was $s_{x_1}^2 = 8{,}281$ so that the variance of the mean was $s_{\bar{x}_1}^2 = 8{,}281/40 = 207.025$.

In 2000 a second inventory was made using the same plot centers. This time, however, the circular plots were only one-tenth acre. The mean volume per plot was $\bar{X}_2 = 122 \text{ ft}^3$. The plot variance was $s_{x_2}^2 = 6{,}084$, so the variance of the mean was $s_{\bar{x}_2}^2 = 152.100$. The covariance of initial and final plot volumes was $s_{x_{1,2}} = 4{,}259$, making the covariance of the means $s_{\bar{x}_1, \bar{x}_2} = 4{,}259/40 = 106.475$.

The net periodic growth per acre would be estimated as

$$G = 10\bar{X}_2 - 4\bar{X}_1 = 10(122) - 4(225) = 320 \text{ cubic feet per acre.}$$

By the rule for linear functions, the Variance of G would be

$$s_G^2 = (10)^2 s_{\bar{x}_2}^2 + (-4)^2 s_{\bar{x}_1}^2 + 2(1)(-4) s_{\bar{x}_1, \bar{x}_2}$$

$$= 100(152.100) + 16(207.025) - 80(106.475)$$

$$= 10{,}004.4$$

In this example, there was a statistical relationship between the 2000 and 1995 means because the same plot locations were used in both samples. The covariance of the means ($s_{\bar{x}_1, \bar{x}_2}$) is a measure of this relationship. If the 2000 plots had been located at random rather than at the 1995 locations, the two means would have been considered statistically independent and their covariance would have been set at zero. In this case, the equation for the variance of the net periodic growth per acre (G) would reduce to

$$s_G^2 = (10)^2 s_{\bar{x}_2}^2 + (-4)^2 s_{\bar{x}_1}^2$$

$$= 100(152.100) + 16(207.025) = 18{,}522.4$$

SAMPLING—MEASUREMENT VARIABLES

SIMPLE RANDOM SAMPLING

Most environmental practitioners are familiar with *simple random sampling*. As in any sampling system, the aim is to estimate some characteristic of a population without measuring all of the population units. In a simple random sample of size n, the units are selected so that every possible combination of n units has an equal chance of being selected. If sampling is without replacement, then at any stage of the sampling each unused unit should have an equal chance of being selected.

Sample estimates of the population mean and total: From a population of $N = 100$ units, $n = 20$ units were selected at random and measured. Sampling was without replacement—once a unit had been included in the sample it could not be selected again. The unit values were:

10	9	10	9	11
16	11	7	12	12
11	3	5	11	14
8	13	12	20	10
Sum of all 20 random units =				214

From this sample we estimate the population mean as

$$\bar{X} = \frac{\sum X}{n} = \frac{214}{20} = 10.7$$

A population of $N = 100$ units having a mean of 10.7 would then have an estimated total of

$$\hat{T} = N\bar{X} = 100(10.7) = 1{,}070$$

Standard Errors

The first step in calculating a standard error is to obtain an estimate of the population variance (σ^2) or standard deviation (σ). As noted in a previous section, the standard deviation for a simple random sample is estimated by

$$s = \sqrt{\frac{\sum X^2 - \frac{\left(\sum X\right)^2}{n}}{n-1}} = \sqrt{\frac{10^2 + 16^2 + \ldots + 10^2 - \frac{214^2}{20}}{19}}$$

$$= \sqrt{13.4842}$$

$$= 3.672$$

For sampling without replacement, the standard error of the mean is

$$s_{\bar{x}} = \sqrt{\frac{s^2}{n}\left(1 - \frac{n}{N}\right)} = \sqrt{\frac{13.4842}{20}\left(1 - \frac{20}{100}\right)}$$

$$= \sqrt{0.539368}$$

$$= 0.734$$

From the formula for the variance of a linear function, we find that the variance of the estimated total is

$$s_{\hat{T}}^2 = N^2 s_{\bar{x}}^2$$

The standard error of the estimated total is the square root of this, or

$$s_{\hat{T}} = N s_{\bar{x}} = 100(0.734) = 73.4$$

Confidence Limits

Sample estimates are subject to variation. How much they vary depends primarily on the inherent variability of the population (Var^2) and on the size of the sample (n) and of the population (N).

The statistical way of indicating the reliability of an estimate is to establish *confidence* limits. For estimates made from normally distributed populations, the confidence limits are given by

$$(\text{Estimate}) \pm (t)(\text{Standard error})$$

For setting confidence limits on the mean and total we already have everything we need except for the value of t, and that can be obtained from a table of the t distribution.

In the previous example, the sample of $n=20$ units had a mean of $\bar{X} = 10.7$ and a standard error of $s_{\bar{x}} = 0.734$. For 95% confidence limits on the mean, we would use a t value (from t table) of 0.05 and (also from a t table) 19 degrees of freedom. As $t_{0.05} = 2.093$, the confidence limits are given by

$$\bar{X} \pm (t)(s_{\bar{x}}) = 10.7 \pm (2.093)(0.734) = 9.16 \text{ to } 12.24$$

This says that unless a 1-in-20 chance has occurred in sampling, the population mean is somewhere between 9.16 and 12.24. It does not say where the mean of future samples from this population might fall. Nor does it say where the mean may be if mistakes have been made in the measurements.

For 99% confidence limits we find $t_{0.01} = 2.861$ (with 19 degrees of freedom), and so the limits are

$$10.7 \pm (2.861)(0.734) = 8.6 \text{ to } 12.8$$

These limits are wider, but they are more likely to include the true population mean.

For the population total the confidence limits are:

95% limits: $1{,}070 \pm (2.093)(73.4) = 916$ to $1{,}224$
99% limits: $1{,}070 \pm (2.861)(73.4) = 860$ to $1{,}280$

For large samples ($n > 60$) the 95% limits are closely approximated by

$$\text{Estimate} \pm (2)(\text{Standard error})$$

and the 99% limits by

$$\text{Estimate} \pm (2.6)(\text{Standard error})$$

Sample Size

Samples cost money. So do errors. The aim in planning a survey should be to take enough observations to obtain the desired precision—no more, no less.

The number of observations needed in a simple random sample will depend on the precision desired and the inherent variability of the population being sampled. Since sampling precision is often expressed in terms of confidence interval on the mean, it is not unreasonable in planning a survey to say that in the computed confidence interval

$$\bar{X} \pm t s_{\bar{x}}$$

we would like to have the $t s_{\bar{x}}$ equal to or less than some specified value E, unless a 1-in-20 (or 1-in-100) chance has occurred in sample. That is, we want

$$t s_{\bar{x}} = E$$

or, since $s_{\bar{x}} = \dfrac{s}{\sqrt{n}}$ we want

$$t\left(\dfrac{s}{\sqrt{n}}\right) = E$$

Solving this for n gives the desired sample size.

$$n = \dfrac{t^2 s^2}{E^2}$$

To apply this equation we need to have an estimate (s^2) of the population variance and a value for students t at the appropriate level of probability.

The variance estimate can be a real problem. One solution is to make the sample survey in two stages. In the first state, n_1 random observations are made, and from these, an estimate (s^2) of the variance is computed. Then this value is plugged into the sample size equation

$$n = \dfrac{t^2 s^2}{E^2}$$

where t has $n_1 - 1$ degrees of freedom and is selected from the appropriate tale. The computed value of n is the total size of sample needed. As we have already observed n_1 units, this means that we will have to observe ($n - n_1$) additional units.

If pre-sampling as described above is not feasible then it will be necessary to make a guess at the variance. Assuming our knowledge of the population is such that the guessed variance (s^2) can be considered fairly reliable, then the size of sample (n) needed to estimate the mean to within $\pm E$ units is approximately

$$n = \dfrac{4 s^2}{E^2}$$

for 95% confidence and

$$n = \dfrac{20(s)^2}{3 E^2}$$

for 99% confidence.

Less reliable variance estimates could be doubled (as a safety factor) before applying these equations. In many cases the variance estimate may be so poor as to make the sample size computation just so much statistical window dressing.

When sampling is without replacement (as it is in most forest sampling situations) the sample size estimates given above apply to populations with an extremely large number (N) of units so that the sampling fraction (n/N) is very small. If the sampling fraction is not small (say $n/N=0.05$) then the sample size estimates should be adjusted. This adjusted value of n is

$$n_a = \frac{n}{1+n/N}$$

DID YOU KNOW?

It is important that the specified error (E) and the estimated variance (s^2) be on the same scale of measurement. We could not, for example, use a board-foot variance in conjunction with an error expressed in cubic feet. Similarly, if the error is expressed in volume per acre, the variance must be put on a per acre basis.

Suppose that we plan to use quarter-acre plots in a survey ad estimate the variance among plot volumes to be $s^2=160,000$. If the error limit is $E=5000$ feet per acre, we must convert the variance to an acre basis or the error to a quarter-acre basis. To convert a quarter-acre volume to a per-acre basis we multiply by 4, and to convert a quarter-acre variance to an acre variance we multiply by 16. Thus, the variance would be 2,560,000 and the sample-size formula would be

$$n = \frac{t^2(2,560,000)}{(500)^2} = t^2(10.24)$$

Alternatively, we can leave the variance alone and convert the error statement from an acre to a quarter-acre basis; $E=125$. Then the sample-size formula is

$$n = \frac{t^2(160,000)}{(125)^2} = t^2(10.24), \text{ as before}$$

STRATIFIED RANDOM SAMPLING

In *stratified sampling*, a population is divided into subpopulations (strata) of known size, and a simple random sample of at least two units is selected in each subpopulation. This approach has several advantages. For one thing, if there is more variation between subpopulations than within them, the estimate of the population mean will be more precise than that given by a simple random sample of the same size. Also, it may be desirable to have separate estimates for each subpopulation (e.g., in timber types or administrative subunits). And it may be administratively more efficient to sample by subpopulations.

Example 5.7

Problem:

A 500-acre forested area was divided into three strata on the basis of timber type. A simple random sample of 0.2-acre pots was taken in each stratum, and the means, variances, and standard errors were computed by the formulae for a simple random sample. These results, along with the size (N_h) of each stratum (expressed in number of 0.1-acre plots) are:

Type	Stratum Number (n)	Stratum Size (N_h)	Sample Size (n_h)	Stratum Mean (X_h)	Within Stratum Variance (s_h^2)	Squared Standard Error of the Mean ($s_{\bar{x}_h}^2$)
Pine Upland	1	1,350	30	251	10,860	353.96
Hardwood	2	700	15	164	9,680	631.50
Bottom-land Hardwoods	3	450	10	110	3,020	265.29
	Sum	2,500				

The squared standard error of the mean for stratum h is computed by the formula given for the simple random sample

$$s_{\bar{x}_h}^2 = s_h^2/n_h \left(1 - n_h/N_h\right)$$

Thus, for stratum 1 (pine type),

$$s_{\bar{x}}^2 = \frac{10,860}{30}\left(1 - \frac{30}{1,350}\right) = 353.96$$

where the sampling fraction (n_h/N_h) is small, the fpc can be omitted.
With this data, the population mean is estimated by

$$\bar{X}_{st} = \sum \frac{N_h \bar{X}_h}{N}$$

where $N = N_h$
For this example we have

$$\bar{X}_{st} = \frac{N_1 \bar{X}_1 + N_2 + \bar{X}_2 + N_3 \bar{X}}{N} = \frac{1,350(251) + 700(164) + 450(110)}{2500}$$

$$= 201.26$$

The formula for the standard error of the stratified mean is cumbersome but not complicated.

$$s_{\bar{x}_{st}} = \sqrt{\frac{1}{N^2}\left[\sum N_h^2 s_{\bar{x}_h}^2\right]}$$

$$= \sqrt{\frac{(1,350)^2(353.96) + (700)^2(631.50) + (450)^2(295.29)}{(2,500)^2}}$$

$$= 12.74$$

Statistics Review

If the sample size is fairly large, the confidence limits on the mean are given by

$$95\% \text{ confidence limits} = \bar{X}_{st} \pm 2s_{\bar{x}_{st}}$$

$$99\% \text{ confidence limits} = \bar{X}_{st} \pm 2.6s_{\bar{x}_{st}}$$

There is no simple way of compiling the confidence limits for small samples.

Sample Allocation If a sample of n units is taken, how many units should be selected in each stratum? Among several possibilities, the most common procedure is to allocate the sample in proportion to the size of the stratum; in a stratum having two-fifths of the units of the population we would take two-fifths of the samples. In the population discussed in the previous example the proportional allocation of the 55 sample units would have been (and was) as follows:

Stratum	Relative Size (N_h/N)	Sample Allocation
1	0.54	29.7 or 30
2	0.28	15.4 or 15
3	0.18	9.9 or 10
Sums	1.00	55

Some other possibilities are equal allocation, allocation proportional to estimated value, and optimum allocation. In optimum allocation, an attempt is made to get the smallest standard error possible for a sample of n units. This is done by sampling more heavily in the state having a larger variation. The equation for optimum allocation is

$$n_h = \left(\frac{N_h S_h}{\Sigma N_h S_h} \right)$$

Optimum allocation requires estimated of the within-stratum variances—information that may be difficult to obtain. A refinement of optimum allocation is to take sampling cost differences into account and allocate the sample so as to get the most information per dollar. If the cost per sampling unit in stratum h is c_h, the equation is

$$n_h = \left(\frac{\frac{N_h S_s}{\sqrt{c_h}}}{\Sigma \left(\frac{N_h S_h}{\sqrt{c_h}} \right)} \right) n$$

Sample Size

To estimate the size of sample to take for a specified error at a given level of confidence, it is first necessary to decide on the method of allocation. Ordinarily, proportional allocation is the simplest and perhaps the best choice. With proportional allocation, the size of sample needed to be within $\pm E$ units of the true value at the 0.05 probability level can be approximated by

$$n = \frac{N \left(\Sigma N_h S_h^2 \right)}{\frac{N^2 E^2}{4} + \Sigma N_h S_h^2}$$

For the 0.01 probability level, use 6.76 in place of 4.

Example 5.8

Problem:

Assume that prior to sampling the 500-acre forest we had decided that we wish to estimate the mean volume per acre to within ±100 cubic feet per acre unless a 1-in-20 change occurs in sampling. As we plan to sample with 0.2-acre plots, the error specification should be put on a 0.2-acre basis. Therefore,

$$E = 20$$

From previous sampling, the stratum variances for 0.2-acre volumes are estimated to be

$$s_1^2 = 8,000 \quad s_2^2 = 10,000 \quad s_3^2 = 5,000$$

Therefore,

$$n = \frac{2,500\left[(1,350)(8,000)+(700)(10,000)+(450)(5,000)\right]}{\frac{(2,500)^2 +(20)}{4}+\left[(1,350)(8,000)+(700)(10,000)+(450)(5,000)\right]} = 77.1 \text{ or } 78$$

The 78 sample units would now be allocated to the strata by the formula

$$n_h = \left(\frac{N_h}{N}\right)n$$

giving

$$n_1 = 42, \quad n_2 = 22, \quad n_3 = 14$$

SAMPLING—DISCRETE VARIABLES

RANDOM SAMPLING

The sampling methods discussed in the previous sections apply to data that are on a continuous or nearly continuous scale of measurement. These methods may not be applicable if each unit observed is classified as alive this type may follow what is known as the binomial distribution. They require slightly different statistical techniques.

As an illustration, suppose that a sample of 1,000 seeds was selected at random and tested for germination. If 480 of the seeds germinated, the estimated viability for the lot would be

$$\bar{p} = \frac{480}{1,000} = 0.48 \text{ or } 48\%$$

For large samples (say $n > 250$) with proportions >0.20 but <0.80, approximate confidence limits can be obtained by first computing the standard error of $\bar{p}$ by the equation

$$s_{\bar{p}} = \sqrt{\frac{\bar{p}(1-\bar{p})}{(n-1)}\left(1-\frac{n}{N}\right)}$$

Then, the 95% confidence limits are given by

$$95\% \text{ confidence interval} = \bar{p} \pm \left[2(s_{\bar{p}}) + 1/2n\right]$$

Applying this to the above example we get

$$s_{\bar{p}} = \sqrt{\frac{(0.48)(0.52)}{999}} \text{ (fpc ignored)}$$

$$= 0.0158$$

And,

$$95\% \text{ confidence interval} = 48 \pm \left[2(0.0158) + \frac{1}{2(1,000)}\right]$$

$$= 0.448 \text{ to } 0.512$$

The 99% confidence limits are approximated by

$$99\% \text{ confidence interval} = \bar{p} \pm \left[2.6 s_{\bar{p}} + \frac{1}{2n}\right]$$

SAMPLE SIZE

An appropriate table can be used to estimate the number of units that would have to be observed in a simple random sample in order to estimate a population proportion with some specified precision.

Suppose, for example, that we wanted to estimate the germination percent for a population to within plus or minus 10% (or 0.10) at the 95-prect confidence level. The first step is to guess about what the proportion of seed germination will be. If a good guess is not possible, then the safest course is to guess $\bar{p} = 0.59$ as this will give the maximum sample size.

Next, pick any of the sample sizes given in the appropriate table (e.g., 10, 15, 20, 30, 50, 100, 250, and 1,000) and look at the confidence interval for the specified value of $\bar{p}$. Inspection of these limits will tell whether or not the precision will be met with a sample of this size of if a larger or smaller sample would be more appropriate.

Thus, if we guess $\bar{p} = 0.2$, then in a sample of $n = 50$ we would expect to observe $(0.2)(50) = 10$, and the table says that the 95% confidence limits on $\bar{p}$ would be 0.10 and 0.34. Since the upper limit is not within 0.10 of $\bar{p}$, a larger sample would be needed. For a sample of $n = 100$ the limits are 0.13–0.29. Since both of these values are within 0.10 of $\bar{p}$, a sample of 100 would be adequate.

If the table indicates the need for a sample of over 250, the size can be approximated by

$$n = \frac{4(\bar{p})(1-\bar{p})}{E^2} \text{ for 95\% confidence}$$

or,

$$n = \frac{20(\bar{p})(1-\bar{p})}{3E^2}, \text{ for 99\% confidence}$$

where E = The precision with which $\bar{p}$ is to be estimated (expressed in same for as $\bar{p}$, either percent or decimal).

CLUSTER SAMPLING FOR ATTRIBUTES

Simple random sampling of discrete variables is often difficult or impractical. In estimating tree plantation survival, for example, we could select individual trees at random and examine them, but it wouldn't make much sense to walk down a row of planted trees in order to observe a single member of that row. It would usually be more reasonable to select rows at random and observe all of the trees in the selected row.

Seed viability is often estimated by randomly selecting several lots of 100 or 200 seeds each and recording for each lot the percentage of the seeds that germinate.

These are examples of cluster sampling; the unit of observation is the cluster rather than the individual tree or single seed. The value attached to the unit is the proportion having a certain characteristic rather than the simple fact of having or not having that characteristic.

If the clusters are large enough (say over 100 individuals per cluster) and nearly equal in size, the statistical methods that have been described for measurement variables can often be applied. Thus, suppose that the germination percent of a seed lot is estimated by selecting $n = 10$ sets of 200 seed each and observing the germination percent for each set.

Set	1	2	3	4	5	6	7	8	9	10	Sum
Germination percent (p)	78.5	82.0	86.0	80.5	74.5	78.0	79.0	81.0	80.5	83.5	803.5

then the mean germination percent is estimated by

$$\bar{p} = \frac{\sum p}{n} = \frac{8.03.5}{10} = 80.35\%$$

The standard deviation of p is

$$s_p = \sqrt{\frac{\sum p^2 - \frac{(\sum p)^2}{n}}{n-1}}$$

$$\sqrt{\frac{78.5^2 + \cdots + 83.5^2 - \frac{(803.5)^2}{10}}{9}}$$

$$= \sqrt{10.002778} = 3.163$$

And the standard error for $\bar{p}$ is

$$s_{\bar{p}} = \sqrt{\frac{s_p^2}{n}\left(1 - \frac{n}{N}\right)}$$

$$= \sqrt{\frac{10.0022778}{10}} = 1.000 \quad (\text{fpc ignored})$$

Note that n and N in these equations refer to the number of clusters, not to the number of individuals.

The 95% confidence interval, computed by the procedure for continuous variables:

$$= \bar{p} \pm (t_{0.05})(s_{\bar{p}}), (t \text{ has } (n-1) = 9 \, \text{df})$$

$$= 80.35 \pm 2.262(1.000) = 78.1 \text{ to } 82.6$$

Statistics Review

TRANSFORMATIONS

The above method of computing confidence limits assumes that the individual percentages follow something close to a normal distribution with homogenous variance (i.e., the same variance regardless of the size of the percent). If the clusters are small (say <100 individuals per cluster) or some of the percentages are >80 or <20, the assumptions may not be valid and the computed confidence limits will be unreliable.

In such cases, it may be desirable to compute the transformation

$$y = \arcsin \sqrt{\text{percent}}$$

and to analyze the transformed variable.

CHI-SQUARE TESTS

TEST OF INDEPENDENCE

Individuals are often classified according to two (or more) distinct systems. A tree can be classified as to species and at the same time according to whether or not it is not infected with some disease. A milacre plot can be classified as to whether or not is stocked with adequate reproduction and whether it is shaded or not shaded. Given such a cross-classification, it may be desirable to know whether the classification of an individual according to one system is independent of its classification by the other system. In the species-infection classification, for example, independence of species and infection would be interpreted to mean that there is no difference in infection rate between species (i.e., infection rate does not depend on species).

The hypothesis that two or more systems of classification are independent can be tested by chi-square. The procedure can be illustrated by a test of three termite repellents. A batch of 1,500 wooden stakes was divided at random into three groups of 500 each, and each group received a different termite-repellent treatment. The treated stakes were driven into the ground, with the treatment at any particular stake location being selected at random. Two years later the stakes were examined for termites. The number of stakes in each classification is shown in the following 2 by 3 (two rows and three columns) contingency table:

	Group I	Group II	Group III	Subtotals
Attached by termites	193	148	210	551
Not attacked	307	352	390	949
Subtotals	500	500	500	1500

If the data in the table be symbolized as shown below:

	I	II	III	II
Attacked	a_1	a_2	a_3	A
Not attacked	b_1	b_2	b_3	B
	T_1	T_2	T_3	G

the test of independence is made by computing

$$x^2 = \frac{1}{(A)(B)} \sum_{i=1}^{3} \left(\frac{[a_i B - b_i A]^2}{T_i} \right)$$

$$= \frac{1}{(551)(949)} \left[\frac{((193)(949)-(307)(551))^2}{500} + \cdots + \frac{((210)(949)-(290)(551))^2}{500} \right]$$

$$= 17.66$$

The result is compared to the appropriate tabular accumulative distribution of chi-square value of x^2 with $(c-1)$ degrees of freedom, where c is the number of columns in the table of data. If the computed value exceeds the tabular value given in the 0.05 column, the difference among treatments is said to be significant at the 0.05 level (i.e., we reject the hypothesis that attack classification is independent of termite repellent treatment).

For illustrative purposes, in this example, we say that the computed value of 17.66 (2 degrees of freedom) exceeds the tabular value in the 0.01 column, and so the difference in rate of attack among treatments is said to be significant at the 1% level. Examination of the data suggests that this is primarily due to the lower rate of attack on the Group II stakes.

TEST OF A HYPOTHESIZED COUNT

A geneticist hypothesized that, if a certain cross were made, the progeny would be of four types, in the proportions

$$A = 0.48 \quad B = 0.32 \quad C = 0.12 \quad D = 0.08$$

The actual segregation of 1,225 progeny is shown below, along with the numbers expected according to the hypothesis.

Type	A	B	C	D	Total
Number (X_i)	542	401	164	118	1,225
Expected (m_i)	588	392	147	98	1,225

As the observed counts differ from those expected, we might wonder if the hypothesis is false. Or can departures as large as this occur strictly by chance?

The chi-square test is

$$X^2 = \sum_{i=1}^{k} \left(\frac{(X_i - m_i)^2}{m_i} \right), \text{ with } (k-1) \text{ degrees of freedom}$$

where:
k = The number of groups recognized
X_i = The observed count for the ith
M_i = The count expected in the ith group if the hypothesis is true.

For the above data,

$$X^2_{3df} = \frac{(542-588)^2}{588} + \frac{(401-392)^2}{392} + \frac{(164-147)^2}{147} + \frac{(118-98)^2}{98} = 9.85$$

This value exceeds the tabular x^2 with 3 degrees of freedom at the 0.05 level (i.e., it is >7.81). Hence the hypothesis would be rejected (if the geneticist believed in testing at the 0.05 level).

BARTLETT'S TEST OF HOMOGENEITY OF VARIANCE

Many of the statistical methods described later are valid only if the variance is homogenous (i.e., variance within each of the populations is equal). The t-test of the following section assumes that the variance is the same for each group, and so does the analysis of variance. The fitting of an unweighted regression as described in the last section also assumes that the dependent variable has the same degree of variability (variance) for all levels of the independent variables.

Bartlett's test offers a means of evaluating this assumption. Suppose that we have taken random samples in each of four groups and obtained variances (s^2) of 84.2, 63.8, 88.6, and 72.1 based on samples of 9, 21, 5, and 11 units, respectively. We would like to know if these variances could have come from populations all having the same variance. The quantities needed for Bartlett's test are tabulated here:

Group	Variance (s^2)	($n-1$)	Corrected Sum of Squares SS	$1/n-1$	log s^2	($n-1$) (log s^2)
1	84.2	8	673.6	0.125	1.92531	15.40248
2	63.8	20	1,276.0	0.050	1.80482	36.09640
3	88.6	5	443.0	0.200	1.94743	9.73715
4	72.1	10	721.0	0.100	1.85794	18.57940
$k=4$	groups	Sums 43	3,113.6	0.475		79.81543

where k = The number of groups (= 4).

$$SS = \text{The corrected sum of squares} \left(\sum X^2 - \frac{(\sum X)^2}{n} \right) = (n-1)s^2$$

From this we compute the pooled within-group variance

$$\bar{s}^2 = \frac{\sum SS_i}{\sum (n-1)} = \frac{3113.6}{43} = 72.4093$$

and

$$\log \bar{s}^2 = 1.85979$$

Then the test for homogeneity is

$$X^2_{(k-1)df} = (2.3026*)[(1.85979)(43) - 79.81543]$$

$$= 0.358$$

This value of X^2 is now compared with the value of X^2 in an accumulative distribution of chi-square for the desired probability level. A value greater than that given in the table would lead us to reject the homogeneity assumption.

*Note: The original form of this equation used natural logarithms in place of the common logarithms shown here. The natural log of any number is ~2.3026 times its common log—hence the constant of 2.3026 in the equation. In computations, common logarithms are usually more convenient than natural logarithms.

The X^2 value given by the above equation is biased upward. If X^2 is nonsignificant, the bias is not important. However, if the computed X^2 is just a little above the threshold value for significance, a correction for bias should be applied. The correction is:

$$C = \frac{3(k-1) + \left[\sum\left(\frac{1}{N_i - 1}\right) - \frac{1}{\sum(n_i - 1)}\right]}{3(k-1)}$$

$$= \frac{3(4-1) + (0.475 - 1/43)}{3(4-1)}$$

$$= 1.0502$$

The corrected value of X^2 is then

$$\text{Corrected } X^2 = \frac{\text{Uncorrected } X^2}{C} = \frac{0.358}{1.0502} = 0.341$$

COMPARING TWO GROUPS BY THE t TEST

THE t TEST FOR UNPAIRED PLOTS

An individual unit in a population may be characterized in a number of different ways. A single tree, for example, can be described as alive or dead, hardwood, or softwood, infected or uninfected and so forth. When dealing with observations of this type we usually want to estimate the proportion of a population having a certain attribute. Or, if there are two or more different groups, we will often be interested in testing whether or not the groups differ in the proportions of individuals having the specified attribute. Some methods of handling these problems have been discussed in previous sections.

Alternatively, we might describe a tree by a measurement of some characteristics such as its diameter, height, or cubic volume. For this measurement type of observation, we may wish to estimate the mean for a group as discussed in the section on sampling for measurement variables. If there are two or more groups we will frequently want to test whether or not the group means are different. Often the groups will represent types of treatment which we wish to compare. Under certain conditions, the t or F tests may be used for this purpose.

Both of these tests have a wide variety of applications. For the present, we will confine our attention to tests of the hypothesis that there is no difference between treatment (or group) means. The computational routine depends on how the observations have been selected or arranged. The first illustration of a *t*-test of the hypothesis that there is no difference between the means of two treatments assumes that the treatments have been assigned to the experimental units completely at random. Except for the fact that there are usually (but not necessarily) an equal number of units or "plots" for each treatment, there is no restriction on the random assignment of treatments.

Statistics Review

> **DID YOU KNOW?**
>
> According to Ernst Mayr (1970, 2002), *races* are distinct generally divergent populations within the same species with relatively small morphological and genetic differences. The populations can be described as ecological races if they arise from adaptation to different local habitats or geographic races when they are geographically isolated. If sufficiently different, two or more races can be identified as subspecies, which is an official biological taxonomy unit subordinate to species. If not, they are denoted as races, which means that a formal rank should not be given to the group, or taxonomist are unsure whether or not a formal rank should be given. Again, according to Mayr, "a subspecies is a geographical race that is sufficiently different taxonomically to be worthy of a separate name" (pp. 89–94).

In this example, the "treatments" were two races of white pine which were to be compared based on their volume production over a specified period of time. Twenty-two square one-acre plots were staked out for the study. Eleven of these were selected entirely at random and planted with seedlings of race A. The remaining 11 were planted with seedlings of race V. After the prescribed time period that pulpwood volume (in cords—a stack of wood 4 ft wide by 4 ft high by 8 ft in length) was determined for each plot. The results were as follows:

Race A			Race B		
11	5	9	9		69
8	10	11	9	13	8
10	8	11	6		56
8	8		10	7	
Sum=99			Sum=88		
Average=9.0			Average=8.0		

To test the hypothesis that there is no difference between the race means (sometimes referred to as a null hypothesis—general or default position) we compute

$$t = \frac{\bar{X}_A - \bar{X}_B}{\sqrt{\frac{s^2(n_A + n_B)}{(n_A)(n_B)}}}$$

where $\bar{X}_A$ and $\bar{X}_B$ = The arithmetic means for groups A and B.
  n_A and n_B = The number of observations in groups A and B (n_A and n_B do not have to be the same).
  s^2 = The pooled within-group variance (calculation shown below).

To compute the pooled within-group variance, we first get the corrected sum of squares (*SS*) within the group.

$$SS_A = \sum X_A^2 - \frac{\left(\sum X_A\right)^2}{n_A} = 11^2 + 8^2 + \cdots + 11^2 - \frac{(99)^2}{11} = 34$$

$$SS_B = \sum X_B^2 - \frac{\left(\sum X_B\right)^2}{n_B} = 9^2 + 9^2 + \cdots + 6^2 - \frac{(88)^2}{11} = 54$$

Then the pooled variance is

$$s^2 = \frac{SS_A + SS_B}{(n_A - 1) + (n_B - 1)} = \frac{88}{20} = 4.4$$

Hence,

$$t = \frac{9.0 - 8.0}{\sqrt{4.4\left(\frac{11+11}{(11)(11)}\right)}} = \frac{1.0}{\sqrt{0.800000}} = 1.118$$

This value of t has $(n_A - 1) + (n_B - 1)$ degrees of freedom. If it exceeds the tabular value (from a distribution of t table) at a specified probability level, we will reject the hypothesis. The difference between the two means would be considered significant (larger than would be expected by chance if there is actually no difference).

In this case, tabular t with 20 degrees of freedom at the 0.05 level is 2.086. Since our sample value is less than this, the difference is not significant at the 0.05 levels.

Requirements: One of the unfortunate aspects of the t-test and other statistical methods is that almost any kind of numbers can be plugged into the equations. But if the numbers and methods of obtaining them do not meet certain requirements, the result may be a fancy statistical facade with nothing behind it. In a handbook of this scope, it is not possible to make the reader aware of all of the niceties of statistical usage, but a few words of warning are certainly appropriate.

A fundamental requirement in the use of most statistical methods is that the experimental material be a random sample of the population to which the conclusions are to be applied. In the t-test of white pine races, the plots should be a sample of the sites on which the pine are to be grown, and the planted seedlings should be a random sample representing the particular race. A test conducted in one corner of an experimental forest may yield conclusions that are valid only for that particular area or sites that are about the same. Similarly, if the seedlings of a particle race are the progeny of a small number of parents, their performance may be representative of those parents only, rather than of the race.

In addition to assuming that the observations for a given race are a valid sample of the population of possible observations, the t-test described above assumes that the population of such observations follows the normal distribution. With only a few observations, it is usually impossible to determine whether or not this assumption has been met. Special studies can be made to check on the distribution, but often the question is left to the judgment and knowledge of the research worker.

Finally, the t-test of unpaired plots assumes that each group (or treatment) has the same population variance. Since it is possible to compute a sample variance for each group, this assumption can be checked with Bartlett's test for homogeneity of variance. Most statistical textbooks present variations of the t-test that may be used if the group variances are unequal.

Sample Size

If there is a real difference of D feet between the two races of white pine, how many replicates (plots) would be needed to show that it is significant? To answer this, we first assume that the number of replicates will be the same of each group $(n_A = n_B = n)$. The equation for t can then be written

$$t = \frac{D}{\sqrt{\frac{2s^2}{n}}} \quad \text{or} \quad n = \frac{2t^2 s^2}{D^2}$$

Next we need an estimate of the within-group variance, s^2. As usual, this must be determined from previous experiments, or by the special study of the populations.

Statistics Review

Example 5.9

Problem:

Suppose that we plan to test at the 0.05 level and wish to detect a true difference of $d=1$ cord if it exists. From previous tests, we estimate $s^2 = 5.0$. Thus we have

$$n = \frac{2t^2 s^2}{D^2} 2t^2 \left(\frac{5.0}{1.0}\right)$$

Here we hit a snag. In order to estimate n we need a value for t, but the value of t depends on the number of degrees of freedom, which depends on n. The situation calls for an iterative solution—a mathematical procedure that generates a sequence of improving approximate solutions for a class of problems; in other words, a fancy name for trial and error. We start with a guessed value for n, say $n_o=20$. At t has $(n_A-1)+(n_B-1)=2(n-1)$ degrees of freedom, we'll use $t=2.025 (= t_{0.05}$ with 38 df) and compute

$$n_1 = 2(2.025)^2 \left(\frac{5.0}{1.0}\right) = 41$$

The proper value of n will be somewhere between n_o and n_1—much closer to n_1 than to n_o. We can now make a second guess at n and repeat the process. If we try $n_2=38$, t will have $2(n-1)=74$ df and $t_{0.05}=1.992$. Hence,

$$n_2 = 2(1.992)^2 \left(\frac{5.0}{1.0}\right) = 39.7$$

Thus, n appears to be over 39 and we will use $n=40$ plots for each group or a total of 80 plots.

THE T TEST FOR PAIRED PLOTS

A second test was made of the two races of white pine. It also had 11 replicates of each race, but instead of the two races being assigned completely at random over the 22 plots, the plots were grouped into 11 pairs and a different race was randomly assigned to each member of a pair; The cordwood volumes at the end of the growth period were

Plot Pair	1	2	3	4	5	6	7	8	9	10	11	Sum	Mean
Race A	12	8	8	11	10	9	11	11	13	10	7	110	10.0
Race B	10	7	8	9	11	6	10	11	10	8	9	99	9.0
$D_i = A_i - B_i$	2	1	0	2	-1	3	1	0	3	2	-2	11	1.0

As before, we wish to test the hypothesis that there is no real difference between the race means. The value of t when the plots have been paired is

$$t = \frac{\overline{X}_A - \overline{X}_B}{\sqrt{\frac{s_d^2}{n}}} = \frac{\overline{d}}{\sqrt{s_{\overline{d}}^2}}, \text{ with } (n-1) \text{ degrees of freedom}$$

where $n=$ The number of pairs of plots

s_d^2 = The variance of the individual differences between A and B

$$s_d^2 = \frac{\sum d_i^2 - \frac{\left(\sum d_i\right)^2}{n}}{n-1} = \frac{2^2 + 1^2 + \cdots + (-2)^2 - \frac{11^2}{11}}{10}$$

$$= 2.6$$

So, in this example we find

$$t_{10}dt = \frac{10.0 - 9.0}{\sqrt{2.6/11}} = 2.057$$

When this value of 2.057 is compared to the tabular value of t in a distribution of t table ($t_{0.05}$ with $10_{df} = 2.228$), we find that the difference is not significant at the 0.05 level. That is, a sample means difference of 1 cord of more could have occurred by chance more than one time in 20 even if there is no real difference between the race means. Usually, such an outcome is not regarded as sufficiently strong evidence to reject the hypothesis.

"The method of paired observations is a useful technique. Compared with the standard two-sample t test, in addition to the advantage that we do not have to assume that the two samples are independent, we also need not assume that the variances of the two samples are equal" (Hamburg, 1987, p. 304). Moreover, the paired test will be more sensitive (capable of detecting smaller real differences) than the unpaired test whenever the experimental units (plots in this case) can be grouped into pairs such that the variation between pairs is appreciably larger than the variation within pairs. The basis for paring plots may be geographic proximity of similarity in any other characteristic that is expected to affect the performance of the plot. In animal-husbandry studies, litter mates are often paired, and where patches of human skin are the plots, the left and right arms may constitute the pair. If the experimental units are very homogenous, there may be no advantage in pairing.

Number of Replicates

The number (n) of plot pairs needed to detect a true mean difference of size D is

$$n = \frac{t^2 s_d^2}{D^2}$$

COMPARISON OF TWO OR MORE GROUPS BY ANALYSIS OF VARIANCE

COMPLETE RANDOMIZATION

A planter wanted to compare the effects of five site-preparation treatments on the early height growth of planted pine seedlings. He laid out 25 plots and applied each treatment to 5 randomly selected plots. The plots were then hand-planted and at the end of 5 years the height for all pines was measured and an average height was computed for each plot. The plot averages (in feet) were as follows:

Statistics Review

Treatments

	A	B	C	D	E	Total
	15	16	13	11	14	
	14	14	12	13	12	
	12	13	11	10	12	
	13	15	12	12	10	
	13	14	10	11	11	
Sums	67	72	58	57	59	313
Treatment means	13.4	14.4	11.6	11.4	11.8	12.52

Looking at the data we see that there are differences among the treatment means: A and B have higher averages than C, D, and E. Soils and planting stock are seldom completely uniform, however, so we would expect some differences even if every plot had been given exactly the same site-preparation treatment. The question is, can differences as large as this occur strictly by chance if there is actually no difference among treatments? If we decide that the observed differences are larger than might be expected to occur strictly by chance, the inference is that the treatment means are not equal. Statistically speaking, we reject the hypothesis of no difference among treatment means.

Problems like this are neatly handled by an analysis of variance. To make this analysis, we need to fill in a table like the following:

Source of Variation	Degrees of Freedom	Sums of Squares
Treatments	4	
Error	20	
Total	24	

Source of variation: There are a number of reasons why the height growth of these 25 plots might vary, but only one can be definitely identified and evaluated—that is attributable to treatments. The unidentified variation is assumed to represent the variation inherent in the experimental material and is labeled error. Thus, total variation is divided into two parts: one part attributable to treatments, and the other unidentified and called error.

Degrees of freedom: Degrees of freedom are hard to explain in non-statistical language. In the simpler analyses of variance, however, they are not difficult to determine. For the total, the degrees of freedom are one less than the number of observations: there are 25 plots, so the total has 24 dfs. For the sources, other than error, the dfs are one less than the number of classes or groups recognized in the source. Thus, in the source labeled treatments there are five groups (five treatments), so there will be four degrees of freedom for treatments. The remaining degrees of freedom (24−4=20) are associated with the error term.

Sums of squares: There is a sum of squares associated with every source of variation. These SS are easily calculated in the following steps:

First we need what is known as a "correction term" or C.T. This is simply

$$\text{C.T.} = \frac{\left(\sum_{}^{n} X\right)^2}{n} = \frac{313^2}{25} = 3918.76$$

where $\sum_{}^{n}$ = the sum of n items

Then the total sum of squares is

$$\text{Total } \underset{24 df}{SS} = \sum^n X^2 - \text{C.T.} = \left(15^2 + 14^2 + \cdots + 11^2\right) - \text{C.T.} = 64.24$$

The sum of squares attributable to treatments is

$$\underset{4 df}{\text{Treatment SS}} = \frac{\sum^5 \left(\text{Treatment totals}^2\right)}{\text{No. of plots per treatment}}$$

$$= \frac{67^2 + 72^2 + \cdots + 59^2}{5} - \text{C.T.} = \frac{19767}{5} - \text{C.T.} = 34.64$$

Note that in both SS calculations, the number of items squared and added was one more than the number of degrees of freedom associated with the sum of squares. The number of degrees of freedom just below the SS and the number of items to be squared and added over the n value provided a partial check as to whether the proper totals are being used in the calculation—the degrees of freedom must be one less than the number of items.

Note also that the divisor in the treatment SS calculation is equal to the number of individual items that go to make up each of the totals being squared in the numerator. This was also true in the calculation of total SS, but there the divisor was 1 and hence did not have to be shown. Note further that the divisor times the number over the summation sign ($5 \times 5 = 25$ for treatments) must always be equal to the total number of observations in the test—another check.

The sum of squares for error is obtained by subtracting the treatment SS from total SS. A good habit to get into when obtaining sums of squares by subtraction is to perform the same subtraction using dfs. In the more complex designs, doing this provides a partial check on whether the right items are being used.

Mean squares: The mean squares are now calculated by dividing the sums of squares by the associated degrees of freedom. It is not necessary to calculate the mean square for the total.

The items that have been calculated are entered directly into the analysis table, which at the present stage would look like this:

Source	df	SS	MS (Mean Square)
Treatment	4	34.64	8.66
Error	20	29.60	1.48
Total	25	64.24	

An F test of treatments (used to reject the null hypothesis) is now made by dividing the MS for treatments by the MS for error. In this case

$$F = \frac{8.66}{1.48} = 5.851$$

Fortunately, critical values of the F ratio have been tabulated for frequently used significance levels analogous to the x^2 distribution. Thus, the result, 5.851, is compared to the appropriate value of F in the table. The tabular F for significance at the 0.05 level is 2.87 and that for the 0.01 level is 4.43.

As the calculated value of F exceeds 4.43, we conclude that the difference in height growth between treatments is significant at the 0.01 level. (More precisely, we reject the hypothesis that there is no difference in mean height growth between the treatments.) If F had been smaller than 4.43 but larger than 2.87, we would have said that the difference is significant at the 0.05 level. If F had been <2.87, we would have said that the difference between treatments is not significant at the 0.05 level. The researcher should select his own level of significance (preferably in advance of the study), keeping in mind that significance at the α (alpha) level (for example) means this: if there is actually no difference among treatments, the probability of getting chance differences as large as those observed is α or less.

The *t*-test versus the analysis of variance: If only two treatments are compared, the analysis of variance of a completely randomized design and the *t*-test of unpaired plots lead to the same conclusion. The choice of test is strictly one of personal preference, as may be verified by applying the analysis of variance to the data used to illustrate the *t*-test of unpaired plots. The resulting F value will be equal to the square of the value of t that was obtained (i.e., $F=t^2$).

Like the *t*-test, the F test is valid only if the variable observed is normally distributed and if all groups have the same variance.

MULTIPLE COMPARISONS

In the example illustrating the completely randomized design, the difference among treatments was found to be significant at the 0.01 probability level. This is interesting as far as it goes, but usually, we will want to take a closer look at the data, making comparisons among various combinations of the treatments.

Suppose, for example, that A and B involved some mechanical form of site preparation while C, D, and E were chemical treatments. Then we might want to test whether the average of A and B together differed from the combined average of C, D, and E. Or we might wish to test whether A and B differ significantly from each other. When the number of replications (n) is the same for all treatments, such comparisons are fairly easy to define and test.

The question of whether the average of treatments A and B differs significantly from the average of treatments C, D, and E is equivalent to testing whether the linear contrast

$$\hat{Q} = (3\bar{A} + 3\bar{B}) - (2\bar{C} + 2\bar{D} + 2\bar{E})$$

differs significantly from zero ($\bar{A}$=the mean for treatment A, etc.). Note that the coefficients of this contrast sum to zero ($3+3-2-2-2=0$) and are selected so as to put the two means in the first group on an equal basis with the three means in the second group.

F Test with Single Degree of Freedom

A comparison specified in advance of the study (on logical grounds and before examination of the data) can be tested by an F test with a single degree of freedom. For the linear contrast

$$\hat{Q} = a_1\bar{X}_1 + a_2\bar{X}_2 + a_3\bar{X}_3 + \ldots$$

among means based on the same number (n) of observations, the sum of squares has one degree of freedom and is computed as

$$SS = \frac{n\hat{Q}^2}{\sum a_i^2}\Big|_{1df}$$

This sum of squares divided by the mean square for error provides an F test of the comparison.

Thus, in testing A and B versus C, D, and E we have

$$\hat{Q} = 3(13.4) + 3(14.4) - 2(11.6) - 2(11.4) - 2(11.8) = 13.8$$

and

$$SS_{1df} = \frac{5(13.8)^2}{3^2 + 3^2 + (-2)^2 + (-2)^2 + (-2)^2} = \frac{952.20}{30} = 31.74$$

Then dividing by the error mean square gives the F value for testing the contrast.

$$F = \frac{31.74}{1.48} = 21.446 \text{ with 1 and 20 degrees of freedom}$$

This exceeds the tabular value of F (4.35) at the 0.05 probability level. If this is the level at which we decided to test, we would reject the hypothesis that the mean of treatments A and B does not differ from the mean of treatments C, D, and E.

If Q is expressed in terms of the treatment totals rather than their means so that

$$\hat{Q}_T = a_1(\Sigma X_1) + a_2(\Sigma X_2) + \ldots$$

then the equation for the single degree of freedom sum of squares is

$$SS_{1df} = \frac{\hat{Q}^2 T}{n \sum a_i^2}$$

The results will be the same as those obtained with the means. For the test of A and B versus C, D, and E,

$$\hat{Q}_T = 3(67) + 3(72) - 2(58) - 2(57) - 2(59) = 69$$

And,

$$SS_{1df} = \frac{69^2}{5[3^2 + 3^2 + (-2)^2 + (-2)^2 + (-2)^2]} = \frac{4,761}{150} = 37.74, \text{ as before.}$$

Working with the totals saves the labor of computing means and avoids possible rounding errors.

Scheffe's Test

Quite often we will want to test comparisons that were not anticipated before the data were collected. If the test of treatments was significant, such unplanned comparisons can be tested by the method of Scheffe or Scheffe's Test. Named after the American statistician Henry Scheffe, the Scheffe Test adjusts significant levels in a linear regression analysis to account for multiple comparisons. It is particularly useful in analysis of variance, and in constructing simultaneous bands for regressions involving basic functions. When there are n replications of each treatment, k degrees of freedom for treatment, and v degrees of freedom forever, any linear contrast among the treatment means

$$\hat{Q} = a_1 \bar{X}_1 + a_2 \bar{X}_2 + \ldots$$

is tested by computing

$$F = \frac{nQ^2}{K\left(\sum a_i^2\right)(\text{Error mean square})}$$

This value is then compared to the tabular value of F with k and v degrees of freedom.

For example, to test treatment B against the means of treatments C and E we would have

$$\hat{Q}\left[2\bar{B} - (\bar{C} + \bar{E})\right] = [2(14.4) - 11.6 - 11.8] = 5.4$$

And

$$F = \frac{5(5.4)^2}{(4)\left[2^2 + (-1) + (-1)^2\right](1.48)} = 4.105, \text{ with 4 and 20 degrees of freedom}$$

This figure is larger than the tabular value of F (= 2.87), and so in testing at the 0.05 level we would reject the hypothesis that the mean for treatment B did not differ from the combined average of treatments C and E. For a contrast (Q_T) expressed in terms of treatment totals, the equation for F becomes

$$F = \frac{\hat{Q}_T^2}{nk\left(\Sigma a_i^2\right)(\text{Error mean square})}$$

Unequal Replication

If the number of replications is not the same for all treatments, then for the linear contrast

$$\hat{Q} = a_1 X_1 + a_2 X_2 + \ldots$$

The sum of squares in the single degree of freedom F test is given by

$$SS = \frac{\hat{Q}^2}{(k)\left(\dfrac{a_1^2}{n_1} + \dfrac{a_2^2}{n_2} + \ldots\right)}$$

where n_i = the number of replications on which $\bar{X}_i$ is based.

With unequal replication, the F value in Scheffe's test is computed by the equation

$$F = \frac{\hat{Q}^2}{(k)\left(\dfrac{a_1^2}{n_1} + \dfrac{a_2^2}{n_2} + \ldots\right)(\text{Error mean square})}$$

Selecting the coefficients (a_i) for such contrasts can be tricky. When testing the hypothesis that there is no difference between the means of two groups of treatments, the positive coefficients are usually

$$\text{Positive } a_i = \frac{n}{p}$$

where p = the total number of plots in the group of treatments with positive coefficients.

The negative coefficients are

$$\text{Negative } a_j = \frac{n_i}{m}$$

where m = the total number of plots in the group of treatments with negative coefficients.

To illustrate, if we wish to compare the mean of treatments A, B, and C with the mean of treatments D and E and there are two plots of treatment A, three of B, five of C, three of D, and two of E, then $p = 2 + 3 + 5 = 10$, $m = 3 + 2 = 5$ and the contrast would be

$$\hat{Q} = \left(\frac{2}{10}\bar{A} + \frac{3}{10}\bar{B} + \frac{5}{10}\bar{C}\right) - \left(\frac{3}{5}\bar{D} + \frac{2}{5}\bar{E}\right)$$

RANDOMIZED BLOCK DESIGN

There are two basic types of the two-factor analysis of variance: the *completely randomized design* (discussed in the previous section) and the *randomized block design*. In the completely randomized design, the error mean square is a measure of the variation among plots treated alike. It is in fact an average of the within-treatment variances, as may easily be verified by computation. If there is considerable variation among plots treated alike, the error mean square will be large and the F test for a given set of treatments is less likely to be significant. Only large differences among treatments will be detected as real and the experiment is said to be insensitive.

DID YOU KNOW?

The term "block" derives from experimental design work in agriculture, in which parcels of land are referred to as blocks. In a randomized block design, treatments are randomly assigned to units within each block. In testing the yield of different fertilizers, for example, this design ensures that the best fertilizer is applied to all types of soil, not just the best soil (Hamburg, 1987).

Often the error can be reduced (thus giving a more sensitive test) by use of a randomized block design in place of complete randomization. In this design, similar plots or plots that are close together are grouped into blocks. Usually, the number of plots in each block is the same as the number of treatments to be compared, though there are variations having two or more plots per treatment in each block. The blocks are recognized as a source of variation that is isolated in the analysis. A general rule in randomized block design is to "block what you can, randomize what you can't." In other words, blocking is used to remove the effects of nuisance variables or factors. Nuisance factors are those that may affect the measured result, but are not of primary inters, For example, in applying a treatment, nuisance factors might be the time of day the experiment was run, the room temperature, or might be the specific operator who prepared the treatment (Addelman, 1969, 1970).

As an example, a randomized block design with five blocks was used to test the height growth of cottonwood cuttings from four selected parent trees. The field layout looked like this:

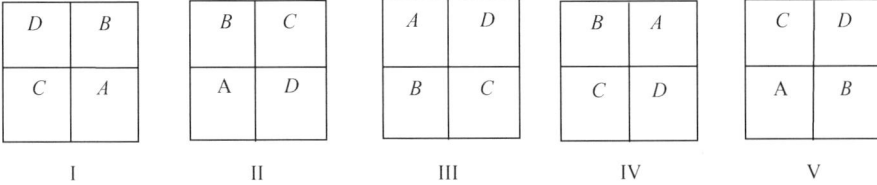

Each plot consisted of a planting of 100 cuttings of the clone assigned to that plot. When the trees were 5 years old the heights of all survivors were measured and an average computed for each plot. The plot averages (in feet) by clones and blocks are summarized below:

Block	Close				Block totals
	A	B	C	D	
I	18	14	12	16	60
II	15	15	16	13	59
III	16	15	8	15	54
IV	14	12	10	12	48
V	12	14	9	14	49
Clone totals	75	70	55	70	270
Clone means	15	14	11	14	

The hypothesis to be tested is that clones do not differ in mean height.

In this design, there are two identifiable sources of variation—that attributable to clones and that associated with blocks. The remaining portion of the total variation is used as a measure of experimental error. The outline of the analysis is therefore as follows:

Source of Variation	df	Sums of Squares	Mean Squares
Blocks	4		
Clones	3		
Error	12		
Total	19		

The breakdown in degrees of freedom and computation of the various sums of squares follow the same pattern as in the completely randomized design. Total degrees of freedom (19) are one less than the total number of plots. Degrees of freedom for clones (3) are one less than the number of clones. With five blocks, there will be four degrees of freedom for blocks. The remaining 12 degrees of freedom are associated with the error term.

Sums-of-squares calculations proceed as follows:

1. The correction term $\text{C.T.} = \dfrac{\left(\sum_{}^{20} X\right)^2}{N} = \dfrac{270^2}{20} = 3{,}645$

2. Total SS $= \sum_{19df}^{20} X^2 - \text{C.T.} = \left(18^2 + 15^2 + \cdots + 14^2\right) - \text{C.T.} = 3,766 - 3,645 = 121$

3. Clone SS $_{3df} = \dfrac{\sum^4 \left(\text{Clone totals}^2\right)}{\text{No. of plots per clone}} - \text{C.T.} = \dfrac{75^2 + 70^2 + 55^2 + 70^2}{5} - \text{C.T.} = 3,690 - 3,645 = 45$

4. Block SS $_{4df} = \dfrac{\sum^5 \left(\text{Block totals}^2\right)}{\text{No. of plots per block}} - \text{C.T.} = \dfrac{60^2 + 59^2 + \cdots + 49^2}{4} - \text{C.T.} = 3,675.5 - 3,645 = 30.5$

5. Error SS $\underset{12\,df}{} = \text{Total SS} \underset{19\,df}{} - \text{Clone SS} \underset{3\,df}{} - \text{Block SS} \underset{4\,df}{} = 45.5$

Note that in obtaining the error SS by subtraction, we get a partial check on ourselves by subtracting clone and block dfs from the total df to see if we come out with the correct number of error df. If these don't check, we have probably used the wrong sums of squares in the subtraction.

Mean squares are again calculated by dividing the sums of squares by the associated number of degrees of freedom.

Tabulating the results of these computations

Source	df	SS	MS
Blocks	4	30.5	7.625
Clones	3	45.0	15.000
Error	12	45.5	3.792
Total	19	121.0	

F for clones is obtained by dividing clone MS by error MS. In this case $F = 15.000/3.792 = 3.956$. As this is larger than the tabular F of 3.49 (obtained from a Distribution of F table) ($F_{0.05}$ with 3 and 12 degrees of freedom) we conclude that the difference between clones is significant at the 0.05 level. The significance appears to be due largely to the low value of C as compared to A, B, and D.

Comparisons among clone means can be made by the methods previously described. For example, to test the prespecified (i.e., before examining the data) hypothesis that there is no difference between the mean of clone C and the combined average A, B, and D we would have:

$$\text{SS for}\left(A + B + D \text{ vs. } C\right)_{1df} = \dfrac{5\left(3\bar{C} - \bar{A} - \bar{B} - \bar{D}\right)^2}{\left[3^2 + (-1)^2 + (-1)^2 + (-1)^2\right]}$$

Then,

$$F = \dfrac{41.667}{3.792} = 10.988$$

Tabular F at the 0.01 level with 1 and 12 degrees of freedom is 9.33. As calculated F is greater than this, we conclude that the difference between C and the average of A, B, and D is significant at the 0.01 level.

The sum of squares for this single-degree-of-freedom comparison (41.667) is almost as large as that for clones (45.0) with three degrees of freedom. This result suggests that most of the clonal variation is attributable to the low value of C, and that comparisons between the other three means are not likely to be significant.

There is usually no reason for testing blocks, but the size of the block mean square relative to the mean square for error does give an indication of how much precision was gained by blocking. If the block mean square is large (at least two or three times as large as the error means square) the test is more sensitive than it would have been with complete randomization. If the block mean square is about equal to or only slightly larger than the error mean square, the use of blocks has not improved the precision of the test. The block mean square should not be appreciably smaller than the error mean square. If it is, the method or conducting the study and the computations should be re-examined.

Assumptions—In addition to the assumption of homogeneous variance and normality, the randomized block design assumes that there is no interaction between treatments and blocks; i.e., that differences among treatments are about the same in all blocks. Because of this assumption, it is not advisable to have blocks that differ greatly—since they may cause an interaction with treatments.

DID YOU KNOW?

With only two treatments, the analysis of variance of a randomized block design is equivalent to the t test of paired replicates. The value of F will be equal to the value of t^2 and the inferences derived from the tests will be the same. The choice of tests is a matter of personal preference.

LATIN SQUARE DESIGN

In the randomized block design, the purpose of blocking is to isolate a recognizable extraneous source of variation. If successful, blocking reduces the error mean square and hence gives a more sensitive test than could be obtained by complete randomization.

In some situations, however, we have a two-way source of variation that cannot be isolated by blocks alone. In an agricultural field, for example, fertility gradients may exist both parallel to and at right angles to plowed rows. Simple blocking isolates only one of these sources of variation, leaving the other to swell the error term and reduce the sensitivity of the test.

When such a two-way source of extraneous variation is recognized or suspected, the Latin square design may be helpful. In this design, the total number of plots or experimental units is made equal to the square of the number of treatments. In forestry and agricultural experiments, the plots are often (but not always) arranged in rows and columns with each row and column having a number of poles equal to the number of treatments being tested. The rows represent different levels of one source of extraneous variation while the columns represent different levels of the other source of extraneous variation. Thus, before the assignment of treatments, the field layout of a Latin square for testing five treatments might look like this:

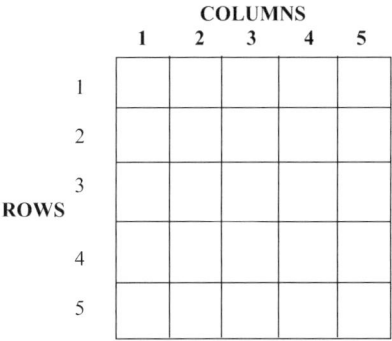

Treatments are assigned to plots at random, but with the very important restriction that a given treatment cannot appear more than once in any row or any column.

An example of a field layout of a Latin square for testing five treatments is given below. The letters represent the assignment of five treatments (which here are five species of hardwoods). The numbers show the average 5-year height growth by plots. The tabulation shows the totals for rows, columns, and treatments.

		COLUMNS			
	1	2	3	4	5
ROWS 1	C 13	A 21	B 16	E 16	D 14
ROWS 2	A 18	B 15	D 17	C 17	E 15
ROWS 3	D 17	C 15	E 15	A 15	B 18
ROWS 4	E 18	D 18	C 16	B 14	A 16
ROWS 5	B 17	E 16	A 25	D 19	C 14

Row, column, and treatment totals

Row		Column		Treatment		$\bar{X}$
1	80	1	83	A	95	19
2	82	2	85	B	80	16
3	80	3	89	C	75	15
4	82	4	81	D	85	16
	= 415		= 415		= 415	16.6

The partitioning of dfs, the calculation of sums of squares, and the subsequent analysis follow much the same pattern illustrated previously for randomized blocks.

$$\text{C.T.} = \frac{\left(\sum_{}^{25} X\right)^2}{n} = \frac{415^2}{25} = \frac{172,225}{25} = 6,889.0$$

$$\text{Total SS}_{24\text{ df}} = \sum_{}^{25} X^2 - \text{C.T.} = 7{,}041 - \text{C.T.} = 152.0$$

$$\text{Row SS}_{4\text{ df}} = \frac{\sum_{}^{5}\left(\text{Row totals}^2\right)}{\text{No. of plots per row}} - \text{C.T.} = \frac{34,529}{5} - \text{C.T.} = 16.8$$

$$\text{Col. SS}_{4\,df} = \frac{\sum_{}^{5}\left(\text{Column totals}^2\right)}{\text{No. of plots per column}} - \text{C.T.} = \frac{34,529}{5} - \text{C.T.} = 16.0$$

$$\text{Species SS}_{4\,df} = \frac{\sum_{}^{5}\left(\text{Column totals}^2\right)}{\text{No. of plots per species}} - \text{C.T.} = \frac{34,675}{5} - \text{C.T.} = 46.0$$

$$\text{Error SS}_{12\,df} = \text{Total SS}_{24\,df} - \text{Species SS}_{4\,df} - \text{Row SS}_{4\,df} - \text{Col. SS}_{4\,df} = 73.2$$

Analysis of variance

Source	df	SS	MS
Rows	4	16.8	4.2
Columns	4	16.0	4.0
Species	4	46.0	11.5
Error	12	73.2	6.1
Total	24	152.0	

$$F(\text{for species}) = \frac{11.5}{6.1} = 1.885$$

As the computed value of F is less than the tabular value of F at the 0.05 level (with 4/12 dfs), the differences among species are considered nonsignificant.

The Latin square design can be used whenever there is a two-way heterogeneity that cannot be controlled simply by blocking. In greenhouse studies, distance from a window could be treated as a row effect while distance from the blower or heater might be regarded as a column effect. Though the plots are often physically arranged in rows or columns, this is not required. In testing, the use of materials in a manufacturing process where different machines and machine operators will be involved, the variation between machines could be treated as a row effect and the variation due to operator could be treated as a column effect.

The Latin square should not be used if an interaction between rows and treatments or columns and treatments is suspected.

FACTORIAL EXPERIMENTS

In environmental practice, knowledge that interactions between elements of the environment occur and an understanding of what their influence or impact on the environment is, or can be, is important. Consider a comparison of corn yields following three rates or levels of nitrogen fertilization indicating that the yields depended on how much phosphorus was used along with the nitrogen. The differences in yield were smaller when no phosphorus was used than when the nitrogen applications were accompanied by 100 pounds per acre of phosphorus. In statistics, this situation is referred to as an interaction between nitrogen and phosphorus. Another example: when leaf litter was removed from the forest floor, the catch of pine seedlings was much greater than when the litter was not removed; but for red oak the reverse was true—the seedling catch was lower where litter was removed. Thus, species and litter treatment were interacting.

Interactions are important in the interpretation of study results. In the presence of an interaction between species and litter treatment it obviously makes no sense to talk about the effects of litter removal without specifying the species. The nitrogen-phosphorus interaction means that it may be misleading to recommend a level of nitrogen without mentioning the associated level of phosphorus.

Factorial experiments are aimed at evaluating known or suspected interactions. In these experiments, each factor to be studied is tested at several levels and each level of a factor is tested at all possible combinations of the levels of the other factors. In a planting test involving three species of trees and four methods of preplanting site preparation, each method will be applied to each species, and the total number of treatment combinations will be 12. In a factorial test of the effects of two nursery treatments on the survival of four species of pine planted by three different methods, there would be 24 ($2 \times 4 \times 3 = 24$) treatment combinations.

The method of analysis can be illustrated by a factorial test of the effects of three levels of nitrogen fertilization (0, 100, and 200 pounds per acre) on the growth of three species (A, B, and C) of planted pine. The nine possible treatment combinations were assigned at random to nine plots in each of three blocks. Treatments were evaluated based on average annual height growth in inches per year over a 3-year period.

Field layout and plot data were as follows (with subscripts denoting nitrogen levels: $0 = 0$, $1 = 100$, $2 = 200$):

Block I

C_2	B_2	C_0
17	18	37
A_0	B_0	C_1
45	24	20
B_1	A_2	A_1
21	24	17

Block II

B_2	C_2	A_1
23	20	18
B_1	A_2	C_0
18	14	43
A_0	C_1	B_0
40	25	35

Block III

C_1	C_2	A_0
19	21	37
B_1	A_2	A_1
19	17	28
B_0	C_0	B_2
29	39	15

The preliminary analysis of the nine combinations (temporarily ignoring their factorial nature) is made just as though this were a straight randomized block design (which is exactly what it is). (*See table below.*)

Statistics Review

Sums of Squares

$$\text{C.T.} = \frac{\left(\sum_{1}^{27} X\right)^2}{27} = \frac{683^2}{27} = 17{,}277.3704$$

$$\text{Total SS}_{24\,df} = \sum_{1}^{27} X^2 - \text{C.T.} = \left(45^2 + 17^2 + \cdots + 21^2\right) - \text{C.T.}$$

$$= 2{,}275.6296$$

$$\text{Block SS}_{2\,df} = \frac{\sum_{1}^{3}\left(\text{Block totals}^2\right)}{\text{No of plots per block}} - \text{C.T.}$$

$$= \frac{\left(223^2 + 236 + 224^2\right)}{9} - \text{C.T.} = 22.6296$$

Species	Nitrogen Level	Blocks I	II	III	Nitrogen Subtotals	Species Totals
A	0	45	40	37	122	
	1	17	18	28	63	
	2	24	14	17	55	
	Block subtotals	86	72	82		240
B	0	24	35	29	88	
	1	21	18	19	58	
	2	18	23	15	56	
	Block subtotals	63	76	63		202
C	0	37	43	39	119	
	1	20	25	19	64	
	2	17	20	21	58	
	Block subtotals	74	88	79		241
All species	0	106	118	105	329	
	1	58	61	66	185	
	2	59	57	53	169	
	Totals	223	236	224	683	

$$\text{Treatment SS}_{8\,df} = \frac{\sum_{1}^{9}\left(\text{Treatment totals}^2\right)}{\text{No. of plots per treatment}} - \text{C.T.}$$

$$= \frac{\left(122^2 + 63^2 + \cdots + 58^2\right)}{3} - \text{C.T.} = 1{,}970.2963$$

Tabulating these in the usual form:

Source	df	SS	MS
Blocks	2	11.6296	5.8148
Treatments	8	1970.2963	246.2870
Error	16	293.7037	18.3565
Totals	26	2,275.6296	

$$\text{Testing treatments}: F_{8/16\,df} = \frac{246.2870}{18.3565}$$

$$= 13.417, \text{ significant at the } 00.1$$

The next step is to analyze the components of the treatment variability. How do the species compare? What is the effect of fertilization? And does fertilization affect all species the same way (i.e., is there a species-nitrogen interaction)? To answer these questions we have to partition the degrees of freedom and sums of squares associate with treatments. This is easily done by summarizing the data for the nine combinations in a two-way table.

Species	Nitrogen Levels			Totals
	0	1	2	
A	122	63	55	240
B	88	58	56	202
C	119	64	58	241
Totals	329	185	169	683

The nine individual values will be recognized as those that entered into the calculation of the treatment SS. Keeping in mind that each entry in the body of the table is the sum of three plot values, and that the species and nitrogen totals are each the sum of 9 plots, the sums of squares for species, nitrogen, and the species-nitrogen interaction can be computed as follows:

$$\text{Treatment SS}_{8\,df} = 1{,}970.2963 \,(\text{as previously calculated})$$

$$\text{Species SS}_{2\,df} = \frac{\sum_{}^{3}(\text{species totals})^2}{\text{No. of plots per species}} - \text{C.T.}$$

$$= \frac{(240^2 + 202^2 + 241^2)}{9} - \text{C.T.}$$

$$= \frac{156{,}485}{9} - \text{C.T.} = 109.8518$$

$$\text{Nitrogen SS}_{\text{3 df}} = \frac{\sum_{}^{3}(\text{Nitrogen totals})^2}{\text{No. of plots per level of nitrogen}} - \text{C.T.}$$

$$= \frac{(329^2 + 185^2 + 169^2)}{9} - \text{C.T.} = \frac{171{,}027}{9} - \text{C.T.}$$

$$= 1{,}725.6296$$

$$\text{Species-nitrogen interaction SS}_{\text{4 df}} = \text{Treatment SS}_{\text{8 df}} - \text{Species SS}_{\text{2 df}} - \text{Nitrogen SS}_{\text{3 df}} = 134.8149$$

The analysis now becomes:

Source	df	SS	MS	F
Blocks	2	11.6296	5.8148	—
Treatments	8	1,970.2963	246.2870	13.417[a]
Species	2[b]	109.8518[b]	54.9259	2.992[c]
Nitrogen	2[b]	1,725.6296[b]	862.8148	47.003[a]
Species-nitrogen	4[b]	134.8149[b]	33.7037	1.836[c]
Error	16	293.7037	18.3565	
Total	26	2,275.6296		

[a] Means significant at the 0.01 level.
[b] Offset figures are a partitioning of the dfs and sum of squares for Treatments and are therefore not included in the total at the bottom of the table.
[c] Nonsignificant.

The degrees of freedom for simple interactions can be obtained in two ways. The first way is by subtracting the dfs associated with the component factors (in this case two for species and two for nitrogen levels) from the dfs associated with all possible treatment combinations (eight in this case). The second way is not to calculate the interaction dfs as the product of the component factor dfs (in this case $2 \times 2 = 4$). Do it both ways as a check.

The F values for species, nitrogen and the species-nitrogen interaction are calculated by dividing their mean squares by the mean square for error.

The analysis indicates a significant difference among levels of nitrogen, but no difference between species and no species-nitrogen interaction.

As before, a prespecified comparison among treatment means can be tested by breaking out the sum of squares associated with that comparison. To illustrate the computations, we will test nitrogen versus no nitrogen and also 100 pounds versus 200 pounds of nitrogen.

$$\text{Nitrogen vs. no nitrogen SS} = \frac{9\left[2\left(\frac{329}{9}\right) - 1\left(\frac{185}{9}\right) - 1\left(\frac{169}{9}\right)\right]^2}{(2^2 + 1^2 + 1^2)}$$

$$= \frac{[2(329) - 185 - 169]^2}{9(6)} = 1{,}711.4074$$

In the numerator, the mean for the zero level of nitrogen is multiplied by 2 to give it equal weight with the mean of levels 1 and 2 with which it is compared. The 9 is the number of plots on which each mean is based. The $(2^2+1^2+1^2)$ in the denominator is the sum of squares of the coefficients used in the numerator.

$$100 \text{ vs. } 200 \text{ pounds } \underset{1\,df}{SS} = \frac{9\left[1\left(\frac{185}{9}\right)-1\left(\frac{169}{9}\right)\right]^2}{\left(1^2+1^2\right)} = \frac{[185-169]^2}{9(2)} = 14.2222$$

Note that these two sums of squares (1,711.4075 and 14.2222), each with 1 df, add up to the sum of squares for nitrogen (1,725.6296) with 2 dfs. This additive characteristic holds true only if the individual df comparisons selected are orthogonal (i.e., independent). When the number of observations is the same for all treatments, the orthogonality of any two comparisons can be checked in the following manner: First, tabulate the coefficients and check to see that for each comparison the coefficients sum to zero.

Comparison		Nitrogen Level		Sum
		1	2	
$2N_0$ vs. N_1+N_2	2	–	–	0
N_1 vs. N_2	0	+	–	0
Product of coefficients	0	–	–	0

For the two comparisons to be orthogonal the sum of the products of corresponding coefficients must be zero. Ay sum of squares can be partitioned in a similar manner, with the number of possible orthogonal individual df comparisons being equal to the total number of degrees of freedom with which the sum of squares is associated.

The sum of squares for species can also be partitioned into two orthogonal single df comparisons. If the comparisons were specified before the data were examined, we might make single df test of the difference between B and the average of A and C and also of the difference between A and C. The method is the same as that illustrated in the comparison of nitrogen treatments. The calculations are as follows:

$$2B \text{ vs. } (A+C) \underset{1\,df}{SS} = \frac{9\left[1\left(\frac{240}{9}\right)+1\left(\frac{241}{9}\right)-2\left(\frac{202}{9}\right)\right]^2}{\left(1^2+1^2+2^2\right)}$$

$$= \frac{[240+241-2(202)]^2}{9(6)} = 109.7963$$

$$A \text{ vs. } C \underset{1\,df}{SS} = \frac{9\left[1\left(\frac{241}{9}\right)-1\left(\frac{240}{9}\right)\right]^2}{\left(1^2+1^2\right)} = \frac{(241-240)^2}{9(2)} = 0.0555$$

These comparisons are orthogonal, so that the sums of squares each with 1 df add up to the species SS with 2 dfs.

Note that in computing the sums of squares for the single-degree-of-freedom comparisons, the equations have been restated in terms of treatment totals rather than means. This often simplifies the computations and reduces the errors due to rounding.

With the partitioning the analysis has become:

Source	df	SS	MS	F
Blocks	2	11.6296	5.8148	—
Species	2	109.8518	54.9259	2.992[a]
2B vs. (A+C)	1	109.7963	109.7963	5.981[b]
A vs. C	1	.0555	.0555	—
Nitrogen	2	1,725.6296	862.8148	47.003[c]
$2N_0$ vs. (N_1+N_2)	1	1,711.4074	1,711.4074	93.232[c]
N_1 vs. N_2	1	14.2222	14.2222	—
Species × nitrogen interaction	4	134.8149	33.7037	1.836[a]
Error	16	293.7037	18.3565	
Total	26	2,275.6296		

[a] Nonsignificant.
[b] Significant at the 0.05 level.
[c] Significant at the 0.01 level.

We conclude that species B is poorer than A or C and that there is no difference in growth between A and C. We also conclude that nitrogen adversely affected growth and that 100 pounds was about as bad as 200 pounds. The nitrogen effect was about the same for all species (i.e., no interaction).

It is worth repeating that the comparisons to be made in an analysis should, whenever possible, be planned and specified prior to an examination of the data. A good procedure is to outline the analysis, putting in all the times that are to appear in the first two columns (source and df) of the table.

The factorial experiment, it well be noted, is not an experimental design. It is, instead, a way of selecting treatments; given two or more factors each at two or more levels, the treatments are all possible combinations of the levels of each factor. If we have three factors with the first at four levels, the second at two levels, and the third at three levels, we will have 4×2×3=24 factorial combinations or treatments. Factorial experiments may be conducted in any of the standard designs. The randomized block and split plot design are the most common for factorial experiments in forest research.

THE SPLIT PLOT DESIGN

When two or more types of treatment are applied in factorial combinations, it may be that one type can be applied on relatively small plots while the other type is best applied to larger plots. Rather than make all plots of the size needed for the second type, a split-plot design can be employed. In this design, the major (large-plot) treatments are applied to a number of plots with replication accomplished through any of the common designs (such as complete randomization, randomized blocks, Latin Square). Each major plot is then split into a number of subplots, equal to the number of minor (small-plot) treatments. Minor treatments are assigned at random to subplots within each major plot.

As an example, a test was to be made of direct seeding of loblolly pine at six different dates, on burned and unburned seedbeds. To get typical burn effects, major plots 6 acres in size were selected.

There were to be four replications of major treatments in randomized blocks. Each major plot was divided into six 1-acre subplots for seeding at six dates. The field layout was somewhat as follows (blocks denoted by Roman numerals, burning treatment by capital letters, day of seeding by small letters):

I

A
f	b	c
a	d	e
e	a	f
d	c	b
B

II

A
c	e	b
a	d	f
a	f	d
e	b	c
B

III
b	f	c
a	d	e
c	b	d
f	a	e

IV
c	a	f
d	b	e
b	e	d
c	f	a

One pound of seed was sowed on each 1-acre subplot. Seedling counts were made at the end of the first growing season. Results were as follows:

Summary of Seedlings Per Acre

	I		II		III		IV		Date subtotals		Date
Date	A	BA	B	A	B	A	B	A		B	Totals
a	900	880	810	1,100	760	960	1,040	1,040	3,510	3,980	7,490
b	880	1,050	1,170	1,240	1,060	1,110	910	1,120	4,020	4,520	8,540
c	1,530	1,140	1,160	1,270	1,390	1,320	1,540	1,080	5,620	4,810	10,430
d	1,970	1,360	1,890	1,510	1,820	1,490	2,140	1,270	7,820	5,630	13,450
e	1,960	1,270	1,670	1,380	1,310	1,500	1,480	1,450	6,420	5,600	12,020
f	830	150	420	380	570	420	760	270	2,580	1,220	3,800
Major Plot Totals	8,070	5,850	7,120	6,880	6,910	6,800	7,870	6,230	29,970	25,760	—
Block Totals	13,920		14,000		13,710		14,100			—	55,730

Calculations

The correction term and total sum of squares are calculated using the 48 subplot values.

$$\text{C.T.} = \frac{(\text{Grand total of all subplots})^2}{\text{Total number of subplots}} = \frac{55,730^2}{48} = 64,704,852$$

$$\text{Total SS}_{47\,df} = \sum^{48}\left(\text{Subplot values}\right)^2 - \text{C.T.}$$

$$= \left(900^2 + 880^2 + \cdots + 270^2\right) - \text{C.T.} = 9{,}339{,}648$$

Before partitioning the total sum of squares into its components, it may be instructive to ignore subplots for the moment and examine the major plot phase of the study. The major phase can be viewed as a straight randomized block design with two burning treatments in each of four blocks. The analysis would be:

Source	df
Blocks	3
Burning	1
Error (major plots)	3
Major plots	7

Now, looking at the subplots, we can think of the major plots as blocks. From this standpoint, we would have a randomized block design with six dates of treatment in each of eight blocks (major plots) for which the analysis is:

Source	df
Major-plots	7
Dates	5
Remainder	35
Subplots (= total)	47

In this analysis, the remained is made up of two components. One of these is the burning-date interaction, with 5 dfs. The rest, with 30 dfs, is called the subplot error. Thus, the complete breakdown of the split-plot design is:

Source	df	
Blocks	3	
Burning	1	
Major plot error	3	**Major plots**: 7 df
Date	5	**Dates**: 5 df
Burning X date	5	**Remainder**: 35 df
Subplot error	30	
Total	47	

The various sums of squares are obtained in an analogous manner. We first compute

$$\text{Major plot SS} \atop 7\,df = \frac{\sum_{}^{8}\left(\text{Major plot totals}^2\right)}{\text{Number of subplots per major plot}} - \text{C.T.}$$

$$= \frac{8{,}070^2 + \cdots + 6{,}230^2}{6} - \text{C.T.} = 647{,}498$$

$$\text{Block SS} \atop 3\,df = \frac{\sum_{}^{4}\left(\text{Block totals}^2\right)}{\text{Subplots per block}} - \text{C.T.}$$

$$= \frac{13{,}920^2 + \cdots + 14{,}100^2}{12} - \text{C.T.} = 6{,}856$$

$$\text{Burning SS} \atop 1\,df = \frac{\sum_{}^{2}\left(\text{Burning treatment totals}^2\right)}{\text{Subplots per burning treatment}} - \text{C.T.}$$

$$= \frac{29{,}970^2 + 25{,}760^2}{24} - \text{C.T.} = 369{,}252$$

$$\text{Major-plot error SS} \atop 3\,df = \text{Major plot SS} \atop 2\,df - \text{Block SS} \atop 3\,df - \text{Burning SS} \atop 1\,df = 271{,}390$$

$$\text{Subplot SS} \atop 40\,df = \text{Total SS} \atop 47\,df - \text{Major plot SS} \atop 7\,df = 8{,}692{,}150$$

$$\text{Date SS} \atop 5\,df = \frac{\sum_{}^{6}\left(\text{Date totals}^2\right)}{\text{Subplots per date}} - \text{C.T.}$$

$$= \frac{7{,}490^2 + \cdots + 3{,}800^2}{8} - \text{C.T.} = 7{,}500{,}086$$

Date-Burning SS

To get the sum of squares for the interaction between date and burning we resort to a factorial experiment device—the two-way table of the treatment combination totals.

Burning	Date						Burning Subtotals
	a	b	c	d	e	f	
A	3,510	4,020	5,620	7,820	6,420	2,580	29,970
B	3,980	4,520	4,810	5,630	5,600	1,220	25,760
Date subtotals	7,490	8,540	10,430	13,450	12,020	3,800	55,730

$$\text{Date-burning subclass SS}_{11\,df} = \frac{\sum_{}^{12}\left(\text{Date-burning combination totals}^2\right)}{\text{Subplots per date – burning combination}}$$

$$= \frac{3{,}510^2 + \cdots + 1{,}220^2}{4} - \text{C.T.} = 8{,}555{,}723$$

$$\text{Date-burning interaction SS}_{5\,df} = \text{Date-burning subclass SS}_{11\,df} - \text{Date SS}_{5\,df} - \text{Burning SS}_{1\,df}$$

$$= 686{,}385$$

$$\text{Subplot error SS}_{30\,df} = \text{Subplot SS}_{40\,df} - \text{Date SS}_{5\,df} - \text{Date-burning interaction SS}_{5\,df}$$

$$= 505{,}679$$

Thus the completed analysis table is

Source	df	SS	MS
Blocks	3	6,856	—
Burning	1	369,252	369,252
Major-plot error	3	271,390	90,463
Date	5	7,500,086	1,500,017
Date-burning	5	686,385	137,277
Subplot error	30	505,679	16,856
Total	47	9,339,648	

The F Test for burning is

$$F_{1/3\,df} = \frac{\text{Burning MS}}{\text{Major-plot error MS}} = \frac{369{,}252}{90{,}463}$$

$$= 4{,}082, \quad \text{not significant at the 0.05 level.}$$

For dates,

$$F_{5/30\,df} = \frac{\text{Date MS}}{\text{Subplot error MS}} = \frac{1{,}500{,}017}{16{,}856}$$

$$= 88{,}99, \quad \text{significant at the 0.01 level.}$$

And for the date-burning interaction,

$$F_{5/30\,df} = \frac{\text{Date-burning MS}}{\text{Subplot error MS}} = \frac{137{,}277}{16{,}856}$$

$$= 8.14, \quad \text{significant at the 0.01 level}$$

Note that the major-plot error is used to test the sources above the dashed line while the subplot error is used for the sources below the line. Because the subplot error is a measure of random variation *within* major plots it will usually be smaller than the major-plot error, which is a measure of the random variation between major plots. In addition to being smaller, the subplot error will generally have more degrees of freedom than the major-plot error, and for these reasons, the sources below the dashed line will usually be tested with greater sensitivity than the sources above the line. This fact is important; in planning a split-plot experiment the designer should try to get the items of greatest interest below the line rather than above.

Rarely will the major-plot error be appreciably smaller than the subplot error. If it is, the conduct of the study and the computations should be carefully examined.

Subplots can also be split: If desired, the subplots can also be split for a third level of treatment, producing a split-split-plot design. The calculations follow the same general pattern but are more involved. A split-split-plot design has three separate error terms.

Comparisons among means in a split-plot design: For comparisons among major- or subplot treatments, F tests with a single degree of freedom may be made in the usual manner. Comparisons among major-plot treatments should be tested against the major-plot error mean square, while subplot treatment comparisons are tested against the subplot error. In addition, it is sometimes desirable to compare the means of two treatment combinations. This can get tricky, for the variation among such means may contain more than one source of error. A few of the more common cases are discussed below.

In general, the t test for comparing two equally replicated treatment means is

$$t = \frac{\text{Mean difference}}{\text{Standard error or the mean difference}} = \frac{\bar{D}}{s_{\bar{D}}}$$

1. For the difference between two major-treatment means:

$$s_{\bar{D}} = \sqrt{\frac{2(\text{Major-plot error MS})}{(m)(R)}}; \quad t \text{ has df equal to the df for the major-plot error.}$$

   where R = Number of replications of major treatments.
   m = Number of subplots per major plot.

2. For the difference between two minor-treatment means:

$$s_{\bar{D}} = \sqrt{\frac{2(\text{Subplot error MS})}{(R)(M)}}; \quad t \text{ has df equal to the df for subplot error;}$$

   where M = Number of major-plot treatments.

3. For the difference between two minor treatments within a single major treatment:

$$s_{\bar{D}} = \sqrt{\frac{2(\text{Subplot error MS})}{R}}; \quad \text{df for } t = \text{df for the subplot error}$$

4. For the difference between the means of two major treatments at a single level of a minor treatment, or between the means of two major treatments at different levels of a minor treatment:

Statistics Review

$$s_{\bar{D}} = \sqrt{2\left[\frac{(m-1)(\text{Subplot error MS}) + \text{Major-plot error MS}}{(m)(R)}\right]}$$

In this case, t will not follow the t distribution. A close approximation to the value of t required for significance at the a level is given by

$$t = \frac{(m-1)(\text{Subplot error MS})t_m + (\text{Major-plot error MS})t_M}{(m-1)(\text{Subplot error MS}) + (\text{Major-plot error MS})}$$

where:
t_m = Tabular value of t at a level for df equal to the df for the subplot error.
T_M = Tabular value of t at a level for df equal to the df for the major-plot error.
Other symbols are as previously defined.

MISSING PLOTS

A mathematician who had developed a complex electronic computer program for analyzing a wide variety of experimental designs was asked how he handled missing plots. His disdainful reply was, "We tell our research workers not to have missing plots."

This is good advice. But it is sometimes hard to follow, and particularly so in forest, environmental and ecological research, where close control over experimental material is difficult and studies may run for several years.

The likelihood of plots being lost during the course of a study should be considered when selecting an experimental design. Lost plots are least troublesome in the simple designs. For this reason, complete randomization and randomized blocks may be preferable to more intricate designs when missing data can be expected.

In the complete randomization design, loss of one or more plots causes no computational difficulties. The analysis is made as though the missing plots never existed. Of course, a degree of freedom will be lost from the total and error terms for each missing plot and the sensitivity of the test will be reduced. If missing plots are likely, the number of replications should be increased accordingly.

In the randomized block design, completion of the analysis will usually require an estimate of the values for the missing plots. A single missing value can be estimated by

$$X = \frac{bB + tT - G}{(b-1)(t-1)}$$

where b = Number of blocks
t = Number of treatments
B = Total of all other units in the block with a missing plot
T = Total of all other units that received the same treatment as the missing plot
G = Total of all observed units

If more than one plot is missing, the customary procedure is to insert guessed values for all but one of the missing units, which is then estimated by the above formula. This estimate is used in obtaining an estimated value for one of the guessed plots, and so on through each missing unit. Then the process is repeated with the first estimates replacing the guessed values. The cycle should be repeated until the new approximations differ little from the previous estimates.

The estimated values are now applied in the usual analysis-of-variance calculations. For each missing unit one degree of freedom is deducted from the total and from the error term.

A similar procedure is used with the Latin square design, but the formula for a missing plot is

$$X = \frac{R(R+C+T)-2G}{(r-1)(r-2)}$$

where r = Number of rows
R = Total of all observed units in the row with the missing plot
C = Total of all observed units in the column with the missing plot
T = Total of all observed unit in the missing plot treatment
G = Grand total of all observed units

With the split-plot design, missing plots can cause trouble. A single missing subplot value can be estimated by the equation

$$X = \frac{rP + m(T_{ij}) - (T_i)}{(r-1)(m-1)}$$

where r = Number of replications of major-plot treatments
P = Total of all observed subplots in the major plot having a missing subplot
m = Number of subplot treatments
T_{ij} = Total of all subplots having the same treatment combination as the missing unit
T_i = Total of all subplots having the same major-plot treatment as the missing unit

For more than one missing subplot the iterative process described for randomized blocks must be used. In the analysis, one df will be deducted from the total and subplot error terms for each missing subplot.

When data for missing plots are estimated, the treatment mean square for all designs is biased upwards. If the proportion of missing plots is small, the bias can usually be ignored. Where the proportion is large, adjustments can be made as described in the standard references on experimental designs.

REGRESSION

SIMPLE LINEAR REGRESSION

An environmental researcher had an idea that she could tell how well a loblolly pine was growing from the volume of the crown. Very simple: big crown—good growth, small crown—poor growth. But she couldn't say how big and how good, or how small and how poor. When she needed regression analysis: it would enable her to express a relationship between tree growth and crown volume in an equation. Given a certain crown volume, she could use the equation to predict what the tree growth was.

To gather data, she ran parallel survey lines across a large tract that was representative of the area in which she was interested. The lines were five chains apart. At each two-chain mark along the lines, she measured the nearest loblolly pine of at least 5.6 inches dbh. (diameter at breast height; i.e., 4.5 ft above the forest floor on the uphill side of the tree) for crown volume and basal area growth over the past 10 years.

A portion of the data is printed below to illustrate the methods of calculation. Crown volume in hundreds of cubic feet is labeled X and basal area growth in square feet is labeled Y. Now, what can we tell the environmental researcher about the relationship?

X Crown Volume	Y Growth	X Crown Volume	Y Growth	X Crown Volume	Y Growth
22	0.36	53	0.47	51	0.41
6	0.09	70	0.55	75	0.66
93	0.67	5	0.07	6	0.18
62	0.44	90	0.69	20	0.21
84	0.72	46	0.42	36	0.29
14	0.24	36	0.39	50	0.56
52	0.33	14	0.09	9	0.13
69	0.61	60	0.54	2	0.10
104	0.66	103	0.74	21	0.18
100	0.80	43	0.64	17	0.17
41	0.47	22	0.50	87	0.63
85	0.60	75	0.39	97	0.66
90	0.51	29	0.30	33	0.18
27	0.14	76	0.61	20	0.06
18	0.32	20	0.29	96	0.58
48	0.21	29	0.38	61	0.42
37	0.54	30	0.53		
67	0.70	59	0.58		
56	0.67	70	0.62		
31	0.42	81	0.66		
17	0.39	93	0.69		
7	0.25	99	0.71		
2	0.06	14	0.14		
Totals				3,050	26.62
Means ($n=62$)				49.1935	0.42935

Often, the first step is to plot the field data on coordinate paper (see Figure 5.3). This is done to provide some visual evidence of whether the two variables are related. If there is a simple relationship, the plotted points will tend to form a pattern (a straight line or curve). If the relationship is very strong, the pattern will generally be distinct. If the relationship is weak, the points will be more spread out and the pattern less definite. If the points appear to fall pretty much at random, there may be no simple relationship or one that is so very poor as to make it a waste of time to fit any regression.

The type of pattern (straight line, parabolic curve, exponential curve, etc.) will influence the regression model to be fitted. In this particular case, we will assume a simple straight-line relationship.

After selecting the model to be fitted, the next step will be to calculate the corrected sums of squares and products. In the following equations, capital letters indicate uncorrected values of the variables; lower-case letters will be used for the corrected values ($y = Y - \bar{Y}$).

$$\text{The corrected sum of squares for } Y: \sum y^2 = \sum_{n} Y^2 - \frac{\left(\sum_{n} y\right)^2}{n}$$

$$= \left(0.36^2 + 0.09^2\right) + \cdots + 0.42^2$$

$$= \frac{26.62^2}{62}$$

$$= 2.7826$$

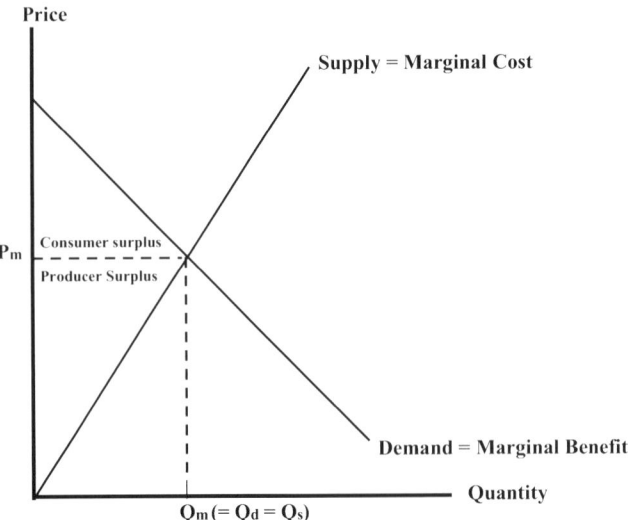

FIGURE 5.3 Market equilibrium.

The corrected sum of squares for X: $\Sigma x^2 = \Sigma X^2 - \dfrac{\left(\sum X\right)^2}{n}$

$$= \left(22^2 + 6^2 + \cdots + 6^2\right) - \dfrac{3{,}050^2}{62}$$

$$= 59{,}397.6778$$

The corrected sum of products: $\sum xy = \sum^{n}(XY) - \dfrac{\left(\sum X\right)\left(\sum^{n} Y\right)}{n}$

$$= \left[220(0.36) + (6)(0.09) + \cdots + (61)(0.42)\right] - \dfrac{(3{,}050)(26{,}62)}{62}$$

$$= 354.1477$$

The general form of equation for a straight line is $Y = a + bX$

In this equation, a and b are constants or regression coefficients that must be estimated. According to the principle of least squares, the best estimates of these coefficients are:

$$b = \dfrac{\sum XY}{\sum X^2} = \dfrac{354.1477}{59{,}397.6775} = 0.005962$$

$$a = \bar{Y} - b\bar{X} = 0.42935 - (0.005962)(49.1935) = 0.13606$$

Substituting these estimates in the general equation gives

$$\bar{Y} = 0.13606 + 0.005962X$$

where $\hat{Y}$ is used to indicated that we are dealing with an estimated value of Y.

With this equation, we can estimate the basal area growth for the past 10 years ($\hat{Y}$) from the measurements of the crown volume X.

Because Y is estimated form a known value of X, it is called the dependent variable and X the independent variable. In plotting on graph paper, the values of Y are usually (purely by convention) plotted along the vertical axis (ordinate) and the values of X along the horizontal axis (abscissa).

HOW WELL DOES THE REGRESSION LINE FIT THE DATA?

A regression line can be thought of as a moving average. It gives an average value of Y associated with a particular value of X. Of course, some values of Y will be above the regression line (moving average) and some below, just as some values of Y are above or below the general average of Y.

The corrected sum of squares for Y (i.e., Σy^2) estimates the amount of variation of individual values of Y about the mean value of Y. A regression equation is a statement that part of the observed variation in Y (estimated by Σy^2) is associated with the relationship of Y to X. The amount of variation in Y that is associated with the regression on X is called the reduction or regression sum of squares.

$$\text{Reduction } SS = \frac{\left(\Sigma xy\right)^2}{\Sigma x^2} = \frac{(34.1477)^2}{(59{,}367.6775)} = 2.1115$$

As noted above, the total variation in Y is estimated by $\Sigma y^2 = 2.7826$ (as previously calculated).

The part of the total variation in Y that is not associated with the regression is called the residual sum of squares. It is calculated by

$$\text{Residual } SS = \Sigma y^2 - \text{Reduction } SS = 2.782 - 2.1115 = 0.6711$$

In analysis of variance we used the unexplained variation as a standard for testing the amount of variation attributable to treatments. We can do the same in regression. What's more, the familiar F test will serve.

Source of Variation	Df[a]	SS	MS[b]
Due to regression $\left[= \frac{(\Sigma xy)^2}{\Sigma X^2} \right]$	1	2.1115	2.1115
Residual (i.e., unexplained)	60	0.6711	0.01118
Total ($= \Sigma y^2$)	61	2.7826	

[a] As there are 62 values of Y, the total sum of squares has 61 df. The regression of Y on X has on df. The residual df are obtained by subtraction.
[b] MS is, as always $=$ SS/df.

The regression is tested by

$$F = \frac{\text{Regression MS}}{\text{Residual MS}} = \frac{2.1115}{0.01118} = 188.86$$

As calculated F is much greater than tabular $F_{0.01}$ with 1/60 df, the regression is deemed significant at the 0.01 level.

Before we fitted a regression line to the data, Y had a certain amount of variation about its mean ($\bar{Y}$). Fitting the regression was, in effect, an attempt to explain part of this variation by the linear association of Y with X. But even after the line had been fitted, some variation was unexplained— that of Y about the regression line. When we tested the recession line above, we merely showed that the part of the variation in Y that is explained by the fitted line is significantly greater than the part that the line left unexplained. The test did not show that the line we fitted gives the best possible description of the data (a curved line might be even better). Nor does it mean that we have found the true mathematical relationship between the two variables. There is a dangerous tendency to ascribe more meaning to a fitted regression than is warranted.

It might be noted that the residual sum of squares is equal to the sum of the squared deviations of the observed values of Y from the regression line. That is,

$$\text{Residual SS} = \Sigma\left(Y - \hat{Y}\right)^2 = \Sigma(Y - a - bX)^2$$

The principle of least squares says that the best estimates of the regression coefficients (a and b) are those that make this sum of squares a minimum.

COEFFICIENT OF DETERMINATION

The coefficient of determination, denoted R^2, is used in the context of statistical models whose main purpose is the prediction of future outcomes on the basis of other related information. Stated differently, the coefficient of determination is a ratio that measures how well a regression fits the sample data.

$$\text{Coefficient of determination}\left(R^2\right) = \frac{\text{Reduction SS}}{\text{Total SS}}$$

$$= \frac{2.1115}{2.7826} = 0.758823$$

When someone says, "76% of variation in Y was associated with X," she means that the coefficient of determination was 0.76. Note that R^2 is most often seen as a number between 0 and 1.0, used to describe how well a regression lien fit a set of data. An R^2 near 1.0 indicates that regression lien fits the data well, while an R^2 Closer to 0 indicates a regression line does not fit the data very well.

The coefficient of determination is equal to the square of the correlation coefficient.

$$\frac{\text{Reduction SS}}{\text{Total SS}} = \frac{(\Sigma xy)^2 / \Sigma x^2}{\Sigma y^2} = \frac{(\Sigma xy)^2}{(\Sigma x^2)(\Sigma y^2)} = R^2$$

In fact, most present-day users of regression refer to R^2 values rather than to coefficients of determination.

CONFIDENCE INTERVALS

Because it is based on sample data, a regression equation is subject to sample variation. Confidence limits (i.e., a pair of numbers used to estimate a characteristics of a population) on the regression line can be obtained by specifying several values over the range of X and computed (refer to Figure 5.4)

Statistics Review

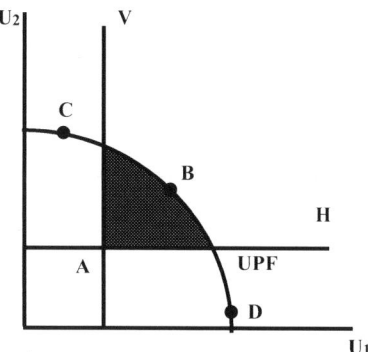

FIGURE 5.4 Utility possibility Frontier.

Where X_0 = a selected value of X, and
Degrees of freedom for t = df for residue MS.
In the example we had:

$$\hat{Y} = 0.13606 + 0.005962X$$

Residual MS = 0.01118 with 60 dfs

$$\Sigma x^2 = 59,397.6775$$

So, if we pick $X_0 = 28$ we have $\bar{Y} = 0.303$, and 95% confidence limits

$$= 0.303 \pm 2.000 \sqrt{0.01118 \left(\frac{1}{62} + \frac{(28 - 49.1935)^2}{59,397.6775} \right)}$$

$$= 0.270 \text{ to } 0.336$$

For other values of X_0 we would get:

X_0	$\hat{Y}$	95% Limits	
		Lower	Upper
8	0.184	0.139	0.229
49.1935	0.429	0.402	0.456
70	0.553	0.521	0.585
90	0.673	0.629	0.717

Note that these are confidence limits on the regression of Y on X. They indicate the limits within which the true mean of Y for a given X will lie unless a 1-in-20 chance has occurred. The limits do not apply to a single predicted value of Y. The limits within which a single Y might lie are given by

$$\hat{Y} \pm t \sqrt{(\text{Residual MS}) \left(1 + \frac{1}{n} + \frac{(X_0 - \bar{X})^2}{\Sigma x^2} \right)}$$

Assumptions: In addition to assuming that the relationship of Y to X is linear, the above method of fitting assumes that the variance of Y about the regression line is the same at all levels of X (the assumption of homogeneous variance or homoscedasticity—that is, the property of having equal variances). The fitting does not assume nor does it require that the variation of Y about the regression line follows the normal distribution. However, the F test does assume normality, and so does the use of t for the computation of confidence limits.

There is also an assumption of independence of the errors (departures from regression) of the sample observations. The validity of this assumption is best insured by selecting the sample units are random. The requirements of independence may not be met if successive observations are made on a single unit or if the units are observed in clusters. For example, a series of observations of tree diameter made by means of a growth band would probably lack independence.

Selecting the sample units so as to get a particular distribution of the X values does not violate any of the regression assumptions, provided the Y values are a random sample of all Y's associated with the selected values of X. Spreading the sample over a wide range of X values will usually increase the precision with which the regression coefficient is estimated. This device must be used with caution however, for if the Y values are not random, the regression coefficients and residual means squares may be improperly estimated.

MULTIPLE REGRESSION

It frequently happens that a variable (Y) in which we are interested is related to more than one independent variable. If this relationship can be estimated, it may enable us to make more precise predictions of the dependent variable than would be possible by a simple linear regression. This brings us up against multiple regression (describes the changes in a dependent variable associated with changes in one or more independent variables), which is a little more work but no more complicated than a simple linear regression.

The calculation methods can be illustrated with the following set of hypothetical data from an environmental study relating the growth of even-aged loblolly-shortleaf pine stands to the total basal area (X_1), the percentage of the basal area in loblolly pine (X_2), and loblolly pine site index (X_3).

Y	X_1	X_2	X_3
65	41	79	75
78	90	48	83
85	53	67	74
50	42	52	61
55	57	52	59
59	32	82	73
82	71	80	72
66	60	65	66
113	98	96	99
86	80	81	90
104	101	78	86
92	100	59	88
96	84	84	93
65	72	48	70
81	55	93	85

(*Continued*)

	Y	X_1	X_2	X_3
	77	77	68	71
	83	98	51	84
	97	95	82	81
	90	90	70	78
	87	93	61	89
	74	45	96	81
	70	50	80	77
	75	60	76	70
	75	68	74	76
	93	75	96	85
	76	82	58	80
	71	72	58	68
	61	46	69	65
Sums	2,206	1,987	2,003	2,179
Means ($n=28$)	78.7857	70.9643	71.5387	77.8214

With this data, we would like to fit an equation of the form

$$Y = a + b_1 X_1 + b_2 X_2 + b_3 X_3$$

According to the principle of least squares, the best estimates of the X coefficients can be obtained by solving the set of least squares normal equations.

b_1 equation: $\left(\Sigma x_1^2\right) b_1 + \left(\Sigma x_1 x_2\right) b_2 + \left(\Sigma x_1 x_3\right) b_3 = \Sigma x_1 y$

b_2 equation: $\left(\Sigma x_1 x_2\right) b_1 + \left(\Sigma x_2^2\right) b_2 + \left(\Sigma x_2 x_3\right) b_2 = \Sigma x_2 y$

b_3 equation: $\left(\Sigma x_1 x_3\right) b_1 + \left(\Sigma x_2 x_3\right) b_2 + \left(\Sigma x_3^2\right) b_3 = \Sigma x_3 y$

$$\text{where}: \Sigma x_i x_j = \Sigma X_i X_j - \frac{(\Sigma X_i)(\Sigma X_j)}{n}$$

Having solved for the X coefficients (b_1, b_2, and b_3), we obtain the constant term by solving

$$a = \bar{Y} - b_1 \bar{X}_1 - b_2 \bar{X}_2 - b_3 \bar{X}_3$$

Derivation of the least squares normal equations requires a knowledge of differential calculus. However, for the general linear mode with a constant term

$$Y = a + b_1 X_1 + b_2 + \cdots + b_k X_k$$

the normal equations can be written quite mechanically once their pattern has been recognized. Every term in the first row contains an x_1, every term in the second row an x_2, and so forth down to the kth row, every term of which will have an x_k. Similarly, every term in the first column has an x_1 and a b_1, every term in the second column has an x_2 and a b_2, and so through the kth column, every term of which has an x_k and a b_k. On the right side of the equations, each term has a y times the x that is appropriate for a particular row. So, for the general linear model given above, the normal equations are:

b_1 equation: $(\Sigma x_1^2)b_1 + (\Sigma x_1 x_2)b_2 + (\Sigma x_1 x_3)b_3 + \cdots + (\Sigma x_1 x_k)b_k = \Sigma x_1 y$

b_2 equation: $(\Sigma x_1 x_2)b_1 + (\Sigma x_2^2)b_2 + (\Sigma x_2 x_3)b_2 + \cdots + (\Sigma x_2 x_k)b_k = \Sigma x_2 y$

b_3 equation: $(\Sigma x_1 x_3)b_1 + (\Sigma x_2 x_3)b_2 + (\Sigma x_3^2)b_3 + \cdots + (\Sigma x_3 x_k)b_k = \Sigma x_3 y$

b_k equation: $(\Sigma x_1 x_k)b_1 + (\Sigma x_2 x_k)b_2 + (\Sigma x_3 x_k)b_3 + \cdots + (\Sigma x_k^2)b_k = \Sigma x_k y$

Given the X coefficients, the constant term can be computed as

$$a = \bar{Y} - b_1 \bar{X}_1 - b_2 \bar{X}_2 - \cdots - b_k \bar{X}_k$$

Note that the normal equations for the general linear model include the solution for the simple linear regression

$$(\Sigma x_1^2)b_1 = \Sigma x_1 y$$

Hence,

$$b_1 = (\Sigma x_1 y)/\Sigma x_1^2 \text{ as previously given.}$$

In fact, all of this section on multiple regression can be applied to the simple linear regression as a special case.

The corrected sums of squares and products are computed in the familiar manner:

$$\Sigma y^2 = \Sigma Y^2 - \frac{(\Sigma Y)^2}{n} = (65^2 + \cdots + 61^2) - \frac{(2,206)^2}{28} = 5,974.7143$$

$$\Sigma x_1^2 = \Sigma X_1^2 - \frac{(\Sigma X_1)^2}{n} = (41^2 + \cdots + 46^2) - \frac{(1,987)^2}{28} = 11,436.9643$$

$$\Sigma x_1 y = \Sigma X_1 Y - \frac{(\Sigma X_1)(\Sigma Y)}{n}$$

$$= (41)(65) + \cdots + (46)(61) - \frac{(1,987)(2,206)}{28}$$

$$= 6,428.7858$$

Similarly

$\Sigma x_1 x_2 = -1,171.4642 \quad \Sigma x_2 y = 2,632.2143$

$\Sigma x_1 x_3 = 3,458.8215 \quad \Sigma x_3^2 = 2,606.1072$

$\Sigma x_2^2 = 5,998.9643 \quad \Sigma x_3 y = 3,327.9286$

$\Sigma x_2 x_3 = 1,789.6786$

Statistics Review

Putting these values in the normal equations gives:

$$11{,}436.9643b_1 - 1{,}171.4642b_2 + 3{,}458.8215b_3 = 6{,}428.7858$$

$$-1{,}171.4642b_1 + 5{,}998.9643b_2 + 1{,}789.6786b_3 = 2{,}632.2143$$

$$3{,}458.8215b_1 + 1{,}789.6786b_2 + 2{,}606.1072b_3 = 3{,}327.9286$$

These equations can be solved by any of the standard procedure for simultaneous equations. One approach (applying the above equations) is as follows:

1. Divide through each equation by the numerical coefficient of b_1.

$$b_1 - 0.102{,}427{,}897b_2 + 0.302{,}424{,}788b_3 = 0.562{,}105{,}960$$

$$b_1 - 5.120{,}911{,}334b_2 + 1.527{,}727{,}949b_3 = -2.246{,}943{,}867$$

$$b_1 + 0.517{,}424{,}389b_2 + 0.753{,}466{,}809b_3 = 0.962{,}156{,}792$$

2. Subtract the second equation form the first and the third from the first so as to leave two equations in b_2 and b_3.

$$5.018{,}483{,}437b_2 + 1.830{,}152{,}737b_3 = 2.809{,}049{,}827$$

$$-0.619{,}852{,}286b_2 - 0.451{,}042{,}021b_3 = -0.400{,}050{,}832$$

3. Divide through each equation by the numerical coefficient of b_2.

$$b^2 + 0.364{,}682{,}430b_3 = 0.559{,}740{,}779$$

$$b^2 + 0.727{,}660{,}494b_2 = 0.645{,}397{,}042$$

4. Subtract the second of these equations from the first, leaving one equation in b_3.

$$-0.362{,}978{,}064b_3 = -0.085{,}656{,}263$$

5. Solve for b_3

$$b_3 = \frac{-0.085{,}656{,}263}{-0.362{,}978{,}064} = 0.235{,}981{,}927$$

6. Substitute this value of b_3 in one of the equations (the first one, for example) of step 3 and solve for b_2.

$$b_2 + (0.364{,}682{,}43)(0.381{,}927) = 0.59{,}740{,}779$$

$$b_2 = 0.473{,}682{,}316$$

7. Substitute the solutions for b_2 and b_3 in one of the equations (the first one, for example) of step 1, and solve for b_1.

$$b_1 - (0.102,427,897)(0.473,682,316) + (0.302,424,788)(0.235,981,927)$$

$$= 0.562,105,960$$

$$b_1 = 0.539,257,459$$

8. As a check, add up the original normal equations and substitute the solutions for b_1, b_2, and b_3.

$$13,724.3216 b_1 + 6,617.1787 b_2 + 7,854.6073 b_3 = 12,388.9287$$

$$12,388.92869 = 12,388.9287, \text{ check}$$

Given the values of b_1, b_2, and b_3 we can now compute

$$a = \bar{Y} - b_1 \bar{X}_1 - b_2 \bar{X}_2 - b_3 \bar{X}_3 = -11.7320$$

Thus, after rounding of the coefficients, the regression equation is

$$\hat{Y} = -11.732 + 0.539 X_1 + 0.474 X_2 + 0.236 X_3$$

It should be noted that in solving the normal equations more digits have been carried than would be justified by the rules for a number of significant digits. Unless this is done, the rounding errors may make it difficult to check the computations.

Tests of Significance

Tests of significance refer to the methods of inference used to support or reject claims based on sample data. To test the significance of the fitted regression, the outline for the analysis of variance is

Source	df
Reduction due to regression on X_1, X_2, and X_3	3
Residuals	24
Totals	27

The degrees of freedom for the total are equal to the number of observations minus 1. The total sum of squares is

$$\text{Total SS} = \Sigma y^2 = 5,974.7143$$

The degrees of freedom for the reduction are equal to the number of independent variables fitted, in this case 3. The reduction sum of squares for any least squares regression is

$$\text{Reduction SS} = \Sigma (\text{Estimated coefficients})(\text{Right side of their normal equations})$$

Statistics Review

In this example, there are three coefficients estimated by the normal equations, and so

$$\text{Reduction SS}_{3\,df} = b_2(\Sigma x_1 y) + b_2(\Sigma x_2 y) + b_3(\Sigma x_3 y)$$

$$= (0.53926)(6,428.7858) + (0.47368)(2,632.2143) + (0.23598)(3,327.9286)$$

$$= 5,498.9389$$

The residual df and sum of squares are obtained by subtraction. Thus the analysis becomes

Source	df	SS	MS
Reduction due to X_1, X_2, and X_3	3	5,498.9389	1,832.9796
Residuals	24	475.7754	19.8240
Total	27	5,974.7143	

To test the regression we compute

$$F_{3/24\,df} = \frac{1,832.9796}{19.8240} = 92.46$$

which is significant at the 0.01 level.

Often we will want to test individual terms of the regression. In the previous example we might want to test the hypothesis that the true value of b_3 is zero. This would be equivalent to testing whether the viable X_3 makes any contribution to the prediction of Y. If we decide a b_3 may be equal to zero, we might rewrite the equation in terms of X_1 and X_2. Similarly, we could test the hypothesis that b_1 and b_3 are both equal to zero.

To test the contribution of any set of the independent variables in the presence of the remaining variables:

1. Fit all independent variables and compute the reduction and residual sums of squares.
2. Fit a new regression that includes only the variables not being tested. Compute the reduction due to this regression.
3. The reduction obtained in the first step minus the reduction in the second step is the gain due to the variables being tested.
4. The mean square for the gain (step 3) is tested against the residual mean square from the first step.

Coefficient of Multiple Determination

As a measure of how well the regression fits the data it is customary to compute the ration of the reduction sum of squares to the total sum of squares. As mentioned earlier, this ratio is symbolized by R^2 and is sometimes called the coefficient of determination:

$$R^2 = \frac{\text{Reduction SS}}{\text{Total SS}}$$

For the regression of Y on X_1, X_2, and X_3,

$$R^2 = \frac{5{,}498.9389}{5{,}974.7143} = 0.92$$

The R^2 value is usually referred to saying that a certain percentage (92 in this case) of the variation in Y is associated with regression. The square root (R) of the ratio is called the multiple correlation coefficient.

The c-Multipliers

Putting confidence limits on a multiple regression requires the computation of the Gauss or c-multipliers. The c-multipliers are the elements of the inverse of the matrix of corrected sums of squares and products as they appear in the normal equations.

CURVILINEAR REGRESSIONS AND INTERACTIONS

Curves

Many forms of curvilinear relationships can be fitted by the regression methods that have been described in the previous sections.

If the relationship between height and age is assumed to be hyperbolic so that

$$\text{Height} = a + \frac{b}{\text{Age}}$$

then we could let $Y=$Height and $X_1=$ 1/Age and fit

$$Y = a + b_1 X_1$$

Similarly, if the relationship between Y and X is quadratic

$$Y = a + bX + cX^2 \quad Y = a + bX + cX^2$$

We can let $X = X_1$ and $X^2 = X_1$ and fit

$$Y = a + b_1 X_1 + b_2 X_2$$

Functions such as

$$Y = aX^b$$

$$Y = a(b^x)$$

$$10^Y = aX^b$$

Which are nonlinear in the coefficients can sometimes be made linear by a logarithmic transformation. The equation

$$Y = aX^b$$

would become

$$\log Y = \log a + b(\log X)$$

which could be fitted by

$$Y' = a' + b_1 X_1$$

where $Y' = \log Y$, and

$$X_1 = \log X.$$

The second equation transforms to

$$\log Y = \log a + (\log b) X$$

The third becomes

$$Y = \log a + b(\log X)$$

The linear model can fit both.

In making these transformations the effect on the assumption of homogeneous variance must be considered. If Y has homogeneous variance, $\log Y$ probably will not have—and vice versa.

Some curvilinear models cannot be fitted by the methods that have been described. Some examples are

$$Y = a + b^x$$

$$Y = a(X - b)^2$$

$$Y = a(X_1 - b)(X_2 - c)$$

Fitting these models requires more cumbersome procedures.

Interactions

Suppose that there is a simple linear relationship between Y and X_1. If the slope (b) of this relationship varies, depending on the level of some other independent variable (X_2), then X_1 and X_2 are said to interact. Such interactions can sometimes be handled by introducing interaction variables.

To illustrate, suppose that we know that there is a linear relationship between Y and X_1

$$Y = a + bX_1$$

Suppose further that we know or suspect that the slope (b) varies linearly with Z

$$b = a' + b'Z$$

This implies the relationship

$$Y = a + (a' + b'Z)X_1$$

or

$$Y = a + a'X_1 + b'X_1Z$$

where $X_2 = X_1Z$, and interaction variable.

If the Y-intercept is also a liner function of Z, then

$$a = a'' + b''Z$$

and the form of relationship is

$$Y = a'' + b''Z + a'X_1 + b'X_1Z$$

Group Regressions

Linear regressions of Y on X were fitted for each of two groups. The basic data and fitted regressions were:

Group A										Sum	Mean
Y	3	7	9	6	8	13	10	12	14	82	9.111
X	1	4	7	7	2	9	10	6	12	58	6.444

$$n = 9 \Sigma Y^2 = 848, \ \Sigma XY = 609 \Sigma X^2 = 480$$

$$\Sigma y^2 = 100.8889, \ \Sigma xy = 80.5556, \ \Sigma x^2 = 106.2222,$$

$$\hat{Y} = 4.224 + 0.7584 X$$

Residual SS = 39.7980, with 7 df.

Group A														Sum	Mean
Y	4	6	12	2	8	7	0	5	9	2	11	3	10	79	6.077
X	4	9	14	6	9	12	2	7	5	5	11	2	13	99	7.616

$$n = 13 \Sigma Y^2 = 653, \ \Sigma XY = 753, \ \Sigma X^2 = 951$$

$$\Sigma y^2 = 172.9231, \ \Sigma xy = 151.3846, \ \Sigma x^2 = 197.0769$$

$$\hat{Y} = 0.228 + 0.7681 X$$

Residual SS = 56.6370, with 11 df.

Statistics Review

Now, we might ask, are these really different regressions? Or could the data be combined to produce a single regression that would be applicable to both groups? If there is no significant difference between the residual mean squares for the two groups (this matter may be determined by Bartlett's test), the test described below helps to answer the question.

Testing for the Common Regressions

Simple linear regressions may differ either in their slope or in their level. In testing for common regressions, the procedure is to test first for common slopes. If the slopes differ significantly, the regressions are different and no further testing is needed. If the slopes are not significantly different, the difference in level is tested. The analysis table is:

Line	Group	df	Σy^2	Σxy	Σx^2	Residuals df	Residuals SS	Residuals MS
1	A	8	100.8889	80.5556	106.2222	7	39.7980	
2	B	12	172.9231	151.3846	97.0769	11	56.6370	
3	Pooled residuals					18	96.4350	5.3575
4	Difference for testing common slopes					1	0.0067	0.0067
5	Common slope	20	273.8120	231.9402	303.2991	19	96.4417	5.0759
6	Difference for testing trends						80.1954	80.1954
7	Single regression	21	322.7727	213.0455	310.5909	20	176.6371	

The first two lines in this table contain the basic data for the two groups. To the left are the total df for the groups (8 for A and 12 for B. In the center are the corrected sums of squares and products. The right side of the table gives the residual sum of squares and df. Since only simple linear regressions have been fitted, the residual df of reach group are one less than the total df. The residual sum of squares is obtained by first computing the reduction sum of squares for each group.

$$\text{Reduction SS} = \frac{\left(\Sigma xy\right)^2}{\Sigma x^2}$$

This reduction is then subtracted from the total sum of squares (Σy^2) to give to residuals.

Line 3 is obtained by pooling the residual df and residual sums of squares for the groups. Dividing the pooled sum of squares by the pooled df gives the pooled mean square.

The left side and center of line (we will skip line 4 for the moment) is obtained by pooling the total df and the corrected sums of square and products for the groups. These are the values that are obtained under the assumption of no difference in the slopes of the group regressions. If the assumption is wrong, the residuals about this common slope regression will be considerably larger than the mean square residual about the separate regressions. The residual df and sum of squares are obtained by fitting a straight line to this pooled data. The residual df are, of course, one less than the total df. The residual sum of squares is, as usual,

$$\text{Residual SS} = 273.8120 - \frac{(231.9402)^2}{303.2991} = 96.4417$$

Now, the difference between these residuals (line 4=5 – line 3) provides a test of the hypothesis of common slopes. The error term for this test is the pooled mean square from line 3.

$$\text{Test of common slopes}: F_{1/18\text{df}} = \frac{0.0067}{5.3575}$$

The difference is not significant.

If the slopes differed significantly, the groups would have different regressions, and we would stop here. Since the slopes did not differ, we now go on to test for a difference in the levels of the regression.

Line 7 is what we would have if we ignored the groups entirely, lumped all the original observations together, and fitted a single linear regression. The combined data are as follows:

$$n = (9+13) = 22 \left(\text{so the df for total} = 21\right)$$

$$\Sigma Y = (82+79) = 161, \quad \Sigma Y^2 = (848+653) = 1,501$$

$$\Sigma y^2 = 1,501 - \frac{(161)^2}{22} = 322.7727$$

$$\Sigma X = (585+99) = 157, \quad \Sigma X^2 = (480+951) = 1,431$$

$$\Sigma x^2 = 1,431 - \frac{(157)^2}{22} = 310.5909$$

$$\Sigma XY = (609+753) = 1,362, \quad \Sigma xy = 1,362 - \frac{(157)(161)}{22} = 213.0455$$

From this we obtain the residual values on the right side of line 7.

$$SS = 322.7727 - \frac{(213.0455)^2}{310.5909} = 176.6371$$

If there is a real difference among the levels of the groups, the residuals about this single regression will be considerably larger than the mean square residual about the regression that assumed the same slopes but different levels. This difference (line 6=line 7 – line 5) is tested against the residual mean square from line 5.

$$\text{Test of levels}: F_{1/18\text{df}} = \frac{80.1954}{5.0759} = 15.80$$

As the levels differ significantly, the groups do not have the same regressions.

The test is easily extended to cover several groups, though there may be a problem in finding which groups are likely to have separate regressions and which can be combined. The test can also be extended to multiple regressions

Analysis of Covariance in a Randomized Block Design

A test was made on the effect of three soil treatments on the height growth of 2-year-old seedlings. Treatments were assigned at random to the three plots within each of 11 blocks. Each plot was made up of 50 seedlings. Average 5-year height growth was the criterion for evaluating treatments. Initial heights and 5-year growths, all in feet, were:

Block	Treatment A		Treatment B		Treatment C		Block Totals	
	Height	Growth	Height	Growth	Height	Growth	Height	Growth
1	3.6	8.9	3.1	10.7	4.7	12.4	11.4	32.0
2	4.7	10.1	4.9	14.2	2.6	9.0	12.2	33.3
3	2.6	6.3	0.8	5.9	1.5	7.4	4.9	19.6
4	5.3	14.0	4.6	12.6	4.3	10.1	14.2	36.7
5	3.1	9.6	3.9	12.5	3.3	6.8	10.3	28.9
6	1.8	6.4	1.7	9.6	3.6	10.0	7.1	26.0
7	5.8	12.3	5.5	12.8	5.8	11.9	17.1	37.0
8	3.8	10.8	2.6	8.0	2.0	7.5	8.4	26.3
9	2.4	8.0	1.1	7.5	1.6	5.2	5.1	20.7
10	5.3	12.6	4.4	11.4	5.8	13.4	15.5	37.4
11	3.6	7.4	1.4	8.4	4.8	10.7	9.8	26.5
Sums	42.0	106.4	34.0	113.6	40.0	104.4	116.0	324.4
Means	3.82	9.67	3.09	10.33	3.64	9.49	3.52	9.83

The analysis of variance of grow this:

Source	df	SS	MS
Blocks	10	132.83	—
Treatment	2	4.26	2.130
Error	20	68.88	3.444
Total	32	205.94	

$$F(\text{for testing treatments})_{2/20\,df} = \frac{2.130}{3.444}$$

Not significant at 0.05 level.

There is no evidence of a real difference in growth due to treatments. This is, however, the reason to believe that, for young seedlings, growth is affected by initial height. A glance at the block totals seems to suggest that plots with greatest initial height had greatest 5-year growth. The possibility that effects of treatment are being obscured by differences in initial heights raises the question of how the treatments would compare if adjusted for differences in initial heights.

If the relationship between height growth and initial height is linear and if the slope of the regression is the same for all treatments, the test of adjusted treatment means can be made by an analysis of covariance as described below. In this analysis, the growth will be labeled Y and initial height X.

Computationally the first step is to obtain total, block, treatment, and error sums of squares of X (SS_x) and sums of products of x and y (SP_{xy}), just as has already been done for Y.

$$\text{For } X \quad \text{C.T.}_{\cdot x} = \frac{116.0^2}{33} = 407.76$$

$$\text{Total SS}_x = \left(3.6^2 + \cdots + 4.8^2\right) - \text{C.T.}_x = 73.2$$

$$\text{Block SS}_x = \left(\frac{11.4^2 + \cdots + 9.8^2}{3}\right) - \text{C.T.}_x = 4.31$$

$$\text{Treatment SS}_x = \left(\frac{42.0^2 + 34.0^2 + 40.0^2}{11}\right) - \text{C.T.}_x = 3.15$$

$$\text{Error SS}_x = \text{Total} - \text{Block} - \text{Treatment} = 15.80$$

$$\text{For } XY: \quad \text{C.T.}_{xy} = \frac{(116.0)(324.4)}{33} = 1{,}140.32$$

$$\text{Total SP}_{xy} = (3.6)(8.9) + \cdots + (4.8)(10.7) - \text{C.T.}_{xy} = 103.99$$

$$\text{Treatment SP}_{xy} = \left(\frac{(11.4)(32.0) + \cdots + (9.8)(26.5)}{3}\right) - \text{C.T.}_{xy} = 82.71$$

$$\text{Treatment SP}_{xy} = \left(\frac{(42.0)(106.4) + (34.0)(113.6) + (40.0)(104.4)}{11}\right) - \text{C.T.}_{xy} = -3.30$$

$$\text{Error SP}_{xy} = \text{Total} - \text{Block} - \text{Treatment} = 24.58$$

These computed terms are arranged in a manner similar that for the test of group regressions (which is exactly what the covariance analysis is). One departure is that the total line is put at the top.

Source	df	SS_y	SP_{xy}	SS_x	Residuals		
					df	SS	MS
Total	32	205.97	103.99	73.26			
Blocks	10	132.83	82.71	54.31			
Treatment	2	4.26	−3.30	3.15			
Error	20	68.88	24.58	15.80	19	30.641	1.613

On the error line, the residual sum of squares after adjusting for a linear regression is

$$SS_y - \frac{(SP_{xy})^2}{SS_x} = 68.88 - \frac{24.58^2}{15.80} = 30.641$$

This sum of squares has 1 df less than the unadjusted sum or squares.

To test treatments we first pool the unadjusted df and sums of squares and products for treatment and error. The residual terms for this pooled line are then computed just as they were for the error line

	df	SSy	SP_{xy}	SS_x	Residuals	
					df	SS
Treatment+error	22	73.14	21.28	18.95	21	49.244

Then to test for a difference among treatments after adjustment for the regression of growth on initial height, we compute the difference in residuals between the error and the treatment+error lines

	df	SS	MS
Difference for testing adjusted treatments	2	18.603	9.302

The mean square for the difference in residual is now tested against the residual mean square for error.

$$F_{2/19df} = \frac{9.302}{1.613} = 5.77$$

Thus, after adjustment, the difference in treatment means is found to be significant at the 0.05 level. It may also happen that differences that were significant before adjustment are not significant afterwards.

If the independent variable has been affected by treatments, the interpretation of a covariance analysis requires careful thinking. The covariance adjustment may have the effect of removing the treatment differences that are being tested. On the other hand, it may be informative to know that treatments are or are not significantly different in spite of the covariance adjustment. The beginner who is uncertain of the interpretations would do well to select as covariates only those that have not been affected by treatments.

The covariance test may be made in a similar manner for any experimental design and, if Desired (and justified), adjustment may be made for multiple or curvilinear regressions.

The entire analysis is usually presented in the following form:

Source	df	SS_y	SP_y	SS_x	df	Adjusted	
						SS	MS
Total	32	205.97	103.99	73.26			
Blocks	10	132.83	82.71	54.31			
Treatment	2	4.26	−3.30	3.15			
Error	20	68.88	24.58	15.80	19	30.641	1.613
Treatment+error	22	73/14	21.28	18.95	21	49.244	—
Difference for testing adjusted treatment means					2	18.603	9.302

Unadjusted treatments: $F_{2/20\,df} = \dfrac{2.130}{3.444}$ not significant.

Adjusted treatments: $F_{2/19\,df} = \dfrac{9.302}{1.613} = 5.77$ significant at the 0.05 level.

Adjusted Means

If we wish to know what the treatment means are after adjustment for regression, the equation is

$$\text{Adjusted } \bar{Y}_i = \bar{Y}_i - b(\bar{X} - X)$$

where
$\bar{Y}$ = Unadjusted mean for treatment i.
b = Coefficient of the linear regression = $\dfrac{\text{Error SP}_{xy}}{\text{Error SS}_x}$
$\bar{X}_i$ = Mean of the independent variable for treatment i.
$\bar{X}$ = Mean of X for all treatments.
In the example we had $\bar{X}_A = 3.82$, $\bar{X}_B = 3.09$, $\bar{X}_C = 3.64$, $\bar{X} = 3.52$, and

$$b = \dfrac{24.58}{15.80} = 1.56$$

So, the unadjusted and adjusted mean growths are

Treatment	Mean Growths	
	Unadjusted	Adjusted
A	9.67	9.20
B	10.33	11.00
C	9.49	9.30

Tests among Adjusted Means

In an earlier section, we encountered methods of making further tests among the means. Ignoring the covariance adjustment, we could for example make an F test for pre-specified comparisons such as $A+C$ vs. B, or A vs. C. Similar tests can also be made after adjustment for covariance, though they involve more labor. The F test will be illustrated for the comparison B vs. $A+C$ after adjustment.

As might be suspected, to make the f test we must first compute sums of squares and products of X and Y for the specified comparison:

$$SS_y = \dfrac{\left[2(\Sigma Y_B) - (\Sigma Y_A + \Sigma Y_c)\right]^2}{\left[2^2 + 1^2 + 1^2\right][11]} = \dfrac{\left[2(113.6) - (106.4 + 104.4)\right]^2}{66} = 4.08$$

$$SS_x = \dfrac{\left[2(\Sigma X_B) - (\Sigma X_A + \Sigma X_C)\right]^2}{\left[2^2 + 1^2 + 1^2\right][11]} = \dfrac{\left[2(34.0) - (42.0 + 40.0)\right]^2}{66} = 2.97$$

$$SP_{xy} = \dfrac{\left[2(\Sigma Y_B) - (\Sigma Y_A + \Sigma Y_C)\right]\left[2(\Sigma X_B) - (\Sigma X_A + \Sigma X_C)\right]}{\left[2^2 + 1^2 + 1^2\right][11]} = -3.48$$

From this point on, the F test of $A+B$ vs. C is made in exactly the same manner as the test of treatments in the covariance analysis.

Source	df	SS_y	SP_{xy}	SS_x	df	Residuals	
						SS	MS
$2B-(A+C)$	1	4.08	−3.48	2.97	—	—	—
Error	20	68.88	24.58	15.80	19	30.641	1.613
Sum	21	72.96	21.10	18.77	20	49.241	—
Difference for testing adjusted comparison					1	18.600	18.600

$F_{1/19df} = 11.531$ significant at the 0.01 level.

NOTE

1. Much of the information in this chapter is modeled after F. Freese (1967) USFS-USDA Agriculture Handbook 317—*Elementary Statistical Methods for Foresters*. Washington, DC.

REFERENCES

Addelman, S. (1969). The generalized randomized block design. The American Statistician **23**(4):35–36.
Addelman, S. (1970). Variability of treatments and experimental units in the design and analysis of experiments. Journal of the American Statistical Association **65**(331):1095–1108.
Freese, F. (1969). *Elementary Statistical Methods for Foresters*. Washington, DC: USDA.
Hamburg, M. (1987). *Statistical Analysis for Decision Making*, 4th ed. New York: Harcourt Brace Jovanovich Publishers.
Mayr, E. (1970). *Populations, Species, and Evolution: An Abridgement of Animal Species and Evolution*. Cambridge, MA: Belknap Press.
Mayr, E. (2002). The biology of race and the concept of equality (http://www.goodurmj.com/Mayr.html). Daedalus, Winter, pp. 89–94.

6 Fundamental Engineering Concepts

DEEP AND GLOOMY WOODS

The mountain, and a deep and gloomy wood, their colours and their forms, were then to me an appetite: a feeling and a love, that had no need of remoter charm ...

W. Wordsworth (1798)

Education can only go so far in preparing the environmental engineer for on-the-job experience. A person who wishes to become an environmental engineer is greatly assisted by two personal factors. First, a well-rounded, broad development of experience in many areas is required, and results in the production of the classic generalist. Although environmental engineers cannot possibly attain great depth in all areas, they must have the desire and the aptitude to do so. They must be interested in—and well informed about—many widely differing fields of study. The necessity for this in the environmental application is readily apparent. Why? Simply because the range of problems encountered is so immense that a narrow education will not suffice; environmental engineers must handle situations that call upon skills as widely diverse as the ability to understand psychological and sociological problems with people to the ability to perform calculations required for mechanics and structures. We have pointed out that the would-be practicing environmental engineer can come from just about any background, and that a narrow education does not preclude students and others from broadening themselves later; however, quite often those who are very specialized lack appreciation for other disciplines, as well as the adaptability necessary for environmental engineering.

A second requirement calls for educational emphasis upon quantitative and logical problem solving. The non-technician whose mathematical ability breaks down at simple algebra is not likely to acquire the necessary quantitative expertise without great effort. Along with mathematics, the environmental engineer must have a good foundation in mechanics and structures. An education that does not include a foundation in the study of forces that act on buildings, machines and processes leaves the environmental practitioner in the same position as a thoracic surgeon with incomplete knowledge of gross anatomy—he or she simply has to feel their way to the target, leaving a lot to be desired (especially for the patient).

Frank R. Spellman (1996)

INTRODUCTION

Though individual learning style is important in your determination of a career choice as an environmental engineer, again, we stress education as the key ingredient in the mix that produces the safety engineer. Along with basic and applied sciences (mathematics, natural, and behavioral sciences—which are applied to the solution of technological, biological, and behavioral problems), education in engineering and technology is a must. Topics including applied mechanics, properties of materials, electrical circuits and machines, and principles of engineering design and computer science fall into this category.

In this chapter, the focus is on applied mechanics and, in particular, forces and the resolution of forces. Why? Because many accidents and resulting injuries are caused by forces of great magnitude for machine, material, or structure. To design and inspect systems, devices, or products to ensure their safety, environmental engineers must account for the forces that act or might act on them. Environmental engineers must also account for forces from objects that may act on the human body (an area of focus that is often overlooked). Important areas that are part of or that interface with applied mechanics are the properties of materials, electrical circuits and machines, and engineering design considerations. We cannot discuss all engineering aspects related to these areas in this text. Instead, our goal is to look at some fundamental concepts and their relationships to environmental engineering.

RESOLUTION OF FORCES

In environmental engineering, we tend to focus our attention on those forces that are likely to cause failure or damage to some device or system, resulting in an occurrence that is likely to produce secondary damage to other devices or systems and harm to individuals. Typically, large forces are more likely to cause failure or damage than small ones.

The environmental engineer must understand force. He or she must understand how a force acts on a body: (1) the direction of force, (2) point of application (location) of force, (3) the area over which force acts, (4) the distribution or concentration of forces that act on bodies, and (5) how essential these elements are in evaluating the strength of materials. For example, a 40-lb. force applied to the edge of a sheet of plastic and parallel to it probably will not break it. If a sledgehammer strikes the center of the sheet with the same force, the plastic will probably break. A sheet metal panel of the same size undergoing the same force will not break.

Practice tells us that different materials have different strength properties. Striking the plastic panel will probably cause it to break, whereas striking a sheet metal panel will cause a dent. The strength of a material and its ability to deform are directly related to the force applied. Important physical, mechanical, and other properties of materials are:

- crystal structure
- strength
- melting point
- density
- hardness
- brittleness
- ductility
- modulus of elasticity
- wear properties
- coefficient of expansion
- contraction
- conductivity
- shape
- exposure to environmental conditions
- exposure to chemicals
- fracture toughness

and many others. Note that all these properties can vary depending on forces that are crushing, corroding, cutting, pulling, or twisting.

The forces an object can encounter are often different from the forces that an object would be able to withstand. The object may be designed to withstand only minimal force before it fails (a toy doll may be designed with either very soft and pliable materials or designed to break or give way in

Fundamental Engineering Concepts

certain places when a child falls on it to prevent injury). Other devices may be designed to withstand the greatest possible load and shock (e.g., a building constructed to withstand an earthquake).

When working with any material to go in an area with a concern for safety, a safety factor (or factor of safety) is often introduced. Safety factor (SF), [as defined by ASSE (1988)], is the ratio allowed for in design, between the ultimate breaking strength of a member, material, structure, or equipment and the actual working stress or safe permissible load placed on it during ordinary use. Simply put, including a factor of safety—into the design of a machine, for example— makes an allowance for many unknowns (inaccurate estimates of real loads or irregularities in materials, for example) related to the materials used to make the machine, related to the machine's assembly, and related to the use of the machine. Safety factor (SF) can be determined in several ways. One of the most commonly used ways is

$$SF = \frac{\text{Failure} - \text{producing load}}{\text{Allowable stress}}. \tag{6.1}$$

Forces on a material or object are classified by the way they act on the material. For example, if the force pulls a material apart, it is called tensile force. Forces that squeeze a material or object are called compression forces. Shear forces cut a material or object. Forces that twist a material or object are called torsional forces. Forces that cause a material or object to bend are called bending forces. A bearing force occurs when one material or object presses against or bears on another material or body.

So, what is force? Force is typically defined as any influence that tends to change the state of rest or the uniform motion in a straight line of a body. The action of an unbalanced or resultant force results in the acceleration of a body in the direction of action of the force, or it may (if the body is unable to move freely) result in its deformation (Hooke's Law). Force is a vector quantity, possessing both magnitude and direction (see Figure 6.1a and b); its SI unit is the newton (equal to 3.6 ounces, or 0.225 lb.).

According to Newton's second law of motion, the magnitude of a resultant force is equal to the rate of change of momentum of the body on which it acts; the force F producing acceleration a m/s^2 on a body of mass kilograms is therefore given by:

$$F = ma \tag{6.2}$$

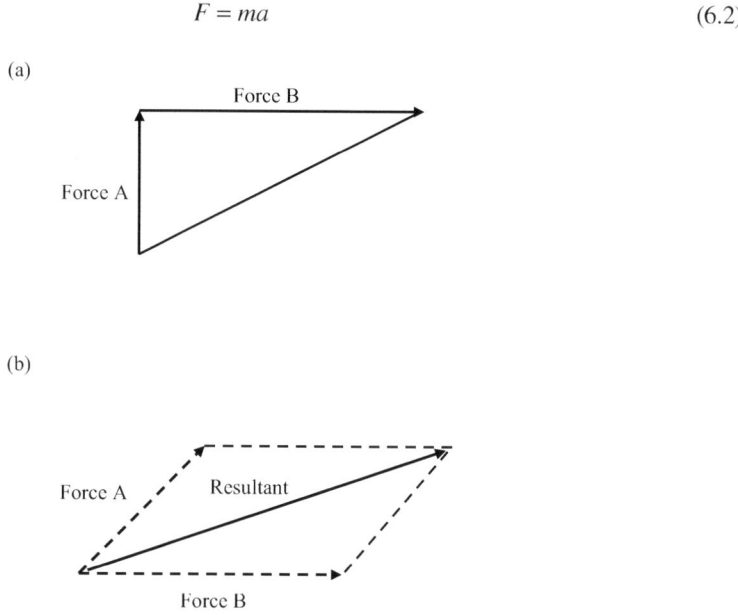

FIGURE 6.1 A-B Force is a vector quantity. Relative change relative to 1980.

With regard to environmental engineering, a key relationship between a force F and the body on which it acts is

$$F = sA \tag{6.3}$$

where

s = force or stress per unit area (e.g., pounds per square inch)
A = area (square inches, square feet, etc.) over which a force acts.

Note: The stress a material can withstand is a function of the material and the type of loading.

Frequently, two or more forces act together to produce the effect of a single force, called a resultant. This resolution of forces can be accomplished in two ways: triangle and/or parallelogram law. The triangle law provides that if two concurrent forces are laid out vectorially with the beginning of the second force at the end of the first, the vector connecting the beginning and the end of the forces represents the resultant of the two forces (see Figure 6.1a). The parallelogram law provides that if two concurrent forces are laid out vectorially, with both forces pointing toward and pointing away from their point of intersection, a parallelogram represents the resultant of the force. The concurrent forces must have both direction and magnitude of their resultant to be determined (see Figure 6.1b). After the triangle or parallelogram has been completed, and if the individual forces are known or one of the individual forces and the resultants are known, the resultant force may be simply calculated by either the trigonometric method (sines, cosines, and tangents), or by the graphic method (which involves laying out the known force, or forces, to an exact scale and exact direction in either a parallelogram or triangle and then measuring the unknown to the same scale).

SLINGS

Let's take a look at a few example problems involving forces that the environmental engineer might be called upon to calculate. In our examples, we use lifting slings under different conditions of loading.

Note: Slings are commonly used between cranes, derricks, and/or hoists and the load, so that the load may be lifted and moved to the desired location. For the safety engineer, knowledge of the properties and limitations of the slings, the type and condition of the material being lifted, the weight and shape of the object being lifted, the angle of the lifting sling to the load being lifted, and the environment in which the lift is to be made are all important considerations to be evaluated—before the transfer of material can take place safely.

Example 6.1

Let us assume a load of 2,000 pounds is supported by a two-legged sling; the legs of the sling make an angle of 60° with the load (see Figure 6.2). What force is exerted on each leg of the sling?
   Note: In solving this type of problem, you should always draw a rough diagram as shown in Figure 6.2.
   A resolution of forces provides the answer. We use the trigonometric method to solve this problem but remember that it may also be solved by using the graphic method. Using the trigonometric method with the parallelogram law, the problem could be solved as follows: (Again, make a drawing which should look like Figure 6.3):
   We could consider the load (2,000 pounds) as being concentrated and acting vertically, which can be indicated by a vertical line. The legs of the slings are at a 60° angle, which can be shown as **ab** and **ac**. The parallelogram can now be constructed by drawing lines parallel to **ab** and **ac**, intersecting at **d**. The point where **cb** and ad intersect can be indicated as **e**. The force on each leg of the sling (**ab** for example) is the resultant of two forces, one acting vertically (**ae**) and the other horizontally (**be**), as shown in the force diagram. Force **ae** is equal to one half of **ad** (the total force acting vertically, 2,000 pounds, so **ae** = 1,000. This value remains constant regardless of the angle

Fundamental Engineering Concepts   213

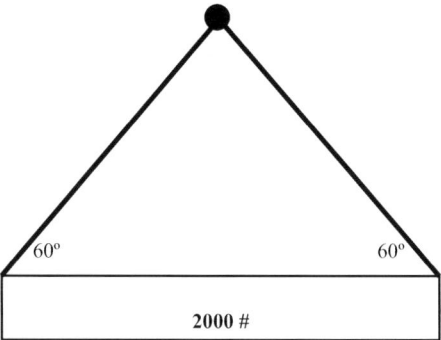

FIGURE 6.2 For Example 6.1.

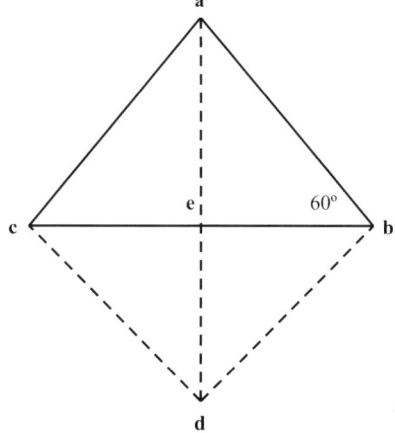

FIGURE 6.3 For Example 6.1.

ab makes with **bd,** because as the angle increases or decreases, **ae** also increases or decreases. But **ae** is always **ad/2**. The force **ab** can be calculated by trigonometry using the right triangle **abe:**

$$\text{Sine of an angle} = \frac{\text{Opposite side}}{\text{Hypotenuse}}$$

$$\text{Therefore,} \quad \sin 60° = \frac{\mathbf{ae}}{\mathbf{ab}}$$

$$\text{Transposing,} \quad \mathbf{ab} = \frac{\mathbf{ae}}{\sin 60°}$$

$$\text{Substituting known value,} \quad \mathbf{ab} = \frac{1,000}{0.866} = 1,155$$

The total weight on each leg of the sling at a 60° angle from the load is 1,155 pounds. Note that the weight is more than half of the load because the load is made up of two forces—one acting vertically and the other horizontally. An important point to remember is—the smaller the angle, the greater the load (force) on the sling. For example, at a 15° angle, the force on each leg of a 2,000-pound load increases to 3,864 pounds.

Let's take a look at what the force would be on each leg of a 2,000-pound load at different angles (angles that are common for lifting slings; see Figure 6.4).

Now let's work on a couple of example problems.

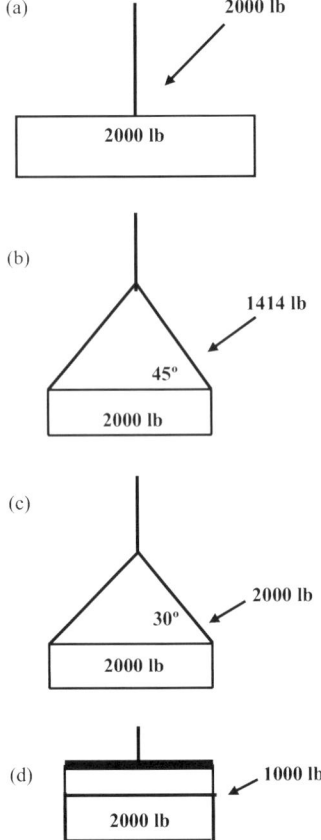

FIGURE 6.4 Sling angle and load examples for Example 6.2.

Example 6.2

Problem:

We have a 3,000-lb. load to be lifted with a 2-leg sling whose legs are at a 30° angle with the load. The load (force) on each leg of the sling is:

Solution:

$$\sin A = \frac{a}{c}$$

$$\sin 30 = 0.500$$

$$a = \frac{3,000 \text{ lb}}{2} = 1,500$$

$$c = \frac{a}{\sin A}$$

$$= \frac{1,500}{0.5}$$

$$= 3000$$

Fundamental Engineering Concepts

Example 6.3

Problem:

Given a two-rope sling supporting 10,000 pounds, what is the load (force) on the left sling? The sling angle to load is 60°.

Solution:

$$\sin A = \frac{a}{c}$$

$$\sin A = \frac{60}{0.866}$$

$$a = \frac{10,000}{2}$$

$$c = \frac{a}{\sin A}$$

$$= \frac{5,000}{0.866}$$

$$= 5,774 \text{ lb}$$

INCLINED PLANE

Another common problem encountered by the environmental engineer involving the resolution of forces occurs in material handling operations in moving a load (a cart, for example) up an inclined plane (a ramp, in our example). The safety implications in this type of work activity should be obvious. The forces acting on an inclined plane are shown in Figure 6.5. Let's take a look at a typical example of how to determine the force needed to pull a fully loaded cart up a ramp (an inclined plane)

Example 6.4

Problem:

Assume that a fully loaded cart weighing 400 pounds is to be pulled up a ramp that has a 5-ft rise for each 12 ft, measured along the horizontal (again, make a rough drawing like Figure 6.6).
    What force is required to pull it up the ramp?

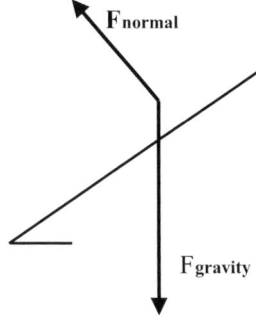

FIGURE 6.5   Forces acting on an inclined plant.

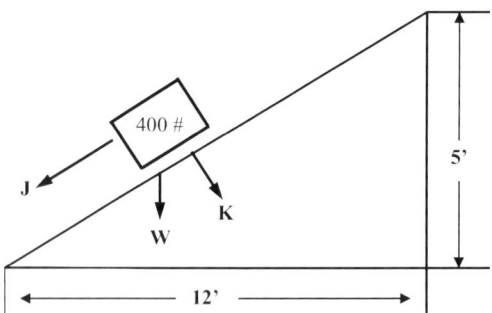

FIGURE 6.6 Inclined plane for Example 6.4.

NOTE: For illustrative purposes, we assume no friction. Without friction, of course, the work done in moving the cart in a horizontal direction would be zero (once the cart was started, it would move with constant velocity—the only work required is that necessary to get it started). However, a force equal to J is necessary to pull the cart up the ramp, or to maintain the car at rest (in equilibrium). As the angle (slope) of the ramp is increased, greater force is required to move it, because the load is being raised as it moves along the ramp, thus doing work (remember, this is not the case when the cart is moved along a horizontal plane without friction—in actual practice, however, friction can never be ignored and some "work" is accomplished in moving the cart).

To determine the actual force involved, we can again use a resolution of forces. The first step is to determine the angle of the ramp. This can be calculated by the formula:

$$\text{Tangent (angle of ramp)} = \frac{\text{Opposite side}}{\text{Adjacent side}} = 5/12 = 0.42$$

Then, arctan $0.42 = 22.8°$

Now you need to draw a force parallelogram and apply the trigonometric method (see Figure 6.7). The weight of the cart W (shown as force acting vertically) can be resolved into two components, one a force J parallel to the ramp, the other a force K perpendicular to the ramp. The component K, being perpendicular to the inclined ramp (plane) does not hinder movement up the ramp. The component J represents a force that would accelerate the cart down the ramp. To pull the cart up the ramp, a force equal to or greater than J is necessary.

Applying the trigonometric method, the angle WOK is the same as the angle of the ramp.

$$OJ = WK \ \& \ OW = 400 \, lb.$$

$$\sin \text{of angle WOK} \ (22.8°) = \frac{\text{Opposite side (WK)}}{\text{Hypotenuse (OW)}}$$

Transposing, $WK = OW \times \sin(22.8°)$

$$= 400 \times 0.388$$

$$= 155.2$$

Thus, a force of 155.2 pounds is necessary to pull the cart up the 22.8° angle of the ramp (friction ignored). Note that the total amount of work is the same, whether the cart is lifted vertically (400 pounds × 5 ft = 2,000 foot-pounds) or pulled up the ramp (155.2 pounds × 13 ft = 2000 foot-pounds). The advantage gained in using a ramp instead of a vertical lift is that less force is required—but through a greater distance.

Fundamental Engineering Concepts

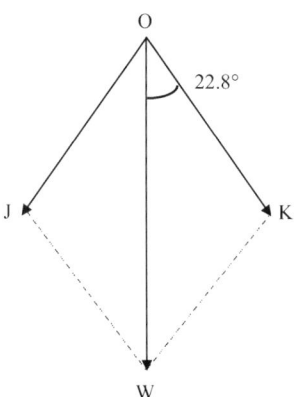

FIGURE 6.7 Force parallelogram.

PROPERTIES OF MATERIALS AND PRINCIPLES OF MECHANICS

To be able to recognize hazards and to select and implement appropriate controls, an environmental engineer must have a good understanding of the properties of materials and principles of mechanics. In this section, we start with the properties of materials then cover the wide spectrum that is mechanics, starting with statics and ending with electrical machines. The intent is to clearly illustrate the wide scope of knowledge required in areas germane to the properties of materials, and the principles of mechanics and those topics on the periphery—all of which blend in the mix—the safety mix—the mix that helps to produce the well-rounded, knowledgeable environmental engineer.

PROPERTIES OF MATERIALS

When we speak of the properties of materials, what are we referring to, and why should we concern ourselves with this topic? The best way to answer this question is to use an example where the environmental engineer, working with design engineers in a preliminary design conference, might typically be exposed to (should be exposed to) data, parameters and specifications related to the properties of a particular construction material to be used in the fabrication of, for example, a large mezzanine in a warehouse. In constructing this particular mezzanine, consideration was given to the fact that it would be used to store large, heavy equipment components. The demands placed on the finished mezzanine create the need for the mezzanine to be built using materials that can support a heavy load.

For illustration, let's say that the design engineers plan to use an aluminum alloy, type structural—No 17ST. Before they decide to include No 17ST and determine the required quantity needed to build the mezzanine, they are concerned with determining its mechanical properties, to ensure that it will be able to handle the intended load (they will also factor in, many times over, for safety—selecting a type of material that will handle a load much greater than expected).

Using a table on the *Mechanical Properties of Engineering Materials in Urquhart's Civil Engineering Handbook*, 4th ed., (1959) they check the following for No 17ST (Table 6.1).

The question is: Is this information important to the environmental engineer? No—not exactly. What is important to the environmental engineer is: (1) that a procedure such as the one just described was actually accomplished; that is, that professional engineers actually took the time to determine the correct materials to use in constructing the mezzanine; and (2) when exposed to this type of information, to specific terms, the environmental engineer must know enough about the "language" used to know what the design engineers are talking about—and to understand its significance. Remember Voltaire: "If you wish to converse with me, define your terms."

TABLE 6.1
Properties of Engineering Materials No. 17ST

Ultimate strength, psi (defined as the ultimate strength in compression for ductile materials, which is usually taken as the yield point)	Tension: 58,000 psi Compression: 35,000 psi, Shear: 35,000 psi
Yield point tension psi	@ 35,000 psi.
Modulus of elasticity, tension or compression, psi:	10,000,000
Modulus of elasticity, shear, and psi	@ 3,750,000
Weight per cu in., lb.:	0.10

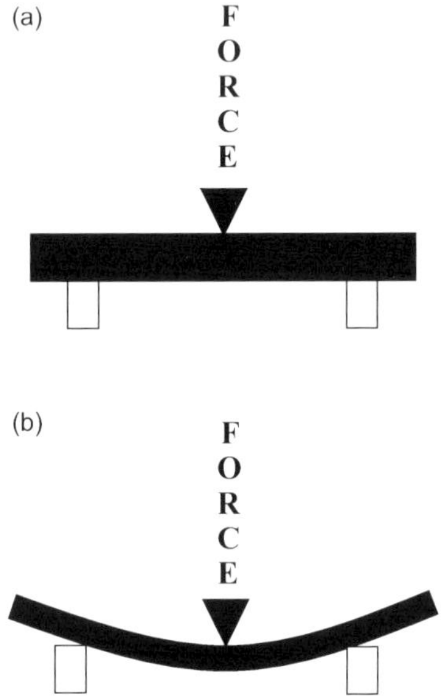

FIGURE 6.8 (a) Stress—measured in terms of the applied load over the area; (b) Strain—expressed in terms of amount per square inch.

Let's take a look at a few other engineering terms and their definitions, so that we will be able to converse. Many of the following engineering terms are from Heisler's *The Wiley Engineer's Desk Reference: A Concise Guide for the Professional Engineer* (1984) and Giachino and Weeks' *Welding Skills* (1985)—which should be standard reference texts for any safety engineer.

Stress: the internal resistance a material offers to being deformed. Measured in terms of the applied load over the area (see Figure 6.8)

Strain: the deformation that results from a stress. Expressed in terms of the amount of deformation per inch.

Intensity of stress: the stress per unit area, usually expressed in pounds per square inch. Due to a force of **P** pounds producing tension, compression, or shear on an area of **A** square inches, over which it is uniformly distributed. The simple term, stress, is normally used to indicate intensity of stress.

Ultimate stress: the greatest stress that can be produced in a body before rupture occurs.

Allowable stress or working stress: the intensity of stress that the material of a structure or a machine is designed to resist.

Fundamental Engineering Concepts

Elastic limit: the maximum intensity of stress to which a material may be subjected and return to its original shape upon the removal of stress (see Figure 6.9).

Yield point: the intensity of stress beyond which the change in length increases rapidly with little (if any) increase in stress.

Modulus of elasticity: the ratio of stress to strain, for stresses below the elastic limit. By checking the modulus of elasticity, the comparative stiffness of different materials can readily be ascertained. Rigidity and stiffness are very important for many machine and structural applications.

Poisson's ratio: the ratio of the relative change of diameter of a bar to its unit change of length under an axial load that does not stress it beyond the elastic limit.

Intensity of stress: the stress per unit area, usually expressed in pounds per square inch. Due to a force of **P** pounds producing tension, compression, or shear on an area of **A** square inches, over which it is uniformly distributed. The simple term, stress, is normally used to indicate intensity of stress.

Tensile strength: the property that resists forces acting to pull the metal apart—a very important factor in the evaluation of a metal (see Figure 6.10).

Compressive strength: the ability of a material to resist being crushed (see Figure 6.11).

Bending strength: that quality that resists forces from causing a member to bend or deflect in the direction in which the load is applied—actually a combination of tensile and compressive stresses (see Figure 6.12).

Torsional strength: the ability of a metal to withstand forces that cause a member to twist (see Figure 6.13).

Shear strength: how well a member can withstand two equal forces acting in opposite directions (see Figure 6.14).

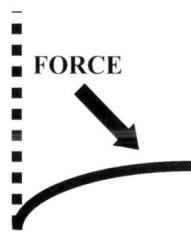

FIGURE 6.9 Elasticity and elastic limit—a metal has the ability to return to its original shape after being elongated or distorted unless it reaches its maximum stress point.

FIGURE 6.10 A metal with tensile strength resists pulling forces.

FIGURE 6.11 Compressive strength—a metal's ability to resist crushing forces.

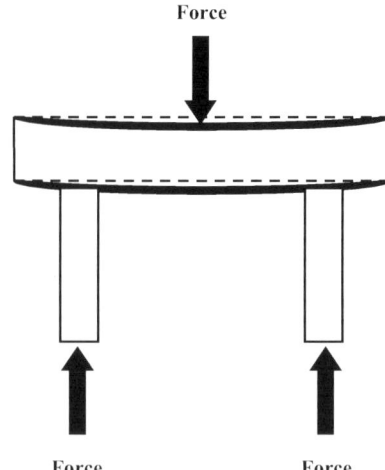

FIGURE 6.12 Bending strength (stress)—a combination of tensile strength and compression stresses.

FIGURE 6.13 Torsional strength—a metal's ability to withstand twisting forces.

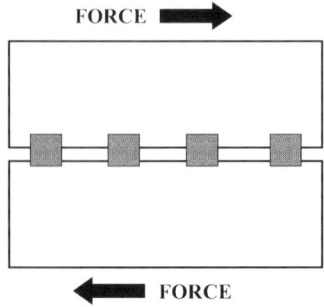

FIGURE 6.14 Shear strength—two equal forces acting in opposite directions.

Fatigue strength: the property of a material to resist various kinds of rapidly alternating stresses.

Impact strength: the ability of a metal to resist loads that are applied suddenly and often at high velocity.

Ductility: the ability of a metal to stretch, bend, or twist without breaking or cracking (see Figure 6.15).

Fundamental Engineering Concepts

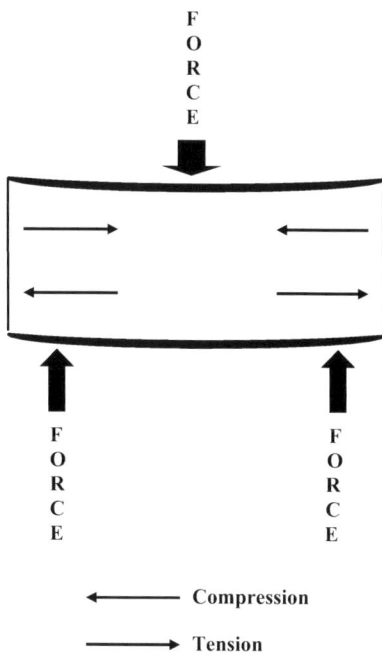

FIGURE 6.15 Distribution of stress in a beam cross section during bending.

Hardness: the property in steel that resists indentation or penetration.
Brittleness: a condition whereby a metal will easily fracture under low stress.
Toughness: may be considered as strength, together with ductility. A tough material can absorb large amounts of energy without breaking.
Malleability: the ability of a metal to be deformed by compression forces without developing defects, such as encountered in rolling, pressing, or forging.

Friction

Earlier, in discussing the principle of the inclined plane, we ignored the effect of friction. In actual use, friction cannot be ignored and you must have some understanding of its characteristics and applications. Friction deals with one body in contact with another that is on the verge of sliding or is sliding. Friction allows us to walk, ski, drive vehicles and power machines (among other things). Whenever one body slides over another, frictional forces opposing the motion are developed between them. Friction force is the force tangent to the contact surface that resists motion. If motion occurs, the resistance is due to kinetic friction, which is normally lower than the value for static friction. Contrary to common perception, the degree of smoothness of a surface area is not responsible for these frictional forces; instead, the molecular structure of the materials is responsible. The coefficient of friction M (which differs among different materials) is the ratio of the frictional force F to the normal force N between two bodies.

$$M = \frac{F}{N} \tag{6.4}$$

For dry surfaces, the coefficient of friction remains constant, even if the weight of an object (i.e., the force N) is changed. The force of friction (F) required to move the block changes proportionally.

Note that the coefficient of friction is independent of the area of contact, which means that pushing a brick across the floor requires the same amount of force, whether it is on end, on edge, or flat. The coefficient of friction is useful in determining the force necessary to do a certain amount of work. Temperature changes only slightly affect friction. Friction causes wear. To overcome this wear problem (to reduce friction), lubricants are used.

Pressure

Pressure, in mechanics, is defined as the force per unit of area or

$$\text{Pressure} = \frac{\text{Total force}}{\text{Area}} \quad (6.5)$$

Pressure is usually expressed in terms of force per unit of area, as in pounds per square inch when dealing with gases, or in pounds per square foot when dealing with weight on a given floor area. The pressure exerted on a surface is the perpendicular force per unit area that acts upon it. Gauge pressure is the difference between total pressure and atmospheric pressure.

Specific Gravity

Specific gravity is the ratio of the weight of a substance to the weight of an equal volume of water, a number that can be determined by dividing the weight of a body by the weight of an equal volume of water. Since the weight of anybody per unit of volume is called density, then:

$$\text{Specific gravity} = \frac{\text{Density of body}}{\text{Density of water}} \quad (6.6)$$

Example: The density of a particular material is 0.24 pounds per cubic inch, and the density of water per cubic inch is 0.0361 pounds per cubic inch,

$$\text{Then: Specific gravity of the material} = \frac{0.24}{0.0361} = 6.6$$

The material is 6.6 times as heavy as water. This ratio does not change, regardless of the units that may be used, which is an advantage for two reasons: (1) it will always be the same for the same material, and (2) it is less confusing than the term density, which changes as the units change.

Force, Mass, and Acceleration

According to Newton's second law of motion:

The acceleration produced by an unbalanced force acting on a mass is directly proportional to the unbalanced force, in the direction of the unbalanced force, and inversely proportional to the total mass being accelerated by the unbalanced force.

If we express Newton's second law mathematically, it is greatly simplified and becomes

$$F = ma \quad (6.7)$$

This equation is extremely important in physics and engineering. It simply relates acceleration to force and mass. Acceleration is defined as the change in velocity divided by the time taken. This definition tells us how to measure acceleration. $F = ma$ tells us what causes the acceleration—an unbalanced force. Mass may be defined as the quotient obtained by dividing the weight of a body by the acceleration caused by gravity. Since gravity is always present, we can, for practical purposes, think of mass in terms of weight, making the necessary allowance for gravitational acceleration.

Fundamental Engineering Concepts

Centrifugal and Centripetal Forces

Two terms the safety engineer should be familiar with are centrifugal and centripetal forces. Centrifugal force is a concept based on an apparent (but not real) force. It may be regarded as a force that acts radially outwards from a spinning or orbiting object (a ball tied to a string whirling about), thus balancing a real force, the centripetal force (the force that acts radially inwards).

This concept is important in safety engineering because many of the machines encountered on the job may involve rapidly revolving wheels or flywheels. If the wheel is revolving fast enough, and if the molecular structure of the wheel is not strong enough to overcome the centrifugal force, it may fracture. Pieces (shrapnel) of the wheel would fly off tangent to the arc described by the wheel. The safety implications are obvious. Any worker using such a device, or near it may be severely injured when the rotating member ruptures. This is what happens when a grinding wheel on a pedestal grinder "bursts." Rim speed determines the centrifugal force, and rim speed involves both the speed (rpm) of the wheel and the diameter of the wheel.

Stress and Strain

In materials, stress is a measure of the deforming force applied to a body. Strain (which is often erroneously used as a synonym for stress) is really the resulting change in its shape (deformation). For perfectly elastic material, stress is proportional to stain. This relationship is explained by Hooke's Law, which states that the deformation of a body is proportional to the magnitude of the deforming force, provided that the body's elastic limit is not exceeded. If the elastic limit is not reached, the body will return to its original size once the force is removed. For example, if a spring is stretched by 2 cm by a weight of 1 N, it will be stretched by 4 cm by a weight of 2 N, and so on; however, once the load exceeds the elastic limit for the spring, Hooke's law will no longer be obeyed, and each successive increase in weight will result in a greater extension until the spring finally breaks. Stress forces are categorized in three ways:

1. Tension (or tensile stress), in which equal and opposite forces that act away from each other are applied to a body; tends to elongate a body.
2. Compression stress, in which equal and opposite forces that act toward each other are applied to a body; tends to shorten it.
3. Shear stress, in which equal and opposite forces that do not act along the same line of action or plane are applied to a body; tends to change its shape without changing its volume.

PRINCIPLES OF MECHANICS

In this section, we discuss mechanical principles: statics, dynamics, soils, beams, floors, columns, electric circuits, and machines. The field safety engineer should have at least some familiarity with all of these. Note: the safety engineer whose function is to verify design specifications (with safety in mind) should have more than just a familiarity with these topics.

Statics

Statics is the branch of mechanics concerned with the behavior of bodies at rest and forces in equilibrium and distinguished from dynamics (concerned with the behavior of bodies in motion). Forces acting on statics do not create motion. Static applications are bolts, welds, rivets, load-carrying components (ropes and chains), and other structural elements. A common example of a static situation is in a bolt and plate assembly. The bolt is loaded in tension and holds two elements together.

Welds

Welding is a method of joining metals to achieve a more efficient use of the materials and faster fabrication and erection. Welding also permits the designer to develop and use new and aesthetically appealing designs and saves weight because connecting plates are not needed and allowances need not be made for reduced load-carrying ability due to holes for rivets, bolts, and so on (Heisler, 1984). Simply put, the welding process joins two pieces of metal together by establishing a metallurgical bond between them. Most processes use a fusion technique; the two most widely used are arc welding and gas welding.

In the welding process, where two pieces of metal are joined together, the mechanical properties of metals are important, of course. The mechanical properties of metals primarily measure how materials behave under applied loads—in other words, how strong a metal is when it comes in contact with one or more forces. The important point is that if you know and use the strength properties of a metal, you can build a structure that is both safe and sound.

In welding, the welder must know the strength of his weld as compared with the base metal to produce a weldment that is strong enough to do the job. Thus, the welder is just as concerned with the mechanical properties of metals as is the engineer.

Dynamics

Dynamics (kinetics in mechanics) is the mathematical and physical study of the behavior of bodies under the action of forces that produce changes of motion in them. In dynamics, certain properties are important: displacement, velocity, acceleration, momentum, kinetic energy, potential energy, work, and power. Environmental engineers work with these properties to determine, for example, if rotating equipment will fly apart and cause injury to workers, or to determine the distance needed to stop a vehicle in motion.

Hydraulics and Pneumatics—Fluid Mechanics

Hydraulics (liquids only) and pneumatics (gases only) make up the study of fluid mechanics, which in turn is the study of forces acting on fluids (liquids and gases are considered fluids). Environmental engineers encounter many fluid mechanics problems and applications of fluid mechanics. In particular, engineers working in chemical industries, or in or around processes using or producing chemicals need an understanding of flowing liquids or gases to be able to predict and control their behavior.

Soil Mechanics

When dealing with nature's nature building material, soil, the engineer should keep the following statement in mind:

> Observe always that everything is the result of change and get used to thinking that there is nothing Nature loves so well as to change existing forms and to make new ones like them.
>
> **Marcus Aurelius,** *Meditations*

Soil for Construction

By the time a student reaches the 3rd or 4th year of elementary school, he or she is familiar with the Leaning Tower of Pisa. Many are also familiar with Galileo's experiments with gravity and the speed of falling objects dropped from the top of the tower. This twelfth century bell tower has been a curiosity for literally millions of people from the time it was first built to the present. Eight stories high and 180 ft tall, with a base diameter of 52 ft, the tower began to lean by the time the third story was completed, and leans about 1/25 in more each year.

How many people know why the tower is leaning in the first place—and who would be more than ordinarily curious about why the Leaning Tower leans? If you are a soil scientist or an engineer, this question has real significance—the question requires an answer. And, in fact, the Leaning Tower of Pisa should never have acquired the distinction of being a "leaning" tower in the first place.

So, why does the Leaning Tower of Pisa lean? —Because it rests on a non-uniform consolidation of clay beneath the structure. This ongoing process may eventually lead to failure of the tower.

Fundamental Engineering Concepts

As you might have guessed, the mechanics of why the Leaning Tower of Pisa leans is what this section is all about. More specifically, it is about the mechanics and physics of the soil—important factors in making the determination as to whether a particular building site is viable for building. Simply put, these two factors are essential in answering the question "Will the soils present support buildings?"

Soil Characteristics

When we refer to the characteristics of soils, we are referring to its mechanical characteristics, physical factors important to safety engineers. Safety engineers focus on the characteristics of the soil related to its suitability as a construction material, and as a substance to be excavated (can the soil in question be safely excavated without caving in on workers?). Simply put, the environmental engineer must understand the response of a particular volume of soil to internal and external mechanical forces. Obviously, to be able to determine the soil's ability to withstand the load applied by structures of various types, and its ability to remain stable when excavated is important.

From a purely engineering point of view, soil is any surficial material of the Earth that is unconsolidated enough to be excavated with tools (from bulldozers to shovels). The engineer takes into consideration both the advantages and disadvantages of using soil for engineering purposes. The obvious key advantage of using soil for engineering is that there is (in many places) no shortage of it—it may already be on the construction site, thus avoiding the expense of hauling it from afar. Another advantage of using soil for construction is its ease of manipulation; it may be easily shaped into almost any desired form. Soil also allows for the passage of moisture, or as needed, it can be made impermeable.

We said that the engineer looks at both the advantages and disadvantages of using soil for construction projects. The most obvious disadvantage of using soil is its variability from place to place and from time to time. Soil is not a uniform material for which reliable data related to strength can be compiled or computed. Cycles of wetting and drying and freezing and thawing affect the engineering properties of soil. A particular soil may be suitable for one purpose, but not for another. Stamford clay in Texas, for example, is rated as "very good" for sealing of farm ponds—and "very poor" for use as base for roads and buildings (Buol et al., 1980).

How does the engineer know if a particular soil is suitable for use as a base for roads or buildings?

The engineer determines the suitability of a particular soil (for whatever purpose) by studying soil survey maps and reports. The engineer also checks with soil scientists and other engineers familiar with the region and the soil types of that region. Any good engineer will also want to conduct field sampling—to ensure that the soil product he or she will be working with possesses the soil characteristics required for its intended purpose.

What are the soil characteristics (the kinds of information) that the engineer is interested in? Important characteristics of soils for engineering purposes include:

- Soil texture
- Kinds of clay present
- Depth to bedrock
- Soil density
- Erodibility
- Corrosivity
- Surface geology
- Plasticity
- Content of organic matter
- Salinity
- Depth to seasonal water table

The engineer will also want to know the soil's density, space and volume or weight-volume relationships, stress-strain, slope stability, and compaction. Because these concepts are of paramount importance to the engineer (and others), these concepts are presented in the following sections.

Soil Weight-Volume or Space and Volume Relationships

All-natural soil consists of at least three primary components or phases, solid (mineral) particles, water and/or air (within void spaces between the solid particles). The physical relationships (for soils in particular) between these phases must be examined.

The proportions of the components vary dramatically between and within various soil types. Why? Because water that is not chemically attached is a void filler, and the relationship between it and void areas are dependent on how much water (moisture) is available.

The volume of the soil mass is the sum of the volumes of the three components, or

$$V_T = V_a + V_w + V_s \tag{6.8}$$

The volume of the voids is the sum of V_a and V_w. However, regarding weight (because the weighing of air in the soil voids would be done within the Earth's atmosphere as with other weighing's), the weight of the solids is determined on a different basis. We consider the weight of air in the soil to be zero and the total weight is expressed as the sum of the weights of the soil solids and the water:

$$W_t = W_s + W_w \tag{6.9}$$

The relationship between weight and volume can be expressed as

$$W_m = V_m G_{m-w} \tag{6.10}$$

where
  W_m = Weight of the material (solid, liquid, or gas).
  V_m = Volume of the material.
  G_m = Specific gravity of the material (dimensionless).
  w = Unit weight of water.

With the relationships described above, a few useful problems can be solved. When an engineer determines that within a given soil, the proportions of the three major components need to be mechanically adjusted this can be accomplished by reorienting the mineral grains by compaction or tilling. The engineer, probably for a specific purpose, may want to blend soil types to alter the proportions, such as increasing or decreasing the percentage of void space.

How do we go about doing this? Relationships between volumes of soil and voids are described by the void ratio (e) and porosity (η). To accomplish this, we must first determine the void ratio (the ratio of the void volume to the volume of solids):

$$e = \frac{V_v}{V_s} \tag{6.11}$$

We must also determine the ratio of the volume of void spaces to the total volume. This can be accomplished by determining the Porosity (η) of the soil, which is the ratio of void volume to total volume. Porosity is usually expressed as a percentage.

$$\eta = \frac{V_v}{V_t} \times 100\% \tag{6.12}$$

Fundamental Engineering Concepts

Two additional relationships, known as the moisture content, (w) and degree of saturation, (S), relate the water content of the soil and the volume of the water in the void space to the total void volume:

$$w = \frac{W_w}{W_s} \times 100\% \tag{6.13}$$

And

$$S = \frac{V_w}{V_v} \times 100\% \tag{6.14}$$

Soil Particle Characteristics

The size and shape of particles in the soil, as well as density and other characteristics, relate to sheer strength, compressibility, and other aspects of soil behavior. Engineers use these index properties to form engineering classifications of soil. Simple classification tests are used to measure index properties (see Table 6.2) in the lab or the field.

From Table 6.2 we see that an important division of soils (from the engineering point of view) is the separation of the cohesive (fine-grained) from the incohesive (coarse-grained) soils. Let's take a closer look at these two important terms.

Cohesion indicates the tendency of soil particles to stick together. Cohesive soils contain silt and clay. The clay and water content makes these soils cohesive, through the attractive forces between individual clay and water particles. The influence of the clay particles makes the index properties of cohesive soils somewhat more complicated than the index properties of cohesionless soils. The resistance of soil at various moisture contents to mechanical stresses or manipulations is called the soil's consistency—the arrangement of clay particles and is the most important characteristic of cohesive soils.

Another important index property of cohesive soils is its *sensitivity*. Simply defined, sensitivity is the ratio of unconfined compressive strength in the undisturbed state to strength in the remolded state (see Equation 6.15). Soils with high sensitivity are highly unstable.

$$S_t = \frac{\text{Strength in undisturbed condition}}{\text{Strength in remolded condition}} \tag{6.15}$$

TABLE 6.2
Index Property of Soils

Soil Type	Index Property
Cohesive (fine-grained)	Water content
	Sensitivity
	Type and amount of clay
	Consistency
	Atterberg limits
Incohesive (coarse-grained)	Relative density
	In-place density
	Particle-size distribution
	Clay content
	Shape of particles

Source: Adaptation from Kehew (1995).

Soil water content (described earlier) is an important factor that influences the behavior of the soil. The water content values of soil are known as the Atterburg limits, a collective designation of so-called limits of consistency of fine-grained soils, which are determined with simple laboratory tests. They are usually presented as the liquid limit (LL) and plastic limit (PL).

Plasticity is exhibited over a range of moisture contents referred to as plasticity limits—and the plasticity index (PI). The plastic limit is the lower water level at which soil begins to be malleable in a semi-solid state, but molded pieces crumble easily when a little pressure is applied. At the point when the volume of the soil becomes nearly constant, with further decreases in water content, the soil reaches the shrinkage state.

The upper plasticity limit (or liquid limit) is the water content at which the soil-water mixture changes from a liquid to a semi-fluid (or plastic) state and tends to flow when jolted. An engineer charged with building a highway or building would not want to choose a soil for the foundation that tends to flow when wet.

The difference between the liquid limit and the plastic limit is the range of water content over which the soil is plastic and is called the plasticity index. Soils with the highest plasticity indices are unstable in bearing loads.

Several systems for classifying the stability of soil materials have been devised, but the best known (and probably the most useful) system is called the Unified System of Classification. This classification gives each soil type (14 classes) a two-letter designation, which are primarily defined based on particle-size distribution, liquid limit, and plasticity index.

Cohesionless coarse-grained soils behave much differently than cohesive soils and are based on (from index properties) the size and distribution of particles in the soil. Other index properties (particle shape, in-place density, and relative density, for example) are important in describing cohesionless soils because they relate to how closely particles can be packed together.

Soil Stress and Strain

If you are familiar with water pressure and its effect as you go deeper into the water (as when diving deep into a lake), the same concept applies to soil and pressure should not surprise you. Like water, the pressure within the soil increases as the depth increases. A soil, for example, that has a unit weight of 75 pounds per cubic feet exerts a pressure of 75 psi at one-foot depth and 225 psi at three feet, etc. As you might expect, as the pressure on a soil unit increases, the soil particles reorient themselves structurally to support the cumulative load. This consideration is important, because the elasticity of the soil sample retrieved from beneath the load may not be truly representative, once it is delivered to the surface. In sampling, the importance of taking representative samples can't be overstated.

The response of a soil to pressure (stress) is similar to what occurs when a load is applied to a solid object; the stress is transmitted throughout the material. The load subjects the material to pressure, which equals the amount of load, divided by the surface area of the external face of the object over which it is applied. The response to this pressure or stress is called displacement or strain. Stress (like pressure), at any point within the object, can be defined as force per unit area.

Soil Compressibility

When a vertical load such as a building or material stockpile is placed above a soil layer, some settlement can be expected. Settlement is the vertical subsidence of the building (or load) as the soil is compressed.

Compressibility refers to the tendency of soil to decrease in volume under load. This compressibility is most significant in clay soils because of the inherent high porosity. Although the mechanics of compressibility and settlement are quite complex and beyond the scope of this text, you should know something about the actual evaluation process for these properties, which is accomplished in the consolidation test. This test subjects a soil sample to an increasing load. The change in thickness is measured after the application of each load increment.

Soil Compaction

The goal of compaction is to reduce void ratio and thus increase the soil density, which, in turn, increases the shear strength. This is accomplished by working the soil to reorient the soil grains into a more compact state. If water content is within a limited range (sufficient enough to lubricate particle movement), efficient compaction can be obtained.

The most effective compaction occurs when the soil placement layer (commonly called lift) is approximately 8 inches. At this depth, the most energy is transmitted throughout the lift. Note that more energy must be dispersed, and the effort required to accomplish maximum density is greatly increased when the lift is >10 in in thickness.

For cohesive soils, compaction is best accomplished by blending or kneading the soil using sheepsfoot rollers and pneumatic tire rollers. These devices work to turn the soil into a denser state.

To check the effectiveness of the compactive effort, the in-place dry density of the soil (weight of solids per unit volume) is tested, by comparing the dry density of field-compacted soil to a standard prepared in an environmental laboratory. Such a test allows a percent compaction comparison to be made.

Soil Failure

The construction and safety engineer must be concerned with soil structural implications involved with natural processes (such as frost heave, which could damage a septic system) and changes applied to soils during remediation efforts (e.g. when excavating to mitigate a hazardous materials spill in soil). Soil failure occurs whenever it cannot support a load. Failure of an overloaded foundation, collapse of the sides of an excavation, or slope failure on the sides of a dike, hill, or similar feature is termed structural failure.

The type of soil structural failure that probably occurs more frequently than any other is slope failure (commonly known in practice as cave-in). In a review of the Bureau of Labor Statistics annual report of on-the-job mishaps, including from 80 to 100 fatalities every year—we see that cave-ins occurring in excavations accomplished in construction are more frequent occurrences than you might think, considering the obvious dangers inherent in excavation.

What is excavation? How deep does an excavation have to be to be considered dangerous? Two good questions—and the answers could save your life, or help you protect others when you become an environmental engineer.

An excavation is any man-made cut, cavity, trench, or depression in the Earth's surface formed by earth removal. This can include excavations for anything from a remediation dig to sewer line installation.

No excavation activity should be accomplished without keeping personnel safety in mind. Any time soil is excavated, care and caution is advised. As a rule of thumb (and as law under 29 CFR 1926.650-652), the Occupational Safety and Health Administration (OSHA) requires trench protection in any excavation 5 ft or more in depth.

Before digging begins, proper precautions must be taken. The responsible party in charge (the competent person, according to OSHA) must:

- Contact utility companies to ensure underground installations are identified and located.
- Ensure underground installations are protected, supported, or removed as necessary to safeguard workers.
- Remove or secure any surface obstacles (e.g., trees, rocks, and sidewalks) that may create a hazard for workers.
- Classify the type of soil and rock deposits at the site as either stable rock, type A, type B, or type C soil. One visual and at least one manual analysis must make the soil classification.

Let's take a closer look at the requirement to classify the type of soil to be excavated. Before excavation can be accomplished, the soil type must be determined. The soil must be classified as: stable rock, type A, type B, or type C soil. Remember, commonly you will find a combination of soil types at an excavation site. Soil classification (used in this manner) is used to determine the need for a protective system.

Exactly what do the soil classifications of stable rock, type A, type B, and type C soil mean?

- **Stable rock**: a natural solid mineral material that can be excavated with vertical sides. Stable rock will remain intact while exposed, but keep in mind that even though solid rock is generally stable, it may become very unstable when excavated (in practice you never work in this kind of rock).
- **Type A soil**: the most stable soil, and includes clay, silty clay, sandy clay, clay loam, and sometimes silty clay loam and sandy clay loam.
- **Type B soil**: moderately stable, and includes silt, silt loam, sandy loam and sometimes silty clay loam and sand clay loam.
- **Type C soil**: the least stable, and includes granular soils like gravel, sand, loamy sand, submerged soil, soil from which water is freely seeping, and submerged rock that is not stable. How is soil tested?

To test and classify soil for excavation, the environmental specialist conducts both visual and manual tests. In visual soil testing, you should look at soil particle size and type. You'll see a mixture of soils, of course. You should check to see if the soil clumps when dug—if so it could be clay or silt. Type B or C soil can sometimes be identified by the presence of cracks in walls and spalling (breaks up in chips or fragments). If you notice layered systems with adjacent hazardous areas—buildings, roads, and vibrating machinery—you may require a professional engineer for classification. Standing water or water seeping through trench walls automatically classifies the soil as type C.

Manual soil testing is required before a protective system (e.g., shoring, or shoring box) is selected. A sample taken from soil dug out into a spoil pile should be tested as soon as possible, to preserve its natural moisture. Soil can be tested either on-site or off-site. Manual soil tests include a sedimentation test, wet shaking test, thread test, and ribbon test.

A sedimentation test determines how much silt and clay are in sandy soil. Saturated sandy soil is placed in a straight-side jar with about 5 in of water. After the sample is thoroughly mixed (by shaking it) and allowed to settle, the percentage of sand is visible. A sample containing 80% sand, for example, will be classified as Type C.

The wet shaking test is another way to determine the amount of sand versus clay and silt in a soil sample. This test is accomplished by shaking a saturated sample by hand to gauge soil permeability based on the following facts: (1) shaken clay resists water movement through it; and (2) water flows freely through sand and less freely through silt.

The thread test is used to determine cohesion (remember, cohesion relates to stability—how well the grains hold together). After a representative soil sample is taken, it is rolled between the palms of the hands to about 1/8″ diameter and several inches in length (any child who has played in dirt has accomplished this at one time or another—nobody said soil science had to be boring). The rolled piece is placed on a flat surface, then picked up. If a sample holds together for 2 in, it is considered cohesive.

The ribbon test is used as a backup for the Thread Test. It also determines cohesion. A representative soil sample is rolled out (using the palms of your hands) to 3/4″ in diameter, and several inches in length. The sample is then squeezed between thumb and forefinger into a flat unbroken ribbon 1/8″ to 1/4″ thick, which is allowed to fall freely over the fingers. If the ribbon does not break off before several inches are squeezed out, the soil is considered cohesive.

Once soil has been properly classified, the correct protective system can be chosen (if necessary). This choice is based on both soil classification and site restrictions. There are two main types of protective systems: (1) sloping or benching and (2) shoring or shielding.

Sloping or benching are excavation protective measures that cut the walls of an excavation back at an angle to its floor.

Fundamental Engineering Concepts

TABLE 6.3
Maximum Safe Side Slopes in Excavations

Soil Type	Side Slope (Vertical to Horizontal)	Side Slope (Degrees from Horizontal)
A	75:1	53°
B	1:1	45°
C	1.5:1	34°

Source: OSHA (1978)

The angle used for sloping or benching is a ratio based on soil classification and site restrictions; the flatter the angle the greater the protection for workers. Reasonably safe side slopes for each of these soil types are presented in Table 6.3.

Shoring and shielding are two protective measures that add support structure to an existing excavation (generally used in excavations with vertical sides but can be used with sloped or benched soil).

Shoring is a system designed to prevent cave-ins by supporting walls with vertical shores called uprights or sheeting. Wales are horizontal members along the sides of a shoring structure. Cross braces are supports placed horizontally between trench walls.

Shielding is a system that employs a trench box or trench shield. They can be pre-manufactured or job-build under the specification of a licensed engineer. Shields are usually portable steel structures placed in the trench by heavy equipment. For deep excavations, trench boxes can be stacked and attached to each other with stacking lugs.

Soil Physics

Soil is a dynamic, heterogeneous body that is non-isotropic (does not have the same properties in all directions). As you might expect, because of these properties, various physical processes are active in soil at all times. This important point is made clear by Windgardner (1996) in the following: "all of the factors acting on a particular soil, in an established environment, at a specified time, are working from some state of imbalance to achieve a balance" (p. 63).

Most soil specialists (and many other people) have little difficulty in understanding why soils are very important to the existence of life on Earth. They know, for example, not only that soil is necessary (in a very direct sense) to sustain plant life (and thus other life forms that depend on plants), but they also know that soil functions to store and conduct water, serves a critical purpose in soil engineering involved with construction, and acts as a sink and purifying medium for waste disposal systems.

The soil practitioner must be well versed in the physical properties of soil. Specifically, he or she must understand those physical processes that are active in soil. These factors include physical interactions related to soil water, soil grains, organic matter, soil gases, and soil temperature. To gain this knowledge, the safety engineer must have training in basic geology, soil science, and/or engineering construction.

REFERENCES

ASSE. (1988). Dictionary of terms used in Safety Profession. San Antonia, TX: American Society of Safety Engineers.
Buol, S.W., Hole, F.D., & McCracken, R.J. (1980). Soil Genesis and Collection. Ames, IA: Iowa State Press.
Heisler, S.I. (1984). *The Wiley Engineer's Desk Reference*. New York: John Wiley & Sons.
Kehew, A.E. (1995). *Geology for Engineers, and Environmental Scientists*. Englewood Cliffs, NJ: Prentice-Hall.
OSHA (1978). Excavation Standard, 29 CFR 1926.650-652.
Spellman, F.R. (1996). Stream Ecology. Lancaster, PA: Technomic Publishing Co.
Windgardner, D.L. (1996). An Introduction to Soils for Environmental Professionals. Boco Raton, FL: CRC Press.
Wordsworth, W. (1798). Tintern Abbey. New York: Everyman's Library.

7 Air Pollution

The difference between science and the fuzzy subjects is that science requires reasoning, while those other subjects merely require scholarship.

R.A. Heinlein (1973, p. 348)

THE 411 ON AIR

Engineers always seem to have a definition for just about anything and everything (most of which cannot be understood by many of us—and maybe that is their intent). An engineer might refer to air as a fluid (because it is; like water, the air is fluid—you can pour it). Engineers are primarily interested in the air as a fluid because he or she deals with fluid mechanics, the study of the behavior of fluids (including air) at rest or in motion. Fluids may be either gases or liquids. You are probably familiar with the physical difference between gases and liquids, as exhibited by air and water, but for the study of fluid mechanics (and the purposes of this text), it is convenient to classify fluids to their compressibility.

- Gases are very readily compressible (you've heard of compressed air).
- Liquids are only slightly compressible (it is unlikely you've heard much about compressed water).

What is air? Air is a mixture of gases that constitutes the Earth's atmosphere. What is the Earth's atmosphere? The atmosphere is that thin shell, veil, an envelope of gases that surrounds Earth like the skin of an apple—thin, very thin—but very, very vital. The approximate composition of dry air is, by volume at sea level, nitrogen 78%, oxygen 21% (necessary for life as we know it) argon 0.93%, and carbon dioxide 0.03%, together with very small amounts of numerous other constituents (see Table 7.1). Because of constant mixing by the winds and other weather factors, the percentages of each gas in the atmosphere are normally constant to 70,000 ft. However, it is important to point out that the water vapor content is highly variable and depends on atmospheric conditions. Air is said to be pure when none of the minor constituents is present in sufficient concentration to be injurious to the health of human beings or animals, to damage vegetation, or to cause loss of amenity (e.g., through the presence of dirt, dust, or odors or by diminution of sunshine).

Where does air come from? Genesis 1:2 states that God separated the water environment into the atmosphere and surface waters on the second day of creation. Many scientists state that 4.6 billion years ago, a cloud of dust and gases forged the Earth and also created a dense molten core enveloped in cosmic gases. This was the *proto-atmosphere* or *proto-air*, composed mainly of carbon dioxide, hydrogen, ammonia, and carbon monoxide, but did not last long before it was stripped away by a tremendous outburst of charged particles from the Sun. As the outer crust of Earth began to solidify a new atmosphere began to form from the gases outpouring from gigantic hot springs and volcanoes. This created an atmosphere of air composed of carbon dioxide, nitrogen oxides, hydrogen, sulfur dioxide, and water vapor. As the Earth cooled, water vapor condensed into highly acidic rainfall, which collected to form oceans and lakes.

For much of Earth's early existence (the first half), only trace amounts of free oxygen were present. But then green plants evolved in the oceans, and they began to add oxygen to the atmosphere as a waste gas and later oxygen increased to about 1% of the atmosphere and with time to its present 21%.

TABLE 7.1
Composition of Air/Earth's Atmosphere

Gas	Chemical Symbol	Volume (%)
Nitrogen	N_2	78.08
Oxygen	O_2	20.94
Carbon dioxide	CO_2	0.03
Argon	Ar	0.093
Neon	Ne	0.0018
Helium	He	0.0005
Krypton	Kr	Trace
Xenon	Xe	Trace
Ozone	O_3	0.00006
Hydrogen	H_2	0.00005

How do we know for sure about the evolution of air on Earth? Are we just guessing, using "voodoo" science? There is no guessing or voodoo involved with the historical geological record. Consider, for example, geological formations that are dated to 2 billion years ago. In these early sediments, there is a clear and extensive band of red sediment ("red bed" sediments)—sands colored with oxidized (ferric) iron. Previously, ferrous formations had been laid down showing no oxidation. But there is more evidence. We can look at the timeframe of 4.5 billion years ago when carbon dioxide in the atmosphere was beginning to be lost in sediments. The vast amount of carbon deposited in limestone, oil, and coal indicates that carbon dioxide concentrations must once have been many times greater than they are today, which stand at only 0.03%. The first carbonated deposits appeared about 1.7 billion years ago, and the first sulfate deposits about 1 billion years ago. The decrease in carbon dioxide was balanced by an increase in the nitrogen content of the air. The forms of *respiration* practiced advanced from fermentation 4 billion years ago to anaerobic *photosynthesis* 3 billion years ago to aerobic photosynthesis 1.5 billion years ago. The aerobic respiration that is so familiar today only began to appear about 500 million years ago.

Fast-forward to the present. The atmosphere itself continues to evolve, but human activities—with their highly polluting effects—have now overtaken nature in determining the changes.

If we cannot live without air—if the air is so precious—so necessary for sustaining life, then two questions arise: (1) Why do we ignore air? (2) Why do we abuse it (pollute it)? We ignore air (like we do water) because it is so common, so accessible (normally), and so unexceptional.

Is air pollution really that big of a deal? Isn't pollution relative? That is, isn't pollution dependent on your point of view—a judgment call? Well, if you could ask the victims of the incidents listed in Table 7.2 below [from U.S. Senate Staff Report, *Air Quality Criteria* (1968)], the answer, simply stated, would be yes; it is.

Beyond the fact that air is one of our essential resources, sustaining life as it stimulates and pleases the senses—though invisible to the human eye, it makes possible such sights as beautiful and dazzling rainbows, heart-pinching sunsets and sun rises, the Northern Lights, and on occasion, a clear view of that high alpine meadow sprinkled throughout with the colors of spring. But the air is more than this. Air is capable of many other wondrous things. For example, have you ever felt the light touch of a cool soothing breeze against your skin? But the air is capable of more, much more: it carries thousands of scents—both pungent and subtle: salty ocean breezes, approaching rain, fragrances from blooming flowers, and others.

It is the "others" that concern us here: the sulfurous gases from industrial processes—that typical rotten egg odor; the stink of garbage, refuse, trash—all part of man's throwaways; the toxic poison remnants from pesticides, herbicides, and all the other "-cides." We are surrounded by it, but we seldom think about it until it displeases us (remember, Man can put up with just about anything until

TABLE 7.2
Mortality Occurring During Air Pollution Events

Location	Year	Deaths Reported as a Result of Pollution Event
Belgium	1930	63
Pennsylvania	1948	17
London	1948	700–800
London	1952	4,000
London	1956	1,000
London	1957	700–800
London	1959	200–250
London	1962	700
London	1963	700
New York	1963	200–400

it displeases him). It is pollution, those discarded, sickening leftovers of the heavy hand of man that causes the problem.

DEFINITION OF KEY TERMS

To work at even the edge of air science and the science disciplines closely related to air science, it is necessary for the reader to acquire a familiarity with the vocabulary used in air pollution control activities.

Definitions

> **Absolute pressure**: the total pressure in a system, including both the pressure of a substance and the pressure of the atmosphere (about 14.7 psi, at sea level).
> **Acid**: any substance that releases hydrogen ions (H^+) when it is mixed with water.
> **Acid precipitation**: rain, snow, or fog that contains higher than normal levels of sulfuric or nitric acid, which may damage forests, aquatic ecosystems, and cultural landmarks.
> **Acid surge**: a period of short, intense acid deposition in lakes and streams as a result of the release (by rainfall or spring snowmelt) of acids stored in soil or snow.
> **Acidic solution**: a solution that contains significant numbers of (H^+) ions.
> **Airborne toxins**: hazardous chemical pollutants that have been released into the atmosphere and are carried by air currents.
> **Albedo**: reflectivity, or the fraction of incident light that is reflected by a surface.
> **Arithmetic mean**: a measurement of average value, calculated by summing all terms and dividing by the number of terms.
> **Arithmetic scale**: a scale is a series of intervals (marks or lines), usually made along the side or bottom of a graph, that represents the range of values of the data. When the marks or lines are equally spaced, it is called an arithmetic scale.
> **Atmosphere**: a 500-km-thick layer of colorless, odorless gases known as the air that surrounds the Earth and is composed of nitrogen, oxygen, argon, carbon dioxide, and other gases in trace amounts.
> **Atom**: the smallest particle of an element that still retains the characteristics of that element.
> **Atomic number**: the number of protons in the nucleus of an atom.
> **Atomic weight**: the sum number of protons and the number of neutrons in the nucleus of an atom.
> **Base**: any substance that releases hydroxyl ions (OH^-) when it dissociates in water.

Chemical bond: the force that holds atoms together within molecules. A chemical bond is formed when a chemical reaction takes place. Two types of chemical bonds are ionic bonds and covalent bonds.

Chemical reaction: a process that occurs when atoms of certain elements are brought together and combined to form molecules, or when molecules are broken down into individual atoms.

Climate: the long-term weather pattern of a particular region.

Covalent bond: a type of chemical bond in which electrons are shared.

Density: the weight of a substance per unit of its volume; e.g., pounds per cubic foot.

Dew point: the temperature at which a sample of air becomes saturated, i.e., has a relative humidity of 100%.

Element: any of more than 100 fundamental substances that consist of atoms of only one kind and that constitute all matter.

Emission standards: the maximum amount of a specific pollutant permitted to be legally discharged from a particular source in a given environment.

Emissivity: the relative power of a surface to reradiate solar radiation back into space in the form of heat, or long-wave infrared radiation.

Energy: the ability to do work, to move matter from place to place, or to change matter from one form to another.

First law of thermodynamics: natural law that dictates that during physical or chemical change energy is neither created not destroyed, but it may be changed in form and moved from place to place.

Global warming: the increase in global temperature predicted to arise from increased levels of carbon dioxide, methane, and other greenhouse gases in the atmosphere.

Greenhouse effect: the prevention of the reradiation of heat waves to space by carbon dioxide, methane, and other gases in the atmosphere. The greenhouse effect makes possible the conditions that enable life to exist on Earth.

Ion: an atom or radical in solution carrying an integral electrical charge either positive (cation) or negative (anion).

Insolation: the solar radiation received by the Earth and its atmosphere — incoming solar radiation.

Lapse rate: the rate of temperature change with altitude. In the troposphere, the normal lapse rate is $-3.5°F$ per 1000 ft.

Matter: anything that exists in time, occupies space, and has mass.

Mesosphere: a region of the atmosphere base on temperature between ~35 and 60 miles in altitude.

Meteorology: the study of atmospheric phenomena.

Mixture: two or more elements, compounds, or both, mixed together with no chemical reaction occurring.

Ozone: the compound O_3. It is found naturally in the atmosphere in the ozonosphere and is also a constituent of photochemical smog.

pH: a means of expressing hydrogen ion concentration in terms of the powers of 10; measurement of how acidic or basic a substance is. The pH scale runs from 0 (most acidic) to 14 (most basic). The center of the range (7) indicates the substance is neutral.

Photochemical smog: an atmospheric haze that occurs above industrial sites and urban areas resulting from reactions, which take place in the presence of sunlight, between pollutants produced in high temperature and pressurized combustion process (such as the combustion of fuel in a motor vehicle). The primary component of smog is ozone.

Photosynthesis: the process of using the Sun's light energy by chlorophyll-containing plants to convert carbon dioxide (CO_2) and water (H_2O) into complex chemical bonds forming simple carbohydrates such as glucose and fructose.

Pollutant: a contaminant at a concentration high enough to endanger the environment.

Pressure: the force pushing on a unit area. Normally, in air applications, measured in atmospheres, Pascal (Pa) or pounds per square inch (psi).

Primary pollutants: pollutants that are emitted directly into the atmosphere where they exert an adverse influence on human health of the environment. The six primary pollutants are carbon dioxide, carbon monoxide, sulfur oxides, nitrogen oxides, hydrocarbons, and particulates. All but carbon dioxide are regulated in the United States.

Raleigh scattering: the preferential scattering of light by air molecules and particles that accounts for the blueness of the sky. The scattering is proportional to $1/\lambda^4$.

Radon: a naturally occurring radioactive gas, arising from the decay of uranium 238, which may be harmful to human health in high concentrations.

Rain shadow effect: the phenomenon that occurs as a result of the movement of air masses over a mountain range. As an air mass rises to clear a mountain, the air cools and precipitation forms. Often, both the precipitation and the pollutant load carried by the air mass will be dropped on the windward side of the mountain. The air mass is then devoid of most of its moisture; consequently, the lee side of the mountain receives little or no precipitation and is said to lie in the rain shadow of the mountain range.

Relative humidity: the concentration of water vapor in the air. It is expressed as the percentage that its moisture content represents of the maximum amount that the air could contain at the same temperature and pressure. The higher the temperature the more water vapor the air can hold.

Secondary pollutants: pollutants formed from the interaction of primary pollutants with other primary pollutants or with atmospheric compounds such as water vapor.

Second law of thermodynamics: natural law that dictates that with each change in form some energy is degraded to a less useful form and given off to the surroundings, usually as low-quality heat.

Solute: the substance dissolved in a solution.

Solution: a liquid containing a dissolved substance.

Specific gravity: the ratio of the density of a substance to a standard density. For gases, the density is compared with the density of air (=1).

Stratosphere: an atmospheric layer extending from 6 to 7 miles to 30 miles above the Earth's surface.

Stratospheric ozone depletion: the thinning of the ozone layer in the stratosphere; occurs when certain chemicals (such as chlorofluorocarbons) capable of destroying ozone accumulate in the upper atmosphere.

Thermosphere: an atmospheric layer that extends from 56 miles to outer space.

Troposphere: the atmospheric layer that extends from the Earth's surface to 6 to 7 miles above the surface.

Weather: the day-to-day pattern of precipitation, temperature, wind, barometric pressure, and humidity.

Wind: horizontal air motion.

COMPONENTS OF AIR: CHARACTERISTICS AND PROPERTIES

It was pointed out that air is a combination of component parts: gases (see Table 9.1) and other matter (suspended minute liquid or particulate matter). In this section, we discuss each of the major components.

Note: Much of the information pertaining to atmospheric gases that follow was adapted from the Compressed Gas Association's *Handbook of Compressed Gases* (1990) and *Environmental Science and Technology: Concepts and Applications* (2006).

Atmospheric Nitrogen

Nitrogen (N_2) makes up the major portion of the atmosphere (78.03% by volume, 75.5% by weight). It is a colorless, odorless, tasteless, nontoxic, and almost totally inert gas. Nitrogen is nonflammable, will not support combustion, and is not life supporting. Nitrogen is part of Earth's atmosphere primarily because, over time, it has simply accumulated in the atmosphere and remained in place and in balance. This nitrogen accumulation process has occurred because, chemically, nitrogen is not very reactive. When released by any process, it tends not to recombine with other elements and accumulates in the atmosphere. And this is a good thing because we need nitrogen. No, we don't need it for breathing, but we need it for other life-sustaining processes.

Although nitrogen in its gaseous form is of little use to us, after oxygen, carbon, and hydrogen, it is the most common element in living tissues. As a chief constituent of chlorophyll, amino acids, and nucleic acids—the "building blocks" of proteins (which are used as structural components in cells)—nitrogen is essential to life. Nitrogen is dissolved in and carried by the blood. Nitrogen does not appear to enter into any chemical combination as it is carried throughout the body. Each time we breathe, the same amount of nitrogen is exhaled as is inhaled. Animals cannot use nitrogen directly but only when it is obtained by eating plant or animal tissues; plants obtain the nitrogen they need when it is in the form of inorganic compounds, principally nitrate and ammonium.

Gaseous nitrogen is converted to a form usable by plants (nitrate ions) chiefly through the process of nitrogen fixation via the nitrogen cycle, shown in simplified form in Figure 7.1.

Via the *nitrogen cycle*, aerial nitrogen is converted into nitrates mainly by microorganisms, bacteria, and blue-green algae. Lightning also converts some aerial nitrogen gas into forms that return to the Earth as nitrate ions in rainfall and other types of precipitation. Ammonia plays a major role in the nitrogen cycle. Excretion by animals and anaerobic decomposition of dead organic matter by bacteria produce ammonia. Ammonia, in turn, is converted by nitrification bacteria into nitrites and then into nitrates. This process is known as nitrification. Nitrification bacteria are aerobic. Bacteria that convert ammonia into nitrites are known as nitrite bacteria (*Nitrosococcus* and *Nitrosomonas*). Although nitrite is toxic to many plants, it usually does not accumulate in the soil. Instead, other bacteria (such as *Nitrobacter*) oxidize the nitrite to form nitrate (NO_3^-), the most common biologically usable form of nitrogen.

Nitrogen reenters the atmosphere through the action of denitrifying bacteria, which are found in nutrient-rich habitats such as marshes and swamps. These bacteria break down nitrates into nitrogen gas and nitrous oxide (N_2O), which then reenter the atmosphere. Nitrogen also reenters the atmosphere from exposed nitrate deposits, and emissions from electric power plants, automobiles, and from volcanoes.

Nitrogen: Physical Properties

The physical properties of nitrogen are noted in Table 7.3.

Nitrogen: Uses

In addition to being the preeminent (in regard to volume) component of Earth's atmosphere and providing an essential ingredient in sustaining life, nitrogen gas has many commercial and technical applications. As a gas, it is used in heat treating primary metals; the production of semi-conductor electronic components, as a blanketing atmosphere; blanketing of oxygen-sensitive liquids and of volatile liquid chemicals; inhibition of aerobic bacteria growth; and the propulsion of liquids through canisters, cylinders, and pipelines.

Nitrogen Oxides

There are six oxides of nitrogen: nitrous oxide (N_2O), nitric oxide (NO), dinitrogen trioxide (N_2O_3), nitrogen dioxide (NO_2), dinitrogen tetroxide (N_2O_4), and dinitrogen pentoxide (N_2O_5).

Air Pollution

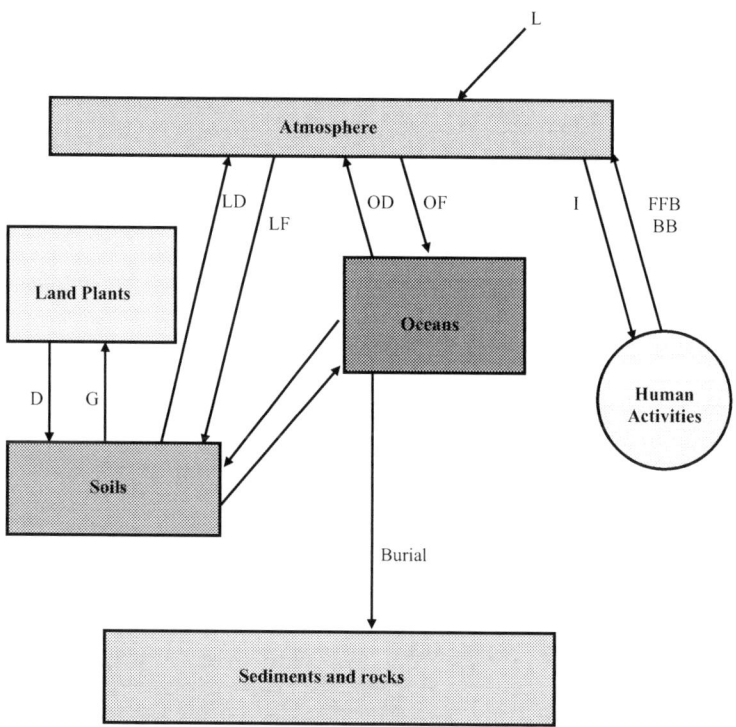

LF = Land Fixation
LD = Land Denitrification
OF = Oceanic Fixation
OD = Oceanic Denitrification
I = Industrial Fixation
FFB = Fossil Fuel Burning
BB = Biomass Burning
L = Lightning
D = Decay
G = Growth

O – L = Ocean-to-Land
L – O = Land-to-Ocean

FIGURE 7.1 Nitrogen cycle.

TABLE 7.3
Nitrogen: Physical Properties

Chemical Formula	N_2
Molecular weight	28.01
Density of gas @ 70°F	0.072 lb/ft³
Specific gravity of gas @ 70°F & 1 atm (air = 1)	0.967
Specific volume of gas @ 70°F & 1 atm	13.89 ft³
Boiling point @ 1 atm	−320.4°F
Melting point @ 1 atm	−345.8°F
Critical temperature	−232.4°F
Critical pressure	493 psia
Critical density	19.60 lb/ft³
Latent heat of vaporization @ boiling point	85.6 Btu/lb
Latent heat of fusion @ melting point	11.1 Btu/lb

TABLE 7.4
Oxygen: Physical Properties

Chemical Formula	O_2
Molecular weight	31.9988
Freezing point	−361.12°F
Boiling point	−297.33°F
Heat of fusion	5.95 Btu/lb
Heat of vaporization	91.70 Btu/lb
Density of gas @ boiling point	0.268 lb/ft^3
Density of gas @ room temperature	0.081 lb/ft^3
Vapor density (air=1)	1.105
Liquid-to-gas expansion ratio	875

Nitric oxide, nitrogen dioxide, and nitrogen tetroxide are fire gases. One or more of them is generated when certain nitrogenous organic compounds (polyurethane) burn. Nitric oxide is the product of incomplete combustion, whereas a mixture of nitrogen dioxide and nitrogen tetroxide is the product of complete combustion.

Nitrogen oxides are usually collectively symbolized by the formula NO_x. USEPA, under the Clean Air Act (CAA), regulates the amount of nitrogen oxides that commercial and industrial facilities may emit into the atmosphere. The primary and secondary standards are the same: The annual concentration of nitrogen dioxide may not exceed 100 µg/m^3 (0.05 ppm).

ATMOSPHERIC OXYGEN

Oxygen (O_2- Greek *oxys* "acid" *genes* "forming") constitutes approximately a fifth (21% by volume and 23.2% by weight) of the air in Earth's atmosphere. Gaseous oxygen (O_2) is vital to life as we know it. On Earth, oxygen is the most abundant element. Most oxygen on Earth is not found in the free-state, but in combination with other elements as chemical compounds. Water and carbon dioxide are common examples of compounds that contain oxygen, but there are countless others.

At ordinary temperatures, oxygen is a colorless, odorless, tasteless gas that not only supports life but also combustion. All the elements except the inert gases combine directly with oxygen to form oxides. However, oxidation of different elements occurs over a wide range of temperatures.

Oxygen is nonflammable but it readily supports combustion; that is, all materials that are flammable in air burn much more vigorously in oxygen. Some combustibles, such as oil and grease, burn with nearly explosive violence in oxygen if ignited.

Oxygen: Physical Properties

The physical properties of oxygen are noted in Table 7.4.

Oxygen: Uses

The major uses of oxygen stem from its life-sustaining and combustion-supporting properties. It also has many industrial applications (when used with other fuel gases such as acetylene) including metal cutting, welding, hardening, and scarfing.

Ozone: Just another Form of Oxygen

Ozone (O_3) is a highly reactive pale-blue gas with a penetrating odor. Ozone is an allotropic modification of oxygen. An allotrope is a variation of an element that possesses a set of physical and

Air Pollution

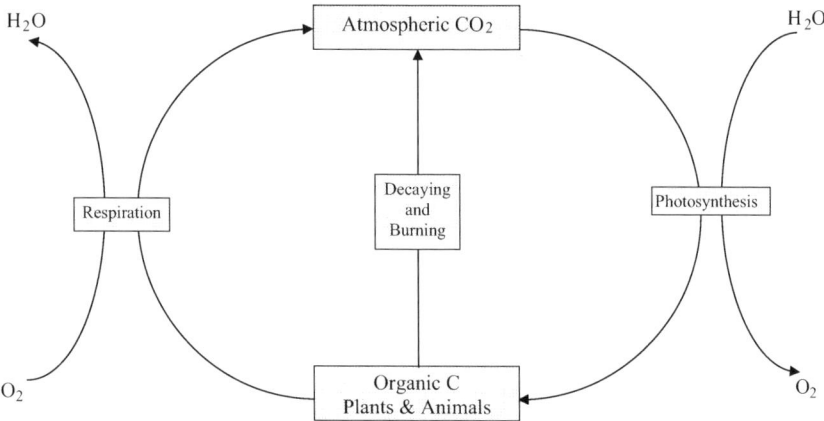

FIGURE 7.2 Carbon cycle.

chemical properties significantly different from the "normal" form of the element. Only a few elements have allotropic forms; oxygen, phosphorous, and sulfur are some of them. Ozone is just another form of oxygen. It is formed when the molecule of the stable form of oxygen (O_2) is split by ultraviolet (UV) radiation or electrical discharge; it has three instead of two atoms of oxygen per molecule. Thus, its chemical formula is represented by O_3.

Ozone forms a thin layer in the upper atmosphere, which protects life on Earth from ultraviolet rays, a cause of skin cancer. At lower atmospheric levels it is an air pollutant and contributes to the greenhouse effect. At ground level, ozone, when inhaled, can cause asthma attacks, stunted growth in plants, and corrosion of certain materials. It is produced by the action of sunlight on air pollutants, including car exhaust fumes, and is a major air pollutant in hot summers. More will be said about ozone and the greenhouse effect later in the text.

ATMOSPHERIC CARBON DIOXIDE

Carbon dioxide (CO_2) is a colorless, odorless gas (although it is felt by some persons to have a slight pungent odor and biting taste), slightly soluble in water and denser than air (one and half times heavier than air), and slightly acid gas. Carbon dioxide gas is relatively non-reactive and nontoxic. It will not burn, and it will not support combustion or life.

CO_2 is normally present in atmospheric air at about 0.035% by volume and cycles through the biosphere (Carbon Cycle) as shown in Figure 7.2. Carbon dioxide, along with water vapor, is primarily responsible for the absorption of infrared energy re-emitted by the Earth and, in turn, some of this energy is reradiated back to the Earth's surface. It is also a normal end product of human and animal metabolism. The exhaled breathe contains up to 5.6% carbon dioxide. In addition, the burning of carbon-laden fossil fuels releases carbon dioxide into the atmosphere. Much of this carbon dioxide is absorbed by ocean water, some of it is taken up by vegetation through photosynthesis in the carbon cycle (see Figure 7.2), and some remains in the atmosphere. Today, it is estimated that the concentration of carbon dioxide in the atmosphere is ~350 parts per million (ppm) and is rising at a rate of ~20 ppm every decade. The increasing rate of combustion of coal and oil has been primarily responsible for this occurrence, which (as we will see later in this text) may eventually have an impact on the global climate.

Carbon Dioxide: Physical Properties

The physical properties of carbon dioxide are noted in Table 7.5.

TABLE 7.5
Carbon Dioxide: Physical Properties

Chemical Formula	CO_2
Molecular weight	44.01
Vapor pressure @ 70°F	838 psig
Density of the gas @ 70°F & 1 atm	0.1144 lb/ft^3
Specific gravity of the gas @ 70°F & 1 atm (air=1)	1.522
Specific volume of the gas @ 70°F & 1 atm	8.741 ft^3/lb
Critical temperature	−109.3°F
Critical pressure	1070.6 psia
Critical density	29.2 lb/ft^3
Latent heat of vaporization @ 32°F	100.8 Btu/lb
Latent heat of fusion @ −69.9°F	5.6 Btu/lb

Carbon Dioxide: Uses

Solid carbon dioxide is used quite extensively to refrigerate perishable foods while in transit. It is also used as a cooling agent in many industrial processes, such as grinding, rubber work, cold-treating metals, vacuum cold traps, and so on.

Gaseous carbon dioxide is used to carbonate soft drinks, for pH control in water treatment, in chemical processing, as a food preservative, and in pneumatic devices.

ATMOSPHERIC ARGON

Argon (Ar - Greek *argos* "idle") is a colorless, odorless, tasteless, nontoxic, nonflammable gaseous element (Noble gas). It constitutes almost 1% of the Earth's atmosphere and is plentiful compared to the other rare atmospheric gases. It is extremely inert and forms no know chemical compounds. It is slightly soluble in water.

Argon: Physical Properties

The physical properties of argon are noted in Table 7.6.

TABLE 7.6
Argon: Physical Properties

Chemical Formula	Ar
Molecular weight	39.95
Density of the gas @ 70°F	0.103 lb/ft^3
Specific gravity of the gas @ 70°F	1.38
Specific volume of the gas @ 70°F	9.71 ft^3/lb
Boiling point at 1 atm	−302.6°F
Melting point at 1 atm	−308.6°F
Critical temperature	−188.1°F
Critical pressure	711.5 psia
Critical density	33.444 lb/ft^3
Latent heat of vaporization @ boiling point and 1 atm	69.8 But/lb
Latent heat of fusion	12.8 Btu/lb

TABLE 7.7
Neon: Physical Properties

Chemical Formula	Ne
Molecular weight	20.183
Density of the gas @ 70°F & 1 atm	0.05215 lb/ft^3
Specific gravity of the gas @ 70°F & 1 atm	0.696
Specific volume of the gas @ 70°F & 1 atm	19.18 ft^3/lb
Boiling point at 1 atm	−410.9°F
Melting point at 1 atm	−415.6°F
Critical temperature	−379.8°F
Critical pressure	384.9 psia
Critical density	30.15 lb/ft^3
Latent heat of vaporization @ boiling point	37.08 But/lb
Latent heat of fusion	7.14 Btu/lb

Argon: Uses

Argon is used extensively in filling incandescent and fluorescent lamps and electronic tubes; to provide a protective shield for growing silicon and germanium crystals; and as a blanket in the production of titanium, zirconium, and other reactive metals.

ATMOSPHERIC NEON

Neon (Ne - Greek *neon* "new") is a colorless, odorless, gaseous, nontoxic, chemically inert element. Air is about 2 parts-per-thousand neon by volume.

Neon: Physical Properties

The physical properties of neon are noted in Table 7.7.

Neon: Uses

Neon is used principally to fill lamp bulbs and tubes. The electronics industry uses neon singly or in mixtures with other gases in many types of gas-filled electron tubes.

ATMOSPHERIC HELIUM

Helium (He - Greek *helios* "Sun") is inert (and as a result, does not appear to have any major effect on, or role in, the atmosphere), nontoxic, odorless, tasteless, non-reactive, forms no compounds, colorless, and makes about 0.00005% (5 ppm) by volume of air in the Earth's atmosphere. Helium, as with neon, krypton, hydrogen, and xenon, is a noble gas. Helium is the second lightest element; only hydrogen is lighter. It is one-seventh as heavy as air. Helium is nonflammable and is only slightly soluble in water.

Helium: Physical Properties

The physical properties of helium are noted in Table 7.8.

ATMOSPHERIC KRYPTON

Krypton (Kr - Greek *kryptos* "hidden") is a colorless, odorless, inert gaseous component of Earth's atmosphere. It is present in very small quantities in the air (about 114 parts per million—ppm).

TABLE 7.8
Helium: Physical Properties

Chemical Formula	He
Molecular weight	4.00
Density of the gas @ 70°F & 1 atm	0.0103 lb/ft^3
Specific gravity of the gas @ 70°F & 1 atm	0.138
Specific volume of the gas @ 70°F & 1 atm	97.09 ft^3/lb
Boiling point @ 1 atm	−452.1°F
Critical temperature	−450.3°F
Critical pressure	33.0 psia
Critical density	4.347 lb/ft^3
Latent heat of vaporization @ boiling point & 1 atm	8.72 But/lb

TABLE 7.9
Krypton: Physical Properties

Chemical Formula	Kr
Molecular weight	83.80
Density of the gas @ 70°F & 1 atm	0.2172 lb/ft^3
Specific gravity of the gas @ 70°F & 1 atm	2.899
Specific volume of the gas @ 70°F & 1 atm	4.604 ft^3/lb
Boiling point @ 1 atm	−244.0°F
Melting point @ 1 atm	−251°F
Critical temperature	−82.8°F
Critical pressure	798.0 psia
Critical density	56.7 lb/ft^3
Latent heat of vaporization @ boiling point	46.2 Btu/lb
Latent heat of fusion	8.41 But/lb

Krypton: Physical Properties

The physical properties of krypton are noted in Table 7.9.

Krypton: Uses

Krypton is used principally to fill lamp bulbs and tubes. The electronics industry uses it singly or in the mixture in many types of gas-filled electron tubes.

ATMOSPHERIC XENON

Xenon (Xe - Greek *xenon* "stranger") is a colorless, odorless, nontoxic, inert, heavy gas that is present in very small quantities in the air (about 1 part in 20 million).

Xenon: Physical Properties

The physical properties of xenon are noted in Table 7.10.

Xenon: Uses

Xenon is used principally to fill lamp bulbs and tubes. The electronics industry uses it singly or in mixtures in many types of gas-filled electron tubes.

TABLE 7.10
Xenon: Physical Properties

Chemical Formula	Xe
Molecular weight	131.3
Density of the gas @ 70°F & 1 atm	0.3416 lb/ft^3
Specific gravity of the gas @ 70°F & 1 atm	4.560
Specific volume of the gas @ 70°F & 1 atm	2.927 ft^3/lb
Boiling point at 1 atm	−162.6°F
Melting point at 1 atm	−168°F
Critical temperature	61.9°F
Critical pressure	847.0 psia
Critical density	68.67 lb/ft^3
Latent heat of vaporization at boiling point	41.4 Btu/lb
Latent heat of fusion	7.57 Btu/lb

TABLE 7.11
Hydrogen: Physical Properties

Chemical formula	H$_2$
Molecular weight	2.016
Density of the gas @ 70°F & 1 atm	0.00521 lb/ft^3
Specific gravity of the gas @ 70° & 1 atm	0.06960
Specific volume of the gas @ 70°F & 1 atm	192.0 ft^3/lb
Boiling point @ 1 atm	−423.0°F
Melting point @ 1 atm	−434.55°F
Critical temperature	−399.93°F
Critical pressure	190.8 psia
Critical density	1.88 lb/ft^3
Latent heat of vaporization @ boiling point	191.7 Btu/lb
Latent heat of fusion	24.97 Btu/lb

ATMOSPHERIC HYDROGEN

Hydrogen (H$_2$- Greek *hydros + gen* "water generator") is a colorless, odorless, tasteless, nontoxic, flammable gas. It is the lightest of all the elements and occurs on Earth chiefly in combination with oxygen as water. Hydrogen is the most abundant element in the universe, where it accounts for 93% of the total number of atoms and 76% of the total mass. It is the lightest gas known, with a density ~0.07 that of air. Hydrogen is present in the atmosphere, occurring in concentrations of only about 0.5 ppm by volume at lower altitudes.

Hydrogen: Physical Properties

The physical properties of hydrogen are noted in Table 7.11.

Hydrogen is used by refineries, petrochemical, and bulk chemical facilities for hydro-treating, catalytic reforming, and hydro-cracking. Hydrogen is used in the production of a wide variety of chemicals. Metallurgical companies use hydrogen in the production of their products. Glass manufacturers use hydrogen as a protective atmosphere in a process whereby molten glass is floated on a surface of molten tin. Food companies hydrogenate fats, oils, and fatty acids to control various physical and chemical properties. Electronic manufacturers use hydrogen at several steps in the complex processes for manufacturing semiconductors.

ATMOSPHERIC WATER

Leonardo da Vinci understood the importance of water when he said: "Water is the driver of nature." da Vinci was actually acknowledging what most scientists and many of the rest of us have come to realize: Water, propelled by the varying temperatures and pressures in Earth's atmosphere, allows life as we know it to exist on our planet (Gradel and Crutzen, 1995).

The water vapor content of the lower atmosphere (troposphere) is normally with a range of 1%–3% by volume with a global average of about 1%. However, the percentage of water in the atmosphere can vary from as little as 0.1% or as much as 5% water, depending upon altitude; water in the atmosphere decreases with increasing altitude. Water circulates in the atmosphere in the hydrologic cycle, as shown in Figure 7.3.

Water vapor contained in Earth's atmosphere plays several important roles: (1) it absorbs infrared radiation; (2) it acts as a blanket at night, retaining heat from the Earth's surface; and (3) it affects the formation of clouds in the atmosphere.

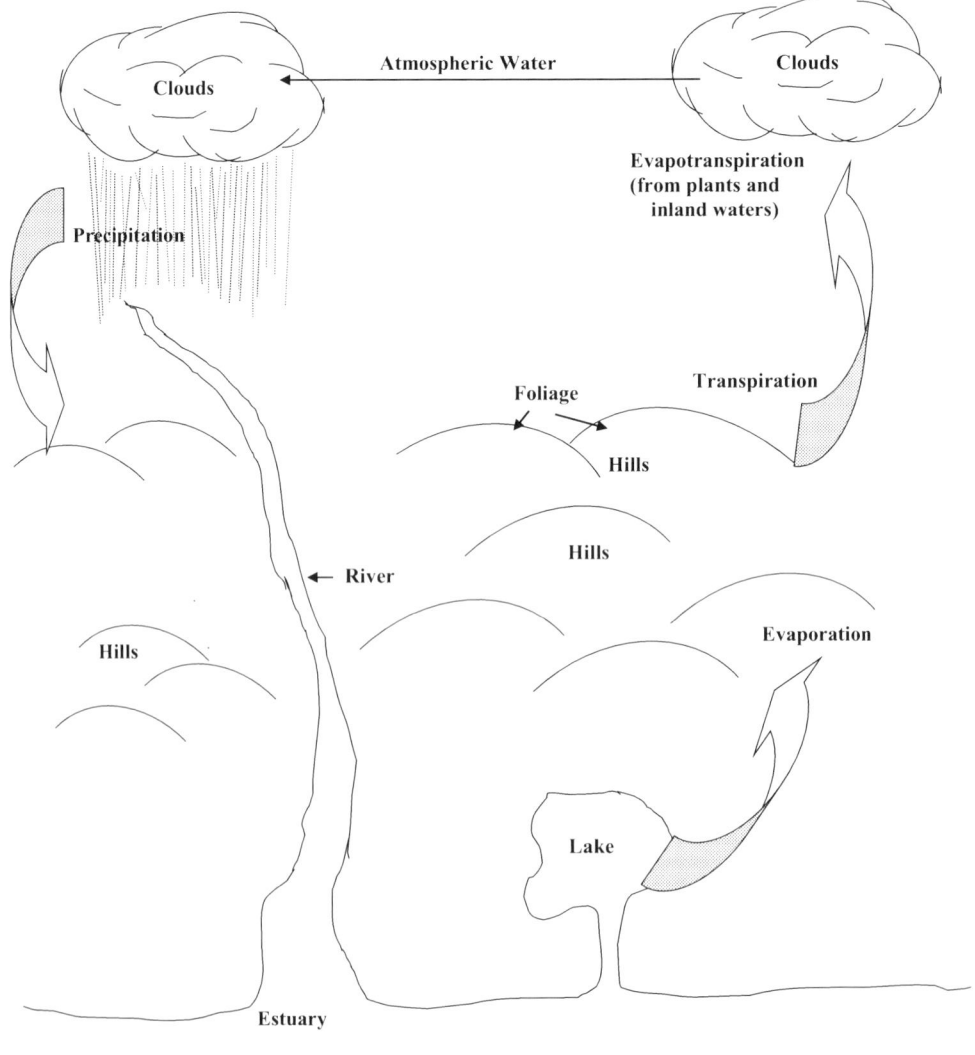

FIGURE 7.3 Water cycle.

Air Pollution

ATMOSPHERIC PARTICULATE MATTER

There are significant number of particles (particulate matter) suspended in the atmosphere, particularly the troposphere. These particles originate in nature from smokes, sea sprays, dust, and the evaporation of organic materials from vegetation. There is also a wide variety of nature's living or semi-living particles—spores and pollen grains, mites and other tiny insects, spider webs, and diatoms. The atmosphere also contains a bewildering variety of anthropogenic (manmade) particles produced by automobiles, refineries, production mills, and many other human activities.

Atmospheric particulate matter varies greatly in size (colloidal-sized particles in the atmosphere are called aerosols—usually <0.1 μm in diameter); the smallest are gaseous clusters and ions and submicroscopic liquids and solids; somewhat larger ones produce the beautiful blue haze in distant vistas; those two to three times larger are highly effective in scattering light; and the largest consist of such things as rock fragments, salt crystals, and ashy residues from volcanoes, forest fires, or incinerators.

The number of which particulates concentrated in the atmosphere varies greatly — ranging from more than 10,000,000/cm^3 to <1/L (0.001/cc). Excluding the particles in gases as well as vegetative material, sizes range from 0.005 to 500 μm, a variation in diameter of 100,000 times.

The largest number of airborne particulates is always in the invisible range. These numbers vary from <1 L to more than a half million per cubic centimeter in heavily polluted air and to at least ten times more than that when a gas-to-particle reaction is occurring (Schaefer and Day, 1981).

Based on particulate level, there are two distinct regions in the atmosphere: very clean and dirty. In the clean parts there are so few particulates that they are almost invisible, making them hard to collect or measure. In the dirty parts of the atmosphere—the air of a large metropolitan area—the concentration of particles includes an incredible variety of particulates from a wide variety of sources.

Atmospheric particulate matter performs a number of functions and undergoes several processes and is involved in many chemical reactions in the atmosphere. Probably the most important function of particulate matter in the atmosphere is their action as nuclei for the formation of water droplets and ice crystals. Much of the work of Vincent J. Schaefer (inventor of cloud seeding) involved using dry ice in early attempts but later evolved around the addition of condensing particles to atmospheres supersaturated with water vapor and the use of silver iodide, which forms huge numbers of very small particles. Another important function of atmospheric particulate matter is that they help determine the heat balance of the Earth's atmosphere by reflecting light. Particulate matter is also involved in many chemical reactions in the atmosphere such as neutralization, catalytic effects, and oxidation reactions. These chemical reactions will be discussed in greater detail later.

AIR FOR COMBUSTION

It is difficult to imagine where Man would be today or how far he would have progressed from his beginning to the present if he had not discovered and developed the use of fire. Today, of course, we are quite familiar with fire. We use the terms fire, combustion, oxidation, and/or burning pretty much in the same light to mean one in the same thing. However, in regard to combustion and oxidation, there is a subtle difference between them. During combustion, two or more substances chemically unite. In practice, one of them is almost always atmospheric oxygen, but combustion reactions are known in which oxygen is not one of the reactants. Thus, it is more correct to describe combustion as a rapid oxidation—or fire.

To state that atmospheric air plays an important role in combustion is to understate its significance; that is, we are stating the obvious —stating something that is readily apparent to most of us. Though the air is important in combustion, it is the actual chemical reaction involved with combustion that most of us give little thought to.

Combustion is a chemical reaction—one in which a fuel combines with air (oxygen) with the evolution of heat: burning. The combustion of fuels containing carbon and hydrogen is said to be complete when these two elements are oxidized to carbon dioxide and water (e.g., the combustion of carbon $C+O_2=CO_2$). In air pollution control, it is incomplete combustion that concerns us. Incomplete combustion may lead to (1) appreciable amounts of carbon remaining in the ash; (2) emission of some of the carbon as carbon monoxide; and (3) reaction of the fuel molecules to give a range of products which are emitted as smoke.

AIR FOR POWER

Along with performing its important function in Earth's atmosphere and its vital role in combustion, most industrial processes use gases to power systems of one type or another. The work is actually performed by a gas under pressure in the system. A gas power system may function as part of a process, such as heating and cooling, or it may be used as a secondary service system, such as compressed air. Compressed air is the gas most often found in industrial applications, but nitrogen and carbon dioxide are also commonly used. A system that uses gas for transmitting force is called a pneumatic system. The word pneumatic is derived from the Greek word for an unseen gas. Originally pneumatic referred only to the flow of air. Now it includes the flow of any gas in a system under pressure.

Pneumatic systems perform work in many ways, including operating pneumatic tools, door openers, linear motion devices, and rotary motion devices. Have you ever watched (heard) an automobile mechanic remove and replace a tire on your car? The device he/she uses to take off and put on tire lug nuts is a pneumatic (air-operated) wrench. Pneumatic hoisting equipment may be found in heavy fabricating environments, and pneumatics conveyors are used in the processing of raw materials. Pneumatic systems are also used to control flow valves in chemical process equipment and in large air-conditioning systems.

The pneumatic system in an industrial plant usually handles compressed air. As pointed out earlier, compressed air is used for operating portable air tools, such as drills, wrenches, and chipping tools, for vises, chucks, and other clamping devices; for movable locating stops; for operating plastic molding machines, and also for supplying air used in manufacturing processes. Although the pieces of pneumatic equipment just described are difference from each other, they all convert compressed air into work. Later, we review some of the laws of force and motion and their relation to pneumatic principles.

GAS PHYSICS

MATERIAL BALANCE

Material balance is one of the most basic and useful tools in the air pollution engineering field. Stated simply, a materials balance means "what goes in, must come out." Matter is neither created nor destroyed in industrial processes (nonradioactive only).

$$\text{Mass}_{(out)} = \text{Mass}_{(in)} \tag{7.1}$$

Material balance is used in a wide variety of air pollution control calculations. For example, it is used to evaluate the following:

- Formation of combustion products in boilers
- Rates of air infiltration into air pollution control systems
- Material requirements for process operations

Air Pollution

- Rate of ash collection in air pollution control systems
- Humidity of exhaust gas streams
- Exhaust gas flow rates from multiple sources controlled by a single air pollution control system
- Gas flow rates from combustion processes

This principle, called the conservation of matter, can be applied in solving problems involving the quantities of matter moving in various parts of a process.

COMPRESSIBILITY

Air, unlike liquids, is readily compressible, and large quantities can be stored in relatively small containers. The more the air is compressed, the higher its pressure becomes. The higher the pressure in a container, the stronger the container must be. Gases are important compressible fluids, not only from the standpoint that a gas can be a pollutant but also because gases convey the particles (particulate matter) and gaseous pollutants (Hesketh, 1991).

GAS LAWS

Gases can be pollutants as well as the conveyors of pollutants. Air (which is mainly nitrogen) is usually the main gas stream. Gas conditions are usually described in two ways: *standard temperature and pressure (STP)*, and *Standard Conditions (SC)*. STP represents 0°C (32°F) and 1 atm. SC is more commonly used and represents typical room conditions of 20°C (70°F) and 1 atm; SC is usually measured in cubic meters, Nm^3, or standard cubic feet (scf).

To understand the physics of air it is imperative to understand various physical laws that govern the behavior of pressurized gases. One of the more well-known physical laws, mentioned earlier, is *Pascal's Law*. In addition to one of its previously mentioned parameters, Pascal's Law also states that a confined gas (fluid) transmits externally applied pressure uniformly in all directions, without change in magnitude. This parameter can be seen in a container that is flexible; it will assume a spherical (balloon) shape. However, you probably have noticed that most compressed gas tanks are cylindrical in shape (which allows the use of thinner sheets of steel without sacrificing safety) with spherical ends to contain the pressure more effectively.

Boyle's Law

Though gases are compressible, note that, for a given mass flow rate, the actual volume of gas passing through the system is not constant within the system due to changes in pressure. This physical property (the basic relationship between the pressure of a gas and its volume) is described by *Boyle's Law* (Named for its discoverer: Irish physicist and chemist Robert Boyle in 1662), which states: "The absolute pressure of a confined quantity of gas varies inversely with its volume, if its temperature does not change." For example, if the pressure of a gas doubles, its volume will be reduced by half and vice versa. That is, *as pressure goes up, volume goes down*, and vice versa. This means, for example, that if 12 ft^3 of air at 14.7 psia is compressed to 1 ft^3, air pressure will rise to 176.4 psia, as long as the air temperature remains the same. This relationship can be calculated as follows:

$$P_1 \times V_1 = P_2 \times V_2 \qquad (7.2)$$

where
 P_1 = original pressure (units for pressure must be absolute)
 P_2 = new pressure (units for pressure must be absolute)

V_1 = original gas volume at pressure P_1
V_2 = new gas volume at pressure P_2
This equation can also be written as:

$$\frac{P_2}{P_1} = \frac{V_1}{V_2} \text{ or } \frac{P_1}{P_2} = \frac{V_2}{V_1} \quad (7.3)$$

To allow for the effects of atmospheric pressure, always remember to convert from gage pressure *before* solving the problem, then convert back to gage pressure *after* solving it:

$$\text{psia} = \text{psig} + 14.7\,\text{psi}$$

and

$$\text{psig} = \text{psia} - 14.7\,\text{psi}$$

Note that in a pressurized gas system where gas is caused to move through the system by the fact that gases will flow from an area of high pressure to that of low pressure, we will always have a greater actual volume of gas at the end of the system than at the beginning (assuming the temperature remains constant).

Let's take a look at a typical gas problem using Boyle's Law.

Example 7.1

Problem:

What is the gage pressure of 12 ft³ of air at 25 psig when compressed to 8 ft³?

Solution:

$$25\,\text{psig} + 14.7\,\text{psi} = 39.7\,\text{psia}$$

$$P_2 = P_1 \times \frac{V_1}{V_2}$$

$$= 39.7 \times \frac{12}{8} = 59.6\,\text{psia}$$

$$\text{psig} = \text{psia} - 14.7\,\text{psi}$$

$$= (59.6\,\text{psia}) - (14.7\,\text{psi}) = 44.9\,\text{psig}$$

The gage pressure is 44.9 psig (remember that the pressures should always be calculated on the basis of absolute pressures instead of gage pressures).

Charles's Law

Another physical law dealing with temperature is *Charles's Law* (discovered by French physicist Jacques Charles in 1787). It states, "the volume of a given mass of gas at constant pressure is directly proportional to its absolute temperature." [The temperature in Kelvin (273 + °C) or Rankine (absolute zero = −460°F, or 0°R).]

This is calculated by using the following equation:

$$P_2 = P_1 \times \frac{T_2}{T_1} \tag{7.4}$$

Charles's Law also states: "If the pressure of a confined quantity of gas remains the same, the change in the volume (V) of the gas varies directly with a change in the temperature of the gas," as given in the equation:

$$V_2 = V_1 \times \frac{T_2}{T_1} \tag{7.5}$$

Ideal Gas Law

The *Ideal Gas Law* combines Boyle's and Charles's Laws because air cannot be compressed without its temperature changing. The Ideal Gas Law is expressed by the equation:

$$\frac{P_1 \times V_1}{T_1} = \frac{P_2 \times V_2}{T_2} \tag{7.6}$$

Note that the Ideal Gas Law is still used as a design equation even though the equation shows that the pressure, volume, and temperature of the second state of a gas are equal to the pressure, volume, and temperature of the first state. In actual practice, however, other factors such as humidity, heat of friction, and efficiency losses all affect the gas. Also, this equation uses absolute pressure (psia) and absolute temperatures (°R) in its calculations.

In air science practice, the importance of the Ideal Gas Law cannot be overstated. It is one of the fundamental principles used in calculations involving gas flow in air pollution-related work. This law is used to calculate actual gas flow rates based on the quantity of gas present at standard pressures and temperatures. It is also used to determine the total quantity of that contaminant in a gas that can participate in a chemical reaction.

- Number of moles of gas
- Absolute temperature
- Absolute pressure

In practical applications, practitioners generally use the following standard ideal gas law equation:

$$V = \frac{nRT}{P} \quad \text{or} \quad PV = nRT \tag{7.7}$$

where
    V = Volume
    n = Number of moles
    R = Universal gas constant
    T = Absolute temperature
    P = Absolute pressure

Example 7.2

What is the volume of 1 pound mole (denoted "lb mole") of combustion gas at an absolute pressure of 14.7 psia and a temperature of 68°F? (These are EPA-defined standard conditions.)

Solution:

$$V = \frac{nRT}{P}$$

1. Convert the temperature form relative to the absolute scale (from °F to °R).

$$T_{\text{Absolute}} = 68°F + 460 = 528°R$$

2. Calculate the gas volume.

$$V = \frac{1\,\text{lb mole} \times \dfrac{10.73(\text{psia})(\text{ft}^3)}{(\text{lb mole})(°R)} \times 528°R}{14.7\ psia}$$

$$= 385.4\ \text{ft}^3$$

FLOW RATE

Gas flow rate is a measure of the volume of gas that passes a point in an industrial system during a given period of time. The ideal gas law tells us that this gas flow rate varies depending on the temperature and pressure of the gas stream and the number of moles of gas moving per unit of time.

When gas flow rates are expressed at actual conditions of temperature and pressure, the actual gas flow rate is used. As you will learn later, gas flow rates can also be expressed at standard conditions of temperature and pressure; this is referred to as the standard gas flow rate.

GAS CONVERSIONS

Gases of interest in air pollution control are usually mixtures of several different compounds. For example, air is composed of three major constituents: nitrogen (N_2) at approximately 78.1%, oxygen (O_2) at ~20.9%, and argon at 0.9%. Many flue gas streams generated by industrial processes consist of the following major constituents: (1) nitrogen, (2) oxygen, (3) argon, (4) carbon dioxide (CO_2), and (5) water vapor (H_2O). Both air and industrial gas streams also contain minor constituents, including air pollutants, present at concentrations that are relatively low compared to these major constituents.

There is a need for ways to express both the concentrations of the major constituents of the gas stream and the concentrations of the pollutants present as minor constituents at relatively low concentrations. There are a variety of ways to express gas phase concentrations, which can easily be converted from on type of units to another. They include the following:

Major Constituents

- Volume percent: is one of the most common formats used to express the concentrations of major gas stream constituents such as oxygen, nitrogen, carbon dioxide, and water vapor. The format is very common partially because the gas stream analysis techniques used in EPA emission testing methods provide data directly in a volume percent format.
- **Partial pressure**: concentrations can also be expressed in terms of partial pressures. This expression refers to the part of the total pressure exerted by one of the constituent gases.

    Gases composed of different chemical compounds such as molecular nitrogen and oxygen behave physically the same as gases composed of a single compound. At any given

Air Pollution

temperature, one mole of gas exerts the same pressure as one mole of any other type of gas. All of the molecules move at a rate that is dependent on the absolute temperature, and they exert pressure. The total pressure is the sum of the pressures of each of the components. The equations below are often Called Dalton's law of partial pressures.

$$P_{Total} = p_i + p_{ii} + p_{iii} + \cdots + p_n \tag{7.8}$$

$$P_{Total} = \sum_{i=1}^{n} P_i \tag{7.9}$$

$$(Gas) = \left[\frac{Volume\,\%(Gas)}{100\%}\right] \times P_{Total} \tag{7.10}$$

Because the partial pressure value is related to the total pressure, concentration data expressed as partial pressure are not the same at actual and standard conditions. The partial pressure values are also different in American Engineering units and Cgs units.

Both Major and Minor Constituents

- **Mole fraction**: is simply an expression of the number of moles of a compound divided by the total number of moles of all the compounds present in the gas.

Minor Constituents

- Parts per million (ppm)
- Milligrams per cubic meter (mg/m^3)
- Micrograms per cubic meter (µg/m^3)
- Nanograms per cubic meter (ng/m^3)

All of the concentration units above can be expressed in a dry format as well as corrected to a standard oxygen concentration. These corrections are necessary because moisture and oxygen concentrations can vary greatly in gas streams, causing variations in pollutant concentrations.

GAS VELOCITY

Gas velocity is one of the fundamental design variables for ventilation systems and air pollution control equipment. Gas streams containing particulate are usually maintained at velocities of 3,000 to 4,500 ft/min n ductwork leading to particulate collectors to minimize particle deposition. The velocity of gas streams without particulate matter is often in the range of 1,500–3,000 ft/min. The gas velocities in air pollution control equipment are usually low to allow for sufficient time to remove the contaminants. For example, gas velocities through electrostatic precipitators are usually in the range of 2.5–6 ft/s. The filtration velocities through pulse jet fabric filters are usually in the range of 2–10 ft/min. Variations in the gas velocity can have direct impact on the contaminant removal efficiency.

The average velocity of a gas stream in an emission testing probe, an industrial duct, or an air pollution control device is a function of the actual gas flow rate and the cross-sectional flow area.

$$v = Gas\,velocity = \frac{Gas\,flow\,rate, actual}{Area} \tag{7.11}$$

Gas Stream Treatment (Residence) Time

The flow rate of the gas stream through an air pollution control system determines the length of time that the pollutants can be removed from the gas stream. This is termed the *treatment time* or *residence time*. These common equipment sizing parameters are defined mathematically in Equation 7.12.

$$\text{Treatment time} = \text{Residence time}$$

$$\frac{\text{Volume of control device}}{\text{Gas flow rate, actual}} \qquad (7.12)$$

Gas Density

Gas density is important primarily because it affects the flow characteristics of the moving gas streams. Gas density affects the velocities of gas through ductwork and air pollution control equipment. It determines the ability to move the gas steam using a fan. Gas density affects the velocities of gases emitted from the stack and thereby influences the dispersion of the pollutants remaining in the stack gases. It affects the ability of particles to move through gases. It also affects emission testing. Gas density data is needed in many of the calculations involved in air pollution control equipment evaluation, emission testing, and other air pollution control-related studies.

As discussed earlier, the volume of a gas increases as the temperature increases due to the motion of the gas molecules. As the volume occupied by the gas increases, its density decreases. Density is the mass per unit volume as indicated in Equation 7.13.

$$P_{(T=i, P=j)} = \frac{m}{V_{(T=I, P=j)}} \qquad (7.13)$$

where

$$P_{(T=I, P=j)} = \text{Density at } T = I, P = j$$

M = Mass of a substance

$$V_{(T=I, P=j)} = \text{Volume at } T = I, P = j$$

T = Absolute temperature
P = Absolute pressure

Heat Capacity and Enthalpy

The heat capacity of a gas is the amount of heat required to change the temperature of a unit-mass of gas by one temperature degree.

Enthalpy represents the total quantity of internal energy, such as heat, measured for a specific quantity of material at a given temperature. Enthalpy data are often represented in units of energy (e.g. Btu, kcal, joule, etc.). The enthalpy content change is often expressed in Btu/unit mass (Btu/lb_m) or Btu/unit time (Btu/SCF). The change in enthalpy of the total quantity of material present in a system is expressed in units of Btu/unit time (Btu/min). The symbols, H and ΔH, denote enthalpy and the change in enthalpy, respectively.

Air Pollution

HEAT AND ENERGY IN THE ATMOSPHERE

In addition to the importance of heat on a particular air stream, it is important to point out that heat also has an impact on Earth's atmosphere and thus on atmospheric science. The Sun's energy is the prime source of Earth's climatic system. From the Sun, energy is reflected, scattered, absorbed, and reradiated within the system but without uniform distribution. Some areas receive more energy than they lose; in some areas, the reverse occurs. If this situation were able to continue for long the areas with an energy surplus would get hotter—too hot, and those with a deficit would get colder—too cold. This does not happen because the temperature differences produced help to drive the wind and ocean currents of the world. They carry heat with them, either in sensible or latent forms, and help to counteract the radiation imbalance. Winds from the tropics are therefore normally warm, carrying excess heat with them. Polar winds are blowing from areas with a deficit of heat and so are cold. Acting together, these energy transfer mechanisms help to produce the present climate on Earth.

Adiabatic Lapse Rate

The atmosphere is restless, always in motion either horizontally or vertically or both. As air rises, pressure on it decreases and in response it expands. The act of expansion to encompass its new and larger dimensions requires an expenditure of energy; since the temperature is a measure of internal energy, this use of energy makes its temperature drop—this is an important point—an important process in physics (especially in air physics).

This phenomenon is known as the *adiabatic lapse rate*. Simply, adiabatic refers to a process that occurs with or without loss of heat, especially the expansion or contraction of a gas in which a change takes place in the pressure or volume, although no heat is allowed to enter or leave.

Lapse rate refers to the rate at which air temperature decreases with height. The normal lapse rate in stationary air is about 3.5°F/1,000 ft (6.5°C/km). This value may vary with latitude and changing atmospheric conditions (e.g., seasonal changes). A parcel of air that is not immediately next to the Earth's surface is sufficiently well insulated by its surroundings that either expansion or compression of the parcel may be assumed to be adiabatic.

The air temperature may be calculated for any height by the general formula

$$T = T_0 - Rh \tag{7.14}$$

where
 T = temperature of the air
 h = height of air
 T_0 = temperature of the air at the level from which the height is measured
 R = lapse rate

Example 7.3

Problem:

If the air temperature of stationary air ($R = 3.5°F/1,000$ ft) at the Earth's surface is 70°F, then at 5,000 ft, the stationary air temperature would be

Solution:

$$T = T_0 - Rh$$

$$= 70°F - (3.5°F/1,000 \text{ ft})(5,000 \text{ ft})$$

$$= 70°F - 17.5°F = 52.5°F$$

The formula simply says that for every 1000 ft of altitude (height), 3.5° is subtracted from the initial air temperature, in this case.

Adiabatic lapse rates have an important relationship with atmospheric stability and will be discussed in greater detail later in the text.

VISCOSITY

All fluids (gases included) resist flow. *Absolute viscosity* is a measure of this resistance to flow. The absolute viscosity of a gas for given conditions may be calculated from the following formula:

$$\mu = 51.12 + 0.372(T) + 1.05 \times 10^{-4}(T)^2 \\ + 53.147(\%O_2/100\%) - 74.143(\%H_2O/100\%) \quad (7.15)$$

where
  μ = Absolute viscosity of gas at the prevailing conditions, micropoise
  T = Gas absolute temperature, °K
  % O_2 = Oxygen concentration, % by volume
  % H_2O = Water vapor concentration, % by Volume

As this equation indicates, the viscosity of a gas increases as the temperature increases. It's harder to push something (e.g. particles) through a hot gas stream than a cooler one due to increased molecular activity as temperature rises, which results in increased momentum transfer between the molecules. For liquids, the opposite relationship between viscosity and temperature holds. The viscosity of a liquid decreases as temperature increases. It's harder to push something through a cold liquid than a hot one because in liquids, hydrogen bonding increases with colder temperatures.

Gas viscosity actually increases very slightly with pressure, but this variation is very small in most air pollution-related engineering calculations.

The absolute viscosity and density of gas are occasionally combined into a single parameter since both of these parameters are found in many common equations describing gas flow characteristics. The combined parameter is termed the kinematic viscosity. It is defined in Equation 7.16.

$$u = \mu/p \quad (7.16)$$

where
  u = Kinematic viscosity, m²/s
  μ = Absolute viscosity, Pa s
  p = Gas density, gm/cm³

The kinematic viscosity can be used in equations describing particle motion through gas streams. The expression for kinematic viscosity is used to simplify these calculations.

FLOW CHARACTERISTICS

When fluids, such as gases, move slowly the bulk material moves as distinct layers in parallel paths. The only movement across these layers is molecular motion, which creates viscosity. This is termed *laminar flow*. As the velocity of the gas stream increases the bulk movement of the gas changes. Eddy currents develop which cause mixing across the flow stream. This is called *turbulent flow* and is essentially the only flow characteristic that occurs in air pollution control equipment and emission testing-related situations.

Air Pollution

A dimensionless parameter called the *Reynolds number* is used to characterize the fluid flow. It is the ratio of the inertial force which is causing gas movement to the viscous force which is restricting movement. The Reynolds number is calculated Using Equation 7.17. Consistent units must be used to ensure that the Reynolds number is dimensionless.

$$N_{Re(g)} = Lvp/\mu \qquad (7.17)$$

where
 L = Linear dimensions (usually duct diameter)
 v = Fluid velocity
 p = Fluid density
 μ = Fluid viscosity

Reynolds numbers <2,000 are associated with laminar flow conditions. Due to the relatively low velocities associated with this type of flow, they are rarely encountered in air pollution field situations.

Reynolds numbers above 10,000 are associated with turbulent flow. In many field situations, the Reynolds numbers exceed 100,000.

Essentially, all gas flow situations are turbulent in air pollution control systems, emission test and monitoring equipment, and dispersion modeling studies. However, this does not mean that the gas stream is entirely well-mixed. In reality, the side-to-side mixing (and even mixing in the direction of flow) can be limited. For this reason, it is possible to have different concentrations of pollutants at different positions in the duct. This is called *pollutant stratification*. It can be caused by a variety of factors: combining two separate gas streams into a single duct, temperature differences in the gas stream, and in-leakage of cold ambition air into the duct.

Stratification does not exist in most industrial gas handling systems. However, it is important to check for this condition prior to the installation of continuous emission monitors or other instruments, which are located at a single sampling or measurement point in the gas stream. These measurements can be in error if stratification is severe.

PARTICLE PHYSICS

Particulates constitute a major class of air pollution. They have a variety of shapes; they include dusts, fumes, mists, and smoke; and they have a wide range of physical and chemical properties. Important particulate matter characteristics include size, size distribution, shape, density, stickiness, corrosivity, reactivity, and toxicity. With a wide range of characteristics, it should be obvious that the type of collection device to be used might be better suited than others for a particular particle. For example, particulate matter typically ranges in size from 0.005 to 100 µm in diameter. This wide range in size distribution must certainly be considered.

CHARACTERISTICS OF PARTICLES

Understanding the characteristics of particles is important in air pollution control technology for the following reasons: The efficiency of the particle collection mechanisms strongly depends on particle size; the particle size distribution of flue gas dictates the manner in which air testing is performed; the particle size distribution of flue gas determines the operating conditions necessary to collect the particles; and particle characteristics are important in determining the behavior of particles in the respiratory tract.

Surface Area and Volume

The size range of particles of concern in air pollution studies is remarkably broad. Some of the droplets collected in the mist eliminators of wet scrubbers and the solid particles collected in

TABLE 7.12
Spherical Particle Diameter, Volume, and Area

Particle Diameter (μm)	Particle Volume (cm³)	Particle Area (cm²)
0.1	5.23×10^{-16}	3.14×10^{-10}
1.0	5.23×10^{-13}	3.14×10^{-8}
10.0	5.23×10^{-10}	3.14×10^{-6}
100.0	5.23×10^{-7}	3.14×10^{-4}
1,000.0	5.23×10^{-4}	3.14×10^{2}

Source: EPA (2007b).

large-diameter cyclones are as large raindrops. Some of the small particles created in high-temperature incinerators and metallurgical processes can consist of <50 molecules clustered together. It is important to recognize the categories since particles behave differently depending on their size. Most particles of primary interest in air pollution emission testing and control have diameters that range from 0.1 to 1,000 μm.

Useful conversions for particle sizes:

$$\text{Micrometer} = \left(\frac{1}{1,000,000}\right) \text{m} \tag{7.18}$$

$$\text{Micrometer} = \left(\frac{1}{10,000}\right) \text{cm} \tag{7.19}$$

$$1,000\,\mu\text{m} = 1.0\,\text{mm} = 0.1\,\text{cm} \tag{7.20}$$

To appreciate the difference in particle sizes commonly found in emission testing and air pollution control, compare the diameters, volumes, and surface areas of particles in Table 7.12.

The data in Table 7.12 indicates that particles of 1,000 μm are 1,000,000,000,000 (one trillion) times larger in volume than 0.1 μm particles.

As an analogy, assume that the 1,000-μm particle is a large domed spots stadium. A basketball in this "stadium" would be equivalent to a 5-μm particle. Approximately 100,000 spherical particles of 0.1 μm diameter would fit into this 5-μm "basketball." The entire 1,000-μm "stadium" is the size of a small raindrop. As previously stated, particles over this entire size range of 0.1–1,000 μm are of interest in air pollution control (EPA, 2007a).

Equations for calculating the volume and surface area of spheres are provided below.

$$\text{Surface area of a sphere} = 4\pi r^2 = \pi D^2 \tag{7.21}$$

$$\text{Volume of a sphere} = 4\pi r^3/3 = \pi D^3/6 \tag{7.22}$$

where
 r = Radius of sphere
 D = Diameter of sphere

Example 7.4

Problem:

Calculate the volumes of three spherical particles (in cm³) given that their actual diameters are 0.6, 6.0, and 60 µm.

Solution:

$$V = 4/3\pi r^3$$
$$= 4.19 r^3$$

For a 0.6 µm particle, $r = 0.3$ µm $= 0.00003$ cm

$$V = 4.19(0.00003 \text{cm})^3$$
$$= 1.13 \times 10^{-13} \text{ cm}^3$$

For a 6.0 µm particle, $4 = 3.0$ µm $= 0.0003$ cm

$$V = 4.19(0.0003 \text{cm})^3$$
$$= 1.13 \times 10^{-10} \text{ cm}^3$$

For a 60-µm particle, $r = 30$ µm $= 0.003$ cm

$$V = 4.19(0.003 \text{ cm})^3$$
$$- 1.13 \times 10^{-7} \text{ cm}^3$$

Aerodynamic Diameter

Particles emitted from air pollution sources and formed by natural processes have a number of different shapes and densities. Defining particle size for spherical particles is easy; it is simply the diameter of the particle. For non-spherical particles, the term "diameter" does not appear to be strictly applicable.

In air pollution control, particle size is based on particle behavior in the Earth's gravitational field. The *aerodynamic equivalent diameter* refers to a spherical particle of unit density that falls at a standard velocity. Particle size is important because it determines atmospheric lifetime, effects on light scattering, and deposition in human lungs. However, it is important to note that when we speak of particulate size we are not referring to particulate shape. This is not only an obvious point but also a very important because, for simplicity and for theoretical and for learning applications, it is common practice to base assumptions on particles being spherical. Particles are not spherical; they are usually quite irregularly shaped. In pollution control technology design applications, the size of the particle is taken into consideration but it is the particle's behavior in a gaseous waste stream that we are most concerned with. To better understand this important point, consider an analogy. If you take a flat piece of regular typing paper and drop it from the top of a six-foot ladder, the paper will settle irregularly to the floor, but crumple the same sheet of paper and drop it from the ladder and see what happens. The crumpled paper will fall rapidly. Thus, size is important but other factors such as particle shape are also important.

The aerodynamic diameter for all particles >0.5 μm can be approximated using the flowing equation. For particles <0.5 μm, refer to aerosol textbooks to determine the aerodynamic diameter.

$$d_{pa} = d_{ps}\sqrt{P_p} \qquad (7.23)$$

where

d_{pa} = Aerodynamic particle diameter, μm
d_{ps} = Stokes diameter, μm
P_p = Particle density, gm/cm³

Along with particle size and shape, particle density must be taken into consideration. For example, a baseball and a whiffle ball have approximately the same diameter but behave quite differently when tossed into the air. Particle density affects the motion of a particle through a fluid and is considered in Equation 7.23. The Stokes diameter for a particle is the diameter of the sphere that has the same density and settling velocity of the particle. It is based on the aerodynamic drag force caused by the difference in velocity of the particle and the surrounding fluid. For smooth, spherical particles, the Stokes diameter is identical to the physical or actual diameter.

Inertial sampling devices such as cascade impactors are used for particle sizing. These sampling devices determine the aerodynamic diameter. The term "aerodynamic diameter" is useful for all particles including fibers and particle clusters. It is not a true size because "non-spherical" particles require more than one dimension to characterize their size.

Note: In this text, the terms particle diameter and particle size refer to the aerodynamic diameter unless otherwise stated.

Particle Size Categories

Since the range of particle sizes of concern for air emission evaluation is quite broad it is beneficial to divide this range into smaller categories. Defining different size categories is useful since particles of different sizes behave differently in the atmosphere and the respiratory system.

EPA has defined four terms for categorizing particles of different sizes Table 7.13 displays the EPA terminology along with the corresponding particle sizes.

Regulated Particulate Matter Categories

In addition to the terminology provided in Table 7.13 EPA also categorizes particles as follows:

- **Total Suspended Particulate Matter (TSP)**: particles ranging in size for 0.1 μm to about 30 μm in diameter are referred to as total suspended particulate matter (TSP). TSP includes a broad range of particle sizes including fine, coarse, and super-coarse particles.
- **PM$_{10}$**: the U.S. EPA defines PM$_{10}$ as particulate matter with a diameter of 10 μm collected with 50% efficiency by a PM$_{10}$ sampling collection device. However, for convenience in this text, the term PM$_{10}$ will be used to include all particles having an aerodynamic diameter of less than or equal to 10 μm.

TABLE 7.13
EPA Terminology for Particle Sizes

EPA Description	Particle Size (μm)
Supercoarse	$d_{pa} > 10$
Coarse	$2.5 < d_{pa} \leq 10$
Fine	$0.1 < d_{pa} \leq 2.5$
Ultrafine	$d_{pa} \leq 0.1$

Air Pollution

PM_{10} is regulated as a specific type of "pollutant" because this size range is considered respirable. In other words, particles less than ~10 μm can penetrate into the lower respiratory tract. The particle size range between 0.1 and 10 μm is especially important in air pollution studies. A major fraction of the particulate matter generated in some industrial sources is in this size range.

- **$PM_{2.5}$**: as with PM_{10}, EPA defines $PM_{2.5}$ as particulate matter with a diameter of 2.5 μm with 50% efficiency by a $PM_{2.5}$ sampling collection device. However, for convenience in this text, the term $PM_{2.5}$ will be used to include all particles having an aerodynamic diameter of less than or equal to 2.5 μm.

  EPA chose 2.5 μm as the partition between fine and coarse particulate matter. Particles less than ~2.5 μm are regulated as **$PM_{2.5}$**. Air emission testing and air pollution control methods for $PM_{2.5}$ particles are different than those for coarse and super coarse particles.

  $PM_{2.5}$ particles settle quite slowly in the atmosphere relative to coarse and super-coarse particles. Normal weather patterns can keep $PM_{2.5}$ particles airborne for several hours to several days and enable these particles to cover hundreds of miles. $PM_{2.5}$ particles can cause health problems due to their potentially long airborne retention time and the inability of the human respiratory system to defend itself against particles of this size.

  In addition, the chemical makeup of $PM_{2.5}$ particles is quite different than for coarse and super-coarse particles. EPA data indicate that $PM_{2.5}$ particles are composed primarily of sulfates, nitrates, organic compounds, and ammonium compounds. The EPA also determined that $PM_{2.5}$ particles often contain acidic materials, metals and other contaminants believed to be associated with adverse health effects.

  Particles <1 μm in diameter are termed submicrometer particles ad can be the most difficult size to collect. Particles in the range of 0.2–0.5 μm are common in many types of combustion, waste incineration, and metallurgical sources. Particles in the range of 0.1–1.0 μm are important because they can represent a significant fraction of the particulate emissions from some types of industrial sources and because they are relatively hard to collect.

- **Particles <0.1 μm**: particles can be much smaller than 0.1 μm. Particles composed of as little as 20–50 molecules clustered together can exist in a stable form. Some industrial processes such as combustion and metallurgical sources generate particles in the range of 0.01–0.1 μm. These sizes are approaching the size of individual gas molecules, which are in the range of 0.0002–0.001 μm. However, particles in the size range of 0.01–0.1 μm tend to agglomerate rapidly to yield particles in the >0.1 μm range. Accordingly, very little of the particulate matter entering an air pollution control device or leaving the stack remains in the very small size range of 0.01–0.1 μm.

- **Condensable particulate matter**: particulate matter that forms from condensing gases or vapors is referred to as *condensable particulate matter*. Condensable particulate matter forms through chemical reactions as well as by physical phenomena.

  Condensable particulate matter is usually formed form material that is not particulate matter at stack conditions but which condenses and/or reacts upon cooling and dilution in the ambient air to form particulate matter. The formation of condensable particulate matter occurs within a few seconds after discharge from the stack.

  From a health standpoint, condensable particulate matter is important because it is almost entirely contained in the $PM_{2.5}$ classification.

These particle categories are important because particulate matter is regulated and tested for under these categories.

Size Distribution

Particulate emissions from both manmade and natural sources do not consist of particles of any one size. Instead, they are composed of particles over a relatively wide size range. It is often necessary to describe this size range. Particulate matter for size distribution evaluation is measured in a variety of ways. The data must be measured in a manner whereby it can be classified into successive particle diameter size categories.

Particle Formation

The range of particle sizes formed in a process is largely dependent on the types of particle formation mechanisms present. The general size range of particles can be estimated by simply recognizing which particle formation mechanisms are most important in the process being evaluated. The most important particle formation mechanisms in air pollution sources include physical attrition/mechanical dispersion, combustion particle burnout, homogeneous and heterogeneous nucleation, and droplet evaporation. Several particle formation mechanisms can be present in an air pollution source. As a result, the particles created can have a wide range of sizes and chemical composition. Particle formation mechanisms are described in detail below.

- **Physical attrition**: generates primarily moderate-to-large-sized particles and occurs when two surfaces rub together. For example, the grinding of a metal rod on a grinding wheel yields small particles that break off from both surfaces. The compositions and densities of these particles are identical to the parent materials.

  In order for fuel to burn, it must be pulverized or atomized so that there is sufficient surface area exposed to oxygen and high temperature. As indicated in Table 7.14, the surface area of particles increases substantially as more and more of the material is reduced in size.

  Accordingly, most industrial-scale combustion processes use one or more types of physical attrition in order to prepare or introduce their fuel into the furnace. For example, oil-fired boilers use pulverizers to reduce the chunks of coal to sizes that can be burned quickly. Oil-fired boilers use atomizers to disperse the oil as fine droplets. In both cases, the fuel particle size range is reduced to primarily the 10- to 1,000-µm range by physical attrition.

- **Combustion particle burnout**—when fuel particles are injected into the hot furnace area of the combustion process, such as in fossil-fuel-fired boilers, most of the organic compounds in the fuel are vaporized and oxidized in the gas stream. Fuel particles become smaller as the volatile matter leaves and they are quickly reduced to only the incombustible matter (ash) and the slow-burning char composed of organic compounds. Eventually, most of the char will also burn to leave primarily the incombustible material.

  As combustion progresses, the fuel particles, which started s 10- to 1,000-µm particles, are reduced to ash and char particles that are primarily in the 1–100 µm range. This mechanism for particle formation can be termed combustion particle burnout.

TABLE 7.14
Surface Area Comparison for Spherical Particles of Different Diameters[a]

Total Mass	Diameter of Particles (µm)	Number of Particles (Approx. in Millions)	Total Surface Area (cm²)	(m²)
1.0 gm	1,000	0.002	60	0.006
	100	2	600	0.06
	10	2,000	6,000	0.6
	1	2,000,000	60,000	6
	0.1	2,000,000,000	600,000	60

Source: EPA (2007b).

[a] Based on density of 1.0 gm/cm³.

- **Homogenous and heterogeneous nucleation**: involve the conversion of vapor phase materials to a particulate form. In both cases, the vapor-containing gas streams must cool to the temperature at which nucleation can occur, which is the dew point. Each vapor phase element and compound has a different dew point. Therefore, some materials nucleate in relatively hot gas zones while others remain as vapor until the gas stream is cold.
  - **Homogeneous nucleation**: is the formation of new particles composed almost entirely of the vapor phase material. The formation of particles by homogeneous nucleation involves only one compound.
  - **Heterogeneous nucleation**: is the accumulation of material on the surfaces of existing particles. In the case of heterogeneous nucleation, the resulting particle consists of more than one compound.

  There are two main categories of vapor phase material that can nucleate air pollution source gas streams: (1) organic compounds, and (2) inorganic metals and metal compounds. In a waste incinerator, waste that volatizes to organic vapor is generally oxidized completely to carbon dioxide and water.

  However, if there is an upset in the combustion process, a portion of the organic compounds of their partial oxidation products remain in the gas stream as it leaves the incinerator. Volatile metals and metal compounds such as mercury, lead, lead oxide, cadmium, cadmium oxide, and arsenic trioxide can also volatilize in the hot incinerator. Once the gas stream passes through the heat exchange equipment (i.e. waste heat boiler) used to produce stream, the organic vapors and metal vapors can condense homogeneously or heterogeneously. In general, the metals and metal compound search their dew point first and begin to nucleate in relatively hot zones of the unit.

  The organic vapors begin to condense in areas downstream form the process where the gas temperatures are cooler. These particles must then be collected in the downstream air pollution control systems. Homogeneous and heterogeneous nucleation generally creates particles that are very small, often between 0.1 and 1.0 µm.

  Heterogeneous nucleation facilitates a phenomenon called enrichment of particles in the submicrometer size range. The elemental metals and metal compounds volatized during high temperature operations (fossil fuel combustion, incineration, industrial furnaces, and metallurgical processes) nucleate preferentially as small particles or on the very small particles produced by these processes. Consequently, very small particles have more potentially toxic materials than the very large particles leaving the processes. Heterogeneous nucleation contributes to the formation of article distributions that have quite different chemical compositions in different size ranges.

  Another consequence of particle formation by heterogeneous nucleation is that a greater variety of chemical reactions may occur in the gas stream than would otherwise happen. During heterogeneous nucleation, small quantities of metals are deposited on the surfaces of many small particles. In this form, the metals are available to participate in catalytic reactions with gases or other vapor phase materials that are continuing to nucleate. Accordingly, heterogeneous nucleation increases the types of chemical reactions that can occur as the particles travel in the gas stream from the process source and through the air pollution control device.
- **Droplet evaporation**: some air pollution control system use solids-containing water recycled from wet scrubbers to cool the gas streams. This practice inadvertently creates another particle formation mechanism that is very similar to fuel burnout. The water streams are atomized during injection into the hot gas streams. As these small droplets evaporate to dryness, the suspended and dissolved solids are released as small particles. The particle size range created by this mechanism has not been extensively studied. However, it probably creates particles that range in size from 0.1 to 20 µm. All of these particles must then be collected in the downstream air pollution control systems.

COLLECTION MECHANISMS

When sunlight steams into a quiet room, particles of many different shapes and sizes can be seen, some appear to float while others slowly settle to the floor. All of these small particles are denser than the room air, but they do not settle very fast. The solid, liquid, and fibrous particles formed in air pollution sources behave in a manner that is very similar to standard household dusts and other familiar particles. What we instinctively understand about these everyday particles can be applied in many respects to the particles from air pollution sources.

There are, however, two major differences between industrially generated particles and those in more familiar settings. Industrial particles are much smaller than most household particles. Also, some industrial particles have complex chemical compositions and include compounds and elements that are known to be toxic.

Emission testing devise and air pollution control systems apply forces to the particles to remove them from the gas stream. These forces include inertial impaction and interception, Brownian diffusion, gravitational settling, electrostatic attraction, thermophoresis, and Diffusiophoresis. These forces are basically the "tools" that can be used for separating particles from the gas stream. All these collection mechanism forces are strongly dependent on particle size.

- **Inertial impaction and interception**: due to inertia, a particle moving in a gas steam can strike slowly moving or stationary obstacles (targets) in its path. As the gas stream deflects around the obstacle, the particle continues toward the object and impacts it. The obstacle may be a solid particle, stationary object, or water droplet.

    Two primary factors affect the probability of an impaction occurring: (1) aerodynamic article size and (2) the difference in velocity between the particle and the obstacle. Larger particles are collected more easily than smaller particles due to their greater inertia. Also, collection efficiency increases as the difference in velocity between the particle in the gas stream and the obstacle (or target) increase.

    Inertial impaction is analogous to a small car riding down an interstate highway at 65 mph and approaching a merle lane where a slowly moving truck is entering. If the care is unable to get into the passing lane to go around the merging truck, there could be an "impaction" incident. Larger cars will have more difficulty going around the tuck than smaller cars. Also, the faster the car is going relative to the truck, the more probable is an impaction.

    The efficiency of impaction is directly proportion to the impaction parameter shown in Equation 7.24 As the value of this parameter increases, the efficiency of inertial impaction increases. This parameter is related to the square of the Stokes particle diameter and the difference in velocity between the particle and the target droplet.

$$K_1 = \frac{C_c \left(d_{ps}\right)^2 P_p}{18\mu D_c} \tag{7.24}$$

where
K_1 = Impaction parameter (dimensionless)
C_c = Cunningham slip correction factor (dimensionless)
d_{ps} = Stokes particle diameter (μm)
v = Difference in velocity (cm/s)
P_p = Particle density (gm/cm^3)
D_c = Diameter of a droplet (cm)
μ = Gas viscosity (gm/cm-s)

The Cunningham slip correction factor (also called Cunningham's correction factor) accounts for molecular slip. Molecular slip occurs when the size of the particle is of the same magnitude as the distance between gas molecules. The particle no longer moves as a continuum in the gas, but as a particle among discrete gas molecules, thereby reducing the drag force. For particles in the air with an actual diameter of 1.0 μm and less, the Cunningham correction factor is significant.

Inertial impaction occurs when obstacles (e.g. water droplets) are directly in the path of the particle moving in the gas stream. Sometimes the obstacle or target is offset slightly from the direct path of the moving particle. In this instance, as the particle approaches the edge of the obstacle, the obstacle may collect the particle through a process called *interception*.

- **Brownian diffusion**: becomes the dominant collection mechanism for particles <0.3 μm and is especially significant for particles in the 0.01–0.1 μm size range.

  Very small particles in a gas stream deflect slightly when gas molecules strike them. Transfer of kinetic energy from the rapidly moving gas molecule to the small particle causes this deflection, called *Brownian diffusion*. These small particles are captured when they impact a target (e.g. liquid droplet) as a result of this random movement.

  Diffusivity is a measure of the extent to which molecular collisions influence very small particles, causing them to move in a random manner across the direction of gas flow. The diffusion coefficient in the equation below represents the diffusivity of a particle at certain gas stream conditions.

$$D_p = \frac{C_c K T}{3\pi d_{pa} \mu} \quad (7.25)$$

where
  D_p = Diffusion coefficient (cm²/s)
  C_c = Cunningham slip correction factor (dimensionless)
  K = Boltzmann constant (gm-cm²/s² °K)
  T = Absolute temperature (°K)
  d_{pa} = Particle aerodynamic diameter (μm)
  μ = Gas viscosity (kg/m-s)

- **Gravitational settling**: particles in still air have two forces acting on them; (1) a gravitational force downward and (2) the air resistance (or drag) force upward. When particles begin to fall, they quickly reach a terminal settling velocity, which represents the constant velocity of a falling particle when the gravitational force downward is balanced by the air resistance (or drag) force upward. The terminal settling velocity can usually be expressed using Equation 7.26. [Note: Equation 7.26 is applicable for particles <80 μm in size (aerodynamic diameter) and having a Reynolds number <2.0 and a low velocity].

$$v_t = \frac{G P_p (d_{ps})^2 C_c}{18\mu} \quad (7.26)$$

where
  v_t = Terminal settling velocity (cm/s)
  g = Gravitational acceleration (cm/s²)
  P_p = Density of particle (gm/cm³)
  d_{ps} = Stokes particle diameter (cm)
  C_c = Cunningham slip correction factor (dimensionless)
  μ = Viscosity of air (gm/cm-s)

- **Electrostatic attraction**—in air pollution control, electrostatic precipitators (ESPs) use electrostatic attraction for particulate collection. Electrostatic attraction of particles is accomplished by establishing a strong electrical field and creating unipolar ions. The particles passing through the electrical field are charged by the ions being driven along the electrical field lines. Several parameters dictate the effectiveness of electrostatic attraction including the particle size, gas flow rate, and resistivity.

    The particles will eventually reach a maximum or saturation charge, which is a function of the particle area. The saturation charge occurs when the localized field created by the already captured ions is sufficiently strong to deflect the electrical field lines. Particles can also be charged by the diffusion of ions in the gas stream. The strength of the electrical charges imposed on the particles by both mechanisms is particle size dependent.
- **Resistivity**: is a measure of the ability of the particle to conduct electricity and is expressed in units of ohm-cm. Particles with low resistivity have a greater ability to conduct electricity (and higher electrostatic attraction) than particles with high resistivity. The following factors influence resistivity:
    - Chemical composition of the gas stream
    - Chemical composition of the particle
    - Gas stream temperature
- **Thermophoresis**: is particle movement caused by thermal differences on two sides of the particle. Gas molecules at higher temperatures have greater kinetic energy than those at lower temperatures. Therefore, when the particle collides with a gas molecule from the hotter side, the particle receives more kinetic energy than when it collides with a gas molecule from the cooler side. Accordingly, particles tend to be deflected toward the colder area.
- **Diffusiophoresis**: is particle movement caused by concentration differences on two sides of the particle. When there is a strong difference in the concentration of gas molecules on two sides of the particle, there is a difference in the number of molecular collisions, which causes an imbalance in the total kinetic energies of the gas molecules. Gas molecules in the high concentration area striking a particle transmit more kinetic energy to the particle than molecules in the lower concentration area. Therefore, particles tend to move toward the area of lower concentration.

ATMOSPHERIC CHEMISTRY AND RELATED PHENOMENA

Note: In the brief discussion that follows describing the chemistry of Earth's atmosphere, keep in mind that the atmosphere as it is at present is what is referred to. The atmosphere previous to this period was chemically quite different. Also, note that "atmospheric chemistry" is a scientific discipline or entity that can stand on its own. The "nuts and bolts" of atmospheric chemistry are beyond the scope of this text. Here certain important atmospheric chemistry phenomena are highlighted.

You should recall that the full range of chemistry (including slow and fast reactions, dissolving crystals, and precipitation of solids) all occur in the atmosphere, and that the atmosphere is as Gradel and Crutzen (1995) described it—a "flask without walls." We also pointed out that (excluding highly variable amounts of water vapor) more than 99% of the molecules constituting Earth's atmosphere are nitrogen, oxygen, and chemically inert gases (noble gases such as argon, etc).

The chemistry (and thus the reactivity) of these natural gases (nitrogen, oxygen, carbon dioxide, argon, and others) is well known. The other chemical reactive chemicals (anthropogenically produced) that are part of Earth's atmosphere are also known, but there are still differing opinions on their "exact" total affect/effect on our environment. For example, methane is by far the most abundant reactive compound in the atmosphere and currently is at a ground-level concentration (in the northern hemisphere) of about 1.7 ppmv. We know significant amounts of information about methane (its generation and fate when discharged) and its influence on the atmosphere; however, we are still conducting research to find out more—as we should.

Many different reactive molecules (other than methane) exist in the atmosphere. We may not be familiar with each of these reactants, but many of us certainly are familiar with their consequences: the ozone hole, greenhouse effect and global warming, smog, acid rain, the rising tide, and so on. It may surprise you to know, however, that the total amount of all these reactants in the atmosphere is seldom more than 10 ppmv volume anywhere in the world at any given time. The significance should be obvious: the atmospheric problems currently occurring on Earth at this time are the result of less than one-thousandth of 1% of all of the molecules in the atmosphere. This indicates that environmental damage (causing global atmospheric problems) can result from far less than the tremendous amounts of reactive substances we imagine are dangerous.

The quality of the air we breathe, the visibility and atmospheric esthetics, and our climate (all of which are dependent upon chemical phenomena that occur in the atmosphere) are important to our health and to our quality of life. Global atmospheric problems, such as the nature and level of air pollutants, concern the air science practitioner the most because they affect our health and of quality of life.

Let's take a look at some of the important chemical species and their reactions within (primarily) the stratosphere of our atmosphere.

PHOTOCHEMICAL REACTION—SMOG PRODUCTION

A *photochemical reaction*, generally, is any chemical reaction in which light is produced or light initiates the reaction. Light can initiate reactions by exciting atoms or molecules and making them more reactive. The light energy becomes converted to chemical energy.

The photochemical reaction we are concerned with here is the action or absorption of electromagnetic solar radiation (light) by chemical species, which causes the reactions. The ability of electromagnetic radiation to cause photochemical reactions to occur is a function is shown in the relationship

$$E = hv \tag{7.27}$$

where
E = energy of a photon
v = frequency
h = Planck' constant, 6.62×10^{-27}

The major photochemical reaction we are concerned with is the one that produces photochemical smog. *Photochemical smog* is initiated by nitrogen dioxide, which absorbs the visible or ultraviolet energy of sunlight, forming nitric oxide to free atoms of oxygen (O), which then combine with molecular oxygen (O_2) to form *ozone* (O_3). In the presence of hydrocarbons (other than methane) and certain other organic compounds, a variety of chemical reactions take place. Some 80 separate reactions have been identified or postulated.

The photochemical reaction that produces the smog we are familiar with are dependent on two factors:

1. Smog concentration is linked to both the amount of sunlight and hydrocarbons present.
2. The amount is dependent on the initial concentration of nitrogen oxides.

In the production of photochemical smog, many different substances are formed in sequence including acrolein, formaldehyde, and PAN (Peroxyacetylnitrate). Photochemical smog (the characteristic haze of minute droplets) is a result of condensed low-volatility organic compounds. The organics irritate the eye, and also, together with ozone, can cause severe damage to leafy plants. Photochemical smog tends to be most intense in the early afternoon when sunlight intensity is greatest. It differs from traditional smog (Los Angeles-type smog), which is most intense in the early morning and is dispersed by solar radiation.

In addition to the photochemical reactions that produce smog, many other chemical reactions take place in the Earth's atmosphere including ozone production, production of free radicals, chain reactions, oxidation processes, acid-base reactions, the presence of electrons and positive ions, and many others.

AIR QUALITY

When undertaking a comprehensive discussion of air, the discussion begins and ends with the Earth's atmosphere—and to a point, we have done just that. In this chapter, concepts are covered that will enable us to better understand the anthropogenic impact of pollution on the atmosphere, which in turn will enable us to better understand the key parameters used to measure air quality. Obviously, having a full understanding of air quality is essential. To set the stage for information to follow in subsequent sections related to air pollution and air pollution control, we must review a few basic concepts.

EARTH'S HEAT BALANCE

The energy expended in virtually all atmospheric processes is originally derived from the Sun. This energy is transferred by radiation of heat in the form of electromagnetic waves. The radiation from the Sun has its peak energy transmission in the visible wavelength range (038–0.78 μm) of the electromagnetic spectrum. However, the Sun also releases considerable energy in the ultraviolet and infrared regions. Ninety-nine percent of the Sun's energy is emitted in wavelengths between 0.15 and 40 μm. Furthermore, wavelengths longer than 2.5 μm are strongly absorbed by water vapor and carbon dioxide in the atmosphere. Radiation at wavelength <0.29 μm is absorbed high in the atmosphere by nitrogen and oxygen. Therefore, solar radiation striking the Earth generally has a wavelength between 0.29 and 2.5 μm (EPA, 2005).

Since energy from the Sun is always entering the atmosphere, the Earth would overheat if all this energy were stored in the Earth-atmosphere system. So, energy must eventually be released back into space. On the whole, this is what happens—approximately 50% of the solar radiation entering the atmosphere reaches Earth's surface, either directly or after being scattered by clouds, particulate matter, or atmospheric gases. The other 50% is either reflected directly back or absorbed in the atmosphere and its energy is reradiated back into space at a later time as infrared radiation. Most of the solar energy reaching the surface is absorbed and must be returned to space to maintain *heat balance (aka radiational balance)*. The energy produced within the Earth's interior (from the hot mantle area via convection and conduction) which reaches the Earth's surface (about 1% of that received from the Sun) must also be lost.

Reradiation of energy from the Earth is accomplished by three energy transport mechanisms: radiation, conduction, and convection. *Radiation* of energy, as stated earlier, occurs through electromagnetic radiation in the infrared region of the spectrum. The crucial importance of the radiation mechanism is that it carries energy away from Earth on a much longer wavelength than the solar energy (sunlight) that brings energy to the Earth and, in turn, works to maintain the Earth's heat balance. The Earth's heat balance is of particular interest to us in this text because it is susceptible to being upset by human activities.

A comparatively smaller but significant amount of heat energy is transferred to the atmosphere by conduction from the Earth's surface. *Conduction* of energy occurs through the interaction of adjacent molecules with no visible motion accompanying the transfer of heat; for example, the whole length of a metal rod will become hot when one end is held in a fire. Because air is a poor heat conductor, conduction is restricted to the layer of air in direct contact with the Earth's surface. The heated air is then transferred aloft by *convection*, the movement of whole masses of air, which may be either relatively warm or cold. Convection is the mechanism by which abrupt temperature variations occur when large masses of air move across an area. Air temperature tends to be greater near the surface of the Earth and decreases gradually with altitude. A large amount of the Earth's surface heat is transported to clouds in the atmosphere by conduction and convection before being lost ultimately by radiation and this redistribution of heat energy plays an important role in weather and climate conditions.

The Earth's average surface temperature is maintained at about 15°C because of the atmospheric greenhouse effect. Greenhouse effect occurs when the gases of the lower atmosphere transmit most of the visible portion of incident sunlight in the same way as the glass of a garden greenhouse. The warmed Earth emits radiation in the infrared region, which is selectively absorbed by the atmospheric gases whose absorption spectrum is similar to that of glass. This absorbed energy heats the atmosphere and helps maintain the Earth's temperature. Without this greenhouse effect, the surface temperature would average around −18°C. Most of the absorption of infrared energy is performed by water molecules in the atmosphere. In addition to the key role played by water molecules, carbon dioxide, although to a lesser extent, also is essential in maintaining the heat balance. Environmentalists and others concerned with environmental issues are concerned that an increase in the carbon dioxide level in the atmosphere could prevent sufficient energy loss, causing damaging increases in the Earth's temperature. This phenomenon, commonly known as anthropogenic greenhouse effect, may occur from elevated levels of carbon dioxide levels caused by increased use of fossil fuels and the reduction in carbon dioxide absorption because of the destruction of the rainforest and other forest areas.

INSOLATION (EPA, 2005)

The amount of **incoming solar radiation** received at a particular time and location in the Earth-atmosphere system is called insolation. Insolation is governed by four factors:

- Solar Constant
- Transparency of the atmosphere
- Daily sunlight duration
- Angle at which the Sun's rays strike the Earth

Solar Constant

The *solar constant* is the average amount of radiation received at a point, perpendicular to the Sun's rays, that is located outside the Earth's atmosphere at the Earth's mean distance from the Sun. The average amount of solar radiation received at the outer edge of the atmosphere would vary slightly depending on the energy output of the Sun and the distance of the Earth relative to the Sun. Due to the eccentricity of the Earth's orbit around the Sun, the Earth is closer to the Sun in January than in July. Also, the radiation emitted from the Sun varies slightly, probably less than a few percent. These slight variations that affect eh solar constant are trivial considering the atmospheric properties that deplete the overall amount of solar radiation reaching the Earth's surface. Transparency of the atmosphere, duration of daylight, and the angle at which the Sun's rays strike the Earth are much more important in influencing the amount of radiation actually received, which in turn includes the weather.

Transparency

Transparency of the atmosphere does have an important bearing on the amount of insolation that reaches the Earth's surface. The emitted radiation is depleted as it passes through the atmosphere. Different atmospheric constituents absorb or reflect energy in different ways and in varying amounts. Transparency of the atmosphere refers to how much radiation penetrates the atmosphere and reaches the Earth's surface without being depleted.

The general reflectivity of the various surfaces of the Earth is referred to as the albedo. *Albedo* is defined as the fraction (or percentage) of incoming solar energy that is reflected back to space. Different surfaces (water, snow, sand, etc.) have different albedo values. For the Earth and atmosphere as a whole, the average albedo is 30% for average conditions of cloudiness over the Earth. This reflectivity is greatest in the visible range of wavelengths.

Some of the gases in the atmosphere (notably water vapor) absorb solar radiation, causing less radiation to reach the Earth's surface. Water vapor, although comprising only about 3% of the atmosphere, on the average absorbs about six times as much solar radiation as all other gases combined. The amount of radiation received at the Earth's surface is therefore considerably less than hat received outside the atmosphere as represented by the solar constant.

Daylight Duration

The duration of daylight also affects the amount of insolation received: the longer the period of sunlight, the greater the total possible insolation. Daylight duration varies with latitude and the seasons. At the equator, day and night are always equal. In the Polar Regions, the daylight period reaches a maximum of 24 hours in summer and a minimum of zero hours in winter.

Angle of Sun's Rays

The angle at which the Sun's rays strike the Earth varies considerably as the Sun "shifts" back and forth across the equator. A relatively flat surface perpendicular to an incoming vertical sun ray receives the largest amount of insolation. Therefore, areas at which the Sun's rays are oblique receive less insolation because the oblique rays must pass through a thicker layer of reflecting and absorbing the atmosphere and are spread over a greater surface area. This same principle also applies to the daily shift of the Sun's rays. At solar noon, the intensity of insolation is greatest. In the morning and evening hours, when the Sun is at a low angle, the amount of insolation is small.

HEAT DISTRIBUTION

The Earth, as a whole, experiences great contrasts in heat and cold at any particular time. Warm, tropical breezes blow at the equator while ice caps are forming in the Polar Regions. In fact, due to the extreme temperature differences at the equator and the poles, the Earth-atmosphere system resembles a giant "heat engine." Heat engines depend on hot-cold contrasts to generate power. As you will see, this global "heat engine" influences the major atmospheric circulation patterns as warm air is transferred to cooler areas. Different parts of the Earth receiving different amounts of insolation account for much of this heat imbalance. As discussed earlier, latitude, the seasons, and daylight duration cause different locations to receive varying amounts of insolation.

Differential Heating

Not only do different amounts of solar radiation reach the Earth's surface, but different Earth surfaces absorb heat energy at different rates. For example, land masses absorb and store heat differently than water masses. Also, different types of land surfaces vary in their ability to absorb and store heat. The color, shape, surface texture, vegetation and presence of buildings can all influence the heating and cooling of the ground. In general, dry surfaces heat and cool faster than moist surfaces. Plowed fields, sandy beaches, and paved roads become hotter than surrounding meadows and wooded areas. During the day, the air over a plowed field is warmer than over a forest or swamp; during the night, the situation is reversed. The property of different surfaces which causes them to heat and cool at different rates is referred to as *differential heating*.

Absorption of heat energy from the Sun is confined to a swallow layer of the land surface. Consequently, land surfaces heat rapidly during the day and cool quickly at night. Water surfaces, on the other hand, heat and cool more slowly than land surfaces for the following reasons:

- Water movement distributes heat
- The Sun's rays are able to penetrate the water surface
- More heat is required to change the temperature of water due to its higher specific heat. (It takes more energy to raise the temperature of water than it does to change the temperature of the same amount of soil.)
- Evaporation of water occurs which is a cooling process

Transport of Heat

Earlier, it was pointed out that in addition to radiation, heat is transferred by conduction, convection, and advection. These processes affect the temperature of the atmosphere near the surface of the Earth. *Conduction* is the process by which heat is transferred through matter without the transfer of matter itself. For example, the handle of an iron skillet becomes hot due to the conduction of heat from the stove burner. Heat is conducted from a warmer object to a cooler one. Heat transfer occurs when the matter is in motion. Air that is warmed by a heated land surface (by conduction) will rise because it is lighter than the surrounding air. This heated air rises, transferring heat vertically. Likewise, cooler air aloft will sink because it is heavier than the surrounding air. This goes hand in hand with rising air and is part of heat transfer by convection. Meteorologists also us the term *advection* to denote heat transfer that occurs mainly by horizontal motion rather than by vertical movement or air (convection).

Global Distribution of Heat

As mentioned before, the world distribution of insolation is closely related to latitude. Total annual insolation is greatest at the equator and decreases toward the poles. The amount of insolation received annually at the equator is over four times that received at either of the poles. As the rays of the sun shift seasonally from on hemisphere to the other, the zone of maximum possible daily insolation moves with them. For the Earth as a whole, the gains in solar energy equal the losses of energy back into space (heat balance). However, since the equatorial region does gain more heat than it loses and the poles lose more heat than they gain, something must happen to distribute heat more evenly around the Earth. Otherwise, the equatorial regions would continue to heat and the poles would continue to cool. Therefore, in order to reach equilibrium, a continuous large-scale transfer of heat (from low to high altitudes) is carried out by atmospheric and oceanic circulations.

The atmosphere drives warm air poleward and brings cold air toward the equator. Heat transfer from the tropics poleward takes place throughout the year, but at a much slower rate in summer than in winter. The temperature difference between low and high latitudes is considerably smaller in summer than in winter (only about half as large in the Northern Hemisphere). As would be expected, the winter hemisphere has a net energy loss and the summer hemisphere a net gain. Most of the summertime gain is stored in the surface layers of land and ocean, mainly in the ocean.

The oceans also play a role in heat exchange. Warm water flows poleward along the western side of an ocean basin and cold-water flows toward the equator on the eastern side. At higher latitudes, warm water moves poleward in the eastern side of the ocean basin and cold-water flows toward the equator on the western side. The oceanic currents are responsible for about 40 percent of the transport of energy from the equator to the poles. The remaining 60 percent is attributed to the movement of air.

AIR QUALITY MANAGEMENT

Proper air quality management includes several different areas related to air pollutants and their control. For example, we can mathematically model to predict where pollutants emitted from a source will be dispersed in the atmosphere and eventually fall to the ground and at what concentration. We have found that pollution control equipment can be added to various sources to reduce the amount of pollutants before they are emitted into the air. We have found that certain phenomena such as acid rain, the greenhouse effect, and global warming are all indicators of adverse effects on the air and other environmental mediums, which result from the excessive amount of pollutants being released into the air. We have found that we must concern ourselves not only with ambient air quality in our local outdoor environment but also with the issue of indoor air quality.

To accomplish air quality management, we have found that managing is one thing—and accomplishing significant change improvement is another. We need to add regulatory authority, regulations, and regulatory enforcement authority to the air quality management scheme; strictly voluntary compliance in ineffective.

We cannot maintain a quality air supply without proper management, regulation, and regulatory enforcement. This section presents the regulatory framework governing air quality management. It provides an overview of the environmental air quality laws and regulations used to protect human health and the environment from the potential hazards of air pollution. New legislation, reauthorizations of Acts and new National Ambient Air Quality Standards (NAAQS) have created many changes in the way government and industry manage their business. Fortunately for our environment and for us, they are management tools that are effective—they are working to manage air quality.

CLEAN AIR ACT (CAA)

When you look at a historical overview of air quality regulations, you might be surprised to discover that most air quality regulations are recent. For example, in the United States, the first attempt at regulating air quality came about through the passage of the Air Pollution Control Act of 1955 (Public Law 84–159). This act was a step forward but that's about all; it did little more than move us toward effective legislation. Revised in 1960 and again in 1962, the act was supplanted by the Clean Air Act (CAA) of 1963 (Public Law 88-206). CAA 1963 encouraged state, local, and regional programs for air pollution control but reserved the right of federal intervention, should pollution from one state endanger the health and welfare of citizens residing in another state. In addition, CAA in 1963 initiated the development of air quality criteria upon which the air quality and emissions standards of the 1970s were based.

The move toward air pollution control gained momentum in 1970 first by the creation of the Environmental Protection Agency (EPA) and second by the passage of the Clean Air Act of 1970 (Public Law 91-604), for which the EPA was given responsibility for implementation. The act was important because it set primary and secondary ambient- air-quality standards. Primary standards (based on air quality criteria), allowed for an extra margin of safety to protect public health, while secondary standards (also based on air quality criteria) were established to protect public welfare—animals, property, plants, and materials.

The Clean Air Act of 1977 (Public Law 95-95) further strengthened the existing laws and set the nation's course toward cleaning up our atmosphere.

In 1990, the President signed the Clean Air Act Amendments of 1990. Specifically, the new law:

- Encourages the use of market-based principles and other innovative approaches, like performance-based standard and emission banking and trading.
- Promotes the use of clean low-sulfur coal and natural gas, as well as the use of innovative technologies to clean high-sulfur coal through the acid rain program.

- Reduces enough energy waste and creates enough of a market for clean fuels derived from grain and natural gas to cut dependency on oil imports by one million barrels/day.
- Promotes energy conservation through an acid rain program that gives utilities the flexibility to obtain needed emission reductions through programs that encourage customers to conserve energy.

Under CAA 1990 several Titles are listed with specific requirements. For example,

- **Title 1**: specifies provisions for the attainment and maintenance of National Ambient Air Quality Standards (NAAQS).
- **Title 2**: specifies provisions relating to mobile sources of pollutants.
- **Title 3**: covers air toxics.
- **Title 4**: covers specifications for acid rain control.
- **Title 5**: addresses permits.
- **Title 6**: specifies stratospheric ozone and global protection measures.
- **Title 7**: discusses provisions relating to enforcement.

TITLE 1: ATTAINMENT AND MAINTENANCE OF NAAQS

The Clean Air Act of 1977 has brought about significant improvements in U.S. air quality, but the urban air pollution problems of smog (ozone), carbon monoxide (CO), and particulate matter (PM-10) still persist. For example, currently, over 100 million Americans live in cities which are out of attainment with the public health standards for ozone.

A new, balanced strategy for attacking the urban smog problem was needed. The Clean Air Act of 1990 created this new strategy. Under these new amendments, states are given more time to meet the air quality standard (e.g., up to 20 years for ozone in Los Angeles) but they must make steady, impressive progress in reducing emissions. Specifically, it requires the federal government to reduce emissions from (1) cars, buses, and trucks; (2) consumer products such as window-washing compounds and hair spray; and (3) ships and barges during loading and unloading of petroleum products. In addition, the federal government must develop the technical guidance that states need to control stationary sources. In urban air pollution problems of smog (ozone), carbon monoxide (CO), and particulate matter (PM-10), the new law clarifies how areas are designated and redesignated "attainment." The EPA is also allowed to define the boundaries of "nonattainment" of areas (geographical areas whose air quality does not meet federal air quality standards designed to protect public health. CAA 1990 also establishes provisions defining when and how the federal government can impose sanctions on areas of the country that have not met certain conditions.

For ozone specifically, the new law established nonattainment area classifications ranked according to the severity of the area's air pollution problem. These classifications are

- marginal
- moderate
- serious
- severe
- extreme

The EPA assigns each nonattainment area one of these categories, thus prompting varying requirements the areas must comply with in order to meet the ozone standard.

Again, nonattainment areas have to implement different control measures, depending on their classifications. Those closest to meeting the standard, for example, are the marginal areas, which are required to conduct an inventory of their ozone-causing emissions and institute a permit program. Various control measures must be implemented by nonattainment areas with more serious air quality problems; that is, the worse the air quality, the more control areas will have to implement.

TABLE 7.15
National Ambient Air Quality Standards (NAAQS)

Pollutant	Standard Value	
Carbon Monoxide (CO)		
8-hour average	9 ppm	10 mg/m³
1-hour average	35 ppm	40 mg/m³
Lead (Pb)		
Quarterly average	1.5 µg/m³	
Nitrogen Dioxide (NO₂)		
Annual arithmetic mean	0.053 ppm	100 µg/m³
1-hour average	0.12 ppm	235 µg/m³
8-hour average	0.08 ppm	157 µg/m³
Particulate Matter (PM-10)		
Annual arithmetic mean	50 µg/m³	
24-hour average 150 µg/m³		
Particulate Matter (PM-2.5)		
Annual arithmetic mean	15 µg/m³	
24-hour average	65 µg/m³	
Annual arithmetic mean	0.03 ppm	80 µg/m³
24-hour average	0.14 ppm	365 µg/m³

Source: USEPA, NAAQS (2007).

For carbon monoxide and particulate matter, CAA 1990 also establishes similar programs for areas that do not meet the federal health standard. Areas exceeding the standards for these pollutants are divided into "moderate" and "serious" classifications. Areas that exceed the carbon monoxide standard (i.e., to the degree to which they exceed it) are required primarily to implement programs introducing oxygenated rules and/or enhanced emission inspection programs. Likewise, areas exceeding the particulate matter standard have to (among other requirements) implement either reasonably available control measures (RACMs) or best available control measures (BACMs).

Title 1 attainment and maintenance of NAAQS requirements have gone a long way toward improving air quality in most locations throughout the United States. However, on November 27, 1996, in an effort to upgrade NAAQS for ozone and particulate matter, USEPA proposed the National Ambient Air Quality Standards. Carol Browner, EPA Administrator, later signed Notice of rulemaking putting into effect the two new NAAQS for ozone and particulate matter smaller than 2.5 µm diameter ($PM_{2.5}$). These rules appear at 62 FR 38651 for particulate matter, and 62 FR 38855 for ozone. They are the first update in 20 years for ozone (smog), and the first in 10 years for particulate matter (soot).

Table 7.15 lists the National Ambient Air Quality Standards including the new requirements (updated as of January 2007). Note that NAAQS is important but is not enforceable by itself. The standards set ambient concentration limits for the protection of human health and environment-related values. However, it is important to remember that it is a very rare case where any one source of air pollutants is responsible for the concentrations in an entire area.

TITLE 2: MOBILE SOURCES

Cars, trucks, and buses account for almost half the emissions (even though great strides have been made since the 1960s in reducing the amounts) of the ozone precursors, volatile organic carbons

(VOCs) and nitrogen oxides, and up to 90% of the CO emissions in urban areas. A large portion of the emission reductions gained from motor vehicle emission controls has been offset by the rapid growth in the number of vehicles on the highways and the total miles driven.

Due to the unforeseen growth in automobile emissions in urban areas, which was further compounded with the serious air pollution problems in many urban areas, Congress made significant changes to the motor vehicle provisions of the 1977 Clean Air Act. The Clean Air Act of 1990 established even tighter pollution standards for emissions from motor vehicles. These standards were designed to reduce tailpipe emissions of hydrocarbons, nitrogen oxides, and carbon monoxide on a phased-in basis which began with the model year 1994. Automobile manufacturers are also required to reduce vehicle emissions resulting from the evaporation of gasoline during refueling.

The latest Clean Air Act (1990 w/1997 amendments for ozone and particulate matter) also require fuel quantity to be controlled. New programs were required for cleaner or reformulated gasoline initiated in 1995 for the cities (nine total) with the worst ozone problems. Other cities were given the option to "buy in" to the reformulated gasoline program. In addition, a clean fuel car pilot program was established in California, which requires the phasing-in of tighter emission limits for several thousand vehicles in model year 1996 and up to 300,000 by model year 1999. The law allows these standards to be met with any combination of vehicle technology and cleaner fuels. Note that the standards became even stricter in 2001.

TITLE 3: AIR TOXICS

Toxic air pollutants (those which are hazardous to human health of the environment—carcinogens, mutagens, and reproductive toxins) were not specifically covered under Clean Air Act 1977. This situation is quite surprising and alarming when you consider that information generated as a result of SARA Title III (Superfund Section 313) indicates that in the U.S. more than 2 billion pounds of toxic air pollutants are emitted annually.

The Clean Air Act of 1990 offered a comprehensive plan for achieving significant reductions in emissions of hazardous air pollutants from major sources. The new law improved EPA's ability to address this problem effectively and dramatically accelerates progress in controlling major toxic air pollutants.

The 1990 law includes a list of 189 toxic air pollutants whose emissions must be reduced. The EPA was required to publish a list of source categories that emit certain levels of these pollutants. The EPA was also required to issue maximum achievable control technology (MACT) standards for each listed source category, and the law also established a Chemical Safety Board to investigate accidental releases of extremely hazardous chemicals.

TITLE 4: ACID DEPOSITION

The purity of rainfall is a major concern for many people, especially regarding the acidity of the precipitation. Most rainfall is slightly acidic because of decomposing organic matter, the movement of the sea, volcanic eruptions, but the principal factor is atmospheric carbon dioxide, which causes carbonic acid to form. *Acid rain* (pH < 5.6) (in the pollution sense) is produced by the conversion of the primary pollutants sulfur dioxide and nitrogen oxides to sulfuric acid and nitric acid, respectively. These processes are complex, depending on the physical dispersion processes and the rates of the chemical conversions.

Contrary to popular belief, acid rain is not a new phenomenon, nor does it result solely from industrial pollution. Natural processes—volcanic eruptions and forest fires, for example—produce and release acid particles into the air. The burning of forest areas to clear land in Brazil, Africa, and other countries also contribute to acid rain. However, the rise in manufacturing which began with the Industrial Revolution literally dwarfs all other contributions to the problem.

The main culprits are emissions of sulfur dioxide from the burning of fossil fuels, such as oil and coal, and nitrogen oxide, formed mostly from internal combustion engine emissions, which is readily transformed into nitrogen dioxide. These mix in the atmosphere to form sulfuric acid and nitric acid.

In dealing with atmospheric acid deposition, the Earth's ecosystems are not completely defenseless; they can deal with a certain amount of acid through natural alkaline substances in soil or rocks that buffer and neutralize acid. The American Midwest and southern England are areas with highly alkaline soil (limestone and sandstone) which provide some natural neutralization. Areas with thin soil and those laid on granite bedrock, however, have little ability to neutralize acid rain.

Scientists continue to study how living beings are damaged and/or killed by acid rain. This complex subject has many variables. We know from various episodes of acid rain that pollution can travel over very long distances. Lakes in Canada and New York are feeling the effects coal-burning in the Ohio Valley. For this and other reasons, the lakes of the world are where most of the scientific studies have taken place. In lakes, the smaller organisms often die off first, leaving the larger animals to starve to death. Sometimes the larger animals (fish) are killed directly; as lake water becomes more acidic, it dissolves heavy metals, leading to concentrations at toxic and often lethal levels. Have you ever wandered up to the local lake shore and observed thousands of fish belly-up? Not a pleasant sight or smell, is it? Loss of life in lakes also disrupts the system of life on the land and the air around them.

In some parts of the United States, the acidity of rainfall has fallen well below 5.6. In the northeastern U.S., for example, the average pH of rainfall is 4.6, and rainfall with a pH of 4.0, which is 1000 times more acidic than distilled water.

Despite intensive research into most aspects of acid rain, scientists still have many areas of uncertainty and disagreement. That is why progressive, forward-thinking countries emphasize the importance of further research into acid rain. And that is why the 1990 Clean Air Act was strengthened to initiate a permanent reduction in SO_2 levels.

One of the interesting features of the 1990 Act is that it allowed utilities to trade allowance within their systems and/or buy or sell allowance to and from other affected sources. Each source must have sufficient allowances to cover its annual emissions. If not, the source is subject to excess emissions fees and a requirement to offset the excess emissions in the following year. The 1990 law also included specific requirements for reducing emissions of nitrogen oxides for certain boilers.

TITLE 5: PERMITS

The 1990 law also introduced an operating permit system similar to the *National Pollution Discharge Elimination System* (NPDES). The permit system has twofold purpose: (1) to ensure compliance with all applicable requirements of the CAA and (2) to enhance the EPA's ability to enforce the Act. Under the Act, air pollution sources must develop and implement the program, and the EPA must issue permit program regulations, review each state's proposed program, and oversee the state's effort to implement any approved program. The EPA must also develop and implement a federal permit program when a state fails to adopt and implement its own program.

TITLE 6: OZONE AND GLOBAL CLIMATE PROTECTION

Ozone is formed in the stratosphere by radiation from the Sun and helps to shield life on Earth from some of the Sun's potentially destructive ultraviolet (UV) radiation.

In the early 1970s, scientists suspected that the ozone layer was being depleted. By the 1980s, it became clear that the ozone shield was indeed thinning in some places, and at times, even has a seasonal hole in it, notably over Antarctica. The exact causes and actual extent of the depletion are not yet fully known, but most scientists believe that various chemicals in the air are responsible.

Most scientists identify the family of chlorine-based compounds, most notably chlorofluorocarbons (CFCs) and chlorinated solvents (carbon tetrachloride and methyl chloroform) as the primary culprits involved in ozone depletion. In 1974, Molina and Rowland hypothesized the CFCs, containing chlorine, were responsible for ozone depletion. They pointed out that chlorine molecules are highly active and readily and continually break apart the three-atom ozone into the two-atom form of oxygen generally found close to the Earth, in the lower atmosphere.

The Interdepartmental Committee for Atmospheric Sciences (1975) estimates that a 5% reduction in ozone could result in nearly a 10% increase in cancer. This already frightening scenario was made even more frightening by 1987 when evidence showed that CFCs destroy ozone in the stratosphere above Antarctica every spring. The ozone hole had become larger, with more than half of the total ozone column wiped out and essentially all ozone disappeared from some regions of the stratosphere (Davis and Cornwell, 1991).

In 1988, Zurer reported that on a worldwide basis, the ozone layer shrunk ~2.5% in the preceding decade. This obvious thinning of the ozone layer, with its increased chances of skin cancer and cataracts, is also implicated in the suppression of the human immune system, and damage to other animals and plants, especially aquatic life, and soybean crops. The urgency of the problem spurred the 1987 signing of the Montreal Protocol by 24 countries, which required signatory countries to reduce their consumption of CFCs by 20% by 1993, and by 50% by 1998, marking a significant achievement in solving a global environmental problem.

The Clean Air Act of 1990 borrowed from EPA requirements already on the books in other regulations and mandated phase-out of the production of substances that deplete the ozone layer. Under these provisions, the EPA was required to list all regulated substances along with their ozone-depletion potential, atmospheric lifetime, and global warming potential.

TITLE 7: ENFORCEMENT

A broad array of authorities is contained within the Clean Air Act to make the law more readily enforceable. The EPA was given new authority to issue administrative penalties with fines, and field citations (with fines) for smaller infractions. In addition, sources must certify their compliance, and the EPA has the authority to issue administrative subpoenas for compliance data.

CLEAN AIR ACT AMENDMENTS (EPA, 2005)

The Clean Air Act amendments require that State Implementation Plans (SIPs) be developed, the impact upon the atmosphere be evaluated for new sources, and air quality modeling analyses be performed. These regulatory programs required knowledge of the air quality in the region around a source, air quality modeling procedures, and the fate and transport of pollutants in the atmosphere. Implicit in air pollution programs is knowledge of the climatology of the area in question.

STATE IMPLEMENTATION PLANS (SIPs)

State Implementation Plans (SIPs) are federally approved plans developed by state (or local) air quality management authorities to attain and maintain the national ambient air quality standards (NAAQS). In general, these SIPs are a state's (local) air quality rules and regulations which are considered an acceptable control strategy once approved by the Environmental Protection Agency (EPA). The purpose of SIPs is to control the amount and types of pollution in any given area of the region of the United States.

In these types of control strategies, emission limits should be based on ambient pollutant concentration estimates for the averaging time that results in the most stringent control requirements. In all cases, these concentrations estimates are assumed to be the sum of the pollutant concentrations contributed by the source and an appropriate background concentration. An air quality model is used to determine which averaging time (e.g., annual, 24-, 8-, 3-, 1-hour) results in the highest ambient impact. For example, if the annual average air quality standard is approached by a greater degree (percentage) than standards for other averaging times, the annual average is considered the restrictive standard. In this case, the sum of the highest estimated annual average concentration and the annual average background concentration provides the concentration which should be used to specify emission limits. However, if a short-term standard is approached by a greater degree and is thus identified as the restrictive standard, other considerations are required because the frequency of occurrence must also be considered.

New Source Review

New major stationary sources or major modifications to existing sources of air pollution are required by the Clean Air Act to obtain an air quality permit before construction is started. This process is called New Source Review (NSR), and it is required for any new major stationary source or major modification to an existing source regardless of whether or not the National Ambient Air Quality Standards (NAAQS) are exceeded. Sources located in areas which exceed the NAAQS (nonattainment areas) would undergo nonattainment New Source Review. New Source Review for major sources in areas where the NAAQS are not violated (attainment areas) would involve the preparation of a Prevention of Significant Deterioration (PSD) permit. Some sources will have the potential to emit pollutants for which their area is in attainment (or unclassifiable) as well as the potential to emit pollutants for which their area is nonattainment. When this is the case, the source's permit will contain terms and conditions to meet both the PSD and nonattainment are major NSR requirements because these requirements are pollutant specific.

In most cases, any new source must obtain a nonattainment NSR permit if it will emit, or has the potential to emit, 100 tons per year or more of any regulated NSR pollutant for which that area is in nonattainment, form "marginal" to "extreme." In areas where air quality problems are more severe, EPA has established lower thresholds for three criteria pollutants: ozone (VOCs), particulate matter (PM_{10}), and carbon monoxide. The "significance levels" are lower for modifications to existing sources.

In general, a new source located in an attainment or unclassifiable must get a PSD permit if it will emit or has the potential to emit, 250 tons per year (tpy) or more of any criteria or NSR-regulated pollutant. If the source is on EPA's list of 28 PSD source categories, a PSD permit is required if it will or may emit 100 tpy or more of any NSR-regulated pollutant. The "significant levels" are lower for modifications to existing sources. In addition, PSD review would be triggered, with respect to a particular pollutant, if a new source or major modification is constructed within 10 km of a Class I area (see below) and would have an impact on such area equal to or >1 mg/m^3, (24-hour average) for the pollutant, even though the emissions of such pollutant would not otherwise be considered "significant."

Some new sources or modifications to sources that are in attainment areas may be required to perform an air quality modeling analysis. This *air quality impact analysis* should determine if the source will cause a violation of the NAAQS or cause air quality deterioration that is greater than the available PSD increments. PSD requirements provide an area classification system based on land use for areas within the United States. These three areas are Class 1, Class 2, and Class III, and each class has an established set of increments that cannot be exceeded. Class I areas consist of national parks and wilderness areas that are only allowed a small amount of air quality deterioration. Due to the pristine nature of these areas, the most stringent limits on air pollution are enforced in the Class I areas. Class II areas consist of normal, well-managed industrial development. Moderate levels of air quality deterioration are permitted in these regions. Class III areas allow the largest amount of air quality deterioration to occur. When a PSD analysis is performed, the PSD increments set forth a maximum allowable increase in pollutant concentrations, which limits the allowable amount of air quality deterioration in an area. This in turn limits the amount of pollution that enters the atmosphere for a given region. To determine if a source of sulfur dioxide, for example, will cause an air quality violation, the air quality analysis uses the highest estimated concentration for annual averaging periods, and the highest, second highest estimated concentration for averaging periods of 24 hours or less. The new NAAQS for PM and Ozone contain specific procedures for determining modeled air quality violations.

For reviews of new or modified sources, the air quality impact analysis should generally be limited to the area where the source's impact is "significant," as defined by regulations. In addition, due to the uncertainties in making concentration estimates of large downwind distances, the air quality impact analysis should generally be limited to a downwind distance of 50 km, unless adverse impacts in a Class I area may occur at greater distances.

AIR QUALITY MONITORING

As mentioned in the previous two sections, air quality modeling is necessary to ensure that a source follows the SIP and New Source Review requirements. When air quality modeling is required, the selection of a model is dependent on the source characteristics, pollutants emitted, terrain, and meteorological parameters. The EPA has compiled the *Guideline on Air Quality Modeling* (40 CFR 1 Appendix W) which summarizes the available models, techniques, and guidance in conducting air quality modeling analyses used in regulatory programs. This document was written to promote consistency among modelers so that all air quality modeling activities would be based on the same procedures and recommendations.

When air quality modeling is required, the specific model used (from a simple screening tool to a refined analysis) will need meteorological data. The data can vary from a few factors such as average wind speed and Pasquill-Gifford stability categories of a mathematical representation of turbulence. Whatever model is chosen to estimate air quality, the meteorological data must match the quality of the model used. For example, the average wind speed used in a simple screening model will not be sufficient for a complex refined model. An air quality modeling analysis incorporates the evaluation of terrain, building dimensions, ambient monitoring data, relevant emissions from nearby sources, and the aforementioned meteorological data.

For a dispersion model to provide useful and valid results the meteorological data used in the model must be representative of the transport and dispersions used in the model must be representative of the transport and dispersions characteristics in the vicinity of the source that the model is trying to simulate. The representativeness of the meteorological data is dependent on the following:

- The proximity of the meteorological monitoring site to the area under consideration
- The complexity of the terrain in the area
- The exposure of the meteorological monitoring site
- The period of time during which the data are collected

In addition, the representativeness of the data can be adversely affected by large distances between the source and the receptor of interest. Similarly, valley/mountain, land/water, and urban/rural characteristics affect the accuracy of the meteorological data for the source under consideration.

For control strategy evaluations and New Source Review, the minimum meteorological data required to describe transport and dispersion of air pollutants in the atmosphere are wind direction, wind speed, mixing height and atmospheric stability (or related indicators of atmospheric turbulence and mixing). Because of the question of the representativeness of meteorological data, site-specific data are preferable to data collected off-site. Typically, one year of on-site data is required. If an off-site database is used (from a nearby airport for example), 5 years of data are normally required. With 5 years of data, the model can incorporate most of the possible variations in the meteorological conditions at the site.

Visibility

Visibility is the distance an observer can see along the horizon. The scattering and absorption of light by air pollutants in the atmosphere impair visibility.

There are generally two types of air pollution which impair visibility. The first type consists of smoke, dust, or gaseous plumes which obscure the sky or horizon and are emitted from a single source or small group of sources. The second type is a widespread area of haze that impairs visibility in every direction over a large area and originates from a multitude of sources. Regardless of the type of air pollution that impairs the visibility at a particular location, any change in the meteorology or source emissions that would increase the pollutant concentration in the atmosphere will result in increased visibility impairment.

PSD Class I areas have the most stringent PSD increments, and therefore, must be protected not only form high pollutant concentrations but also from the additional problems pollutants in the atmosphere can cause. Under the Clean Air Act, PSD Class I areas must be evaluated for visibility impairment. This may involve a visibility impairment analysis. According to EPA regulations visibility impairment is defined as any humanly perceptible change in visibility (visual range, contrast, or coloration) from natural conditions. Therefore, any location is susceptible to visibility impairment due to air pollution sources. Since PSD Class I areas (national parks and wilderness areas) are known for their aesthetic quality any change or alteration in the visibility of the area must be analyzed.

Pollutant Dispersion

Pollutant dispersion is the process of pollutants being removed from the atmosphere and deposited onto the surface of the Earth. Stack plumes contain gases and a small number of particles that are not removed from the gas stream. When the plume emerges from the stack, these particles are carried with it. Once airborne, the particles begin to settle out and become deposited on the ground and on surface objects. There are two ways the particles can be deposited: dry deposition (gravitational settling) or wet deposition (precipitation scavenging). Depending on the meteorological conditions during the time of pollutant emission, these may:

1. Settle out quickly due to their weight and the effect of gravity
2. Be transported further downwind of the source due to buoyancy and wind conditions; or
3. Be washed out of the atmosphere by precipitation or clouds (wet deposition).

In any case, the deposition of these pollution particles is important to understand and quantify since pollutants deposited upon the ground can impact human health, vegetation, and wildlife.

Pollutant deposition concentrations must be predicted to minimize the risk to human health. To quantify the amount of pollutant deposition which occurs from stack emissions, air quality models can be used. These models determine pollution deposition based on the chemical reactivity and solubility of various gases and by using detailed data on precipitation for the areas in question.

Vapor Plume Induced ICING

Vapor plumes are emitted from cooling towers and stacks and consist mainly of water vapor. Although pollutant concentrations are not a major concern with vapor plumes, other problems arise when vapor plume sources are located close to frequently traveled roads and populated areas. Vapor emitted from a stack is warm and mist. When meteorological conditions are favorable, the moisture in the vapor plume condenses out and settles on cooler objects (e.g., road surfaces). This phenomenon is similar to the moisture that collects on the sides of a glass of water on a warm day. If temperatures are at or below freezing when the moisture condenses, road surfaces can freeze rapidly creating hazardous driving conditions. In addition, light winds can cause the plume to remain

Air Pollution

stagnant creating a form of ground fog that can cause low visibilities as well. Water vapor plumes that lower visibility can create hazards for aircraft, especially during critical phases of flight including landings and takeoffs.

AIR POLLUTION MECHANICS

In the past, the sight of belching smokestacks was a comforting sight to many people: more smoke meant more business, which indicated that the economy was healthy. But many of us are now troubled by evidence that indicates that polluted air adversely affects our health. Many toxic gases and fine particles entering the air pose health hazards: cancer, genetic defects, and respiratory disease. Nitrogen and sulfur oxides, ozone, and other air pollutants from fossil fuels are inflicting damage on our forests, crops, soils, lakes, rivers, coastal waters, and buildings. Chlorofluorocarbons (CFCs) and other pollutants entering the atmosphere are depleting the Earth's protective ozone layer, allowing more harmful ultraviolet radiation to reach the Earth's surface. Fossil fuel combustion is increasing the amount of carbon dioxide in the atmosphere which can have a severe long-term environmental impact.

It is interesting to note that when the ambient air is considered, the composition of "unpolluted" air is unknown to us. Humans have lived on the planet for thousands of years and influenced the composition of the air through their many activities before it was possible to measure the constituents of the air.

EPA (2006) points out that in theory the air has always been polluted to some degree. Natural phenomena such as volcanoes, wind storms, the decomposition of plants and animals, and even the aerosols emitted by the ocean "pollute" the air. However, the pollutants we usually refer to when we talk about air pollution are those generated as a result of human activity. An *air pollutant* can be considered as a substance in the air that, in high enough concentrations, produces a detrimental environmental effect. These effects can be either health effects or welfare effects. A pollutant can affect the health of humans, as well as the health of plants and animals. Pollutants can also affect non-living materials such as paints, metals, and fabrics. An *environmental effect* is defined as a measurable or perceivable detrimental change resulting from contact with an air pollutant.

Human activities have had a detrimental effect on the makeup of the air. Activities such as driving cars and trucks, burning of coal, oil and other fossil fuels, and manufacturing chemicals have changed the composition of the air by introducing many pollutants. There are hundreds of pollutants in the ambient air. Ambient air is the air to which the general public has access, i.e. any unconfined portion of the atmosphere. The two basic physical forms of air pollutants are particulate matter and gases. Particulate matter includes small solid and liquid particles such as dust, smoke, sand, pollen, mist, and fly ash. Gases include substances such as carbon monoxide (CO), sulfur dioxide (SO_2), nitrogen oxides (NO_2) and volatile organic compounds (VOCs).

Historically, many felt that the air renewed itself, through interaction with vegetation and the oceans, in sufficient quantities to make up for the influx into our atmosphere of anthropogenic pollutants. Today, however, this kind of thinking is being challenged by evidence that clearly indicates that increased use of fossil fuels, expanding industrial production, and growing use of motor vehicles are having a detrimental effect on the atmosphere, air, and the environment. In this section, we discuss air pollution mechanics: pollutant dispersal, transformation, and deposition mechanisms; and we examine the types and sources of air pollutants that are related to these concerns.

ATOMPHERIC DISPERSON, TRANSPORTATION AND DEPOSITION (EPA, 2005)

A source of air pollution is any activity that causes pollutants to be emitted into the air.

There have always been natural sources of air pollution, also known as biogenic sources. For example, volcanoes have spewed articulate matter and gases into our atmosphere for millions of years. Lightning strikes have caused forest fires, with their resulting contribution of gases and particles, for as long as storms and forests have existed. Organic matter in swaps decay and wind storms whip up dust. Trees and other vegetation contribute large amounts of pollen and spores to our atmosphere. These natural pollutants can be problematic generated pollutants or anthropogenic sources.

The quality of daily life depends on many modern conveniences. People enjoy the freedom to drive cars and travel in airplanes for business and pleasure. They expect their homes to have electricity and their water to be heated for bathing and cooking. They use a variety of products such as clothing, pharmaceuticals, and furniture made of synthetic materials. At times, they rely on services that use chemical solvents, such as the local dry cleaner and print shop. Yet the availability of these everyday conveniences comes at a price because they all contribute to air pollution.

Air pollutants are released from both stationary and mobile sources. Scientists have gathered much information that is available on the sources, quantity, and toxicity levels of these pollutants. The measurement of air pollution is an important scientific skill, and practitioners of this skill are usually well-founded in the pertinent related sciences, in modeling aspects applicable to their studies and analyses of air pollutants in the ambient atmosphere. However, to get at the very heart of air pollution, the practitioner must also be well versed in how to determine the origin of the pollutants and understand the mechanics of the pollutant dispersal, transport, and deposition.

Air pollution practitioners must constantly deal with one basic fact: air pollutants rarely stay at their release location. Instead, wind flow conditions and turbulence, local topographic features, and other physical conditions work to disperse these pollutants. So, along with having a thorough knowledge and understanding of the pollutants in question, the air pollution practitioner has a definite need for detailed knowledge of the atmospheric processes that govern their subsequent dispersal and fate.

Conversion of precursor substances to secondary pollutants such as ozone is an example of chemical transformation in the atmosphere. Transformations are (both physical and chemical) and affect the ultimate impact of originally emitted air pollutants.

Pollutants emitted into the atmosphere do not remain there forever. Two common deposition (depletion) mechanisms are *dry deposition* (the removal of both particles and gases as they come into contact with the Earth's surface) and *washout* (the uptake of particles and gases by water droplets and snow and their removal from the atmosphere as precipitation that falls to the ground). Acid deposition (acid rain) is a form of pollution depletion from the atmosphere.

The following sections discuss the atmospheric dispersion of air pollutants in greater detail and the main factors associated with this phenomenon, including weather, turbulence, air parcels, buoyancy factors, lapse rates, mixing, topography, inversions, plume behavior, and transport.

Weather

The air contained in Earth's atmosphere is not still. Constantly in motion, air masses warmed by solar radiation rise at the equator and spread toward the colder poles where they sink, and flow downward, eventually returning to the equator. Near the Earth's surface, as a result of the Earth's rotation, major wind patterns develop. During the day the land warms more quickly than the sea; at night, the land cools more quickly. Local wind patterns are driven by this differential warming and cooling between the land and adjacent water bodies. Normally, onshore breezes bring cooler, denser air from over the land masses out over the waters during the night. Precipitation is also affected by wind patterns. Warm, moisture-laden air rising from the oceans is carried inland, where the air masses eventually cool, causing the moisture to fall as rain, hail, sleet, or snow.

Even though pollutant emissions may remain relatively constant, air quality varies tremendously from day to day. The determining factors have to do with the weather.

Weather conditions have a significant impact on air quality and air pollution, both favorable and unfavorable, especially on local conditions. For example, on hot, Sun-filled days, when the weather is calm with stagnating high-pressure cells, air quality suffers, because these conditions allow the buildup of pollutants on the ground level. When local weather conditions include cool, windy, stormy weather with turbulent low-pressure cells and cold fronts, these conditions allow the upward mixing and dispersal of air pollutants.

Air Pollution

Weather has a direct impact on pollution levels in both mechanical and chemical ways. Mechanically, precipitation works to cleanse the air of pollutants (transferring the pollutants to rivers, streams, lakes, or the soil). Winds transport pollutants from one place to another. Winds and storms often dilute pollutants with cleaner air, making pollution levels less annoying in the area of their release. Air (and its accompanying pollution in a low-pressure cell) is also carried aloft by air heated by the Sun. When wind accompanies this rising air mass, the pollutants are diluted with fresh air. In a high-pressure cell, the opposite occurs with air and the pollutants it carries sinking toward the ground. With no wind, these pollutants are trapped and concentrated near the ground where serious air pollution episodes may occur.

Chemically, weather can also affect pollution levels. Winds and turbulence mix pollutants together in a sort of giant chemical broth in the atmosphere. Energy from the Sun, moisture in the clouds and the proximity of highly reactive chemicals may case chemical reactions, which lead to the formation of secondary pollutants. Many of these secondary pollutants may be more dangerous than the original pollutants.

Turbulence

In the atmosphere, the degree of turbulence (which results from wind speed and convective conditions related to the change of temperature with height above the Earth's surface) is directly related to stability (a function of the vertical distribution of atmospheric temperature). The stability of the atmosphere refers to the susceptibility of rising air parcels to vertical motion (attributed to high- and low-pressure systems, air lifting over terrain or fronts and convection), consideration of atmospheric stability or instability is essential in establishing the dispersion rate of pollutants. When specifically discussing the stability of the atmosphere, we are referring to the lower boundary of the Earth where air pollutants are emitted. The degree of turbulence in the atmosphere is usually classified by stability class. Ambient and adiabatic lapse rates are a measure of atmospheric stability.

Stability is divided into three classes: stable, unstable, and neutral. A *stable atmosphere* is marked by air cooler at the ground than aloft, by low wind speeds, and consequently, by a low degree of turbulence. A plume of pollutants released into a stable lower layer of the atmosphere can remain relatively intact for long distances. Thus, we can say that stable air discourages the dispersion and dilution of pollutants. An *unstable atmosphere* is marked by a high degree of turbulence. A plume of pollutants released into an unstable atmosphere may exhibit a characteristic looping appearance produced by turbulent eddies. A *neutrally stable atmosphere* is an intermediate class between stable and unstable conditions. A plume of pollutants released into a neutral stability condition is often characterized by a coning appearance as the edges of the plume spread out in a V-shape.

The importance of the "state of the atmosphere" and stability's effects cannot be overstated. The ease with which pollutants can disperse vertically into the atmosphere is mainly determined by the rate of change of air temperature with height (altitude). Therefore, air stability is a primary factor in determining where pollutants will travel, and how long they will remain aloft. Stable air discourages the dispersion and dilution of pollutants. Conversely, in unstable air conditions, rapid vertical mixing takes place, encouraging pollutant dispersal, which increases air quality.

Air Parcels

With regard to air parcels described in this section, think of air inside a balloon as an analogy for the air parcel. This theoretically infinitesimal parcel is a relatively well-defined body of air (a constant number of molecules) that acts as a whole. Self-contained, it does not readily mix with the surrounding air. The exchange of heat between the parcel and its surroundings is minimal, and the temperature within the parcel is generally uniform.

Buoyancy Factors

Atmospheric temperature and pressure influence the buoyancy of air parcels. When holding other conditions constant, the temperature of air (a fluid) increases as atmospheric pressure increases and conversely decreases as pressure decreases. With respect to the atmosphere, where air pressure decreases with rising altitude, the normal temperature profile of the troposphere is one where temperature decreases with height.

An air parcel that becomes warmer than the surrounding air (for example, by heat radiating from the Earth's surface) begins to expand and cool. As long as the parcel's temperature is greater than the surrounding air, the parcel is less dense than the cooler surrounding air. Therefore, it rises or is buoyant. As the parcel rises, it expands thereby decreasing its pressure and, therefore, its temperature decreases as well. The initial cooling of an air parcel has the opposite effect. In short, warm air rises and cools, while cool air descends and warms.

The extent to which an air parcel rises of falls depends on the relationship of its temperature to that of the surrounding air. As long as the parcel's temperature is cooler, it will descend. When the temperatures of the parcel and the surrounding air are the same, the parcel will neither rise nor descend unless influenced by wind flow.

Lapse Rates

The *lapse rate* is defined as the rate of temperature change with height. With an increase in altitude in the troposphere, the temperature of the ambient air usually decreases. On the average, temperature decreases −6 to −7°C per km. This is the normal lapse rate but it varies widely depending on location and time of day. We define a temperature *decrease* with height as a negative elapse rate and a temperature *increase* with height as a positive elapse rate.

In a dry environment, when a parcel of warm dry air is lifted in the atmosphere, it undergoes adiabatic expansion and cooling. For the most part, a parcel of air does no exchange heat across its boundaries. Therefore, an air parcel that is warmer than the surrounding air does not transfer heat to the atmosphere. Any temperature changes that occur within the parcel are caused by increases or decreases of molecular activity within the parcel. Such changes, occur adiabatically, and are due only to the change in atmospheric pressure as a parcel moves vertically. The term "adiabatic" literally means impassable from, corresponding in this instance to an absence of heat transfer. That is, an adiabatic process is one in which there is no transfer of heat or mass across the boundaries of the air parcel. In an adiabatic process, compression results in heating and expansion results in cooling. A dry air parcel rising in the atmosphere cools at the dry adiabatic rate of 9.8°C/1,000 m and has a lapse rate of −9.8°C/1,000 m. Likewise, a dry air parcel sinking in the atmosphere heats up at the dry adiabatic rate of 9.8°C/1,000 m and has a lapse rate of 9.8°C/1,000 m. Air is considered dry, in this context, as long as any water in it remains in a gaseous state.

The *dry adiabatic lapse rate* is a fixed rate, entirely independent of ambient air temperature. A parcel of dry air moving upward in the atmosphere, then, will always cool at the rate of 9.8°C/1,000 m, regardless of its initial temperature or the temperature of the surrounding air.

When the ambient lapse rate exceeds the adiabatic lapse rate, the ambient rate is said to be *super adiabatic*, and the atmosphere is highly unstable. When the two lapse rates are exactly equal, the atmosphere is said to be *neutral*. When the ambient lapse rate is less than the dry adiabatic lapse rate, the ambient lapse rate is termed *subadiabatic*, and the atmosphere is stable.

The cooling process within a rising parcel of air is assumed to be adiabatic (occurring without the addition or loss of heat). A rising parcel of air (under adiabatic conditions) behaves like a rising balloon, with the air in that distinct parcel expanding as it encounters air of lesser density until its own density is equal to that of the atmosphere which surrounds it. This process is assumed to occur with no heat exchange between the rising parcel and the ambient air (Peavy et al., 1985).

A rising parcel of dry air containing water vapor will continue to cool at the dry adiabatic lapse rate until it reaches its condensation temperature, or dew point. At this point, the pressure of the water vapor equals the saturation vapor pressure of the air, and some of the water vapor begins to condense. Condensation releases latent heat in the parcel, and thus the cooling rate of the parcel slows. This new rate is called the *wet adiabatic lapse rate*. Unlike the dry adiabatic lapse rate, the wet adiabatic lapse rate is not constant but depends on temperature and pressure. In the middle troposphere, however, it is assumed to be approximately $-6°C$ to $-7°C/1,000\,m$.

Mixing

Within the atmosphere, for effective pollutant dispersal to occur, turbulent mixing is important. Turbulent mixing, the result of the movement of air in the vertical dimension, is enhanced by vertical temperature differences. The steeper the temperature gradient and the larger the vertical air column in which the mixing takes place, the more vigorous the convective and turbulent mixing of the atmosphere.

Topography

On a local scale, the topography may affect air motion. In the United States, most large urban centers are located along sea and lake coastal areas. Contained within these large urban centers is much heavy industry. Local air flow patterns in these urban centers have a significant impact on pollution dispersion processes. Topographic features also affect local weather patterns, especially in large urban centers located near lakes, seas, and open land. Breezes from these features affect vertical mixing and pollutant dispersal. Seasonal differences in heating and cooling land and water surfaces may also precipitate the formation of inversions near the sea or lake shore.

River valley areas are also geographical locations that routinely suffer from industry-related pollution. Many early settlements began in river valleys because of the readily available water supply and the ease of transportation afforded to settlers by river systems within such valleys. Along with settlers came industry—the type of industry that invariably produces air pollutants. These air pollutants, because of the terrain and physical configuration of the valley, are not easily removed from the valley.

Winds that move through a typical river valley are called slope winds. Slope winds, like water, flow downhill into the valley floor. At valley floor level, slope winds transform to valley winds which flow down-valley with the flow of the river. Down-valley winds are lighter than slope winds. The valley floor becomes flooded with a large volume of air which intensifies the surface inversion that is normally produced by radiative cooling. As the inversion deepens over the course of the night, it often reaches its maximum depth just before sunrise with the height of the inversion layer dependent on the depth of the valley and the intensity of the radiative cooling process.

Hills and mountains can also affect local air flow. These natural topographical features tend to decrease wind speed (because of their surface roughness) and form physical barriers preventing the air movement.

Inversions

An inversion occurs when the air temperature increases with altitude. Temperature inversions (extreme cases of atmospheric stability) create a virtual lid on the upward movement of atmospheric pollution. This situation occurs frequently but is generally confined to a relatively shallow layer. Plumes emitted into air layers that are experiencing an inversion (inverted layer) do not disperse very much as they are transported with the wind. Plumes that are emitted above or below an inverted layer do not penetrate that layer; rather these plumes are trapped either above or below that inverted layer. High concentrations of air pollutants are often associated with inversions since they inhibit plume dispersions. Two types of inversions are important from an air quality standpoint: radiation and subsidence inversions.

Radiation inversions are the most common form of surface inversion and occur when the Earth's surface cools rapidly. They prompt the formation of fog, and simultaneously trap gases and particulates, creating a concentration of pollutants. They are characteristically a nocturnal phenomenon caused by cooling of the Earth's surface. On a cloudy night, the Earth's radiant heat tends to be absorbed by water vapor in the atmosphere. Some of this is re-radiated back to the surface.

However, on clear winter nights, the surface more readily radiates energy to the atmosphere and beyond, allowing the ground to cool more rapidly. The air in contact with the cooler ground also cools, and the air just above the ground becomes cooler than the air above it, creating an inversion close to the ground, lasting for only a matter of hours. These radiation inversions usually begin to form at the worst time of the day for human concerns in large urban areas—during the late afternoon rush hour, trapping automobile exhaust at ground level and causing elevated concentrations of pollution for commuters. During evening hours, photochemical reactions cannot take place, so the biggest problem can be the accumulation of carbon monoxide. At sunrise, the Sun warms the ground and the inversion begins to break up. Pollutants that have been trapped in the stable air mass are suddenly brought back to Earth in a process known as *fumigation*, which can cause a short-lived, high concentration of pollution at ground level (Masters, 1991).

The second type of inversion is *subsidence inversion*, usually associated with anticyclones (high-pressure systems); they may significantly affect the dispersion of pollutants over large regions. A subsidence inversion is caused by the characteristic sinking motion of air in a high-pressure cell. Air in the middle of a high-pressure zone descends slowly. As the air descends, it is compressed and heated. It forms a blanket of warm air over the cooler air below, thus creating an inversion (located anywhere from several hundred meters above the surface to several thousand meters) that prevents further vertical movement of air.

Plume Behavior

One way to quickly determine the stability of the lower atmosphere is to view the shape of a smoke trail or *plume* from a tall stack located on flat terrain. Visible plumes usually consist of pollutants emitted from a smoke-stack into the atmosphere. The formation and fate of the plume itself depend on a number of related factors: (1) the nature of the pollutants, (2) meteorological factors (combination of vertical air movement and horizontal air flow), (3) source obstructions, and (4) local topography, especially downwind. Overall, maximum ground-level concentrations will occur in a range from the vicinity of the smokestack to some distance downwind.

Air quality problems associated with the dispersion of city plumes are compounded by the presence of an already contaminated environment. Even though conventional processes normally work to disperse emissions from point sources do occur within the city plume, because of micro-climates within the city and the volume of pollutants they must handle, the conventional processes often cannot disperse effectively. Other compounding conditions present in areas where city plumes are generated (topographical barriers, surface inversions, and stagnating anticyclones) work to intensify the city plume and result in high pollutant concentrations.

Plume Rise Equation

Plume rise has been studied over the years. The most common plume rise formulas are those developed by Briggs, which have been extensively validated with stack plume observations (EPA, 2005). One of these that applies to buoyancy-dominated plumes is included in Equation 7.28. Plume rise formulas are to be used on plumes with temperatures greater than the ambient air temperature. The *Briggs' plume rise formula* is as follows:

$$\Delta h = \frac{1.6 F^{1/3} x^{2/3}}{u} \tag{7.28}$$

where Δh = plume rise (above stack)
F = Buoyancy Flux (see below; 7.28)
u = average wind speed
x = downwind distance from the stack/source
g = acceleration due to gravity (9.8 m/s^2)

V = volumetric flow rate of tack gas
T_s = temperature of stack gas
T_a = temperature of ambient air

$$\text{Buoyancy flux} = F = \frac{g}{\pi} V \left(\frac{T_s - T_a}{T_s} \right) \quad (7.29)$$

Transport

Those people living east of the Mississippi River would be surprised to find out that they are breathing air contaminated by pollutants from various sources many miles from their location. Most people view pollution under the old cliché "out of sight out of mind." As far as they are concerned, if they don't see it, it doesn't exist. For example, assume that a person on a farm heaps together a huge pile of assorted rubbish to be burned. The person preparing this huge bonfire probably gives little thought about the long-range transport (and consequences) of any contaminants that might be generated from that bonfire. This person simply has trash he or she no longer wants, and an easy solution is to burn it.

This pile of rubbish contains various elements: discarded rubber tires, old, compressed gas bottles, assorted plastic containers, paper, oils and greases, wood and old paint cans. The person burning it doesn't consider this hazardous material, though, just household trash. The truth is when the pile of rubbish burns, a huge plume of smoke forms and is carried away by a westerly wind. The fire-starter looks downwind and notices that the smoke disappears just a few miles over the property line. The dilution processes and the enormity of the atmosphere work together to dissipate and move away the smoke plume; the fire-starter doesn't give it a second thought. However, elevated levels of pollutants from many such fires may occur hundreds to thousands of miles downwind of the combination point sources producing such plumes. The result is that people living many miles from such pollution generators end up breathing contaminated air, transported over distance to their location.

Transport or dispersion estimates are determined by using distribution equations and/or air quality models. These dispersion estimates are typically valid for the layer of the atmosphere closet to the ground where frequent changes occur in the temperature and distribution of the winds. These two variables have an enormous effect on how plumes are dispersed.

DISPERSION MODELS

Air quality dispersion models consist of a set of mathematical equations that interpret and predict pollutant concentration due to plume dispersal and impaction. They are essentially used to predict or describe the fate of airborne gases, particulate matter, and ground-level concentrations downwind of point sources. To determine the significance of air quality impact on a particular area, the first consideration is normal background concentrations, those pollutant concentrations from natural sources and/or distant, unidentified man-made sources. Each particular geographical area has a "signature" or background level of contamination considered an annual mean background concentration level of certain pollutants. An area, for example, might normally have a particulate matter reading of 30–40 $\mu g/m^3$. If particulate matter readings are significantly higher than the background level, this suggests an additional source. To establish background contaminations for a particular source, air quality data related to that site and its vicinity must be collected and analyzed.

The USEPA recognized that in calculating the atmospheric dispersion of air pollutants, some means by which consistency could be maintained in air quality analysis had to be established. Thus, the USEPA promulgated two guidebooks to assist in modeling for air quality analyses: *Guidelines on Air Quality Models (Revised) (1986)* and *Industrial Source Complex (ISC) Dispersion Models User's Guide (1986).*

In performing dispersion calculations, particularly for health effect studies, the USEPA and other recognized experts in the field recommend following a four-step procedure:

1. Estimate the rate, duration, and location of the release into the environment.
2. Select the best available model to perform the calculations.
3. Perform the calculations and generate downstream concentrations, including lines of constant concentration (isopleths) resulting from the source emission(s).
4. Determine what effect, if any, the resulting discharge has on the environment, including humans, animals, vegetation, and materials of construction. These calculations often include estimates of the so-called vulnerability zones—that is, regions that may be adversely affected because of the emissions (Holmes et al., 1993).

Before beginning any dispersion determination activity, you must first determine the acceptable ground-level concentration of the waste pollutant(s). Local meteorological conduits and local topography must be considered and having an accurate knowledge of the constituents of the waste gas and its chemical and physical properties is paramount.

Air quality models provide a relatively inexpensive means of determining compliance and predicting the degree of emission reduction necessary to attain ambient air-quality standards. Under the 1977 Clean Air Act Amendments, the use of models is required for the evaluation of permit applications associated with permissible increments under the so-called Prevention of Significant Deterioration (PSD) requirements, which require localities "to protect and enhance" air that is not contaminated (Godish, 1997).

Several dispersion models have been developed. Really equations, these models are mathematical descriptions of the meteorological transport and dispersion of air contaminants in a particular area, which permit estimates of contaminant concentrations, either in plume from a ground-level or an elevated source (Carson and Moses, 1969). User-friendly modeling programs are available now that produce quick, accurate results from the operator's pertinent data.

There are four generic types of models: Gaussian, numerical, statistical, and physical. The *Gaussian* models use the Gaussian distribution equation and are widely used to estimate the impact of nonreactive pollutants. *Numerical* models are more appropriate than Gaussian models for area sources in urban locations that involve reactive pollutants, but numerical models require extreme.

Detailed source and pollutant information and are not widely used. *Statistical* models are used when scientific information about the chemical and physical processes of a source is incomplete or vague and therefore make the use of either Gaussian or numerical models impractical. Lastly, *physical* models required fluid modeling studies or wind tunneling. This approach involves the construction of scaled model and observing fluid flow around these models. This type of modeling is very complex and requires expert technical support. However, for large areas with complex terrain, stack downwash, complex flow conditions, or large buildings, this type of modeling may be the best choice.

As mentioned, selection of an air quality model for a particular air quality analysis is dependent on the type of pollutants being emitted, the complexity of the source, and the type of topography surrounding the facility. Some pollutants are formed by the combination of precursor pollutants. For example, ground-level ozone is formed when volatile organic compounds (VOCs) and nitrogen oxides (NO_x) react in the presence of sunlight. Models to predict ground-level ozone concentrations would use the emission rate of VOCs and NO_x as inputs. Also, some pollutants readily react to one emitted into the atmosphere. These reactions deplete the concentrations of these pollutants and may need to be accounted for in the model. Source complexity also plays a role in model selection. Some pollutants may be emitted from short stacks that are subject to aerodynamic downwash. If this is the case, a model must be used that is capable of accounting for this phenomenon. Again,

Air Pollution

topography plays a major role in the dispersal of plumes and their air pollutants and must be considered in the selection of an air quality model. Elevated plumes may impact areas of high terrain. Elevated terrain heights may experience higher pollutant concentrations since they are closer to the plume centerline. A model which considers terrain heights should be used when elevated terrain exists.

This handbook's intent is not to develop each dispersion model in detail but rather to recommend the one with the greatest applicability today. Probably the best atmospheric dispersion workbook for modeling published to date is that by D.B. Turner in his *Workbook of Atmospheric Dispersion Estimates* for the EPA, and most of the air dispersion models used today are based on the Pasquill-Gifford Model.

MAJOR AIR POLLUTANTS

The most common and widespread anthropogenic pollutants currently emitted are sulfur dioxide (SO_2), nitrogen oxides (NO_x), carbon monoxide (CO), carbon dioxide (CO_2), volatile organic compounds (hydrocarbons), particulates, lead, and several toxic chemicals. Table 7.16 lists important air pollutants and their sources.

Recall that in the United States, the Environmental Protection Agency (EPA) regulates air quality under the Clean Air Act (CAA) and amendments which charged the federal government to develop uniform National Ambient Air-Quality Standards (NAAQS). These were to include a dual standard requirement of primary standards (covering criteria pollutants) designed to protect health and secondary standards to protect public welfare. Primary standards were to be achieved by July 1975, and secondary standards in "a reasonable period of time." Pollutant levels of protective of public welfare take priority (and are more stringent) than those for public health; achievement of the primary health standard had immediate priority. In 1971 the USEPA promulgated NAAQS for six classes of air pollutants. Later, in 1978, an air-quality standard was also promulgated for lead and the photochemical oxidant standard was revised to an ozone (O_3) standard (the ozone permissible level was increased). The PM standard was revised and redesignated PM_{10} standard in 1987. This revision reflected the need for a PM standard based on particle sizes ($\leq 10\,\mu m$) that have the potential for entering the respiratory tract and affecting human health.

Thus, air pollutants were categorized into two groups: primary and secondary. Primary pollutants are emitted directly into the atmosphere where they exert an adverse influence on human health or the environment. Of particular concern are primary pollutants emitted in large quantities: carbon dioxide, carbon monoxide, sulfur dioxide, nitrogen dioxides, hydrocarbons, and particulate matter (PM). Once in the atmosphere, primary pollutants may react with other primary pollutants or atmospheric compounds such as water vapor to form secondary pollutants. A secondary pollutant that has received a lot of press and attention otherwise is acid precipitation, which is formed when sulfur or nitrogen oxides react with water vapor in the atmosphere.

TABLE 7.16
Important Air Pollutants and Their Sources

Pollutant	Source
Sulfur and nitrogen oxides	From fossil fuel combustion
Carbon monoxide	Mostly from motor vehicles
Volatile organic compounds	From vehicles and industry
Ozone	From atmospheric reactions between nitrogen oxides and organic compounds

Source: USEPA (1988a).

Sulfur Dioxide (SO_2)

Sulfur enters the atmosphere in the form of corrosive *sulfur dioxide* (SO_2) gas. Sulfur dioxide is a colorless gas possessing the sharp, pungent odor of burning rubber. On a global basis, nature and anthropogenic activities produce sulfur dioxide in roughly equivalent amounts. Its natural sources include volcanoes, decaying organic matter, and sea spray, while anthropogenic sources include combustion of sulfur-containing coal and petroleum products and smelting of nonferrous ores. In industrial areas much more sulfur dioxide comes from human activities than from natural sources. Sulfur-containing substances are often present in fossil fuels; SO_2 is a product of combustion that results from the burning sulfur-containing materials. The largest single source (65%) of sulfur dioxide is from the burning of fossil fuels to generate electricity. Thus, near major industrialized areas, it is often encountered as an air pollutant.

In the air, sulfur dioxide converts to sulfur trioxide (SO^3) and sulfate particles (SO_4). Sulfate particles restrict visibility and, in the presence of water, form sulfur acid (H_2SO_4), a highly corrosive substance that also lowers visibility. According to McKenzie and El-Ashry (1988), global output of sulfur dioxide has increased six-fold since 1900. Most industrial nations, however, have since (1975–1985) lowered sulfur dioxide levels by 20%–60% by shifting away from heavy industry and imposing stricter emission standards. Major sulfur dioxide reductions have come from burning coal with lower sulfur content and from using less coal to generate electricity.

Two major environmental problems have developed in highly industrialized regions of the world, where the atmospheric sulfur dioxide concentration has been relatively high: sulfurous smog and acid rain. Sulfurous smog is the haze that develops in the atmosphere when molecules of sulfuric acid accumulate, growing in size as droplets until they become sufficiently large to serve as light scatterers. The second problem, acid rain, is precipitation contaminated with dissolved acids like sulfuric acid. Acid rain has posed a threat to the environment by causing certain lakes to become void of aquatic life.

Nitrogen Oxides (NO_x)

There are seven oxides of nitrogen that are known to occur—NO, NO_2, NO_3, N_2O, N_2O_3, N_2O_4, and N_2O_5—but only two are important in the study of air pollution: nitric oxide (NO) and nitrogen dioxide (NO_2). Nitric oxide is produced by both natural and human actions. Soil bacteria are responsible for the production of most of the nitric oxide that is produced naturally and released to the atmosphere. Within the atmosphere, nitric oxide readily combines with oxygen to form nitrogen dioxide, and together, those two oxides of nitrogen are usually referred to as NO_x (nitrogen oxides). NO_x is formed naturally by lightning and by decomposing organic matter. Approximately 50% of anthropogenic NO_x is emitted by motor vehicles and about 30% comes from power plants, with the other 20% produced by industrial processes.

Scientists distinguish between two types of NO_x—thermal and fuel—depending on its mode of formation. Thermal NO_x is created when nitrogen and oxygen in the combustion air, such as those within internal combustion engines, are heated to a high enough temperature (above 1,000 K) to cause nitrogen (N_2) and oxygen (O_2) in the air to combine. Fuel NO_x results from the oxidation (i.e., it combines with oxygen in the air) of nitrogen contained within a fuel such as coal. Both types of NO_x generate nitric oxide first, and then when vented and cooled, a portion of nitric oxide is converted to nitrogen dioxide. Although both thermal NO_x can be significant contributors to the total NO_x emissions, fuel NO_x is usually the dominant source, with ~50% coming from power plants (stationary sources) and the other half is released by automobiles (mobile sources).

Nitrogen dioxide is more toxic than nitric oxide and is a much more serious air pollutant. Nitrogen dioxide, at high concentrations, is believed to contribute to heart, lung, liver, and kidney damage. In addition, because nitrogen dioxide occurs as a brownish haze (giving smog its reddish-brown color), it reduces visibility. When nitrogen dioxide combines with water vapor in the atmosphere, it forms

TABLE 7.17
United States Emission Estimates, 1986 (10^{12} g/yr)

Source	SO_x	NO_x	VOC	CO	Lead	PM
Transportation	0.9	8.5	6.5	42.6	0.0035	1.4
Stationary source fuel	17.2	10.0	2.3	7.2	0.0005	1.8
Industrial processes	3.1	0.6	7.9	4.5	0.0019	2.5
Solid waste disposal	0.0	0.1	0.6	1.7	0.0027	0.3
Miscellaneous	0.0	0.1	2.2	5.0	0.0000	0.8
Total	28.4	18.1	19.5	60.9	0.0086	6.8

Source: USEPA (1988b).

nitric acid (HNO_3), a corrosive substance that, when precipitated out as acid rain, causes damage to plants and corrosion of metal surfaces.

NO_x rose in several countries and then leveled off or declined during the 1970s. During this same timeframe, levels of nitrogen oxide have not dropped as dramatically as those of sulfur dioxide, primarily because a large part of total NO_x emissions comes from millions of motor vehicles, while most sulfur dioxide is released by a relatively small number of emission-controlled, coal-burning power plants.

CARBON MONOXIDE (CO)

Carbon monoxide is a colorless, odorless, tasteless gas formed when carbon in fuel is not burned completely; it is by far the most abundant of the primary pollutants, as Table 7.17 indicates. When inhaled, carbon monoxide gas restricts the blood's ability to absorb oxygen, causing angina, impaired vision, and poor coordination. Carbon monoxide has a little direct effect on ecosystems but has an indirect environmental impact by contributing to the greenhouse effect and depletion of the Earth's protective ozone layer.

The most important natural source of atmospheric carbon monoxide is the combination of oxygen with methane (CH_4), which is a product of the anaerobic decay of vegetation. (Anaerobic decay takes place in the absence of oxygen.) At the same time, however, carbon monoxide is removed from the atmosphere by the activities of certain soil microorganisms, so the net result is a harmless average concentration that is <0.12 to 15 ppm, in the northern hemisphere. Because stationary source combustion facilities are under much tighter environmental control than are mobile sources, the principal source of carbon monoxide that is caused by human activities is motor vehicle exhaust, which contributes to about 70% of all CO emissions in the United States.

VOLATILE ORGANIC COMPOUNDS (VOCS—HYDROCARBONS)

Volatile Organic Compounds (VOCs) (also listed under the general heading of hydrocarbons) encompass a wide variety of chemicals that contain exclusively hydrogen and carbon. Emissions of volatile hydrocarbons from human resources are primarily the result of incomplete combustion of fossil fuels. Fires and the decomposition of matter are natural sources. Of the VOCs that occur naturally in the atmosphere, methane (CH_4) is present at highest concentrations (~1.5 ppm). But even at relatively high concentrations methane does not interact chemically with other substances and causes no ill health effects. However, in the lower atmosphere, sunlight causes VOCs to combine with other gases, such as NO_2, oxygen, and CO to form secondary pollutants such as formaldehyde, ketones, ozone, peroxyacetyl nitrate (PAN) and other types of photochemical oxidants. These active chemicals can irritate the eyes and damage the respiratory system and damage vegetation.

Ozone and Photochemical Smog

By far the most damaging photochemical air pollutant is ozone (each ozone molecule contains three atoms of oxygen and thus is written O_3). Other photochemical oxidants [peroxyacetyl nitrate (PAN), hydrogen peroxide (H_2O_2), and aldehydes] play minor roles. All of these are secondary pollutants because they are not emitted but are formed in the atmosphere by photochemical reactions involving sunlight and emitted gases, especially NO_x and hydrocarbons.

Ozone is a bluish gas, about 1.6 times heavier than air, and relatively reactive as an oxidant. Ozone is present in a relatively large concentration in the stratosphere and is formed naturally by ultraviolet radiation. At ground level, ozone is a serious air pollutant; it has caused serious air pollution problems throughout the industrialized world, posing threats to human health, and damaging foliage and building material.

According to MacKenzie and El-Ashry (1988), ozone concentrations in industrialized countries of North America and Europe are up to three times higher than the level at which damage to crops and vegetation begins. Ozone harms vegetation by damaging plant tissues, inhibiting photosynthesis, and increasing susceptibility to disease, drought, and other air pollutants.

In the upper atmosphere, where "good" (vital) ozone is produced, ozone is being depleted by anthropogenic emission of ozone-depleting chemicals has increased on the ground. With this increase, concern has been raised over a potential upset of the dynamic equilibria among stratospheric ozone reactions, with a consequent reduction in ozone concentration. This is a serious situation because stratospheric ozone absorbs much of the incoming solar ultraviolet (UV) radiation. As a UV shield, ozone helps to protect organisms on the Earth's surface from some of the harmful effects of this high-energy radiation. If not interrupted, UV radiation could cause serious damage, as disruption of genetic material, which could lead to increased rates of skin cancers and heritable problems.

In the mid-1980s a serious problem with ozone depletion became apparent. A springtime decrease in the concentration of stratospheric ozone (ozone holes) had been observed at high latitudes, most notably over Antarctica between September and November. Scientists strongly suspected that chlorine atoms or simple chlorine compounds may play a key role in this ozone depletion problem.

On rare occasions, it is possible for upper stratospheric ozone (good ozone) to enter the lower atmosphere (troposphere). In general, this phenomenon only occurs during an event of great turbulence in the upper atmosphere. On rare incursions, atmospheric ozone reaches ground level for a short period of time. Most of the tropospheric ozone is formed and consumed by endogenous photochemical reactions, which are the result of the interaction of hydrocarbons, oxides of nitrogen, and sunlight, which produces a yellowish-brown haze commonly called smog (Los Angeles-type smog).

Although the incursion of stratospheric ozone into the troposphere can cause smog formation, the actual formation of Los Angeles-type smog involves a complex group of photochemical interactions. These interactions are between anthropogenically emitted pollutants (NO and hydrocarbons) and secondarily produced chemicals (PAN, aldehydes, NO_2, and ozone). Note that the concentrations of these chemicals exhibit a pronounced diurnal pattern, depending on their rate of emission, and on the intensity of solar radiation and atmospheric stability at different times of the day (Freedman, 1989).

Historical records show that the presence of various air pollutants in the atmosphere of Los Angeles include NO (emitted as NO_x), which has a morning peak of concentration at 0600–0700, largely due to emissions from morning rush-hour vehicles. Hydrocarbons are emitted both from vehicles and refineries; they display a similar pattern to that of NO except that their peak concentration is slightly later. In bright sunlight the NO is photochemically oxidized to NO_2, resulting in a decrease in NO concentration and a peak of NO_2 at 0700–0900. Photochemical reactions involving NO_2 produce O atoms, which react with O_2 to form O_3. These result in a net decrease in NO_2 concentration and an increase in O_3 concentration; peaks between 1,200 and 1,500. Aldehydes also formed photochemically, peak earlier than O_3. As the day proceeds, the various gases decrease in concentration as they are diluted by fresh air masses or are consumed by photochemical reactions. This cycle is typical of an area that experiences photochemical smog and is repeated daily (Urone, 1976).

TABLE 7.18
Tropospheric Ozone Budget (Northern Hemisphere) (kg/ha-year)

Transport from Stratosphere	13–20
Photochemical production	48–78
Destruction at ground	18–35
Photochemical destruction	48–55

Source: Adapted from Hov (1984).

A tropospheric ozone budget for the northern hemisphere is shown in Table 7.18. The considerable range of the estimates reflects uncertainty in the calculation of the ozone fluxes. On average, stratospheric incursions account for about 18% for the total ozone influx to the troposphere, while endogenous photochemical production accounts for the remaining 82%. About 31% of the tropospheric ozone is consumed by oxidative reactions at vegetative and inorganic suffocates at ground level, while the other 69% is consumed by photochemical reactions in the atmosphere (Freedman, 1989).

CARBON DIOXIDE

Carbon-laden fuels, when burned, release carbon dioxide (CO_2) into the atmosphere. Much of this carbon dioxide is dissipated and then absorbed by ocean water, some are taken up by vegetation through photosynthesis, and some remains in the atmosphere. Today, the concentration of carbon dioxide in the atmosphere is ~350 ppm and is rising at a rate of ~20 ppm every decade. The increasing rate of combustion of coal and oil has been primarily responsible for this occurrence, which may eventually have an impact on global climate.

PARTICULATE MATTER

Atmospheric particulate matter is defined as any dispersed matter, solid or liquid, in which the individual aggregates are larger than single small molecules, but smaller than about 500 μm. Particulate matter is extremely divers and complex, since size and chemical composition, as well as atmospheric concentrations, are important characteristics (Masters, 1991).

A number of terms are used to categorize particulates, depending on their size and phase (liquid or solid). These terms are listed and described in Table 7.19.

Dust, spray, forest fires, and the burning of certain types of fuels are among the sources of particulates in the atmosphere. Even with the implementation of stringent emission controls, which have worked to reduce particulates in the atmosphere, the U.S. Office of Technology Assessment (Postel, 1987) estimates that current levels of particulates and sulfates in ambient air may cause the premature death of 50,000 Americans every year.

LEAD

Lead is emitted to the atmosphere primarily from human sources, such as burning leaded gasoline, in the form of inorganic particulates. In high concentrations, lead can damage human health and the environment. Once lead enters an ecosystem, it remains there permanently. In humans and animals, lead can affect the neurological system and cause kidney disease. In plants, lead can inhibit respiration and photosynthesis as well as block the decomposition of microorganisms. Since the 1970s, stricter emission standards have caused a dramatic reduction in lead output.

TABLE 7.19
Atmospheric Particulates

Term	Description
Aerosol	General term for particles suspended in air
Mist	Aerosol consisting of liquid droplets
Dust	Aerosol consisting of solid particles that are blown into the air or are produced from larger particles by grinding them down
Smoke	Aerosol consisting of solid particles or a mixture of solid and liquid particles produced by chemical reactions such as fires
Fume	Generally means the same as smoke, but often applies specifically to aerosols produced by condensation of hot vapors, especially of metals
Plume	The geometrical shape or form of the smoke coming out of a stack or chimney
Fog	Aerosol consisting of water droplets
Haze	Any aerosol, other than fog, that obscures the view through the atmosphere
Smog	Popular term originating in England to describe a mixture of smoke and fog; implies photochemical pollution

AIR POLLUTION CONTROL TECHNOLOGY

There are two primary motivations behind the utilization of industrial air pollution control technologies. These are

1. They must be used because of legal or regulatory requirements
2. They are integral to the economical operation of an industrial process

Although economists would point out that both of these motivations are really the same, that is, it is less expensive for an industrial user to operate with air pollution control than without the distinction in application type is an important one. In general, air pollution control is used to describe those applications that are driven by regulations and/or health considerations, while applications that deal with product recovery are considered process applications. Nevertheless, the technical issues, equipment design, operation, etc., will be similar if not identical. What differs between these uses is that the economics that affects the decision-making process will often vary to some degree (Heumann, 1997).

Air pollution control begins with regulation. Regulations (for example, to clean up, reduce, or eliminate a pollutant emission source) in turn, are generated because of certain community concerns. Buonicore et al. (1992) pointed out regulations usually evolve around three considerations:

1. Legal limitations imposed for the protection of public health and welfare.
2. Social limitations imposed by the community in which the pollution source is or is to be located.
3. Economic limitations imposed by marketplace constraints

The environmental engineer assigned to mitigate an air pollution problem must ensure that the design control methodology used will bring the source into full compliance with applicable regulations. To accomplish this feat, environmental engineers must first understand the problem(s) and then rely heavily on technology to correct the situation. Various air pollution control technologies are available to environmental engineers or air pollution control practitioners working to mitigate air pollution source problems. By analyzing the problem carefully and applying the most effective method for the situation, the engineer or practitioner can ensure that a particular pollution source is brought under control and the responsible parties are in full compliance with regulations.

Air Pollution

In this section, we discuss the various air pollution control technologies available to environmental engineers and air pollution control practitioners in mitigating air pollution source problems.

AIR POLLUTION CONTROL: CHOICES

Assuming that the design engineer has complete knowledge of the contaminant and the source, all available physical and chemical data on the effluent from the source, and the regulations of the control agencies involved, he or she must then decide which control methodology to employ. Since only a few control methods exist, the choice is limited. Control of atmospheric emissions from a process will generally consist of one of four methods depending on the process, types, fuels, availability of control equipment, etc. The four general control methods are (1) elimination of the process entirely or in part, (2) modification of the operation to a fuel which will give the desired level of emission, (3) installation of control equipment between the pollutant source and the receptor, and (4) relocation of the operation.

Tremendous costs are involved with eliminating or relocating a complete process, which makes either of these choices the choice of last resort, let's take a look at the first and last control methods first. Eliminating a process is no easy undertaking, especially when the "process" to be eliminated is the process for which the facility exists. Relocation is not always an answer, either.

The second pollution control method—modification of the operation to a fuel which will give the desired level of emission—often looks favorable to those who have weighed the high costs associated with air pollution control systems. Modifying the process to eliminate as much of the pollution problem as possible at the source is generally the first approach to be examined.

Again, often the easiest way to modify a process for air pollution control is to change the fuel. If a power plant, for example, emits large quantities of sulfur dioxide and fly ash, conversion to cleaner-burning natural gas is cheaper than installing the necessary control equipment to reduce the pollutant emissions to permitted values.

Changing from one fuel to another, however, causes its own problems related to costs, availability, and competition. Today's fuel prices are high, and no one counts on the trend reversing. Finding a low-sulfur fuel isn't easy, especially since many industries own their own dedicated supplies (which are not available for use in other industries). With regulation compliance threatening everyone, everyone wants their share of any available low cost, low sulfur fuel. With limited supplies available, the law of supply and demand takes over and prices go up.

Some industries employ other process modification techniques. These may include the evaluation of alternative manufacturing and production techniques, substitution of raw materials, and improved process control methods (Buonicore et al., 1992).

When elimination of the process entirely or in part, or when relocation of the operation, or when modification of the operation to a fuel which will give the desired level of emission is not possible the only alternative control method left is installation of control equipment between the pollutant source and the receptor (the purpose of installing pollution control equipment or a control system, obviously, is to remove the pollution from the polluted carrier gas). To accomplish this, the polluted carrier gas must pass through a control device or system, which collects or destroys the pollutant and releases the cleaned carrier gas into the atmosphere (Boubel et al., 1994). The rest of this section will focus on these air pollution control equipment devices and systems.

AIR POLLUTION CONTROL EQUIPMENT AND SYSTEMS

Several considerations must be factored into any selection decision for air pollution control equipment or systems. Careful consideration must be given to costs. No one ever said air pollution equipment/systems were inexpensive—they are not. The equipment/system must be designed to comply with applicable regulatory emission limitations. The operational and maintenance history/record

TABLE 7.20
Factors in Selecting Air Pollution Control Equipment/Systems

(1)	Best available technology (BAT)
(2)	Reliability
(3)	Lifetime and salvage value
(4)	Power requirements
(5)	Collection efficiency
(6)	Capital cost, including operation and maintenance costs
(7)	Track record of equipment/system and manufacturer
(8)	Space requirements and weight
(9)	Power requirements
(10)	Availability of space parts and manufacturer's representatives

(costs of energy, labor, and repair parts should also be factored in) of each equipment/system must be evaluated. Remember, emission control equipment must be operated on a continual basis, without interruptions. Any interruption could be subject to severe regulatory penalties, which could again be quite costly.

Probably the major factor to consider in the equipment/system selection process is what type of pollutant or pollutant stream is under consideration. If the pollutant is conveyed in a carrier gas, for example, factors such as carrier gas pressure, temperature, viscosity, toxicity, density, humidity, corrosiveness, and inflammability must all be considered before any selection is made. Other important factors must also be considered when selecting air pollution control equipment. Many of the general factors are listed in Table 7.20.

In addition to those factors listed in Table 7.20, process considerations dealing with gas flow rate and velocity, pollutant concentration, allowable pressure drop, and the variability of gas and pollutant flow rates, including temperature must all be considered.

The type of pollutant is also an important factor that must be taken into consideration: gaseous or particulate. Certain pertinent questions must be asked and answered. If the pollutant, for example, is gaseous, how corrosive, inflammable, reactive, and toxic is it? After these factors have been evaluated, the focus shifts to the selection of the best air pollution control equipment/system—affordable, practical, and permitted by regulatory requirements—depending, of course, on the type of pollutant to be removed.

In the following sections, two types of pollutants (dry particulates and gaseous pollutants) and the various air pollution control equipment/processes available for their removal are discussed.

REMOVAL OF DRY PARTICULATE MATTER

Constituting a major class of air pollutants, particulates have a variety of shapes and sizes, and as either liquid droplet or dry dust, they have a wide range of physical and chemical characteristics. Dry particulates are emitted from a variety of different sources, including both combustion and non-combustion sources in industry, mining, construction activities, incinerators, and internal combustion engines. Dry particulates are also emitted from natural sources—volcanoes, forest fires, pollen, and windstorms.

All particles and particulate matter exhibit certain important characteristics, which, along with process conditions, must be considered in any engineering strategy to separate and remove them from a stream of carrier gas. Particulate size range and distribution, particle shape, corrosiveness, agglomeration tendencies, abrasiveness, toxicity, reactivity, inflammability, and hygroscopic tendencies must all be examined in light of equipment limitations.

Air Pollution

In an air pollution control system, particulates are separated from the gas stream by the application of one or more forces, in gravity settlers, centrifugal settles, fabric filters, electrostatic precipitators, or wet scrubbers. The particles are then collected and removed from the system.

As mentioned earlier in the text, when a flowing fluid (engineering and science applications consider both liquid and gaseous states as a fluid) approaches a stationary object such as a metal plate, a fabric thread, or a large water droplet, the fluid flow will diverge around that object. Particles in the fluid (because of inertia) will not follow stream flow exactly but will tend to continue in their original directions. If the particles have enough inertia and are located close enough to the stationary object, they will collide with the object and can be collected by it. This is an important phenomenon.

Particles are collected by impaction, interception, and diffusion. *Impaction* occurs when the center of mass of a particle that is diverging from the fluid strikes a stationary object. *Interception* occurs when the particle's center of mass closely misses the object, but, because of its finite size, the particle strikes the object. *Diffusion* occurs when small particulates happen to "diffuse" toward the object while passing near it. Particles that strike the object by any of these means are collected—if short-range forces (chemical, electrostatic, and so forth) are strong enough to hold them to the surface (Cooper and Alley, 1990).

Control technologies for particles focus on capturing the particles emitted by a pollution source. Several factors must be considered before choosing a particulate control device. Typically, particles are collected and channeled through a duct or stack. The characteristics of the particulate exhaust stream affect the choice of the control device. These characteristics include the range of particle sizes, the exhaust flow rate, the temperature, the moisture content, and various, chemical properties such as explosiveness, acidity, alkalinity, and flammability.

The most commonly used control devices for controlling particulate emissions include:

gravity settlers, cyclones, electrostatic precipitators, wet (Venturi) scrubbers, and baghouse (fabric filters). In many cases, more than one of these devices is used in a series to obtain the desired removal efficiencies. For example, a settling chamber can be used to remove larger particles before a pollutant stream enters an electrostatic precipitator. In this section, we will briefly introduce each of the major types of particulate control equipment and point out their advantages and disadvantages.

Gravity Settlers

Gravity settlers (or settling chambers) have long been used by industry for removing solid and liquid waste materials from gaseous streams. Simply constructed, a gravity settler is actually nothing more than an enlarged chamber in which the horizontal gas velocity is reduced, allowing large particles to settle out of the gas by gravity and be recollected in hoppers. Gravity settlers have the advantage of having low initial cost and are relatively inexpensive to operate—there's not a lot to go wrong. However, because settling chambers are effective in removing only larger particles, they have relatively low efficiency, especially for the removal of small particles (<50 µm). Thus, gravity settlers are used in conjunction with a more efficient control device. In addition, although simple in design, gravity settlers require a large space for installation.

Cyclone Collectors

The cyclone (or centrifugal) collector provides a low-cost, low-maintenance method of removing larger particulates from a gas stream. The cyclone removes particles by inertia separation causing the entire gas stream to flow in a spiral pattern inside a tube and is the collector of choice for removing particles >10 µm in diameter. By centrifugal force, the larger particles more outward and collide with the narrowing wall of the tube. The particles slide down the wall and fall to the bottom of the cone, where they are removed. The cleaned gas flows out the top of the cyclone. Along with their relatively low construction costs, cyclones need relatively small space requirements for installation. However, cyclones are efficient in removing large particles but are not as efficient with smaller particles, especially on particles below 10 µm in size—and they do not handle sticky materials well. For this reason, they are used with other particulate control devices. The most serious problems encountered with cyclones are with air flow equalization, and their tendency to plug. Cyclones have been used successfully at feed and grain mills, cement plants, fertilizer plants, petroleum refineries, and other applications involving large quantities of gas containing relatively large particles.

Electrostatic Precipitators (ESPs)

An electrostatic precipitator (ESP) is a particle control device that uses electrical forces to move the particles out of the flowing gas stream and onto collector plates. ESPs are usually used to remove small particles from moving gas streams at high collection efficiencies. Widely used in power plants for removing fly ash from the gases prior to discharge, an electrostatic precipitator applies electrical force to separate particles from the gas stream. A high voltage drop is established between electrodes, and particles passing through the resulting electrical field acquire a charge. The charged particles are attracted to and collected on an oppositely charged plate, and the cleaned gas glows through the device. Periodically, the plates are cleaned by rapping to shake off the layer of dust that accumulates, and the dust is collected in hoppers at the bottom of the device. Although electrostatic precipitators have the advantages of low operating costs, capability for operation a high-temperature applications (to 1,300°F), low-pressure drop—and extremely high particulate (coarse and fine) collection efficiencies can be attained, they have the disadvantages of high capital costs and space requirements. The removal efficiencies for ESPs are highly variable; however, for very small particles alone the removal efficiency is about 99%. Typical ESP applications include use in industrial and utility boilers, cement plants, steel mills, petroleum refineries, municipal waste incinerators, hazardous waste incinerators, Kraft pulp and paper mills, and Lead, zinc, and copper smelters.

Wet (Venturi) Scrubbers

Wet scrubbers (or collectors) have found widespread use in cleaning contaminated gas streams (e.g., foundry dust emissions, acid mists, and furnace fumes) because of their ability to effectively remove particulate and gaseous pollutants. Wet scrubbers vary in complexity from simple spray chambers to remove coarse particles to high-efficiency systems (venturi-types) to remove fine particles. Whichever system is used, operation employs the same basic principles of inertial impingement or impaction and interception of dust particles by droplets of water. The larger, heavier water droplets are easily separated from the gas by gravity. The solid particles can then be independently separated from the water, or the water can be otherwise treated before reuse or discharge. Increasing either the gas velocity or the liquid droplet velocity in a scrubber increases the efficiency because of the greater number of collisions per unit time. For the ultimate in wet scrubbing, where high collection efficiency is desired, the venturi scrubber is used. The venturi operates at extremely high gas and liquid velocities with a very high-pressure drop across the venturi throat. The reduced velocity at the expanded section of the throat allows the droplets of water containing the particles to drop out of the gas stream. Venturi scrubbers are most efficient for removing particulate matter in the size range of 0.5–5 μm, with removal efficiencies of up to 99%, which makes them especially effective for the removal of sub-micron particulates associated with smoke and fumes.

Although wet scrubbers require relatively small space requirements, can remove both gases and particles, can neutralize corrosive gases, have low capital cost, and can handle high-temperature, high-humidity gas streams, their power and maintenance costs are relatively high, they may create wastewater disposal problems, their corrosion problems are more severe than dry systems, and the final product they produce is collected wet. Table 7.21 summarizes these advantages and disadvantages. Wet scrubbers have been used in a variety of industries such as acid plants, fertilizer plants, steel mills, asphalt plants, and larger power plants.

Baghouse (Fabric) Filters

Baghouse filters (or fabric filters) are the most commonly used air pollution control filtration system. In much the same manner as the common vacuum cleaner, filter fabric filter material, capable of removing most particles as small as 0.5 μm and substantial quantities of particles as small as 0.1 μm,

TABLE 7.21
Advantages and Disadvantages of Wet Scrubbers

Advantages	Disadvantages
Small space requirements	Corrosion problems
No secondary dust sources	High power requirements
Handles high-temperature, high-humidity gas streams	Water-disposal problems
Minimal fire and explosion hazards	Difficult product recovery
Ability to collect both gases and particles	Meteorological problems (plume=fog)

Source: EPA (2006).

is formed into cylindrical or envelope bags and suspended in the baghouse. The particulate-laden gas stream is forced through the porous fabric filter, and as the air passes through the fabric, particulates accumulate on the cloth, providing a clean air stream. As particulates build up on the inside surfaces of the bags, the pressure drop increases. Before the pressure drop becomes too severe, the bags must be relieved of some of the particulate layers. The particulates are periodically removed from the cloth by shaking or by reversing the air flow.

The selection of the fiber material and fabric construction is important to baghouse performance. The fiber material from which the fabric is made must have adequate strength characteristics at the maximum gas temperature expected and adequate chemical compatibility with both the gas and the collected dust. One disadvantage of the fabric filter is that high-temperature gases often have to be cooled before contacting the filter medium.

Fabric filters are used in the power generation, incineration, chemical, steel, cement, food, pharmaceutical, metal working, aggregate, and carbon black industries.

Fabric filters are relatively simple to operate, provide high overall collection efficiencies up to 99+%, and are very effective in controlling sub-micrometer particles, but they do have limitations. These include relatively high capital costs, high maintenance requirements (bag replacement, etc), high space requirements, and flammability hazards for some dusts.

REMOVAL OF GASEOUS POLLUTANTS: STATIONARY SOURCES

In the removal of gaseous air pollutants, the principal gases of concern are the sulfur oxides (SO_x), carbon oxide (CO_x), nitrogen oxides (NO_x), organic and inorganic acid gases, and hydrocarbons (HC). The most common method for controlling gaseous pollutants is the addition of add-on control devices to recover or destroy a pollutant. Four major treatment processes (add-ons) currently available for control of these and other gaseous emissions: absorption, adsorption, condensation, and combustion (incineration).

The decision of which single or combined air pollution control technique to use for stationary sources is not always easy. Gaseous pollutants can be controlled by a wide variety of devices, and choosing the most cost-effective, most efficient units requires careful attention to the particular operation for which the control devices are intended. Specifically, the choice of control technology depends on the pollutant(s) to be removed, the removal efficiency required, pollutant and gas stream characteristics, and specific characteristics of the site (EPA, 2006). Absorption, adsorption, and condensation all are recovery techniques while incineration involves the destruction of the pollutant.

In making the difficult and often complex decision on which air pollution control technology to employ, it is helpful to follow guidelines based on experience and set forth by Buonicore et al. (1992) in the prestigious engineering text: *Air Pollution Engineering Manual*. Table 7.22 summarizes these.

TABLE 7.22
Comparison of Air Pollution Control Technologies

Treatment Technology	Concentration and Efficiency	Comments
Incineration	(<100 ppmv) 90%–95% efficient (>100 ppmv) 95%–99% efficient	Incomplete combustion may require additional controls
Carbon adsorption	(>200 ppmv) 90+% efficiency (>1,000 ppmv) 95+% efficiency	Recovered organics may need additional treatment—can increase cost
Absorption	(<200 ppmv) 90%–95% efficiency (>200 ppmv) 95+% efficiency	Can blowdown stream be accommodated at site?
Condensation	(>2,000 ppmv) 80+% efficiency	Must have low temperature or high pressure for efficiency

Note: Typically, only incineration and absorption technologies can achieve >99% gaseous pollutant removal consistently.

Absorption

Absorption (or scrubbing) is a major chemical engineering unit operation that involves bringing contaminated effluent gas into contact with a liquid absorbent so that one or more constituents of the effluent gas are selectively dissolved into a relatively nonvolatile liquid. Absorption units are designed to transfer the pollutant from a gas phase to a liquid phase (water is the most commonly used absorbent liquid). The absorption unit accomplishes this by providing intimate contact between the gas and the liquid, providing optimum diffusion of the gas into the solution. The actual removal of a pollutant from the gas stream takes place in three steps: (1) diffusion of the pollutant gas to the surface of the liquid; (2) transfer across the gas/liquid interface; and (3) diffusion of the dissolved gas away from the interface into the liquid (Davis and Cornwell, 1991). Absorption is commonly used to recover products or to purify gas streams that have high concentrations of organic compounds. Absorption equipment is designed to get as much mixing between the gas and liquid as possible.

Several types of absorbers are available, including spray chambers (and towers or columns), plate or tray towers, packed towers, and venturi scrubbers. Pollutant gases commonly controlled by absorption include sulfur dioxide, hydrogen sulfide, hydrogen chloride, chlorine, ammonia, and oxides of nitrogen.

Absorbers are often referred to as scrubbers, and there are various types of absorption equipment. The principal types of gas absorption equipment include spray towers, packed columns, spray chambers, and venture scrubbers. The two most common absorbent units in use today are the plate and packed tower systems. Plate towers contain perforated horizontal plates or trays designed to provide large liquid-gas interfacial areas. The polluted air stream is usually introduced at one side of the bottom of the tower or column and rises up through the perforations in each plate; the rising gas prevents the liquid from draining through the openings rather than through a downpipe. During continuous operation, contact is maintained between air and liquid, allowing gaseous contaminants to be removed, with clean air emerging from the top of the tower.

The packed tower scrubbing system is by far the most commonly used for the control of gaseous pollutants in industrial applications, where it typically demonstrates a removal efficiency of 90%–95%. Usually configured in vertical fashion the packed tower is literally "packed" with devices that have a large surface-to-volume ratio and a large void ratio that offers minimum resistance to gas flow. In addition, packing should provide even distribution of both fluid-phases, be sturdy enough to support them in the tower, and be low cost, available, and easily handled (Hesketh, 1991).

The flow through a packed tower is typically countercurrent, with gas entering at the bottom of the tower and liquid entering at the top. Liquid flows over the surface of the packing in a thin film, affording continuous contact with the gases.

Though highly efficient for the removal of gaseous contaminants, packed towers may create wastewater disposal problems (converting an air pollution problem to a water pollution problem), become easily clogged when gases with high particulate loads are introduced and have relatively high maintenance costs.

Adsorption

Adsorption is a mass transfer process that involves passing a stream of effluent gas through the surface of prepared porous solids (adsorbents). The surfaces of the porous solid substance attract and hold (bind) the gas (the adsorbate) by either physical or chemical adsorption. In physical adsorption (a readily reversible process), a gas molecule adheres to the surface of the solid because of an imbalance of electron distribution. In chemical adsorption (not readily reversible) once the gas molecule adheres to the surface, it reacts chemically with it.

Several materials possess adsorptive properties. These materials include activated carbon, alumina, bone char, magnesia, silica gel, molecular sieves, strontium sulfate, and others. The most important adsorbent for air pollution control is activated charcoal. Activated carbon s the universal standard for the purification and removal of trace organic contaminants form liquid and vapor streams. The surface area of activated charcoal will preferentially adsorb hydrocarbon vapors and odorous organic compounds from an airstream.

In an adsorption system (in contrast to the absorption system where the collected contaminant is continuously removed by flowing liquid), the collected contaminant remains in the adsorption bed. The most common adsorption system is the fixed bed adsorber which can be contained in either a vertical or horizontal cylindrical shell. The adsorbent (usually activated carbon) is arranged in beds or trays in layers about 0.5 in. thick. In multiple bed systems, one or more beds are adsorbing vapors, while the other bed is being regenerated.

The efficiency of most adsorbers is near 100% at the beginning of the operation and remains high until a breakpoint or breakthrough occurs. When the adsorbent becomes saturated with adsorbate, contaminant begins to leak out of the bed, signaling that the adsorber should be renewed or regenerated. By regenerating the carbon bed, the same activated carbon particles can be used again and again.

Although adsorption systems are high-efficiency devices that may allow recovery of product, have excellent control and response to process changes, and have the capability of being operated unattended, they also have some disadvantages, including the need for exotic, expensive extraction schemes if product recovery is required, relatively high capital cost, and gas stream prefiltering needs (to remove any particulate capable of plugging the adsorbent bed).

Condensation

Condensation is a process by which volatile gases are removed from the contaminant stream and changed into a liquid. In air pollution control, a condenser can be used in two ways: either for pretreatment to reduce the load problems with other air pollution control equipment or for effectively controlling contaminants in the form of gases and vapors.

Condensers condense vapors to liquid phase by either increasing the system pressure without a change in temperature, or by decreasing the system temperature to its saturation temperature without a pressure change. Condensation is affected by the composition of the contaminant gas stream. When gases are present in the stream which condense under different conditions, condensation is hindered.

Condensers are widely used to recover valuable products in a waste stream. Condensers are simple, relatively inexpensive devices that normally use water or air to cool and condense a vapor stream. Condensers are typically used as pretreatment devices. They can be used ahead of adsorbers, absorbers, and incinerators to reduce the total gas volume to be treated by more expensive control equipment.

There are two basic types of condensation equipment used for pollution control—surface and contact condensers. A surface condenser is normally a shell-and-tube heat exchanger. A surface condenser uses a cooling medium of air or water where the vapor to be condensed is separated from the cooling medium by a metal wall. Coolant flows through the tubes, while the vapor is passed over and condenses on the outside of the tubes, and drains off to storage (EPA, 2006).

In a contact condenser (which resembles a simple spray scrubber), the vapor is cooled by spraying liquid directly on the vapor stream. The cooled vapor condenses, and the water and condensate mixture are removed, treated, and disposed of.

In general, contact condensers are less expensive, more flexible, and simpler than surface condensers, but surface condensers require much less water and produce many times less wastewater that must be treated than do contact condensers. Removal efficiencies of condensers typically range from 50 percent to more than 95 percent, depending on design and applications. Condensers are used in a wide range of industrial applications including, petroleum refining, petrochemical manufacturing, basic chemical manufacturing, and dry cleaning, and degreasing.

Combustion (Incineration)

Even though combustion (or incineration) is a major source of air pollution, it is also, if properly operated, a beneficial air pollution control system in which the objective is to convert certain air contaminants (usually CO and hydrocarbons) to innocuous substances such as carbon dioxide and water (EPA, 2006).

Combustion is a chemical process defined as rapid, high-temperature gas-phase oxidation. The combustion equipment used to control air pollution emissions is designed to push these oxidation reactions as close to complete combustion as possible, leaving a minimum of unburned residue. The operation of any combustion operation is governed by four variables: oxygen, temperature, turbulence, and time. For complete combustion to occur, oxygen must be available and put into contact with sufficient temperature (turbulence) and held at this temperature for a sufficient time. These four variables are not independent—changing one affects the entire process.

Depending upon the contaminant being oxidized, equipment used to control waste gases by combustion can be divided into three categories: direct-flame combustion (or flaring), thermal combustion (afterburners), or catalytic combustion. Choosing the proper device depends on many factors, including the type of hazardous contaminants in the waste stream, the concentration of combustibles in the stream, process flowrate, control requirements, and an economic evaluation.

Direct-Flame Combustion (Flaring)

Direct-flame combustion devices (flares) are the most commonly used air pollution control devices by which waste gases are burned directly (with or without the addition of a supplementary fuel). Common flares include steam-assisted, air-assisted, and pressure head types. Studies conducted by EPA have shown that the destruction efficiency of a flare is about 98%. Flares are normally elevated from 100 to 400 ft to protect the surroundings from heat and flames. Often designed for steam injection at the flare top flares commonly use steam in this application because it provides sufficient turbulence to ensure complete combustion, which prevents the production of visible smoke or soot. Flares are also noisy, which can cause problems for adjacent neighborhoods, and some flares produce oxides of nitrogen, thus creating a new air pollutant.

TABLE 7.23
Advantages of Catalytic over Thermal Incinerators

(1)	Catalytic incinerators have lower fuel requirements
(2)	Catalytic incinerators have lower operating temperatures
(3)	Catalytic incinerators have little or no insulation requirements
(4)	Catalytic incinerators have reduced fire hazards
(5)	Catalytic incinerators have reduced flashback problems

Source: Adaptation from Buonicore et al. (1992).

Thermal Combustion (Afterburners)

The thermal incinerator or afterburner is usually the unit of choice in cases where the concentration of combustible gaseous pollutants is too low to make flaring practical. Widely used in industry, typically, the thermal combustion system operates at high temperatures. Within the thermal incinerator, the contaminant airstream passes around or through a burner and into a refractory-line residence chamber where oxidation occurs. Residence time is the amount of time the fuel mixture remains in the combustion chamber. Flue gas from a thermal incinerator (which is relatively clean) is at high temperature and contains recoverable heat energy. Thermal incinerators can destroy gaseous pollutants at efficiencies of greater than 99 percent when operated correctly.

Catalytic Combustion

Catalytic combustion operates by passing a preheated contaminant-laden gas stream through a catalyst bed (usually thinly-coated platinum mesh mat, honeycomb of other configurations designed to increase surface area), which promotes the oxidization reaction at lower temperatures. The metal catalyst is used to initiate and promote combustion at much lower temperatures than those required for thermal combustion (metals in the platinum family are recognized for their ability to promote combustion at low temperature). Catalytic incineration may require 20 to 50 times less residence time than thermal incineration (see Table 7.23) for other advantages of catalytic incinerators over thermal incinerators). Catalytic incinerators normally operate at 700°F–900°F. At this reduced temperature range, a saving in fuel usage and cost is realized; however, this may be offset by the cost of the catalytic incinerator itself. Destruction efficiencies >95% are possible using a catalytic incinerator. Higher efficiencies are possible if larger catalyst volumes or higher temperatures are used.

A heat exchanger is an option for systems with heat transfer between two gas streams (recuperative heat exchange). The need for dilution air, combustion air, and/or flue gas treatment is based on site-specific conditions. Catalysts are subject to both physical and chemical deterioration and their usefulness is suppressed by sulfur-containing compounds. For best performance, catalyst surfaces must be clean and active.

Catalytic incineration is used in a variety of industries to treat effluent gases, including emissions from paint and enamel bake ovens, asphalt oxidation, coke ovens, formaldehyde manufacturing and varnish cooking. Catalytic incinerators are best suited for emission streams with low VOC content.

INDOOR AIR QUALITY

The quality of the air we breathe and the attendant consequences for human health are influenced by a variety of factors. These include hazardous material discharges indoors and outdoors, meteorological and ventilation conditions, and pollutant decay and removal processes. Over 80% of our time is spent in indoor environments so that the influence of building structures, surfaces, and ventilation are important considerations when evaluating air pollution exposures (Wadden and Scheff, p. 1, 1983).

For those familiar with *Star Trek*, *Trekees* (and for those who are not), consider a quotable quote: "The air is the air." In regard to the air we breathe, according to USEPA (2001), few of us realize that we all face a variety of risks to our health as we go about our day-to-day lives. Driving our cars, flying in planes, engaging in recreational activities, and being exposed to environmental pollutants all pose-varying degrees of risk. Some risks are simply unavoidable. Some we choose to accept because to do otherwise would restrict out the ability to lead our lives the way we want. And some are risks we might decide to avoid if we had the opportunity to make informed choices. Indoor air pollution is one risk that we can do something about.

In the last several years, a growing body of scientific evidence has indicated that the air within homes and other buildings can be more seriously polluted than the outdoor air in even the largest and most industrialized cities. Other research indicates that people spend ~90% of their time indoors. A type of climate we don't often think about (if at all) is the micro-climates we spend 80% of our time in: the office and/or the home (indoors) (Wadden and Scheff, 1983). Thus, for many people, the risks to health may be greater due to exposure to air pollution indoors than outdoors (USEPA, 2001).

Not much attention was given to indoor micro-climates until after two events took place a few years ago. The first event had to do with Legionnaires' disease and the second with Sick Building Syndrome. In addition, people who may be exposed to indoor air pollutants for the longest periods of time are often those most susceptible to the effects of indoor air pollution. Such groups include the young, the elderly, and the chronically ill, especially those suffering from respiratory or cardio-vascular disease.

The impact of energy conservation on inside environments may be substantial, particularly with respect to decreases in ventilation rates (Hollowell et al., 1979a) and "tight" buildings constructed to minimize infiltration of outdoor air (Woods, 1980; Hollowell et al., 1979b). The purpose of constructing "tight buildings" is to save energy—to keep the heat or air conditioning inside the structure. The problem is indoor air contaminants within these tight structures are not only trapped within but also can be concentrated, exposing inhabitants to even more exposure.

These topics and others along with causal factors leading to indoor air pollution are covered in this section. What about indoor air quality problems in the workplace? In this section, we also discuss this pervasive but often overlooked problem. In this regard, we discuss the basics of IAQ (as related to the workplace environment) and the major contaminants that currently contribute to this problem. Moreover, mold and mold remediation, although not new to the workplace, are the new buzzwords attracting attention these days. Contaminants such as asbestos, silica, lead, and formaldehyde contamination are also discussed. Various related remediation practices are also discussed.

LEGIONNAIRES' DISEASE

Since that infamous event that occurred in Philadelphia in 1976 at the Belleview Stratford Hotel during a convention of American Legion members, which included 182 cases and 29 deaths, *Legionella pneumophila* (the deadly bacterium) has become synonymous with the term *Legionnaires' disease*. The deaths were attributed to colonized bacteria in the air conditioning system cooling tower.

Let's take a look at this deadly killer—a killer that inhabits the micro-climate we call office, hotel, and other indoor spaces.

Organisms of the genus *Legionella* are ubiquitous in the environment and are found in natural fresh water, potable water, as well as in closed-circuit systems, such as evaporative condensers, humidifiers, recreational whirlpools, air handling systems, and, of course, in cooling tower water.

The potential for the presence of *Legionella* bacteria is dependent on certain environmental factors: moisture, temperature (50°F–140°F), oxygen, and a source of nourishment such as slime or algae.

Not all the ways in which Legionnaires' disease can be spread are known to us at this time; however, we do know that it can be spread through the air. Centers for Disease Control (CDC) states (in its *Questions and Answers on Legionnaires' disease*, CDC No. 28L0343779) that there is no evidence that Legionnaires' disease is spread person-to-person.

Air-conditioning cooling towers and evaporative condensers have been the source of most outbreaks to date and the bacterium is commonly found in both. Unfortunately, we do not know if this is an important means of spreading of Legionnaires' disease because other outbreaks have occurred in buildings that did not have air-conditioning.

Not all people are at risk of contracting Legionnaires' disease. The people most at risk include persons:

1. with lowered immunological capacity
2. who smoke cigarettes and abuse alcohol
3. who are exposed to high concentrations of *Legionella pneumophila*

Most commonly recognized as a form of pneumonia, the symptoms of Legionnaires' disease usually become apparent 2–10 days after known or presumed exposure to airborne Legionnaires' disease bacteria. A sputum-free cough is common, but sputum production is sometimes associated with the disease. Within less than a day, the victim can experience rapidly rising fever and the onset of chills. Mental confusion, chest pain, abdominal pain, impaired kidney function, and diarrhea are associated manifestations of the disease. CDC estimates that around 25,000 people develop Legionnaires' disease annually.

The obvious question becomes: How do we prevent or control Legionnaires' disease? Good question.

The controls presently being used are targeted at cooling towers and air handling units (condensate drain pans).

Cooling tower procedures used to control bacterial growth vary somewhat on the various regions in a cooling tower system. However, control procedures usually include a good maintenance program, including repair/replacement of damaged components, routine cleaning, and sterilization.

In sterilization, a typical protocol calls for the use of chlorine in a residual solution at about 50 ppm combined with a detergent that is compatible to produce the desired sterilization effect. It is important to ensure that even those spaces that are somewhat inaccessible are properly cleaned of slime and algae accumulations.

Control measures for air handling units: condensate drain pans typically involve keeping the pans clean and checked for proper drainage of fluid—this is important to prevent stagnation and the buildup of slime/algae/bacteria. A cleaning and sterilization program is required anytime algae or slime are found in the unit.

SICK BUILDING SYNDROME

The second event that got the Public's attention regarding micro-climates and the possibility of unhealthy environments contained therein was actually spawned by the first, the Legionnaires' event and other incidents or complaints that followed. What we are referring to here is *Sick Building Syndrome*.

The term Sick Building Syndrome was coined by an international working group under the *World Health Organization* (WHO) in 1982. The WHO working group studied the literature about indoor climate problems and found that these micro-climates in buildings are characterized by the same set of frequently appearing complaints and symptoms. WHO came up with five categories of symptoms exemplified by some complaints reported by occupants supposed to suffer from sick building syndrome (SBS). These categories are listed in the following:

1. **Sensory irritation in eyes, nose, and throat**: Pain, sensation of dryness, smarting feeling, stinging, irritation, hoarseness, and voice problems.
2. **Neurological or general health symptoms**: Headache, sluggishness, mental fatigue, reduced memory, reduced capability to concentrate, dizziness, intoxication, nausea and vomiting, tiredness.

3. **Skin irritation**: Pain, reddening, smarting or itching sensations, dry skin.
4. **Nonspecific hypersensitivity reactions**: Running nose and eyes, asthma-like symptoms among nonasthmatics, sounds from the respiratory system.
5. **Odor and taste symptoms**: Changed sensitivity of olfactory or gustatory sense, unpleasant olfactory or gustatory perceptions.

In the past similar symptoms had been used to define other syndromes such as the building disease, the building illness syndrome, building-related illness, or the tight-fitting office syndrome, which in many cases appear to be synonyms for the sick building syndrome; thus, the WHO definition of the SBS worked to combine these syndromes into one general definition or summary. A summary compiled by WHO (1982, 1984) and Molhave (1986) of this combined definition includes the five categories of symptoms listed earlier and also:

1. Irritation of mucous membranes in the eye, nose, and throat is among the most frequent symptoms
2. Other symptoms, e.g., from lower airways or from internal organs, should be infrequent.
3. A large majority of occupants report symptoms.
4. The symptoms appear especially frequently in one building or in part of it.
5. No evident causality can be identified in relation either to exposures or to occupant sensitivity.

The WHO group suggested the possibility that the SBS symptoms have a common causality and mechanism (WHO, 1982). However, the existence of SBS is still a postulate because the descriptions of the symptoms in the literature are anecdotal and unsystematic.

INDOOR AIR POLLUTION

Why is indoor air pollution a problem? As indicated above, the recognition that the indoor air environment may be a health problem is a relatively recent emergence. The most significant of indoor air quality are the impact of cigarette smoking, stove and oven operation, and emanations from certain types of particleboard, cement, and other building materials (Wadden and Scheff, 1983).

The significance of the indoor air quality problem became apparent not only because of the Legionnaires' Incident of 1976 and the WHO study of 1982 but also because of another factor that came to the forefront in the mid-1970s: The need to conserve energy. In the early 1970s when hundreds of thousands of people were standing in line to obtain gasoline for their automobiles, it was not difficult to drive home the need to conserve energy supplies.

The resulting impact of energy conservation on inside environments has been substantial. This is especially the case in regard to building modifications that were made to decrease ventilation rates and new construction practices that were incorporated to ensure "tight" buildings to minimize infiltration of outdoor air.

There is some irony in this development, of course. While there is a need to ensure proper building design, construction, and ventilation guidelines to avoid the exposure of inhabitants to unhealthy environments, what really resulted in this mad dash to reduce ventilation rates and "tighten" buildings from infiltration was a tradeoff: energy economics versus air quality.

According to Byrd (2003), Indoor Air Quality (IAQ) refers to the effect, good or bad, of the contents of the air inside a structure, on its occupants. Stated differently, indoor air quality (IAQ), in this text, refers to the quality of the air inside workplaces as represented by concentrations of pollutants and thermal (temperature and relative humidity) conditions that affect the health, comfort and performance of employees. Usually, temperature (too hot or too cold), humidity (too dry or too damp), and air velocity (draftiness or motionlessness) are considered "comfort" rather than indoor air quality issues. Unless they are extreme, they may make someone uncomfortable, but they won't

Air Pollution

make a person ill. Other factors affecting employees, such as light and noise, are important indoor environmental quality considerations, but are not treated as core elements of indoor air quality problems. Nevertheless, most industrial hygienists must take these factors into account in investigating environmental quality situations.

Byrd (2003) further points out that "good IAQ is the quality of air, which has no unwanted gases or particles in it at concentrations, which will adversely affect someone. Poor IAQ occurs when gases or particles are present at an excessive concentration so as to affect the satisfaction of health of occupants."

In the workplace, poor IAQ may only be annoying to one person, however, at the extreme, it could be fatal to all the occupants in the workplace.

The concentration of the contaminant is crucial. Potentially infectious, toxic, allergenic, or irritating substances are always present in the air. Note that there is nearly always a threshold level below which no effect occurs.

COMMON INDOOR AIR POLLUTANTS IN THE HOME

This section takes a brief source-by-source look at the most common indoor air pollutants, their potential health effects, and ways to reduce their levels in the home.

Radon

Radon is a noble, nontoxic, colorless, odorless gas produced in the decay of radium-226 and is found everywhere at very low levels. Radon is ubiquitously present in the soil and air near to the surface of the Earth. As radon undergoes radioactive decay, it releases an alpha particle, gamma ray, and progeny that quickly decay to release alpha and beta particles and gamma rays. Because radon progeny are electrically charged, they readily attach to particles, producing a radioactive aerosol. It is when radon becomes trapped in buildings and concentrations build up in indoor air that exposure to radon becomes of concern. This is the case because aerosol radon-contaminated particles may be inhaled and deposited in the bifurcations of respiratory airways. Irradiation of tissue at these sites poses a significant risk of lung cancer (depending on exposure dose).

How does radon enter a house?

The most common way in which radon enters a house is through the soil or rock upon which the house is built. The most common source of indoor radon is uranium, which is common to many soils and/or rocks. As uranium breaks down, it releases soil or radon gas, and radon gas breaks down into radon decay products or progeny (commonly called *radon daughters*). Radon gas is transported into buildings by pressure-induced convective flows.

There are other sources of radon, for example, from well water and masonry materials.

Radon levels in a house vary in response to temperature-dependent and wind-dependent pressure differentials and to changes in barometric pressures. When the base of a house is under significant negative pressure, radon transport is enhanced.

Studies by the USEPA (1987, 1988a) indicate that as many as 10% of all American homes, or about 9 million homes, may have elevated levels of radon, and the percentage may be higher in geographic areas with certain soils and bedrock formations.

According to USEPA's booklets, *A Citizen's Guide to Radon*, *Radon Reduction Methods: A Homeowner's Guide*, and *Radon Measurement Proficiency Report* (for each state) exposure to radon in the home can be reduced by the following steps:

1. Measure levels of radon in the home.
2. The state radiation protection office can provide you with information on the availability of detection devices or services.
3. Refer to EPA guidelines in deciding whether and how quickly to act based on test results.
4. Learn about control measures.

5. Take precautions not to draw larger amounts of radon into the house.
6. Select a qualified contractor to draw up and implement a radon mitigation plan.
7. Stop smoking and discourage smoking in your home.
8. Treat radon-contaminated well water by aerating or filtering through granulated activated charcoal.

Environmental Tobacco Smoke

The use of tobacco products by ~45 million smokers in the United States results in significant indoor contamination from combustion by-products that pose significant exposures to millions of others who do not smoke but who must breathe contaminated indoor air. Composed of side-stream smoke (smoke that comes from the burning end of a cigarette) and smoke that is exhaled by the smoker, it contains a complex mixture of over 4,700 compounds, including both gases and particles.

According to reports issued in 1986 by the Surgeon general and the National Academy of Sciences, environmental tobacco smoke is a cause of disease, including lung cancer, in both smokers and healthy nonsmokers. Environmental tobacco smoke may also increase the lung cancer risk associated with exposure to radon.

The following steps can reduce exposure to environmental tobacco smoke in the office and/or home:

1. Give up smoking and discourage smoking in your home and place of work or require smokers to smoke outdoors.
2. A common method of reducing exposure to indoor air pollutants such as environmental tobacco smoke is ventilation which works to reduce but not eliminate exposure.

Biological Contaminants

A variety of biological contaminants can cause significant illness and health risks. These include mold and mildew, viruses, animal dander and cat saliva, mites, cockroaches, pollen, and infections form airborne exposures to viruses that cause colds and influenza and bacteria that cause Legionnaires' disease and tuberculosis (TB).

The following steps can reduce exposure to biological contaminants in the home and/or office.

1. Install and use exhaust fans that are vented to the outdoors in kitchens, bathrooms, and vent clothes dryers outdoors.
2. Ventilate the attic and crawl spaces to prevent moisture buildup.
3. Keep water trays in cool mist or ultrasonic humidifiers clean and filled with fresh distilled water daily.
4. Water-damaged carpets and building materials should be thoroughly dried and cleaned within 24 hours.
5. Main good housekeeping practices both in the home and office.

Combustion by-Products

Combustion byproducts are released into indoor air from a variety of sources. These include unvented kerosene and gas space heaters, woodstoves, fireplaces, gas stoves, and hot water heaters. The major pollutants released from these sources are carbon monoxide, nitrogen dioxide, and particles.

The following steps can reduce exposure to combustion products in the home (and/or office):

1. Fuel-burning unvented space heaters should only be operated using great care and special safety precautions.

Air Pollution

2. Install and use exhaust fans over gas cooking stoves and ranges and keep the burners properly adjusted.
3. Furnaces, flues, and chimneys should be inspected annually, and any needed repairs should be made promptly.
4. Woodstove emissions should be kept to a minimum.

Household Products

A large variety of organic compounds are widely used in household products because of their useful characteristics, such as the ability to dissolve substances and evaporate quickly. Cleaning, disinfecting, cosmetic, degreasing, and hobby products all contain organic solvents, as do paints, varnishes, and waxes. All of these products can release organic compounds while using them, and when they are stored.

The following steps can reduce exposure to household organic compounds:

1. Always follow label instructions carefully.
2. Throw away partially full containers of chemicals safely.
3. Limit the amount you buy.

Pesticides

Pesticides represent a special case of chemical contamination of buildings where the EPA estimates 80%–90% of most people's exposure to the air occurs. These products are extremely dangerous if not used properly.

The following steps can reduce exposure to pesticides in the home:

1. Read the label and follow directions.
2. Use pesticides only in well-ventilated areas.
3. Dispose of unwanted pesticides safely.

Asbestos in the Home

Asbestos became a major indoor air-quality concern in the U.S. in the late 1970s. Asbestos is a mineral fiber commonly used in a variety of building materials and has been identified as having the potential (when friable) to cause cancer in humans.

The following steps can reduce exposure to asbestos in the home (and/or office):

1. Do not cut, rip, or sand asbestos-containing materials.
2. When you need to remove or clean up asbestos, use a professional, trained contractor.

BUILDING FACTORS AFFECTING INDOOR AIR QUALITY

Building factors affecting indoor air quality can be grouped into two factors: Factors affecting indoor climate and factors affecting indoor air pollution.

- **Factors affecting indoor climate**: The thermal environment (temperature, relative humidity, and airflow) are important dimension of indoor air quality for several reasons. First, many complaints of poor indoor air may be resolved by simply altering the temperature or relative humidity. Second, people that are thermally uncomfortable will have a lower tolerance to other building discomforts. Third, the rate at which chemicals are released from building materials is usually higher at higher building temperatures. Thus, if occupants are too warm, it is also likely that they are being exposed to higher pollutant levels.

- **Factors affecting indoor air pollution**: Much of the building fabric, its furnishings and equipment, its occupants and their activities produce pollution. In a well-functioning building, some of these pollutants will be directly exhausted to the outdoors and some will be removed as outdoor air enters that building and replaces the air inside. The air outside may also contain contaminants which will be brought inside in this process. This air exchange is brought about by the mechanical introduction of outdoor air (outdoor air ventilation rate), the mechanical exhaust of indoor air, and the air exchanged through the building envelope (infiltration and exfiltration).

Pollutants inside can travel through the building as air flows from areas of higher atmospheric pressure to areas of lower atmospheric pressure. Some of these pathways are planned and deliberate so as to draw pollutants away from occupants, but problems arise when unintended flows draw contaminants into occupied areas. In addition, some contaminants may be removed from the air through natural processes, as with the adsorption of chemicals by surfaces or the settling of particles onto surfaces. Removal processes may also be deliberately incorporated into the building systems. Air filtration devices, for example, are commonly incorporated into building ventilation systems.

Thus, the factors most important to understanding indoor pollution are (1) indoor sources of pollution, (2) outdoor sources of pollution, (3) ventilation parameters, (4) airflow patterns and pressure relationships, and (5) air filtration systems.

TYPES OF WORKPLACE AIR POLLUTANTS

Common pollutants or pollutant classes of concern in commercial buildings along with common sources of these pollutants are provided in Table 7.24.

SOURCES OF WORKPLACE AIR POLLUTANTS

Air quality is affected by the presence of various types of contaminants in the air. Some are in the form of gases. These would be generally classified as toxic chemicals. The types of interest are combustion products (carbon monoxide, nitrogen dioxide), volatile organic compounds (formaldehyde,

TABLE 7.24
Indoor Pollutants and Potential Sources

Pollutant or Pollutant Class	Potential Sources
Environmental Tobacco Smoke	Lighted cigarettes, cigars, pipes
Combustion Contaminants	Furnaces, generators, gas or kerosene space heaters, tobacco products, outdoor air, vehicles
Biological Contaminants	Wet or damp materials, cooling towers, humidifiers, cooling coils or drain pans, damp duct insulation or filters, condensation, re-entrained sanitary exhausts, bird droppings, cockroaches or rodents, dust mites on upholstered furniture or carpeting, body odors
Volatile Organic Compounds (VOCs)	Paints, stains, varnishes, solvents, pesticides, adhesives, wood preservatives, waxes, polishes, cleansers, lubricants, sealants, dyes, air fresheners, fuels, plastics, copy machines, printers, tobacco products, perfumes, dry cleaned clothing
Formaldehyde	Particle board, plywood, cabinetry, furniture, fabrics
Soil gases (radon, sewer gas, VOCs, methane)	Soil, and rock (radon), sewer drain leak, dry drain traps, leaking underground storage tanks, land fill
Pesticides	Termiticides, insecticides, rodenticides, fungicides, disinfectants, herbicides

Air Pollution

solvents, perfumes, fragrances, etc.), and semi-volatile organic compounds (pesticides). Other pollutants are in the form of animal dander, etc.); soot; particles from buildings, furnishings, and occupants such as fiberglass, gypsum powder, paper dust, lint from clothing, carpet fibers, etc.; dirt (sandy and earthy material), etc.

Burge and Hoyer (1998) point out many specific sources for contaminants that result in adverse health effects in the workplace, including the workers (contagious diseases, carriage of allergens, and other agents on clothing); building compounds (VOCs, particles, and fibers); contamination of building components (allergens, microbial agents, pesticides); and outdoor air (microorganisms, allergens, and chemical air pollutants).

When workers complain of IAQ problems, the Industrial Hygienist is called upon to determine if the problem really is an IAQ problem. If he/she determines that some form of contaminant is present in the workplace, proper remedial action is required. This usually includes removing the source of the contamination.

Tables 7.25 and 7.26 identify indoor and outdoor sources (respectively) of contaminants commonly found in the workplace and offer some measures for maintaining control of these contaminants.

Table 7.26 identifies common sources of contaminants that are introduced from outside buildings. These contaminants frequently find their way inside through the building shell, openings, or other pathways to the inside.

INDOOR AIR CONTAMINANT TRANSPORT

Air contaminants reach worker breathing-zones by traveling from the source to the worker by various pathways. Normally, the contaminants travel with the flow of air. Air moves from areas of high

---

TABLE 7.25
Indoor Sources of Contaminants

Category/Common Sources	Mitigation and Control
Housekeeping and Maintenance	
• Cleanser	• Use low-emitting products
• Waxes and polishes	• Avoid aerosols and sprays
• Disinfectants	• Dilute to proper strength
• Air fresheners	• Do not overuse; use during unoccupied hours
• Adhesives	• Use proper protocol when diluting and mixing
• Janitor's/storage closets	• Store properly with containers closed and lid tight
• Wet mops	• Use exhaust ventilation for storage spaces (eliminate return air)
• Drain cleaners	• Clean mops, store mop top up to dry
• Vacuuming	• Avoid "air fresheners"—clean and exhaust instead
• Paints and coatings	• Use high efficiency vacuum bags/filters
• Solvents	• Use integrated Pest Management
• Pesticides	
• Lubricants	
Occupant-Related Sources	
• Tobacco products	• Smoking policy
• Office equipment (printers/copiers)	• Use exhaust ventilation with pressure control for major local sources
• Cooking/microwave	
• Art supplies	• Low emitting art supplies/marking pens
• Marking pens	• Avoid paper clutter
• Paper products	• Education material for occupants and staff
• Personal products (e.g., perfume)	
• Tracked in dirt/pollen	

(Continued)

TABLE 7.25 (*Continued*)
Indoor Sources of Contaminants

Category/Common Sources	Mitigation and Control
Building Uses as Major Sources	
• Print/photocopy shop	• Use exhaust ventilation and pressure control
• Dry cleaning	• Use exhaust hoods where appropriate; check hood airflows
• Science laboratory	
• Medical office	
• Hair/nail salon	
• Cafeteria	
• Pet store	
Building-Related Sources	
• Plywood/compressed wood	• Use low emitting sources
• Construction adhesives	• Air out in an open/ventilated area before installing
• Asbestos products	• Increase ventilation rates during and after installing
• Insulation	• Keep material dry prior to enclosing
• Wall/floor coverings (vinyl/plastic)	
• Carpets/carpet adhesives	
• Wet building products	
• Transformers	
• Upholstered furniture	
• Renovation/remodeling	
HVAC System	
• Contaminated filters	• Perform HVAC preventive maintenance
• Contaminated duct lining	• Change filter
• Dirty drain pans	• Clean drain pans; proper slope and drainage
• Humidifiers	• Use portable water for humidification
• Lubricants	• Keep duct lining dry; move lining outside of duct if possible
• Refrigerants	• Fix leaks/clean spills
• Mechanical room	• Maintain spotless mechanical room (not a storage area)
• Maintenance activities	• Avoid back drafting
• Combustion appliances boilers/furnaces/stoves/ generators	• Check/maintain flues from boiler to outside
	• Keep combustion appliances properly tuned
	• Disallow unvented combustion appliances
	• Perform polluting activities during unoccupied hours
Moisture	
• Mold	• Keep building dry
Vehicles	
• Underground/attached garage	• Use exhaust ventilation
	• Maintain garage under negative pressure relative to the building
	• Check air flow patterns frequently
	• Monitor CO

pressure to areas of low pressure. That is why controlling workplace air pressure is an integral part of controlling pollution and enhancing building IAQ performance.

Air movements should be from occupants, toward a source, and out of the building rather than from the source to the occupants and out of the building. Pressure differences will control the direction of air motion and the extent of occupant exposure.

TABLE 7.26
Outdoor Sources of Contaminants

Category/Common Sources	Mitigation and Control

Ambient Outdoor Air
- Air quality in the general area
  - Filtration or air cleaning of intake air

Vehicular sources
- Local vehicular traffic
- Vehicle idling areas
- Loading dock
  - Locate air intake away from source
  - Require engines shot off at loading dock
  - Pressurize building/zone
  - Add vestibules/sealed doors near source

Commercial/Manufacturing Sources
- Laundry or dry cleaning
- Paint shop
- Restaurant
- Photo-processing
- Automotive shop/gas station
- Electronics manufacture/assembly
- Various industrial operations
  - Locate air intake away from source
  - Pressurize building relative to outdoors
  - Consider air cleaning options for outdoor air intake
  - Use landscaping to block or redirect flow of contaminants

Utilities/Public Works
- Utility power plant
- Incinerator
- Water treatment plant

Agricultural
- Pesticide spraying
- Processing or packing plants
- Ponds

Construction/Demolitions
- Pressurize building
- Use walk-off mats

Building Exhaust
- Bathrooms exhaust
- Restaurant exhaust
- Air handler relief vent
- Exhaust from major tenant (e.g., dry cleaner)
  - Separate exhaust or relief from air intake
  - Pressurize building

Water Sources
- Pools of water on roof
- Cooling tower mist
  - Proper roof drainage
  - Separate air intake from source of water
  - Treat and maintain cooling tower water

Birds and Rodents
- Fecal contaminants
- Bird nesting
  - Bird proof intake grills
  - Consider vertical grills

Building Operations and Maintenance
- Trash and refuse area
- Chemical/fertilizer/grounds keeping storage
- Painting/roofing/sanding
  - Separate source from air intake
  - Keep source area clean/lids on tight
  - Isolate storage area from occupied areas

Ground Sources
- Soil gas
- Sewer gas
- Underground fuel storage tanks
  - Depressurize soil
  - Seal foundation and penetrations to foundations
  - Keep air ducts away from ground sources

TABLE 7.27
Major Driving Forces

Driving Force	Effect
Wind	Positive pressure is created on the windward side causing infiltration, and negative pressure on the leeward side causing exfiltration, though wind direction can be varied due to surrounding structures
Stack effect Stairwells This buoyant force on the higher floors floors and a neutral between	When the air inside is warmer than outside, it rises sometimes creating a column of rising—up elevator shafts, vertical pipe chases etc of the air results in positive pressure and negative pressure on the lower pressure plane somewhere
HVAC/fans	Fans are designed to push air in a directional flow and create positive pressure in front, and negative pressure behind the fan
Flues and exhaust	Exhausted air from a building will reduce the building air pressure relative to the outdoors. Air exhausted will be replaced either through infiltration or through planned outdoor air intake vent
Elevators	The pumping action of a moving elevator can push air out of or draw air into the elevator shaft as it moves

Driving forces change pressure relationships and create airflow. Common driving forces are identified in Table 7.27.

INDOOR AIR DISPERSION PARAMETERS

Several parameters (some characterize observed patterns of contaminant distribution and others characterize flow features, such as stability) are used to characterize the dispersion of a contaminant inside a room.

PARAMETERS

- **Contaminant concentration**: indicator of contaminant distribution in a room, i.e. the mass of contaminant per unit volume of air (measured in kg/m^3).
- **Local mean age of air**: is the average time it takes for air to travel from the inlet to any point P in the room (Di Tommaso et al., 1999).
- **Purging effectiveness of inlets**: is a quantity that can be used to identify the relative performance of each inlet in a room where there are multiple inlets.
- **Local specific contaminant-accumulating index**: a general index capable of reflecting the interaction between the ventilation flow and a specific contaminant source.
- **Air change efficiency (ACE)**: is a measure of how effectively the air present in a room is replaced by fresh air from the ventilation system (Di Tommaso et al., 1999). It is the ratio of room mean age that would exist if the air in the room were completely missed to the average time of replacement of the room.
- **Ventilation effectiveness factor (VEF)**: is defined as the ratio of two contaminant concentration differentials: the contaminant concentration in the supply air (typically zero) and the contaminant concentration in the room under complete mixing conditions (Zhang et al., 2001).
- **Relative ventilation efficiency**: is the ratio of the local mean age that would exist if the air in the room were completely mixed to the local mean age that is actually measured at a point.

Air Pollution

- **Air diffusion performance index (ADPI)**: is primarily a measure of occupant comfort rather than an indicator of contaminant concentrations. It expresses the percentage of locations in an occupied zone that meet air movement and temperature specifications for comfort.
- **Temperature effectiveness**: is similar in concept to ventilation effectiveness and reflects the ability of a ventilation system to remove heat.
- **Effective draft temperature**: indicates the feeling of coolness due to air motion.
- **Reynolds number**: expresses the ratio of the inertial forces to viscous forces.
- **Rayleigh number**: characterizes natural convection flows.
- **Grashof number**: is equivalent to the Rayleigh number divided by the Prandtl number (a dimensionless number approximating the ratio of viscosity and thermal diffusivity).
- **Froude number**: a dimensionless number used to characterize flow through corridors and doorways and in combined displacement and wind ventilation cases.
- **Richardson number**: characterizes the importance of buoyancy.
- **Flux Richardson number**: is used to characterize the stabilizing effect of stratification on turbulence.
- **Buoyancy flux**: used to characterize buoyancy-driven flows.
- **Archimedes number**: conditions of the supplied air are often characterized by the discharge Archimedes number, which expresses the ration of the buoyancy forces to momentum forces or the strength of natural convection to forced convection.

COMMON AIRFLOW PATHWAYS

Contaminants travel along pathways—sometimes over great distances. Pathways may lead from an indoor source to an indoor location or from an outdoor source to an indoor location.

The location experiencing a pollution problem may be close by, in the same or an adjacent area, but it may be a great distance from, and/or on a different floor from a contaminant source.

Knowledge of common pathways helps to track down the source and/or prevent contaminants from reaching building occupants (see Table 7.28).

TABLE 7.28
Common Airflow Pathways for Contaminants

Common Pathway	Comment
Indoors	
Stairwell/elevator shaft vertical electrical or plumbing chases	The stack effect brings about air flow by drawing air toward these chases on the lower floors and elevator shaft away from these chases on the higher floors, affecting the flow of contaminants
Receptacles, outlets, openings	Contaminants can easily enter and exit building cavities and thereby move from space to space
Duct or plenum	Contaminants are commonly carried by the HVAC system throughout the occupied spaces
Duct or plenum leakage	Duct leakage accounts for significant unplanned air flow and energy loss in buildings
Flue or exhaust leakage	Leaks from sanitary exhausts or combustion flues can cause serious health problems
Room spaces	Air and contaminants move within a room or through doors and corridors to adjoining spaces
Outdoors to Indoors	
Indoor air intake	Polluted outdoor air or exhaust air can enter the building through the air intake
Windows/doors	A negatively pressurized building will draw air and outside pollutants into the building through any cracks and crevices available opening
Substructures/slab penetrations	Radon and other soil gases and moisture laden air or microbial contaminated air often travel through crawlspaces and other substructures into the building

MAJOR IAQ CONTAMINANTS

Environmental engineers and industrial hygienists spend a large portion of their time working with and mitigating air contaminant problems in the workplace. The list of potential contaminants workers might be exposed to while working is extensive. There are, however, a few major chemical-/material-derived air contaminants (other than those poisonous gases and materials that are automatically top priorities for the engineer and industrial hygienist to investigate and mitigate) that are considered extremely hazardous. These too garner the industrial hygienist's immediate attention and remedial action(s). In this text, we focus on asbestos, silica, formaldehyde, and lead as "those" hazardous contaminants (keeping in mind that there are others) requiring the immediate attention of the industrial hygienist.

ASBESTOS EXPOSURE (OSHA, 2002)

Asbestos is the name given to a group of naturally occurring minerals widely used in certain products, such as building materials and vehicle brakes, to resist heat and corrosion. Asbestos includes chrysotile, amosite, crocidolite, tremolite asbestos, anthophyllite asbestos, actinolite asbestos, and any of these materials that have been chemically treated and/or altered. Typically, asbestos appears as a whitish, fibrous material which may release fibers that range in texture from coarse to silky; however, airborne fibers that can cause health damage may be too small to be seen with the naked eye.

An estimated 1.3 million employees in the construction and general industry face significant asbestos exposure on the job. Heaviest exposures occur in the construction industry, particularly during the removal of asbestos during renovation or demolition (abatement). Employees are also likely to be exposed during the manufacture of asbestos products (such as textiles, friction products, insulation, and other building materials) and automotive brake and clutch repair work.

The inhalation of asbestos fibers by workers can cause serious diseases of the lungs and other organs that may not appear or manifest themselves until years after the exposure has occurred. For instance, asbestosis can cause a buildup of scar-like tissue in the lungs and result in loss of lung function that often progresses to disability and death. As mentioned, asbestos fibers associated with these health risks are too small to be seen with the naked eye, and smokers are at higher risk of developing some asbestos-related diseases. For example, exposure to asbestos can cause asbestosis (scarring of the lungs resulting in loss of lung function that often progresses to disability and to death); mesothelioma (cancer affecting the membranes lining the lungs and abdomen); lung cancer; and cancers of the esophagus, stomach, colon, and rectum.

OSHA has issued the following three standards to assist industrial hygienists with compliance and to protect workers from exposure to asbestos in the workplace:

- 29 CFR 1926.1101 covers construction work, including alteration, repair, renovation, and demolition of structures containing asbestos.
- 29 CFR 1915.1001 covers asbestos exposure during work in shipyards.
- 29 CFR 1910.1001 applies to asbestos exposure in general industry, such as exposure during brake and clutch repair, custodial work, and manufacture of asbestos-containing products.

The standards for the construction and shipyard industries classify the hazards of asbestos work activities and prescribe particular requirements for each classification:

- **Class I**: is the most potentially hazardous class of asbestos jobs and involves the removal of thermal system insulation and sprayed-on or troweled-on surfacing asbestos-containing materials or presumed asbestos-containing materials.
- **Class II**: includes the removal of other types of asbestos-containing materials that are not thermal systems insulation, such as resilient flooring and roofing materials containing asbestos.

- **Class III**: focuses on repair and maintenance operations where asbestos-containing materials are disturbed.
- **Class IV**: pertains to custodial activities where employees clean up asbestos-containing waste and debris.

There are equivalent regulations in states with OSHA-approved state plans.

Permissible Exposure Limits

Employee exposure to asbestos must not exceed 0.1 fibers per cubic centimeter (f/cc) of air, averaged over an 8-hour work shift. Short-term exposure must also be limited to not more than 1 f/cc, averaged over 30 minutes. Rotation of employees to achieve compliance with either permissible exposure limit (PEL) is prohibited.

Exposure Monitoring

In construction and shipyard work, unless the industrial hygienist is able to demonstrate that employee exposures will be below the PELs (a "negative exposure assessment"), it is generally a requirement that monitoring for workers in Class I and II regulated areas be conducted. For workers in other operations where exposures are expected to exceed one of the PELs, periodic monitoring must be conducted. In general industry, for workers who may be exposed above a PEL or above the excursion limit, initial monitoring must be conducted. Subsequent monitoring at reasonable intervals must be conducted, and in no case at intervals greater than 6 months for employees exposed above a PEL.

Competent Person

In all operations involving asbestos removal (abatement), employers must name a "competent person" qualified and authorized to ensure worker safety and health, as required by Subpart C., "General Safety and Health Provisions for Construction" (29 CFR 1926.20). Under the requirements for safety and health prevention programs, the competent person must frequently inspect job sites, materials, and equipment. A fully trained and licensed industrial hygienist often fills this role.

In addition, for Class I jobs the competent person must inspect onsite at least once during each work shift and upon employee request. For Class II and III jobs, the competent person must inspect often enough to assess changing conditions and upon employee request.

Regulated Areas

In general industry and construction, regulated areas must be established where the 8-hour TWA or 30-minute excursion values for airborne asbestos exceed the PELs. Only authorized persons wearing appropriate respirators can enter a regulated area. In regulated areas, eating, smoking, drinking, chewing tobacco or gum, and applying cosmetics are prohibited. Warning signs must be displayed at each regulated area and must be posted at all approaches to regulated areas.

Methods of Compliance

In both general industry and construction, employers must control exposures using engineering controls, to the extent feasible. Where engineering controls are not feasible to meet the exposure limit, they must be used to reduce employee exposures to the lowest levels attainable and must be supplemented by the use of respiratory protection.

Respirators

In general industry and construction, the level of exposure determines what type of respirator is required; the standards specify the respirator to be used. Keep in mind that respirators must be used during all Class I asbestos jobs. Refer to 29 CFR 1926.103 for further guidance on when respirators must be worn.

TABLE 7.29
Deaths from Silica in the Workplace

(a) Annual Morbidity rates from silicosis

Occupation	PMR
Miscellaneous metal and plastic machine operators	168.44
Hand molders and shapers, except jewelers	64.12
Crushing and grinding machine operators	50.97
Hand molding, casting, and forming occupations	35.70
Molding and casting machine operators	30.60
Mining machine operators	19.61
Mining occupations (not elsewhere classified)	15.33
Construction trades (not elsewhere classified)	14.77
Grinding, abrading, buffing, and polishing machine operators	8.47
Heavy equipment mechanics	7.72
Miscellaneous material moving equipment operators	6.92
Millwrights	6.56
Crane and tower operators	6.02
Brick masons and stonemasons	4.71
Painters, construction, and maintenance	4.50
Furnace, kiln, oven operators, except food	4.10
Laborers, except construction	3.79
Operating engineers	3.56
Welders and cutters	3.01
Machine operators, not specified	2.86
Not specified mechanics and repairers	2.84
Supervisors, production occupations	2.73
Construction laborers	2.14
Machinists	1.79
Janitors and cleaners	1.78

(b) Lung Disease—Work Based (2002)

Industry	PMR
Metal mining	69.51
Miscellaneous nonmetallic mineral and stone products	55.31
Nonmetallic mining and quarrying, except fuel	49.77
Iron and steel foundries	31.15
Pottery and related products	30.73
Structural clay products	27.82
Coal mining	9.26
Blast furnaces, steelworks, rolling and finishing mills	6.49
Miscellaneous fabricated metal products	5.87
Miscellaneous retail stores	4.63
Machinery, except electrical, (not elsewhere classified)	3.96
Other primary metal industries	3.63
Industrial and miscellaneous chemicals	2.72
Not specified manufacturing industries	2.67
Construction	1.82

Source: Work-Related Lung Disease Surveillance Report (2002).

Air Pollution

Labels
Caution labels must be placed on all raw materials, mixtures, scrap, waste, debris, and other products containing asbestos fibers.

Protective Clothing
For any employee exposed to airborne concentrations of asbestos that exceed the PEL, the employer must provide and require the use of protective clothing such as coveralls or similar full-body clothing, head coverings, gloves, and foot covering. Wherever the possibility of eye irritation exists, face shields, vented goggles, or other appropriate protective equipment must be provided and worn.

Training
For employees involved in each identified work classification, training must be provided. The specific training requirements depend upon the particular class of work being performed. In general industry, training must be provided to all employees exposed above a PEL. Asbestos awareness training must also be provided to employees who perform housekeeping operations covered by the standard. Warning labels must be placed on all asbestos products, containers, and installed construction materials when feasible.

Recordkeeping
The employer must keep an accurate record of all measurements taken to monitor employee exposure to asbestos. This record is to include: the date of measurement, operation involving exposure, sampling and analytical methods used, and evidence of their accuracy; number, duration, and results of samples taken; type of respiratory protective devices worn; name, social security number, and the results of all employee exposure measurements. This record must be kept for 30 years.

Hygiene Facilities and Practices
Clean change rooms must be furnished by employers for employees who work in areas where exposure is above the TWA and/or excursion limit. Two lockers or storage facilities must be furnished and separated to prevent contamination of the employee's street clothes from protective work clothing and equipment. Showers must be furnished so that employees may shower at the end of the work shift. Employees must enter and exit the regulated area through the decontamination area.

The equipment room must be supplied with impermeable, labeled bags and containers for the containment and disposal of contaminated protective clothing and equipment.

Lunchroom facilities for those employees must have a positive pressure, filtered air supply and be readily accessible to employees. Employees must wash their hands and face prior to eating, drinking, or smoking. The employer must ensure that employees do not enter lunchroom facilities with protective work clothing or equipment unless surface fibers have been removed from the clothing or equipment.

Employees may not smoke in work areas where they are occupationally exposed to asbestos.

Medical Exams
In general industry, exposed employees must have a preplacement physical examination before being assigned to an occupation exposed to airborne concentrations of asbestos at or above the action level or the excursion level. The physical examination must include chest X-ray, medical and work history, and pulmonary function tests. Subsequent exams must be given annually and upon the termination of employment, though chest X-rays are required annually only for older workers whose first asbestos exposure occurred more than 10 years ago.

In construction, examinations must be made available annually for workers exposed above the action level or excursion limit for 30 or more days per year or who are required to wear negative pressure respirators; chest X-rays are at the discretion of the physician.

Silica Exposure

Crystalline silica (SiO_2) is a major component of the Earth's crust. In pure, natural form, SiO_2 crystals are minute, very hard, translucent, and colorless. Most mined minerals contain some SiO_2. "Crystalline" refers to the orientation of SiO_2 molecules in a fixed pattern as opposed to a nonperiodic, random molecular arrangement defined as amorphous (e.g., diatomaceous earth). Therefore, silica exposure occurs in a wide variety of settings, such as mining, quarrying, and stone cutting operations; ceramics and vitreous enameling; and in the use of filters for paints and rubber. The wide use and multiple applications of silica in industrial applications combine to make silica a major occupational health hazard (silicosis), which can lead to death.

Silicosis is a disabling, nonreversible and sometimes fatal lung disease caused by overexposure to respirable crystalline silica. More than one million U.S. workers are exposed to crystalline silica, and each year more than 250 die from silicosis (see 29a and b). There is no cure for the disease, but it is 100% preventable if employers, workers, and health professionals work together to reduce exposure.

The first column is the occupation title. The second column (PMR) is the observed number of deaths from silicosis per occupation divided by the expected number of deaths. Therefore, a value of one indicates no additional risk. A value of ten would indicate a risk ten times greater than the normal risk of silicosis. The first table provides risk by occupation and the second provides risk by industry.

Guidelines for Control of Occupational Exposure to Silica

In accordance with the Occupational Safety and Health Administration's (OSHA) standard for air contaminants (29 CFR 1910.1000), employee exposure to airborne crystalline silica shall not exceed an 8-hour time-weighted average limit (variable) as stated in 29 CFR 1910.1000, Table Z-3, or a limit set by a state agency whenever a state-administered Occupational Safety and Health Plan is in effect.

As mandated by OSHA, the first mandatory requirement is that employee exposure is eliminated through the implementation of feasible engineering controls (e.g., dust suppression and ventilation). After all, such controls are implemented and they do not control the permissible exposure, each employer must rotate its employees to the extent possible in order to reduce exposure. Only when all engineering or administrative controls have been implemented, and the level of respirable silica still exceeds permissible exposure limits, may an employer rely on a respirator program pursuant to the mandatory requirements of 29 CFR1910.134. Generally where working conditions or other practices constitute recognized hazards likely to cause death or serious physical harm, they must be corrected.

Formaldehyde (HCHO) Exposure

Formaldehyde (HCHO) is a colorless, flammable gas with a pungent suffocating odor. Formaldehyde is common in the chemical industry. It is the most important aldehyde produced commercially and is used in the preparation of urea-formaldehyde and phenol-formaldehyde resins. It is also produced during the combustion of organic materials and is a component of smoke.

The major sources in workplace settings are in manufacturing processes (used in the paper, photographic and clothing industries) and building materials. Building materials may contain phenol, urea, thiourea, or melamine resins which contain HCHO. Degradation of HCHO resins can occur when these materials become damp from exposure to high relative humidity, or if the HCHO materials are saturated with water during flooding, or when leaks occur. The release of HCHO occurs when the acid catalysts involved in the resin formulation are reactivated. When temperatures and relative humidity increase, out-gassing increases (DOH, Wash., 2003).

Formaldehyde exposure is most common through gas-phase inhalation. However, it can also occur through liquid-phase skin absorption. Workers can be exposed during direct production, treatment of materials, and production of resins. Health care professionals; pathology and histology technicians; and teachers and students who handle preserved specimens are potentially at high risk.

Air Pollution

Studies indicate that formaldehyde is a potential human carcinogen. Airborne concentrations above 0.1 ppm can cause irritation of the eyes, nose, and throat. The severity of irritation increases as concentrations increase; at 100 ppm it is immediately dangerous to life and health. Dermal contact causes various skin reactions including sensitization, which might force persons thus sensitized to find other work.

OSHA requires that the employer conduct initial monitoring to identify all employees who are exposed to formaldehyde at or above the action level or short-term-exposure-limit (STEL) and to accurately determine the exposure of each employee so identified. If the exposure level is maintained below the STEL and the action level, employers may discontinue exposure monitoring, until such there is a change which could affect exposure levels. The employer must also monitor employee exposure promptly, upon receiving reports of formaldehyde-related signs and symptoms.

In regard to exposure control, the best prevention is provided by source control (if possible). The selection of HCHO-free or low-emitting products such as exterior grade plywood which use phenol HCHO resins for indoor use is the best starting point.

Secondary controls include filtration, sealants, and fumigation treatments. Filtration can be achieved using selected adsorbents. Sealants involve coating the materials in question with two or three coats of nitro-cellulose vanish, or water-based polyurethane. Three coats of these materials can reduce out-gassing by as much as 90%.

Training is required at least annually for all employees exposed to formaldehyde concentrations of 0.1 ppm or greater. The training will increase employees' awareness of specific hazards in their workplace and of the control measures employed. The training also will assist in successful medical surveillance and medical removal programs. These provisions will only be effective if employees know what signs or symptoms are related to the health effects of formaldehyde if they are periodically encouraged to do so.

LEAD EXPOSURE

Lead has been poisoning workers for thousands of years. Most occupational overexposures to lead have been found in the construction trades, such as plumbing, welding, and painting. In plumbing, soft solder (banned for many uses in the U.S.), used chiefly for soldering tinplate and copper pipe joints, is an alloy of lead and tin. Although the use of lead-based paint in residential applications has been banned, since lead-based paint inhibits the rusting and corrosion of iron and steel, it is still used on construction projects. Significant lead exposures can also arise from removing paint from surfaces previously coated with lead-based paint. According to OSHA 93-47 (2002), the operations that generate lead dust and fume include the following:

Flame-torch cutting, welding, the use of heat guns, sanding, scraping and grind of lead-painted surfaces in repair, reconstruction, dismantling, and demolition work

- Abrasive blasting of structures containing lead-based paints
- Use of torches and heat guns, and sanding, scraping, and grinding lead-based paint surfaces during remodeling or abating lead-based paint
- Maintaining process equipment or exhaust duct work.

Health Effects of Lead

There are several routes of entry in which lead enters the body. When absorbed into the body in certain doses, lead is a toxic substance. Lead can be absorbed into the body by inhalation and ingestion. Except for certain organic lead compounds not covered by OSHA's Lead Standard (29 CFR 1926.62), such as tetraethyl lead, lead, when scattered in the air as a dust, fume, or mist, can be absorbed into the body by inhalation.

A significant portion of the lead that can be inhaled or ingested gets into the blood stream. Once in the blood stream, lead is circulated throughout the body and stored in various organs and tissues.

Some of this lead is quickly filtered out of the body and excreted, but some remains in the blood and other tissues. As exposure to lead continues, the amount stored in the body will increase if more lead is being absorbed that is being excreted. Cumulative exposure to lead, which is typical in construction settings, may result in damage to the blood, nervous system, kidneys, bones, heart, and reproductive system and contributes to high blood pressure. Some of the symptoms of lead poisoning include the following:

- Poor appetite
- Dizziness
- Pallor
- Headache
- Irritability/anxiety
- Constipation
- Sleeplessness
- Weakness
- Insomnia
- "Lead line" in gums
- Fine tremors
- Hyperactivity
- "Wrist drop" (weakness of extensor muscles)
- Excessive tiredness
- Numbness
- Muscle and joint pain or soreness
- Nausea
- Reproductive difficulties

Lead Standard Definitions

According to OSHA's Lead Standard, the terms listed below have the following meanings:

- **Action level**: means employee exposure, without regard to the use of respirators, to an airborne concentration of lead of 30 micrograms per cubic meter of air (30 µg/m^3), averaged over an 8-hour period.
- **Permissible exposure limit (PEL)**: means the concentration of airborne lead to which an average person may be exposed without harmful effects. OSHA has established a PEL of 50 micrograms per cubic meter of air (50 µg/m^3) averaged over an 8-hour period. If an employee is exposed to lead for more than 8 hours in any work day, the permissible exposure limit, a time-weighted average (TWA) for that day, shall be reduced according to the following formula:

$$\text{Maximum permissible limit}\left(\text{in μg/m}^3\right) = 400 \times \text{hours worked in the day}.$$

When respirators are used to supplement engineering and administrative controls to comply with the PEL and all the requirements of the lead standard's respiratory protection rules have been met, employee exposure, to determine whether the employer has complied with the PEL, may be considered to be at the level provided by the protection factor of the respirator for those periods the respirator is worn. Those periods may be averaged with exposure levels during periods when respirators are not worn to determine the employee's daily TWA exposure.

- **µg/m^3**: means micrograms per cubic meter of air. A microgram is one-millionth of a gram. There are 454 g in a pound.

Worker Lead Protection Program

The employer is responsible for the development and implementation of a worker lead protection program. This program is essential in minimizing worker risk of lead exposure.

The most effective way to protect workers is to minimize exposure through the use of engineering controls and good work practices.

At the minimum, the following elements should be included in the employer's worker protection program for employees exposed to lead:

- Hazard determination, including exposure assessment
- Engineering and work practice controls
- Respiratory Protection
- PPE (protective work clothing and equipment)
- Housekeeping
- Hygiene facilities and practices
- Medical surveillance and provisions for medical removal
- Employee information and training
- Signs
- Recordkeeping

MOLD CONTROL

Molds can be found almost anywhere; they can grow on virtually any organic substance, as long as moisture and oxygen are present. The earliest known writings that appear to discuss mold infestation and remediation (removal, cleaning up) are found in Leviticus, Chapter 14, Old Testament.

Where is mold "typically" found?

As mentioned, name the spot or place; they have been found growing in office buildings, schools, automobiles, in private homes, and other locations where water and organic matter are left unattended. Mold is not a new issue—just one which, until recently, has received little attention by regulators in the United States. That is, there are no state or federal statutes or regulations regarding molds and IAQ.

Molds reproduce by making spores that usually cannot be seen without magnification. Mold spores waft through the indoor and outdoor air continually. When mold spores land on a damp spot indoors, they may begin growing and digesting whatever they are growing on in order to survive. Molds generally destroy the things they grow on (USEPA, 2001).

The key to limiting mold exposure is to prevent the germination and growth of mold. Since mold requires water to grow, it is important to prevent moisture problems in buildings. Moisture problems can have many causes, including uncontrolled humidity. Some moisture problems in workplace buildings have been linked to changes in building construction practices during the 1970s, 1980s, and 1990s. Some of these changes have resulted in buildings that are tightly sealed, but may lack adequate ventilation, potentially leading to moisture buildup. Building materials, such as drywall, may not allow moisture to escape easily. Moisture problems may include roof leaks, landscaping, or gutters that direct water into or under the building, and unvented combustion appliances. Delayed maintenance or insufficient maintenance are also associated with moisture problems in buildings. Moisture problems in temporary structures have frequently been associated with mold problems.

Building maintenance personnel, architects, and builders need to know effective means of avoiding mold growth which might arise from maintenance and construction practices. Locating and cleaning existing growths are also paramount to decreasing the health effects of mold contamination. Using proper cleaning techniques is important because molds are incredibly resilient and adaptable (Davis, 2001).

Molds can elicit a variety of health responses in humans. The extent to which an individual may be affected depends upon his or her state of health, susceptibility to disease, the organisms with which he or she came in contact, and the duration and severity of exposure (Ammann, 2000). Some people experience temporary effects that disappear when they vacate infested areas (Burge, 1997). In others, the effects of exposure may be long-term or permanent (Yang, 2001).

It should be noted that systemic infections caused by molds are not common. Normal, healthy individuals can resist systemic infection from airborne molds.

Those at risk for system fungal infection are severely immunocompromised individuals such as those with HIV/AIDs, individuals who have had organ or no marrow transplants, and persons undergoing chemotherapy.

In 1994, an outbreak of *Stachybotrys chartarum* in Cleveland, Ohio was believed by some to have caused pulmonary hemorrhage in infants. Sixteen of the infants died. CDC sponsored a review of the cases and concluded that the scientific evidence provided did not warrant the conclusion that inhaled mold was the cause of the illnesses in infants. However, the panel also stated that further research was warranted, as the study design for the original research appeared to be flawed (CDC, 1999).

Below is a list of mold components known to elicit a response in humans.

- **Volatile organic compounds (VOCs)**: "Molds produce a large number of volatile organic compounds. These chemicals are responsible for the musty odors produced by growing molds" (McNeel and Kreutzer, 1996). VOCs also provide the odor in cheeses and the "off" taste of mold infested foods. Exposure to high levels of volatile organic compounds, affect the CNS, producing such symptoms as headaches, attention deficit, inability to concentrate, and dizziness (Ammann, 2000). According to McNeel, at present, the specific contribution of mole volatile organic compounds in building-related health problems has not been studied. Also, mold volatile organic compounds are likely responsible for only a small fraction of total VOCs indoors (Davis, 2001).
- **Allergens**: all molds, because of the presence of allergen son spores, have the potential to cause an allergic reaction in susceptible humans (Rose, 1999). Allergic reactions are believed to be the most common exposure reaction to molds. These reactions can range from mile, transitory response, such as runny eyes, runny nose, throat irritation, coughing, and sneezing; to severe, chronic illnesses such as sinusitis and asthma (Ammann, 2000).
- **Mycotoxins**: are natural organic compounds that are capable of initiating a toxic response in vertebrates (McNeel and Kreutzer, 1996). Some molds are capable of producing mycotoxins. Molds known to potentially produce mycotoxins, and which have been isolated in infestations causing adverse health effects include certain species of *Acremonium, Alternaria, Aspergillus,* Chaetomium, *Cladosporium, Fusarium, Paecilomyces, Penicillium, Stachybotrys,* and Trichoderma (Yang, 2001).

While a certain type of mold or mold strain type may have the genetic potential for producing mycotoxins; specific environmental conditions are believed to be needed for the mycotoxins to be produced. In other words, although a given mold might have the potential to produce mycotoxins, it will not produce them if the appropriate environmental conditions are not present (USEPA, 2001).

Currently, the specific conditions that cause mycotoxin production are not fully understood. The USEPA recognizes that mycotoxins tend to concentrate in fungal spores and that there is limited information currently available regarding the process involved in fungal spore release. As a result, USEPA is currently conducting research in an effort to determine "the environmental conditions required for sporulation, emission, aerosolization, dissemination and transport of [Stachybotrys] into the air (USEPA, 2001).

Mold Prevention

As mentioned, the key to mold control is moisture control. Solve moisture problems before they become mold problems. Several mold prevention tips are listed below.

- Fix leaky plumbing and leaks in the building envelope as soon as possible.
- Watch for condensation and wet spots. Fix sources(s) of moisture problem(s) as soon as possible.
- Prevent moisture due to condensation by increasing surface temperature or reducing the moisture level in air (humidity). To increase surface temperature, insulate or increase air circulation. To reduce the moisture level in the air, repair leaks, increase ventilation (if outside air is cold and dry), or dehumidify (if outdoor air is warm and humid).
- Keep heating, ventilation, and air conditioning (HVAC) drip pans clean, flowing properly, and unobstructed.
- Vent moisture-generating appliances, such as dryers, to the outside where possible.
- Perform regular build/HVAC inspections and maintenance as scheduled.
- Maintain low indoor humidity, below 60% relative humidity (RH), ideally 30%–50%, if possible.
- Clean and dry wet or damp spots within 48 hours.
- Don't let foundations stay wet. Provide drainage and slope the ground away from the foundation.

Mold Remediation

At present, there are no standardized recommendations for mold remediation; however, USEPA is working on guidelines. There are certain aspects of mold cleanup, however, which are agreed upon by many practitioners in the field.

- A common-sense approach should be taken when assessing mold growth. For example, it is generally believed those small amounts of growth, like those commonly found on shower walls pose no immediate health risk to most individuals.
- Persons with respiratory problems, a compromised immune system, or fragile health, should not participate in cleanup operations.
- Cleanup crews should be properly attired. Mold should not be allowed to touch bare skin. Eyes and lungs should be protected from aerosol exposure.
- Adequate ventilation should be provided while, at the same time, containing the infestation in an effort to avoid spreading mold to other areas.
- The source of moisture must be stopped and all areas infested with mold thoroughly cleaned. If thorough cleaning is not possible due to the nature of the material (porous versus semi- and non-porous), al contaminated areas should be removed.

Safety tips that should be followed when remediating moisture and mold problems include:

- Do not touch mold or moldy items with bare hands.
- Do not get mold or mold spores in your eyes.
- Do not breathe in mold or mold spores.
- Consider using PPE when disturbing mold. The minimum PPE is a respirator, gloves, and eye protection.

Mold Cleanup Methods

A variety of mold cleanup methods are available for remediating damage to building materials and furnishings caused by moisture control problems and mold growth. These include wet vacuum, damp wipe, HEPA vacuum, and the removal of damaged materials and sealing off in plastic bags. The specific method or group of methods used will depend on the type of material affected.

REFERENCES

Ammann, H. (2000). Is Indoor Mold Contamination a Threat? Washington State Department of Health. Accessed August 9, 2003. Posted at: http://www.doh.wa.gov/ehp/ocha/mold.html.

Boubel, R.W., Fox, D.L., Turner, D.B., & Stern, A.C. (1994). *Fundamentals of Air Pollution*. New York: Academic Press.

Buonicore, A.J., Theodore, L. & Davis, W.T. (1992). *Air Pollution Engineering Manual*. New York: Van Nostrand Reinhold.

Burge, H.A. (1997). The fungi; how they grow and their effects on human health. Heating, Piping, Air Conditioning, **July**, pp. 69–70.

Burge, H.A. & Hoyer, M.E. (1998). Indoor air quality. In: *The Occupational Environment: Its Evaluation and Control*, pp. 218–222. DiNardi, S.R., (Ed.). Fairfax, VA: American Industrial Hygiene Association.

Byrd, R.R. (2003). *IAQ FAG Part 1*. Glendale, CA: Machado Environmental Corporation.

Carson, J.E. & Moses, H. (1969). The validity of several plume rise formulas. *Journal of the Air Pollution Control Association* 19(11):862.

CDC (1999). *Reports of Members of the CDC External Expert Panel on Acute Idiopathic Pulmonary Hemorrhage in Infants: A Synthesis*. Washington, DC: Centers for Disease Control.

Compressed Gas Association, Inc. (1990). *Handbook of Compressed Gases, 3rd ed*. New York: Van Nostrand Reinhold.

Cooper, C.D., & Alley, F.C. (1990). *Air Pollution Control: A Design Approach, Prospect Heights*. Long Grove, IL: Waveland Press, Inc.

Davis, P.J. (2001). *Molds, Toxic Molds, and Indoor Air Quality*. Sacramento, CA: California Research Bureau, California State Library.

Davis, M.L. & Cornwell, D.A. (1991). *Introduction to Environmental Engineering*. New York: McGraw-Hill, Inc.

Di Tommaso, R.M., Nino, E., & Fracastro, G.V. (1999). Influence of the boundary thermal conditions on the air change efficiency indexes. *Indoor Air* **9**:63–69.

DOH, Wash (2003). Formaldehyde. Washington State Department of Health, Office of Environmental Health, and Safety. Accessed August 2003 at http://www.doh.wa.gov/ehp/ts/IAQ/Formaldehyde.HTM.

EPA (2005). Basic Air Pollution Meteorology. Accessed January 12, 2008 @ www.epa.gov/apti.

EPA (2006). Air Pollution Control Orientation Course. Accessed January 26, 2008 @ www//www.epa.gov/air/oaqps/eop/course422/ce6c.html.

EPA (2007a). Basic Concepts in Environmental Sciences. Accessed January 2, 2008 @ www.epa.gov/eogapti1/module3/collect/collect.htm.

EPA (2007b). Introduction to Air Pollution. United States Environmental Protection Agency. Accessed December 25, 2007 @ http://www.epa.gov/air/oaqps/eog/course42/ap.html.

Freedman, B. (1989). *Environmental Ecology*. New York: Academic Press.

Godish, T. (1997). *Air Quality*, 3rd ed. Boca Raton, FL: Lewis Publishers.

Graedel, T.E. & Crutzen, P.J. (1995). *Atmosphere, Climate, and Climate*. New York: Scientific American Library.

Heinlein, R.A. (1973). *Time Enough for Love*. New York: G.P. Putnam Sons.

Hesketh, H.E. (1991). *Air Pollution Control: Traditional and Hazardous Pollutants*. Lancaster, PA: Technomic Publishing Company.

Heumann, W.L. (1997). *Industrial Air Pollution Control Systems*. New York: McGraw-Hill, Inc.

Hollowell, C.D. et al. (1979a). Impact of infiltration and ventilation on indoor air quality. *ASHRAE Journal* (**12**): 49–53.

Hollowell, C.D. et al. (1979b). Impact of energy conservation in buildings on health. In: *Changing Energy Use Futures*, pp. 638–647. Razzolare, R.A. and Smith, C.B. (Eds.) New York: Pergamon.

Holmes, G., Singh, B.R., & Theodore, L. (1993). *Handbook of Environmental Management & Technology*. New York: John Wiley & Sons.

Hov, O. (1984). Ozone in the troposphere: High level pollution. *Ambio* **13**:73–79.

Interdepartmental Committee for Atmospheric Science (1975). *Ozone*. Washington, DC: Department of Commerce.

MacKenzie, J.J. & El-Ashry, T. (1988). *Ill Winds: Airborne Pollutant's Toll on Trees and Crops*. Washington, DC: World Resource Institute.

Masters, G.M. (1991). *Introduction to Environmental Engineering and Science*. Englewood Cliffs, NJ: Prentice-Hall.

McNeel, S. & Kreutzer, R. (1996). Fungi & indoor air quality. *Health & Environment Digest* 10(2): 9–12.

McKenzie, J.J., & El-Ashry, T. (1988). *Ill Winds: Airborne Pollutants Toll on Trees and Crops*. Washington, DC: World Resources Institute.

Molhave, L. (1986). Indoor air quality in relation to sensory irritation due to volatile organic compounds. *ASHRAE Transactions* **92**(1):306–316.

OSHA 93-47 (2002). Lead Exposure in Construction. Washington, DC: U.S. Department of Labor, Occupational Safety & Health Administration. Accessed August 2022 at http://www.osha-slc.gov/pls/oshaweb/owadisp.show_document?p_table=FACT_Sheet.

Peavy, H.S., Rowe, D.R., & Tchobanoglous, G. (1985). *Environmental Engineering*. New York: McGraw-Hill, Inc.

Postel, S. (1987). Stabilizing chemical cycles. In: *State of the World*, pp. 157–178. Brown, L.R. (Ed.), New York: Norton.

Rose, C.F. (1999). Antigens (Cincinnati: American Conference of Governmental Industrial Hygienists 1999) ACGIH Bioaerosols Assessment and Control, Ch. 25, 25-1 to 25-11.

Schaefer, V.J. & Day, J.A. (1981). *Atmosphere: Clouds, Rain, Snow, Storms*. Boston, MA: Houghton Mifflin Company.

Urone, P. (1976). The primary air pollutants—gaseous: Their occurrence, sources, and effects. In: *Air Pollution*, pp. 23–75. Stern, A.C. (Ed.), Vol. 1. New York: Academic Press.

USEPA (1987). Indoor Air Facts No. 1, EPA, and Indoor Air Quality.

USEPA (1988a). *Environmental Progress & Challenges*. Collingdale, PA: DIANE Publishing, p. 13.

USEPA (1988b). National Air Pollutant Emission Estimates 1940–1986, Environmental Protection Agency, Washington, D.C.

USEPA (2001). Indoor Air Facts No. 4 (revised) Sick Building Syndrome. Accessed August 9, 2022. Posted at http://www.epa.gov/iaq/pubs/sbs.html.

Wadden, R.A. & Scheff, P.A. (1983). *Indoor Air Pollution: Characteristics, Predications, and Control*. New York: John Wiley & Sons.

WHO (1984). *Indoor Air Quality Research. Euro Reports and Studies no. 103*. Copenhagen: World Health Organization.

Woods, J.E. (1980). Environmental Implications of Conservation and Solar Space Heating. Engineering Research Institute, Iowa State University, Ames, Iowa, BEUL 80-3. Meeting of the New York Academy of Sciences, New York, January 16.

Work-Related Lung Disease Surveillance Report (2002). National Institute for Occupational Safety and Health, U.S. Department of Health and Human Services, Public Health Service, Centers for Disease Control and Prevention, Table 3-8; DHHA (NIOSH) Publication No. 96-134.Publications Dissemination, EID, National Institute for Occupational Safety and Health, 4676 Columbia Parkway Cincinnati, OH.

World Health Organization (WHO) (1982). Indoor Air Pollutants, Exposure, and Health Effects Assessment. Euro-Reports and Studies No.78: Working Group Report. Copenhagen: WHO Regional Office.

Yang, C.S. (2001). Toxic Effects of Some Common Indoor Fungi. Accessed August 6, 2022. Posted at http://www.envirovillage.com/Newsletters/Enviros/No4_09.htm.

Zhang, Y., Wang, X., Riskowski, G.L., & Christianson, L.L. (2001). Quantifying ventilation effectiveness for air quality control. *Transactions of the American Society of Agricultural and Biological engineers (ASABE)* **44**(2): 385–390.

Zurer, P.S. (1988). Studies on Ozone Destruction Expand Beyond Antarctic, *C&E News*, pp. 18–25, May.

8 Water Pollution

INTRODUCTION

You may have heard the old saying, idiom, phrase, or expression that water is the elixir of life on Earth. Having spent the majority of my life studying water, analyzing water, drinking water, watering my plants and animals, practicing personal hygiene using water, writing about water, and teaching environmental health and engineering students about water, I have come to realize, early on, that water is indeed the elixir of life on Earth.

Now the irony in this statement about water being the elixir of life on Earth is that when explained to humans, who do not think about water unless they are thirsty or dying of thirst, they agree about the true value of water. But only to a point.

Only to a point?

Yes. The average person learns, usually early on, that water is a must, a "have to have," a crucial ingredient to maintain his or her good health and life. The problem, the irony, is manifested when we forget that we are not the only life on Earth. For example, trees, flowers, grass, and weeds all require water to survive. And then there are the animals. They are living organisms too; they must have water to survive. The problem with the animals on Earth is that most are only visible if they can be seen. How about the animals and other lifeforms on Earth that are not normally seen? For example, we do not normally observe, in their natural environment, fishes, marine algae, lichens, birds, and invertebrates such as crabs, lobsters, sea stars, urchins, mussels, clams, barnacles, snails, limpets, sea squirts, sea anemones, and so many more lifeforms in the sea. These listed sea life forms not only depend on water to survive but also their survival is important to maintaining life and good health in humans. So, the question is: why would we pollute marine environments?

It is a gross understatement to say that the marine environment is of pronounced significance to humankind. Why? Well, consider that more than 20% of the world's population (~1.3 billion people) lives within 62.1 miles (100 km) of the coast, a figure that is likely to rise exponentially by 2050. In addition, the marine environment supports nearly half of the universal primary production, a great share of which drives global fisheries.

> The bottom line: even though human welfare is intricately linked, interconnected with the sea and its nature resources humans have substantially altered the face of the ocean within only a few centuries. The face of today's sea is quite apparent; it floats.
>
> **Frank R. Spellman (2007)**

The average American uses more than 180 gallons of fresh, clean water a day, while many rural villagers in the Third World nations spend up to six hours a day obtaining their water from distant, polluted streams. Nearly 10 million people die each year as a result of intestinal diseases caused by unsafe water.

A. Boyce (1997)

THE 411 ON WATER

Let's look at some basic concepts or facts about water that most people are familiar with.

- Life, as we know it, cannot exist without water. A lack of water first enervates, then prostrates, and then kills. Most of us are familiar with this fact.

- When literally dying from lack of water, any fear of drowning in the first river, lake, stream, or deep depression containing water that we stumble upon leaves us instantly. Why? Simply put, we do not fear drowning when we intend to drink it all—every last drop.
- The leading cause of accidental death in desert regions is drowning.
- In a rainstorm, we know that when lightning streaks like gunfire through clouds dense as wool, and volleys of thunder roar like cannonballs tumbling down a stone staircase, shaking air permeated with the smell of ozone, that rain, hard rain, rain that splatters on mighty rock like pellets, knocking berries off junipers, leaves off trees, limbs off oak, is on its way—driven by wind perhaps capable of knocking down the oak itself.

These are things we know or imagine about water, about rain, and about the results of water or rain events. We know these things—some through instinct, some through experience. We are aware of and have knowledge of many facts concerning water. But, of course, there is much more that we do not know about water—things or facts that we should know. Most importantly, as our Native American friends would tell us if we bothered to ask, man needs to learn to respect water. Why? Simply put, water is the essence of life.

Now let's take a look at some other facts about water—facts that we might not be familiar with at all, but we should be.

- If we took all the available supplies of water on Earth from every ocean, every lake, every stream, every river, every brook, every well, every tiny puddle, and from the atmosphere itself, the quantity would all add up to an overwhelming volume of more than 326 million cubic miles, making water the most common substance on Earth.
- Of these 326 million cubic miles of water, about 97% (or 317 million cubic miles) is salty, residing mostly in the oceans, but also in glaciers and ice caps. Note that although saltwater is more plentiful than freshwater, freshwater is what concerns us in this text. Saltwater fails its most vital test: it fails to provide us with the vital refreshment we require.
- Of the global total of water, freshwater makes up only a small fraction of it, <1%. Just over 2 million cubic miles of freshwater are contained in lakes, rivers, streams, bays, the atmosphere, and under the ground.
- Most living organisms depend on water for survival. Human bodies contain about 60% water. To survive, we need to take in about two and a half quarts a day, with one and a half quarts coming from the liquids we drink, and the rest from the water content in our foods. We put our lives in jeopardy if we lose more than 12% of our body water.

Water covers almost three-quarters of the Earth's surface. Although it is our most abundant resource, water is only present on Earth in a thin film, and almost 97% of that film is salt. Earth's water performs many functions essential to life on Earth, including helping to maintain the climate, diluting environmental pollution, and, of course, supporting all life. Fresh or salt, without water, life as we define it would never have begun on Earth. Without fresh water, we would have no agriculture, no manufacturing, no transportation, and no human life as we know it.

Freshwater sources include groundwater and surface water. First, we focus on surface water—rivers, lakes, streams, wetlands, coastal waters, and oceans and secondly, on groundwater. More specifically, we focus our attention on surface and groundwater pollution.

Most human settlements evolved and continue to develop along the shores of many freshwater bodies, mainly rivers. The obvious reason for this is threefold: accessibility; a plentiful source of drinking water; and, later, a source of energy (waterpower) for our earliest machines.

When human populations began to spread out and leave the watercourses, we found that some areas had too little water and others too much. Human beings (being the innovative and destructive creatures we are) have, with varying degrees of success, attempted to correct these imbalances by capturing fresh water in reservoirs behind dams, transferring fresh water in rivers and streams from one area to another, tapping underground supplies, and endeavoring to reduce water use, waste, and contamination. We define it in a general way as the presence of unwanted substances in water beyond levels acceptable for health or aesthetics. In some of these efforts, we have been successful; in others, when gauged against the present condition of our water supplies, we are still learning—and we have a lot more to learn.

Unfortunately, only a small proportion (about 0.5%) of all the water on Earth is found in lakes, rivers, streams, or the atmosphere. Obviously, this seems like a small amount, relative to Earth's total water supply. But the irony is that it is more than enough—even this small amount, if it were kept free of pollution and distributed evenly—to provide for the drinking, food preparation, and agricultural needs of all of Earth's people. We simply need to learn how to better manage and conserve the fresh water readily available to us.

SURFACE WATER

Precipitation that does not infiltrate into the ground or return to the atmosphere is called surface water. When it is freshly fallen and still mobile, not having yet reached a body of water, we call it runoff—water that flows into nearby lakes, wetlands, streams, rivers, and reservoirs.

Before we continue our discussion of surface water, we need to review the basic concepts of the hydrologic cycle.

Actually a manifestation of an enormous heat engine, the water cycle raises water from the oceans in warmer latitudes by a prodigious transformation of solar energy. Transferred through the atmosphere by the winds, the water is deposited far away over sea or land. Water taken from the Earth's surface to the atmosphere (either by evaporation from the surface of lakes, rivers, streams, and oceans or through transpiration of plants), forms clouds that condense to deposit moisture on the land and sea as rain or snow. The water that collects on land flows back to the oceans in streams and rivers.

The water that we see is surface water. The USEPA defines surface water as all water open to the atmosphere and subject to runoff. Surface freshwater can be broken down into four components.

Lakes—Rivers and Streams—Estuaries—Wetlands

Limnology is the study of surface or open freshwater bodies. Specifically, limnology is the study of bodies of open freshwater (lakes, rivers, and streams), and of their plant and animal biology and physical properties. Freshwater systems are grouped or classified as either lentic or lotic. Lentic (lenis=calm) systems are represented by lakes, ponds, impoundments, reservoirs, and swamps—standing water systems. Lotic (lotus=washed) systems are represented by rivers, streams, brooks, and springs—running water systems. On occasion, distinguishing between these two different systems is difficult. In old, wide, and deep rivers where water velocity is quite low, for example, the system becomes similar to that of a pond.

Surface water (produced by melting snow or ice or from rainstorms and falling to Earth's surface) always follows the path of least resistance. In other words, water doesn't run uphill. Beginning with droplets and ever-increasing, runoff is carried by rills, rivulets, brooks, creeks, streams, and rivers from elevated land areas that slope down toward one primary water course—a topographically defined drainage area. Drainage areas, known as watersheds or drainage basins, are surrounded by a ridge of high ground called the watershed divide. Watershed divides separate drainage areas from each other.

DID YOU KNOW?

LENTIC (STANDING OR STILL) WATER SYSTEMS

Natural lentic water systems include lakes, ponds, bogs, marshes, and swamps. Other standing freshwater bodies, including reservoirs, oxidation ponds, and holding basins, are usually constructed.

LOTIC (FLOWING) WATER SYSTEMS

The human circulatory system can be compared to the Earth's water circulation system. The hydrological cycle pumps water as our hearts pump blood, continuously circulating water through air, water bodies, and various vessels. As our blood vessels are essential to the task of carrying blood throughout our bodies, water vessels (rivers) carry water, fed by capillary creeks, brooks, streams, rills, and rivulets.

SURFACE WATER POLLUTANTS

Before discussing surface water pollution, we must define several important terms related to water pollution. One of these is point source. A point source is, as defined by the Clean Water Act (CWA), any discernible, confined, and discrete conveyance, including but not limited to any pipe, ditch, channel, tunnel, conduit, well, discrete fissure, container, rolling stock, concentrated animal feeding operation, or vessel or other floating craft, from which pollutants are discharged. For example, the outfalls of industrial facilities or wastewater treatment plants are *point sources*. A *non-point source*, in contrast, is more widely dispersed. An example of nonpoint source of pollution is rainwater carrying topsoil (sediments) and chemical contaminants into a stream. Nonpoint pollution includes water runoff from farming, urban areas, forestry, and construction activities. Nonpoint sources comprise the largest source of water pollution, contributing an estimated 70% or more of the contamination in quality-impaired surface waters. Note that atmospheric deposition of pollutants is also a nonpoint source of acids, nutrients, metals, and other airborne pollutants.

Another important term associated with nonpoint sources is runoff. *Runoff* means a nonpoint source that originated on land. The USEPA considers polluted runoff the most serious water pollution problem in the U.S. Why? —What exactly is runoff? Runoff occurs because of human intervention in landscapes. When land is disturbed by parking lots, tarmac, roads, factories, homes, and buildings, rainwater is not free to percolate through the soil, which absorbs and detoxifies many pollutants. Instead, in disturbed areas, there is little if any soil, and contaminated rainwater runs off into area water bodies, polluting them.

Surface water pollutants can harm aquatic life, threaten human health, or result in the loss of recreational or aesthetic potential. Surface water pollutants come from industrial sources, nonpoint sources, municipal sources, background sources, and other/unknown sources. The eight chief pollutants are biochemical oxygen demand, nutrients, suspended solids, pH, oil and grease, pathogenic microorganisms, toxic pollutants, and nontoxic pollutants.

Biochemical Oxygen Demand (BOD)

Organic matter (dead plants and animal debris, and wild animal and bird feces), human sewage, food-processing wastes, chemical plant wastes, slaughterhouse wastes, pulp- and paper-making operations wastes and tannery wastes discharged to a water body are degraded by oxygen-requiring microorganisms. The amount of oxygen consumed during microbial utilization of organics is called the biochemical oxygen demand (BOD). Although some natural BOD is almost always present, BOD is often an indication of the presence of sewage and other organic waste. High levels of BOD can deplete the oxygen in water. Fish and other aquatic organisms present in such waters with low oxygen levels may die.

Nutrients

Nutrients are elements essential to the growth and reproduction of plants and animals, and aquatic species depend on the surrounding water to provide their nutrients. However, just as too much of any good thing can have serious side effects for all of us, so is the case with too many nutrients in water. For example, when fertilizers composed of nutrients enter surface water systems, over-enrichment with nitrogen and phosphorus may result. A rich supply of such nutrients entering a lake may hasten eutrophication, which the USEPA defines as a process during which a lake evolves into a bog or marsh and eventually disappears. Excess nutrients can also stimulate a very abundant and dense growth of aquatic plants (bloom), especially algae.

pH

pH refers to the acidity or alkalinity of water. A low pH may mean a water body is too acidic to support life optimally. Some water bodies are naturally acidic, but others are made so by acidic deposition or acid runoff from mining operations.

Suspended Solids

Suspended solids are physical pollutants, and may consist of inorganic or organic particles, or of immiscible liquids. Inorganic solids such as clay, silt, and other soil constituents are common in surface water. Organic materials—plant fibers and biological solids—are also common constituents of surface waters. These materials are often natural contaminants resulting from the erosive action of water flowing over surfaces. Fine particles from soil runoff can remain suspended in water and increase its turbidity or cloudiness. This can stunt the growth of aquatic plants by limiting the amount of sunlight reaching them. Effluents from wastewater treatment plants, from industrial plants, and runoff from forestry and agricultural operations are sources of suspended solids. Note that because of the filtering capacity of the soil, suspended solids are seldom a constituent of groundwater.

Oil and Grease

Oil spills in or near surface water bodies that eventually reach the water body can have a devastating effect on fish, other aquatic organisms, and birds and mammals. Note that spills are not the only source of oil in water: oil leaking from automobiles and other vehicles or released during accidents is washed off roads with rainwater and into water bodies. Improper disposal of used oil from vehicles is another source; motor and other recreational boats release unburned fuel into water bodies.

Pathogenic Organisms

From the perspective of human use and consumption, the biggest concern associated with microorganisms is infectious disease. Microorganisms are naturally found in water (and elsewhere in the environment) and can cause infections. However, organisms that are not native to aquatic systems are of greatest concern—native or not, they can be transported by natural water systems. These organisms usually require an animal host for growth and reproduction. Nonpoint sources of these microorganisms include runoff from livestock operations and stormwater runoff. Point sources include improperly operating sewage treatment plants. When the surface water body functions to provide drinking water to a community, the threat of infectious microorganism contamination is very real and may be life-threatening. Those who live in industrial nations with generally safe water supplies think of pathogenic contamination as a third world problem. However, several recent problems in industrial nations (Sydney, Australia's local water supply, for example, had serious problems over the summer of 1998) have alerted us to the very real possibility of dangerous contamination in our own water supplies.

Toxic Pollutants

There are hundreds of potentially toxic water pollutants. Of these, the USEPA, under the Clean Water Act, regulates more than 100 pollutants of special concern. These include arsenic and the metals mercury, lead, cadmium, nickel, copper, and zinc. Organic toxic pollutants include benzene, toluene, and many pesticides.

Nontoxic Pollutants

Nontoxic pollutants include chemicals such as chlorine, phenols, iron, and ammonia. Color and heat are also nontoxic pollutants regulated under CWA. Pure water is colorless, but water in natural water systems is often colored by foreign substances. For example, many facilities discharge colored effluents into surface water systems. However, colored water is not aesthetically acceptable to the general public; thus, the intensity of the color that can be released is regulated by law. Heat or thermal nontoxic pollution can cause problems, but is not ordinarily a serious pollutant, although in localized situations, it can cause problems with temperature-sensitive organism populations.

EMERGING CONTAMINANTS

It has been suggested that humans and domestic and wildlife species have suffered adverse health consequences resulting from exposure to environmental chemicals that interact with the endocrine system. However, considerable uncertainty exists regarding the relationship(s) between adverse health outcomes and exposure to environmental contaminants. Collectively, chemicals with the potential to interfere with the function of endocrine systems are called endocrine disrupting chemicals (EDCs). EDCs have been broadly defined as exogenous agents that interfere with the production, release, transport, metabolism, binding, action, or elimination of the natural hormones in the boy responsible for the maintenance of homeostasis and the regulation of developmental processes.

It is important to point out that EDCs along with, in a broader sense, PPCPs (pharmaceutical and personal care products) fall into the category of "emerging contaminants." Emerging contaminants can fall into a wide range of groups defined by their effects, uses, or by their key chemical or microbiological characteristics. These compounds are found in the environment, often as a result of human activities.

We are quickly approaching a timeline in which we will enter the fifth generation of people exposed to toxic chemicals from before conception to adulthood. In a few cases we have identified the hazards of certain chemicals and their compounds and have implemented restrictions. One well known chemical compound that comes to mind in regard to its environmental harm and subsequent banning is dichloro-diphenyl-trichloroethane (DDT). Let's take a look at DDT.

The insecticide DDT was first produced in the 1920s and was developed as the first of the modern synthetic insecticides in the 1940s. It was extensively used between 1945 and 1965 with great effect to control and eradicate insects that were responsible for malaria, typhus, and the other insect-borne human diseases among both military and civilian populations and for insect control in crop and livestock production, institutions, homes, and gardens. DDT was an excellent insecticide because it was very effective at killing a wide variety of insects at low levels. DDT's quick success as a pesticide and broad use in the United States and other countries led to the development of resistance by many insect pest species. Moreover, the chemical properties that made this a good pesticide also made it persist in the environment for a long time.

This persistence led to accumulation of the pesticide in non-target species, especially raptorial birds (e.g., falcons, eagles, etc.). Due to the properties of DDT, the concentration of DDT in birds could be much higher than concentrations in insects or soil. Birds at the top of the food chain (e.g., pelicans, falcons, eagles, and grebes) had the highest concentrations of DDT. Although the amount of DDT did not kill the birds, it interfered with calcium metabolism, which led to thin eggshells. As a result, eggs would crack during development, allowing bacteria to enter, which killed the developing embryos. This had a great impact on the population levels of these birds. Peregrine falcons and brown pelicans were placed on the endangered species list in the US, partially due to declining reproductive success of the birds from DDT exposure.

Rachel Carson, that unequaled environmental journalist of profound vision and insight, published *Silent Spring* in 1962, which helped to draw public attention to this problem and to the need for better pesticide controls. This was the very beginning of the environmental movement in the US and is an excellent example of reporting by someone affiliated with the media that identified a problem and warned of many similar problems that could occur unless restrictions were put in place related to chemical pesticide use. Partially as a result of Carson's flagship book, scientists documented the link between DDT and eggshell thinning. This led to the U.S. Department of Agriculture, the federal agency with responsibility of regulating pesticides before the formation of the U.S. Environmental Protection Agency in 1970, initiating regulatory actions in the later 1950s and 1960s to prohibit many of DDT's uses because of mounting evidence of the pesticide's declining benefits and environmental and toxicological effects.

In 1972, EPA issued a cancellation order for DDT based on adverse environmental effects of its use, such as those to wildlife (the known effects on other species, such as raptors), as well as DDT's potential human health risks. Since then, studies have continued, and a causal relationship between DDT exposure and reproductive effects is suspected. Today, DDT is classified as a probable human carcinogen by the U.S. and international authorities. This classification is based on animal studies in which some animals developed liver tumors.

DDT is known to be very persistent in the environment, will accumulate in fatty tissues, and can travel long distances in the upper atmosphere. Since the use of DDT was discontinued in the United States, its concentration in the environment and animals has decreased, but because of its persistence, residues of concern from historical use still remain. Moreover, DDT is still used in developing countries because it is inexpensive and highly effective. Other alternatives are too expensive for these other countries to use (USEPA, 2012).

There are ~13,500 chemical manufacturing facilities in the United States owned by more than 9,000 companies. There are 84,000 chemical in use in the United States with ~700 new ones added each year. Manufacturers generally manufacture chemicals classified into two groups: commodity chemicals and specialty chemicals. Commodity chemical manufacturers produce large quantities of basic and relatively inexpensive compounds in large plants, often built specially to make one chemical. Commodity plants often run continuously, typically shutting down only a few weeks a year for maintenance. Specialty-batch or performance chemical manufacturers produce smaller quantities of more expensive chemical on an "as needed" basis that are used less frequently. Facilities are located all over the country, with many companies in Texas, Ohio, New Jersey, Illinois, Louisiana, Pennsylvania, and the Carolinas.

In the U.S., under the Toxic Substances Control Act, five chemicals were banned—only five. This may seem odd and difficult to comprehend because when TSCA was passed in 1976, there were 60,000 chemicals on the inventory of existing chemicals. Since that time, EPA has only successfully restricted or banned the five just mentioned and has only required testing on another 200 existing chemicals. An additional 24,000 chemicals have entered the marketplace for a total of more than 84,000 chemicals on the TSCA inventory. The chemical industry is an essential contributor to the U.S. economy, with shipments valued at about $555 billion per year.

So, the obvious question might be: If we know several chemicals are dangerous or harmful to us and our environment, why have only five been banned? Good question. Best answer is that under TSCA it is difficult to ban a chemical that predated the Rule and thus has been grandfathered.

Next question: which five chemicals has TSCA banned? If you were a contestant on the TV show Jeopardy and you thought you knew the correct answer you might respond with the question: What are chemicals PCBs, chlorofluorocarbons, dioxin, hexavalent chromium, and asbestos? A no brainer, right? An easy question with a straightforward answer, right? Well, not so fast. Technically, you would be incorrect with that answer. Why? Well, even though EPA did initially ban most asbestos-containing products in the United States, in 1991, the rule was vacated and remanded by the Fifth Circuit Court of Appeals. As a result, most of the original bans on the manufacture, importation, processing, or distribution in commerce for most asbestos-containing product categories originally covered in the 1989 final rule were overturned. Only the bans on corrugated paper, roll board, commercial paper, specialty paper, and flooring felt, and any new uses of asbestos remained banned under the 1989 rule. Although most asbestos containing products can still legally be manufactured, imported, processed and distributed in the U.S., this is still the case even though more than 45,000 Americans have died from asbestos exposure in the past 33 years. In spite of that, according to the U.S. Geological Survey, the production and use of asbestos have declined significantly (USEPA, 2011).

Endocrine Disruptors

A growing body of evidence suggests that humans and wildlife species have suffered adverse health effects after exposure to endocrine-disrupting chemicals (aka environmental endocrine disruptors).[1] In this book, environmental endocrine disruptors are defined as exogenous agents that interfere with the production, release, transport, metabolism binding, action, or elimination of natural hormones in the body, which are responsible for the maintenance of homeostasis and the regulation of developmental processes. The definition reflects a growing awareness that the issue of endocrine disruptors in the environment extends considerably beyond that of exogenous estrogens and includes antiandrogens and agents that act on other components of the endocrine system such as the thyroid and pituitary glands (Kavlock et al., 1996). Disrupting the endocrine system can occur in various ways. Some chemicals can mimic a natural hormone, fooling the body into over-responding to the stimulus (e.g., a growth hormone that results in increased muscle mass) or responding at inappropriate times (e.g., producing insulin when it is not needed). Other endocrine-disrupting chemicals can block the effects of a hormone from certain receptors. Still others can directly stimulate or inhibit the endocrine system, causing overproduction or underproduction of hormones. Certain drugs are used to intentionally cause some of these effects, such as birth control pills. In many situations involving environmental chemicals, an endocrine effect may not be desirable.

In recent years, some scientists have proposed that chemicals might inadvertently be disrupting the endocrine system of humans and wildlife. Reported adverse effects include declines in populations, increases in cancers, and reduced reproductive function. To date, these health problems have been identified primarily in domestic or wildlife species with relatively high exposures to organochlorine compounds, including DDT and its metabolites, polychlorinated biphenyls (PCBs) and dioxides, or to naturally occurring plant estrogens (phytoestrogens). However, the relationship of human diseases of the endocrine system and exposure to environmental contaminants is poorly understood and scientifically controversial.

Although domestic and wildlife species have demonstrated adverse health consequences from exposure to environment that interact with the endocrine system, it is not known if similar affects are occurring in the general human population, but again there is evidence of adverse effects in populations with relatively high exposures. Several reports of declines in the quality and decrease in the quantity of sperm production in humans over the last five decades and the reported increase in incidences of certain cancers (breast, prostate, testicular) that may have an endocrine-related basis have led to speculation about environmental etiologies (Kavlock et al., 1996). For example, Carlson et al. (1992) point to the increasing concern about the impact of the environment on public health, including reproductive ability. They also point out that controversy has arisen from some reviews which have claimed that the quality of human semen has declined. However, only little notice has been paid to these warnings, possibly because the suggestions were based on data on selected groups of men recruited from infertility clinics, from among semen donors, or from candidates for vasectomy. Furthermore, the sampling of publications used for review was not systematic, thus implying a risk of bias. As a decline in semen quality may have serious implications for human reproductive health, it is of great importance to elucidate whether the reported decrease in sperm count reflects a biological phenomenon or, rather, is due to methodological errors.

Data on semen quality collected systematically from reports published worldwide indicate clearly that sperm density had declined appreciably between 1938 and 1990, although we cannot conclude whether or not this decline is continuing. Concomitantly, the incidence of some genitourinary abnormalities, including testicular cancer and possibly also maldescent (faulty descent of the testicle into the scrotum) and hypospadias (an abnormally placed urinary meatus opening in the male penis), has increased. Such remarkable changes in semen quality and the occurrence of genitourinary abnormalities over a relatively short period are more likely due to environmental than genetic factors. Some common prenatal influences could be responsible both for the decline in sperm density and for the increase in cancer of the testis, hypospadias, and cryptorchidism (where one or both testicles fail to move to the scrotum). Whether estrogens, compounds with estrogen-like activity, or other environmental or endogenous factors actually damage testicular function remains to be determined (Carlson et al., 1992). Even though we stated that we do not know what we do not know about endocrine disruptors, it is known that the normal functions of all organ systems are regulated by endocrine factors, and small disturbances in endocrine function, especially during certain stages of the life cycle such as development, pregnancy, and lactation, can lead to profound and lasting effects. The critical issue is whether sufficiently high levels of endocrine-disrupting chemicals exist in the ambient environment to exert adverse health effects on the general population.

Current methodologies for assessing, measuring, and demonstrating human and wildlife health effects (e.g., the generation of data in accordance with testing guideline) are in their infancy. EPA has developed testing guidelines and the Endocrine Disruption Screening Program (EDSP), which is mandated to use validated methods for screening the testing chemicals to identify potential endocrine disruptors, determine adverse effects, dose-response, assess risk and ultimately manage risk under current laws.

Biological Effects

In this section, we discuss what we know about the carcinogenic effects of endocrine-disrupting agents in humans and wildlife; the major classes of chemicals thought to be responsive for these effects; and the uncertainties associated with the reported effects.

Numerous field studies of teleost (bony) fishes in localized, highly contaminated areas (i.e., "hot spots") have shown high prevalences of liver tumors (Baumann et al., 1990; Harshbarger and Clark, 1990; Meyers et al., 1994). The predominant risk factor that has been associated with these liver tumors is exposure to poly-aromatic hydrocarbons (PAHs) and to a lesser degree, PCBs and DDT. Certain species, such as carp and fathead minnows, are more resistant, while trout are more sensitive. There has been no indication that the liver tumors in fish involve an endocrine modulation mechanism. Other than for localized areas of high contamination, field studies have shown no increasing trends for tumors of any type in fish. Two tumor registries for wildlife species exist in the United States (the Smithsonian Registry of Tumors in Lower Animals, and the armed Forces Institute of Pathology's Registry of Comparative Pathology). A variety of dose-related tumors can be produced in fish given carcinogens under experimental laboratory conditions (Couch and Harshbarger, 1985). Again, let's be clear: there is no specific evidence that the development of these tumors involves a hormonal disruption mechanism. Research has shown that estradiol and certain hormone precursors (e.g., dehydroepiandrosterone [DHEA]) act as promoters after treatment of fishes with carcinogenic substances such as aflatoxin and N-methyl-N-nitroso-N'-nitroguanidine

(MNNG). Toxicopathic liver lesions have been associated with contaminant exposure in some marine fish (Meyers et al., 1994).

There is a paucity of carcinogenicity data for other forms of wildlife. One study of beluga whales in the St. Lawrence seaway found that ~50% of dead whales examined had neoplasms (i.e., abnormal masses of tissue), of which about 25% were malignant (Beland et al., 1993; DeGuise et al., 1995).

In 1996, Theo Colborn, together with science writers Dianne Dumanoski and John Peterson Myers, compiled various observations concerning chemical exposure and wildlife health into the book *Our Stolen Future* (often compared to Rachel Carson's *Silent Spring*) and drew a straight line between the effects observed in wild animals and human health effects, including breast and prostate cancer and decreasing male fertility caused by decreasing sperm counts, cryptorchidism (where one or both testicles fail to descend from the body) and hypospadias (deformation of the phallus) (Breithaupt, 2004).

Is this animal-to-humankind endocrine disrupter exposure connection to ill health effects an old-style fishing expedition or is it food for thought? Is it enough to drive maximum effort by scientific researchers? Conversely it can be asked: When it comes to researching and determining the causal factors related to human disease is there a limiting factor, or should there be one?

Kavlock et al. (1996) pointed out the hypothesis that endocrine disruption can cause cancer in humans is based on the causal associated between DES exposure of pregnant women and clear-cell adenocarcinoma of the vagina and cervix in their female offspring, hormone-related risk factors for breast and uterine cancer, and limited evidence of an association between body burden levels of 1,1-dichloro-2.2-bis(p-chlorophenyl)ethylene (DDE) or PCBs and breast cancer risk. Young women who developed cancer of the vagina were more likely to have had mothers who used DES during pregnancy to avoid miscarriage form mothers who did not use the drug (Herbst et al., 1971). The finding has led to a number of important conclusions. First, maternal exposures during gestation can lead to cancer in offspring, and second, it demonstrates that a synthetic estrogen can cause cancer. Some of the male offspring of women who took DES display pseudohermaphroditism (Kaplan, 1959) and genital malformations, including epididymal cysts, testicular abnormalities such as small testes and microphallus, and reduced semen quality (Gill et al., 1979; Driscoll and Taylor, 1980; Penny, 1982). Follow-up surveys of DES-exposed male offspring, however, have not shown impairment in fertility or sexual function (Leary et al., 1984; Wilcox et al., 1995), nor is there evidence of increased risk of testicular cancer (Leary et al., 1984).

According to CDC (2013) and based on U.S. cancer statistics: 1999–2009 incidence and mortality data the three most common cancers among woman of all races and Hispanic origins in the United States are:

- Breast cancer (123.1 per 100,000)
- Lung cancer (54.1 per 100,000) (second among white, black, and American Indian/Alaska Native women; third among Asian/Pacific Islander and Hispanic Women)
- Colorectal cancer (37.1 per 100,000) Second among Asian/Pacific Islander and Hispanic women; third among white, black, and American Indian/Alaska Native women

The leading causes of cancer death among women are:

- Lung cancer (38.6 per 100,000 women of all races); first among white, black, Asian/Pacific Islander, and the American Indian/Alaska Native women; second among Hispanic women.
- Breast cancer (22.2 per 100,00 women of all races); first among Hispanic women; second among white, black, Asian/Pacific Islander, and American Indian/Alaska Native
- Colorectal cancer (13.1 per 100,000 women of all races); third among women of all races and Hispanic origin populations.

As mentioned, the most common cancer among women in the United States is breast cancer. A number of epidemiological studies have examined the risk factors for breast cancer. Identified risk factors include several that relate to hormonal activity: decreased parity, age at first delivery, age at menarche (the beginning of the menstrual function), age, race, and unopposed estrogen therapy. In addition, breast tumors can be characterized as to their degree of estrogen-receptor positivity resulting in relevant prognostic information. The evidence supports a causal relationship between female breast cancer and hormonal activity.

A number of organochlorine pesticides or pesticidal metabolites are found in beast mile and human adipose tissue (Jensen and Slorach, 1991; Sonawane, 1995). Several recent cross-sectional studies suggest a possible relationship between levels of some organohalide residues in human tissues and breast cancer risk, although the observations are not entirely consistent across studies, and no clear relationship has been established (Unger et al., 1984; Mussalo-Rauhamaa et al., 1990; Austin et al., 1989; Falck et al., 1992; Wolff et al., 1993; DeWally et al., 1994; Krieger, et al., 1994; Henderson et al., 1995). In general, these studies suggested that levels of p, p'-DDE and total PCBs were higher in the fat or serum of woman who had breast cancer than in comparison groups. The meaning of these findings is unclear, in part, because op, p'-DDE and the few PCB congeners that have been tested have little or no discernible estrogenic activity, while the short-lived forms of DDT, o, p'-DDT and o, p'-DDE, have only very weak estrogenic properties; further, a 1994 case-control study with historical data from serum DDE and PCBs conflicts with earlier findings (Krieger et al., 1994). This study showed no overall effect of serum residue levels on breast cancer risk, although subcategorical analysis did suggest a possible increase in risk among black women with higher levels of serum p, p'-DDE. The women in these studies, except those in the study by Henderson et al. (1995), were not exposed to high levels of PCBs or DDE and the actual differences in levels measured between cases and controls were not large. Studies of women occupationally exposed high levels of PCBs have not demonstrated an excess risk of breast center mortality (Brown, 1987; Sinks et al., 1992). The results of these studies, therefore, are equivocal and further research is needed (including examination of effects in subsequent generations for parental exposures).

The bottom line: Examination of the U.S. EPA database on pesticide registration or organochlorines showed no correlation of the spectrum of tumor types observed in laboratory animals with the assertion that organochlorines are related to human breast cancer. Organochlorines frequently increased the incidence of liver tumors in rates but did not increase the incidence of mammary tumors (Kavlock et al., 1996). One subclass of herbicides, the chloro-S-triazines, produces an earlier onset of mammary tumors in Sprague-Dawley rats (Wetzel et al., 1994; Stevens et al., 1994), but there are no epidemiological studies that suggest a relationship between exposure to triazine herbicide and human breast cancer. An examination of the National Toxicology Program (NTP) database involving ~450 animal studies showed increased incidences of mammary tumors in ~10% of the studies. However, based on the evaluation of chemical structures and other available information, this subset of test substances in not likely to be estrogenic (Dunnick et al., 1995). This analysis only considered the possibility that the chemicals were direct estrogen agonists. Other possible mechanisms exist for the induction of endocrine-mediated tumors. Moreover, good animal or cellular models do not yet exist for the study of some endocrine-mediated tumors (e.g., testicular), which are reported to be on the increase in the human population, and such models would be useful in testing cause-and-effect relationships.

In the preceding, we provided a detailed discussion of the unproven but potential for endocrine disruptors causing breast cancer in women. What about men? What is the potential for men to contract various cancers via exposure to certain endocrine disruptors? Well, the jury is still out on this hypothesis, too, but there have been many studies conducted on the subject and many researchers attempt to show or explain a possible straight-line connection between the two.

For example, the endocrine disruptor and prostate cancer risk have been studied extensively.

Reproductive Effects

Currently, some of the major questions being asked about the reproductive effects of endocrine disruptors include: What do we know about the reproductive and developmental effects of endocrine-disrupting agents in humans and wildlife? What are the major classes of chemicals though responsible for these effects? What are the uncertainties associated with the reported effects? In this section, we provide both speculative opinions and authoritative answers to these questions.

Field and laboratory studies of wildlife populations and individuals have revealed effects in offspring that appear to be the result of endocrine disruption. Examples include reproductive problems in wood ducks (*Aix sponsa*) from Bayou Meto, Arkansas, which is downstream form a hazardous waste site in central Arkansas (White, 1992); wasting and embryonic deformities in Great Lakes colonial fish-eating birds (i.e., herring gulls, common terns and double-crested cormorants) (Peakall and Fox, 1987; Hoffman et al., 1987; Colborn, 1991; Gilbertson et al., 1991; Fox et al., 1991a,b; Giesy et al., 1995; Bowerman et al., 1995); feminization and demasculinization of gulls (Boss and Witschi, 1943; Fry and Toone, 1981; Fry et al., 1987; Fox et al., 1978); developmental effects in Great lakes snapping turtles (*Chelydra serpentina*) (Bishop et al., 1991); embryonic mortality and developmental dysfunction in lake trout and other salminiods in the Great Lakes (Mac and Edsall, 1991; Mac et al., 1993; Leatherland, 1993); abnormalities of sexual development in Lake Apopka alligators (Guillette et al., 1994, 1995); reproductive failure in mink from the Great Lakes area (Wren, 1991); and reproductive impairment in the Florida Panther (Facemire et al., 1995). In each case, detectable concentrations of chemicals with known endocrine-disruption effects have been reported in the animals or in their environment, but an etiological link has been established for only a few of these observations. In ecological studies, these effects were not recognized until the populations began to decline. However, the observation that a population is stable is not an assurance that endocrine-disrupting chemicals are not affecting the reproduction, development, and/or growth of individuals.

Neurological Effects

Currently, some of the major questions being asked about the neurological effects of endocrine disruptors include: What do we know about the neurological effects of endocrine-disrupting agents in humans and wildlife? What are the major classes of chemicals thought to be responsible for these effects? What are the uncertainties associated with the reported effects? In this section, we provide both speculative opinions and authoritative answers for these questions.

Current knowledge indicates that neuroendocrine disruption can be induced by multiple mechanisms. Direct effects on endocrine glands (e.g., the thyroid) may alter the hormonal milieu, which in turn can affect the nervous system, resulting in neurotoxicity. Conversely, EDCs may initially act on the central nervous system (CNS) (e.g., neuroendocrine disruptors), which in turn can influence the endocrine system. Simply, exposure to chemicals can adversely affect the structure and function of the nervous system without any endocrine system involvement. Moreover, alterations in the following would be indicative of neuroendocrine disruption: reproductive behaviors mediated by alterations in the hypothalamic-pituitary axis (e.g., courtship and parental behavior in avian species); alterations in metabolic rate, which could indirectly affect behavior; altered sexual differentiation in the brain, which could affect sexually dimorphic reproductive and nonreproductive neural end points; and some types of neuroteratogenic effects. There are clear examples in the human and animal literature in which exposure to endocrine disruptors had occurred and effects on behavior, learning and memory, attention, sensory function, and psychomotor development were observed (Fox et al., 1978; McArthur et al., 1983; Colborn et al., 1993). Some of these effects, however, can also be produced by developmental neurotoxicants having little or no known endocrine-disrupting properties and, therefore, cannot be regarded as specific to the endocrine-disrupting class of chemicals. It is known that exposures to a number of nonchemical factors (e.g., food or oxygen deprivation, infections, and temperatures) could also adversely affect the nervous system resulting in effects similar to those produced by endocrine disruptors. These nonchemical factors may also interact in as yet unpredictable ways with chemical stressors. Therefore, considerable care should be taken to eliminate nonchemical causes before concluding that neurotoxicity is causally related to the effects of a chemical acting on the endocrine system.

There are several examples of chemicals or classes of chemicals that produce neurotoxicity by an endocrine mechanism. Environmental toxicologist should consider the dose at which neuroendocrine dysfunctions are produced relative to the concentrations existing in the environment and relative to dose levels at which other toxic effects occur, the relationship between exposure and effect, and the role of naturally occurring chemicals with endocrine-mimicking properties. With these caveats in mind, examples of directly or indirectly acting neuroendocrine disruptors include some PCBs, dioxins, DDT and related chlorinated pesticides and their metabolites, some metals (methylmercury, lead, and organotins), insect growth regulators, dithiocarbamates, synthetic steroids, tamoxifen, phytoestrogen, and triazine herbicides. Identification of chemicals as neuroendocrine disruptors should be based on mechanistic information at the cellular or molecular level in the endocrine system or defined functionally in terms of activity on response known to be mediated by or dependent on hormones. All definitions of neuroendocrine disruptors should be interpreted specifically with respect to gender, hormonal status, and developmental stage since the expression of toxicities for chemicals may change significantly depending on these variables.

A number of uncertainties critical to understand the significance and effects of neuroendocrine disruptors have been identified, including:

a. Chemicals occur as mixtures in the environment, thereby making it difficult to assign cause and effect for specific agents. It is possible that the parent chemical may not affect the endocrine system but is metabolized to an active form. The toxicokinetics of and relative tissue distribution into the nervous system are generally unknown for most chemicals; little is known about the metabolic interaction between chemicals in mixtures.
b. There are ranges of possible specific and nonspecific effects that could be measured. Research to date has used only a small number of techniques and methods, and it is likely that many neuroendocrine effects may be subtle and not easily detected with currently available procedures. It is also a concern that the functions most sensitive to chemically induced alterations in neuroendocrine function are the most difficult to measure in the field.
c. It is critical to know when exposure occurred relative to when the effects are measured. Observed effects could be dependent on a number of intrinsic factors such as seasonal variability and intrinsic facts such as hormonal status. In addition, the nervous system is known to be differently sensitive to chemical perturbation at various stages of development. A chemical may have a significant effect on neuroendocrine function if exposure occurs at a critical period of development but have little or no effect at other states of maturation.
d. Several issues related to extrapolation are critical to understanding neuroendocrine disruptors. For example, it is difficult to evaluate the significance to human health of a chemically induced change in a behavior that does not naturally occur in humans, i.e., there are concerns about the appropriateness of some animal models for toxicologic studies. In addition, there are uncertainties about extrapolating from species to species and from experiments conducted in the laboratory to those performed in the field.
e. There are uncertainties about the shape of the respective dose-response curves for many neuroendocrine effects. It is likely that some chemicals may have multiple effects occurring at different points on the dose-response curve.
f. Basic information concerning the mechanism of action of chemicals on the developing nervous system and the neurological role of hormones during development would greatly reduce uncertainties about the risk of exposure to neuroendocrine disruptors. Furthermore, it is also important to understand the consistency of the effects relative to the hypothesis that chemicals are affecting the nervous system.

Immunological Effects

Currently, some of the questions being asked about the immunological effects of endocrine disrupters include: What do we know about the immunological effects of endocrine-disrupting agents in humans and wildlife? What are the major classes of chemicals though responsible for these effects? What are the uncertainties associated with the reported effects? In this section, both speculative opinions and authoritative answers are provided for these questions.

Published studies have demonstrated associations between autoimmune syndromes and DES exposures (Noller et al., 1988). A relationship is well established between physiological estrogen levels and autoimmune diseases in women (Schuurs and Veheul, 1990; Homo-Delarche et al., 1991; Grossman, 1985). The observations that exposure of humans to DES, TCDD, PCBs, carbamates, organochlorines, organometals, and certain heavy metals alters immune phenotypes or function are suggestive of immunosuppression (Davis and Safe, 1990; McKinney et al., 1976; Luster et al., 1990; Dean et al., 1994a). Experimental animal studies support these observations (e.g., Loose et al., 1977), although dose-response information is needed to clarify whether these are directly-or indirect-acting agents. With respect to fish and wildlife, it was also noted that several of the agents listed above induce immune suppression or hyperactivity similar to that reported in experimental animals and humans. Embryonic exposure of trout to aflatoxin has led to alterations in adult immune capacity (Arkoosh and Kaattari, 1987; Arkoosh, 1989; Kaattari et al., 1994). With regard to disease susceptibility and exposure, there have been examples such as the dolphin epizootic of 1987–1988 (Aguilar and Raga, 1983). In this case there was an association between PCBs and DDT in the blood, decreased immune function, and increase the incidence of infections among affected individuals (Swenson et al., 1994). Impairment in immune function has been reported in bottlenose dolphins exposure to PCBs and DDT (Lahvis et al., 1995) and in harbor seals fed fish from polluted waters (de Swart et al., 1994; Ross et al., 1995). From 1991 to 1993, specific immune functions and general hematologic parameters were measured in herring gull and Caspian tern chicks from a number of study sites in the great Lakes chosen across a wide range of organochlorine contamination (primarily PCBs). As the hepatic activity of ethoxinyresorufin-O-deethylase EROD), an index of exposure, increased, thymus mass decrease. At highly contaminated sites both gull and tern chicks showed marked reductions in T-cell-mediated immunity as measured by the phytohemagglutinin skin test (Grassman, 1995).

A variety of immunoassays have been used to demonstrate effects in experimental laboratory animals, humans, fish, and wildlife. These include modulation of antibody response (both in vivo and in vitro), the phytohemagglutinin skin test, mitogenesis, phagocytosis, levels of complement or lack of acute phase reactants, cytotoxic T-Lymphocyte reactivity, and natural killer cell activity (Luster et al., 1988; Schrank et al., 1990; Zelikoff et al., 1991; Bowser et al., 1994; Dean et al., 1994b; Grassman et al., 1995; Zelikoff et al., 1995).

Evidence of an increased rate of autoimmunity associated with prenatal DES exposure suggests the possibility that other endocrine-disrupting chemicals (EDCs) may induce a similar pathologic state. Studies are needed to determine if there has been an increase in cases of immune dysregulation in areas or sites where EDC exposures have occurred. Evidence indicates that the incidences of allergy and asthma (which are forms of hypersensitivity) are increasing in humans (Buist and Vollmer, 1990; Weiss and Wagener, 1990; Gergen and Weiss, 1990). It is not known whether EDC exposures are responsible for some part of this development. Alteration of sex-steroid balance has been shown to lead to increase or accelerated onset of autoimmune syndromes in mice (Homo-Delarche et al., 1991). In rats and mice, heavy metals such as lead, mercury, and gold enhance autoimmune syndromes (Zelikoff et al., 1994; Pelletier et al., 1994). There have also been reports of exposures of fish to EDCs in the environment that lead to immune enhancement. Although autoantibodies have been reported in sharks and trout

(Gonzalez et al., 1989; Michel et al., 1990; Marcholonis et al., 1993), no attempts have been made to correlate exposure to EDCs with incidences of autoantibodies. In trout, embryonic exposure to aflatoxin B_1 can lead to immune stimulation or suppression in the adult, depending upon the immune parameter analyzed (Kaattari et al., 1994). Other data suggest that small changes in physiologic levels of estrogens can affect the immune system, and studies in gull and tern chicks in the great Lakes clearly indicate that the findings are associated with developmental exposures (Grassman, 1995).

Concerning direct-acting EDCs, although it would appear that these agents directly affect the immune system, it is unknown whether there may be disruptive effects on the endocrine-immune axis. Since the immune and endocrine systems are linked via various cytokine signaling process (IL-1, ACTH, catecholamines, prolactin, and endorphins), it is likely that EDC effects on the immune system modulate elements of the endocrine or nervous systems or vice versa. Too little is known about the dose-response curves for immunotoxicity, neurotoxicity, or endocrine effects to decipher the independent or interactive effects of these systems. Because of the high degree of intercommunication between these systems, there is a need for coordinated and cooperative studies among laboratories in all these disciplines.

Although the most forceful arguments for the overall consequences of immune dysfunction would be increased disease incidence, this is difficult to assess in humans or wildlife populations. Furthermore, only certain subpopulations (the very young or elderly) may be affected. Disease may only be manifested as a population decline. Disease trials can be undertaken, but they require controlled laboratory experiments employing populations of wild animals or fish that can easily be maintained in the laboratory. In humans, the variability within a population makes it difficult to decipher exogenously triggered effects.

The fact that the employment of a variety of *in vitro* assays has been successful in correlative exposure studies leads to the question of whether those immune parameters can be correlated with the increased risk of disease. A number of immune parameters operate independently (i.e., lysozyme levels, complement activity, phagocytosis, induction of cytotoxic T lymphocytes, plaque-forming cells, etc.). Which combination of these assessments would make for an optimal predictive suite of assays?

Knowledge of normal baseline values for wildlife species, and in most cases humans, is lacking. If these populations are to be screened for perturbations in immune function, control populations must be defined, and standardized control values obtained. Also, the types of exposures must be well documented (i.e., dose, length of exposure, timing).

PPCPs

According to USEPA (2010), pharmaceuticals and personal care products were first called PPCPs only a few years ago, but these bioactive chemicals (substances that have an effect on living tissue) have been around for decades. Their effect on the environment is now recognized as an important area of research.[2] PPCPs include:

Prescription and over-the-counter therapeutic drugs
Veterinary drugs
Fragrances
Cosmetics
Sun-screen products
Diagnostic agents
Nutraceuticals (e.g., vitamins)

Sources of PPCPs:

   Human activity
   Residues from pharmaceutical manufacturing
   Residues from hospitals
   Illicit drugs
   Veterinary drug use, especially antibiotics and steroids
   Agribusiness

USEPA (2010) points out the significance of individuals directly contributing to the combined load of chemicals in the environment and that these personal actions have been largely unrecognized. PPCPs in the environment illustrate the immediate connection of the actions/activities of individuals with their environment.

Individuals add PPCPs to the environment through excretion (the elimination of waste material from the body) and bathing, and disposal of unwanted medications to sewers and trash. Some PPCPs are easily broken down and processed by the human body or degrade quickly in the environment, but others are not easily broken down and processed, so they enter domestic sewers. Excretion of biologically unused and unprocessed drugs depends on:

- Individual drug composition (certain excipients—i.e., inert ingredients—can minimize absorption and therefore maximize excretion)
- The ability of individual bodies to break down drugs (this ability depends on age, sex, health, and individual idiosyncrasies)

Because they dissolve easily and don't evaporate at normal temperatures or pressure,

PPCPS make their way into the soil and into aquatic environments via sewage, treated sewage biosolids (sludge), and irrigation with reclaimed water. Figure 8.1 shows the origins and fate of PPCPS in the environment.

For the purposes of this discussion, pharmaceutical (and veterinary and illicit) drugs (and the ingredients in cosmetics, food supplements, and other personal care products), together with their respective metabolites and transformation products, will collectively be referred to as PPCPs (pharmaceuticals and personal care products) and more appropriately called micropollutants. PPCPs are commonly infused into the environment via sewage treatment facilities, outhouses, septic tanks, cesspools, concentrated animal feeding operations (CAFOs), human and animal excretion into the environment (water and soil), and wet weather runoff (i.e., stormwater runoff). In many instances, untreated sewage is discharged into receiving waters (e.g., flood overload events, domestic 'straight-piping,' bypassing due to interceptor and pumping failures, or sewage waters lacking municipal treatment). In the United States alone, possibly more than a million homes do not have sewage systems but instead rely on direct discharge or raw sewage into streams by straight-piping or by outhouses not connected to leach fields (Pressley, 1999).

Note that even with wastewater treatment, many of the micropollutants in the sewage waste stream remain in the effluent discharged into the receiving waters. This is the case because many treatment processes are not designed to remove the low concentrations of PPCP micropollutants discussed in this text. The big unknown (remember, the warp and woof woven throughout this text is that we do not know what we do not know about PPCPs and EDCs) is whether the combined low concentrations from each of the numerous PPCPs and their transformation products have any significance with respect to ecologic function while recognizing that immediate effects could escape detection if they are subtle and that long-term cumulative consequences could be insidious. Another question is whether the pharmaceuticals remaining in water used for domestic purposes poses long-term risks for human health after lifetime ingestion via potable waters multiple times a day of very low, subtherapeutic doses of numerous pharmaceuticals; this issue, however, is not addressed in this text.

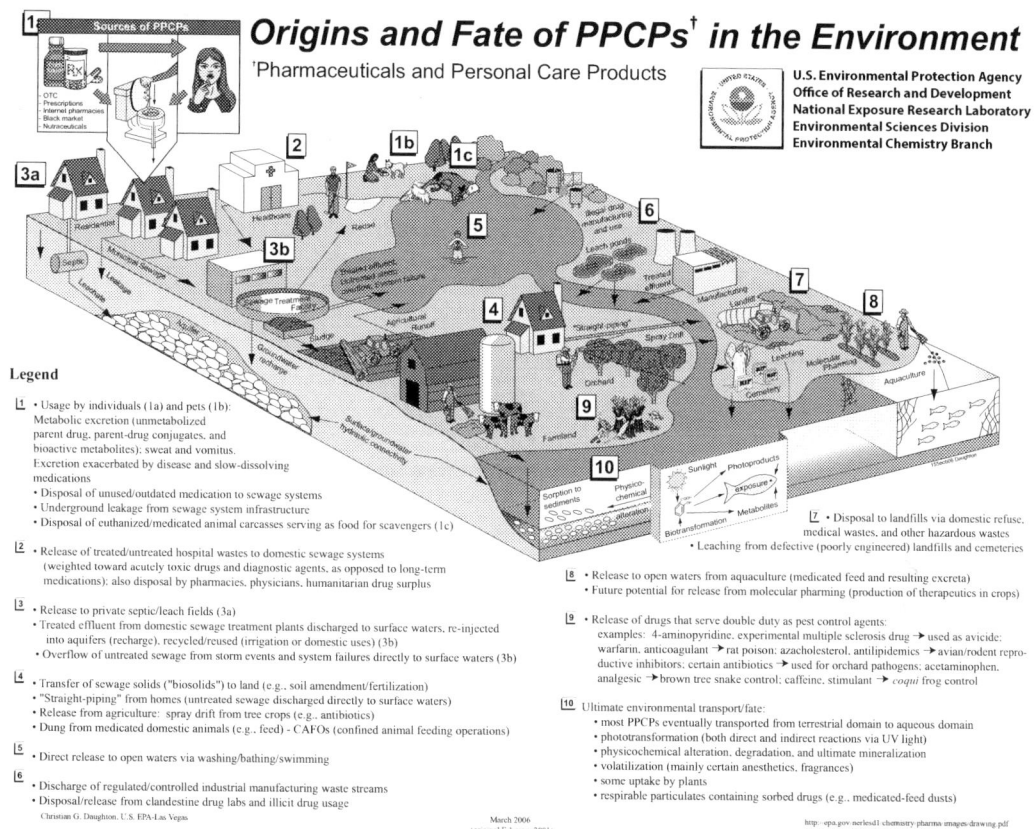

FIGURE 8.1 Origins and fate of PPCPs in the environment. 2010. C.G. Daughton, US. EPA. NERL, Las Vegas, NV.

The questions continue and are further complicated by the fact that while the concentration of individual drugs in the aquatic environment could be low (sub-parts per billion or sub-nanomolar, again, often referred to as micropollutants), the presence of numerous drugs sharing a specific mode of action could lead to significant effects through additive exposures. It is also significant that drugs, at least until recently, unlike pesticides, have not been subject to the same scrutiny regarding possible adverse environmental effects. They have therefore enjoyed several decades of unrestricted discharge to the environment, mainly via wastewater treatment works. This is surprising especially since certain pharmaceuticals are designed to modulate endocrine and immune systems and cellular signal transduction and as such (as opposed to pesticides and other industrial chemicals already undergoing scrutiny as endocrine disruptors) have obvious potential as endocrine disruptors in the environment, especially for aquatic organisms, may differ from that of pesticides and other industrial chemicals in one significant respect—exposures may be of a more chronic nature because PPCPs are constantly infused into the environment wherever humans live or visit, whereas pesticide fluxes are more sporadic and have greater spatial heterogeneity. At the present time, it is quite apparent that little information exists from which to construct comprehensive risk assessments for the vast majority of PPCPs having the potential to enter the environment.

Although little is known of the occurrence and effects of pharmaceuticals in the environment, more data exist for antibiotics than for any other therapeutic class. This is the result of their extensive use in both human therapy and animal husbandry, their more easily deterred effects end points (e.g., via microbial and immunoassays), and their greater chances of introduction into the environment, not just by wastewater treatment plants, but also by run-off and groundwater contamination, especially from confined animal feeding operations (CAFOs). The literature on antibiotics is much more developed because of the obvious issues of direct effects on native microbiota (and consequent alteration of microbial community structure) and development of resistance in potential human pathogens. Because of the considerably larger literature on antibiotics, this text only touches on the issue; for the same reason, this presentation only touches on steroidal drugs (those purposefully designed to modulate endocrine systems).

Pharmaceuticals in the Environment

The fact that pharmaceuticals have been entering the environment from a number of different routes and possibly causing untoward effects in biota has been noted in the scientific literature for several decades, but, until recently, its significance has gone largely unnoticed. This results in a large part from the international regulation of drugs by human health agencies, which usually have limited expertise in environmental issues. In the past, drubs were rarely viewed as potential environmental pollutants; there was seldom serious consideration as to their fates once they were excreted from the user. Then again, until the 1990s, any concerted efforts to look for drugs in the environment would have met with limited success because the requisite chemical analysis tools with sufficiently high separatory efficiencies, to resolve the drugs from the plethora of other substances—native and anthropogenic alike, and low detection limits (i.e., nanograms per liter or parts per trillion), were not commonly available. Other obstacles, which still exist to a large degree, are that many pharmaceuticals and cosmetic ingredients and their metabolites are not available in the widely used environmentally oriented mass spectral libraries. There are available in speciality libraries such as Pfleger (e.g., Pfleger, 1999a,b), which are not frequently used by environmental chemists. Analytical reference standards when available are often difficult to acquire and are quite costly. The majority of drugs are also highly water soluble. This precludes the application of straightforward, conventional sample clean-up/preconcentration methods, coupled with direct gas chromatographic separation, that have been used for years for "conventional" pollutants, which tend to be less polar and more volatile.

Drugs in the environment did not capture the attention of the scientific or popular press until the last couple of years, with some early overviews and reviews presented by Halling-Sorenson et al. (1998), Montague (1998), Raloff (1998), Roembke et al. (1996), Ternes et al. (1999; 1998), Velagaleti (1998), and more recently by USEPA's Christian, G. Daughton (2010).

The evidence supports the case that PPCPs refractory to degradation and transformation does indeed have the potential to reach the environment (Halling-Sorenson et al., 1998). What is not known, however, is whether these chemicals and their transformation products can elicit physiologic effects on biota at the low concentrations (ng-µg/L) which they are observed to occur. Another unknown is the actual quantity of each of the numerous commercial drugs that is ingested or disposed. With respect to determining the potential extent of the problem, this contrasts sharply with pesticides in which usage is much better documented and controlled.

In discussing the disposal or wastage of pharmaceuticals (drugs), one thing seems certain; namely, we must understand the terminology currently used in discussing drug disposal or drug wastage. Daughton (2010) points out that discussions of drug disposal are complicated enough, but sometimes it is not even clear as to what is meant by the various terms used to describe drugs that are subject to disposal. Terms used in the literature include: unused, unwanted, unneeded, expired, wasted, and leftover. The distinctions between these can be subtle or ambiguous. For instance, "unused" and "expired" are not good descriptors as they only comprise subsets of the total spectrum of medications that can require disposal. "Unused" omits those medications requiring disposal but which have indeed already been used (such as used medical devices). The point is that just because a medication's container or package has been opened does not necessarily mean it has been "used." "Unused" can also mean to the patient that they are literally no longer using the medication (for its intended purpose), despite the fact that many patients continue using medications on a self-medicating basis (administering the medication for a condition or during not originally intended—one of many forms of non-compliance). The term "expired" omits the preponderance of drugs that are discarded before expiry—often soon after they are dispensed. The term "leftover" is sufficiently expansive, as it includes all medications no longer begin used for the original prescribed condition or intended use—or even unintended purpose.

Another term often used to refer to unused consumer pharmaceuticals is "home-generated pharmaceuticals" (or home-generated pharmaceutical waste). But this too is not a rigorous term, as many drugs from consumer use are not kept in the home but are dispersed in countless locations throughout society (Ruhoy and Daughton, 2008).

A major obstacle in any discussion of drug wastage is what exactly is meant by "wastage." A definition of wastage is notoriously difficult—especially since the topic involves countless variables and perspectives. A simple definition for drug waste is medications dispensed to—or purchased by—a consumer that is never used for the original intended purpose. But, on closer examination, this is not as straightforward as it might first appear. A better term might be "leftover" medications, as this avoids any inference of whether the medications were actually "wasted" (that is, served no purpose). The term "leftovers" does not infer a reason for why medications accumulated unused or unwanted. Would a medication intended for emergency contingency purposes (and now expired) be considered "wasted"? After all, such medications severed their purpose of being available for possible emergencies. How about medications intended for unscheduled consumption "as the situation arises" or "as needed" (PRN: "pro re nata," Latin for "in the circumstances" or "as the circumstance arises"). These scenarios show that it would not be possible to completely eliminate leftover medications—only to reduce them to a necessary minimum.

It could be argued whether the basic premise that medications experience undue wastage is even valid. No one really knows how much drug waste occurs in commerce (at the consumer level or in the healthcare setting) in terms of either the total quantity or the cost. In one review of medication wastage, it was estimated that in the UK 1%–10% of the total cost of medications are wasted; but estimates in the UK are usually based on the quantities of medications returned to pharmacies by consumers, omitting the quantities that are disposed of at home, stored indefinitely, or shared with others.

Many statements regarding drug wastage are based on rates of patient compliance, which is an enormously complex and controversial topic by itself. But non-compliance rates include not just the frequency with which drugs go unused, but also the frequency with which prescriptions are NOT filled or with which they are consumed incorrectly. Neither of the latter contributes to any need for disposal. Failure to fill a prescription may even reduce the need for disposal; so, noncompliance does not necessarily lead to leftover drugs. Few make this distinction in the literature.

Table 8.1 lists the PPCPs covered in this text, together with their chemical names, use/origin, and some representative environmental occurrence/effects data. These chemicals, together with their synthetic precursors and transformation products, are continually released into the environment in enormous quantities as a result of their manufacture, use (via excretion, mainly in urine and feces), and disposal of unused/unwanted drugs and those that have expired both directly into the domestic sewage system and via burial in landfills. Although largely unknown, there is evidence that large quantities of prescription and nonprescription, "over-the-counter" (OTC) drugs are never consumed (for any number of reasons) (Bosch, 1998), and many of these are undoubtedly eventually disposed down toilets or via domestic refuse.

A striking difference between pharmaceuticals and pesticides with respect to environmental release is that pharmaceuticals have the potential for ubiquitous direct release into the environment worldwide—anywhere that humans live or visit. Even areas considered relatively pristine (e.g., national parks) are subject to pharmaceutical exposures, especially given that some parks have very large, aging sewage treatment systems some of which discharge into park surface waters and some which overflow during wet weather events and infrastructure failures (e.g., Yellowstone National Park) (Milstein, 1999). Other possible sources include the disposal of unwanted illicit drugs and synthesis byproducts into domestic sewage systems by clandestine drug operations; disposal of

TABLE 8.1
PPCPs Identified in Environmental Samples—or Having Significance with Respect to Aquatic Life

Compound	Use/Origin	Environmental Occurrence
Acetaminophen	Analgesic/anti-inflammatory	Efficiently removed by POTW; POTW max. effluent: 6.0 µg/L; not detected in surface waters (Ternes, 1998)
Acetylsalicylic acid	Analgesic/anti-inflammatory	Ubiquitous. One of first pharmaceuticals identified in sewage influent/effluent; POTW removal efficiency 81%; POTW max. Effluent: 1.5 µg/L; max. in surface waters: 0.34 µg/L (Ternes, 1998; Richardson and Bowron, 1985)
Betaxolol	Beta-blocker	POTW max. effluent: 0.19 µg/L; max. in surface waters: 0.028 µg/L (Hirsch et al., 1996)
Bezafibrate	Lipid regulator	Loading of ~300 g/day in Germany POTW (Ternes, 1998); POTW removal efficiency 83% (Ternes, 1996); POTW max. effluent: 4.6 µg/L; max. in surface waters: 3.1 µg/L influent concentration of 1.2: µg/L in Brazilian STWs (Stumpf et al., 1999) with removal efficiencies ranging from 27% to 50%
Biphenylol	Antiseptic, fungicide	POTWs in Germany biphenylol routinely found in both influents (up to 2.6 µg/L) and effluents (Ternes et al., 1998), but removal was extensive
Bisoprolol	Beta-blocker	POTW max. effluent: 0.37 µg/L; max. in surface waters: 2.9 µg/L (Hirsch et al., 1996)
Carazolol	Beta-blocker	POTW max. effluent: 0.12 µg/L; max. in surface waters: 0.11 µg/L (Hirsch et al., 1996)
Carbamazepine	Analgesic; antiepileptic	Loading of over 100 g/day in German POTW (Ternes, 1998); but load in effluent can be 114 g/day; POTW removal efficiency 7% (Ternes, 1998); POTW max. effluent: 6.3 µg/L; max. in surface waters: 1.1 µg/L
4-Chlor-3,5-xylenol	Antiseptic	POTWs in Germany: 4-chloroxylenol occasionally found in both influents and effluents (<0.1 µg/L) (Ternes et al., 1998)
Chlorophene	Antiseptic	POTWs in Germany: chlorophene routinely found in both influents (up to 0.71 µg/L) and effluents (Ternes et al., 1998); removal not as extensive as for biphenylol
Clenbuterol	Bronchodilator	POTW max. effluent: 0.08 µg/L; max. in surface waters: 0.05 µg/L (Ternes, 1998)
Clofibrate	Lipid regulator	Not detected in POTW effluent (Ternes, 1998); not detected in surface waters. River water: ~40 ng/L (Richardson and Bowron, 1985)
Clofibric acid	Polar, active metabolite of liquid regulators	One of the first prescription drugs/metabolites ever reported in sewage influent/effluent: Missouri STW effluent avg. 2.1 kg/day (Hignite and Azarnoff, 1977); 0.8–2.0 µg/L in raw sewage and activated sludge effluent (Garrison et al., 1976). Loading of over 50 g/day in German POTW (Ternes, 1998); POTW removal efficiency 51% (Ternes, 1998); POTW max. effluent: 1.6 µg/L; mas. in surface waters: 0.55 µg/L. Swiss rural/urban lakes: 1–9 ng/L (ppt); North Sea (up to 7.8 ng/L) (Ternes, 1998) in Brazilian STWs (Buser et al., 1998b) with removal efficiencies ranging from 15% to 34%. up to 270 ng/L in German tap waters (Stumpf et al., 1999; Heberer et al., 1998)
Cyclophosphamide	Antineoplastic	POTW max. effluent: 0.02 µg/L; Not detected in surface waters (Ternes, 1998). Hospital sewage 146 ng/L (Steger-Hartmann et al., 1996); POTW receiving hospital waste: influent up to 143 ng/L effluent up to 17 ng/L
Diatrizoate (Na)	X-ray contrast media	Resistant to biodegradation and refractory, unidentified metabolites (Kalsch, 1999). In German surface waters, median concentration of 0.23 µg/L (Ternes and Hirsch, 2000; isolated maximum values above 100 µg/L indicate that locally very high concentrations can occur, especially in small streams containing a high percentage of STW discharges

(Continued)

TABLE 8.1 (Continued)
PPCPs Identified in Environmental Samples—or Having Significance with Respect to Aquatic Life

Compound	Use/Origin	Environmental Occurrence
Diazepam	Psychiatric drug	POTW mas. effluent: 0.04 µg/L; not detected in surface waters (Ternes, 1998). Groundwater from a Super-fund site near Atlantic City, NJ: 10–40 µg/L (Genicola, 1999).
Diclofenac-Na	Analgesic/anti-inflammatory	Loading of ~100 g/day in German POTW (Ternes, 1998); POTW removal efficiency 69% (Ternes, 1998); POTW mas. effluent: 2.1 µg/L; max. in surface waters: 1.2 µg/L. Influent to Swiss STWs 500–1,800 ng/L and effluents more than 505 as much: Swiss lakes/rivers 1–12 ng/L with lower order streams 11–310 ng/L (Buser et al., 1998b). Influent concentration of 0.8 µg/L in Brazilian STWs (Stumpf et al., 1999) with removal efficiencies ranging from 9% to 75%
Aminopyrine	Analgesic/anti-inflammatory	Loading of over 50 g/day in German POTW (Ternes, 1998); POTW removal efficiency 38% (Ternes, 1998); POTW max. effluent: 1.0 µg/L; max. in surface waters: 0.34 µg/L
17 α-ethinyl estradiol	Oral contraceptive	Up to 7 ng/L in POTW effluent (Routledge et al., 1998). Not detected in German surface water above 0.5 ng/L (Ternes et al., 1999), but found in Dutch Rhine water up to 4.3 ng/L (Belfroid et al., 1999)
Etofibrate	Lipid regulator	Not detected in POTW effluent (Ternes, 1998); not detected in surface waters
Fenfluramine	Sympathomimetic amine	While no one has looked for fenfluramine in sewage, it is known to enhance the release of serotonin (5-HT), and in the crayfish, 5-HT in turn triggers the release of ovary-stimulating hormone—resulting in larger oocytes with enhanced amounts of vitellin (consequences unknown) (Kulkarni et al., 1992). Similarly, in fiddler crabs, fenfluramine (dose of 125 nmol) stimulates (through 5-HT) the production of gonad-stimulating hormone—accelerating testicular maturate (Sarojini et al., 1993
Fenofibrate	Lipid regulator	Efficiently removed by POTW (Ternes, 1998); POTW max. effluent: 0.03 µg/L; not detected in surface waters
Fenofibric acid	Polar, active metabolite of fenofibrate	Loading of over 50 g/day in German POTW (Ternes, 1998); POTW removal efficiency 64% (Ternes, 1998); POTW max. effluent: 1.2 µg/L; max.in surface waters: 0.28 µg/L in Brazilian STWs (Stumpf et al., 1999) with removal efficiencies ranging from 6% to 45%
Fenoprofen	Analgesic/anti-inflammatory	Not detected in POTW effluent or surface waters (Ternes, 1998; Stumpf et al., 1996)
Fenoterol	Bronchodilator	POTW max. effluent: 0.06 µg/L; max.in surface waters: 0.061 µg/L (Ternes, 1998).
Fluoroquinolone carboxylic acids	Antibiotics	As one of many classes of pharmaceuticals, antibiotics in general have been investigated for their occurrence in the environment more than any other class of PPCPs. Their ubiquitous occurrence in the environment is a leading proposed cause of the rise in resistance among pathogenic bacteria. Strongly sorbs to soil (Burhenne et al., 1997; Hartmann et al., 1999). Highly active in hospital wastewaters (Hartmann et al., 1998; Hartmann et al., 1999)
Fluxetine	Antidepressant	No data available
Fluvoxamine	Antidepressant	No data available
Gemfibrozil	Lipid regulator	Loading of over 50 g/day in German POTW (Ternes, 1998); POTW removal efficiency 69% (Ternes, 1998); POTW max. effluent: 1.5 µg/L; max. in surface waters: 0.51 µg/L. Influent concentration of 0.3 µg/L in Brazilian STWs (Stumpf et al., 1999) with removal efficiencies ranging from 16% to 46%

(Continued)

TABLE 8.1 (Continued)
PPCPs Identified in Environmental Samples—or Having Significance with Respect to Aquatic Life

Compound	Use/Origin	Environmental Occurrence
Gentisic acid	Hydroxylated metabolite of acetylsalicylic acid	Efficiently removed by POTW (Ternes, 1998); POTW max. effluent: 0.59 µg/L; max. in surface waters: 1.2 µg/L. Average gentisic acid concentrations in POTW influents of 4.6 µg/L (Ternes et al., 1998) with no detectable amounts in the effluents
o-Hydroxyhippuric acid	Metabolite of acetylsalicylic acid	Efficiently removed by POTW Ternes (1998); not detected in POTW effluent or surface waters (Ternes, 1998); average o-hydroxyhippuric acid concentrations in POTW influent of 6.8 µg/L; no detectable amounts in effluents (Ternes et al., 1998)
Ibuprofen	Analgesic/anti-inflammatory	Loading of over 200 g/day in German POTW (Ternes, 1998); POTW removal efficiency 90% (Ternes, 1998); POTW max. effluent: 3.4 µg/L; max. in surface waters: 0.53 µg/L influent concentration of 0.3 µg/L in Brazilian STWs (Stumpf et al., 1999) with removal efficiencies ranging from 22% to 75%. STW influents up to 3.3 µg/L. POTW removal >95%, surface waters up to 8 ng/L; one of few studies to at metabolites (Buser et al., 1999)
Ifosfamide	Antineoplastic	POTW max. effluent: 2.9 µg/L; not detected in surface waters (Ternes, 1998). Hospital sewage 24 ng/L (Steger-Hartmann et al., 1996). Hospital effluent: max. 1.91 µg/L, median 109 ng/L; POTW influent/effluent max 43 ng/L median 6.5–9.3 ng/L (Kummerer et al., 1997). Found to be totally refractory to removal by POTW (Kummerer et al. 1997)
Indomethacine German	Analgesic/anti-inflammatory	Loading of ~10 g/day in POTW (Ternes, 1998); POTW removal efficiency 75% (Ternes, 1998); POTW max. effluent: 0.60 µg/L: max. in surface waters: 0.20 µg/L. Influent concentration of 0.95 µg/L in Brazilian STWs (Stumpf et al., 1999) with removal efficiencies ranging from 71% to 83%
Iohexol	X-ray contrast	Very low toxicity reported by Steger-Hartmann et al. (1998)
Iopamidol	X-ray contrast	Concentrations as high as 15 µg/L in municipal STW effluents (Ternes and Hirsch, 2000), and median concentration of 0.49 µg/L
Iopromide	X-ray contrast	Resistant to biodegradation and yields refractory, unidentified metabolites (Kalsch, 1999). Reported by Ternes et al. (1998) in rivers. Concentrations as high as 11 µg/L in municipal STW effluents (Ternes et al., 1998)
Iotrolan	X-ray contrast	Very low aquatic toxicity reported by Steger-Hartmann et al. (1998).
Ketoprofen	Analgesic/anti-inflammatory	POTW max. effluent: 0.38 µg/L:max. in surface waters: 0.12 µg/L (Ternes, 1998). Influent concentration of 0.5 µg/L in Brazilian STWs (Stumpf et al., 1999) with removal efficiencies ranging from 48% to 69%
Meclofenamic acid	Analgesic/anti-inflammatory	Not detected in POTW effluent or surface waters (Ternes, 1998; Stumpf et al., 1999)
Metoprolol	Beta-blocker	Loading of nearly 400 g/day in German POTW (Ternes, 1998); POTW removal efficiency 83% (Ternes, 1998); POTW max. effluent: 2.2 µg/L; max. in surface waters: 2.2 µg/L
Musk ambrette	Nitro musk	Synthetic musks first began to be identified in environmental samples almost 30 years ago
Musk xylene	Nitro musk	
Musk ketone	Nitro musk	
Musk moskene	Nitro musk	
Musk tibetene	Nitro musk	

(Continued)

TABLE 8.1 (Continued)
PPCPs Identified in Environmental Samples—or Having Significance with Respect to Aquatic Life

Compound	Use/Origin	Environmental Occurrence
Galaxolide	Polycyclic musk	Widely used in a wide array of fragrances for cosmetics and other personal care products. Introduced to commerce in 1950s
Tonalide	Polycyclic musk	
Celestolide	Polycyclic musk	
Musk xylene influent/effluent	Transformation products of nitro musks	Identified in sewage (Behechi et al., 1998; Gatermann et al., 1998)
Nadolol	Beta-blocker	POTW max. effluent: 0.06 µg/L; not detected in surface waters (Ternes, 1998)
Naproxen	Analgesic/anti-inflammatory	Loading of over 50 g/day in German POTW (Ternes, 1998); POTW removal efficiency 66% (Ternes, 1998); POTW max. effluent: 0.52 µg/L; max. in surface waters: 0.39 µg/L. Influent concentration of 0.6 µg/L in Brazilian STWs (Stumpf et al., 1999) with removal efficiencies ranging from 15% to 78%
Paroxetine	Antidepressant	No data
Phenazone	Analgesic	Loading of ~10 g/day in German POTW (Ternes, 1998); POTW removal efficiency 33% (Ternes, 1998); POTW max. effluent: 0.41 µg/L; max.in surface waters: 0.95 µg/L
Propranolol	Beta-blocker	Loading of over 500 g/day in German POTW (Ternes, 1998); POTW removal efficiency 96% (Ternes, 1998); POTW max. effluent: 029 µg/L; max in surface waters: 059 µg/L
Propyphenazone	Analgesic/anti-inflammatory	General (Denmark) landfill leachates: 0.3–4.0 mg/L directly beneath and declined depending on depth and distance along pole (Holm et al., 1995); prevalent in Berlin waters (Heberer et al., 1998)
Salbutamol	Bronchodilator	POTW max. influent: 0.17 µg/L;max. in surface waters: 0.035 µg/L (Ternes, 1998)
Salicylic acid		Primary hydrolytic metabolite Of acetylsalicylic acid keratolytic, dermatice preservative of food Up to 54 µg/L in POTW effluent but efficiently removed in effluent (Ternes, 1998); POTW max. effluent: 0.14 µg/L; max. in surface Waters: 4.1 µg/L. Average salicylic acid concentrations in POTW influents of 55 µg/L and in effluents of 0.5 µg/L (Ternes et al., 1998)
Sulfonamides	Antibiotics	Grinstead (Denmark) landfill leachates: 0.04–6.47 mg/L directly beneath and declining depending on depth and distance along plume (Holm et al., 1995)
Terbutaline	Bronchodilator	POTW max. effluent: 0.12 µg/L; not detected in surface waters (Ternes, 1998)
3,4,5,6-Tetrabrom-	Antiseptic, fungicide	POTWs in Germany: tetrabromo-o-cresol found in both influent and effluents (<0.1 µg/L (Ternes, et al. 1998)
Timolol	Beta-blocker	POTW max. effluent: 0.07 µg/L; max. in surface waters: 0.01 µg/L (Ternes, 1998)
Triclosan	Antiseptic	0.05–0.15 µg/L in water (Okurura and Nishikawa, 1996). Antibacterial widely used for 30 years in a vast array of consumer products. Its usage as a preservative and disinfectant continues to grow. Tricolosan's use in commercial products spans footwear (in hosiery and insoles of shoes called "odor-eaters"), hospital hand soap, acne creams (e.g., Clearasil), and rather recently as a slow-release product called Microban, which is incorporated in a wide variety of plastic products (from children's toys to kitchen utensils, such as cutting boards)
Verapamil	Cardiac drug	No data available

raw products and intermediates (e.g., ephedrine) via toilets is not uncommon in illegal laboratories. Also, in contrast to pesticides, pharmaceuticals in any stage of clinical testing (not yet approved for dispensing by the FDA) are subject to release into the environment, although their overall concentrations would be very low.

Some drugs are excreted essentially unaltered in their free form (e.g., methotrexate and platinum antineoplastics), often with the help of active cellular "multidrug transporters" for moderately lipophilic drugs. Others are metabolized to various extents, which is partly a function of the individual patient and the circadian timing of the dose (the P450 microsomal oxidase system is a major route of formation of more polar, more easily excreted metabolites). Still others are converted to more soluble forms by the formation of conjugates (with sugars or peptides). The subsequent transformation products—metabolites and conjugates from eukaryotic and prokaryotic metabolism, and form physicochemical alteration—add to the already complex picture of thousands of highly bioactive chemicals. The FDA refers to all metabolites and physicochemical transformation products, for example, those that range from the dissociated parent compound to photolysis products, for a given drug as structurally related substances (SRSs), which can have greater or lesser physiological activity than the parent drug.

As in mammals, the metabolic disposition of lipophilic xenobiotics, such as numerous drugs, in vertebrate aquatic species is largely governed by what is referred to as Phase I and Phase II reactions (James, 1986); less is known about invertebrate metabolism. Phase I makes use of monooxygenases (e.g., cytochrome P450), reductases, and hydrolases (for esters and epoxides) to add reactive functional groups to the molecule. Phase II uses covalent conjugation (glucuronidation) to make the molecule hydrophilic and more excretable. These reactions are catalyzed by glycosyltransferases and sulfotransferases (for hydroxy aromatics and carboxy groups), glutathione S-transferases (for electrophilic functional groups such as halogens, nitro groups, or unsaturated/conjugated sites), acetyltransferases (for primary amines or hydrazine), and amino acyltransferases (for forming peptides from carboxy groups using free amino acids). This metabolic strategy creates metabolites successively more polar than the parent compound, thereby enhancing excretion. Considerable interspecies and intraspecies diversity, however, can be observed in actual metabolic potentials.

The introduction of drugs into the environment is partly a function of the quantity of drugs manufactured, the dosage frequency and amount [the 200 most frequently prescribed drugs, representing about two-thirds of all prescriptions filled in the United States for the most recently documented year, are listed in RxList (1999)], the excretion efficiency of the parent compound and metabolites, propensity of the drug to sorb to solids, and the metabolic transformation capability of subsequent sewage treatment (or landfill) microorganisms. Publicly owned wastewater treatment plants (POTWs) receive influent from both domestic, municipal, and industrial (including pharmaceutical manufacture) sewage systems. The processed liquid effluents from primary and secondary treatments are then discharged to surface waters and the residual solids (biosolids) to landfills/farms; land disposal, including manure from treated animals at CAFOs, creates the potential for introduction into groundwaters or surface waters (via wet weather run-off). Theoretically, PPCPs in sewage biosolids applied to crop lands could be taken up by plants.

Compounds surviving the various phases of metabolism and other degradative or sequestering actions (i.e., display environmental persistence) can then pose an exposure risk for organisms in the environment. Even the less/nontoxic conjugates (glucuronides) can later be converted back to the original bioactive compounds via enzymatic (β-glucuronidases) or chemical hydrolysis (e.g., acetylsalicylic acid can be hydrolyzed to the free salicylic acid).

Some degradation products can even be more bioactive than the parent compound. Therefore, conjugates can essentially act as storage reservoirs from which the free drugs can later be released into the environment. Up to 90% of certain drugs can become conjugated (Ternes, 1998), conjugations varying as a function of chemical class. These pathways of introduction into the environment have been summarized by Velagaleti (1998).

Wastewater Treatment and PPCPs

Treatment facilities, primarily POTWs or wastewater treatment plants (WWTPs), which include privately owned works as well, play a key role in the introduction of pharmaceuticals into the environment. WWTPs were designed to handle the human waste of mainly natural origin, primarily via the acclimated degradative action of microorganisms (the efficiency of metabolism of a given drug can increase with the duration of treatment because of enzyme induction and cellular adaptation) and the coagulation/flocculation of suspended solids; sometimes, tertiary treatment (e.g., chemical/ultraviolet [UV] oxidations) is used. Most anthropogenic chemicals introduced along with this normal waste suffer unknown fates. Two primary mechanisms remove substances from the incoming waste stream: (1) microbial degradation to lower molecular weight products, leading sometimes to complete mineralization—CO_2 and H_2O; and (2) sorption to filterable solids which are later removed with the biosolids.

Although the microbiota of wastewater treatment systems may have been exposed to many PPCPs for a number of years, two factors work against the effective microbial removal of these substances for WWTPs. First, the concentrations of most drugs are probably so low that the lower limits for enzyme affinities may not be met. For example, the daily loadings of PPCPs into WWTPs are largely a function of the serviced human population, the dosages/duration of medications consumed, and the metabolic/excretory half-lives, which are all large variables. As an example, the daily load of a subset of pharmaceuticals to a particular POTW near Frankfurt/Main, Germany, ranged from tens to hundreds of grams, with approximate individual removal efficiencies varying widely from 10% to 100% but trending to around 60% (Ternes, 1998). This particular POTW serviced about a third of a million people at a flow rate of roughly 60,000 m^3/day. Despite the number of studies on treatment efficiencies, a widespread investigation is still lacking for the differences in removal efficiencies for distinct types of WWTPs as well as for individual treatment techniques. The extent to which a particular plant uses primary, secondary, and tertiary technologies will greatly influence removal efficiencies; the technologies employed vary widely among cities. The biodegradative fate of most compounds in WWTPs is governed by nongrowth-limiting (enzyme-saturating) substrate concentrations (copiotrophic metabolism; they thrive in nutrient-rich environments). In contrast, PPCPs are present in WWTPs at concentrations at enzyme-subsaturating levels, which necessitates oligotrophic metabolism (nutrient-poor environments). These micropollutants might be handled by only a small subset of specialist oligotrophic organisms whose occurrence is probably more prevalent in native environments (e.g., lakes) characterized by low-carbon fluxes (e.g. sediments and associated pore waters, where desorption mass transfer is limiting) than in WWTPS.

For clarity of the preceding discussion and for what follows, let's look at *Oligotropic lakes* (meaning few foods): They are young, deep, crystal-clear water, nutrient-poor lakes with little biomass productivity. Only a small quantity of organic matter grows in an oligotrophic lake; the phytoplankton, the zooplankton, the attached algae, the macrophytes (aquatic weeds), the bacteria, and the fish are all present as small populations. "It's like planting corn in sandy soil, not much growth" (Kevern et al., 1999). Lake Superior is an example from the Great Lakes.

The bottom line on degradation of PPCPs in WWTPs is that their degradation may occur more prevalently in the receiving waters/sediments (especially in the case of running waters such as streams and rivers) than in WWTPs.

Many new drugs are introduced to the market each year; some of these drugs are from entirely new classes never seen before by the microbiota of a WTTP. Each of these presents a new challenge to biodegradation. A worst-case scenario may not be unusual—the concentration of a drug leaving a WWTP in the effluent could essentially be the same as that entering. Only the several-fold to multiple order of magnitude dilution when the effluent is mixed into the receiving water, assuming a sufficiently high natural flow, serves to reduce the concentration; obviously, smaller streams have increased potential for having higher concentrations of any PPCP that has been introduced. In general, most pharmaceuticals resist extensive microbial degradation (e.g., mineralization) (Velagaleti,

1998). Although some parent drugs often show poor solubility in water (Velagaleti, 1998), leading to preferential sorption of suspended particles, they can thereby sorb to colloids and therefore be discharged in the aqueous effluent. Metabolites, including breakdown products and conjugates, will partition mainly to the aqueous effluent. Some published data demonstrate that many parent drugs do make their way into the environment.

In a 2004–2009 study, USGS (2013) scientists found and reported that pharmaceutical manufacturing facilities can be a significant source of pharmaceuticals to the environment. Effluents from two wastewater treatment plants that receive the discharge from pharmaceutical manufacturing facilities (PMFs) had 10–1,000 times higher concentrations of pharmaceuticals than effluents from 4 WWTPs across the nation that do not receive PMF discharge. The effluents from these two WWTPs are discharged to streams where the measured pharmaceuticals were traced downstream, and as far as 30 km (18 miles) from one plant's outfall. This is the first study that assesses PMFs as a potential source of pharmaceuticals to the environment. The PMFs investigated are pharmaceutical formulation facilities, where ingredients are combined to form final drug products and products are packaged for distribution. While pharmaceuticals have been measured in many streams and aquifers across the nation, levels are generally lower than one part per billion (1 ppb). As mentioned, however, concerns persist other 23 plants that higher levels may occur in environmental settings where wastewaters are released into the environment.

In this study, 35–38 effluent samples were collected from each of three WWTPs in New York State and one effluent sample was collected from each of 23 strategically selected WWTPs across the nation. The samples were analyzed for seven target pharmaceuticals including opioids and muscle relaxants, some of which have not been previously studied in the environment. Pharmaceutical concentrations in effluents from two of the three WWTPs in New York State, which both receive more than 20% of their discharge form PMFs, were compared to the measurements made the third plant in New York State and at across the nation, which all do not receive the discharge from PMFs. Maximum pharmaceutical concentrations in effluent samples from the 24 WWTPs that do not receive the discharge from PMFs rarely (about 1%) exceeded one part per billion. By contrast, maximum concentrations in effluents from the two WWTPs receiving PMF discharge were as high as 3,888 ppb of metaxalone (a muscle relaxant), 1,700 ppb or oxycodone (an opioid prescribed for pain relief), >400 ppb of methadone (an opioid prescribed for pain relief and drug withdrawal), 160 ppb of butalbital (a barbiturate), and >40 ppb of both phendimetrazine (a stimulant prescribed for obesity) and carisoprodol (a muscle relaxant).

The pharmaceuticals investigated in this study were identified using a forensic approach that identified pharmaceuticals present in samples and subsequently developed methods to quantify these pharmaceuticals at a wide range of concentrations. Additional pharmaceuticals which may be formulated at these sites also were identified as present in the effluents of these two WWTPs. Ongoing studies are documenting the levels at which these additional pharmaceuticals occur in the environment (USGS, 2013).

The efficiency of the removal of pharmaceuticals by WWTPs is largely unknown. Currently, the most extensive study of treatment efficiency (Ternes, 1998) reports the removal from German WWTPs of 14 drugs representing five broad physiologic categories. Removal of the parent compound (keep in mind that possible subsequent metabolites were not accounted for) ranged from 7% (carbamazepine an antiepileptic) to 96% (propranolol, a beta-blocker); most removal efficiencies averaged about 60%. Fenofibrate, acetaminophen, and salicylic acid, o-hydroxyhippuric acid, and gentisic acid (acetylsalicylic acid metabolites) could not be detected in effluent; salicylic acid was found in the influent at concentrations up to 54 µg/L. It is important to understand that absent the stoichiometric accounting of metabolic products, one cannot distinguish between the three major fates of a substance: (1) degradation to lower molecular weight compounds, (2) physical sequestration by solids (and subsequent removal as sludge), and (3) conjugates that can later be hydrolyzed to yield the parent compound (e.g., clofibric and Fenofibric acid conjugates) (Ternes, 1998). Therefore, by simply following the disappearance (removal) of a substance, one cannot conclude that it was structurally altered or destroyed—it may simply reside in another state or form. Identifying metabolic products is difficult not only because of the number of metabolites (sometimes several per parent compound) but also because standard reference materials are difficult to obtain commercially and can be costly.

Despite high removal rates in WWTPs for some drugs, upsets in the homeostasis of a treatment plant can result in higher-than-normal discharges. For example, Ternes (1998) found that wet weather runoff dramatically reduce the removal rates for certain drugs (e.g., several nonsteroidal anti-inflammatory drugs [NSAIDS] and lip regulators) in a facility located close to Frankfurt/Main. During the increased period of influent flow, the removal rate dropped to below 5% from over 60% previously; several days were required for the removal rates to recover. Clearly, even for drugs efficiently removed, the operational state of the WWTP can have a dramatic effect on the removal efficiencies. Other transients that could affect removal include transitions between seasons and sporadic plug-flow influx of toxicants from various sources. Overflows from WWTP failure or overcapacity events (e.g., floods, excessive water use) lead to direct, untreated introduction of sewage into the environment. In efforts to improve tributary conditions (by increasing stream flow); some cities, for example, Portland, Oregon, have considered increasing the percentage of annual overflow events. The highest concentration in a WWTP effluent reported by Ternes (1998) was for Bezafibrate (4.6 µg/L); the highest concentration in surface water also was for Bezafibrate (3.1 µg/L ppb).

Landfills

In most instances, we have been trained to get rid of expired or unwanted medications by "disposing of them in a manner that children cannot get access to them." In practical terms, this usually involves flushing them down the toilet or putting them into the household trash. In the former instance they end up in the sewer, whereas in the latter instance, they end up in the landfill (Jjemba, 2008). PPCPs can also be introduced to landfills via domestic industrial routes and indirectly via sewage sludge (biosolids). Holm et al. (1995) first reported leachates carrying pharmaceuticals from a landfill. Large amounts of numerous sulfonamides (antibiotics) and barbiturates from domestic waste and from a pharmaceutical manufacturer were disposed of at a Danish landfill over a 45-year period. High concentrations (ppm) of many of these drugs were found in leachates close to the landfill; these compounds even accounted for 5% of the total nonvolatile organic carbon found in the leachate. It was also found that the concentration dropped off dramatically tens of meters down gradient, presumably a result of microbial attenuation.

Drinking Water

In the 1990–1995 timeframe, few pharmaceuticals were identified in domestic drinking water, probably because of the dearth of monitoring efforts and because the required detection limits are too low for current routine analytical technology. In Germany, however, clofibric acid concentrations up to 165 ng/L (Stan et al., 1994) and 270 ng/L (Heberer et al., 1998) have been measured in tap water; the presumed source was from recharged groundwaters that had been contaminated by sewage. Stumpf et al. (1996) and Ternes et al. (1999) found several pharmaceutics in German drinking water in the lower nanograms-per-liter range, with a maximum of 70 ng/L for clofibric acid. Additionally, these investigators found that diclofenac, Bezafibrate, Phenazone, and carbamazepine were sometimes present. In the memory of the samples analyzed, however, no drugs were observed. The investigations performed to date therefore indicate that contamination of drinking water does not appear to be a general problem. Depending on the water source for drinking/water production, however, certain facilities can experience contamination, especially if the source is polluted groundwater and if polishing technology does not remove the PPCP (Heberer et al., 1998; Stumpf et al., 1999). A major unaddressed issue regarding human health is the long-term effects of ingesting via potable waters very low, subtherapeutic doses of numerous pharmaceuticals multiple times a day for many decades. This concern especially relates to infants, fetuses, and people suffering from certain enzyme deficiencies (which can even be food-induced, e.g., microsomal oxidase inhibition by grapefruit juice).

Domestic Animals

Whereas the concentration of many drugs is greatly attenuated through sewage treatment plants, larger quantities of many pharmaceuticals are used in various animal husbandry operations, especially CAFOs. With aquaculture in particular, which uses many anti-infectives and anesthetics, the chance for introduction into the immediate environment is greatly enhanced, and the possibility of direct human consumption of therapeutic quantities is correspondingly heightened. Even in the United States extremely large populations of pet dogs and cars are recipients of numerous drugs (e.g., tranquillizers and antidepressants)—some prescribed by veterinarians and others intended for their owners' use as pet owners sometimes administer medications to their pets to test off-label uses for themselves. PPCPs (both veterinary drugs and OTC products) used with terrestrial domestic animals can be dispersed into the environment through the same routes as those PPCPs used for humans, with the added major route of run-off/leaching of on-ground fecal material.

DRUG CLASSES AND ENVIRONMENTAL OCCURRENCES

Hormones/Mimics

An excellent overview of hormone systems is given by the Endocrine Disruptor Screening and Testing Advisory Committee (EDSTAC) (1998). Steroids were the first physiologic compounds to be reported in sewage effluent (Aherne and Briggs, 1989; Shore et al., 1993; Tabak and Bunch, 1970; Tabak et al., 1981) and as such were the first pharmaceuticals to capture the attention of environmental scientists. Estrogenic drugs, primarily synthetic xenoestrogens, are used extensively in estrogen-replacement therapy and in oral contraceptives, in veterinary medicine for growth enhancement, and in athletic performance enhancement. A special issue of *The Science of the Total Environment* (Ternes and Wilkens, 1998) is devoted to drugs (especially hormones) as pollutants in the aquatic environment.

Although the synthetic oral contraceptive (17α-ethynylestradiol) occurs generally at low concentration (<7 ng/L) in POTW effluent, it is still suspected, in combination with the steroidal estrogens 17β-estradiol and estrone (Desbrow et al., 1998), of causing vitellogenin production (feminization) in male fish. Feminization is a phenomenon first observed for fish in sewage treatment lagoons in the mid-1980s (Routledge et al., 1998). An overview of pharmaceutical hormones in the environment is presented by Arcand-Hoy et al. (1998). The estrogenic activity of various waters (from sewage to drinking water) has been shown to vary dramatically, spanning six orders of magnitude. Some other widely used synthetic hormone modulators include Proscar/Propecia (finasteride: an androgen hormone inhibitor) and various thyroxine analogs (thyroid hormones); noting is known of the environmental fates of these compounds. In general, the lipophilicity of these hormones is sufficiently great that a least a large portion are removed via sorptive processes in sewage treatment (Johnson et al., 1999; Routledge et al., 1999) and therefore partition to the sludge; but even the low concentrations that remain in the effluents may be capable of exerting physiologic effects in aquatic biota.

In addition to these synthetic steroids and xenoestrogens is a suite of naturally occurring estrogen hormones, for example, phytoestrogens such as the complex series of leguminous isoflavonoids, including genistein, daidzein, and glycitein in soy. Further complicating the picture are a host of newly suspected endocrine-disrupting compounds (EDCs), more recently referred to as hormonally active agents (HAAs) by the NRC (1999), which have gained attention in the last few years, primarily as a result of the 1996 publication *Our Stolen Future* by Colborn et al. (1997). These inadvertent EDCs include such commonly recognized industrial pollutants and products as halogenated dioxins/furans, PCBs, organohalogen pesticides, phthalates, and bisphenol A.

The issue of screening many of the major commercial chemicals (over 87,000 total) for endocrine disruption potential has been formalized with the creation of the EDSTAC, which had been charged by the U.S. EPA with the task of implementing a screening and testing program by August 1999 (EDSTAC, 1998). The Chemical Manufacturers Association (CMA) also has launched an intensive health effects investigation for over 3,000 high-volume chemicals (called the Health and Environmental Research Initiative) (CMA, 1999). It is significant, however, and should be noted, that pharmaceuticals are not specifically targeted by the EDSTAC (or the CMA) in its tiered screening program that focuses on pesticides, commodity chemicals, naturally occurring nonsteroidal estrogens (Phytoestrogens and mycotoxins), food additives, cosmetics, nutritional supplements, and representative mixtures (for possible synergistic effects). Even though the strategy gives top priority to "chemicals with widespread exposure at the national level" (CMA, 1999), PPCPs are not specially targeted. It is also significant that the screening strategy will initially focus on only the three primary hormone systems—estrogen, androgen, and thyroid—hormone systems of relatively unknown importance to invertebrates (EDSTAC, 1998).

A controversial hypothesis regarding multiple toxicants (sharing a common mode of action), when each is present at a low level, is that of synergism. Evidence of synergism among estrogenic mimics (where the effect can be elicited at orders-of-magnitude lower concentration than predicted by additive action) was reported by Arnold et al. (1996) and others rebutted this hypothesis. They did not find any evidence of synergism in mixtures of mild estrogenic pollutants.

Another subclass of hormonelike substances includes those that are being purposefully designed to mimic the activity of therapeutically significant hormones. A long-sought objective has been to obviate the need for hormone-replacement therapy (e.g., insulin) by designing small synthetic (non-peptidyl) molecules that mimic the hormone's effect yet can be ingested orally, taken up the gut, and remain stable for a sufficiently long period of time in the blood. The first report of a "designer" hormone mimic (Tian et al., 1998; Baringa, 1998), a polybenzimidazole that activates the receptor for a cytokine that regulates white blood cell production, perhaps portends the advent of many synthetic hormone mimics in therapeutic medicine. If the finding can be generalized, it could mean that the possible routes of hormone disruption by simple molecules could extend beyond that of the estrogen/androgen system.

With the exception of estrogenic mimics, the possibility of disrupting the activity of proteinaceous hormones by lower molecular weight anthropogenic chemicals has been held in low regard. This view has been based on the fact that a relatively large, complex proteinaceous molecule (the hormone) neatly "fits" within the complex three-dimensional domain of its target receptor, whereas in contrast a much smaller nonproteinaceous molecule would have little to offer in terms of recognition specificity. It has been believed that the complexity of larger proteins such as insulin was required to enable recognition by the corresponding receptors; smaller compounds simply did not convey enough three-dimensional information to have high-binding constraints for one or multiple receptors.

The report by Tian et al. (1998) demonstrates for the first time that a relatively small non-peptide molecule can bind to a receptor normally dedicated to a proteinaceous hormone. While this has high therapeutic significance (this research might catalyze concerted attempts to develop the first protein-mimicking and therefore perhaps hormone-mimicking low molecular weight drugs), it also alludes to the possibility that existing anthropogenic compounds might have a greater chance of interacting with hormone receptors than was previously believed. Although the synthetic substance was three to six orders of magnitude less potent, its ability to bind to the receptor was undisputed (in the mouse *in vitro* and, more importantly, *in vivo*).

Antibiotics

A large body of literature exists on antibiotics in the environment. Veterinary and animal husbandry, especially aquaculture, usage plays a major role in their introduction into the environment. In one study of hospital effluent, fluoroquinolones was the chemical class contributing the major portion to overall DNA toxicity (Hartmann et al., 1998); ciprofloxacin, for example was identified at 3–87 µg/L. Hirsch et al. (1999) analyzed German WWTP effluents and groundwaters/surface waters for 18 antibiotics representing macrolides, sulfonamides, penicillins, and tetracyclines. Although the penicillins (susceptible to hydrolysis) and the tetracyclines (can precipitate with calcium and similar cations) were not found, the others were detected in the microgram per liter range. Indeed, the rampant, widespread (and sometimes indiscriminate) use of antibiotics, coupled with their subsequent release into the environment, is the leading proposed cause of accelerated/spreading resistance among bacterial pathogens, which is exacerbated by the fact that resistance is maintained even in the absence of continued selective pressure (an irreversible occurrence). Sufficiently high concentrations could also have acute effects on bacteria. Such exposures could easily lead to altered microbial community structures in nature and thereby affect the higher food chain. Their use in aquaculture results in eventual human consumption.

In 1999, a number of stream surveys documented the significant prevalence of native bacteria that display resistance to a wide array of antibiotics including vancomycin (Ash et al., 1999). Isolates from wild geese near Chicago, Illinois, are reported to be resistant to ampicillin, tetracycline, penicillin, and erythromycin (Eichorst et al., 1999). All these reports could simply indicate that the natural occurrence of antibiotic resistance in native bacterial populations is much higher than expected or that these bacteria are being selected for by the uncontrolled release of antibiotics into the environment. If the latter is true then excluding the significance of antibiotics themselves in the environment, their occurrence can be reviewed as marking or indicating the possible presence of other PPCPs.

Blood Lipid Regulators

Clofibric acid was the first prescription drug (actually an SRS) reported in a sewage effluent (Garrison et al., 1976; Stan and Heberer, 1997), and it continues to be one of the most frequently reported PPCPs in monitoring studies. Clofibric acid (2-[4]-chlorophenoxy-2-methyl propanoic acid), the active metabolite from a series of widely used blood lipid regulators, and which also happens to be structurally related to the phenylalkanoic acid herbicide mecoprop (the methylphenoxy structural analog), has captured much attention from investigators in Europe. Stan et al. (1994) first reported clofibric acid in Berlin tap water at concentrations between 10 and 165 ng/L. Heberer and Stan (1997) found clofibric acid at level up to 4 µg/L in groundwater under a sewage treatment farm; they also found clofibric acid concentrations up to 270 ng/L in drinking water samples. They concluded that it is not removed by sewage/water treatment processes.

Buser et al. (1998a) report finding clofibric acid in various Swiss waters ranging from rural to urban lakes. Concentrations ranged from 1 to 9 ng/L (ppt), whereas the parallel concentrations for mecoprop where higher at 8–45 ng/L; little of either compound was found in a relatively remote mountain lake, indicating no atmospheric deposition. Because this drug is not manufactured in Switzerland, its route of introduction into the environment had to be through medical use and subsequent excretion/disposal. Although these concentrations are very low, they are significant in that they are similar to the concentrations found for any of the conventional ubiquitous and persistent pollutants, sometimes referred to as persistent organic pollutants (POPs) or persistent bioaccumulative toxicants (PBTs) such as lindane. In one of the lakes studied, Buser et al. (1998a) calculated steady-state amounts of clofibric acid to be roughly 19 kg (with export and import amounts balancing each other). Perhaps more significantly, they also found amounts of clofibric acid up to 7.8 ng/L in the North Sea; the parallel concentrations of mecoprop in the same North Sea samples were lower, up to only 2.7 ng/L, indicating that mecoprop was less persistent than clofibric acid.

Stumpf et al. (1996) and Ternes (1998) reported Bezafibrate, Gemfibrozil, and clofibric/fenofibric acids in river waters at the nanograms per liter level. Stumpf et al. (1999) reported that the removal efficiencies from Brazilian WWTPs for clofibric/Fenofibric acids, Bezafibrate, and gemfibrozil ranged from only 6% to 50%, verifying extremely limited degradation for these compounds. This chemical class is ubiquitous because the daily human dosages are generally high (grams per day). Buser et al. (1998a) concluded that the concentrations seen in urban Swiss and German rivers, coupled with essentially the same concentrations in the North Sea, lead to an annual input of 50–100 tons of clofibric acid into the North Sea. The concentration of clofibric acid in the environment is more a function of dilution than of degradation. Clofibric acid is the most widely and routinely reported drug found in open waters. It would be expected that its occurrence in other parts of the world would parallel these studies.

Nonopiod Analgestics/Nonsteroidal Anti-Inflammatory Drugs

Stumpf et al. (1996) were the first to identify diclofenac, ibuprofen, acetylsalicylic acid, and ketoprofen in sewage and river water. Ternes (1998) reported levels of diclofenac, indometacine, ibuprofen, naproxen, ketoprofen, and Phenazone in POTW effluent exceeding 1 µg/L; all these except Ketoprofen were also found in surface waters at concentrations several-fold lower. In another study, Ternes et al. (1998) reported average concentrations of acetylsalicylic acid generally <1 µg/L in most POTW effluents as well as <0.14 µg/L in rivers. They also reported salicylic acid concentrations of 54 µg/L in POTW influents, with two other acetylsalicylic metabolites, gentisic acid (4.6 µ/L) and o-hydroxyhippuric acid (6.8 µg/L). While low levels (0.5 µg/L) of salicylic acid appeared in the effluents, no detectable amounts of the metabolites could be found. Ternes et al. (1998) also found naproxen (a nonsteroidal anti-inflammatory drug) in all POTW effluents examined and in river waters (~0.05 to 0.4 µg/L); two veterinary NSAIDs, meclofenamic and tofenamic acids, were not detectable in any river sample. In their screening of waters in Berlin, Heberer et al. (1998) found that the most prevalent drugs, other than clofibric acid, were the NSAIDs diclofenac, ibuprofen, and propyphenazone. In groundwater from a drinking water plant, they found diclofenac, ibuprofen, and N-methylphenacetin (from phenacetin) (Heberer et al., 1998). In the influent to Swiss WWTPs, Buser et al. (1998b) found diclofenac at concentrations of 0.5–1.8 µg/L, whereas the concentrations in the respective effluents were only moderately reduced (at most 50%). In the receiving water (Swiss lakes/rivers), they found 11–310 ng/L but only 1–12 ng/L in exiting waters. They concluded that photolysis was the major cause of the diminished concentrations of diclofenac in surface waters (Buser et al., 1998a). Buser et al. (1999) showed that ibuprofen, while present in influents at 1–3.3 µg/L, was easily degraded to yield low effluent concentrations (nanograms/liter) in contrast to the NSAIDs, which were more refractory.

Beta-Blockers/β_2-Sympathomimetics

Hirsch et al. (1996) and Ternes (1998) identified the beta-blockers metoprolol and propranolol, with lesser amounts of Betaxolol, bisoprolol, and nadolol, in POTW effluent. Only metoprolol and propranolol were found in surface waters at concentrations just about the limit of detection. The β_2-sympatnomimetics (bronchodilators) terbutalin and salbutarnol (albuterol in the United States), but rarely clenbuterol and fenoterol, were detected in POTW effluent and only at a low concentrations, <0.2 µg/L. They were rarely seen in surface waters. It may be significant to note that medications delivered by inhalers could result in portions of the dose being deposited externally because of improper dosing technique.

Fenfluramine (N-ethyl-α-methyl-3-[trifluoromethyl] benzene ethanamine hydrochloride), known as Pondimin in addition to other brand names, is a sympathomimetic amine, which was used as popular diet (anorectic) drug and was removed from the U.S. market in 1998 by the FDA because of heart valve damage. Although no one has looked for fenfluramine in sewage, it is known to enhance the release of serotonin (3-(2-aminoetyl)indol-5-ol or 5-hydroxytryptamine creatinine sulfate [5-HT]); in the crayfish, 5-HT in turn triggers the release of ovary-stimulating hormone, resulting in larger oocytes with enhanced amounts of vitellin (consequences unknown) (Kulkarni et al., 1992). Similarly, in fiddler crabs, fenfluramine at a dose of 125 nmol stimulates (thought 5-HT) the production of gonad-stimulating hormone, which accelerates testicular maturation (Sarojini et al., 1993).

Antidepressants/Obsessive-Compulsive Regulators

Selective serotonin reuptake inhibitors (SSRIs) are a major class of widely prescribed antidepressants that includes Prozac, Zoloft, Luvox, and Paxil. These drugs enjoy widespread and heavy use. One of the few series of studies reported in the literature that addresses the effects of drugs on nontarget organisms (albeit not the intent of the studies) was performed in a quest for more effective spawning inducers for economically important bivalves (Fong, 1998). Fong's studies and those of other physiologists studying the function of serotonin in a wide array of aquatic creatures could prove highly significant function of serotonin in a wide array of aquatic creatures could prove highly significant in any discussion of the importance of low levels of pharmaceuticals in the environment. Fong's work is perhaps the most significant to date for showing the potential for dramatic physiologic effects on nontarget species (in this case invertebrates) by low (ppb) concentrations of pharmaceuticals.

Serotonin is a biogenic amine common in both vertebrate and invertebrate nervous systems. SSRIs increase serotonin neurotransmission by inhibiting its reuptake at the synapse by inhibiting the transporter enzymes. In addition to playing a key role in mammalian neurotransmission, serotonin is involved in a wide array of physiologic regulatory roles in mollusks, among most other creatures. For bivalves, reproductive functions including spawning, oocyte maturation, and parturition are regulated by serotonin (Fong, 1998). Serotonin controls a wide spectrum of additional behaviors and reflexes in mollusks, including heartbeat rhythm, feeding/biting, swimming motor patterns, beating cilia, and induction of larval metamorphosis (Couper and Leise, 1996). It also stimulates the release of various neurohormones in crustaceans (hyperglycemic hormone, red pigment-dispersing hormone, neurodepressing hormone, and molt-inhibiting hormone) and ovarian maturation (Sarojini et al., 1995).

It has long been known that serotonin at concentrations of 10^{-4} to 10^{-3} M (~0.18 to 1.8 g/L) induces spawning in bivalves. Some commercial farmers make use of this by adding serotonin to induce spawning. Fong (1998) found that Prozac (fluoxetine) and Luvox (fluvoxamine) are the most potent inducers ever found, eliciting spawning behavior in zebra mussels at aqueous concentrations many orders of magnitude lower than serotonin. Fluoxetine elicited significant spawning in male mussels at concentrations of 10^{-7} M (~150 µg/L); females were an order of magnitude less sensitive at 10^{-6} M. Fluvoxamine was the most potent of the SSRIs, eliciting significant spawning in male mussels, at 10^{-9} M (~0.318 µg/L); females were two orders of magnitude less sensitive, at 10^{-7} M. In males, spawning was complete in the first hour, while females were slower (within 2 hours). Paxil (paroxetine) was the least potent of these three SSRIs, eliciting male spawning, but to a less degree, at 10^{-6} M, and having no inducing effect on females at any concentration. It should be noted that Fong states that the evidence is not clear whether these compounds are indeed acting as SSRIs, or via some other mechanism. It is also unknown how these compounds are taken up by mollusks (Fong, 1998).

In another study, Fong et al. (1998) showed that fluvoxamine induces significant parturition in fingernail clams at 1 nM; 1 nM fluvoxamine also potentiated the effect of 10 µM 5-HT by almost 5-fold. Paroxetine was less potent, requiring a concentration of 10 µM to bring about significant parturition. In contrast, even at concentrations of 100 µM, fluoxetine displayed no effect, although it was capable at 5 µM of potentiating 5-HT at concentrations that were otherwise subthreshold. It is interesting that the order of potency for inducing parturition in clams differs from the order for induction of spawning in mussels (above). This points out the complexity of considering any approach involving extrapolations from one species to another or from one drug to another within a given class.

In crustaceans, Kulkarni et al. (1992) found that fluoxetine significantly potentiates the effect of 5-HT in crayfish, enhancing the release of ovary-stimulating hormone, which results in larger oocytes with enhanced amounts of vitellin; any ecologic consequences of higher vitellin protein levels are unknown. Similarly, in fiddler crabs, fluoxetine at a dose of 125 nmol stimulates (through 5-HT) the production of gonad-stimulating hormone, which accelerates testicular maturation (Sarojini et al., 1993).

It is clear that aquatic life can be exquisitely sensitive to at least some of this class of compounds. Although some SSRIs are extremely potent, others have almost no effect, which possibly makes that approach of assessing ecologic risk on a class-by-class basis infeasible.

Concentration of SSRIs plays a complicated role with respect to effects. For example, Couper and Leise (1996) found that while injected fluoxetine induced significant metamorphosis in a gastropod, 10^{-4}M induced less metamorphosis than 10^{-6}M. Simple extrapolations of effects from higher concentrations do not necessarily have any relevance to effects at lower concentrations.

The potential for SSRIs to elicit subtle effects on aquatic life is further extended by serotonin reuptake mechanisms that also are a factor in snails and squids (Fong, 1998), particularly in the regulation of aggression (Huber et al., 1997). Yet another example of a subtle effect that would go unnoticed is the fighting behavior of lobsters, in which serotonin causes behavior reversal by stimulating subordinates to engage in fighting against dominants by reducing their propensity to retreat (Huber et al., 1997).

Antiepileptics

Antiepileptics (used to treat epileptic seizures) are ubiquitous and prevalent due to poor WWTP removal. Carbamazepine was the drug detected most frequently and in highest concentrations during a study by Ternes (1998). This drug was detected in all POTWs and receiving waters, with a maximum concentration of 6.3 μg/L. Ternes hypothesized that the ubiquitous occurrence resulted from the very low removal efficiency from POTWs, which was calculated to be only 7%, Sacher et al. (1998) found carbamazepine levels in the river Rhine in Germany up to 0.90 μg/L and always above 0.1 μg/L.

Antineoplastics

Antineoplastics are highly [geno] toxic compounds, primarily from hospitals, with poor removal WWTPs. These agents, antitumor agents primarily used only within hospitals for chemotherapy, are found sporadically and in a range of concentrations, probably because only small amounts are introduced to WWTPs via domestic sewage because of their long-lived physiologic retention. These compounds act as nonspecific alkylating agents (i.e., specific receptors are not involved) and therefore have the potential to act as either acute or log-felt stressors (mutagens/carcinogens/teratogens/embryotoxins) in any organism. The fact that two oxazaphosphorines, Ifosfamide and cyclophosphamide, were found in certain effluents in the low microgram-per-liter range indicates that these highly toxic compounds, which are probably refractory to microbial degradation at POTWs (Steger-Hartmann et al., 1997), can find their way into the environment. Indeed, Steger-Hartmann et al. (1997) found levels of cyclophosphamide in sewage influent from servicing hospitals ranging from undetectable to 143 ng/L; the levels in the effluent reached 17 ng/L.

Additional evidence pointing to the refractory nature of ifosfamide is presented by Kummerer et al. (1997), who found that concentrations of Ifosfamide in hospital effluent matched the predicted values of up to 1.91 μg/L; also, the concentrations in the influent and effluent of POTWs that serviced chemotherapy hospitals were essentially unchanged (influent/effluent maximum, 43 ng/L; median, 6.5–9.3 ng/L). Kummerer et al. (1997) found Ifosfamide to be totally refractory to removal by POTWs and to totally resist alteration during a 2-month bench-scale POTW simulations.

Another class of antineoplastics, the palatinate, includes carboplatin and cisplatin. Although the stability of these compounds in sewage systems is unknown, Kummerer et al. (1999) calculated that if they were present in hospital sewage effluents as the intact parent compound, they could be present at daily average concentrations of up to 600 ng/L (based on total platinum). Although the majority of the dose for these compounds is excreted in the urine in the first day, a large amount (~30%) resides in the body and is slowly excreted over a period of years and therefore could be excreted to residential sewage systems. Falter and Wilken (1999) showed that while these compounds are difficult to determine analytically, their potential to remain in the aqueous phase after sewage treatment is high.

In the most detailed overview to date on the genotoxicity of wastewaters, elaboration is made that while the genotoxic potency of industrial wastewaters is often the highest, the overall loading of genotoxic compounds to surface waters is far greater, up to several orders of magnitude, from municipal treatment plants. They present a striking correlation between the occurrence of direct-acting mutagens in surface waters and the human population served by the discharging WWTPs. This correlation points to the activities/metabolism of humans, not industrial activities, as the origin for these mutagens. A number of possible sources for the mutagens are discussed, an obvious one of which is antineoplastic drugs.

These data point to antineoplastics as a class of drugs of potential concern for environmental effects, not just for their acute toxicity but perhaps more for their ability to effect subtle genetic changes, the cumulative impact of which over time can lead to more profound ecologic change. Hospitals are the major source of genotoxic drugs. POTWs that service hospitals, especially multiple hospitals, are likely candidates for releasing these chemicals into surface waters.

Impotence Drugs

This class of drugs displays widespread use, new modes of action, and unknown effects on nontarget organisms. It is interesting to note that even though a number of drugs from various chemical classes have been used over the years for treating importance, the emergence of Viagra (sildenafil citrate) has focused tremendous attention on this market. The significance of this therapeutic class of drugs, with new ones awaiting FDA approval, is that they all tend to have distinct modes of action, most of which differ from those of traditional drugs. While potential effects on wildlife are totally unknown, the fact that Viagra, for example, works by inhibiting a phosphodiesterase responsible for regulating the concentration of cyclic guanosine monophosphate, which indirectly relaxes muscles and increases blood flow (Wilson, 1998) gives causes for concern regarding the disruption of this common phosphodiesterase in unintended target species. Impotence drugs will prove to have very high usage rates, especially since they are one of the most common drugs available without prescription over the Internet, yielding high potential for environmental exposure and possibly nontarget effects.

Tranquilizers

Little is known about the possible occurrence of tranquilizers; however, Ternes (1998) reported diazepam in almost half of the POTWs but only in low concentrations of <0.04 µg/L; it could not be deterred in surface waters. Genicola (1999) reported diazepam in the groundwater from a monitoring well at a Superfund site near Atlantic City, New Jersey. Concentrations were ~10 to 40 µg/L and probably originated in a landfill in which pharmaceutical manufacturers disposed of chemicals.

Retinoids

Retinoids, low molecular weight lipophilic derivatives of vitamin A, can have profound effects upon the development of various embryonic systems (Maden, 1996) especially amphibians in which retinoic acid receptors have been hypothesized to play a role in frog deformities. Although naturally occurring, retinoids have been used for a number of years for a wide array of medical conditions including skin disorders (e.g., Accutane [isotretinoin] for acne), antiaging treatments (e.g., Retin-A [tretinoin] for skin wrinkles), and cancer (e.g., Vesanoid [tretinoin] for leukemia). Isotretinoin (13-cisretinoic acid) is related to both retinoic acid and retinol (vitamin A). Tretinoin is among the top 200 prescribed drugs in the United States. Methoprene, an insecticidal synthetic retinoic acid mimic, is photolabile and yields numerous photo-products, some of which also elicit strong retinoic acid activity. Although retinoic acids would also be expected to be photolabile (and therefore not persistent), their products may also still possess receptor activity.

Diagnostic Contrast Media

X-ray images of soft tissues are routinely captured by the use of contrast media. Some of the more widely used members of contrast media are highly substituted and sterically hindered amidated, iodinated aromatics such as diatrizoate and Iopromide (Kalsch, 1999) which are used worldwide at annual rates exceeding 3,000 tons. Kalsch (1999) found these compounds to be quite resistant to transformation in WWTPs and in river waters. When transformations were effected, they merely terminated with unidentified resistant metabolites. Ternes et al. (2000) recently reported significant amounts of iopromide in rivers.

In municipal WWTP effluents, Ternes et al. (1999) found concentrations as high as 15 µg/L (Iopamidol) and 11 µg/L (Iopromide). In a WWTP close to Frankfurt/Main, they found two other contrast agents, diatrizoate and iomeprol, at concentrations up to 8.7 µg/L, as well as iothalmic acid and ioxithalamic acid in the nanogram-per-liter range. In rivers and streams, five iodinated diagnostics were repeatedly detected, with median values up to 0.49 µg/L for Iopamidol and up to 0.23 µg/L for diatrizoate. Isolated maximum values above 100 µg/L for diatrizoate indicated that relatively high local concentrations can occur, especially in small streams containing a high percentage of WWTP discharges. Maximum groundwater concentrations for iodinated contrast agents ranged up to 2.4 µg/L and may well represent a worst case with respect to the occurrence of pharmaceuticals in native waters. In Germany alone, individual contrast agents can experience annual usage rates of 100 tons. Such high usage, coupled with inefficient human metabolism (95% unmetabolized) and ineffective elimination of iodinated contrast agents by WWTPs, can lead to very high environmental accumulations and persistence. Despite these negative attributes, contrast agents have no bioaccumulation potential and low toxicity (Steger-Hartmann et al., 1998); Steger-Hartmann et al. (1998) also found no acute toxicity for bacteria (*Vibrio fisheri*), algae (*Scenedesmus subspicatus*), crustaceans (*Daphnia*), and fish (*Danio rerio, Lueciscus idus meanotus*) exposed to no more than 10 g/L of Iohexol, Iotrolan, diatrizoate, or iopromide.

Personal Care Products in the Environment

For the purposes of this text, personal care products are defined as chemicals marketed for direct use by the consumer (excluding OTC medication with documented physiologic effects) and having intended end uses primarily on the human body (products not intended for ingestion, with the exception of food supplements). In general, these chemicals are directed at altering odor, appearance, touch, or taste while not displaying significant biochemical activity. Most of these chemicals are used as active ingredients or preservatives in cosmetics, toiletries, or fragrances. They are not used for the treatment of disease, but some may be intended to prevent diseases (e.g., sunscreen agents). In contrast to drugs, almost no attention has been given to the environmental fate or effects of personal care products—the focus has traditionally been on the effects of intended use on human health. Many of these substances are used in very large quantities frequently more than recommended.

Personal care products differ from pharmaceuticals in that large amounts can be directly introduced to the environment. For example, these products can be released directly into—recreational waters or volatilized into the air (e.g., musks). Because of this release they can bypass possible degradation in POTWs. Also, in contrast to pharmaceuticals, less is known about the effects of this broad and diverse class of chemicals on nontarget organisms, especially aquatic organisms. Data are also limited on the unexpected effects on humans. For example, common sunscreen ingredients, 2-phenylbenzimidazole-5-sulfonic acid and 2-phenybenzimidazole, can effect DNA breakage when exposed to UV-B (USEPA, 2010).

The quantities of personal care products commercially can be very large. For example, in Germany alone the combined annual output for eight separate categories has been estimated (Statistisches Bundesamt, 1993) at 559,000 tons for 1993. A few examples are given below of common personal care products that are ubiquitous pollutants and that may possess substantial bioactivity.

Fragrances (Musks)

Fragrances (musks) are ubiquitous, persistent, bioaccumulative pollutants that are sometimes highly toxic; amino musk transformation products are toxicologically significant. Synthetic musks comprise a series of structurally similar chemicals (which emulate the odor but not the structure of the expensive, natural product from the Asian musk deer) used in a broad spectrum of fragranced consumer items, both as fragrance and as fixative. Included are the older, synthetic nitro musks (e.g., ambrette, musk ketone, musk xylene, and the lesser known musks moskene and tibetene) and a variety of new, synthetic polycyclic musks that are best known by their individual trade names or acronyms. The polycyclic musks (substituted indanes and tetralins are the major musks used today, accounting for almost two-thirds of worldwide production) and especially the inexpensive nitro musks (nitrated aromatics accounting for about one-third of worldwide production) are used in nearly every commercial fragrance formulation (cosmetics, detergents, toiletries) and most other personal care products with fragrance; they are also used as food additives and in cigarettes and fish baits (Gatermann et al., 1998).

The nitro musks are under scrutiny in a number of countries because of their persistence and possible adverse environmental impacts and therefore are beginning to be phased out in some countries. Musk xylol has proved carcinogenic in a rodent bioassay and is significantly absorbed through human skin; from exposure to combined sources, a person could absorb 240 µg/day (Bronaugh et al., 1998). The human lipid concentration of various musks parallels that of other bioaccumulative pollutants such as PCBs (Muller et al., 1996). Worldwide production of synthetic musks in 1988 was 7,000 tons (Gatermann et al., 1998); worldwide production for nitro musks in 1993 was 1,000 tons, two-thirds of which were musk xylene (Kafferlein et al., 1998).

Synthetic musks first began to be identified in environmental samples almost 20 years ago (Yamagishi et al., 1981; Yamagishi et al., 1983). By 1981, Yamagishi et al. (1981) had identified musk xylene and musk ketone in goldfish (*Carassius auratus langsdorfii*) in Japanese rivers and not much later (Yamagishi et al., 1983) in river water, sewage, marine mussels (*Mytilus edulis*), and oysters (*Crassosterea gigas*). Yamagishi's studies comprised the first comprehensive monitoring efforts, identifying musk xylene and musk ketone in freshwater fish, marine shellfish, river water, and WWTP waters. Musk xylene was found in all samples, and musk ketone was found in 80% of the 74 samples analyzed. Concentrations in WWTP effluents ranged from 25 to 36 ng/L for musk xylene and from 140 to 410 ng/L for musk ketone. Concentrations of musk xylene in fish muscle were in the tens of parts per billion, whereas those for musk ketone were <10 µg/kg, with highest values occurring in fish downstream of STWs. In contrast, for shellfish, the concentrations were lower, between 1 and 5.3 µg/kg, presumably because of their lower lipid contents. In river water, musk xylene occurred in all samples, whether upstream or downstream of WWTPs and ranged between 1 and 23 ng/L; those of musk ketone were generally in the same range, but in distinct contrast, they were not detectable in upstream samples.

Geyer et al. (1994) have published an excellent review on residues of nitro musk fragrances in fish and mussels as well as in breast milk and human lipids and the current ecotoxicologic and toxicologic knowledge for these personal care products. Residues of musk xylene and musk ketone found in the fillet of freshwater fish (e.g., pike, eel, brass, Zander, rainbow trout) from rivers of North Germany were between 10 and 350 µg/kg lipid and 10 and 380 µg/kg lipid for musk xylene and musk ketone, respectively. In mussels (*Mytilus edulis*) 10–30 µg/kg lipid of both fragrances were detected. In human breast milk from German women, musk xylene and musk ketone were detected between 10 and 240 mg/kg lipid (Geyer et al., 1994). The literature has a number of additional publications from Europe, especially Germany and Switzerland. Rimkus et al. (1997) give a brief overview of the occurrence of musks in the environment. Kafferlein et al. (1998) and Geyer et al. (1994) published the most thorough reviews to date on the occurrence (in the environment and in personal care products), transformation, and toxicology of the ubiquitous musk xylene; these reviews summarize many more occurrence studies (for musk xylene) than mentioned here.

Musks are refractory to biodegradation (other than reduction of nitro musks to amino derivatives), which explains why they have been measured in water bodies throughout the world (Gatermann et al., 1998). They also are very lipophilic [octanol-water partition coefficients are similar to those for DDT and hexacholorocyclohexane (Winkler et al., 1998)] and therefore can bioconcentrate/bioaccumulate (Geyer et al., 1994; Rimkus et al., 1997). Concern has been expressed regarding developmental toxicity in aquatic organisms. Musk ambrette (2,6-dinitro-3-methoyxy-4-ter-butyl toluene) may play a role in damaging the nervous system (Kirschner, 1997).

Draisci et al. (1998) examined freshwater fish in Italy and identified two of five targeted polycyclic musks in most fish samples; a hexa-hydro-hexamethylcyclopental-benzopyran (HHCB, trade name Galaxolide) and an acetylhexamethyltetralin (AHTN, trade name Tonalide) were identified at levels ranging from <4 ng/g (ppb) to 105 ng/g in fish muscle tissue. In the Swiss river Glatt, Muller et al. (1996) identified Galaxolide, Tonalide, and Celestolie (ADBI, 4-acetyl-6-tert-buryl-1,1-demethylindane) at concentrations of 136, 75, and 3.2 ng/L, respectively; they also found the nitro musks tibetene ambrette, moskene, ketone, and xylene at concentrations of 0.04, <0.03, 0.08, 8.3, and 0.62 ng/L, respectively. Eschke et al. (cited in Mersch-Sundermanno et al., 1998) identified Galaxolide, Tonalide, and Celestolide in the fatty tissue of bream and perch from the Ruhr River, Germany, at average concentrations between 2.5 and 4.6 mg/kg (ppm), illustrating the extreme bioaccumulation potential for these compounds. In the late 1990s, Heberer et al. (1999) investigated the contamination of surface waters in Berlin, Germany, (and vicinity) receiving high percentages of treated sewage and found maximum concentrations above 10 µg/L for the polycyclic musks Galaxolide, Tonalide, and Celestolide.

Winkler et al. (1998) measured musks in 31 particulate matter and water samples from the Elbe River, Germany. In all particulate matter samples, concentrations for musk ketone were 4–22 ng/g, for Galazolide 148–736 ng/g, and for Tonalide 194–770 ng/g; Celestolide was found at concentrations of 4–43 ng/g in 23 of the particulate matter samples. The values for the three most prevalent musks were within the same order of magnitude as those for 15 polycyclic aromatic hydrocarbons (PAHs) and exceeded those for 14 common polychlorinated organic pollutants (only hexachlorobiphenyl [HCB] and p, p'-DDT were of similar concentration).

It is not surprising that musks have been detected in air, Kallenborn et al. (1999) detected three polycyclic musks and two nitro musks in Norwegian outdoor air samples. The polycyclic musks were more prevalent. Concentrations of all these musks ranged from low pictograms per cubic meter to hundreds of pictograms per cubic meter. The most common was the polycyclic musk Galaxolide, but the relative ratios among the musks are a function of usage (which varies among countries) and photolability.

Although the significance of the aquatic toxicity of the nitro and polycyclic musks is debatable (genotoxicity from the polycyclic seems to not be a concern) (Kevekordes et al., 1998), the aminobenzene (reduced) versions of the nitro musks can be highly toxic; these reduced derivatives are undoubtedly created under the anaerobic conditions of sewage sludge digestion. Behechi et al. (1998) tested the acute toxicity of four reduced analogs of musk xylene on *Daphnia magna*. The p-aminodinitro compound exhibited the most toxicity of the four, with extremely low median effective concentration (EC_{50}) values averaging 0.25 µg/L (0.25 ppb).

In 1998, the amino transformation products of nitro musks were identified in sewage treatment effluent and in the Elbe River, German. Gatermann et al. (1998) identified musk xylene and musk ketone together with their amino derivatives 4- and 2-amino musk xylenes and 2-amino musk ketone. In sewage treatment influent, the concentrations of musk xylene and musk ketone were 150 and 550 ng/L, respectively. In the effluent, their concentrations dropped to 10 and 6 ng/L, respectively. In contrast, although the amino derivatives could not be detected in the influent, their concentrations in the effluents dramatically increase, showing the extensive transformation of the parent nitro musks: 2-amino musk xylene (10 ng/L), 4-amino musk xylene (34 ng/L), and 2-amino musk ketone (250 ng/L). It was concluded that the amino derivatives could be expected in sewage effluent at concentrations more than an order of magnitude higher than the patent nitro musks. In the Elbe, 4-amino musk xylene was found at higher concentrations (1–9 ng/L) than the parent compound.

Given that the amino nitro musk transformation products (1) are more water soluble than the parent musks, (2) still have significant octanol-water partition coefficients (high bioconcentration potential), and (3) are more toxic than the parent nitro musks, more attention should be focused on these compounds. Because synthetic musks are ubiquitous, used in large quantities, introduced into the environment almost exclusively via treated sewage effluent, and are persistent and bioconcentratable, they are prime candidates for monitoring in both water and biota as indicators for the presence of other PPCPs. Their analysis, especially in biota, has been thoroughly discussed by Gatermann et al. (1998) and by Rimkus et al. (1997).

Disinfectants/Antiseptics
Triclosan (Irgasan DP 300, is chlorinated diphenyl ether: 2.4.4′-trichloro-2′-hydroxy-dephenyl ether) is an antiseptic agent that has been widely used for almost 30 years in a vast array of consumer products. Its use as a preservative and disinfectant continues to grow; for example, it is incorporated at <1% in Colgate's Total toothpaste, the first toothpaste approved by the FDA to fight gingivitis; however, there is no evidence according to the food and Drug Administration (FDA) that triclosan provides an extra benefit to health beyond its anti-gingivitis effect in toothpaste. Nevertheless, triclosan has many other uses. For example, triclosan is registered with the U.S. EPA as a pesticide and it is freely available OTC. Triclosan's use in commercial products includes footwear (in hosiery and insoles of shoes called Odor-Eaters), hospital handsoap, acne creams (e.g., Clearasil), and as a slow-release product called Microban, which is incorporated into a wide variety of plastic products from children's toys to kitchen utensils such as cutting boards. Many of these uses can result in direct discharge of triclosan to wastewater treatment systems, and as such, this compound can find its way into receiving waters depending on its resistance to microbial degradation. Okumura and Nishikawa (1996) found traces of triclosan ranging from 0.05 to 0.15 µg/L in water. Although triclosan has long been regarded as a biocide, a toxicant having a wide-ranging, nonspecific mechanism(s) of action—in this case, gross membrane disruption, McMurry et al. (1998) report that triclosan is rather an antibacterial having particular enzymatic targets (lipid synthesis). As such, bacteria could develop resistance to triclosan. As with all antibiotics in the environment, this could lead to the development of resistance and change in microbial community structure (diversity).

A wide array of disinfectants is used in rather large amounts not just by hospitals, but also by households and livestock breeders. These compounds are often substituted phenolics as well as others such as triclosan. Biphenylol, 4-chlorocrsol, chlorophene, bromophene, 4-chloroxylenol, and tetrabromo-o-cresol (Ternes et al., 1998) are some of the active ingredients, at percentage volumes of <1%–20%. A survey of 49 WWTPs in Germany (Ternes et al., 1998) routinely found biphenylol and chlorophene in both influents, up to 2.6 µg/L for biphenylol and up to 0.71 µg/L for chlorophene, and effluents. The removal of chlorophene from the effluent was less sensitive than for biphenylol, with surface waters having concentrations similar to that of the effluents.

Preservatives
Parabens (alkyl-p-hydroxybenzoates) are one of the most widely and heavily used suites of antimicrobial preservatives by cosmetic and pharmaceutical industries for skin creams, tanning lotions, toiletries, pharmaceuticals, and foodstuffs. Parabens are effective preservatives in many types of formulas. Although the acute toxicity of these compounds is very low, Routledge et al. (1998) report that these compounds (methyl through butyl homologs) display weak estrogenic activity in several assays. Although the risk from dermal application in humans is unknown, the probable continual introduction of these benzoates into wastewater treatment systems and directly to recreational waters from the skin leads to the question of risk to aquatic organisms. Butylparaben showed the most competitive binding to the rat estrogen receptor at concentration one to two orders of magnitude higher than that of nonylphenol and showed estrogenic activity in a yeast estrogen screen at 10^{-6} M.

Sunscreen Agents
The occurrence of sunscreen agents (UV filters) in the German lake Meerfelder Maar was investigated by Nagtegaal et al. (1998). The combined concentrations of six sunscreen agents (SSAs)

identified in perch (*Perca fluviatilis*) in the summer of 1991 were as high as 2.0 mg/kg lipid and in roach (*Rutilus rutilus L*) in the summer of 1993, as high as 0.50 mg/kg lipid. Methylbenzylidene camphor (MBC) was *detected* in roach from three other German lakes. These lipophilic SSAs seem to occur widely in fish from small lakes used for recreational swimming. Both fish species had body burdens of SSA on par with PCBs and DDT. The bioaccumulation factor, calculated as quotient of the MBC concentration in the whole fish (21 µg/kg) versus that in the water (0.004 µg/L), exceeded 5,200, indicating high lipophilicity. The fact that SSAs (e.g., 2-hydroxy-4-methoxybenzophenone [oxybenzone] and 2-ethylhexyl-4-methoxycinamate) can be deterred in human breast milk [16 and 417 ng/g lipid, respectively (Hany and Nagel, 1995)] shows the potential for dermal absorption and bioconcentration in aquatic species.

Nutraceuticals/Herbal Remedies

Nutraceutical, a portmanteau of the words "nutrition" and "pharmaceutical," lumps together a group of nutritional supplements of highly bioactive food supplements. The term is applied to products that range from isolated nutrients, dietary supplements and herbal products, specific diets, genetically modified food, and processed foods such as cereals, soups, and beverages that are intended as supplements to the diet. Nutraceuticals and many herbal remedies can have potent physiologic effects. These are a mainstay of alternative medicine and have enjoyed explosive growth in use in the United States and other countries during the last 30 years. Nutraceuticals are not classified as drugs by the FDA, primarily because a given botanical usually has not one but an array of distinct compounds whose assemblage elicits the putative effect and because these arrays cannot be easily standardized. As such they are not regulated and are available OTC (heavily promoted via the Internet). Even in those cases in which the natural product is identical to a prescription pharmaceutical (e.g., the Chinese red-yeast product Cholestin newly introduced to the United States contains lovastatin, an active ingredient in the approved prescription drug Mevacor used to lower cholesterol levels), a 1999 ruling (Pharmanex, 1999) prevented the FDA from regulation.

With the accelerating inverted age structure of our society, coupled with the U.S. 1994 Dietary Supplement Health and Education Act (DSHEA, 1994) (which eases regulations on the introduction and marketing of supplements), the use of Nutraceuticals could greatly escalate. The significance of dietary supplements in the United States was epitomized by the creation of the Office of Dietary Supplements (ODS) via the DSHEA in 1995 under the National Institutes of Health (NIH) (DSHEA, 1994). The ODS maintains a searchable database (International Bibliographic Information on Dietary Supplements [IBIDS]) of published scientific literature on dietary supplements (NIH, 1999). The NIH was also mandated to create the National Center for complementary and Alternative Medicine (NCCAM, 1999) to facilitate the evaluations of alternative medical treatment modalities to determine their effectiveness.

There are countless Nutraceuticals, both new and ancient, experiencing vigorous consumption. There are several unknowns regarding the Nutraceuticals being excreted, surviving wastewater treatment, and then eliciting effects on aquatic organisms. Nutraceuticals and herbal remedies would have the same potential fate in the environment as pharmaceuticals, with the added dimension that their usage rates could be much higher, as they are readily available and taken without the controls of prescription medication. Because these compounds are natural products, however, they would be expected to be more easily biodegraded.

Although the argument can be made that naturally occurring compounds would not pose an ecologic risk, this ignores that (1) the concentrations of these compounds in effluents could be higher than they are in the environment in which they occur naturally, and (2) many of these substances/mixtures come only from isolated parts of the world (e.g., Kava, huperzine), and their use/dispersal in other parts of the world would essentially make them anthropogenic. The use of these compounds serves to redistribute their normal occurrence in the environment, and even though they might be naturally occurring, this promotes exposure to organisms that normally would never occur.

Illicit Drugs in Wastewater

Screening for Big Bucks

In wastewater treatment, the purpose of *screening* is to remove large solids such as rags, cans, rocks, branches, leaves, roots, etc. from the flow before the flow moves on to downstream processes.

Note: Typically, a treatment plant will remove anywhere from 0.5 to 12 ft^3 of screenings for each million gallons of influent received.

A *bar screen* traps debris as wastewater influent passes through. Typically, a bar screen consists of a series of parallel, evenly spaced bars or a perforated screen placed in a channel. The waste stream passes through the screen and the large solids (*screenings*) are trapped on the bars for removal.

Note: The screenings must be removed frequently enough to prevent accumulation which will block the screen and cause the water level in front of the screen to build up.

The bar screen may be coarse (2–4-inch openings) or fine (0.75–2.0-inch openings). The bar screen may be manually cleaned (bars or screens are placed at an angle of 30° for easier solids removal) or mechanically cleaned (bars are placed at 45°–60° angle to improve mechanical cleaner operation).

The screening method employed depends on the design of the plant, the amount of solids expected and whether the screen is for constant, or emergency use only. Manually cleaned screens are cleaned at least once per shift (or often enough to prevent buildup which may cause reduced flow into the plant) using a long tooth rake. Solids are manually pulled to the drain platform and allowed to drain before storage in a covered container. The area around the screen should be cleaned frequently to prevent a buildup of grease or other materials, which can cause odors, slippery conditions, and insect and rodent problems. Because screenings may contain organic matter as well as large amounts of grease they should be stored in a covered container. Screenings can be disposed of by burial in approved landfills or by incineration. Some treatment facilities grind the screenings into small particles, which are then returned to the wastewater flow for further processing and removal later in the process.

Mechanically cleaned screens use a mechanized rake assembly to collect the solids and move them (carries them) out of the wastewater flow for discharge to a storage hopper. The screen may be continuously cleaned or cleaned on a time or flow-controlled cycle. As with the manually cleaned screen, the area surrounding the mechanically operated screen must be cleaned frequently to prevent buildup of materials, which can cause unsafe conditions. As with all mechanical equipment, operator vigilance is required to ensure proper operation and that proper maintenance is performed. Maintenance includes lubricating equipment and maintaining it in accordance with manufacturer's recommendations or the plant's O&M Manual (Operations & Maintenance Manual). Screenings from mechanically operated bar screens are disposed of in the same manner as screenings from manually operated screen: landfill disposal, incineration, or ground into smaller particles for return to the wastewater flow.

As stated earlier, bar screens are designed to remove large solids such as rags, cans, rocks, branches, leaves, roots, and so forth. Based on actual experience, during my walk-around inspections of wastewater treatment plants that always begin at the plant's headworks (bar screen area where raw influent enters the plant) I have found that the "and so forth" can include not-often-thought-about or normally mentioned items such as small guns of all makes and models; ammunition, spent and live; aborted fetuses; human tissues, adult diapers, mop heads and towels, illicit drugs and drug paraphernalia; and one hundred dollar bills.

One-hundred-dollar bills? Yes. Absolutely.

I have had a few plant operators over the years show me their recent acquisition of paper money, including one-hundred-dollar bills that they have removed from the screenings after the bills had entered the plant with the raw sewage flow.

Apparently, the money in the sewage influent is the result of the police knocking at some illicit drugger's hangout (typically a residential home) and the drugger depositing drugs and money down the toilet in order to do away with incriminating evidence.

This book is not about illicit drug money, however. The interest is in whether the disposal of illicit drugs into the environment is harming the environment. At this point and time, we do not know what we do not know about this problem; whether they have adverse effects is completely unknown (Daughton, 2001). Not knowing the impact of illicit drugs on the environment is not the point at this point of time; what is important today is our awareness of the potential issues related to illicit drug disposal and our need to investigate this problem.

SURFACE WATER POLLUTION: THE IMPACT

Unless you notice water bodies best described as cesspools, and/or experience water that smells foul and tastes worse (and ultimately might make you ill) you may think that water pollution is relative and find it hard to define. Once you come up with a definition (it might have something to do with physical characteristics and negative impact) you may also consider the idea that freshwater pollution is not a new phenomenon. Only the issue of freshwater pollution as a major public concern is relatively new.

Natural forms of pollutants have always been present in surface waters. Many of the pollutants we have discussed in this section were being washed from the air, eroded from land surfaces, or leached from the soil, and ultimately found their way into surface water bodies long before humans evolved to walk on Earth. Floods and dead animals pollute, but their effects are local and generally temporary. In prehistoric times (and even in more recent times) natural disasters have contributed to surface water pollution. Cataclysmic events—earthquakes, volcanic eruptions, meteor impact, transition from the ice age to inter glacial to ice age—have all contributed to surface water pollution. Natural purification processes—over time—were able to self-clean surface water bodies. We can accurately say that without these self-purifying processes, the water-dependent life on Earth could not have developed as it did.

But the natural problems, the ones the environment could eventually self-clean, are augmented by anthropogenic ones. Made-made problems piled on top of the natural pollutants present us with greater risks—and greater challenges.

GROUNDWATER

Of the ~3 ft of water that falls each year on every square foot of Earth (on average), ~6 inches returns to the sea. Evaporation takes another 2 ft. The last 6 inches infiltrate through the Earth's interstices, voids, hollows, and cavities, filtering into the sponge-like soil. In traveling down into and through the soil, the course that water follows may carry it only a few inches, a few feet—or several hundred feet—before it joins the subterranean water stores that comprise the Earth's groundwater supply. This water supply (one people are often oblivious to) contains an estimated 1,700,000 cubic miles of water, hidden underground. Enough water, if you could spread over the Earth's surface, to blanket all Earth's land surfaces with one thousand feet of water. U.S. groundwater sources constitute a freshwater supply greater than all the surface water in the U.S.—and that is including the Great Lakes.

This enormous reservoir, our groundwater supply, feeds all the natural fountains and springs of Earth. These natural exits to groundwater allow it to bubble up in cool, blue pools from springs. In more unusual circumstances, from places too deep within the Earth to imagine, groundwater heats up, forms steam and bursts from the surface in geysers and hot springs. Though we make use of many different groundwater sources, not all groundwater supplies can be tapped for use. In some places, the water is not accessible, because of pumping costs and drilling difficulties. Groundwater supplies, too, are not always pure. Contaminated groundwater supplies have become a significant pollution problem.

However, most of the Earth's groundwater supply lies within reach of the surface, accessible by drilling a borehole or well down to the water table. Humans have obtained water this way for millennia, and as more and more people inhabit Earth, more use is made of our groundwater supplies. Currently, groundwater serves as a reliable source of potable water for millions of Earth's inhabitants, and if used with moderation, groundwater should remain a viable source for years to come.

GROUNDWATER USES AND SOURCES

The water we use, by population, breaks down to roughly 50% groundwater sources and 50% surface water sources. Large cities rely primarily on surface water for their supplies, but 95% of small communities and rural areas use groundwater. A larger percent of the U.S. population is supplied by surface water—but four times the total number of communities supplied by surface water is supplied by groundwater.

As a water supply source, groundwater has several desirable characteristics: (1) natural storage, eliminating the need for man-made impoundments; (2) usually available at point of demand, so transmission cost is reduced significantly; and (3) filtration through the natural geologic strata means groundwater usually appears clearer to the eye than surface water (McGhee, 1991). For these reasons, groundwater is generally preferred as a municipal and industrial water source.

For many years, we believed that groundwater was safe from contamination, and naturally cleansed by travelling through the soil. Groundwater was considered safe to drink, and many water utilities delivered it to their customers with no further treatment. We know better now.

We have discovered that groundwater is not automatically safe to use as a potable water supply. Discoveries of contaminated groundwater have led to the closure of thousands of potable water wells across the U.S. The USEPA reported that in the mid-1980s, more than 8,000 drinking water wells in areas all over the nation were no longer usable—because of contamination. Monitoring the complex groundwater situation nationwide is fraught with difficulty, because of the vast number of potential and possible contamination sources, including contamination by toxic or hazardous materials leaking from waste treatment facilities, natural sources, or landfills that may not be evident to either the public or regulatory agencies, as well as from many other sources. Groundwater contamination's biggest problem is twofold: monitoring its condition is difficult, and when contaminated, restoring it is difficult (and expensive)—if possible, at all.

Aquifers

An aquifer performs two important functions: storage and transport. Expressed simply, the subsurface is charged with the water that then becomes groundwater when surface water seeps down from the rain-soaked surface, and sinks until it reaches an impermeable layer, where it collects and fills all the pores and cracks of the permeable portions. The top of this saturated zone is called the water table (see Figure 8.2).

In reality, the groundwater system is a bit more complicated than that shown in Figure 8.2. Groundwater occurs in two different zones in unconfined aquifers (an aquifer not overlain by an impermeable layer is unconfined). These zones are distinguished by whether or not water fills all the cracks and pores between particles of soil and rock. The unsaturated zone, which lies just beneath the land surface, is characterized by crevices that contain both air and water. While the unsaturated zone contains water (vadose water), this water is essentially unavailable for use. Water flow in a confined aquifer (a water-bearing layer sandwiched between two less permeable layers) is restricted to vertical movement only (see Figure 8.3). An unconfined aquifer allows water to flow with more freedom of movement and resembles flow in an open channel.

Water Pollution   371

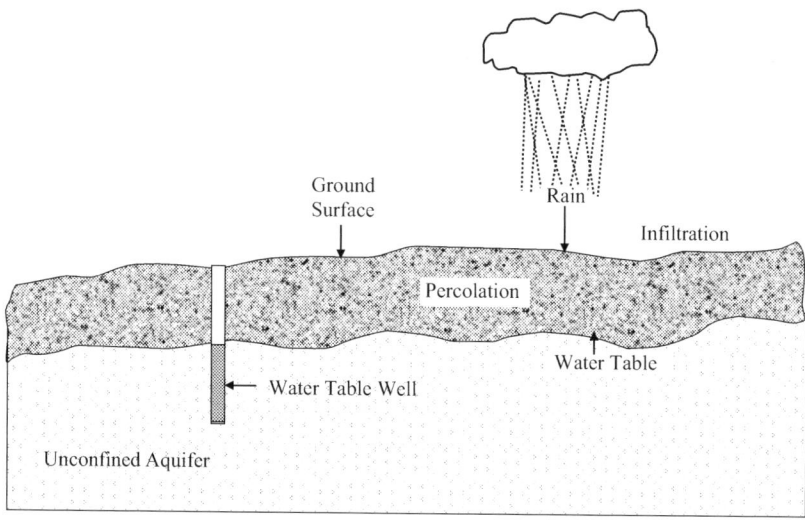

FIGURE 8.2 Unconfined aquifer.

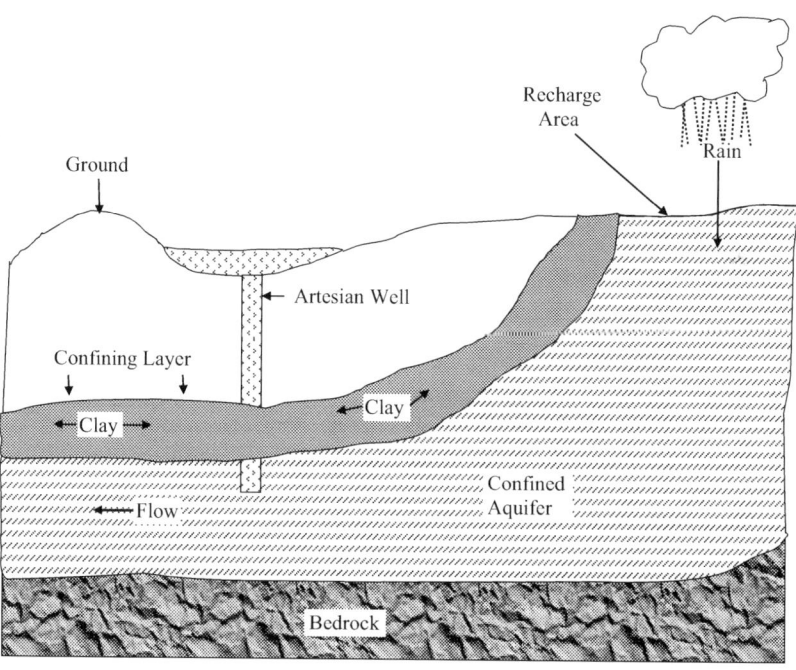

FIGURE 8.3 Confined aquifer.

Groundwater Flow

To have any flow at all, a hydraulic gradient must exist, whether groundwater flow occurs in an unconfined aquifer's open channel-like flow or a confined aquifer's vertical-only (pipe-like) flow. The hydraulic gradient is the difference in hydraulic head divided by the distance along the fluid flow path. For our applications of the concept, you should know that groundwater moves through an aquifer in the direction of the hydraulic gradient, at a rate proportional to the gradient (the direction of the slope of the water table), inversely related to the aquifer's permeability. The more permeable the substrate and the steeper the slope, the more rapidly the water flows.

Groundwater, of course, contrary to popular belief, does not flow like a river. Percolating downward, groundwater moves from high elevations to lower elevations at a variable rate that is dependent on underground conditions. Sometimes slow-moving, it can sometimes move surprisingly quickly, from less than an inch to a several feet a day.

Groundwater aquifers, as we said previously, supply a large portion of the U.S. population—and almost all of the rural population. They form important sources of water. Groundwater use demand continues to increase, which threatens both the quantity and quality of this critically important resource. You should remember two important points that concern groundwater: (1) The groundwater supply is not inexhaustible; and (2) groundwater is not exempt from surface contamination. It is not completely purified as it percolates through the ground—even though the interconnectedness of the hydrological cycle in self-purification works to our advantage. The processes integral to the water cycle can trap toxins, complicating efforts to clean them up—of special concern for persistent pollutants. The natural processes that clean our water as it travels through the hydrological cycle worked well for centuries, but now, in many places, humans have overloaded the capacity of the water cycle to self-purify. While we are now cleaning up problems created by past environmental abuse and ignorance, inevitably, we create problems that future generations will have to clean up. The solutions we try now will present future generations with problems we did not foresee—but re-creating past mistakes is foolish and foolhardy. Our water system is too valuable for us to risk.

Groundwater Pollution

Groundwater pollution can be a very serious problem. We know, through experience and study, that any pollutant that contacts the ground holds the possibility for contaminating groundwater. As water enters the ground, it filters naturally through the soil, and in some soils, that process quite effectively removes many substances, including suspended solids and bacteria—and pollutants. Some chemicals are removed as they bind themselves to the surface of soil particles (phosphates). In some areas, though, industrial, and municipal wastes are sprayed on the ground surface, and employing the natural self-purification process, wastewater filters through the soil, becomes purified in the process, and recharges the groundwater reservoir. Though a beneficial process, natural purification of water as it passes through the soil is a slow process, because the water has no access to air and is not readily diluted.

Drainage-basin activities that pollute surface waters also cause groundwater contamination. Problems can occur from sources as diverse as septic tanks, agriculture, industrial waste lagoons, underground injection wells, underground storage tanks, and landfills. Waste disposal sites located in unsuitable soils (or even directly over fractured dolomites and limestones) can cause major problems. Disposal sites located directly on top of such rock allow polluted water to travel into wells. At least 25% of the usable groundwater (from wells) is already contaminated in some areas (Draper, 1987).

The causes of groundwater contamination vary. Increasingly, groundwater contamination from saltwater, microbiological contaminants and toxic organic and inorganic chemicals has occurred and are now being observed. The major source of groundwater contamination in the U.S. comes from the improper disposal of toxic industrial wastes. The levels of and problems related to contamination from these wastes are increased significantly when waste disposal sites are not protected by some type of lining; when disposal sites lie in permeable materials above usable water aquifers; and when these sites are located close to water supply wells. In 1982, the Conservation Foundation reported that groundwater contamination was responsible for closing hundreds of U.S. wells.

WETLANDS

Wetlands routinely replenish and purify groundwater supplies. By absorbing excess nutrients and immobilizing pesticides, heavy metals, and other toxins, wetland plants prevent them from moving

Water Pollution

up the food chain. Wetlands have been used to treat sewage in some locations. However, wetland ecosystems are relatively fragile, and their capacity to cleanse polluted water is limited. Many have been overwhelmed by pollution, though more of our natural wetland areas have been destroyed by anthropogenic activities. In the United States, for example, half our wetlands have been lost to urban and agricultural development (Goldsmith and Hildyard, 1988).

WATER TREATMENT

In this section, the focus is on water treatment operations and the various unit processes currently used to treat raw source water before it is distributed to the user. In addition, the reasons for water treatment and the basic theories associated with individual treatment unit processes are discussed. Water treatment systems are installed to remove those materials that cause disease and/or create nuisances. At its simplest level, the basic goal of water treatment operations is to protect public health, with a broader goal to provide potable and palatable water. The bottom line is that the water treatment process functions to provide water that is safe to drink and is pleasant in appearance, taste, and odor.

Water treatment is defined as any unit process that changes/alters the chemical, physical, and/or bacteriological quality of water with the purpose of making it safe for human consumption and/or appealing to the customer. Treatment also is used to protect the water distribution system components from corrosion.

Many water treatment unit processes are commonly used today. Treatment processes used depend upon the evaluation of the nature and quality of the particular water to be treated and the desired quality of the finished water. In water treatment unit processes employed to treat raw water, one thing is certain: as new USEPA regulations take effect, many more processes will come into use in the attempt to produce water that complies with all current regulations, despite source water conditions.

Small water systems tend to use a smaller number of the wide array of unit treatment processes available, in part because they usually rely on groundwater as the source, and in part because small makes many sophisticated processes impractical (i.e., too expensive to install, too expensive to operate, too sophisticated for limited operating staff). This section concentrates on those individual treatment unit processes usually found in conventional water treatment systems, corrosion control methods, and fluoridation. A summary of basic water treatment processes (many of which are discussed in this section) is presented in Table 8.2.

Purpose of Water Treatment

The purpose of water treatment is to condition, modify and/or remove undesirable impurities, to provide water that is safe, palatable, and acceptable to users. This may seem an obvious, expected purpose of treating water, but various regulations also require water treatment. Some regulations state that if the contaminants listed under the various regulations are found in excess of maximum contaminant levels (MCLs), the water must be treated to reduce the levels. If a well or spring source is surface influenced, treatment is required, regardless of the actual presence of contamination. Some impurities affect the aesthetic qualities (taste, odor, color, and hardness) of the water; if they exceed secondary MCLs established by USEPA and the state, the water may need to be treated.

If we assume that the water source used to feed a typical water supply system is groundwater (usually the case in the U.S.), a number of common groundwater problems may require water treatment. Keep in mind that water that must be treated for any one of these problems may also exhibit several other problems:

- Bacteriological contamination
- Hydrogen sulfide odors

TABLE 8.2
Basic Water Treatment Processes

Process	Purpose
Screening	Removes large debris (leaves, rocks, fish, branches) that can foul or damage plant equipment
Chemical pretreatment	Conditions the water for removal of algae and other aquatic nuisances
Presedimentation	Removes gravel, sand, silt, and other gritty materials
Microstraining	Removes algae, aquatic plants, and small debris
Chemical feed/rapid mix	Adds chemical (coagulants, pH adjusters, etc.) to water
Coagulation/flocculation	Converts nonsettleable or settable particles
Sedimentation	Removes settleable particles
Softening	Removes hardness-causing chemicals from water
Filtration	Removes particles of solid matter which can include biological contamination and turbidity
Disinfection	Kills disease-causing organisms
Adsorption using granular activated carbon	Removes radon and many organic chemicals such as pesticides, solvents, and trihalomethanes
Aeration	Removes volatile organic chemicals (VOCs), radon H2S, and other dissolved gases; oxidizes iron and manganese
Corrosion control	Prevents scaling and corrosion
Reverse osmosis, electrodialysis	Removes nearly all inorganic contaminants
Ion exchange	Removes some inorganic contaminants including hardness-causing chemicals
Activated alumina	Removes some inorganic contamination
Oxidation filtration	Removes some inorganic contaminants (e.g., iron, manganese, radium)

Source: Adapted from AWWA (1984).

- Hard water
- Corrosive water
- Iron and manganese

STAGES OF WATER TREATMENT

Earlier it was stated that the focus of our discussion in this text is on the conventional model of water treatment. Figure 8.4 presents the very basic conventional model discussed in this text. The figure clearly illustrates that water treatment is made up of various stages, unit processes, or a train of processes combined to form one treatment system. Note that a given waterworks may contain all the unit processes discussed in the following or any combination of them. One or more of these stages

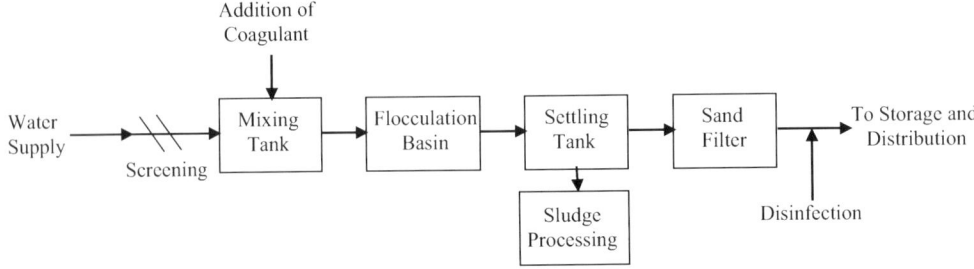

FIGURE 8.4 Conventional water treatment model.

may be used to treat any one or more of the source water problems listed above. Also, note that the model shown in Figure 8.4 does not necessarily apply to very small water systems. In some small systems, water treatment may consist of nothing more than the removal of water via pumping from a groundwater source to storage to distribution. In some small water supply operations, disinfection may be added because it is required. Although it is likely that the basic model shown in Figure 8.4 does not reflect the type of treatment process used in most small systems, we use it in this handbook for illustrative and instructive purposes.

Pretreatment

Simply stated, water pretreatment (also called preliminary treatment) is any physical, chemical, or mechanical process used before main water treatment processes. It can include screening, presedimentation, and chemical addition. Pretreatment in water treatment operations usually consists of oxidation or other treatment for the removal of tastes and odors, iron, and manganese, trihalomethane precursors, or entrapped gases (like hydrogen sulfide). Unit processes may include chlorine, potassium permanganate or ozone oxidation, activated carbon addition, aeration, and presedimentation. Pretreatment of surface water supplies accomplishes the removal of certain constituents and materials that interfere with or place an unnecessary burden on conventional water treatment facilities.

Based on our experience and according to the Texas Water Utilities Association's Manual of Water Utility Operations, 8th edition, typical pretreatment processes include the following:

1. Removal of debris from water from rivers and reservoirs that would clog pumping equipment.
2. Destratification of reservoirs to prevent anaerobic decomposition that could result in reducing iron and manganese from the soil to a state that would be soluble in water. This can cause subsequent removal problems in the treatment plant. The production of hydrogen sulfide and other taste- and odor-producing compounds also results from stratification.
3. Chemical treatment of reservoirs to control the growth of algae and other aquatic growths that could result in taste and odor problems.
4. Presedimentation to remove excessively heavy silt loads prior to the treatment processes.
5. Aeration to remove dissolved odor-causing gases such as hydrogen sulfide and other dissolved gases or volatile constituents, and to aid in the oxidation of iron and manganese, although manganese or high concentrations of iron are not removed in detention provided in conventional aeration units.
6. Chemical oxidation of iron and manganese, sulfides, taste- and odor-producing compounds, and organic precursors that may produce trihalomethanes upon the addition of chlorine.
7. Adsorption for removal of tastes and odors.

Note: An important point to keep in mind is that in small systems, using groundwater as a source, pretreatment may be the only treatment process used.

Note: Pretreatment may be incorporated as part of the total treatment process or may be located adjacent to the source before the water is sent to the treatment facility.

Aeration

Aeration is commonly used to treat water that contains trapped gases (such as hydrogen sulfide) that can impart an unpleasant taste and odor to the water. Just allowing the water to rest in a vented tank will (sometimes) drive off much of the gas, but usually some form of forced aeration is needed. Aeration works well (about 85% of the sulfides may be removed) whenever the pH of the water is <6.5. Aeration may also be useful in oxidizing iron and manganese, oxidizing humic substances that might form trihalomethanes when chlorinated, eliminating other sources of taste and odor, or imparting oxygen to oxygen-deficient water.

Note: Iron is a naturally occurring mineral found in many water supplies. When the concentration of iron exceeds 0.3 mg/L, red stains will occur on fixtures and clothing. This increases customer costs for cleaning and replacement of damaged fixtures and clothing. Manganese, like iron, is a naturally occurring mineral found in many water supplies. When the concentration of manganese exceeds 0.05 mg/L, black stains occur on fixtures and clothing. As with iron, this increases customer costs for cleaning and replacement of damaged fixtures and clothing. Iron and manganese are commonly found together in the same water supply. We discuss iron and manganese later.

Screening

Screening is usually the first major step in the water pretreatment process (see Figure 8.4). It is defined as the process whereby relatively large and suspended debris is removed or retained from the water before it enters the plant. River water, for example, typically contains suspended and floating debris varying in size from small rocks to logs. Removing these solids is important, not only because these items have no place in potable water, but also because this river trash may cause damage to downstream equipment (clogging and damaging pumps, etc.), increase chemical requirements, impede hydraulic flow in open channels or pipes, or hinder the treatment process. The most important criteria used in the selection of a particular screening system for water treatment technology are the screen opening size and flow rate; they range in size from microscreens to trash racks. Other important criteria include: costs related to operation and equipment; plant hydraulics; debris handling requirements; and operator qualifications and availability. Large surface water treatment plants may employ a variety of screening devices including rash screens (or trash rakes), traveling water screens, drum screens, bar screens, or passive screens.

Chemical Addition

Much of the procedural information presented in this section applies to both water and wastewater operations. Two of the major chemical pretreatment processes used in treating water for potable use are iron and manganese and hardness removal. Another chemical treatment process that is not necessarily part of the pretreatment process, but is also discussed in this section, is corrosion control. Corrosion prevention is affected by chemical treatment—not only in the treatment process but is also in the distribution process. Before discussing each of these treatment methods in detail, however, it is important to describe chemical addition, chemical feeders, and chemical feeder calibration.

When chemicals are used in the pretreatment process, they must be the proper ones, fed in the proper concentration and introduced to the water at the proper locations. Determining the proper amount of chemical to use is accomplished by testing. The operator must test the raw water periodically to determine if the chemical dosage should be adjusted. For surface supplies, checking must be done more frequently than for groundwater (remember, surface water supplies are subject to change on short notice, while groundwaters generally remain stable). The operator must be aware of the potential for interactions between various chemicals and how to determine the optimum dosage (e.g., adding both chlorine and activated carbon at the same point will minimize the effectiveness of both processes, as the adsorptive power of the carbon will be used to remove the chlorine from the water).

Note: Sometimes using too many chemicals can be worse than not using enough.

Prechlorination (distinguished from chlorination used in disinfection at the end of treatment) is often used as an oxidant to help with the removal of iron and manganese. However, currently, concern for systems that prechlorinate is prevalent because of the potential for the formation of total trihalomethanes (TTHMs), which form as a by-product of the reaction between chlorine and naturally occurring compounds in raw water. USEPA's TTHM standard does not apply to water systems that serve <10,000 people, but operators should be aware of the impact and causes of TTHMs. Chlorine dosage or application point may be changed to reduce problems with TTHMs.

Note: TTHMs such as chloroform are known or suspected to be carcinogenic and are limited by water and state regulations.

Note: To be effective, pretreatment chemicals must be thoroughly mixed with the water. Short-circuiting or plug flows of chemicals that do not come in contact with most of the water will not result in proper treatment.

All chemicals intended for use in drinking water must meet certain standards. Thus, when ordering water treatment chemicals, the operator must be assured that they meet all appropriate standards for drinking water use.

Chemicals are normally fed with dry chemical feeders or solution (metering) pumps. Operators must be familiar with all of the adjustments needed to control the rate at which the chemical is fed to the water (wastewater). Some feeders are manually controlled and must be adjusted by the operator when the raw water quality or the flow rate changes; other feeders are paced by a flow meter to adjust the chemical feed, so it matches the water flow rate. Operators must also be familiar with chemical solution and feeder calibration.

As mentioned, a significant part of waterworks operator's important daily operational functions includes measuring quantities of chemicals and applying them to water at preset rates. Normally accomplished semiautomatically by use of electro-mechanical-chemical feed devices, waterworks operators must still know what chemicals to add, how much to add to the water (wastewater), and the purpose of the chemical addition.

Chemical Solutions

A water solution is a homogeneous liquid made of the solvent (the substance that dissolves another substance) and the solute (the substance that dissolves in the solvent). Water is the solvent (see Figure 8.5). The solute (whatever it may be) may dissolve up to a certain limit. This is called its solubility—that is, the solubility of the solute in the particular solvent (water) at a particular temperature and pressure.

Note: Temperature and pressure influence stability of solutions but not by filtration because only suspended material can be eliminated by filtration or by sedimentation.

Remember, in chemical solutions, the substance being dissolved is called the solute, and the liquid present in the greatest amount in a solution (that does the dissolving) is called the solvent. Environmental practitioners should also be familiar with another term, concentration—the amount of solute dissolved in a given amount of solvent. Concentration is measured as:

$$\% \text{ Strength} = \frac{\text{Weight of solute}}{\text{Weight of solution}} \times 100 = \frac{\text{Weight of solute}}{\text{Weight of solute} + \text{solvent}} \times 100 \qquad (8.1)$$

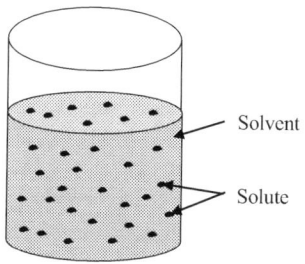

FIGURE 8.5 Solution with two components: solvent and solute.

Example 8.1

Problem:

If 30 lb of chemical is added to 400 lb of water, what is the percent strength (by weight) of the solution?

Solution:

$$\% \text{ Strength} = \frac{30 \text{ lb solute}}{400 \text{ lb water}} \times 100 = \frac{30 \text{ lb solute}}{30 \text{ lb solute} + 400 \text{ lb water}} \times 100$$

$$= \frac{30 \text{ lb solute}}{430 \text{ lb solute/water}} \times 100$$

$$= 7.0 \text{ (rounded)}$$

Important to the process of making accurate computations of chemical strength is a complete understanding of the dimensional units involved. For example, operators should understand exactly what milligrams per liter (mg/L) signify.

$$\text{Milligrams per Liter}(\text{mg/L}) = \frac{\text{Milligrams of solute}}{\text{Liters of solution}} \qquad (8.2)$$

Another important dimensional unit commonly used when dealing with chemical solutions is parts per million (ppm).

$$\text{Parts per Million}(\text{ppm}) = \frac{\text{Parts of solute}}{\text{Million parts of solution}} \qquad (8.3)$$

Note: "Parts" is usually a weight measurement.
An example is:

$$9 \text{ ppm} = \frac{9 \text{ lb solids}}{1,000,000 \text{ lb solution}}$$

or

$$9 \text{ ppm} = \frac{9 \text{ mg solids}}{1,000,000 \text{ mg solution}}$$

This leads us to two important parameters that environmental professionals should commit to memory:

$$1\text{-mg/L} = 1 \text{ ppm}$$

$$1\% = 10,000 \text{ mg/L}$$

When working with chemical solutions, you should also be familiar with two chemical properties we briefly described earlier: density and specific gravity. Density is defined as the weight of a substance per a unit of its volume; for example, pounds per cubic foot or pounds per gallon. Specific gravity is defined as the ratio of the density of a substance to a standard density.

$$\text{Density} = \frac{\text{Mass of substance}}{\text{Volume of substance}} \qquad (8.4)$$

Here are a few key facts about density:

- measured in units of lb/cf, lb/gal, or mg/l
- density of water = 62.5 lbs/cf = 8.34 lb/gal

- other densities: concrete = 130 lb/cf, alum (liquid, @ 60°F) = 1.33, and hydrogen peroxide (35%) = 1.132

$$\text{Specific gravity} = \frac{\text{Density of substance}}{\text{Density of water}} \qquad (8.5)$$

Here are a few facts about specific gravity:

- Specific gravity has no units
- Specific gravity of water = 1.0
- Specific gravity of concrete = 2.08, alum (liquid, @ 60°F) = 1.33, and hydrogen peroxide (355) = 1.132

Chemical Feeders

Simply put, a chemical feeder is a mechanical device for measuring a quantity of chemical and applying it to water at a preset rate. Two types of chemical feeders are commonly used: solution (or liquid) feeders and dry feeders. Liquid feeders apply chemicals in solutions or suspensions, and dry feeders apply chemicals in granular or powdered forms. In a solution feeder, chemical enters feeder and leaves feeder in a liquid state; in a dry feeder, chemical enters and leaves feeder in a dry state.

- **Solution feeders**: are small, positive displacement metering pumps of three types: (1) reciprocating (piston-plunger or diaphragm types); (2) vacuum type (e.g., gas chlorinator); or (3) gravity feed rotameter (e.g., drip feeder). Positive displacement pumps are used in high pressure, low flow applications; they deliver a specific volume of liquid for each stroke of a piston or rotation of an impeller.
- **Dry feeders**: two types of dry feeders are volumetric and gravimetric, depending on whether the chemical is measured by volume (volumetric-type) or weight (gravimetric-type). Simpler and less expensive than gravimetric pumps, volumetric dry feeders are also less accurate. Gravimetric dry feeders are extremely accurate, deliver high feed rates, and are more expensive than volumetric feeders.

Chemical Feeder Calibration

Chemical feeder calibration ensures effective control of the treatment process. Obviously, chemical feed without some type of metering and accounting of chemical used adversely affects the water treatment process. Chemical feeder calibration also optimizes economy of operation; it ensures the optimum use of expensive chemicals. Finally, operators must have accurate knowledge of each individual feeder's capabilities at specific settings. When a certain dose must be administered, the operator must rely on the feeder to feed the correct amount of chemical. Proper calibration ensures chemical dosages can be set with confidence. At a minimum, chemical feeders must be calibrated on an annual basis. During operation, when the operator changes chemical strength or chemical purity or makes any adjustment to the feeder, or when the treated water flow changes, the chemical feeder should be calibrated. Ideally, any time maintenance is performed on chemical feed equipment, calibration should be performed.

What factors affect chemical feeder calibration (i.e., feed rate)? For solution feeders, calibration is affected any time solution strength changes, any time a mechanical change is introduced in the pump (change in stroke length or stroke frequency), and/or whenever flow rate changes. In the dry chemical feeder, calibration is affected any time chemical purity changes, mechanical damage occurs (e.g., belt change), and/or whenever flow rate changes. In the calibration process, calibration charts are usually used or made up to fit the calibration equipment. The calibration chart is also affected by certain factors, including change in chemical, change in flow rate of water being treated, and/or a mechanical change in the feeder.

Example 8.2

To demonstrate that performing a chemical feed procedure is not necessarily as simple as opening a bag of chemicals and dumping the contents into the feed system, we provide a real-world example below.

Problem: *Consider the chlorination dosage rates below.*

Setting		Dosage
100%	111/121	0.93 mg/L
70%	78/121	0.66 mg/L
50%	54/121	0.45 mg/L
20%	20/121	0.16 g/L

Solution:

This is not a good dosage setup for a chlorination system. Maintenance of a chlorine residual at the ends of the distribution system should be within 0.5–1.0 ppm. At 0.9 ppm, dosage will probably result in this range—depending on the chlorine demand of the raw water and detention time in the system. However, the pump is set at its highest setting. We have room to decrease the dosage, but no ability to increase the dosage without changing the solution strength in the solution tank. In this example, doubling the solution strength to 1% provides the ideal solution, resulting in the following chart changes.

Setting		Dosage
100%	222/121	1.86 mg/L
70%	154/121	1.32 mg/L
50%	108/121	0.90 mg/L
20%	40/121	0.32 mg/L

This is ideal, because the dosage we want to feed is at the 50% setting for our chlorinator. We can now easily increase or decrease the dosage whereas the previous setup only allowed the dosage to be decreased.

Iron and Manganese Removal

Iron and manganese are frequently found in groundwater and in some surface waters. They do not cause health-related problems but are objectionable because they may cause aesthetic problems. Severe aesthetic problems may cause consumers to avoid an otherwise safe water supply in favor of one of unknown or of questionable quality or may cause them to incur unnecessary expense for bottled water. Aesthetic problems associated with iron and manganese include the discoloration of water (iron = reddish water, manganese = brown or black water); staining of plumbing fixtures; imparting a bitter taste to the water; and stimulating the growth of microorganisms.

Although there is no direct health concerns associated with iron and manganese, although the growth of iron bacteria slimes may cause indirect health problems. Economic problems include damage to textiles, dye, paper, and food. Iron residue (or tuberculation) in pipes increases pumping head, decreases carrying capacity, may clog pipes, and may corrode through pipes.

Note: Iron and manganese are secondary contaminants. Their secondary maximum contaminant levels (SMCLs) are: iron = 0.3 mg/L; manganese = 0.05 mg/L.

Iron and manganese are most likely found in groundwater supplies, industrial waste, and acid mine drainage, and as by-products of pipeline corrosion. They may accumulate in lake and reservoir sediments, causing possible problems during lake/reservoir turnover. They are not usually found in running waters (streams, rivers, etc.).

Iron and Manganese Removal Techniques

Chemical precipitation treatments for iron and manganese removal are called deferrization and demanganization. The usual process is aeration; dissolved oxygen in the chemical causing precipitation; chlorine or potassium permanganate may be required.

Precipitation

Precipitation (or pH adjustment) or iron and manganese from water in their solid forms can be effected in treatment plants by adjusting the pH of the water by adding lime or other chemicals. Some of the precipitate will settle out with time, while the rest is easily removed by sand filters. This process requires pH of the water to be in the range of 10–11.

Note: Although the precipitation or pH adjustment technique for treating water containing iron and manganese is effective, note that the pH level must be adjusted higher (10–11 range) to cause the precipitation, which means that the pH level must then also be lowered (to the 8.5 range or a bit lower) to use the water for consumption.

Oxidation

One of the most common methods of removing iron and manganese is through the process of oxidation (another chemical process), usually followed by settling and filtration. Air, chlorine, or potassium permanganate can oxidize these minerals. Each oxidant has advantages and disadvantages, an each operates slightly differently:

1. **Air**: To be effective as an oxidant, the air must come in contact with as much of the water as possible. Aeration is often accomplished by bubbling diffused air through the water by spraying the water up into the air, or by trickling the water over rocks, boards, or plastic packing materials in an aeration tower. The more finely divided the drops of water, the more oxygen comes in contact with the water and the dissolved iron and manganese.
2. **Chlorine**: This is one of the most popular oxidants for iron and manganese control because it is also widely used as a disinfectant; iron and manganese control by prechlorination can be as simple as adding a new chlorine feed point in a facility already feeding chlorine. It also provides a pre-disinfecting step that can help control bacterial growth through the rest of the treatment system. The downside to chorine use, however, is that when chlorine reacts with the organic materials found in surface water and some groundwaters, it forms TTHMs. This process also requires that the pH of the water be in the range of 6.5–7; because many groundwaters are more acidic than this, pH adjustment with lime, soda ash, or caustic soda may be necessary when oxidizing with chlorine.
3. **Potassium permanganate**: This is the best oxidizing chemical to use for manganese control removal. An extremely strong oxidant, it has the additional benefit of producing manganese dioxide during the oxidation reaction. Manganese dioxide acts as an adsorbent for soluble manganese ions. This attraction for soluble manganese provides removal to extremely low levels.

The oxidized compounds form precipitates that are removed by a filter. Note that sufficient time should be allowed from the addition of the oxidant to the filtration step. Otherwise, the oxidation process will be completed after filtration, creating insoluble iron and manganese precipitates in the distribution system.

Ion Exchange
The ion exchange process is used primarily to soften hard waters, but it will also remove soluble iron and manganese. The water passes through a bed of resin that adsorbs undesirable ions from the water, replacing them with less troublesome ions. When the resin has given up all its donor ions, it is regenerated with strong salt brine (sodium chloride); the sodium ions from the brine replace the adsorbed ions and restore the ion exchange capabilities.

Sequestering
Sequestering or stabilization may be used when the water contains mainly low concentration of iron, and the volumes needed are relatively small. This process does not actually remove the iron or manganese from the water but complexes (binds it chemically) it with other ions in a soluble form that is not likely to come out of solution (i.e., not likely oxidized).

Aeration
The primary physical process uses air to oxidize the iron and manganese. The water is either pumped up into the air or allowed to fall over an aeration device. The air oxidizes the iron and manganese that is then removed by use of a filter. The addition of lime to raise the pH is often added to the process. While this is called a physical process, removal is accomplished by chemical oxidation.

Potassium Permanganate Oxidation and Manganese Greensand
The continuous regeneration potassium greensand filter process is another commonly used filtration technique for iron and manganese control. Manganese greensand is a mineral (gluconite) that has been treated with alternating solutions manganous chloride and potassium permanganate. The result is a sand-like (zeolite) material coated with a layer of manganese dioxide—an adsorbent for soluble iron and manganese. Manganese greensand has the ability to capture (adsorb) soluble iron and manganese that may have escaped oxidation, as well as the capability of physically filtering out the particles of oxidized iron and manganese. Manganese greensand filters are generally set up as pressure filters, totally enclosed tanks containing the greensand. The process of adsorbing soluble iron and manganese "uses up" the greensand by converting the manganese dioxide coating to manganic oxide, which does not have the adsorption property. The greensand can be regenerated in much the same way as ion exchange resins, by washing the sand with potassium permanganate.

Hardness Treatment
Hardness in water is caused by the presence of certain positively charged metallic irons in solution in the water. The most common of these hardness-causing ions are calcium and magnesium; others include iron, strontium, and barium. As a general rule, groundwaters are harder than surface waters, so hardness is frequently of concern to the small water system operator. This hardness is derived from contact with soil and rock formations such as limestone. Although rainwater itself will not dissolve many solids, the natural carbon dioxide in the soil enters the water and forms carbonic acid (HCO), which is capable of dissolving minerals. Where soil is thick (contributing more carbon dioxide to the water) and limestone is present, hardness is likely to be a problem. The total amount of hardness in water is expressed as the sum of its calcium carbonate ($CaCO_3$) and its magnesium hardness. However, for practical purposes, hardness is expressed as calcium carbonate. This means that regardless of the amount of the various components that make up hardness, they can be related to a specific amount of calcium carbonate (e.g., hardness is expressed as mg/l as $CaCO_3$—milligrams per liter as calcium carbonate).

Note: The two types of water hardness are temporary hardness and permanent hardness. Temporary hardness is also known as carbonate hardness (hardness that can be removed by boiling); permanent hardness is also known as noncarbonate hardness (hardness that cannot be removed by boiling).

Water Pollution

TABLE 8.3
Classification of Hardness

Classification	mg/L CaCO$_3$
Soft	0–75
Moderately hard	75–150
Hard	150–300
Very hard	Over 300

Hardness is of concern in domestic water consumption because hard water increases soap consumption, leaves a soapy scum in the sink or tub, can cause water heater electrodes to burn out quickly, can cause discoloration of plumbing fixtures and utensils, and is perceived as less desirable water. In industrial water use, hardness is a concern because it can cause boiler scale and damage to industrial equipment.

The objection of customers to hardness is often dependent on the amount of hardness they are used to. People familiar with water with a hardness of 20 mg/L might think that a hardness of 100 mg/L is too much. On the other hand, a person who has been using water with a hardness of 200 mg/L might think that 100 mg/L was very soft. Table 8.3 lists the classifications of hardness.

Hardness Calculation

Recall that hardness is expressed as mg/L as CaCO$_3$. The mg/L of Ca and Mg must be converted to mg/L as CaCO$_3$ before they can be added. The hardness (in mg/L as CaCO$_3$) for any given metallic ion is calculated using the formula:

$$\text{Hardness}\left(\text{mg/L as CaCO}_3\right) = M\left(\text{mg/L}\right) \times \frac{50}{\text{Eq. Wt. of } M} \tag{8.6}$$

where:

M = metal ion concentration (mg/L)
Eq. Wt. = equivalent weight = gram molecular weight ÷ valence

Treatment Methods

Two common methods are used to reduce hardness: ion exchange and cation exchange:

- **Ion exchange process**: The ion exchange process is the most frequently used process for softening water. Accomplished by charging a resin with sodium ions, the resin exchanges the sodium ions for calcium and/or magnesium ions. Naturally occurring and synthetic cation exchange resins are available. Natural exchange resins include such substances as aluminum silicate, zeolite clays [zeolites are hydrous silicates found naturally in the cavities of lavas (greensand); glauconite zeolites; or synthetic, porous zeolites], humus, and certain types of sediments. These resins are placed in a pressure vessel. Salt brine is flushed through the resins. The sodium ions in the salt brine attach to the resin. The resin is now said to be charged. Once charged, water is passed through the resin and the resin exchanges the sodium ions attached to the resin for calcium and magnesium ions, thus removing them from the water. The zeolite clays are most common because they are quite durable, can tolerate extreme ranges in pH, and are chemically stable. They have relatively limited exchange capacities, however, so they should be used only for water with a moderate total hardness. One of the results is that the water may be more corrosive than before. Another concern is that addition of sodium ions to the water may increase the health risk of those with high blood pressure.

- **Cation exchange process**: The cation exchange process takes place with little or no intervention from the treatment plant operator. Water containing hardness-causing cations (Ca^{++}, Mg^{++}, Fe^{+3}) is passed through a bed of cation exchange resin. The water coming through the bed contains hardness near zero, although it will have elevated sodium content. (The sodium content is not likely to be high enough to be noticeable, but it could be high enough to pose problems to people on highly restricted salt-free diets.) The total lack of hardness in the finished water is likely to make it very corrosive, so normal practice bypasses a portion of the water around the softening process. The treated and untreated waters are blended to produce an effluent with a total hardness around 50–75 mg/L as $CaCO_3$

Corrosion

Water operators add chemicals (e.g., lime or sodium hydroxide) to water at the source or at the waterworks to control corrosion. Using chemicals to achieve slightly alkaline chemical balance prevents the water from corroding distribution pipes and consumers' plumbing. This keeps substances like lead from leaching out of plumbing and into the drinking water. For our purposes, we define corrosion as the conversion of a metal to a salt or oxide with a loss of desirable properties such as mechanical strength. Corrosion may occur over an entire exposed surface or may be localized at micro- or macroscopic discontinuities in metal. In all types of corrosion, a gradual decomposition of the material occurs, often due to an electrochemical reaction. Corrosion may be caused by (1) stray current electrolysis, (2) galvanic corrosion caused by dissimilar metals, or (3) differential concentration cells. Corrosion starts at the surface of a material and moves inward.

The adverse effects of corrosion can be categorized according to health, aesthetics, economic effects, and/or other effects. The corrosion of toxic metal pipe made from lead creates a serious health hazard. Lead tends to accumulate in the bones of humans and animals. Signs of lead intoxication include gastrointestinal disturbances, fatigue, anemia, and muscular paralysis. Lead is not a natural contaminant in either surface waters or groundwaters, and the MCL of 0.005 mg/l in source waters is rarely exceeded. It is corrosion by-product from high lead solder joints in copper and lead piping. Small dosages of lead can lead to developmental problems in children. The USEPA's Lead and Copper Rule addresses the matter of lead in drinking water exceeds specified action levels.

Note: The USEPA's Lead and Copper Rule requires that a treatment facility achieve optimum corrosion control. Since lead and copper contamination generally occurs after water has left the public water system, the best way for the water system operator to find out if customer water is contaminated is to test water that has come from a household faucet.

With regard to the Lead and Copper Rule, it is important to note that EPA made minor changes in 1999 to the original Rule. These minor revisions (also known as the Lead and Copper Rule Minor Revisions or LCRMR) streamline requirements, promote consistent nation implementation, and in many cases, reduce burden for water systems. The LCRMR do not change the action levels of 0.015 mg/L for lead and 1.3 mg/L, for copper, or Maximum Contaminant Level Goals established by the 1991 Lead and Copper Rule ("the rule"), which are 0 mg/L of lead and 1.3 mg/L for copper. They also do not affect the rule's basic requirements to optimize corrosion control and, if appropriate, treat source water, deliver public education, and replace lead service lines.

Cadmium is the only other toxic metal found in samples from plumbing systems. Cadmium is a contaminant found in zinc. Its adverse health effects are best known for being associated with severe bone and kidney syndrome in Japan. The PMCL for cadmium is 0.01 mg/L.

Note: Water systems should try to supply water free of lead and has no more than 1.3 milligrams of copper (mg/L). This is a non-enforceable health goal.

Aesthetic effects that are a result of corrosion of iron are characterized by "pitting" and are a consequence of the deposition of ferric hydroxide and other products and the solution of iron—tuberculation. Tuberculation reduces the hydraulic capacity of the pipe. Corrosion of iron can cause customer complaints of reddish or reddish-brown staining of plumbing fixtures and laundry. Corrosion of copper lines can cause customer complaints of bluish or blue-green stains on plumbing fixtures.

Water Pollution

Sulfide corrosion of copper and iron lines can cause a blackish color in the water. The by-products of microbial activity (especially iron bacteria) can cause foul tastes and/or odors in the water.

The economic effects of corrosion may include the need for water main replacement, especially when tuberculation reduces the flow capacity of the main. Tuberculation increases pipe roughness, causing an increase in pumping costs and reducing distribution system pressure. Tuberculation and corrosion can cause leaks in distribution mains and household plumbing. Corrosion of household plumping may require extensive treatment, public education, and other actions under the Lead and Copper Rule.

Other effects of corrosion include short service life of household plumbing caused by pitting. Buildup of mineral deposits in the hot water system may eventually restrict hot water flow. Also, the structural integrity of steel water storage tanks may deteriorate, causing structural failures. Steel ladders in clear wells or water storage tanks may corrode, introducing iron into the finished water. Steel parts in flocculation tanks, sedimentation basins, clarifiers, and filters may also corrode.

Types of Corrosion

Three types of corrosion occur in water mains: galvanic, tuberculation, and pitting:

- Galvanic occurs when two dissimilar metals are in contact and are exposed to a conductive environment, a potential exists between them and current flows. This type of corrosion is the result of an electrochemical reaction when the flow of electric current itself is an essential part of the reaction.
- Tuberculation refers to the formation of localized corrosion products scattered over the surface in the form of knob-like mounds. These mounds increase the roughness of the inside of the pipe, increasing resistance to water flow and decreasing the C-factor of the pipe.
- Pitting is localized corrosion is generally classified as pitting when the diameter of the cavity at the metal surface is the same or less than the depth.

Factors Affecting Corrosion

The primary factors affecting corrosion are pH, alkalinity, hardness (calcium), dissolved oxygen, and total dissolved solids. Secondary factors include temperature, velocity of water in pipes, and carbon dioxide (CO_2).

Determination of Corrosion Problems

To determine if corrosion is taking place in water mains, materials removed from the distribution system should be examined for sings of corrosion damage. A primary indicator of corrosion damage is pitting. (Note: measure depth of pits to gauge the extent of damage.) Another common method used to determine if corrosion or scaling is taking place in distribution lines is by inserting special steel specimens of known weight (called coupons) in the pipe and examining them for corrosion after a period of time. Detecting evidence of leaks, conducting flow tests and chemical tests for dissolved oxygen and toxic metals, as well as customer complains (red or black water and/or laundry and fixture stains) are also used to indicate corrosion problems.

Formulas can also be used to determine corrosion (to an extent). The Langlier Saturation Index (L.I.) and Aggressive Index (A.I.) are two of the commonly used indices. The L.I. is a method used to determine if water is corrosive. A.I refers to waters that have low natural pH, are high in dissolved oxygen, are low in total dissolved solids, and have low alkalinity and low hardness. These waters are very aggressive and can be corrosive. Both the Langlier Saturation and Aggressive Indices are typically used as starting points in determining the adjustments required to produce a film:

L.I. ~0.5
A.I. value of 12 or higher

Note: LI and AI are based on the dissolving of and precipitation of calcium carbonate; therefore, the respective indices may not actually reflect the corrosive nature of the particular water for a specific pipe material. However, they can be useful tools in selecting materials or treatment options for corrosion control.

Corrosion Control

As mentioned, one method used to reduce the corrosive nature of water is chemical addition. Selection of chemicals depends on the characteristics of the water, where the chemicals can be applied, how they can be applied and mixed with water, and the cost of the chemicals.

If the product of the calcium hardness times the alkalinity of the water is <100, treatments may be required. Both lime and CO_2 may be required for proper treatment of the water. If the calcium hardness and alkalinity levels are between 100 and 500, either lime or soda ash (Na_2CO_3) will be satisfactory. The decision regarding which chemical to use depends on the cost of the equipment and chemicals. If the product of the calcium hardness times the alkalinity is >500, either lime or caustic (NaOH) may be used. Soda ash will be ruled out because of the expense.

The chemicals chosen for treatment of public drinking water supplies modify the water characteristics, making the water less corrosive to the pipe. Modification of water quality can increase the pH of the water, reducing the hydrogen ions available for galvanic corrosion, as well as reducing the solubility of copper, zinc, iron, lead, and calcium, and increasing the possibility of forming carbonate protective films.

Calcium carbonate stability is the most effective means of controlling corrosion. Lime, caustic soda, or soda ash is added until the pH and the alkalinity indicates the water is saturated with calcium carbonate. Saturation does not always assure non-corrosiveness. Utilities should also exercise caution when applying sodium compounds, since high sodium content in water can be a health concern for some customers. By increasing the alkalinity of the water, the bicarbonate and carbonate available to form protective carbonate film increase. By decreasing the dissolved oxygen of the water, the rate of galvanic corrosion is reduced, along with the possibility of iron tuberculation.

Inorganic phosphates used include the following:

a. Zinc phosphates, which can cause algae blooms on open reservoirs.
b. Sodium silicate, which is used by individual customers, such as apartments, houses, and office buildings.
c. Sodium polyphosphates (tetrasodium pyrophosphate or sodium hexametaphosphate), which control scale formation in supersaturated waters and are known as sequestering agents.
d. Silicates (SiO_2), which form a film; an initial dosage of 12–16 mg/L for about 30 days will adequately coat the pipes, and a 1.0 mg/L concentration should be maintained thereafter.

Caution: Great care and caution must be exercised any time feeding corrosion control chemicals into a public drinking water system!

Another corrosion control method is aeration. Aeration works to remove carbon dioxide (CO_2); it can be reduced to about 5 mg/L. Cathodic protection, often employed to control corrosion, is achieved by applying an outside electric current to the metal to reverse the electromechanical corrosion process. The application of D-C current prevents normal electron flow. Cathodic protection uses a sacrificial metal electrode (a magnesium anode) that corrodes instead of the pipe or tank. Linings, coatings, and paints can also be used in corrosion control. Slip-line with plastic liner, cement mortar, zinc or magnesium, polyethylene, epoxy, and coal tar enamels are some of the materials that can be used.

Caution: Before using any protective coatings, consult the district engineer first!

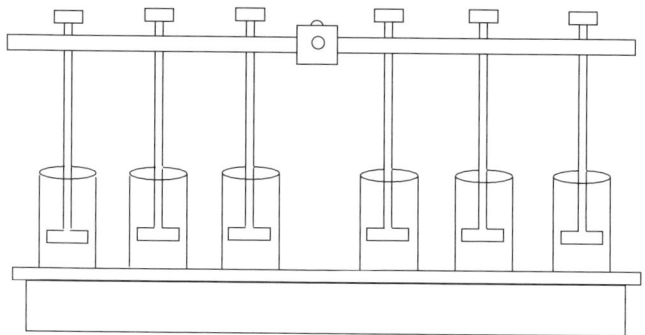

FIGURE 8.6 Variable-speed paddle mixer used in jar testing procedure.

Several *corrosive resistant pipe materials* are used to prevent corrosion:

1. PVC plastic pipe
2. Aluminum
3. Nickel
4. Silicon
5. Brass
6. Bronze
7. Stainless steel
8. Reinforced concrete

In addition to internal corrosion problems, waterworks operators must also be concerned with external corrosion problems. The primary culprit involved with external corrosion of distribution system pipe is soil. The measure of corrosivity of the soil is the soil resistivity. If the soil resistivity is >5,000 Ω/cm, serious corrosion is unlikely. Steel pipe may be used under these conditions. If soil resistivity is <500 Ω/cm, plastic PVC pipe should be used. For intermediate ranges of soil resistivity (500–5,000 Ω/cm), use ductile iron pipe, lining, and coating.

Several operating problems associated are commonly associated with corrosion control:

1. **$CaCO_3$ not depositing a film**: usually a result of poor pH control (out of the normal range of 6.5–8.5). This may also cause excessive film deposition.
2. **Persistence of red water problems**: most probably a result of poor flow patterns, insufficient velocity, tuberculation of pipe surface, and presence of iron bacteria.
    a. **Velocity**: Chemicals need to contact pipe surface. Dead ends and low-flow areas should have flushing program; dead ends should be looped.
    b. **Tuberculation**: the best approach is to clean with "pig." In extreme cases, clean pipe with metal scrapers and install cement-mortar lining.
    c. **Iron bacteria**: Slime prevents film contact with pipe surface. Slime will grow and lose coating. Pipe cleaning and disinfection program are needed.

Coagulation

The primary purpose in surface-water treatment is chemical clarification by coagulation and mixing, flocculation, sedimentation, and filtration. These units processes, along with disinfection, work to remove particles, naturally occurring organic matter (NOM—i.e., bacteria, algae, zooplankton, and organic compounds), microbes from water and to produce water that is non-corrosive. Specifically, coagulation/flocculation work to destabilize particles and agglomerate dissolved and particulate matter. Sedimentation removes solids and provides ½ log Giardia and 1 log virus removal. Filtration removes solids and provides 2 log Giardia and 1 log virus removal. Finally, disinfection provides microbial inactivation and ½ Giardia and 2 log Virus removal.

From Figure 8.4, it can be seen that following screening and the other pretreatment processes, the next unit process in a conventional water treatment system is a mixer where chemicals are added in what is known as coagulation. The exception to this unit process configuration occurs in small systems using groundwater, when chlorine or other taste and odor control measures are introduced at the intake and are the extent of treatment.

Materials present in raw water may vary in size, concentration, and type. Dispersed substances in the water may be classified as suspended, colloidal, or solution. Suspended particles may vary in mass and size and are dependent on the flow of water. High flows and velocities can carry larger material. As velocities decrease, the suspended particles settle according to size and mass.

Other material may be in solution, for example, salt dissolves in water. Matter in the colloidal state does not dissolve, but the particles are so small they will not settle out of the water. Color (as in tea-colored swamp water) is mainly due to colloids or extremely fine particles of matter in suspension. Colloidal and solute particles in water are electrically charged. Because most of the charges are alike (negative) and repel each other, the particles stay dispersed and remain in the colloidal or soluble state.

Suspended matter will settle without treatment, if the water is still enough to allow it to settle. The rate of settling of particles can be determined, as this settling follows certain laws of physics. However, much of the suspended matter may be so slow in settling that the normal settling processes become impractical, and if colloidal particles are present, settling will not occur. Moreover, water drawn from a raw water source often contains many small unstable (un-sticky) particles; therefore, sedimentation alone is usually an impractical way to obtain clear water in most locations, and another method of increasing the settling rate must be used: coagulation, which is designed to convert stable (unsticky) particles to unstable (sticky) particles.

The term coagulation refers to the series of chemical and mechanical operations by which coagulants are applied and made effective. These operations are comprised of two distinct phases: (1) rapid mixing to disperse coagulant chemicals by violent agitation into the water being treated and (2) flocculation to agglomerate small particles into well-defined floc by gentle agitation for a much longer time.

Coagulation results from adding salts of iron or aluminum to the water. The coagulant must be added to the raw water and perfectly distributer into the liquid; such uniformity of chemical treatment is reached through rapid agitation or mixing. Common coagulants (salts) include:

- Alum—aluminum sulfate
- Sodium aluminate
- Ferric sulfate
- Ferrous sulfate
- Ferric chloride
- Polymers

Coagulation is the reaction between one of these salts and water. The simplest coagulation process occurs between alum and water. Alum or aluminum sulfate is made by a chemical reaction of bauxite ore and sulfuric acid. The normal strength of liquid alum is adjusted to 8.3%, while the strength of dry alum is 17%.

When alum is placed in water, a chemical reaction occurs that produces positively charged aluminum ions. The overall result is the reduction of electrical charges and the formation of a sticky substance—the formation of floc, which when properly formed, will settle. These two destabilizing factors are the major contributions that coagulation makes to the removal of turbidity, color, and microorganisms.

Water Pollution

Liquid alum is preferred in water treatment because it has several advantages over other coagulants, including the following:

1. Ease of handling
2. Lower costs
3. Less labor required to unload, store, and convey
4. Elimination of dissolving operations
5. Less storage space required
6. Greater accuracy in measurement and control provided
7. Elimination of the nuisance and unpleasantness of handling dry alum
8. Easier maintenance

The formation of floc is the first step of coagulation; for greatest efficiency, rapid, intimate mixing of the raw water and the coagulant must occur. After mixing, the water should be slowly stirred so that the very small, newly formed particles can attract and enmesh colloidal particles, holding them together to form larger floc. This slow mixing is the second stage of the process (flocculation), covered later.

A number of factors influence the coagulation process—pH, turbidity, temperature, alkalinity, and the use of polymers. The degree to which these factors influence coagulation depends upon the coagulant use. The raw water conditions, optimum pH for coagulation and other factors must be considered before deciding which chemical is to be fed and at what levels.

To determine the correct chemical dosage, a Jar Test or Coagulation Test is performed. Jar tests (widely used for many years by the water treatment industry) simulate full-scale coagulation and flocculation processes to determine optimum chemical dosages. It is important to note that jar testing is only an attempt to achieve a ballpark approximation of correct chemical dosage for the treatment process. The test conditions are intended to reflect the normal operation of a chemical treatment facility. The test can be used to:

- Select the most effective chemical
- Select the optimum dosage
- Determine the value of a flocculant aid and the proper dose

The testing procedure requires a series of samples to be placed in testing jars (see Figure 8.6) and mixed at 100 ppm. Varying amounts of the process chemical or specified amounts of several flocculants are added (one volume/sample container). The mix is continued for one minute. Next, the mixing is slowed to 30 rpms to provide gentle agitation, and then the floc is allowed to settle. The flocculation period and settling process is observed carefully to determine the floc strength, settleability, and clarity of the supernatant liquor (defined: the water that remains above the settled floc). Additionally, the supernatant can be tested to determine the efficiency of the chemical addition for removal of TSS, BOD_5, and phosphorus.

The equipment required for the jar test includes a six-position, variable speed paddle mixer (see Figure 8.6), six two-quart wide-mouthed jars, an interval timer, and assorted glassware, pipettes, graduates, and so forth. The jar testing procedure follows:

1. Place an appropriate volume of water sample in each of the jars (250–1,000 mL samples may be used, depending upon the size of the equipment being used). Start mixers and set for 100 rpm.
2. Add previously selected amounts of the chemical being evaluated. (Initial tests may use wide variations in chemical volumes to determine the approximate range. This is then narrowed in subsequent tests).
3. Continue mixing for 1 minute.

4. Reduce the mixer speed to a gentle agitation (30 rpm) and continue mixing for 20 minutes. Again, time and mixer speed may be varied to reflect the facility.
    Note: During this time, observe the floc formation—how well the floc holds together during the agitation (floc strength).
5. Turn off the mixer and allow solids to settle for 20–30 minutes. Observe the settling characteristics, the clarity of the supernatant, the settleability of the solids, the flocculation of the solids, and the compactibilty of the solids.
6. Perform phosphate tests to determine removals.
7. Select the dose that provided the best treatment based on the observations made during the analysis.

Note: After initial ranges and/or chemical selections are completed, repeat the test using a smaller range of dosages to optimize performance.

Flocculation

As can be seen in Figure 8.4, flocculation follows coagulation in the conventional water treatment process. Flocculation is the physical process of slowly mixing the coagulated water to increase the probability of particle collision—unstable particles collide and stick together to form fewer larger flocs. Through experience, we see that effective mixing reduces the required amount of chemicals and greatly improves the sedimentation process, which results in longer filter runs and higher quality finished water.

The goal of flocculation is to form a uniform feather-like material similar to snowflakes—a dense, tenacious floc that entraps the fine, suspended, and colloidal particles and carries them down rapidly in the settling basin. Proper flocculation requires from 15 to 45 minutes. The time is based on water chemistry, water temperature, and mixing intensity. Temperature is the key component in determining the amount of time required for floc formation. To increase the speed of floc formation and the strength and weight of the floc, polymers are often added.

Sedimentation

After raw water and chemicals have been mixed and the floc formed, the water containing the floc (because it has a higher specific gravity than water) flows to the sedimentation or settling basin (see Figure 8.4). Sedimentation is also called clarification. Sedimentation removes settleable solids by gravity. Water moves slowly through the sedimentation tank/basin with a minimum of turbulence at entry and exit points with minimum short-circuiting. Sludge accumulates at bottom of tank/basin. Typical tanks or basins used in sedimentation include conventional rectangular basins, conventional center-feed basins, peripheral-feed basins, and spiral-flow basins.

In conventional treatment plants, the amount of detention time required for settling can vary from 2 to 6 hours. Detention time should be based on the total filter capacity when the filters are passing 2 gpm per square foot of superficial sand area. For plants with higher filter rates, the detention time is based on a filter rate of 3–4 gpm per square foot of sand area. The time requirement is dependent on the weight of the floc, the temperature of the water, and how quiescent (still) the basin.

A number of conditions affect sedimentation: (1) uniformity of flow of water through the basin; (2) stratification of water due to difference in temperature between water entering and water already in the basin; (3) release of gases that may collect in small bubbles on suspended solids, causing them to rise and float as scum rather than settle as sludge; (4) disintegration of previously formed floc; and (5) size and density of the floc.

Filtration

In the conventional water treatment process, filtration usually follows coagulation, flocculation, and sedimentation (see Figure 8.4). At present, filtration is not always used in small water systems. However, recent regulatory requirements under USEPA's Interim Enhanced Surface Water Treatment rules may make water filtering necessary at most water supply systems. Water filtration is a physical process of separating suspended and colloidal particles from water by passing water through a granular material. The process of filtration involves straining, settling, and adsorption. As floc passes into the filter, the spaces between the filter grains become clogged, reducing this opening, and increasing removal. Some material is removed merely because it settles on a media grain. One of the most important processes is the adsorption of the floc onto the surface of individual filter grains. This helps collect the floc and reduces the size of the openings between the filter media grains. In addition to removing silt and sediment, floc, algae, insect larvae, and any other large elements, filtration also contributes to the removal of bacteria and protozoans such as Giardia lamblia and Cryptosporidium. Some filtration processes are also used for iron and manganese removal.

Types of Filter Technologies

The Surface Water Treatment Rule (SWTR) specifies four filtration technologies, although SWTR also allows the use of alternate filtration technologies (e.g., cartridge filters). The specified technologies are (1) slow sand filtration/rapid sand filtration, (2) pressure filtration, (3) diatomaceous Earth filtration, and (4) direct filtration. Of these, all but rapid sand filtration is commonly employed in small water systems that use filtration. Each type of filtration system has advantages and disadvantages. Regardless of the type of filter, however, filtration involves the processes of straining (where particles are captured in the small spaces between filter media grains), sedimentation (where the particles land on top of the grains and stay there), and adsorption (where a chemical attraction occurs between the particles and the surface of the media grains).

Slow Sand Filters

The first slow sand filter was installed in London in 1829 and was used widely throughout Europe, though not in the U.S. By 1900, rapid sand filtration began taking over as the dominant filtration technology, and a few slow sand filters are in operation today. However, with the advent of the Safe Drinking Water Act and its regulations (especially the Surface Water Treatment Rule) and the recognition of the problems associated with Giardia lamblia and Cryptosporidium in surface water, the water industry is reexamining slow sand filters. Because of low technology requirements may prevent many state water systems from using this type of equipment.

On the plus side, slow sand filtration is well suited for small water systems. It is a proven, effective filtration process with relatively low construction costs and low operating costs (it does not require constant operator attention). It is quite effective for water systems as large as 5,000 people, beyond that, surface area requirements and manual labor required to recondition the filters make rapid sand filters more effective. The filtration rate is generally in the range of 45–150 gallons per day per square foot. Components making up a slow sand filter include the following:

- A covered structure to hold the filter media
- An underdrain system
- Graded rock that is placed around and just above the underdrain
- The filter media, consisting of 30–55 inches of sand with a grain size of 0.25–0.35 mm
- Inlet and outlet piping to convey the water to and from the filter, and the means to drain filtered water to waste

The area above the top of the sand layer is flooded with water to a depth of 3–5 ft, and the water is allowed to trickle down through the sand. An overflow device prevents excessive water depth. The filter must have provisions for filling it from the bottom up, and it must be equipped with a loss-of-head gauge, a rate-of-flow control device (such as an orifice or butterfly valve), a weir or effluent pipe that assures that the water level cannot drop below the sand surface, and filtered waste sample taps.

When the filter is first placed in service, the head loss through the media caused by the resistance of the sand is about 0.2 ft (i.e., a layer of water 0.2 ft deep on top of the filter will provide enough pressure to push the water downward through the filter). As the filter operates, the media becomes clogged with the material being filtered out of the water, and the head loss increase. When it reaches about 4–5 ft, the filter needs to be cleaned.

For efficient operation of a slow sand filter, the water being filtered should have a turbidity average <5 TU, with a maximum of 30 TU. Slow sand filters are not backwashed the way conventional filtration units are. The 1–2 inches of material must be removed on a periodic basis to keep the filter operating.

Rapid Sand Filters

The rapid sand filter, which is similar in some ways to slow sand filter, is one of the most widely used filtration units. The major difference is in the principle of operation; that is, in the speed or rate at which water passes through the media. In operation, water passes downward through a sand bed that removes the suspended particles. The suspended particles consist of the coagulated matter remaining in the water after sedimentation, as well as a small amount of uncoagulated suspended matter.

Some significant differences exist in construction, control, and operation between slow sand filters and rapid sand filters. Because of the construction and operation of the rapid sand filtration with its higher filtration the land area needed to filter the same quantity of water is reduced. Components of a rapid sand filter include:

- Structure to house media
- Filter media
- Gravel media support layer
- Underdrain system
- Valves and piping system
- Filter backwash system
- Waste disposal system

Usually 2–3 ft deep, the filter media is supported by ~1 foot of gravel. The media may be fine sand or a combination of sand, anthracite coal, and coal (dual-multimedia filter). Water is applied to a rapid sand filter at a rate of 1.5 to gallons per minute per square foot of filter media surface. When the rate is between 4 and 6 gpm/ft^2, the filter is referred to as a high-rate filter; at a rate over 6 gpm/ft^2, the filter is called ultra-high-rate. These rates compare to the slow sand filtration rate of 45–150 gallons per day per square foot. High-rate and ultra-high-rate filters must meet additional conditions to assure proper operation.

Generally, raw water turbidity is not that high. However, even if raw water turbidity values exceed 1,000 TU, properly operated rapid sand filters can produce filtered water with turbidity or well under 0.5 TU. The time the filter is in operation between cleanings (filter runs) usually lasts from 12 to 72 hours, depending on the quality of the raw water; the end of the run is indicated by the head loss approaching 6–8 ft. Filter breakthrough (when filtered material is pulled through the filter into the effluent) can occur if the head loss becomes too great. Operation with head loss too high can also cause air binding (which blocks part of the filter with air bubbles), increasing the flow rate through the remaining filter area.

Rapid sand filters have the advantage of lower land requirements, and they have other advantages, too. For example, rapid sand filters cost less, are less labor-intensive to clean, and offer higher efficiency with highly turbid waters. On the downside, operation and maintenance costs of rapid sand filters are much higher because of the increased complexity of the filter controls and backwashing system.

When backwashing a rapid sand filter, the filter is cleaned by passing treated water backwards (upwards) through the filter media and agitating the top of the media. The need for backwashing is determined by a combination of filter run time (i.e., the length of time since the last backwashing), effluent turbidity, and head loss through the filter. Depending on the raw water quality, the run time varies from one filtration plant to another (and may even vary from one filter to another in the same plant).

Note: Backwashing usually requires 3%–7% of the water produced by the plant.

Pressure Filter Systems

When raw water is pumped or piped from the source to a gravity filter, the head (pressure) is lost as the water enters the floc basin. When this occurs, pumping the water from the plant clear well to the reservoir is usually necessary. One way to reduce pumping is to place the plant components into pressure vessels, thus maintaining the head. This type of arrangement is called a pressure filter system. Pressure filters are also quite popular for iron and manganese removal and for filtration of water from wells. They may be placed directly in the pipeline from the well or pump with little head loss. Most pressure filters operate at a rate of about 3 gpm/ft^2.

Operationally the same and consisting of components similar to those of a rapid sand filter, the main difference between a rapid sand filtration system and a pressure filtration system is that the entire pressure filter is contained within a pressure vessel. These units are often highly automated and are usually purchased as self-contained units with all necessary piping, controls, and equipment contained in a single unit. They are backwashed in much the same manner as the rapid sand filter.

The major advantage of the pressure filter is its low initial cost. They are usually prefabricated, with standardized designs. A major disadvantage is that the operator is unable to observe the filter in the pressure filter and so is unable to determine the condition of the media. Unless the unit has an automatic shutdown feature on high effluent turbidity, driving filtered material through the filter is possible.

Diatomaceous Earth Filters

Diatomaceous earth is a white material made from the skeletal remains of diatoms. The skeletons are microscopic, and in most cases, porous. There are different grades of diatomaceous earth, and the grade is selected based on filtration requirements. These diatoms are mixed in water slurry and fed onto a fine screen called a septum, usually of stainless steel, nylon, or plastic. The slurry is fed at a rate of 0.2 lb per square foot of filter area. The diatoms collect in a pre-coat over the septum, forming an extremely fine screen. Diatoms are fed continuously with the raw water, causing the buildup of a filter cake ~1/8 to 1/5 inch thick. The openings are so small that the fine particles that cause turbidity are trapped on the screen. Coating the septum with diatoms gives it the ability to filter out very small, microscopic material. The fine screen and the buildup of filtered particles cause a high head loss through the filter; when the head loss reaches a maximum level (30 psi on a pressure-type filter or 15 inches or mercury on a vacuum-type filter), the filter cake must be removed by backwashing.

A slurry of diatoms is fed with raw water during filtration in a process called body feed. The body feed prevents premature clogging of the septum cake. These diatoms are caught on the septum, increasing the headloss and preventing the cake from clogging too rapidly by the particles being filtered. While the body feed increases headloss, headloss increases are more gradual than if body feed were not used.

Diatomaceous earth filters are relatively low in cost to construct, but they have high operating costs and can give frequent operating problems if not properly operated and maintained. They can be used to filter raw surface waters or surface-influenced groundwaters, with low turbidity (<5 NTU), low coliform concentrations (no more than 50 coliforms/100 mL) and may also be used for iron and manganese removal following oxidation. Filtration rates are between 1.0 and 1.5 gpm/ft^2.

Direct Filtration
Direct filtration is a treatment scheme that omits the flocculation and sedimentation steps prior to filtration. Coagulant chemicals are added, and the water is passed directly onto the filter. All solids removal takes place on the filter, which can lead to much shorter filter runs, more frequent back-washing, and a greater percentage of finished water used for backwashing. The lack of a flocculation process and sedimentation basin reduces construction cost but increases the requirement for skilled operators and high-quality instrumentation. Direct filtration must be used only where the water flow rate and raw water quality are fairly consistent and where the incoming turbidity is low.

Alternative Filters
A cartridge filter system can be employed as an alternate filtering system to reduce turbidity and remove Giardia. A cartridge filter is made of a synthetic media contained in a plastic or metal housing. These systems are normally installed in a series of three or four filters. Each filter contains a media that is successively smaller than the previous filter. The media sizes typically range from 50μ to 5μ or less. The filter arrangement is dependent on the quality of the water, the capability of the filter, and the quantity of water needed. The USEPA and state agencies have established criteria for the selection and use of cartridge filters. Generally, cartridge filter systems are regulated in the same manner as other filtration systems.

Because of new regulatory requirements and the need to provide more efficient removal of pathogenic protozoans (e.g., Giardia and Cryptosporidium) from water supplies, membrane filtration systems are finding increased application in water treatment systems. A membrane is a thin film separating two different phases of a material acting as a selective barrier to the transport of matter operated by some driving force. Simply, a membrane can be regarded as a sieve with very small pores. Membrane filtration processes are typically pressure, electrically, vacuum, or thermally driven. The types of drinking water membrane filtration systems include microfiltration, ultrafiltration, nanofiltration, and reverse osmosis. In a typical membrane filtration process, there is one input and two outputs. Membrane performance is largely a function of the properties of the materials to be separated and can vary throughout the operation.

Filtration and Compliance with Turbidity Requirements
Under the1996 Safe Drinking Water Act (SDWA) Amendments, USEPA must supplement the existing 1989 Surface Water Treatment Rule (SWTR) with the Interim Enhanced Surface Water Treatment Rule (IESWTR) to improve protection against waterborne pathogens. Key provisions established in the IESWTR include (USEPA, 1998).

- A maximum contaminant level goal (MCLG) of zero for Cryptosporidium; 2-log (99%) Cryptosporidium removal requirements for systems that filter
- Strengthened combined filter effluent turbidity performance standards
- Individual filter turbidity monitoring provisions
- Disinfection benchmark provisions to assure continued levels of microbial protection while facilities take the necessary steps to comply with new disinfection byproduct standards
- Inclusion of Cryptosporidium in the definition of groundwater under the direct influence of surface water (GWUDI) and in the watershed and in the watershed control requirements for unfiltered public water systems
- Requirements for covers on new finished water reservoirs
- Sanitary surveys for all surface water systems regardless of size

Additional Compliance Issues
The following section outlines additional compliance issues associated with the IESWTR.

Water Pollution

Schedule

The IESWTR was published on December 16, 1998, and became effective on February 16, 1999. The SDWA requires, within 24 months following the promulgation of a rule, that the Primacy Agencies adopt any State regulations necessary to implement the rule. Under Sec. 14.13, these rules must be at least as stringent as those required by USEPA. Thus, primary agencies must promulgate regulations which are at least as stringent as the IESWTR by December 17, 2000. Beginning December 17, 2001, systems serving at least 10,000 people must meet the turbidity requirements in Section 141.173.

Individual Filter Follow-up Action

Based on the monitoring results obtained through continuous filter monitoring, a system may have to conduct one of the following follow-up actions due to persistently high turbidity levels at an individual filter:

- Filter profile
- Individual filter self-assessment
- Comprehensive Performance Evaluation

These specific requirements are found in section 141.175(b)(1)-(4).

Abnormal Filter Operations—Filter Profile

A filter profile must be produced if no obvious reason for abnormal filter performance can be identified. A filter profile is a graphical representation of individual filter performance based on continuous turbidity measurements or total particle counts versus time for an entire filter run, from startup to backwash inclusively that includes an assessment of filter performance while another filter is being backwashed. The run length during this assessment should be representative of typical plant filter runs. The profile should include an explanation of the cause of any filter performance spikes during the run. Examples of possible abnormal filter operations which may be obvious to operators include the following:

- Outages or maintenance activities at processes within the treatment train
- Coagulant feed pump or equipment failure
- Filters being run at significantly higher loading rates than approved

It is important to note that while the reasons for abnormal filter operation may appear obvious, they could be masking other reasons which are more difficult to identify. These may include situations such as:

- Distribution in filter media
- Excessive or insufficient coagulant dosage
- Hydraulic surges due to pump changes or other filters being brought on/offline.

Systems need to use the best professional judgment and discretion when determining when to develop a filter profile. Attention at this stage will help systems avoid the other forms of follow-up action described below.

Individual Filter Self-Assessment

A system must conduct an individual filter self-assessment for any individual filter that has a measured turbidity level of greater than 1.0 NTU in two consecutive measurements taken 15 minutes apart in each of three consecutive months. The system must report the filter number, the turbidity measurement, and the dates on which the exceedance occurred.

Comprehensive Performance Evaluation

A system must conduct a comprehensive performance evaluation (CPE) if any individual filter has a measured turbidity level of greater than 2.0 NTU in two consecutive measurements taken 15 minutes apart in two consecutive months. The system must report the filter number, the turbidity measurement, and the date(s) on which the exceedance occurred. The system shall contact the State or a third party approved by the State to conduct a comprehensive performance evaluation.

Note: USEPA has developed a guidance document called, Handbook: Optimizing Water Treatment Plant Performance Using the Composite Correction Program (EPA/625/6-91/027) (1998).

Notification

The IESWTR contains two distinct types of notification: state and public. It is important to understand the differences between each and the requirements of each.

- **State notification**: Systems are required to notify States under Section 141.31. Systems must report to the State within 48 hours, the failure to comply with any national primary drinking water regulation. The system within 10 days of completion of each public notification required pursuant to Section 141.32 must submit to the State a representative copy of each type of notice distributed, published, posted, and/or made available to persons served by the system and/or the media. The water supply system must also submit to the State (within the time stated in the request made by the State) copies of any records required to be maintained under Section 141.33 or copies of any documents then in existence which the State or the Administrator is entitled to inspect pursuant to the authority of Section 1445 of the Safe Drinking Water Act or the equivalent provisions of the State Law.

Public Notification

The IESWTR specifies that the public notification requirements of the Safe Drinking Water Act (SDWA) and the implementation regulations of 40 CFR Section 141.32 must be followed. These regulations divide public notification requirements into two tiers. These tiers are defined as follows:

- TIER 1
    Failure to comply with MCL
    Failure to comply with the prescribed treatment technique
    Failure to comply with a variance or exemption schedule
- TIER 2
    Failure to comply with monitoring requirements
    Failure to comply with a testing procedure prescribed by a NPDWR
    Operating under a variance/exemption. This is not considered a violation but public notification is required.

Certain general requirements must be met by all public notices must meet. All notices must provide a clear and readily understandable explanation of the violation, any potential adverse health effects, the population at risk, the steps the system is taking to correct the violation, the necessity of seeking alternate water supplies (if any) and any preventative measures the consumer should take. The notice must be conspicuous, and not contain any unduly technical language, unduly small print, or similar problems. The notice must include the telephone number of the owner or operator or designee of the public water system as a source of additional information concerning the violation where appropriate. The notice must be bi- or multilingual if appropriate.

Tier 1 Violations In addition, the public notification rule requires that when providing notification on potential adverse health effects in Tier 1 public notices and in notices on the granting and continued existence of a variance or exemption, the owner-operator of a public water system must include certain mandatory health effects language. For violations of treatment technique requirements for filtration and disinfection, the mandatory health effects language is:

> The USEPA sets drinking water standards and has determined that the presence of microbiological contaminants is a health concern at certain levels of exposure. If water is inadequately treated, microbiological contaminants in that water cause disease. Disease symptoms may include diarrhea, cramps, nausea, and possibly jaundice, and any associated headaches and fatigue. These symptoms, however, are not just associated with disease-causing organisms in drinking water, but also may be caused by a number of factors other than your drinking water. USEPA has set enforceable requirements for treating drinking water to reduce the risk of these adverse health effects. Treatment such as filtering and disinfection the water removes or destroys microbiological contaminants. Drinking water which is treated to meet UEPA requirements is associated with little to none of this risk and should be considered safe.

Further, the owner or operator of a community water system must give a copy of the most recent notice for any Tier 1 violations to all new billing units or hookups prior to or at the time service begins.

The medium for performing public notification and the time period in which notification must be sent varies with the type of violation and is specified in Section 141.32. For Tier 1 violations, the owner or operator of a public water system must give notice:

> By publication in a local daily newspaper as soon as possible but in no case later than 14 days after the violation or failure. If the area does not have a daily newspaper, then notice shall be given by publication in a weekly newspaper of general circulation in the area, and
>
> By either direct mail delivery or hand delivery of the notice, either by itself or with the water bill no later than 45 days after the violation or failure. The Primacy Agency may waive the requirement if it determines that the owner or operator has corrected the violation with 45 days.

Although the IESWTR does not specify any acute violations, the Primacy Agency may specify some Tier 1 violations as posing an acute risk to human health; examples might include:

- A waterborne outbreak in an unfiltered supply
- Turbidity of a filtered water exceeds 1.0 NTU at any time
- Failure to maintain a disinfectant residual of at least 0.2 mg/L in the water being delivered to the distribution system.

For these violations or any others defined by the Primacy Agency as "acute" violations, the system must furnish a copy of the notice to the radio and television stations serving the area as soon as possible but in no case later than 72 hours after the violation. Depending on the circumstances particular to the system, as determined by the Primacy Agency, the notice may instruct that all water be boiled prior to consumption.

Following the initial notice, the owner or operator must give notice at least once every three months by mail delivery (either with the water bill), or by hand delivery, for as long as the violation or failures exist.

There are two variations on these requirements. First, the owner or operator of a community water system in an area not served by a daily or weekly newspaper must give notice within 14 days after the violation by hand delivery or continuous posting of a notice of the violation. The notice must continue for as long as the violation exists. Notice by hand delivery must be repeated at least every 3 months for the duration of the violation. Secondly, the owner or operator of a noncommunity water system (i.e., one serving a transitory population) may give notice by hand delivery or continuous posting of the notice in conspicuous places in the area served by the system. The notice must be given within 14 days after the violation. If notice is given by posting, then it must continue as long as the violations exist. Notice given by hand delivery must be repeated at least every 3 months for as long as the violation exists.

Tier 2 Violations For Tier 2 violations (i.e., violations of 40 CFR Sections 141.74 and 141.174) notice must be given within 3 months after the violation by publication in a daily newspaper of general circulation, or if there is no daily newspaper, then in a weekly newspaper. In addition, the owner or operator shall give notice by mail (either by itself or with the water bill) or by hand delivery at least once every 3 months for as long as the violation exists. Notice of a variance or exemption must be given every 3 months from the date it is granted for as long as it remains in effect.

If a daily or weekly newspaper does not serve the area, the owner or operator of a community water system must give notice by continuous posting in conspicuous places in the area served by the system. This must continue as long as the violation exists or the variance or exemption remains in effect. Notice by hand delivery must be repeated at least every 3 months for the duration of the violation or the variance or exemption.

For noncommunity water systems, the owner or operator may give notice by hand delivery or continuous posting in conspicuous places; beginning within 3 months of the violation or the variance or exemption. Posting must continue for the duration of the violation or variance or exemption and notice by hand delivery must be repeated at least every 3 months during this period.

The Primacy Agency may allow for owner or operator to provide less frequent notice for minor monitoring violations (as defined, by the Primacy Agency if EPA has approved the Primacy Agency's substitute requirements contained in a program revision application).

Variances and Exemptions

As with the SWTR, no variances from the requirements in Section 141 are permitted for subpart H systems. Under Section 1416(a), USEPA or a State may exempt a public water system from any requirements related to an MCL or treatment technique of a National Primary Drinking Water Regulation (NPDWR) if it finds that (1) due to compelling factors (which may include economic factors such as qualifications of the PWS as serving a disadvantaged community), the Public Water System (PWS) is unable to comply with the requirement or implement measures to develop an alternative source of water supply; (2) the exemption will not result in an unreasonable risk to health; and (3) the PWS was in operation on the effective date of the NPDWR, or for a system that was not in operation by that date, only if no reasonable alternative source of drinking water is available to the new systems; and (4) management or restructuring changes (or both) cannot reasonably result in compliance with the Act or improve the quality of drinking water.

Disinfection

Disinfection is a unit process used both in water and wastewater treatment. Many of the terms, practices, and applications discussed in this section apply to both water and wastewater treatment. There are also some differences—mainly in the types of disinfectants used and applications—between the use of disinfection in water and wastewater treatment. Thus, in this section, we discuss disinfection as it applies to water treatment. Later we cover disinfection as it applies to wastewater treatment. Much of the information presented in this section is based on personal experience.

To comply with the SDWA regulations, the majority of PWSs use some form of water treatment. The 1995 Community Water System Survey reports that in the United States, 99% of surface water systems provide some treatment to their water, with 99% of these treatment systems using disinfection/oxidation as part of the treatment process. Although 45 perfect of groundwater systems provide no treatment, 92% of those groundwater plants that do provide some form of treatment include disinfection/oxidation as part of the treatment process (USEPA, 1997). In regard to groundwater supplies, why is the Public Health concern? According to USEPA's Bruce Macler [in What is the Ground Water Disinfection Rule, @ www.groc.org/winter96/gwdr.htm.],

> There are legitimate concerns for public health from microbial contamination of groundwater systems. Microorganisms and other evidence of fecal contamination have been detected in a large number of wells tested, even those wells that had been previously judged not vulnerable to such contamination.

The scientific community believes that microbial contamination of groundwater is real and widespread. Public health impact from this contamination while not well quantified, appear to be large. Disease outbreaks have occurred in many groundwater systems. Risk estimates suggest several million illnesses each year. Additional research is underway to better characterize the nature and magnitude of the public health problem.

The most commonly used disinfectants/oxidants (in no particular order) are chlorine, chlorine dioxide, chloramines, ozone, and potassium permanganate.

As mentioned, the process used to control waterborne pathogenic organisms and prevent waterborne disease is called disinfection. The goal in proper disinfection in a water system is to destroy all disease-causing organisms. Disinfection should not be confused with sterilization. Sterilization is the complete killing of all living organisms. Waterworks operators disinfect by destroying organisms that might be dangerous, they do not attempt to sterilize water.

Disinfectants are also used to achieve other specific objectives in drinking water treatment. These other objectives include nuisance control (e.g., for zebra mussels and Asiatic clams), oxidation of specific compounds (i.e., taste and odor-causing compounds, iron, and manganese), and use as a coagulant and filtration aid. The goals of this section are to:

- Provide a brief overview of the need for disinfection in water treatment
- Provide basic information that is common to all disinfectants
- Discuss other uses for disinfectant chemicals (i.e., as oxidants)
- Describe trends in DBP formation and the health effects of DBPs found in water treatment
- Discuss microorganisms of concern in water systems, their associated health impact, and the inactivation mechanisms and efficiencies of various disinfectants
- Summarize current disinfection practices in the Unites States, including the use of chlorine as a disinfectant and an oxidant

In water treatment, disinfection is almost always accomplished by adding chlorine or chlorine compounds after all other treatment steps (see Figure 8.7); although in the United States ultraviolet (UV) light and potassium permanganate and ozone processes may be encountered.

The effectiveness of disinfection in a drinking water system is measured by testing for the presence or absence of coliform bacteria. Coliform bacteria found in water are generally not pathogenic, though they are good indicators of contamination. Their presence indicates the possibility of contamination, and their absence indicates the possibility that the water is potable—if the source is adequate, the waterworks history is good, and acceptable chlorine residual is present.

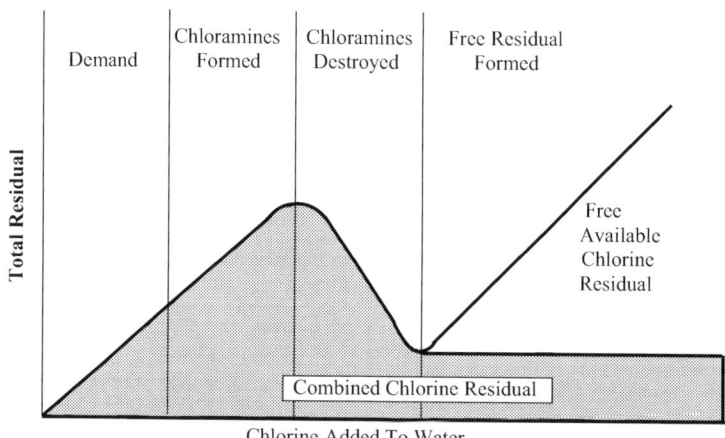

FIGURE 8.7 Breakpoint chlorination curve.

Desired characteristics of a disinfectant include the following:

- It must be able to deactivate or destroy any type or number of disease-causing microorganisms that may be in a water supply, in a water supply, in reasonable time, within expected temperature ranges, and despite changes in the character of the water (pH, for example).
- It must be nontoxic.
- It must not add unpleasant taste or odor to the water.
- It must be readily available at a reasonable cost and be safe and easy to handle, transport, store, and apply.
- It must be quick and easy to determine the concentration of the disinfectant in the treated water.
- It should persist within the disinfected water at a high enough concentration to provide residual protection through the distribution.

Need for Disinfection in Water Treatment

Although the epidemiological relation between water and disease had been suggested as early as the 1850s, it was not until the establishment of the germ theory of disease by Pasteur in the mid-1880s that water as a carrier of disease-producing organisms was understood. And in the 1850s, while London experienced the "Broad Street Well" cholera epidemic, Dr. John Snow conducted his now-famous epidemiological study. Dr. Snow concluded that the well had become contaminated by a visitor, with the disease, who had arrived in the vicinity. Cholera was one of the first diseases to be recognized as capable of being waterborne. Also, this incident was probably the first reported disease epidemic attributed to the direct recycling of non-disinfected water. Now, over 100 years later, the list of potential waterborne disease due to pathogens is considerably larger, and includes bacterial, viral, and parasitic microorganisms, as shown in Tables 8.4–8.6, respectively.

A major cause for the number of disease outbreaks in potable water is contamination of the distribution system from cross connections and back siphonage with non-potable water. However, outbreaks resulting from distribution system contamination are usually quickly contained and result in relatively few illnesses compared to contamination of the source water or a breakdown in the treatment system, which typically produces many cases of illnesses per incident. When considering

TABLE 8.4
Waterborne Diseases from Bacteria

Causative Agent	Disease
Salmonella typhosa	Typhoid fever
S. paratyphi	Paratyphoid fever
S. schottinulleri	
S. hirschfeldi C.	
Shigella flexneri	Bacillary dysentery
Sh. Dysenteriae	
Sh. Sonnel	
Sh. Paradysinteriae	
Vibrio comma	Cholera
V. cholerae	
Pasteurella tularensis	Tularemia
Brucella melitensis	Brucellosis
Leptospira icterchaemorrihagiae	Leptospirosis
Enteropathogenic *E. coli*	Gastroenteritis

TABLE 8.5
Waterborne Diseases from Human Enteric Viruses

Group	Subgroup
Eternovirus	Poliovirus
	Echovirus
	Coxsackie-virus
	A
	B
Reovirus	
Adenovirus	
Hepatitis	

TABLE 8.6
Waterborne Diseases from Parasites

Causative Agent	Symptoms
Ascario lumbricoides (round worm)	Ascariasis
Cryptosporidium muris and parvum	Cryptosporidiosis
Entamoeba histolytica	Amebiasis
Giardia lamblia	Giardiasis
Naegleria gruberi	Amoebic meningoecephalitis
Schistosoma mansoni	Schistosomiasis
Taenis saginata (beef tapeworm)	Taeniasis

the number of cases, the major causes of disease outbreaks are source water contamination and treatment deficiencies (White, 1992). Historically, about 46% of the outbreaks in public water systems are found to be related to deficiencies in source water and treatment systems with 92% of the causes of illness due to these two particular problems.

All natural waters support biological communities. Because some microorganisms can be responsible for public health problems, biological characteristics of the source water are one of the most important parameters in water treatment. In addition to public health problems, microbiology can also affect the physical and chemical water quality and treatment plant operation.

Pathogens of Primary Concern

There are three groups of pathogens of concern in water treatment, namely bacteria, viruses, and protozoa.

1. **Bacteria**: Recall that bacteria are single-celled organisms typically ranging in size from 0.1 to 10 µm. Shape, components, size, and the manner in which they grow can characterize the physical structure of the bacterial cell. Most bacteria can be grouped by shape into four general categories: spheroid, rod, curved rod or spiral, and filamentous. Cocci, or spherical bacteria, are ~1 to 3 µm in diameter. Bacilli (rod-shaped bacteria) are variable in size and range from 0.3 to 1.5 µm in width (or diameter) and from 1.0 to 10.0 µm in length. Vibrio, or curved rod-shaped bacteria, typically vary in size from 0.6 to 1.0 µm in width (or diameter) and from 2 to 6 µm in length. Spirilla (spiral bacteria) can be found in lengths up to 50 µm whereas filamentous bacteria can occur in length in excess of 100 µm.

2. **Viruses**: Viruses are microorganisms composed of the genetic material deoxyribonucleic acid (DNA) or ribonucleic acid (RNA) and a protective protein coat (either single, double, or partially double-stranded). All viruses are obligate parasites, unable to carry out any form of metabolism and are completely dependent upon host cells for replication. Viruses are typically 0.01–0.1 μm in size and are very species specific with respect to infection, typically attacking only one type of host. Although the principal modes of transmission for the hepatitis B virus and poliovirus are through food, personal contact, or exchange of body fluids, these viruses can be transmitted through potable water. Some viruses, such as the retroviruses (including the HIV group), appear to be too fragile for water transmission to be a significant danger to public health (Spellman, 2007).
3. **Protozoa**: are single-cell eukaryotic microorganisms without cell walls that utilize bacteria and other organisms for food. Most protozoa are free-living in nature and can be encountered in water; however, several species are parasitic and live on or in host organisms. Host organisms can vary from a primitive organism such as algae to highly complex organisms such as human beings. Several species of protozoa are known to utilize human beings as hosts.

Recent Waterborne Disease Outbreaks

Within the past 40 years, several pathogenic agents never before associated with documented waterborne outbreaks have appeared in the United States. Enteropathogenic *E. coli* and Giardia lamblia were first identified to be the etiological agent responsible for waterborne outbreaks in the 1960s. The first recorded Cryptosporidium infection in humans occurred in the mid-1970s. Also during that time was the first recorded outbreak of pneumonia caused by Legionella pneumophila. Recently, there have been numerous documented waterborne disease outbreaks that have been caused by *E. coli*, *G. lamblia*, Cryptosporidium, and *L. pneumophila*.

Escherichia coli

The first documented case of waterborne disease outbreaks in the United States associated with enteropathogenic *E. coli* occurred in the 1960s. Various serotypes of *E. coli* have been implicated as the etiological agent responsible for disease in newborn infants, usually the result of cross contamination in nurseries. Now, there have been several well-documented outbreaks of *E. coli* associated with adult waterborne disease. In 1975, the etiologic agent of a large outbreak at Crater Lake National Park was *E. coli* serotype 06:H16 (Craun, 1981).

Giardia lamblia

Similar to Escherichia coli, Giardia lamblia was first identified in the 1960s to be associated with waterborne outbreaks in the United States. Recall that *G. lamblia* is a flagellated protozoan that is responsible for Giardiasis, a disease that can range from being mild to extremely debilitating. Giardia is currently one of the most commonly identified pathogens responsible for waterborne disease outbreaks. The life cycle of Giardia includes a cyst stage when the organism remains dormant and is extremely resilient (i.e., the cyst can survive some extreme environmental conditions). Once ingested by a warm-blooded animal, the life cycle of Giardia continues with excystation.

The cysts are relatively large (8–14 μm) and can be removed effectively by filtration using diatomaceous earth, granular media, or membranes. Giardiasis can be acquired by ingesting viable cysts from food or water or by direct contact with fecal material. In addition to humans, wild and domestic animals have been implicated as hosts. Between 1972 and 1981, 50 waterborne outbreaks of Giardiasis occurred with about 20,000 reported cases (Craun and Jakubowski, 1996). Currently, no simple and reliable method exists to assay Giardia cysts in water samples. Microscopic methods for detection and enumeration are tedious and require examiner skill and patience. Giardia cysts are relatively resistant to chlorine, especially at higher pH and low temperatures.

Cryptosporidium

Cryptosporidium is a protozoan similar to Giardia. It forms resilient oocysts as part of its life cycle. The oocysts are smaller than Giardia cysts, typically about 4–6 µm in diameter. These oocysts can survive under adverse conditions until ingested by a warm-blooded animal and then continue with excystation. Due to the increase in the number of outbreaks of Cryptosporidiosis, a tremendous amount of research has focused on Cryptosporidium within the last 10 years. Medical interest has increased because of its occurrence as a life-threatening infection to individuals with depressed immune systems. As previously mentioned, in 1993, the largest documented waterborne disease outbreak in the U.S. occurred in Milwaukee and was determined to be caused by Cryptosporidium. An estimated 403,000 people became ill, 4,400 people were hospitalized, and 100 people died. The outbreak was associated with deterioration in raw water quality and a simultaneous decrease in the effectiveness of the coagulation-filtration process, which led to an increase in the turbidity of treated water and inadequate removal of Cryptosporidium oocysts.

Legionella pneumophila

An outbreak of pneumonia occurred in 1976 at the annual convention of the Pennsylvania American Legion. A total of 221 people were affected by the outbreak, and 35 of those afflicted died. The cause of pneumonia was not determined immediately despite an intense investigation by the Centers for Disease Control. Six months after the incident, microbiologists were able to isolate a bacterium from the autopsy lung tissue of one of the Legionnaires. The bacterium responsible for the outbreak was found to be distinct from other known bacterium and was named Legionella pneumophila (Witherell et al., 1988). Following the discovery of this organism, other Legionella-like organisms were discovered. Legionnaires' disease does not appear to be transferred person-to-person. Epidemiological studies have shown that the disease enters the body through the respiratory system. Legionella can be inhaled in water particles <5 µm in size from facilities such as cooling towers, hospital hot water systems, and recreational whirlpools.

Mechanism of Pathogen Inactivation

The three primary mechanisms of pathogen inactivation are:

- Destroy or impair cellular structural organization by attacking major cell constituents, such as destroying the cell wall or impairing the functions of semi-permeable membranes;
- Interfere with energy-yielding metabolism through enzyme substrates in combination with prosthetic groups of enzymes, thus rendering non-functional; and
- Interfere with biosynthesis and growth by preventing the synthesis of normal proteins, nucleic acids, coenzymes, or the cell wall.

Depending on the disinfectant and microorganism type, combinations of these mechanisms can also be responsible for pathogen inactivation. In water treatment, it is believed that the primary factors controlling disinfection efficiency are: (1) the ability of the disinfectant to oxidize or rupture the cell wall; and (2) the ability of the disinfectant to diffuse into the cell and interfere with cellular activity (Montgomery, 1985). In addition, it is important to point out that disinfection is effective in reducing waterborne diseases because most pathogenic organisms are more sensitive to disinfection than are nonpathogens. However, disinfection is only as effective as the care used in controlling the process and assuring that all of the water supply is continually treated with the amount of disinfectant required producing safe water.

Other Uses of Disinfectants in Water Treatment

Disinfectants are used for more than just disinfection in drinking water treatment. While the inactivation of pathogenic organisms is a primary function, disinfectants are also used as oxidants in drinking water treatment for several other functions:

- Minimization of DBP formation
- Control of nuisance Asiatic clams and zebra mussels
- Oxidation of iron and manganese
- Prevention of regrowth in the distribution system and maintenance of biological stability
- Removal of taste and odors through chemical oxidation
- Improvement of coagulation and filtration efficiency
- Prevention of algal growth in sedimentation basins and filters
- Removal of color

A brief discussion of these additional oxidant uses follows.

Minimization of DBP Formation

Strong oxidants may play a role in disinfection and DBP control strategies in water treatment. Several strong oxidants, including potassium permanganate and ozone, may be used to control DBP precursors.

Note: Potassium permanganate can be used to oxidize organic precursors at the head of the treatment plant, thus minimizing the formation of byproducts at the downstream disinfection stage of the plant. The use of ozone for the oxidation of DBP precursors is currently being studied. Early work has shown that the effects of ozonation, prior to chlorination, were highly site-specific and unpredictable. The key variables that seem to determine the effect of ozone are dose, pH, alkalinity, and the nature of the organic material Ozone has been shown to be effective for DBP precursor reduction at low pH. However, at higher pH (i.e., above 7.5), ozone may actually increase the amount of chlorination byproduct precursors.

Control of Nuisance Asiatic Clams and Zebra Mussels

The Asiatic clam (Corbicula fluminea) was introduced to the U.S from Southeast Asia in 1938 and now inhabits almost every major river system south of 40° latitude. Asiatic clams have been found in the Trinity River, TX; the Ohio River at Evansville, IN; New River at Narrows and Glen Lyn, VA; and the Catawba River in Rock Hill, SC. This animal has invaded many water utilities, clogging source water transmission systems and valves, screens, and meters; damaging centrifugal pumps; and causing taste and odor problems (Britton and Morton, 1982; Sinclair, 1964; Cameron et al., 1989).

Cameron et al. (1989) investigated the effectiveness of several oxidants to control the Asiatic clam in both the juvenile and adult phases. As expected, the adult clam was found to be much more resistant to oxidants than the juvenile form. In many cases, the traditional method of control, free chlorination, cannot be used because of the formation of excessive amounts of THMs. As shown in Table 8.7, Cameron et al. (1989) compared the effectiveness of four oxidants for controlling the

TABLE 8.7
The Effects of Various Oxidants on Mortality of the Asiatic Clam

Chemical	Residual (mg/L)	Temperature	pH (°C)	Life Stage	LT50 (days)
Free chlorine	0.5	23	8.0	Adult	8.7
	4.8	21	7.9	Adult	5.9
	4.7	16	7.8	Juvenile	4.8
Potassium	1.1	17	7.6	Juvenile	7.9
Permanganate	4.8	17	7.6	Juvenile	8.6
Chlorine	1.2	24	6.9	Juvenile	0.7
dioxide	4.7	22	6.6	Juvenile	0.6

Source: Adaptation from Cameron et al. (1989)

juvenile Asiatic clam in terms of the LT50 (time required for 50% mortality). Monochloramine was found to be the best for controlling juvenile clams without forming THMs. The effectiveness of monochloramine increased greatly as the temperature increased. Clams can tolerate temperatures between 2°C and 35°C.

The zebra mussel (Dresissmena polymorpha) is a recent addition to the fauna of the Great Lakes. It was first found in Lake St. Clair in 1988, though it is believed that this native of the Black and Caspian seas was brought over from Europe in ballast water around 1985. The zebra mussel population in the Great Lakes has expanded very rapidly, both in size and geographical distribution (Herbert et al., 1989; Roberts, 1990). Lang (1994) reported that zebra mussels have been found in the Ohio River, Cumberland River, Arkansas River, Tennessee River, and the Mississippi River south to New Orleans.

Klerks and Fraleigh (1991) evaluated the effectiveness of hypochlorite, permanganate, and hydrogen peroxide with iron for their effectiveness in controlling adult zebra mussels. Both continuous and intermittent 28-day static renewal tests were conducted to determine the impact or intermittent dosing. Intermittent treatment proved to be much less effective than continuous dosing. The hydrogen peroxide-iron combination (1–5 mg/L with 25% iron) was less effective in controlling the zebra mussel than either permanganate or hypochlorite. Permanganate (0.5–2.5 mg $KnnO_4$/L) was usually less effective than hypochlorite (0.5–10 mg Cl_2/L).

Van Benschoten et al. (1995) developed a kinetic model to predict the rate of mortality of the zebra mussel in response to chlorine. The model shows the relationship between chlorine residual and temperature on the exposure time required to achieve 50% and 95% mortality. Data were collected for chlorine residuals between 0.5 and 3.0 mg Cl_2/L and temperatures from 0.3°C to 24°C. The results show a strong dependence on temperature and required contact times ranging from 2 days to more than a month, depending on environmental factors and mortality required.

Brady et al. (1996) compared the efficiency of chlorine to control the growth of zebra mussel and quagga mussel (Dreissenda bugensis). The quagga mussel is a newly identified mollusk within the Great Lakes that is similar in appearance to the zebra mussel. Full-scale chlorination treatment found a significantly higher mortality for the quagga mussel. The required contact time for 100% mortality for quagga and zebra mussels was 23 and 37 days, respectively, suggesting that chlorination programs designed to control zebra mussels should also be effective for controlling populations of quagga mussels.

Matisoff et al. (1996) evaluated chlorine dioxide (ClO_2) to control adult zebra mussels using simple, intermittent, and continuous exposures. A single 30-minute exposure to 20 mg/L chlorine dioxide or higher concentration induced at least 50% mortality, while sodium hypochlorite produced only 26% mortality, and permanganate and hydrogen peroxide were totally ineffective when dose at 30 mg/L for 30 minutes under the same conditions. These high dosages, even though only used for a short period, may not allow application directly in water for certain applications due to byproducts that remain in the water. Continuous exposure to chlorine dioxide for 4 days was effective at concentrations above 0.5 mg/L (LC50=0.35 mg/L), and 100% mortality was achieved at chlorine dioxide concentrations above 1 mg/L.

These studies all show that the dose required to induce mortality to these nuisance organism is extremely high, both in terms of chemical dose and contact time. The potential impact on DBPs is significant especially when the water is high in organic content with a high propensity to form THMs and other DBPs.

Oxidation of Iron and Manganese

Iron and manganese occur frequently in ground waters but are less problematic in surface waters. Although not harmful to human health at the low concentrations typically found in water, these compounds can cause staining and taste problems. These compounds are readily treated by oxidation to produce a precipitant that is removed in subsequent sedimentation and filtration processes.

TABLE 8.8
Oxidant Doses Required for Oxidation of Iron and Manganese

Oxidant	Iron (II) (mg/mg Fe)	Manganese (II) (mg/mg Mn)
Chorine	0.62	0.77
Chlorine dioxide	1.21	2.45
Ozone	0.43	0.85
Oxygen	0.14	0.29
Potassium permanganate	0.94	1.92

Source: Adapted from Culp and Culp (1974b).

Almost all the common oxidants except chloramines will convert ferrous (2+) iron to the ferric (3+) state and manganese (2+) to the (4+) state, which will precipitate as ferric hydroxide and manganese dioxide, respectively (AWWA, 1984). The precise chemical composition of the precipitate will depend on the nature of the water, temperature, and pH. Table 8.8 shows that oxidant doses for iron and manganese control are relatively low. In addition, the reactions are relatively rapid, on the order or seconds while DBP formation occurs over hours. Therefore, with proper dosing, residual chlorine during iron and manganese oxidation is therefore relatively low and short-lived. These factors reduced the potential for DBP formation as a result of oxidation for iron and manganese removal.

Prevention of Regrowth in the Distribution System and Maintenance of Biological Stability
Biodegradable organic compounds and ammonia in treated water can cause microbial growth to the distribution system. Biological stability refers to a condition wherein the water quality does not enhance biological growth in the distribution system. Biological stability can be accomplished in several ways:

- Removing nutrients from the water prior to distribution
- Maintaining a disinfectant residual in the treated water
- Combining nutrient removal and disinfectant residual maintenance

To maintain biological stability in the distribution system, the Total Coliform Rule (TCR) requires that treated water have a residual disinfectant of 0.2 mg/L when entering the distribution system. A measurable disinfectant residual must be maintained in the distribution system, or the utility must show through monitoring that the heterotrophic plate count (HPC) remains <500/100mL. A system remains in compliance as long as 95% of samples meet these criteria. Chlorine, monochloramine, and chlorine dioxide are typically used to maintain a disinfectant residual in the distribution system. Filtration can also be used to enhance biological stability by reducing the nutrients in the treated water.

The level of secondary disinfectant residual maintained is low, typically in the range of 0.1–0.3 mg/L, depending on the distribution system and water quality. However, because the contact times in the system are quite long, it is possible to generate significant amounts of DBPs in the Distribution system, even at low disinfectant does. Distribution system problems associated with the use of combined chlorine residual (chloramines), or no residual, have been documented in several instances. The use of combined chlorine is characterized by an initial satisfactory phase in which chloramine residuals are easily maintained throughout the system and bacterial counts are very low. However, problems may develop over a period of years including increased bacterial counts, reduced combined chlorine residual, increased taste and odor complaints, and reduced transmission main carrying capacity. Conversion of the system to free-chlorine residual produces an initial increase in consumer complaints of taste and odors resulting from oxidation of accumulated organic material. Also, it is difficult to maintain a free-chlorine concentration at the ends of the distribution system (AWWA, 1984).

Removal of Taste and Odors through Chemical Oxidation

Taste and odors in drinking water are caused by several sources, including microorganisms, decaying vegetation, hydrogen sulfide, and specific compounds of municipal, industrial, or agricultural origin. Disinfectants themselves can also create taste and odor problems. In addition to a specific taste-and odor-causing compound, the sanitary impact is often accentuated by a combination of compounds.

Oxidation is commonly used to remove taste and odor-causing compounds. Because many of these compounds are very resistant to oxidation, advanced oxidation processes (ozone/hydrogen peroxide, ozone/UV, etc.) and ozone by itself are often used to address taste and odor problems. The effectiveness of various chemicals to control taste and odors can be site-specific. Suffet et al. (1986) found that ozone is generally the most effective oxidant for use in taste and odor treatment. They found ozone is generally the most effective oxidant for use in taste and odor treatment. They found ozone dose of 2.5–2.7 mg/L and 10 minutes of contact time (residual 0.2 mg/L) significantly reduce levels of taste and odors. Lalezary et al. (1986) used chlorine, chlorine dioxide, ozone, and permanganate to treat earthy-musty smelling compounds. In that study, chlorine dioxide was found most effective, although none of the oxidants was able to remove Geosmin and MIB by more than 40%–60%. Potassium permanganate has been used in doses of 0.25–20 mg/L.

Prior experiences with taste and odor treatment indicate that oxidant does are dependent on the source or the water and causative compounds. In general, small doses can be effective for many taste and odor compounds, but some of the difficult-to-treat compounds. In general, small doses can be effective for many taste and odor compounds, but some of the difficult-to-treat compounds require strong oxidants such as ozone and/or advanced oxidation processes or alternative technologies such as granular activated carbon (GAC) adsorption.

Improvement of Coagulation and Filtration Efficiency

Oxidants, specifically ozone, have been reported to improve coagulation and filtration efficiency. Others, however, have found no improvement in effluent turbidity from oxidation. Prendiville (1986) collected data from a large treatment plant showing that preozonation was more effective than prechlorination to reduce filter effluent turbidities. The cause of and improved coagulation is not clear, but several possibilities have been offered, including (Gurol and Pidatella, 1983; Reckhow et al., 1986):

- Oxidation of organics into more polar forms
- Oxidation of metal ions to yield insoluble complexes such as ferric iron complexes
- Change in the structure and size of suspended particles.

Prevention of Algal Growth in Sedimentation Basins and Filters

Prechlorination is often used to minimize operational problems associated with biological growth in water treatment plants (AWWA, 1984). Prechlorination will prevent slime formation on filters, pipes, and tanks, and reduce potential taste and odor problems associated with such slimes. Many sedimentation and filtration facilities operate with a small chlorine residual to prevent the growth of algae and bacteria in laundries (in like equipment or machinery) and on the filter surfaces. This practice has increased in recent years as utilities take advantage of additional contact time in the treatment units to meet disinfection requirements under the SWTR.

Removal of Color
Free chlorine is used for color removal. A low pH is favored. Humic compounds which have a high potential for DBP formation cause color. The chlorine dosage and kinetics for color removal are best determined through bench studies.

Types of Disinfection Byproducts and Disinfection Residuals
Table 8.9 provides a list compiled by USEPA of DBPs and disinfection residuals that may be of health concern. The table includes both the disinfectant residuals and the specific byproducts produced by the disinfectants of interest in drinking water treatment. These contaminants of concern are grouped into four distinct categories and include disinfectant residuals, inorganic byproducts, organic oxidation byproducts, and halogenated organic byproducts.

The production of DBPs depends on the type of disinfectant, the presence of organic material (e.g., TOC), bromide ion, and other environmental factors as discussed in this section. By removing DBP precursors, the formation of DBPs can be reduced. The health effects of DBPs and disinfectants are generally evaluated with epidemiological studies and/or toxicological studies using laboratory animals. Table 8.10 indicates the cancer classifications of both disinfectants and DBPs as of January 1999. The classification scheme used by USEPA is shown at the bottom of Table 8.10. The USEPA classification scheme for carcinogenicity weighs both animal studies and epidemiologic studies, but places greater weight on the evidence of carcinogenicity in humans.

TABLE 8.9
List of Disinfection Byproducts and Disinfection Residuals

Disinfectant Residuals	**Halogenated Organic Byproducts**
Free chlorine	Trihalomethanes
Hypochlorous acid	Chloroform
Hypochlorite ion	Bromodichloromethane
Chloramines	Dibromochloromethane
Monochloramine	Bromoform
Chlorine dioxide	Haloacetic acids
Inorganic Byproducts	Monochloroacetic acid
Chlorate ion	Dichloroacetic acid
Chlorite ion	Trichloroacetic acid
Bromate ion	Monobromoacetic acid
Iodate ion	Dibromoacetic acid
Hydrogen peroxide	Haloacetonitriles
Ammonia	Dichloroacetonitrile
	Bromochloroacetonitrile
	Dibromoacetonitrile
Organic Oxidation Byproducts	Trichloroacetonitrile
Aldehydes	Haloketones
Formaldehyde	1,1-Dichloropropanone
Acetaldehyde	1,1,1-Trichloropropanone
Glyoxal	Chlorophenols
Hexanal	2-Chlorophenol
Heptanal	2,4-Dichlorophenol
Carboxylic acids	2,4,6-Trichlorophenol
Hexanoic acid	Chloropicrin
Heptanoic acid	Chloral hydrate
Oxalic acid	Cyanogen chloride
Assimilable organic carbon	N-Organochloramines

TABLE 8.10
Status of Health Information for Disinfectants and DBPs

Contaminant	Cancer Classification
Chloroform	Probable human carcinogen
Bromodichloromethane	Probable human carcinogen
Dibromochloromethane	Possible human carcinogen
Bromoform	Probable human carcinogen
Monochloroacetic acid	–
Dichloroacetic acid	Probable human carcinogen
Trichloroacetic acid	Possible human carcinogen
Dichloroacetonitrile	Possible human carcinogen
Bromochloroacetonitrile	–
Dibromoacetonitrile	Possible human carcinogen
Trichloroacetonitrile	–
1,1-dichloropropanone	–
1,1,1-trichloropropanone	–
2-chlorophenol	Not classifiable
2,4-dichlorphenol	Not classifiable
2,4,6-trichlorophenol	Probable human carcinogen
Chloropicrin	–
Chloral hydrate	Possible human carcinogen
Cyanogen chloride	–
Formaldehyde	Probable human carcinogen
Chlorate	–
Chlorite	Not classifiable
Bromate	Probable human carcinogen
Ammonia	Not classifiable
Hypochlorous acid	–
Hypochlorite	–
Monochloramine	–
Chlorine dioxide	Not classifiable

Source: USEPA (1996).

Disinfection Byproduct Formation

Halogenated organic byproducts are formed when natural organic matter (NOM) reacts with free chlorine or free bromine. Free chlorine can be introduced to water directly as a primary or secondary disinfectant, with chlorine dioxide, or with chloramines. Free bromine results from the oxidation of the bromide ion in source water. Factors affecting the formation of halogenated DBPs include the type and concentration of natural organic matter, oxidant type and dose, time, bromide ion concentration, pH, organic nitrogen concentration, and temperature. Organic nitrogen significantly influences the formation of nitrogen-containing DBPs such as the haloacetonitriles, halopicrins, and cyanogen halides. The parameter TOX represents the concentration of total organic halides in a water sample (calculated as chloride). In general, <50% of the TOX content has been identified, despite evidence that several of these unknown halogenated byproducts of water chlorination may be harmful to humans (Reckhow et al., 1990; Singer and Chang, 1989).

Nonhalogenated DBPs are also formed when strong oxidants react with organic compounds found in water. Ozone and peroxone oxidation of organics leads to the production of aldehydes, aldo- and keto-acids, organic acids, and, when bromide ion is present, brominated organics. Many of the oxidation byproducts are biodegradable and appear as biodegradable dissolved organic carbon (BDOC) and assimilable organic carbon (AOC) in treated water.

Bromide ion plays a key role in DBP formation. Ozone or free chlorine oxidizes bromide ions to hypobromate ions/hypobromous acid, which subsequently forms brominated DBPs. Brominated organic byproducts include compounds such as bromoform, brominated acetic acids and acetonitriles, bromopicrin, and cyanogen bromide. Only about one-third of the bromide ions incorporated into byproducts have been identified.

DBP Precursors

Numerous researchers have documented that NOM is the principal precursor of organic DBP formation. Chlorine reacts with NOM to produce a variety of DBPs, including THMs, haloacetic acids (HAAs), and others. Ozone reacts with NOM to produce aldehydes, organic acids, and aldo- and keto-acids; many of these are produced by chlorine as well (Stevens, 1976; Singer and Harrington, 1993). Natural waters contain mixtures of both humic and nonhumic organic substances. NOM can be subdivided into a hydrophobic fraction composed of primarily humic material, and a hydrophilic fraction composed of primarily fulvic material. The type and concentration of NOM are often assessed using surrogate measures. Although surrogate parameters have limitations, they are used because they may be measured more easily, rapidly, and inexpensively than the parameter of interest, often allowing online monitoring of the operation and performance of water treatment plants. Surrogates used to assess NOM include:

- Total and dissolved organic carbon (TOC and DOC)
- Specific ultraviolet light absorbance (SUVA), which is the absorbance at 254 nm wavelength (UV-254) divided by DOC (SUVA=(UV-254/DOC)100 in L/mg-m)
- THM formation potential (THMFP)—a test measuring the quantity of THMs formed with a high dosage of free chlorine and a long reaction time
- TTHM Stimulated Distribution System (SDS)—a test to predict the TTHM concentration at some selected point in a given distribution system, where the conditions of the chlorination test simulate the distribution system at the point desired.

On average, about 90% of the TOC is dissolved. DOC is defined as the TOC able to pass through a 0.45 μm filter. UV absorbance is a good technique for assessing the presence of DOC through a 0.45 μm filter. UV absorbance is a good technique for assessing the presence of DOC because DOC primarily consists of humic substances, which contain aromatic structures that absorb light in the UV spectrum. Oxidation of DOC reduces the UV absorbance of the water due to oxidation of some of the organic bonds that absorb UV absorbance. Complete mineralization of organic compounds to carbon dioxide usually does not occur underwater treatment conditions; therefore, the overall TOC concentration usually is constant.

Concentrations of DBPs vary seasonally and are typically greatest in the summer and early fall for several reasons:

- The rate of DBP formation increases with increasing temperature
- The nature of organic DBP precursors varies with the season
- Due to warmer temperatures, chlorine demand may be greater during summer months requiring higher dosages to maintain disinfection.

If the bromide ion is present in source waters, it can be oxidized to hypobromous acid that can react with NOM to form brominated DBPs, such as bromoform. Furthermore, under certain conditions, ozone may react with the hypobromite ion to form bromate ion (Singer, 1992).

The ratio of bromide ion to the chlorine dose affects THM formation and bromine substitution of chlorine. Increasing the bromide ion to chlorine dose ratio shifts the speciation of THMs to produce more brominated forms. In the Krasner (1989) study, the chlorine dose was roughly proportional to TOC concentration. As TOC was removed through the treatment train, the chlorine dose decrease and TTHM formation declined. However, at the same time, the bromide ion to chlorine dose increased, thereby shifting TTHM concentrations to the more brominated THMs. Therefore, improving the removal of NOM prior to chlorination can shift the speciation of halogenated byproducts toward more brominated forms.

Chloropicrin is produced by the chlorination of humic materials in the presence of nitrate ion. Thibaud et al. (1988) chlorinated humic compounds in the presence of bromide ion to demonstrate the formation of brominated analogs to chloropicrin.

Impacts of pH on DBP Formation

The pH of water being chlorinated has an impact on the formation of halogenated byproducts. THM formation increases with increasing pH. Trichloroacetic acid, dichloroacetonitile, and trichloropropanone formation decrease with increased pH. Overall TOX formation decreases with increasing pH. Based on chlorination studies of humic material in model systems, high pH tends to favor chloroform formation over the formation of trichloroacetic acid and other organic halides. Accordingly, water treatment plants practicing precipitative softening at pH values >9.5 to10 are likely to have a higher fraction of TOX attributable to THMs than plants treating surface waters by conventional treatment in pH ranges of 6–8.

Because the application of chlorine dioxide and chloramines may introduce free chlorine into water, chlorination byproducts that may be formed would be influenced by pH as discussed above. Ozone application to bromide ion-containing waters at high pH favors the formation of bromate ion, while application at low pH favors the formation of brominated organic byproducts.

The pH also impacts enhanced coagulation (i.e., for ESWTR compliance) and Lead and Copper Rule Compliance. These issues are addressed in USEPA's Microbial and Disinfection Byproduct Simultaneous Compliance Guidance Manual.

Organic Oxidation Byproducts

Organic oxidation byproducts are formed by reactions between NOM and all oxidizing agents added during drinking water treatment. Some of these byproducts are halogenated, as discussed in the previous section, while others are not. The types and concentrations of organic oxidation byproducts produced depend on the type and dosage of the oxidant being used, chemical characteristics and concentration of the NOM being oxidized, and other factors such as the pH and temperature.

Inorganic Byproducts and Disinfectants

Table 8.11 shows some of the inorganic DBPs that are produced or remain as residual during disinfection. As discussed earlier, bromide ion reacts with strong oxidants to form bromate ion and other organic DBPs. Chlorine dioxide and chloramines leave residuals that are of concern for health considerations, as well as for taste and odor.

DBP Control Strategies

In 1983, the USEPA identified technologies, treatment techniques, and plant modifications that community water systems could use to comply with the maximum contaminant level for TTHMs.

TABLE 8.11
Inorganic DBPs Produced During Disinfection

Disinfectant	Inorganic byproduct or disinfectant residual discussed
Chlorine dioxide	Chlorine dioxide, chlorite ion, chlorate ion, bromate ion
Ozone	Bromate ion, hydrogen peroxide
Chloramination	Monochloramine, dichloramine, trichloramine, ammonia

The principal treatment modifications involved moving the point of chlorination downstream in the water treatment plant, improving the coagulation process to enhance the removal of DBP precursors, and using chloramines to supplement or replace the use of free chlorine (Singer and Harrington, 1993). Moving the point of chlorination downstream in the treatment train often is very effective in reducing DBP formation, because it allows the NOM precursor concentration to be reduced during treatment prior to chlorine addition. Replacing prechlorination by preoxidation with an alternate disinfectant that produces less DBPs is another option for reducing formation of chlorinated byproducts. Other options to control the formation of DBPs include; source water quality control, DBP precursor removal, and disinfection strategy selection. An overview of each is provided below.

Source Water Quality Control

Source water control strategies involve managing the source water to lower the concentrations of NOM and bromide ion in the source water. Research has shown that algal growth leads to the production of DBP precursors (Oliver and Shindler, 1980). Therefore, nutrient and algal management is one method of controlling DBP formation potential of source waters. Control of bromide ion in source waters may be accomplished by preventing brine or saltwater intrusion into the water source.

DBP Precursor Removal

Raw water can include DBP precursors in both dissolved and particulate forms. For the dissolved precursors to be removed in conventional treatment, they must be converted to particulate form for subsequent removal during settling and filtering. The THM formation potential generally decreases by about 50% through conventional coagulation and settling, indicating the importance of moving the point of chlorine application after coagulation and settling (and even filtration) to control TOX as well as TTHM formation (Singer and Chang, 1989). Conventional systems can lower the DBP formation potential of water prior to disinfection by further removing precursors with enhanced coagulation, GAC adsorption, or membrane filtration prior to disinfection. Precursor removal efficiencies are site-specific and vary with different source waters and treatment techniques.

Aluminum (alum) and iron (ferric) salts can remove variable amounts of NOM. For alum, the optimal pH for NOM removal is in the range of 5.5–6.0. The addition of alum decreases pH and may allow the optimal pH range to be reached without acid addition. However, waters with very low or very high alkalinities may require the addition of base or acid to reach the optimal NOM coagulation pH (Singer, 1992).

Granular activated carbon adsorption can be used following filtration to remove additional NOM. For most applications, empty bed contact times in excess of 20 minutes are required, with regeneration frequencies of about 2–3 months. These long control times and frequent regeneration requirements make GAC an expensive treatment option. In cases where prechlorination is practiced, the chlorine rapidly degrades GAC. The addition of a disinfectant to the GAC bed can result in specific reactions in which previously absorbed compounds leach into the treated water.

Membrane filtration has been shown effective in removing DBP precursors in some instances. In pilot studies, ultrafiltration (UF) with a molecular weight cutoff (MWCO) of 100,000 Daltons was ineffective for controlling DBP formation. However, when little or no bromide ion was present in source water, nanofiltration (NF) membranes with MWCOs of 400–800 Daltons effectively controlled DBP formation (Laine, 1993).

In waters containing bromide ions, higher bromoform concentrations were observed after chlorination of membrane permeate (compared with raw water). This occurs as a result of filtration removing NOM while concentrating bromide ions in permeate thus providing a higher ratio of bromide ions to NOM than in raw water. This reduction in chlorine demand increases the ratio of bromide to chlorine, resulting in higher bromoform concentrations after chlorination of NF membrane permeate (compared with the raw water). TTHMs were lower in chlorinated permeate than chlorinated raw water. However, due to the shift in speciation of THMs toward more brominated forms, bromoform concentrations were actually greater in chlorinated treated water than in chlorinated

TABLE 8.12
Required Removal of TOC by Enhanced Coagulation for Surface Water Systems Using Convention Treatment (Percent Reduction)

Source Water TOC (mg/L)	Source Water Alkalinity (mg/L as CaCO$_3$)		
	0–60	>60–120	>120
>2.0–4.0	35.0	25.0	15.0
>4.0–8.0	45.0	35.0	25.0
>8.0	50.0	40.0	30.0

raw water. The use of spiral-wound NF membranes (200–300 Daltons) more effectively controlled the formation of brominated THMs, but pretreatment of the water was necessary. Significant limitations in the use of membranes are disposal of the waste brine generated, fouling of membranes, cost of membrane replacement, and increasing energy cost.

Disinfection byproduct regulations require enhanced coagulation as an initial step for the removal of DBP precursors. In addition to meeting MCLs and MRDLs, some water suppliers also must meet treatment requirements to control the organic material (DBP precursors) in the raw water that combines with disinfectant residuals to form DBPs. Systems using conventional treatment are required to control precursors (measured as TOC) by using enhanced coagulation or enhanced softening. A system must remove a specified percentage of TOC (based on raw water quality) prior to the point of continuous disinfection (Table 8.12).

Systems using ozone followed by biologically active filtration or chlorine dioxide that meet specific criteria would be required to meet the TOC removal requirements prior to the addition of a residual disinfectant. Systems able to reduce TOC by a specified percentage level have met the DBP treatment technique requirements. If the system does not meet the percent reduction, it must determine its alternative minimum TOC removal level. The primacy agency approves the alternative minimum TOC removal possible for the system based on the relationship between coagulant dose and TOC in the system based on the results of bench or pilot-scale testing. Enhanced coagulation is determined in part as the coagulant does where an incremental addition of 10 mg/L of alum (or an equivalent amount of ferric salt) results in a TOC removal below 0.3 mg/L.

Disinfection Strategy Selection

In addition to improving the raw or predisinfectant water quality, alternative disinfection strategies can be used to control DBPs. These strategies include the following:

- Use an alternative or supplemental disinfectant or oxidant such as chloramines or chlorine dioxide that will produce fewer DBPs.
- Move the point of chlorination to reduce TTHM formation and, where necessary, substitute chloramines, chlorine dioxide, or potassium permanganate for chlorine as a preoxidant
- Use two different disinfectants or oxidants at various points in the treatment plant to avoid DBP formation at locations where precursors are still present in high quantities
- Use of powdered activated carbon for THM precursor or TTHM reduction seasonally or intermittently
- Maximize precursor removal

CT Factor

One of the most important factors for determining or predicting the germicidal efficiency of any disinfectant is the CT factor, a version of the Chick-Watson law (1908). The CT factor is defined as the product of the residual disinfectant concentration, C, in mg/L, and the contact time, T, in

TABLE 8.13
CT Values for Inactivation of Viruses

Disinfectant (@ 10°C)	Units	Inactivation	
		2-log	3-log
Chlorine	mg·min/L	4	4
Chloramine	mg·min/L	643	1,067
Chlorine dioxide	mg·min/L	4.2	12.8
Ozone	mg·min/L	0.5	0.8
UV	mW·s/cm^2	21	36

Source: Modified from AWWA (1991).

TABLE 8.14
CT Values for Inactivation of Giardia Cysts

Disinfectant (@ 10°C)	Inactivation (mg·min/L)		
	1-log	2-log	3-log
Chlorine	35	69	104
Chloramine	615	1,240	1,850
Chlorine dioxide	7.7	15	23
Ozone	0.48	0.95	1.43

Source: Modified from AWWA (1991).

minutes that residual disinfectant is in contact with the water. USEPA developed CT values for the inactivation of Giardia and viruses under the SWTR. Table 8.13 compares the CT values for virus inactivation using chlorine, chlorine dioxide, ozone, chloramine, and ultraviolet light disinfection under specified conditions. Table 8.14 shows the CT values for the inactivation of Giardia cyst using chlorine, chloramine, chlorine dioxide, and ozone under specified conditions. The CT values shown in Tables 8.13 and 8.14 are based on water temperatures of 10°C and pH values in the range of 6–9. CT values for chlorine disinfection are based on free chlorine residual. Note that chlorine is less effective as pH increases from 6 to 9. In addition, for a given CT value, a low C and a high T are more effective than the reverse (i.e., a high C and a low T). For all disinfectants, as temperature increases, effectiveness increases.

Disinfectant Residual Regulatory Requirements

One of the most important factors for evaluating the merits of alternative disinfectants is their ability to maintain the microbial quality in the water distribution system. Disinfectant residuals may serve to protect the distribution system against regrowth. The Surface Water Treatment Rule (SWTR) requires that filtration and disinfection must be provided to ensure that the total treatment of the system achieves at least a 3-log removal/inactivation of Giardia cysts and a 4-log removal/inactivation of viruses. In addition, the disinfection process must demonstrate by continuous monitoring and recording that the disinfection residual in the water entering the distribution system is never <0.2 mg/L for more than 4 hours (Spellman, 2007).

Several of the alternative disinfectants examined in the handbook cannot be used to meet the residual requirements stated in the SWTR. For example, if either ozone or ultraviolet light disinfection is used as the primary disinfectant, a secondary disinfectant such as chlorine or chloramines should be utilized to obtain a residual in the distribution system.

Disinfectant byproduct formation continues in the distribution system due to reactions between the residual disinfectant and organics in the water. Koch et al. (1991) found that with a chlorine dose of 3–4 mg/L, THM and HAA concentrations increase rapidly during the first 24 hours in the distribution system. After the initial 48 hours, the subsequent increase in THMs is very small. Chloral hydrate concentrations continued to increase after the initial 24 hours, but at a reduced rate. Haloketones actually decreased in the distribution system.

Nieminski et al. (1993) evaluated DBP formation in the simulated distribution systems of treatment plants in Utah. Finished water chlorine residuals ranged from 0.4 to 2.8 mg/L. Generally, THM values in the distribution system studies increased by 50%–100% (range of 30%–200%) of the plant effluent value after 24-hour contact time. The 24-hour THM concentration was essentially the same as the 7-day THM formation potential. HAA concentrations in the simulated distribution system were about 100% (range of 30%–200%) of the HAA in the plant effluent. The 7-day HAA formation potential was sometimes higher, or below the distribution system values. If chlorine is used as a secondary disinfectant, one should therefore anticipate a 100% increase in the plant effluent THMs or plan to reach the 7-day THM formation level in the distribution system.

Summary of Current National Disinfection Practices

Most water treatment plants disinfect water prior to distribution. The 1995 Community Water Systems Survey reports that 81% of all community water systems provide some form of treatment on all or a portion of their water sources. The survey also found that virtually all-surface water systems provide some treatment of their water. Of those systems reporting no treatment, 80% rely on groundwater as their only water source.

The most commonly used disinfectants/oxidants are chlorine, chlorine dioxide, chloramines, ozone, and potassium permanganate. Chlorine is predominately used in surface and groundwater disinfection treatment systems; more than 60% of the treatment systems use chlorine as disinfectant/oxidant. Potassium permanganate, on the other hand, is used by many systems, but its application is primarily for oxidation, rather than for disinfection.

Permanganate will have some beneficial impact on disinfection since it is a strong oxidant that will reduce the chemical demand for the ultimate disinfection chemical. Chloramine is used by some systems and is more frequently used as a post-treatment disinfectant.

The International Ozone Association (IOA, 1997) conducted a survey of ozone facilities in the United States. The survey documented the types of ozone facilities, size, the objective of ozone application, and year of operation. The International Ozone Association (1997) summarizes its findings of its survey of water treatment plants in the U.S. in Table 8.15. The most common use of ozone is for the oxidation of iron and manganese, and for taste and odor control. Twenty-four of the 158 ozone facilities used GAC following ozonation. In addition to the 158 operating ozone facilities, the survey identified 19 facilities under construction and another 30 under design. The capacity of the systems range from <25 gpm to exceeding 500 mg. Nearly half of the operating facilities have a capacity exceeding 1 mg. Rice et al. (1998) found that as of May 1998, 264 drinking water plants in the United States are using ozone.

TABLE 8.15

Ozone Application in Water Treatment Plants in the U.S

Ozone Objective	Number of Plants	% Plants
THM control	50	32
Disinfection	63	40
Iron/manganese, taste and odor control	92	58
Total	158	–

The methods of disinfection include:

Heat: possibly the first method of disinfection. Disinfection is accomplished by boiling water for 5–10 minutes. Good, obviously, only for household quantities of water when bacteriological quality is questionable.

Ultraviolet (UV) light: while a practical method of treating large quantities, adsorption of UV light is very rapid, so the use of this method is limited to nonturbid waters close to the light source.

Metal ions: silver, copper, mercury.

Alkalis and acids

pH adjustment: to under 3.0 or over 11.0
Oxidizing agents: bromine, ozone, potassium permanganate, and chlorine

The vast majority of drinking water systems in the U.S. use chlorine for disinfection (Spellman, 2007). Along with meeting the desired characteristics listed above, chlorine has the added advantage of a long history of use—it is fairly well understood. Although some small water systems may use other disinfectants, we focus on chlorine in this handbook and provide only a brief overview of other disinfection alternatives.

Note: One of the recent developments in chlorine disinfection is the use of multiple and interactive disinfectants. In these applications, chlorine is combined with a second disinfectant to achieve improved disinfection efficiency and/or effective DBP control.

Chlorination

The addition of chlorine or chlorine compounds to water is called chlorination. Chlorination is considered to be the single most important process for preventing the spread of waterborne disease. Chlorine has many attractive features that contribute to its wide use in industry. Five of the key attributes of chlorine are:

- It damages the cell wall.
- It alters the permeability of the cell (the ability to pass water in and out through the cell wall).
- It alters the cell protoplasm.
- It inhibits the enzyme activity of the cell so it is unable to use its food to produce energy.
- It inhibits cell reproduction.

Some concerns regarding chlorine usage that may inspect its uses include:

- Chlorine reacts with many naturally occurring organic and inorganic compounds in water to produce undesirable DBPs
- Hazards associated with using chlorine, specifically chlorine gas, require special treatment and response programs
- High chlorine doses can cause taste and odor problems.

Chlorine is used in water treatment facilities primarily for disinfection. Because of chlorine's oxidizing powers, it has been found to serve other useful purposes in water treatment, such as (White, 1992):

- Taste and odor control
- Prevention of algal growths

Water Pollution

- Maintenance of clear filter media
- Removal of iron and manganese
- Destruction of hydrogen sulfide
- Bleaching of certain organic colors
- Maintenance of distribution system water quality by controlling slime growth
- Restoration and preservation of pipeline capacity
- Restoration of well capacity, water main sterilization
- Improved coagulation

Chlorine is available in a number of different forms: (1) as pure elemental gaseous chlorine (a greenish-yellow gas possessing a pungent and irritating odor that is heavier than air, nonflammable, and nonexplosive), when released to the atmosphere, this form is toxic and corrosive; (2) as solid calcium hypochlorite (in tablets or granules); or (3) as a liquid sodium hypochlorite solution (in various strengths).

The selection of one form of chlorine over the others for a given water system depends on the amount of water to be treated, configuration of the water system, the local availability of the chemicals, and the skill of the operator.

One of the major advantages of using chlorine is the effective residual that it produces. A residual indicates that disinfection is completed, and the system has an acceptable bacteriological quality. Maintaining a residual in the distribution system provides another line of defense against pathogenic organisms that could enter the distribution system and helps to prevent the regrowth of those microorganisms that were injured but not killed during the initial disinfection stage.

Chlorine Terminology

Often it is difficult for new waterworks operators to understand the terms used to describe the various reactions and processes used in chlorination. Common chlorination terms include the following:

- **Chlorine reaction**: regardless of the form of chlorine used for disinfection, the reaction in water is basically the same. The same amount of disinfection can be expected, provided the same amount of available chlorine is added to the water. The standard term for the concentration of chlorine in water is milligrams per liter (mg/L) or parts per million (ppm); these terms indicate the same quantity.
- **Chlorine dose**: the amount of chlorine added to the system. It can be determined by adding the desired residual for the finished water to the chlorine demand of the untreated water. Dosage can be either milligrams per liter (mg/L) or pounds per day. The most common is mg/L.
- **Chlorine demand**: the amount of chlorine used by iron, manganese, turbidity, algae, and microorganisms in the water. Because the reaction between chlorine and microorganisms is not instantaneous, demand is relative to time. For instance, the demand 5 minutes after applying chlorine will be less than the demand after 20 minutes. Demand, like dosage, is expressed in mg/l. The chlorine demand is as follows:

$$Cl_2 \text{ demand} = Cl_2 \text{ dose} - Cl_2 \text{ residual} \qquad (8.7)$$

- **Chlorine residual**: the amount of chlorine (determined by testing) that remains after the demand is satisfied. Residual, like demand, is based on time. The longer the time after dosage, the lower the residual will be, until all of the demand has been satisfied. Residual, like dosage and demand, is expressed in mg/L. The presence of a free residual of at least 0.2–0.4 ppm usually provides a high degree of assurance that the disinfection of the water is complete. Combined residual is the result of combining free chlorine with nitrogen compounds. Combined residuals are also called chloramines. Total chlorine residual is the

mathematical combination of free and combined residuals. Total residual can be determined directly with standard chlorine residual test kits.
- **Chorine contact time**: one of the key items in predicting the effectiveness of chlorine on microorganisms. It is the interval (usually only a few minutes) between the time when chlorine is added to the water and the time the water passes by the sampling point, contact time is the "T" in CT. CT is calculated based on the free chlorine residual prior to the first customer, times the contact time in minutes.

$$CT = \text{Concentration} \times \text{Contact time} = \text{mg/L} \times \text{minutes} \tag{8.8}$$

A certain minimum time period is required for the disinfecting action to be completed. The contact time is usually a fixed condition determined by the rate of flow of the water and the distance from the chlorination point to the first consumer connection. Ideally, the contact time should not be <30 minutes, but even more time is needed at lower chlorine doses, in cold weather, or under other conditions.

Pilot studies have shown that specific CT values are necessary for the inactivation of viruses and Giardia. The required CT value will vary depending on pH, temperature, and the organisms to be killed. Charts and formulae are available to make this determination. USEPA has set a CT value of three-log ($CT_{99.9}$) inactivation to assure the water is free of Giardia. State drinking water regulations include charts giving this value for different pH and temperature combinations. Filtration, in combination with disinfection, must provide a 3-log removal/inactivation of Giardia. Charts in the USEPA Surface Water Treatment Rule Guidance manual list the required CT values for various filter systems. Under the 1996 Interim Enhanced Surface Water Treatment (IESWT) rules, the USEPA requires systems that filter to remove 99% (2 log) of Cryptosporidium oocysts. To be assured that the water is free of viruses, a combination of filtration and disinfection to provide a 4-log removal of viruses has been judged the best for drinking water safety—99.99% removal. Viruses are inactivated (killed) more easily than cysts or oocysts.

Chlorine Chemistry
The reactions of chlorine with water and the impurities that might be in the water are quite complex, but a basic understanding of these reactions can aid the operator in keeping the disinfection process operating at its highest efficiency. When dissolved in pure water, chlorine reacts with the H$^+$ ions and the OH$^-$ radicals in the water. Two of the products of this reaction (the actual disinfecting agents) are hypochlorous acid, HOCl, and the hypochlorite radical, OCl$^-$. If microorganisms are present in the water, the HOCl and the OCl$^-$ penetrate the microbe cells and react with certain enzymes. This reaction disrupts the organisms' metabolism and kills them. The chemical equation for hypochlorous acid is as follows:

$$\frac{Cl_2}{(\text{chlorine})} + \frac{H_2O}{(\text{water})} \leftrightarrow \frac{HOCl}{(\text{hypochlorous acid})} + \frac{HCl}{(\text{hydrochloric acid})} \tag{8.9}$$

Note: The symbol $\leftrightarrow$ indicates that the reactions are reversible.

Hypochlorous acid (HOCl) is a weak acid—meaning it dissociates slightly into hydrogen and hypochlorite ions–but a strong oxidizing and germicidal agent. Hydrochloric acid (HCl) in the above equation is a strong acid and retains more of the properties of chlorine. HCl tends to lower the pH of the water, especially in swimming pools where the water is recirculated and continually chlorinated. The total hypochlorous acid and hypochlorite ions in water constitute the free available chlorine. Hypochlorites act in a manner similar to HCl when added to water, because hypochloric acid is formed.

When chlorine is first added to water containing some impurities, the chlorine immediately reacts with the dissolved inorganic or organic substances and is then unavailable for disinfection. The amount of chlorine used in this initial reaction is the chlorine demand of the water. If dissolved

Water Pollution

ammonia (NH_3) is present in the water, the chlorine will react with it to form compounds called chloramines. Only after the chlorine demand is satisfied and the reaction with all the dissolved ammonia is complete is the chlorine actually available in the form of HOCl and OCl^-. The equation for the reaction of hypochlorous acid (HOCl) and ammonia (NH_3) is as follows:

$$\underset{\text{(hypochlorous acid)}}{HOCl} + \underset{\text{(ammonia)}}{NH_3} \leftrightarrow \underset{\text{(monochloramine)}}{NH_2Cl} + \underset{\text{(Water)}}{H_2O} \tag{8.10}$$

Note: The chlorine as hypochlorous acid and hypochlorite ions remaining in the water after the above reactions are complete is known as free available chlorine, and it is a very active disinfectant.

Breakpoint Chlorination

To produce a free chlorine residual, enough chlorine must be added to the water to produce what is referred to as breakpoint chlorination, which is the point at which near complete oxidation of nitrogen compounds is reached; any residual beyond breakpoint is mostly free chlorine (see Figure 8.7). When chlorine is added to natural waters, the chlorine begins combining with and oxidizing the chemicals in the water before it begins disinfecting. Although residual chlorine will be detectable in the water, the chlorine will be in the combined form with a weak disinfecting power. As we see in Figure 8.7, adding more chlorine to the water at this point actually decreases the chlorine residual as the additional chlorine destroys the combined chlorine compounds. At this stage, water may have a strong swimming pool or medicinal taste and odor. To avoid this taste and odor, add still more chlorine to produce a free residual chlorine. Free chlorine has the highest disinfecting power. The point at which most of the combined chlorine compounds have been destroyed and the free chlorine starts to form is the breakpoint.

The chlorine breakpoint of water can only be determined by experimentation. This simple experiment requires 20 1,000-mL breakers and a solution of chlorine. Place the raw water in the beakers and dose with progressively larger amounts of chlorine. For instance, you might start with zero in the first beaker, then 0.5 and 1.0 mg/L, and so on. After a period of time, say 20 minutes, test each beaker for total chlorine residual and plot the results.

Breakpoint Chlorination Curve

Refer to Figure 8.7 for the following explanation. Where the curve starts, no residual exists, even though there was a dosage. This is called the initial demand and is microorganisms and interfering agents using the result of the chlorine. After the initial demand, the curve slopes upward. Chlorine combining to form chloramines produces this part of the curve. All of the residual measured on this part of the curve is combined residual. At some point, the curve begins to drop back toward zero; this portion of the curve results from a reduction in combined residual, which occurs because enough chlorine has been added to destroy (oxidize) the nitrogen compounds used to form combined residuals. The breakpoint is the point where the downward slope of the curve breaks upward. At this point, all of the nitrogen compounds that could be destroyed have been destroyed. After breakpoint, the curve starts upward again, usually at a 45° angle. Only on this part of the curve can free residuals be found. The distance that the breakpoint is above zero is a measure of the remaining combined residual in the water. This combined residual exists because some of the nitrogen compounds will not have been oxidized by chlorine. If the irreducible combined residual is more than 15% of the total residual, chlorine odor, and taste complaints will be high.

Gas Chlorination

Gas chlorine is provided in 100-lb or 1-ton containers. Chlorine is placed in the container as a liquid. The liquid boils at room temperature, reducing to a gas and building pressure in the cylinder. At room temperature of 70°F, a chlorine cylinder will have a pressure of 85 psi; 100/150-lb cylinders should be maintained in an upright position and chained to the wall. To prevent a chlorine cylinder from rupturing in a fire, the cylinder valves are equipped with special fusible plugs that melt between 158°F and 164°F.

Chlorine gas is 99.9% chlorine. A gas chlorinator meters the gas flow and mixes it with water, which is then injected as a water solution of pure chlorine. As the compressed liquid chlorine is withdrawn from the cylinder, it expands as a gas, withdrawing heat from the cylinder. Care must be taken not to withdraw the chlorine at too fast a rate: if the operator attempts to withdraw more than about 40 pounds of chlorine per day from a 150-pound cylinder, it will freeze up.

Note: All chlorine gas feed equipment sold today is vacuum operated. This safety feature ensures that if a break occurs in one of the components in the chlorinator, the vacuum will be lost, and the chlorinator will shut down without allowing gas to escape.

Chlorine gas is a highly toxic lung irritant, and special facilities are required for storing and housing it. Chlorine gas will expand to 500 times its original compressed liquid volume at room temperature (one gallon of liquid chlorine will expand to about 67 ft^3). Its advantage as a drinking water disinfectant is the convenience afforded by a relatively large quantity of chlorine available for continuous operation for several days or weeks without the need for mixing chemicals. Where water flow rates are highly variable, the chlorination rate can be synchronized with the flow.

Chlorine gas has a very strong, characteristic odor that can be detected by most people at concentrations as low as 3.5 ppm. Highly corrosive in moist air, it is extremely toxic and irritating in concentrated form. Its toxicity ranges from throat irritation 15 ppm to rapid death at 1,000 ppm. Although chlorine does not burn, it supports combustion, so open flames should never be used around chlorination equipment.

When changing chlorine cylinders, an accidental release of chlorine may occasionally occur. To handle this type of release, an approved (NIOSH-approved) Self-Contained Breathing Apparatus (SCBA) must be worn. Special emergency repair kits are available from the Chlorine Institute for use by emergency response teams to deal with chlorine leaks. Because chlorine gas is 2.5 times heavier than air, exhaust and inlet air ducts should be installed at floor level. A leak of chlorine gas can be found by using the fumes from a strong ammonia mist solution. A white cloud develops when ammonia mist and chlorine combine.

Hypochlorination

Combining chlorine with calcium or sodium produces hypochlorites. Calcium hypochlorites are sold in powder or tablet forms and can contain chlorine concentrations up to 67%. Sodium hypochlorite is a liquid (bleach, for example) and is found in concentrations up to 16%. Chlorine concentrations of household bleach range from 4.75% to 5.25%. Most small system operators find using these liquid or dry chlorine compounds more convenient and safer than chlorine gas.

The compounds are mixed with water and fed into the water with inexpensive solution feed pumps. These pumps are designed to operate against high system pressures but can also be used to inject chlorine solutions into tanks, though injecting chlorine into the suction side of a pump is not recommended as the chlorine may corrode the pump impeller.

Calcium hypochlorite can be purchased as tablets or granules with ~65% available chlorine (10 pounds of calcium hypochlorite granules contain only 6.5 pounds of chlorine). Normally, 6.5 pounds of calcium hypochlorite will produce a concentration of 50-mg/L chlorine in 10,000 gallons of water. Calcium hypochlorite can burn (at 350°F) if combined with oil or grease. When mixing calcium hypochlorite, operators must wear chemical safety goggles, a cartridge breathing apparatus, and rubberized gloves. Always place the powder in the water. Placing the water into the dry powder could cause an explosion.

Sodium hypochlorite is supplied as a clear, greenish-yellow liquid in strengths from 5.25% to 16% available chlorine. Often referred to as "bleach," it is, in fact, used for bleaching. As we stated earlier, common household bleach is a solution of sodium hypochlorite containing 5.25% available chlorine. The amount of sodium hypochlorite needed to produce a 50-mg/l-chlorine concentration in 10,000 gallons of water can be calculated using the solutions equation:

$$C_1V_1 = C_2V_2 \tag{8.11}$$

where:
   C = the solution concentration in mg/L or %
   V = the solution volume in the liters, gal., qt., etc.
   1.0% = 10,000 mg/L

In this example, C_1 and V_1 and associated with the sodium hypochlorite and C_2 and V_2 are associated with the 10,000 gallons of water with a 50 mg/L chlorine concentration. Therefore:

$$C_1 = 5.25\%$$

$$C_1 = \frac{(5.25\%)(10,000\,\text{mg/L})}{1.0\%} = 52,500\,\text{mg/L}$$

V_1 = unknown volume of sodium hypochlorite
C_2 = 50 mg/L
V_2 = 10,000 gallons

$$(52,500\,\text{mg/L})(V_1) = (50\,\text{mg/l})(10,000\,\text{gal})$$

$$V_1 = \frac{(50\,\text{mg/L})(10,000\,\text{gal})}{52,500\,\text{mg/L}}$$

$$V_1 = 9.52\,\text{gal of sodium hypochlorite}$$

Sodium hypochlorite solutions are introduced to the water in the same manner as calcium hypochlorite solutions. The purchased stock "bleach" is usually diluted with water to produce a feed solution that is pumped into the water system.

Hypochlorites stored properly to maintain their strengths. Calcium hypochlorite must be stored in airtight containers in cool, dry, dark locations. Sodium hypochlorite degrades relatively quickly even when properly stored; it can lose more than half of its strength in 3–6 months. Operators should purchase hypochlorites in small quantities to assure they are used while still strong. Old chemicals should be discarded safely.

The pumping rate of a chemical metering pump is usually manually adjusted by varying the length of the piston or diaphragm stroke. Once the stroke is set, the hypochlorinator feeds accurately at that rate. However, chlorine measurements must be made occasionally at the beginning and end of the well pump cycle to assure the correct dosage. A metering device may be used to vary the hypochlorinator feed rate, synchronized with the water flow rate. Where a well pump is used, the hypochlorinator is connected electrically with the pump's on-off controls to assure that chlorine solution is not fed into the pipe when the well is not pumping.

Determining Chlorine Dosage
Proper disinfection requires the calculation of the amount of chlorine that must be added to the water to produce the required dosage. The type of calculation used depends on the form of chlorine being used. The basic chlorination calculation used is the same one used for all chemical addition calculations—the pounds formula. The pounds formula is

$$\text{Pounds} = \text{mg/L} \times 8.34 \times \text{MG} \tag{8.12}$$

where
   Pounds = pounds of available chlorine required
   mg/L = desired concentration in milligrams per liter
   8.34 = conversion factor
   MG = millions of gallons of water to be treated

Example 8.3

Problem:

Calculate the number of pounds of gaseous chlorine needed to treat 250,000 gallons of water with 1.2 mg/l of chlorine.

Solution:

$$\text{Pounds} = 1.2\,\text{mg/L} \times 8.34 \times 0.25\,\text{MG}$$

$$= 2.5\,\text{lb}$$

Note: Hypochlorites contain <100% available chlorine. Thus, we must use more hypochlorite to get the same number of pounds of chlorine into the water.

If we substitute calcium hypochlorite with 65% available chlorine in our example, 2.5 lb of available chlorine is still needed, but more than 2.5 pounds of calcium hypochlorite is needed to provide that much chlorine. Determine how much of the chemical is needed by dividing the pounds of chlorine needed by the decimal form of the percent available chlorine. Since 65% is the same as 0.65, we need to add:

$$(\text{OCl})\frac{2.5\,\text{lb}}{0.65\,\text{available chlorine}} = 3.85\,\text{lb Ca} \qquad (8.13)$$

to get that much chlorine.

In practice, because most hypochlorites are fed as solutions, we often need to know how much chlorine solution we need to feed. In addition, the practical problems faced in day-to-day operation are never so clearly stated as the practice problems we work. For example, small water systems do not usually deal with water flow in million gallons per day. Real-world problems usually require a lot of intermediate calculations to get everything ready to plug into the pounds formula.

Example 8.4

Problem:

We have raw water with a chlorine demand 2.2 mg/L. We need a final residual of 1.0 mg/L at the entrance to the distribution system. We can use sodium hypochlorite or calcium hypochlorite granules as the source of chlorine. If well output is 65 gallons per minute (gpm) and the chemical feed pump can inject 100 milliliters per minute (mL/min) at the 50% setting. What is the required strength of the chlorine solution we will feed? What volume of 5.20% sodium hypochlorite will be needed to produce one gallon of the chlorine feed solution? How many pounds of 65% calcium hypochlorite will be needed to mix each gallon of solution?

Solution:

- **Step 1**: Determine the amount of chlorine to be added to the water (the chlorine dose). The dose is defined as the chlorine demand of the water, plus the desired residual, or in this case:

$$Q_1 = 65\,\text{gal/min} \qquad Q_t = Q_1 + Q_2$$

$$C_L = 0.0\,\text{mg/L} \qquad C_t = 3.2\,\text{mg/L}$$

Well pump →→→ Treated water

↑

Chlorination metering pump

$$Q_2 = 100\,\text{mL/min}$$

$$C_2 = \text{unknown}$$

Water Pollution

Chlorination system problem

$$\text{Dose} = \text{Demand} + \text{Residual} = 2.2\,\text{mg/L} + 1.0\,\text{mg/L} = 3.2\,\text{mg/L}$$

To obtain a 1.0-mg/L residual when the water enters the distribution system, we must add 3.2 mg/L of chlorine to the water.

- **Step 2**: Determine the strength of the chlorine feed solution that would add 3.2 mg/L of chlorine to 65 gallons/minute of water when fed at a rate of 100 mL/ minute. The well pump is producing water at a flowrate (Q_1) of 65 gpm, with a chlorine concentration (C_1) of 0.0 mg/L. The metering pump will add 100-mL/min (Q_2) of chlorine solution, but we do not know its concentration (C_2) yet. The finished water will have been dosed with a chlorine concentration of 3.2 mg/L (C_t) and will enter the distribution system at a rate (Q_t) of 65 gallons + 100 mL per minute, with a free residual chlorine concentration of 1.0 mg/L after the chlorine contact time.
- **Step 3**: Convert the metering pump flow (100 mL/min) to gpm so we can calculate Q_1. To do this, we use standard conversion factors:

$$\frac{100\,\text{mL}}{\text{min}} \times \frac{1\,\text{L}}{1{,}000\,\text{mL}} \times \frac{1\,\text{gal}}{3.785\,\text{L}} = 0.026\,\text{gal/min}$$

So, now we know that Q_2 is 0.026 gpm and that Q_t is 65 gpm + 0.026 gpm = 65.026 gpm.

- **Step 4**: We must now use what is known as the stand mass balance equation. The mass balance equation says that the flow rate times the concentration of the output is equal to the flow rate times the concentration of each of the inputs added together, or in equation form,

$$(Q_1 \times C_1) + (Q_2 \times C_2) = (Q_1 \times C_1) \tag{8.14}$$

Substituting the numbers given and those we have calculated so far gives us:

$$(65\,\text{gpm} \times 0.0\,\text{mg/L}) + (0.026\,\text{gpm} \times \text{mg/L})$$

$$[(65 + 0.026)\,\text{gpm} \times 3.2\,\text{mg/L}]$$

$$0 + 0.026x = 65.026 \times 3.2$$

$$0.026x = 208.1$$

$$x = \frac{208.1}{0.026}$$

$$x = 8{,}004\,\text{mg/L}$$

This is the answer to the first part of the question, the required strength of the chlorine feed solution. Because a 1% solution is equal to 10,000 mg/L, solution strength of 8,004 mg/L is approximately a 0.80% solution.

To determine the required volume of bleach per gallon to produce this 0.80% solution, we go back to the pounds formula:

$$\text{Pounds} = \text{mg/L} \times 8.34 \times \text{MG}$$

$$= 8{,}004\,\text{mg/L} \times 8.34 \times 0.000001\,\text{MG}$$

$$= 0.067\,\text{lb chlorine/gal solution}$$

Note: Remember to convert gallons to MG

Recalling that the bleach, like water, weighs 8.34 lb/gal and contains ~5.20% (0.0520) available chlorine,

$$1\,\text{gal bleach} = 8.34\,\text{lb/gal} \times 0.0520$$

$$= 0.43\,\text{lb available chlorine per gallon of bleach}$$

If one gallon of bleach contains 0.43 lb of available chlorine, how many gallons of bleach do we need to provide the 0.067-lb of chlorine we need for each gallon of chlorine feed solution? To determine the gallons of bleach needed, we use the simple ration equation:

$$\frac{1 \text{ gal}}{0.443 \text{ lb Cl}_2} = \frac{x \text{ gal}}{0.067 \text{ lb Cl}_2}$$

$$x = \frac{1 \text{ gal} \times 0.067 \text{ lb}}{0.43 \text{ lb}}$$

$$= 0.16\text{-gal bleach per gal solution}$$

Or, to determine the required volume of bleach per gallon to produce this 0.80% solution, we use the solutions equation and calculate it this way:

$$C_1 V_1 = C_2 V_2$$

$C_1 = 0.80\%$
$V_1 = 1.0 \text{ gal}$
$C_2 = 5.20\%$
$V_2 = \text{unknown}$

$$0.80\% \text{ solution} \times 1.0 \text{ gal} = 5.20\% \text{ solution} \times \text{gal}$$

$$0.80\% \text{ solution} \times 1.0 \text{ gal} = 5.20\% \text{ solution} \times \text{gal}$$

$$x = \frac{0.80\% \times 1.0 \text{ gal}}{5.20\%}$$

$$= 0.15\text{-gal bleach per gal solution}$$

To summarize, 1 gallon of household bleach:

- Contains 0.85-gal water
- Contains 5.20% or 52,000 mg/L available chlorine
- Contains 0.15-gal available chlorine
- Contains 0.43 lb available chlorine
- Weighs 8.34 lb

The third part of the problem requires that we determine the pounds of calcium hypochlorite needed for each gallon of feed solution. We know that we need 0.066 lb of chlorine for each gallon of solution and that HTH contains 65% available chlorine; that is, 1.0 lb of HTH contains 0.65 lb of available chlorine. Using the ratio equation,

$$\frac{1 \text{ lb HTH}}{0.65 \text{ lb Cl}_2} = \frac{x \text{ lb HTH}}{0.067 \text{ lb Cl}_2}$$

$$x = \frac{1 \text{ lb} \times 0.067 \text{ lb}}{0.65 \text{ lb}}$$

$$x = \frac{0.067}{0.65}$$

$$= 0.1 \text{ lb HTH per gal solution}$$

Chlorine Generation
Onsite generation of chlorine has recently become practical. These generation systems, using only salt and electric power, can be designed to meet disinfection and residual standards and to operate

unattended at remote sites. Considerations for chlorine generation include cost, concentration of the brine produced, and availability of the process.

Chlorine Gas Chlorine gas can be generated by a number of processes including the electrolysis of alkaline brine or hydrochloric acid, the reaction between sodium chloride and nitric acid, or the oxidation of hydrochloric acid. About 70% of the chlorine produced in the U.S. is manufactured from the electrolysis of salt brine and caustic solutions in a diaphragm cell (White, 1992). Because chlorine is a stable compound, it is typically produced off-site by a chemical manufacturer. Once produced, chlorine is packaged as a liquefied gas under pressure for delivery to the site in railcars, tanker trucks, or cylinders.

Sodium Hypochlorite Dilute sodium hypochlorite solutions (<1%) can be generated electrochemically on-site from salt brine solution. Typically, sodium hypochlorite solutions are referred to as liquid bleach or Javelle water. Generally, the commercial or industrial-grade solutions produced have hypochlorite strengths of 10%–16%. The stability of sodium hypochlorite solution depends on the hypochlorite concentration, the storage temperature, the length of storage (time), the impurities of the solution, and exposure to light. Decomposition of hypochlorite over time can affect the feed rate and dosage, as well as produce undesirable byproducts such as chlorite ions or chlorate (Spellman, 2007). Because of the storage problems, many systems are investigating onsite generation of hypochlorite in lieu of its purchase from a manufacturer or vendor.

Calcium Hypochlorite To produce calcium hypochlorite, hypochlorous acid is made by adding chlorine monoxide to water and then neutralizing it with lime slurry to create a solution of calcium hypochlorite. Generally, the final product contains up to 70% available chlorine and 4%–6% lime. Storage of calcium hypochlorite is a major safety consideration. It should never be stored where it is subject to heat or allowed to contact with any organic material of an easily oxidized nature (Spellman, 2007).

Primary Uses and Points of Application of Chlorine
Uses The main usage of chlorine in drinking water treatment is for disinfection. However, chlorine has also found application for a variety of other water treatment objectives such as the control of nuisance organisms, oxidation of taste and odor compounds, oxidation of iron and manganese, color removal, and as a general treatment aid to filtration and sedimentation processes (White, 1992). Table 8.16 presents a summary of chlorine uses and doses.

Points of Application At conventional surface water treatment plants, chlorine is typically added for prechlorination at either the raw water intake or flash mixer, for intermediate chlorination ahead of the filters, for postchlorination at the filter clear well, or for rechlorination of the distribution system. Table 8.17 summarizes typical points of application.

Typical Doses
Table 8.18 shows the typical dosages for the various forms of chlorine. The wide range of chlorine gas dosages most likely represents its use as both an oxidant and a disinfectant. While sodium hypochlorite and calcium hypochlorite can also serve as both an oxidant and a disinfectant, their higher cost may limit their use.

Disinfection by chlorination is a rather straightforward process, but several factors (interferences) can affect the ability of chlorine to perform its main function: disinfection. Turbidity is one such interference. As we described earlier, turbidity is a general term that describes particles suspended in the water. Water with a high turbidity appears cloudy. Turbidity interferes with disinfection when microorganisms "hide" from the chlorine within the particles causing turbidity. This problem is magnified when turbidity comes from organic particles, such as those from sewage effluent.

TABLE 8.16
Chlorine Uses and Doses

Application	Typical Dose	Optimal pH	Reaction Time	Effectiveness	Other Considerations
Iron	0.62 mg/mg Fe	7.0	>1 hour	Good	
Manganese	0.77 mg/mg Mn	7–8	1–3-hour	Slow Kinetics	Reaction time increases at lower pH
	Lower pH	9.5	Minutes		
Biological growth	1–2 mg/L	6–8	NA	Good	DBP Formation
Taste/odor	Varies	6–8	Varies	Varies	Effectiveness depends on compound
Color removal	Varies	4.0–6.8	Minutes	Good	DBP formation
Zebra	2–5 mg/L		Shock level	Good	DBP formation
Mussels	0.2–0.5 mg/L		Maintenance level		
Asiatic clams	0.3–0.5 mg/L		Continuous	Good	DBP formation

Source: Adapted in part from White (1992) and Culp et al. (1986).

TABLE 8.17
Typical Chlorine Points of Application

Point of Application

Raw water intake

Flash mixer (prior to sedimentation)

Filter influent

Filter clearwell

Distribution system

Sources: Adapted in part from White (1992).

TABLE 8.18
Typical Chlorine Dosages at Water Treatment Plants

Chlorine Compound	Range of Doses (mg/L)
Calcium hypochlorite	0.6–5
Sodium hypochlorite	0.2–2
Chlorine gas	1–16

Source: Science Applications International Corporation (SAIC), 1998, San Diego, CA, as adapted from USEPA's review of public water systems' initial Sampling Plans which were required by USEPA's Information Collection Rule (ICR).

To overcome this, the length of time the water is exposed to the chlorine or the chlorine dose must be increased, although highly turbid waters may still shield some microorganisms from the disinfectant. USEPA took this type of problem into consideration in its Surface Water Treatment Rule and the 1996 Amendments (the Interim Enhanced Surface Water Treatment) (IESWT) rule that tightens controls on DBPs and turbidity and regulates Cryptosporidium. The rule requires continuous turbidity monitoring of individual filters and tightens allowable turbidity limits for combined filter effluent, cutting the maximum from 5 NTU to 1 NTU and the average monthly limit from 0.5 NTU to 0.3 NTU.

Note: Recall that the IESWT rule applies to large (those serving more than 10,000 people) public water systems that use surface water or groundwater directly influenced by surface water and is the first to directly regulate Cryptosporidium.

Temperature affects the solubility of chlorine, the rate at which disinfecting ions are produced, and the proportion of highly effective forms of chlorine that will be present in the water. More importantly, however, temperature affects the rate at which the chlorine reacts with the microorganisms themselves. As water temperature decreases, the rate at which the chlorine reacts with the microorganisms themselves. As water temperature decreases, the rate at which the chlorine can pass through the microorganisms' cell wall decreases, making the chlorine less effective as a disinfectant. Along with the presence of turbidity-causing agents such as suspended solids and organic matter, chemical compounds in the water may influence chlorination: high alkalinity, nitrates, manganese, iron, and hydrogen sulfide.

Measuring Chlorine Residual

During normal operations, waterworks operators perform many operating checks and tests on unit processes throughout the plant. One of the most important and most frequent operating checks is chlorine residual test. This test must be performed whenever a distribution sample is collected for microbiological analysis and should be done frequently where the treatment facility discharges to the water distribution system to ensure that the disinfection system is working properly.

To test for residual chlorine, several methods are available. The most common and most convenient is the DPD Color Comparator Method. This method uses a small portable test kit with prepared chemicals that produce a color reaction indicating the presence of chlorine. By comparing the color produced by the reaction with a standard, we can determine the approximate chlorine residual concentration of the sample. DPD color comparator chlorine residual test kits are available from the manufacturers of chlorination equipment.

Note: The color comparator method is acceptable for most groundwater systems and for chlorine residual measurements in the distribution system. However, keep in mind that the methods used to take chlorine residual measurements for controlling the disinfection process in surface water systems and some groundwater systems where adequate disinfection is essential must be approved by standard methods.

Pathogen Inactivation and Disinfection Efficacy

Inactivation Mechanisms

Research has shown that chlorine is capable of producing lethal events at or near the cell membrane as well as affecting DNA. In bacteria, chlorine was found to adversely affect cell respiration, transport, and possibly DNA activity. Chlorination was found to cause an immediate decrease in oxygen utilization in both Escherichia coli and Candida parapsilosis studies. The results also found that chlorine damages the cell wall membrane, promotes leakage through the cell membrane, and produces lower levels of DNA synthesis for Escherichia coli, Candida parapsilosis, and Mycobacterium fortuitum bacteria. This study also showed that chlorine inactivation is rapid and does not require bacteria reproduction. These observations rule out mutation or lesions as the principal inactivation mechanisms since these mechanisms require at least one generation of replication for inactivation to occur.

Environmental Effects

Several environmental factors influence the inactivation efficiency of chlorine, including water temperature, pH, contact time, mixing, turbidity, interfering substances, and the concentration of available chlorine. In general, the highest levels of pathogen inactivation are achieved with high chlorine residuals, long contact times, high water temperature, and good mixing, combined with a low pH, low turbidity, and the absence of interfering substances. Of the environmental factors, pH and temperature have the most impact on pathogen inactivation by chlorine. The effect of pH and temperature on pathogen inactivation is discussed below.

- **pH**: The germicidal efficiency of hypochlorous acid (HOCl) is much higher than that of the hypochlorite ion (OCl$^-$). The distribution of chlorine species between HOCl and OCl$^-$. The distribution of chlorine species between HOCl and OCl$^-$ is determined by pH, as discussed above. Because HOCl dominates at low pH, chlorination provides more effective disinfection at low pH. At high pH, OCl$^-$ dominates, which causes a decrease in disinfection efficiency. The inactivation efficiency of gaseous chlorine and hypochlorite is the same at the same pH after chlorine addition. Note, however, that the addition of gaseous chlorine will decrease the pH while the addition of hypochlorite will increase the pH of the water. Therefore, without pH adjustment to maintain the same treated water pH, gaseous chlorine will have greater disinfection efficiency than hypochlorite.

    The impact of pH on chlorine disinfection has been demonstrated in the field. For example, virus inactivation studies have shown that 50% more contact time is required at pH 7.0 than at pH 6.0 to achieve comparable levels of inactivation. These studies also demonstrated that a rise in pH from 7.0 to 8.8 or 9.0 requires six times the contact time to achieve the same level of virus inactivation (Culp and Culp, 1974a). Although these studies found a decrease in inactivation with increasing pH, some studies have shown the opposite effect. A 1972 study reported that viruses were more sensitive to free chorine at high pH than at low pH (Scarpino et al., 1972).
- **Temperature**: For typical drinking water treatment temperatures, pathogen inactivation increases with temperature. Virus studies indicated that the contact time should be increased by two to three times to achieve comparable inactivation levels when the water temperature is lowered by 10°C (Clarke et al., 1962).
- **Disinfection efficacy**: Since its introduction, numerous investigations have been made to determine the germicidal efficiency of chlorine. Although there are widespread differences in the susceptibility of various pathogens the general order of increasing chlorine disinfection difficulty are bacteria, viruses, and then protozoa.

    Bacteria inactivation: Chlorine is an extremely effective disinfectant for inactivating bacteria. A study conducted during the 1940s investigated the inactivation levels as a function of time for *E. coli*, Pseudomonas aeruginosa, Salmonella typhi, and Shigella dysenteriae (Butterfield et al., 1943). Study results indicated that HOCl is more effective than OCl$^-$ for the inactivation of these bacteria. These results have been confirmed by several researchers that concluded that HOCl is 70–80 times more effective than OCl$^-$ for inactivating bacteria (Culp et al., 1986).

    Virus inactivation: Chlorine has been shown to be a highly effective viricide. One of the most comprehensive virus studies was conducted in 1971 using treated Potomac estuary water (Liu et al., 1971). The tests were performed to determine the resistance of 20 different enteric viruses to free chlorine under constant conditions of 0.5 mg/L free chlorine and a pH and temperature of 7.8°C and 2°C, respectively. In this study, the least resistant virus was found to be reovirus and required 2.7 minutes to achieve 99.99% inactivation (4 log removal). The most resistant virus was found to be a poliovirus, which required more than 60 minutes for 99.99 inactivation. The corresponding CT range required to achieve 99.99% inactivation for all 20 viruses was between 1.4 to over 30-mg·min/L. All of the virus inactivation tests in this study were performed at a free chlorine residual of 0.4 mg/L, a pH of 7.0, a temperature of 5°C, and contact times of either 10, 100, or 1,000 minutes. Test results showed that of the 20 cultures tested only two poliovirus strains reached 99.99% inactivation after 10 minutes (CT=4 mg·min/L), six poliovirus strains reached

99.99% inactivation after 100 minutes (CT=40 mg·min/L), and 11 of the 12 polioviruses plus one Coxsackievirus strain (12 out of a total of 20 viruses) reached 99.99% inactivation after 1,000 minutes (CT=400 mg·min/L).

Protozoa inactivation: Chlorine has been shown to have limited success inactivating protozoa. Data obtained during a 1984 study indicated that the resistance of Giardia cysts is two orders of magnitude higher than that of enteroviruses and more than three orders of magnitude higher than the enteric bacteria (Hoff et al., 1984). CT requirements for Giardia cysts inactivation when using chlorine as a disinfectant has been determined for various pH and temperature conditions. The CT values increase at low temperatures and high pH. Chlorine has little impact on the viability of Cryptosporidium oocysts when used at the relatively low doses encountered in water treatment (e.g., 5 mg/L). Approximately 40% removals (0.2 log) of Cryptosporidium were achieved at CT values of both 30 and 3,600 mg·min/L (Finch et al., 1994). Another study determined that "no practical inactivation was observed" when oocysts were exposed to free chlorine concentrations ranging from 5 to 80 mg/L at pH 8, a temperature of 22°C, and contact times of 48–245 minutes (Gyurek et al., 1996). CT values ranging from 3,000 to 4,000 mg·min/L were required to achieve 1-log of Cryptosporidium inactivation at pH 6.0 and temperature of 22°C. During this study, one trial in which oocysts were exposed to 80 mg/L of free chlorine for 120 minutes was found to produce greater than 3-logs of inactivation.

Disinfection By-Products (DBPs)

Halogenated organics are formed when natural organic matter (NOM) reacts with free chlorine or free bromine. Free chlorine is normally introduced into water directly as a primary or secondary disinfectant. Free bromine results from the oxidation by chlorine of the bromide ion in the source water. Factors affecting the formation of these halogenated DBPs include type and concentration of NOM, chlorine form and dose, time, bromide ion concentration, pH, organic nitrogen concentration, and temperature. Organic nitrogen significantly influenced the formation of nitrogen containing DBPs, including haloacetonitriles (Reckhow et al., 1990), halopicrins, and cyanogen halides. Because most water treatment systems have been required to monitor for total trihalomethanes (TTHM) in the past, most water treatment operators are probably familiar with some of the requirements that the D/DBP Rule involves. The key points of the DBP Rule and some of the key changes water supply systems are required to comply with are summarized below:

- **Chemical limits and testing**: Testing requirements will include total trihalomethanes (TTTHM) and five haloacetic acids (HAA5). The Maximum Contaminant Level (MCL) for TTHM is 0.080 mg/L for surface water systems. In addition, a new MCL of 0.060 mg/L has been established for haloacetic acids (HAA5). New MCLs have been established for Bromate (0.010 mg/L and Chlorite (1.0 mg/L). Bromate monitoring is required for systems which use ozone. Chlorite monitoring only will be required of systems which use chlorine dioxide (i.e., sodium- and calcium hypochlorite are not included). Maximum Residual Disinfectant Levels (MRDLs) will be established for Total Chlorine (4.0 mg/L) and Chlorine Dioxide (0.8 mg/L).
- **Operational requirements**: Analytical requirements for measuring chlorine residual have been changed to require digital equipment (i.e., no color wheels or analog test kits). The test kit must have a detection limit of at least 0.1 mg/L.
- **Monitoring and reporting**: Individual state requirements will differ, but at the minimum, the following requirements are listed.
    - Surface water system monitoring requirements include four quarterly samples per treatment plant (Source Treatment Unit or STU) for TTHM and HAA5. One of these quarterly samples, or 25% of the total samples, must be collected at the maximum residence time location. The remaining samples must be collected at representative

locations throughout the entire distribution system. Compliance is based on a running annual average computed quarterly.
- For those surface water systems using conventional filtration or lime softening, a D/DBP monthly operating report for Total Organic Carbon (TOC) removal will be required to be completed and filed with state EPA. This report will include TOC, alkalinity, and Specific Ultraviolet Absorption (SUVA) parameters. There will be an additional monthly operating report for bromate chlorite, chlorine dioxide, and chlorine residual. More information will be forthcoming closer to the compliance date.
- TTHM monitoring results may indicate the possible need for additional treatment to include the best available technology for the reduction of DBP. This may include granular activated carbon, enhanced coagulation (for surface water systems using conventional filtration), or enhanced softening (for systems using lime softening).
- Operators are required to develop and implement a sample-monitoring plan for disinfectant residual and disinfection byproducts. The plan will be required to be submitted to and approved by state EPA. Disinfection residual monitoring compliance for total chlorine, including chloramines, will be based upon a running annual average, computed quarterly, of the monthly average of all samples collected under this rule. Disinfectant residual monitoring compliance for chlorine dioxide will be based on consecutive daily samples. Disinfectant residual monitoring will be required at the same distribution point and time as total coliform monitoring. In addition, if the operator feeds ozone or chlorine dioxide, a sample monitoring plan for bromate or chlorite, respectively, must be submitted to and approved by state EPA.

ARSENIC REMOVAL FROM DRINKING WATER

Much of the following information is based on USEPA (2000). Operators may be familiar with the controversy created when newly elected President George W. Bush placed the pending Arsenic Standard on temporary hold. The President prevented implementation of the Arsenic Standard to give scientists time to review the Standard, to take a closer look at the possible detrimental effects on health and well-being of consumers in certain geographical areas of the U.S., and to give economists time to determine the actual cost of implementation.

President Bush's decision caused quite a stir, especially among environmentalists, the media, and others who felt that the Arsenic Standard should be enacted immediately to protect affected consumers. The President, who understood the emotional and political implications of shelving the Arsenic Standard, also understood the staggering economic implications involved in implementing the new, tougher standard. Many view the President's decision as the wrong one. Others view his decision as the right one; they base their opinion on the old adage, "It is best to make scientific judgments based on good science instead of on "feel good" science." Whether the reader shares the latter view or not, the point is that arsenic levels in potable water supplies are required to be reduced to a set level and in the future will have to be reduced to an even lower level. Accordingly, water treatment plants affected by the existing arsenic requirements and the pending tougher arsenic requirements should be familiar with the technologies for removal of arsenic from water supplies. In this section, we describe a number of these technologies.

ARSENIC EXPOSURE

Arsenic (As) is a naturally occurring element present in food, water, and air. Known for centuries to be an effective poison, some animal studies suggest that arsenic may be an essential nutrient at low concentrations. Non-malignant skin alterations, such as keratosis and hypo- and hyperpigmentation, have been linked to arsenic ingestion, and skin cancers have developed in some patients. Additional studies indicate that arsenic ingestion may result in internal malignancies, including cancers of the kidney, bladder, liver, lung, and other organs. Vascular system effects have also been observed, including peripheral vascular disease, which in its most severe form, results in gangrene or Blackfoot Disease. Other potential effects include neurologic impairment.

The primary route of exposure to arsenic for humans is ingestion. Exposure via inhalation is considered minimal, though there are regions where elevated levels of airborne arsenic occur periodically (Hering and Chiu, 1998). Arsenic occurs in two primary forms; organic and inorganic. Organic species of arsenic are predominately found in foodstuffs, such as shellfish, and include such forms as monomethyl arsenic acid (MMAA), dimethyl arsenic acid (DMAA), and arseno-sugars. Inorganic arsenic occurs in two valence states, arsenite and arsenate. In natural surface waters arsenate is the dominant species.

ARSENIC REMOVAL TECHNOLOGIES

Some of the arsenic removal technologies discussed in this section are traditional treatment processes which have been tailored to improve the removal of arsenic from drinking water. Several treatment techniques discussed here are at the experimental stage with regard to arsenic removal, and some have not been demonstrated at full scale. Although some of these processes may be technically feasible, their cost may be prohibitive. Technologies discussed in this section are grouped into four broad categories: prescriptive processes, adsorption processes, ion exchange processes, and separation (membrane) processes. Each category is discussed here, with at least one treatment technology described in each category.

Prescriptive Processes

Coagulation/Filtration

Coagulation/flocculation (C/F) is a treatment process by which the physical or chemical properties of dissolved chemical properties of dissolved colloidal or suspended matter are altered such that agglomeration is enhanced to an extent that the resulting particles will settle out of solution by gravity or will be removed by filtration. Coagulants change surface charge properties of solids to allow agglomeration and/or enmeshment of particles into a flocculated precipitate. In either case, the final products are larger particles, or floc, which more readily filter or settle under the influence of gravity.

The coagulation/filtration process has traditionally been used to remove solids from drinking water supplies. However, the process is not restricted to the removal of particles. Coagulants render some dissolved species [e.g., natural organic matter (NOM), inorganics, and hydrophobic synthetic organic compounds (SOCs)] insoluble and the metal hydroxide particles produced by the addition of metal salt coagulants (typically aluminum sulfate, ferric chloride, or ferric sulfate) can adsorb other dissolved species. Major components of a basic coagulation/filtration facility include chemical feed systems, mixing equipment, basins for rapid mix, flocculation, settling, filter media, sludge handling equipment, and filter backwash facilities. Settling may not be necessary in situations where the influent particle concentration is very low. Treatment plants without settling are known as direct filtration plants.

Iron/Manganese Oxidation

Iron/Manganese (Fe/Mn) oxidation is commonly used by facilities treating groundwater. The oxidation process used to remove iron and manganese leads to the formation of hydroxides that remove soluble arsenic by precipitation or adsorption reactions. Arsenic removal during iron precipitation is fairly efficient. Removal of 2 mg/L of iron achieved a 92.5% removal of arsenic from a 10 µg/L arsenate initial concentration by adsorption alone. Even removal of 1 mg/L of iron resulted in the removal of 83% of influent arsenic from a source with 22-µg/L arsenate. Indeed, field studies of iron removal plants have indicated that this treatment can feasibly reach 3 g/L.

The removal efficiencies achieved by iron removal are not as high or as consistent as those realized by activated alumina or ion exchange (Edwards, 1994). Note, however, that arsenic removal during manganese precipitation is relatively ineffective when compared to iron even when removal by both adsorption and coprecipitation is considered. For instance, precipitation of 3-mg/L manganese removed only 69% of arsenate of a 12.5-µg/L arsenate influent concentration.

Oxidation filtration technologies may be effective arsenic removal technologies. Research of oxidation filtration technologies has primarily focused on greensand filtration. As a result, the following discussion focuses on the effectiveness of greensand filtration as an arsenic removal technology. Substantial arsenic removal has been seen using greensand filtration (Subramanian et al., 1997). The active material in "greensand" is glauconite, a green, iron-rich, clay-like mineral that has ion exchange properties. Glauconite often occurs in nature as small pellets mixed with other sand particles, giving a green color to the sand. The glauconite sand is treated with $KMnO_4$ until the sand grains are coated with a layer of manganese oxides, particularly manganese dioxide. The principle behind this arsenic removal treatment is multi-faceted and includes oxidation, ion exchange, and adsorption. Arsenic compounds displace species from the manganese oxide (presumably OH^- and H_2O), becoming bound to the greensand surface—in effect an exchange of ions. The oxidative nature of the manganese surface converts arsenite to arsenate and arsenate is adsorbed to the surface. As a result of the transfer of electrons and adsorption of arsenate, reduced manganese (Mn) is released from the surface.

The effectiveness of greensand filtration for arsenic filtration for arsenic removal is dependent on the influent water quality. Subramanian et al. (1997) showed a strong correlation between influent Fe concentration and arsenic percent removal. Removal increased from 41% to more than 80% as the Fe/As ratio increased from 0 to 20 when treating tap water with a spiked arsenite concentration of 200 mg/L. The tap water contained 366-mg/L sulfate and 321 mg/L TDS; neither constituent seemed to affect arsenic removal. The authors also point out that the influent manganese concentration may play an important role. Divalent ions, such as calcium, can also compete with arsenic for adsorption sites. Water quality would need to be carefully evaluated for applicability for treatment using greensand. Other researchers have also reported substantial arsenic removal using this technology, including arsenic removals of >90% for treatment of groundwater (Subramanian et al., 1997).

As with other treatment media, greensand must be regenerated when its oxidative and adsorptive capacity has been exhausted. Greensand filters are regenerated using a solution of excess potassium permanganate ($KMnO_4$). Like other treatment media, the regeneration frequency will depend on the influent water quality in terms of constituents which will degrade the filter capacity. Regenerant disposal for greensand filtration has not been addressed in previous research.

Coagulation-Assisted Microfiltration

Arsenic is removed effectively by the coagulation process. Microfiltration is used as a membrane separation process to remove particulates, turbidity, and microorganisms. In coagulation-assisted microfiltration technology, microfiltration is used in a manner similar to a conventional gravity filter. The advantages of microfiltration over conventional filtration are outlined below (Muilenberg, 1997):

- More effective microorganism barrier during the coagulation process upsets
- Smaller floc sizes can be removed (smaller amounts of coagulants are required)
- Increased total plant capacity

Vickers et al. (1997) reported that microfiltration exhibited excellent arsenic removal capability. This report is corroborated by pilot studies conducted by Clifford, which found that coagulation-assisted microfiltration could reduce arsenic levels below 2 g/L in waters with a pH of between 6 and 7, even when the influent concentration of Fe is ~2.5 mg/L (Clifford et al., 1997). These studies also found that the same level of arsenic removal could be achieved by this treatment process even if source water sulfate and silica levels were high. Further, coagulation-assisted microfiltration can reduce arsenic levels to an even greater extent at a slightly lower pH (~5.5).

The addition of a coagulant did not significantly affect the membrane-cleaning interval, although the solids level in the membrane system increased substantially. With an iron and manganese removal system, it is critical that all of the iron and manganese removal system, it is critical that all of the iron and manganese be fully oxidized before they reach the membrane to prevent fouling (Muilenberg, 1997).

Enhanced Coagulation

The Disinfectant/Disinfection By-Product (D/DBP) Rule requires the use of enhanced coagulation treatment for the reduction of disinfection byproduct (DBP) precursors for surface water systems which have sedimentation capabilities. The enhanced process involves modifications to the existing coagulation process such as increasing the coagulant dosage, reducing the pH, or both. Cheng et al. (1994) conducted bench, pilot, and demonstration scale studies to examine arsenate removals during enhanced coagulation. The enhanced coagulation conditions in these studies included the increase of alum and ferric chloride coagulant dosage from 10 to 30 mg/L, decrease of pH from 7 to 5.5, or both. Results from these studies indicated the following:

- Greater than 90% arsenate removal can be achieved under enhanced coagulation conditions. Arsenate removals >90% were easily attained under all conditions when ferric chloride was used.
- Enhanced coagulation using ferric salts is more effective for arsenic removal than enhanced coagulation using alum. With an influent arsenic concentration of 5 µg/L, ferric chloride achieved 96% arsenate removal with a dosage of 10 mg/L and no acid addition. When alum was used, 90% arsenate removal could not be achieved without reducing the pH.
- Lowering pH during enhanced coagulation improved arsenic removal by alum coagulation. With ferric coagulation pH does not have a significant effect between 5.5 and 7.0.

Note: Post-treatment pH adjustment may be required for corrosion control when the process is operated at a low pH.

Lime Softening

Recall that hardness is predominately caused by calcium and magnesium compounds in solution. Lime softening removes this hardness by creating a shift in the carbonate equilibrium. The addition of lime to water raises the pH. Bicarbonate is converted to carbonate as the pH increases, and as a result, calcium is precipitated as calcium carbonate. Soda ash (sodium carbonate) is added if insufficient bicarbonate is present in the water to remove hardness to the desired level. Softening for calcium removal is typically accomplished at a pH range of 9–9.5. For magnesium removal, excess lime is added beyond the point of calcium carbonate precipitation. Magnesium hydroxide precipitates at pH levels >10.5. Neutralization is required if the pH of the softened water is excessively high (above 9.5) for potable use. The most common form of pH adjustment in softening plants is recarbonation with carbon dioxide.

Lime softening has been widely used in the U.S. for reducing hardness in large water treatment systems. Lime softening, excess lime treatment, split lime treatment, and lime-soda softening are all common in municipal water systems. All of these treatment methods are effective in reducing arsenic. Arsenite or arsenate removal is pH-dependent. Oxidation of arsenite is the predominant form. Considerable amounts of sludge are produced in a lime softening system and its disposal is expensive. Large capacity systems may find it economically feasible to install recalcination equipment to recover the reuse the lime sludge and reduce disposal problems. Construction of a new lime softening plant for the removal of arsenic would not generally be recommended unless hardness must also be reduced.

Adsorptive Processes
Activated Alumina
Activated alumina is a physical/chemical process by which ions in the feed water are sorbed to the oxidized activated alumina surface. Activated alumina is considered an adsorption process, although the chemical reactions involved are actually an exchange of ions. Activated alumina is prepared through dehydration of Al (OH)$_3$ at high temperatures and consists of amorphous and gamma alumina oxide (Clifford and Lin, 1985). Activated alumina is used in packed beds to remove contaminants such as fluoride, arsenic, selenium, silica, and NOM. Feed water is continuously passed through the bed to remove contaminants. The contaminant ions are exchanged with the surface hydroxides on the alumina. When adsorption sites on the activated alumina surface become filled, the bed must be regenerated. Regeneration is accomplished through a sequence of rinsing with regenerant, flushing with water, and neutralizing with acid. The regenerant is a strong base, typically sodium hydroxide; the neutralizer is a strong acid, typically sulfuric acid. Many studies have shown that activated alumina is an effective treatment technique for arsenic removal. Factors such as pH, arsenic oxidation state, competing ions, empty bed contact time, and regeneration have significant effects on the removals achieved with activated alumina. Other factors include spent regenerant disposal, alumina disposal, and secondary water quality.

Ion Exchange
Ion exchange is a physical/chemical process by which an ion in the solid phase is exchanged for an ion in the feed water. This solid phase is typically a synthetic resin which has been chosen to preferentially adsorb the particular contaminant of concern. To accomplish this exchange of ions, feed water is continuously passed through a bed of ion exchange resin beads in a downflow or upflow mode until the resin is exhausted. Exhaustion occurs when all sites on the resin beads have been filled by contaminant ions. At this point, the bed is regenerated by rinsing the ion exchange column with a regenerant—a concentrated solution of ions initially exchanged from the resin. The number of bed volumes that can be treated before exhaustion varies with resin type and influent water quality. Typically, from 300 to 60,000 bed volume (BV) can be treated before regeneration is required. In most cases, regeneration of the bed can be accomplished with only 1–5 BV of regenerant followed by 2–20 BV of rinse water. Important considerations in the applicability of the ion exchange process for the removal of a contaminant include water quality parameters such as pH, competing ions, resin type, alkalinity, and influent arsenic concentration. Other factors include the affinity of the resin for the contaminant, spent regenerant and resin disposal requirements, secondary water quality effects, and design operating parameters.

Membrane Processes
Membranes are a selective barrier, allowing some constituents to pass while blocking the passage of others. The movement of constituents across a membrane requires a driving force (i.e., a potential difference between the two sides of the membrane). Membrane processes are often classified by the type of driving force, including pressure, concentration, electrical potential, and temperature. The processes discussed here include only pressure-driven and electrical potential-driven types.

Pressure-driven membrane processes are often classified by pore size into four categories: microfiltration (MF), ultrafiltration (UF), nanofiltration (NF), and reverse osmosis (RO). High-pressure processes (i.e., NF and UF). Typical pressure ranges for these processes are given in Table 8.19. NF and RO primarily remove constituents through chemical diffusion. MF and UF primarily remove constituents through physical sieving. An advantage of high-pressure processes is that they tend to remove a broader range of constituents than low-pressure processes. However, the drawback to broader removal is the increase in energy required for high-pressure processes (Aptel and Buckley, 1996).

TABLE 8.19
Typical Pressure Ranges for Membrane Processes

Membrane Process	Pressure Range (psi)
MF	5–45
UF	7–100
NF	50–150
RO	100–150

Electrical potential-driven membrane processes can also be used for arsenic removal. These processes include, for the purposes of this document, only electrodialysis reversal (EDR). In terms of achievable contaminant removal, EDR is comparable to RO. The separation process used in EDR, however, is ion exchange.

Alternative Technologies

Iron Oxide Coated Sand

Iron-oxide-coated sand is a rare process which has shown some tendency for arsenic removal. Iron oxide-coated sand consists of sand grains coated with ferric hydroxide which are used in fixed bed reactors to remove various dissolved metal species. The metal ions are exchanged with the surface hydroxides on the iron oxide-coated sand. Iron oxide-coated sand exhibits selectivity in the adsorption and exchange of ions present in the water. Like other processes, when the bed is exhausted it must be regenerated by a sequence of operations consisting of rinsing with regenerant, flushing with water, and neutralizing with strong acid. Sodium hydroxide is the most common regenerant, and sulfuric acid is the most common neutralizer. Several studies have shown that iron oxide-coated sand is effective for arsenic removal. Factors such as pH, arsenic oxidation state, competing ions, and regeneration have significant effects on the removes achieved with iron oxide-coated sand.

Sulfur-Modified Iron

A patented Sulfur-Modified Iron (SMI) process for arsenic removal has recently been developed. The process consists of three components: (1) finely divided metallic iron; (2) powdered elemental sulfur, or other sulfur compounds; and (3) an oxidizing agent. The powdered iron, powdered sulfur, and the oxidizing agent (H_2O_2 in preliminary tests) are thoroughly mixed and then added to the water to be treated. The oxidizing agent serves to convert arsenite to arsenate. The solution is then mixed and settled. Using the sulfur-modified iron process on several water types, high adsorptive capacities were obtained with final arsenic concentration of 0.050 mg/L. Arsenic removal was influenced by pH. Approximately 20 mg/L arsenic per gram of iron was removed at pH 8, and 50 mg arsenic per gram of iron was removed at pH 7. Arsenic removal seems to be very dependent on the iron-to-arsenic ratio.

Packed-bed column tests demonstrated significant arsenic removal at residence times of 5–15 minutes. Significant removal of both arsenate and arsenite was measured. The highest adsorption capacity measured was 11 mg arsenic removed per gram of iron. Flow distribution problems were evident, as several columns became partially plugged and better arsenic removal was observed with reduced flow rates.

Spent media from the column tests were classified as nonhazardous waste. Projected operating costs for sulfur-modified iron, when the process is operated below a pH of 8, are much lower than alternative arsenic removal technologies such as ferric chloride addition, reverse osmosis, and activated alumina. Cost savings would increase proportionally with increased flow rates and increased arsenic concentrations.

Possible treatment systems using sulfur-modified iron include continuously stirred tank reactors, packed bed reactors, fluidized bed reactors, and passive in situ reactors. Packed bed and fluidized bed reactors appear to be the most promising for successful arsenic removal in pilot-scale and full-scale treatment systems based on present knowledge of the sulfur-modified iron process.

Granular Ferric Hydroxide

A new removal technique for arsenate, which has recently been developed at the Technical University of Berlin (Germany), Department of Water Quality Control, is adsorption on granular ferric hydroxide in fixed bed reactors. This technique combines the advantages of the coagulation-filtration process, efficiency, and small residual mass, with the fixed bed adsorption on activated alumina, and sample processing. Demers and Renner (1992) reported that the application of granular ferric hydroxide in test adsorbers showed a high treatment capacity of 30,000–40,000 bed volumes with an effluent arsenate concentration never exceeding 10 µ/L. The typical residual mass was in the range of 5–25 g/m^3 treated water. The residue was a solid with an arsenate content of 1–10 g/kg.

The competition of sulfate on arsenate adsorption was not very strong. Phosphate, however, competed strongly with arsenate, which reduced arsenate removal with GFH. Arsenate adsorption decreases with pH, which is typical for anion adsorption. At high pH values GFH our-performs alumina. Below a pH of 7.6, the performance is comparable.

A field study reported by Simms et al. (2000) confirms the efficacy of GFH for arsenic removal. Over the course of this study, a 5.3 mgd GHF located in the United Kingdom was found to reduce average influent arsenic concentrations or 20 to <10 g/L reliably and consistently for 200,000 BV (over a year of operation) at an empty bed contact time (EBCT) of 3 minutes. Despite insignificant headloss, routine backwashing was conducted on a monthly basis to maintain media condition and to reduce the possibility of bacterial growth. The backwash was not hazardous and could be recycled or disposed to a sanitary sewer. At the time of replacement, arsenic loading on the media was 2.3%. Leachate tests conducted on the spent media found that arsenic did not leach from the media.

The most significant weakness of this technology appears to be its cost. Currently, GFH media costs ~$4,000 per ton. However, if a GFH bed can be used several times longer than an alumina bed, for example, it may prove to be the more cost-effective technology. Indeed, the system profiled in the field study presented above tested activated alumina as well as GFH and found the GFH was sufficiently more efficient that smaller adsorption vessels and less media could be used to achieve the same level of arsenic removal (reducing costs). In addition, unlike activated alumina, GFH does not require preoxidation.

A treatment for leaching arsenic from the media to enable regeneration of GFH seems feasible, but it results in the generation of an alkaline solution with high levels of arsenate, which requires further treatment to obtain a solid waste. Thus, direct disposal of spent GFH should be favored.

Iron Filings

Iron filings and sand may be used to reduce inorganic arsenic species to iron co-precipitates, mixed precipitates and, in conjunction with sulfates, arsenopyrites. This type of process is essential a filter technology, much like greensand filtration, wherein the source water is filtered through a bed of sand and iron filings. Unlike some technologies, ion exchange, for example, sulfate is actually introduced in this process to encourage arsenopyrite precipitation. This arsenic removal method was originally developed as a batch arsenic remediation technology. It appears to be quite effective in this use. Bench-scale tests indicate an average removal efficiency of 81% with much higher removals at lower influent concentrations. This method was tested to arsenic levels of 20,000 ppb, and at 2000-ppb consistently reduced arsenic levels to <50 ppb (the current MCL). While it is quite effective in this capacity, its use as a drinking water treatment technology appears to be limited. In batch tests, a residence time of ~7 days was required to reach the desired arsenic removal. In flowing conditions, even though removals averaged 81% and reached >95% at 2,000 ppb arsenic, there is no indication that this technology can reduce arsenic levels below ~25 ppb, and there are no data to indicate how the technology can reduce arsenic levels below ~25 ppb, and there are no data to indicate how the technology performs at normal source water arsenic levels. This technology needs to be further evaluated before it can be recommended as an approved arsenic removal technology for drinking water.

Photooxidation

Researchers at the Australian Nuclear Science and Technology Organization (ANSTO) have found that in the presence of light and naturally occurring light-absorbing materials, the oxidation rate of arsenite by oxygen can be increased ten-thousandfold. The oxidized arsenic, now arsenate, can then be effectively removed by co-precipitation. ANSTO evaluated both UV lamp reactors and sunlight-assisted-photo-oxidation using acidic, metal-bearing water from an abandoned gold, silver, and lead mine. Air sparging was required for sunlight-assisted oxidation due to the high initial arsenate concentration (12 mg/L). Tests demonstrated that near complete oxidation of arsenite could be achieved using the photochemical process. Analysis of process waters 97% of the arsenic of the arsenic in the process stream was present as arsenate. Researchers also concluded that arsenite was preferentially oxidized in the presence of excess dissolved Fe (22:1 iron to arsenic mole ratio). This is a contrast to conventional plants where dissolved Fe represents an extra chemical oxidant demand.

Photooxidation of the mine water followed by co-precipitation was able to reduce arsenic concentrations to as low as 17 g/L, which meets the current MCL for arsenic. Initial total arsenic concentrations were unknown, though the arsenite concentration was given as ~12 mg/L, which is considerably higher than typical raw water arsenic concentrations. ANSTO reported residuals from this process are environmentally stable and passed the Toxicity Characteristic Leaching Procedure (TCLP) test necessary to declare waste non-hazardous suitable for landfill disposal. Based on the removals achieved and residual characteristics, it is expected that photo-oxidation followed by co-precipitation would be an effective arsenic removal technology. However, this technology is still largely experimental and should be further evaluated before recommendations as an approved arsenic removal technology for drinking water.

WASTEWATER TREATMENT

According to the Code of Federal Regulations (CFR) 40 CFR Part 403, regulations were established in the late 1970s and early 1980s to help Publicly Owned Treatment Works (POTW) control industrial discharges to sewers.[3] These regulations were designed to prevent pass-through and interference at the treatment plants and interference in the collection and transmission systems.

Pass-through occurs when pollutants literally "pass through" a POTW without being properly treated and cause the POTW to have an effluent violation or increase the magnitude or duration of a violation. Interference occurs when a pollutant discharge causes a POTW to violate its permit by inhibiting or disrupting treatment processes, treatment operations, or processes related to sludge use or disposal.

Unit operations (unit processes), which are the components that are linked together to form a process train (as shown in Figure 8.8; keep in mind the caboose attached to this train is treated and cleaned wastewater, which when outfalled is usually cleaner than the water in the receiving body), are commonly divided based on the fundamental mechanisms acting with them (i.e., physical, chemical, and biochemical). Physical operations are those, such as sedimentation, that are governed by the laws of physics (gravity). Chemical operations are those in which strictly chemical reactions occur, such as precipitation. Biochemical operations are those that use living microorganisms to destroy or transform pollutants through enzymatically catalyzed chemical reactions (Grady et al., 2011).

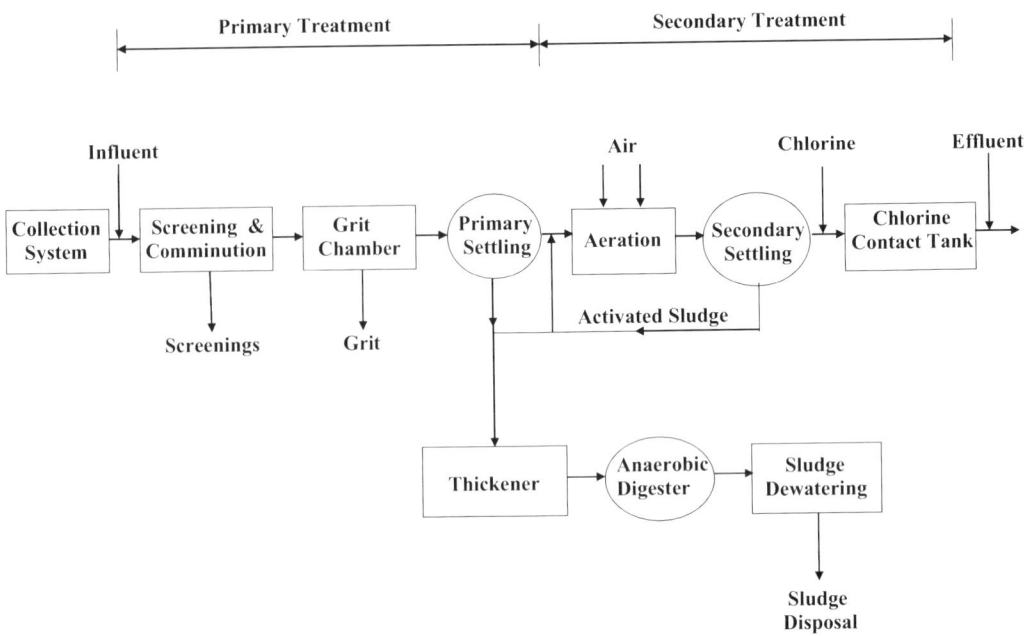

FIGURE 8.8 Schematic of an example wastewater treatment process providing primary and secondary treatment using the activated sludge process.

THE WASTEWATER TREATMENT MODEL

Figure 8.8 shows a basic schematic of an example wastewater treatment process providing primary and secondary treatment using the *Activated sludge process*. This is the model, the prototype, the paradigm used in this handbook. Though it is true that in secondary treatment (which provides BOD removal beyond what is achievable by simple sedimentation) there are actually three commonly used approaches--trickling filter, activated sludge, and oxidation ponds—we focus, for instructive and illustrative purposes, on the activated sludge process throughout this handbook. The purpose of Figure 8.8 is to allow the reader to follow the treatment process step-by-step as it is presented (and as it is actually configured in the real world) and to assist understanding of how all the various unit processes sequentially follow and tie into each other. Therefore, we begin certain sections (which discuss unit processes) with frequent reference to Figure 8.8. It is important to begin these sections in this manner because wastewater treatment is a series of individual steps (unit processes) that treat the wastestream as it makes its way through the entire process. Thus, it logically follows that a pictorial presentation along with pertinent written information enhances the learning process. It should also be pointed out, however, that even though the model shown in Figure 8.8 does not include all unit processes currently used in wastewater treatment we do not ignore the other major processes: trickling filters, rotating biological contactors (RBCs), and oxidation ponds.

WASTEWATER TERMINOLOGY AND DEFINITIONS

Wastewater treatment technology, like many other technical fields, has its own unique terms with their own meaning. Though some of the terms are unique, many are common to other professions. Remember that the science of wastewater treatment is a combination of engineering, biology, mathematics, hydrology, chemistry, physics, and other disciplines. Therefore, many of the terms used in engineering, biology, mathematics, hydrology, chemistry, physics, and others are also used in wastewater treatment. Those terms not listed or defined in the following section will be defined as they appear in the text.

Activated sludge: the solids formed when microorganisms are used to treat wastewater using the activated sludge treatment process. It includes organisms, accumulated food materials and waste products from the aerobic decomposition process.

Advanced waste treatment: treatment technology to produce an extremely high-quality discharge.

Aerobic: conditions in which free, elemental oxygen is present. Also used to describe organisms, biological activity, or treatment processes, which require free oxygen.

Anaerobic: conditions in which no oxygen (free or combined) is available. Also used to describe organisms, biological activity or treatment processes which function in the absence of oxygen.

Anoxic: conditions in which no free, elemental oxygen is present, the only source of oxygen is combined oxygen such as that found in nitrate compounds. Also used to describe biological activity or treatment processes, which function only in the presence of combined oxygen.

Average monthly discharge limitation: the highest allowable discharge over a calendar month.

Average weekly discharge limitation: the highest allowable discharge over a calendar week.

Biochemical oxygen demand, BOD_5: the amount of organic matter which can be biologically oxidized under controlled conditions (5 days @ 20°C in the dark).

Biosolids: From *Merriam-Webster's Collegiate Dictionary, Tenth Edition* (1998): biosolid *n* (1977)–solid organic matter recovered from a sewage treatment process and used especially as fertilizer–usually used in plural.

  Note: In this text, biosolids are used in many places (activated sludge being the exception) to replace the standard term sludge. The author (along with others in the field) views the term sludge as an ugly four-letter word that is inappropriate to use in describing biosolids. Biosolids is a product that can be reused; it has some value. Because biosolids have some value, they should not be classified as a "waste" product and when biosolids for beneficial reuse is addressed, it is not.

Buffer: a substance or solution which resists changes in pH.

Carbonaceous biochemical oxygen demand, $CBOD_5$: the amount of biochemical oxygen demand which can be attributed to carbonaceous material.

Chemical oxygen demand (COD): the amount of chemically oxidizable materials present in the wastewater.

Clarifier: a device designed to permit solids to settle or rise and be separated from the flow. Also known as a settling tank or sedimentation basin.

Coliform: a type of bacteria used to indicate possible human or animal contamination of water.

Combined sewer: a collection system which carries both wastewater and stormwater flows.

Comminution: a process to shred solids into smaller, less harmful particles.

Composite sample: a combination of individual samples taken in proportion to flow.

Daily discharge: the discharge of a pollutant measured during a calendar day or any 24-hour period that reasonably represents a calendar day for the purposes of sampling. Limitations expressed as weight is total mass (weight) discharged over the day.

  Limitations expressed in other units are average measurements of the day.

Daily maximum discharge: the highest allowable value for a daily discharge.

Detention time: the theoretical time water remains in a tank at a given flow rate.

Dewatering: the removal or separation of a portion of water present in a sludge or slurry.

Discharge Monitoring Report (DMR): the monthly report required by the treatment plant's NPDES discharge permit.

Dissolved oxygen (DO): free or elemental oxygen, which is dissolved in water.

Effluent: the flow leaving a tank, channel, or treatment process.
Effluent limitation: any restriction imposed by the regulatory agency on quantities, discharge rates, or concentrations of pollutants which are discharged from point sources into state waters.
Facultative: organisms that can survive and function in the presence or absence of free, elemental oxygen.
Fecal coliform: a type of bacteria found in the bodily discharges of warm-blooded animals. Used as an indicator organism.
Floc: solids which join together to form larger particles which will settle better.
Flume: a flow rate measurement device.
Food-to-microorganism ratio (F/M): an activated sludge process control calculation based upon the amount of food (BOD_5 or COD) available per pound of mixed liquor volatile suspended solids.
Grab sample: an individual sample collected at a randomly selected time.
Grit: heavy inorganic solids such as sand, gravel, eggshells, or metal filings.
Industrial wastewater: wastes associated with industrial manufacturing processes.
Infiltration/inflow: extraneous flows in sewers; defined by Metcalf and Eddy in *Wastewater Engineering: Treatment, Disposal, Reuse*, 3rd. Ed., New York: McGraw-Hill, Inc., pp. 29–31, 1991 as follows:
- **Infiltration**: water entering the collection system through cracks, joints, or breaks.
- **Steady inflow**: water discharged from cellar and foundation drains, cooling water discharges, and drains from springs and swampy areas. This type of inflow is steady and is identified and measured along with infiltration.
- **Direct flow**: those types of inflow that have a direct stormwater runoff connection to the sanitary sewer and cause an almost immediate increase in wastewater flows. Possible sources are roof leaders, yard and areaway drains, manhole covers, cross connections from storm drains and catch basins, and combined sewers.
- **Total inflow**: the sum of the direct inflow at any point in the system plus any flow discharged from the system upstream through overflows, pumping station bypasses, and the like.
- **Delayed inflow**: stormwater that may require several days or more to drain through the sewer system. This category can include the discharge of sump pumps from cellar drainage as well as the slowed entry of surface water through manholes in ponded areas.

Influent: the wastewater entering a tank, channel, or treatment process.
Inorganic: mineral materials such as salt, ferric chloride; iron, sand, gravel, etc.
License: a certificate issued by the State Board of Waterworks/Wastewater Works Operators authorizing the holder to perform the duties of a wastewater treatment plant operator.
Mean Cell Residence Time (MRCT): the average length of time a mixed liquor-suspended solids particle remains in the activated sludge process. May also be known as sludge retention time.
Mixed liquor: the combination of return activated sludge and wastewater in the aeration tank.
Mixed Liquor Suspended Solids (MLSS): the suspended solids concentration of the mixed liquor.
Mixed Liquor Volatile Suspended Solids (MLVSS): the concentration of organic matter in the mixed liquor suspended solids.
Milligrams/Liter (mg/L): a measure of concentration. It is equivalent to parts per million (ppm).
Nitrogenous Oxygen Demand (NOD): a measure of the amount of oxygen required to biologically oxidize nitrogen compounds under specified conditions of time and temperature.

NPDES Permit: National Pollutant Discharge Elimination System permit that authorizes the discharge of treated wastes and specifies the condition, which must be met for discharge.

Nutrients: substances required to support living organisms. Usually refers to nitrogen, phosphorus, iron, and other trace metals.

Organic: materials which consist of carbon, hydrogen, oxygen, sulfur, and nitrogen. Many organics are biologically degradable. All organic compounds can be converted to carbon dioxide and water when subjected to high temperatures.

Pathogenic: disease causing. A pathogenic organism is capable of causing illness.

Point Source: any discernible, defined, and discrete conveyance from which pollutants is or may be discharged.

Part per million: an alternative (but numerically equivalent) unit used in chemistry is milligrams per liter (mg/L). As an analogy think of a ppm as being equivalent to a full shot glass in a swimming pool.

Return Activated Sludge Solids (RASS): the concentration of suspended solids in the sludge flow being returned from the settling tank to the head of the aeration tank.

Sanitary wastewater: wastes discharged from residences and from commercial, institutional, and similar facilities, which include both sewage and industrial wastes.

Scum: the mixture of floatable solids and water, which is removed from the surface of the settling tank.

Septic: a wastewater which has no dissolved oxygen present. Generally characterized by black color and rotten egg (hydrogen sulfide) odors.

Settleability: a process control test used to evaluate the settling characteristics of the activated sludge. Readings taken at 30–60 minutes are used to calculate the settled sludge volume (SSV) and the sludge volume index (SVI).

Settled sludge volume: the volume in percent occupied by an activated sludge sample after 30–60 minutes of settling. Normally written as SSV with a subscript to indicate the time of the reading used for calculation (SSV_{60}) or (SSV_{30}).

Sewage: wastewater containing human wastes.

Sludge: the mixture of settleable solids and water, which is removed from the bottom of the settling tank.

Sludge Retention Time (SRT): See Mean Cell Residence Time.

Sludge Volume Index (SVI): a process control calculation, which is used to evaluate the settling quality of the activated sludge. Requires the SSV_{30} and mixed liquor suspended solids test results to calculate.

Storm sewer: a collection system designed to carry only stormwater runoff.

Stormwater: runoff resulting from rainfall and snowmelt.

Supernatant: in a digester, it is the amber-colored liquid above the sludge.

Wastewater: the water supply of the community after it has been soiled by use.

Waste Activated Sludge Solids (WASS): the concentration of suspended solids in the sludge, which is being removed from the activated sludge process.

Weir: a device used to measure wastewater flow.

Zoogleal Slime: the biological slime which forms on fixed film treatment devices. It contains a wide variety of organisms essential to the treatment process.

MEASURING PLANT PERFORMANCE

To evaluate how well a plant or treatment unit process is operating; *performance efficiency* or *percent (%) removal* is used. The results can be compared with those listed in the plant's operation and maintenance manual (O&M) to determine if the facility is performing as expected. In this section sample calculations often used to measure plant performance/efficiency are presented.

Plant Performance and Efficiency

The calculation used for determining the performance (percent removal) for a digester is different from that used for performance (percent removal) for other processes. Care must be taken to select the right formula.

$$\% \text{Removal} = \frac{[\text{Influent concentration} - \text{Effluent concentration}] \times 100}{\text{Influent concentration}} \quad (8.15)$$

Example 8.5

Problem:

The influent BOD_5 is 247 mg/L and the plant effluent BOD is 17 mg/L. What is the percent removal?

Solution:

$$\% \text{Removal} = \frac{(247 \text{ mg/L} - 17 \text{ mg/L}) \times 100}{247 \text{ mg/L}} = 93\%$$

Equation 8.15 is used again to determine unit process efficiency. The concentration entering the unit and the concentration leaving the unit (i.e., primary, secondary, etc.) are used to determine the unit performance.

$$\% \text{Removal} = \frac{[\text{Influent concentration} - \text{Effluent concentration}] \times 100}{\text{Influent concentration}}$$

Example 8.6

Problem:

The primary influent BOD is 235 mg/L and the primary effluent BOD is 169 mg/L. What is the percent removal?

$$\% \text{Removal} = \frac{(235 \text{ mg/L} - 169 \text{ mg/L}) \times 100}{235 \text{ mg/L}} = 28\%$$

Percent Volatile Matter Reduction in Sludge

The calculation used to determine *percent volatile matter reduction* is more complicated because of the changes occurring during sludge digestion.

$$\% \text{V.M. Reduction} = \frac{(\% \text{V.M.}_{\text{in}} - \% \text{V.M.}_{\text{out}}) \times 100}{[\% \text{V.M.}_{\text{in}} - (\% \text{V.M.}_{\text{in}} \times \% \text{V.M.}_{\text{out}})]} \quad (8.16)$$

V.M. = Volatile Matter.

Example 8.7

Problem:

Using the digester data provided below, determine the % Volatile Matter Reduction for the digester.

Data:
Raw Sludge Volatile Matter = 74%
Digested Sludge Volatile Matter = 54%

$$\% \text{Volatile matter reduction} = \frac{(0.74 - 0.54) \times 100}{[0.74 - (0.74 \times 0.54)]} = 59\%$$

Hydraulic Detention Time

The term, *detention time* or *hydraulic detention time (HDT)*, refers to the average length of time (theoretical time) a drop of water, wastewater, or suspended particles remains in a tank or channel. It is calculated by dividing the water/wastewater in the tank by the flow rate through the tank. The units of flow rate used in the calculation are dependent on whether the detention time is to be calculated in seconds, minutes, hours, or days. Detention time is used in conjunction with various treatment processes, including sedimentation and coagulation-flocculation. Generally, in practice, detention time is associated with the amount of time required for a tank to empty. The range of detention time varies with the process. For example, in a tank used for sedimentation, detention time is commonly measured in minutes. The calculation methods used to determine detention time are illustrated in the following sections.

Hydraulic Detention Time in Days

$$\text{HDT, Days} = \frac{\text{Tank volume, ft}^3 \times 7.48 \text{ gal/ft}^3}{\text{Flow, gallon/day}} \tag{8.17}$$

Example 8.8

Problem:

An anaerobic digester has a volume of 2,400,000 gal. What is the detention time in days when the influent flow rate is 0.07 MGD?

Solution:

$$\text{D.T., Days} = \frac{2,400,000 \text{ gal}}{0.07 \text{ MGD} \times 1,000,000 \text{ gal/MG}}$$

$$= 34 \text{ days}$$

Hydraulic Detention Time in Hours

$$\text{HDT, Hours} = \frac{\text{Tank volume, ft}^3 \times 7.48 \text{ gal/ft}^3 \times 24 \text{ hours/day}}{\text{Flow, gal/day}} \tag{8.18}$$

Example 8.9

Problem:

A settling tank has a volume of 44,000 ft³. What is the detention time in hours when the flow is 4.15 MGD?

$$\text{D.T., Hours} = \frac{44,000 \text{ ft}^3 \times 7.48 \text{ gal/ft}^3 \times 24 \text{ hours/day}}{4.15 \text{ MGD} \times 1,000,000 \text{ gal/MG}}$$

$$= 1.9 \text{ hours}$$

Detention Time in Minutes

$$\text{HDT, Minutes} = \frac{\text{Tank volume, ft}^3 \times 7.48\,\text{gal/ft}^3 \times 1{,}440\,\text{minutes/day}}{\text{Flow, gal/day}} \quad (8.19)$$

Example 8.10

Problem:

A grit channel has a volume of 1,340 ft^3. What is the detention time in minutes when the flow rate is 4.3 MGD?

Solution:

$$\text{D.T., Minutes} = \frac{1{,}340\,\text{ft}^3 \times 7.48\,\text{gal/ft}^3 \times 1{,}440\,\text{min/day}}{4{,}300{,}000\,\text{gal/day}}$$

$$= 3.36\,\text{minutes}$$

Note: The tank volume and the flow rate must be in the same dimensions before calculating the hydraulic detention time.

WASTEWATER SOURCES AND CHARACTERISTICS

Wastewater treatment is designed to use the natural purification processes (self-purification processes of streams and rivers) to the maximum level possible. It is also designed to complete these processes in a controlled environment rather than over many miles of stream or river. Moreover, the treatment plant is also designed to remove other contaminants, which are not normally subjected to natural processes, as well as treating the solids, which are generated through the treatment unit steps. The typical wastewater treatment plant is designed to achieve many different purposes:

- Protect public health
- Protect public water supplies
- Protect aquatic life
- Preserve the best uses of the waters
- Protect adjacent lands

Wastewater treatment is a series of steps. Each of the steps can be accomplished using one of more treatment processes or types of equipment. The major categories of treatment steps are:

- **Preliminary treatment**: removes materials that could damage plant equipment or would occupy treatment capacity without being treated.
- **Primary treatment**: removes settleable and floatable solids (may not be present in all treatment plants).
- **Secondary treatment**: Removes BOD$_5$ and dissolved and colloidal suspended organic matter by biological action; organics are converted to stable solids, carbon dioxide and more organisms.
- **Advanced waste treatment**: uses physical, chemical, and biological processes to remove additional BOD$_5$, solids and nutrients (not present in all treatment plants).
- **Disinfection**: removes microorganisms to eliminate or reduce the possibility of disease when the flow is discharged.
- **Sludge treatment**: stabilizes the solids removed from wastewater during treatment, inactivates pathogenic organisms and/or reduces the volume of the sludge by removing water.

The various treatment processes described above are discussed in detail later.

Wastewater Sources

The principal sources of domestic wastewater in a community are residential areas and commercial districts. Other important sources include institutional and recreational facilities and stormwater (runoff) and groundwater (infiltration). Each source produces wastewater with specific characteristics. In this section wastewater sources and the specific characteristics of wastewater are described.

Wastewater is generated by five major sources: human and animal wastes, household wastes, industrial wastes, stormwater runoff, and groundwater infiltration.

1. **Human and animal wastes**: contain the solid and liquid discharges of humans and animals and are considered by many to be the most dangerous from a human health viewpoint. The primary health hazard is presented by the millions of bacteria, viruses, and other microorganisms (some of which may be pathogenic) present in the wastestream.
2. **Household wastes**: are wastes, other than human and animal wastes, discharged from the home. Household wastes usually contain paper, household cleaners, detergents, trash, garbage, and other substances the homeowner discharges into the sewer system.
3. **Industrial wastes**: includes industry-specific materials, which can be discharged from industrial processes into the collection system. Typically contains chemicals, dyes, acids, alkalis, grit, detergents, and highly toxic materials.
4. **Stormwater runoff**: many collection systems are designed to carry both the wastes of the community and stormwater runoff. In this type of system when a storm event occurs, the wastestream can contain large amounts of sand, gravel, and other grit as well as excessive amounts of water.
5. **Groundwater infiltration**: groundwater will enter older improperly sealed collection systems through cracks or unsealed pipe joints. Not only can this add large amounts of water to wastewater flows but also additional grit.

Classification of Wastewater

Wastewater can be classified according to the sources of flows:

1. **Domestic (sewage) wastewater**: mainly contains human and animal wastes, household wastes, small amounts of groundwater infiltration and small amounts of industrial wastes.
2. **Sanitary wastewater**: consists of domestic wastes and significant amounts of industrial wastes. In many cases, industrial wastes can be treated without special precautions. However, in some cases, the industrial wastes will require special precautions or a pretreatment program to ensure the wastes do not cause compliance problems for the wastewater treatment plant.
3. **Industrial wastewater**: industrial wastes only. Often the industry will determine that it is safer and more economical to treat its waste independently of domestic waste.
4. **Combined wastewater**: is the combination of sanitary wastewater and stormwater runoff. All the wastewater and stormwater of the community is transported through one system to the treatment plant.
5. **Stormwater**: a separate collection system (no sanitary waste) that carries stormwater runoff including street debris, road salt, and grit.

Wastewater Characteristics

Wastewater contains many different substances, which can be used to characterize it. The specific substances and amounts or concentrations of each will vary, depending on the source. Thus, it is difficult to "precisely" characterize wastewater. Instead, wastewater characterization is usually based on and applied to an average domestic wastewater. Wastewater is characterized in terms of its physical, chemical, and biological characteristics.

Note: Keep in mind that other sources and types of wastewater can dramatically change the characteristics.

Physical Characteristics

The *physical characteristics* of wastewater are based on color, odor, temperature, and flow.

- **Color**: fresh wastewater is usually a light brownish-gray color. However, typical wastewater is gray and has a cloudy appearance. The color of the wastewater will change significantly if allowed to go septic (if travel time in the collection system increases). Typical septic wastewater will have a black color.
- **Odor**: odors in domestic wastewater usually are caused by gases produced by the decomposition of organic matter or by other substances added to the wastewater. Fresh domestic wastewater has a musty odor. If the wastewater is allowed to go septic, this odor will change significantly—to a rotten egg odor associated with the production of hydrogen sulfide (H_2S).
- **Temperature**: the temperature of wastewater is commonly higher than that of the water supply because of the addition of warm water from households and industrial plants. However, significant amounts of infiltration or stormwater flow can cause major temperature fluctuations.
- **Flow**: the actual volume of wastewater is commonly used as a physical characterization of wastewater and is normally expressed in terms of gallons per person per day. Most treatment plants are designed using an expected flow of 100–200 gallons per person per day. This figure may have to be revised to reflect the degree of infiltration or storm flow the plant receives. Flow rates will vary throughout the day. This variation, which can be as much as 50%–200% of the average daily flow is known as the *diurnal flow variation*.

Note: *Diurnal* means occurs in a day or each day; daily.

Chemical Characteristics

When describing the chemical characteristics of wastewater, the discussion generally includes topics such as organic matter, the measurement of organic matter, inorganic matter, and gases. For the sake of simplicity, in this handbook we specifically describe chemical characteristics in terms of alkalinity, biochemical oxygen demand (BOD), chemical oxygen demand (COD), dissolved gases, nitrogen compounds, pH, phosphorus, solids (organic, inorganic, suspended and dissolved solids), and water.

- **Alkalinity**: is a measure of the wastewater's capability to neutralize acids. It is measured in terms of bicarbonate, carbonate, and hydroxide alkalinity. Alkalinity is essential to buffer (hold the neutral pH) of the wastewater during the biological treatment processes.
- **Biochemical Oxygen Demand (BOD)**: a measure of the amount of biodegradable matter in wastewater. Normally measured by a 5-day test conducted at 20°C. The BOD_5 domestic waste is normally in the range of 100–300 mg/L.
- **Chemical Oxygen Demand (COD)**: a measure of the amount of oxidizable matter present in the sample. The COD is normally in the range of 200–500 mg/L. The presence of industrial wastes can increase this significantly.
- **Dissolved gases**: gases that are dissolved in wastewater. The specific gases and normal concentrations are based on the composition of the wastewater. Typical domestic wastewater contains oxygen in relatively low concentrations, carbon dioxide, and hydrogen sulfide (if septic conditions exist).

- **Nitrogen compounds**: the type and amount of nitrogen present will vary from the raw wastewater to the treated effluent. Nitrogen follows a cycle of oxidation and reduction. Most of the nitrogen in untreated wastewater will be in the forms of organic nitrogen and ammonia nitrogen. Laboratory tests exist for determination of both of these forms. The sum of these two forms of nitrogen is also measured and is known as *Total Kjeldahl Nitrogen (TKN)*. Wastewater will normally contain between 20 and 85 mg/L of nitrogen. Organic nitrogen will normally be in the range of 8–35 mg/L and ammonia nitrogen will be in the range of 12–50 mg/L.
- **pH**: a method of expressing the acid condition of the wastewater. pH is expressed on a scale of 1–14. For proper treatment, wastewater pH should normally be in the range of 6.5–9.0 (ideal–6.5–8.0).
- **Phosphorus**: essential to biological activity and must be present in at least minimum quantities or secondary treatment processes will not perform. Excessive amounts can cause stream damage and excessive algal growth. Phosphorus will normally be in the range of 6–20 mg/L. The removal of phosphate compounds from detergents has had a significant impact on the amounts of phosphorus in wastewater.
- **Solids**: most pollutants found in wastewater can be classified as solids. Wastewater treatment is generally designed to remove solids or to convert solids to a form which is more stable or can be removed. Solids can be classified by their chemical composition (organic or inorganic) or by their physical characteristics (settleable, floatable, and colloidal). Concentration of total solids in wastewater is normally in the range of 350–1,200 mg/L.
  - **Organic solids**: consist of carbon, hydrogen, oxygen, and nitrogen and can be converted to carbon dioxide and water by ignition at 550°C. Also known as fixed solids or loss on ignition.
  - **Inorganic solids**: mineral solids which are unaffected by ignition. Also known as fixed solids or ash.
  - **Suspended solids**: will not pass through a glass fiber filter pad. Can be further classified as Total Suspended Solids (TSS), Volatile Suspended Solids and/or Fixed Suspended Solids. Can also be separated into three components based on settling characteristics. Settleable solids, floatable solids, and colloidal solids. Total suspended solids in wastewater are normally in the range of 100–350 mg/L.
  - **Dissolved solids**: will pass through a glass fiber filter pad. Can also be classified as Total Dissolved Solids (TDS), volatile dissolved solids, and fixed dissolved solids. Total dissolved solids are normally in the range of 250–850 mg/L.
- **Water**: always the major constituent of wastewater. In most cases, water makes up 99.5%–99.9% of the wastewater. Even in the strongest wastewater, the total amount of contamination present is <0.5% of the total and in average strength wastes it is usually <0.1%.

Biological Characteristics and Processes

After undergoing physical aspects of treatment (i.e., screening, grit removal, and sedimentation) in preliminary and primary treatment, wastewater still contains some suspended solids and other solids that are dissolved in the water. In a natural stream, such substances are a source of food for protozoa, fungi, algae, and several varieties of bacteria. In secondary wastewater treatment, these same microscopic organisms (which are one of the main reasons for treating wastewater) are allowed to work as fast as they can to biologically convert the dissolved solids to suspended solids which will physically settle out at the end of secondary treatment.

Raw wastewater influent typically contains millions of organisms. The majority of these organisms are non-pathogenic; however, several pathogenic organisms may also be present (these may include the organisms responsible for diseases such as typhoid, tetanus, hepatitis, dysentery, gastroenteritis, and others). Many of the organisms found in wastewater are microscopic (microorganisms); they include algae, bacteria, protozoans (such as amoeba, flagellates, free-swimming ciliates,

TABLE 8.20
Typical Domestic Wastewater Characteristics

Characteristic	Typical Characteristic
Color	Gray
Odor	Musty
Dissolved oxygen	>1.0 mg/L
pH	6.5–9.0
TSS	100–350 mg/L
BOD_5	100–300 mg/L
COD	200–500 mg/L
Flow	100–200 gallons/person/day
Total nitrogen	20–85 mg/L
Total phosphorus	6–20 mg/L
Fecal coliform	500,000–3,000,000 MPN/100 mL

and stalked ciliates), rotifers, and virus. Table 8.20 is a summary of typical domestic wastewater characteristics.

WASTEWATER COLLECTION SYSTEMS

Wastewater collection systems collect and convey wastewater to the treatment plant. The complexity of the system depends on the size of the community and the type of system selected. Methods of collection and conveyance of wastewater include gravity systems, force main systems, vacuum systems, and combinations of all three types of systems.

Gravity Collection System

In a gravity collection system, the collection lines are sloped to permit the flow to move through the system with as little pumping as possible. The slope of the lines must keep the wastewater moving at a velocity (speed) of 2–4 ft per second. Otherwise, at lower velocities, solids will settle out causing clogged lines, overflows, and offensive odors. To keep collection systems lines at a reasonable depth, wastewater must be lifted (pumped) periodically so that it can continue flowing downhill to the treatment plant. Pump stations are installed at selected points within the system for this purpose.

Force Main Collection System

In a typical force main collection system, wastewater is collected to central points and pumped under pressure to the treatment plant. The system is normally used for conveying wastewater long distances. The use of the force main system allows the wastewater to flow to the treatment plant at the desired velocity without using sloped lines. It should be noted that the pump station discharge lines in a gravity system are considered to be force mains since the content of the lines is under pressure.

Note: Extra care must be taken when performing maintenance on force main systems since the content of the collection system is under pressure.

Vacuum System

In a vacuum collection system, wastewaters are collected to central points and then drawn toward the treatment plant under vacuum. The system consists of a large amount of mechanical equipment and requires a large amount of maintenance to perform properly. Generally, the vacuum-type collection systems are not economically feasible.

Water Pollution

Pumping Stations

Pumping stations provide the motive force (energy) to keep the wastewater moving at the desired velocity. They are used in both the force main and gravity systems. They are designed in several different configurations and may use different sources of energy to move the wastewater (i.e., pumps, air pressure or vacuum). One of the more commonly used types of pumping station designs is the wet well/dry well design.

Wet-Well/Dry-Well Pumping Stations

The Wet Well/Dry Well pumping station consists of two separate spaces or sections separated by a common wall. Wastewater is collected in one section (wet well section) and the pumping equipment (and in many cases, the motors, and controllers) are located in a second section known as the dry well. There are many different designs for this type of system, but in most cases, the pumps selected for this system are of a centrifugal design. There are a couple of major considerations in selecting centrifugal design: (1) allows for the separation of mechanical equipment (pumps, motors, controllers, wiring, etc.) from the potentially corrosive atmosphere (sulfides) of the wastewater; and (2) this type of design is usually safer for workers because they can monitor, maintain, operate, and repair equipment without entering the pumping station wet well.

Note: Most pumping station wet wells are confined spaces. To ensure safe entry into such spaces compliance with OSHA's 29 CFR 1910.146 (Confined Space Entry Standard) is required.

Wet-Well Pumping Stations

Another type of pumping station design is the *wet well* type. This type consists of a single compartment, which collects the wastewater flow. The pump is submerged in the wastewater with motor controls located in the space or has a weatherproof motor housing located above the wet well. In this type of station, a submersible centrifugal pump is normally used.

Pneumatic Pumping Stations

The pneumatic pumping station consists of a wet well and a control system, which controls the inlet and outlet value operations and provides, pressurized air to force or "push" the wastewater through the system. The exact method of operation depends on the system design. When operating, wastewater in the wet well reaches a predetermined level and activates an automatic valve, which closes the influent line. The tank (wet well) is then pressurized to a predetermined level. When the pressure reaches the predetermined level, the effluent line valve is opened and the pressure pushes the wastestream out of the discharge line.

Pumping Station Wet Well Calculations

Calculations normally associated with pumping station wet well design (determining design lift or pumping capacity, etc.) are usually left up to design and mechanical engineers. However, on occasion, wastewater operators or interceptor's technicians may be called upon to make certain basic calculations. Usually, these calculations deal with determining either pump capacity without influent (e.g., to check the pumping rate of the station's constant speed pump) or pump capacity with influent (e.g., to check how many gallons per minute the pump is discharging). In this section, we use examples to describe instances of how and where these two calculations are made.

Example 8.11

Problem:

A pumping station wet well is 10 ft by 9 ft. The operator needs to check the pumping rate of the station's constant speed pump. To do this, the influent valve to the wet well is closed for a 5-minutes test, the level in the well dropped 2.2-ft. What is the pumping rate in gallons per minute?

Solution:

Using the length and width of the well, we can find the area of the water surface:

$$10\,\text{ft} \times 9\,\text{ft} = 90\,\text{ft}^2$$

The water level dropped 2.2 ft. From this we can find the volume of water removed by the pump during the test:

$$\text{Area} \times \text{Depth} = \text{Volume} \tag{8.20}$$

$$90\,\text{ft}^2 \times 2.2\,\text{ft} = 198\,\text{ft}^3$$

One cubic foot of water holds 7.48 gal. We can convert this volume in cubic feet to gallons.

$$198\,\text{ft}^3 \times \frac{7.48\,\text{gal}}{1\,\text{ft}^3} = 1{,}481\,\text{gal}$$

The test was done for 5 minutes. From this information, a pumping rate can be calculated.

$$\frac{1{,}481\,\text{gal}}{5\,\text{minutes}} = \frac{296.2}{1\,\text{minutes}} = 296.2\,\text{gpm}$$

PRELIMINARY WASTEWATER TREATMENT

The initial stage in the wastewater treatment process (following collection and influent pumping) is *preliminary treatment*. Raw influents entering the treatment plant may contain many kinds of materials (trash). The purpose of preliminary treatment is to protect plant equipment by removing these materials which could cause clogs, jams, or excessive wear to plant machinery. In addition, the removal of various materials at the beginning of the treatment process saves valuable space within the treatment plant.

Preliminary treatment may include many different processes; each designed to remove a specific type of material, which is a potential problem for the treatment process. Processes include: wastewater collections–influent pumping, screening, shredding, grit removal, flow measurement, preaeration, chemical addition, and flow equalization–the major processes are shown in Figure 8.8. In this section, we describe and discuss each of these processes and their importance in the treatment process.

Note: As mentioned, not all treatment plants will include all of the processes shown in Figure 8.8. Specific processes have been included to facilitate discussion of major potential problems with each process and its operation; this is information that may be important to environmental engineers and wastewater operators.

Screening

The purpose of *screening* is to remove large solids such as rags, cans, rocks, branches, leaves, roots, etc. from the flow before the flow moves on to downstream processes (see Figure 8.8).

Note: Typically, a treatment plant will remove anywhere from 0.5 to 12 ft^3 of screenings for each million gallons of influent received.

A *bar screen* traps debris as wastewater influent passes through. Typically, a bar screen consists of a series of parallel, evenly spaced bars or a perforated screen placed in a channel (see Figure 8.9). The wastestream passes through the screen and the large solids (*screenings*) are trapped on the bars for removal.

Note: The screenings must be removed frequently enough to prevent accumulation which will block the screen and cause the water level in front of the screen to build up.

Water Pollution

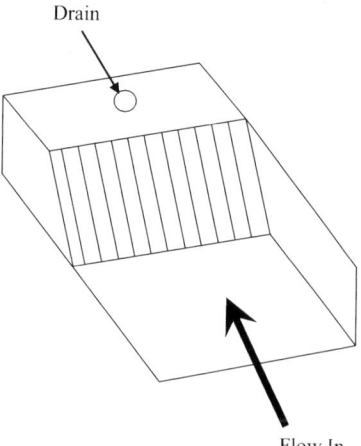

FIGURE 8.9 Basic bar screen.

The bar screen may be coarse (2 to 4-inch openings) or fine (0.75 to 2.0-inch openings). The bar screen may be manually cleaned (bars or screens are placed at an angle of 30° for easier solids removal—see Figure 8.9) or mechanically cleaned (bars are placed at 45° to 60° angle to improve mechanical cleaner operation).

The screening method employed depends on the design of the plant, the amount of solids expected and whether the screen is for constant or emergency use only.

Manually Cleaned Screens

Manually cleaned screens are cleaned at least once per shift (or often enough to prevent buildup which may cause reduced flow into the plant) using a long tooth rake. Solids are manually pulled to the drain platform and allowed to drain before storage in a covered container. The area around the screen should be cleaned frequently to prevent a buildup of grease or other materials, which can cause odors, slippery conditions, and insect and rodent problems. Because screenings may contain organic matter as well as large amounts of grease they should be stored in a covered container. Screenings can be disposed of by burial in approved landfills or by incineration. Some treatment facilities grind the screenings into small particles, which are then returned to the wastewater flow for further processing and removal later in the process.

Manually cleaned screens require a certain amount of operator attention to maintain optimum operation. Failure to clean the screen frequently can lead to septic wastes entering the primary; surge flows after cleaning; and/or low flows before cleaning. On occasion, when such operational problems occur, it becomes necessary to increase the frequency of the cleaning cycle. Another operational problem is excessive grit in the bar screen channel. Improper design or construction or insufficient cleaning may cause this problem. The corrective action required is either to correct the design problem or increase cleaning frequency and flush channel regularly. Another common problem with manually cleaned bar screens is their tendency to clog frequently. This may be caused by excessive debris in the wastewater or the screen is too fine for its current application. The operator should locate the source of the excessive debris and eliminate it. If the screen is the problem, a coarser screen may need to be installed. If the bar screen area is filled with obnoxious odors, flies, and other insects, it may be necessary to dispose of screenings more frequently.

Mechanically Cleaned Screens

Mechanically cleaned screens use a mechanized rake assembly to collect the solids and move them (carries them) out of the wastewater flow for discharge to a storage hopper. The screen may be continuously cleaned or cleaned on a time or flow-controlled cycle. As with the manually cleaned screen, the area surrounding the mechanically operated screen must be cleaned frequently to prevent buildup of materials, which can cause unsafe conditions. As with all mechanical equipment, operator vigilance is required to ensure proper operation and that proper maintenance is performed. Maintenance includes lubricating equipment and maintaining it in accordance with manufacturer's recommendations or the plant's O&M Manual (Operations & Maintenance Manual). Screenings from mechanically operated bar screens are disposed of in the same manner as screenings from manually operated screen: landfill disposal, incineration, or ground into smaller particles for return to the wastewater flow.

Many of the operational problems associated with mechanically cleaned bar screens are the same as those for manual screens: septic wastes entering the primary; surge flows after cleaning; excessive grit in the bar screen channel; and/or the screen clogs frequently. The same corrective actions employed for manually operated screens would be applied for these problems in mechanically operated screens. In addition to these problems, however, mechanically operated screens also have other problems, including the cleaner will not operate at all; and the rake does not operate but the motor does. Obviously, these are mechanical problems that could be caused by jammed cleaning mechanism, broken chain, broken cable, or a broken shear pin. Authorized and fully trained maintenance operators should be called in to handle these types of problems.

Screenings Removal Computations

Operators responsible for screenings disposal are typically required to keep a record of the amount of screenings removed from the wastewater flow. To keep and maintain accurate screenings' records, the volume of screenings withdrawn must be determined. Two methods are commonly used to calculate the volume of screenings withdrawn.

$$\text{Screenings removed, cu ft/day} = \frac{\text{Screenings, cu ft}}{\text{Days}} \quad (8.21)$$

$$\text{Screenings removed, cu ft/MG} = \frac{\text{Screenings, cu ft}}{\text{Flow, MG}} \quad (8.22)$$

Example 8.12

Problem:

A total of 65 gal of screenings are removed from the wastewater flow during a 24-hour period. What is the screenings removal reported as cu/ft/day?

Solution:

First, convert gallons screenings to cu ft:

$$\frac{65 \text{ gal}}{7.48 \text{ gal/cu ft}} = 8.7 \text{ cu ft screenings}$$

Next, calculate screenings removed as cu ft/day:

$$\text{Screenings removed (cu ft/day)} = \frac{8.7 \text{ cu ft}}{1 \text{ day}} = 8.7 \text{ cu ft/day}$$

Example 8.13

Problem:

For one week, a total of 310 gal of screenings were removed from the wastewater screens. What is the average screening removal in cu ft/day?

Solution:

First, gallons of screenings must be converted to cubic feet of screenings:

$$\frac{310\,\text{gal}}{7.48\,\text{gal/cu ft}} = 41.4\,\text{cu ft screenings}$$

$$\text{Screenings removed, cu ft/day} = \frac{41.4\,\text{cu ft}}{7} = 5.9\,\text{cu ft/day}$$

Shredding

As an alternative to screening, shredding can be used to reduce solids to a size, which can enter the plant without causing mechanical problems or clogging. Shredding processes include comminution (comminute means to cut up) and barminution devices.

Comminution

The *comminutor* is the most common shredding device used in wastewater treatment. In this device, all the wastewater flow passes through the grinder assembly. The grinder consists of a screen or slotted basket, a rotating or oscillating cutter and a stationary cutter. Solids pass through the screen and are chopped or shredded between the two cutters. The comminutor will not remove solids, which are too large to fit through the slots, and it will not remove floating objects. These materials must be removed manually. Maintenance requirements for comminutors include aligning, sharpening, and replacing cutters and corrective and preventive maintenance performed in accordance with plant O&M Manual.

Common operational problems associated with comminutors include output containing coarse solids. When this occurs, it is usually a sign that the cutters are dull or misaligned. If the system does not operate at all, the unit is either clogged, jammed, a shear pin or coupling is broken or electrical power is shut off. If the unit stalls or jams frequently, this usually indicates cutter misalignment, excessive debris in influent, or dull cutters.

Note: Only qualified maintenance operators should perform maintenance of shredding equipment.

Barminution

In barminution, the *barminutor* uses a bar screen to collect solids, which are then shredded and passed through the bar screen for removal at a later process. In operation each device cutter alignment and sharpness are critical factors in effective operation. Cutters must be sharpened or replaced and alignment must be checked in accordance with manufacturer's recommendations. Solids, which are not shredded, must be removed daily, stored in closed containers, and disposed of by burial or incineration. Barminutor operational problems are similar to those listed above for comminutors. Preventive and corrective maintenance as well as lubrication must be performed by qualified personnel and in accordance with the plant's O&M Manual. Because of higher maintenance requirements the barminutor is less frequently used.

Grit Removal

The purpose of grit removal is to remove the heavy inorganic solids, which could cause excessive mechanical wear. Grit is heavier than inorganic solids and includes, sand, gravel, clay, eggshells, coffee grounds, metal filings, seeds, and other similar materials. There are several processes or devices used for grit removal. All of the processes are based on the fact that grit is heavier than the organic solids, which should be kept in suspension for treatment in the following processes. Grit removal may be accomplished in grit chambers (see Figure 8.8) or by the centrifugal separation of sludge. Processes use gravity/velocity, aeration, or centrifugal force to separate the solids from the wastewater.

Gravity/Velocity-Controlled Grit Removal

Gravity/velocity-controlled grit removal is normally accomplished in a channel or tank where the speed or the velocity of the wastewater is controlled to about 1 foot per second (ideal), so that grit will settle while organic matter remains suspended. As long as the velocity is controlled in the range of 0.7–1.4 feet per second (fps) the grit removal will remain effective. Velocity is controlled by the amount of water flowing through the channel, the depth of the water in the channel, by the width of the channel, or by cumulative width of channels in service.

Process Control Calculations

Velocity of the flow in a channel can be determined either by the float and stopwatch method or by channel dimensions.

Example 8.14: Velocity by Float and Stopwatch

$$\text{Velocity, feet/second} = \frac{\text{Distance traveled (ft)}}{\text{Time required (seconds)}} \quad (8.23)$$

Problem:

A float takes 25 seconds to travel 34 ft in a grit channel. What is the velocity of the flow in the channel?

Solution:

$$\text{Velocity, fps} = \frac{34 \text{ ft}}{25 \text{ seconds}} = 1.4 \text{ fps}$$

Example 8.15: Velocity by Flow and Channel Dimensions

This calculation can be used for a single channel or tank or multiple channels or tanks with the same dimensions and equal flow. If the flows through each unit of the unit dimensions are unequal, the velocity for each channel or tank must be computed individually.

$$\text{Velocity, fps} = \frac{\text{Flow, MGD} \times 1.55 \text{ cfs/MGD}}{\text{\# Chan. in ser.} \times \text{chan width, ft} \times \text{water D, ft}} \quad (8.24)$$

Problem:

The plant is currently using two grit channels. Each channel is 3 ft wide and has a water depth of 1.2 ft. What is the velocity when the influent flow rate is 3.0 MGD?

Solution:

$$\text{Velocity, fps} = \frac{3.0 \text{ MGD} \times 1.55 \text{ cfs/MGD}}{2 \text{ channels} \times 3 \text{ ft} \times 1.2 \text{ ft}}$$

$$\text{Velocity, fps} = \frac{4.65 \text{ cfs}}{7.2 \text{ ft}^2} = 0.65 \text{ fps}$$

Note: The channel dimensions must always be in feet. Convert inches to feet by dividing by 12 inches per foot.

Example 8.16: Required Settling Time

This calculation can be used to determine the time required for a particle to travel from the surface of the liquid to the bottom at a given settling velocity. In order to compute the

Water Pollution

settling time, the settling velocity in fps must be provided or determined experimentally in a laboratory.

$$\text{Settling time (seconds)} = \frac{\text{Liquid depth (ft)}}{\text{Settling, velocity (fps)}} \quad (8.25)$$

Problem:

The plant's grit channel is designed to remove sand, which has a settling velocity of 0.085 fps. The channel is currently operating at a depth of 2.2 ft. How many seconds will it take for a sand particle to reach the channel bottom?

Solution:

$$\text{Settling time, seconds} = \frac{2.2 \text{ ft}}{0.085 \text{ fps}} = 25.9 \text{ seconds}$$

Example 8.17: Required Channel Length

This calculation can be used to determine the length of channel required to remove an object with a specified settling velocity.

$$\text{Required channel length} = \frac{\text{Channel depth, ft} \times \text{flow velocity, fps}}{\text{Settling velocity, fps}} \quad (8.26)$$

Problem:

The plant's grit channel is designed to remove sand, which has a settling velocity of 0.070 fps. The channel is currently operating at a depth of 3 ft. The calculated velocity of flow through the channel is 0.80 fps. The channel is 35 ft long. Is the channel long enough to remove the desired sand particle size?

Solution:

$$\text{Required channel length, ft} = \frac{3 \text{ ft} \times 0.80 \text{ fps}}{0.070 \text{ fps}} = 34.3 \text{ ft}$$

Yes, the channel is long enough to ensure all of the sand will be removed.

Cleaning

Gravity-type systems may be manually or mechanically cleaned. Manual cleaning normally requires that the channel be taken out of service, drained, and manually cleaned. Mechanical cleaning systems are operated continuously or on a time cycle. Removal should be frequent enough to prevent grit carry-over into the rest of the plant.

Note: Before and during cleaning activities always ventilate the area thoroughly.

Aeration

Aerated grit removal systems use aeration to keep the lighter organic solids in suspension while allowing the heavier grit particles to settle out. Aerated grit removal may be manually or mechanically cleaned; however, the majority of the systems are mechanically cleaned. During normal operation, adjusting the aeration rate produces the desired separation. This requires observation of mixing and aeration and sampling of fixed suspended solids. Actual grit removal is controlled by the rate of aeration. If the rate is too high, all of the solids remain in suspension. If the rate is too low, both grit and organics will settle out. The operator observes the same kinds of conditions as those listed for the gravity/velocity-controlled system but must also pay close attention to the air distribution system to ensure proper operation.

Cyclone Degritter

The cyclone degritter uses a rapid spinning motion (centrifugal force) to separate the heavy inorganic solids or grit from the light organic solids. This unit process is normally used on primary sludge rather than the entire wastewater flow. The critical control factor for the process is the inlet pressure. If the pressure exceeds the recommendations of the manufacturer, the unit will flood, and grit will carry through with the flow. Grit is separated from flow, washed, and discharged directly to a storage container. Grit removal performance is determined by calculating the percent removal for inorganic (fixed) suspended solids. The operator observes the same kinds of conditions listed for the gravity/velocity-controlled and aerated grit removal systems, with the exception of the air distribution system. Typical problems associated with grit removal include mechanical malfunctions and rotten egg odor in the grit chamber (hydrogen sulfide formation), which can lead to metal and concrete corrosion problems. Low recovery rate of grit is another typical problem. Bottom scour, overaeration, or not enough detention time normally causes this. When these problems occur, the operator must make the required adjustments or repairs to correct the problem.

Grit Removal Calculations

Wastewater systems typically average 1–15 ft³ of grit per million gallons of flow (sanitary systems: 1–4 ft³/million gal; combined wastewater systems average from 4 to 15 ft³/million gals of flow), with higher ranges during storm events. Generally, grit is disposed of in sanitary landfills. Because of this practice, for planning purposes, operators must keep accurate records of grit removal. Most often, the data is reported as cubic feet of grit removed per million gallons of flow:

$$\text{Cubic removed, ft}^3/\text{MG} = \frac{\text{Grit volume, ft}^3}{\text{Flow, MG}} \qquad (8.27)$$

Over a given period, the average grit removal rate at a plant (at least a seasonal average) can be determined and used for planning purposes. Typically, grit removal is calculated as cubic yards because excavation is normally expressed in terms of cubic yards.

$$\text{Grit}(\text{yd}^3) = \frac{\text{Total grit}(\text{ft}^3)}{27 \text{ ft}^3/\text{yd}^3} \qquad (8.28)$$

Example 8.18

Problem:

A treatment plant removes 10 ft³ of grit in one day. How many cubic feet of grit are removed per million gallons if the plant flow was 9 MGD?

Solution:

$$\text{Grit removed, ft}^3/\text{MG} = \frac{\text{Grit volume, ft}^3}{\text{Flow, MG}}$$

$$= \frac{10 \text{ ft}^3}{9 \text{ MG}} = 1.1 \text{ ft}^3/\text{MG}$$

Example 8.19

Problem:

The total daily grit removed for a plant is 250 gal. If the plant flow is 12.2 MGD, how many cubic feet of grit are removed per MG flow?

Solution:

First, convert gallon grit removed to cubic ft:

$$\frac{250 \text{ gal}}{7.48 \text{ gal/ft}^3} = 33 \text{ ft}^3$$

Next, complete the calculation of ft³/MG:

$$\text{Grit removal, ft}^3/\text{MG} = \frac{\text{Grit voloume, ft}^3}{\text{Flow, MG}}$$

$$= \frac{33 \text{ ft}^3}{12.2 \text{ MGD}} = 2.7 \text{ ft}^3/\text{MGD}$$

Example 8.20

Problem:

The monthly average grit removal is 2.5 ft³/MG. If the monthly average flow is 2,500,000 gpd, how many cu yards must be available for grit disposal pit is to have a 90-day capacity?

Solution:

First, calculate the grit generated each day:

$$\frac{\left(2.5 \text{ ft}^3\right)}{\text{MG}}\left(2.5 \text{ MGD}\right) = 6.25 \text{ ft}^3 \text{ eachday}$$

The ft³ grit generated for 90 days would be

$$\frac{\left(6.25 \text{ ft}^3\right)}{\text{day}}\left(90 \text{ days}\right) = 562.5 \text{ ft}$$

Convert ft³ grit to yd³ grit:

$$\frac{562.5 \text{ ft}^3}{27 \text{ ft}^3/\text{yd}^3} = 21 \text{ yd}^3$$

Preaeration

In the preaeration process (diffused or mechanical), we aerate wastewater to achieve and maintain an aerobic state (to freshen septic wastes), strip off hydrogen sulfide (to reduce odors and corrosion), agitate solids (to release trapped gases and improve solids separation and settling), and to reduce BOD_5. All of this can be accomplished by aerating the wastewater for 10–30 minutes. To reduce BOD_5, preaeration must be conducted from 45 to 60 minutes.

In preaeration grit removal systems, the operator is concerned with maintaining proper operation and must be alert to any possible mechanical problems. In addition, the operator monitors dissolved oxygen levels and the impact of preaeration on influent.

Chemical Addition

Chemical addition to the wastestream is done (either via dry chemical metering or solution feed metering) to improve settling, reduce odors, neutralize acids or bases, reduce corrosion, reduce BOD_5, improve solids and grease removal, reduce loading on the plant, add or remove nutrients, add organisms, and/or aid subsequent downstream processes. The particular chemical and amount used depend on the desired result. Chemicals must be added at a point where sufficient mixing will occur to obtain maximum benefit. Chemicals typically used in wastewater treatment include chlorine, peroxide, acids and bases, miner salts (ferric chloride, alum, etc.), and bioadditives and enzymes.

When adding chemicals to the wastestream to remove grit, the operator monitors the process for evidence of mechanical problems and takes proper corrective actions when necessary. The operator also monitors the current chemical feed rate and dosage. The operator ensures that mixing at the point of addition is accomplished in accordance with Standard Operating Procedures and monitors the impact of chemical addition on influent.

Equalization

The purpose of flow equalization (whether by surge, diurnal, or complete methods) is to reduce or remove the wide swings in flow rates normally associated with wastewater treatment plant loading; it minimizes the impact of storm flows. The process can be designed to prevent flows above maximum plant design hydraulic capacity; to reduce the magnitude of diurnal flow variations; and to eliminate flow variations. Flow equalization is accomplished using mixing or aeration equipment, pumps, and flow measurement. Normal operation depends on the purpose and requirements of the flow equalization system. Equalized flows allow the plant to perform at optimum levels by providing stable hydraulic and organic loading. The downside to flow equalization is in additional costs associated with the construction and operation of the flow equalization facilities.

During normal operations, the operator must monitor all mechanical systems involved with flow equalization and must watch for mechanical problems and take the appropriate corrective action. The operator also monitors dissolved oxygen levels, the impact of equalization on influent, and water levels in equalization basins, and makes necessary adjustments.

PRIMARY WASTEWATER TREATMENT (SEDIMENTATION)

The purpose of primary treatment (primary sedimentation or primary clarification) is to remove settleable organic and flotable solids. Normally, each primary clarification unit can be expected to remove 90%–95% settleable solids, 40%–60% total suspended solids and 25%–35% BOD_5.

Note: Performance expectations for settling devices used in other areas of plant operation is normally expressed as overall unit performance rather than settling unit performance.

Sedimentation may be used throughout the plant to remove settleable and flotable solids. It is used in primary treatment, secondary treatment, and advanced wastewater treatment processes. In this section, we focus on primary treatment or primary clarification, which uses large basins in which primary settling is achieved under relatively quiescent conditions (see Figure 8.8). Within these basins, mechanical scrapers collect the primary settled solids into a hopper, from which they are pumped to a sludge-processing area. Oil, grease, and other floating materials (scum) are skimmed from the surface. The effluent is discharged over weirs into a collection trough.

In primary sedimentation, wastewater enters a settling tank or basin. Velocity is reduced to ~1 foot per minute.

Note: Notice that the velocity is based on *minutes* instead of seconds, as was the case in the grit channels. A grit channel velocity of 1 ft/seconds would be 60 ft/minutes.

Solids, which are heavier than water, settle to the bottom while solids which are lighter than water float to the top. Settled solids are removed as sludge and floating solids are removed as scum. Wastewater leaves the sedimentation tank over an effluent weir and on to the next step in treatment. Detention time, temperature, tank design and condition of the equipment control the efficiency of the process.

- Primary treatment reduces the organic loading on downstream treatment processes by removing a large amount of settleable, suspended, and floatable materials.
- Primary treatment reduces the velocity of the wastewater through a clarifier to ~1 to 2 ft/minutes, so that settling and floatation can take place. Slowing the flow enhances the removal of suspended solids in wastewater.

Water Pollution

- Primary settling tanks remove floated grease and scum, remove the settled sludge solids, and collect them for pumped transfer to disposal or further treatment.
- Clarifiers used may be rectangular or circular. In rectangular clarifiers, wastewater flows from one end to the other, and the settled sludge is moved to a hopper at the one end, either by flights set on parallel chains or by a single bottom scraper set on a traveling bridge. Floating material (mostly grease and oil) is collected by a surface skimmer.
- In circular tanks, the wastewater usually enters the middle and flows outward. Settled sludge is pushed to a hopper in the middle of the tank bottom, and a surface skimmer removes floating material.
- Factors affecting primary clarifier performance include:
  - Rate of flow through the clarifier
  - Wastewater characteristics (strength; temperature; amount and type of industrial waste; and the density, size, and shapes of particles)
  - Performance of pretreatment processes
  - Nature and amount of any wastes recycled to the primary clarifier
- Key factors in primary clarifier operation include the following concepts:

$$\text{Retention time (hours)} = \frac{(\text{Volume, gal})(24 \text{ hours/day})}{\text{Flow, gal per day}}$$

$$\text{Surface loading rate (gal/day/ft}^2) = \frac{Q(\text{gal/day})}{\text{Surface area (ft}^2)}$$

$$\text{Solids loading rate (lb/day/ft}^2) = \frac{\text{Solids into clarifier (lb/day)}}{\text{Surface area (ft}^2)}$$

$$\text{Weir overflow rate (gal/day/lineal ft)} = \frac{Q(\text{gal/day})}{\text{Weir length (lineal ft)}}$$

Types of Sedimentation Tanks

Sedimentation equipment includes septic tanks, two-story tanks and plain settling tanks or clarifiers. All three devices may be used for primary treatment while plain settling tanks are normally used for secondary or advanced wastewater treatment processes.

Septic Tanks

Septic tanks are prefabricated tanks that serve as a combined settling and skimming tank and as an unheated-unmixed anaerobic digester. Septic tanks provide long settling times (6–8 hours or more) but do not separate decomposing solids from the wastewater flow. When the tank becomes full, solids will be discharged with the flow. The process is suitable for small facilities (i.e., schools, motels, homes, etc.) but, due to the long detention times and lack of control, it is not suitable for larger applications.

Two-Story (Imhoff) Tank

The two-story or Imhoff tank, named for German engineer Karl Imhoff (1876–1965), is similar to a septic tank in the removal of settleable solids and the anaerobic digestion of solids. The difference is that the two-story tank consists of a settling compartment where sedimentation is accomplished, a lower compartment where settled solids and digestion takes place, and gas vents. Solids removed from the wastewater by settling pass from the settling compartment into the digestion compartment through a slot in the bottom of the settling compartment. The design of the slot prevents solids from returning to the settling compartment. Solids decompose anaerobically in the digestion section. Gases produced as a result of the solids decomposition are released through the gas vents running along each side of the settling compartment.

Plain Settling Tanks (Clarifiers)

The plain settling tank or clarifier optimizes the settling process. Sludge is removed from the tank for processing in other downstream treatment units. Flow enters the tank, is slowed, and distributed evenly across the width and depth of the unit, passes through the unit and leaves over the effluent weir. Detention time within the primary settling tank is from 1 to 3 hours (2-hour average). Sludge removal is accomplished frequently on either continuous or intermittent basis. Continuous removal requires additional sludge treatment processes to remove the excess water resulting from removal of sludge which contains <2% to 3% solids. Intermittent sludge removal requires the sludge be pumped from the tank on a schedule frequent enough to prevent large clumps of solids rising to the surface but infrequent enough to obtain 4%–8% solids in the sludge withdrawn.

Scum must be removed from the surface of the settling tank frequently. This is normally a mechanical process but may require manual start-up. The system should be operated frequently enough to prevent excessive buildup and scum carryover but not so frequent as to cause hydraulic overloading of the scum removal system. Settling tanks require housekeeping and maintenance. Baffles (prevent flotable solids, scum, from leaving the tank), scum troughs, scum collectors, effluent troughs and effluent weirs require frequent cleaning to prevent heavy biological growths and solids accumulations. Mechanical equipment must be lubricated and maintained as specified in the manufacturer's recommendations or in accordance with procedures listed in the plant O&M Manual.

Process control sampling and testing are used to evaluate the performance of the settling process. Settleable solids, dissolved oxygen, pH, temperature, total suspended solids and BOD_5, as well as sludge solids and volatile matter testing are routinely carried out.

Sedimentation Calculations

As with many other wastewater treatment plant unit processes, process control calculations aid in determining the performance of the sedimentation process. Process control calculations are used in the sedimentation process to determine

- Percent removal
- Hydraulic detention time
- Surface loading rate (surface settling rate)
- Weir overflow rate (weir loading rate)
- Sludge pumping
- Percent total solids (% TS)

In the following sections, we take a closer look at a few of these process control calculations and example problems.

Note: The calculations presented in the following sections allow you to determine values for each function performed. Keep in mind that an optimally operated primary clarifier should have values in an expected range.

Percent Removal

The expected range of % removal for a primary clarifier is

- Settleable solids 90%–95%
- Suspended solids 40%–60%
- BOD_5 25%–35%

Water Pollution

Detention Time

The primary purpose of primary settling is to remove settleable solids. This is accomplished by slowing the flow down to ~1 ft/minutes. The flow at this velocity will stay in the primary tank from 1.5 to 2.5 hours. The length of time the water stays in the tank is called the hydraulic detention time.

Surface Loading Rate (Surface Settling Rate/Surface Overflow Rate)

Surface loading rate is the number of gallons of wastewater passing over 1 ft² of tank per day. This can be used to compare actual conditions with design. Plant designs generally use a surface-loading rate of 300–1,200 gallons/day/ft². Other terms used synonymously with surface loading rate include *surface overflow rate* and *surface settling rate*.

$$\text{Surface settling rate, gpd/ft}^2 = \frac{\text{Flow, gal/day}}{\text{Settling tank area, ft}^2} \tag{8.29}$$

Example 8.21

Problem:

The settling tank is 120 ft in diameter and the flow to the unit is 4.5 MGD. What is the surface loading rate in gallons/day/ft²?

Solution:

$$\text{Surface loading rate} = \frac{4.5\,\text{MGD} \times 1{,}000{,}000\,\text{gal/MGD}}{0.785 \times 120\,\text{ft} \times 120\,\text{ft}} = 398\,\text{gpd/ft}^2$$

Example 8.22

Problem:

A circular clarifier has a diameter of 50 ft. If the primary effluent flow is 2,150,000 gpd, what is the surface overflow rate in gpd/ft²?

Solution:

$$\text{Area} = (0.785)(50\,\text{ft})(50\,\text{ft})$$

$$\text{Surface overflow rate} = \frac{\text{Flow, gpd}}{\text{Area, ft}^2}$$

$$= \frac{2{,}150{,}000\,\text{gpd}}{(0.785)(50\,\text{ft})(50\,\text{ft})}$$

$$= 1{,}096\,\text{gpd/ft}^2$$

Weir Overflow Rate (Weir Loading Rate)

The weir overflow rate or weir loading rate is the amount of water leaving the settling tank per linear foot of weir. The result of this calculation can be compared with design. Normally weir overflow rates of 10,000–20,000 gallon/day/foot are used in the design of a settling tank.

$$\text{Weir overflow rate, gpd/ft} = \frac{\text{Flow, gal/day}}{\text{Weir length, ft}} \tag{8.30}$$

Example 8.23

Problem:

The circular settling tank is 90 ft in diameter and has a weir along its circumference. The effluent flow rate is 2.55 MGD. What is the weir overflow rate (gpd/ft)?

Solution:

$$\text{Weir overflow, gpd/ft} = \frac{2.55 \text{ MGD} \times 1,000,000 \text{ gal/MG}}{3.14 \times 90 \text{ ft}}$$

$$= 9,023 \text{ gpd/ft}$$

Sludge Pumping

Determination of sludge pumping (the quantity of solids and volatile solids removed from the sedimentation tank) provides accurate information needed for process control of the sedimentation process.

$$\text{Sol. pumped, lb/day} = \text{Pump rate} \times \text{pump time} \times 8.34 \text{ lbs/gal} \times \% \text{ solids} \quad (8.31)$$

$$\text{Vol. solids lb/day} = \text{Pump rate} \times \text{pump time} \times 8.34 \times \% \text{ solids} \times \% \text{ volume} \quad (8.32)$$

Example 8.24

Problem:

The sludge pump operates 20 minutes/hours. The pump delivers 20 gal/minutes of sludge. Laboratory tests indicate that the sludge is 5.2% solids and 66% volatile matter. How many pounds of volatile matter are transferred from the settling tank to the digester?

Solution:

Pump time = 20 minutes/hours
  Pump rate = 20 gpm
  % Solids = 5.2%
  % V.M. = 66%

$$\text{Vol. solids, lb/day} = 20 \text{ gpm} \times (20 \text{ minutes/hours} \times 24 \text{ hours/day}) \times 8.34 \text{ lbs/gal} \times 0.052 \times 0.66$$

$$= 2,748 \text{ lb/day}$$

Percent Total Solids (% TS)

Example 8.25

Problem:

A settling tank sludge sample is tested for solids. The sample and dish weight 74.69 g. The dish alone weighed 21.2 g. After drying, the dish with dry solids weighed 22.3 g.
  What is the percent total solids (% TS) of the sample?

Solution:

Sample + Dish	74.69 g	Dish + Dry solids	22.3 g
Dish alone	−21.2 g	Dish alone	−21.2 g
Sample weight	53.49 g	Dry solids weight	1.1 g

$$\frac{1.1 \text{ g}}{53.49 \text{ g}} \times 100\% = 2\%$$

BOD and SS Removal

To calculate the pounds of BOD or suspended solids removed each day, you need to know the mg/L BOD or SS removed and the plant flow. Then, you can use the mg/L to lb/day equation.

$$\text{SS Removed} = \text{mg/L} \times \text{MGD} \times 8.34 \text{ lb/gal} \tag{8.33}$$

Example 8.26

Problem:

If 120 mg/L suspended solids are removed by a primary clarifier, how many lb/day suspended solids are removed when the flow is 6,230,000 gpd?

Solution:

$$\text{SS Removed} = 120 \text{ mg/L} \times 6.25 \text{ MGD} \times 8.34 \text{ lb/gal} = 6,255 \text{ lb/day}$$

Example 8.27

Problem:

The flow to a secondary clarifier is 1.6 MGD. If the influent BOD concentration is 200 mg/L and the effluent BOD concentration is 70 mg/L, how many pounds of BOD are removed daily?

$$\text{lb/day BOD removed} = 200 \text{ mg/L} - 70 \text{ mg/L} = 130 \text{ mg/L}$$

After calculating mg/L BOD removed, calculate lb/day BOD removed:

$$\text{BOD removed, lb/day} = (130 \text{ mg/L})(1.6 \text{ MGD})(8.34 \text{ lb/gal}) = 1,735 \text{ lb/day}$$

Effluent from Settling Tanks

Upon completion of screening, degritting, and settling in sedimentation basins, large debris, grit and many settleable materials have been removed from the wastestream. What is left is referred to as *Primary Effluent*. Usually cloudy and frequently gray in color, primary effluent still contains large amounts of dissolved food and other chemicals (nutrients). These nutrients are treated in the next step in the treatment process (Secondary Treatment) which is discussed in the next section.

Note: Two of the most important nutrients left to remove are phosphorus and ammonia. While we want to remove these two nutrients from the wastestream, we do not want to remove too much. Carbonaceous microorganisms in secondary treatment (biological treatment) need both phosphorus and ammonia.

SECONDARY TREATMENT

The main purpose of *secondary treatment* (sometimes referred to as biological treatment) is to provide biochemical oxygen demand (BOD) removal beyond what is achievable by primary treatment (see Figure 8.8). There are three commonly used approaches, all of which take advantage of the ability of microorganisms to convert organic wastes (via biological treatment), into stabilized, low-energy compounds. Two of these approaches, the *trickling filter* [and/or its variation, the *rotating biological contactor (RBC)*] and the *activated sludge* process, sequentially follow normal primary treatment. The third, *ponds* (oxidation ponds or lagoons), however, can provide equivalent results without preliminary treatment. In this section, we present a brief overview of the secondary treatment process followed by a detailed discussion of wastewater treatment ponds (used primarily in smaller treatment plants), trickling filters and RBCs. We then shift focus to the activated sludge process—the secondary treatment process, which is used primarily in large installations and is the main focus of the handbook.

Secondary treatment refers to those treatment processes which use biological processes to convert dissolved, suspended, and colloidal organic wastes to more stable solids which can either be removed by settling or discharged to the environment without causing harm. Exactly what is secondary treatment? As defined by the Clean Water Act (CWA), secondary treatment produces an effluent with no more than 30 mg/L BOD_5 and 30 mg/L total suspended solids.

Note: The CWA also states that ponds and trickling filters will be included in the definition of secondary treatment even if they do not meet the effluent quality requirements continuously.

Most secondary treatment processes decompose solids aerobically producing carbon dioxide, stable solids, and more organisms. Since solids are produced, all of the biological processes must include some form of solids removal (settling tank, filter, etc.). Secondary treatment processes can be separated into two large categories: fixed film systems and suspended growth systems.

Fixed film systems are processes, which use a biological growth (biomass or slime) which is attached to some form of media. Wastewater passes over or around the media and the slime. When the wastewater and slime are in contact, the organisms remove and oxidize the organic solids. The media may be stone, redwood, synthetic materials, or any other substance that is durable (capable of withstanding weather conditions for many years), provides a large area for slime growth while providing open space for ventilation and is not toxic to the organisms in the biomass. Fixed film devices include trickling filters and rotating biological contactors (RBCs). *Suspended growth systems* are processes, which use a biological growth, which is mixed with the wastewater. Typical suspended growth systems consist of various modifications of the activated sludge process.

Treatment Ponds

Wastewater treatment can be accomplished using *ponds* (aka, *lagoons*). Ponds are relatively easy to build, to manage, they accommodate large fluctuations in flow, and they can also provide treatment that approaches conventional systems (producing a highly purified effluent) at much lower cost. It is the cost (the economics) that drives many managers to decide on the pond option. The actual degree of treatment provided depends on the type and number of ponds used. Ponds can be used as the sole type of treatment or they can be used in conjunction with other forms of wastewater treatment; that is, other treatment processes followed by a pond or a pond followed by other treatment processes.

Stabilization ponds (treatment ponds) have been used for treatment of wastewater for over 3,000 years. The first recorded construction of a pond system in the U.S. was at San Antonio, Texas, in 1901. Today, over 8,000 wastewater treatment ponds are in place, involving more than 50% of the wastewater treatment facilities in the U.S (CWNS, 2000). Facultative ponds account for 62%, aerated ponds 25%, anaerobic 0.04% and total containment 12% of the pond treatment systems. They treat a variety of wastewaters from domestic wastewater to complex industrial wastes, and they function under a wide range of weather conditions, form tropical to arctic. Ponds can be used alone or in combination with other wastewater treatment processes. As our understanding of pond operating mechanisms has increased, different types of ponds have been developed for application in specific types of wastewater under local environmental conditions. This handbook focuses on municipal wastewater treatment pond systems.

While the tendency in the U.S. has been for smaller communities to build ponds, in other parts of the world, including Australia, New Zealand, Mexico and Latin America, Asia and Africa, treatment ponds have been built for large cities. As a result, our understanding of the biological, biochemical, physical, and climatic factors that interact to transform the organic compounds, nutrients sand pathogenic organisms found in sewage into less harmful chemicals and unviable organisms (i.e., dead or sterile) has grown since 1983. A wealth of experience has been built up as civil, sanitary, or environmental engineers, operators, public works managers and public health and environmental agencies have gained more experience with these systems. While some of this information makes its way into technical journals and textbooks, there is a need for a less formal presentation of the subject for those working in the field every day (USEPA, 2011).

Ponds are designed to enhance the growth of natural ecosystems that are either anaerobic (providing conditions for bacteria that grow I the absence of oxygen (O_2) environments), aerobic (promoting the growth of O_2 producing and/or requiring organs, such as algae and bacteria), or facultative, which is a combination of the two. Ponds are managed to reduce concentrations of biochemical oxygen demand (BOD), TSS and coliform numbers (fecal or total) to meet water quality requirements.

Ponds can be classified based on their location in the system, by the type wastes they receive, and by the main biological process occurring in the pond. First, we look at the types of ponds according to their location and the type wastes they receive: Raw Sewage Stabilization ponds (see Figure 8.10), oxidation ponds, and polishing ponds. Then, in the following section, we look at ponds classified by the type of processes occurring within the pond: *Aerobic Ponds, Anaerobic Ponds, Facultative Ponds*, and *Aerated Ponds*.

Ponds Based on Location and Types of Wastes They Receive

Raw Sewage Stabilization Pond—The raw sewage stabilization pond is the most common type of pond (see Figure 8.10). With the exception of screening and shredding, this type of pond receives no prior treatment. Generally, raw sewage stabilization ponds are designed to provide a minimum of 45 days detention time and to receive no more than 30 pounds of BOD_5 per day per acre. The quality of the discharge is dependent on the time of the year. Summer months produce high BOD_5 removal but excellent suspended solids removals. The pond consists of an influent structure, pond berm or walls and an effluent structure designed to permit selection of the best quality effluent. Normal operating depth of the pond is 3–5 ft. The process occurring in the pond involves bacteria decomposing the organics in the wastewater (aerobically and anaerobically) and algae using the products of the bacterial action to produce oxygen (photosynthesis). Because this type of pond is the most commonly used in wastewater treatment, the process that occurs within the pond is described in greater detail in the following text.

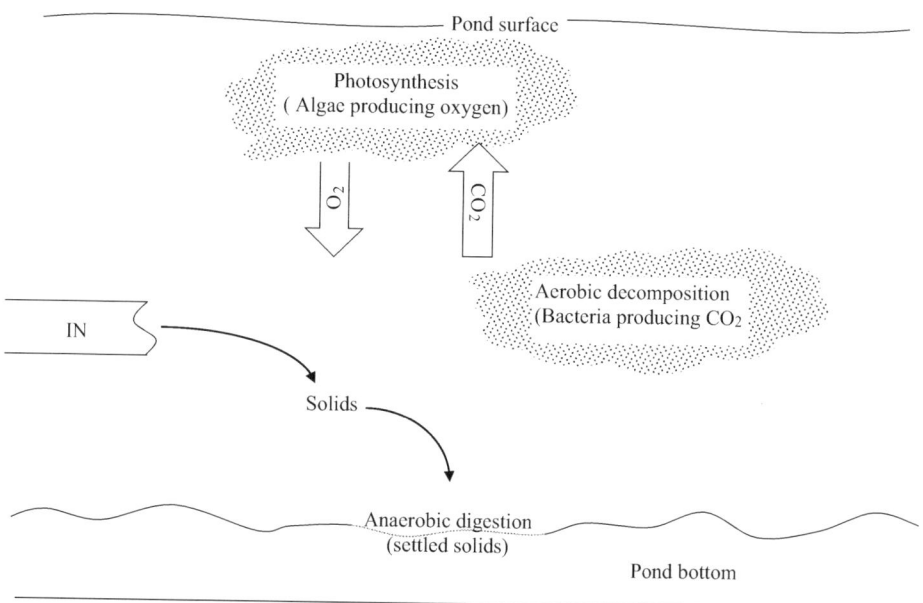

FIGURE 8.10 Stabilization pond processes.

When wastewater enters the stabilization pond several processes begin to occur. These include settling, aerobic decomposition, anaerobic decomposition, and photosynthesis (see Figure 8.10). Solids in the wastewater will settle to the bottom of the pond. In addition to the solids in the wastewater entering the pond, solids, which are produced by the biological activity, will also settle to the bottom. Eventually this will reduce the detention time and the performance of the pond. When this occurs (20–30 years normal) the pond will have to be replaced or cleaned.

Bacteria and other microorganisms use the organic matter as a food source. They use oxygen (aerobic decomposition), organic matter and nutrients to produce carbon dioxide, water, and stable solids, which may settle out, and more organisms. The carbon dioxide is an essential component of the photosynthesis process occurring near the surface of the pond. Organisms also use the solids that settled out as food material, however, the oxygen levels at the bottom of the pond are extremely low so the process used is anaerobic decomposition. The organisms use the organic matter to produce gases (hydrogen sulfide, methane, etc.) which are dissolved in the water, stable solids, and more organisms. Near the surface of the pond a population of green algae will develop which can use the carbon dioxide produced by the bacterial population, nutrients, and sunlight to produce more algae and oxygen which is dissolved into the water. The dissolved oxygen is then used by organisms in the aerobic decomposition process.

When compared with other wastewater treatment systems involving biological treatment, a stabilization pond treatment system is the simplest to operate and maintain. Operation and maintenance activities include collecting and testing samples for dissolved oxygen (D.O), and pH, removing weeds and other debris (scum) from the pond, mowing the berms, repairing erosion, and removing burrowing animals.

Note: Dissolved oxygen and pH levels in the pond will vary throughout the day. Normal operation will result in very high D.O. and pH levels due to the natural processes occurring.

Note: When operating properly the stabilization pond will exhibit a wide variation in both dissolved oxygen and pH. This is due to the photosynthesis occurring in the system.

> **Oxidation pond**: An oxidation pond, which is normally designed using the same criteria as the stabilization pond, receives flows that have passed through a stabilization pond or primary settling tank. This type of pond provides biological treatment, additional settling, and some reduction in the number of fecal coliform present.
>
> **Polishing pond**: A polishing pond, which uses the same equipment as a stabilization pond, receives flow from an oxidation pond or from other secondary treatment systems. Polishing ponds remove additional BOD_5, solids and fecal coliform and some nutrients. They are designed to provide 1–3 days detention time and normally operate at a depth of 5–10 ft. Excessive detention time or too shallow a depth will result in alga growth, which increases influent, suspended solids concentrations.

Ponds Based on the Type of Processes Occurring Within

The type of processes occurring within the pond may also classify ponds. These include the aerobic, anaerobic, facultative, and aerated processes.

> **Aerobic ponds**: In aerobic ponds, also known as oxidation ponds or high-rate aerobic ponds, which are not widely used, oxygen is present throughout the pond. All biological activity is aerobic decomposition. They are usually 30–45 cm deep, which allows light to penetrate throughout the pond. Mixing is often provided, keeping algae at the surface to maintain maximum rates of photosynthesis and O_2 production and to prevent algae from settling and producing an anaerobic bottom layer. The rate of photosynthetic production of O_2 may be enhanced by surface re-aeration; O_2 and aerobic bacteria biochemically stabilize the waste. Detention time is typically 2–6 days.
>
> These ponds are appropriate for treatment in warm, sunny climates. They are used where a high degree of BOD_5 removal is desired but land area is limited. The chief advantage of these ponds is

that they produce a stable effluent during short detention times with low land and energy requirements. However, their operation is somewhat more complex than that of facultative ponds and, unless the algae are removed, the effluent will contain high TSS. While the shallow depths allow penetration of ultra-violet (UV) light that may reduce pathogens, shorter detention times vary work against effective coliform and parasite die-off. Since they are shallow, bottom paving or veering is usually necessary to prevent aquatic plants form colonizing the ponds. The Advanced Integrated Wastewater Pond System® (AIWPS®) uses the high-rate pond to maximize the growth of microalgae using a low-energy paddle-wheel (USEPA, 2011).

Anaerobic ponds: Anaerobic ponds are normally used to treat high strength industrial wastes; that is, they receive heavy organic loading, so much so that there is no aerobic zone—no oxygen is present and all biological activity is anaerobic decomposition. They are usually 2.5–4.5 m in depth and have detention times of 5–50 days. The predominant biological treatment reactions are bacterial acid formation and methane fermentation.

Anaerobic ponds are usually used for treatment of strong industrial and agricultural (food processing) wastes, as pretreatment step in municipal systems, or where an industry is a significant contributor to a municipal system. The biochemical reactions in an anaerobic pond produce hydrogen sulfide (H_2S) and other odorous compounds. To reduce odors, the common practice is to recirculate water form a downstream facultative or aerated pond. This provides a thin aerobic layer at the surface of the anaerobic pond, which prevents odors from escaping into the air. A cover may also be used to contain odors. The effluent from anaerobic ponds usually requires further treatment prior to discharge (USEPA, 2011).

Facultative pond: The facultative pond, which may also be called an oxidation or photosynthetic pond, is the most common type pond (based on processes occurring). Oxygen is present in the upper portions of the pond and aerobic processes are occurring. No oxygen is present in the lower levels of the pond where processes occurring are anoxic and anaerobic. Facultative ponds are usually 0.9–2.4 m deep or deeper, with an aerobic layer overlying an anaerobic layer. Recommended detention times vary form 5–50 days in warm climates and 90–180 days in colder climates (NEIWPCC, 1998). Aerobic treatment processes in the upper layer provide odor control, nutrient, and BOD removal. Anaerobic fermentation processes, such as sludge digestion, denitrification, and some BOD removal, occur in the lower layer. The key to successful operation of this type of pond is O_2 production by photosynthetic algae and/or re-aeration at the surface.

Facultative ponds are used to treat raw municipal wastewater in small communities and for primary or secondary effluent treatment for small or large cities. They are also used in industrial applications, usually in the process line after aerated or anaerobic ponds, to provide additional treatment prior to discharge. Commonly achieve effluent BOD values, as measured in the BOD_5 test, range from 20 to 60 mg/L, and TSS levels may range from 30 to 150 mg/L. The size of the pond needed to treat BOD loadings depends on specific conditions and regulatory requirements.

Aerated ponds: Facultative ponds overloaded due to unplanned additional sewage volume or higher strength influent from a new industrial connection may be modified by the addition of mechanical aeration. Ponds originally designed for mechanical aeration are generally 2–6 m deep with detention times of 3–10 days. For colder climates, 20–40 days is recommended. Mechanically aerated ponds require less land area but have greater energy requirements.

When aeration is used, the depth of the pond and/or the acceptable loading levels may increase. Mechanical or diffused aeration is often used to supplement natural oxygen production or to replace it.

Pond Organisms

Although our understanding of wastewater pond ecology is far from complete, general observations about the interactions of macro- and microorganisms in these biologically driven systems support our ability to design, operate and maintain them.

Bacteria

Bacteria found in ponds help to decompose complex, organic constituents in the influent simple, non-toxic compounds. Certain pathogen bacteria and other microbial organism (viruses, protozoa) associated with human waste enter in that system with the influent; the wastewater treatment process is designed so that nether numbers will be reduced adequately to meet public health standards.

- **Aerobic bacteria** are found in the aerobic zone of a wastewater pond and are primarily the same type as those found in an activated sludge process or in the Zoogleal mass of a trickling filter. The most frequently isolated bacteria include *Beggiatoa alba*, *Sphaerotilus natans*, *Achromobacter*, *Alcaligenes*, *Flavobacterium*, *Pseudomonas* and *Zoogoea spp.* (Spellman, 2000; Lynch and Poole, 1979; Pearson, 2005). These organisms decompose the organic materials present in the aerobic zone into oxidized end products.
- **Anaerobic bacteria** are hydrolytic bacteria that convert complex organic material into simple alcohols and acids, primarily amino acids, glucose, fatty acid and glycerols (Spellman, 2000; Brockett, 1976; Pearson, 2005; Paterson and Curtis, 2005). Acidogenic bacteria convert the sugars and amino acids into acetate, ammonia (NH_3), hydrogen (H), and carbon dioxide (CO_2). Methanogenic bacteria break down these products further to methane (CH_4) and CO_2 (Gallert and Winter, 2005).
- **Cyanobacteria**, formerly classified as blue-green algae, are autotrophic organism that are able to synthesize organic compound using CO_2 as the major carbon source. Cyanobacteria produce O_2 as a by-product of photosynthesis, providing an O_2 source for other organisms in the ponds. They are found in very large numbers as blooms when environmental conditions are suitable (Gaudy and Gaudy, 1980). Commonly encountered cyanobacteria include *Oscillatoria, Arthrospira, Spirulina,* and *Microcystis* (Vasconcelos and Pereira, 2001; Spellman, 2000).
- **Purple sulfur bacteria** (Chromatiaceae) may grow in any aquatic environment to which light of the required wavelength penetrates, provided that CO_2, nitrogen (N), and a reduced form of sulfur (S) or H are available. Purple surf bacteria occupy the anaerobic layer below the algae, cyanobacteria, and other aerobic bacteria in a pond. They are commonly found at a specific depth, in a thin layer where light and nutrient conditions are at an optimum (Gaudy and Gaudy, 1980; Pearson, 2005). Their biochemical conversion of odorous sulfide compounds to elemental S or sulfate (SO_4) helps to control odor in facultative and anaerobic ponds.

Algae

Algae constitute a group of aquatic organisms that may be unicellular or multicellular, motile, or immotile, and, depending on the phylogenetic family, have different combinations of photosynthetic pigments. As autotrophs, algae need only inorganic nutrient, such as N, phosphorus (P) and a suite of microelements, to fix CO_2 and grow in the present of sunlight. Algae do no fix atmospheric N; they require an external source of inorganic N in the form of nitrate (NO_3) or NH_3. Some algal species are able to use amino acids and other organic N compounds. Oxygen is a by-product of these reactions.

Algae are generally divided into three major groups, based on the color reflected from the cells by the chlorophyll and other pigments involved in photosynthesis. Green and brown algae are common to wastewater ponds; red algae occur infrequently. The algal species that is dominant at any particular time is thought to be primarily a function of temperature, although the effects of predation, nutrient availability, and toxins are also important.

Green algae (Chlorophyta) include unicellular, filamentous, and colonial forms. Some green algal genera commonly found in facultative and aerobic ponds are *Euglena, Phacus, Chlamydomonas, Ankistrodesmus, Chlorella, Micractinium, Scenedesmus, Selenastrum, Dictyosphaerium* and *Volvox*.

Chrysophytes, or brown algae, are unicellular and may be flagellated, and include the diatoms. Certain brown algae are responsible for toxic red blooms. Brown algae found in wastewater ponds include the diatoms *Navicula* and *Cyclotella*.

Red algae (Rhodophyta) include a few unicellular forms, but are primarily filamentous (Gaudy and Gaudy, 1980; Pearson, 2005).

With regard to the importance of interactions between bacteria and algae it is generally accepted that the presence of both algae and bacteria is essential for the proper functioning of a treatment pond. Bacteria break down the complex organic waste components found in anaerobic and aerobic pond environments into simple compounds, which are then available for uptake by the algae. Algae, in turn, produce the O_2 necessary for the survival of aerobic bacteria.

In the process of pond reactions of biodegradation and mineralization of waste material by bacteria and the synthesis of new organic compounds in the form of algal cells a pond effluent might contain a higher than acceptable TSS. Although this form of TSS does not contain the same constituents as the influent TSS, it does contribute to turbidity and needs to be removed before the effluent is discharged. Once concentrated and removed, depending on regulatory requirements, algal TSS may be used as a nutrient for use in agriculture or as a feed supplement.

Invertebrates

Invertebrates although bacteria and algae are the primary organisms through which waste stabilization is accomplished; predator life forms do play a role in wastewater pond ecology. It has been suggested that the planktonic invertebrate *Cladocera* spp and the benthic invertebrate family Chironomidae are the most significant fauna in the pond community in terms of stabilizing organic material. The cladocerans feed on the algae and promote flocculation and settling of particulate matter. This in turn results in better light penetration and algal growth at greater depths. Settled matter is further broken down and stabilized by the benthic feeding Chironomidae. Predators, such as rotifers, often control the population levels of certain of the smaller life forms in the pond, thereby influencing the succession of species throughout the seasons.

Mosquitoes can present a problem in some ponds. Aside from their nuisance characteristics, certain mosquitoes are also vectors for such diseases as encephalitis, malaria, and yellow fever, and constitute a hazard to public health which must be controlled. *Gambusia*, commonly called mosquito fish, have been introduced to eliminate mosquito problems in some ponds in warm climates (Ullrich, 1967; Pipes, 1961; Pearson, 2005), but their introduction has been problematic as they can out-compete native fish that also feed on mosquito larvae. There are also biochemical controls, such as the larvicides *Bacillus thuringiensis* israelensis (Bti), and Abate®, which may be effective if the product is applied directly to the area containing mosquito larvae. The most effective means of control of mosquitoes in ponds is the control of emergent vegetation (USEPA, 2011).

Biochemistry in a Pond

Photosynthesis

Photosynthesis is the process whereby organisms use solar energy to fix CO_2 and obtain the reducing power to convert it to organic compounds. In wastewater ponds, the dominant photosynthetic organisms include algae, cyanobacteria, and purple sulfur bacteria (Pipes, 1961; Pearson, 2005). Photosynthesis may be classified as oxygenic or anoxygenic, depending on the source of reducing power used by a particular organism. In oxygenic photosynthesis, water serves as the source of reducing power, with O_2 as a by-product. The equation representing oxygenic photosynthesis is:

$$H_2O + sunlight \rightarrow 1/2 O_2 + 2H^+ + 2e^- \tag{8.34}$$

Oxygenic photosynthetic algae and cyanobacteria convert CO_2 to organic compounds, which serve as the major source of chemical energy for other aerobic organisms. Aerobic bacteria need the

O_2 produced to function in their role as primary consumers in degrading complex organic waste material.

Anoxygenic photosynthesis does not produce O_2 and, in fact, occurs in the complete absence of O_2. The bacteria involved in anoxygenic photosynthesis are largely strict anaerobes, unable to function in the presence of O_2. They obtain energy by reducing inorganic compounds. Many photosynthetic bacteria utilize reduced S compounds or element S in anoxygenic photosynthesis according to the following equation:

$$H_2S \to S^0 + 2H^+ + 2e^- \tag{8.35}$$

Respiration

Respiration is a physiological process by which organic compounds are oxidized into CO_2 and water. Respiration is also an indicator of cell material synthesis. It is a complex process that consists of many interrelated biochemical reactions (Pearson, 2005). Aerobic respiration, common to species of bacteria, algae, protozoa, invertebrates and higher plants and animals, may be represented by the following equation:

$$_2H_{12}O_6 + 6O_2 + Enzymes \to 6CO_2 + 6H_2O + New\,cells \tag{8.36}$$

The bacteria involved in aerobic respiration are primarily responsible for degradation of waste products. In the presence of light, respiration and photosynthesis can occur simultaneously in algae. However, the respiration rate is low compared to the photosynthesis rate, which results in a net consumption of CO_2 and production of O_2. In the absence of light, on the other hand, algal respiration continues while photosynthesis stops, resulting in a net consumption of O_2 and production of CO_2 (USEPA, 2011).

Nitrogen Cycle

The N cycle occurring in a wastewater treatment pond consists of a number of biochemical reactions mediated by bacteria. A schematic representation of the changes in N speciation in wastewater ponds over a year is represented in Figure 8.11. Organic N and NH_3 enter with the influent wastewater. Organic N in fecal matter and other organic materials undergo conversion to NH_3 and ammonium ion NH_4^+ by microbial activity. The NH_3 may volatilize into the atmosphere. The rate of gaseous NH_3 losses to the atmosphere is primarily a function of pH, surface-to-volume ratio, temperature, and mixing conditions. An alkaline pH shifts the equilibrium of NH_3 gas and NH_4^+ towards gaseous NH_3 production, while the mixing conditions affect the magnitude of the mass transfer coefficient.

Ammonium is nitrified to nitrite (NO_2^-) by the bacterium *Nitrosomonas* and then to NO_3^- by *Nitrobacter*. The overall nitrification reaction is:

$$NH_4^+ + 2O_2 \to NO_3^- + 2H^+ + H_2O \tag{8.37}$$

The NO_3^- produced in nitrification process, as well as a portion of the NH_4^- produced from ammonification, can be assimilated by organisms to produce cell protein and other N-containing compounds. The NO_3^- may also be denitrified to form NO_2^- and then N gas. Several species of bacteria may be involved in the denitrification process, including *Pseudomonas, Micrococcus, Achromobacter,* and *Bacillus*. The overall denitrification reaction is

$$6NO_3^- + 5CH_3OH \to 3N_2 + 5CO_2 + 7H_2O + 6OH^- \tag{8.38}$$

Nitrogen gas may be fixed by certain species of cyanobacteria when N is limited. This may occur in N-poor industrial ponds, but rarely in municipal or agricultural ponds (USEPA, 1975, 1993).

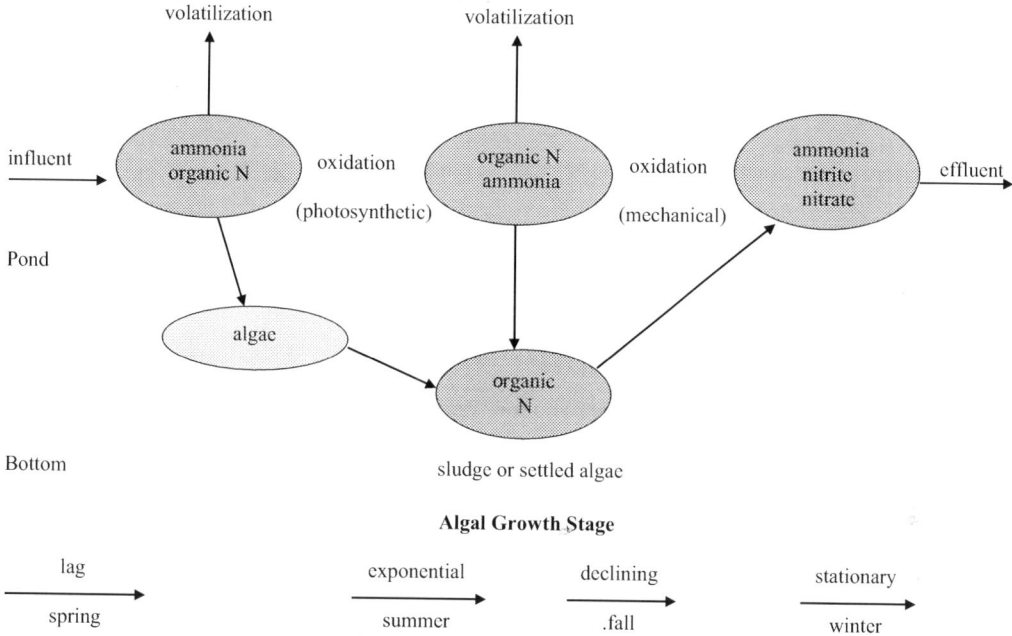

FIGURE 8.11 The nitrogen cycle in wastewater pond system.

Nitrogen removal in facultative wastewater ponds can occur through any of the following processes: (1) gaseous NH_3 stripping to the atmosphere, (2) NH_4^- assimilation in algal biomass, (3) NO_3^- uptake by floating vascular plants and algae, and (4) biological nitrification-denitrification. Whether NH_4^- is assimilated into algal biomass depends on the biological activity in the system and is affected by several factors such as temperature, organic load, detention time, and wastewater characteristics.

Dissolved Oxygen (DO)
Oxygen is a partially soluble gas. Its solubility varies in direct proportion to the atmospheric pressure at any given temperature. DO concentrations of ~8 mg/L are generally considered to be maximum available under local ambient conditions. In mechanically aerated ponds, the limited solubility of O_2 determines its absorption rate (Sawyer et al., 1994). The natural sources of DO in ponds are photosynthetic oxygenation and surface re-aeration. In areas of low wind activity, surface re-aeration may be relatively unimportant, depending on the water depth. Where surface turbulence is created by excessive wind activity, surface re-aeration can be significant. Experiments have shown that DO in wastewater ponds varies almost directly with the level of photosynthetic activity, which is low at night and early morning and rises during daylight hours to a peak in the early afternoon. At increased depth, the effects of photosynthetic oxygenation and surface re-aeration decrease, as the distance from the water-atmosphere interface increases and light penetration decreases. This can result in the establishment of a vertical gradient. The microorganisms in the pond will segregate along the gradient.

pH and Alkalinity
In wastewater ponds, the H ion concentration, expressed as pH, is controlled through the carbonate buffering system represented by the following equations:

$$CO_2 + H_2O \leftrightarrow H_2CO_3 \leftrightarrow HCO_3^- + H^+ \tag{8.39}$$

$$HCO_3^- \leftrightarrow CO_3^{-2} + H^+ \qquad (8.40)$$

$$CO_3^{-2} + H_2O \leftrightarrow HCO_3^- + OH^- \qquad (8.41)$$

$$OH^- + H^+ \leftrightarrow H_2O \qquad (8.42)$$

The equilibrium of this system is affected by the rate of algal photosynthesis. In photosynthetic metabolism, CO_2 is removed from the dissolved phase, forcing the equilibrium of the first expression (8.32) to the left. This tends to decrease the hydrogen ion (H^+) concentration and the bicarbonate (HCO_3^-) alkalinity. The effect of the decrease in HCO_3^- concentration is to force the third Equation (8.41) to the left and the fourth (8.42) to the right, both of which decrease total alkalinity. The decreased alkalinity associated with photosynthesis will simultaneously reduce the carbonate hardness present in the waste. Because of the close correlation between pH and photosynthetic activity, there is a diurnal fluctuation in pH when respiration is the dominant metabolic activity.

Physical Factors

Light The intensity and spectral composition of light penetrating a pond surface significantly affect all resident microbial activity. In general, activity increases with increasing light intensity until the photosynthetic system becomes light saturated. The rate at which photosynthesis increases in proportion to an increase in light intensity, as well as the level at which an organism's photosynthetic system becomes light-saturated, depends upon the particular biochemistry of the species (Lynch and Poole, 1979; Pearson, 2005). In ponds, photosynthetic O_2 production has been shown to be relatively constant with the range of 5,380–53,800 lumens/m^2 light intensity with a reduction occurring at higher and lower intensities (Pipes, 1961; Paterson and Curtis, 2005).

The spectral composition of available light is also crucial in determining photosynthetic activity. The ability of photosynthetic organisms to utilize available light energy depends primarily upon their ability to absorb the available wavelengths. This absorption ability is determined by the specific photosynthetic pigment of the organism. The main photosynthetic pigments are chlorophylls and phycobilins. Bacterial chlorophyll differs from algal chlorophyll in both chemical structure and absorption capacity. These differences allow the photosynthetic bacteria to live below dense algal layers where they can utilize light not absorbed by the algae (Lynch and Poole, 1979; Pearson, 2005).

The quality and quantity of light penetrating the pond surface to any depth depend on the presence of dissolved and particulate matter as well as the water absorption characteristics. The organisms themselves contribute to water turbidity, further limiting the depth of light penetration. Given the light penetration interferences, photosynthesis is significant only in the upper pond layers. This region of net photosynthetic activity is called the euphotic zone (Lynch and Poole, 1979; Pearson, 2005). Light intensity from solar radiation varies with the time of day and difference in latitudes. In cold climates, light penetration can be reduced during the winter by ice and snow cover. Supplementing the treatment ponds with mechanical aeration may be necessary in these regions during that time of year.

Temperature Temperature at or near the surface of the aerobic environment of a pond determines the succession of predominant species of algae, bacteria, and other aquatic organisms. Algae can survive at temperatures of 5°C–40°C. Green algae show most efficient growth and activity at temperatures of 30°C–35°C. Aerobic bacteria are viable within a temperature range of 10°C–40°C;

35°C–40°C is optimum for cyanobacteria (Anderson and Zwieg, 1962; Gloyna, 1976; Paterson and Curtis, 2005; Crites et al., 2006).

As the major source of heat for these systems in solar radiation, a temperature gradient can develop in a pond with depth. This will influence the rate of anaerobic decomposition of solids that have settled at the bottom of the pond. The bacteria responsible for anaerobic degradation are active in temperatures from 15°C to 65°C. When they are exposed to lower temperatures, their activity is reduced.

The other major source of heat is the influent water. In sewerage systems with no major inflow or infiltration problems, the influent temperature is higher than that of the pond contents. Cooling influences are exerted by evaporation, contact with cooler groundwater and wind action. The overall effect of temperature in combination with light intensity is reflected in the fact that nearly all investigators report improved performance during summer and autumn months when both temperature and light are at their maximum. The maximum practical temperature of wastewater ponds is likely <30°C, indicating that most ponds operate at less than optimum temperature for anaerobic activity (Oswald, 1996; Paterson and Curtis, 2005; Crites et al., 2006; USEPA, 2011).

During certain times of the year, cooler, denser water remains at depth, while the warmer water stays at the surface. Water temperature differences may cause ponds to stratify throughout their depth. As the temperature decreased during the fall and the surface water cools, stratification decreases and the deeper water mix with the cooling surface water. This phenomenon is call *mixis*, or pond or lake overturn. As the density of water decreases and the temperature falls below 4°C, winter stratification can develop. When the ice cover breaks up and the water warms, a spring overturn can also occur (Spellman, 1996).

Pond overturn, which releases odorous compounds into the atmosphere, can generate complaints from property owners living downwind of the pond. The potential for pond overturn during certain times of the year is the reason why regulations may specify that ponds be located downwind, based on prevailing winds during overturn periods, and away from dwellings.

Wind Prevailing and storm-generated wind should be factored into pond design and siting as they influence performance and maintenance in several significant ways:

- **Oxygen transfer and dispersal**: By producing circulatory flows, winds provide the mixing needed for O_2 transfer and diffusion below the surface of facultative ponds. This mixing action also helps disperse microorganisms and augments the movement of algae, particularly green algae.
- **Prevention of short-circuiting and reduction of odor events**: Care must be taken during design to position the pond inlet/outlet axis perpendicular to the direction of prevailing winds to reduce short-circuiting, which is the most common cause of poor performance. Consideration must also be made for the transport and fate of odors generated by treatment by-products in anaerobic and facultative ponds.
- **Disturbance of pond integrity**: Waves generated by strong prevailing or storm winds are capable of eroding or overtopping embankments. Some protective material should extend one or more feet above and below the water level to stabilize earthen berms.
- **Hydraulic detention time**: wind effects can reduce hydraulic retention time.

Pond Nutritional Requirements

In order to function as designed, the wastewater pond must provide sufficient macro- and micro-nutrients for the microorganisms to grow and populate the system adequately. It should be understood that a treatment pond system should be neither overloaded nor underloaded with wastewater nutrients.

Nitrogen Nitrogen can be a limiting nutrient for primary productivity in a pond. The conversion of organic N to various other N forms results in a total net loss (Assenzo and Reid, 1966; Pano and Middlebrooks, 1982; Middlebrooks et al., 1982; Middlebrooks and Pano, 1983; Craggs, 2005). This N loss may be due to algal uptake or bacterial action. It is likely that both mechanisms contribute to the overall total N reduction. Another factor contributing to the reduction of total N is the removal of gaseous NH_3 under favorable environmental conditions. Regardless of the specific removal mechanism involved, NH_3 under favorable environmental conditions. Regardless of the specific removal mechanism involved, NH_3 removal in facultative wastewater ponds has been observed at levels >90%, with the major removal occurring in the primary cell of a multicell pond system (Middlebrooks et al., 1982; Shilton, 2005; Crites et al., 2006; USEPA, 2011).

Phosphorus Phosphorus is most often the growth-limiting nutrient in aquatic environments. Municipal wastewater in the United States is normally enriched in P even though restrictions on P-containing compound sin laundry detergents in some states have resulted in reduced concentrations since the 1970s. As of 1999, 27 states and the District of Columbia had pass laws prohibiting the manufacture and use of laundry detergents containing P. However, phosphate (PO_4^{-3}) content limits in automatic dishwashing detergents and other household cleaning agents containing P remain unchanged in most states. With a contribution of ~15%, the concentration of P from wastewater treatment plants is still adequate to promote growth in aquatic organisms.

In aquatic environments, P occurs in three forms: (1) particulate P, (2) soluble organic P, and (3) inorganic P. Inorganic P, primarily in the form of orthophosphate ($OP(OR)_3$), is readily utilized by aquatic organisms. Some organism may store excess P as polyphosphate. At the same time, some PO_4^{-3} is continuously lost to sediments, where it is locked up in insoluble precipitates (Lynch and Poole, 1979; Craggs, 2005; Crites et al., 2006). Phosphorus removal in ponds occurs via physical mechanisms such as adsorption, coagulation, and precipitation. The uptake of P by organisms in metabolic function as well as for storage can also contribute to its removal. Removal in wastewater ponds has been reported to range from 30% to 95% (Assenzo and Reid, 1966; Pearson, 2005; Crites et al., 2006). Algae discharged in the final effluent may introduce organic P to receiving waters. Excessive algal "afterblooms" observed in waters receiving effluents have, in some cases, been attributed to N and P compounds remaining in the treated wastewater.

Sulfur Sulfur is a required nutrient for microorganisms, and it is usually present in sufficient concentration in natural waters. Because S is rarely limiting, its removal from wastewater is usually not considered necessary. Ecologically, S compounds such as hydrogen sulfide (H_2S and sulfuric acid (H_2SO_4) are toxic, while the oxidation of certain S compound is an important energy source for some aquatic bacteria (Lynch and Poole, 1979; Pearson, 2005).

Carbon The decomposable organic C content of waste is traditionally measured in terms of its BOD_5, or the amount of O_2 required under standardized conditions for the aerobic biological stabilization of the organic matter over a certain period of time. Since complete treatment by biological oxidation can take several weeks, depending on the organic material and the organism present, standard practice is to use the BOD_5 as an index of the organic carbon content or organic strength of a waste. The removal of BOD_5 is a primary criterion by which treatment efficiency is evaluated.

BOD_5 reduction in wastewater ponds ranging from 50% to 95% has been reported in the literature. Various factors affect the rate of reduction of BOD_5. A very rapid reduction occurs in a wastewater pond during the first 5–7 days. Subsequent reductions take place at a sharply reduced rate. BOD_5 removals are generally much lower during winter and early spring than in summer and early fall. Many regulatory agencies recommend that pond operations do not include discharge during cold periods.

Process Control Calculations for Stabilization Ponds

Process control calculations are an important part of wastewater treatment operations, including pond operations.

Determining Pond Area in Acres

$$\text{Area, acres} = \frac{\text{Area, ft}^2}{43,560 \text{ ft}^2/\text{acre}} \qquad (8.43)$$

Determining Pond Volume in ac-ft

$$\text{Volume, ac-ft} = \frac{\text{Volume, ft}^3}{43,560 \text{ ft}^3/\text{ac-ft}} \qquad (8.44)$$

Determining Flow Rate in Acre Feet/Day

$$\text{Flow, ac-ft/day} = \text{Flow, MGD} \times 3.069 \text{ ac-ft/MG} \qquad (8.45)$$

Note: Acre-feet (ac-ft) is a unit that can cause confusion, especially for those not familiar with pond or lagoon operations. The 1 ac-ft is the volume of a box with a 1-acre top and 1 ft of depth—but the top doesn't have to be an even number of acres in size to use acre-feet.

Determining Flow Rate in ac-inches/day

$$\text{Flow, acre-inches/day} = \text{Flow, MGD} \times 36.8 \text{ ac-inches/MG} \qquad (8.46)$$

Hydraulic Detention Time in Days

$$\text{Hydraulic detention time, days} = \frac{\text{Pond volume, ac-ft}}{\text{Influent flow, ac-ft/day}} \qquad (8.47)$$

Note: Normally, hydraulic detention time ranges from 30 to 120 days for stabilization ponds.

Example 8.28

Problem:

A stabilization pond has a volume of 53.5 ac-ft. What is the detention time in days when the flow is 0.30 MGD?

Solution:

$$\text{Flow, ac-ft/day} = 0.30 \text{ MGD} \times 3.069$$

$$= 0.92 \text{ ac-ft/day}$$

$$\text{D.T. Days} = \frac{53.5 \text{ acre}}{0.92 \text{ ac-ft/day}} = 58.2 \text{ days}$$

Hydraulic Loading, Inches/Day (Overflow Rate)

$$\text{Hydraulic loading, inches/say} = \frac{\text{Influent flow, ac-inch/day}}{\text{Pond area, ac}} \qquad (8.48)$$

$$\text{Pop. loading, people/acre/day} = \frac{\text{Population that system serves (people)}}{\text{Pond area (ac)}} \qquad (8.49)$$

Note: Population loading normally ranges from 50 to 500 people per acre.

Organic Loading

Organic loading can be expressed as pounds of BOD_5 per acre per day (most common), pounds BOD_5 per acre-foot per day or people per acre per day.

$$\text{Organic } L, \text{ lb } BOD_5/\text{acre/day} = \frac{BOD_5, \text{mg/L} \times \text{infl. flow, MGD} \times 8.34}{\text{Pond area (ac)}} \qquad (8.50)$$

Note: Normal range is 10–50 lb BOD_5 per day per acre.

Example 8.29

Problem:

A wastewater treatment pond has an average width of 380 ft and an average length of 725 ft. The influent flowrate to the pond is 0.12 MGD with a BOD concentration of 160 mg/L. What is the organic loading rate to the pond in pounds per day per acre (lb/d/ac)?

Solution:

$$725\,\text{ft} \times 380\,\text{ft} \times \frac{1\,\text{ac}}{43,560\,\text{ft}^2} = 6.32\,\text{acre}$$

$$0.12\,\text{MGD} \times 160\,\text{mg/L} \times 8.34\,\text{lb/gal} = 160.1\,\text{lb/day}$$

$$\frac{160.1\,\text{lb/day}}{6.32\,\text{acre}} = 25.3\,\text{lb/day/acre}$$

Trickling Filters

Trickling filters have been used to treat wastewater since the 1890s. It was found that if settled wastewater was passed over rock surfaces, slime grew on the rocks and the water became cleaner. Today we still use this principle but, in many installations, instead of rocks we use plastic media. In most wastewater treatment systems, the *trickling filter* follows primary treatment and includes a secondary settling tank or clarifier as shown in Figure 8.12. Trickling filters are widely used for the treatment of domestic and industrial wastes. The process is a fixed film biological treatment method designed to remove BOD_5 and suspended solids.

A trickling filter consists of a rotating distribution arm that sprays and evenly distributes liquid wastewater over a circular bed of fist-sized rocks, other coarse materials, or synthetic media (see Figure 8.13). The spaces between the media allow air to circulate easily so that aerobic conditions can be maintained. The spaces also allow wastewater to trickle down through, around and over the media. A layer of biological slime that absorbs and consumes the wastes trickling through the bed covers the media material. The organisms aerobically decompose the solids producing more

Water Pollution

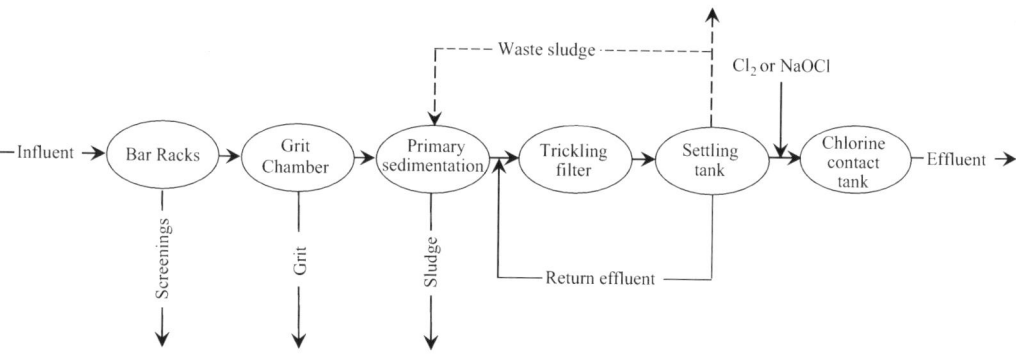

FIGURE 8.12 Simplified flow diagram of trickling filter used for wastewater treatment.

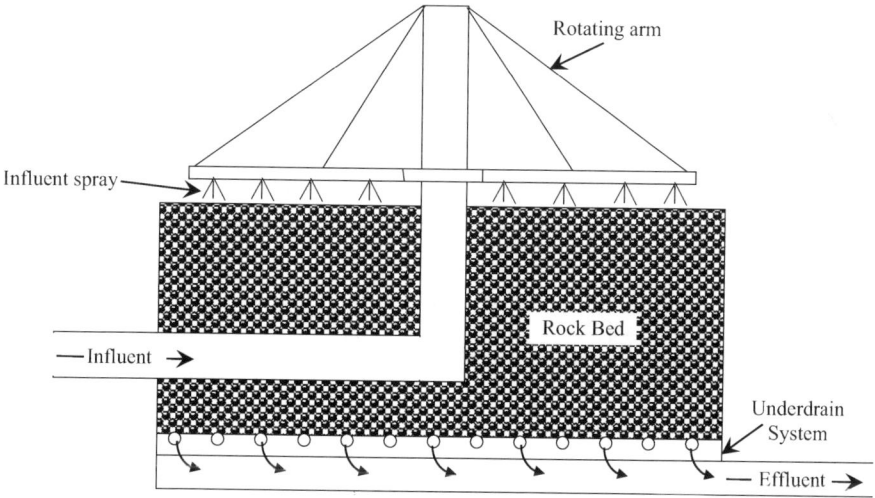

FIGURE 8.13 Schematic of cross-section of a trickling filter.

organisms and stable wastes, which either become part of the slime or are discharged back into the wastewater flowing over the media. This slime consists mainly of bacteria, but it may also include algae, protozoa, worms, snails, fungi, and insect larvae. The accumulating slime occasionally sloughs off (*sloughings*) individual media materials (see Figure 8.14) and is collected at the bottom of the filter, along with the treated wastewater, and passed on to the secondary settling tank where it is removed. The overall performance of the trickling filter is dependent on hydraulic and organic loading, temperature, and recirculation.

Trickling Filter Definitions

To clearly understand the correct operation of the trickling filter, the operator must be familiar with certain terms.

Note: The following list of terms applies to the trickling filter process. We assume that other terms related to other units within the treatment system (plant) are already familiar to operators.

- **Biological towers**: a type of trickling filter that is very deep (10–20 ft). Filled with a lightweight synthetic media, these towers are also known as oxidation or roughing towers or (because of their extremely high hydraulic loading) super-rate trickling filters.

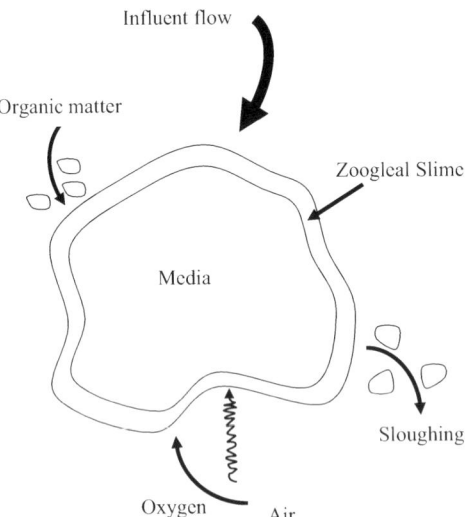

FIGURE 8.14 Filter media showing biological activities that take place on the surface area.

- **Biomass**: the total mass of organisms attached to the media. Similar to solids inventory in the activated sludge process, it is sometimes referred to as the *zoogleal slime*.
- **Distribution arm**: the device most widely used to apply wastewater evenly over the entire surface of the media. In most cases, the force of the wastewater being sprayed through the orifices moves the arm.
- **Filter underdrain**: the open space provided under the media to collect the liquid (wastewater and sloughings) and to allow air to enter the filter. It has a sloped floor to collect the flow to a central channel for removal.
- **Hydraulic loading**: the amount of wastewater flow applied to the surface of the trickling filter media. It can be expressed in several ways: flow per square foot of surface per day (gpd/ft^2); flow per acre per day (MGAD); or flow per acre foot per day (MGAFD). The hydraulic loading includes all flow entering the filter.
- **High-rate trickling filters**: a classification in which the organic loading is in the range of 25–100 pounds of BOD$_5$ per 1,000 ft^3 of media per day. The standard rate filter may also produce a highly nitrified effluent.
- **Media**: an inert substance placed in the filter to provide a surface for the microorganism to grow on. The media can be filed stone, crushed stone, slag, plastic, or redwood slats.
- **Organic loading**: the amount of BOD$_5$ or chemical oxygen demand (COD) applied to a given volume of filter media. It does not include the BOD$_5$ or COD contributed to any recirculated flow and is commonly expressed as pounds of BOD$_5$ or COD per 1,000 ft^3 of media.
- **Recirculation**: the return of filter effluent back to the head of the trickling filter. It can level flow variations and assist in solving operational problems, such as ponding, filter flies, and odors.
- **Roughing filters**: a classification of trickling filters (see Table 8.21) in which the organic is in excess of 200 pounds of BOD$_5$ per 1,000 ft^3 of media per day. A roughing filter is used to reduce the loading on other biological treatment processes to produce an industrial discharge that can be safely treated in a municipal treatment facility.
- **Sloughing**: the process in which the excess growths break away from the media and wash through the filter to the underdrains with the wastewater. These sloughings must be removed from the flow by settling.

TABLE 8.21
Trickling Filter Classification

Filter Class	Standard	Intermediate	High-Rate	Super High-Rate	Roughing
Hydraulic loading gpd/ft^2	25–90	90–230	230–900	350–2,100	>900
Organic loading BOD per 1,000 ft^3	5–25	15–30	25–300	Up to 300	>300
Sloughing frequency	Seasonal	Varies	Continuous	Continuous	Continuous
Distribution	Rotary	Rotary fixed	Rotary fixed	Rotary	Rotary fixed
Recirculation	No	Usually	Always	Usually	Not usually
Media depth, ft	6–8	6–8	3–8	Up to 40	3–20
Media type	Rock	Rock	Rock	Plastic	Rock
	Plastic	Plastic	Plastic	Plastic	
	Wood	Wood	Wood	Wood	
Nitrification	Yes	Some	Some	Limited	None
Filter flies	Yes	Variable	Variable	Very few	Not usually
BOD removal	80%–85%	50%–70%	65%–80%	65%–85%	40%–65%
TSS removal	80%–85%	50%–70%	65%–80%	65%–85%	40%–65%

- **Staging**: the practice of operating two or more trickling filters in series. The effluent of one filter is used as the influent of the next. This practice can produce a higher quality effluent by removing additional BOD$_5$ or COD.

Trickling Filter Equipment

The trickling filter distribution system is designed to spread wastewater evenly over the surface of the entire media. The most common system is the rotary distributor which moves above the surface of the media and sprays the wastewater on the surface. The force of the water leaving the orifices drives the rotary system. The distributor arms usually have small plates below each orifice to spread the wastewater into a fan-shaped distribution system. The second type of distributor is the fixed nozzle system. In this system, the nozzles are fixed in place above the media and are designed to spray the wastewater over a fixed portion of the media. This system is used frequently with deep bed synthetic media filters.

Note: Trickling filters that use ordinary rock are normally only about 3 m in depth because of structural problems caused by the weight of rocks—which also requires the construction of beds that are quite wide, in many applications, up to 60 ft in diameter. When synthetic media is used, the bed can be much deeper.

No matter which type of media is selected, the primary consideration is that it must be capable of providing the desired film location for the development of the biomass. Depending on the type of media used and the filter classification, the media may be 3–20 or more feet in depth.

The underdrains are designed to support the media, collect the wastewater and sloughings and carry them out of the filter and to provide ventilation to the filter.

Note: In order to ensure sufficient airflow to the filter the underdrains should never be allowed to flow more than 50% full of wastewater.

The effluent channel is designed to carry the flow from the trickling filter to the secondary settling tank. The secondary settling tank provides 2–4 hours of detention time to separate the sloughing materials from the treated wastewater. Design, construction, and operation are similar to that of the primary settling tank. Longer detention times are provided because the sloughing materials are lighter and settle more slowly.

Recirculation pumps and piping are designed to recirculate (and thus improve the performance of the trickling filter or settling tank) a portion of the effluent back to be mixed with the filter influent. When recirculation is used, obviously, pumps and metering devices must be provided.

Filter Classifications

Trickling filters are classified by hydraulic and organic loading. Moreover, the expected performance and the construction of the trickling filter are determined by the filter classification. Filter classifications include: standard rate, intermediate rate, high rate, supers high rate (plastic media) and roughing rate types. Standard rate, high rate, and roughing rate are the filter types most commonly used. The *standard rate filter* has a hydraulic loading (gpd/ft^3) of from 25 to 90; a seasonal sloughing frequency; does not employ recirculation; and typically has an 80%–85% BOD$_5$ removal rate and 80%–85% TSS removal rate. The *high-rate filter* has a hydraulic loading (gpd/ft^3) of 230–900; a continuous sloughing frequency; always employs recirculation; and typically has a 65%–80% BOD$_5$ removal rate and 65%–80% TSS removal rate. The *roughing filter* has a hydraulic loading (gpd/ft^3) of >900; a continuous sloughing frequency; does not normally include recirculation; and typically has a 40%–65% removal rate and 40%–65% TSS removal rate.

Standard operating procedures for trickling filters include sampling and testing, observation, recirculation, maintenance, and expectations of performance. Collection of influent and process effluent samples to determine performance and monitor process condition of trickling filters is required. Dissolved oxygen, pH and settleable solids testing should be collected daily. BOD$_5$ and suspended solids testing should be done as often as practical to determine the per cent removal.

The operation and condition of the filter should be observed daily. Items to observe include the distributor movement, uniformity of distribution, evidence of operation or mechanical problems, and the presence of objectionable odors. In addition to the items above the normal observation for a settling tank should also be performed.

Recirculation is used to reduce organic loading, improve sloughing, reduce odors, and reduce or eliminate filter fly or ponding problems. The amount of recirculation is dependent on the design of the treatment plant and the operational requirements of the process. Recirculation flow may be expressed as a specific flow rate (i.e., 2.0-MGD). In most cases, it is expressed as a ratio (3:1, 0.5: 1.0, etc). The recirculation is always listed as the first number and the influent flow listed as the second number. Because the second number in the ratio is always 1.0, the ratio is sometimes written as a single number (dropping the: 1.0)

Flows can be recirculated from various points following the filter to various points before the filter. The most common form of recirculation removes flow from the filter effluent or settling tank and returns it to the influent of the trickling filter as shown in Figure 8.15.

Maintenance requirements include lubrication of mechanical equipment, removal of debris from the surface and orifices, as well as adjustment of flow patterns and maintenance associated with the settling tank.

The trickling filter process involves spraying wastewater over a solid media such as rock, plastic, or redwood slats (or laths). As the wastewater trickles over the surface of the media, the growth of microorganisms (bacteria, protozoa, fungi, algae, helminthes or worms, and larvae) develops. This growth is visible as a shiny slime very similar to the slime found on rocks in a stream. As the wastewater passes over this slime, the slime adsorbs the organic (food) matter. This organic matter is used for food by microorganisms. At the same time, air moving through the open spaces in the filter transfers oxygen to the wastewater. This oxygen is then transferred to the slime to keep the outer layer aerobic. As the microorganisms use the food and oxygen, they produce more organisms, carbon dioxide, sulfates, nitrates, and other stable by-products; these materials are then discarded from the slime back into the wastewater flow and are carried out of the filter:

$$\text{Organics} + \text{Organisms} + O_2 = \text{More Organisms} + CO_2 + \text{Solid Wastes} \tag{8.51}$$

The growth of the microorganisms and the buildup of solid wastes in the slime make it thicker and heavier. When this slime becomes too thick, the wastewater flow breaks off parts of the slime. These must be removed in the final settling tank. In some trickling filters, a portion of the filter

Water Pollution

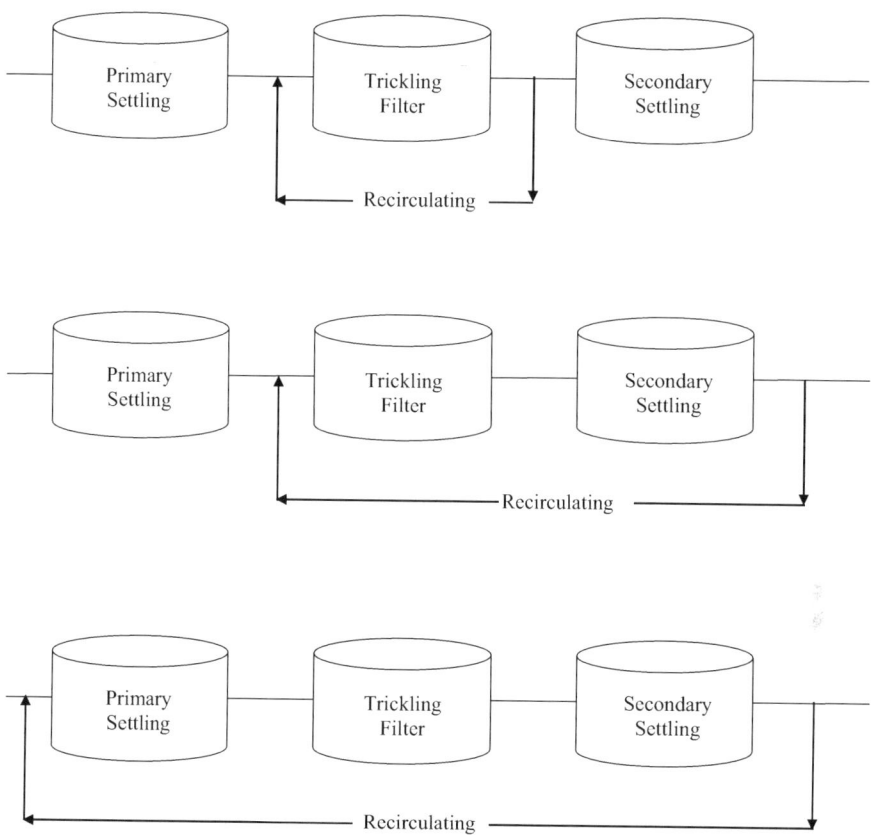

FIGURE 8.15 Common forms of recirculation.

effluent is returned to the head of the trickling filter to level out variations in flow and improves operations (recirculation).

A trickling filter consists of a bed of coarse media, usually rocks or plastic, covered with microorganisms.

Note: Trickling filters that use ordinary rock are normally only about 10 ft in depth because of structural problems caused by the weight of rocks, which also requires the construction of beds that are quite wide—in many applications, up to 60 ft in diameter. When synthetic media are used, the bed can be much deeper.

- The wastewater is applied to the media at a controlled rate, using a rotating distributor arm or fixed nozzles. Organic material is removed by contact with the microorganisms as the wastewater trickles down through the media openings. The treated wastewater is collected by an underdrain system.

    Note: To ensure sufficient airflow to the filter, the underdrains should never be allowed to flow more than 50% full of wastewater.
- The trickling filter is usually built into a tank that contains the media. The filter may be square, rectangular, or circular.
- The trickling filter does not provide any actual filtration. The filter media provides a large amount of surface area that the microorganisms can cling to and grow in a slime that forms on the media as they feed on the organic material in the wastewater.

- The slime growth on the trickling filter media periodically sloughs off and is settled and removed in a secondary clarifier that follows the filter.
- Key factors in trickling filter operation include the following concepts:
  1. Hydraulic loading rate:

  $$\frac{\text{gal/day}}{\text{sq.ft}} = \frac{\text{Flow, gal/day (including recirculation)}}{\text{Media top surface, sq.ft}}$$

  2. Organic loading rate:

  $$\frac{\text{lb/day}}{1{,}000\,\text{ft}^3} = \frac{\text{BOD in filter, lb/day}}{\text{Media volume, }1{,}000\,\text{ft}^3}$$

  3. Recirculation:

  $$\text{Recirculation ratio} = \frac{\text{Recirculation flow, MGD}}{\text{Ave. influent flow, MGD}}$$

Process Calculations

Several calculations are useful in the operation of a trickling filter, these include: total flow, hydraulic loading, and organic loading.

Total Flow If the recirculated flow rate is given, total flow is

$$\text{Total flow, MGD} = \text{Influent flow, MGD} + \text{Recirculation flow, MGD}$$
$$\text{Total flow, gpd} = \text{Total flow, MGD} \times 1{,}000{,}000\,\text{gallons/MG}$$

(8.52)

Note: The total flow to the tricking filter includes the influent flow and the recirculated flow. This can be determined using the recirculation ratio.

$$\text{Total flow, MGD} = \text{Influent flow} \times (\text{recirculation rate} + 1.0)$$

Example 8.30

Problem:

The trickling filter is currently operating with a recirculation rate of 1.5. What is the total flow applied to the filter when the influent flow rate is 3.65 MGD?

Solution:

$$\text{Total flow, MGD} = 3.65\,\text{MGD} \times (1.5 + 1.0)$$
$$= 9.13\,\text{MGD}$$

Hydraulic Loading Calculating the hydraulic loading rate is important in accounting for both the primary effluent as well as the recirculated trickling filter effluent. Both of these are combined before being applied to the surface of the filter. The hydraulic loading rate is calculated based on the surface area of the filter.

Example 8.31

Problem:

A trickling filter 90-ft in diameter is operated with a primary effluent of 0.488 MGD and a recirculated effluent flowrate of 0.566 MGD. Calculate the hydraulic loading rate on the filter in units gpd/ft².

Solution:

The primary effluent and recirculated trickling filter effluent are applied together across the surface of the filter; therefore,

$$0.488\,MGD + 0.566\,MGD = 1.054\,MGD = 1{,}054{,}000\,gpd$$

$$\text{Circular surface area} = 0.785 \times (\text{diameter})^2$$

$$= 0.785 \times (90\,ft)^2$$

$$= 6{,}359\,ft^2$$

$$\frac{1{,}054{,}000\,gpd}{6{,}359\,ft^2} = 165.7\,gpd/ft^2$$

Organic Loading Rate As mentioned earlier, trickling filters are sometimes classified by the Organic Loading Rate applied. The organic loading rate is expressed as a certain amount of BOD applied to a certain volume of media.

Example 8.32

Problem:

A trickling filter, 50 ft in diameter, receives a primary effluent flow rate of 0.445 MGD. Calculate the organic loading rate in units of pounds of BOD applied per day per 900 ft³ of media volume. The primary effluent BOD concentration is 85 mg/L. The media depth is 9 ft.

Solution:

$$0.445\,MGD \times 85\,mg/L \times 8.34\,lb/gal = 315.5\,BOD\,applied/day$$

$$\text{Surface area} = 0.785 \times (50)^2$$

$$= 1962.5\,ft^2$$

$$\text{Area} \times \text{Depth} = \text{Volume}$$

$$1962.5\,ft^2 \times 9\,ft = 17{,}662.5\,(\text{TF Volume})$$

Note: To determine the pounds of BOD per 1,000 ft³ in a volume of thousands of cubic feet, we must set up the equation as shown below.

$$\frac{315.5\,lb\,BOD/day}{17{,}662.5} \times \frac{1{,}000}{1{,}000}$$

Regrouping the numbers and the units together:

$$\frac{315.5 \text{lb} \times 1,000}{17,662.5} \times \frac{\text{lb BOD/day}}{1,000 \text{ft}^3}$$

$$= 17.9 \frac{\text{lb BOD/day}}{1,000 \text{ft}^3}$$

Settling Tank In the operation of settling tanks that follow trickling filters, various calculations are routinely made to determine detention time, surface settling rate, hydraulic loading, and sludge pumping.

Rotating Biological Contactors (RBCs)

The rotating biological contactor (RBC) is a biological treatment system (see Figure 7.19) and is a variation of the attached growth idea provided by the trickling filter. Still relying on microorganisms that grow on the surface of a medium, the RBC is instead a **fixed film** biological treatment device—the basic biological process, however, is similar to that occurring in the trickling filter. An RBC consists of a series of closely spaced (mounted side by side), circular, plastic (synthetic) disks, that are typically about 3.5 m in diameter and attached to a rotating horizontal shaft (see Figure 8.16). Approximately 40% of each disk is submerged in a tank containing the wastewater to be treated. As the RBC rotates, the attached biomass film (zoogleal slime) that grows on the surface of the disk move into and out of the wastewater. While submerged in the wastewater, the microorganisms absorb organics; while they are rotated out of the wastewater, they are supplied with needed oxygen for aerobic decomposition. As the zoogleal slime reenters the wastewater, excess solids and waste products are stripped off the media as sloughings. These sloughings are transported with the wastewater flow to a settling tank for removal.

Modular RBC units are placed in series (see Figure 8.17). Simply because a single contactor is not sufficient to achieve the desired level of treatment; the resulting treatment achieved exceeds conventional secondary treatment. Each individual contactor is called a stage and the group is known as a train. Most RBC systems consist of two or more trains with three or more stages in each. The key advantage in using RBCs instead of trickling filters is that RBCs are easier to operate under varying load conditions, since it is easier to keep the solid medium wet at all times. Moreover, the level of nitrification, which can be achieved by a RBC system, is significant—especially when multiple stages are employed.

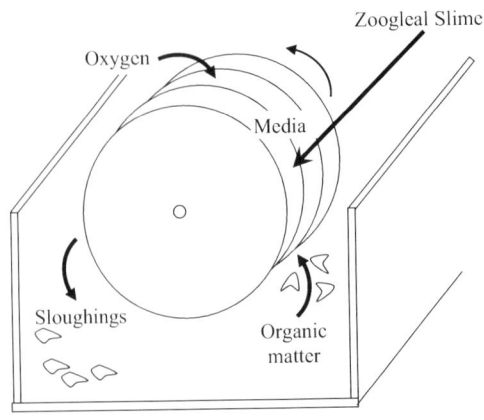

FIGURE 8.16 Rotating biological contactor (RBC) cross-section and treatment system.

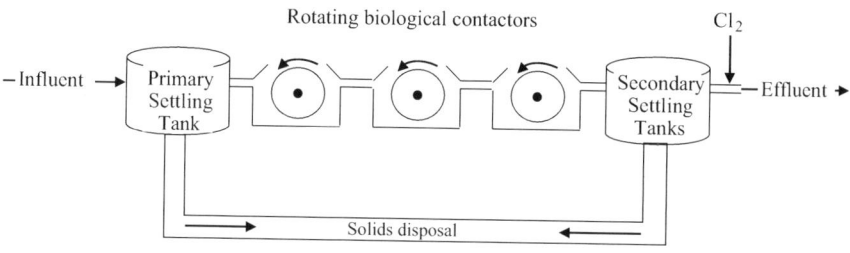

FIGURE 8.17 Rotating biological contactor (RBC) treatment system.

RBC Equipment

The equipment that makes up an RBC includes the rotating biological contactor (the media: either standard or high density), a center shaft, drive system, tank, baffles, housing or cover, and a settling tank. The *rotating biological contactor* consists of circular sheets of synthetic material (usually plastic) which are mounted side by side on a shaft. The sheets (media) contain large amounts of surface area for the growth of the biomass. The *center shaft* provides the support for the disks of media and must be strong enough to support the weight of the media and the biomass. Experience has indicated a major problem has been the collapse of the support shaft. The *drive system* provides the motive force to rotate the disks and shaft. The drive system may be mechanical or air driven or a combination of each. When the drive system does not provide uniform movement of the RBC, major operational problems can arise.

The *tank* holds the wastewater that the RBC rotates in. It should be large enough to permit variation of the liquid depth and detention time. *Baffles* are required to permit proper adjustment of the loading applied to each stage of the RBC process. Adjustment can be made to increase or decrease the submergence of the RBC. RBC stages are normally enclosed in some type of protective structure (*cover*) to prevent the loss of biomass due to severe weather changes (snow, rain, temperature, wind, sunlight, etc.). In many instances, this housing greatly restricts access to the RBC. The *settling tank* is provided to remove the sloughing material created by the biological activity and is similar in design to the primary settling tank. The settling tank provides 2-to-4-hour detention times to permit the settling of lighter biological solids.

During normal operation, operator vigilance is required to observe the RBC movement, slime color, and appearance. However, if the unit is covered, observations may be limited to that portion of the media, which can be viewed through the access door. Slime color and appearance can indicate process condition, for example:

- Gray, shaggy slime growth–indicates normal operation
- Reddish brown, golden shaggy growth–indicates nitrification
- White chalky appearance–indicates high sulfur concentrations
- No slime–indicates severe temperature or pH changes

Sampling and testing should be conducted daily for dissolved oxygen content and pH. BOD_5 and suspended solids testing should also be accomplished to aid in assessing performance.

The RBC normally produces a high-quality effluent with BOD_5 at 85%–95% and Suspended Solids Removal at 85%–95%. The RBC treatment process may also significantly reduce (if designed for this purpose) the levels of organic nitrogen and ammonia nitrogen.

RBC: Process Control Calculations

Several process control calculations may be useful in the operation of an RBC. These include soluble BOD, total media area, organic loading rate, and hydraulic loading rate. Settling tank calculations and sludge pumping calculations may be helpful for the evaluation and control of the settling tank following the RBC.

RBC Soluble BOD

The soluble BOD_5 concentration of the RBC influent can be determined experimentally in the laboratory, or it can be estimated using the suspended solids concentration and the "K" factor. The "K" factor is used to approximate the BOD_5 (particulate BOD) contributed by the suspended matter. The K factor must be provided or determined experimentally in the laboratory. The K factor for domestic wastes is normally in the range in the range of 0.5–0.7.

$$\text{Soluble BOD}_5 = \text{Total BOD}_5 - \left(\text{K factor} \times \text{total suspended solids}\right) \quad (8.53)$$

Example 8.33

Problem:

The suspended solids concentration of a wastewater is 250 mg/L. If the normal K-value at the plant is 0.6, what is the estimated particulate biochemical oxygen demand (BOD) concentration of the wastewater?

Solution:

Note: The K-value of 0.6 indicates that about 60% of the suspended solids are organic suspended solids (particulate BOD).

$$(250\,\text{mg/L})(0.6) = 150\,\text{mg/L (Particulate BOD)}$$

Example 8.34

Problem:

A rotating biological contactor receives a flow of 2.2 MGD with a BOD content of 170 mg/L and suspended solids (SS) concentration of 140 mg/L. If the K-value is 0.7, how many pounds of soluble BOD enter the RBC daily?

Solution:

$$\text{Total BOD} = \text{Particulate BOD} + \text{Soluble BOD}$$

$$170\,\text{mg/L} = (140\,\text{mg/L})(0.7) + x\,\text{mg/L}$$

$$170\,\text{mg/L} = 98\,\text{mg/L} + x\,\text{mg/L}$$

$$170\,\text{mg/L} - 98\,\text{mg/L} = x$$

$$x = 72\,\text{mg/L Soluble BOD}$$

Now, we can determine lb/day soluble BOD:

$$(\text{mg/L Soluble BOD})(\text{MGD Flow})(8.34\,\text{lb/gal}) = \text{lb/day}$$

$$(72\,\text{mg/L})(2.2\,\text{MGD})(8.34\,\text{lb/gal}) = 1{,}321\,\text{lb/day soluble BOD}$$

RBC Total Media Area

Several process control calculations for the RBC use the total surface area of all the stages within the train. As was the case with the soluble BOD calculation, plant design information or information

Water Pollution

supplied by the unit manufacturer must provide the individual stage areas (or the total train area), because the physical determination of this would be extremely difficult.

$$\text{Total area} = 1^{st} \text{ stage area} + 2^{nd} \text{ stage area} + \cdots + n\text{th stage area} \tag{8.54}$$

RBC Organic Loading Rate

If the soluble BOD concentration is known, the organic loading on an RBC can be determined. Organic loading on an RBC based on soluble BOD concentration can range from 3 to 4 lb/day/1,000 ft².

Example 8.35

Problem:

An RBC has a total media surface area of 102,500 ft² and receives a primary effluent flow rate of 0.269 MGD. If the soluble BOD concentration of the RBC influent is 159 mg/L, what is the organic loading rate in lbs/1,000 ft²?

Solution:

$$0.269 \text{ MGD} \times 159 \text{ mg/L} \times \frac{8.34 \text{ lb}}{1 \text{ gal}} = 356.7 \text{ lb/day}$$

$$\frac{356.7 \text{ lb/day}}{102,500 \text{ ft}^2} \times \frac{1,000(\text{number})}{1,000(\text{unit})} = 3.48 \text{ lb/day/1,000 ft}^2$$

RBC Hydraulic Loading Rate

The manufacturer normally specifies the RBC media surface area and the hydraulic loading rate is based on the media surface area, usually in square feet (ft²). Hydraulic loading on an RBC can range from 1 to 3 gpd/ft².

Example 8.36

Problem:

An RBC treats a primary effluent flowrate of 0.233 MGD. What is the hydraulic loading rate in gpd/ft² if the media surface area is 96,600 ft²?

Solution:

$$\frac{233,000 \text{ gpd}}{96,600 \text{ ft}^2} = 2.41 \text{ gpd/ft}^2$$

ACTIVATED SLUDGE

The biological treatment systems discussed to this point [ponds, trickling filters, and rotating biological contactors (RBCs)] have been around for years. The trickling filter, for example, has been around and successfully used since the late 1800s. The problem with ponds, trickling filters and RBCs is that they are temperature sensitive, remove less BOD and, trickling filters, for example, cost more to build than the activated sludge systems that were later developed.

Note: Although trickling filters and other systems cost more to build than activated sludge systems, it is important to point out that activated sludge systems cost more to operate because of the need for energy to run pumps and blowers.

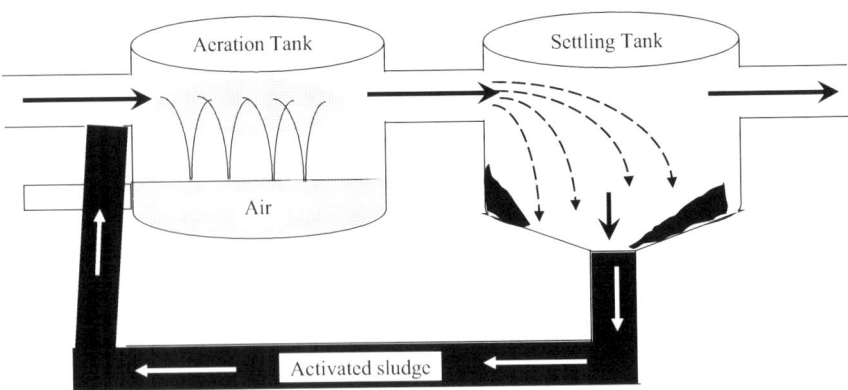

FIGURE 8.18 Activated sludge process.

As shown in Figure 8.8, the activated sludge process follows primary settling. The basic components of an activated sludge sewage treatment system include an aeration tank and a secondary basin, settling basin, or clarifier (see Figure 8.18). Primary effluent is mixed with settled solids recycled from the secondary clarifier and is then introduced into the aeration tank. Compressed air is injected continuously into the mixture through porous diffusers located at the bottom of the tank, usually along one side.

Wastewater is fed continuously into an aerated tank, where the microorganisms metabolize and biologically flocculate the organics. Microorganisms (activated sludge) are settled from the aerated mixed liquor under quiescent conditions in the final clarifier and are returned to the aeration tank. Left uncontrolled, the number of organisms would eventually become too great; therefore, some must periodically be removed (wasted). A portion of the concentrated solids from the bottom of the settling tank must be removed from the process (waste-activated sludge or WAS). Clear supernatant from the final settling tank is the plant effluent.

Activated Sludge Terminology

To better understand the discussion of the activated sludge process presented in the following sections, you must understand the terms associated with the process. Some of these terms have been used and defined earlier in the text, but we list them here again to refresh your memory. Review these terms and remember them. They are used throughout the discussion.

- **Adsorption**: taking in or reception of one substance into the body of another by molecular or chemical actions and distribution throughout the absorber.
- **Activated**: to speed up reaction. When applied to sludge, it means that many aerobic bacteria and other microorganisms are in the sludge particles.
- **Activated sludge**: a floc or solid formed by microorganisms. It includes organisms, accumulated food materials, and waste products from the aerobic decomposition process.
- **Activated sludge process**: a biological wastewater treatment process in which a mixture or influent and activated sludge is agitated and aerated. The activated sludge is subsequently separated from the treated mixed liquor by sedimentation and is returned to the process as needed. The treated wastewater overflows the weir of the settling tank in which separation from the sludge takes place.
- **Adsorption**: the adherence of dissolved, colloidal, or finely divided solids to the surface of solid bodies when they are brought into contact.
- **Aeration**: mixing air and a liquid by one of the following methods: spraying the liquid in the air; diffusing air into the liquid; or agitating the liquid to promote surface adsorption of air.

- **Aerobic**: a condition in which "free" or dissolved oxygen is present in the aquatic environment. Aerobic organisms must be in the presence of dissolved oxygen to be active.
- **Bacteria**: single-cell plants that play a vital role in the stabilization of organic waste.
- **Biochemical oxygen demand (BOD)**: a measure of the amount of food available to the microorganisms in a particular waste. It is measured by the amount of dissolved oxygen used up during a specific time period (usually 5 days, expressed as BOD_5).
- **Biodegradable**: from "degrade" (to wear away or break down chemically) and "bio" (by living organisms). Put it all together, and you have a "substance, usually organic, which can be decomposed by biological action."
- **Bulking**: a problem in activated sludge plants that results in poor settleability of sludge particles.
- **Coning**: a condition that may be established in a sludge hopper during sludge withdrawal, when part of the sludge moves toward the outlet while the remainder tends to stay in place. Development of a cone or channel of moving liquids surrounded by relatively stationary sludge.
- **Decomposition**: generally, in waste treatment, decomposition refers to the changing of waste matter into simpler, more stable forms that will not harm the receiving stream.
- **Diffuser**: a porous plate or tube through which air is forced and divided into tiny bubbles for distribution in liquids. Commonly made of carborundum, aluminum, or silica sand.
- **Diffused air aeration**: a diffused air-activated sludge plant takes air, compresses it, and then discharges the air below the water surface to the aerator through some type of air diffusion device.
- **Dissolved oxygen**: atmospheric oxygen dissolved in water or wastewater, usually abbreviated as DO.
  Note: The typical required DO for a well-operated activated sludge plant is between 2.0 and 2.5 mg/L.
- **Facultative**: facultative bacteria can use either molecular (dissolved) oxygen or oxygen obtained from food materials. In other words, facultative bacteria can live under aerobic or anaerobic conditions.
- **Filamentous bacteria**: organisms that grow in thread or filamentous form.
- **Food-to-microorganisms ratio**: a process control calculation used to evaluate the amount of food (BOD or COD) available per pound of mixed liquor volatile suspended solids. This may be written as F/M ratio.

$$\frac{\text{Food}}{\text{Microorganisms}} = \frac{\text{BOD}(\text{lb/day})}{\text{MLVSS}(\text{lb})}$$

$$= \frac{\text{Flow}(\text{MGD}) \times \text{BOD}(\text{mg/L}) \times 8.34\,\text{lb/gal}}{\text{Vol.}(\text{MG}) \times \text{MLVSS}(\text{mg/L}) \times 8.34\,\text{lb/gal}}$$

- **Fungi**: multicellular aerobic organisms.
- **Gould sludge age**: a process control calculation used to evaluate the amount of influent suspended solids available per pound of mixed liquor suspended solids.
- **Mean cell residence time** (MCRT): the average length of time mixed liquor-suspended solids particle remains in the activated sludge process. This is usually written as MCRT and may also be referred to as *sludge retention rate* (STR).

$$\text{MCRT, days} = \frac{\text{Solids in activated sludge process, lbs}}{\text{Solids removed from process, lb/day}}$$

- **Mixed liquor**: the contribution of return activated sludge and wastewater (either influent or primary effluent) that flows into the aeration tank.
- **Mixed liquor suspended solids (MLSS)**: the suspended solids concentration of the mixed liquor. Many references use this concentration to represent the amount of organisms in the liquor. Many references use this concentration to represent the amount of organisms in the activated sludge process. This is usually written MLSS.
- **Mixed liquor volatile suspended solids (MLVSS)**: the organic matter in the mixed liquor suspended solids. This can also be used to represent the amount of organisms in the process. This is normally written as MLVSS.
- **Nematodes**: microscopic worms that may appear in biological waste treatment systems.
- **Nutrients**: substances required to support plant organisms. Major nutrients are carbon, hydrogen, oxygen, sulfur, nitrogen, and phosphorus.
- **Protozoa**: single-cell animals that are easily observed under the microscope at a magnification of 100×. Bacteria and algae are prime sources of food for advanced forms of protozoa.
- **Return activated sludge**: the solids returned from the settling tank to the head of the aeration tank. This is normally written as RAS.
- **Rising sludge**: rising sludge occurs in the secondary clarifiers or activated sludge plant when the sludge settles to the bottom of the clarifier, is compacted, and then rises to the surface in relatively short time.
- **Rotifiers**: multicellular animals with flexible bodies and cilia near their mouths used to attract food. Bacteria and algae are their major source of food.
- **Secondary treatment**: a wastewater treatment process used to convert dissolved or suspended materials into a form that can be removed.
- **Settleability**: a process control test used to evaluate the settling characteristics of the activated sludge. Readings taken at 30–60 minutes are used to calculate the settled sludge volume (SSV) and the sludge volume index (SVI).
- **Settled sludge volume**: the volume of ml/L (or percent) occupied by an activated sludge sample after 30 or 60 minutes of settling. Normally written as SSV with a subscript to indicate the time of the reading used for calculation (SSV_{30} or SSV_{60}).
- **Shock load**: the arrival at a plant of a waste toxic to organisms, in sufficient quantity or strength to cause operating problems, such as odor or sloughing off of the growth of slime on the trickling filter media. Organic overloads also can cause a shock load.
- **Sludge volume index**: a process control calculation used to evaluate the settling quality of the activated sludge. Requires the SSV_{30} and mixed liquor suspended solids test results to calculate.

$$\text{Sludge vol. index (SVI), mL/g} = \frac{(30 \text{ minutes settled vol., mL/L})(1{,}000 \text{ mg/g})}{\text{Mixed liquor suspended solids, mg/L}}$$

- **Solids**: material in the solid state
- **Dissolved**: solids present in solution. Solids that will pass through a glass fiber filter.
    Fixed: also known as the inorganic solids. The solids that are left after a sample is ignited at 550°C for 15 minutes.
- **Floatable solids**: solids that will float to the surface of still water, sewage, or other liquid. Usually composed of grease particles, oils, light plastic material, etc. Also called *scum*.
- **Non-settleable**: finely divided suspended solids that will not sink to the bottom in still water, sewage, or other liquid in a reasonable period, usually 2 hours. Non-settleable solids are also known as colloidal solids.

- **Suspended**: the solids that will not pass through a glass fiber filter.
- **Total**: the solids in water, sewage, or other liquids; it includes the suspended solids and dissolved solids.
- **Volatile**: the organic solids. Measured as the solids that are lost on ignition of the dry solids at 550°C.
- **Waste activated sludge**: the solids being removed from the activated sludge process. This is normally written as WAS.

Activated Sludge Process

The equipment requirements for the activated sludge process are more complex than other processes discussed. Equipment includes an *aeration tank, aeration, system-settling tank, return sludge*, and *waste sludge system*. These are discussed in the following.

Aeration Tank: the *aeration tank* is designed to provide the required detention time (depending on the specific modification) and ensure that the activated sludge and the influent wastewater are thoroughly mixed. Tank design normally attempts to ensure no dead spots are created.

Aeration: can be mechanical or diffused. Mechanical aeration systems use agitators or mixers to mix air and mixed liquor. Some systems use sparge ring to release air directly into the mixer. Diffused aeration systems use pressurized air released through diffusers near the bottom of the tank. Efficiency is directly related to the size of the air bubbles produced. Fine bubble systems have a higher efficiency. The diffused air system has a blower to produce large volumes of low-pressure air (5–10 psi), air lines to carry the air to the aeration tank, headers to distribute the air to the diffusers which release the air into the wastewater.

Settling tank: activated sludge systems are equipped with plain *settling tanks* designed to provide 2 to 4 hours hydraulic detention time.

Return sludge: the return sludge system includes pumps, a timer or variable speed drive to regulate pump delivery and a flow measurement device to determine actual flow rates.

Waste sludge: in some cases, the *waste activated sludge* withdrawal is accomplished by adjusting valves on the return system. When a separate system is used it includes pump(s), timer or variable speed drive and a flow measurement device.

The activated sludge process is a treatment technique in which wastewater and reused biological sludge full of living microorganisms are mixed and aerated. The biological solids are then separated from the treated wastewater in a clarifier and are returned to the aeration process or wasted. The microorganisms are mixed thoroughly with the incoming organic material, and they grow and reproduce by using the organic material as food. As they grow and are mixed with air, the individual organisms cling together (flocculate). Once flocculated, they more readily settle in the secondary clarifiers.

The wastewater being treated flows continuously into an aeration tank where air is injected to mix the wastewater with the returned activated sludge and to supply the oxygen needed by the microbes to live and feed on the organics. Aeration can be supplied by injection through air diffusers in the bottom of tank or by mechanical aerators located at the surface. The mixture of activated sludge and wastewater in the aeration tank is called the "mixed liquor." The mixed liquor flows to a secondary clarifier where the activated sludge is allowed to settle.

The activated sludge is constantly growing, and more is produced than can be returned for use in the aeration basin. Some of this sludge must, therefore, be wasted to a sludge handling system for treatment and disposal. The volume of sludge returned to the aeration basins is normally 40%–60% of the wastewater flow. The rest is wasted.

Factors Affecting the Operation of the Activated Sludge Process

Several factors affect the performance of an activated sludge system. These include the following:

- Temperature
- Return rates
- Amount of oxygen available
- Amount of organic matter available
- pH
- Waste rates
- Aeration time
- Wastewater toxicity

To obtain the desired level of performance in an activated sludge system, a proper balance must be maintained between the amount of food (organic matter), organisms (activated sludge), and oxygen (dissolved oxygen, DO). The majority of problems with the activated sludge process result from an imbalance between these three items.

To fully appreciate and understand the biological process taking place in a normally functioning activated sludge process, the operator must have knowledge of the key players in the process: the organisms. This makes a certain amount of sense when you consider that the heart of the activated sludge process is the mass of settleable solids formed by aerating wastewater containing biological degradable compounds in the presence of microorganisms. Activated sludge consists of organic solids plus bacteria, fungi, protozoa, rotifers, and nematodes.

Microorganism Growth Curve

To understand the microbiological population and its function in an activated sludge process, the operator must be familiar with the microorganism *growth curve;* (see Figure 8.19). In the presence of excess organic matter, the microorganisms multiply at a fast rate. The demand for food and oxygen is at its peak. Most of this is used for the production of new cells. This condition is known as the *log growth phase* (see Figure 8.19). As time goes on, the amount of food available for the organism declines. Floc begins to form while the growth rate of bacteria and protozoa begins to decline. This is referred to as the *declining growth phase* (see Figure 8.19). The *endogenous respiration* phase occurs as the food available becomes extremely limited and the organism mass begins to decline (see Figure 8.19). Some of the microorganisms may die and break apart, thus releasing organic matter that can be consumed by the remaining population.

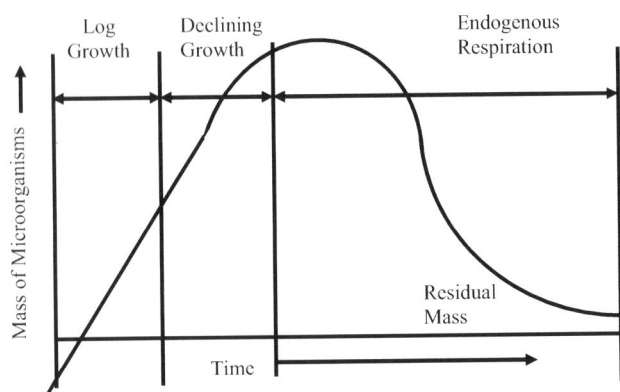

FIGURE 8.19 Microorganism growth curve.

Water Pollution

The actual operation of an activated-sludge system is regulated by three factors: (1) the quantity of air supplied to the aeration tank, (2) the rate of activated-sludge recirculation, and (3) the amount of excess sludge withdrawn from the system. Sludge wasting is an important operational practice because it allows the operator to establish the desired concentration of MLSS, food/microorganism ratio, and sludge age.

Note: Air requirements in an activated sludge basin are governed by (1) biological oxygen demand (BOD) loading and the desired removal effluent; (2) volatile suspended solids concentration in the aerator; and (3) suspended solids concentration of the primary effluent.

Activated Sludge Formation

The formation of activated sludge is dependent on three steps on three steps. The first step is the transfer of food from wastewater to organism. Second is the conversion of wastes to a usable form. Third is the flocculation step.

1. **Transfer**: Organic matter (food) is transferred from the water to the organisms. Soluble material is absorbed directly through the cell wall. Particulate and colloidal matter is adsorbed to the cell wall, where it is broken down into simpler soluble forms, then absorbed through the cell wall.
2. **Conversion**: Food matter is converted to cell matter by synthesis and oxidation into end products such as CO_2, H_2O, NH_3, stable organic waste, and new cells.
3. **Flocculation**: Flocculation is the gathering of fine particles into larger particles. This process begins in the aeration tank and is the basic mechanism for removal of suspended matter in the final clarifier. The concentrated *bio-floc* that settles and forms the sludge blanket in the secondary clarifier is known as activated sludge.

To maintain the working organisms in the activated sludge process, the operator must ensure that a suitable environment is maintained by being aware of the many factors influencing the process and by monitoring them repeatedly "Control" is defined as maintaining the proper solids (floc mass) concentration in the aerator for the incoming water (food) flow by adjusting the return and waste sludge pumping rate and regulating the oxygen supply to maintain a satisfactory level of dissolved oxygen in the process.

Aeration: The activated sludge process must receive sufficient aeration to keep the activated sludge in suspension and to satisfy the organism's oxygen requirements. Insufficient mixing results in dead spots, septic conditions, and/or loss of activated sludge.

Alkalinity: The activated sludge process requires sufficient alkalinity to ensure that pH remains in the acceptable range of 6.5–9.0. If organic nitrogen and ammonia are being converted to nitrate (nitrification), sufficient alkalinity must be available to support this process, as well.

Nutrients: The microorganisms of the activated sludge process require nutrients (nitrogen, phosphorus, iron, and other trace metals) to function. If sufficient nutrients are not available, the process will not perform as expected. The accepted minimum ratio of carbon to nitrogen, phosphorus, and iron is 100 parts carbon to five parts nitrogen, one part phosphorus, and 0.5 parts iron.

pH: The pH of the mixed liquor should be maintained within the range of 6.5–9.0 (6.0–8.0 is ideal). Gradual fluctuations within this range will normally not upset the process. Rapid fluctuations or fluctuations outside this range can reduce organism activity.

Temperature: As temperature decreases, the activity of the organisms will also decrease. Cold temperatures also require longer recovery time for systems that have been upset. Warm temperatures tend to favor denitrification and filamentous growth.

Note: The activity level of bacteria within the activated sludge process increases with rise in temperature.

Toxicity: Sufficient concentrations of elements or compounds that enter a treatment plant that have the ability to kill the microorganisms (the activated sludge) are known as toxic waste (shock level). Common to this group are cyanides and heavy metals.

Note: A typical example of a toxic substance added by operators is the uninhabited use of chlorine for odor control or control or control of filamentous organisms (prechlorination). Chlorination is for disinfection. Chlorine is a toxicant and should not be allowed to enter the activated sludge process; it is not selective with respect to type of organisms damaged or killed. If may kill the organisms that should be retained in the process as workers. Chlorine is very effective in disinfecting the plant effluent after treatment by the activated sludge process, however.

Hydraulic loading: Hydraulic loading is the amount of flow entering the treatment process. When compared with the design capacity of the system, it can be used to determine if the process is hydraulically overloaded or underloaded. If more flow is entering the system than it was designed to handle, the system is hydraulically overloaded. If less flow is entering the system than it was designed for, the system is hydraulically underloaded. Generally, the system is more affected by overloading than by underloading. Overloading can be caused by stormwater, infiltration of groundwater, excessive return rates, or many other causes. Underloading normally occurs during periods of drought or in the period following initial startup when the plant has not reached its design capacity. Excess hydraulic flow rates through the treatment plant will reduce the efficiency of the clarifier by allowing activated sludge solids to rise in the clarifier and pass over the effluent weir. This loss of solids in the effluent degrades effluent quality and reduces the amount of activated sludge in the system, in turn, reducing process performance.

Organic loading: Organic loading is the amount of organic matter entering the treatment plant. It is usually measured as biochemical oxygen demand (BOD). An organic overload occurs when the amount of BOD entering the system exceeds the design capacity of the system. An organic underload occurs when the amount of BOD entering the system is significantly less than the design capacity of the plant. Organic overloading may occur when the system receives more waste than it was designed to handle. It can also occur when an industry or other contributor discharges more wastes to the system than originally planned. Wastewater treatment plant processes can also cause organic overloads returning high-strength wastes from the sludge treatment processes.

Regardless of the source, an organic overloading of the plant results in increased demand for oxygen. This demand may exceed the air supply available from the blowers. When this occurs, the activated sludge process may become septic. Excessive wasting can also result in a type of organic overload. The food available exceeds the number of activated sludge organisms, resulting in increased oxygen demand and very rapid growth.

Organic underloading may occur when a new treatment plant is initially put into service. The facility may not receive enough waste to allow the plant to operate at its design level. Underloading can also occur when excessive amounts of activated sludge are allowed to remain in the system. When this occurs, the plant will have difficulty in developing and maintaining a good, activated sludge.

Activated Sludge Modifications

First developed in 1913, the original activated sludge process has been modified over the years to provide better performance for specific operating conditions or with different influent waste characteristics.

1. *Conventional Activated Sludge*
   - Employing the conventional activated sludge modification requires primary treatment.
   - Conventional activated sludge provides excellent treatment; however, large aeration tank capacity is required, and construction costs are high.
   - In operation, initial oxygen demand is high. The process is also very sensitive to operational problems (e.g., bulking).

2. *Step Aeration*
   - Step aeration requires primary treatment.
   - It provides excellent treatment.
   - Operation characteristics are similar to conventional.
   - It distributes organic loading by splitting influent flow.
   - It reduces oxygen demand at the head of the system.
   - It reduces solids loading on settling tank.
3. *Complete Mix*
   - May or may not include primary treatment.
   - Distributes waste, return, and oxygen evenly throughout tank.
   - Aeration may be more efficient.
   - Maximizes tank use.
   - Permits a higher organic loading.

   Note: During the complete mix, activated sludge process organisms are in declining phase on growth curve.
4. *Pure Oxygen*
   - Requires primary treatment.
   - Permits higher organic loading.
   - Uses higher solids levels.
   - Operates at higher F:M ratios.
   - Uses covered tanks.
   - Potential safety hazards (pure oxygen).
   - Oxygen production is expensive.
5. *Contact Stabilization*
   - Contact stabilization does not require primary treatment.
   - During operation, organisms collect organic matter (during contact).
   - Solids and activated sludge are separated from flow via settling.
   - Activated sludge and solids are aerated for 3–6 hours (stabilization).

   Note: *Return* sludge is aerated before it is mixed with influent flow.
   - The activated sludge oxidizes available organic matter.
   - While the process is complicated to control, it requires less tank volume than other modifications and can be prefabricated as a *package* unit for flows of 0.05–1.0 MGD.
   - A disadvantage is that common process control calculations do not provide usable information.
6. *Extended Aeration*
   - Does not require primary treatment.
   - Used frequently for small flows such as schools and subdivisions.
   - Uses 24-hour aeration.
   - Produces low BOD effluent.
   - Produces the least amount of waste-activated sludge.
   - Process is capable of achieving 95% or more removals of BOD.
   - Can produce effluent low in organic and ammonia nitrogen.
7. *Oxidation Ditch*
   - Does not require primary treatment.
   - The oxidation ditch process is similar to the extended aeration process.

Table 8.22 lists the process parameters for each of the four most commonly used activated sludge modifications.

Extended Aeration: Package Plants

One of the most common types of modified active sludge processes which provides biological treatment for the removal of biodegradable organic waste under aerobic conditions is the extended

TABLE 8.22
Activated Sludge Modifications

Parameter	Conventional	Contact Stabilization	Extended Aeration	Oxidation Ditch
Aeration time (hours)	4–8	0.5–1.5 (contact) 3–6 (reaeration)	24	24
Settling time (hours)	2–4	2–4	2–4	2–4
Return rate, % of influent flow	25–100	25–100	25–100	25–100
MLSS (mg/L)	1,500–4,000	1,000–3,000 3,000–8,000	2,000–6,000	2,000–6,000
D.O. (mg/L)	1–3	1–3	1–3	1–3
SSV_{30} (ml/L)	400–700	400–700 (contact)	400–700	400–700
Food: mass ratio lbs BOD_5/lb MLVSS	02–0.5	02–0.6 (contact)	0/05–0.15	0.05–0.15
MCRT (whole system) days	5%–15	N/A	20–30	20–30
% Removal BOD_5	85%–95%	85%–95%	85%–95%	85%–95%
% Removal TSS	85%–95%	85%–95%	85%–95%	85%–95%
Primary treatment	Yes	No	No	No

aeration process called the package plant. Package plants are pre-manufactured treatment facilities used to treat wastewater in small communities or on individual properties. According to manufacturers, package plants can be designed to treat flows as low as 0.002 MGD or as high as 0.5 MGD, although they more commonly treat flows between 0.01 and 0.25 MGD (Metcalf and Eddy, 1991).

In operation, air may be supplied to the extended aeration package plant by mechanical or diffused aeration to provide the oxygen required to sustain the aerobic biological process. Mixing must be provided by aeration or mechanical means to maintain the microbial organisms in contact with the dissolved organics. In addition, the pH must be controlled to optimize the biological process and essential nutrients must be present to facilitate biological growth and the continuation of biological degradation.

As depicted in Figure 8.20, wastewater enters the treatment system and is typically screened immediately to remove large suspended, settleable, or floating solids that could interfere with or

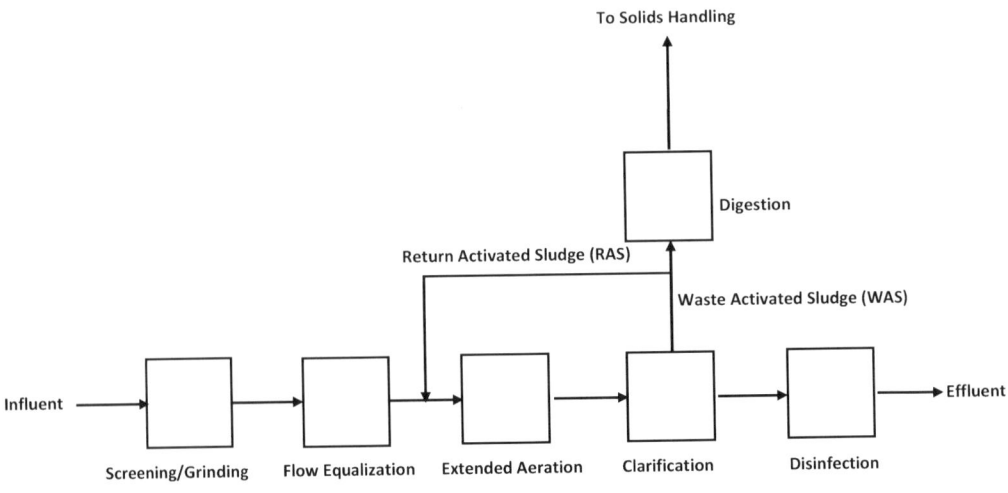

FIGURE 8.20 A typical extended aeration plant.

Water Pollution

damage equipment downstream in the process. Wastewater may then pass through a grinder to reduce large particles that are not captured in the screening process. If the plant requires the flow to be regulated, the effluent with then flow into equalization basins which regulate peak wastewater flow rates. Wastewater then enters the aeration chamber, where it is mixed and oxygen is provided to the microorganisms. The mixed liquor then flows to a clarifier or settling chamber where most microorganism settle to the bottom of the clarifier and a portion are pumped back to the incoming wastewater at the beginning of the plant. This returned material is the retune activated sludge (RAS). The material that is not returned, the waste activated sludge (WAS), is removed for treatment and disposal. The clarified wastewater then flows over a weir and into a collection channel before being diverted to the disinfection system (USEPA, 2007b).

Package plants are typically located in small municipalities, suburban subdivisions, apartment complexes, highway rest areas, trailer parks, small institutions, and other sites where follow rates are below 0.1 MGD. Extended aeration package plants consist of a steel tank that is compartmentalized into flow equalization, aeration, clarification, disinfection, and aerated sludge holding/digestion segments. Extended aeration systems are typically manufactured to treat wastewater flow rates between 0.002 and 0.1 MGD. Use of concrete tanks may be preferable for large sizes (Spellman, 2000).

Extended aeration plants are usually started up using "seed sludge" from another sewage plant. It may take as many as 2–4 weeks form the time it is seeded for the plant to stabilize (Spellman, 2000). These systems are also useful for areas requiring nitrification. Key internal components of extended aeration treatment package plants consist of the following: transfer pumps to move wastewater between the equalization and aeration zones; a bar screen and/or grinder to decrease the size of large solids; an aeration system consisting of blowers and diffusers for the equalization, aeration, and sludge holding zones; transfer pumps to move wastewater between the equalization and aeration zones; an airlift pump for returning sludge; a skimmer and effluent weir for the clarifier; and UV, liquid hypochlorite, or tablet modules used in the disinfection zone. Blowers and the control panel containing switches, lights, and motor starters are typically attached to either the top or one side of the package plant (Sloan, 1999).

Advantages and Disadvantages
Advantages:

- Plants are easy to operate, as many are manned for a maximum of 2 or 3 hours per day.
- Extended aeration processes are often better at handling organic loading and flow fluctuations, as there is a greater detention time for the nutrients to be assimilated by microbes.
- Systems are easy to install, as they are shipped in one or two pieces and then mounted on an onsite concrete pad, above or below grade.
- Systems are odor free, can be installed in most locations, have a relatively small footprint, and can be landscaped to match the surrounding area.
- Extended aeration systems have a relatively low sludge yield due to long sludge ages, can be designed to provide nitrification, and do not require a primary clarifier.

Disadvantages:

- Extended aeration plants do not achieve denitrification of phosphorus removal without additional unit processes.
- Flexibility is limited to adapt to changing effluent requirements resulting from regulatory changes.
- A longer aeration period requires more energy.
- Systems require a larger amount of space and tankage that other "higher rate" processes, which have shorter aeration detention times.

Oxidation Ditches

An oxidation ditch is modified extended aeration activated sludge biological treatment process that utilizes longs solids retention times (SRTs) to remove biodegradable organics. Oxidation ditches are typically complete mix systems, but they can be modified to approach plug flow conditions. (Note: as conditions approach plug flow, diffused air must be used to provide enough mixing. The system will also no longer operate as an oxidation ditch). Typical oxidation ditch treatment systems consist of a single or multi-channel configuration within a ring, oval, or horseshoe-shaped basin. As a result, oxidation ditches are called "racetrack type" reactors. Horizontally or vertically mounted aerators provide circulation, oxygen transfer, and aeration in the ditch.

Preliminary treatment, such as bar screens and grit removal, normally precedes the oxidation ditch. Primary settling prior to an oxidation ditch is sometimes practiced but is not typical in this design. Tertiary filters may be required after clarification, depending on the effluent requirements. Disinfection is required and reaeration may be necessary prior to final discharge. Flow to the oxidation ditch is aerated and mixed with return sludge from a secondary clarifier. A typical process flow diagram for an activated sludge plant using an oxidation ditch is shown in Figure 8.21.

Surface aerators, such as brush rotors, disc aerators, draft tube aerators, or find bubble diffusers are used to circulate the mixed liquor. The mixing process entrains oxygen into the mixed liquor to foster microbial growth and the motive velocity ensures contact of microorganisms with the incoming wastewater. The aeration sharply increases the dissolved oxygen (DO) concentration but decreases as biomass uptake oxygen as the mixed liquor travels through the ditch. Solids are maintained in suspension as the mixed liquor travels through the ditch. Solids are maintained in suspension as the mixed liquor circulated around the ditch. If design SRT's are selected for nitrification, a high degree of nitrification will occur. Oxidation ditch effluent is usually settled in a separate secondary clarifier. An anaerobic tank may be added prior to the ditch to enhance biological phosphorus removal.

An oxidation ditch may also be operated to achieve partial denitrification. One of the most common design modifications for enhanced nitrogen removal is known as the Modified Ludzack-Ettinger (MLE) process. In this process, illustrated in Figure 8.22, an anoxic tank is added upstream of the

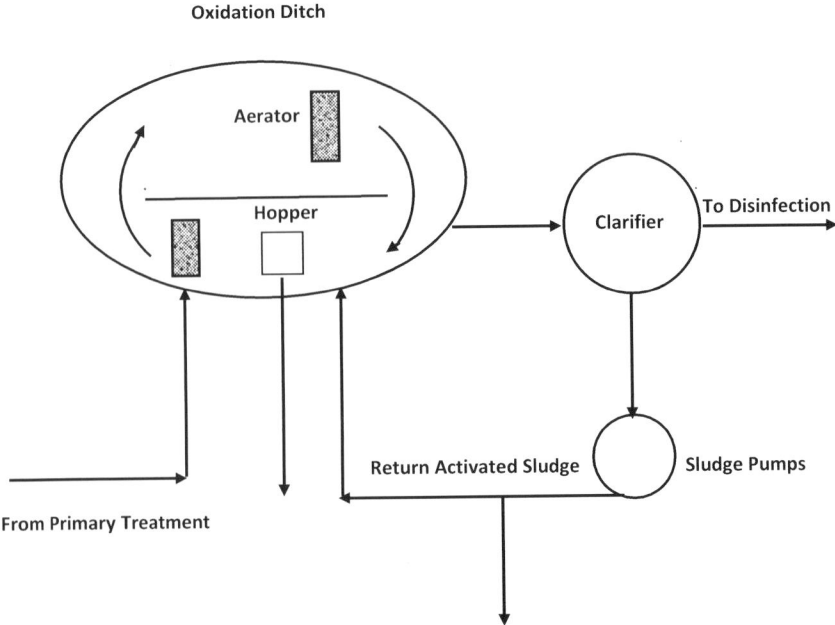

FIGURE 8.21 Typical oxidation ditch-activated sludge system.

Water Pollution

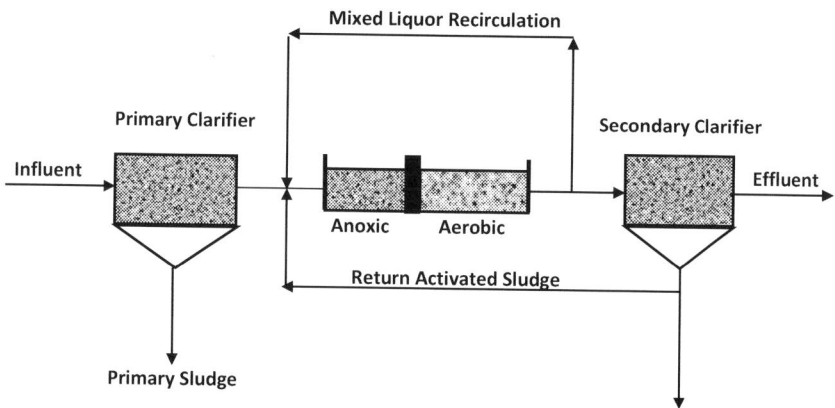

FIGURE 8.22 The Modified Ludzack-Ettinger (MLE) process.

ditch along with mixed liquor recirculation for the aerobic zone to the tank to achieve higher levels of denitrification. In the aerobic basin, autotrophic bacteria (nitrifiers) convert ammonia-nitrogen to nitrite-nitrogen and then to nitrate-nitrogen. In the anoxic zone, heterotrophic bacteria convert nitrate-nitrogen to nitrogen gas which is released to the atmosphere. Some mixed liquor from the aerobic basin is recirculated to the anoxic zone to provide mixed liquor with a high concentration of nitrate-nitrogen to the anoxic zone.

Several manufacturers have developed modifications to the oxidation ditch design to remove nutrient sin conditions cycled or phased between the anoxic and aerobic states. While the mechanics of operation differ by manufacturer, in general, the process consists of two separate aeration basins, the first anoxic and the second aerobic. Wastewater and return activated sludge (RAS) are introduced into the first reactor which operates under anoxic conditions. Mixed liquor then flows into the second reactor operating under aerobic conditions. The process is then reversed and the second reactor begins to operate under anoxic conditions.

With regard to applicability, the oxidation ditch process is a fully demonstrated secondary wastewater treatment technology, applicable in any situation where activated sludge treatment (conventional or extend aeration) is appropriate. Oxidation ditches are applicable in plants that require nitrification because the basins can be sized using an appropriate SRT to achieve nitrification at the mixed liquor minimum temperature. This technology is very effective in small installations, small communities, and isolated institutions, because it requires more land than conventional treatment plants.

There are currently more than 9,000 municipal oxidation ditch installations in the United States (Spellman, 2007). Nitrification to <1 mg/L ammonia nitrogen consistently occurs when ditches are designed and operated for nitrogen removal. An excellent example of an upgrade to the MLE process is to provide in the following case. Keep in mind that the motivation for this upgrade was 2-fold: to increase optimal plant operation (DO optimization) and to conserve energy.

Advantages and Disadvantages
Advantages:

The main advantage of the oxidation ditch is the ability to achieve removal performance objectives with low operational requirements and operation and maintenance costs. Some specific advantages of oxidation ditches include:

- An added measure of reliability and performance over other biological processes owing to a constant water level and continuous discharge which lowers the weir overflow rate and eliminates the periodic effluent surge common to other biological processes, such as SBRS.

- Long hydraulic retention time and complete mixing minimize the impact of a shock load or hydraulic surge.
- Produces less sludge than other biological treatment processes owing to extended biological activity during the activated sludge process.
- Energy efficient operations result in reduced energy costs compared with other biological treatment processes.

Disadvantages:

- Effluent suspended solids concentrations are relatively high compared to other modifications of the activated sludge process.
- Requires a larger land area than other activated sludge treatment options. This can prove costly, limiting the feasibility of oxidation ditches in urban, suburban, or other areas where land acquisition costs are relatively high.

Activated Sludge Process Control Parameters

When designing and/or operating an activated sludge process, the environmental engineer and/or the operator must be familiar with the many important process control parameters, which must be monitored frequently and adjusted occasionally to maintain optimal performance.

Alkalinity: Monitoring alkalinity in the aeration tank is essential to control of the process. Insufficient alkalinity will reduce organism activity and may result in low effluent pH and, in some cases, extremely high chlorine demand in the disinfection process.

Dissolved Oxygen (DO): The activated sludge process is an aerobic process that requires some dissolved oxygen be present at all times. The amount of oxygen required is dependent on the influent food (BOD), the activity of the activated sludge, and the degree of treatment desired.

pH: Activated sludge microorganisms can be injured or destroyed by wide variations in pH. The pH of the aeration basin will normally be in the range of 6.5–9.0. Gradual variations within this range will not cause any major problems; however, rapid changes of one or more pH units can have a significant impact on performance. Industrial waste discharges, septic wastes, or significant amounts of stormwater flows may produce wide variations in pH. pH should be monitored as part of the routine process control-testing schedule. Sudden changes or abnormal pH values may indicate an industrial discharge of strongly acidic or alkaline wastes. Because these wastes can upset the environmental balance of the activated sludge, the presence of wide pH variations can result in poor performance. Processes undergoing nitrification may show a significant decrease in effluent pH.

Mixed Liquor Suspended Solids, Mixed Liquor Volatile Suspended Solids and Mixed Liquor Total Suspended Solids

The mixed liquor suspended solids (MLSS) or mixed liquor volatile suspended solids MLVSS can be used to represent the activated sludge or microorganisms present in the process. Process control calculations, such as sludge age and sludge volume index, cannot be calculated unless the MLSS is determined. Adjust the MLSS and MLVSS by increasing or decreasing the waste sludge rates. The mixed liquor total suspended solids or MLTSS is an important activated sludge control parameter. To increase the MLTSS, for example, the operator must decrease the waste rate and/or increase the MCRT. The MCRT must be decreased to prevent the MLTSS from changing when the number of aeration tanks in service is reduced.

Note: In performing the Gould Sludge Age Test, assume that the source of the MLTSS in the aeration tank is influent solids.

Water Pollution

Return Activated Sludge Rate and Concentration

The sludge rate is a critical control variable. The operator must maintain a continuous return of activated sludge to the aeration tank or the process will show a drastic decrease in performance. If the rate is too low, solids remain in the settling tank, resulting in solids loss and a septic return. If the rate is too high, the aeration tank can become hydraulically overloaded, causing reduced aeration time and poor performance. The return concentration is also important because it may be used to determine the return rate required to maintain the desired MLSS.

Waste-Activated Sludge Flow Rate

Because the activated sludge contains living organisms that grow, reproduce, and produce waste matter, the amount of activated sludge is continuously increasing. If the activated sludge is allowed to remain in the system too long, the performance of the process will decrease. If too much activated sludge is removed from the system, the solids become very light and will not settle quickly enough to be removed in the secondary clarifier.

Temperature

Because temperature directly affects the activity of the microorganisms, accurate monitoring of temperature can be helpful in identifying the causes of significant changes in organization populations or process performance.

Sludge Blanket Depth

The separation of solids and liquid in the secondary clarifier results in a blanket of solids. If solids are not removed from the clarifier at the same rate they enter, the blanket will increase in depth. If this occurs, the solids may carry over into the process effluent. The sludge blanket depth may be affected by other conditions, such as temperature variation, toxic wastes, or sludge bulking. The best sludge blanket depth is dependent upon such factors as hydraulic load, clarifier design, sludge characteristics, and many more. The best blanket depth must be determined on an individual basis by experimentation.

Note: In measuring sludge blanket depth, it is general practice to use a 15–20 ft long clear plastic pipe marked at 6-inch intervals, the pipe is equipped with a ball valve at the bottom.

DISINFECTION OF WASTEWATER

Like drinking water, liquid wastewater effluent is disinfected. Unlike drinking water, wastewater effluent is disinfected not directly (direct end-of-pipe connection) to protect drinking water supplies, but instead is treated to protect public health in general. This is particularly important when the secondary effluent is discharged into a body of water used for swimming or water supply for a downstream water supply. In the treatment of water for human consumption, treated water is typically chlorinated (although ozonation is also currently being applied in many cases). Chlorination is the preferred disinfection in potable water supplies because of chlorine's unique ability to provide a residual. This chlorine residual is important because when treated water leaves the waterworks facility and enters the distribution system; the possibility of contamination is increased. The residual works to continuously disinfect water right up to the consumer's tap.

Considering the previously provided in-depth discussion of the disinfection of water, this section continues that discussion with description of basic chlorination and dechlorination as applied to wastewater. In addition, we describe ultraviolet (UV) irradiation, ozonation, bromine chlorine, and no disinfection. Again, keep in mind that much of the chlorination material presented in the following is similar to the chlorination information presented earlier in the water disinfection section.

Chlorine Disinfection

Chlorination for disinfection, as shown in Figure 8.8, follows all other steps in conventional wastewater treatment. The purpose of chlorination is to reduce the population of organisms in the wastewater to levels low enough to ensure that pathogenic organisms will not be present in sufficient quantities to cause disease when discharged.

Note: Chlorine gas is heavier than (vapor density of 2.5). Therefore, exhaust from a chlorinator room should be taken from floor level.

Note: The safest action to take in the event of a major chlorine container leak is to call the fire department.

Note: You might wonder why it is that chlorination of critical waters such as natural trout streams is not normal practice. This practice is strictly prohibited because chlorine and its by-products (i.e., chloramines) are extremely toxic to aquatic organisms.

Chlorination Terminology

Remember that there are several terms used in discussion of disinfection by chlorination. Because it is important for the operator to be familiar with these terms, we repeat key terms again.

- **Chlorine**: a strong oxidizing agent which has strong disinfecting capability. A yellow-green gas which is extremely corrosive and is toxic to humans in extremely low concentrations in air.
- **Contact time**: the length of time the disinfecting agent and the wastewater remain in contact.
- **Demand**: the chemical reactions, which must be satisfied before a residual or excess chemical will appear.
- **Disinfection**: refers to the selective destruction of disease-causing organisms. All the organisms are not destroyed during the process. This differentiates disinfection from sterilization, which is the destruction of all organisms.
- **Dose**: the amount of chemical being added in milligrams/liter.
- **Feed rate**: the amount of chemical being added in pounds per day.
- **Residual**: the amount of disinfecting chemical remaining after the demand has been satisfied.
- **Sterilization**: the removal of all living organisms.

Wastewater Chlorination: Facts and Process Description

Chlorine Facts

- Elemental chlorine (Cl_2—gaseous) is a yellow-green gas, 2.5 times heavier than air.
- The most common use of chlorine in wastewater treatment is for disinfection. Other uses include odor control and activated sludge bulking control. Chlorination takes place prior to the discharge of the final effluent to the receiving waters.
- Chlorine may also be used for nitrogen removal, through a process called *breakpoint chlorination*. For nitrogen removal, enough chlorine is added to the wastewater to convert all the ammonium nitrogen gas. To do this, ~10 mg/L of chlorine must be added for every 1 mg/L of ammonium nitrogen in the wastewater.
- For disinfection, chlorine is fed manually or automatically into a chlorine contact tank or basin, where it contacts flowing wastewater for at least 30 minutes to destroy disease-causing microorganisms (pathogens) found in treated wastewater.
- Chorine may be applied as a gas, a solid, or in liquid hypochlorite form.
- Chorine is a very reactive substance. It has the potential to react with many different chemicals (including ammonia), as well as with organic matter. When chlorine is added to wastewater, several reactions occur:
    1. Chlorine will react with any reducing agent (i.e., sulfide, nitrite, iron, and thiosulfate) present in wastewater. These reactions are known as *chlorine demand*. The chlorine used for these reactions is not available for disinfection.

2. Chlorine also reacts with organic compounds and ammonia compounds to form chlor-organics and chloramines. Chloramines are part of the group of chlorine compounds that have disinfecting properties and show up as part of the chlorine residual test.
3. After all of the chlorine demands are met, addition of more chlorine will produce free residual chlorine. Producing free residual chlorine in wastewater requires very large additions of chlorine.

Hypochlorite Facts: Hypochlorite, though there are some minor hazards associated with its use (skin irritation, nose irritation, and burning eyes), is relatively safe to work with. It is normally available in dry form as a white powder, pellet, or tablet or in liquid form. It can be added directly using a dry chemical feeder or dissolved and fed as a solution.

Note: In most wastewater treatment systems, disinfection is accomplished by means of combined residual.

Wastewater Chlorination Process Description

Chlorine is a very reactive substance. Chlorine is added to wastewater to satisfy all chemical demands, that is, to react with certain chemicals (such as sulfide, sulfite, ferrous iron, etc.). When these initial chemical demands have been satisfied, chlorine will react with substances such as ammonia to produce chloramines and other substances which, although not as effective as chlorine, have disinfecting capability. This produces a combined residual, which can be measured using residual chlorine test methods. If additional chlorine is added, free residual chlorine can be produced. Due to the chemicals normally found in wastewater, chlorine residuals are normally combined rather than free residuals. Control of the disinfection process is normally based upon maintaining total residual chlorine of at least 1.0 mg/L for a contact time of at least 30 minutes at design flow.

Note: Residual level, contact time, and effluent quality affect Disinfection. Failure to maintain the desired residual levels for the required contact time will result in lower efficiency and increased probability that disease organisms will be discharged.

Based on water quality standards, *total residual limitations* on chlorine are:

- **Freshwater**: <11 ppb total residual chlorine.
- **Estuaries**: Less than 7.5-ppb for halogen-produced oxidants.
- **Endangered species**: Use of chlorine is prohibited.

Hypochlorite Systems

Dependent on the form of hypochlorite selected for use, special equipment, which will control the addition of hypochlorite to the wastewater, is required. Liquid forms require the use of metering pumps, which can deliver varying flows of hypochlorite solution. Dry chemicals require the use of a feed system designed to provide variable doses of the form used. The tablet form of hypochlorite requires the use of a tablet chlorinator designed specifically to provide the desired dose of chlorine. The hypochlorite solution or dry feed systems dispenses the hypochlorite, which is then mixed with the flow. The treated wastewater then enters the contact tank to provide the required contact time.

Chlorine Systems

Because of the potential hazards associated with the use of chlorine, the equipment requirements are significantly greater than those associated with hypochlorite use. The system most widely used is a *solution feed system*. In this system, chlorine is removed from the container at a flow rate controlled by a variable orifice. Water moving through the chlorine injector creates a vacuum, which draws the chlorine gas to the injector and mixes it with the water. The chlorine gas reacts with the water to form hypochlorous and hydrochloric acid. The solution is then piped to the chlorine contact tank and dispersed into the wastewater through a diffuser. Larger facilities may withdraw the liquid form of chlorine and use evaporators (heaters) to convert to the gas form. Small facilities will normally draw the gas form of chlorine from the cylinder. As gas is withdrawn, liquid will be converted to the gas form. This requires heat energy and may result in chlorine line freeze-up if the withdrawal rate exceeds the available energy levels.

In either type of system, normal operation requires adjustment of feed rates to ensure the required residual levels are maintained. This normally requires chlorine residual testing and adjustment based upon the results of the test. Other activities include removal of accumulated solids from the contact tank, collection of bacteriological samples to evaluate process performance and maintenance of safety equipment (respirator-air pack, safety lines, etc.). Hypochlorite operation may also include make-up solution (solution feed systems), adding powder or pellets to the dry chemical feeder or tablets to the tablet chlorinator.

Chlorine operations include adjustment of chlorinator feed rates, inspection of mechanical equipment, testing for leaks using ammonia swab (white smoke means leaks), changing containers (requires more than one person for safety) and adjusting the injector water feed rate when required. Chlorination requires routine testing of plant effluent for total residual chlorine and may also require collection and analysis of samples to determine the Fecal Coliform concentration in the effluent.

Chlorination Process Calculations

Several calculations may be useful in operating a chlorination system. Many of these calculations are discussed and illustrated in this section.

Chlorine Demand Chlorine demand is the amount of chlorine in milligrams per liter that must be added to the wastewater to complete all of the chemical reactions that must occur prior to producing a residual.

$$\text{Chlorine demand} = \text{Chlorine dose, mg/L} - \text{Chlorine residual, mg/L} \tag{8.55}$$

Example 8.37

Problem:

The plant effluent currently requires a chlorine dose of 7.1 mg/L to produce the required 1.0 mg/L chlorine residual in the chlorine contact tank. What is the chlorine demand in milligrams per liter?

Solution:

$$\text{Chlorine demand mg/L} = 7.1\,\text{mg/L} - 1.0\,\text{mg/L} = 6.1\,\text{mg/L}$$

Chlorine Feed Rate The chlorine feed rate is the amount of chlorine added to the wastewater in pounds per day.

$$\text{Chlorine feed rate} = \text{Dose, mg/L} \times \text{Flow, MGD} \times 8.34\,\text{lb/mg/L/MG} \tag{8.56}$$

Example 8.38

Problem:

The current chlorine dose is 5.55 mg/L. What is the feed rate in pounds per day if the flow is 22.89 MGD?

Solution:

$$\text{Feed, lbs/day} = 5.55\,\text{mg/L} \times 22.89\,\text{MGD} \times 8.34\,\text{lb/mg/L/MG} = 1{,}060\,\text{lb/day}$$

Water Pollution

Chlorine Dose Chlorine dose is the concentration of chlorine being added to the wastewater. It is expressed in milligrams per liter.

$$\text{Dose, mg/L} = \frac{\text{Chlorine feed rate in pounds/day}}{\text{Flow in million gal/day} \times 8.34 \, \text{lb/mg/L/MG}} \tag{8.57}$$

Example 8.39

Problem:

A total of 320 pounds of chlorine are added per day to a wastewater flow of 5.60 MGD. What is the chlorine dose in milligrams per liter?

$$\text{Dose, mg/L} = \frac{320 \, \text{lb/day}}{5.60 \, \text{MGD} \times 8.34 \, \text{lb/mg/L/MG}} = 6.9 \, \text{mg/L}$$

Available Chlorine When hypochlorite forms of chlorine are used, the available chlorine is listed on the label. In these cases, the amount of chemical added must be converted to the actual amount of chlorine using the following calculation.

$$\text{Available chlorine} = \text{Amount of hypochlorite} \times \% \, \text{available chlorine} \tag{8.58}$$

Example 8.40

Problem:

The calcium hypochlorite used for chlorination contains 62.5% available chlorine. How many pounds of chlorine are added to the plant effluent if the current feed rate is 30 pounds of calcium hypochlorite per day?

Solution:

$$\text{Quantity of chlorine} = 30 \, \text{pounds} \times 0.625 = 18.75 \, \text{lb Chlorine}$$

Required Quantity of Dry Hypochlorite This calculation is used to determine the amount of hypochlorite needed to achieve the desired dose of chlorine:

$$\text{Hypochlorite quantity, lb/day} = \frac{\text{Required chlorine dose, mg/L} \times \text{flow, MGD} \times 8.34 \, \text{lb/mg/L/MG}}{\% \, \text{Available chlorine}} \tag{8.59}$$

Example 8.41

Problem:

The chlorine dose is 8.8 mg/L and the flow rate is 3.28 MGD. The hypochlorite solution is 71% available chlorine and has a specific gravity of 1.25. How many pounds of hypochlorite must be used?

Solution:

$$\text{Hypochlorite quantity} = \frac{8.8 \, \text{mg/L} \times 3.28 \, \text{MGD} \times 8.34 \, \text{lb/mg/L/MG}}{0.71 \times 8.34 \, \text{lb/gal} \times 1.25} = 32.5 \, \text{gal/day}$$

Ordering Chlorine ∥Because disinfection must be continuous, the supply of chlorine must never be allowed to run out. The following calculation provides a simple method for determining when additional supplies must be ordered. The process consists of three steps:

Step 1: Adjust the flow and use variations if projected changes are provided.
Step 2: If an increase in flow and/or required dosage is projected, current flow rate and/or dose must be adjusted to reflect the projected change.
Step 3: Calculate the projected flow and dose:

$$\text{Projected flow} = \text{Current flow, MGD} \times (1.0 + \%\text{Change})$$
$$\text{Projected dose} = \text{Current dose, mg/L} \times (1.0 + \%\text{Change})$$

(8.60)

Example 8.42

Problem:

Based on the available information for the past 12 months the operator projects that the effluent flow rate will increase by 7.5% during the next year. If the average daily flow has been 4.5 MGD, what will be the projected flow for the next 12 months?

Solution:

$$\text{Projected flow, MGD} = 4.5\,\text{MGD} \times (1.0 + 0.075) = 4.84\,\text{MGD}$$

To determine the amount of chlorine required for a given period:

$$\text{Chlorine required} = \text{Feed rate, lb/day} \times \text{No. of days required.}$$

Example 8.43

Problem:

The plant currently uses 90 lb of chlorine per day. The Town wishes to order enough chlorine to supply the plant for 4 months (assume 31 days/month). How many pounds of chlorine should be ordered to provide the needed supply?

Solution:

$$\text{Chlorine required} = 90\,\text{lb/day} \times 124\,\text{days} = 11{,}160\,\text{lb}$$

Note: In some instances, projections for flow or dose changes are not available but the plant operator wishes to include an extra amount of chlorine as a safety factor. This safety factor can be stated as a specific quantity or as a percentage of the projected usage. Safety Factor as a specific quantity can be expressed as:

$$\text{Total required Cl}_2 = \text{Chlorine required, lbs} + \text{safety factor}$$

Note: Because chlorine is only shipped in full containers unless asked specifically for the amount of chlorine actually required or used during a specified period, all decimal parts of a cylinder are rounded up to the next highest number of full cylinders.

ULTRAVIOLET IRRADIATION

Although ultraviolet (UV) disinfection was recognized as a method for achieving disinfection in the late 19th century, its application virtually disappeared with the evolution of chlorination technologies. However, in recent years, there has been resurgence in its use in the wastewater field, largely as a consequence of concern for discharge of toxic chlorine residual. Even more recently, UV has gained more attention because of the tough new regulations on chlorine use imposed by both OSHA and USEPA. Because of this relatively recent increased regulatory pressure, many facilities are actively engaged in substituting chlorine for other disinfection alternatives. Moreover, UV technology itself has made many improvements, which now makes UV attractive as a disinfection alternative. Ultraviolet light has very good germicidal qualities and is very effective in destroying microorganisms. It is used in hospitals, biological testing facilities, and many other similar locations. In wastewater treatment, the plant effluent is exposed to ultraviolet light of a specified wavelength and intensity for a specified contact period. The effectiveness of the process is dependent on:

- UV light intensity
- Contact time
- Wastewater quality (turbidity)
- For any one treatment plant, disinfection success is directly related to the concentration of colloidal and particulate constituents in the wastewater.

The Achilles' heel of UV for disinfecting wastewater is turbidity. If the wastewater quality is poor, the ultraviolet light will be unable to penetrate the solids and the effectiveness of the process decreases dramatically. For this reason, many states limit the use of UV disinfection to facilities that can reasonably be expected to produce an effluent containing ≤30 mg/L, or less of BOD_5 and total suspended solids. The main components of a UV disinfection system are mercury arc lamps, a reactor, and ballasts. The source of UV radiation is either the low-pressure or medium-pressure mercury arc lamp with low or high intensities. Note that in the operation of UV systems, UV lamps must be readily available when replacements are required. The best lamps are those with a stated operating life of at least 7,500 hours and those that do not produce significant amounts of ozone or hydrogen peroxide. The lamps must also meet technical specifications for intensity, output, and arc length. If the UV light tubes are submerged in the wastestream, they must be protected inside quartz tubes, which not only protect the lights but also make cleaning and replacement easier.

Contact tanks must be used with UV disinfection. They must be designed with the banks of UV lights in a horizontal position, either parallel or perpendicular to the flow or with banks of lights placed in a vertical position perpendicular to the flow.

Note: The contact tank must provide, at a minimum, 10-second exposure time.

We stated earlier that turbidity problems have been the problem with using UV in wastewater treatment—and this is the case. However, if turbidity is its Achilles' heel, then the need for increased maintenance (as compared to other disinfection alternatives) is the toe of the same foot. UV maintenance requires that the tubes be cleaned on a regular basis or as needed. In addition, periodic acid washing is also required to remove chemical buildup.

Routine monitoring is required. Monitoring to check on bulb burnout, buildup of solids on quartz tubes, and UV light intensity is necessary.

Note: UV light is extremely hazardous to the eyes. Never enter an area where UV lights are in operation without proper eye protection. Never look directly into the ultraviolet light.

Advantages and Disadvantages

Advantages:

- UV disinfection is effective at inactivating most viruses, spores, and cysts.
- UV disinfection is a physical process rather than a chemical disinfectant, which eliminates the need to generate, handle, transport, or store toxic/hazardous or corrosive chemicals.
- There is no residual effect that can be harmful to humans or aquatic life.
- UV disinfection is user-friendly for operators.
- UV disinfection has a shorter contact time when compared with other disinfectants (~20 to 30 seconds with low-pressure lamps).
- UV disinfection equipment requires less space than other methods.

Disadvantages:

- Low dosages may not effectively inactivate some viruses, spores, and cysts.
- Organisms can sometimes repair and reverse the destructive effects of UV through a "repair mechanism," known as photo reactivation, or in the absence of light known as "dark repairs."
- A preventive maintenance program is necessary to control fouling of tubes.
- Turbidity and total suspended solids (TSS) in the wastewater can render UV disinfection ineffective. UV disinfection with low-pressure lamps is not as effective for secondary effluent with TSS levels above 30 mg/L.
- UV disinfection is not as cost-competitive when chlorination dechlorination is used and fire codes are met (USEPA, 1999).

When choosing a UV disinfection system, there are three critical areas to be considered. The first is primarily determined by the manufacturer; the second, by design and Operation and Maintenance (O&M); and the third has to be controlled at the treatment facility. Choosing a UV disinfection system depends on three critical factors listed below:

- **Hydraulic properties of the reactor**: Ideally, a UV disinfection system should have a uniform flow with enough axial motion (radial mixing) to maximize exposure to UV radiation. The path that an organism takes in the reactor determines the amount of UV radiation it will be exposed to before inactivation. A reactor must be designed to eliminate short-circuiting and/or dead zones, which can result in inefficient use of power and reduced contact time.
- **Intensity of the UV radiation**: Factors affecting the intensity are the age of the lamps, lamp fouling, and the configuration and placement of lamps in the reactor.
- **Wastewater characteristics**: These include the flow rate, suspended and colloidal solids, initial bacterial density, and other physical and chemical parameters. Both the concentration of TSS and the concentration of particle-associated microorganism determine how much UV radiation ultimately reaches the target organism. The higher these concentrations, the lower the UV radiation absorbed by the organisms. UV disinfection can be used in plants of various sizes that provide secondary or advanced levels of treatment.

The proper O&M of a UV disinfection system ensures that sufficient UV radiation is transmitted to the organisms to render them sterile. All surfaces between the UV radiation and the target organism must be clean, and the ballasts, lamps, and reactors must be functioning at peak efficiency. Inadequate cleaning is one of the most common causes of a UV system's ineffectiveness. The quartz sleeves or Teflon tubes need to be cleaned regularly by mechanical wipes, ultrasonics, or chemicals. The cleaning frequency is very site-specific; some systems need to be cleaned more often than others.

Chemical cleaning is most commonly done with citric acid. Other cleaning agents include mild vinegar solutions and sodium hydrosulfite. A combination of cleaning agents should be tested to find the agent most suitable for the wastewater characteristics without producing harmful or toxic by-products. Noncontact reactor systems are most effectively cleaned by using sodium hydrosulfite.

Any UV disinfection should be pilot-tested prior to full-scale platform to ensure that it will meet discharge permit requirements for a particular site. The average lamp life ranges from 8,760 to 14,000 working hours, and the lamps are usually replaced after 12,000 hours of use. Operating procedures should be set to reduce the on/off cycles of the lamps since efficacy is reduced with repeated cycles. The ballast must be compatible with the lamps and should be ventilated to protect it from excessive heating, which may shorten its life or even result in fires. Although the life cycle of ballasts is ~10 to 15 years, they are usually replaced every 10 years. Quartz sleeves will last about 5–8 years but are generally replaced every 5 years (USEPA, 1999).

Ozonation

Ozone is a strong oxidizing gas that reacts with most organic and many inorganic molecules. It is produced when oxygen molecules separate, collide with other oxygen atoms, and form a molecule consisting of three oxygen atoms. For high-quality effluents, ozone is a very effective disinfectant. Current regulations for domestic treatment systems limit the use of ozonation to filtered effluents unless the system's effectiveness can be demonstrated prior to installation.

Note: Effluent quality is the key performance factor for ozonation.

For ozonation of wastewater, the facility must have the capability to generate pure oxygen along with an ozone generator. A contact tank with ≥10-minute contact time at design average daily flow is required. Off-gas monitoring for process control is also required. In addition, safety equipment capable of monitoring ozone in the atmosphere and a ventilation system to prevent ozone levels exceeding 0.1 ppm is required.

The actual operation of the ozonation process consists of monitoring and adjusting the ozone generator and monitoring the control system to maintain the required ozone concentration in the off-gas. The process must also be evaluated periodically using biological testing to assess its effectiveness.

Note: Ozone is an extremely toxic substance. Concentrations in the air should not exceed 0.1 ppm. It also has the potential to create an explosive atmosphere. Sufficient ventilation and purging capabilities should be provided.

Note: Ozone has certain advantages over chlorine for disinfection of wastewater: (1) Ozone increases DO in the effluent; (2) ozone has a briefer contact time; (3) ozone has no undesirable effects on marine organisms; and (4) ozone decreases turbidity and odor.

Advantages and Disadvantages

Advantages:

- Ozone is more effective than chlorine in destroying viruses and bacteria.
- The ozonation process utilizes a short contact time (~10 to 30 minutes).
- There are no harmful residuals that need to be removed after ozonation because ozone decomposes rapidly.
- After ozonation, there is no regrowth of microorganisms, except for those protected by the particulates in the wastewater stream.
- Ozone is generated onsite, and thus, there are fewer safety problems associated with shipping and handling.
- Ozonation elevates the dissolved oxygen (DO) concentration of the effluent. The increase in DO can eliminate the need for reaeration and also raise the level of DO in the receiving stream.

Disadvantages:

- Low dosage may not effectively inactivate some viruses, spores, and cysts.
- Ozonation is a more complex technology than is chlorine or UV disinfection, requiring complicated equipment and efficient contacting systems.
- Ozone is very reactive and corrosive, thus requiring corrosion-resistant material such as stainless steel.
- Ozonation is not economical for wastewater with high levels of suspended soils (SS), biochemical oxygen demand (BOD), chemical oxygen demand, or total organic carbon.
- Ozone is extremely irritating and possibly toxic, so off-gases from the contactor must be destroyed to prevent worker exposure.
- The cost of treatment can be relatively high in capital and in power intensiveness.

Applicability

Ozone disinfection is generally used at medium to large-sized plants after at least secondary treatment. In addition to disinfection, another common use for ozone in wastewater treatment is odor control. Ozone disinfection is the least used method in the U.S. although this technology has been widely accepted in Europe for decades. Ozone treatment has the ability to achieve higher levels of disinfection than either chlorine or UV; however, the capital costs as well as maintenance expenditures are not competitive with available alternatives. Ozone is therefore used only sparingly, primarily in special cases where alternatives are not effective (USEPA, 2000).

Operation and Maintenance

Ozone generation uses a significant amount of electrical power. Thus, constant attention must be given to the system to ensure that power is optimized for controlled disinfection performance.

There must be no leaking connections in or surrounding the ozone generator. The operator must on a regular basis monitor the appropriate subunits to ensure that they are not overheated. Therefore, the operator must check for leaks routinely, since a very small leak can cause unacceptable ambient ozone concentrations. The ozone monitoring equipment must be tested and calibrated as recommended by the equipment manufacturer.

Like oxygen, ozone has limited solubility and decomposes more rapidly in water than in the air. This factor, along with ozone reactivity, requires that the ozone contactor be well covered and that the ozone diffuses into the wastewater as effectively as possible. Ozone in gaseous form is explosive once it reaches a concentration of 240 g/m^3. Since most ozonation systems never exceed a gaseous ozone concentration of 50–200 g/m^3, this is generally not a problem. However, ozone in gaseous form will remain hazardous for a significant amount of time thus, extreme caution is needed when operating the ozone gas systems.

It is important that the ozone generator, distribution, contracting, off-gas, and ozone destructor inlet piping be purged before opening the various systems or subsystems. When entering the ozone contactor, personnel must recognize the potential for oxygen deficiencies or trapped ozone gas in spite of best efforts to purge the system. The operator should be aware of all emergency operating procedures required if a problem occurs. All safety equipment should be available for operators to use in case of an emergency. Key O&M parameters include:

- Clean feed gas with a dew point of $-60°C$ ($-76°F$), or lower, must be delivered to the ozone generator. If the supply gas is moist, the reaction of the ozone and the moisture will yield a very corrosive condensate on the inside of the ozonator. The output of the generator could be lowered by the formation of nitrogen oxides (such as nitric acid).
- Maintain the required flow of generator coolant (air, water, or other liquid).
- Lubricate the compressor or blower in accordance with the manufacturer's specifications. Ensure that all compressor sealing gaskets are in good condition.

- Operate the ozone generator within its design parameters. Regulatory inspect and clean the ozonator, air supply, and dielectric assemblies, and monitor the temperature of the ozone generator.
- Monitor the ozone gas-feed and distribution system to ensure that the necessary volume comes into sufficient contact with the wastewater.
- Maintain ambient levels of ozone below the limits of applicable safety regulations.

BROMINE CHLORIDE

Bromine chloride is a mixture of bromine and chlorine. It forms hydrocarbons and hydrochloric acid when mixed with water. Bromine chloride is an excellent disinfectant that reacts quickly and normally does not produce any long-term residuals.

Note: Bromine chloride is an extremely corrosive compound in the presence of low concentrations of moisture.

The reactions occurring when bromine chloride is added to the wastewater are similar to those occurring when chlorine is added. The major difference is the production of bromamine compounds rather than chloramines. The bromamine compounds are excellent disinfectants but are less stable and dissipate quickly. In most cases, the bromamines decay into other, less toxic compounds rapidly and are undetectable in the plant effluent. The factors that affect performance are similar to those affecting the performance of the chlorine disinfection process. Effluent quality, contact time, etc. have a direct impact on the performance of the process.

NO DISINFECTION

In a very limited number of cases, treated wastewater discharges without disinfection is permitted. These are approved on a case-by-case basis. Each request must be evaluated based upon the point of discharge, the quality of the discharge, the potential for human contact, and many other factors.

ADVANCED WASTEWATER TREATMENT

Advanced wastewater treatment is defined as the method(s) and/or process (es) that remove more contaminants (suspended and dissolved substances) from wastewater than are taken out by conventional biological treatment. Put another way, advanced wastewater treatment is the application of a process or system that follows secondary treatment or that includes phosphorus removal or nitrification in conventional secondary treatment.

Advanced wastewater treatment is used to augment conventional secondary treatment because secondary treatment typically removes only between 85% and 95% of the biochemical oxygen demand (BOD) and total suspended solids (TSS) in raw sanitary sewage. Generally, this leaves 30 mg/L or less of BOD and TSS in the secondary effluent. To meet stringent water-quality standards, this level of BOD and TSS in secondary effluent may not prevent violation of water-quality standards—the plant may not make permit. Thus, advanced wastewater treatment is often used to remove additional pollutants from treated wastewater.

In addition to meeting or exceeding the requirements of water-quality standards, treatment facilities use advanced wastewater treatment for other reasons as well. For example, sometimes, conventional secondary wastewater treatment is not sufficient to protect the aquatic environment. In a stream, for example, when periodic flow events occur, the stream may not provide the amount of dilution of effluent needed to maintain the necessary dissolved oxygen (DO) levels for aquatic organism survival.

Secondary treatment has other limitations. It does not significantly reduce the effluent concentration of nitrogen and phosphorus (important plant nutrients) in sewage. An over-abundance of these nutrients can over-stimulate plant and algae growth such that they create water quality problems.

For example, if discharged into lakes, these nutrients contribute to algal blooms and accelerated Eutrophication (lake aging). Also, the nitrogen in the sewage effluent may be present mostly in the form of ammonia compounds. If in high enough concentration, ammonia compounds are toxic to aquatic organisms. Yet another problem with these compounds is that they exert a *nitrogenous* oxygen demand in the receiving water, as they convert to nitrates. This process is called nitrification.

Note: The term *tertiary treatment* is commonly used as a synonym for advanced wastewater treatment. However, these two terms do not have precisely the same meaning. Tertiary suggests a third step that is applied after primary and secondary treatment.

Advanced wastewater treatment can remove more than 99% of the pollutants from raw sewage and can produce an effluent of almost potable (drinking) water quality. However, obviously, advanced treatment is not cost-free. The cost of advanced treatment, for operation and maintenance as well as for retrofit of present conventional processes, is very high (sometimes doubling the cost of secondary treatment). Therefore, a plan to install advanced treatment technology calls for careful study—the benefit-to-cost ratio is not always big enough to justify the additional expense.

Even considering the expense, the application of some form of advanced treatment is not uncommon. These treatment processes can be physical, chemical, or biological. The specific process used is based on the purpose of the treatment and the quality of the effluent desired.

CHEMICAL TREATMENT

The purpose of chemical treatment is to remove:

- Biochemical oxygen demand (BOD)
- Total suspended solids (TSS)
- Phosphorus
- Heavy metals
- Other substances that can be chemically converted to a settleable solid

Chemical treatment is often accomplished as an "add-on" to existing treatment systems or by means of separate facilities specifically designed for chemical addition. In each case, the basic process necessary to achieve the desired results remains the same:

- Chemicals are thoroughly mixed with the wastewater.
- The chemical reactions that occur form solids (coagulation).
- The solids are mixed to increase particle size (flocculation).
- Settling and/or filtration (separation) then remove the solids.

The specific chemical used depends on the pollutant to be removed and the characteristics of the wastewater. Chemicals may include the following:

- Lime
- Alum (aluminum sulfate)
- Aluminum salts
- Ferric or ferrous salts
- Polymers
- Bioadditives

MICROSCREENING

Microscreening (also called *microstraining*) is an advanced treatment process used to reduce suspended solids. The microscreens are composed of specially woven steel wire fabric mounted around the perimeter of a large revolving drum. The steel wire cloth acts as a fine screen, with openings as small as 20 µm (or millionths of a meter)—small enough to remove microscopic organisms and debris. The rotating drum is partially submerged in the secondary effluent, which must flow into the drum and then outward through the microscreen. As the drum rotates, captured solids are carried to the top where a high-velocity water spray flushes them into a hopper or backwash tray mounted on the hollow axle of the drum. Backwash solids are recycled to plant influent for treatment. These units have found the greatest application in the treatment of industrial waters and final polishing filtration of wastewater effluents. Expected performance for suspended solids removal is 95%–99%, but the typical suspended-solids removal achieved with these units is about 55%. The normal range is from 10% to 80%.

According to Metcalf and Eddy (2003), the functional design of the microscreen unit involves the following considerations: (1) The characterization of the suspended solids with respect to the concentration and degree of flocculation; (2) the selection of unit design parameter values that will not only ensure capacity to meet maximum hydraulic loadings with critical solids characteristics but also provide desired design performance over the expected range of hydraulic and solids loadings; (3) the provision of backwash and cleaning facilities to maintain the capacity of the screen.

FILTRATION

The purpose of *filtration* processes used in advanced treatment is to remove suspended solids. The specific operations associated with a filtration system are dependent on the equipment used. In operation, wastewater flows to a filter (gravity or pressurized). The filter contains single, dual, or multimedia. Wastewater flows through the media, which removes solids. The solids remain in the filter. Backwashing the filter as needed removes trapped solids. Backwash solids are returned to the plant for treatment. Processes typically remove 95%–99% of the suspended matter.

MEMBRANE BIOREACTORS

The use of microfiltration membrane bioreactors (MBRs), a technology that has become increasingly used in the past 10 years, overcomes many of the limitations of conventional systems. These systems have the advantage of combining a suspended growth biological reactor with solids removal via filtration. The membranes can be designed for and operated in small spaces and with high removal efficiency of contaminants such as nitrogen, phosphorus, bacteria, biochemical oxygen demand, and total suspended solids. The membrane filtration system in effect can replace the secondary clarifier and sand filters in a typical activated sludge treatment system. Membrane filtration allows a higher biomass concentration to be maintained, thereby allowing smaller bioreactors to be used (USEPA, 2007a).

For new installations, the use of MBR systems allows for higher wastewater flow or improved treatment performance in a smaller space than a conventional design i.e., a facility using secondary clarifiers and sand filters. Historically, membranes have been used for smaller-flow systems due to the high capital cost of the equipment and high operation and maintenance (O&M) costs. Today, however, they are receiving increased use in larger systems. MBR systems are also well suited for some industrial and commercial applications. The high-quality effluent produced by MBRs makes them particularly applicable to reuse applications for surface water discharge applications requiring extensive nutrient (nitrogen and phosphorus) removal (USEPA, 2007a).

Advantages

The advantages of MBR systems over conventional biological system include better effluent quality, smaller space requirements, and ease of automation. Specifically, MBRs operate at higher volumetric loading rates which result in lower hydraulic retention times. The low retention times mean that less space is required compared to a conventional system. MBRs have often been operated with longer solids residence times (SRTs), which results in lower sludge production; but this is not a requirement, and more conventional SRTs have been used (Crawford et al., 2000). The effluent from MBRs contains low concentrations of bacteria, total suspended solids (TSS), biochemical oxygen demand (BOD), and phosphorus. This facilitates high-level disinfection. Effluents are readily discharged to surface streams or can be sold for reuse, such as irrigation (USEPA, 2007a).

Disadvantages

The primary disadvantage of MBR systems is the typically higher capital and operating costs than the conventional system for the same throughput. O&M costs include membrane cleaning and fouling control, and eventual membrane replacement. Energy costs are also higher because of the need for air scouring to control bacterial growth on the membranes. In addition, the waste sludge from such a system might have a low settling rate, resulting in the need for chemicals to produce biosolids acceptable for disposal (Hermanowicz et al., 2006). Fleischer et al. (2005) have demonstrated that waste sludges from MBRs can be processed using standard technologies used for activated sludge processes.

BIOLOGICAL NITRIFICATION

Biological *nitrification* is the first basic step of *biological nitrification-denitrification*.

In nitrification, the secondary effluent is introduced into another aeration tank, trickling filter, or biodisc. Because most of the carbonaceous BOD has already been removed, the microorganisms that drive in this advanced step are the nitrifying bacteria *Nitrosomonas* and *Nitrobacter*. In nitrification, the ammonia nitrogen is converted to nitrate nitrogen, producing a *nitrified effluent*. At this point, the nitrogen has not actually been removed, only converted to a form that is not toxic to aquatic life and that does not cause an additional oxygen demand. The nitrification process can be limited (performance affected) by alkalinity (requires 7.3 parts alkalinity to 1.0-part ammonia nitrogen); pH; dissolved oxygen availability; toxicity (ammonia or other toxic materials); and process mean cell residence time (sludge retention time). As a general rule, biological nitrification is more effective and achieves higher levels of removal during the warmer times of the year.

BIOLOGICAL DENITRIFICATION

Biological denitrification removes nitrogen from the wastewater. When bacteria come in contact with a nitrified element in the absence of oxygen, they reduce the nitrates to nitrogen gas, which escapes the wastewater. The denitrification process can be done in either an anoxic-activated sludge system (suspended growth) or in a column system (fixed growth). The denitrification process can remove up to 85% or more of nitrogen. After effective biological treatment, little oxygen-demanding material is left in the wastewater when it reaches the denitrification process. The denitrification reaction will only occur if an oxygen demand source exists when no dissolved oxygen is present in the wastewater. An oxygen demand source is usually added to reduce the nitrates quickly. The most common demand source added is soluble BOD or methanol. Approximately 3 mg/L of methanol is added for every 1 mg/L of nitrate-nitrogen. Suspended growth denitrification reactors are mixed mechanically, but only enough to keep the biomass from settling without adding unwanted oxygen. Submerged filters of different types of media may also be used to provide denitrification. A fine media downflow filter is sometimes used to provide both denitrification and effluent filtration. A fluidized sand bed where wastewater flows upward through a media of sand or activated carbon at a rate to fluidize the bed may also be used. Denitrification bacteria grow on the media.

CARBON ADSORPTION

The main purpose of *carbon adsorption* used in advanced treatment processes is the removal of refractory organic compounds (non-BOD_5) and soluble organic material that are difficult to eliminate by biological or physical/chemical treatment. In the carbon adsorption process, wastewater passes through a container filled either with carbon powder or carbon slurry. Organics adsorb onto the carbon (i.e., organic molecules are attracted to the activated carbon surface and are held there) with sufficient contact time. A carbon system usually has several columns or basins used as contactors. Most contact chambers are either open concrete gravity-type systems or steel pressure containers applicable to either upflow or downflow operation. With use, carbon loses its adsorptive capacity. The carbon must then be regenerated or replaced with fresh carbon. As head loss develops in carbon contactors, they are backwashed with clean effluent in much the same way the effluent filters are backwashed. Carbon used for adsorption may be in a granular form or in a powdered form.

Note: Powdered carbon is too fine for use in columns and is usually added to the wastewater, then later removed by coagulation, and settling.

LAND APPLICATION

The application of secondary effluent onto a land surface can provide an effective alternative to the expensive and complicated advanced treatment methods discussed previously and the biological nutrient removal (BNR) system discussed later. A high-quality polished effluent (i.e., effluent with high levels of TSS, BOD, phosphorus, and nitrogen compounds as well as refractory organics are reduced) can be obtained by the natural processes that occur as the effluent flows over the vegetated ground surface and percolates through the soil. Limitations are involved with land application of wastewater effluent. For example, the process needs large land areas. Soil type and climate are also critical factors in controlling the design and feasibility of a land treatment process.

Three basic types or modes of land application or treatment are commonly used: Irrigation (slow rate), overland flow, and infiltration-percolation (rapid rate). The basic objectives of these types of land applications and the conditions under which they can function vary. In *irrigation* (also called slow rate), wastewater is sprayed or applied (usually by ridge and-furrow surface spreading or by sprinkler systems) to the surface of the land. Wastewater enters the soil. Crops growing on the irrigation area utilize available nutrients. Soil organisms stabilize organic content of the flow. Water returns to hydrologic (water) cycle through evaporation or by entering the surface water or groundwater.

The irrigation land application method provides the best results (compared with the other two types of land application systems) with respect to advanced treatment levels of pollutant removal. Not only are suspended solids and BOD significantly reduced by filtration of the wastewater, but also biological oxidation of the organics in the top few inches of soil occurs. Nitrogen is removed primarily by crop uptake, and phosphorus is removed by adsorption within the soil. Expected performance levels for irrigation include:

- BOD_5—98%
- Suspended solids—98%
- Nitrogen—85%
- Phosphorus—95%
- Metals—95%

The overland flow application method utilizes physical, chemical, and biological processes as the wastewater flows in a thin film down the relatively impermeable surface. In the process, wastewater sprayed over sloped terraces flows slowly over the surface. Soil and vegetation remove suspended solids, nutrients, and organics. A small portion of the wastewater evaporates. The remainder flows to collection channels. Collected effluent is discharged to surface waters. Expected performance levels for overflow flow include:

- BOD_5—92%
- Suspended solids—92%
- Nitrogen—70%–90%
- Phosphorus—40%–80%
- Metals—50%

In the infiltration-percolation (rapid rate) land application process, wastewater is sprayed/pumped to spreading basins (a.k.a. recharge basins or large ponds). Some wastewater evaporates. The remainder percolates/infiltrates into the soil. Solids are removed by filtration. Water recharges the groundwater system. Most of the effluent percolates to the groundwater; very little of it is absorbed by vegetation. The filtering and adsorption action of the soil removes most of the BOD, TSS, and phosphorous from the effluent; however, nitrogen removal is relatively poor. Expected performance levels for infiltration-percolation include:

- BOD_5—85%–99%
- Suspended solids—98%
- Nitrogen—0%–50%
- Phosphorus—60%–95%
- Metals—50%–95%

BIOLOGICAL NUTRIENT REMOVAL (BNR)

Nitrogen and phosphorus are the primary causes of cultural Eutrophication (i.e., nutrient enrichment due to human activities) in surface waters. The most recognizable manifestations of this Eutrophication are algal blooms that occur during the summer. Chronic symptoms of over-enrichment include low dissolved oxygen, fish kills, murky water, and depletion of desirable flora and fauna. In addition, the increase in algae and turbidity increases the need to chlorinate drinking water, which, in turn, leads to higher levels of disinfection by-products that have been shown to increase the risk of cancer (USEPA, 2007c). Excessive amounts of nutrients can also stimulate the activity of microbes, such as *Pfisteria*, which may be harmful to human health (USEPA, 2007d).

Approximately 25% of all water body impairments are due to nutrient-related causes (e.g., nutrients, oxygen depletion, algal growth, ammonia, harmful algal blooms, biological integrity, and turbidity) (USEPA, 2007d). In efforts to reduce the number of nutrient impairments, many point source discharges have received more stringent effluent limits for nitrogen and phosphorus. To achieve these new, lower effluent limits, facilities have begun to look beyond traditional treatment technologies.

Recent experience has reinforced the concept that biological nutrient removal (BNR) systems are reliable and effective in removing nitrogen and phosphorus. The process is based upon the principle that, under specific conditions, microorganisms will remove more phosphorus and nitrogen than is required for biological activity; thus, treatment can be accomplished without the use of chemicals. Not having to use and therefore having to purchase chemicals to remove nitrogen and phosphorus potentially has numerous cost-benefit implications. In addition, because chemicals are not required to be used, chemical waste products are not produced, reducing the need to handle and dispose of waste. Several patented processes are available for this purpose. Performance depends on the biological activity and the process employed.

There are a number of BNR process configurations available. Some BNR systems are designed to remove only TN or TP, while others remove both. The configuration most appropriate for any particular system depends on the target effluent quality, operator experience, influent quality, and existing treatment processes, if retrofitting an existing facility. BNR configuration varies based on the sequencing of environmental conditions (i.e., aerobic, anaerobic, and anoxic) and timing (USEPA, 2007c). Common BNR system configurations include:

- **Modified Ludzack-Ettinger (MLE) process**: continuous-flow suspended-growth process with an initial anoxic stage followed by an aerobic stage; used to remove
- **A²/O process**: MLE process preceded by an initial anaerobic stage; used to remove both TN and TP
- **Step feed process**: alternating anoxic and aerobic stages; however, influent flow is split to several feed locations and the recycle sludge stream is sent to the beginning of the process; used to remove TN
- **Bardenpho process (four-stage)**: continuous-flow suspended-growth process with alternating anoxic/aerobic/anoxic/aerobic stages; used to remove TN
- **Modified bardenpho process**: Bardenpho process with the addition of an initial anaerobic zone; used to remove both TN and TP
- **Sequencing Batch Reactor (SBR) process**: suspended-growth batch process sequenced to simulate the four-stage process; used to remove TN (TP removal is inconsistent)
- **Modified University of Cape town (UCT) process**: A²/O Process with a second anoxic stage where the internal nitrate recycle is returned; used to remove both TN and TP
- **Rotating Biological Contactor (RBC) process**: continuous-flow process using RBCs with sequential anoxic/aerobic stages; used to remove TN
- **Oxidation ditch**: continuous-flow process using looped channels to create time-sequenced anoxic, aerobic, and anaerobic zones; used to remove both TN and TP.

Although the exact configurations of each system differ, BNR systems designed to remove TN must have an aerobic zone for nitrification and anoxic zone for denitrification, and BNR systems designed to remove TP must have an anaerobic zone free of dissolved oxygen and nitrate. Often, sand, or other media filtration is used as a polishing step to remove particulate matter when low TN and TP effluent concentrations are required.

Choosing which system is most appropriate for a particular facility primarily depends on the target effluent concentrations, and whether the facility will be constructed as new or retrofit with BNR to achieve more stringent effluent limits. New plants have more flexibility and options when deciding which BNR configuration to implement because they are not constrained by existing treatment units and sludge handling procedures.

Retrofitting an existing plant with BNR capabilities should involve consideration of the following factors (Park, 2012):

- Aeration basin size and configuration
- Clarifier capacity
- Type of aeration system
- Sludge processing units
- Operator skills

The aeration basin size and configuration dictate which BNR configurations are the most economical and feasible. Available excess capacity reduces the need for additional basins and may allow for a more configuration (e.g., five-stage Bardenpho versus four-state Bardenpho configuration). The need for additional basins could result in the need for more land if the space needed is not available. If land is not available, another BNR process configuration may have to be considered.

Clarifier capacity influences the return activated sludge (RAS) rate and effluent suspended solids, which in turn, affects effluent TN and TP levels. If the existing facility configuration does not allow for a preanoxic zone so that nitrates can be removed prior to the anaerobic zone, then the clarifier should be modified to have sludge blanket just deep enough to prevent the release of phosphorus to the liquid.

The aeration system will most likely need to be modified to accommodate an anaerobic zone and to reduce the DO concentration in the return sludge. Such modifications could be as simple as removing aeration equipment from the zone designated for anaerobic conditions or changing the type of pump used for the recycled sludge stream (to avoid introduction oxygen).

The manner in which sludge is processed at a facility is important in designing nutrient removal systems. Sludge is recycled within the process to provide the organisms necessary for the TN and TP removal mechanism to occur. The content and volume of sludge recycled directly impacts the system's performance. Thus, sludge handling processes may be modified to achieve optimal TN and TP removal efficiencies. For example, some polymers in sludge dewatering could inhibit nitrification when recycled. Also, because aerobic digestion of sludge process nitrates, denitrification and phosphorus uptake rates may be lowered when the sludge is recycled (USEPA, 2007c).

Operators should be able to adjust the process to compensate for constantly varying conditions. BNR processes are very sensitive to influent conditions which are influenced by weather events, sludge processing, and other treatment processes (e.g., recycling after filter backwashing). Therefore, operator skills and training are essential for achieving target TN and TP effluent concentrations (USEPA, 2007c).

ENHANCED BIOLOGICAL NUTRIENT REMOVAL (EBNR)

Removing phosphorus from wastewater in secondary treatment processes has evolved into innovative *enhanced biological nutrient removal* (EBNR) technologies. An ENBR treatment process promotes the production of phosphorus-accumulating organisms which utilize more phosphorus in their metabolic processes than a conventional secondary biological treatment process (USEPA, 2007b). The average total phosphorus concentrations in raw domestic wastewater are usually between 6 to 8 mg/L and the total phosphorus concentration in municipal wastewater after conventional secondary treatment is routinely reduced to 3 or 4 mg/L. Whereas, EBNR incorporated into the secondary treatment system can often reduce total phosphorus concentrations to 0.3 mg/L and less. Facilities using EBNR significantly reduced the amount of phosphorus to be removed through the subsequent chemical addition and tertiary filtration process. This improved the efficiency of the tertiary process and significantly reduced the costs of chemicals used to remove phosphorus. Facilities using EBNR reported that their chemical dosing was cut in half after EBNR was installed to remove phosphorus (USEPA, 2007b).

Treatment provided by these WWTPs also removes other pollutants which commonly affect water quality to very low levels (USEPA, 2007b). Biochemical oxygen demand (BOD) and total suspended solids are routinely <2 mg/L and fecal coliform bacteria <10 fcu/100 mL. Turbidity of the final effluent is very low which allows for effective disinfection using ultraviolet light, rather than chlorination. Recent studies report that wastewater treatment plants using EBNR also significantly reduced the amount of pharmaceuticals and personal healthcare products from municipal wastewater, as compared to the removal accomplished by conventional secondary treatment.

WASTEWATER SOLIDS (SLUDGE/BIOSOLIDS) HANDLING

The wastewater treatment unit processes described to this point remove solids and BOD from the wastestream before the liquid effluent is discharged to its receiving waters. What remains to be disposed of is a mixture of solids and wastes, called *process residuals*—more commonly referred to as *sludge* or *biosolids*.

Note: *Sludge* is the commonly accepted name for wastewater solids. However, if wastewater sludge is used for beneficial reuse (e.g., as a soil amendment or fertilizer), it is commonly called *biosolids*.

The most costly and complex aspect of wastewater treatment can be the collection, processing, and disposal of sludge. This is the case because the quantity of sludge produced may be as high as 2% of the original volume of wastewater, depending somewhat on the treatment process being used.

Because sludge can be as much as 97% water content, and because the cost of disposal will be related to the volume of sludge being processed, one of the primary purposes or goals (along with stabilizing it so it is no longer objectionable or environmentally damaging) of sludge treatment is to separate as much of the water from the solids as possible. Sludge treatment methods may be designed to accomplish both of these purposes.

Note: Sludge treatment methods are generally divided into three major categories: thickening, stabilization, and dewatering. Many of these processes include complex sludge treatment methods (i.e., heat treatment, vacuum filtration, incineration, and others).

BACKGROUND INFORMATION ON SLUDGE

When we speak of *sludge* or *biosolids*, we are speaking of the same substance or material; each is defined as the suspended solids removed from wastewater during sedimentation, and then concentrated for further treatment and disposal or reuse. The difference between the terms *sludge* and *biosolids* is determined by the way they are managed.

Note: The task of disposing, treating, or reusing wastewater solids is called *sludge* or *biosolids management*.)

Sludge is typically seen as wastewater solids that are "disposed" of. Biosolids is the same substance managed for reuse—commonly called beneficial reuse (e.g., for land application as a soil amendment, such as biosolids compost). Note that even as wastewater treatment standards have become more stringent because of increasing environmental regulations, so has the volume of wastewater sludge increased. Also note that before sludge can be disposed of or reused, it requires some form of treatment to reduce its volume, to stabilize it, and to inactivate pathogenic organisms.

Sludge initially forms as a 3%–7% suspension of solids, and with each person typically generating about 4 gallons of sludge per week, the total quantity generated each day, week, month, and year is significant. Because of the volume and nature of the material, sludge management is a major factor in the design and operation of all water pollution control plants.

Note: Wastewater solids treatment, handling, and disposal account for more than half of the total costs in a typical secondary treatment plant.

SOURCES OF SLUDGE

Wastewater sludge is generated in primary, secondary, and chemical treatment processes. In primary treatment, the solids that float or settle are removed. The floatable material makes up a portion of the solid waste known as scum. Scum is not normally considered sludge; however, it should be disposed of in an environmentally sound way. The settleable material that collects on the bottom of the clarifier is known as *primary sludge*. Primary sludge can also be referred to as raw sludge because it has not undergone decomposition. Raw primary sludge from a typical domestic facility is quite objectionable and has a high percentage of water, two characteristics that make handling difficult.

Solids not removed in the primary clarifier are carried out of the primary unit. These solids are known as *colloidal suspended solids*. The secondary treatment system (i.e., trickling filter, activated sludge, etc.) is designed to change those colloidal solids into settleable solids that can be removed. Once in the settleable form, these solids are removed in the secondary clarifier. The sludge at the bottom of the secondary clarifier is called *secondary sludge*. Secondary sludges are light and fluffy and more difficult to process than primary sludges—in short, secondary sludges do not de-water well.

The addition of chemicals and various organic and inorganic substances prior to sedimentation and clarification may increase the solids capture and reduce the amount of solids lost in the effluent. This *chemical addition* results in the formation of heavier solids, which trap the colloidal solids or convert dissolved solids to settleable solids. The resultant solids are known as *chemical sludges*. As chemical usage increases, so does the quantity of sludge that must be handled and disposed of. Chemical sludges can be very difficult to process; they do not de-water well and contain lower percentages of solids.

Sludge Characteristics

The composition and characteristics of sewage sludge vary widely and can change considerably with time. Notwithstanding these facts, the basic components of wastewater sludge remain the same. The only variations occur in the quantity of the various components as the type of sludge and the process from which it originated changes. The main component of all sludges is *water*. Prior to treatment, most sludge contains 95% to 99+% water (see Table 8.23). This high-water content makes sludge handling and processing extremely costly in terms of both money and time. Sludge handling may represent up to 40% of the capital cost and 50% of the operation cost of a treatment plant. As a result, the importance of optimum design for handling and disposal of sludge cannot be overemphasized. The water content of the sludge is present in a number of different forms. Some forms can be removed by several sludge treatment processes, thus allowing the same flexibility in choosing the optimum sludge treatment and disposal method.

The various forms of water and their approximate percentages for a typical activated sludge are shown in Table 8.24. The forms of water associated with sludges include:

- **Free water**: water that is not attached to sludge solids in any way. This can be removed by simple gravitational settling.
- **Floc water**: water that is trapped within the floc and travels with them. Its removal is possible by mechanical de-watering.
- **Capillary water**: water that adheres to the individual particles and can be squeezed out of shape and compacted.
- **Particle water**: water that is chemically bound to the individual particles and can't be removed without inclination.

From a public health view, the second and probably more important component of sludge is the *solids matter*. Representing from 1% to 8% of the total mixture, these solids are extremely unstable.

TABLE 8.23
Typical Water Content of Sludges

Water Treatment Process	% Moisture of Sludge	lb Water/lb Sludge Solids Generated
Primary sedimentation	95	19
	Trickling Filter	
Humus—low rate	93	13.3
Humus—high rate	97	32.3
Activated sludge	99	99

Source: USEPA (1978).

Water Pollution

TABLE 8.24
Distribution of Water in an Activated Sludge

Water Type	% Volume
Free water	75
Floc water	20
Capillary water	
Particle water	2.5
Solids	0.5
Total	100

Source: USEPA (1978).

Wastewater solids can be classified into two categories based on their origin—organic and inorganic. *Organic solids* in wastewater, simply put, are materials that are or were at one time alive and that will burn or volatilize at 550°C after 15 minutes in a muffle furnace. The percent organic material within sludge will determine how unstable it is.

The inorganic material within sludge will determine how stable it is. The *inorganic solids* are those solids that were never alive and will not burn or volatilize at 550°C after 15 minutes in a muffle furnace. Inorganic solids are generally not subject to breakdown by biological action and are considered stable. Certain inorganic solids, however, can create problems when related to the environment, for example, heavy metals such as copper, lead, zinc, mercury, and others. These can be extremely harmful if discharged.

Organic solids may be subject to biological decomposition in either an aerobic or anaerobic environment. Decomposition of organic matter (with its production of objectionable by-products) and the possibility of toxic organic solids within the sludge compound the problems of sludge disposal.

The pathogens in domestic sewage are primarily associated with insoluble solids. Primary wastewater treatment processes concentrate these solids into sewage sludge, so untreated or raw primary sewage sludges have higher quantities of pathogens than the incoming wastewater. Biological wastewater treatment processes such as lagoons, trickling filters, and activated sludge treatment may substantially reduce the number of pathogens in wastewater (USEPA, 1989). These processes may also reduce the number of pathogens in sewage sludge by creating adverse conditions for pathogen survival.

Nevertheless, the resulting biological sewage sludges may still contain sufficient levels of pathogens to pose a public health and environmental concern. Moreover, insects, birds rodents and domestic animals may transport sewage sludge and pathogens from sewage sludge to humans and to animals. Vectors are attracted to sewage sludge as food source, and the reduction of the attraction of vectors to sewage sludge to prevent the spread of pathogens is a focus of current regulations. Sludge-borne pathogens and vector attraction are discussed in the following section.

Sludge Pathogens and Vector Attraction

As discussed earlier, a pathogen is an organism capable of causing disease. Pathogens infect humans through several different pathways including ingestion, inhalation, and dermal contact. The infective dose, or the number of pathogenic organisms to which a human must be exposed to become infected, varies depending on the organism and on the health status of the exposed individual. Pathogens that propagate in the enteric or urinary system so humans and are discharged in feces or urine pose the greatest risk to public health with regard to the use and disposal of sewage sludge. Pathogens are also found in the urinary and enteric systems of other animals and may propagate in non-enteric settings.

The four major types of human pathogenic (disease-causing) organisms (bacteria, viruses, protozoa, and helminths) all may be present in domestic sewage. The actual species and quantity of pathogens present in the domestic sewage from a particular municipality (and the sewage sludge produced when treating the domestic sweater) depend on the health status of the local community and may vary substantially at different times. The level of pathogens presents in treated sewage sludge (biosolids) also depends on the reductions achieved by the wastewater and sewage sludge treatment processes.

If improperly treated sewage sludge were illegally applied to land or placed on a surface disposal site, humans and animals could be exposed to pathogens directly by coming into contact with sewage sludge, or indirectly by consuming drinking water or food contaminated by sewage sludge pathogens, insects, birds, rodents, and even farm workers could contribute to these exposure routes by transporting sewage sludge and sewage sludge pathogens away from the site. Potential routes of exposure include the following:

Direct Contact

- Touching the sewage sludge.
- Walking through an area—such as a filed, forest, or reclamation area—shortly after sewage sludge application.
- Handling soil from fields where sewage sludge has been applied.
- Inhaling microbes that become airborne (via aerosols, dust, etc.) during sewage sludge spreading or by strong winds, plowing or cultivating the soils after application.

Indirect Contact

- Consumption of pathogen-contaminated crops grown on sewage sludge-amended soil or of other food products that have been contaminated by contact with these crops or field workers, etc.
- Consumption of pathogen-contaminated milk or other food products from animals contaminated by grazing in pastures or fed crops grown on sewage sludge-amended fields.
- Ingestion of drinking water or recreational waters contaminated by runoff from nearby land application sites or by organisms from sewage sludge migrating into ground-water aquifers.
- Consumption of inadequately cooked or uncooked pathogen-contaminated fish from water contaminated by runoff from a nearby sewage sludge application site.
- Contact with sewage sludge or pathogens transported away from the land application or surface disposal site by rodents, insects, or other vectors, including grazing animals or pets.

One of the lesser impacts on public health can be from inhalation of airborne pathogens. Pathogens may become airborne via the spray of liquid biosolids from a splash plate or high-pressure hose, or in fine particulate dissemination as dewatered biosolids are applied or incorporated. While high-pressure spray applications may result in some aerosolization of pathogens, this type of equipment is generally used on large, remote sites such as forests, where the impact on the public is minimal. Fine particulates created by the application of dewatered biosolids or the incorporation of biosolids into the soil may cause very localized fine particulate/dusty conditions, but particles in dewatered biosolids are too large to travel far, and the fine particulates do not spread beyond the immediate area. The activity of applying and incorporating biosolids may create dusty conditions. However, the biosolids are moist materials and do not add to the dusty condition, and by the time biosolids have dried sufficiently to create fine particulates, the pathogens have been reduced (Yeager and Ward, 1981).

With regard to vector attraction reduction, it can be accomplished in two ways: by treating the sewage sludge to the point at which vectors will no longer be attracted to the sewage sludge and by placing a barrier between the sewage sludge and vectors.

Note: Before moving on to a discussion of the fundamentals of sludge treatment methods, it is important to begin by covering sludge pumping calculations. It is important to point out that it is difficult (if not impossible) to treat the sludge unless it is pumped into the specific sludge treatment process.

SLUDGE PUMPING CALCULATIONS

Environmental engineers who design wastewater unit processes and managers and operators who oversee and operate wastewater treatment facilities are often called upon to make various process control calculations. An important calculation involves sludge pumping. The sludge pumping calculations the operator may be required to make during plant operations are covered in this section.

Estimating Daily Sludge Production

The calculation for *estimation of the required sludge-pumping rate* provides a method to establish an initial pumping rate or to evaluate the adequacy of the current withdrawal rate:

$$\text{Est. pump rate} = \frac{(\text{Influent TSS conc.} - \text{Effluent TSS conc.}) \times \text{Flow} \times 8.34}{\% \text{ Solids in sludge} \times 8.34 \times 1,440 \text{ minutes/day}} \quad (8.61)$$

Example 8.44

Problem:

The sludge withdrawn from the primary settling tank contains 1.4% solids. The unit influent contains 285 mg/L TSS and the effluent contains 140 mg/L TSS. If the influent flow rate is 5.55 MGD, what is the estimated sludge withdrawal rate in gallons per minute (assuming the pump operates continuously)?

Solution:

$$\text{Sludge rate, gpm} = \frac{(285 \text{ mg/L} - 140 \text{ mg/L}) \times 5.55 \times 8.34}{0.014 \times 8.34 \times 1,440 \text{ minutes/day}} = 40 \text{ gpm}$$

Sludge Pumping Time

The *Sludge Pumping Time* is the total time the pump operates during a 24 period in minutes.

$$\text{Pump op. time} = \text{Time, (cycle/minutes)} \times \text{Frequency, (cycles/day)} \quad (8.62)$$

Note: The following information is used for examples 8.45–8.49.

Operating Time	15 minutes/cycle
Frequency	24 times/day
Pump Rate	120 gpm
Solids	3.70%
Volatile Matter	66%

Example 8.45

Problem:

What is the pump operating time?

Solution:

$$\text{Pump operating time} = 15 \text{ minutes/hour} \times 24 \text{ (cycles)/day} = 360 \text{ minutes/day}$$

Gallons of Sludge Pumped per Day

$$\text{Sludge, gpd} = \text{Operating time, minutes/day} \times \text{Pump rate, gpm} \tag{8.63}$$

Example 8.46

Problem:

What is the sludge pumped/Day in gallons?

Solution:

$$\text{Sludge, gpd} = 360 \text{ minutes/day} \times 120 \text{ gpm} = 43,200 \text{ gpd}$$

Pounds Sludge Pumped per Day

$$\text{Sludge, lb/day} = \text{Gallons of sludge pumped} \times 8.34 \text{ lb/gal} \tag{8.64}$$

Example 8.47

Problem:

What is the sludge pumped per day in gallons?

Solution:

$$\text{Sludge, lb/day} = 43,200 \text{ gal/day} \times 8.34 \text{ lb/gal} = 360,300 \text{ lbs/day}$$

Pounds Solids Pumped per Day

$$\text{Solids pumped, lb/day} = \text{Sludge pumped, gpd} \times \% \text{ Solids} \tag{8.65}$$

Example 8.48

Problem:

What are the solids pumped per day?

Solution:

$$\text{Solids pumped lb/day} = 360,300 \text{ lb/day} \times 0.0370 = 13,331 \text{ lb/day}$$

Pounds Volatile Matter (VM) Pumped per Day

$$\text{Vol. matter (lb/day)} = \text{Solids pumped, lb/day} \times \% \text{ Volatile matter} \tag{8.66}$$

Water Pollution

Example 8.49

Problem:

What is the volatile matter in pounds per day?

Solution:

$$\text{Volatile matter, lb/day} = 13{,}331\,\text{lb/day} \times 0.66 = 8{,}798\,\text{lb/day}$$

Note: If we wish to calculate the pounds of solids or the pounds of volatile solids removed per day, the individual equations demonstrated above can be combined into a single calculation.

Solids, lb/day = Pump time, min/cycle × Frequency, cycles/day × Rate, gpm × 8.34 lb/gal × Solids vol.

Matter, lb/day = Time, min/cycle × Frequency, cycles/day × Rate, gpm × 8.34 × % Solids × % V.M.

(8.67)

Sludge Production in Pounds/Million Gallons

A common method of expressing sludge production is in pounds of sludge per million gallons of wastewater treated.

$$\text{Sludge, lb/MG} = \frac{\text{Total sludge production, lb}}{\text{Total wastewater flow, MG}} \tag{8.68}$$

Example 8.50

Problem:

Records show that the plant has produced 85,000 gallons of sludge during the past 30 days. The average daily flow for this period was 1.2 MGD. What was the plant's sludge production in pounds per million gallons?

Solution:

$$\text{Sludge, lb/MG} = \frac{85{,}000\,\text{gallons} \times 8.34\,\text{lb/gallon}}{1.2\,\text{MGD} \times 30\,\text{days}} = 19{,}692\,\text{lb/MG}$$

Sludge Production in Wet Tons/Year

Sludge production can also be expressed in terms of the amount of sludge (water and solids) produced per year. This is normally expressed in wet tons per year.

$$\text{Sludge, wet tons/year} = \frac{\text{Sludge prod., lb/MG} \times \text{Ave. daily flow, MGD} \times 365\,\text{days/year}}{2{,}000\,\text{lb/ton}} \tag{8.69}$$

Example 8.51

Problem:

The plant is currently producing sludge at the rate of 16,500 lb/MG. The current average daily wastewater flow rate is 1.5 MGD. What will be the total amount of sludge produced per year in wet tons per year?

Solution:

$$\text{Sludge, wet tons/year} = \frac{16{,}500\,\text{lb/MG} \times 1.5\,\text{MGD} \times 365\,\text{days/year}}{2{,}000\,\text{lb/ton}}$$

$$= 4{,}517\,\text{wet tons/year}$$

Important Note: Release of wastewater solids without proper treatment could result in severe damage to the environment. Obviously, we must have a system to treat the volume of material removed as sludge throughout the system. Release without treatment would defeat the purpose of environmental protection. A design engineer can choose from many processes when developing sludge treatment systems. No matter what the system or combination of systems chosen, the ultimate purpose will be the same: the conversion of wastewater sludges into a form that can be handled economically and disposed of without damage to the environment or creating nuisance conditions. Leaving either condition unmet will require further treatment. The degree of treatment will generally depend on the proposed method of disposal. Sludge treatment processes can be classified into a number of major categories. In this handbook, we discuss the processes of thickening, digestion (or stabilization), de-watering, incineration, and land application. Each of these categories has then been further subdivided according to the specific processes that are used to accomplish sludge treatment. As mentioned, the importance of adequate, efficient sludge treatment cannot be overlooked when designing wastewater treatment facilities. The inadequacies of a sludge treatment system can severely affect a plant's overall performance capabilities. The inability to remove and process solids as fast as they accumulate in the process can lead to the discharge of large quantities of solids to receiving waters. Even with proper design and capabilities in place, no system can be effective unless it is properly operated. Proper operation requires proper engineering design and operator performance. Proper operator performance begins and ends with proper training.

SLUDGE THICKENING

The solids content of primary, activated, trickling filter, or even mixed sludge (i.e., primary plus activated sludge) varies considerably, depending on the characteristics of the sludge. Note that the sludge removal and pumping facilities and the method of operation also affect the solids content. *Sludge thickening* (or *concentration*) is a unit process used to increase the solids content of the sludge by removing a portion of the liquid fraction. By increasing the solids content, more economical treatment of the sludge can be affected. Sludge thickening processes include:

- Gravity Thickeners
- Flotation Thickeners
- Solids Concentrators

Gravity Thickening

Gravity thickening is most effective on primary sludge. In operation, solids are withdrawn from primary treatment (and sometimes-secondary treatment) and pumped into the thickener. The solids buildup in the thickener forms a solid blanket on the bottom. The weight of the blanket compresses the solids on the bottom and "squeezes" the water out. By adjusting the blanket thickness, the percent solids in the underflow (solids withdrawn from the bottom of the thickener) can be increased or decreased. The supernatant (clear water) which rises to the surface is returned to the wastewater flow for treatment. Daily operations of the thickening process include pumping, observation, sampling and testing, process control calculations, maintenance, and housekeeping.

Note: The equipment employed in thickening depends on the specific thickening processes used.

Equipment used for gravity thickening consists of a thickening tank, which is similar in design to the settling tank used in primary treatment. Generally, the tank is circular and provides equipment for continuous solids collection. The collector mechanism uses heavier construction than that in a settling tank because the solids being moved are more concentrated. The gravity thickener pumping facilities (i.e., pump and flow measurement) are used for the withdrawal of thickened solids.

Solids concentrations achieved by gravity thickening are typically 8%–10% solids from primary underflow, 2%–4% solids from waste-activated sludge, 7%–9% solids from trickling filter residuals and 45–9% from combined primary and secondary residuals. The performance of gravity thickening processes depends on various factors, including:

- Type of sludge
- Condition of influent sludge
- Temperature
- Blanket depth
- Solids loading
- Hydraulic loading
- Solids retention time
- Hydraulic detention time

Flotation Thickening

Flotation thickening is used most efficiently for waste sludges from the suspended growth biological treatment process, such as the activated sludge process. In operation, recycled water from the flotation thickener is aerated under pressure. During this time the water absorbs more air than it would under normal pressure. The recycled flow together with chemical additives (if used) is mixed with the flow. When the mixture enters the flotation thickener, the excess air is released in the form of fine bubbles. These bubbles become attached to the solids and lift them toward the surface. The accumulation of solids on the surface is called the *float cake*. As more solids are added to the bottom of the float cake it becomes thicker and water drains from the upper levels of the cake. The solids are then moved up an inclined plane by a scraper and discharged. The supernatant leaves the tank below the surface of the float solids and is recycled or returned to the wastestream for treatment. Typically, flotation thickener performance is 3%–5% solids for waste-activated sludge with polymer addition and 2%–4% solids without polymer addition.

The flotation thickening process requires pressurized air, a vessel for mixing the air with all or part of the process residual flow, a tank for the flotation process to occur, solids collector mechanisms to remove the float cake (solids) from the top of the tank and accumulated heavy solids from the bottom of the tank. Since the process normally requires chemicals to be added to improve separation, chemical mixing equipment, storage tanks, and metering equipment to dispense the chemicals at the desired dose are required. The performance of the dissolved air-thickening process depends on various factors:

- Bubble size
- Solids loading
- Sludge characteristics
- Chemical selection
- Chemical dose

Solids Concentrators

Solids concentrators (belt thickeners) usually consist of a mixing tank, chemical storage and metering equipment and a moving porous belt. In operation, the process residual flow is chemically treated and then spread evenly over the surface of the moving porous belt. As the flow is carried down the belt (similar to a conveyor belt) the solids are mechanically turned or agitated and water drains through the belt. This process is primarily used in facilities where space is limited.

Process Calculations (Gravity/Dissolved Air Flotation)

Sludge thickening calculations are based on the concept that the solids in the primary or secondary sludge are equal to the solids in the thickened sludge. Assuming a negligible amount of solids are lost in the thickener overflow, the solids are the same. Note that the water is removed to thicken the sludge which results in higher percent solids.

Estimating Daily Sludge Production

Equation 8.70 provides a method to establish an initial pumping rate or to evaluate the adequacy of the current pump rate.

$$\text{Est. Pump Rate} = \frac{\left(\text{Infl. TSS Conc.} - \text{Eff. TSS Conc.}\right) \times \text{Flow} \times 8.34}{\%\text{Solids in Sludge} \times 8.34 \times 1{,}440 \text{ min/day}} \quad (8.70)$$

Example 8.52

Problem:

The sludge withdrawn from the primary settling tank contains 1.5% solids. The unit influent contains 280 mg/L TSS, and the effluent contains 141 mg/L. If the influent flow rate is 5.55 MGD, what is the estimated sludge withdrawal rate in gallons per minute (assuming the pump operates continuously)?

Solution:

$$\text{Sludge rate, gpm} = \frac{(280 \text{ mg/L} - 141 \text{ mg/L}) \times 5.55 \text{ MGD} \times 8.34}{0.015 \times 8.34 \times 1{,}440 \text{ min/day}} = 36 \text{ gpm}$$

Surface Loading Rate (gpd/ft²)

The surface loading rate (surface settling rate) is hydraulic loading—the amount of sludge applied per square foot of gravity thickener:

$$\text{Surface loading, gal/day/ft}^2 = \frac{\text{Sludge applied to the thickener, gpd}}{\text{Thickener area, ft}^2} \quad (8.71)$$

Example 8.53

Problem:

The 70-ft-diameter gravity thickener receives 32,000 gpd of sludge. What is the surface loading in gallons per square foot per day?

Solution

$$\text{Surface loading} = \frac{32{,}000 \text{ gpd}}{0.785 \times 70 \text{ ft} \times 70 \text{ ft}} = 8.32 \text{ gpd/ft}^2$$

Solids Loading Rate, lb/day/ft²

The solids loading rate is the pounds of solids per day being applied to 1 square foot of tank surface area. The calculation uses the surface area of the bottom of the tank. It assumes the floor of the tank is flat and has the same dimensions as the surface.

$$\text{Solids loading rate, lb/day/ft} = \frac{\%\text{Sludge solids} \times \text{sludge flow, gpd} \times 8.34 \text{ lb/gal}}{\text{Thickener area, ft}^2} \quad (8.72)$$

Example 8.54

Problem:

The thickener influent contains 1.6% solids. The influent flow rate is 39,000 gpd. The thickener is 50 ft in diameter and 10 ft deep. What is the solid loading in pounds per day?

Solution:

$$\text{Solids loading rate, lb/day/ft}^2 = \frac{0.016 \times 39{,}000 \text{ gpd} \times 8.34 \text{ lb/gal}}{0.785 \times 50 \text{ ft} \times 50 \text{ ft}} = 2.7 \text{ lb/ft}^2$$

Concentration Factor

The concentration factor (CF) represents the increase in concentration resulting from the thickener:

$$\text{CF} = \frac{\text{Thickened sludge concentration, \%}}{\text{Influent sludge concentration, \%}} \tag{8.73}$$

Example 8.55

Problem:

The influent sludge contains 3.5% solids. The thickened sludge solids concentration is 7.7%. What is the concentration factor?

Solution:

$$\text{CF} = \frac{7.7\%}{3.5\%} = 2.2$$

Air-to-Solids Ratio

The air-to-solids ratio is the ratio of air being applied to the pounds of solids entering the thickener:

$$\text{Air:solids ratio} = \frac{\text{Air flow, ft}^3/\text{min} \times 0.075 \text{ lb/ft}^3}{\text{Sludge flow, gpm} \times \% \text{ Solids} \times 8.34 \text{ lb/gal}} \tag{8.74}$$

Example 8.56

Problem:

The sludge pumped to the thickener is 0.85% solids. The airflow is 13 cfm. What is the air-to-solids ratio if the current sludge flow rate entering the unit is 50 gpm?

Solution:

$$\text{Air:solids ratio} = \frac{13 \text{ cfm} \times 0.075 \text{ lb/ft}}{50 \text{ gpm} \times 0.0085 \times 8.34 \text{ lb/gal}} = 0.28$$

Recycle Flow in Percent

The amount of recycle flow expressed as a percent:

$$\text{Recycle, \%} = \frac{\text{Recycle flow rate, gpm} \times 100}{\text{Sludge flow, gpm}} = 175\% \tag{8.75}$$

Example 8.57

Problem:

The sludge flow to the thickener is 80 gpm. The recycle flow rate is 140 gpm. What is the % recycle?

Solution:

$$\% \text{ Recycle} = \frac{140 \text{ gpm} \times 100}{80 \text{ gpm}} = 175\%$$

SLUDGE STABILIZATION

The purpose of sludge stabilization is to reduce volume, stabilize the organic matter, and eliminate pathogenic organisms to permit reuse or disposal. The equipment required for stabilization depends on the specific process used. Sludge stabilization processes include:

- Aerobic Digestion
- Anaerobic Digestion
- Composting
- Lime Stabilization
- Wet Air Oxidation (Heat Treatment)
- Chemical Oxidation (Chlorine Oxidation)
- Incineration

Aerobic Digestion

Equipment used for *aerobic digestion* consists of an aeration tank (digester) which is similar in design to the aeration tank used for the activated sludge process. Either diffused or mechanical aeration equipment is necessary to maintain the aerobic conditions in the tank. Solids and supernatant removal equipment are also required. In operation, process residuals (sludge) are added to the digester and aerated to maintain a dissolved oxygen (D.O.) concentration of 1.0 mg/L. Aeration also ensures that the tank contents are well mixed. Generally, aeration continues for ~20 days retention time. Periodically, aeration is stopped and the solids are allowed to settle. Sludge and the clear liquid supernatant are withdrawn as needed to provide more room in the digester. When no additional volume is available, mixing is stopped for 12–24 hours before solids are withdrawn for disposal. Process control testing should include alkalinity, pH, % Solids, % Volatile solids for influent sludge, supernatant, digested sludge, and digester contents. Normal operating levels for an aerobic digester are listed in Table 8.25. A typical operational problem associated with an aerobic digester is pH control. When pH drops, for example, this may indicate normal biological activity or low influent alkalinity. This problem is corrected by adding alkalinity (lime, bicarbonate, etc.).

TABLE 8.25
Aerobic Digester Normal Operating Levels

Parameter	Normal Levels
Detention time, days	10–20
Volatile solids loading lbs/ft³/day	0.1–0.3
D.O. mg/L	1.0
pH	5.9–7.7
Volatile solids reduction	40%–50%

Water Pollution

Environmental engineers who design treatment plant unit processes and wastewater operators who operate aerobic digesters are required to make certain process control calculations to ensure that proper operational parameters are engineered into the system and the parameters are operated correctly. These process control calculations are explained in the following sections.

Volatile Solids Loading

Volatile solids loading for the aerobic digester is expressed in pounds of volatile solids entering the digester per day per cubic foot of digester capacity:

$$\text{Volatile solids loading} = \frac{\text{Volatile solids added, lb/day}}{\text{Digester volume, ft}^3} \quad (8.76)$$

Example 8.58

Problem:

The aerobic digester is 25 ft in diameter and has an operating depth of 24 ft. The sludge added to the digester daily contains 1,350 lb of volatile solids. What is the volatile solids loading in pounds per day per cubic foot?

Solution:

$$\text{Volatile solids loading} = \frac{1{,}350 \text{ lb/day}}{.785 \times 25 \text{ ft} \times 25 \text{ ft} \times 24 \text{ ft}}$$

$$= 0.11 \text{ lb/day/ft}^3$$

Digestion Time, Days

Digestion time is the theoretical time the sludge remains in the aerobic digester:

$$\text{Digestion time, days} = \frac{\text{Digester volume, gallons}}{\text{Sludge added, gpd}} \quad (8.77)$$

Example 8.59

Problem:

Digester volume is 240,000 gal. Sludge is being added to the digester at the rate of 13,500 gpd. What is the digestion time in days?

Solution:

$$\text{Digestion time, days} = \frac{240{,}000 \text{ gal}}{13{,}500 \text{ gpd}} = 17.8 \text{ days}$$

Digester Efficiency (% Reduction)

To determine digester efficiency or the % of reduction, a two-step procedure is required. First the % Volatile Matter Reduction must be calculated and then the % moisture reduction.

Step 1: Calculate Volatile Matter

Because of the changes occurring during sludge digestion, the calculation used to determine percent volatile matter reduction is more complicated.

$$\% \text{ Reduction} = \frac{(\% \text{ Volatile matter}_{in} - \% \text{ Volatile matter}_{out}) \times 100}{[\% \text{Vol. matter}_{in} - (\% \text{ Vol. matter}_{in} \times \% \text{ Vol. matter}_{out})]} \quad (8.78)$$

Example 8.60

Problem:

Using the digester data provided below, determine the % Volatile Matter Reduction for the digester.
  Raw Sludge Volatile Matter 71%
  Digested Sludge Volatile Matter 53%

Solution:

$$\% \text{ Vol. matter reduction} = \frac{(0.71-0.53)\times 100}{[0.71-(0.71\times 0.53)]} = 53.9 \text{ or } 54\%$$

Step 2: Calculate Moisture Reduction

$$\% \text{ Moisture reduction} = \frac{(\% \text{Moisture}_{in} - \% \text{Moisture}_{out})\times 100}{[\% \text{Moisture}_{in} - (\% \text{Moisture}_{in} \times \% \text{Moisture}_{out})]} \quad (8.79)$$

Example 8.61

Problem:

Using the digester data provided below, determine the % Moisture Reduction for the digester.
  Note: Percent moisture = 100% − percent solids

Solution:

	% Solids	6%
Raw sludge	% Moisture	94% (100% − 6%)
Digested sludge	% Solids	15%
	% Moisture	85% (100% − 15%)

$$\% \text{ Reduction} = \frac{(0.94-0.85)\times 100}{[0.94-(0.94\times 0.85)]} = 64\%$$

pH Adjustment

Occasionally, the pH of the aerobic digester will fall below the levels required for good biological activity. When this occurs, the operator must perform a laboratory test to determine the amount of alkalinity required to raise the pH to the desired level. The results of the lab test must then be converted to the actual quantity of chemical (usually lime) required by the digester.

$$\text{Chem. required, lb} = \frac{\text{Chemical used in lab test, mg}}{\text{Sample volume, liters}} \times \text{Dig. vol, MG} \times 8.34 \quad (8.80)$$

Example 8.62

Problem:

The lab reports that it took 225 mg of lime to increase pH of a 1-L sample of the aerobic digester contents to pH 7.2. The digester volume is 240,000 gallons. How many pounds of lime will be required to increase the digester pH to 7.2?

Solution:

$$\text{Chemical required, lb} = \frac{225 \text{ mg} \times 240{,}000\text{-gal} \times 3.785 \text{ L/gal}}{1 \text{ L} \times 454 \text{ g/lb} \times 1{,}000 \text{ mg/g}} = 450 \text{ lb}$$

Anaerobic Digestion

Anaerobic digestion is the traditional method of sludge stabilization. It involves using bacteria that thrive in the absence of oxygen and is slower than aerobic digestion but has the advantage that only a small percentage of the wastes are converted into new bacterial cells. Instead, most of the organics are converted into carbon dioxide and methane gas.

Note: In an anaerobic digester, the entrance of air should be prevented because of the potential for air mixed with the gas produced in the digester which could create an explosive mixture.

Equipment used in anaerobic digestion includes a sealed digestion tank with either a fixed or a floating cover, heating and mixing equipment, gas storage tanks, solids and supernatant withdrawal equipment and safety equipment (e.g., vacuum relief, pressure relief, flame traps, explosion proof electrical equipment).

In operation, process residual (thickened or unthickened sludge) is pumped into the sealed digester. The organic matter digests anaerobically through a two-stage process. Sugars, starches, and carbohydrates are converted to volatile acids, carbon dioxide and hydrogen sulfide. The volatile acids are then converted to methane gas. This operation can occur in a single tank (single stage) or in two tanks (two stages). In a single-stage system, supernatant and/or digested solids must be removed whenever flow is added. In a two-stage operation, solids, and liquids from the first stage flow into the second stage each time fresh solids are added. Supernatant is withdrawn from the second stage to provide additional treatment space. Periodically, solids are withdrawn for dewatering or disposal. The methane gas produced in the process may be used for many plant activities.

Note: The primary purpose of a secondary digester is to allow for solids separation.

Various performance factors affect the operation of the anaerobic digester. For example, % Volatile Matter in raw sludge, digester temperature, mixing, volatile acids/alkalinity ratio, feed rate, % solids in raw sludge and pH are all important operational parameters that the operator must monitor. Along with being able to recognize normal/abnormal anaerobic digester performance parameters, wastewater operators must also know and understand normal operating procedures. Normal operating procedures include sludge additions, supernatant withdrawal, sludge withdrawal, pH control, temperature control, mixing, and safety requirements. Important performance parameters are listed in Table 8.26.

TABLE 8.26
Anaerobic Digester—Sludge Parameters

Raw Sludge Solids	Impact
<4% solids	Loss of alkalinity
	Decreased sludge retention time
	Increased heating requirements
	Decreased volatile acid: Alk ratio
4%–8% solids	Normal operation
>8% solids	Poor mixing
	Organic overloading
	Decreased volatile acid: Alk ratio

Sludge Additions

Sludge must be pumped (in small amounts) several times each day to achieve the desired organic loading and optimum performance.

Note: Keep in mind that in fixed cover operations additions must be balanced by withdrawals. If not, structural damage occurs.

Supernatant Withdrawal

Supernatant withdrawal must be controlled for maximum sludge retention time. When sampling, sample all draw off points and select the level with the best quality.

Sludge Withdrawal

Digested sludge is withdrawn only when necessary—always leave at least 25% seed.

pH Control

pH should be adjusted to maintain 6.8–7.2 pH by adjusting feed rate, sludge withdrawal or alkalinity additions.

Note: The buffer capacity of an anaerobic digester is indicated by the volatile acid/alkalinity relationship. Decreases in alkalinity cause a corresponding increase in ratio.

Temperature Control

If the digester is heated, the temperature must be controlled to a normal temperature range of 90°F–95°F. Never adjust the temperature by more than 1°F per day.

Mixing

If the digester is equipped with mixers, mixing should be accomplished to ensure organisms are exposed to food materials.

Safety

Anaerobic digesters are inherently dangerous—several catastrophic failures have been recorded. To prevent such failures, safety equipment such as pressure relief and vacuum relief valves, flame traps, condensate traps, and gas collection safety devices are installed. It is important that these critical safety devices be checked and maintained for proper operation.

Note: Because of the inherent danger involved with working inside anaerobic digesters, they are automatically classified as permit-required confined spaces. Therefore, all operations involving internal entry must be made in accordance with OSHA's confined space entry standard.

Anaerobic Digester: Process Control Calculations

Process control calculations involved with anaerobic digester operation include determining the required seed volume, volatile acid-to-alkalinity ratio, sludge retention time, estimated gas production, volatile matter reduction, and percent moisture reduction in digester sludge. Examples of how to make these calculations are provided in the following sections.

Required Seed Volume in Gallons

$$\text{Seed volume (gallons)} = \text{Digester volume} \times \% \text{Seed} \tag{8.81}$$

Example 8.63

Problem:

The new digester requires a 25% seed to achieve normal operation within the allotted time. If the digester volume is 266,000 gal, how many gallons of seed material will be required?

Water Pollution

Solution:

$$\text{Seed volume} = 266{,}000 \times 0.25 = 66{,}500 \text{ gal}$$

Volatile Acids to Alkalinity Ratio

The volatile acids to alkalinity ratio can be used to control operation of an anaerobic digester.

$$\text{Ratio} = \frac{\text{Volatile acids concentration}}{\text{Alkalinity concentration}} \qquad (8.82)$$

Example 8.64

Problem:

The digester contains 240 mg/L volatile acids and 1,860-mg/L alkalinity. What is the volatile acids-to-alkalinity ratio?

$$\text{Ratio} = \frac{240 \text{ mg/L}}{1{,}860 \text{ mg/L}} = 0.13$$

Note: Increases in the ratio normally indicate a potential change in the operation condition of the digester as shown in Table 8.27.

Sludge Retention Time

Sludge retention time (SRT) is the length of time the sludge remains in the digester.

$$\text{SRT, days} = \frac{\text{Digester vol (gal)}}{\text{Sludge vol. added per day (gpd)}} \qquad (8.83)$$

Example 8.65

Problem:

Sludge is added to a 525,000-gal digester at the rate of 12,250 gal per day.

Solution:

$$\text{SRT} = \frac{525{,}000 \text{ gal}}{12{,}250 \text{ gpd}} = 42.9 \text{ days}$$

Estimated Gas Production in Cubic Feet/Day

The rate of gas production is normally expressed as the volume of gas (ft^3) produced per pound of volatile matter destroyed. The total cubic feet of gas a digester will produce per day can be calculated by:

$$\text{Gas prod.}(ft^3) = \text{Vol. matter in, lb/day} \times \% \text{ Vol. matter red.} \times \text{Prod. rate, } ft^3/\text{lb} \qquad (8.84)$$

TABLE 8.27

Operating Condition	V.A./Alkalinity Ratio
Optimum	≤0.1
Acceptable range	0.1–0.3
% carbon dioxide in gas increases	≥0.5
pH decreases	≥0.8

Example 8.66

Problem:

The digester receives 11,450 lb of volatile matter per day. Currently the volatile matter reduction achieved by the digester is 52%. The rate of gas production is 11.2 ft³ of gas per pound of volatile matter destroyed.

Solution:

$$\text{Gas prod.} = 11{,}450 \, \text{lb/day} \times 0.52 \times 11.2 \, \text{ft}^3/\text{lb} = 66{,}685 \, \text{ft}^3/\text{day}$$

Percent Volatile Matter Reduction

Because of the changes occurring during sludge digestion, the calculation used to determine percent volatile matter reduction is more complicated.

$$\% \text{VM reduction} = \frac{(\% \text{VM}_{in} - \% \text{VM}_{out}) \times 100}{[\% \text{VM}_{in} - (\% \text{VM}_{in} \times \% \text{VM}_{out})]} \quad (8.85)$$

Example 8.67

Problem:

Using the data provided below, determine the % Volatile Matter Reduction for the digester.

Raw sludge volatile matter   74%

Digested sludge volatile matter   55%

$$\% \text{Volatile matter reduction} = \frac{(0.74 - 0.55) \times 100}{[0.74 - (0.74 \times 0.55)]} = 57\%$$

Percent Moisture Reduction in Digested Sludge

$$\% \text{Moisture reduction} = \frac{(\% \text{Moisture}_{in} - \% \text{Moisture}_{out}) \times 100}{[\% \text{Moisture}_{in} - (\% \text{Moisture}_{in} \times \% \text{Moisture}_{out})]} \quad (8.86)$$

Example 8.68

Problem:

Using the digester data provide below, determine the % Moisture Reduction and % Volatile Matter Reduction for the digester.

Solution:

Raw sludge   % Solids 6%

Digested sludge   % Solids 14%

Note: Percent Moisture = 100% − Percent Solids

$$\% \text{Moisture reduction} = \frac{(0.94 - 0.86) \times 100}{[0.94 - (0.94 \times 0.86)]} = 61\%$$

Composting

The purpose of composting sludge is to stabilize the organic matter, reduce volume, and to eliminate pathogenic organisms. In a *composting operation* dewatered solids are usually mixed with a bulking agent (i.e., hardwood chips) and stored until biological stabilization occurs. The composting mixture is ventilated during storage to provide sufficient oxygen for oxidation and to prevent odors. After the solids are stabilized, they are separated from the bulking agent. The composted solids are then stored for curing and applied to farmlands or other beneficial uses. Expected performance of the composting operation for both percent volatile matter reduction and percent moisture reduction ranges from 40% to 60%.

Three methods of composting wastewater biosolids are common. Each method involves mixing dewatered wastewater solids with a bulking agent to provide carbon and increase porosity. The resulting mixture is piled or placed in a vessel where microbial activity cases the temperatures of the mixture to rise during the "active composing" period. The specific temperatures that must be achieved and maintained for successful composing vary based on the method and use of the end product. After active composting the material is cured and distributed. Again, there are three commonly employed composting methods but we only described the aerated static pile (ASP) method because is commonly used.

The ASP Model Composting Facility uses the homogenized mixture of bulking agent (coarse hardwood wood chips) and dewatered biosolids piled by front-end loaders onto a large concrete composting pad where it is mechanically aerated via PVC plastic pipe embedded within the concrete slab. This ventilation procedure is part of the 26-day period of "active" composting when adequate air and oxygen is necessary to support aerobic biological activity in the compost mass and to reduce the heat and moisture content of the compost mixture. Keep in mind that a compost pile without a properly sized air distribution system can lead to the onset of anaerobic conditions and the appearance of putrefactive odors.

For illustration and discussion purposes, we assume a typical overall composting pad area is ~200 ft by 240 ft consisting of eleven (11) blowers and twenty-four (24) pipe troughs (troffs). Three blowers are 20 hp 2,400 cfm, variable speed drive units capable of operating in either the positive or negative aeration mode. Blowers A, B, & C are each connected to two (2) piping troughs that run the full length of the pad. The two troughs are connected at the opposite end of the composting pad to create an "aeration pipe loop." The other eight (8) blowers are rated at 3 hp 1,200 cfm and are arranged 1 blower per 6 troughs at half-length feeding 200 cfm per trough. These blowers can be operated in the positive or negative aeration mode. Aeration piping within the six (6) pipe troughs is perforated PVC plastic pipe, 6 inches inside diameter and 1/4-inch wall thickness. Perforation holes/orifices vary in size from 7/32 to 1/2 inch, increasing in diameter as the distance from the blower increases.

The variable speed motor drives installed with blowers A, B, & C are controlled by five thermal probes mounted at various depths in the compost pile and various parameters are fed back to the recorder whereas the other eight blowers are constant speed, controlled by a timer that cycles them on and off. To ensure optimum composting operations it is important to verify that these thermal probes are calibrated on a regular basis. In the constant speed system, thermal probes are installed but all readings are taken and recorded manually.

For water and leachate drainage purposes, all aeration piping within the troughs slopes downward with the highest point at the center of the composting pad. Drain caps located at each end of the pipe length are manually removed on a regular basis so that any build-up of debris or moisture will not interfere with the airflow.

The actual construction process involved in building the compost pile will be covered in detail later, but for now, a few key points should be made. For example, prior to the piling of the mixture on the composting pad, an 18-inch layer of wood chips is used as a base material. The primary purpose of the wood chips base is to keep the composting mixture clear of the aeration pipes, which reduces clogging of the air distribution openings in the pipes and allows free air circulation.

A secondary benefit is that the wood chips insulate the composting mixture from the pad. The compost pad is like a heat sink and this insulating barrier improves the uniformity of heat distribution within the composting mixture.

Lime or alkaline stabilization can achieve the minimum requirements for both Class A (no detectable pathogens) and Class B (a reduced level of pathogens) biosolids with respect to pathogens, depending on the amount of alkaline material added and other processes employed. Generally, alkaline stabilization meets the Class B requirements when the pH of the mixture of wastewater solids and alkaline material is at 12 or above after 2 hours of contact.

Class A requirements can be achieved when the pH of the mixture is maintained at or above 12 for at least 72 hours, with a temperature of 52°C maintained for at least 12 hours during this time. In one process, the mixture is air-dried to over 50% solids after the 72-hour period of elevated pH. Alternatively, the process may be manipulated to maintain temperatures at or above 70°F for 30 or more minutes, while maintaining the pH requirement of 12. This higher temperature can be achieved by overdosing with lime (that is, adding more than is needed to reach a pH of 12), by using a supplemental heat source, or by using a combination of the two. Monitoring for fecal coliform or *Salmonella* sp. is required prior to release by the generator for use.

Materials that may be used for alkaline stabilization include hydrated lime, quicklime (calcium oxide), fly ash, lime and cement kiln dust, and carbide lime. Quicklime is commonly used because it has a high heat of hydrolysis (491 British thermal units) and can significantly enhance pathogen destruction. Fly ash, lime kiln dust, or cement kiln dust are often used for alkaline stabilization because of their availability and relatively low cost.

The alkaline stabilized product is suitable for application in many situations, such as landscaping, agriculture, and mine reclamation. The product serves as a lime substitute, source of organic matter, and a specialty fertilizer. The addition of alkaline stabilized biosolids results in more favorable conditions for vegetative growth by improving soil properties such as pH, texture, and water-holding capacity. Appropriate applications depend on the needs of the soil and crops that will be grown and the pathogen classification. For example, a Class B material would not be suitable for blending in a topsoil mix intended for use in home landscaping but is suitable for agriculture, mine reclamation, and landfill cover where the potential for contact with the pulse is lower and access can be restricted. Class A alkaline stabilized biosolids are useful in agriculture and as a topsoil blend ingredient. Alkaline stabilized biosolids provide pH adjustment, nutrients, and organic matter, reducing reliance on other fertilizers.

Alkaline stabilized biosolids are also useful as daily landfill cover. They satisfy the federal requirement that landfills must be covered with soil or soil-like material at the end of each day (40 CFR 258). In most cases, lime stabilized biosolids are blended with other soil to achieve the proper consistency for daily cover.

As previously mentioned, alkaline stabilize biosolids are excellent for land reclamation in degraded areas, including acid mine spills or mine tailings. Soils conditions at such sites are very unfavorable for vegetative growth often due to acid content, lack of nutrients, elevated levels of heavy metals, and poor soil texture. Alkaline stabilized biosolids help to remedy these problems, making conditions more favorable for plant growth and reducing erosion potential. In addition, once a vegetative cover is established, the quality of mine drainage improves.

Thermal treatment (or wet air oxidation) subjects sludge to high temperature and pressure in a closed reactor vessel. The high temperature and pressure rupture the cell walls of any microorganisms present in the solids and causes chemical oxidation of the organic matter. This process substantially improves dewatering and reduces the volume of material for disposal. It also produces a very high-strength waste, which must be returned to the wastewater treatment system for further treatment.

Chlorine oxidation also occurs in a closed vessel. In this process chlorine (100–1,000 mg/L) is mixed with a recycled solids flow. The recycled flow and process residual flow are mixed in the reactor. The solids and water are separated after leaving the reactor vessel. The water is returned

Sludge Dewatering

Digested sludge removed from the digester is still mostly liquid. The primary objective of dewatering biosolids is to reduce moisture and consequently volume to a degree that will allow for economical disposal or reuse. Probably one of the best summarizations of the various reasons why it is important to dewater biosolids is given by Metcalf and Eddy (1991) in the following: (1) The costs of transporting biosolids to the ultimate disposal site is greatly reduced when biosolids volume is reduced; (2) dewatered biosolids allow for easier handling; (3) dewatering biosolids (reduction in moisture content) allows for more efficient incineration; (4) if composting is the beneficial reuse choice, dewatered biosolids decrease the amount and therefore the cost of bulking agents; (5) with the USEPA's new 503 rule, dewatering biosolids may be required to render the biosolids less offensive; and (6) when landfilling is the ultimate disposal option, dewatering biosolids is required to reduce leachate production.

Again, the point being made here is that the importance of adequately dewatering biosolids for proper disposal/reuse can't be overstated.

The unit processes that are most often used for dewatering biosolids are: (1) vacuum filtration, (2) pressure filtration, (3) centrifugation, and (4) drying beds. Solids content achievable by various dewatering techniques is shown in Table 8.28. The biosolids cake produced by common dewatering processes has a consistency similar to dry, crumbly, bread pudding (Spellman, 1996). This nonfluid dewatered dry, crumbly cake product is easily handled, non-offensive, and can be land applied manually and by conventional agricultural spreaders (Outwater, 1994).

Dewatering processes are usually divided into natural air drying and mechanical methods. Natural dewatering methods include those methods in which moisture is removed by evaporation and gravity or induced drainage such as sand beds, biosolids lagoons, paved beds, Phragmites reed beds, vacuum-assisted beds, Wedgewater beds, and dewatering via freezing. These natural dewatering methods are less controllable than mechanical dewatering methods but are typically less expensive. Moreover, these natural dewatering methods require less power because they rely on solar energy, gravity, and biological processes as the source of energy for dewatering. Mechanical dewatering processes include pressure filters, vacuum filters, belt filters, and centrifuges. The aforementioned air drying and mechanical dewatering processes will be discussed in greater detail later in this text.

Sand Drying Beds

Sand beds have been used successfully for years to dewater sludge. Composed of a sand bed (consisting of a gravel base, underdrains and 8–12 inches of filter grade sand), drying beds include

TABLE 8.28
Solids Content of Dewatered Biosolids

Dewatering Method	Approximate Solids Content (%)
Lagoons/ponds	30
Drying beds	40
Filter press	35–45
Vacuum filtration	25
Standard centrifuge	20–25
High G/high solids centrifuge	25–40

an inlet pipe, splash pad containment walls and a system to return filtrate (water) for treatment. In some cases, the sand beds are covered to provide drying solids protection from the elements.

In operation, solids are pumped to the sand bed and allowed to dry by first draining off excess water through the sand and then by evaporation. This is the simplest and cheapest method for dewatering sludge. Moreover, no special training or expertise is required. However, there is a downside; namely, drying beds require a great deal of manpower to clean beds; they can create odor and insect problems; and they can cause sludge buildup during inclement weather.

According to Metcalf and Eddy (1991), four types of drying are commonly used in dewatering biosolids: (1) sand, (2) paved, (3) artificial media, and (4) vacuum-assisted. In addition to these commonly used dewatering methods, a few of the innovative methods of natural dewatering will also be discussed in this section. The innovative natural dewatering methods to be discussed include experimental work on biosolids dewatering via freezing. Moreover, dewatering biosolids with aquatic plants, which have been tested and installed in several sites throughout the U.S., is also discussed.

Drying beds are generally used for dewatering well-digested biosolids. Attempting to air dry raw biosolids is generally unsuccessful and may result in odor and vector control problems. Biosolids drying beds consist of perforated or open joint drainage system in a support media, usually gravel, covered with a filter media, usually sand but can consist of extruded plastic or wire mesh. Drying beds are usually separated into workable sections by wood, concrete, or other materials. Drying beds may be enclosed or open to the weather. They may rely entirely on natural drainage and evaporation processes or may use a vacuum to assist the operation (both types are discussed in the following sections).

Traditional Sand Drying Beds

This is the oldest biosolids dewatering technique and consists of 6–12 inches of coarse sand underlain by layers of graded gravel ranging from 1/8 to 1/4 inches at the top and 3/4 to 1–1/2 inches at the bottom. The total gravel thickness is typically about 1 ft. Graded natural earth (4–6 inches) usually makes up the bottom with a web of drain tile placed on 20- to 30-ft centers. Sidewalls and partitions between bed sections are usually of wooden planks or concrete and extend about 14 inches above the sand surface (McGhee, 1991).

Large open areas of land are required for sand-drying biosolids. For example, it is not unusual to have drying beds that are up to 125+ ft long and from 20 to 35 ft in width. Even at the smallest wastewater treatment plants, it is normal practice to provide at least two drying beds.

The actual dewatering process occurs as a result of two different physical processes: evaporation and drainage. The liquor which drains off the biosolids goes to a central sump which pumps it back into the treatment process to undergo further treatment. The operation is very much affected by climate. In wet climates, it may be necessary to cover the beds with a translucent material that will allow at least 85% of the Sun's ultraviolet radiation to pass through.

Typical loading rates for primary biosolids in dry climates range up to 200 kg/(square meter × year) and from 60 to 125 kg/(square meter × year) for mixtures of primary and waste-activated biosolids.

When a drying bed is put into operation, it is generally filled with digested biosolids to a depth ranging from 8 to 12 inches. The actual drying time is climate-sensitive; that is, drying can take from a few weeks to a few months, depending on the climate and the season. After dewatering, the biosolids solids content will range from about 20% to 35% and, more importantly, the volume will have been reduced up to 85%. Upon completion of the drying process, the dried biosolids are generally removed from the bed with handheld forks or front-end loaders. It is important to note that in the dried biosolids removal process a small amount of sand is lost and the bed must be refilled and graded periodically. Dried solids removed from a biosolids drying bed can be either incinerated or land-filled.

Paved Drying Beds

The main reason for using paved drying beds is that they alleviate the problem of mechanical biosolids removal equipment damaging the underlain piping networks. The beds are paved with concrete or asphalt and are generally sloped toward center where a sump-like area with underlain pipes

Water Pollution

is arranged. These dewatering beds, like biosolids lagoons, depend on evaporation for dewatering of the applied solids. Paved drying beds are usually rectangular in shape with a center drainage strip. They can be heated via buried pipes in the paved section and generally are covered to prevent rain incursion.

In this type of natural dewatering, solids contents of 45%–50% can be achieved within 35 days in dry climates under normal conditions (McGhee, 1991). The operation of paved drying beds involves applying the biosolids to a depth of about 12 inches. The settled surface area is routinely mixed by a special vehicle-mounted machine which is driven through the bed. Mixing is important because it breaks up the crust and exposes wet surfaces to the environment. Supernatant is decanted in a manner similar to biosolids lagoons. Biosolids loadings in relatively dry climates range from about 120 to 260 kg/(square meter/year). High capital cost and larger land requirements than for sand beds are the two major disadvantages of paved drying beds.

In attempting to determine the bottom area dimensions of a paved drying bed, Metcalf and Eddy (1991) recommend computation by trial using the following equation:

$$A = \frac{1.04 S\left[(1-S_d)/S_d - (1-S_e)/S_e\right] + (62.4)(P)(A)}{(62.4)(K_e)(E_p)}$$

where

A = bottom area of paved bed, square ft.
S = annual biosolids production, dry solids, lb
S_d = percent dry solids in the biosolids after decanting, as a decimal
S_e = percent dry solids required from final disposal, as a decimal
P = annual precipitation, ft
K_e = reduction factor for evaporation from biosolids versus a free water surface (use 0.6 for a preliminary estimate; pilot test to determine factor for final design)
E_p = free water pan evaporation rate for the area, ft/year

Although the construction and operation methodologies for biosolids drying beds are well-known and widely accepted; this is not to say that the wastewater industry has not attempted to incorporate further advances into their construction and operation. For example, in an attempt to reduce the amount of dewatered biosolids that must be manually removed from drying beds, attempts have been made to construct drying beds in a specific manner whereby they can be planted with reeds; namely, the *Phragmites communis* variety (to be covered in greater detail later). The intent of augmenting the biosolids drying bed with reeds is to effect further desiccation. Moreover, tests have shown that the plants extend their root systems into the biosolids mass. This extended root system has the added benefit of helping to establish a rich microflora which eventually feeds upon the organic content of the biosolids. It is interesting to note that normal plant activity works to keep the system aerobic.

Artificial Media Drying Beds

The first artificial media drying beds, developed in England in 1970, used a stainless-steel medium called Wedgewire. Later, as the technology advanced, the stainless-steel fine wire screen mesh (Wedgewire) was replaced with a high-density polyurethane medium (Wedgewater). Polyurethane is less expensive than the stainless-steel medium but has a shorter life expectancy.

Wedgewire beds are similar in concept to vacuum-assisted drying beds (to be described later). The medium used in Wedgewire beds consists of a septum with wedge-shaped slots about 0.01 in. wide. Initially, the bed is filled with water to a level above the wire screen. Chemically conditioned biosolids are then added and, after a brief holding period, are allowed to drain through the screen and since excess water cannot return to the biosolids through capillary action, the biosolids dewaters faster with this process (McGhee, 1991).

Vacuum-Assisted Drying Beds

For small plants which process small quantities of biosolids and have limited land area, vacuum-assisted drying beds may be the preferred method of dewatering biosolids. Vacuum-assisted drying beds normally employ the use of a small vacuum to accelerate dewatering of biosolids applied to a porous medium plate. This porous medium is set above an aggregate-filled support underdrain which, as the name implies, drains to a sump. A small vacuum is applied to this underdrain, which works to extract free water from the biosolids; with biosolids loadings of about <10 kg/m^2 per cycle, the time required to dewater conditioned biosolids is about 1 day (McGhee, 1991). Using this method of dewatering it is possible to achieve a solids content of >30%, although 20% solids is a more normal expectation.

Removal of dewatered biosolids is usually accomplished with mechanized machinery such as front-end loaders. Once the solids have been removed it is important to wash the surface of the bed with high-pressure hoses to ensure residuals are removed. The main advantage cited for this dewatering method is the reduced amount of time that is needed for dewatering, which reduces the effects of inclement climatic conditions on biosolids drying. The main disadvantage of this type of dewatering may be its dependence on adequate chemical conditioning for successful operation (Metcalf and Eddy, 1991).

Natural Methods of Dewatering Biosolids

Two of the innovative methods of natural biosolids dewatering are discussed in this section: (1) Dewatering via freezing and (2) dewatering using aquatic plants.

Note: Dewatering via freezing is primarily in the pilot study stage. Experimental work in this area has led to pilot plants and has yielded various mathematical models, but no full-scale operations (Outwater, 1994).

Freeze-Assisted Drying Beds   In freeze-assisted drying, low winter temperatures accelerate the dewatering process. Freezing biosolids works to separate the water from the solids. The free water drains quickly when the granular mass is thawed. It is not unusual to attain a solids concentration >25% when the mass thaws and drains.

Determining the feasibility of freezing biosolids in a particular area is dependent on the depth of frost penetration. The maximum depth of frost penetration for an area can be found in published sources or local records (Reed, 1987).

In attempting to calculate the depth of frost penetration it may be helpful to use the following equation for 3-inch (75-mm) layers (McGhee, 1991):

$$Y = 1.76 F_p - 101$$

where
  Y = total depth of biosolids, cm
  F_p = the maximum depth of frost penetration, cm

In this example, the biosolids are applied in 75-mm layers. As soon as the first layer is frozen, another layer is applied. The goal is to fill the bed with biosolids by the end of winter. It must be pointed out that in using this layered method of freeze-dewatering biosolids it is important to ensure that each layer is frozen before the next is applied. Moreover, any snow or debris which falls should be removed from the surface, if not it will serve to insulate the biosolids. Outwater (1994) points out that "to ensure successful performance at all times, the design should be based on the warmest winter in the past 20 years and on a layer thickness which will freeze in a reasonable amount of time if freeze-thaw cycles occur during the winter" (p. 86). Another rule of thumb to use in deciding whether or not biosolids freezing is a viable dewatering option is that biosolids freezing is unlikely to be a practical concept unless more than 100 cm frost penetration is assured (Reed, 1987).

Dewatering Using Aquatic Plants   Using aquatic plants to dewater biosolids was developed in Germany in the 1960s. In this first project, reed beds were constructed in submerged wetlands. In the reed bed system, a typical sand drying bed for biosolids is modified. Instead of removing the dewatered biosolids from the beds after each application, *Phragmites communis* (reeds) are planted in the sand. For the next several years (5–10 years), biosolids are added and then the beds are emptied.

In the reed bed operation, biosolids is spread on the surface of the bed via troughs or gravity-fed pipes. When the bed is filled to capacity, about 4 inches of standing liquid will remain on the surface until it evaporates or drains down through the bed, where tile drains return it to the treatment process.

In an aquatic reed bed, the reeds perform the important function of developing (as described earlier), near the root zone, a rich microflora which feeds on the organic material in biosolids. *Phragmites* reeds are particularly suitable for this application because they are resistant to biosolid contaminants. Although the roots penetrate into the finer gravel and sand, they do not penetrate through lower areas where the larger stones or fragments are located. This is important because root penetration to the lower bed levels could interfere with free drainage.

Another advantage of using *Phragmites* reeds in the aquatic plant drying bed is their growing pattern. *Phragmites* roots grow and extend themselves through rhizomes (Rhizomes defined: an underground horizontal stem, often thickened and tuber-shaped, and possessing bud, nodes, and scale-like leaves). From each rhizome, several plants branch off and grow vertically; this vertical growth aids in dewatering by providing channels through which the water drains. The reeds also absorb some of this water which is then given off to the atmosphere through evapotranspiration.

Phragmites reed beds are operated year-round. In the fall the reeds are harvested, leaving their root systems intact. The harvested reeds, depending on contaminant concentrations, can be incinerated, land-filled, or composted.

It takes about 8 years to fill an average reed bed to capacity. When this occurs, it must be taken out of service and allowed to stand fallow for 6 months to a year. This fallow period allows for the stabilization of the top surface layer. The resulting biosolids product is dry and crumbles in the hands (it is friable). If contamination levels are within acceptable limits as per EPA's 503 Rule, the dewatered biosolids product can be land-applied.

Operation of a Phragmites reed bed has limitations. For example, in order to ensure a successful dewatering operation, it is prudent to hire the services of an agronomist who is familiar with plant growth, care, and control of plant pests such as aphids. Moreover, this dewatering system is designed for those regions that are subject to four distinct seasons, i.e., northern exposures. This is the case because phragmites require a dormancy, wintering-over, and period for proper root growth. Additionally, reed beds are not suitable for large-scale operations for operational reasons and also may be cost-prohibitive due to the cost of land.

The jury is still out on how effective reed bed dewatering systems are. This is the case because the technology is relatively new and none of the beds has been emptied yet; thus, it is difficult to predict the quality of the end-product.

Rotary Vacuum Filtration

Rotary vacuum filters have also been used for many years to dewater sludge. The vacuum filter includes filter media (belt, cloth, or metal coils), media support (drum), vacuum system, chemical feed equipment and conveyor belt(s) to transport the dewatered solids. In operation, chemically treated solids are pumped to a vat or tank in which a rotating drum is submerged. As the drum rotates, a vacuum is applied to the drum. Solids collect on the media and are held there by the vacuum as the drum rotates out of the tank. The vacuum removes additional water from the captured solids. When solids reach the discharge zone, the vacuum is released and the dewatered solids are discharged onto a conveyor belt for disposal. The media is then washed prior to returning to the start of the cycle.

Types of Rotary Vacuum Filters

The three principal types of rotary vacuum filters are rotary drum, coil, and belt. The *rotary drum* filter consists of a cylindrical drum rotating partially submerged in a vat or pan of conditioned sludge. The drum is divided length-wise into a number of sections that are connected through internal piping to ports in the valve body (plant) at the hub. This plate rotates in contact with a fixed valve plate with similar parts, which are connected to a vacuum supply, a compressed air supply, and an atmosphere vent. As the drum rotates, each section is thus connected to the appropriate service.

The *coil type* vacuum filter uses two layers of stainless-steel coils arranged in corduroy fashion around the drum. After a de-watering cycle, the two layers of springs leave the drum bed and are separated from each other so that the cake is lifted off the lower layer and is discharged from the upper layer. The coils are then washed and reapplied to the drum. The coil filter is used successfully for all types of sludges; however, sludges with extremely fine particles or ones that are resistant to flocculation de-water poorly with this system.

The media on a *belt filter* leaves the drum surface at the end of the drying zone and passes over a small diameter discharge roll to aid cake discharge. Washing of the media occurs next. Then the media are returned to the drum and to the vat for another cycle. This type of filter normally has a small-diameter curved bar between the point where the belt leaves the drum and the discharge roll. This bar primarily aids in maintaining belt dimensional stability.

Filter Media

Drum and belt vacuum filters use natural or synthetic fiber materials. On the drum filter, the cloth is stretched and secured to the surface of the drum. In the belt filter, the cloth is stretched over the drum and through the pulley system. The installation of a blanket requires several days. The cloth will (with proper care) last several hundred to several thousand hours. The life of the blanket depends on the cloth selected, the conditioning chemical, backwash frequency, and cleaning (i.e., acid bath) frequency.

Filter Drum

The filter drum is a maze of pipe work running from a metal screen and wooden skeleton and connecting to a rotating valve port at each end of the drum. The drum is equipped with a variable speed drive to turn the drum from 1/8 to 1 rpm. Normally, solids pickup is indirectly related to the drum speed. The drum is partially submerged in a vat containing the conditioned sludge. Normally, submergence is limited to 1/5 or less of the filter surface at a time.

Chemical Conditioning

Sludge dewatered using vacuum filtration is normally chemically conditioned just prior to filtration. Sludge conditioning increases the percentage of solids captured by the filter and improves the de-watering characteristics of the sludge. However, conditional sludge must be filtered as quickly as possible after chemical addition to obtain these desirable results.

Process Control Calculations

Filter Yield (lb/hr/ft^2): Vacuum Filter

Probably the most frequent calculation vacuum filter operators have to make is for determining filter yield. Example 8.68 illustrates how this calculation is made.

Water Pollution

Example 8.69

Problem:

Thickened, thermally condition sludge is pumped to a vacuum filter at a rate of 50 gpm. The vacuum area of the filter is 12 ft wide with a drum diameter of 9.8 ft. If the sludge concentration is 12%, what is the filter yield in lb/hour/ft^2? Assume the sludge weighs 8.34 lb/gal.

Solution:

First calculate the filter surface area.

$$\text{Area of a cylinder side} = 3.14 \times \text{Diameter} \times \text{Length}$$

$$= 3.14 \times 9.8\,\text{ft} \times 12\,\text{ft} = 369.3\,\text{ft}^2$$

Next calculate the pounds of solids per hour:

$$\frac{50\,\text{gpm}}{1\,\text{minutes}} \times \frac{60\,\text{minutes}}{1\,\text{hours}} \times \frac{8.34\,\text{lb}}{1\,\text{gal}} \times \frac{12\%}{100\%} = 3{,}002.4\,\text{lb/hours}$$

Dividing the two:

$$\frac{3{,}002.4\,\text{lb/hours}}{369.3\,\text{ft}^2} = 8.13\,\text{lb/hours/ft}^2$$

Pressure Filtration

Pressure filtration differs from vacuum filtration in that the liquid is forced through the filter media by a positive pressure instead of a vacuum. Several types of presses are available, but the most commonly used types are plate and frame presses and belt presses. *Filter presses* include the belt or plate and frame types. The belt filter includes two or more porous belts, rollers, and related handling systems for chemical makeup and feed, and supernatant and solids collection and transport.

The plate-and-frame filter consists of a support frame, filter plates covered with porous material, hydraulic or mechanical mechanism for pressing plates together, and related handling systems for chemical makeup and feed, and supernatant and solids collection and transport. In the plate-and-frame filter, solids are pumped (sandwiched) between plates. Pressure (200–250 psi) is applied to the plates and water is "squeezed" from the solids. At the end of the cycle, the pressure is released and as the plates separate the solids drop out onto a conveyor belt for transport to storage or disposal. Performance factors for plate and frame presses include feed sludge characteristics, type and amount of chemical conditioning, operating pressures, and the type and amount of precoat.

The belt filter uses a coagulant (polymer) mixed with the influent solids. The chemically treated solids are discharged between two moving belts. First water drains from the solids by gravity. Then, as the two belts move between a series of rollers, pressure "squeezes" additional water out of the solids. The solids are then discharged onto a conveyor belt for transport to storage/disposal. Performance factors for the belt press include sludge feed rate, belt speed, belt tension, belt permeability, chemical dosage, and chemical selection.

Filter presses have lower operation and maintenance costs than vacuum filters or centrifuges. They typically produce a good quality cake and can be batch operated. However, construction and installation costs are high. Moreover, chemical addition is required and the presses must be operated by skilled personnel.

Filter Press Process Control Calculations

As part of the operating routine for filter presses, operators are called upon to make certain process control calculations. The process control calculation most commonly used in operating the belt filter press determines the hydraulic loading rate on the unit. The most commonly used process control calculation used in operation of plate and filter presses determines the pounds of solids pressed per hour. Both of these calculations are demonstrated below.

Example 8.70

Problem:

A belt filter press receives a daily sludge flow of 0.30 gal. If the belt is 60 inches wide, what is the hydraulic loading rate on the unit in gallons per minute for each foot of belt width (gpm/ft)?

Solution:

$$\frac{0.30\,\text{MG}}{1\,\text{day}} \times \frac{1{,}000{,}000\,\text{gal}}{1\,\text{MG}} \times \frac{1\,\text{day}}{1{,}400\,\text{minutes}} = \frac{208.3\,\text{gal}}{1\,\text{minutes}}$$

$$60\,\text{inches} \times \frac{1\,\text{ft}}{12\,\text{inches}} = 5\,\text{ft}$$

$$\frac{208.3\,\text{gal}}{5\,\text{ft}} = 41.7\,\text{gpm/ft}$$

Example 8.71

Problem:

A plate and frame filter press can process 850 gal of sludge during its 120-minute operating cycle. If the sludge concentration is 3.7%, and if the plate surface area is 140 ft^2, how many pounds of solids are pressed per hour for each square foot of plate surface area?

Solution:

$$850\text{-gal} \times \frac{3.7\%}{100\%} \times \frac{8.34\,\text{lb}}{1\,\text{gal}} = 262.3\,\text{lb}$$

$$\frac{262.3\,\text{lb}}{120\,\text{minutes}} \times \frac{60\,\text{minutes}}{1\,\text{hour}} = 131.2\,\text{lb/hours}$$

$$\frac{131.2\,\text{lb/hours}}{140\,\text{ft}^2} = 0.94\,\text{lb/hours/ft}^2$$

Centrifugation

Centrifuges of various types have been used in dewatering operations for at least 30 years and appear to be gaining in popularity. Depending on the type of centrifuge used, in addition to centrifuge pumping equipment for solids feed and centrate removal, chemical makeup and feed equipment and support systems for removal of dewatered solids are required. Centrifuge operation is dependent upon various performance factors.

- Bowl design: length/diameter ratio; flow pattern
- Bowl speed
- Pool volume
- Conveyor design

- Relative conveyor speed
- Type/condition of sludge
- Type and amount of chemical conditioning
- Operating pool depth
- Relative conveyor speed (if adjustable)

Centrifuge operators often find that the operation of centrifuges can be simple, clean, and effluent. In most cases, chemical conditioning is required to achieve optimum concentrations. Operators soon discover that centrifuges are noisemakers; units run at very high speed and produce high-level noise, which can cause loss of hearing with prolonged exposure. Therefore, when working in an area where a centrifuge is in operation, special care must be taken to provide hearing protection.

Actual operation of a centrifugation unit requires the operator to control and adjust chemical feed rates; to observe unit operation and performance; to control and monitor centrate returned to treatment system; and to perform required maintenance as outlined in the manufacturer's technical manual.

Sludge Incineration

Not surprisingly, incinerators produce the maximum solids and moisture reductions. The equipment required depends on whether the unit is a multiple hearth or fluid-bed incinerator. Generally, the system will require a source of heat to reach ignition temperature, a solids feed system and ash handling equipment. It is important to note that the system must also include all required equipment (e.g., scrubbers) to achieve compliance with air pollution control requirements. Solids are pumped to the incinerator. The solids are dried and then ignited (burned). As they burn the organic matter is converted to carbon dioxide and water vapor and the inorganic matter is left behind as ash or "fixed" solids. The ash is then collected for reuse of disposal.

Multiple Hearth Furnace   The *multiple hearth furnace* consists of a circular steel shell surrounding a number of hearths. Scrappers (rabble arms) are connected to a central rotating shaft. Units range from 4.5 to 21.5 ft in diameter and have from 4 to 11 hearths. In operation, de-watered sludge solids are placed on the outer edge of the top hearth. The rotating rabble arms move them slowly to the center of the hearth. At the center of the hearth, the solids fall through ports to the second level. The process is repeated in the opposite direction. Hot gases are generated by burning on lower hearths dry solids. The dry solids pass to the lower hearths. The high temperature on the lower hearths ignites the solids. Burning continues to completion. Ash materials discharge to lower cooling hearths where they are discharged for disposal. Air flowing inside center column and rabble arms continuously cools internal equipment.

Fluidized Bed Furnace   The *fluidized bed* incinerator consists of a vertical circular steel shell (reactor) with a grid to support a sand bed and an air system to provide warm air to the bottom of the sand bed. The evaporation and incineration process takes place within the super-heated sand bed layer. In operation, the air is pumped to the bottom of the unit. The airflow expands (fluidize) the sand bed inside. The fluidized bed is heated to its operating temperature (1,200°F–1,500°F). Auxiliary fuel is added when needed to maintain operating temperature. The sludge solids are injected into the heated sand bed. Moisture immediately evaporates. Organic matter ignites and reduces to ash. Residues are ground to fine ash by the sand movement. Fine ash particles flow up and out of unit with exhaust gases. Ash particles are removed using common air pollution control processes. Oxygen analyzers in the exhaust gas stack control the airflow rate.

Note: Because these systems retain a high amount of heat in the sand, the system can be operated as little as 4 hours per day with little or no reheating.

LAND APPLICATION OF BIOSOLIDS

The purpose of land application of biosolids is to dispose of the treated biosolids in an environmentally sound manner by recycling nutrients and soil conditioners. In order to be land-applied, wastewater biosolids must comply with state and federal biosolids management/disposal regulations. Biosolids must not contain materials that are dangerous to human health (i.e., toxicity, pathogenic organisms, etc.) or dangerous to the environment (i.e., toxicity, pesticides, heavy metals, etc.). Treated biosolids are land applied by either direct injection or application and plowing in (incorporation).

Land application of biosolids requires precise control to avoid problems. The quantity and the quality of biosolids applied must be accurately determined. For this reason, the operator's process control activities include biosolids sampling/testing functions. Biosolids sampling and testing includes the determination of % solids, heavy metals, organic pesticides and herbicide, alkalinity, total organic carbon (TOC), organic nitrogen, and ammonia nitrogen.

Process Control Calculations

Process control calculations include determining disposal cost, plant available nitrogen (PAN), application rate (dry tons and wet tons/acre), metals loading rates, maximum allowable applications based upon metals loading, and site life based on metals loading.

Disposal Cost

The cost of disposal of biosolids can be determined by

$$\text{Cost} = \text{Wet tons/year} \times \% \text{solids} \times \text{cost/dry ton} \tag{8.87}$$

Example 8.72

Problem:

The treatment system produces 1,925 wet tons of biosolids for disposal each year. The biosolids are 16% solids. A contractor disposes of the biosolids for $28.00 per dry ton. What is the annual cost for sludge disposal?

Solution:

$$\text{Cost} = 1{,}925 \text{ wet tons/year} \times 0.16 \times \$28.00/\text{dry ton} = \$8{,}624$$

Plant Available Nitrogen (PAN)

One factor considered when land applying biosolids is the amount of nitrogen in the biosolids available to the plants grown on the site. This includes ammonia nitrogen and organic nitrogen. The organic nitrogen must be mineralized for plant consumption. Only a portion of the organic nitrogen is mineralized per year. The mineralization factor (f_1) is assumed to be 0.20. The amount of ammonia nitrogen available is directly related to the time elapsed between applying the biosolids and incorporating (plowing) the sludge into the soil. We provide volatilization rates based on the example below.

$$\text{Pan, lb/dry ton} = \left[\left(\text{Org. nit., mg/kg} \times f_1\right) + \left(\text{Amm. nit., mg/kg} \times V_1\right)\right] \times 0.002 \text{ lb/dry ton} \tag{8.88}$$

Where
  f_1 = Mineral rate for organic nitrogen (assume 0.20)
  V_1 = Volatilization rate ammonia nitrogen
  V_1 = 1.00 if biosolids are injected

Water Pollution

$V_1 = 0.85$ if biosolids are plowed in within 24 hours
$V_1 = 0.70$ if biosolids are plowed in within 7 days

Example 8.73

Problem:

The biosolids contain 21,000 mg/kg of organic nitrogen and 10,000 mg/kg of ammonia nitrogen. The biosolids are incorporated into the soil within 24 hours after application. What is the plant available nitrogen (PAN) per dry ton of solids?

Solution:

$$\text{PAN, lb/dry ton} = \left[(21,000\,\text{mg/kg} \times 0.20) + (10,000 \times 0.85)\right] \times 0.002$$

$$= 25.4\,\text{lb PAN/dry ton}$$

Application Rate Based on Crop Nitrogen Requirement

In most cases, the application rate of domestic biosolids to crop lands will be controlled by the amount of nitrogen the crop requires. The biosolids application rate based on the nitrogen requirement is determined by the following:

1. Using an agriculture handbook to determine the nitrogen requirement of the crop to be grown
2. Determining the amount of sludge in dry tons required to provide this much nitrogen

$$\text{Dry tons/acre} = \frac{\text{Plant nitrogen requirement, lb/acre}}{\text{Plant available nitrogen, lb/dry ton}} \tag{8.89}$$

Example 8.74

Problem:

The crop to be planted on the land application site requires 150 lb of nitrogen per acre. What is the required biosolids application rate if the PAN of the biosolids is 25 lb/dry ton?

Solution:

$$\text{Dry tons/acre} = \frac{150\,\text{lb nitrogen/acre}}{26\,\text{lb/dry ton}} = 6\,\text{dry tons per acre}$$

Metals Loading

When biosolids are land applied, metals concentrations are closely monitored and their loading on land application sites is calculated.

$$\text{Loading, lb/acre} = \text{Metal conc., mg/kg} \times 0.002\,\text{lb/dry ton} \times \text{Appl. rate, dry tons/acre} \tag{8.90}$$

Example 8.75

Problem:

The biosolids contain 14 mg/kg of lead. Biosolids are currently being applied to the site at a rate of 10 dry tons per acre. What is the metals loading rate for lead in pounds per acre?

Solution:

$$\text{Loading rate, lb/acre} = 14\,\text{mg/kg} \times 0.002\,\text{lb/dry ton} \times 10\,\text{dry tons} = 0.28\,\text{lb/acre}$$

Maximum Allowable Applications Based upon Metals Loading

If metals are present, they may limit the total number of applications a site can receive. Metals loading is normally expressed in terms of the maximum total amount of metal that can be applied to a site during its use.

$$\text{Applications} = \frac{\text{Max. allowable cumulative load for the metal, lb/ac}}{\text{Metal loading, lb/acre/application}} \quad (8.91)$$

Example 8.76

Problem:

The maximum allowable cumulative lead loading is 48.0 lb/acre. Based upon the current loading of 0.30 lb/acre, how many applications of biosolids can be made to this site?

Solution:

$$\text{Applications} = \frac{48.0 \, \text{lb/acre}}{0.30 \, \text{lb/acre}} = 160 \, \text{applications}$$

Site Life Based on Metals Loading

The maximum number of applications based on metals loading and the number of applications per year can be used to determine the maximum site life.

$$\text{Site life, years} = \frac{\text{Maximum allowable applications}}{\text{Number of applications planned per year}} \quad (8.92)$$

Example 8.77

Problem:

Biosolids are currently applied to a site twice annually. Based on the lead content of the biosolids, the maximum number of applications is determined to be 120 applications. Based upon the lead loading and the application rate, how many years can this site be used?

Solution

$$\text{Site life} = \frac{120 \, \text{applications}}{2 \, \text{applications/year}} = 60 \, \text{years}$$

Note: When more than one metal is present, the calculations must be performed for each metal. The site life would then be the lowest value generated by these calculations.

 Note: The following case study illustrates how Clark County Water Reclamation Facility, Las Vegas, Nevada, developed a *Process Today's Sludge Today* policy. For those seeking a more in-depth treatment of this case study it can be obtained from USEPA (2008) *Municipal Nutrient removal Technologies Reference Document Volume 2—Appendices*. Washington, DC: Environmental Protection Agency.

NOTES

1. Based on material from F.R. Spellman (2014) in *Personal Care Products and Pharmaceuticals in Wastewater and the Environment*. Lancaster, PA: DEStech Publishers.

2. Based on and adapted from the materials by Daughton and Ternes (1999). Pharmaceuticals and Personal Care Products in the Environment: Agents of Subtle Change? *Environmental Health Perspectives*. Vol 107, (Suppl 6): 907938; Daughton, C.G., (2010). *Drugs and the Environmental: Stewardship & Sustainability*. USEPA, Las Vegas.
3. Material in this section is from F.R. Spellman (2014). *Handbook of Water and Wastewater Treatment Plant Operations*, 3rd ed. Boca Raton, FL: CRC Press.

REFERENCES

Aguilar, A., & Raga, J.A. (1983). The striped dolphin epizootic in the Mediterranean Sea. *Ambio* **22**:524–528.

Aherne, G.W., & Briggs, R. (1989). The relevance of the presence of certain synthetic steroids in the aquatic environment. *Journal of Pharmacy and Pharmacology* **41**:735–736.

American Water Works Association (AWWA) (1984). *Introduction to Water Treatment: Principles and Practices of Water Supply Operations*. Denver, CO: AWWA.

AWWA (1991). *Guidance Manual for Compliance with the Filtration and Disinfection Requirements for Public Works Systems Using Surface Water Sources*. Denver, CO: AWWA.

Anderson, J.B., & Zwieg, H.P. (1962). Biology of waste stabilization ponds. *Southwest Water Works Journal* **44**(2):15–18.

Aptel, P. & Buckley, C.A. (1996). Categories of membrane operations. In: *Water Treatment Membrane Processes*, Chapter 2, pp. 87–96. New York: McGraw-Hill.

Arcand-Hoy, L.D., Nimrod, A.C., & Benson, W.H. (1998). Endocrine-modulating substances in the environment: Estrogenic effects of pharmaceutical products. *International Journal of Toxicology* **17**(21):139–158.

Arkoosh, M.R. (1989). Development of immunological memory in rainbow trout (Oncorhynchus mykiss) and aflatoxin modulation of the response [PhD dissertation]. Oregon State University, Corvallis, OR.

Arkoosh, M.R., & Kaattari, S. (1987). The effect of early aflatoxin B1 exposure on in vivo and in vitro antibody response in rainbow trout, Salmo gairdneri. *Journal of Fish Biology* **31**(Suppl A):19–22.

Arnold, S.F. Klotz, D.M., Collins, B.M., Vonier, PM., Guillette, L.G. Jr., & MaLachlan, J.A. (1996). Synergistic activation of estrogen receptor with combinations of environmental chemicals. *Science* **272**:1489–1492.

Ash, R.J., Mauch, B., Moulder, W., & Morgan, M. (1999). Antibiotic-resistant bacteria in U.S. rivers. Abstract no Q-383. In: *Proceedings of the Conference of the American Society for Microbiology 99th Annual Meeting*. June 1999, Chicago, IL and Herndon, VA: ASM Press.

Assenzo, J.R., & Reid, G.W. (1966). Removing nitrogen and phosphorus by bio-oxidation ponds in central Oklahoma. *Water and Sewage Works* **13**(8):294–299.

Austin, H., Keil, J.E., & Cole, P. (1989). A prospective follow-up of cancer mortality in relation to serum DDT. *American Journal of Public Health* **79**:43–46.

Barinaga, M. (1998). Small molecule fills hormone's shoes. *Science* **281**:149–151.

Baumann, P.C., Harshbarger, J.C., & Harman, K.J. (1990). Relationship between liver tumors and age in brown bullhead populations from two Lake Erie tributaries. *Science of the Total Environment* **94**:71–87.

Behechi, A., Schramm, K.-W., Attar, A., Niederfellner, J., & Kettrup, A. (1998). Acute aquatic toxicities of four musk xylene derivatives on Daphnia magna. *Water Research* **32**(5):1704–1707.

Beland, P., DeGuise, S., Girard, C., Lagace, A., Martineau, D., Michaud, R., Muir, E.C.G, Norstrom, R.J., Pelletier, E., Ray, S., & Shugart, L.R. (1993). Toxic compounds and health and reproductive effects in St. Lawrence beluga whales. *Journal of Great Lakes Research* **19**:766–775.

Belfroid, A.C., Van der Horst, A., Vethaak, A.D., Schafer, A.J., Rijs, G.B.J., Wegner, J., & Cofino, W.P. (1999). Analysis and occurrence of estrogenic hormones and their glucuronides in surface water and wastewater in the Netherlands. *Science of the Total Environment* **225**(1–2):101–108.

Bishop, C.A., Brooks, R.J., Carey, J.H., Ng, P., Norstrom, R.J., & Lean D.R.S. (1991). The case for a cause-effect linkage between environmental contamination and development in eggs of the common snapping turtle (Chelydra S. serpentina) from Ontario, Canada. *Journal of Toxicology and Environmental Health* **33**:521–547.

Bosch, X. (1998). Household antibiotic storage. *Science* **281**:785.

Boss, W.R., & Witschi, E. (1943). The permanent effects of early stilbestrol injections on the sex organs of the herring gull (Larus argentatus). *Journal of Experimental Zoology* **94**:181–209.

Bowerman, W.W., et al. (1995). A review of factors affecting productivity of bald eagles in the Great Lakes region: Implications for recovery. *Envir Health Perspectives* **103**(Supp. 4):51–59.

Bowser, D., Frenkel, K., & Zelikoff, J.T. (1994). Effects of in vitro nickel exposure on macrophage-mediate immunity in rainbow trout. *Bulletin of Environmental Contamination and Toxicology* **52**:367–373.

Brady, T.J., et al. (1996). Chlorination effectiveness for zebra and quagga mussels. *Journal AWWA* **88**(1):107–110.
Breithaupt, H. (2004). A cause without a disease. *EMBO Reports* **5**(1):16–18.
Britton, J.C., & Morton, B.A. (1982). Dissection guide, field, and laboratory manual for the introduced bivalve Corbicula fluminea. *Malacological Review* **3**:1–82.
Brockett, O.D. (1976) Microbial reactions in facultative ponds-1. The anaerobic nature of oxidation pond sediments. *Water Research* **10**(1):45–49.
Bronaugh, R.L., Yourick, J.J., & Havery, D.C. (1998). Dermal exposure assessment for the fragrances musk xylol. Abstract No 274. In: *Proceedings of the Society of Toxicology* 1998 *Annual Meeting*, Guildford.
Brown, D.P. (1987). Mortality of workers exposed to polychlorinated biphenyl: An update. *Archives of Environmental Health* **42**:333–339.
Buist, A.S., & Vollmer, W.M. (1990). Reflections of the rise in asthma morbidity and mortality. *JAMA* **264**:1719–1720.
Burhenne, J., Ludwig, M., Nikoloudis, P., & Spiteller, M. (1997). Photolytic degradation of fluoroquinolone carboxylic acids in aqueous solution. Part I: Primary photoproducts and half-lives. *Environmental Science and Pollution Research* **4**(1):10–15.
Buser, H-R., Muller, M.D., & Theobald, N. (1998a). Occurrence of the pharmaceutical drug clofibric acid and the herbicide mecoprop in various Swiss lakes and in the North Sea. *Environmental Science & Technology* **32**:188–192.
Buser, H-R., Poiger, T., & Muller, M.D. (1998b). Occurrence and fate of the pharmaceutical drug diclofenac in surface waters: Rapid photodegradation in a lake. *Environmental Science & Technology* **32**:3449–3456.
Buser, H-R., Poiger, T., & Muller, M.D. (1999). Occurrence and environmental behavior of the pharmaceutical drug Ibuprofen in surface waters and in wastewater. *Environmental Science & Technology* **33**:2529–2535.
Butterfield, C.T., et al. (1943). Chlorine vs. hypochlorite. *Public Health Report* **58**:1837.
Cameron, G.N., Symons, J.M., Spencer, S.R., & Ja, J.Y. (1989). Minimizing THM formation during control of the Asiatic Clam: A comparison of biocides. *Journal of AWWA* **81**(10):53–62.
Carlson, E., Giwercam, A., Keiding, N., & Skakkebaek, N.E. (1992). Evidence for decreasing quality of semen during past 50 years. *BMJ* **305**:609–612.
CDC (2013). Cancer among women. Accessed 03/29/13 @ http://www.cdc.gov/canger/depc/dat/women.htm.
Chemical Manufacturers Association (CMA) (1999). The Chemical Industry's Health and Environmental Effects Research Initiative. Available: http://www.cmahq.com/CMAwevstie3.nstf/pages/healthre-search.
Cheng, R.C., et al. (1994). Enhanced coagulation for arsenic removal. *Journal of AWWA* **9**:79–90.
Clarke, N.A., et al. (1962). Human enteric viruses in water, source, survival, and removability. *International Conference on Water Pollution Research*, New York.
Clifford, D.A., & Lin, C.C. (1985). Arsenic (arsenite) and arsenic (arsenate) removal from drinking water in San Ysidro, New Mexico. University of Houston, Texas.
Clifford, D.A., et al. (1997). Final report: Phases 1 & 2 City of Albuquerque Arsenic Study Field studies on Arsenic Removal in Albuquerque, New Mexico using the University of Houston/EPA Mobile Drinking Water Treatment Research Facility, Houston Texas: University of Houston.
Colborn, T. (1991). Epidemiology of great lakes bald eagles. *Journal of Toxicology and Environmental Health* **33**:395–453.
Colborn, T., vom Saal, F.S, & Soto, A.M. (1993). Developmental effects of endocrine-disrupting chemicals in wildlife and humans. *Environmental Health Perspectives* **101**(5):378–384.
Colborn, T., Dumanoski, D., & Myers, J.P. (1997). *Our Stolen Future*. New York: Dutton.
Couch, J.A., & Harshbarger, J.C. (1985). Effects of carcinogenic agents on aquatic animals: An environmental and experimental overview. *Environmental Carcinogenesis Reviews* **3**:63–105.
Couper, J.M., & Leise, E.M. (1996). Serotonin injections induce metamorphosis in larvae of the gastropod mollusk Ilyanassa obsolete. *The Biological Bulletin* **191**:178–86.
Craggs, R. (2005). Nutrients. In: *Pond Treatment Technology*, pp. 211–266. Hilton, A. (Ed.), London, UK: IWA Publishing.
Craun, G.F. (1981). Outbreaks of waterborne disease in the United States. *Journal AWWA* **73**(7):360.
Craun, G.F., & Jakubowski, W. (1996). Status of waterborne giardiasis outbreaks and monitoring methods. American Water Resources Association, *Water Related Health Issue Symposium*, Atlanta, GA.
Crawford, G., Thompson, D., Lozier, J., Daigger, G., & Fleischer, E. (2000). Membrane bioreactors: A designer's perspective. In: *Proceedings of the water Environment Federation 73rd Annual Conference & Exposition on Water Quality and Wastewater Treatment*, Anaheim, CA, CD-ROM, October 14–18, 2000.
Crites, R.W., Middlebrooks, E.J., & Reed, S.C. (2006). *Natural Wastewater Treatment Systems*. Boca Raton, FL: CRC, Taylor & Francis Group.
Culp, G.L. et al. (1986). *Handbook of Public Water Systems*. New York: Van Nostrand Reinhold.

Culp, G.L., & Culp, R.L. (1974a). Outbreaks of waterborne disease in the United States. *Journal of AWWA* **73**(7):360.

Culp, G.L., & Culp, R.L. (1974b). *New Concepts in Water Purification*. New York: Van Nostrand Reinhold Company.

CWNS (Clean Watersheds Needs Survey) (2000). Report to Congress, EPA-832-R-10-002. Washington, DC.

Daughton, C.G. (2001). Illicit drugs in municipal sewage. Accessed 05/12/13 @ ttp://www.epa.gov/nerlesd1/chemisity/phama/book-post.

Daughton, C.G. (2010). *Drugs and the Environment: Stewardship & Sustainability*. Las Vegas, NV: United States Environmental Protection Agency.

Davis, D., & Safe, S. (1990). Immunosuppressive activities of polychlorinated biphenyls in C57BL/6N mice: Structure-activity relationships as Ah report agonists and partial antagonists. *Toxicology* **63**:97–111.

de Swart, R.L., Ross, P.S., Vedder, L.J., Timmerman, H.H., Heisterkamp, S., Van Loveren, H., Vos, J.G., Reijnders, P.J.H., & Osterhaus, A.D.M.E. (1994). Impairment of immune function in harbor seals (Phoca vitulina) feeding on fish from polluted waters. *Ambio* 23:155–159.

Dean, J.H., Luster, M.I., Munson, A., & Kimoor, I, Eds. (1994a). *Immunotoxicology and Immunopharmacology*, 2nd ed. Target Organ Toxicology Series. New York: Raven Press.

Dean, J.H., Cornacoff, G.F., Rosenthal, G.J., & Luster, M.I. (1994b). Immune system: Evaluation of injury. In: *Principles and Methods of Toxicology*, Hayes, A.W. (Ed.), New York: Raven Press, pp. 1065–1090.

DeGuise, S., Martineau, D., Beland, P., & Fournier, M. (1995). Possible mechanisms of action of environmental contamination on St. Lawrence beluga whales (Delphinapterus leucas). *Environmental Health Perspectives* **103**:73–77.

Demers, L.D. & Renner, R.C. (1992). *Alternative Disinfection Technologies for Small Drinking Water Systems*, Denver, CO: AWWA and AWWART.

Desbrow, C., Routledge, E.J., Brightly, G.C., Sumpter, J.P., & Waldock, M. (1998). Identification of estrogenic chemicals in STW effluent. 1: Chemical Fractionation and in vitro biological screening. *Environmental Science & Technology* **32**(11):1549.

DeWally, E., Dodin, S., Verreault, R., Ayotte, Pl, Sauve, L., Morin, J., & Brisson, J. (1994). High organochlorine body burden in women with estrogen receptor-positive breast cancer. *Journal of the National Cancer Institute* **86**:232–234.

Draisci, R., Marchiafava, C., Ferretti, E., Palleschi, L., Catellani, G., & Anastasi, A. (1988). Evaluation of musk contamination of freshwater fish in Italy by accelerated solvent extraction and gas chromatography with mass spectrometric detection. *Journal of Chromatography A* **814**:187–197.

Draper, E. (1987). Groundwater Protection, Clean Water Action News, p. 4, Fall.

Driscoll, S.G., & Taylor, S.H. (1980). Effects of prenatal estrogen on the male urogenital system. *Obstetrics & Gynecology* **56**:537–542.

DSHEA (1994). Dietary supplement health and education act of 1994. Pub L no 103-417, 108 Stat. 4325. Available: http://odp.od.nih.gov/ods/about/law.html.

Dunnick, J.K., Elwell, M.R., Huff, J., & Barrett, J.C. (1995). Chemically-induced mammary gland cancer in the National Toxicology Program's carcinogenesis bioassay. *Carcinogenesis* **16**:173–179.

EDSTAC (1998). Endocrine disruptor screening and testing advisory committee final report. Available http://www.epa.gov/opptintr/opptendo/finalpt.htm.

Edwards, M.A. (1994). Chemistry of arsenic removal during coagulation and Fe-Mn oxidation. *Journal of AWWA* **86**:64–77.

Eichorst, S., Pfeifer, A., Magill, N.G., & Tischler, M.L. (1999). Antibiotic resistance among bacteria isolated from wild populations of resident Canada Geese in a suburban setting Abstract no Q-402. In: *Proceedings of the American Society for Microbiology 99th Annual Meeting*, Chicago, Illinois.

Facemire, C.F., Gross, T.S., & Gillette, L.H, Jr. (1995). Reproductive impairment in the Florida panther: Nature or nurture? *Environmental Health Perspectives* **103**:79–86.

Falck, F., Ricci, A., Wolff, M.S., Godbold, J., & Deckers, P. (1992). Pesticides and polychlorinated biphenyl residues in human breast lipids and their relation to breast cancer. *Archives of Environmental Health* **47**:143–146.

Falter, R., & Wilken, R-D. (1999). Determination of carboplatinum and cis-platinum by interfacing HPLC with ICP-MS using ultrasonic nebulisation. *Science of the Total Environment* 225(1–2):167–176.

Finch, G.R., et al. (1994). Ozone and chlorine inactivation of cryptosporidium. Conference proceedings, *Water Quality Technology Conference, Part II*, San Francisco, CA.

Fleischer, E.J., Broderick, T.A., Daigger, G.T., Fonseca, A.D., Holbrook, R.D., & Murthy, S.N. (2005). Evaluation of membrane bioreactor process capabilities to meet stringent effluent nutrient discharge requirements. *Water Environment Research* **77**:162–178.

Fong, P.P. (1998). Zebra mussel spawning is induced in lower concentrations of putative serotonin reuptake inhibitors. *The Biological Bulletin* **194**:143–149.

Fong, P.P., Huminski, P.T., & d'Uros, L.M. (1998). Induction of potentiation of parturition in fingernail calms (Sphaerium striatinum) by selective serotonin re-uptake inhibitors (SSRIs). *Journal of Experimental Zoology* **280**(3):260–264.

Fox, G.A., Gilman, A.P., Peakall, D.B., & Anderka, F.W. (1978). Aberrant behavior of nesting gulls. *Journal of Wildlife Management* **42**:477–483.

Fox, G.A., Collins, B., Hayakawa, E., Weseloh, D.V., Ludwig, J.P., Kubiak, T.J., & Erdman, T.C. (1991a). Reproductive outcomes in colonial fish-eating birds: a biomarker for developmental toxicants in Great Lakes food chains. *Journal of Great Lakes Research* **17**:158–167.

Fox, G.A., Gilbertson, M., Gilman, A.P., & Kubiak, T.J. (1991b). A rationale for the use of colonial fish-eating birds to monitor the presence of developmental toxicants in Great Lakes fish. *Journal of Great Lakes Research* **17**:151–152.

Fry, D.M., & Toone, C.K. (1981). DDT-induced feminization of gull embryos. *Science* **213**:922–924.

Fry, D.M., Toone, C.K., Speich, S.M., & Peard, R.J. (1987). Sex ratio skew and breeding patterns of gulls: Demographic and toxicological considerations. *Studies in Avian Biology* **10**:26–43.

Gallert, C., & Winter, J. (2005). Bacterial metabolism in wastewater treatment systems. In: *Environmental Biotechnology*, pp. 211–266, Jordening, H.J. & Winter, J. (Eds). Weinheim, Germany: Wiley VCH.

Garrison, A.W., Pope, J.D., & Allen, F.R. (1976). GC/MS analysis of organic compound sin domestic wastewaters. In: *Identification and Analysis of Organic Pollutants in Water.* Keith, C.H. (Ed). Ann Arbor, MI: Ann Arbor Science Publishers, pp. 517–556.

Gatermann, R., Huhnerfuss, H., Rimkus, G., Attar, A., & Kettrup, A. (1998). Occurrence of musk xylene and musk ketone metabolites in the aquatic environment. *Chemosphere* **36**(11):2535–2547.

Gaudy, A.F., Jr., & Gaudy, E.T. (1980). *Microbiology for Environmental Scientists and Engineers.* McGraw Hill, New York, NY.

Genicola, F.A. (1999). New Jersey Dept. Environmental Protection Office of Quality Assurance. Communication, 3 June 1999.

Gergen, P.J., & Weiss, K.B. (1990). Changing pattern of asthma hospitalization among children: 1979–1987. *JAMA* **264**:1688–1692.

Geyer, H.J. Rimkus, G., Wolf, M., Attar, A., Steinberg, C., & Kettrup, A. (1994). Synthetic nitro musk fragrances and bromocycien: New environmental chemicals in fish and mussels as well as in breast milk and human lipids. *Z Umweltchem Okotox* **6**(1):9–17.

Giesy, D.A., et al. (1995). Contaminants in fisheries in Great Lakes: Influenced sections and above dams of three Michigan rivers: In implications for health of bald eagles. *Archives of Environmental Contamination and Toxicology* **29**:309–321.

Gilbertson, M., Kubiak, T., Ludwig, J., & Fox, G.A. (1991). Great Lakes embryo mortality, edema, and deformities syndrome (Glemeds) in colonial fish-eating birds: Similarity to chick-edema disease. *Journal of Toxicology and Environmental Health* **33**:455–520.

Gill, W.H., Schumacher, F.B., Bibbo, M., Straus, F.H., & Schoenberg, H.W. (1979). Association of diethylstilbestrol exposure in utero with cryptorchidism, testicular hypoplasia, and semen abnormalities. *The Journal of Urology* **122**:36–39.

Gloyna, E.F. (1976). Facultative waste stabilization pond design. In: *Ponds as a Waste treatment Alternative*, Gloyna, E.F., Malina, J.F., Jr., & Davis, E.M. (Eds.), Water Resources Symposium No. 9, Austin, TX: University of Texas Press.

Goldsmith, E. & Hildyard, N. (Eds.) (1988). *The Earth Report: The Essential Guide to Global Ecological Issues.* Los Angeles, CA: Price Stern Sloan.

Gonzalez, R., Matsiola, P, Torchy, C., de Kinkelin, P.L., & Avrameas, S. (1989). Natural anti-TNP antibodies from rainbow trout interfere with viral infection *in vitro*. *Journal of Immunology Research* **140**:675–684.

Grady, C.P.L., Jr., Daigger, G.T., Lover, N.G., & Filipe, C.D.M. (2011). *Biological Wastewater Treatment*, 3rd ed. Boca Raton, FL: CRC Press.

Grossman, C.J. (1985). Interactions between the gonadal steroids and the immune system. *Science* **227**(4684):257–261.

Grassman, K.A. (1995). Immunological and hematological biomarkers from contaminants in fish-eating birds of the great Lakes (PhD dissertation). Virginia Polytechnic University, Blacksburg, VA.

Grassman, K.A., & Scanlon, P.F. (1995). Effects of acute lead ingestion and diet on antibody and T-cell-mediated immunity in Japanese quail. *Archives of Environmental Contamination and Toxicology* **28**:161–167.

Guillette, L.J., Gross, T.S., Masson, G.R., Matter, J.M., Percival, H.F., & Woodward, A.R. (1994). Developmental abnormalities of the gonad and abnormal sex hormone concentrations in juvenile alligators form contaminated and control lakes in Florida. *Environmental Health Perspectives* **102**:680–688.

Guillette, L.J., Gross, T.S., Gross, D.A., Rooney, A.A., & Percival, H.F. (1995). Gonadal steroidogenesis in vitro form juvenile alligators obtained from contaminated or control lakes. *Environmental Health Perspectives* **103**:31–36.

Gurol, M.D. & Pidatella, M.A. (1983). Study on ozone-induced coagulation. In: *ASCE Environmental Engineering Division Specialty Conference*. Medicine, A. & Anderson, M. (Ed), Boulder, CO.

Gyurek, L.L., et al. (1996). Disinfection of cryptosporidium parvum using single and sequential application of ozone and chlorine species. *Conference Proceedings, AWWA Water Quality Technology Conference*, Boston, MA.

Halling-Sorenson, B., Nors Nielsen, S., Lanzky, P.F., Ingerlev, F., Holten Lutzhof, H.C., & Jergensen, S.E. (1998). Occurrence fate and effects of pharmaceutical substances in the environment: A review. *Chemosphere* **36**(2):357–393.

Hany, J., & Nagel, R. (1995). Detection of UV-sunscreen agents in breast milk. *Deutsche Lebensmittel* **91**:341–245.

Harshbarger, J.C., & Clark, J.B. (1990). Epizootiology of neoplasms in boy fish of North America. *Science of the Total Environment* **94**:1–32.

Hartmann, A., Alder, A.C., Killer, T., & Widmer, R.M. (1998). Identification of fluoroquinolone antibiotics as the main source of umuC genotoxicity in native hospital wastewater. *Environmental Toxicology and Chemistry* **17**(3):377–382.

Hartmann, A., Golet, E.M., Gartiser, S., Alder, A.C., Killer, T., & Widmer, R.M. (1999). Primary DNA damage but not mutagenicity correlates with ciprofloxacin concentrations in German Hospital wastewaters. *Archives of Environmental Contamination and Toxicology* **36**:115–119.

Heberer, T., & Stan, H-J. (1997). Determination of clofibric acid and N-(phenylsulfonyl)-sarcosine in sewage river and drinking water. *International Journal of Environmental Analytical Chemistry* **67**:113–124.

Heberer, T., Schmidt-Baumier, K., & Stand, H-J. (1998). Occurrence and distribution of organic contaminants in the aquatic system in Berlin. Part I: Drug residues and other polar contaminants in Berlin surface and ground water. *Acta Hydrochimica et Hydrobiologica* **26**(5):272–278.

Heberer, T., Gramer, S., & Stan, H-J. (1999). Occurrence and distribution of organic contaminants in the aquatic system in Berlin. Pat iii: Determination of synthetic musks in Berlin surface water applying solid-phase microextraction (SPME) and gas chromatography-mass spectrometry (GC/MS). *Acta Hydrochimica et Hydrobiologica* **27**:150–156.

Henderson, A.K., Rosen, D., Miller, G.I., Figgs, L.W., Zahm, S.H., Sieber, S.M., Humphrey, H.B., Sinks, T. (1995). Breast cancer among women exposed to polybrominated biphenyls. *Epidemiology* **6**(5):544–546.

Herbert, P.D.N., et al. (1989). Ecological and genetic studies on dresissmena polymorpha (Pallas): A new mollusc in the great Lakes. *Canadian Journal of Fisheries and Aquatic Sciences* **46**:187.

Herbst, A.L., Ulfelder, H., & Poskanzer, D.C. (1971). Adenocarcinoma of the vagina. Association of maternal diethylstilbestrol therapy with tumor appearance in young women. *The New England Journal of Medicine* **284**:878–881.

Hering, J.G., & Chiu, V.Q. (1998). The chemistry of arsenic: Treatment and implications of arsenic speciation and occurrence. *AWWA Inorganic Contaminants Workshop*, San Antonio, TX.

Hermanowicz, S.W., Jenkins, D., Merlo, R.P., & Trussell, R.S. (2006). Effects of biomass properties on Submerged Membrane Bioreactor (SMBR) performance and solids processing. Document no. 01-CTS-19UR. Water Environment Federation.

Hignite, C., & Azarnoff, D.L. (1977). Drugs and drug metabolites as environmental contaminants: chlorophenoxyisobutyrate and salicylic acid in sewage water effluent. *Life Sciencs* **20**:337–342.

Hirsch, R., Ternes, T.A., Haberer, K., & Kratz, K-L. (1996). Determination of betablockers and –sympathomimetics in the aquatic environment. *Vom Wasser* **87**:263.

Hirsch, R., Ternes, T., Haberer, K., & Kratz, K-L. (1999). Occurrence of antibiotics in the aquatic environment. *Science of the Total Environment* **225**(1–2):109–118.

Hoff, J.C., et al. (1984). Disinfection and the control of waterborne giardiasis. *ASCE Specialty Conference*, Chicago.

Hoffman, D.J., Rattner, B.A., Sileo, L, Docherty, D., & Kubiak, T.J. (1987). Embryotoxicity, teratogenicity and arylhydrocarbon hydroxylase activity in Foster's terns on Green Bay, Lake Michigan. *Environmental Research* **42**:176–184.

Holm, J.V., Rugge, K., Bjerg, P.L., & Christensen, TH. (1995). Occurrence and distribution of pharmaceutical organic compounds in the groundwater downgradient of a landfill (Grindsted, Denmark). *Environmental Science & Technology* **29**(5):1415–1420.

Homo-Delarche, F., Fitzpatrick, F., Christeff, N. Nunez, E.A., Bach, J.F., & Dardenne, M. (1991). Sex steroids, glucocorticoids, stress, and autoimmunity. *The Journal of Steroid Biochemistry and Molecular Biology* **40**:619–637.

Huber, R., Smith, K., Delago, A., Isaksson, K., & Kravitz, E.A. (1997). Serotonin and aggressive motivation in crustaceans: Altering the decision to retreat. *Proceedings of the National Academy of Sciences of the United States of America* **94**:5939–5942.

IOA (1997). Survey of Water Treatment Plants. Stanford, CT: International Ozone Association.

James, M.O. (1986). Overview of in vivo metabolism of drugs by aquatic species. *Veterinary and Human Toxicology* **28**(suppl. 1):2–8.

Jensen A.A., & Slorach, S.A. (1991). *Chemical Contaminants in Human Milk*. Boston, MA: CRC Press.

Jjemba, P.K. (2008). *Pharma-Ecology.* New York: Wiley & Sons.

Johnson, A.C., Williams, R.J., & Ulahannan, T. (1999). Comment on identification of estrogenic chemicals in STW effluent. 1: Chemical fractionation and in vitro biological screening. *Environmental Science & Technology* **38**(2):369–370.

Kaattari, S.L., Adkison, M., Shapiro, D., & Arkoosh, A.R. (1994). Mechanisms of immunosuppression by aflatoxin B1. In: *Modulators of Fish Immune Response*, Vol. 1. Stolen, J. (Ed). Fair Haven, NJ: SOS Publication, pp. 151–156.

Kafferlein, H.U., Goen, T., & Angerer, J. (1998). Musk xylene: Analysis, occurrence, kinetics, and toxicology. *Critical Reviews in Toxicology* **28**(5):431–476.

Kallenborn, R., Gatermann, R., Planting, S., Rimkus, G.G., Lund, M., Schlabach, M., & Burkow, I.C. (1999). Gas chromatographic determination of synthetic musk compounds in Norwegian air samples. *Journal of Chromatography A* **846**(1–2):295–306.

Kalsch, W. (1999). Biodegradation of the iodinated X-ray contrast media diatrizoate and iopromide. *Science of the Total Environment* 225(1–2):143–153.

Kaplan, N.M. (1959). Male pseudohemaphrodism: Report of a case with observations on pathogenesis. *The New England Journal of Medicine* **261**:641–644.

Kavlock, R.J., Daston, G.P., DeRosa, C., Fenner-Crisp, P., Gray, L.E., Kaattari, S., Lucier, G., Luster, M., Mac, M.J., Maczka, C., Miller, R., Moore, J., Rolland, R., Scott, G., Sheehan, D.M., Sinks, T., & Tilson, H.A. (1996). Research needs for the risk assessment of health and environmental effects of endocrine disruptors: A report of the U.S. EPA-sponsored workshop. *Environmental Health Perspectives* **104**(Suppl 4):715–740.

Kevekordes, S., Mersch-Sundermann, V., Diez, M., Bolter, C., & Dunkelberg, H. (1998). Genotoxicity of polycyclic musk fragrances in the sister-chromatid exchange test. *Anticancer Research* **18**:449–452.

Kevern, N.R., King, D.L., & Ring, R. (1999). Lake classification systems, part I. The Michigan Riparian, p. 2.

Kirschner, E.M. (1997). Boomers quest for agelessness. *Chemical & Engineering News* **75**(16):19–25.

Klerks, P.L., & Fraleigh, P.C. (1991). Controlling adult zebra mussels with oxidants. *Journal AWWA* **83**(12):92–100.

Koch, B., et al. (1991). Predicting the formation of DBPs by the simulate distribution system. *Journal AWWA* **83**(10):62–70.

Krasner, S.W. (1989). The occurrence of disinfection byproducts in US drinking water. *Journal AWWA* **81**(8):41–53.

Krieger, N., Wolff, M.S., Hiatt, R.A., Rivera, M., Vogelman, J., & Orentreich, N. (1994). Brest cancer and serum organochlorines: A prospective study among white, black, and Asian women. *Journal of the National Cancer Institute* **86**(9):589–599.

Kulkarni, G.K., Nagabhushanam, R., Amaldoss, G., Jaiswal, R.G., & Fingerman, M. (1992). In vivo simulation of ovarian development in the red swamp crayfish Procambarus clarkia (Girad) by 5-hydroxytyptamine. *Invertebrate Reproduction and Development* **21**(3):231–240.

Kummerer, K., Steger-Hartmann, T., & Meyer, M. (1997). Biodegradability of the anti-tumour agent ifosfamide and its occurrence in hospital effluents and communal sewage. *Water Research* **31**(11):2673–2934.

Kummerer, K., Helmers, E., Hubner, P., Mascart, G., Milandri, M., Reinthaler, F., & Zwakenberg, M. (1999). European hospital as a source for platinum in the environment in comparison with other sources. *Science of the Total Environment* 225(1–2):155–165.

Lahvis, G.P., Wells, R.S., Kuehl, D.W., Stewart, J.L., Rhinehart, H.L., & Via, C.S. (1995). Decreased lymphocyte response in free-ranging bottlenose dolphins (Tursiops truncates) are associated with increased concentrations of PCBs and DDT in peripheral blood. *Environmental Health Perspectives* **103**:67–72.

Laine, J.M. (1993). Influence of bromide on low-pressure membrane filtration for controlling DBPs in surface waters. *Journal AWWA* **85**(6):87–99.

Lalezary, S., et al. (1986). Oxidation of five earthy-musty taste and odor compounds. *Journal AWWA* **78**(3):62.

Lang, C.L. (1994). The impact of the freshwater macrofouling zebra mussel (dretssena polymorpha) on drinking water supplies. *Conference Proceedings AWWA Water Quality Technology Conference Part II*, San Francisco, CA.

Leary, F.J., Resseguie, L.J., Kurland, L.T., O'Brien, P.C., Emslander, R.F., & Noller, K.L. (1984). Males exposed in utero to diethylstilbestrol. *JAMA* **252**:2984–2989.

Leatherland, J.F. (1993). Field observations on reproduce and developmental dysfunction in introduced and native salmonids from the Great Lakes. *Journal of Great Lakes Research* **19**:737–751.

Liu, O.C., et al. (1971). Relative resistance of twenty human enteric viruses to free chlorine. Virus and water a quality: Occurrence and control. *Conference Proceedings, Thirteenth Water Quality Conference*. Urban-Champaign, Illinois: University of Illinois.

Loose, L.D., Pittman, K.A., Benitz, K.F., & Silkworth, J.B. (1977). Polychlorinated biphenyl and hexchlorobenzene induced humoral immunosuppression. *Journal of the Reticuloendothelial Society* **22**:253–271.

Luster, M.I., Munson, A.E., Thomas, P.T., Holsapple, M.P., Fenters, J.D., White, K.L., Lauer, L.D., Germolec, D.R., Rosenthal, G.J., & Dean, J.H. (1988). Development of a testing battery to assess chemical-induced immunotoxicity: Nation toxicology program's guidelines for immunotoxicity evaluation in mice. *Fundamental and Applied Toxicology* **10**:2–19.

Luster, M.I., Germolec, D.R., & Rosenthal, G.J. (1990). Immunotoxicology: Review of current status. *Annals of Allergy, Asthma & Immunology* **64**:427–432.

Lynch, J.M., & Poole, N.J. (1979). *Microbial Ecology: A Conceptual Approach*. New York, NY: John Wiley & Sons.

Mac, M.J., & Edsall, C.C. (1991). Environmental contaminants and the reproductive success of lake trout in the great Lakes: An epidemiological approach. *Journal of Toxicology and Environmental Health* 33:375–394.

Mac, M.J., Schwartz, T.R., Edsall, C.C., & Frank, A.M. (1993). Polychlorinated biphenyls in Great Lakes lake trout and their eggs: Relations to survival and congener composition 1979–1988. *Journal of Great Lakes Research* **19**:752–765.

Maden, M. (1996). Retinoic acid in development and regeneration. *Journal of Biosciences* **21**(3):299–312.

Marcholonis, J.J., Hohman, V.S., Thomas, C., & Schulter, S.F. (1993). Antibody production in sharks and humans: A role for natural antibodies. *Developmental and Comparative Immunology* **17**:41–53.

Matisoff, G., et al. (1996). Toxicity of chlorine dioxide to adult zebra mussels. *Journal AWWA* **88**(8):93–106.

McArthur, M.L.B., Fox, G.A., Peakall, D.B., & Philogene, B.J.R. (1983). Ecological significance of behavior and hormonal abnormalities in breeding ring doves fed an organochlorine chemical mixture. *Archives of Environmental Contamination and Toxicology* **12**:343–353.

McGhee, T.J. (1991). *Water Supply and Sewerage*. New York: McGraw-Hill, Inc. McKinney, J.D., Chae, K., Gupia, B.N., Moore, J.A., & Goldstein, J.A. (1976). Toxicological assessment of hexachlorobiphenyl isomers and 2,3,7,8-tetrachlorodibenzofuran in chicks. *Toxicology and Applied Pharmacology* **36**:65–80.

McMurry, L.M., Oethinger, M., & Levy, S.B. (1998). Triclosan targets lipid synthesis. *Nature* **394**(6693):531–532.

Mersch-Sundermanno, V., Kevekordeso, S., & Jentero, C. (1998). Lack of mutagenicity of polycyclic musk fragrances in Salmonella typhimurium. *Toxicology in Vitro* **12**(4):389–393.

Metcalf & Eddy, Inc. (1991). *Wastewater Engineering: Treatment, Disposal, Reuse*, 3rd ed. New York: McGraw-Hill.

Metcalf & Eddy, Inc. (2003). *Wastewater Engineering: Treatment, Disposal, Reuse*, 4th ed. New York: McGraw-Hill.

Meyers, M.S., Stehr, C.M., Olson, A.P., Jonson, L.L., McCain, B.B., Chan, S-L., & Varanasi, U. (1994). Relationships between toxicopathic hepatic lesions and exposure to chemical contaminants in English sole (Pleuoronectes vetulus), starry flounder (Platichtys stellatus), and white croaker (Genyonemus lineatus) from selected marine sites on the Pacific coast. USA. *Environmental Health Perspectives* **102**:200–215.

Michel, C., Gonzalez, R., Bonjour, E., & Avrameas, S. (1990). A concurrent increasing of natural antibodies and enhancement to furunculosis in rainbow trout. *Annales de recherches veterinaires* **21**:211–218.

Middlebrooks, E.J., & Pano, A. (1983). Nitrogen removal in aerated lagoons. *Water Research* **17**(10):1369–1378.

Middlebrooks, E.J., Middlebrooks, C.H., Reynolds, J.H., Watters, G.Z., Reed, S.C. & George, D.B. (1982). *Wastewater Stabilization Lagoon Design, Performance and Upgrading*. New York, NY: Macmillan Publishing Co. Inc.

Milstein, M. (1999). Park sewage systems on the verge of failure internal report states. Available: http://www.billigsgazetter.comainhtm.

Montague, P. (1998). Drugs in the water. Rachel's Environment & Health Weekly #614. Available: gopher://ftp.std.com:70/11/periodicals/rachel [cited 3 September 1998].

Montgomery, J.M. (1985). *Water Treatment Principles and Design*. New York: John Wiley & Sons.

Muilenberg, T. (1997). Microfiltration basics: Theory and practice. *Proceedings Membrane Technology Conference*, New Orleans, LA.

Muller, S., Schmid, P., & Schlatter, C. (1996). Occurrence of nitro and non-nitro benzenoid musk compounds in human adipose tissue. *Chemosphere* 33(1):17–28.

Mussalo-Rauhamaa, H., Hasanen, E., Pyysalo, H., Antervo, K., Kauppila, R., & Pantzar, P. (1990). Occurrence of β-hexachlorocyclohexane in breast cancer patients. *Cancer* 66:2124–2128.

Nagtegaal, M., Ternes, T.A., Baumann, W., & Nagel, R. (1998). Detection of UV-sunscreen agents in water and fish of the Meerfelder Maar the Eifel Germany. *UWSF-Z fur Umweltchem Okotox* 9(2):79–86.

NCCAM (1999). National Center for complementary and Alternative Medicine. Home Page. Available: http://altmed.od.nih.gov/nccam.

NEIWPCC. (1988). Guides for the design of wastewater treatment works TR-16, New England Interstate Water Pollution Control Commission, Wilmington, MA.

Nieminski, E.C., et al. (1993). The occurrence of DBPs in Utah drinking waters. *Journal AWWA* 85(9):98–105.

NIH (1999). Office of dietary supplements. The International Bibliographic Information on Dietary Supplements (IBIDS) Database. Available: http://odp.od.nih.gov/ods/databases/ibids.html.

Noller, K.L., Blair, P.B., O'Brien, P.C., Melton, L., III, Offord, J.R., Kaufman, R.H., & Colton, T. (1988). Increased occurrence of autoimmune disease among women exposed in utero to diethylstilbestrol. *Fertility and Sterility* 49(6):1080–1082.

NRC (1999). *Hormonally Active Agents in the Environment*. Washington, DC: National Academy Press.

Okumura, T., & Nishikawa, Y. (1996). Gas chromatography: Mass spectrometry determination of triclosans in water sediment and fish samples via methylation with diazomethane. *Analytica Chimica Acta* 325(3):175–184.

Oliver, B.G., & Shindler, D.B. (1980). Trihalomethanes for chlorination of aquatic algae. *Environmental Science & Technology* 1492:1502.

Oswald, W.J. (1996). *A Syllabus on Advanced Integrated Pond Systems®*. Berkeley, CA: University of California.

Outwater, A.B. (1994). *Reuse of Sludge and Minor Wastewater Residuals*. Boca Raton, FL: Lewis Publishers.

Pano, A., & Middlebrooks, E.J. (1982). Ammonia nitrogen removal in facultative wastewater stabilization ponds. *Journal of WPCF* 54(4):2148.

Park, J. (2012). Biological nutrient removal theories and design. Online at http://www.dnr.state.wius/org/water/wm/ww/biophos/bnr_remvoal.htm.

Paterson, C., & Curtis, T. (2005). Physical and chemical environments. In: *Pond Treatment Technology*, pp. 49–65, Shilton, A. (Ed.). London, UK: IWA Publishing.

Peakall, D.B., & Fox, G.A. (1987). Toxicological investigations of pollutant-related effects in Great Lakes gulls. *Environmental Health Perspectives* 77:187–193.

Pearson, H. (2005). Microbiology of waste stabilization ponds. In: *Pond Treatment Technology*, pp. 14–48, Shilton, A. (Ed.). London, UK: IWA Publishing.

Pelletier, L., Castedo, M., Bellon, B., & Druet, P. (1994). Mercury and autoimmunity. In: *Immunotoxicology and Immunopharmacology*, Dean, J., Luster, M., Munsen, A.E, & Kimber, I., (Eds). New York: Raven Press, p. 539.

Penny, R. (1982). The effects of DES on male offspring. *Western Journal of Medicine* 136:329–330.

Pfleger, K. (1999a). *Mass Spectral and GC Data of Drugs, Poison, Pesticides, Pollutants, and Their Metabolites*, 2nd rev ed, vols. 1–3. Weinheim, Germany: VCH.

Pfleger, K. (1999b). *Mass Spectral and GC Data Drugs, Poisons, Pesticides, Pollutants and Their Metabolites*, vol. 4, 2nd rev ed. Weinheim, Germany: VCH.

Pharmanex, Inc. v. Donna Shalala. (1999). 2:97 CV 0262 K. Available http://wwwljx.com/LJXfiles/fda/cholestin.html.

Pipes, W.O., Jr. (1961). Basic biology of stabilization ponds. *Water and Sewage Works* 108(4):131–136.

Prendiville, P.W. (1986). Ozonation at the 900cfs Los Angeles Water Purification Plant. *Ozone: Science & Engineering* 8:77.

Pressley, S.A. (1999). N. Carolina effort seeks to wipe out outhouses. *Washington Post*, p. A03. Sunday, 25 April, Available http://serach.washingtonpost.com[cited 3].

Raloff, J. (1998). Drugged waters: Does it matter that pharmaceuticals are turning up in water supplies? *Science News* 153:187–189.

Reckhow, D.A., et al. (1986). Ozone as a coagulant aid. *Seminar Proceedings, Ozonation, Recent Advances and Research Needs. AWWA Annual Conference*, Denver, CO.

Reckhow, D.A., et al. (1990). Chlorination of humic materials: Byproduct formation and chemical interpretations. *Environmental Science & Technology* **24**(11):1655.

Reed, S.C. (1987). *USEPA Design Manual: Dewatering Municipal Wastewater Sludges*. Washington, DC: EPA.

Rice, R.G., et al. (1998). Ozone treatment for small water systems. *Presented at First International Symposium on Safe Drinking Water in Small Systems*, NSF International, Arlington, VA.

Richardson, M.L., & Bowron, J.M. (1985). The fate of pharmaceutical chemicals in the aquatic environments. *Journal of Pharmacy and Pharmacology* **37**:1–12.

Rimkus, G.G., Butte, W., & Geyer, H.J. (1997). Critical considerations on the analysis and bioaccumulation of musk xylene and other synthetic nitro musks in fish. *Chemosphere* **35**(7):1497–1507.

Roberts, R. (1990). Zebra mussel invasion threatens US waters. *Science* **249**:1370.

Roembke, J., Kacker, Th., & Stahlschmidt-Allner, P. (1996). Studie uber Umweltprobleme im/Zusammernhang mit Arzneimittein. [Study about environmental problems in context with drugs.] F+E Vorhabens N. 106 04 121 Umweltbundesamt Berlin. German Report of Research and Development project no 106 04 121 of Federal ministry of Research and Development, Berlin, Germany.

Ross, P.S., deSwart, R.L., Reijnders, P.J.H, Van Lovern, H., Vos, J.G., & Osterhaus, A.D.M.E. (1995). Contaminant-related suppression of delayed-type hypersensitivity and antibody response in harbor seals fed herring form the Baltic Sea. *Environmental Health Perspectives* **103**:162–167.

Routledge, E.J., Sheahan, D., Desbrow, C., Brightly, G.C., Waldock, M., & Sumpter, J.P. (1998). Identification of estrogenic chemicals in STW effluent. 2: In vivo responses in trout and roach. *Environmental Science & Technology* **32**:1559–1565.

Routledge, E.J., Waldock, M., & Sumpter, J.P. (1999). Response to comment on identification of estrogenic chemicals in STW effluent. 1. Chemical fractionation and in vitro biological screening. *Environmental Science & Technology* **33**(2):371.

Ruhoy, I.S. & Daughton, C.G. (2008). Beyond the medicine cabinet: An analysis of where and why medications accumulate. *Environment International* **34**(8):1157–1169.

RxList (1999). The internet Drug Index (The Top 2000 Prescriptions: 1998 US Prescriptions based on More than 2.4 Billion U.S. Prescriptions). Available via: http://www.rxlist.com.

Sacher, F., Lochow, E., Bethmann, D., & Brauch, H.-J. (1998). Occurrence of drugs in surface waters. *VOM Wasser* **90**:233.

Sarojini, R., Nagabhushanam, R., & Fingerman, M. (1993). In vivo evaluation of 5-hydroxytryptameine stimulation of the testes in the fiddler crab Uca pugilator: A presumed action on the neuroendocrine system. *Comparative Biochemistry and Physiology* **106C**(2):321–325.

Sarojini, R., Nagabhushanam, R., & Fingerman, M. (1995). Mode of action of the neurotransmitter 5-hydorytryptamine in stimulating ovarian maturation in the red swamp Crayfish Procambarus clarkia: An in vivo and in vitro study. *Journal of Experimental Zoology* **271**:395–400.

Sawyer, C.N., McCarty, P.L. & Parkin, G.F. (1994). *Chemistry for Environmental Engineering*. New York: McGraw Hill.

Scarpino, P.V., et al. (1972). A comparative study of the inactivation of viruses in water by chlorine. *Water Research* **6**:959.

Schrank, C.S., Cook, M.E., & Hansen, W.R. (1990). Immune response of mallard ducks treated with immunosuppressive agents: Antibody response to erythrocytes and in vivo response to phytohemagglutinin. *Journal of Wildlife Diseases* **26**:307–315.

Schuurs, A.H.W.M., & Veheul, H.A.M. (1990). Effects of gender and sex steroids on the immune response. *The Journal of Steroid Biochemistry and Molecular Biology* **35**:157–172.

Shilton, A., (Ed.). (2005). *Pond Treatment Technology*. London: IWA Publishing.

Shore, L.S., Gurevita, M., & Shemesh, M. (1993). Estrogen as an environmental pollutant. *Bulletin of Environmental Contamination and Toxicology* **51**:361–366.

Simms, J., et al. (2000). Arsenic removal studies and the design of a 20,000 m3 per day plant in U.K. AWWA Inorganic Contaminants Workshop, Albuquerque, NM.

Sinclair, R.M. (1964). Clam pests in tennessee water supplies. *Journal of AWWA* **56**(5):592.

Singer P.C. (1992). Formation and characterization of disinfection byproducts. *Presented at the First International Conference on the Safety of Water Disinfection: Balancing Chemical and Microbial Risks*, Washington, DC.

Singer, P.C., & Chang, S.D. (1989). Correlations between trihalomethanes and total organic halides formed during water treatment. *Journal AWWA* **81**(8):61–65.

Singer, P.C., & Harrington, G.W. (1993). Coagulation of DBP precursors: Theoretical and practical considerations. *AWWA Water Quality Technology Conference*, Miami, FL.

Sinks, T., Steele, G., Smith, A.B., Rinsky, R., & Watkins, K. (1992). Risk factors associated with excess mortality among polychlorinated biphenyl exposed workers. *American Journal of Epidemiology* **136**:389–398.

Sloan Equipment (1999). *Aeration Products*. Maryland: Owings Mills.

Sonawane, B.R. (1995). Chemical contaminants in human milk: An overview. *Environ Health Perspct* **103**(Suppl 6):197–205.

Spellman, F.R. (1996). *Stream Ecology and Self-Purification*. Boca Raton, FL: CRC Press.

Spellman, F.R. (2000). *Microbiology for Water and Wastewater Operators*. Boca Raton, FL: CRC Press.

Spellman, F.R. (2007). *The Science of Water*, 2nd ed. Boca Raton, FL: CRC Press.

Stan, H.J., & Heberer, T. (1997). Pharmaceuticals in the aquatic environment. *Analytics India Magazine* **25**(7):20–23.

Stan, H.J., Heberer, T., & Linkerhagner, M. (1994). Occurrence of clofibric acid in the aquatic system—is the use in human medical care the source of the contamination of surface ground and drinking water? *Vom Wasser* **83**:57–68.

Statistisches Bundesamt (1993). *Monthly Production Report: Production of Selected Personal Care Products in Germany*. Wiesbaden, Germany: Statistische Bundesamt.

Steger-Hartmann, T., Kummerer, K., & Schecker, J. (1996). Trace analysis of the antineoplastics ifosfamide and cyclophosphamide in sewage water by two-step solid-phase extraction and gas chromatography-mass spectrometry. *Journal of Chromatography A* **726**:179–184.

Steger-Hartmann, T., Kummerer, K., & Hartmann, A. (1997). Biological degradation of cyclophosphamide and its occurrence in sewage water. *Ecotoxicology and Environmental Safety* **36**:174–179.

Steger-Hartmann, T., Lange, R., & Schweinfurth, H. (1998). Environmental behavior and ecotoxicological assessment. *Vom Wasser* **91**:185–194.

Stevens, A.A. (1976). Chlorination of organics in drinking water. *Journal AWWA* **8**(11):615.

Stevens, J.T., Breckenridge, C.B., Wetzel, L.T., Gillis, H.H., Luempert L.G. III, & Eldridge, J.C. (1994). Hypothesis for mammary tumorigenesis in Sprague-Dawley rates exposed to certain triazine herbicides. *Journal of Toxicology and Environmental Health* **43**:139–153.

Stumpf, M., Ternes, T.A., Haberer, K., Seel, P., & Baumann, W. (1996). Nachweis von Arzneimittleruckstanden in Klaranlagen und Fliebgewassern [Determination of drugs in sewage treatment plants and river water]. *Vom Wasser* **86**:291–303.

Stumpf, M., Ternes, T.A., Wilken, R-D., Rodrigues, S.V., & Baumann, W. (1999). Polar drug residues in sewage and natural waters in the state of Rio de Janeiro Brazil. *Science of the Total Environment* **225**(1):135–141.

Subramanian, K.D. et al. (1997). Manganese greensand for removal of arsenic in drinking water. *Water Quality Research Journal Canada* **32**(3):551561.

Suffet, I.H., et al. (1986). Removal of tastes and odors by ozonation. *Proceeding of AWWA Seminar on Ozonation: Recent Advances and Research Needs*, Denver, CO.

Swenson, B.G., Hallberg, T., Nilsson, A., Schutz, A., & Hagmar, L (1994). Parameters of immunological competence in subjects with high consumption of fish contaminated with persistent organochlorine compounds. *Archives of Environmental and Occupational Health* **65**:351–358.

Tabak, H.H., & Bunch, R.L. (1970). Steroid hormones as water pollutants. Metabolism of natural and synthetic ovulation-inhibiting hormones by microorganisms of activated sludge and primary settled sewage. *Developments in Industrial Microbiology* **11**:367–376.

Tabak, H.H., Bloomhuff, R.N., & Bunch, R.L. (1981). Steroid hormones as water pollutions. Studies on the persistence and stability of natural urinary and synthetic ovulation-inhibiting hormones and treated wastewaters. *Developments in Industrial Microbiology* **22**:497–519.

Ternes, T.A. (1998). Occurrence of drugs in German sewage treatment plants and rivers. *Waste Resources* **32**(11):3245–3260.

Ternes, T.A., & Hirsch, R. (2000). Occurrence and behavior of iodinated contrast media in the aquatic environment. *Environmental Science & Technology* **34**:2741–2748.

Ternes, T.A., & Wilkens, R-D. (1998). Drugs and hormones as pollutants of the aquatic environment: Determination and ecotoxicological impacts. *Science of the Total Environment* **225**(1–2):176 pp.

Ternes, T.A., Stumpf, M., Schuppert, B., & Haberer, K. (1998). Simultaneous determination of antiseptics and acid drugs in sewage and river. *Vom Wasser* **90:**295–309.

Ternes, T.A., Hirsch, R., Stumpf, M., Eggert, T., Schuppert, B., & Haberer, K. (1999). Identification and screening of pharmaceuticals diagnostics and antiseptics in the aquatic environment. Bundesministerium fur Bildung Wissenschaft Forshung Technologie (BMBF)/53170. German Report of the Federal Ministry of Education, Science, research and technology. (Ref no 02WU9567/3). Bonn, Germany.

Thibaud, H., et al. (1988). Effects of bromide concentration on the production of chloropicrin during chlorination of surface waters: Formation of brominated trihalonitromethanes. *Water Reserch* **22**(3):381.
Tian, S.-S., Lamb, P., King, A.G., Miller, S.G., Kessler, L., Luengo, JI., Averill, L, Johnson, R.K., Gleason, J.G., Pelus, L.M., Dillon. S.G., & Rosen, J. (1998). A small nonpeptidyl mimic of granulocyte-colony-stimulating factor. *Science* **281**:257–259.
Ullrich, A.H. (1967). Use of wastewater stabilization ponds in two different systems. *JWPCF* **39**(6):965–977.
Unger, M., Olsen, J., & Clausen, J. (1984). Organochlorine compouns in the adipose tissue of deceased persons with and without cancer: A statistical survey of some potential confounders. *Environmental Research* **29**:371–376.
USEPA (1975). Process design manual for nitrogen control, EPA-625/1-75-007, Center for Environmental Research Information, Cincinnati, OH.
USEPA (1978). Operational manual: Sludge handling and conditioning. EPA-430/9-78-002.
USEPA (1989). Technical support document for pathogen reducing in sewage sludge. NTIS No. PB89-136618. Springfield, VA: National Technical Information Service.
USEPA (1993). Manual: Nitrogen control, EPA-625/R-93/010, Cincinnati, OH.
USEPA (1996). Drinking water regulations and health advisories. EPA 822-B-96-002.
USEPA. (1997). Community water system survey: Volumes I and II; overview. EPA 815-R-97-001a.
USEPA (1998). *National Primary Deinking Water regulation: Interim Enhanced Surface Water treatment Final Ule*. Washington, DC: United States Environmental Protection Agency.
USEPA. (1999). *Turbidity Requirements: OESWTR Guidance Manual: Turbidity Provisions*. Washington, DC: U.S. Environmental Protection Agency.
USEPA. (2000). Technologies and costs for the removal of arsenic from drinking water, EPA-815-R-00-028. Washington, DC: US Environmental Protection Agency.
USEPA (2007a). *Wastewater Management Fact Sheet: Membrane Bioreactors*. Washington, DC: United States Environmental Protection Agency.
USEPA (2007b). *Advanced Wastewater Treatment to Achieve Low Concentration of Phosphorus*. Washington, DC: Environmental Protection Agency.
USEPA (2007c). *Biological Nutrient Removal Processes and Costs*. Washington, DC: United States Environmental Protection Agency.
USEPA (2007d). National section 303(d) list fact sheet. Online at http://iaspub.epa.gov/waters/national_rept.control.
USEPA (2008) *Municipal Nutrient Removal Technologies Reference Document Volume 2—Appendices*. Washington, DC: United States Environmental Protection Agency.
USEPA (2010). Pharmaceuticals and Personal Care Products (PPCPs). Accessed 04/23/13 @ http://www.epa.gov/ppcp/basic2.html.
USEPA (2011). *Principles of Design and Operations of Wastewater Treatment Pond Systems for Plant Operators, Engineers, and Managers*. Washington, DC: U.S. Environmental Protection Agency.
USEPA (2012). Eagles, other birds thrive after EPA's 1972 DDT Ban. Accessed 03/20/12 @ http://www.epa.gov/aboutepa/hiistory/photos/h13.html.
USGS (2013). Manufacturing facilities release pharmaceuticals to the environment.
Van Benschoten, J.E., et al. (1995). Zebra mussel mortality with chorine. *Journal AWWA* **87**(5):101–108.
Vasconcelos, V.M., & Pereira, E. (2001). Cyanobacteria diversity and toxicity in a wastewater treatment plant (Portugal). *Water Research* **35**(5):1354–1357.
Velagaleti, R. (1998). Behavior of pharmaceutical drugs (human and animal health) in the environment. *Drug Information Journal* **32**:715–722.
Vickers, J.C., et al. (1997). Bench scale evaluation of microfiltration for removal of particles and natural organic matter. *Proceedings Membrane Technology Conference*, New Orleans, LA.
Weiss, K.B., & Wagener, D.K. (1990). Changing patterns of asthma mortality. Identifying target populations at high risk. *JAMA* **264**:1688–1692.
Wetzel, L.T., Leumpert, L.G. III, Breckenridge, C.B., Tisdel, M.O., Stevens, J.T., Thakur, A.K., Extrom, P.J., & Eldridge, J.C. (1994). Chronic effects of atrazine on estrus and mammary tumor formation in female Sprague-Dawley and Fischer 344 rats. *Journal of Toxicology and Environmental Health* **43**:169–182.
Wilson, K. (1998). Impotence drugs: more than Viagra. *Chemical and Engineering News* **29**:29–33.
White, G.C. (1992). *Handbook of Chlorination and Alternative Disinfectants*. New York: Van Nostrand Reinhold.
Wilcox, A.J., Baird, D.D., Weinberg, C.R., Hornsby, P.P, & Herbst, A.L. (1995). Fertility in mend exposed prenatally to diethylstilbestrol. *The New England Journal of Medicine* **332**:1411–1415.

Winkler, M., Kopf, G., Hauptvogel, C., & Neu, T. (1998). Fate of artificial musk fragrances associated with suspended particulate matter (SPM) from the river Elbe (Germany) in comparison to other organic contaminants. *Chemosphere* **37**(6):1139–1156.

Witherell, L.E. et al. (1988). Investigation of Legionella Pneumophila in drinking water. *Journal AWWA* **80**(2):88–93.

Wolff, M.S., Toniolo, P.G., Lee, E.W.., Rivera, M., & Dubin, N. (1993). Blood levels of organochlorine residues and risk of breast cancer. *Journal of the National Cancer Institute* **85**:648–652.

Wren, C.D. (1991). Cause-effect linkages between chemicals and populations of mink (Mustela vison) and otter (Utra canadensis) in the great Lakes basin. *Journal of Toxicology and Environmental Health* **33**:549–585.

Yamagishi, T., Miyazaki, T., Horii, S., & Kaneko, S. (1981). Identification of musk xylene and musk ketone in freshwater fish collected from the Tama River, Tokyo. *Bulletin of Environmental Contamination and Toxicology* **26**:656–662.

Yamagishi, T., Miyazaki, T., Horii, S., & Akiyama, K. (1983). Synthetic musk residues in biota and water from Tama River and Tokyo Bay (Japan). *Archives of Environmental Contamination and Toxicology* **12**:83–89.

Yeager, J.G. & Ward, R.I. (1981). Effects of moisture content on long-term survival and regrowth of bacteria in wastewater sludge. *Applied and Environmental Microbiology* **41**(5):1117–1122.

Zelikoff, J.T., Enane, N.A., Bowser, D, Squibb, K., & Frenkel, K. (1991). Development of fish peritoneal macrophages as a model for higher vertebrates in immunotoxicological studies. *Fundamental and Applied Toxicology* **16**:576–589.

Zelikoff, J.T., Smialowicz, R., Bigazzim, P.E., Goyer, R.A., Lawrence, D. A., Maibach, & Gardner, D. (1994). Immunomodulation by metals. *Fundamental and Applied Toxicology* **22**:1–8.

Zelikoff, J.T., Bowser, D., Squibb, K.S., & Frenkel, K. (1995). Immunotoxicity of low cadmium exposure in fish: Alternative animal models for immunotoxicological studies. *Journal of Toxicology and Environmental Health* **45**:235–248.

9 Soil Quality

One can make a day of any size and regulate the rising and setting of his own sun and the brightness of its shining.

John Muir, 1875

The sediments are a sort of epic poem of the Earth. When we are wise enough, perhaps we can read in them all of past history.

Rachel Carson, *The Sea Around Us*

When we talk about learning something from practical experience we say we need to get in there and get our hands dirty. Increasing urbanization means increasing separation from practical knowledge in many areas of land use—from getting our hands dirty. American society has reached the point where "Fresh Air Kids" are sent from New York to what amounts to smaller cities for a week in the "country." Many members of our society have never grown anything but houseplants potted in soil that comes packaged in plastic from a store—a packaged, sterile experience of what soil is and does.

But soil doesn't begin wrapped in plastic, any more than does the meat and produce people buy at the supermarket. When we forget that we are reliant on what our earth produces out of the fertility of that thin, fine layer of topsoil, we become wasteful—and we put ourselves at risk. We underestimate the value of our soil. We clear-cut it, we pave over it, we expand our communities needlessly on it, through carelessness and unconcern, we poison it—in short, we waste it.

As awareness of the serious soil pollution problems, we must now mitigate and remediate grows, as we work to develop effective methods to reuse and recycle contaminated soil, we still tend to think of soil pollution only as it affects our water supply. Again, we undervalue soil's worth. We should not lose sight of the mountains of stone and eternity of time that went into making the soil under our feet.

In many respects, soil pollution is to environmental science in the 1990s what water and air pollution were to earlier decades—the pressing environmental problem at hand. While developing methods to control air and water was difficult, socially and politically, once the regulations were in place and working effectively, the problems confronting the environmentalists were relatively easy to locate.

Soil pollution, however, presents us with a new problem. These contamination sites, especially those from old underground sources (USTs, for example), create a difficult game of contamination "hide and seek." While new techniques for handling the contamination show signs of promise, and while we have no shortage of sites in need of remediation, we must also remember that hidden sites are still affecting us beneath our feet.

Soil pollution has come to our attention in a time when regulation to begin remediation is in place, and when the public knows the importance of mitigating the contaminated areas—however, at this time, political attacks have weakened regulatory agencies' ability to clean up contaminated sites.

Frank R. Spellman (1997)

INTRODUCTION[1]

If modern man were transported back in time, he would have instantly recognized the massive structure before him, even though he might have been taken aback at what he saw: A youthful mountain range with considerable mass, steep sides, and a height that certainly reached beyond any cloud. He would instantly relate to one particular peak—the tallest, most massive one. The polyhedron-shaped object, with its polygonal base and triangular faces culminating in a single sharp-tipped apex would have looked familiar—comparable in shape, though larger in size, to the largest of the Great Egyptian Pyramids, though the Pyramids were originally covered in a sheet of limestone, not the thick, perpetual sheet of solid ice and snow that covered the mountain peak.

But if man walked this same site in modern times and knew what had once stood upon this site, the changes would be obvious and startling—and entirely relative to time. What stood as an incomparable mountain peak eons ago, today's man cannot see in its ancient majesty. In fact, he wouldn't give it a second thought as he walked across its remnants and through the vegetation that grows from its pulverized and amended remains.

300 million years ago, the pyramid-shaped mountain peak stood in full, unchallenged splendor above the clouds, wrapped in a cloak of ice, a mighty fortress of stone, seemingly vulnerable to nothing, standing tallest of all—higher than any mountain ever stood—or will ever stand—on earth.

And so, it stood, for millions upon millions of passings of the earth around the sun. Born when Mother Earth took a deep breath, the pyramid-shaped peak stood tall and undisturbed until millions of years later, when Mother Earth yawned and stretched her mighty limbs. Today we would call this stretch a massive earthquake—humans have never witnessed one of such magnitude. Rather than registering on the Richter scale, it would have destroyed it.

But when this massive earthquake shattered the earth's surface, nothing we would call intelligent life lived on Earth—and that's a good thing.

During this massive upheaval, the peak shook to its very foundations, and after the initial shockwave and the hundred plus aftershocks, the solid granite structure had fractured. This immense fracture was so massive that each aftershock widened it and loosened the base foundation of the pyramid-shaped peak itself. Only 10,000 years later (a few seconds relative to geologic time), the fracture's effects totally altered the shape of the peak forever. During a horrendous windstorm of an intensity known only in earth's earliest days, a sharp tremor (emanating from deep within the earth and shooting up the spine of the mountain itself, up to the very peak) widened the gaping wound still more.

Decades of continued tremors and terrible windstorms passed (no present-day structure could withstand a blast from such a wind), and finally, the highest peak of that time, of all time, fell. It broke off completely at its base and, following the laws of gravity (as effective and powerful a force then as today, of course), tumbled from its pinnacle position and fell more than 20,000 ft, straight down. It collided with the expanding base of the mountain range, the earth-shattering impact destroying several thousand acres. It finally came to rest (what remained intact) on a precipitous ledge, at 15,000 ft in elevation. The pyramid-shaped peak, much smaller now, sat precariously perched on the precipitous ledge for about 5 million years.

Nothing, absolutely nothing, is safe from time. The most inexorable natural law is that of entropy. Time and entropy mean change and decay—harsh, sometimes brutal, but always inevitable. The bruised, scarred, truncated, but still massive rock form, once a majestic peak, was now a victim of nature's way. Nature, with its chief ally, time, at its side, works to degrade anything and everything that has substance and form. For better or for worse, in doing so, nature is ruthless, sometimes brutal, and always inevitable—but never without purpose.

While resting on the ledge, the giant rock, over the course of 5 million years, was exposed to constantly changing conditions. For several thousand years, the earth's climate was unusually warm—almost tropical—everywhere. Throughout this unusually warm era, the rock was not covered with ice and snow, but instead baked in intense heat, steamed in hot rain, and seared in the gritty, heavy windstorms that arose and released their abrasive fury, sculpting the rock's surface each day for more than ten thousand years.

Then came a pause in the endless windstorms and upheavals of the young planet; a span of time when the weather wasn't furnace-hot or arctic-cold, but moderate. The rock was still exposed to sunlight, but at lower temperatures; to rainfall at increased levels; and to fewer windstorms of increased fury. The climate remained so for some years—then the cycle repeated itself—arctic cold, moderately warm, furnace hot—and the cycle continued.

During the last of these cycles, the rock, considerably affected by physical and chemical exposure, was reduced in size and shape even more. Considerably smaller now than when it landed on the ledge, and a mere pebble compared to its former size, it fell again, this time 8,000 ft to the base of the mountain range, coming to rest on a bed of talus. Reduced in size still more, it remained on its sloping talus bed for many more thousand years.

Somewhere around 15,000 B.C., the rock form, continuously exposed to chemical and mechanical weathering, its physical structure weakened by its long-ago falls, fractured, split—broke into ever-decreasing size rocks, until the largest intact fragment left from the original rock was no bigger than a four-bedroom house. But change did not stop, and neither did time. It rolled on until, about the time the Egyptians were building their pyramids, the rock was reduced, by this long, slow decaying process, to roughly 10 ft^2.

Over the next thousand years, the rock continued to decrease in size, wearing, crumbling, and flaking away, surrounded by fragments of its former self, until it was about the size of a beach ball. Covered with moss and lichen, a web of fissures, tiny crevices, and fractures was now woven through the entire mass.

Over the next thousand or so years, via *bare rock succession*, what had once been the mother of all mountain peaks, the highest point on earth, had been reduced to nothing more than a handful of soil.

How did this happen? What is "bare rock succession"?

If a layer of soil is completely stripped off land by natural means (water, wind, etc.), by anthropogenic means (tillage plus erosion), or by cataclysmic occurrence (a massive landslide or earthquake), only after many years can a soil-denuded area return to something approaching its original state, or can a bare rock be converted to soil. But given enough time—perhaps a millennium—the scars heal over, and a new, virgin layer of soil forms where only bare rock once existed. We call the series of events that take place in this restoration process "bare rock succession". It is indeed a true "succession" with identifiable stages. Each stage in the pattern dooms the existing community as it succeeds the state that existed before.

Bare rock, however, is laid open to view, is exposed to the atmosphere. The geologic processes that cause weathering begin, breaking down the surface into smaller and smaller fragments. Many forms of weathering exist, and all effectively reduce the bare rock surface to smaller particles or chemicals in solution.

Lichens appear to cover the bare rock first. These hardy plants grow on the rock itself. They produce weak acids that assist in the slow weathering of the rock surface. The lichens also trap wind-carried soil particles, which eventually produce a very thin soil layer—a change in environmental conditions that gives rise to the next stage in bare rock succession.

Mosses replace lichens, growing in the meager soil the lichens and weathering provide. They produce a larger growing area and trap even more soil particles, providing a moister bare rock surface. The combination of more soil and moisture establishes abiotic conditions that favor the next succession stage.

Now the seeds of herbaceous plants invade what was once bare rock. Grasses and other flowering plants take hold. Organic matter provided by the dead plant tissue is added to the thin soil, while the rock still weathers from below. More and more organisms join the community as it becomes larger and more complex.

By this time, the plant and animal communities are fairly complicated. The next major invasion is by weedy shrubs that can survive with the amount of soil and moisture present. As time passes, the process of building soil speeds up as more and more plants and animals invade the area. Soon trees take root, and forest succession is evident. Many years are required, of course, before a climax forest will grow here, but the scene is set for that to occur (Tomera, 1989).

Today, only the remnants of the former, incomparable pyramid-shaped peak are left. Soil—soil packed full of organic humus, soil that looks like mud when wet, and that, when dry, most people would think was just a handful of dirt.

SOIL: WHAT IS IT?

In any discussion about soil (the third environmental medium), we must initially describe, explain, and define exactly what soil is and why it is so important to us. Having said the obvious, we must also clear up a major misconception about soil. As the chapter's introduction indicates, people often confuse soil with dirt. Soil is not dirt. Dirt is misplaced soil—soil where we don't want it, contaminating our hands or clothes, tracked in on the floor. We try to clean up and to keep out of our environment.

But *soil* is special—almost mysterious, critical to our survival, and whether we realize it or not, essential to our existence. We have relegated soil to an ignoble position. We commonly degrade it—we consider only feces to be a worse substance. But the soil deserves better.

Before we move on, let's take another look at that handful of "dirt" our modern man is holding after the mountain peak was crafted into soil by the sure hand of Mother Nature over millions and millions of years.

What does someone really have in hand when he or she reaches down and grabs a handful of "dirt?" We make the point that it isn't dirt, it is soil. But what is soil?

Perhaps no question causes more confusion in communication between various groups of "lay persons, " and "professionals"—environmental scientists, environmental engineers, specialized groups of earth scientists, and engineers in general—than does the word "soil." Why? From the professional's perspective, the problem lies in the reasons why different groups study soils.

Pedologists (soil scientists) constitute a group interested in soils as a medium for plant growth. A corresponding branch of engineering soils specialists (*soil engineers*) looks at soil as a medium that can be excavated with tools. A *geologists'* view of soil falls between pedologists and soil engineers—they are interested in soils and the weathering processes as past indicators of climatic conditions, and in relation to the geologic formation of useful materials ranging from clay deposits to metallic ores.

How do we clear up the confusion? To answer, let's view that handful of soil from a different—but much more basic and revealing perspective. Consider the following descriptions of soil to better understand what soil is, and why it is critically important to us all:

1. A handful of soil is alive, a delicate living organism—as lively as an army of migrating caribou and as fascinating as a flock of egrets. Literally teeming with life in incomparable forms, soil deserves to be classified as an independent ecosystem, or, more correctly stated, as many ecosystems.
2. When we reach down and pick up a handful of soil, exposing Earth's stark bedrock surface, it should remind us, maybe startle some of us—the realization that without its thin living soil layer, Earth is a planet as lifeless as our own moon.

If you still prefer to call soil dirt, that's okay. Maybe you view dirt in the same way as 1996 Newbery Award Winner E. L. Konigsburg's character Ethan does:

> The way I see it, the difference between farmers and suburbanites is the difference in the way we feel about dirt. To them, the earth is something to be respected and preserved, but dirt gets no respect. A farmer likes dirt. Suburbanites like to get rid of it. Dirt is the working layer of the earth and dealing with dirt is as much a part of farm life as dealing with manure: neither is user-friendly, but both are necessary (p. 64).

Soil Quality

SOIL BASICS

Soil is the layer of bonded particles of sand, silt, and clay that covers the land surface of the earth. Most soils develop multiple layers. The topmost layer (*topsoil*) is the layer in which plants grow. This topmost layer is actually an ecosystem composed of both biotic and abiotic components—inorganic chemicals, air, water, decaying organic material that provides vital nutrients for plant photosynthesis, and living organisms. Below the topmost layer (usually no more than a meter in thickness) is the *subsoil*, which is much less productive, partly because it contains much less organic matter. Below that is the *parent material*, the bedrock or other geologic material from which the soil is ultimately formed. The general rule of thumb is that it takes about 30 years to form 1 in of topsoil from subsoil; it takes much longer than that for subsoil to be formed from parent material, the length of time depending on the nature of the underlying matter (Franck & Brownstone, 1992).

Soil Properties

From the environmental scientist's view (regarding land conservation and remediation methodologies for contaminated soil remediation through reuse and recycling), four major properties of soil are of interest: soil texture, slope, structure, and organic matter. *Soil texture* (see Figure 9.1) is a given and cannot be easily or practically changed in any significant way. It is determined by the size of the rock particles (sand, silt, and clay particles) within the soil. The largest soil particles are gravel, which consists of fragments larger than 2.0 mm in diameter. Particles between 0.05 and 2.0 mm are classified as sand. Silt particles range from 0.002 to 0.05 mm in diameter and the smallest particles (clay particles) are less than 0.002 mm in diameter. Though clays are composed of the smallest particles, those particles have stronger bonds than silt or sand, though once broken apart, they erode more readily. Particle size has a direct impact on erodibility. Rarely does a soil consist of only one single size of particle—most are a mixture of various sizes.

The *slope* (or steepness of the soil layer) is another given, important because the erosive power of runoff increases with the steepness of the slope. Slope also allows runoff to exert increased force on soil particles, which breaks them apart more readily, and carries them farther away.

Soil structure (tilth) should not be confused with soil texture—they are different. In fact, in the field, the properties determined by soil texture may be considerably modified by soil structure. Soil structure refers to the way various soil particles clump together. The size, shape, and arrangement

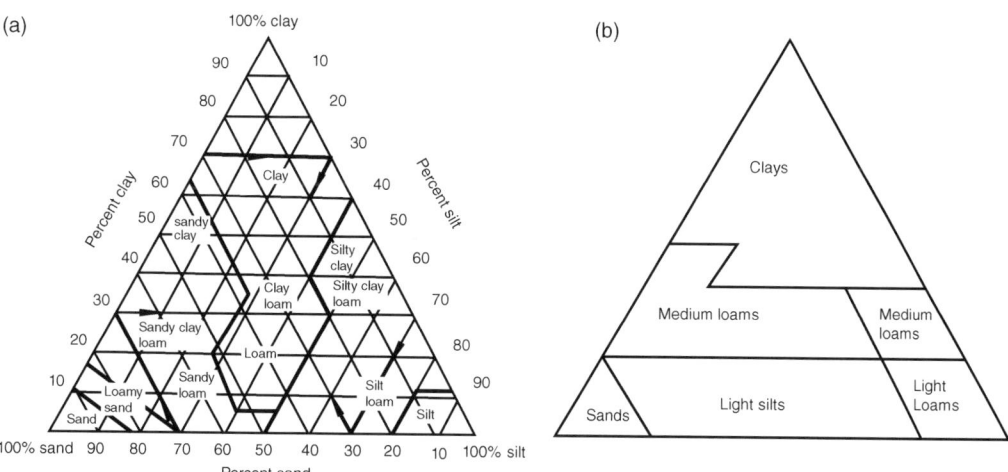

FIGURE 9.1 (a) Textural triangle similar to U.S. Department of Agriculture model; (b) broad groups of textural classes. Adapted from Briggs, 1997.

of clusters of soil particles called *aggregates* to form larger clumps called *peds*. Sand particles do not clump—sandy soils lack structure. Clay soils tend to stick together in large clumps. Good soil develops small *friable* (crumble easily) clumps. Soil develops a unique, fairly stable structure in undisturbed landscapes, but agricultural practices break down the aggregates and peds, lessening erosion resistance.

The presence of decomposed or decomposing remains of plants and animals (*organic matter*) in the soil helps not only fertility, but also soil structure, and especially the soil's ability to store water. Live organisms—protozoa, nematodes, earthworms, insects, fungi, and bacteria are typical inhabitants of soil. These organisms work to either control the population of organisms in the soil or to aid in the recycling of dead organic matter. All soil organisms, in one way or another, work to release nutrients from the organic matter, changing complex organic materials into products that can be used by plants.

Soil Formation

Soil is formed as a result of physical, chemical, and biological interactions in specific locations. Just as vegetation varies among biomes, so do the soil types that support that vegetation. The vegetation of the tundra and rain forest differs vastly from each other and from vegetation of the prairie and coniferous forest; soils differ in a like manner.

In the *soil-forming process*, two related but fundamentally different processes occur simultaneously. The first is the *formation of soil parent materials* by weathering of rocks, rock fragments and sediments. This set of processes is carried out in the *zone of weathering*. The end point is to produce parent material for the soil to develop in, which is referred to as C horizon material (see Figure 9.2) and applies in the same way for glacial deposits as for rocks. The second set of processes

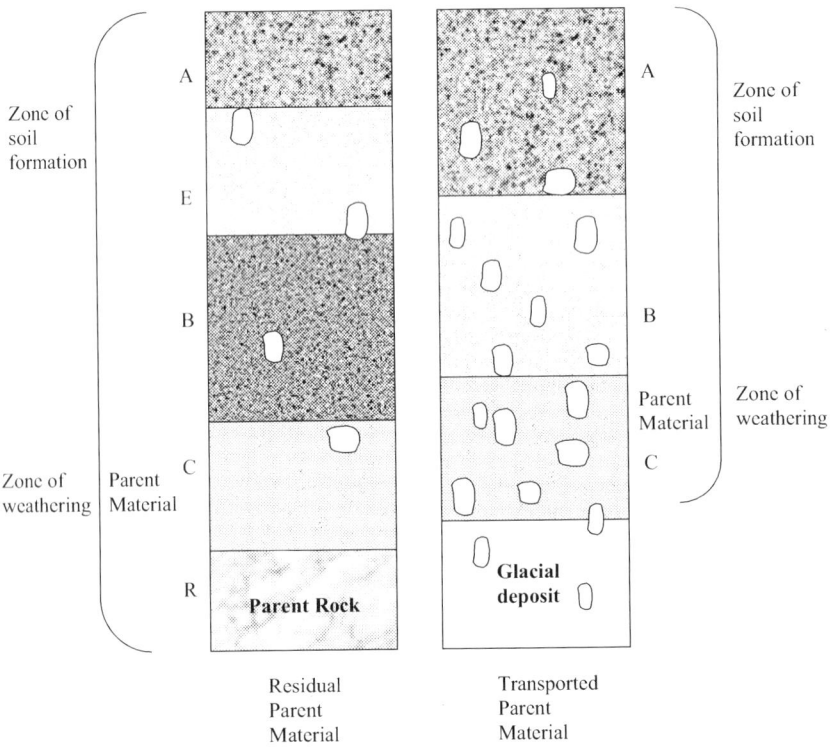

FIGURE 9.2 Soil profiles on residual and transported parent materials.

is the *formation of the soil profile* by *soil forming processes*, which changed the C horizon material into A, E, and B horizons. Figure 9.2 illustrates two soil profiles, one on hard granite, and one on a glacial deposit.

Soil development takes time and is the result of two major processes: weathering and morphogenesis. *Weathering,* the breaking down of bedrock and other sediments that have been deposited on the bedrock by wind, water, volcanic eruptions, or melting glaciers, happens physically, chemically, or a combination of both.

Physical weathering involves the breaking down of rock primarily by temperature changes and the physical action of water, ice, and wind. When a geographical location is characterized as having an arid desert biome, the repeated exposure to very high temperatures during the day, followed by low temperatures at night, causes rocks to expand and contract, and eventually to crack and shatter. At the other extreme, in cold climates, rock can crack and break as a result of repeated cycles of expansion of water in rock cracks and pores during freezing, and contraction during thawing. Another example of physical weathering occurs when various vegetation types spread their roots and grow, and the roots exert enough pressure to enlarge cracks in solid rock, eventually splitting the rock. Plants such as mosses and lichens also penetrate rocks and loosen particles.

In addition to physical weathering, bare rocks are subjected to *chemical weathering,* which involves chemical attack and dissolution of rock. Chemical weathering, accomplished primarily through oxidation via exposure to oxygen gas in the atmosphere, acidic precipitation (after having dissolved small amounts of carbon dioxide gas from the atmosphere), and acidic secretions of microorganisms (bacteria, fungi, and lichens), speeds up in warm climates and slows down in cold ones.

Physical and chemical weathering do not always (if ever) occur independently of each other. Instead, they normally work in combination, and the results can be striking. A classic example of the effect, the power of their simultaneous actions can be seen in the ecological process known as *bare rock succession.* The final stages of soil formation consist of the processes of *morphogenesis,* or the production of a distinctive *soil profile* with its constituent layers or *horizons* (see Figure 9.2). The soil profile (the vertical section of the soil from the surface through all its horizons, including C horizons) gives the environmental scientist critical information. When properly interpreted, soil horizons can provide warning about potential problems in using the land and tell much about the environment and history of a region. The soil profile allows us to describe, sample, and map soils.

Soil horizons are distinct layers, roughly parallel to the surface, which differ in color, texture, structure, and content of organic matter (see Figure 9.2). The clarity with which horizons can be recognized depends upon the relative balance of the migration, stratification, aggregation and mixing processes that take place in the soil during morphogenesis. In *podzol-type soils,* striking horizonation is quite apparent; in *Vertisol-type soils,* the horizons are less distinct. When horizons are studied, they are each given a letter symbol to reflect the genesis of the horizon (see Figure 9.2).

Certain processes work to create and destroy clear soil horizons. Formation of soil horizons that tend to create clear horizons by vertical redistribution of soil materials includes the leaching of ions in the soil solutions, movement of clay-sized particles, upward movement of water by capillary action, and surface deposition of dust and aerosols. Clear soil horizons are destroyed by mixing processes that occur because of organisms, cultivation practices, creep processes on slopes, frost heave, and by swelling and shrinkage of clays—all part of the natural soil formation process.

Soil Fertility

Soil fertility is a major concern, not only throughout the United States but worldwide. The impact on soil fertility from agricultural practices (erosion, salination, and waterlogging) are well known, well studied, and well documented. Remediation practices are also known, and actually in place in many locations throughout the globe. Indeed, solving problems related to soil fertility has received considerable attention, driven not only by a growing and hungry worldwide population but also by pocketbook issues—economics. However, one major problem related to soil fertility has only recently become apparent, important, and critical in the human population's continuing fight to maintain soil for its primary purpose (as pointedly and correctly stated by The World Resource Institute—"its ability to process nutrients into a form usable by plants."). This "recent" problem is soil contamination or pollution.

SOIL POLLUTION

Soil pollution generated by industrial contamination, management of Superfund sites, exploration and production, mining and nuclear industrial practices, among others, are all having impacts on soil quality that we have only recently begun to comprehend. Complicating the problem is that soil pollution remains difficult to assess. However, some evidence clearly indicates the impact of a few industrial practices related to soil pollution. For example, we know that petroleum-contaminated soil affects the largest number of sites and the largest total volume of contaminated material. However, the volume of petroleum-contaminated soil that is either discovered or is generated each year is not consistently tracked on a local basis, so this total is unknown. We also know, from the evidence (for example, in Oklahoma, contaminated soil accounts for about 90% of the waste generated as a one-time occurrence), that the overall amount of contaminated soil generated can be staggering (Testa, 1997).

Pollution of soil and water is a problem common to all human societies. Throughout the history of civilization, man has probably had little problem recognizing surface-water contamination. Treatment of surface water for drinking became common in the late nineteenth century, and health problems linked to impure drinking water in developed countries are now rare. Underdeveloped countries, however, are still faced with a lack of safe drinking water.

Only in the past several decades has a new problem come to light, literally and figuratively: contamination of the soil and its underground environment. In developed countries, this problem is much more serious because of their history of industrialization and the wide range of hazardous materials and other chemicals that have been introduced, either by design or accident, to the underground environment. Ignorance—more than intent—is the culprit. We were ignorant in the sense that we did not comprehend the degree to which contaminants could migrate through the soil, or the damage they could do to the soil medium and the groundwater under its "protective" surface—or the difficulty we would encounter in tracing and removing most contaminants after discovery.

Starting with the most contaminated sites, the response (in developed countries) to underground contamination has been a massive effort to define the extent of contamination and to remediate the subsurface. This response has been driven by governmental regulations dealing with waste handling and disposal, and many other potentially contaminating activities.

The range of activities that cause underground contamination is much larger than most environmental scientists would have guessed even a few years ago. We briefly discuss these activities in the sections that follow.

Soil quality problems originating on the surface include natural atmospheric deposition of gaseous and airborne particulate pollutants; infiltration of contaminated surface water; land disposal of solid and liquid waste materials; stockpiles, tailing, and spoil; dumps; salt spreading on roads; animal feedlots; fertilizers and pesticides; accidental spills; and composting of leaves and other yard wastes.

Though we do not discuss them in detail in this text, note that other sources of soil contamination relate to petroleum products. These other sources include direct disposal of used oils on the ground by individuals or industries; seepage from landfills, illegal dumps, unlined pits, ponds and lagoons; and spills from transport accidents. Even auto accidents make contributions to the soil burden (Tucker, 1989).

In this section, we focus both on the *surface origins* of soil contaminants and the *industrial practices* that can contaminate soil and in the next section we focus on the concepts of remediation and resource recovery.

NOTE: The following discussion focuses on contamination originating on the land surface. However, note that soil and subsurface contamination may also originate below ground but above the water table from septic tanks, landfills, sumps and dry wells, graveyards, USTs, leakage from underground pipelines, and other sources. In addition, soil, the subsurface, and groundwater contamination may also originate below the water table from mines, test holes, agricultural drainage wells and canals, and others.

Gaseous and Airborne Pollutants

We don't commonly associate soil with being a prominent member of the biogeochemical cycles—carbon, nitrogen, and sulfur cycles—but we should, because they are. Not only is soil a prominent part of the rapid, natural cycles of carbon, nitrogen, and sulfur, but along with these cycles, soil has a strong and important interface with the atmosphere. Consider the nitrogen cycle, where nitrates and ammonium ions in rainwater are absorbed by plant roots and soil microorganisms and converted to amino acids or to gaseous N_2 and N_2O, which diffuse back to the atmosphere. N_2 uptake and conversion to amino acids (*nitrogen fixation*) by *symbiotic* and free-living soil microorganisms balance this loss of gaseous nitrogen. NO, NO_2, and NH_3 (other nitrogen gases) are also emitted and absorbed by soils. Soil reactions are major determinants of trace gas concentrations in the atmosphere.

Air pollutants—sulfur dioxide, hydrogen sulfide, hydrocarbons, carbon monoxide, ozone, and atmospheric nitrogen gases—are absorbed by soil. Because the reactions are subtle, they have often been discounted in importance when environmentalists assess air pollution damage. Sulfur dioxide in arid regions is probably the most obvious example of direct soil absorption. The basicity of arid soils makes them an active sink for sulfur dioxide and other acidic compounds from the atmosphere.

Two classic examples of *airborne particulate soil contamination* can be seen in the accumulation of heavy metals around smelters, and in soils in urban areas, contaminated by exhaust fumes associated with auto emissions. These two soil polluters are serious in localized areas, but otherwise are generally thought to be minor.

Infiltration of Contaminated Surface Water

Often wells are intentionally installed near streams and rivers to induce recharge from the water body and to provide high yield with low drawdowns. Occasionally, if the stream or river is polluted, contamination of the soil-water well field can result. This process normally occurs when a shallow water supply well draws water from the alluvial aquifer adjacent to the stream. The cone of depression imposed by pumping the well or well field creates a gradient on the water table directed toward the well, pulling or drawing the polluted water through and contaminating the well field and well.

Land Disposal of Solid and Liquid Waste Materials

Land disposal, stockpiling, or land-applying wastes or materials, including liquid and sludge (biosolids) wastes from sewage treatment plants (nearly half of the municipal sewage biosolids produced in the U.S. is applied to the soil, either for agricultural purposes or to remediate land disturbed by mining and other industrial activities), food processing companies, and other sources, has become common practice. The purpose of this practice is twofold: it serves as a means of disposal and provides beneficial use/reuse of such materials as fertilizers for agricultural lands, golf courses, city parks, and other areas. The objective is to allow biological and chemical processes in the soil, along with plant uptake, to break down the waste products into harmless substances. In many cases, such practices are successful. However, a contamination problem may arise if any of the wastes are water-soluble and mobile, which could allow them to be carried deep into the subsurface. If the drainage or seepage area is underlain by shallow aquifers, a groundwater contamination problem may arise.

STOCKPILES, TAILINGS, AND SPOILS

Stockpiles of certain chemical products can contribute to soil and subsurface contamination. Stockpiling road salt, for example, is a common practice used by many local highway departments and some large industries as a precautionary measure to treat snow- and ice-covered surfaces in winter. *Tailings* are usually produced in mining activities and commonly contain materials (asbestos, arsenic, lead, and radioactive substances) that present a health threat to humans and other living organisms. Remember that tailings from mining operations may contain contaminants, including sulfide, which, when mixed with precipitation, forms sulfuric acid. As the chemically altered precipitation runs off or is leached from the tailing piles, it infiltrates the surface layer, contaminates soil, and ultimately may reach groundwater. *Spoil* is generally the result of excavations such as road-building operations, where huge amounts of surface cover are removed, excavated, and piled, and then moved somewhere else. Problems with spoil are similar to the tailing problems—precipitation removes materials in solution from the spoil by percolating waters (leaching) any contaminants from the spoil. Those contaminants find their way into the soil and ultimately into shallow aquifers.

DUMPS

Until recently a common waste disposal practice was to take whatever not wanted and dump it somewhere, out of sight. Today, uncontrolled dumping is prohibited in most industrialized countries, but the "old" dumping sites can contain just about anything and may still constitute a threat of subsurface contamination. Another problem that is still with us is "*midnight dumping*." Because dumping today is controlled and regulated, many disposers attempt to find ways to "get rid of junk." Unfortunately, much of this "junk" consists of hazardous materials and toxins that end up finding their way into and through soil to aquifers. In addition to the "midnight dumping" problem, another illegal disposal practice has developed in some industries because of the high cost involved with proper disposal. This practice, commonly called "Immaculate Conception," occurs when workers in industrial facilities discover unmarked drums or other containers of unknown wastes that suddenly appear on loading docks or elsewhere in the facility. Then, of course, these immaculately conceived vessels of toxic junk end up being thrown out with the common trash, and their contents eventually percolate through the soil to an aquifer.

SALT SPREADING ON ROADS

In northern climates, especially in urban areas, spreading deicing salts on highways is widespread. In addition to causing deterioration of automobiles, bridges, and the roadway itself, and adversely affecting plants growing alongside a treated highway or sidewalk, salt contamination quickly leaches below the land surface. Because most plants cannot grow in salty soils, the productivity of the land decreases. Continued use can lead to the contamination of wells used for drinking water.

ANIMAL FEEDLOTS

Animal feedlots are a primary source of nonpoint surface water pollution. Animal feedlots are also significant contributors to groundwater pollution. Because animal waste in feedlots literally piles up and is stationary (sometimes for extended periods), runoff containing contaminants may not only enter the nearest surface water body but may also seep into the soil, contaminating it. If the contaminated flow continues unblocked through the subsurface, the flow may eventually make its way into shallow aquifers.

Soil Quality

FERTILIZERS AND PESTICIDES

Fertilizers and pesticides have become the mainstays of high-yield agriculture. They have also had a significant impact on the environment, with each yielding different types of contaminants.

When we apply fertilizers and pesticides to our soil, are we treating the soil—or poisoning it? This question is relatively new to us—and one we are still trying to definitively answer. One thing is certain; with fertilizer and pesticide application, and the long-term effects of such practices, the real quandary is that we do not know what we do not know. We are only now starting to see and understand the impact of using these chemicals. We have a lot to learn. Let's take a look at a few of the known problems with using chemical fertilizers and pesticides.

Nitrogen *fertilizers* are applied to stimulate plant growth, but often in greater quantities than plants can use at the time of application. Nitrate, the most common chemical form of these fertilizers, can easily leach below the plant root zone by rainfall or irrigation. Once it moves below the root zone, it usually continues downward to the water table.

When we place total dependency on chemical fertilizers, they can change the physical, chemical, and biotic properties of the soil—another serious problem.

Pesticides (any chemical used to kill or control populations of "unwanted" animals, fungi, or plants) are not as mobile as nitrate, but they are toxic at much lower concentrations. The perfect pesticide would be inexpensive; affect only the target organism; would have a short half-life; and would break down into harmless substances. However, as of yet, we have not developed the perfect pesticide, and herein lies a multi-faceted problem. Pesticides used in the past—and some of those being used today—are very stable, persistent, and can become long-term problems. They may also be transported from the place of the original application to other parts of the world by wind or ocean currents.

Another problem associated with persistence in pesticide use is that they may accumulate in the bodies of organisms in the lower trophic levels. Recall that when a lower trophic level animal receives small quantities of certain pesticides in its food and cannot eliminate them, the concentration within the organism increases. Remember that when lower trophic level organisms accumulate higher and higher amounts of materials within their bodies, *bioaccumulation* takes place. Eventually, this organism may pass on its accumulation to higher trophic level organisms, and that bioaccumulated toxins pass up the food chain to the highest levels.

ACCIDENTAL SPILLS

Accidental spills of chemical products can be extremely damaging to any of the three environmental mediums—air, water, and soil. Disturbingly common, chemical spills in the soil media that are not discovered right away may allow the contaminant to migrate into and through the soil (contaminating it) to the water table. As an oversimplified general rule of thumb, we can say that the impact of a chemical spill in soil (or any other medium) is directly related to the concentration present at the point and time of release, the extent to which the concentration increases or decreases during exposure, and the time over which the exposure continues.

COMPOSTING OF LEAVES AND OTHER WASTES

Composting of leaves and other wastes, a common practice for many homeowners (especially gardeners), has proven its worth as an environmentally friendly way to dispose of, or beneficially reuse, common waste products. When the feed materials (leaves, twigs, and other organics) have been treated with chemical pesticides and some fertilizers, however, composting this material may be harmful to the soil. In the composting process, the organic material is degraded via a curing process that occurs over time. When water is intentionally added with a garden hose or by precipitation, any chemicals present can be washed or leached from the decaying organic material and drain into the soil, contaminating it.

Industrial Practices and Soil Contamination

Industrial practices that can contaminate soil include the use of underground storage tanks (USTs); contamination from oil field sites; contamination from chemical sites; contamination from geothermal sites; contamination from manufactured gas plants; contamination from mining sites, from many other industrial activities, and, as a result of the Persian Gulf War, contamination from environmental terrorism.

Contamination from Oil Field Sites

People often repeat two cliches—"the past has a way of catching up with us", and "if we don't learn from past mistakes we are doomed to repeat them." But clichés become cliché by being true. And they are certainly true when we consider the problems with soil contamination from oil field sites, a source of large volumes of hydrocarbon-contaminated soil resulting from past and existing oil fields. The extent of the impact of this problem is location-specific. For example, past and present-day oil exploration and production activities located in remote parts of Oklahoma and Texas are not highly visible and thus not subject to public scrutiny. In these remote locations, disposing of hydrocarbon-contaminated soil is easy and inexpensive.

However, in highly urbanized locations (Los Angeles County, CA, for example, where more than 3,000 acres of prime real estate is being or has been exploited for petroleum), developers sit back and eagerly wait until existing fields reach their productive ends. When this occurs, the developers move right in and redevelop the prime real estate (prime in the sense of location, and shortage of available real estate). The ill-informed developer quickly finds out, however, that just because the well runs dry does not mean that the land can be immediately developed for marketing. In these areas, disposal of contaminated soils has emerged as a serious and expensive undertaking.

On or near petroleum-producing properties, the primary sources of soil contamination include oil wells, sumps, pits, dumps leakage from aboveground storage tanks, and leakage and/or spillage. Secondary sources include USTs, transformers, piping ratholes, well cellars, and pumping stations. In addition, the large stationary facilities used for the refining of petroleum have the potential to cause chronic pollution through the discharge of hydrocarbon-laden wastewaters, and by frequent small spills. The primary hazardous constituents associated with oil field properties include drilling mud and constituents, methane, and crude oil. When crude contains certain constituents above maximum contaminant levels—arsenic, chloride, chromium, lead, polychlorinated biphenols (PCBs)—and has a flash point less than the minimum standard as set by American Society for Testing and Materials (ASTM) for recycled products, it may be considered a hazardous waste.

Close to many oil fields exist complete handling and processing ancillaries—refineries, terminals, and pipelines—which also contribute to the overall volume of contaminated soil generated. Of primary concern are contaminants such as crude oil, refined products, and volatile organic compounds.

Contamination from Chemical Sites

The 1979 PEDCO-Eckhardt Survey (commonly referred to as the Eckhardt Survey) of more than 50 of the largest manufacturing companies in the U.S., reported 16,843 tons of organic generated wastes were disposed. Of this total, more than 10 million tons were untreated (residing in landfills, ponds, lagoons, and injection wells). Approximately 0.5 million tons were incinerated, and approximately 0.5 million tons were either recycled or reused. The volume of contaminated soil as a result of one-time occurrence was not addressed in this survey.

Soil contamination from organic chemicals is a serious matter. Some of these organic compounds are biologically damaging even in small concentrations. When they do find their way into the soil, certain organic chemicals may kill, or inhibit sensitive soil organisms, which can undermine the balance of the soil community. Once in the soil, the contaminate may be transported from the soil to the air, water, or vegetation, where it may be inhaled, ingested, or contacted, over a wide area, by a number of organisms. Because of their potential harm, controlling the release of organic chemicals, and understanding their fate and effects in the soil is imperative.

Contamination from Geothermal Sites

Geothermal energy is natural heat generated beneath the earth's surface. The earth's mantle, 15–30 miles below the earth's crust, is composed of a semi-molten rock layer. Beneath the mantle, intense pressure, caused by a molten rock of iron and nickel and decaying radioactive elements, helps warm the earth's surface. Geothermal energy generally lies too deep to be harnessed, but in certain areas, where the molten rock has risen closer to the earth's surface through massive fractures in the crust, underground reservoirs of dry steam, wet steam, and hot water are formed. As with oil deposits, these deposits can be drilled, and their energy is used to heat water, drive industrial processes, and generate electricity.

Generally, geothermal resources are more environmentally friendly than nuclear energy or fossil fuels. However, several drawbacks to geothermal energy use adversely impact the environment. As with oil field operations, geothermal operations provide another example of the close relationship between site usage and the potential for adverse environmental impact. The two constituents associated with geothermal plants that may be considered hazardous are brine and lead-mind scale (Testa, 1997).

Disposal of wastewater from geothermal wells containing *brine* (geothermal mineralizing fluids composed of warm to hot saline waters containing sodium, potassium, chloride, calcium, and minor amounts of other elements that may be harmful to plants and animals) is a major problem. The problem with *lead-mine scale* is more directly related to process equipment failures—developing from scale-buildup in pipes and other process equipment—than environmental problems. However, indirectly, scale buildup may lead to equipment failure (pipe rupture, for example), which, in turn, may lead to geothermal fluid spills with the ultimate result of soil, air, and/or water contamination.

Contamination from Manufactured Gas Plants

The manufacture of gas is not a new process. Since the late 1890s, manufactured gas plants (approximately 3,000 of them located in the U.S.) have been in operation, have been upgraded, or have been completely redeveloped in one way or another. The environmental soil pollution problem associated with manufactured gas plants is the production and disposal of tarry substances, primarily produced in the coal gasification processes—coal carbonization, carbureted waste gas, natural gas, or combination processes.

Other than the obvious mess that the production of any tar-like substance can produce, the main environmental problem with tars is that they are known to contain organic and inorganic compounds that are known or suspected carcinogens. The average volume of tar-contaminated soil averages 10,000 yd^3 per site (Testa, 1997).

Contamination from Mining Sites

According to the U.S. Departments of Interior and Agriculture, since the mid-1860s, more than 3 million acres of land in the United States have been surface-mined for various commodities. Leading the list of commodities related to acreage is mined coal, followed by sand and gravel, stone, gold, phosphate rock, iron ore, and clay, respectively.

Mining operations can give rise to land and water pollution. Sediment pollution via erosion is the most obvious problem associated with surface mining. Sediment pollution to natural surface water bodies is well documented. The Chesapeake Bay, for example, is not the fertile oyster-producing environment it was in the past. Many environmentalists initially blamed the Bay's decline on nutrient-rich substances and chemical pollutants. Recent studies of the Bay's tributaries and the Bay itself, however, indicate that oysters may be suffering (literally suffocating) from sedimentation rather than nutrient contamination.

Less known because less studied is the affect mining sediments and mining wastes (from mining, milling, smelting, and leftovers) have and are having on soil. Typical mining wastes include acid produced by oxidation of naturally occurring sulfides in mining waste; asbestos produced in asbestos mining and milling operations; cyanide produced in precious metal heap-leaching operations; *leach liquors* produced during copper-dump leaching operations; metals from mining and milling operations; and radionuclides (radium) from uranium and phosphate mining operations.

One soil contaminant source is well known and well documented: *acid mine drainage*. Recall that acid formations occur when oxygen from the air and water react with sulfur-bearing minerals to form sulfuric acid and iron compounds. These compounds may directly affect the plant life that absorbs them or have an indirect effect on the flora of a region by affecting the soil minerals and microorganisms.

Another problem with mining is solid waste. Metals are always mixed with material removed from a mine. These materials usually have little commercial value and thus must be disposed somewhere. The piles of rock and rubble are not only unsightly, but they are also prone to erosion, and leaching releases environmental poisons into the soil.

Contamination from Environmental Terrorism

Many human activities that have resulted in environmental contamination have been the result of accidents or poor planning, poor decision-making, inferior design, shoddy workmanship, ignorance, or faulty equipment. Whenever the public reads, hears about, or witnesses environmental contamination, they could assume that those factors were behind the contamination—some fact of human error—at least prior to 1991.

Certainly, the 1st Persian Gulf War changed this perception. After the Gulf War, almost half of Kuwait's 1,500 oil wells were releasing oil into the environment. An estimated 11 million barrels of oil were either being burned or spilled each day, in 600 wells. After the well capping operation got underway and more than 200 wells had been capped, this amount was reduced to approximately 6 million barrels by the end of the summer. The harmful effects to the atmosphere and Persian Gulf were only part of the problem, however. Numerous pools of oil formed, some of them up to 4 ft deep, collectively containing an estimated 20 million barrels of oil (Andrews, 1992).

The larger long-term problem of this blatant act of terrorism is twofold: the presence of oil pools and huge volumes of petroleum-contaminated soil.

USTs: THE PROBLEM

Soil or *subsurface remediation* is a still-developing branch of environmental science and engineering. Because of the regulatory programs of CERCLA (Comprehensive Environmental Response, Compensation, and Liabilities Act of 1980—better known as *superfund*) and RCRA (Resource Conservation and Recovery Act of 1976—better known as the *"cradle to grave act"*) remediation has not only been added to the environmental vocabulary but also has become common and widespread. Just how common and widespread? To best answer this question, follow the response of venture capitalists in their attempts to gain a foothold in this new technological field. Regarding this issue, MacDonald (1997) points out:

> In the early 1990s, venture capitalists began to flock to the market for groundwater and soil cleanup technologies, seeing it as offering significant new profit potential. The market appeared large; not only was $9 billion per year being spent on contaminated site cleanup, but existing technologies were incapable of remediating many serious contamination problems (p. 560, 1997).

Since CERCLA and RCRA, numerous remediation technologies (also commonly known as *Innovative Cleanup Technologies*) have been developed and become commercially available. In this chapter, we discuss these technologies, especially those designed and intended to cleanup sources of subsurface contamination caused by underground storage tanks (USTs). We focus on the technology used in contamination from failed USTs primarily because these units have been the cause of the majority of contamination events and remediation efforts to date. As a result, enormous volumes of information have been recorded on this remediation practice, both from the regulators and the private industries involved with their cleanup.

Keep in mind that (though it likely is the goal of the regulatory agency monitoring a particular UST cleanup effort) no matter what the contaminant, removing every molecule of contamination and restoring the landscape to its natural condition is highly unlikely.

No one knows for sure the exact number of underground storage tank (UST) systems installed in the U.S. However, all present-day estimates range in the millions. Several thousands of these tanks—including ancillaries such as piping—are currently leaking.

Why are so many USTs leaking? USTs leak for several reasons: (1) corrosion, (2) faulty tank construction, (3) faulty installation, (4) piping failure, (5) overfills and spills, or (6) incompatibility of UST contents.

CORROSION PROBLEMS

The most common cause of tank failure is corrosion. Many older tanks were constructed of single shell, unprotected bare steel, and have leaked in the past (and have been removed, hopefully), are leaking at present, or (if not removed or rehabilitated) will leak in the future. If undetected or ignored, such a leak (even a small one) can cause large amounts of petroleum product to be lost to the subsurface.

FAULTY CONSTRUCTION

As with any material item, USTs are only as good as their construction, workmanship, and the materials used in their construction. If a new washing machine is improperly assembled, it will likely fail, sooner than later. If a ladder is made of a material not suited to handle load bearing, it may fail, and the consequences may result in injury. You can probably devise your own list of possible failures resulting from poor or substandard construction or poor workmanship. We all have such a list—some lengthier than others. USTs are no different than any other manufactured item. If they are not constructed properly, and if workmanship is poor, they will fail. It's that simple   although the results of such failure are not so simple.

FAULTY INSTALLATION

Probably the most important step in tank installation is to ensure that adequate backfilling is provided to ensure that no possible movement of the tank can occur after it is placed in the ground. Any such movement might not only damage the tank (especially FRP tanks) but also could also jar loose pipe connections or separate pipe joints. In our experience, failure to use special care in this installation process results in leaks.

Care must also be taken to ensure that underground leak detection devices are carefully and correctly installed. Obviously, if a tank is leaking, knowing it as soon as possible is best—so that remediation can be initiated quickly before a minor spill turns into a nasty environmental contamination incident.

PIPING FAILURES

We have mentioned tank-piping *failure* resulting from improper installation, but piping can fail in other ways as well. Before we discuss them, note that the EPA and other investigators clearly indicate that piping failure is one of the most common causes of "larger" UST spills. If metal piping is used for connecting tanks together, to delivery pumps, to fill drops, or for whatever reason, the danger of corrosion from rust or from electrolytic action is always present. Electrolytic action occurs because threaded pipes (or other metal parts made electrically active by threading) have a strong tendency to corrode if not properly coated or otherwise protected. To prevent electrolytic action, usually *cathodic protection* is installed to negate the electrolytic action. Piping failures are caused equally by poor workmanship, which usually appears around improperly fitted piping joints (both threaded and PVC types), incomplete tightening of joints, construction accidents, and improper installation of cover pad.

Spills and Overfills

All UST facilities are subject to environmental pollution occurring as the result of spills and overfill—usually the result of human error. Though the USEPA has promulgated tank filling procedures in its 40 CFR 280 regulations, and the National Fire Protection Association (NFPA) has issued its NFPA-385 Tank Filling Guidelines, spills from overfilling still frequently occur. Overfilling a UST is bad enough in itself, but the environmental contamination problem is further compounded when such actions occur repeatedly. Petroleum products or hazardous wastes can literally saturate the spill area and can intensify the corrosiveness of soils (Blackman, 1993).

Compatibility of Contents and UST

Obviously, materials must be stored in containers that will contain and hold them. Although it seems obvious, we must not forget that placing highly corrosive materials into containers not rated to contain them is asking for trouble. New chemicals (including fuels) are being developed all the time. Usually, the motive for developing such fuels is to achieve improved air quality but improving air quality does little good at the expense of the other two environmental mediums (water and soil)—if a new fuel that is incompatible with a particular storage tank threatens them.

Many of the USTs presently in use are FRP tanks put in place to replace the old, unprotected, bare steel tanks. FRPs are rated (or can be modified using a different liner) to safely store the fuel products now in common use. The problem occurs when a new, exotic blend of fuel is developed and then placed in an incompatible FRP-type tank.

Common incompatibility problems that have been observed include blistering, internal stress, cracking, or corrosion of the under film. To help prevent FRP-constructed or lined tank problems, the American Petroleum Institute (1980) has put together a standard that should be referred to whenever existing tanks are to be used for different fuel products.

RISK ASSESSMENT

Hydrocarbon spillage or disposal problems are complex. The risk assessment process, which enables scientists, regulatory officials and industrial managers to evaluate the public health risks associated with the hydrocarbon releases (or any other toxic chemical release) to soil and groundwater, can ease the problem's complexity. The risk assessment process consists of the following four steps.

1. **Toxicological evaluation** (hazard identification): should answer the question "Does the chemical have an adverse effect?" The factors that should be considered during the toxicological evaluation for each contaminant include routes of exposure (ingestion, absorption, and inhalation), types of effects, reliability of data, dose, mixture effects, and the strength of evidence supporting the conclusions of the toxicological evaluation.
2. **Dose-response evaluation**: once a chemical is toxicologically evaluated and the result indicates it is likely to cause a particular adverse effect, the next step is to determine the potency of the chemical. The *dose-response curve* is used to describe the relationship that exists between the degree of exposure to a chemical (dose) and the magnitude of the effect (response) in the exposed organism.

3. **Exposure assessment**: conducted to estimate the magnitude of actual and/or potential human exposures, the frequency and duration of these exposures, and the pathways by which humans are potentially exposed.
4. **Risk characterization**: the final step in risk assessment, risk characterization is the process of estimating the incidence of an adverse health effect under the conditions of exposure found and described in the exposure assessment (Ehrhardt et al., 1986; ICAIR, 1985; Blackman, 1993).

EXPOSURE PATHWAYS

Along with performing and evaluating the findings from the four steps involved in risk assessment, determining exposure pathways resulting from the performance of the remediation option chosen to mitigate a particular UST leak or spill is also important. Exposure pathways may be encountered during site excavation, installation, operations, maintenance, and monitoring. They consist of two categories: (1) direct human exposure pathways and (2) environmental exposure pathways. These two categories are subdivided into primary and secondary exposure pathways.

Primary exposure pathways directly affect site operations and personnel (skin contact during soil sampling, for example) or directly affect cleanup levels, which must be achieved by the remedial technology (for example, when soil impact is the principal issue at a site, soil impact sets the cleanup level and the corresponding period when cleanup ceases).

Secondary exposure pathways occur as a minor component during site operations (wind-blown dust, for example), and exhibit significant decreases with time as treatment progresses (EPRI & EEI, 1988).

REMEDIATION OF UST CONTAMINATED SOILS

Before petroleum-contaminated soil from a leaking UST can be remediated, preliminary steps must be taken. *Soil sampling* is important, not only to confirm that a tank is actually leaking, but also to determine the extent of contamination. Any petroleum product remaining within a UST should be pumped out into above ground holding tanks or containers before the tank area is excavated and the tank removed. Removing any residual fuel before excavation is undertaken because of potential damage to the tank during removal.

When site sampling is completed, the range and extent of contamination determined, and the UST is removed, the type of remediation technology to employ in the actual cleanup effort must be determined.

Various organizations, environmental industries, and regulatory agencies have performed technical investigations and evaluations of the various aspects of remediation methods for petroleum hydrocarbons in soil, fate and behavior of petroleum hydrocarbons in soil, and economic analyses. Certainly, one of those industries in the forefront in conducting such studies is the electric utility industry. This particular industry owns and operates many USTs, as well as facilities for using, storing, or transferring petroleum products, primarily motor and heating fuels.

The U.S. Environmental Protection Agency (EPA) developed Federal regulations for reducing and controlling environmental damage from UST leakage, and many states and localities have developed and implemented strict regulations governing USTs and remedial actions for product releases to soil and groundwater. As a result, the Electric Power Research Institute (EPRI), the Edison Electric Institute (EEI), and the Utility Solid Waste Activities Group (USWAG), in a cooperative effort, conducted a technical investigation. From their findings, they developed a report entitled *Remedial Technologies for Leaking Underground Storage Tanks*. This 1988 report focuses on one of the major components of the technical investigation, which, as its title suggests, described and evaluated available technologies for remediating soil and groundwater that contain petroleum products released from an underground storage tank leak.

The EPRI-EEI/USWAG report provides a general introduction to state-of-the-art cleanup technology and serves as a reference in determining the feasible methods, a description of their basic elements, and discussion of the factors to be considered in their selection and implementation for a remedial program.

The available technologies for remediating soil and groundwater containing petroleum products listed by EPRI-EEI/USWAG are divided into two categories: *in situ* treatment and *non-in situ* treatment–the treatment of soil in place and treatment of soil removed from the site. Each of these remedial technologies is briefly described in the following sections, using information adapted from the 1988 EPRI-EEI/USWAG study (the standard reference since 1988).

IN SITU TECHNOLOGIES

In *in situ* technologies, because no excavation is required, exposure pathways are minimized.

In Situ Volatilization (ISV)

In situ volatilization (ISV)—or in situ air stripping—uses forced or drawn air currents through in-place soil to remove volatile compounds. ISV has a successful track record for both effectiveness and cost efficiency. A common ISV system used to enhance subsurface ventilation and volatilization of volatile organic compounds consist of the following operations:

1. A pre-injection air heater warms the influent air to raise subsurface temperatures and increase the volatilization rate.
2. Injection and/or induced draft forces establish airflow through the unsaturated zone.
3. Slotted or screened pipe allows airflow through the system, and to restrict the entrainment of soil particles.
4. A treatment unit (usually activated carbon) recovers volatilized hydrocarbon, minimizing air emissions.
5. Miscellaneous air flow meters, bypass and flow control valves, and sampling ports incorporated into the design facilitate air flow balancing and system efficiency assessment.

Certain factors influence the volatilization of hydrocarbon compounds from soils. These factors fall into four categories: soil, environment, chemical, and management (Jury, 1986).

Soil Factors include water content, porosity/permeability, clay content, and adsorption site density.

1. Water content influences the rate of volatilization by affecting the rates at which chemicals diffuse through the vadose zone. An increase in soil water content decreases the rate at which volatile compounds are transported to the surface via vapor diffusion.
2. Soil porosity and permeability factors relate to the rate at which hydrocarbon compounds volatize and are transported to the surface. A function of the travel distance and cross-sectional area available for flow, diffusion distance increases and cross-sectional flow area decreases with decreasing porosity.
3. Clay content affects soil permeability and volatility. Increased clay content decreases soil permeability, which inhibits volatilization.
4. Adsorption site density refers to the concentration of sorptive surface available from the mineral and organic contents of soils. An increase in adsorption sites indicates an increase in the ability of the soils to immobilize hydrocarbon compounds in the soil matrix.

Environmental Factors include temperature, wind, evaporation, and precipitation.

1. Temperature increase increases the volatilization of hydrocarbon compounds.
2. Wind increase decreases the boundary layer of relatively stagnant air at the ground/air interface, which can assist volatilization.

3. Evaporation of water at the soil surface is a factor controlling the upward flow of water through the unsaturated zone, which can assist volatilization.
4. Precipitation provides water for infiltration into the vadose zone.

Chemical Factors are critical players in affecting the way in which various hydrocarbon compounds interact with the soil matrix. Solubility, concentration, octanol-water participating coefficient, and vapor pressure are the primary chemical properties that affect the susceptibility of chemicals to the in-situ volatilization process.

Management Factors -related to soil management techniques (fertilization, irrigation) decrease leaching, increase soil surface contaminant concentrations, assist volatilization, or maximize soil aeration.

Site-specific conditions (soil porosity, clay content, temperature, and so forth) drive the effectiveness of in situ volatilization techniques. Pilot studies and actual experience confirm:

- In situ volatilization has been successful for remediation in unsaturated zones containing highly permeable sandy soils with little or no clay.
- Recovery periods are typically about 6–12 months.
- Gasoline (which is light and volatile) has the greatest recovery rate.
- In situ volatilization can be used in conjunction with product recovery systems.
- Because ultimate cleanup levels are site-dependent and cannot be predicted, they are usually set by regulatory agencies.

In Situ Biodegradation

In situ biodegradation uses naturally occurring microorganisms in the soil to degrade contaminants into another form. Most petroleum hydrocarbons can be degraded to carbon dioxide and water by microbial processes (Grady, 1985). For hydrocarbon removal, stimulating their growth and activities (primarily through the addition of oxygen and nutrients) enhances this process. Factors such as temperature and pH influence their rate of growth.

Based upon documentation and significant background information related to the successful land treatment of refinery waste, biodegradation has proven its worth as an efficient and cost-effective method for the reduction of hydrocarbons in soil.

Heyse et al. (1986) described the biodegradation process:

1. A submersible pump transports groundwater from a recovery well to a mixing pump.
2. Nutrients, including nitrogen, phosphorous and trace metals, are added to the water in a mixing tank. These nutrients are then transported by the water to the soil, supporting microbial activity.
3. Hydrogen peroxide is added to the conditioned groundwater from the mixing tank just prior to reintroduction to the soil. As hydrogen peroxide decomposes, it provides the needed oxygen for microbial activity.
4. Groundwater pumped to an **infiltration gallery** and/or injection well reintroduces the conditioned water to the aquifer or soils.
5. Groundwater flows from the infiltration galleries or injection wells through the affected area, then back to the recovery wells. The flow of the water should contact all soils containing degradable petroleum hydrocarbons.
6. The water is drawn to the recovery well and pumped to the mixing tank to complete the treatment loop.
7. Groundwater in which hydrocarbon concentrations have been reduced to very low levels is often sent through a carbon adsorption process for the removal of the residual hydrocarbons.

Environmental Factors

The environmental factors that influence biodegradation in soils are temperature and microbial community.

1. **Temperature** is important in biodegradation of contaminants in soils. In general, biodegradation of petroleum fraction increases as temperatures increase (up to 104°F) from increased biological activity (Bossert & Bartha, 1984).
2. A **microbial community** capable of degrading the target compound is important in the biodegradation process. Most in situ biodegradation schemes make use of existing microbial populations; however, attempts have been made to supplement these populations with additional organisms or engineered organisms.

Chemical Factors

While biodegradation is impossible if substrate concentrations are too high, biodegradation relies on a substantial substrate (target compound) presence to ensure microbes metabolize the target compound. Biodegradation is also limited by the solubility of a compound in water because most microbes need moisture to acquire nutrients and avoid desiccation.

Soil Factors

The degradation of hydrocarbons in soil requires proper aerobic conditions. Moisture is also essential for microbial life—however, too much moisture (saturation) limits oxygen levels and can hinder biological activity. Bossert and Bartha (1984) reported that moisture content between 50% and 80% of the water-holding capacity is considered optimal for aerobic activities. In the in-situ biodegradation processes, oxygen transfer is a key factor; soils must be fairly permeable to allow this transfer to occur.

Another important soil factor is soil pH, which directly affects the microbial population supported by the soil. Biodegradation is usually greater in a soil environment with a pH of 7.8. Optimal biodegradation of petroleum hydrocarbons requires nutrients (nitrogen and phosphorus) in the proper amounts.

Environmental Effectiveness

The effectiveness of in situ biodegradation depends upon the same site-specific factors as other in situ technologies, but the historical record for this technology is limited. However, several case studies suggest:

- In situ biodegradation is most effective for situations involving large volumes of subsurface soils.
- Significant degradation of petroleum hydrocarbons normally occurs in the range of 6–18 months (Brown et al., 1986).
- In situ biodegradation has most often been used for the remediation of groundwaters impacted by gasoline.
- Research suggests limited biodegradation of benzene or toluene may occur under anaerobic conditions (Wilson et al., 1986).
- In soils, the remedial target level for in situ biodegradation could be in the low mg/L (ppm) level for total hydrocarbons (Brown et al., 1986).

In Situ Leaching and Chemical Reaction

The *in situ leaching and chemical reaction* process uses water mixed with a surfactant (a surface-active substance—soap) to increase the effectiveness of flushing contaminated soils in the effort to leach the contaminants into the groundwater. The groundwater is then collected downstream of the leaching site, through a collection system for treatment and/or disposal.

Environmental Effectiveness

The in situ leaching and chemical reaction process is not commonly practiced. Little performance data on its environmental effectiveness exists.

In Situ Vitrification

The *in-situ vitrification* process employs electrical current passed through electrodes (driven into the soil in a square configuration), which produces extreme heat and converts soil into a durable glassy material. The organic constituents are pyrolyzed in the melt and migrate to the surface where they combust in the presence of oxygen. Inorganics in the soil are effectively bound in the solidified glass (Johnson & Cosmos, 1989).

Environmental Effectiveness

Organic materials are combusted and/or destroyed by the high temperatures encountered during the vitrification process. The in-situ vitrification process is a developing technology. The jury is still out in determining its environmental effectiveness.

In Situ Passive Remediation

The *in-situ passive remediation process* is the easiest to implement and the least expensive, mainly because it involves no action at the site; however, it is generally unacceptable to the regulatory agencies. It relies upon several natural processes to destroy the contaminant. These natural processes include biodegradation, volatilization, photolysis, leaching, and adsorption.

Environmental Effectiveness

Because passive remediation depends upon a variety of site-specific and constituent-specific factors, the environmental effectiveness of passive remediation must be decided on a case-by-case basis.

In Situ Isolation/Containment

As the name implies, isolation/containment methods are directed toward preventing the migration of liquid contaminant or leachates containing contaminants. Accomplished by separating the contamination area from the environment, and by installation of impermeable barriers to retain liquid containments within the site, successful application of these methods is usually contingent on the presence of an impervious layer beneath the contaminant to be contained, and the attainment of a good seal at the vertical and horizontal surfaces.

The containment devices discussed in this section adequately isolate the contamination. However, the destruction of the contaminant is not accomplished.

Containment Methods

- **Slurry walls**: fixed underground physical barriers formed in an excavated trench by pumping slurry, usually a bentonite or cement and water mixture.
- **Grout curtains**: (similar to slurry walls) suspension grouts composed of Portland cement or grout are injected under pressure to form a barrier.
- **Sheet-piling**: construction involves physically driving rigid sheets, pilings of wood, steel, or concrete into the ground to form a barrier.

Environmental Effectiveness

Isolation/containment systems are effective in physically preventing or impeding migration, but the contaminant is not removed or destroyed.

Non-*In Situ* Technologies

Unlike in-situ techniques, *non-in situ techniques* require the removal (usually by excavation) of contaminated soils. These soils can be either treated on-site or hauled off-site for treatment. Another difference that must be taken into consideration when employing non-in situ techniques is the exposure pathways associated with the handling and/or transport of contaminated soil. The non-in situ technologies for soils discussed in this section include land treatment, thermal treatment, asphalt incorporation, solidification/stabilization, chemical extraction and excavation.

Land Treatment

Land treatment or land farming is the process by which affected soils are removed and spread over an area to enhance naturally-occurring processes, including volatilization, aeration, biodegradation, and photolysis. The land treatment process involves tilling and cultivating soils to enhance the biological degradation of hydrocarbon compounds.

- The area used for land treatment is prepared by removing surface debris, large rocks and brush.
- The area is graded to provide positive drainage and surrounded by a soil berm to contain run-off within the land treatment area.
- The pH is adjusted with lime (if necessary) to provide a neutral pH.
- If the site is deficient in nutrients, fertilizer is added.
- The petroleum-contaminated soil is spread uniformly over the surface of the prepared area.
- The contaminated material is incorporated into the top 6–8 in of soil (to increase contact with microbes) with a tiller, disc harrow, or other plowing devices.
- More soils that contain petroleum products are applied at proper intervals to replenish hydrocarbon supply.
- Hydrocarbon and nutrient levels, and soil pH are monitored to assure that the hydrocarbons are properly contained and treated in the land treatment area.

Environmental Effectiveness

The effectiveness of land treatment or land farming is highly dependent on site-specific conditions. Several years of experience with treating petroleum compounds using this technology confirm:

- Land treatment is an effective means of degrading hydrocarbon compounds.
- Continuous treatment of petroleum-laden soils can result in the accumulation of metals in the soil matrix.
- Ultimate degradation rates are site-dependent and cannot be predicted.

Thermal Treatment

Thermal treatment of contaminated soils requires special equipment but is capable of providing complete destruction of the petroleum-laden contaminant. Affected soils are removed from the ground and exposed to excessive heat in one of the various types of incinerators currently available. These include rotating kilns, fluidized bed incinerators, fixed kilns or hearths, rotating lime or cement kilns, and asphalt plants.

Environmental Effectiveness

High-temperature incineration for the destruction of petroleum product-laden soil is well documented. Destruction and removal efficiencies of 99% can be expected.

Asphalt Incorporation and Other Methods

Asphalt incorporation is a recently developed remedial technology that goes beyond remediation, in the sense that the asphalt incorporation technique is actually a *reuse* and/or *recycling technology*, whereby the containment entrained in soil is used in *beneficial reuse* (to make asphalt, cement products, and bricks), not just destroyed or disposed.

Asphalt incorporation and other reuse/recycling technologies involve the assimilation of petroleum-laden soils into hot or cold asphalt processes, wet or dry cement production processes, or brick manufacturing. During these processes, the petroleum-laden soils are mixed with other constituents to make the final product. In turn, the petroleum contaminants are either volatilized during some treatments or trapped within the substance, thereby limiting contaminant migration.

The conversion of asphalt into asphalt concrete or bituminous concrete involves producing a material that is plastic when being worked and that sets up to a specified hardness sufficient for its end use. The incorporation of contaminated soil into bituminous end products is accomplished by two conventional processes: Cold-Mix Asphalt Processes (CMA) and Hot-Mix Asphalt processes (HMA) (Testa, 1997).

The cold-mix asphalt process (commonly referred to as environmentally processed asphalt) is a mobile or in-place process. It uses soils contaminated with a variety of contaminants (including petroleum hydrocarbons) to serve as the fine-grained component in the mix, along with asphalt emulsion and specific aggregates to produce a wide range of cold-mix asphaltic products. The mix is usually augmented with lime, Portland cement, or fly ash to enhance the stability of the end product. The mixing or incorporation method is accomplished physically by either mixed-in-place methods for large quantities or windrowing for smaller quantities.

The CMA process has several advantages: (1) a variety of contaminants can be processed; (2) large volumes of contaminated soil can be incorporated; (3) possesses flexible mix design and specifications; (4) is a mobile process; (5) has minimal weather restrictions; (6) is cost-effective; (7) can be stockpiled and used when needed; and (8) processing can occur on site. The limitations of CMA include: (1) any volatiles present must be controlled and (2) small volumes of contaminated soil may not be economically viable for mobile plants.

The *hot-mix asphalt process* involves the incorporation of petroleum-laden soils into hot asphalt mixes as a partial substitute for aggregate. This mixture is most often applied to pavement. HMA is conventionally produced using either the batch or drum mixing processes. In either of these processes, both mixing and heating are used to produce pavement material.

During the incorporation process, the mixture, including the contaminated soils (usually limited to 5% of the total aggregate feed at any one time), is heated. This causes volatilization of the more-volatile hydrocarbon compounds at various temperatures. Compound migration is limited by incorporating the remainder of the compounds into the asphalt matrix during cooling.

The advantages associated with using the HMA process are: (1) the time required to dispose of hydrocarbon-laden material is limited only by the size of the batching plant (material may be excavated and stored until it can be used); and (2) can process small volumes of affected soil easily. The disadvantages include: (1) the compound must be applied immediately after processing; (2) has potential for elevated emissions; (3) has emission restrictions; and (4) incomplete burning of light-end hydrocarbons can affect the quality of end product.

Raw materials such as limestone, clay, and sand are incorporated into the *cement production process*. Once incorporated, these materials are usually fed into a rotary kiln. Contaminated soil may be introduced along with the raw materials or dropped directly into the hot part of the kiln. The mix is then heated to up to 2,700°F. Petroleum-laden soil chemically breaks apart during this process, whereas the inorganic compounds recombine with the raw materials and are incorporated into a clinker—dark, hard, golf ball-sized nodules of rapidly formed Portland cement—which are mixed with gypsum and ground to a fine powder (Testa, 1997).

Advantages of the cement production process include: (1) the technology is in place and has been tested; (2) raw materials are readily available; (3) relatively low water solubility and low water permeability; and (4) can accommodate a wide variety of contaminants and material. The disadvantages include: (1) odorous material limitations; (2) wide range of volume increase; and (3) material restrictions both technically and aesthetically.

Petroleum-laden soil has been used as an ingredient in the production of bricks. The contaminated soil replaces either the shale and/or firing clay normally used in the *brick manufacturing process*. Generally, clay and shale are incorporated into a plasticized mixture, then extruded and molded into brick. When dried, the brick is fired in a kiln with temperatures ranging up to 2,000°F during a 3-day residence period. When contaminated soil is added to the process, it is mixed with clay and shale, molded into brick, dried, and preheated. Then the brick is fired at 1,700°F–2000°F for approximately 12 hours in the kiln. While in the kiln, high temperature and residence time destroy organics and incorporate inorganics into the vitrified end product.

Advantages to the brick manufacturing process to reuse or recycle contaminated soils are: (1) fine-grained, low permeability soils can be accommodated; (2) the technology is in place and has been tested; and (3) processing can occur on-site. The disadvantage is that this process is restricted primarily to petroleum hydrocarbons and fly ash.

Solidification/Stabilization

Solidification/stabilization of petroleum-laden soils is used to immobilize contaminants by either encapsulating or converting them but does not change the physical nature of the contaminant. This is not a commonly used practice for soils—because the ultimate destruction of the contaminants does not occur.

Solidification/stabilization processes can be performed either on- or off-site. Various stabilizers and additives are mixed with the material to be disposed. For example, one procedure consists of a generalized process for the manufacture of pozzolanic material (burnt shale or clay resembling volcanic dust that will chemically react with calcium hydroxide at ordinary temperature to form compounds possessing cementitious properties) using fly ash (Mehta, 1983; Transportation Research Board, 1976).

More commonly used to stabilize oily wastes and sludges contained in surface impoundments, solidification/stabilization processes accomplish this in two ways. In situ surface impoundments, the stabilizing agent is added directly to the impoundment and thoroughly mixed. Treated in sections, as each solidifies, it is used as a base that allows the equipment to reach further out into the impoundment.

The second method involves excavation of the sludges contained in the impoundment, following this procedure:

- Earth-moving machines level piles of kiln dust into 6- to 12-in-deep layers;
- A machine lifts the sludge from the impoundment and places it on top of the kiln dust;
- Machines then mix the two materials, and a pulverizing mixer is driven over the mixture until homogeneity is achieved;
- The mixture is allowed to dry for about 24 hours, then compacted and field tested (Musser & Smith, 1984).

Usually, the layers are then stacked to build an in-place landfill, or the semi-solidified sludge can be trucked to another landfill location.

Chemical Extraction

Chemical extraction is the process in which excavated contaminated soils are washed to remove the contaminants of concern. This washing process typically is accomplished in a washing plant that uses a water/surfactant or a water/solvent mixture to remove the contaminants. This method is very similar to the in-situ leaching process described later. The primary difference is that by removing the soil from the ground, wash mixtures can be used that do not expose the environment to further contamination. This process increases product recovery and is a proven method for the removal of hydrocarbon contaminants from the soil.

Excavation

Excavation involves the safe physical removal (using trench boxes, for example) of the contaminated soil for disposal at a hazardous waste or other disposal landfill site. This process has been the mainstay of site remediation for several decades, but recently has been discouraged by newer regulations that favor alternative waste treatment technologies at the contaminated site. Today, excavation is generally considered a storage and not a treatment, process and raises issues of future liability for the responsible parties regarding the ultimate disposal of the soils. One of the factors contributing to the regulators pushing for on-site treatment methodologies versus landfilling is that landfills are quickly reaching their fill limits, with fewer and fewer new landfilling sites authorized for construction and operation.

USEPA (1985) pointed out the positive and negative aspects of excavation. On the positive side, excavation takes little time to complete and allows for complete cleanup of the site. The negative aspects are the necessary worker/operator safety considerations, the production of dust and odor, and the relatively high costs associated with the excavation, transportation, and ultimate disposal of the soil.

NOTE

1. Much of the material in this chapter is from Spellman, F.R. (2009). *The Science of Environmental Pollution*. Boca Raton, FL: CRC Press.

REFERENCES

American Petroleum Institute (1980). *Landfarming: An Effective and Safe Way to Treat/Dispose of Oily Refinery Wastes*. Solid Waste Management Committee.

Andrews, J.S., Jr. (1992). The cleanup of Kuwait. In *Hydrocarbon Contaminated Soils*, Vol. II, pp. 21–23. Kostecki, P.T., et al. (Ed.). Boca Raton, FL: CRC/Lewis Publishers.

Blackman, W.C., Jr. (1993). *Basic Hazardous Waste Management*. Boca Raton, FL: Lewis Publishers.

Bossert, I. & Bartha, R. (1984). The fate of petroleum in soil ecosystems. In *Petroleum Microbiology*, pp. 69–71. Atlas, R.M. (Ed.). New York: Macmillan Co.

Briggs, D. (1997). *Fundamentals of the Physical Environment*. Boca Raton, FL: Routledge Publishers, p. 323.

Brown, R.S., Norris, R.D., & Estray, M.S. (1986). *In Situ Treatment of Groundwater*. Baltimore, Maryland: HazPro 86: Professional Certification Symposium and Exposition.

Ehrhardt, R.F., Stapleton, P.J., Fry, R.L., & Stocker, D.J. (1986). *How clean is clean?—Cleanup Standards for Groundwater and Soil*. Washington, DC: Edison Electric Institute.

EPRI & EEI (1988). *Remedial Technologies for Leaking Underground Storage Tanks*. Chelsea, MI: Lewis Publishers.

Franck, I. & Brownstone, D. (1992). *The Green Encyclopedia*, New York: Prentice-Hall.

Grady, P.C. (1985). Biodegradation: Its measurement and microbiological basis. *Biotechnology and Bioengineering* **27**:660–674.

Heyse, E., James, S.C., & Wetzel, R. (1986). In situ aerobic biodegradation of aquifer contaminants at Kelly air force base. *Environmental Progress* **5**:207–211.

ICAIR, Life Systems, Inc. (1985). *Toxicology Handbook*. Washington, DC: USEPA.

Johnson, N.P. & Cosmos, M.G. (1989). Thermal treatment technologies for HazWaste remediation. *Pollution Engineering* **21**:16.

Jury, W.A. (1986). Volatilization from soil, *Guidebook for Field Testing Soil Fate and Transport Models-Final Report*, Washington, DC: USEPA.

Konigsburg, E.L. (1996). *The View from Saturday*. New York: Scholastic Books.

MacDonald, J.A. (1997). Hard times for innovation cleanup technology. *Environmental Science & Technology* **31**(12):560–563.

Mehta, P.K. (1983). Pozzolanic and cementitious by-products as miner admixtures for concrete—A critical review. In *Fly Ash, Silica Fume, Slag, and Other Mineral By-Products in Concrete*, Vol. 1. Malhotra, V.M. (Ed.). American Concrete Institute.

Muir, J. (1875). *110 John Muir Quotes on Nature, Moutnains are the beauty of the Great Outdoors*. Accessed 12/12/21 @ https://brightdrops.com/john-muir-quotes#.

Musser, D.T. & Smith, R.L. (1984). Case study: In situ solidification/fixation of oil field production fluids—A novel approach. In *Proceedings of the 39th Industrial Waste Conference*, Purdue University, West Lafayette, Indiana.

PEDCO (1979). *PEDCO Analysis of Eckhardt Committee Survey for Chemical Manufacturer's Association*. Washington, DC: PEDCO Environmental Inc.

Spellman, F.R. (1997). *Safe Work Practices for Water/Wastewater Treatment Plants*. Lancaster, PA: Technomic Publishing Co.

Testa, S.M. (1997). *The Reuse and Recycling of Contaminated Soil*. Boca Raton, FL: CRC/Lewis Publishers.

Tomera, A.N. (1989). *Understanding Basic Ecological Concepts*. Portland: Maine: J. Weston Walch, Publisher.

Transportation Research Board (1976). *Lime-Fly Ash: Stabilized Bases and Subbases*. TRB-NCHRP Synthesis Report 37.

Tucker, R.K. (1989). Problems dealing with petroleum contaminated soils: A New Jersey perspective. In *Petroleum Contaminated Soils*, Vol. I, pp. 121–124. Kostecki, P.T. & Calabrese, E.J. (Eds.). Boca Raton, FL: CRC/Lewis Publishers.

Wilson, J.T., Leach, L.E., Benson, M., & Jones, J.N. (1986). In situ biorestoration as a ground water remediation technique. *Ground Water Monitoring Review* **6**:56–64.

USEPA (1985). *Remedial Action at Waste Disposal Sites* (revised). Washington, DC: United States Environmental Protection Agency.

10 Solid and Hazardous Waste

... Unfortunately, man is in the woods, and waste and pure destruction are already making rapid headway.

John Muir (1877)

When we "throw away" waste, it is not gone. Dealing with the waste has only been postponed. Sometimes this postponement means that when we go back, the wastes are rendered helpful and harmless (as with some biodegradable wastes) but more often, it means that the problems we must face will be worse—increased by chemistry and entropy--that 55-gallon drum of toxic waste was easier to handle before it rusted out.

Amid the cries of "Not In My Backyard" and "Pick up the Trash, but Don't Put It Down," we need to hear a more realistic, and environmentally kinder truth: **There's no such thing as a free lunch**.

We pay, somehow, for what we get or use, whether we see the charges or not. The price for our solid waste habits will soon be charged to us. In some places (big cities, for example), the awareness of the size of the bill is sinking in.

Environmentally, what does that mean? In short, if we, as a society, are going to consume as we do, build as we do, grow as we do, we have to pay the price for our increase. And that, sometimes, is going to mean that our waste is going to be "in our backyard." We will have to increase the amount of solid waste we reuse and recycle; we will have to spend tax dollars to solve the problems with landfills and trash incineration, we will have to seriously look at how we live, how the goods we buy are packaged, how our industries deal with their wastes—because if we don't, the bill will be more than we can afford to pay.

Advancements in technology have made our lives more comfortable, safer, healthier, and in many cases more enjoyable. Some would say that progress is not without cost. This statement is correct—however, what costs do they refer to? Can we afford the consequences if these costs include more Bhopal's, Times Beaches, Love Canals, or another Exxon Valdez? If such disasters are "to be included as a cost of progress," then we must say that the cost outweighs the gain.

What we must do to ensure a balance between technological progress and its environmental results is to use technological advances to ensure that "progress" is not too costly—or life-threatening—to both our environment and to ourselves.

RCRA's Waste Management Hierarchy sums up what could/ should/ would happen with waste—any kind of waste—in a "best of all possible worlds." But though it is idealistic and too simple to say we "should" follow these standards, in practical terms, we benefit in the long term by striving to achieve them.

Regulating problem wastes, developing safe and environmentally friendly ways to dispose of them, and using the technologies we develop to control the future of such wastes is in the best interests for us all.

Frank R. Spellman (1995)

With regard to teaching college level courses dealing with solid and hazardous wastes, the portrait and portrayal of American Society that I often use in my lectures presented to environmental science/health/engineering students is not flattering: I describe America as The Throwaway Society. I explain that the American Throwaway Society is one that displays and underscores a characteristic that might be described as habit, trend, custom or practice—the tendency we have to discard those objects we no longer want. When we don't want it anymore, we discard it. When it offends us, we dispose of it. Throughout history, a common practice was, "I don't want it any longer. Take it down to the river and dump it." When mile-long caravans of wagon trains crossed the great American prairies, the pioneers discarded anything and everything that over-burdened the wagon; this was a common practice for survival—homemaking materials, furnishings that made the trip more difficult—some of them necessary to successful homesteading, but too heavy to carry were left along the trail west. Lifelong treasures once so important that they couldn't be left behind now became burdens (and waste) to the American pioneering Waggoner's . . . things to be disposed of to make their lives easier. This pioneering throwaway tendency has been called the Frontier Mentality. That is, "we can throw away anything we no longer want out here or in the Frontier (the prairie) . . . there is so much endless space, trash placed in the vast, endless prairie won't hurt a thing . . . it's like urinating in the ocean . . . won't hurt a thing!" This, of course, ignores the effect of accumulation, which inexorably clogs up all that 'empty' space eventually.

Again, when something is no longer of value because it is broken, worn out, out of style, or no longer needed for whatever reason, we feel discarding it is not a big issue. But it is—particularly when the item we throw away is a hazardous substance that is persistent, nonbiodegradable, and poisonous.

What is the magnitude of the problem with hazardous substance/waste disposal? Let's take a look at a few facts.

- Hazardous substances—including industrial chemicals, toxic waste, pesticides, and nuclear waste—are entering the marketplace, the workplace, and the environment in unprecedented quantities.
- The United States produces almost 300+ million metric tons of hazardous waste each year—with a present population of 320,000,000+, this amounts to more than one ton for every person in the country.
- Through pollution of air, soil and water supplies, hazardous wastes pose both short- and long-term threats to human health and environmental quality.

Frank R. Spellman (1996)

INTRODUCTION

In this chapter, we discuss a growing and significant problem facing not only all practitioners of environmental practice but also all of humanity: Anthropogenically produced wastes. Specifically, we are faced with daunting questions: What are we going to do with all the waste we generate? What are the alternatives? What are the technologies available to us at present to mitigate the "waste problem"—a problem that grows with each passing day.

Before beginning our discussion, we focus on an important question: When we throw waste away, is it really gone? Remember, though we are faced today and in the immediate future with growing mountains of wastes that we produce (and we are running out of places on earth to dispose of them), an even more pressing twofold problem is approaching: the waste's toxicity and persistence.

Later, we discuss waste and the toxicity problem, but for now, think about the persistence of the wastes that we dispose. For example, when we excavate a deep trench and place within it several 55-gallon drums of liquid waste, then bury the entire sordid mess, are we really disposing of the waste in an earth-friendly way? Are we disposing of it permanently? What happens a few years later when the 55-gallon drums corrode and leak? Where does the waste go? Where does the waste end up? What are the consequences of such practices? Are they insignificant to us today because they are tomorrow's problems? Does anyone really care?

We need to ask ourselves these questions and determine the answers now. If we are uncomfortable with the answers we come up with, shouldn't we feel the same about the answers someone else (our grandchildren) will have to come up with later—later, when it is far too late?

Waste is not easily disposed. We can hide it. We can mask it. We can move it from place to place. We can take it to the remotest corners of the earth. But because of its persistence, waste is not always gone when we think it is. It has a way of coming back, a way of reminding us—a way of persisting. How persistent is waste? It is very persistent, as thousands of documented cases make clear.

In this section, we define and discuss solid wastes. In particular, we focus on a significant portion of solid wastes, *Municipal Solid Wastes (MSW)*, because people living in urban areas where many of the problems associated with solid waste occur generate these wastes. In addition, we discuss another significant waste problem: Hazardous wastes. We also discuss waste control technologies related to waste minimization, treatment, and disposal.

SOLID WASTE REGULATORY HISTORY (UNITED STATES)

For most of the nation's history, municipal ordinances (rather than federal regulatory control) were the only solid waste regulations in effect. These local urban governments controlled solid waste almost from the beginning of each settlement—because of the inherent severe health consequences derived from street disposal. Along with prohibiting the dumping of waste in the streets, municipal regulations usually stipulate requirements for proper disposal in designated waste dump sites and mandate owners to remove their waste piles from public property.

The federal government did not begin regulating solid waste dumping until the nation's harbors and rivers were either overwhelmed with raw wastes or headed in that direction. The federal government used its constitutional powers under the *Interstate Commerce Clause* of the constitution to enact the *Rivers and Harbors Act* in 1899. The U.S. Army Corps of Engineers was empowered to regulate and, in some cases, prohibit private and municipal dumping practices.

Not until 1965 did Congress finally get into the picture (as a result of strong public opinion) by adopting the *Solid Waste Disposal Act* of 1965, which became the responsibility of the U.S. Public Health Service to enforce. The intent of this act was to:

1. Promote the demonstration, construction, and application of solid waste management and resource recovery systems that preserve and enhance the quality of air, water, and land resources.
2. Provide technical and financial assistance to state and local governments and interstate agencies in the planning and development of resource recovery and solid waste disposal programs.
3. Promote a national research and development program for improved management techniques; more effective organizational arrangements; new and improved methods of collection, separation, recovery, and recycling of solid wastes; and the environmentally safe disposal of nonrecoverable residues.
4. Provide for the promulgation of guidelines for solid waste collection, transport, separation, recovery, and disposal systems.
5. Provide for training grants in occupations involving the design, operation, and maintenance of solid waste disposal systems (Tchobanoglous et al., 1993).

After Earth Day 1970, Congress became more sensitive to waste issues. In 1976, Congress passed solid waste controls as part of the *Resource Conservation and Recovery Act* (RCRA). "Solid waste" was defined as any garbage, refuse, sludge, or other discarded material from a waste treatment plant, a water supply treatment plant, an air pollution control facility, or other source.

In 1980, Public Law 96-510, 42 U.S.C. Article 9601, the *Comprehensive Environmental Response, Compensation, and Liability Act* (CERCLA) was enacted to provide a means of directly responding to and funding the activities of responding to problems at uncontrolled hazardous waste disposal sites. Uncontrolled MSW landfills are facilities that have not operated or are not operating under RCRA (USEPA, 1989).

Many other laws that apply to the control of solid waste management problems are now in effect. Federal legislation and associated regulations have encouraged solid waste management programs to be implemented at the state level of government. Apparently, legislation will continue to be an important part of future solid waste management.

SOLID WASTE CHARACTERISTICS

Solid waste (also called *refuse, litter, rubbish, waste, trash,* and (incorrectly) *garbage*) refers to any of a variety of materials that are rejected or discarded as being spent, useless, worthless, or in excess. Table 10.1 provides a useful waste classification system.

Solid waste is probably more correctly defined as "any material thing that is no longer wanted." O'Reilly (1992) points out that defining solid waste is tricky because solid waste is a series of paradoxes:

- personal in the kitchen trash can—but impersonal in a landfill;
- what one individual may deem worthless (an outgrown or out-of-fashion coat, for example) and fit only for the trash can—another individual may find valuable;
- of little cost concern to many Americans—yet very costly to our society in the long term;
- an issue of serious federal concern—yet a very localized problem from municipality to municipality.

The popular adage is accurate—everyone wants waste to be picked up, but no one wants it to be put down. It goes almost without saying that the other adage, "Not in My Back Yard" (NIMBY) is also accurate. The important point, though, is that whenever a material object is thrown away, regardless of its actual or potential value, it becomes a solid waste.

Garbage (with its tendency to decompose rapidly and create offensive odors) is often used as a synonym for solid waste, but actually refers strictly to animal or vegetable wastes resulting from handling, storage, preparation, or consumption of food.

TABLE 10.1
Classification of Solid Waste

Type	Principal Components
Trash	Highly combustible wastepaper, wood, cardboard cartons, including up to 10% treated papers, plastic or rubber scraps; commercial and industrial sources.
Rubbish	Combustible waste, paper, cartons, rags, wood scraps, combustible floor sweepings; domestic, commercial, and industrial sources.
Refuse	Rubbish and garbage; residential sources
Garbage	Animal and vegetable wastes, restaurants, hotels, markets; institutional, commercial, and club sources

Source: Adapted from Davis and Cornwell (1991).

Solid and Hazardous Waste

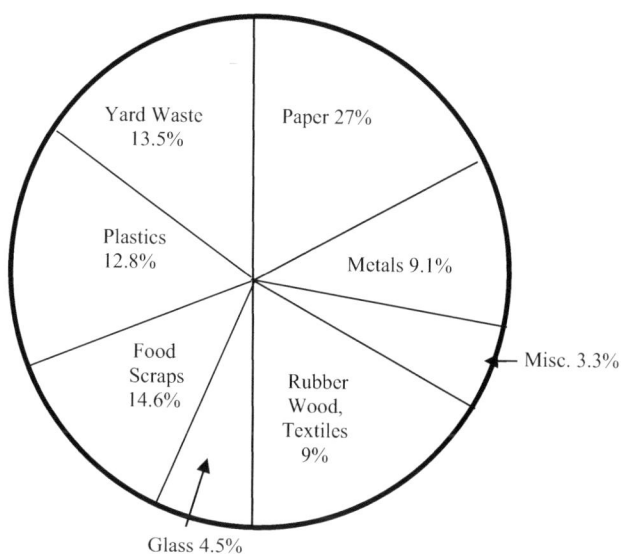

FIGURE 10.1 Composition of municipal solid waste discarded in a typical day by each American. (EPA, Meeting the Environmental Challenge, 2013; Characterization of Municipal Solid Waste in U.S., 2013 Update, EPA/530-5-019.)

The collective and continual production of all refuse (the sum of all solid wastes from all sources) is referred to as the *solid waste stream*. As stated previously, an estimated 6 billion metric tons of solid waste are produced in the United States each year. The two largest sources of solid wastes are agriculture (animal manure, crop residues, and other agricultural by-products) and mining (dirt, waste rock, sand and slag, the material separated from metals during the smelting process). About 10% of the total waste stream is generated by industrial activities (plastics, paper, fly ash, slag, scrap metal, and sludge or biosolids from treatment plants).

From Figure 10.1, we see that paper and paperboard account for the largest percentage (about 29%) of refuse materials by volume of MSW. Yard wastes account for almost 14%. Glass and metals make up almost 14% of MSW, food wastes just under 15%, and plastics about 12.3%.

USEPA (2009) points out that approximately 243 million metric tons of MSW were generated in the U.S. in 2009, equivalent to a bit more than 5 pounds per person per day. By the year 2010, the EPA estimates that waste generation in the U.S. will rise to more than 250 million metric tons annually, almost 5.5 pounds per person per day.

SOURCES OF MUNICIPAL SOLID WASTES (MSW)

Sources of municipal solid wastes in a community are generally related to land use and zoning. MSW sources include residential, commercial, institutional, construction and demolition, municipal services, and treatment plants.

Residential Sources of MSW

Residential sources of MSW are generated by single and multifamily detached dwellings and apartment buildings. The types of solid wastes generated include food wastes, textiles, paper, cardboard, glass, wood, ashes, tin cans, aluminum, street leaves, and special bulky items including yard wastes collected separately, white goods (refrigerators, washers, dryers, etc.), batteries, oil, tires and household hazardous wastes.

COMMERCIAL SOURCES OF MSW

Commercial sources of MSW are generated in restaurants, hotels, stores, motels, service stations, repair shops, markets, office buildings and print shops. The types of solid wastes generated include paper, cardboard, wood, plastics, glass, special wastes such as white goods and other bulky items, and hazardous wastes.

INSTITUTIONAL SOURCES OF MSW

Institutional sources of MSW are generated in hospitals, schools, jails and prisons, and government centers. The types of solid wastes generated by institutional sources are the same as those generated by commercial sources.

CONSTRUCTION AND DEMOLITION SOURCES OF MSW

Construction and demolition sources of MSW are generated at new construction sites, the razing of old buildings, road repair/renovation sites, and broken pavement. The types of solid wastes generated by construction and demolition sources are made up of standard construction materials such as wood, steel, plaster, concrete, and soil.

MUNICIPAL SERVICES SOURCES OF MSW

Municipal services (excluding treatment plants) *sources of MSW* are generated in street cleaning, landscaping, parks and beaches, recreational areas, and catch basin maintenance and cleaning activities. The types of solid wastes generated by municipal services are made up of rubbish, street sweepings, general wastes from parks, beaches, and recreational areas, and catch basin debris.

TREATMENT PLANT SITE SOURCES OF MSW

Treatment plant site sources of MSW are generated in water, wastewater, and other industrial treatment processes (for example, incineration). The principal types of solid wastes generated at treatment plant sites are sludges or biosolids, fly ash, and general plant wastes.

CASE STUDY 10.1 Problem Wastes—Tire Disposal

Since the invention of the automobile, what to do with worn-out tires has presented a disposal problem. America's love affair with cars means that hundreds of millions of tires are discarded every year. Stockpiles across the country store billions of tires. Some of these stockpiles are legal, others are not, but all present us with problems, including the risk of catastrophic fire and the creation of prime breeding habitat for mosquitoes, some varieties of which carry encephalitis or West Nile Virus.

In recent years, tire fires have received national attention. Tire fires are particularly difficult to extinguish. Water used as an extinguishing agent can cause oily run-off, and burning tires emit toxic black smoke, causing pollution problems for air, surface and groundwater supplies and soil. Smothering with dirt and sand appears to be the most cost-effective and efficient way to control burning tire stockpiles. However, sometimes more unusual problems occur with tire fires.

In January 1996, a major road in Ilwaco, Washington, began to heat up, and 2 months later, a major oil leak occurred as the result of a massive underground fire. While response teams immediately contained the oil, they were forced to allow the fire to smolder while they figured out how to even get to it.

Using scrap tires as sub-grade road base is probably the most successful effort at recycling used tires, along with using shredded tires as supplemental fuel for modern, scrubber-equipped boilers.

However, the chances of further episodes of burning roads, which could create contamination problems in all environmental media, mean that the risks for the use of scrap tires are great. As more states enact legislation prohibiting the disposal of tires in landfills, recycling, stockpiling and waste tire dumps will increase, until we can properly manage, store and process these wastes.

With the advent of West Nile Virus, another serious problem with tires has surfaced, one related to international shipping. In the past, shipping techniques and reliance on human labor meant that items shipped internationally by boat were in route and in port for long periods of time. Mechanized loading and unloading practices and modern shipping containers have drastically shortened shipping times—in general, a positive. Serious speculation on how the virus arrived in the US suggests that a likely culprit is an international market in used tires: Mosquito larvae (infected or not) living in water in the tires can survive container shipment—the trip too short to kill off the disease-carrying insects. Whether or not West Nile Virus came to the US this way, shipping used tires poses a recognized threat to public health.

WHAT IS A HAZARDOUS SUBSTANCE?

Hazardous wastes can be informally defined as a subset of all solid and liquid wastes that are disposed of on land rather than being shunted directly into the air or water, and which have the potential to adversely affect human health and the environment. We have the tendency to think of hazardous wastes as resulting mainly from industrial activities, but households also play a role in the generation and improper disposal of substances that might be considered hazardous wastes. Hazardous wastes (as a result of Bhopal and other disastrous episodes) have been given much attention, but surprisingly little is known of their nature and the actual scope of the problem. In this section, we examine definitions of hazardous materials, substances, wastes and so forth, and attempt to bring hazardous wastes into perspective as a major environmental concern.

Unfortunately, defining a *hazardous substance* is largely a matter of "pick and choose," with various regulatory agencies and pieces of environmental legislation defining that term, and related terms somewhat differently. Many of the terms are used interchangeably. Even experienced professionals in environmental health and safety fields, like environmental engineers with the Certified Hazardous Materials Manager (CHMM certification), sometimes interchange these terms, though the terms are generated by different Federal agencies, by different pieces of legislation, and have somewhat different meanings, depending upon the nature of the problem addressed. To understand the scope of the dilemma we face in defining *hazardous substance*, let's take a look at the terms commonly used today, used interchangeably, and often thought to mean the same thing.

HAZARDOUS MATERIALS

A *hazardous material* is a substance (gas, liquid, or solid) capable of causing harm to people, property, and the environment. The United States Department of Transportation (DOT) uses the term *hazardous materials* to cover nine categories identified by the *United Nations Hazard Class Number System*, including:

- Explosives
- Gases (compressed, liquefied, dissolved)
- Flammable Liquids
- Flammable Solids
- Oxidizers
- Poisonous Materials
- Radioactive Materials
- Corrosive Materials
- Miscellaneous Materials

HAZARDOUS SUBSTANCES

The term *hazardous substance* is used by the USEPA for chemicals that, if released into the environment above a certain amount, must be reported, and depending on the threat to the environment, federal involvement in handling the incident can be authorized. USEPA lists hazardous substances in its 40 CFR Part 302, Table 302.4.

The Occupational Safety and Health Administration (OSHA) uses the term *hazardous substance* in 29 CFR 1910.120 (which resulted from Title I of SARA and covers *emergency response*) differently than does the EPA. Hazardous substances (as defined by OSHA) cover every chemical regulated by both DOT and the EPA.

EXTREMELY HAZARDOUS SUBSTANCES

Extremely hazardous substance is a term used by the EPA for chemicals that must be reported to the appropriate authorities if released above the *threshold reporting quantity* (RQ). The list of extremely hazardous substances is identified in Title III of the *Superfund Amendments and Reauthorization Act* (SARA) of 1986 (40 CFR Part 355). Each substance has a threshold reporting quantity.

TOXIC CHEMICALS

EPA uses the term *toxic chemical* for chemicals whose total emissions or releases must be reported annually by owners and operators of certain facilities that manufacture, process, or otherwise use listed toxic chemicals. The list of toxic chemicals is identified in Title III of SARA.

HAZARDOUS WASTES

> The most alarming of all man's assaults upon the environment is the contamination of air, earth, rivers, and sea with dangerous and even lethal materials. This pollution is for the most part irrecoverable; the chain of evil it initiates not only in the world that must support life but in living tissues is for the most part irreversible. In this now universal contamination of the environment, chemicals are the sinister, and little-recognized partners of radiation in changing the very nature of the world—the very nature of life.
>
> **Rachel Carson (1962)**

Rachel Carson was able to combine the insight and sensitivity of a poet with the realism and observations of science more adeptly than anyone before her. Famous for her classic and highly influential book *Silent Spring*, to us today that such a visionary as Rachel Carson was (after the publication of her magnum opus) ostracized, vilified, laughed at, lambasted, and disregarded seems strange. To those guilty of the sins that she revealed, Rachel Carson was an enemy to be discredited—and silenced. She was not, however, disregarded by those who understood. To these concerned folks with conscience, her message was clear: waste, if not properly treated and handled, threatens not only human life in the short term, but the environment as a whole in the long term. Her plea was also clear: stop poisoning the earth.

Examined with a clear vision of retrospect, the environmental missionary Rachel Carson was well ahead of her time. The fears she expressed in 1962 were based on limited data—but have since been confirmed. Rachel Carson was right.

EPA uses the term *hazardous wastes* for chemicals regulated under the Resource, Conservation and Recovery Act (RCRA-40 CFR Part 261.33). Hazardous wastes in transportation are regulated by DOT (49 CFR Parts 170-179).

For the purposes of this text, we define hazardous waste as any hazardous substance that has been spilled or released into the environment. For example, chlorine gas is a hazardous material.

Solid and Hazardous Waste

When chlorine is released into the environment, it becomes a hazardous waste. Similarly, when asbestos is in place and undisturbed, it is a hazardous material. When it is broken, breached, or thrown away, it becomes a hazardous waste.

HAZARDOUS CHEMICALS

OSHA uses the term hazardous chemical to denote any chemical that poses a risk to employees if they are exposed to it in the workplace. Hazardous chemicals cover a broader group of chemicals than the other chemical lists.

AGAIN, WHAT IS A HAZARDOUS SUBSTANCE?

To form the strongest foundation for understanding the main topic of this chapter (hazardous waste), and because RCRA's definition for a hazardous substance can also be used to describe a *hazardous waste*, we use RCRA's definition. RCRA defines something as a *hazardous substance* if it possesses any of the following four characteristics: *reactivity, ignitability, corrosiveness,* or *toxicity*. Briefly,

- **Ignitability** refers to the characteristic of being able to sustain combustion and includes the category of flammability (ability to start fires when heated to temperatures less than 140°F or less than 60°C).
- **Corrosive** substances (or wastes) may destroy containers, contaminate soils and groundwater, or react with other materials to cause toxic gas emissions. Corrosive materials provide a specific hazard to human tissue and aquatic life where the pH levels are extreme.
- **Reactive** substances may be unstable or have tendency to react, explode, or generate pressure during handling. Pressure-sensitive or water-reactive materials are included in this category.
- **Toxicity** is a function of the effect of hazardous materials (or wastes) that may come into contact with water or air and be leached into the groundwater or dispersed in the environment.

Toxic effects that may occur to humans, fish, or wildlife are the principal concerns here. Toxicity, until 1990, was tested using a standardized laboratory test, called the *extraction procedure* (EP Toxicity Test). The EP Toxicity test was replaced in 1990 by the *Toxicity Characteristics Leaching Procedure (TCLP)* because the EP test failed to adequately stimulate the flow of toxic contaminants to drinking water. The TCLP test is designed to identify wastes likely to leach hazardous concentrations of particular toxic constituents into the surrounding soils or groundwater as a result of improper management.

TCLP extracts constituents from the tested waste in a manner designed to simulate leaching actions that occur in landfills. The extract is then analyzed to determine if it possesses any of the toxic constituents listed in Table 10.2. If the concentrations of the toxic constituents exceed the levels listed in the table, the waste is classified as hazardous.

WHAT IS A HAZARDOUS WASTE?

Recall our general rule of thumb that states that any hazardous substance spilled or released to the environment is no longer classified as a hazardous substance, but as a hazardous waste. The EPA uses the same definition for hazardous waste as it does for hazardous substance. The four characteristics described in the previous section (reactivity, ignitability, corrosivity, or toxicity) can be used to identify hazardous substances as well as hazardous wastes.

TABLE 10.2
Maximum Concentration of Contaminants for TCLP Toxicity Test

Contaminant	Regulatory Level (mg/L)
Arsenic	5.0
Barium	100.0
Benzene	0.5
Cadmium	1.0
Carbon tetrachloride	0.5
Chlordane	0.03
Chlorobenzene	100.0
Chloroform	6.0
Chromium	5.0
Cresol	200.0
2,4-D	10.0
1,4-Dichlorobenzene	7.5
1,5-Dichloroethane	0.5
2,4-Dinitrololuene	0.13
Endrin	0.02
Heptachlor	0.008
Hexachlorobenzene	0.13
Hexachloroethane	3.0
Lead	5.0
Lindane	0.4
Mercury	0.2
Methoxychlor	10.0
Methyl ethyl ketone	200.0
Nitrobenzene	2.0
Pentachlorophenol	100.0
Pyridine	5.0
Selenium	1.0
Silver	5.0
Tetrachloroethylene	0.7
Toxaphene	0.5
Trichloroethylene	0.5
2,4,5-Trchlorophenol	400.0
2,4,6-Trchlorophenol	2.0
2,4,5-TP (Silvex)	1.0
Vinyl chloride	0.2

Source: USEPA (1990), 40 CFR 261.24.

Note that the EPA lists substances that it considers hazardous wastes. These lists take precedence over any other method used to identify and classify substances as hazardous (i.e., if a substance is listed in one of the EPA's lists described below, legally, it is a hazardous substance, no matter what).

EPA Lists of Hazardous Wastes

EPA-listed hazardous wastes are organized into three categories: Nonspecific source wastes, specific source wastes, and commercial chemical products; all listed wastes are presumed hazardous regardless of their concentrations. EPA developed these lists by examining different types of wastes and chemical products to determine whether they met any of the following criteria:

- Exhibits one or more of the four characterizations of hazardous waste.
- Meet the statutory definition of hazardous waste.
- Are acutely toxic or acutely hazardous.
- Are otherwise toxic.

These lists are described briefly as:

- **Nonspecific source wastes**: generic wastes, commonly produced by manufacturing and industrial processes. Examples from this list include spent halogenated solvents used in degreasing, and wastewater treatment sludge from electroplating processes, as well as dioxin wastes, most of which are "acutely hazardous" wastes because of the danger they present to human health and the environment.
- **Specific source wastes**: wastes from specially identified industries such as wood preserving, petroleum refining, and organic chemical manufacturing. These wastes typically include sludge, still bottoms, wastewaters, spent catalysts, and residues, for example, wastewater treatment sludge from pigment production.
- **Commercial chemical products** (also called "P" or "U" list wastes because their code numbers begin with these letters): wastes from specific commercial chemical products or manufacturing chemical intermediates. This list includes chemicals such as chloroform and creosote, acids such as sulfuric and hydrochloric, and pesticides such as DDT and kepone (40 CFR 261.31, .32, .33).

Note that the EPA ruled that any waste mixture containing a listed hazardous waste is also considered a hazardous waste—and must be managed accordingly. This applies regardless of what percentage of the waste mixture is composed of listed hazardous wastes. Wastes derived from hazardous wastes (residues from the treatment, storage, and disposal of a listed hazardous waste) are considered hazardous waste as well (USEPA, 1990).

WHERE DO HAZARDOUS WASTES COME FROM?

Hazardous wastes are derived from several waste generators. Most of these waste generators are in the manufacturing and industrial sectors and include chemical manufacturers, the printing industry, vehicle maintenance shops, leather products manufacturers, the construction industry, metal manufacturing, and others. These industrial waste generators produce a wide variety of wastes, including strong acids and bases, spent solvents, heavy metal solutions, ignitable wastes, cyanide wastes, and many more.

WHY ARE WE CONCERNED ABOUT HAZARDOUS WASTES?

From the environmental scientist's perspective, any hazardous waste release that could alter the environment in any way is of major concern. The specifics of their concern lie in acute and chronic toxicity to organisms, bioconcentration, biomagnification, genetic change potential, etiology, pathways, change in climate and/or habitat, extinction, persistence, and esthetics (visual impact).

Remember that when a hazardous substance or hazardous material is spilled or released into the environment, it becomes a hazardous waste. Since specific regulatory legislation is in place regarding hazardous wastes, responding to hazardous waste leak/spill contingencies, and for proper handling, storage, transportation, and treatment of hazardous wastes, this distinction is important—the goal being, of course, protecting the environment—and ultimately, protecting ourselves.

Why so much concern about hazardous substances and hazardous wastes? This question is relatively easy to answer because of the hard lessons we have learned in the past. Our answers are based on experience—actual hazardous materials incidents that we know of, and that we have witnessed—have resulted in tragic consequences, not only to the environment but also to human life. Consider an example from *Surviving an OSHA Audit* (1998); maybe it will provide a better explanation of why the control of hazardous substances and wastes is critically important to us all.

HAZARDOUS WASTE LEGISLATION

A few people (Rachel Carson for one) could have predicted that a disaster on the scale of Bhopal was ripe to occur—but humans are strange in many ways. We may know that a disaster is possible, is likely, could happen, and is predictable. We predict it—but do we act? Do we act before someone dies? No. Not often. Not often enough. We don't think about the human element. We forget the victims of hazardous materials spills—that is, until they suffer or after they die, after we can no longer help them.

Is it fair? Don't mouth that platitude about "life not being fair"—we all know it isn't, but ideally, we strive to make it more so. Is it right? So, what do we do about it? We legislate, of course.

Because of Bhopal and other similar (but less catastrophic) chemical spill events, the United States Congress (pushed by public concern) developed and passed certain environmental laws and regulations to regulate hazardous substances/wastes in the U.S. This section focuses on the two regulatory acts most crucial to the current management programs for hazardous wastes. The first (mentioned several times throughout the text) is the Resource Conservation and Recovery Act (RCRA). Specifically, RCRA provides guidelines for the prudent management of new and future hazardous substances/wastes. The second act (more briefly mentioned) is the Comprehensive Environmental Response, Compensation, and Liability Act (CERCLA), otherwise known as Superfund, which deals primarily with mistakes of the past: inactive and abandoned hazardous waste sites.

RESOURCE CONSERVATION AND RECOVERY ACT

The Resource Conservation and Recovery Act (RCRA) is the U.S.'s single most important law dealing with the management of hazardous waste. RCRA and its amendment *Hazardous and Solid Waste Act* (HSWA-1984) deal with the ongoing management of solid wastes throughout the country—with emphasis on hazardous waste. Keyed to the waste side of hazardous materials, rather than broader issues dealt with in other acts, RCRA is primarily concerned with land disposal of hazardous wastes. The goal is to protect groundwater supplies by creating a "cradle-to-grave" management system with three key elements: *a tracking system, a permitting system, and control of disposal*.

1. A **tracking system**: a **manifest** document accompanies any waste that is transported from one location to another.
2. A **permitting** system: helps assure the safe operation of facilities that treat, store, or dispose of hazardous wastes.
3. A **disposal control** system: controls and restrictions governing the **disposal** of hazardous wastes onto, or into, the land (Masters, 1991).

The RCRA regulates five specific areas for the management of hazardous waste (with a focus on *treatment, storage,* and *disposal*). These are:

1. Identifying what constitutes a hazardous waste and providing a classification of each.
2. Publishing requirements for generators to identify themselves, which includes notification of hazardous waste activities and standards of operation for generators.

3. Adopting standards for transporters of hazardous wastes.
4. Adopting standards for treatment, storage, and disposal facilities.
5. Providing for enforcement of standards through a permitting program and legal penalties for noncompliance (Griffin, 1989).

Arguably, the RCRA is our single most important law dealing with the management of hazardous waste—it certainly is the most comprehensive piece of legislation that the EPA has promulgated to date.

CERCLA

The mission of the Comprehensive Environmental Response, Compensation, and Liabilities Act of 180 (Superfund or SARA) is to clean up hazardous waste disposal mistakes of the past and to cope with emergencies of the present. More often referred to as the *Superfund Law*, as a result of its key provisions a large trust fund (about $1.6 billion) was created. Later, in 1986, when the law was revised, this fund was increased to almost $9 billion. The revised law is designated as the *Superfund Amendments and Reauthorization Act of 1986 (SARA)*. The key requirements under CERCLA include:

1. CERCLA authorizes the EPA to deal with both short-term (emergency situations triggered by a spill or release of hazardous substances), as well as long-term problems involving abandoned or uncontrolled hazardous waste sites for which more permanent solutions are required.
2. CERCLA has set up a remedial scheme for analyzing the impact of contamination on sites under a hazard ranking system. From this hazard ranking system, a list of prioritized disposal and contaminated sites is compiled. This list becomes the National Priorities List (NPL) when promulgated. The NPL identifies the worst sites in the nation, based on such factors as the quantities and toxicity of wastes involved, the exposure pathways, the number of people potentially exposed, and the importance and vulnerability of the underlying groundwater.
3. CERCLA also forces those parties who are responsible for hazardous waste problems to pay the entire cost of cleanup.
4. Title III of SARA requires federal, state, and local governments and industries to work together in developing emergency response plans and reporting on hazardous chemicals. This requirement is commonly known as the *Community Right-To-Know Act*, which allows the public to obtain information about the presence of hazardous chemicals in their communities and the releases of these chemicals into the environment.

WASTE CONTROL TECHNOLOGY

How to handle society's toxic chemical waste now ranks among the top environmental issues in most industrial countries. Without concerted efforts to reduce, recycle, and reuse more industrial waste, the quantities produced will overwhelm even the best treatment and disposal systems.

Sandra Postel (1987, p. 37)

One of the most challenging and pressing current environmental concerns (dilemmas) confronting environmental scientists (and many others) is what to do with all of the solid and hazardous wastes our throwaway society produces. In simple (and simplistic) terms, we could say that we should shift from a throwaway society to a recycling one, which would help restore a gain in our living standards. We could also say that since disposing of hazardous waste is so expensive and risky, to make the situation better we should follow RCRA's Waste Management Hierarchy (in descending order of desirability) to (1) stop producing waste in the first place; (2) if we cannot avoid producing it, then produce only minimum quantities; (3) recycle it; (4) if it must be produced, but cannot be recycled, then treat it: (5) if it cannot be rendered non-hazardous, dispose of it in a safe manner; and (6) once it is disposed, continuously monitor it to ensure no adverse effects to the environment.

All of these statements have merit. The question is are they realistic? To a point, yes. We have developed several different strategies to curb the spread of hazardous substances/wastes. One approach is the treatment of hazardous wastes to neutralize them or make them less toxic. However, again, a better strategy would be to reduce or eliminate the use of toxic substances and the generation of hazardous waste. To a degree, we can accomplish this, but to think that we can simply do away with all our hazardous materials, processes that use hazardous materials, and processes that produce hazardous materials is, at the present time, wishful thinking.

What we need to do is refine our waste reduction programs as much as possible and develop technologies that will better treat waste products that we are not able to replace, do away with, or reduce. We have such technologies and/or practices available to us today. Environmental science and technology can be put to work to develop and use measures and practices by which hazardous chemical wastes can be minimized, recycled, treated, and disposed. We review these measures, practices and technologies in this section.

WASTE MINIMIZATION

Waste minimization (or source reduction measures) is accomplished in a variety of ways and includes feedstock or input substitution, process modifications, and good operating practices. Note that before any source reduction measure can be put into place, considerable amounts of information must be gathered.

One of the first steps to be taken in the information-gathering process is determining the exact nature of the waste produced. The waste must initially be characterized and categorized by type, composition, and quantity, a task accomplished by performing a **chemical process audit** or **survey** of the chemical process. Keep in mind that during this information-gathering survey, looking closely for any off-specification input materials that might produce defective outputs, any inadvertent contamination of inputs, process chemical, and outputs, and any obsolete chemicals (which should be properly disposed of) is important.

During the survey, particular attention should be given to problem areas—excessive waste amounts per unit of production, excessive process upsets or bad batches, or frequent off-specification inputs. The effect of process variables on the waste stream created, and the relationship of waste stream composition to the input chemicals and process methods used should be examined. For example, determine exactly how much process water is used. Can the amount of water used be reduced? Can process water be reused? Questions like these should be addressed during the chemical process survey (Lindgren, 1989).

To determine the feasibility of reuse, recycling, materials recovery, waste transfer or proper methods of waste disposal, the exact nature of the waste must also be determined. Usually accomplished through sampling the wastestream, and then analyzing the sample in the laboratory, the nature of the waste can yield valuable information about the industrial process and the condition of process equipment.

Substitution of Inputs

After completing the chemical process survey, the information gathered may suggest or justify the substitution of certain chemicals, process materials or feedstock to enable the process hazardous wastes to be reduced in volume or no longer produced. Note that input substitutions are often inseparable from process modifications. A few specific examples of possible input substitutions include:

- use of synthetic coolants in place of emulsified oil coolants;
- use of water-based paints instead of solvent-based paints;
- use of noncyanide-based electroplating solutions;
- use of cartridge filters in lieu of earth filters.

Process Modifications

One of the key benefits derived from performing a chemical process audit or survey is that audits often point to or suggest modifications to production systems that work to minimize hazardous waste stream production. Whenever a chemical process can be made more efficient, a reduction in the volume and toxicity of the residuals usually results.

Good Operating Practices

Reducing wastage, preventing inadvertent releases of chemicals, and increasing the useful lifetime of process chemicals are all directly related to **good operating practices**. Ensuring good operating practices by workers can only be accomplished through effective worker training. This training should not only include proper process operations, but also effective spill response training.

Recycling

>Use it up, wear it out, make it do, or do without.
>
>**New England Proverb**

If waste generation is unavoidable in a process, then strategies that minimize the waste to the greatest extent possible should be pursued, such as recycling. Recycling is the process of collecting and processing materials that would otherwise be thrown away as trash and turning them into new products. Recycling can benefit communities and the environment. Benefits of recycling include:

- Reduces the amount of waste sent to landfills and incinerators
- Conserves natural resources such as timber, water, and minerals
- Prevents pollution by reducing the need to collect new raw materials
- Saves energy
- Reduces greenhouse gas emissions that contribute to global climate change
- Helps sustain the environment for future generations
- Helps create new well-paying jobs in the recycling and manufacturing industries in the United States

STEPS TO RECYCLING MATERIALS

Recycling includes three steps, collection and processing, manufacturing and purchasing new products made from recycled materials, which create a continuous loop.

Collection and Processing

There are several methods for collecting recyclables, including curbside collection, drop-off centers, and deposit or refund programs. After collection, recyclables are sent to a recovery facility to be sorted, cleaned and processed into materials that can be used in manufacturing. Recyclables are bought and sold just like raw materials would, and prices go up and down depending on supply and demand in the United States and around the world.

Manufacturing

More and more of today's products are being manufactured with recycled content. Common household items that contain recycled materials include:

- newspapers and paper towels
- aluminum, plastic, and glass soft drink containers
- steel cans
- plastic laundry detergent bottles

Recycled materials are also used in new ways such as recovered glass in asphalt to pave roads or recovered plastic in carpeting and park benches.

Purchasing New Products Made from Recycled Materials

By buying new products made from recycled materials, you help close the recycling loop. There are thousands of products that contain recycled content. When you go shopping, look for:

- Products that can be easily recycled
- Products that contain recycled content

Here are some of the terms used:

- **Recycled-content product**: This means the product was manufactured with recycled materials, either collected from a recycling program or from waste recovered during the normal manufacturing process. Sometimes the label will tell you how much of the content was from recycled materials.
- **Postconsumer content**: This is very similar to recycled content, but the material comes only from recyclables collected from consumers or businesses through a recycling program.
- **Recyclable product**: These are products that can be collected, processed, and manufactured into new ones after they have been used. These products do not necessarily contain recycled materials.

Some of the common products you can find that can be made with recycled content include:

- Aluminum cans
- Car bumpers
- Carpenters
- Cereal boxes
- Comic books
- Egg cartons
- Glass containers
- Laundry detergent bottles
- Motor oil
- Nails
- Newspapers
- Paper towels
- Steel products
- Trash bags

Recycling Hazardous Wastes

With regard specifically to hazardous wastes, various strategies have been developed to *recycle* (and thus minimize) the volume of hazardous wastes to dispose of. These strategies recover or recycle resources, either materials or energy, from the waste stream. The key point to note in chemical process recycling is that the product must receive some processing before re-use. Wastes generally recognized as having components of potential value include:

- flammable and combustible liquids
- oils
- slags and sludge
- precious metal wastes
- catalysts
- acids
- solvents

From the list above, we can see that one such recycling or recovery effort involves the reclamation of organic solvents. Usually accomplished by using highly effective distillation techniques, solvents contaminated with metals and organics are heated to produce a liquid phase and a vapor phase. Lighter components with high volatiles rise to the top of the liquid phase and begin to vaporize. By carefully controlling the waste mixture's temperature, the desired substance can be vaporized and recovered by condensation, leaving the heavier contaminants behind. What remains is a concentrated, highly toxic mixture (far reduced in volume) referred to as still bottoms. Bottoms may contain usable metals and other solvents. As distillation technology improves, more of these bottom materials will be recovered and possibly reused.

TREATMENT TECHNOLOGIES

Because of the 1984 and 1991 amendments to RCRA, hazardous wastes must be treated prior to ultimate disposal in a landfill. Even with process modifications, material substitution, and recycling, some portions of some waste streams may still be hazardous—and must be properly contained. These hazardous waste components require additional treatment. Such treatment takes place in vessels (tanks), reactors, incinerators, kilns, boilers, or impoundments.

At present, several technologies are available for the treatment of hazardous waste streams. In this section, we discuss a few examples, including biological treatment, thermal treatment, activated carbon sorption, electrolytic recovery techniques, air stripping, stabilization and solidification, and filtration and separation treatment systems.

Note: some of these treatment techniques were covered in greater detail earlier—for example, in situ and non-in situ soil contamination treatment; some technologies combine two of more of these basic technologies.

BIOLOGICAL TREATMENT

Several *biological treatment processes* are available for treating "liquid" hazardous waste streams (contaminated soils and solids are more difficult to treat), including activated sludge, aerobic lagoons, anaerobic lagoons, spray irrigation, trickling filters, and waste stabilization ponds. These processes are normally associated with the biological treatment of municipal and industrial wastewater and are generally used for the removal of organic pollutants from wastewater. Generally effective on wastewaters with low-to-moderate concentrations of simple organic compounds and lower concentrations of complex organics, these are generally ineffective in attacking mineral components and useless against heavy metals.

Biological treatment of toxic organic components requires considerably more sophisticated operational control (including pretreatment) than is necessary with nontoxic wastewaters. Microorganisms used in biological treatment processes can easily be destroyed by rapid increases in rate of feed. Acclimation and development of a functional population of biota may require considerable time, and the system is continuously subject to upset (Blackman, 1993).

The two biological processes used for the treatment of toxic waste are the **aerobic processes** (treatment in the presence of oxygen—conventional aeration) and **anaerobic processes** (treatment in the absence of oxygen—in a simple septic tank).

In aerobic treatment, organisms require both an energy source and a carbon source for growth, and both affect what type of organisms will grow in a particular environment. Many hazardous-waste streams satisfy both basic requirements, and if appropriate nutrients are present, a thriving organic population for waste treatment can exist. Under these conditions, if pH and temperature are controlled, substances that are toxic to the active organisms can be eliminated.

The most important aspect limiting the applicability of aerobic biological treatment of hazardous waste is the biodegradability of the waste—its conversion by biological processes to simple inorganic molecules and to biological materials. Biodegradability of a particular waste is very system-specific, and the correct conditions for successful treatment (detoxification or biological conversion of a toxic substance to one less toxic) must be maintained to encourage the correct microbe mixture.

Anaerobic treatment of toxic waste streams has been effectively practiced on many different types of toxic waste streams. This form of treatment is a fermentation process in which organic waste is both oxidized and reduced.

THERMAL PROCESSES

Thermal treatment processes (incineration) are commonly used to treat both liquids and solids to either destroy the hazardous components or allow the disposal of the process residue or treated waste in an EPA-approved hazardous waste landfill.

During *incineration*, carbon-based (organic) materials are burned at high temperatures—typically ranging from 1500°F to 3000°F—to break them down, chiefly into hydrogen, carbon, sulfur, nitrogen, and chlorine. These constituent elements then combine with oxygen to form inorganic gases like water vapor, carbon dioxide, and nitrogen oxides. After combustion, the gases pass through a pollution control system to remove acidic gases and particulate matter prior to being released into the atmosphere.

The advantages of hazardous wastes incineration are twofold: (1) it permanently reduces or eliminates the hazardous character of the waste; and (2) it substantially reduces the volume of the waste being disposed.

Waste characteristics and treatment requirements determine the incinerator design to accommodate liquid or solid wastes. Temperature, turbulence, and retention time (commonly known as the 3 T's of incineration) are the prime factors determining incineration treatment design for both solid and liquid wastes.

Hazardous waste incinerators are regulated by the EPA and require a permit for operation. To receive an operating permit, an incineration facility must demonstrate a 99.99% destruction and removal efficiency (DRE) for each principal organic hazardous constituent in the feed material.

Non-in situ thermal processes (primarily incinerators) include designs such as liquid injection and boilers, rotary kilns, fluidized beds and catalytics. More sophisticated and less common types of thermal treatment systems include wet oxidation, pyrolytic, and plasma processes. Some of these processes can be conducted in situ—with steam injection, radio frequency heating, and vitrification (molten glass treatment) processes.

ACTIVATED CARBON SORPTION

Organic substances may be removed from aqueous hazardous waste streams with activated carbon by sorption. Sorption is the transfer of a substance from a solution to a solid phase. In adsorption (a final chemical reaction that forms a cementitious precipitated sludge, not to be confused with absorption, which is defined as a physical process that does not chemically stabilize a waste material), chemical substances are removed from the waste stream onto a carbon matrix. The carbon may be used in granular of powdered form, depending on the application.

The effectiveness of activated carbon in removing hazardous constituents from aqueous streams is directly proportional to the amount of surface area of the activated carbon; in some cases, it is adequate for complete treatment. It can also be applied to the pretreatment of industrial hazardous waste streams prior to follow-up treatment. Activated carbon sorption is most effective for removing from water those hazardous waste materials that are not water-soluble.

ELECTROLYTIC RECOVERY TECHNIQUES

The *electrolytic recovery technique* (used primarily for the recovery of metals from process streams to clean process waters, or to treat wastewaters prior to discharge) is based on the *oxidation-reduction* reaction, where electrode surfaces are used to collect the metals from the waste stream.

Typically, an electrolytic recovery system consists of a treatment vessel (tank, etc.) with electrodes, an electrical power supply, and a gas handling and treatment system. Recovered metal must be removed from the electrodes periodically when the design thickness is achieved for the recovered metal.

AIR STRIPPING

The *air stripping* technique for removing hazardous constituents from waste streams, although not particularly effective, has been used for many years. *Stripping* is a means of separating volatile components from less volatile ones in a liquid mixture by partitioning the more volatile materials to a gas phase of air or steam. In air stripping, the moving gas is usually ambient air, which is used to remove volatile dissolved organic compounds from liquids including groundwaters and wastewaters. Additional treatment must be applied to the exhaust vapors to destroy and/or capture the separated volatiles. The process is driven by the concentration gradient between air and liquid phase equilibrium for particular molecules according to *Henry's Law*, which states that at constant temperature, the weight of gas absorbed by a given volume of a liquid is proportional to the pressure at which the gas is supplied.

STABILIZATION AND SOLIDIFICATION

Stabilization and *Solidification* are techniques used to convert hazardous waste from the original form to a physically and chemically more stable material. Accomplished by reducing the mobility of hazardous compounds in the waste prior to its land disposal, stabilization and solidification are particularly useful when recovery, removal, or converting hazardous components (as required by RCRA) from a waste prior to disposal in a landfill is not possible.

A wide variety of stabilization and solidification treatment processes use Portland cement as a binding agent. Waste/concrete composites can be formed that have exceptional strength and excellent durability, and that retain wastes very effectively (Blackman, 1993). Stabilization and solidification treatment processes improve handling and physical characteristics, and result in a reduction of solubility or limit the leachability of hazardous components with a waste.

FILTATION AND SEPARATION

Filtration and separation hazardous waste treatment processes are physical processes. Filtration (the separation of solid particles from a liquid stream through the use of semi-porous media) is driven by a pressure difference across the media. This pressure difference is caused by gravity, centrifugal force, vacuum, or elevated pressure.

Filtration applied to hazardous waste treatment falls into two categories—clarification and dewatering. Clarification takes place when liquids of less than 200 ppm are placed within a clarifier, and the solids are allowed to settle out, producing a cleaner effluent. Dewatering is performed on slurries and sludge. The goal of dewatering is to concentrate the solids into a semi-solid form for further treatment or land disposal.

ULTIMATE DISPOSAL

Most of us are familiar with open dumps. However, you might not be familiar with some of the dumping practices that occurred because of environmental legislation of the 1970s, which placed increasingly stringent controls on releases to the atmosphere and to the nation's waterways. To protect our atmosphere and our waterways, the 1970s legislative mindset drove us to dump hazardous materials into open dumps. Why not? Land disposal is safer and proper. Isn't it?

No, it is not. But we didn't realize this until the tragic consequences of these practices became apparent to us later—and today, we're still cleaning up the resulting mess.

We are now well aware that the land is not a bottomless sink that can be used to absorb all of our discards. We've learned that we must pretreat our wastes to detoxify them, to degrade them, to make them less harmful, to make them more earth-friendly—before we deposit them on or in the ground, the soil, the land—our Earth.

Regardless of the treatment, destruction, and immobilization techniques used, some residue(s) that must be contained somewhere will always remain from hazardous wastes. This "somewhere" is burial in land, deep-well injection, surface impoundments, waste piles, and landfills. In this section, we discuss each of these ultimate disposal methods.

DEEP-WELL INJECTION

The practice of *deep-well injection* is not new; it was used in the 1880s by the petroleum industry to dispose of salt water produced when drilling for oil. However, disposing of hazardous materials by deep-well injection is a relatively recent development. The EPA estimates that about 9 billion gallons of all the hazardous waste produced in the United States (about 22% of the total produced) is injected deep into the ground. Most of the deep-well injection sites are located in the Great Lakes region and along the Gulf Coast.

Deep-well injection involves the injection of liquid waste under pressure to underground strata isolated by the impermeable rock where geologists believe they will be contained permanently, isolated from aquifers, typically at a depth of more than 700 m below the surface. A high-pressure pump forces the hazardous liquids into pores in the underground rock, where they displace the water, oil and gases originally present. Sandstone and other sedimentary rock formations are used because they are porous and allow the movement of liquids.

In theory, when properly constructed, operated, and monitored, deep-well injection systems may be the most environmentally sound disposal method for toxic and hazardous wastes currently available. However, as with anything else that in theory is "perfect" or affords us the "best available technology," deep-well injection has its problems. For example, though constructed at a depth below the groundwater table, fractures in the underground geology could allow waste to go where it is not wanted; namely, into the groundwater. The biggest problem with deep-well injection concerns the unknown. We are not certain of the exact fate of hazardous substances after injection—another example of the "we don't know what we do not know" syndrome.

Because of our uncertainty about the results of our hazardous waste disposal practices, the 1984 amendments to RCRA ban unsafe, untreated wastes from land disposal. For those land disposal facilities allowed to accept hazardous substances, USEPA (1986) implemented restrictions requiring:

- Banning liquids from landfills;
- Banning underground injection of hazardous waste within 1/4-mile of a drinking water well;
- Requiring more stringent structural and design conditions for landfills and surface impoundments, including two or more liners, leachate collection systems above and between the liners, and groundwater monitoring;
- Requiring cleanup or corrective action if hazardous waste leaks from a facility;
- Requiring information from disposal facilities on pathways of potential human exposure to hazardous substances;
- Requiring location standards that are protective of human health and the environment.

Surface Impoundments

Surface impoundments are diked or excavated areas used to store liquid hazardous wastes. Because most surface impoundments are temporary, relatively cheap to construct, and allow easy access for treatment, they have been popular for many years.

Unfortunately, in the past, surface impoundments were poorly constructed (literally quickly dug out or diked and put into operation), poorly sited (built on a thin layer of permeable soil that allowed leachate to infiltrate into groundwater), located too close to sources of high-quality drinking water (wells or running water sources), and either not monitored at all or poorly monitored. In 1984, the USEPA estimated that of the more than 180,000 surface impoundments surveyed, that prior to 1980 only about 25% were lined, and fewer than 10% had monitoring systems.

Because of the problems associated with poor siting, construction, and management of the early surface impoundments, EPA regulations have toughened the requirements for the construction of new surface impoundments. Under the *Hazardous and Solid Waste Amendments (HSWA)* of 1984, for example, the EPA now requires new surface impoundments to include

- The installation of two or more liners;
- A leachate collection system between liners; and
- Groundwater monitoring.

Provisions must also ensure the prevention of liquid escaping from overfilling or run-on, and the prevention of erosion of dams and dikes. During construction and installation, liners must be inspected for uniformity, damage, and imperfections. These liners must also meet permit specifications for materials and thickness.

Waste Piles

Waste piles are normally associated with industrial sites, where it was common practice for years to literally pile up industrial waste, and later, when the pile became "too large," dispose it in a landfill. Industrial practice has been to list such piles as "treatment" piles, and even 40 CFR 264/265 subpart L refers to such piles as treatment or storage units.

The environmental problem with such piles is similar to the problems we discussed related to mining waste. Like mining waste, industrial waste piles are subject to weather exposure, including evaporation of volatile components to the atmosphere, and wind and water erosion. The most significant problem related to industrial waste piles is related to precipitation—leaching of contaminants (producing leachate), which may percolate into the subsurface.

RCRA specifications for waste piles are similar to those for landfills (to be discussed in the next section) and are listed in 40 CFR 264/265 subpart L. Under the RCRA guidelines, the owner or operator of a waste pile used for storage or treatment of noncontainerized solid hazardous wastes is given a choice between compliance with the waste pile or landfill requirements. If the waste pile is used for disposal, it must comply with landfill requirements. The waste pile must be placed on an impermeable surface, and if leachate is produced, a control and monitor system must be in place. Waste piles must also be protected from wind dispersion.

LANDFILLING

Landfilling wastes has a history of causing environmental problems—including fires, explosions, production of toxic fumes, and storage problems when incompatible wastes are commingled. Landfills also have a history of contaminating surface and groundwaters (EPA, 1990).

Sanitary landfills are designed and constructed to dispose municipal solid wastes only. Not designed, constructed, or allowed to be operated for the disposal of bulk liquids and/or hazardous wastes, landfills that can legally receive hazardous wastes are known as *secure landfills*.

Under RCRA, the design and operation of hazardous waste landfills have become much more technically sophisticated. Instead of the past practice of gouging out a huge maw from the subsurface and then dumping countless truckloads of assorted waste materials (including hazardous materials) into it until it was full, a hazardous waste landfill is now designed as a modular series of three-dimensional control cells. Design and operating procedures have evolved to include elaborate safeguards against leakage and migration of leachates.

Secure landfills for hazardous waste disposal are equipped with double liners. Leakage detection, leachate collection and monitoring, and groundwater monitoring systems are required (see Figure 10.2). Liners used in secure landfills must meet regulatory specifications. For example, the upper liner must consist of a 10–100-mil thick *flexible-membrane liner* (FML), usually made of sheets of rubber or plastic. The lower liner is usually FML, but recompacted clay at 3 ft thick is also acceptable.

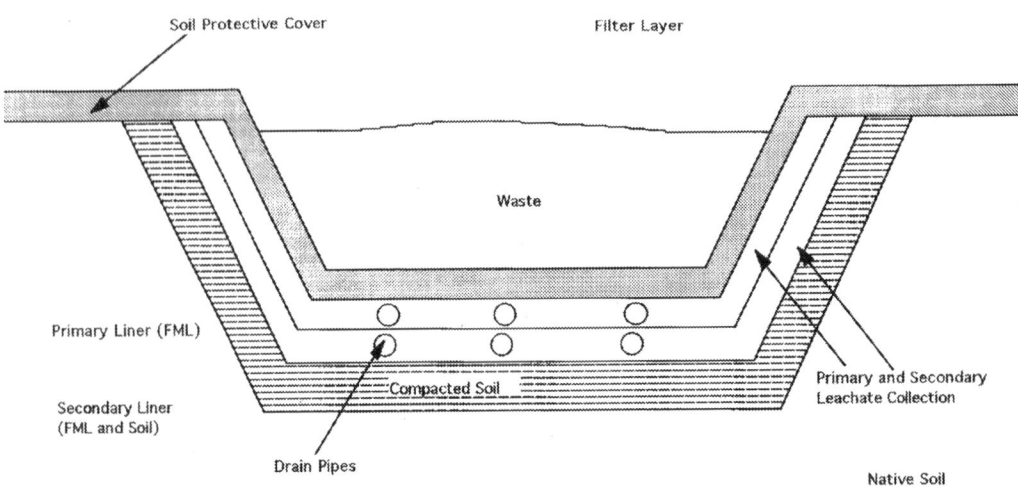

FIGURE 10.2 Cross section of a secure landfill double liner system.

Secure landfills must be constructed to allow the collection of leachate (usually via perforated drainage pipes with an attached pumping system) that accumulates above each liner. Leachate control is critical. To aid in this control process (especially from leachate produced by precipitation) a low permeability cap must be placed over completed cells. When the landfill is finally closed, a cap that will prevent leachate formation via precipitation must be put in place. This cap should be sloped to allow drainage away from the wastes.

When a landfill is filled and capped, it cannot be completely abandoned, ignored, or forgotten. The site must be monitored to ensure that leachate is not contaminating the groundwater. This is accomplished by installing test wells downgradient to assure the detection of any leakage from the site.

REFERENCES

Blackman, W.C. (1993). *Basic Hazardous Waste Management*. Boca Raton, FL: Lewis Publishers.
Carson, R. (1962). *Silent Spring*. Boston, MA: Houghton Mifflin Company.
Davis, M.L. & Cornwell, D.A. (1991). *Introduction to Environmental Engineering*, 2nd ed. New York: McGraw and Hill, p. 585.
DOT 49 CFR-170-179, U.S. Department of Transportation.
40 CFR 261.24, USEPA, (1990).
40 CFR 261.31, .32, .33.
40 CFR 302.4.
EPA (1990). MSW landfill criteria manual. Accessed 12/12/21 @ https://archive.epa.gov/epawaste/nonhaz/municipal/web/html/index-html.
Griffin, R.D. (1989). *Principles of Hazardous Materials Management*, Chelsea, MI: Lewis Publishers.
Lindgren, G.F. (1989). *Managing Industrial Hazardous Waste*. Chelsea, MI: Lewis Publishers.
Masters, G.M. (1991). *Introduction to Environmental Engineering and Science*. New York: Prentice-Hall.
Muir, J. (1877). Change to a man in the woods. Accessed 12/12/21 @ https://medium.com/@yennie.jun/muir-woods-118clb35bb16.
O'Reilly, J.T. (1992). *State & Local Government Solid Waste Management*. Deerfield, IL: Clark, Boardman, Callahan.
Postel, S. (1987). *Defusing the Toxics Threat: Controlling Pesticides and Industrial Wastes, Worldwatch Paper 79*, Washington, DC: Worldwatch Institute, pp. 36–37.
Spellman, F.R. (1995). *Laboratory Safe Work Practices*. Lancaster, PA: Technomic Publishing Company.
Spellman, F.R. (1996). *Process Safety Management*. Lancaster, PA: Technomic Publishing Company.
Spellman, F.R. (1998). *Surviving an OSHA Audit: A Manager's Guide*, Lancaster, PA: Technomic Publishing Company.
Tchobanoglous, G., Theisen, H., and Vigil, S. (1993). *Integrated Solid Waste Management: Engineering Principles and Management Issues*. New York: McGraw-Hill.
USEPA (1986). *Solving the Hazardous Waste Problem: EPA's RCRA Program*. Washington, DC: EPA Office of Solid Waste.
USEPA (1989). *Decision-Makers Guide to Solid Waste Management*, Washington, DC: EPA/530-SW89-072.
USEPA (1990). *RCRA Orientation Manual*. Washington, DC: United States Environmental Protection Agency.
USEPA (2009). *Characterization of Municipal Solid Waste in U.S.: 1992 Update*, Washington, DC: EPA/530-5-92-019.

11 Industrial Hygiene

Industrial hygiene is an area of specialization within the broader field of industrial health and safety. This chapter provides prospective and practicing health and safety professionals with the information they need to know about this area of specialization.

D.L. Goetsch

WHAT IS INDUSTRIAL HYGIENE?

According to the Occupational Safety and Health Administration (OSHA), industrial hygiene is the science of anticipating, recognizing, evaluating, and controlling workplace conditions that may cause workers' injury or illness. Industrial hygienists use environmental monitoring and analytical methods to detect the extent of worker exposure and employee engineering, administrative controls, and other methods, such as personal protective equipment (PPE), to control potential health hazards (OSHA, 1998).

INDUSTRIAL HYGIENE TERMINOLOGY

Every branch of science, every profession, and every engineering process has its own language for communication. Industrial hygiene is no different. In this section, we define many of the terms or "tools" (concepts and ideas) used by industrial hygienists in applying their skills to make our dynamic technological world safer. The practicing industrial hygienist and/or student of industrial hygiene should know these concepts—without them, it is difficult (if not impossible) to practice industrial hygiene. In addition, environmental engineers who become involved in designing and/or maintaining workplace safety and who are responsible for ensuring the environmental health of workers in the workplace must be knowledgeable of all aspects of industrial hygiene practice. To work effectively even at the margin of industrial hygiene, the practitioner must be familiar with industrial hygiene terminology. In this section, key industrial hygiene terms and concepts and related terms and concepts are identified and defined. As Voltaire said: *"If you wish to converse with me, define your terms."*

INDUSTRIAL HYGIENE TERMS AND CONCEPTS

Abatement period: The amount of time given to an employer to correct a hazardous condition that has been cited.

Abrasive blasting: A process for cleaning surfaces by means of materials such as sand, alumina, or steel grit in a stream of high-pressure air.

Absorption: The taking up of one substance by another, such as a liquid by a solid or a gas by a liquid.

Accuracy: The exactness of an observation obtained from an instrument or analytical technique with the true value.

ACGIH: American Conference of Governmental Industrial Hygienists.

Acid: Any chemical with a low pH that in water solution can burn the skin or eyes. Acids turn litmus paper red and have pH values of 0–6.

Accident: An accident is an unplanned and sometimes injurious or damaging event which interrupts the normal progress of an activity and is invariably preceded by an unsafe act or unsafe condition thereof. An accident may be seen as resulting from a failure to identify a hazard or from some inadequacy in an existing system of hazard controls. Based on applications in casualty insurance, an accident is an event that is definite in point of time and place but unexpected as to either its occurrence or its results.

Accident analysis: (see accident investigation) A comprehensive, detailed review of the data and information compiled from an accident investigation. An accident analysis should be used to determine causal factors only, and not to point the finger of blame at anyone. Once the causal factors have been determined; corrective measures should be prescribed to prevent recurrence.

Accident prevention: The act of prevention of a happening which may cause loss or injury to a person.

Accommodation: The ability of the eye to become adjusted after viewing the VDT so as to be able to focus on other objects, particularly objects at a distance.

Acoustics: In general, the experimental and theoretical science of sound and its transmission; in particular, that branch of the science that has to do with the phenomena of sound in a particular space such as a room or theater. Industrial hygiene is concerned with the technical control of sound and involves architecture and construction, studying control of vibration, soundproofing, and the elimination of noise —to engineer out the noise hazard.

Action level: Term used by OSHA and NIOSH (National Institute for Occupational Safety and Health—a federal agency that conducts research on safety and health concerns) and is defined in the Code of Federal Regulations (CFR), Title 40, Protection of Environment. Under OSHA, action level is the level of toxicant, which requires medical surveillance, usually 50% of the PEL (Personal Exposure Level). Note that OSHA also uses the action level in other ways besides setting the level of "toxicant." For example, in its hearing conservation standard, 29 CFR 1910.95, OSHA defines the action level as an 8-hour time-weighted average (TWA) of 85 decibels measured on the A-scale, slow response, or equivalently, a dose of 50%. Under CFR 40 §763.121, action level means an airborne concentration of asbestos of 0.1 fiber per cubic centimeter (f/cc) of air calculated as an 8-hour time-weighted average.

Activated charcoal: Charcoal is an amorphous form of carbon formed by burning wood, nutshells, animal bones, and other carbonaceous material. Charcoal becomes activated by heating it with steam to 800°C–900°C. During this treatment, a porous, submicroscopic internal structure is formed which gives it an extensive internal surface area. Activated charcoal is commonly used as a gas or vapor adsorbent in air-purifying respirators and as a solid sorbent in air-sampling.

Acute: Health effects, which show up a short length of time after exposure. An acute exposure runs a comparatively short course, and its effects are easier to reverse than those of chronic exposure.

Acute toxicity: The discernible adverse effects induced in an organism with a short period of time (days) of exposure to an agent.

Administrative controls: Methods of controlling employee exposures by job rotation, work assignment, time periods away from the hazard, or training in specific work practices designed to reduce exposure.

Adsorption: The taking up of a gas or liquid at the surface of another substance, usually a solid (e.g., activated charcoal adsorbs gases).

Aerosols: Liquid or solid particles that are so small they can remain suspended in the air long enough to be transported over a distance.

Air: The mixture of gases that surrounds the Earth; its major components are as follows: 78.08% nitrogen, 20.95% oxygen, 0.03% carbon dioxide, and 0.93% argon. Water vapor (humidity) varies.

Air cleaner: A device designed to remove atmospheric airborne impurities, such as dusts, gases, vapors, fumes, and smoke.

Air contamination: The result of introducing foreign substances into the air so as to make the air contaminated.

Air-line respirator: A respirator that is connected to a compressed breathing air source by a hose of small inside diameter. The air is delivered continuously or intermittently in a sufficient volume to meet the wearer's breathing.

Air monitoring: The sampling for and measurement of pollutants in the atmosphere.

Air-purifying respirator: A respirator that uses chemicals to remove specific gases and vapors from the air or that uses a mechanical filter to remove particulate matter. An air-purifying respirator must only be used when there is sufficient oxygen to sustain life and the air contaminant level is below the concentration limits of the device.

Air pollution: Contamination of the atmosphere (indoor or outdoor) caused by the discharge (accidental or deliberate) of a wide range of toxic airborne substances.

Air sampling: Industrial hygienists are interested in knowing what contaminants workers are exposed to, and the contaminant concentrations. Determining the quantities and types of atmospheric contaminants is accomplished by measuring and evaluating a representative sample of air. The types of air contaminants that occur in the workplace depend upon the raw materials used and the processes employed. Air contaminants can be divided into two broad groups, depending upon physical characteristics: (1) gases and vapors and (2) particulates.

Air-supplied respirator: Respirator that provides a supply of breathable air form a clean source outside of the contaminated work area.

Allergens: Due to the presence of allergens on spores, all molds studied to date have the potential to cause allergic reactions in susceptible humans. Allergic reactions are believed to be the most common exposure reaction to molds (Rose, 1999).

Alpha particle: A small, positively charged particle made up of two neutrons and two protons of very high velocity, generated by many radioactive materials, including uranium and radium.

Alveoli: Tiny air sacs I the lungs, located at the ends of bronchioles. Through the thin walls of the alveoli, blood takes in oxygen and gives up carbon dioxide in respiration.

Ambient: Descriptive of any condition of the environment surrounding a given point. For example, ambient air means that portion of the atmosphere, external to buildings, to which the general public has access. Ambient sound is the sound generated by the environment.

Amorphous: noncrystalline.

ANSI: The American National Standards Institute is a voluntary membership organization (run with private funding) that develops consensus standards nationally for a wide variety of devices and procedures.

Aromatic: Applied to a group of hydrocarbons and their derivatives characterized by the presence of the benzene nucleus.

Asbestosis: A disease of the lungs caused by inhalation of fine airborne asbestos fibers.

Asphyxiant: A vapor gas which can cause unconsciousness or death by suffocation (lack of oxygen). Asphyxiation is one of the principal potential hazards of working in confined spaces.

Asphyxiation: Suffocation from lack of oxygen. A substance (e.g., carbon monoxide), that combines with hemoglobin to reduce the blood's capacity to transport oxygen produces chemical asphyxiation. Simple asphyxiation is the result of exposure to a substance (such as methane) that displaces oxygen.

ASTM: American Society for Testing and Materials.

Atmosphere: In physics, a unit of pressure whereby 1 atmosphere (atm) equals 14.7 pounds per square inch (psi).

Atmosphere-supplying respirator: A respirator that provides breathing air from a source independent of the surrounding atmosphere. There are two types: air-line and self-contained breathing apparatus.

Atmospheric pressure: The pressure exerted in all directions by the atmosphere. At sea level, mean atmospheric pressure is 29.92 in. Hg, 14.7 psi, or 407 in. wg.

Attenuation: The reduction of the intensity at a designated first location as compared with intensity at a second location, which is farther from the source (reducing the level of noise by increasing distance from the source is a good example).

Atomic weight: The atomic weight is approximately the sum of the number of protons and neutrons found in the nucleus of an atom. This sum is also called the mass number. The atomic weight of oxygen is approximately 16, with most oxygen atoms containing 8 neutrons and 8 protons.

Attenuate: To reduce in amount. Used to refer to noise or ionizing radiation.

Audible range: The frequency range over which normal hearing occurs—approximately 20–20,000 Hz. Above the range of 20,000 Hz, the term ultrasonic is used. Below 20 Hz, the term subsonic is used.

Audiogram: A record of hearing loss or hearing level measured at several different frequencies—usually 500–6000 Hz. The audiogram may be presented graphically or numerically. Hearing level is shown as a function of frequency.

Audiometric testing: Objective measuring of a person's hearing sensitivity. By recording the response to a measured signal, a person's level of hearing sensitivity can be expressed in decibels, as related to an audiometric zero, or no-sound base.

Authorized person: (see competent or qualified person) A person designated or assigned by an employer or supervisor to perform a specific type of duty or duties, to use specified equipment, and/or to be present in a given location at specified times (e.g., an authorized or qualified person is used in confined space entry).

Auto-ignition temperature: The lowest temperature at which a vapor-producing substance or a flammable gas will ignite even without the presence of a spark or flame.

Avogadro's Number: The number of molecules in a mole of any substance (6.02217×10^3). Named after Italian physicist Amedeo Avogadro (1776–1856). At 0°C and 29.92 in. HG, 1 mole occupies 22.4 liters of volume.

Background noise: The radiation coming from sources other than the particular noise sources being monitored.

Baghouse: Term commonly used for the housing containing bag filters for recovery of fumes from arsenic, lead, sulfa, etc. Many different trade meanings, however.

Base: A compound that reacts with an acid to form a salt. It is another term for alkali.

Baseline data: Data collected prior to a project for later use in describing conditions before the project began. Also commonly used to describe the first audiogram given (within 6 months) to a worker after he or she has been exposed to the action level (85 dBA)—to establish his or her baseline for comparison to subsequent audiograms for comparison.

Bel: A unit equal to 10 decibels (see decibel).

Benchmarking: A process for rigorously measuring company performance vs. "best-in-class" companies, and using analysis to meet and exceed the best in class.

Behavior-based management models: A management theory, based on the work of B.F. Skinner, explains behavior in terms of stimulus, response, and consequences.

Benign: Not malignant. A benign tumor does not metastasize or invade tissue. Benign tumors may still be lethal, due to pressure on vital organs.

Beta particle: Beta radiation. A small electrically charged particle thrown off by many radioactive materials, identical to the electron. Beta particles emerge from radioactive material at high speeds.

Bioaerosols: Mold spores, pollen, viruses, bacteria, insect parts, animal dander, etc.

Biohazard: (biological hazard) Organisms or products of organisms that present a risk to humans.

Boiler code: ANSI/ASME Pressure Vessel Code whereby a set of standards prescribing requirements for the design, construction, testing, and installation of boilers and unfired pressure vessels.

Boiling point: The temperature at which the vapor pressure of a liquid equals atmospheric pressure.
Boyle's Law: States that the product of a given pressure and volume is constant with a constant temperature.
Breathing zone: A hemisphere-shaped area from the shoulders to the top of the head.
Cancer: a cellular tumor formed by mutated cells.
Capture velocity: Air velocity at any point in front of an exhaust hood is necessary to overcome opposing air currents and to capture the contaminated air by causing it to flow into the hood.
Carbon monoxide: A colorless, odorless toxic gas produced by any process that involves the incomplete combustion of carbon-containing substances. It is emitted through the exhaust of gasoline-powered vehicles.
Carcinogen: A substance or agent capable of causing or producing cancer in mammals, including humans. A chemical is considered to be a carcinogen if: (1) it has been evaluated by the International Agency for Research on Cancer (IARC) and found to be a carcinogen or potential carcinogen; or (2) it is listed as a carcinogen or potential carcinogen in the Annual Report on Carcinogens published by the National Toxicology Program (NTP) (latest edition); or (3) it is regulated by OSHA as a carcinogen.
Carpal tunnel syndrome: An injury to the median nerve inside the wrist.
CAS: Chemical Abstracts Service is an organization under the American Chemical Society. CAS abstracts and indexes chemical literature from all over the world in "Chemical Abstracts." "CAS Numbers" are used to identify specific chemicals or mixtures.
Catalyst: A substance that alters the speed of, or makes possible, a chemical or biochemical reaction, but remains unchanged at the end of the reaction.
Catastrophe: A loss of extraordinarily large dimensions in terms of injury, death, damage, and destruction.
Casual factor: (accident cause) A person, thing, or condition that contributes significantly to an accident or to a project outcome.
Ceiling limit (C): An airborne concentration of a toxic substance in the work environment, which should never be exceeded.
CERCLA: Comprehensive Environmental Response, Compensation and Liability Act of 1980. Commonly known as "Superfund."
Charles's law: Law stating that the volume of a given mass of gas at constant pressure is directly proportional to its absolute temperature (temperature in kelvin).
Chemical cartridge respirator: A respirator that uses various chemical substances to purify inhaled air of certain gases and vapors. This type of respirator is effective for concentrations no more than ten times the TLV of the contaminant, if the contaminant has warning properties (odor or irritation) below the TLV.
Chemical change: Change that occurs when two or more substances (reactants) interact with each other, resulting in the production of different substances (products) with different chemical compositions. A simple example of chemical change is the burning of carbon in oxygen to produce carbon dioxide.
Chemical hazards: Include mist, vapors, gases, dusts, and fumes.
Chemical spill: An accidental dumping, leakage, or splashing of a harmful or potentially harmful substance.
CHEMTREC: Chemical transportation Emergency Center. Public service of the Chemical Manufacturers Association that provides immediate advice for those at the scene of hazardous materials emergencies. CHEMTREC has a 24-hour toll-free telephone number (800-424-9300) to help respond to chemical transportation emergencies.
Chromatograph: An instrument that separates and analyzes mixtures of chemical substances.
Chronic: *Persistent, prolonged, repeated*. Chronic exposure occurs when repeated exposure to or contact with a toxic substance occurs over a period of time, the effects of which become evident only after multiple exposures.

Coefficient of friction: A numerical correlation of the resistance of one surface against another surface.

Colorimetry: A term used for all chemical analyses involving reactions in which a color is developed when a particular contaminant is present in the sample and reacts with the collection medium. The resultant stain length or color intensity is measured to determine the actual concentration.

Combustible gas indicator: An instrument which samples the air and indicates whether an explosive mixture is present, and the percentage of the lower explosive limit (LEL) of the air-gas mixture that has been reached.

Combustible liquid: Liquids having a flash point at or above 37.8°C (100°F).

Combustion: Burning, defined in chemical terms as the rapid combination of a substance with oxygen, accompanied by the evolution of heat and usually light.

Concentration: The amount of a given substance in a stated unit of measure. Common methods of stating concentration are percent by weight or by volume, weight per unit volume, normality, etc.

Conductive hearing loss: Type of hearing loss is usually caused by a disorder affecting the middle or external ear.

Contact dermatitis: Dermatitis caused by contact with a substance.

Competent person: As defined by OSHA, one who is capable of recognizing and evaluating employee exposure to hazardous substances or to unsafe conditions, and who is capable of specifying protective and precautionary measures to be taken to ensure the safety of employees as required by particular OSHA regulations under the conditions to which such regulations apply.

Confined space: A vessel, compartment, or any area having limited access and (usually) no alternate escape route, having severely limited natural ventilation or an atmosphere containing less than 19.5% oxygen, and having the capability of accumulating a toxic, flammable, or explosive atmosphere, or of being flooded (engulfing a victim).

Containment: In fire terminology, restricting the spread of fire. For chemicals, restricting chemicals to an area that is diked or walled off to protect personnel and the environment.

Contingency plan: (commonly called the emergency response plan) Under CFR 40 § 260.10), a document that sets forth an organized, planned, and coordinated course of action to be followed in the event of an emergency that could threaten human health or the environment.

Convection: The transfer of heat from one location to another by way of a moving medium.

Corrosive Material: Any material that dissolves metals or other materials, or that burns the skin.

Coulometer: A chemical analysis instrument that determines the amount of a substance released in electrolysis by measurement of the quantity of electricity used. The number of electrons transferred in terms of coulombs in an indication of the contaminant concentration.

CFR: Code of Federal Regulations. A collection of the regulations that have been promulgated under United States Law.

Cumulative trauma disorder (CTD): A disorder of a musculoskeletal or nervous system component caused by repeated and or forceful movements of the said musculoskeletal system.

Cumulative injury: A term used to describe any physical or psychological disability that results from the combined effects of related injuries or illnesses in the workplace.

Cumulative trauma disorder: A disorder caused by the highly repetitive motion required of one or more parts of a worker's body, which in some cases, can result in moderate to total disability.

Cutaneous: Pertaining to or affecting the skin.

Cyclone device: A dust-collecting instrument that has the ability to separate particles by size.

Dalton's Law of Partial Pressures: States that in a mixture of theoretically ideal gases, the pressure exerted by the mixture is the sum of the pressure exerted by each component gas of the mixture.

Decibel (dB): A unit of measure used originally to compare sound intensities and subsequently electrical or electronic power outputs; now also used to compare voltages. In hearing conservation, a logarithmic unit is used to express the magnitude of a change in the level of sound intensity.

Decontamination: The process of reducing or eliminating the presence of harmful substances such as infectious agents, to reduce the likelihood of disease transmission from those substances.

Degrees celsius (Centigrade): The temperature on a scale in which the freezing point of water is 0°C and the boiling point is 100°C. To convert to degrees Fahrenheit, use the following formula: °F=(°C×1.8)+32.

Degrees fahrenheit: The temperature on a scale in which the boiling point of water is 212°F and the freezing point is 32°F.

Density: A measure of the compactness of a substance; it is equal to its mass per unit volume and is measured in kg per cubic meter/LB per cubic foot (**D = mass/Volume**).

Dermatitis: Inflammation or irritation of the skin from any cause. Industrial dermatitis is an occupational skin disease.

Dermatosis: A broader term than dermatitis; it includes any cutaneous abnormality, thus encompassing folliculitis, acne, pigmentary changes, and nodules and tumors.

Design load: The weight, which can be safely supported by a floor, equipment or structure, as defined by its design characteristics.

Dike: An embankment or ridge of either natural or man-made materials used to prevent the movement of liquids, sludges, solids, or other materials.

Dilute: Adding material to a chemical by the user or manufacturer to reduce the concentration of active ingredient in the mixture.

Direct reading instruments: Devices that provide an immediate indication of the concentration of aerosols, gases, and vapors by means of a color change in colorimetric devices, or a register on a meter or instrument.

Dose: An exposure level. Exposure is expressed as weight or volume of test substance per volume of air (mg/l), or as parts per million (ppm).

Dose-response relationship: Correlation between the amount of exposure to an agent or toxic chemical and the resulting effect on the body.

Dosimeter: Provides a time-weighted average over a period of time such as one complete work shift.

Dry bulb temperature: An ordinary thermometer. The temperature of a gas or mixture of gases indicated on a thermometer after correction for radiation.

Duct: A conduit used for moving air at low pressures.

Duct velocity: Air velocity through the duct cross section.

Dusts: Solid particles generated by handling, crushing, grinding, rapid impact, detonation, and decrepitation of organic or inorganic materials, such as rock ore, metal, coal, wood and grain. Dusts do not tend to flocculate, except under electrostatic forces; they do not diffuse in air but settle under the influence of gravity.

Dyspnea: Shortness of breath, difficult or labored breathing.

Electrical grounding: Precautionary measures designed into an electrical installation to eliminate dangerous voltages in and around the installation, and to operate protective devices in case of current leakage from energized conductors to their enclosures.

Emergency plan: See contingency plan.

Emergency response: The response made by firefighters, police, health care personnel, and/or other emergency service upon notification of a fire, chemical spill, explosion, or other incident in which human life and/or property may be in jeopardy.

Energized ("live"): The conductors of an electrical circuit. Having voltage applied to such conductors and to surfaces which a person might touch; having voltage between such surfaces and other surfaces might complete a circuit and allow current to flow.

Energy: The capacity for doing work. Potential energy (PE) is energy deriving from position; thus, a stretched spring has elastic PE, and an object raised to a height above the Earth's surface, or the water in an elevated reservoir has gravitational PE. A lump of coal and a tank of oil, together with oxygen needed for their combustion, have chemical energy. Other sorts of energy include electrical and nuclear energy, light, and sound. Moving bodies possess kinetic energy (KE). Energy can be converted from one form to another, but the total quantity stays the same (in accordance with the conservation of energy principle). For example, as an orange falls, it loses gravitational PE, but gains KE.

Engineering: The application of scientific principles to the design and construction of structures, machines, apparatus, manufacturing processes, and power generation and utilization, to satisfy human needs. Industrial hygiene is concerned with the control of environment and humankind's interface with it, especially safety interaction with machines, hazardous materials and radiation.

Engineering controls: Methods of controlling employee exposures by modifying the source of reducing the quantity of contaminants released into the workplace environment.

Epidemiological Theory: Holds that the models used for studying and determining epidemiological relationships can also be used to study causal relationships between environmental factors and accidents or diseases.

Ergonomics: A multidisciplinary activity dealing with interactions between man and his total working environment, plus stresses related to such environmental elements as atmosphere, heat, light, and sound, as well as all tools and equipment of the workplace.

Etiology: The study or knowledge of the cases of disease.

Exhaust ventilation: A mechanical device used to remove air from any space.

Exposure: Contact with a chemical, biological, or physical hazard.

Exposure ceiling: Refers to the concentration level of a given substance that should not be exceeded at any point during an exposure period.

Evaporation: The process by which a liquid is changed into the vapor state.

Evaporation rate: The ratio of the time required to evaporate a measured volume of a liquid to the time required to evaporate the same volume of a reference liquid (butyl acetate, ethyl ether) under ideal test conditions. The higher the ratio, the slower the evaporation rate. The evaporation rate can be useful in evaluating the health and fire hazards of a material.

Face velocity: Average air velocity into an exhaust system measured at the opening of the blood or booth.

Fall arresting system: A system consisting of a body harness, a lanyard or lifeline, and an arresting mechanism with built-in shock absorber, designed for use by workers performing tasks in locations from which falls would be injurious or fatal, or where other kinds of protection are not practical.

Federal Register: Publication of U.S. government documents officially promulgated under the law, documents whose validity depends upon such publication. It is published on each day following a government working day. It is, in effect, the daily supplement to the Code of Federal Regulations, CFR.

Fiber: Particle with an aspect ratio greater than 3:1 (NIOSH 5:1).

Fire: A chemical reaction between oxygen and a combustible fuel.

Fire point: The lowest temperature at which a material can evolve vapors fast enough to support continuous combustion.

Flame ionization detector: A direct reading instrument that ionize gases and vapors with an oxyhydrogen flame and measures the differing electrical currents thus generated.

Flammable liquid: Any liquid having a flash point below 37.8°C (100°F), except any mixture having components with flashpoints of 100°F or higher, the total of which make up 99% or more of the total volume of the mixture.

Flammable range: The difference between the lower and upper flammable limits, expressed in terms of percentage of vapor or gas in the air by volume, and is also often referred to as the "explosive range."

Flammable solid: A non-explosive solid liable to cause fire through friction, absorption of moisture, spontaneous chemical change, or heat retained from a manufacturing process, or that can be ignited readily and when ignited, burns so vigorously and persistently as to create a serious hazard.

Flash point: The lowest temperature at which a liquid gives off enough vapor to form ignitable moisture with air and produce a flame when a source of ignition is present. Two tests are used—open cup and closed cup.

Foot-candle: A unit of illumination. The illumination at a point on a surface which is one foot from, and perpendicular to, a uniform point source of one candle.

Fume: Airborne particulate matter formed by the evaporation of solid materials, e.g., metal fume emitted during welding. Usually less than one micron in diameter.

Gage pressure: Pressure measured with respect to atmospheric pressure.

Gamma rays: High energy X-rays.

Gas: A state of matter in which the material has very low density and viscosity, can expand and contract greatly in response to changes in temperature and pressure, easily diffuses into other gases, and readily and uniformly distributes itself throughout any container. Examples include sulfur dioxide, ozone, and carbon monoxide.

Gas chromatography: A detection technique that separates a gaseous mixture by passing it through a column, enabling the components to be released at various times depending on their molecular structure. Used as an analytical tool for air sampling gases and vapors.

Geiger-muller counter: A gas-filled electrical instrument that indicates the presence of an atomic particle or ray by detecting the ions produced.

General ventilation: A ventilation system using natural or mechanically generated make-up air to mix and dilute contaminants in the workplace.

Globe thermometer: A thermometer set in the center of a black metal sphere to measure radiant heat.

Grab sample: A sample taken within a short duration to quantify or identify air contaminants.

Gram (g): A metric unit of weight. Once ounce equals 28.4 grams.

Grounded system: A system of conductors in which at least one conductor or point is intentionally grounded, either solidly or through a current-limiting (current transformer) device

Ground-fault circuit interrupter (GFCI): A sensitive device intended for shock protection, which functions to de-energize an electrical circuit or portion thereof within a fraction of a second, in case of leakage to ground of current sufficient to be dangerous to persons but less than that required to operate the overcurrent protective device of the circuit.

Half-life: For a single radioactive decay process, the time required for the activity to decrease to half its value by that process.

Hazard: The potential for an activity, condition, circumstance, or changing conditions or circumstances to produce harmful effects. Also, it is an unsafe condition.

Hazard analysis: A systematic process for identifying hazards and recommending corrective action.

Hazard assessment: A qualitative evaluation of potential hazards in the interrelationships between and among the elements of a system, upon the basis of which the occurrence probability of each identified hazard is rated.

Hazard communication standard (HazCom): An OSHA workplace standard found in 29 CFR 1910.1200 that requires all employers to become aware of the chemical hazards in their workplace and relay that information to their employees. In addition, a contractor conducting work at a client's site must provide chemical information to the client regarding the chemicals that are brought onto the work site.

Hazard and operability (HAZOP) analysis: A systematic method in which process hazards and potential operating problems are identified, using a series of guide words to investigate process deviations.

Hazard identification: The pinpointing of material, system, process and plant characteristics that can produce undesirable consequences through the occurrence of an accident.

Hazard control: A means of reducing the risk from exposure to a hazard.

Hazardous material: Any material possessing a relatively high potential for harmful effects upon persons.

Hazardous substance: Any substance which has the potential for causing injury by reason of its being explosive, flammable, toxic, corrosive, oxidizing, irritating, or otherwise harmful to personnel.

Hazardous waste: A solid, liquid, or gaseous waste that may cause or significantly contribute to serious illness or death, or that poses a substantial threat to human health or the environment when the waste is improperly managed. When a hazardous material is spilled, it becomes a hazardous waste.

Heat cramps: Muscle spasms resulting from exposure to excessive heat.

Heat exhaustion: A condition usually caused by loss of water due to exposure to excessive heat. Symptoms include nausea, headache, and fainting.

Heat rash: A rash caused by sweating and inadequate hygiene practices.

Heat stroke: A serious disorder resulting from exposure to excessive heat. Caused by sweat suppression and increased storage of body heat. Symptoms include hot, dry skin, high temperature, mental confusion, convulsions, and coma.

HEPA filter: (High-Efficiency Particulate Air Filter) A disposable, extended medium, dry type filter with a particle removal efficiency of no less than 99.97% for 0.3 m particles.

Hearing conservation: The prevention of, or minimizing of noise-induced deafness through the use of hearing protection devices, the control of noise through engineering controls, annual audiometric tests and employee training.

Heat cramps: A type of heat stress that occurs as a result of salt and potassium depletion.

Heat exhaustion: A condition usually caused by loss of body water from exposure to excess heat. Symptoms include headache, tiredness, nausea, and sometimes fainting.

Heatstroke: A serious disorder resulting from exposure to excess heat. It results from sweat suppression and increased storage of body heat, characterized by high fever, collapse, and sometimes convulsions or coma.

Homeland security: Federal cabinet-level department created to protect America as a result of 9/11. The new Department of Homeland Security (DHS) has three primary missions: Prevent terrorist attacks within the United States, reduce American's vulnerability to terrorism, and minimize the damage from potential attacks and natural disasters.

Hood entry loss: The pressure loss from turbulence and friction as air enters a ventilation system.

Hot work: Work involving electric or gas welding, cutting, brazing, or similar flame or spark-producing operations.

Human factor engineering: (used in the United States) **ergonomics**: (used in Europe)—for practical purposes, the terms are synonymous, and focus on human beings and their interaction with products, equipment, facilities, procedures and environments used in work and everyday living. The emphasis is on human beings (as opposed to engineering, where the emphasis is more strictly on technical engineering considerations) and how the design of things influences people. Human factors, then, seek to change the things people use and the environments in which they use these things to better match the capabilities, limitations, and needs of people (Sanders & McCormick, 1993).

IDLH: Immediately Dangerous to Life and Health. An atmospheric concentration of any toxic, corrosive or asphyxiant substance that poses an immediate threat to life or would cause irreversible or delayed adverse health effects or would interfere with an individual's ability to escape from a dangerous atmosphere.

Ignition source: Anything that provides heat, spark or flame sufficient to cause combustion or an explosion.

Ignition temperature: The temperature at which a given fuel can burst into flame.

Illumination: The amount of light flux a surface receives per unit area. May be expressed in lumens per square foot or in foot-candles.

Impaction: Forcibly lodging particles into matter.

Impervious: A material that does not allow another substance to pass through or penetrate it. Frequently used to describe gloves or chemical clothing.

Impingement: The process of collecting particles by pulling contaminated air through a device filled with water or reagent (particles remain in the liquid).

Impulse noise: A noise characterized by rapid rise time, high peak value, and rapid decay.

Inches of mercury column: A unit used in measuring pressures. One inch of mercury column equals a pressure of 1.66 kPa (0.491 psi).

Inches of water column. A unit used in measuring pressures. One inch of water column equals a pressure of 0.25 kPa (0.036 psi).

Incident: An undesired event that, under slightly different circumstances, could have resulted in personal harm or property damage; any undesired loss of resources.

Incompatible: Materials which could cause dangerous reactions from direct contact with one another.

Indoor air quality (IAQ): refers to the effect, good or bad, of the contents of the air inside a structure, on its occupants. Usually, temperature (too hot and cold), humidity (too dry or too damp), and air velocity (draftiness or motionless) are considered "comfort" rather than indoor air quality issues. Unless they are extreme, they may make someone unhappy, but they won't make a person ill. Nevertheless, most IAQ professionals will take these factors into account in investigating air quality situations.

Industrial hygiene: The American Industrial Hygiene Association (AIHA) defines industrial hygiene as "that science and art devoted to the anticipation, recognition, evaluation, and control of those environmental factors or stresses—arising in the workplace—which may cause sickness, impaired health and well-being, or significant discomfort and inefficiency among workers or among citizens of the community."

Ingestion: Taking in by the mouth.

Inhalation: Breathing of a substance in the form of a gas, vapor, fume, mist, or dust.

Injury: A wound or other specific damage.

Insoluble: Incapable of being dissolved in a liquid.

Interlock: A device that interacts with another device or mechanism to govern succeeding operations. For example: an interlock on an elevator door prevents the car from moving unless the door is properly closed.

Ionizing radiation: Radiation that becomes electrically charged or changed into ions.

Irritant: A substance that produces an irritating effect when it contacts the skin, eyes, nose, or respiratory system.

Job hazard analysis: (also called job safety analysis) The breaking down into its component parts of any method or procedure, to determine the hazards connected therewith and the requirements for performing it safely.

Kinetic energy: The energy resulting from a moving object.

Kelvin: A temperature scale, also called absolute temperature, where the temperature is measured on the average kinetic energy per molecule of a perfect gas.

Laboratory safety standard: A specific hazard communication program for laboratories, found in 29 CFR 1910.1450. These regulations are essentially a blend of hazard communication and emergency response for laboratories. The cornerstone of the Lab Safety Standard is the requirement for a written Chemical Hygiene Plan.

Laser: Light amplification by stimulated emission of radiation.

Latent period: The time that elapses between exposure and the first manifestation of damage.

LC_{50}: Lethal concentration that will kill 50% of the test animals within a specified time.

LD_{50}: The dose required to produce the death in 50% of the exposed species within a specified time.

Liter (L): A measure of capacity—one quart equals 0.9 L.

Local exhaust ventilation: A ventilation system that captures and removes contaminants at the point of generation before escaping into the workplace.

Lockout/tagout procedure: An OSHA procedure found in 29 CFR 1910.147. A tag or lock is used to tag out or log out a device, so that no one can inadvertently actuate the circuit, system, or equipment that is temporarily out of service.

Log and summary of occupational injuries and illnesses (OSHA-200 Log): A cumulative record that employers (generally of more than 10 employees) are required to maintain, showing essential facts of all reportable occupational injuries and illnesses.

Loss: The degradation of a system or component. Loss is best understood when related to dollars lost. Examples include death or injury to a worker, destruction or impairment of facilities or machines, destruction or spoiling of raw materials, and creation of delay. In the insurance business, loss connotes dollar loss, and we have seen underwriters who write it as LO$$ to make that point.

Lower explosive limit (LEL): The minimum concentration of a flammable gas in air required for ignition in the presence of an ignition source. Listed as a percent by volume in air.

Makeup air: Clean, tempered outdoor air supplied to a workplace to replace air removed by exhaust ventilation.

Malignant: As applied to a tumor. Cancerous and capable of undergoing metastasis, or invasion of surrounding tissue.

Material safety data sheet (MSDS): (now known as SDS) Chemical information sheets provided by the chemical manufacturer that include information such as chemical and physical characteristics; long- and short-term health hazards; spill control procedures; personal protective equipment (PPE) to be used when handling the chemical; reactivity with other chemicals; incompatibility with other chemicals; and manufacturer's name, address and phone number. Employee access to and understanding of MSDS are important parts of the HazCom Program.

Medical monitoring: The initial medical exam of a worker, followed by periodic exams. The purpose of medical monitoring is to assess workers' health, determine their fitness to wear personal protective equipment and maintain records of their health.

Mesothelioma: Cancer of the membranes that line the chest and abdomen, almost exclusively associated with asbestos exposure.

Metabolic heat: Produced within a body as a result of activity that burns energy.

Metastasis: Transfer of the causal agent (cell or microorganisms) of a disease form a primary focus to a distant one through the blood or lymphatic vessels. Also, the spread of malignancy from the site of primary cancer to secondary sites.

Meter: A metric unit of length, equal to about 39 inches.

Micron (micrometer, m): A unit of length equal to one-millionth of a meter, approximately 1/25,000 of an inch.

Milligram (mg): A unit of weight in the metric system. One thousand milligrams equals one gram.

Milligrams per cubic meter (mg/m^3): Unit used to measure air concentrations of dusts, gases, mists and fumes.

Milliliter (mL): A metric unit used to measure volume. One milliliter equals one cubic centimeter.

Millimeter or mercury (mmHg): The unit of pressure equal to the pressure exerted by a column of liquid mercury one millimeter high at a standard temperature.

Mists: Tiny liquid droplets suspended in the air.

Molds: These are the most typical form of fungus found on Earth, comprising approximately 25% of the Earth's biomass (McNeel & Kreutzer, 1996)

Monitoring: Periodic or continuous surveillance or testing to determine the level of compliance with statutory requirements and/or pollutant levels, in various media, or in humans, animals or other living things.

Mucous membranes: Lining of the hollow organs of the body, notably the nose, mouth, stomach, intestines, bronchial tubes, and urinary tract.

Mutagen: A substance or material that causes a change in the genetic material of a cell.

Mycotoxins: Some molds are capable of producing mycotoxins, natural organic compounds that are capable of initiating a toxic response in vertebrates (McNeel & Kreutzer, 1996).

NFPA: The National Fire Protection Association is a voluntary membership organization whose aim is to promote and improve fire protection and prevention. The NFPA publishes 16 volumes of codes known as the National Fire Codes.

NIOSH: The National Institute for Occupational Safety and Health is a federal agency. It conducts research on health and safety concerns, texts and certifies respirators, and trains occupational health and safety professionals.

NTP: National Toxicology Program. The NTP publishes an Annual Report on carcinogens.

Nuisance dust: Have a long history of little adverse effect on the lungs and do not produce significant organic disease or toxic effects when exposures are kept under reasonable control.

Nonionizing radiation: That radiation on the electromagnet spectrum that has a frequency of 10^{15} or less and a wavelength in meters of 3×10^{-7}.

Occupational safety and health act (OSH Act): A federal law passed in 1970 to assure, so far as possible, every working man and woman in the nation of safe and healthful working conditions. To achieve this goal, the Act authorizes several functions, such as encouraging safety and health programs in the workplace and encouraging labor-management cooperation in health and safety issues.

OSHA form 300: Log and Summary of Occupational Injuries and Illnesses.

Oxidation: When a substance either gains oxygen or loses hydrogen or electrons in a chemical reaction. One of the chemical treatment methods.

Oxidizer: Also known as an oxidizing agent, a substance that oxidizes another substance. Oxidizers are a category of hazardous materials that may assist in the production of fire by readily yielding oxygen.

Oxygen deficient atmospheres: The legal definition of an atmosphere where the oxygen concentration is less than 19.5% by volume of air.

Oxygen-enriched atmosphere: An atmosphere containing more than 23.5% oxygen by volume.

Particulate matter: Substances (such as diesel soot and combustion products resulting from the burning of wood) released directly into the air; any minute, separate particle of liquid or solid material.

Performance standards: A form of OSHA regulation standards that lists the ultimate goal of compliance, but does not explain exactly how compliance is to be accomplished. Compliance is usually based on accomplishing the act or process in the safest manner possible, based on experience (past performance).

Permissible exposure limit (PEL): The time-weighted average concentration of an airborne contaminant that a healthy worker may be exposed to 8-hours per day or 40-hours per week without suffering any adverse health effects. Established by legal means and enforceable by OSHA.

Personal protective equipment (PPE): Any material or device is worn to protect a worker from exposure to or contact with any harmful substance or force.

pH: Means used to express the degree of acidity or alkalinity of a solution with neutrality indicated as seven.

Pitot tube: A device used for measuring static pressure within ventilation ducts.

Polymerization: A chemical reaction in which two or more small molecules (monomers) combine to form larger molecules (polymers) that contain repeating structural units of the original molecules. A hazardous polymerization is an above reaction, with an uncontrolled release of energy.

Ppm: Parts per million parts of air by volume of vapor or gas or other contaminants. Used to measure air concentrations of vapors and gases.

Precision: The degree of exactness of repeated measurements.

Preliminary assessment: A quick analysis to determine how serious the situation is, and to identify all potentially responsible parties. The preliminary assessment uses readily available information; for instance, forms, records, aerial photographs, and personnel interviews.

Pressure: The force exerted against an opposing fluid or thrust distributed over a surface.

Psi: Pounds per square inch (MSDS purposes) is the pressure a material exerts on the walls of a confining vessel or enclosure. For technical accuracy, pressure must be expressed as psig (pounds per square inch gauge) or psia (pounds per square absolute; that is, gauge pressure plus sea level atmospheric pressure, or psig plus approximately 14.7 pounds per square inch).

Radiant heat: The result of electromagnetic nonionizing energy that is transmitted through space without the movement of matter within that space.

Raynaud's syndrome: An abnormal constriction of blood vessels in the fingers when exposure to cold temperature. Caused by vibrating hand tools.

Radiation: Consists of energetic nuclear particles and includes alpha rays, beta rays, gamma rays, X-rays, neutrons, high-speed electrons, and high-speed protons.

RCRA: Resource Conservation and Recovery Act of 1976.

Reactivity (chemical): A substance that reacts violently by catching on fire, exploding, or giving off fumes when exposed to water, air, or low heat.

Reactivity hazard: The ability of a material to release energy when in contact with water. Also, the tendency of a material, when in its pure state or as a commercially produced product, to vigorously polymerize, decompose, condense, or otherwise self-react and undergo violent chemical change.

Reportable quantity (RQ): The minimum amount of a hazardous material that if spilled while in transport, must be reported immediately to the National Response Center. Minimum reportable quantities range from 1 to 5,000 pounds per 24-hour day.

Respirable size particulates: Particulates in the size range that permits them to penetrate deep into the lungs upon inhalation.

Respirator (approved): A device which has met the requirements of 30 CFR Part 11 and is designed to protect the wearer for inhalation of harmful atmospheres and has been approved by the National Institute for Occupational Safety and Health (NIOSH) and the Mine Safety and Health Administration (MSHA).

Respiratory system: Consists of (in descending order)—the nose, mouth, nasal passages, nasal pharynx, pharynx, larynx, trachea, bronchi, bronchioles, air sacs (alveoli) of the lungs, and muscles of respiration.

Risk: The combination of the expected frequency (event/year) and consequence (effects/event) of a single accident or a group of accidents; the result of a loss-probability occurrence and the acceptability of that loss.

Risk assessment: A process that uses scientific principles to determine the level of risk that actually, exists in a contaminated area.

Risk characterization: The final step in the risk assessment process involves determining a numerical risk factor. This step ensures that exposed populations are not at significant risk.

Risk management: The professional assessment of all loss potentials in an organization's structure and operations, leading to the establishment and administration of a comprehensive loss control program.

Rotameters: A small, tapered tube with a solid ball (float) inside is used to measure the flow rate of air sampling equipment.

Safety: A general term denoting an acceptable level of risk of relative freedom from, and low probability of harm.

Safety factor: Based on experimental data, the amount added (e.g., 1,000-fold) to ensure worker health and safety.

Safety standard: A set of criteria specifically designed to define a safe product, practice, mechanism, arrangement, process or environment, produced by a body representative of all concerned interests, and based upon currently available scientific and empirical knowledge concerning the subject or scope of the standard.

SARA: Superfund Amendments and Reauthorization Act of 1986.

SCBA: Self-contained breathing apparatus.

Secondary containment: A method using two containment systems so that if the first is breached, the second will contain all of the fluid in the first. For USTs, secondary containment consists of either a double-walled tank or a liner system.

Sensitizers: Chemicals that in very low doses trigger an allergic response.

Short term exposure limit (STEL): The time-weighted average concentration to which workers can be exposed continuously for a short period of time (typically 15 minutes) without suffering irritation, chronic or irreversible tissue damage, or impairment for self-rescue.

Silica: Crystalline silica (SiO_2) is a major component of the Earth's crust and a responsible for causing silicosis.

"Skin": A notation (sometimes used with PEL or TLV exposure data) which indicates that the stated substance may be absorbed by the skin, mucous membranes, and eyes—either airborne or by direct contact—and that this additional exposure must be considered part of the total exposure to avoid exceeding the PEL or TLV for that substance.

Solubility in water: A term expressing the percentage of a material (by weight) that will dissolve in water at ambient temperature. Solubility information can be useful in determining spill cleanup methods and re-extinguishing agents and methods for a material.

Solvent: A substance, usually a liquid, in which other substances are dissolved. The most common solvent is water.

Sorbent: (1) A material that removes toxic gases and vapors from the air inhaled through a canister or cartridge. (2) Material used to collect gases and vapors during air-sampling.

Specific gravity: The ratio of the densities of a substance to water.

Stability: An expression of the ability of a material to remain unchanged. For MSDS purposes, a material is stable if it remains in the same form under expected and reasonable conditions of storage or us. Conditions which may cause instability (dangerous change) are stated. Examples are temperatures above 150°F, shock from dropping.

Synergism: Cooperative action of substances whose total effect is greater than the sum of their separate effects.

Systemic: Spread throughout the body, affecting all body systems and organs, not localized in one spot or area.

Temporary threshold shift (TTS): Temporary hearing loss due to noise exposure. Many be partially or completely recovered when exposure ends.

Tendonitis: Inflammation of a tendon.

Tenosynovitis: Inflammation of the connective tissue sheath of a tendon.

Threshold: The lowest dose or exposure to a chemical at which a specific effect is observed.

Threshold limit value (TLV): The same concept as PEL except that TLVs do not have the force of governmental regulations behind them, but are based on recommended limits established and promoted by the American Conference of Governmental Industrial Hygienists.

Time-weighted average (TWA): A mathematical average [(exposure in ppm×time in hours) ¼ time in hrs.=time weighted average in ppm] of exposure concentration over a specific time.

Total quality management (TQM): A way of managing a company that revolves around a total and willing commitment of all personnel at all levels to quality.

Toxicity: The relative property of a chemical agent with reference to a harmful effect on some biologic mechanism and the condition under which this effect occurs. The quality of being poisonous.

Toxicology: The study of poisons, which are substances that can cause harmful effects on living things.

Toxin: A poison.

Unsafe condition: Any physical state that deviates from that which is acceptable, normal, or correct in terms of past production or potential future production of personal injury and/or damage to property; any physical state that results in a reduction in the degree of safety normally present.

Upper explosive limit (UEL): The maximum concentration of flammable gas in the air required for ignition in the presence of an ignition source.

Vapor: The gaseous form of substances that are normally solid or liquid at room temperature.

Vapor pressure: Pressure (measured in pounds per square inch absolute—psia) exerted by a vapor. If a vapor is kept in confinement over its liquid so that the vapor can accumulate above the liquid (the temperature being held constant), the vapor pressure approaches a fixed limit called the maximum (or saturated) vapor pressure, dependent only on the temperature and the liquid.

Vapors: The gaseous form of substances that are normally in the solid or liquid state (at room temperature and pressure). The vapor can be changed back to the solid or liquid state either by increasing the pressure or decreasing the temperature alone. Vapors also diffuse. Evaporation is the process by which a liquid is changed into the vapor state and mixed with the surround air. Solvents with low boiling points will volatize readily. Examples include benzene, methyl alcohol, mercury, and toluene.

Velometer: A device used in ventilation to measure air velocity.

Viscosity: The property of a fluid that resists internal flow by releasing counteracting forces.

Vulnerability assessment: A vulnerability assessment is a very regulated, controlled, cooperative, and documented evaluation of an organization's security posture from outside-in and inside-out, for the purpose of defining or greatly enhancing security policy.

Volatility: The tendency or ability or a liquid to vaporize. Such liquids as alcohol and gasoline, because of their well-known tendency to evaporate rapidly, are called volatile liquids.

Water column: A unit used in measuring pressure.

Wet bulb globe thermometer: Temperature as determined by the wet bulb thermometer or a standard sling psychrometer or its equivalent. Influenced by the evaporation rate of water which, in turn, depends on relative air humidity. A wet bulb thermometer consists of a bulb covered with a cloth saturated with water.

Workers' compensation: A system of insurance required by state law and financed by employers, which provides payments to employees and their families for occupational illnesses, injuries, or fatalities incurred while at work and resulting in loss of wage income, usually regardless of the employer's or employee's negligence.

Zero energy state: The state of equipment in which every power source that can produce movement of a part of the equipment, or the release of energy, has been rendered inactive.

Zoonoses: Diseases communicable from animals to humans under natural conditions.

HISTORY OF INDUSTRIAL HYGIENE (OSHA, 1998)

There has been an awareness of industrial hygiene since antiquity. The environment and its relation to worker health were recognized as early as the fourth century BC when Hippocrates noted lead toxicity in the mining industry. In the first century AD, Pliny the-elder, a Roman scholar, perceived health risks to those working with zinc and sulfur. He devised a face mask made from an animal bladder to protect workers from exposure to dust and lead fumes. In the second century AD, the Greek physician, Galen, accurately described the pathology of lead poisoning and also recognized the hazardous exposures of copper miners to acid mists.

In the Middle Ages, guilds worked at assisting sick workers and their families. In 1556 the German scholar, Agricola, advanced the science of industrial hygiene even further when, in his book *De Re Metallica* (On the Nature of Metals), he described the disease of miners and prescribed preventive measures. The book included suggestions for mine ventilation and worker protection, discussed mining accidents, and described diseases associated with mining occupations such as silicosis.

Industrial Hygiene

Industrial hygiene gained further respectability in 1700 when Bernardo Ramazzini, known as the "father of industrial medicine," published in Italy the first comprehensive book on industrial medicine, *De Morbis Artificum Diatriba* (The Diseases of Workmen). The book contained accurate descriptions of the occupational diseases of most of the workers of his time. Ramazzini greatly affected the future of industrial hygiene because he asserted that occupational diseases should be studied in the work environment rather than in hospital wards.

Industrial hygiene received another major boost in 1743 when Ulrich Ellenborg published a pamphlet on occupational diseases and injuries among gold miners. Ellenborg also wrote about the toxicity of carbon monoxide, mercury, lead, and nitric acid.

In England in the eighteenth century, Percival Pott, as a result of his findings on the insidious effects of soot on chimney sweepers, was a major force in getting the British Parliament to pass the *Chimney-Sweepers Act of 1788*. The passage of the English Factory Acts beginning in 1833 marked the first effective legislative acts in the field of industrial safety. The Acts, however, were intended to provide compensation for accidents rather than to control their causes. Later, various other European nations developed workers' compensation acts, which stimulated the adoption of increased factory safety precautions and the establishment of medical services within industrial plants.

In the early twentieth century in the U.S., Dr. Alice Hamilton led efforts to improve industrial hygiene. She observed industrial conditions firsthand and startled mine owners, factory managers, and state officials with evidence that there was a correlation between worker illness and their exposure to toxins. She also presented definitive proposals for eliminating unhealthful working conditions.

At about the same time, U.S. federal and state agencies began investigating health conditions in industry. In 1908, the public's awareness of occupationally related diseases stimulated the passage of compensation acts for certain civil employees. States passed the first workers' compensation laws in 1911. And in 1913, the New York Department of Labor and the Ohio Department of Health established the first state industrial hygiene programs. All states enacted such legislation by 1948. In most states, there is some compensation coverage for workers contracting the occupational disease.

The U.S. Congress has passed three landmark pieces of legislation reliant to safeguarding workers' health: (1) *the Metal and Nonmetallic Mines Safety Act of 1966*, (2) *the Federal Coal Mine Safety and Health Act of 1969*, and (3) *the Occupational Safety and Health Act of 1970*. Today, nearly every employer is required to implement the elements of industrial hygiene and safety, occupational health, or hazard communication program and to be responsive to the Occupational Safety and Health Administration (OSHA) and the Act and its regulations.

OSHA/NIOSH AND INDUSTRIAL HYGIENE

The principal piece of federal legislation relating to industrial hygiene is the Occupational Safety and Health Act of 1970 (OSH Act) as amended. Under the Act, the Occupational Safety and Health Administration (OSHA) develops and sets mandatory occupational safety and health requirements applicable to the more than 6 million workplaces in the U.S. OSHA relies on, among others, industrial hygienists to evaluate jobs for potential health hazards. Developing and setting mandatory occupational safety and health standards involves determining the extent of employee exposure to hazards and deciding what is needed to control these hazards, thereby protecting the workers. As mentioned, industrial hygienists are trained to anticipate, recognize, evaluate, and recommend controls for environmental and physical hazards that can affect the health and well-being of workers. More than 40% of the OSHA compliance officers who inspect America's workplaces are industrial hygienists. Industrial hygienists also play a major role in developing and issuing OSHA standards to protect workers from health hazards associated with toxic chemicals, biological hazards, and harmful physical agents. They also provide technical assistance and support to the agency's national and regional offices. OSHA also employs industrial hygienists who assist in setting up field enforcement procedures, and who issue technical interpretations of OSHA regulations and standards. Industrial hygienists analyze, identify, and measure workplace hazards or stressors that can cause sickness, impaired health, or significant discomfort in workers through chemical, physical, ergonomic, or biological exposures. Two roles of the OSHA industrial hygienist are to spot those conditions and help eliminate or control them through appropriate measures (OSHA, 1998).

The OSH Act sets forth the following requirements relating to industrial hygiene:

- Use of warning labels and other means to make employees aware of potential hazards, symptoms of exposure, precautions, and emergency treatment.
- Prescription of appropriate personal protective equipment and other technological preventive measures (29 CFR Subpart I, 1910.133 and 1910.134).
- Provision of medical tests to determine the effect on employees of exposure to environmental stressors.
- Maintenance of accurate records of employee exposures to environmental stressors that are required to be measured or monitored.
- Making monitoring tests and measurement activities open to the observation of employees.
- Making records of monitoring tests and measurement activities available to employees on request.
- Notification of employees who have been exposed to environmental stressors at a level beyond the recommended threshold and corrective action begin taken (Goetsch, 1996).

Effective management of worker safety and health protection is a decisive factor in reducing the extent and severity of work-related injuries and illnesses and their related costs. To assist employers and employees in developing effective safety and health program, OSHA published recommended safety and Health Program Management Guidelines (Federal Register 54 (18):3908–3916, January 26, 1989). These voluntary guidelines apply to all places of employment covered by OSHA. The guidelines identify four general elements that are critical to the development of a successful safety and health management program:

- management commitment and employee involvement
- worksite analysis
- hazard prevention and control
- safety and health training

The National Institute for Occupational Safety and Health (NIOSH) is part of the Department of Health and Human Services (HHS). This agency is important to industrial hygiene professionals. The main focus of the agency's research is on toxicity levels and human tolerance levels of hazardous substances. NIOSH prepares recommendations for OSHA standards dealing with hazardous substances, and NIOSH studies are made available to employers.

INDUSTRIAL HYGIENE: WORKPLACE STRESSORS

The industrial hygienist focuses on evaluating the healthfulness of the workplace environment, either for short periods or for a work-life of exposure. When required, the industrial hygienist recommends corrective procedures to protect health, based on solid quantitative data, experience, and knowledge. The control measures he or she often recommends include: isolation of a work process, substitution of a less harmful chemical or material, and/or other measures designed solely to increase the healthfulness of the work environment.

To ensure a healthy workplace environment and associated environs, the industrial hygienist focuses on the recognition, evaluation, and control of chemical, physical, or biological and ergonomic *stressors* that can cause sickness, impaired health, or significant discomfort to workers.

The key word just mentioned was *stressors*, or simply, *stress*—the stress caused by the workplace external environment demands placed upon a worker. Increases in external stressors beyond a worker's tolerance level affect his or her on-the-job performance and overall health.

The industrial hygienist must not only understand that workplace stressors exist, but also that they are sometimes cumulative (additive). For example, studies have shown that some assembly line processes are little affected by neither low illumination nor vibration; however, when these two stressors are combined, assembly-line performance deteriorates.

Other cases have shown just the opposite effect. For example, the worker who has had little sleep and then is exposed to a work area where noise levels are high actually benefits (to a degree, depending on the intensity of the noise level and the worker's exhaustion level) from increased arousal level; a lack of sleep combined with a high noise level is compensatory.

To recognize environmental stressors and other factors that influence worker health, the industrial hygienist must be familiar with work operations and processes. An essential part of the new industrial hygienist' employee orientation process should include an overview of all pertinent company work operations and processes. The newly hired industrial hygienist who has not been fully indoctrinated on company work operations and processes not only is not qualified to study the environmental effects of such processes but also suffers from another disability—lack of credibility with supervisors and workers. This point cannot be emphasized strongly enough—know your organization and what it is all about.

What are the workplace stressors the industrial hygienist should be concerned with? The industrial hygienist should be concerned with those workplace stressors that are likely to accelerate the aging process, cause significant discomfort and inefficiency, or may be immediately dangerous to life and health (Spellman, 1998). Several stressors fall into these categories; the most important health stressors include:

- **Chemical stressors**: gases, dusts, fumes, mists, liquids, or vapors.
- **Physical stressors**: noise, vibration, extremes of pressure and temperature, and electromagnetic and ionizing radiation.
- **Biological stressors**: bacteria, fungi, molds, yeasts, insects, mites, and viruses.
- **Ergonomic stressors**: repetitive motion, work pressure, fatigue, body position in relation to work activity, monotony/boredom, and worry.

OSHA lists these health stressors as major job risks. Each is explained further in the following (OSHA, 1998).

Chemical stressors: harmful chemical compounds in the form of solids, liquids, gases, mists, dusts, fumes, and vapors exert toxic effects by inhalation (breathing), absorption (though direct contact with the skin), or ingestion (eating or drinking). Airborne chemical hazards exist as concentrations of mists, vapors, gases, fumes, or solids. Some are toxic through inhalation and some of them irritate the skin on contact; some can be toxic by absorption through the skin or through ingestion, and some are corrosive to living tissue.

The degree of worker risk from exposure to any given substance depends on the nature and potency of the toxic effects and the magnitude and duration of exposure.

Information on the risk to workers from chemical hazards can be obtained from Material Safety Data Sheet (MSDS) that OSHA's Hazard Communication Standard requires to be supplied by the manufacturer or importer to the purchaser of all hazardous materials. The MSDS is a summary of the important health, safety, and toxicological information on the chemical or the mixture's ingredients. Other provisions of the Hazard Communication Standard require that all containers of hazardous substances in the workplace have appropriate warning and identification labels.

Physical stressors: These include excessive levels of ionizing and nonionizing electromagnetic radiation, noise, vibration, illumination, and temperature. In occupations where there is exposure to ionizing radiation, time, distance, and shielding are important tools in ensuring worker safety.

Danger from radiation increases with the amount of time one is exposed to it; hence, the shorter the time of exposure the smaller the radiation danger. Distance also is a valuable tool in controlling exposure to both ionizing and non-ionizing radiation. Radiation levels from some sources can be estimated by comparing the squares of the distances between the work and the source. For example, at a reference point of 10 feet from a source, the radiation is 1/100 of the intensity at 1 foot from the source. Shielding also is a way to protect against radiation. The greater the protective mass between a radioactive source and the worker, the lower the radiation exposure. Nonionizing radiation also is dealt with by shielding workers form the source. Sometimes limiting exposure times to nonionizing radiation or increasing the distance is not effective. Laser radiation, for example, cannot be controlled effectively by imposing time limits. An exposure can be hazardous that is faster than the blinking of an eye. Increasing the distance from a laser source may require miles before the energy level reaches a point where the exposure would not be harmful.

Noise, another significant physical hazard, can be controlled by various measures. Noise can be reduced by installing equipment and systems that have been engineered, designed, and built to operate quietly; by enclosing or shielding noisy equipment; by making certain that equipment is in good repair and properly maintained with all worn or unbalanced parts replaced; by mounting noisy equipment on special mounts to reduce vibration; and by installing silencers, mufflers, or baffles. Substituting quiet work methods for noisy ones is another significant way to reduce noise, for example, welding parts rather than riveting them. Also, treating floors, ceilings, and walls with acoustical material can reduce reflected or reverberant noise. In addition, erecting sound barriers at adjacent workstations around noisy operations will reduce worker exposure to noise generated at adjacent work stations.

It is also possible to reduce noise exposure by increasing the distance between the source and the receiver, by isolating workers in acoustical booths, limiting workers' exposure time to noise, and by providing hearing protection. OSHA requires that workers in noisy surroundings be periodically tested as a precaution against hearing loss.

Another physical hazard, radiant heat exposure in factories such as steel mills, can be controlled by installing reflective shields and by providing protective clothing.

Biological stressors: These include bacteria, viruses, fungi, and other living organisms that can cause acute and chronic infections by entering the body either directly or through breaks in the skin. Occupations that deal with plants or animals or their products or with food and food processing may expose workers to biological hazards. Laboratory and medical personnel also can be exposed to biological hazards. Any occupations that result in contact with bodily fluids pose a risk to workers from biological hazards.

In occupations where animals are involved, biological hazards are dealt with by preventing and controlling diseases in the animal population as well as proper care and handling of infected animals. Also, effective personal hygiene, particularly proper attention to minor cuts and scratches, especially those on the hands and forearms, helps keep worker risks to a minimum.

In occupations where there is potential exposure to biological hazards, workers should practice proper personal hygiene, particularly hand washing. Hospitals should provide proper ventilation, proper personal protective equipment such as gloves and respirators, adequate infectious waste disposal systems, and appropriate controls including isolation in instances of particularly contagious diseases such as tuberculosis.

Ergonomic stressors: The science of ergonomics studies and evaluates a full range of tasks including, but not limited to, lifting holding, pushing, walking, and reaching. Many ergonomic problems result from technological changes such as increased assembly line speeds, adding specialized tasks, and increased repetition; some problems arise from poorly designed job tasks. Any of those conditions can cause ergonomic hazards such as excessive vibration and noise, eye strain, repetitive motion, and heavy lifting problems. Improperly designed tools or work areas also can be ergonomic hazards. Repetitive motions or repeated shocks over prolonged periods of time as in jobs involving sorting, assembling, and data entry can often cause irritation and inflammation of the tendon sheath of the hands and arms, a condition known as carpal tunnel syndrome.

Ergonomic hazards are avoided primarily by the effective design of a job or jobsite and better-designed tools or equipment that meet workers' needs in terms of physical environment and job tasks. Through thorough worksite analyses, employers can set up procedures to correct or control ergonomic hazards by using the appropriate engineering controls (e.g., designing or redesigning work stations, lighting, tools, and equipment); teaching correct work practices (e.g., proper lifting methods); employing proper administrative controls (e.g., shifting workers among several different tasks, reducing production demand, and increasing rest breaks); and, if necessary, providing and mandating personal protective equipment. Evaluating working conditions from an ergonomics standpoint involves looking at the total physiological and psychological demands of the job on the worker.

Overall, industrial hygienists point out that the benefits of a well-designed, ergonomic work environment can include increased efficiency, fewer accidents, lower operating costs, and more effective use of personnel.

In the workplace, the industrial hygienist should review the following to anticipate potential health stressors:

- Raw materials
- Support materials
- Chemical reactions
- Chemical interactions
- Products
- By-products
- Waste products
- Equipment
- Operating procedures

INDUSTRIAL HYGIENE: AREAS OF CONCERN

From the list of health stressors above, it can be seen that the industrial hygienist has many areas of concern related to protecting the health of workers on the job. In this section, we focus on the major areas that the industrial hygienist typically is concerned with in the workplace. We also discuss the important areas of industrial toxicology and industrial health hazards. Later we cover industrial noise, vibration, and environmental control. All of these areas are important to the industrial hygienist (and to the worker, of course), but they are not all-inclusive; the industrial hygienist also is concerned with other areas—ionizing and non-ionizing radiation, for example, and many others.

INDUSTRIAL TOXICOLOGY

The practice of Industrial Toxicology (especially in the area of industrial poisons) owes its genesis in the United States to the work of Alice Hamilton (1869–1970). In 1919 Hamilton became the first woman to be appointed to the staff at the Harvard Medical School. She also did studies on industrial pollution for the federal government and the United Nations. She also authored several books including Industrial *Poisons in the United States* (1925), *Industrial Toxicology* (1934) and *Exploring the Dangerous Trades* (1943).

Currently, we have witnessed unprecedented industrialization, explosive population growth, and a massive introduction of new chemical agents into the workplace. Unfortunately, we lag far behind in our understanding of the impact that many of these new chemicals, particularly mixtures, have on the health or workers and other members of our ecosystem.

The industrial hygienist must be well-versed and knowledgeable in toxicology. He or she must be a student/practitioner who constantly studies the nature and effects of poison and their treatment in the workplace. Thus, the need and importance of the industrial hygienist possessing a full understanding of industrial toxicology can't be overstated.

Consider this, normally, we give little thought to the materials (chemical substances, for example) that we are exposed to on a daily (almost constant) basis, unless they interfere with our lifestyles, irritate us, or noticeably physically affect us. However, keep in mind that all chemical substances have the potential for being injurious at some sufficiently high concentration and level of exposure. Again, the industrial hygienist understands this, and to prevent the lethal effects of over-exposure for workers must have an adequate understanding and knowledge of general toxicology.

What Is Toxicology?

Toxicology is a very broad, interdisciplinary science, which studies the adverse effects of chemicals on living organisms, using knowledge and research methods drawn from virtually all areas of the biomedical sciences. It deals with chemicals used in industry, drugs, food, and cosmetics, as well as those occurring naturally in the environment. Toxicology is the science that deals with the poisonous or toxic properties of substances. The primary objective of industrial toxicology is the prevention of adverse health effects in workers exposed to chemicals in the workplace. The industrial hygienist's responsibility is to consider all types of exposure and the subsequent effects on workers. Following the prescribed precautionary measures and limitations placed on exposure to certain chemical substances is the industrial toxicologist is worker's responsibility. The industrial hygienist uses toxicity information to prescribe safety measures for protecting workers.

To gain better appreciation for what industrial toxicology is all about, you must understand some basic terms and factors—many of which contribute to determining the degree of hazard particular chemicals present. You must also differentiate between *toxicity* and *hazard*. *Toxicity* is the intrinsic ability of a substance to produce an unwanted effect on humans and other living organisms when the chemical has reached a sufficient concentration at a certain site in the body. *Hazard* is the probability that a substance will produce harm under specific conditions. The industrial hygienist and other safety professionals employ the opposite of hazard—safety—that is, the probability that harm will not occur under specific conditions. A toxic chemical—used under safety conditions—may not be hazardous.

All toxicological considerations are based on the *dose-response relationship*, another toxicological concept important to the industrial hygienist. In its simplest terms, the dose of a chemical to the body resulting from exposure is directly related to the degree of harm. This relationship means that the toxicologist is able to determine a *threshold level* of exposure for a given chemical—the highest amount of a chemical substance to which one can be exposed with no resulting adverse health effects. Stated differently, chemicals present a threshold of effect or a no-effect level.

Threshold levels are critically important parameters. For instance, under the OSHA Act, threshold limits have been established for the air contaminants most frequently found in the workplace. The contaminants are listed in three tables in 29 CFR 1910 subpart Z—Toxic and Hazardous Substances. The threshold limit values listed in these tables are drawn from values published by the American Conference of Governmental Industrial Hygienists (ACGIH) and from the "Standards of Acceptable Concentrations of Toxic Dusts and Gases," issued by the American National Standards Institute (ANSI).

An important and necessary consideration when determining levels of safety for exposure to contaminants is their effect over a period of time. For example, during an 8-hour work shift, a worker may be exposed to a concentration of Substance A [with a 10 ppm (parts per million—analogous to a full shot glass of water taken from a standard-size in-ground swimming pool) TWA (time-weighted average), 25 ppm ceiling and 50 ppm peak] above 25 ppm (but never above 50 ppm) only for a maximum period of 10 minutes. Such exposure must be compensated by exposures to concentrations less than 10 ppm, so that the cumulative exposure for the entire 8-hour work shift does not

Industrial Hygiene

exceed a weighted average of 10 ppm. Formulas are provided in the regulations for computing the cumulative effects of exposures in such instances. Note that the computed cumulative exposure to a contaminant may not exceed the limit value specified for it.

Air Contaminants

One of the primary categories of industrial health hazards that the industrial hygienist must deal with is airborne contaminants. Air contaminants are commonly classified as either particulate or gas and vapor contaminants. The most common particulate contaminants include dusts, fumes, mists, aerosols, and fibers.

Dusts are solid particles that are formed or generated from solid organic or inorganic materials by reducing their size through mechanical processes such as crushing, grinding, drilling, abrading or blasting.

Industrial atmospheric contaminants exist in virtually every workplace. Sometimes they are readily apparent to workers, because of their odor, or because they can actually be seen. Industrial hygienists, however, can't rely on odor or vision to detect or measure airborne contaminants. They must rely on measurements taken by monitoring, sampling or detection devices.

Fumes are formed when material from a volatilized solid condenses in cool air. In most cases, the solid particles resulting from the condensation react with air to form an oxide.

The term *mist* is applied to a finely divided liquid suspended in the atmosphere. Mists are generated by liquids condensing from a vapor back to a liquid or by breaking up a liquid into a dispersed state such as by splashing, foaming or atomizing. *Aerosols* are also a form of a mist characterized by highly respirable, minute liquid particles.

Fibers are solid particles whose length is several times greater than their diameter.

Gases are formless fluids that expand to occupy the space or enclosure in which they are confined. Examples are welding gases such as acetylene, nitrogen, helium, and argon; and carbon monoxide generated from the operation of internal combustion engines or by its use as a reducing gas in a heat-treating operation. Another example is hydrogen sulfide which is formed wherever there is decomposition of materials containing sulfur under reducing conditions.

Liquids change into vapors and mix with the surrounding atmosphere through evaporation. *Vapors* are the volatile form of substances that are normally in a solid or liquid state at room temperature and pressure. Vapors are the gaseous form of substances which are normally in the solid or liquid state at room temperature and pressure. They are formed by evaporation from a liquid or solid and can be found where parts cleaning and painting take place and where solvents are used.

Although air contaminant values are useful as a guide for determining conditions that may be hazardous and may demand improved control measures, the industrial hygienist must recognize that the susceptibility of workers varies.

Even though it is essential not to permit exposures to exceed the stated values for substances, note that even careful adherence to the suggested values for any substance will not assure an absolutely harmless exposure. Thus, the air contaminant concentration values should only serve as a tool for indicating harmful exposures, rather than the absolute reference on which to base control measures.

Routes of Entry

For a chemical substance to cause or produce a harmful effect, it must reach the appropriate site in the body (usually via the bloodstream) at a concentration (and for a length of time) sufficient to produce an adverse effect. Toxic injury can occur at the first point of contact between the toxicant and the body, or in later, systemic injuries to various organs deep in the body. Common routes of entry are ingestion, injection, skin absorption and inhalation. However, entry into the body can occur by more than one route (e.g., inhalation of a substance that is absorbed through the skin).

Ingestion of toxic substances is not a common problem in industry—most workers do not deliberately swallow substances they handle in the workplace. However, ingestion does sometimes occur either directly or indirectly. Industrial exposure to harmful substance through ingestion may occur when workers eat lunch, drink coffee, chew tobacco, apply cosmetics, or smoke in a contaminated work area. The substances may exert their toxic effect on the intestinal tract or at specific organ sites.

Injection of toxic substances may occur just about anywhere in the body where a needle can be inserted but is a rare event in the industrial workplace.

Skin absorption or contact is an important route of entry in terms of occupational exposure. While the skin (the largest organ in the human body) may act as a barrier to some harmful agents, other materials may irritate or sensitize the skin and eyes, or travel through the skin into the bloodstream, thereby impacting on specific organs.

Inhalation is the most common route of entry for harmful substances in industrial exposures. Nearly all substances that are airborne can be inhaled. Dusts, fumes, mists, gases, vapors, and other airborne substances may enter the body via the lungs and may produce local effects on the lungs or may be transported by the blood to specific organs in the body.

Upon finding a route of entry into the body, chemicals and other substances may exert their harmful effects on specific organs of the body, such as the lungs, liver, kidneys, central nervous system, and skin. These specific organs are termed *target organs* and will vary with the chemical of concern (see Table 11.1).

The toxic action of a substance is divided into *short-term (acute)* and long-term (*chronic*) effects. Short-term adverse (acute) effects are usually related to an accident where exposure symptoms (effects) may occur within a short time period following either a single exposure or multiple exposures to a chemical. Long-term adverse (chronic) effects usually occur slowly after a long period of time, following exposures to small quantities of a substance (as lung disease may follow cigarette smoking). Chronic effects may sometimes occur following short-term exposures to certain substances.

INDUSTRIAL HEALTH HAZARDS

NIOSH and OSHA's Occupational Health Guidelines for Chemical Hazards, DHHS (NIOSH) Publication No. 81-123 (Washington, DC: Superintendent of Documents, U.S. Government Printing Office, current edition) summarizes information on permissible exposure limits, chemical and physical properties, and health hazards. It provides recommendations for medical surveillance, respiratory protection, and personal protection and sanitation practices for specific chemicals that have Federal occupational safety and health regulations. These recommendations reflect good industrial hygiene practices, and their implementation will assist development and maintenance of an effective occupational health program. The practicing industrial hygienist should maintain a current copy of this important and useful document on his or her bookshelf, within easy reach. It is also available in digital format.

Generally, determining if a substance is hazardous or not is simple, if the following is known: (1) what the agent is and what form it is in; (2) the concentration; and (3) the duration and form of exposure.

It should be noted, however, that because of the dynamic (ever changing) characteristics of the chemical and product industries, the practicing industrial hygienist comes face to face with trying to determine the uncertain toxicity of new chemical products that are frequently introduced into the workplace each year. Another related problem occurs when manufacturers develop chemical products with unfamiliar trade names and do not properly label them to indicate the chemical constituents of the compounds (of course, under OSHA's Hazard Communication Program, 29 CFR 1910.1200, this practice is illegal). Many different commercially available instruments permit the detection and concentration evaluation of many different contaminants. Some of these instruments

TABLE 11.1
Selected Toxic Contaminants and the Target Organs They Endanger

Target Organs	Toxic Contaminants
Blood	Benzene
	Carbon monoxide
	Arsenic
	Aniline
	Toluene
Kidneys	Mercury
	Chloroform
Heart	Aniline
Brain	Lead
	Mercury
	Benzene
	Manganese
	Acetaldehyde
Eyes	Cresol
	Acrolein
	Benzyl chloride
	Butyl alcohol
Skin	Nickel
	Phenol
	Trichloroethylene
Lungs	Asbestos
	Chromium
	Hydrogen sulfide
	Mica
	Nitrogen dioxide
Liver	Chloroform
	Carbon tetrachloride
	Toluene

Source: Spellman (1998).

are so simple that nearly any worker can learn to properly operate them. A note of caution, however; the untrained worker may receive an instrument reading that seems to indicate a higher degree of safety than may actually exist. Thus, the qualitative and quantitative measurement of atmospheric contaminants generally is the job of the industrial hygienist. Any samples taken should also be representative—samples should be taken of the actual air the workers breathe, at the point, they inhale them, in their breathing zone—between the top of the head and the shoulders.

Environmental Controls

Industrial hygienists recognize that engineering controls, work practice controls, administrative controls, and personal protective equipment (PPE) are the primary means of reducing employee exposure to occupational hazards. Workplace exposure to toxic materials and physical hazards can be reduced or controlled by a variety of these control methods, or by a combination of methods.

Engineering Controls

Engineering controls are methods of environmental control whereby the hazard is "engineered out," either by initial design specifications or by applying methods of substitution (e.g., replacing toxic chlorine used in disinfection processes with relatively non-toxic sodium hypochlorite). Engineering control may entail the utilization of isolation methods. For example, an operating diesel generator producing noise levels in excess of 120 decibels (120 dBA) could be controlled by enclosing it inside a soundproofed enclosure—effectively isolating the noise hazard. Another example of hazard isolation can be seen in the use of tightly closed enclosures that isolate an abrasive blasting operation. This method of isolation is typically used in conjunction with local exhaust ventilation. Ventilation is one of the most widely used and effective engineering controls (because it is so crucial in controlling workplace atmospheric hazards).

Work Practice Controls

Work practice controls alter the manner in which a task is performed. Some fundamental and easily implemented work practice controls include (1) following proper procedures that minimize exposures while operating production and control equipment; (2) inspecting and maintaining process and control equipment on a regular basis; (3) implementing good housekeeping procedures; (4) providing good supervision and (5) mandating that eating, drinking, smoking, chewing tobacco or gum, and applying cosmetics in regulated areas be prohibited.

Administrative Controls

After the design, construction and installation phase, installing engineering controls to control a workplace hazard or hazards often becomes difficult and expensive. A question industrial hygienists face on almost an on-going basis is "If I can't engineer out the hazard, what can I do?"

This question would not arise, of course, if the industrial hygienist had been invited to sit in and participate in the design, construction and installation phases. However, based on experience, this is the exception to the rule—this is not "good engineering practice"—but happens more often than not.

As a remedial action (a third line of defense)—after determining that engineering controls and work practice controls can't be accomplished for technological or budgetary reasons (or for any other reason)—administrative controls might be the alternative.

What are administrative controls? Simply, administrative controls include controlling workers' exposure by scheduling production and workers' tasks, or both, in ways that minimize exposure levels. The employer might schedule operations with the highest exposure potential during periods when the fewest workers are present. For example, a worker who is required to work in an extremely high noise area where engineering controls and work practice controls are not possible would be rotated from the high noise area to a quiet area when the daily permissible noise exposure limit is reached.

It should be noted that reducing exposures by limiting the duration of exposure (basically by modifying the work schedule) must be carefully managed (most managers soon find that attempting to properly manage this procedure takes a considerable amount of time, effort and "imagination"). When practiced, reducing worker exposure is based on limiting the amount of time a worker is exposed, ensuring that OSHA Permissible exposure Limits (PELs) are not exceeded.

Because administrative controls are not easy to implement and manage, many practicing industrial hygienists don't particularity like the practice; furthermore, they feel that such a strategy merely spreads the exposure out and does nothing to control the source. Experience has shown that in many instances this view is correct. Nevertheless, work schedule modification is commonly used for exposures to such stressors as noise and lead.

Though mentioned under work practice controls, *good housekeeping practices* are also an administrative control. Think about it. If dust and spilled chemicals are allowed to accumulate in the work area, workers will be exposed to these substances. This is of particular importance for flammable and toxic materials (i.e., in the prevention of fires, explosions or poisoning). Housekeeping practices that prevent toxic or hazardous materials from being dispersed into the air are also an important concern.

Administrative controls implemented at work can also reach beyond the workplace. For example, if workers abate asbestos all day, they should only wear approved protective suits and other required personal protective equipment (PPE; discussed in detail later). After the work assignment is completed each day, the workers must be decontaminated, following the standard protocol. Moreover, these workers should be prohibited from wearing personal clothing while removing asbestos. They should be required to decontaminate and remove their contaminated protective clothing and change it into uncontaminated personal clothing before leaving the job site. The proper procedure is to leave any contaminated clothing at work and not to take it into the household where family members could be exposed. The bottom line: leave asbestos (and any other contaminant) at the worksite.

Implementation of standardized *materials handling or transferring procedures* is another administrative control often used to protect workers. In handling chemicals, any transfer operation taken should be closed-system (i.e., the transfer of chemicals directly from a storage container to the application point) or should have adequate exhaust systems to prevent worker exposure to contamination of the workplace air. This practice should also include the use of spill trays to collect overfill spills or leaking materials between transfer points.

Administrative controls that involve visual inspection and automatic sensor services (leak detection programs) allow not only for quick detection but also for quick repair and minimal exposure. When automatic system sensors and alarms are deployed as administrative controls, tying the alarm system into an automatic shutdown system (close a valve, open an electrical circuit, et.) allows the sensor to detect a leak, sound the alarm, and initiate corrective action (e.g., immediate shutdown of the system).

Two other administrative control practices include *training* and *personal hygiene*. For workers to best protect themselves from workplace hazards (to reduce the risk of injury or illness), they must be made aware of the hazards; they must be trained. OSHA puts great emphasis on the worker training requirement. This emphasis is well-placed. No worker can be expected to know about every workplace process or equipment hazard unless he or she has been properly trained on the hazards and/or potential hazards. Thus, an important part of the training process is worker awareness. Legally (and morally) workers have the right to know what they are working with, what they are exposed to while on the job; they must be made aware of the hazards. They must also be trained on what actions to take when they are exposed to specific hazards.

Personal hygiene practices are an important part of worker protection. The industrial hygienist must ensure that appropriate cleaning agents and facilities such as soap, sinks, showers, and toilets are available to workers. In addition, such equipment as emergency eyewashes and deluge showers, and changing rooms must be made available and conveniently located for worker use.

Personal Protective Equipment (PPE)

Personal protective equipment (PPE) is the workers' last line of defense against injury on the job. Industrial hygienists prefer to incorporate engineering, work practice, and administrative controls whenever possible. However, when the work environment can't be made safe by any other method, PPE is used as the last resort. PPE imposes a barrier between the work and the hazard but does nothing to reduce or eliminate the hazard. Typical PPE includes safety goggles, helmets, face shields, gloves, safety shoes, hearing protection devices, full-body protective clothing, barrier creams, and respirators (Olishifski, 1988). To be effective, PPE must be individually selected, properly fitted and periodically refitted; it must be conscientiously and properly worn; regularly maintained; and replaced as necessary.

HAZARD COMMUNICATION

The Bhopal Incident in India, the ensuing chemical spill, and the resulting tragic deaths and injuries are well known. However, not all of the repercussions—the lessons learned—from this incident are as well known. After Bhopal arose a worldwide outcry. "How could such an incident occur? Why wasn't something done to protect the inhabitants? Weren't there pre-event safety measures taken or in place to prevent such a disaster from occurring?"

In the U.S., these questions and others were bandied around and about by the Press and Congress, and Congress took the first major step to prevent such incidents from occurring in the U.S. What Congress did was to direct OSHA to take a close look at chemical manufacturing in the U.S. to see if a Bhopal-type Incident could occur in this country. OSHA did a study and then reported to Congress that a Bhopal-type incident in the U.S. was very unlikely. Within a few months of OSHA's report to Congress, however, a chemical spill occurred, similar to Bhopal, but fortunately, not deadly (no deaths, 100+ people became ill), in Institute, West Virginia.

Needless to say, Congress was upset. Because of Bhopal and the Institute, West Virginia fiascoes, in 1984 OSHA implemented its Hazard Communication Standard (HCS) which mandates the Hazard Communication Program (HazCom), 29 CFR 1910.1200. Later, other programs like SARA (Superfund) Title III reporting requirements for all chemical users, producers, suppliers, and storage entities were mandated by USEPA.

There is no all-inclusive list of chemicals covered by the HAZCOM Standard; however, the regulation refers to "any chemical which is a physical or health hazard." Those specifically deemed hazardous include:

- Chemicals regulated by OSHA in 29 CFR Part 1910, Subpart Z, Toxic and Hazardous Substances
- Chemicals included in the American Conference of Governmental Industrial Hygienists' (ACGIH) latest edition of Threshold Limit Values (TLVs) for *Chemical Substances and Physical Agents in the Work Environment*
- Chemicals found to be suspected or confirmed carcinogens by the National Toxicology Program in the *Registry of Toxic Effects of Chemicals Substances* published by NIOSH or appearing in the latest edition of the *Annual Report on Carcinogens,* or by the International Agency for Research on Cancer in the latest editions of its IARC *Monographs*

Congress decided that those personnel involved with working with or around hazardous materials "had a right to know" about the hazards near them, or the ones they worked with. Thus, OSHA's Hazard Communication Standard was created. The Hazard Communication Standard is, without a doubt, the regulation most important to the communication of hazards to employees.

Under its Hazard Communication Standard (more commonly known as "HAZCOM" or the "Right to Know Law"), OSHA requires employers who use or produce chemicals on the worksite to inform all employees of the hazards that might be involved with those chemicals. HAZCOM says that employees have the right to know what chemicals they are handling or could be exposed to. HAZCOM's intent is to make the workplace safer. Under the HAZCOM Standard, the employer is required to fully evaluate all chemicals on the worksite for possible physical and health hazards. All information relating to these hazards must be made available to the employee 24 hours each day. The standard is written in a performance manner, meaning that the specifics are left to the employer to develop.

HAZCOM AND THE ENVIRONMENTAL PROFESSIONAL

The environmental professional responsible for the implementation of or oversight of the Hazard Communication Standard in the workplace must take a personal interest in ensuring that his/her facility is in full compliance with the HCS for three major reasons:

1. it is the law;
2. it is consistently the number one cause of citations issued by OSHA for noncompliance
3. compliance with the standard goes a long way toward protecting workers.

Major elements of the Hazard Communication Standard, which the environmental professional must ensure are part of the organization's HAZCOM Program, include hazard determination, written hazard communication program, labels and other forms of warning, safety data sheets, and employee training.

MODIFICATION OF THE HAZARD COMMUNICATION STANDARD (HCS)[1]

To conform to the United Nations' (UN) Globally Harmonized System (GHS) of Classification and Labeling of Chemicals OSHA has adopted the international approach to hazard communication. The original HAZCOM Standard required the *employer* to ensure proper labeling of each chemical, including chemicals that might be produced by a *process* (process hazards). For example, in the wastewater industry, deadly methane gas is generated in the waste stream. Another common wastewater hazard (off gas) is the generation of hydrogen sulfide (which produces the characteristic rotten-egg odor) during degradation of organic substances in the wastestream and can kill quickly. OSHA's HAZCOM requires the employer to label methane and hydrogen sulfide hazards so that workers are warned, and safety precautions are followed.

In the original HAZCOM Standard, labels are required to be designed to be clearly understood by all workers. Employers are required to provide both training and written materials to make workers aware of what they are working with and what hazards they might be exposed to. Employers are also required to make Safety Data Sheets (SDS) available to all employees. An SDS is a fact sheet for a chemical posing a physical or health hazard at work. SDS must be in English and contain the following information:

- Identity of the chemical (label name)
- Physical hazards
- Control measures
- Health hazards
- Is it a carcinogen?
- Emergency and first aid procedures
- Date of preparation of the latest revision
- Name, address, and telephone number of manufacturer, importer, or other responsible party

Blank spaces are not permitted on an SDS. If relevant information in any one of the categories is unavailable at the time of preparation, the SDS must indicate no information was available. Every facility must have an SDS for each hazardous chemical it uses. Copies must be made available to other companies working on the worksite (outside contractors, for example), and they must do the same for the host site. The facility Hazard Communication Program was and is required to be in writing and, along with SDS made available to all workers 24 hours each day/each shift.

In the Globally Harmonized System, all the above requirements still stand; they are still required. The difference is (i.e., the new requirements) that under GHS the employer must now abide by agreed criteria for the classification of chemical hazards and incorporate into their labeling procedure a standardized approach to label elements and safety data sheets. The GHS was negotiated in a multi-year process by hazard communication experts from many different countries, internal organizations, and stakeholder groups. It is based on major existing systems around the world, including OHSA's Hazard Communication Standard and the chemical classification and labeling systems of other US agencies.

The result of this negotiation process is the United Nations' document entitled "Globally Harmonized System of Classification and Labeling of Chemical," commonly referred to as The Purple book. This document provides harmonized classification criteria for health, physical, and environmental hazards of chemicals. It also includes standardized label elements that are assigned to these hazard classes and categories, and provide the appropriate signal words, pictograms, and hazard and precautionary statements to convey the hazards to users. A standardized order of information for safety data sheets is also provided. These recommendations can be used by regulatory authorities such as OSHA to establish mandatory requirements for hazard communication, but do not constitute a model regulation

OSHA has modified the Hazard Communication Standard (HCS) to adopt the GHS to improve the safety and health of workers through more effective communications on chemical hazards. Since it was first promulgated in 1983, the HCS has provided employers and employees with extensive information about the chemicals in their workplaces. As mentioned, the original standard is performance-oriented, allowing chemical manufacturers and importers to convey information of labels and material safety data sheets in whatever format they choose. While the available information has been helpful in improving employee safety and health, a more standardized approach to classifying the hazards and conveying the information will be more effective and provide further improvements in American workplaces. The GHS provides such a standardized approach, including detailed criteria for determining what hazardous effects a chemical poses, as well as standardized label elements assigned by hazard class and category. This will enhance both employer and worker comprehension of the hazards, which will help to ensure appropriate handling and safe use of workplace chemicals. In addition, the safety data sheet requirements establish an order of information that is standardized. The harmonized format of the safety data sheets will enable employers, workers, health professionals, and emergency responders to access the information more efficiently and effectively, thus increasing their utility.

The specific changes to the Hazard Communication Standard are in the hazard classification, labels, and safety data sheets.

- **Hazard classification**: The definitions of hazard have been changed to provide specific criteria for the classification of health and physical hazards, as well as the classification of mixtures. These specific criteria will help to ensure that evaluations of hazardous effects are consistent across manufacturers and that labels and safety data sheets are more accurate as a result.
- **Labels**: Chemical manufacturers and importers will be required to provide a label that includes a harmonized signal word, pictogram, and hazard statement for each hazard class and category. Precautionary statements must also be provided.
- **Safety data sheets**: Will now have a specified 16-section format.

While the revised Hazard Communication Standard (HCS) is a modification of the existing standard, the parts of the standard that did not relate to the GHS (such as the basic framework, scope, and exemptions) remained largely unchanged. There have been some modifications to terminology in order to align the revised HCS with the language used in the GHS. For example, the term "hazard determination" has been changed to "hazard classification" and "material safety data sheet" was changed to "safety data sheet."

Under both the current Hazard Communication Standard (HCS) and the revised HCS, an evaluation of chemical hazards must be performed considering the available scientific evidence concerning such hazards. Under the current HCS, the hazard determination provisions have definitions of hazard and the evaluator determines whether or not the data on a chemical meets those definitions. It is a performance-oriented approach that provides parameters for the evaluation, but not specific, detailed criteria. The hazard classification approach in the revised HCS is quite different. The revised HCS has specific criteria for each health and physical hazard, along with detailed instructions for hazard evaluation and determinations as to whether mixtures or substances are covered. It also establishes both hazard classes and hazard categories—for most of the effects; the classes are divided into categories that reflect the relative severity of the effect. The current HCS does not include categories for most of the health hazards covered, so this new approach provides additional information that can be related to the appropriate response to address the hazard. OSHA has included the general provisions for hazard classification in paragraph (d) of the revised rule and added extensive appendixes that address the criteria for each health or physical effect.

Adoption of the GHS in the US and around the world will also help to improve information received from other countries—since the US is both a major importer and exporter of chemicals, American workers often see labels and safety data sheets from other countries. The diverse and sometimes conflicting national and international requirements can create confusion among those who seek to use hazard information effectively. For example, labels and safety data sheets may include symbols and hazard statements that are unfamiliar t readers or not well understood. Containers may be labeled with such a large volume of information that important statements are not easily recognized. Given the differences in hazard classification criteria, labels may also be incorrect when used in other countries. If countries around the world adopt the GHS, these problems will be minimized, and chemicals crossing borders with have consistent information, thus improving communication globally.

Table 11.2 summarizes the phase-in dates required under the revised Hazard communication Standard (HCS):

It is important to point out that even though full compliance with HCS modifications is to be phased-in and GHS does not include harmonized training provisions, OSHA is requiring that employees be trained on the new label elements (i.e., pictograms, hazard statement, precautionary statements, and signal words) and SDS format by December 1, 2013, while full compliance

TABLE 11.2
Phase-in Dates Required under the Revised Hazard Communication Standard (HCS)

Effective Completion Date	Requirement(s)	Who
December 1, 2013	Train employees on the new label on the new label elements and safety data sheet (SDS) format.	Employers
June 1, 2015[a]	Compliance with all modified provisions of this final rule, except:	Chemical manufacturers, importers, distributors and employers
December 1, 2015	The Distributor shall not ship containers labeled by the chemical manufacturer or importer unless it is a GHS label	
June 1, 2016	Update alternative workplace labeling and hazard communication program as necessary and provide or health hazards	Employers
Transition Period to the effective completion dates noted above	May comply with either 29 CFR 1910.1200 (the final standard), or the current standard, or both	Chemical manufacturers, importers, distributors, and employers

Note: During the phase-in period, employers would be required to be following either the existing HCS or the revised HCS, or both. OSHA recognizes that hazard communication programs will go through a period of time where labels and SDSs under both standards will be present in the workplace. This will be considered acceptable, and employers are not required to maintain two sets of labels and SDSs for compliance purposes.

[a] This date coincides with the EU implementation date for the classification of mixtures.

with the final rule will begin in 2015. OSHA believes that American workplaces will soon begin to receive labels and SDSs that are consistent with the GHS since many American and foreign chemical manufactures have already begun to produce HAZCOM 2012/GHS-compliant labels and SDSs. It is important to ensure that when employees begin to see the new labels and SDSs in the workplace, they will be familiar with them, understand how to use them, and access the information effectively.

Under the current Hazard Communication Standard (HCS), the label preparer must provide the identity of the chemical, and the appropriate hazard warnings. This may be done in a variety of ways, and the method to convey the information is left to the preparer. Under the revised HCS, once the hazard classification is completed, the standard specifies what information is to be provided for each hazard class and category. Labels will require the following elements:

- **Pictogram**: a symbol plus other graphic elements, such as a border, background pattern, or color that is intended to convey specific information about the hazards of a chemical. Each pictogram consists of a different symbol on a white background within a red square frame set on a point (i.e., a red diamond). There are nine pictograms under the GHS. However, only eight pictograms re required under the HCS.
- **Signal words**: a single word used to indicate the relative level of severity of hazard and alert the reader to a potential hazard on the label. The signal words used are "danger" and "warning." "Danger is used for the more sever hazards, while "Warming" is used for less severe hazards.
- **Hazard statement**: a statement assigned to a hazard class and category that describes the nature of the hazard(s) of a chemical, including, where appropriate, the degree of hazard.
- **Precautionary statement**: a phrase that describes recommended measures to be taken to minimize or prevent adverse effects resulting from exposure to a hazardous chemical, or improper storage or handling of a hazardous chemical.

The required pictograms required under GHS are shown in Figure 11.1.

Under the revised Hazard Communication Standard, the old Material Safety Data Sheet (MSDS) has not only been replaced by the Safety Data Sheet (SDS) but the SDS has also been slightly modified. Essentially the information required on the safety data sheet (SDS) will remain the same as that in the current standard (HAZCOM, 1994). HAZCOM 1994 indicates what information has to be included on an SDS but does not specify a format for presentation or order of information. The revised Hazard Communication Standard (HAZCOM, 2012) requires that the information of the SDS be presented using specific headings in a specified sequence. Paragraph (g) of the final rule provides the headings of information to be included on the SDS and the order in which they are to be provided. In addition, Appendix D provides the information to be included under each heading.

Section 1. Identification
Section 2. Hazard(s) identification
Section 3. Composition/information on ingredients
Section 4. First-Aid measures
Section 5. Fire-fighting measures
Section 6. Accidental release measures
Section 7. Handling and storage
Section 8. Exposure controls/personal protection
Section 9. Physical and chemical properties
Section 10. Stability and reactivity
Section 11. Toxicological information
Section 12. Ecological information

Section 13. Disposal considerations
Section 14. Transport information
Section 15. Regulatory information
Section 16. Other information, including the date of preparation or last revision

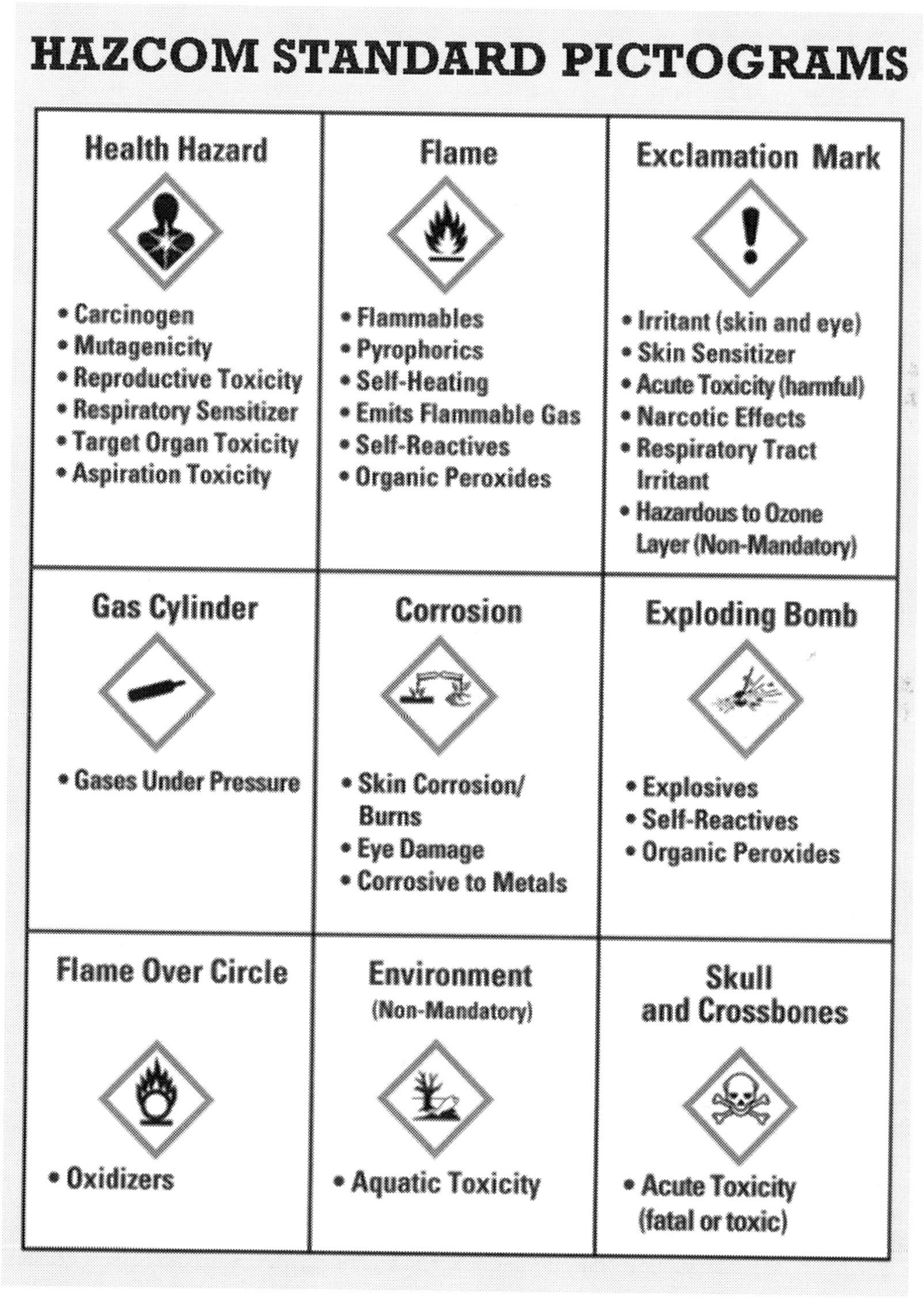

HAZCOM STANDARD PICTOGRAMS

Health Hazard	Flame	Exclamation Mark
• Carcinogen • Mutagenicity • Reproductive Toxicity • Respiratory Sensitizer • Target Organ Toxicity • Aspiration Toxicity	• Flammables • Pyrophorics • Self-Heating • Emits Flammable Gas • Self-Reactives • Organic Peroxides	• Irritant (skin and eye) • Skin Sensitizer • Acute Toxicity (harmful) • Narcotic Effects • Respiratory Tract Irritant • Hazardous to Ozone Layer (Non-Mandatory)
Gas Cylinder	**Corrosion**	**Exploding Bomb**
• Gases Under Pressure	• Skin Corrosion/Burns • Eye Damage • Corrosive to Metals	• Explosives • Self-Reactives • Organic Peroxides
Flame Over Circle	**Environment** (Non-Mandatory)	**Skull and Crossbones**
• Oxidizers	• Aquatic Toxicity	• Acute Toxicity (fatal or toxic)

FIGURE 11.1 HCS pictograms and hazards. OSHA Hazard Communication (2012). http://www.osha.gov/dsg/hazcom/hazcome-gaq.html.

OCCUPATIONAL ENVIRONMENTAL LIMITS (OELS)

Many processes and procedures generate hazardous air contaminants that can get into the air people breathe. Normally, the body can take in limited amounts of hazardous air contaminants, metabolize them, and eliminate them from the body without producing harmful effects. Safe levels of exposure to many hazardous materials have been established by governmental agencies after much research on their short term (acute) and cumulative (chronic) health effects using available human exposure data (usually from industrial sources) and animal testing. When the average air concentrations repeatedly exceed certain thresholds, called *exposure limits*, adverse health effects are more likely to occur. Exposure limits do change with time as more research is conducted and more occupational data is collected.

OELs

A fairly standard terminology has come to be used with regard to Occupational Environmental Limits (OELs).

Let's pause right here and talk about Permissible Exposure Limits (PELs) and threshold Limit Values (TLVs).

A workplace exposure level, such as PEL or TLV, is expressed as the concentration of the air contaminant in a volume of air. It is important to know what they are and what significance they play in the industrial hygienist's daily activities. Let's begin with TLVs.

Threshold Limit Values (TLVs) are published by the American Conference of Governmental Industrial Hygienists (ACGIH) (an organization made up of physicians, toxicologists, chemists, epidemiologists, and industrial hygienists) in its *Threshold Limit Values for Chemical Substances and Physical Agents in the Work Environment*. These values are used in assessing the risk of a worker exposed to a hazardous chemical vapor; concentrations in the workplace can often be maintained below these levels with proper controls. The substances listed by ACGIH are evaluated annually, limits are revised as needed, and new substances are added to the list, as information becomes available. The values are established from the experience of many groups in industry, academia, and medicine, and from laboratory research.

The chemical substance exposure limits listed under both ACGIH and OSHA are based strictly on airborne concentrations of chemical substances in terms of milligrams per cubic meter (mg/m^3), parts per million (ppm)(the number of "parts" of air contaminant per million parts of air), and fibers per cubic centimeters ($fibers/cm^3$). The smaller the concentration number, the more toxic the substance is by inhalation. The ACGIH has established some "rules of thumb" in regard to exposure limits. Substances with exposure limits below 100 ppm are considered highly toxic by inhalation. Those substances with exposure limits of 100–500 ppm are considered moderately toxic by inhalation. Substances with exposure limits greater than 500 ppm are considered slightly toxic by inhalation. Allowable limits are based on three different time periods of average exposure: (1) 8-hour work shifts known as TWA (time-weighted average), (2) short terms of 15 minutes or STEL (short-term exposure limit), and (3) instantaneous exposure of "C" (ceiling). Unlike OSHA's PELs, TLVs are recommended levels only and do not have the force of regulation to back them up.

OSHA has promulgated limits for personnel exposure in workplace air for approximately 400 chemicals listed in Tables Z1, Z2, and Z3 in Part 1910.1000 of the Federal Occupational Safety and Health Standard. These limits are defined as *permissible exposure limits* (PEL) and like TLVs are based on 8-hour time-weighted averages or ceiling limits when preceded by a "C" (Exposure limits expressed in terms other than ppm must be converted to ppm before comparing to the guidelines). Keeping within the limits in the Subpart Z Tables is the only requirement specified by OSHA for these chemicals. The significance of OSHA's PELs is that they have the force of regulatory law behind them to back them up—compliance with OSHA's PELs is the law.

Evaluation of personnel exposure to physical and chemical stresses in the industrial workplace requires the use of the guidelines provided by TLVs and the regulatory guidelines of

Industrial Hygiene

PELs. For the Industrial hygienist to carry out the goals of recognizing, measuring, and effecting controls (of any type) for workplace stresses, such limits are a necessity and have become the ultimate guidelines in the science of industrial hygiene. A word of caution is advised, however. These values are set only as guides for the best practice and are not to be considered absolute values. What are we saying here? These values provide reasonable assurance that occupational disease will not occur if exposures are kept below these levels. On the other hand, occupational disease is likely to develop in some people—if the recommended levels are exceeded on a consistent basis.

Time-Weighted Average (TWA) is the fundamental concept of most OELs. It is usually presented as the average concentration over an 8-hour workday for a 40-hour workweek.

Eight-hour *Threshold Limit Values-Time Weighted Averages* (TLV-TWA) exist for some four hundred plus chemical agents commonly found in the workplace. NIOSH list sampling and analytical methods for most of these agents.

Short-Term Exposure Limits (STELs) are recommended when exposures of even short duration to high concentrations of a chemical are known to produce acute toxicity.

The STEL is the concentration to which workers can be exposed continuously for a short period of time without suffering form (1) irritation, (2) chronic or irreversible tissue damage, or (3) narcosis of sufficient degree to increase the likelihood of accidental injury, impaired self-rescue, or reduced work efficiency.

The STEL is defined as a 15-minute TWA exposure that should not be exceeded at any time during a workday, even if the overall 8-hour TWA is within limits, and it should not occur more than four times per day. Note: There should be at least 60 minutes between successive exposures in this range. If warranted, an averaging period other than 15 minutes can also be used. STELS are not available for all substances.

Ceiling (C) is the concentration that should not be exceeded during any part of the working exposure, assessed over a 15-minute period.

The *Action Level* is the concentration or level of an agent at which it is deemed that some specific action should be taken. Action levels are found only in certain substance-specific standards by OSHA. In practice, the action level is usually set at one-half of the TLV.

Skin notation denotes the possibility that dermal absorption may be a significant contribution to the overall body burden of the chemical. A "SKIN" notation that follows the exposure limit indicates that a significant exposure can be received if the skin is in contact with the chemical in the gas, vapor, or solid form.

Airborne particulate matter is divided into three classes based on the likely deposition with the respiratory tract. While the past practice was to provide TLVs in terms of total particulate mass, the recent approach is to consider the aerodynamic diameter of the particle and its site of action. The three classes of airborne particulate matter are described below.

1. *Inhalable Particulate Mass (IPM) TLVs* are designated for compounds that are toxic if deposited at any site within the respiratory tract. The typical size for these particles can range from submicron size to approximately 100 microns.
2. *Thoracic Particulate Mass (TPM) TLVs* are designated for compounds that are toxic if deposited either within the airways of the lung or the gas-exchange region. The typical size for these particles can range from approximately 5 to 15 microns.
3. *Respirable Particulate Mass (RPM) TLVs* are designated for those compounds that are toxic if deposited within the gas-exchange region of the lung. The typical size for these particles is approximately 5 microns or less.

Nuisance dust is no longer used since all dusts have biological effects at some dose. The term *particulates* not otherwise classified is now being used in place of nuisance dusts.

Biological Exposure Limits (BEl) covers nearly 40 chemicals. A BEl has been defined as a level of a determinant that is likely to be observed in a specimen (such as blood, urine, or air) collected from a worker who was exposed to a chemical and who has similar levels of the determinant as if he or she had been exposed to the chemical at the TLV.

AIR MONITORING/SAMPLING

Air monitoring is widely used to measure human exposure and to characterize emission sources. It is often employed within the context of the general survey, investigating a specific complaint, or simply for regulatory compliance. It is also used for more fundamental purposes, such as in confined space entry operations. While it is true that just about any confined space entry team member can be trained to properly calibrate and operate air monitors for safe confined space entry, it is also true that a higher level of knowledge and training is often required in the actual evaluation of confined spaces for possible oxygen deficiency and/or air contaminant problems. After an overview of the basics of air monitoring/sampling, we discuss air monitoring requirements for permit-required confined space entry.

AIR SAMPLE VOLUME

OSHA's *Analytical Method* and NIOSH's *Manual of Analytical Methods* list a range of air sample volumes, minimum volume (VOL-MIN) to maximum volume (VOL-MAX) that should be collected for an exposure assessment. The volume is based on the sampler's sorptive capacity and assumes that the measured exposure is the OSHA PEL.

The range of volumes listed may not collect a sufficient mass for accurate laboratory analysis if the actual contaminant concentration is less than the PEL or the Time Weighted average-Threshold Limit Value (TWA-TLV).

If a collection method recommends sampling 10 liters of air, the industrial hygienist may not be sure if an 8.5-liter air sample will collect sufficient mass for the lab to quantify. This makes an important point. That is, finding that an insufficient air volume (i.e., too little mass) was sampled after the lab results are returned can turn out to be an enormous waste of resources. To avoid this situation, it is essential to understand the restrictions faced by analytical labs: Limit of Detection (LOD) and Limit of Quantification (LOQ).

To compute a minimum air sample volume to provide useful information for the evaluation of airborne contaminant concentration in the workplace, the industrial hygienist must understand how to correctly manipulate the LOD and the LOQ. Knowledge of these limits will provide increased flexibility in sampling.

LIMIT OF DETECTION (LOD)

The *limit of detection* (LOD) has many definitions in the literature. For example, the American Chemical Society (ACS) Committee on Environmental Analytical Chemistry (2002) defines the LOD as the lowest concentration level than can be determined to be statistically different from a blank sample (a blank sample is a sample of a carrying agent—a gas, liquid, or solids—that is normally used to selectively capture a material of interest, and that is subjected to the usual analytical or measurement process to establish a zero baseline or background value, which is used to adjust or correct routine analytical results). For our purpose, the ACS definition of LOD is used in this handbook.

The analytical instrument output signal produced by the sample must be three to five times the instrument's background noise level to be at the limit of detection; that is, the signal-to-noise ratio (S/N) is three-to-one (3/1). The S/N ratio is the analytical method's lower limit of detection.

Industrial Hygiene

LIMIT OF QUANTIFICATION (LOQ)

The *limit of quantification* (LOQ) is the concentration level above which quantitative results may be obtained with a certain degree of confidence. That is, LOQ is the minimum mass of the analyte above which the precision of the reported result is better than a specified level. The recommended value of the LOQ is the amount of analyte that will give rise to a signal that is 10 times the standard deviation of the signal from a series of media blanks.

Precision, Accuracy, and Bias

Sample results are only as good as the sampling technique and equipment used. Thus, in any type of air monitoring operation (air, water, or soil), it is important that the industrial hygienist factor in the precision, accuracy and any possible bias involved in the monitoring process.

Precision is the reproducibility of replicate analyses of the same sample (mass or concentration). For example, how close to each other is a target shooter able to place a set of shots anywhere on the target?

Accuracy is the degree of agreement between measured values and the accepted reference value. The investigator must carefully design his sampling program and use certain statistical tools to evaluate his data before making any inferences from the data. In our target shooting analogy, accuracy can be equated to how close does a target shooter come to the bull's eye?

Bias is the error introduced into sampling that causes estimates of parameters to be inaccurate. More specifically, bias is the difference between the average measured mass or concentration and a reference mass or concentration, expressed as a fraction of the reference mass or concentration. For example, how far from the bull's eye is the target shooter able to place a cluster of shots?

CALIBRATION REQUIREMENTS

The American National Standards Institute (2014) defines calibration as the set of operations which establishes, under specified conditions (i.e., instrument manufacturer's guidelines or regulator's protocols), the relationship between values indicated by a measuring instrument or measuring system, and the corresponding standard or known values derived from the standard. Note: Before any air-monitoring device can be relied on as accurate, it must be calibrated. Calibration procedures can be found in OSHA's Personal Sampling for Air Contaminants [www.osha.gov].

TYPES OF AIR SAMPLING

Although the information provided in the following discussion is area-specific (*area sampling*), it is important to point out that one of the most important air sampling operations is personal sampling. *Personal sampling* puts the sample detection device on the worker. This is done to obtain samples that represent the worker's exposure while working. As the preferred method of evaluating worker exposure to airborne contaminants, personal sampling, as mentioned, requires the worker to wear the detection device on their person in the breathing zone area. A small air pump and associated tubing connected to the detector is also worn by the worker. Personal sampling allows the industrial hygienist to define a potential hazard, check compliance with specific regulations, and determine the worker's daily time-weighted-average (TWA) exposure.

ANALYLTICAL METHODS FOR GASES AND VAPORS

In routine practice, the industrial hygienist will collect air samples to determine the concentration of a known contaminant or group of contaminants and will request or conduct analysis for these compounds. The first step in this procedure is to develop a sampling plan. When developing a sampling plan or strategy, the sampler should review the specific sampling and analytical methods available for the contaminants of interest. Several organizations have compiled and published collections of sampling and analytical methods for gases and vapors (see Table 11.3).

TABLE 11.3
Available Publications on Sampling/Analytical Methods for Gases/Vapors

NIOSH Manual of Analytical Methods, 4th ed. Centers for Disease Control and Prevention National Institute for
 Occupational Safety and Health
4676 Columbia Parkway
Cincinnati, OH 45226

Annual Book of ASTM Standards
American Society for Testing and Materials
100 Barr Harbor Dr.
West Conshohocken, PA 19428

OSHA Analytical Methods Manual
Occupational Safety and Health Administration
OSHA Salt Lake Technical Center
P.O Box 65200
1781 South 300 West
Salt Lake City, UT 84165

Methods of Air Sampling and Analysis
Lewis Publishers/CRC Press Inc.
2000 Corporate Blvd. NW
Boca Raton, FL 33431

For discussion purposes, analytical methods for gases and vapors (vapors are the gaseous phase of a substance that is liquid or solid at normal temperature and pressure; vapors diffuse) are grouped into chromatographic, volumetric, and optical methods. In the following we briefly discuss chromatography.

The primary type of analytical equipment used in *chromatography* for the analysis of gases and vapors in air samples is the *gas chromatograph* (GC). The GC is a powerful tool for the analysis of low-concentration air contaminants. It is generally a reliable analytical instrument. GC analysis is applicable to compounds with sufficient vapor pressure and thermal stability to dissolve in the carrier gas and pass through the chromatographic column in sufficient quantity to be detectable. Air samples to be analyzed by GC are typically collected on sorbent tubes and desorbed into a liquid for analysis. It should be noted that the GC instrument cannot be used for the reliable identification of specific substances. Because of this limitation, the GS is often married to the mass spectrometer (MS) instrument to provide specific results.

As mentioned, the GC is limited in its analysis for contaminant specificity. To correct this limitation, the GC/MS combination is used. When the industrial hygienist or engineering analyst uses the GC instrument to separate compounds before analysis with an MS instrument, a complementary relationship exists. The technician has access to both the retention times and mass spectral data. Many environmental professionals consider GC/MS analysis as a tool for conclusive proof of identity— the 'gold standard" in scientific analysis.

Some common applications of GC/MS include:

1. Evaluation of complex mixtures
2. Identification of pyrolysis and combustion products from fries
3. Analysis of insecticides and herbicides—conventional analytical methods frequently cannot resolve or identify the wide variety of industrial pesticides currently in use, but GC/MS can both identify and quantify these compounds.

Industrial Hygiene

AIR MONITORING VERSUS AIR SAMPLING

In the practice of industrial hygiene, the terms air monitoring and air sampling are often used interchangeably to mean the same thing. But are they the same? This depends on your choice and use of the vernacular.

In reality, they are different; that is, air monitoring and air sampling are separate functions. The difference is related to time: real time versus time-integration.

Air monitoring is real-time monitoring and generally includes monitoring with hand-held, direct-reading units such as portable gas chromatographs (GC), photoionization detectors (PIDs), flame ionization detectors (FIDs), dust monitors, and colorimetric tubes. Real-time air monitoring instrumentation is generally easily portable and allows the user to collect multiple samples in a relatively short sample period—ranging from a few seconds to a few minutes. Most portable real-time instruments measure low parts per million (ppm) of total volatile organics.

Real-time monitoring methods have higher detection limits than time-integrated sampling methods, react with entire classes of compounds and, unless real-time monitoring is conducted continuously, provide only a "snapshot" of the monitored ambient air concentration. Air monitoring instruments and methods provide results that are generally used for the evaluation of short-term exposure limits and can be useful in providing timely information to those engaged in various activities such as confined space entry operations. That is, in confined space operations, proper air monitoring can detect the presence or absence of life-threatening contaminants and/or insufficient oxygen levels within the confined space, alerting the entrants not to enter before making the space safe (e.g., by using forced air ventilation, etc.) for entry.

On the other hand, time-integrated *air sampling* is intended to document actual exposure for comparison to long-term exposure limits. Air sampling data is collected at "fixed" locations along the perimeter of the sample area (work area) and at locations adjacent to other sensitive receptors. Because most contaminants are (or will be) present in ambient air at relatively low levels, some type of sample concentrating is necessary to meet detection limits normally required in evaluating long-term health risks. Air sampling is accomplished using air-monitoring instrumentation designed to continuously sample large volumes of air over extended periods of time (typically from 8 to 24 hours).

Air sampling methods involve collecting air samples on sampling media designed specifically for the collection of the compounds of interest or as whole air samples. Upon completion of the sampling period the sampling media is collected, packaged, and transported for subsequent analysis. Analysis of air samples usually requires a minimum of 48 hours to complete.

Now you should have a basic understanding of air monitoring and air sampling and the difference, though in some cases subtle, between the two. Both procedures are important and significant tools in the industrial hygienist's toolbox.

To effectively evaluate a potentially hazardous worksite, an industrial hygienist must obtain objective and quantitative data. To do this, the IH must perform some form of air sampling, dependent upon, of course, the airborne contaminant in question. Moreover, sampling operations involve the use of instruments to measure the concentration of the particulate, gas, or vapor of interest. Many instruments perform both sampling and analysis. The instrument of choice in conducting sampling and analysis typically is a direct-reading-type instrument. The IH must be familiar with the uses, advantages, and limitations of such instruments. In addition, the IH must use math calculations to calculate sample volumes, sample times, TLVs, air concentrations from vapor pressures, and determine the additive effects of chemicals when multiple agents are used in the workplace. These calculations must consider changing conditions, such as temperature and pressure change in the workplace. Finally, the IH must understand how particulates, gases, and vapors are generated, how they enter the human body, how they impact worker's health, and how to evaluate particulate-, gas-, and vapor-laden workplaces.

Because air sampling is integral to just about everything the industrial hygienist does and is about, in the following we include important sections focusing on air sampling principles, dealing with airborne particulates, airborne gases and vapors, direct-reading instruments and basic air sampling calculations.

Air Sampling for Airborne Particulates

As mentioned, airborne particulates (or particulate matter, PM) includes solid and liquid matter such as:

- dusts
- fumes
- mists
- smokes
- bioaerosols

Inhalation of particulates is a major cause of occupational illness and disease. Pneumoconiosis (Gr. "dusty lung") is a lung disease caused by inhalation. Four critical factors influence the health impact of airborne particulates. Each of these four factors is interrelated in such a way that on one factor can be considered independently of the others:

- The size of the particles
- The duration of exposure time
- The nature of the dust in question
- The airborne concentration of the dust, in the breathing zone of the exposure person.

Note: The *breathing zone* of the worker is described by a hemisphere bordering the shoulders to the top of the head.

Dusts

Dusts are generated by mechanical processes such as grinding or crushing. Dusts range in size from 0.5 to 50 µm in size. Note that dust is a relatively new term used to describe dust that is hazardous when deposited anywhere in the respiratory tree including the nose and mouth. It has a 50% cut-point of 10 microns and includes the big and the small particles. The cut-point describes the performance of cyclones and other particle size selective devices. For personal sampling, the 50% cut point is the size of the dust that the device collects with 50% efficiency.

Alpaugh (1988) points out that common workplace dusts are either inorganic or organic. Inorganic dusts are derived from metallic and non-metallic sources. Non-metallic dusts can be silica bearing; that is, in combined or free silica as crystalline or amorphous form. Organic dusts are either synthetic or natural. Natural organic dust can be animal or vegetable-derived.

Examples of organic and inorganic dust would be:

- Sand (inorganic, non-metallic, silica-bearing, free silica, crystalline; Beryllium (inorganic, metallic)
- Cotton (organic, naturally occurring, vegetable)

Dusts have been classified based on their health effects. He identifies dusts as being:

- Innocuous – iron oxide, limestone (may also be considered as nuisance dusts)
- Acute respiratory hazards – cadmium fume

Industrial Hygiene

- Chronic respiratory hazards – airborne asbestos fibers
- Sensitizers – many hardwood dusts

DURATION OF EXPOSURE

The duration of exposure may be *acute* (short term) or it may be *chronic* (long-term). Some airborne particulates, for example beryllium, may exert a toxic effect after a single acute exposure or metal fume fever may occur following acute exposure to metal fumes. Other particulates, such as lead or manganese, may exert a toxic effect following a longer period of exposure, maybe several days to several weeks. Such exposures could be termed *sub-chronic*. Chronic lung conditions, such as pneumoconiosis or mesothelioma, may follow prolonged exposure to silica dusts or asbestos (crocidolite or blue asbestos) respectively.

PARTICLE SIZE

Particle size is critical in determining where particulates will settle in the lung. Smaller particles outnumber larger ones but vary widely in size. Larger particles will settle in the upper respiratory tract in the bronchi and the bronchioles and will not tend to penetrate the smaller airways found in the alveolar (air sac) region. These are termed *inspirable* particles. Those smaller-sized particles that can penetrate the alveolar (the gas exchange) region of the lungs are termed *respirable* particles.

Particle size is expressed as "aerodynamic" or "equivalent" diameter. This is equal to the diameter of spherical particles of unit density that have the same falling velocity (terminal velocity; settling velocity is the same as terminal velocity) in the air as the particle in question. The terminal velocity is proportional to the specific gravity of the particle, p, and the square of its diameter, d.

Particles with an aerodynamic diameter greater than approximately 20 microns (µm) will be trapped in the nose and upper airways.

Particles in the region of 7–20 microns will penetrate the bronchioles and are inspirable, while particles in the size range 0.5–7 microns are respirable. Particles smaller than this will not settle out as their terminal velocity is so small that there is insufficient time for them to be deposited in the alveolus and they are exhaled out again.

An understanding of aerodynamic diameter is important when calculating terminal settling rates of particulates. Constants for these calculations include:

- 1 gm/cm^3 = unit density
- Gravity = 32.2 ft/s^2 or 98 cm/s^2

Stoke's Law

Stoke's Law is the relationship that relates the "settling rate" to a particle's density and diameter. Stoke's Law applies to the fate of particulates in the atmosphere. Stoke's Law is given as:

$$u = \frac{gd^2 \rho_1 - \rho_2}{18\eta} \tag{11.1}$$

where:

u = settling velocity in cm/s (settling velocity = 0.006 ft/min (specific gravity)d^2)
g = acceleration due to gravity in cm/s^2
d^2 = diameter of particle squared in cm^2
ρ_1 = particle density in g/cm^3
ρ_2 = air density in g/cm^3
η = air viscosity in poise g/cm – s

What Stoke's Law tells us is that all other things being constant, dense particles settle faster, larger particles settle faster, and denser, more viscous air causes particles to settle slower.

Stoke's Law is used in several ways. We can predict the settling rate for a given particle if its diameter and density are known. Another way in which Stoke's Law is used is to estimate particle diameters (called "Stoke's diameter") from observed settling rates.

Airborne Dust Concentration

The concentration of dust to which a person is exposed is a critical factor to the impact on the health of the worker exposed. This is measured in the breathing zone of the worker. Airborne concentrations of dust are usually assessed by collecting dust on a pre-weighed filter. A known volume of air is drawn through the filter, which is then re-weighed. The difference in weight is the mass of dust, usually in milligrams (mg) or micrograms (μg), and the volume is expressed as cubic meters of air (m^3). Hence, the overall concentration of dust in the air is measured in mg/m^3 or $\mu g/m^3$.

PARTICULATE COLLECTION

To evaluate the workplace atmosphere for particulates a sample must be taken and analyzed. To obtain a sample the following collection mechanisms are used:

- Impaction
- Sedimentation
- Diffusion
- Direct interception
- Electrostatic attraction

The *filter* is the most common particulate collection device. The mixed cellulose ester membrane filter is the most commonly used type of filter. This type of filter is used for collecting asbestos and metals. Other filter types include polyvinyl chloride, silver, glass fiber, and Teflon filters.

ANALYSIS OF PARTICULATES

There are several methods for the analysis of particulates, including:

- **Gravimetric**: coal dust, free silica, total dust
- **Instrumental**: atomic absorption (AA) for metals
- **Optical microscopy**: fibers, asbestos, dust
- **Direct-reading instruments**: aerosol photometers, piezo-electric instruments
- **Wet chemical**: lead, isocyanates, free silica

HEALTH AND ENVIRONMENTAL IMPACTS OF PARTICULATES

Particulates cause a wide variety of health and environmental impacts. Many scientific studies have linked breathing particulates to a series of significant health problems, including:

- aggravated asthma
- increases in respiratory symptoms like coughing and difficult or painful breathing
- chronic bronchitis
- decreased lung function
- premature death

Industrial Hygiene

Particulate matter is the major cause of reduced visibility (haze) in parts of the United States, including many of our national parks.

Particles can be carried over long distances by wind and then settle on ground or water. The effects of this settling include:

- making lakes and streams acidic
- changing the nutrient balance in coastal waters and large river basins
- depleting the nutrients in soil
- damaging sensitive forests and farm crops
- affecting the diversity of ecosystems

Soot particulate stains and damages stone and other materials, including culturally important objects such as monuments and statues.

CONTROL OF PARTICULATES

As with all other methods of industrial hygiene hazard control, control of particulates is most often accomplished through the use of engineering controls, administrative controls, and PPE.

The best method for controlling particulates is the use of ventilation—an engineering control. To be most effective, the particulate generating process should be totally enclosed with a negative pressure exhaust ventilation system in place.

Administrative controls include using wet methods of housekeeping and wet cleanup methods to minimize dust regeneration and prohibiting the use of compressed air to clean work surfaces.

PPE, used as a last resort, includes equipping workers with proper respiratory protection for the prevention of inhalation of particulates and protective clothing to protect the worker from contacting particulates.

AIR SAMPLING FOR GASES AND VAPORS

Gases and *vapors* are "elastic fluids," so-called because they take the shape and volume to their containers. A *fluid* is generally termed as gas if its temperature is very far removed from that required for liquefaction; it is called a vapor if its temperature is close to that of liquefaction.

In the industrial hygiene field, a substance is considered a gas if this is its normal physical state at room temperature and atmospheric pressure. It is considered a vapor if, under the existing environmental conditions, conversion of its liquid or solid form to the gaseous state results from its vapor pressure affecting its volatilization or sublimation into the atmosphere of the container, which may be the process equipment or the worksite. Our chief interest in distinguishing between gases and vapors lies in our need to assess the potential occupational hazards associated with the use of specific chemical agents, an assessment which requires knowledge of the physical and chemical properties of these substances (NIOSH, 2004). The type of air sampling for gases and vapors employed depends on the purpose of sampling, environmental conditions, equipment available and the nature of the contaminant.

TYPES OF AIR SAMPLES

No matter the type of air sampling used, workplace samples must be obtained that represents the worker's exposure (i.e., a representative sample). In taking a *representative sample*, a sampling plan should be used that specifies the following:

- Where to sample
- Whom to sample
- How long to sample
- How many samples to take
- When to sample

Generally, as mentioned previously, two methods of sampling for airborne contaminants are used: Personal air sampling and area sampling. *Personal air sampling* (the worker wears a sampling device that collects an air sample) is the preferred method of evaluating worker exposure to airborne contaminants. *Area monitoring* (e.g., in confined spaces) is used to identify high-exposure areas.

Methods of Sampling

Standardized sampling methods provide the information needed to sample air for specific contaminants. Standard air sampling methods specify procedures, collection media, sample volume, flowrate, and chemical analysis to be used. For example, NIOSH's *Manual of Analytical Methods* and OSHA's *Chemical Information Manual* provide the information necessary to sample for air of specific contaminants.

Generally, two methods of sampling are used in sampling for airborne contaminants: grab sampling and continuous (or integrated) sampling. *Grab sampling* (i.e., instantaneous sampling) is conducted using a heavy-walled evacuated (air removed) flask. The flask is placed in the work area and a valve is opened to allow air to fill the flask. The sample represents a "snapshot" of an environmental concentration at a particular point in time. The sample is analyzed either in the laboratory or with suitable field instruments. *Continuous sampling* is the preferred method for determining time-weighted average (TWA) exposures. The sample is taken for a sample air stream.

Air Sampling Collection Processes

Airborne contaminants are collected on media or in liquid media through absorption or adsorption processes. *Absorption* is the process of collecting gas or vapor in a liquid (dissolving gas/vapor in a liquid). Absorption theory states that gases and vapors will go into solution up to an equilibrium concentration. Samplers include gas washing bottles (impingers), fritted bubblers, spiral and helical absorbers, and glass-bead columns. *Gas adsorbents* (gas onto a solid) typically use activated charcoal, silica gel, or other materials to collect gases. *Diffusive samplers* (passive samplers) depend on the flow of contaminant across a quiescent (uses no pump to draw air across adsorbent) layer of air or a membrane. Diffusion depends on well-established rules from physical chemistry, known as Fick's law.

Calibration of Air Sampling Equipment

In order to gather accurate sampling data, the equipment used must be properly calibrated. The calibration of any instrument is an absolute necessity if the data are to have any meaning. Various devices are used to calibrate air sampling equipment. Calibration is based on primary or secondary calibration standards. Primary calibrations standards include:

- soap bubble meter (or frictionless piston meter)
- spirometer (measures displaced air)
- Mariotti bottle (measures displaced water)
- electronic calibrators (provide instantaneous airflow readings)

Industrial Hygiene

Secondary calibration standards include:

- wet test meter
- dry gas meter
- rotameters

DIRECT READING INSTRUMENTS FOR AIR SAMPLING

Direct reading instruments are used for on-site evaluations for a number of reasons, including:

- To find the sources of emission of hazardous contaminants on the spot
- To ascertain if select OSHA air standards are being exceeded
- To check the performance of control equipment
- As continuous monitors at fixed locations,
  - to trigger an alarm system in the event of breakdown in a process control which could result in the accidental release of copious amounts of harmful substances to the workroom atmosphere
  - To obtain permanently recorded documentation of the concentrations of a contaminant in the atmospheric environment for future use in epidemiological and other types of occupational studies, in legal actions, to inform employees as to their exposure, and for information required for improved design of control measures.

Such on-site evaluations of the atmospheric concentrations of hazardous substances make possible the immediate assessment of undesirable exposures and enable the industrial hygienist to make an immediate correction of an operation, in accordance with his/her judgment of the seriousness of a situation, without permitting further risk of injury to the workers (NIOSH, 2004).

Types of Direct Reading Instruments

There are two types of direct reading instruments used in air sampling: direct reading physical instruments and direct reading colorimetric devices.

1. **Direct Reading Physical Instruments**

   The physical properties of gases, aerosols, and vapors are used in the design of direct reading physical instruments for quantitative estimations of these types of contaminants in the atmosphere. The various types of these instruments, the principle of operation, and a brief description of application are presented in the following discussion.
   - **Aerosol photometry**: measures, records, and controls particulates continuously in areas requiring sensitive detection of aerosol levels; detection of 0.05–40 µm diameter particles. Computer interface equipment is available.
   - **Chemiluminescence**: measurement of NO in ambient air selectivity and NO_x after conversion to NO by hot catalyst. Specific measurement of O_2. No atmospheric interferences.
   - **Colorimetry**: measure and separate recording of NO_2^-, NO_x, SO_2, total oxidants, H_2S, HF, NH_3, Cl_2 and aldehydes in ambient air.
   - **Combustion**: detects and analyzes combustible gases in terms of percent LEL (Lower Explosive Limit) on a graduated scale. Available with alarm set at 1/3 LEL.
   - **Conductivity, electrical**: records SO_2 concentrations in ambient air. Some operate off a 12-volt car battery. Operate unattended for periods of up to 30 days.
   - **Coulometry**: continuous monitoring of NO, NO_2, O_x, and SO_2 in ambient air. Provided with strip chart recorders. Some require attention only once a month.
   - **Flame ionization (with gas chromatograph)**: continuous determination and recording of methane, total hydrocarbons, and carbon monoxide. Catalytic conversion of SO

to CH_4. Operates for up to 3 days unattended. Separate model for continuous monitoring of SO_2, H_2S and total sulfur in the air. Unattended operation up to 3 days.
- **Flame ionization (hydrocarbon analyzer)**: continuous monitoring of total hydrocarbons in ambient air; potentiometric or optional current outputs compatible with any recorded. Electronic stability from 32° to 110°F.
- **Gas chromatograph, portable**: on site determination of fixed gases, solvent vapors, nitro and halogenated compounds and light hydrocarbons. Instruments available with a choice of flame ionization, electron capture or thermal conductivity detectors and appropriate columns for desired analyses. Rechargeable batteries.
- **Infrared analyzer (photometry)**: continuous determination of a given component in a gaseous or liquid stream by measuring the amount of infrared energy absorbed by component of interest using pressure sensor technique. A wide variety of applications include CO, CO_2, Freons, hydrocarbons, nitrous oxide, NH_3, SO_2, and water vapor.
- **Photometry, ultraviolet (tuned to 253.7 mµ)**—direct readout of mercury vapor; calibration filter is built into the meter. Other gases or vapors which interfere include acetone, aniline, benzene, ozone, and others with absorb radiation at 253.7 mµ.
- **Photometry, visible (narrow-centered 394 mµ band pass)**: continuous monitoring of SO_2, SO_3, H_2S, mercaptans and total sulfur compounds in ambient air. Operates for more than 3 days unattended.
- **Particle counting (near forward scattering)**: Reads and directly prints particle concentrations at 1 of 3 preset time intervals of 100, 1,000 or 10,000 seconds, corresponding to 0.01, 0.1 and 1 cubic foot of sampled air.
- **Polarography**: monitor gaseous oxygen in flue gases, auto exhausts, hazardous environments and in food storage atmospheres and dissolved oxygen in wastewater samples. Battery operated, portable, sample temperature 32°F–110°F, up to 95% relative humidity. Potentiometric recorder output. The maximum distance between sensor and amplifier is 1,000 ft.
- **Radioactivity**: Continuous monitoring of ambient gamma and X-radiation by measurement of ion chamber currents, averaging or integrating over a constant recycling time interval, sample temperature limits 32°F–120°F; 0%–95% relative humidity (weatherproof detector); up to 1,000 ft remote sensing capability. Recorder and computer outputs. Complete with alert, scram, and failure alarm systems. All solid-state circuitry.
- **Radioactivity**: Continuous monitoring of beta or gamma-emitting radioactive materials within gaseous or liquid effluents; either a thin wall Geiger-Mueller tube or a gamma scintillation crystal detector is selected depending on the isotope of interest; gaseous effluent flow – 4 cfm; effluent sample temperature limits 32°F–120°F using scintillation detector and −65°F to 165°F using G-M detector. Complete with high radiation, alert, and failure alarms.
- **Radioactivity**: Continuous monitoring of radioactive airborne particulates collected on a filter tape transport system; rate or air flow – 10 SCFM; scintillation and G-M detectors, optional but a beta-sensitive plastic scintillator is provided to reduce shielding requirements and offer greater sensitivity. Air sample temperature limits 32°F–120°F; weight 550 pounds. Complete with high and low flow alarm and a filter failure alarm.

2. **Direct-Reading Colorimetric Devices**

Direct-reading colorimetric devices are widely used, easy to operate, and inexpensive. They utilize the chemical properties of an atmospheric contaminant for the reaction of that substance with a color-producing reagent, revealing stain length or color intensity. Stain lengths or color intensities can be read directly to provide an instantaneous value of the concentration accurately within ±25%. Reagents used in detector kits may be in either a liquid or a solid phase or provided in the form of chemical treated papers. The liquid and solid reagents are generally supported in sampling devices through which a measured amount of contaminated air is drawn. On the other hand, chemically treated papers are usually exposed to the atmosphere and the reaction time noted for a color change to occur (NIOSH, 2014).

Calibration of Direct-Reading Instruments

Two common methods used for calibrating direct-reading instruments are:

- **The static method**: is easy to use and efficient. A known volume of gas is introduced into the instrument and sampling is performed for a limited period of time.
- **The dynamic method**: the instrument is used to monitor a known concentration of the contaminant to test its accuracy.

AIR MONITORING: CONFINED SPACE ENTRY

When a confined space is to be entered, it is important to remember that one can never rely on his/her senses to determine if the air in the confined space is safe. You cannot see or smell many toxic gases and vapors, nor can you determine the level of oxygen present (Spellman, 1998).

One of the most common and important functions that an environmental professional responsible for safety and health is called upon to perform is in regard to the evaluation of confined spaces for safe entry. A *confined space* is defined as a space large enough and so configured that an employee can bodily enter and perform assigned work; has limited or restricted means for entry or exit; and is not designed for continuous employee occupancy. A *permit-required confined space* (a permit is a written or printed document provided by the employer to allow and control entry into a permit space) has one or more of the following characteristics: (1) contains or has the potential to contain a hazardous atmosphere; (2) contains a material that has the potential for engulfing an entrant; (3) has a configuration such that an entrant could be trapped or asphyxiated by inwardly converging walls or by a floor that slopes downward and tapers to a smaller cross section; or (4) contains any recognized serious safety or health hazard.

To ensure the safety and health of a confined space entrant, atmospheric monitoring (testing) is required for two distinct purposes: evaluation of the hazards of the permit space and verification that acceptable entry conditions for entry into that space exist.

1. **Evaluation testing**: The atmosphere of a confined space should be analyzed using equipment of sufficient sensitivity and specificity to identify and evaluate any hazardous atmospheres that may exist or arise, so that appropriate permit entry procedures can be developed and acceptable entry conditions stipulated for that space. Evaluation and interpretation of these data, and development of entry procedure, should be done by, or reviewed by, a technically qualified industrial hygienist based on evaluation of all serious hazards.
2. **Verification testing**: The atmosphere of a permit space that may contain a hazardous substance should be tested for residues of all contaminants identified by evaluation testing using permit-specified equipment to determine that residual concentrations at the time of testing and entry are within the range of acceptable entry conditions. Results of testing (i.e., actual concentrations, etc.) should be recorded on the permit in the space provided adjacent to the acceptable entry condition.
3. **Duration of testing**: Measurement of values for each atmospheric parameter should be made for at least the minimum response time of the test instrument specified by the manufacturer.

4. **Testing stratified atmospheres**: When monitoring for entries involving a descent into the atmosphere that may be stratified, the atmospheric envelope should be tested at a distance of approximately 4 ft in the direction of travel and to each side. If a sampling probe is used, the entrant's rate of progress should be slowed to accommodate the sampling speed and detector response.

NOISE & VIBRATION

High noise levels in the workplace are a hazard to employees. High noise levels are physical stress that may produce psychological effects by annoying, startling, or disrupting the worker's concentration, which can lead to accidents. High levels can also result in damage to worker's hearing, resulting in hearing loss. In this section, we discuss the basics of noise, including those elements the industrial hygienist needs to know to ensure that his/her organization's hearing conservation program follows OSHA. We also discuss the basics of vibration and its control; vibration is closely related to noise.

OSHA Noise Control Requirements

In 1983, OSHA adopted a Hearing Conservation Amendment to OSHA 29 CFR 1910.95 requiring employers to implement *hearing conservation programs* in any work setting where employees are exposed to an 8-hour time-weighted average of 85 dBA and above. Employers must monitor all employees whose noise exposure is equivalent to or greater than a noise exposure received in 8 hours where the noise level is constantly 85 dB. The exposure measurement must include all continuous, intermittent, and impulsive noise within an 80–130 dB range and must be taken during a typical work situation. This requirement is performance-oriented because it allows employers to choose the monitoring method that best suits each individual situation (OSHA, 1998).

The basic requirements of OSHA's *Hearing Conservation Standard* are explained here:

- **Monitoring noise levels**: Noise levels should be monitored on a regular basis. Whenever a new process is added, an existing process is altered, or new equipment is purchased, special monitoring should be undertaken immediately.
- **Medical surveillance**: The medical surveillance component of the regulation specifies that employees who will be exposed to high noise levels be tested upon being hired and again at least annually.
- **Noise controls**: The regulation requires that steps be taken to control noise at the source. Noise controls are required in situations where the noise level exceeds 90 dBA. Administrative controls are sufficient until noise levels exceed 100 dBA. Beyond 100 dBA engineering, controls must be used.
- **Personal protection**: Personal protective devices are specified as the next level of protection when administrative and engineering controls do not reduce noise hazards to acceptable levels. They are to be used in addition to rather than instead of administrative and engineering controls.
- **Education and training**: The regulation requires the provision of education and training to do the following: ensure that employees understand (1) how the ear works, (2) how to interpret the results of audiometric tests, (3) how to select personal protective devices that will protect them against the types of noise hazards to which they will be exposed, and (4) how to properly use personal protective devices.

Industrial Hygiene

NOISE AND HEARING LOSS TERMINOLOGY

There are many specialized terms used to express concepts in noise, noise control and hearing loss prevention. The environmental practitioner responsible for ensuring compliance with OSHA's Hearing Conservation Program requirements must be familiar with these terms. The NIOSH definitions below were written in as non-technical a fashion as possible.

Acoustic trauma: A single incident which produces an abrupt hearing loss. Welding sparks (to the eardrum), blows to the head, and blast noise are examples of events capable of providing acoustic trauma.

Action level: The sound level which when reached or exceeded necessitates the implementation of activities to reduce the risk of noise-induce hearing loss. OSHA currently uses an 8-hour time-weighted average of 85 dBA as the criterion for implementing an effective hearing conservation program.

Attenuate: To reduce the amplitude of sound pressure (noise).

Attenuation:

Real Ear Attenuation at Threshold (REAT): A standardized procedure for conducting psycho-acoustic tests on human subjects designed to measure sound protection features of hearing protective devices. Typically, these measures are obtained in a calibrated sound field and represent the difference between subjects' hearing thresholds when wearing a hearing protector vs. not wearing the protector.

Attenuation:

Real-world: Estimated sound protection provided by hearing protective devices as worn in "real-world" environments.

Audible range: The frequency range over which normal ears hear: approximately 20–20,000 Hz.

Audiogram: A chart, graph, or table resulting from an audiometric test showing an individual's hearing threshold levels as a function of frequency.

Audiologist: A professional, specializing in the study and rehabilitation of hearing, who is certified by the American Speech-Language-Hearing Association or licensed by a state board of examiners.

Background noise: Noise coming from sources other than the particular noise sources being monitored.

Baseline audiogram: A valid audiogram against which subsequent audiograms are compared to determine if hearing thresholds have changed. The baseline audiogram is preceded by a quiet period so as to obtain the best estimate of the person's hearing at that time.

Continuous noise: Noise of a constant level as measured over at least 1 second using the "slow" setting on a sound level meter. Note, that a noise which is intermittent, e.g., on for over a second and then off for a period would be both variable and continuous.

Controls:

Administrative: Efforts, usually by management, to limit workers' noise exposure by modifying workers' schedule or location, or by modifying the operating schedule of noisy machinery.

Controls:

Engineering: Any use of engineering methods to reduce or control the sound level of a noise source by modifying or replacing equipment, making any physical changes at the noise source or along the transmission path (with the exception of hearing protectors).

Criterion sound level: A sound level of 90 decibels.

dB (Decibel): The unit used to express the intensity of sound. The decibel was named after Alexander Graham Bell. The decibel scale is a logarithmic scale in which 0 dB approximates the threshold of hearing in the mid frequencies for young adults and in which the threshold of discomfort is between 85- and 95-dB SPL and the threshold for pain is between 120 and 140 dB SPL.

Double hearing protection: A combination of both ear plug and earmuff type hearing protection devices is required for employees who have demonstrated Temporary Threshold Shift during the audiometric examination and for those who have been advised to wear double protection by a medical doctor in work areas that exceed 104 dBA.

Dosimeter: When applied to noise, refers to an instrument that measures sound levels over a specified interval, stores the measures, and calculates the sound as a function of sound level and sound duration and describes the results in terms of, dose, time-weighted average and (perhaps) other parameters such as peak level, equivalent sound level, sound exposure level, etc.

Equal-energy rule: The relationship between sound level and sound duration based upon a 3 dB exchange rate, i.e., the sound energy resulting from doubling or halving a noise exposure's duration is equivalent to increasing or decreasing the sound level by 3 dB, respectively.

Exchange rate: The relationship between intensity and dose. OSHA uses a 5-dB exchange rate. Thus, if the intensity of exposure increases by 5 dB, the dose doubles. Sometimes, this is also referred to as the doubling rate. The U.S. Navy uses a 4-dB exchange rate; the U.S. Army and Air Force use a 3-dB exchange rate. NIOSH recommends a 3-dB exchange rate. Note that the equal-energy rule is based on a 3 dB exchange rate.

Frequency: Rate in which pressure oscillations are produced. Measured in hertz (Hz).

Hazardous noise: Any sound for which any combination of frequency, intensity, or duration is capable of causing permanent hearing loss in a specified population.

Hazardous task inventory: A concept based on using work tasks as the central organizing principle for collecting descriptive information on a given work hazard. It consists of a list(s) of specific tasks linked to a database containing the prominent characteristics relevant to the hazard(s) of interest which are associated with each task.

Hearing conservation record: Employee's audiometric record. Includes name, age, job classification, TWA exposure, date of audiogram, and name of audiometric technician. To be retained for the duration of employment for OSHA. Kept indefinitely for Workers' Compensation.

Hearing damage risk criteria: A standard which defines the percentage of a given population expected to incur a specified hearing loss as a function of exposure to a given noise exposure.

Hearing handicap: A specified amount of permanent hearing loss usually averaged across several frequencies which negatively impacts employment and/or social activities. Handicap is often related to an impaired ability to communicate. The degree of handicap will also be related to whether the hearing loss is in one or both ears, and whether the better ear has normal or impartial hearing.

Hearing loss: Hearing loss is often characterized by the area of the auditory system responsible for the loss. For example, when injury or a medical condition affects the outer ear or middle ear (i.e., from the pinna, ear canal, and ear drum to the cavity behind the ear drum—which includes the ossicles) the resulting hearing loss is referred to as a *conductive* loss. When an injury or medical condition affects the inner ear or the auditory nerve that connects the inner ear to the brain (i.e., the cochlea and the VIIIth cranial nerve) the resulting hearing loss is referred to as a *sensorineural* loss. Thus, a welder's spark which damaged the ear drum would cause a conductive hearing loss. Because noise can damage the tiny hair cells located in the cochlea, it causes a sensorineural hearing loss.

Hearing loss prevention program audit: An assessment performed prior to putting a hearing loss prevention program into place or before changing an existing program. The audit should be a top-down analysis of the strengths and weaknesses of each aspect of the program.

HTL (Hearing Threshold Level): The hearing level, above a reference value, at which a specified sound or tone is heard by an ear in a specified fraction of the trials. Hearing threshold levels have been established so that 0 dB HTL reflects the best hearing of a group of persons.

Hz (HERTZ): The unit of measurement for audio frequencies. The frequency range for human hearing lies between 20 HA and approximately 20,000 Hz. The sensitivity of the human ear drops off sharply below about 500 Hz and above 4,000 Hz.

Impulsive noise: Used to generally characterize impact or impulse noise which is typified by a sound which rapidly rises to a sharp peak and then quickly fades. The sound may or may not have a "ringing" quality (such as a striking a hammer on a metal plate or a gunshot in a reverberant room). Impulsive noise may be repetitive or may be a single event (as with a sonic boom). Note: If impulses occur in very rapid succession (such as with some jack hammers), the noise would not be described as impulsive.

Loudness: The subjective attribute of a sound by which it would be characterized along a continuum from "soft" to "loud." Although this is a subjective attribute, it depends primarily upon sound pressure level, and to a lesser extent, the frequency characteristics and duration of the sound.

Material hearing impairment: As defined by OSHA, a material hearing impairment is an average hearing threshold level of 25 dB HTL as the frequencies of 1,000, 2,000, and 3,000 Hz.

Medical pathology: A disorder or disease. For purposes of this program, a condition or disease affecting the ear, which a physician specialist should treat.

Noise: Noise is any unwanted sound.

Noise dose: The noise exposure expressed as a percentage of the allowable daily exposure. For OSHA, a 100% dose would equal an 8-hour exposure to a continuous 90 dBA noise; a 50% dose would equal an 8-hour exposure to an 85 dBA noise or a 4-hour exposure to a 90 dBA noise. If 85 dBA is the maximum permissible level, then an 8-hour exposure to a continuous 85 dBA noise would equal a 100% dose. If a 3 dB exchange rate is used in conjunction with an 85 dBA maximum permissible level, a 50% dose would equal a 2-hour exposure to 88 dBA or an 8-hour exposure to 82 dBA.

Noise dosimeter: An instrument that integrates a function of sound pressure over a period of time to directly indicate a noise dose.

Noise hazard area: Any area where noise levels are equal to or exceed 85 dBA. OSHA requires employers to designate work areas, post warning signs, and warn employees when work practices exceed 90 dBA as a "Noise Hazard Area." Hearing protection must be worn whenever 90 dBA is reached or exceeded.

Noise hazard work practice: Performing or observing work where 90 dBA is equaled or exceeded. Some work practices will be specified, however, as a "Rule of Thumb, " whenever attempting to hold a normal conversation with someone who is one foot away and shouting must be employed to be heard, one can assume that a 90 dBA noise level or greater exists and hearing protection is required. Typical examples of work practices where hearing protection is required are jack hammering, heavy grinding, heavy equipment operations, and similar activities.

Noise-induced hearing loss: A sensorineural hearing loss that is attributed to noise and for which no other etiology can be determined.

Noise level measurement: Total sound level within an area. Includes workplace measurements indicating the combined sound levels of tool noise (from ventilation systems, cooling compressors, circulation pumps, etc.).

NRR (Noise Reduction Rating): The NRR is a single-number rating method which attempts to describe a hearing protector based on how much the overall noise level is reduced by the hearing protector. When estimating A-weighted noise exposures, it is important to remember to first subtract 7 dB from the NRR and then subtract the remainder from the A-weighted noise level. The NRR theoretically provides an estimate of the protection that should be met or exceeded by 98% of the wearers of a given device. In practice, this does not prove to be the case, so a variety of methods for "de-rating" the NRR have been discussed.

Ototoxic: A term typically associated with sensorineural hearing loss resulting from the therapeutic administration of certain prescription drugs.

Ototraumatic: A broader term than ototoxic. As used in hearing loss prevention, refers to any agent (e.g., noise, drugs, or industrial chemicals) which has the potential to cause permanent hearing loss subsequent to acute or prolong exposure.

Presbycusis: The gradual increase in hearing loss that is attributable to the effects of aging, and not related to medical causes or noise exposure.

Sensori-neural hearing loss: A hearing loss resulting from damage to the inner ear (from any source).

Sociacusis: A hearing loss related to non-occupational noise exposure.

Sound Intensity (I): Sound intensity at a specific location is the average rate at which sound energy is transmitted through a unit area normal to the direction of sound propagation.

Sound Level Meter (SLM): A device which measures sound and provides a readout of the resulting measurement. Some provide only A-weighted measurements, others provide A- and C-weighted measurements, and some can provide weighted, linear, and octave (or narrower) ban measurements. Some SLMs are also capable of providing time-integrated measurements.

Sound power: Is the total sound energy radiated by a source per unit time. Sound power cannot be measured directly.

SPL (Sound Pressure Level): A measure of the ratio of the pressure of a sound wave relative to a reference sound pressure. Sound pressure level in decibels is typically referenced to 20 mPa. When used alone, (e.g., 90 dB APL) a given decibel level implies an unweighted sound Pressure level.

STS:
Standard threshold shift: OSHA uses the term to describe a change in hearing threshold relative to the baseline audiogram of an average of 10 dB or more in 2,000, 3,000 and 4,000 Hz in either ear. Used by OSHA to trigger additional audiometric testing and related follow-up.

OR

Significant threshold shift: NIOSH uses this term to describe a change of 15 dB or more at any frequency, 5,000–6000 Hz, from baseline levels that are present on an immediate retest in the same ear and at the same frequency. NIOSH recommends a confirmation audiogram within 30 days with the confirmation audiogram preceded by a quiet period of at least 14 hours.

Threshold shift: Audiometric monitoring programs will encounter two types of changes in hearing sensitivity, i.e., threshold shifts: permanent threshold shift (PTS) and temporary threshold shift (TTS). As the names imply, any change in hearing sensitivity which is persistent is considered a PTS. Persistence may be assumed if the change is observed on a 30-day follow-up exam. Exposure to loud noise may cause a temporary worsening in hearing sensitivity (i.e., a TTS) that may persist for 14 hours (or even longer in cases where the exposure duration exceeded 12–16 hours). Hearing health professionals need to recognize that not all threshold shifts represent decreased sensitivity, and not all temporary or permanent threshold shifts are due to noise exposure. When a permanent threshold shift can be attributable to noise exposure, it may be referred to as a noise-induced permanent threshold shift (NIPTS).

Velocity (c): Is the speed at which the regions of sound producing pressure changes move away from the sound source.

Wavelength (λ): This term refers to the distance required for one complete pressure cycle to be completed (1 wavelength) and is measured in feet or meters.

Weighted measurements: Two weighting curves are commonly applied to measures of sound levels to account for the way the ear perceives the "loudness" of sounds.

> **A-weighting**: A measurement scale that approximates the "loudness" of tones relative to a 40 dB SPL 1,000 Hz reference tone. A-weighting has the added advantage of being correlated with annoyance measures and is most responsive to the mid frequencies, 500–4,000 Hz.

Industrial Hygiene

C-weighting: A measurement scale that approximates the "loudness" of tones relative to a 90 dB SPL 1,000 Hz reference tone. C-weighting has the added advantage of providing a relatively "flat" measurement scale which includes very low frequencies.

OCCUPATIONAL NOISE EXPOSURE

As mentioned above, *noise* is commonly defined as any unwanted sound. Noise literally surrounds us every day and is with us just about everywhere we go. However, the noise we are concerned with here is that produced by industrial processes. Excessive amounts of noise in the work environment (and outside it) cause many problems for workers, including increased stress levels, interference with communication, disrupted concentration, and most importantly, varying degrees of hearing loss. Exposure to high noise levels also adversely affects job performance and increases accident rates.

One of the major problems with attempting to protect workers' hearing acuity is the tendency of many workers to ignore the dangers of noise. Because hearing loss, like cancer, is insidious, it's easy to ignore. It sort of sneaks up slowly and is not apparent (in many cases) until after the damage is done. Alarmingly, hearing loss from occupational noise exposure has been well documented since the eighteenth century, yet since the advent of the industrial revolution, the number of exposed workers has greatly increased. However, today the picture of hearing loss is not as bleak as it has been in the past, as a direct result of OSHA's requirements. Now that noise exposure must be controlled in all industrial environments, that well-written and well-managed hearing conservation programs must be put in place, and that employee awareness must be raised to the dangers of exposure to excessive levels of noise, job-related hearing loss is coming under control.

DETERMINING WORKPLACE NOISE LEVELS

The unit of measurement for sound is the decibel. *Decibels* are the preferred unit for measuring sound, derived from the bel, a unit of measure in electrical communications engineering. A decibel is a dimensionless unit used to express the logarithm of the ratio of a measured quantity to a reference quantity.

With regards to noise control in the workplace, the industrial hygienist's primary concern is first to determine if any "noise-makers" in the facility exceed the OSHA limits for worker exposure—exactly which machines or processes produce noise at unacceptable levels. Making this determination is accomplished by conducting a noise level survey of the plant or facility. Sound measuring instruments are used to make this determination. These include noise dosimeters, sound level meters, and octave-band analyzers. The uses and limitations of each kind of instrument are discussed below.

1. **Noise dosimeter**: The noise dosimeters used by OSHA meet the American National Standards Institute (ANSI) Standard S1.25-1978, "Specifications for Personal Noise Dosimeter," which set performance and accuracy tolerances. For OSHA use, the dosimeter must have a 5-dB exchange rate, use a 90-dBA criterion level, be set at slow response, and use either an 80 or 90-dBA threshold gate, or a dosimeter that has both capabilities, whichever is appropriate for evaluation.
2. **Sound Level Meter (SLM)**: When conducting the noise level survey, the industrial hygienist should use an ANSI-approved Sound-level meter (SLM)—a device used most commonly to measure sound pressure. The SLM measures in decibels. One decibel is one-tenth of a bel and is the minimum difference in loudness that is usually perceptible.
3. **Octave-band noise analyzers**: Several Type 1 sound level meters (such as the GenRad 1982 and 1983 and the Quest 155) used by OSHA have built-in octave band analysis capability. These devices can be used to determine the feasibility of controls for individual

noise sources for abatement purposes and to evaluate hearing protectors. Octave-band analyzers segment noise into its component parts. The octave-band filter sets provide filters with the following center frequencies: 31.5; 63; 125; 250; 500; 1,000; 2,000; 4,000; 8,000; and 16,000 Hz.

The special signature of a given noise can be obtained by taking sound level meter readings at each of these settings (assuming that the noise is fairly constant over time). The results may indicate those octave-bands that contain the majority of the total radiated sound power.

Octave-band noise analyzers can assist industrial hygienists in determining the adequacy of various types of frequency-dependent noise controls. They also can be used to select hearing protectors because they can measure the amount of attenuation offered by the protectors in the octave-bands responsible for most of the sound energy in a given situation.

ENGINEERING CONTROL FOR INDUSTRIAL NOISE

When the environmental practitioner investigates the possibility of using engineering controls to control noise, the first thing he or she recognizes is that reducing and/or eliminating all noise is virtually impossible. And this should not be the focus in the first place, eliminating or reducing the "hazard" is the goal. While the primary hazard may be the possibility of hearing loss, the distractive effect (or its interference with communication) must also be considered. The distractive effect or excessive noise can certainly be classified as hazardous whenever the distraction might affect the attention of the worker. The obvious implication of noise levels that interfere with communications is emergency response. If ambient noise is at such a high level that workers can't hear fire or other emergency alarms, this is obviously an unacceptable situation.

So, what does all this mean? The environmental practitioner must determine the "acceptable" level of noise. Then he or she can look into applying the appropriate noise control measures. These include making alterations in engineering design (obviously this can only be accomplished in the design phase) or making modifications after installation. Unfortunately, this latter method is the one the environmental practitioner is usually forced to apply—and also the most difficult, depending upon circumstances.

Let's assume that the industrial hygienist is trying to reduce noise levels generated by an installed air compressor to a safe level. The first place to start is at the *source*: the air compressor. Several options are available for the environmental practitioner to employ at the source. First, the industrial hygienist would look at the possibility of modifying the air compressor to reduce its noise output. One option might be to install resilient vibration mounting devices. Another might be to change the coupling between the motor and the compressor.

If the options described for use at the source of the noise are not feasible or are only partially effective, the next component the environmental practitioner would look at is the *path* along which the sound energy travels. Increasing the distance between the air compressor and the workers could be a possibility. (NOTE: Sound levels decrease with distance). Another option might be to install acoustical treatments on ceilings, floors, and walls. The best option available (in this case) probably is to enclose the air compressor, so that the dangerous noise levels are contained within the enclosure, and the sound leaving the space is attenuated to a lower, safety level. If total enclosure of the air compressor is not practicable, then erecting a barrier or baffle system between the compressor and the open work area might be an option.

The final engineering control component the environmental practitioner might incorporate to reduce the air compressor's noise problem is to consider the *receiver* (the worker/operator). An attempt should be made to isolate the operator by providing a noise reduction or soundproof enclosure or booth for the operator.

Industrial Hygiene

AUDIOMETRIC TESTING (OSHA, 1998)

Audiometric testing monitors an employee's hearing acuity over time. It also provides an opportunity for employers to educate employees about their hearing and the need to protect it.

The employer must establish and maintain an audiometric testing program. The important elements of the program include baseline audiograms, annual audiograms, training, and follow-up procedures. Employers must make audiometric testing available at no cost to all employees who are exposed to an action level of 85 dB or above, measured as an 8-hour TWA.

The audiometric testing program follow-up should indicate whether the employer's hearing conservation program is preventing hearing loss. A licensed or certified audiologist, otolaryngologist, or other physician must be responsible for the program. Both professionals and trained technicians may conduct audiometric testing. The professional in charge of the program does not have to be present when a qualified technician conducts tests. The professional's responsibilities include overseeing the program and the work of the technicians, reviewing problem audiograms, and determining whether referral is necessary.

The employee needs a referral for further testing when test results are questionable or when related medical problems are suspected. If additional testing is necessary or if the employer suspects a medical pathology of the ear that is caused or aggravated by wearing hearing protectors, the employer must refer the employee for a clinical audiological evaluation or ontological exam, as appropriate. There are two types of audiograms required in the hearing conservation program: baseline and annual audiograms.

The *baseline audiogram* is the reference audiogram against which future audiograms are compared. Employers must provide baseline audiograms within 6 months of an employee's first exposure at or above an 8-hour TWA of 85 dB. An exception is allowed when the employer uses a mobile test van for audiograms. In these instances, baseline audiograms must be completed within 1 year after an employee's first exposure to workplace noise at or above a TWA of 85 dB. Employees, however, must be fitted with, issued, and required to wear hearing protectors whenever they are exposed to noise levels above a TWA of 85 dB or any period exceeding 6 months after their first exposure until the baseline audiogram is conducted.

Baseline audiograms taken before the hearing conservation program took effect in 1983 are acceptable if the professional supervisor determines that the audiogram is valid. Employees should not be exposed to workplace noise for 14 hours before the baseline test or be required to wear hearing protectors during this time period.

Employers must provide *annual* audiograms within 1 year of the baseline. It is important to test workers' hearing annually to identify deterioration in their hearing acuity as early as possible. This enables employers to initiate protective follow-up measures before hearing loss progresses. Employers must compare annual audiograms to baseline audiograms to determine whether the audiogram is valid and whether the employee has lost hearing acuity or experienced a standard threshold shift (STS). An STS is an average shift in either ear of 10 dB or more at 2,000, 3,000, and 4,000 Hz.

The employer must fit or refit any employee showing an STS with adequate hearing protectors, show the employee how to use them, and require the employee to wear them. Employers must notify employees within 21 days after the determination that their audiometric test results show an STS. Some employees with an STS may need further testing if the professional determines that their test results are questionable or if they have an ear problem thought to be caused or aggravated by wearing hearing protectors. If the suspected medical problem is not thought to be related to wearing hearing protection, the employer must advise the employee to see a physician. If subsequent audiometric tests show that the STS identified on a previous audiogram is not persistent, employees whose exposure to noise is less than a TWA of 90 dB may stop wearing hearing protectors.

The employer may substitute an annual audiogram for the original baseline audiogram if the professional supervising the audiometric program determines that the employee's STS is persistent.

TABLE 11.4
Permissible Noise Exposures[a]

Duration Per Day, hours	Sound Level dBA Slow Response
8	90
6	92
4	95
3	97
2	100
1.5	102
1	105
1/2	110
1/4 or less	115

Source: 29 CFR 1910.95, OSHA.

[a] When the daily noise exposure is composed of two or more periods of noise exposure of different levels, their combined effect should be considered, rather than the individual effect of each. If the sum of the following fractions $C_1/T_1 + C_2/T_2 + C_n/T_n$ exceeds unity, then, the mixed exposure should be considered to exceed the limit value. C_n indicates the total time of exposure at a specified noise level, and T_n indicates the total time of exposure permitted at that level. Exposure to impulsive or impact noise should not exceed 140-dB peak sound pressure level.

The employer must retain the original baseline audiogram, however, for the length of the employee's employment. This substitution will ensure that the same shift is not repeatedly identified. The professional also may decide to revise the baseline audiogram if the employee's hearing improves. This will ensure that the baseline reflects actual hearing thresholds to the extent possible. Employers must conduct audiometric tests in a room meeting specific background levels and with calibrated audiometers that meet American National Standard Institute (ANSI) specifications of SC-1969.

Included in any company's written Hearing Conservation Program the effects of noise exposure must be provided when the sound levels exceed those shown in Table 11.4 (when measured on the A scale of a standard sound level meter at slow response).

NOISE UNITS, RELATIONSHIPS & EQUATIONS

A number of noise units, relationships and equations that are important to the industrial hygienist involved with controlling noise hazards in the workplace are discussed below.

1. **Sound Power** (w): Sound power of a source is the total sound energy radiated by the source per unit time. It is expressed in terms of the sound power level (L_w) in decibels referenced to 10^{-12} watts (w_0). The relationship to decibels is shown below:

$$L_w = 10 \log w/w_0$$

   where:
   L_w = sound power level (decibels)
   w = sound power (watts)
   w_0 = reference power (10^{-12} watts)
   log = a logarithm to the base 10

2. Units used to describe **sound pressures** are

$$1 \, \mu bar = 1 \, dyne/cm^2 = 0.1 \, N/cm^2 = 0.1 \, Pa$$

Industrial Hygiene

3. **Sound pressure level** or SPL = $10 \log p^2/p_0$
   where:
   SPL = sound pressure level (decibels)
   p = measured root-mean-square (rms) sound pressure (N/m², μbars). Root-mean-square (rms) value of a changing quantity, such as sound pressure, is the square root of the mean of the squares of the instantaneous values of the quantity.
   p_0 = reference rms sound pressure (20 μPa, N/m², μbars)

4. **Speed of sound** $(c) = c = f\lambda$
5. **Wavelength** $(\lambda) = c/f$
6. Calculation of **frequency of octave bands** can be calculated using the following formulae:

   Upper frequency band: $f_2 = 2f_1$
   where:
   f_2 = upper frequency band
   f_1 = lower frequency band
   One-half octave band: $f_2 = \sqrt{2}(f_1)$
   where:
   f_2 = ½ octave band
   f_1 = lower frequency band
   One-third octave band: $f_2 = \sqrt[3]{2}(f_1)$
   where:
   f_2 = 1/3 octave band
   f_1 = lower frequency band

7. Formula for **adding noise sources** when sound power is known:

$$L_w = 10 \log(w_1 + w_2)/(w_0 + w_0)$$

where:
L_w = sound power in watts
w_1 = sound power of noise source 1 in watts
w_2 = sound power of noise source 2 in watts
w_0 = reference sound power (reference 10^{-12}) watts

8. Formula for **sound pressure additions** when sound pressure is known:

$$SPL = 10 \log p^2 / p_0^2$$

where:
$p^2/p_0^2 = 10^{SPL/10}$
and:
SPL = sound pressure level (decibels)
p = measured root-mean-square (rms) sound pressure (N/m², μbars)
p_0 = reference rms sound pressure (20 μPa, N/m², μbars)
For three sources, the equation becomes:

$$SPL = 10 \log \left(10^{SPL_1/10}\right) + \left(10^{SPL_2/10}\right) + \left(10^{SPL_3/10}\right)$$

When adding any number of sources, whether the sources are identical or not, the equation becomes:

$$SPL = 10\log\left(10^{SPL_1/10} + \cdots + 10^{SPL_n/10}\right)$$

Determining the sound pressure level from multiple identical sources:

$$SPL_f = SPL_i + 10\log n$$

where:
SPL_f = total sound pressure level (dB)
SPL_i = individual sound pressure level (dB)
n = number of identical sources

9. The equation for determining **noise levels in a free field** is expressed as:

$$SPL = L_w - 20\log r - 0.5$$

where:
SPL = sound pressure (reference 0.00002 N/m^2)
L_w = sound power (reference 10^{-12} watts)
r = distance in feet

10. Calculation for noise levels with directional characteristics is expressed as:

$$SPL = L_w - 20\log r - 0.5 + \log Q$$

where:
SPL = sound pressure (reference 0.00002 N/m^2)
L_w = sound power (reference 10^{-12} watts)
r = distance in feet
Q = directivity factor
$Q = 2$ for one reflecting plane
$Q = 4$ for two reflecting planes
$Q = 8$ for three reflecting planes

11. Calculating the **noise level at a new distance** from the noise source can be computed as follows:

$$SPL = SPL_1 + 20\log(d_1)/(d_2)$$

where:
SPL = sound pressure level at new distance (d_2)
SPL_1 = sound pressure level at d_1
d_n = distance from source

12. **Calculating Daily Noise Dose** can be accomplished using the following formula, which combines the effects of different sound pressure levels and allowable exposure times.

$$\text{Daily Noise Dose} = \frac{C_1 + C_2 + C_3 + \cdots + C_n}{T_1 + T_2 + T_3 + \cdots + T_n}$$

where:
C_i = number of hours exposed at given SPL_i

T_i = number of hours exposed at given SPL_i

13. Calculating OSHA Permissible Noise Levels using the formula below:

$$T_{SPL} = 8 / 2^{(SPL-90)/5}$$

where:
T_{SPL} = time in hours at given SPL
SPL = sound pressure level (dBA)

14. Formula for converting noise dose measurements to the **equivalent eight-hour TWA:**

$$TWA_{eq} = 90 + 16.61 \log(D) / (100)$$

where:
TWA_{eq} = 8-hour equivalent TWA in dBA
D = noise dosimeter reading in %

INDUSTRIAL VIBRATION CONTROL

Vibration is often closely associated with noise but is frequently overlooked as a potential occupational health hazard. Vibration is defined as the oscillatory motion of a system around an equilibrium position. The system can be in a solid, liquid, or gaseous state, and the oscillation of the system can be periodic or random, steady state or transient, continuous, or intermittent (NIOSH, 1973). Vibrations of the human body (or parts of the human body) are not only annoying, but they also affect worker performance, and sometimes causing blurred vision and loss of motor control. Excessive vibration can cause trauma, which results when external vibrating forces accelerate the body or some part so that amplitudes and restraining capacities by tissues are exceeded.

Vibration results in the mechanical shaking of the body or parts of the body. These two types of vibration are called *whole-body vibration* (affects vehicle operators) and *segmental vibration* (occurs in foundry operations, mining, stonecutting, and a variety of assembly operations, for example). Vibration originates from mechanical motion, generally occurring at some machine or series of machines. This mechanical vibration can be transmitted directly to the body or body part or it may be transmitted through solid objects to a worker located at some distance away from the actual vibration.

The effect of vibration on the human body is not totally understood; however, we do know that vibration of the chest may create breathing difficulties, and that an inhibition of tendon reflexes is a result of vibration. Excessive vibration can cause reduced ability on the part of the worker to perform complex tasks, and indications of potential damage to other systems of the body also exist.

More is known about the results of segmental vibration (typically transmitted through hand to arm), and a common example is the vibration received when using a pneumatic hammer—jackhammer. One recognized indication of the effect of segmental vibration is impaired circulation to the appendage, a condition known as *Raynaud's Syndrome*, also known as "dead fingers" or "white fingers." Segmental vibration can also result in the loss of the sense of touch in the affected area. Some indications that decalcification of the bones in the hand can result from vibration transmitted to that part of the body exist. In addition, muscle atrophy has been identified as a result of segmental vibration.

As with noise, the human body can withstand short-term vibration, even though this vibration might be extreme. The dangers of vibration are related to certain frequencies that are resonant with various parts of the body. Vibration outside these frequencies is not nearly as dangerous as vibration that results in resonance.

Control measures for vibration include substituting some other device (one that does not cause vibration) for the mechanical device that causes the vibration. An important corrective measure (often overlooked) that helps in reducing vibration is proper maintenance of tools, or support mechanisms for tools, including coating the tools with materials that attenuate vibrations. Another engineering control often employed to reduce vibration is the application of balancers, isolators, and damping devices/materials that help to reduce vibration.

RADIATION

The type of radiation that most of us are familiar with is *ionizing radiation*. Very few people have difficulty in recognizing the potential destructive power of this type of radiation. However, fewer individuals are aware of another type of radiation, *non-ionizing radiation*, which we are exposed to each day. Even fewer people can differentiate between the two types. The industrial hygienist must be familiar with the nature of radiation and understand the detection of radiation, permissible exposure limits, biological effects of radiation, monitoring techniques, control measures and procedures.

RADIATION SAFETY PROGRAM ACRONYMS AND DEFINITIONS

[*Note*: The following listing of acronyms and definitions that are typically included in radiation safety programs is adapted from the U.S. Department of Health and Human Services Public Health Service Centers for Disease Control and prevention, Atlanta, GA, August 1999 *Radiation Safety Manual*].

Abbreviations Typically Used in Radiation Safety Programs

- ALARA: As Low as Reasonably Achievable
- ALI: Annual Limit on Intake
- AU: Authorized User
- CDC: Centers for Disease Control and Prevention
- Ci: Curie
- cm^2: square centimeters
- cpm: counts per minute
- DAC: Derived Air Concentration
- dpm: disintegrations per minute
- GM: Geiger-Muller
- NaI: Sodium Iodide
- kg: kilogram
- lfm: linear feet per minute
- LSC: Liquid Scintillation Counter
- mCi: milliCurie
- ml: milliliters
- MeV: mega electron-volts
- mrem: millirem (0.001 rem)
- NRC: Nuclear Regulatory Commission
- OHC: Occupational Health Clinic
- OHS: Office of Health and safety
- PSA: Physical Security Activity
- RIA: Radioimmunoassay
- RSC: Radiation Safety Committee
- RSO: Radiation safety Officer
- TLD: Thermoluminescent Dosimeter
- ^3H: Tritium (hydrogen-3)
- ^{14}C: Carbon-14

Industrial Hygiene

- ^{32}P: Phosphorous-32
- ^{33}P: Phosphorous-33
- ^{35}S: Sulfur-35
- ^{51}Cr: Chromium-51
- ^{60}Co: Cobalt-60
- ^{125}I: Iodine-125
- ^{129}I: Iodine-129
- ^{131}I: Iodine-131
- ^{137}CS: Iodine-137
- 10 CFR 19: NRC's Title 10, Chapter 1, Code of Federal Regulations, Part 19
- 10 CFR 20: NRC's Title 10, Chapter 1, Code of Federal Regulations, Part 20

Typical Radiation Program Definitions

Absorbed dose: is the energy imparted by ionizing radiation per unit mass of irradiated material. The units of absorbed dose are the rad and the gray (Gy).

Activity: is the rate of disintegration (transformation) or decay of radioactive material. The units of activity are the curie (Ci) and the Becquerel (Bq).

Alpha particle: is a strongly ionizing particle emitted form the nucleus of an atom during radioactive decay, containing two protons and neutrons and having a double positive charge.

Alternate Authorized User: serves in the absence of the Authorized user and can assume any duties as assigned.

Authorized User: an employee who is approved by the RSO and RSC and is ultimately responsible for the safety of those who use radioisotopes under his/her supervision.

Beta particle: is an ionizing charge particle emitted form the nucleus of an atom during radioactive decay, equal in mass and charge to an electron.

Bioassay: means the determination of kinds, quantities, or concentrations, and, in some cases, the locations of radioactive material in the human body, whether by direct measurement (in vivo counting) or by analysis and evaluation of materials excreted or removed from the human body.

Biological half-life: is the length of time required for on-half of a radioactive substance to be biologically eliminated from the body.

Bremsstrahlung: is electromagnetic (X-ray) radiation associated with the deceleration of charged particles passing through matter.

Contamination: is the deposition of radioactive material in any place where it is not wanted.

Controlled area: means an area, outside of a restricted area but inside the site boundary, access to which can be limited by the licensee for any reason.

Counts per minute (cpm): is the number of nuclear transformations from radioactive decay able to be detected by a counting instrument in a 1-minute time interval.

Curie (Ci): is a unit of activity equal to 37 billion disintegrations per second.

Declared pregnant woman: means a woman who has voluntary informed her employer, in writing, of her pregnancy and the estimated date of conception.

Disintegrations per minute (dpm): is the number of nuclear transformation from radioactive decay in a 1-minute time interval.

Dose equivalent: is a quantity of radiation dose expressing all radiation on a common scale for calculating the effective absorbed dose. The units of dose equivalent are the rem and sievert (SV).

Dosimeter: is a device used to determine the external radiation dose a person has received.

Effective half-life: is the length of time required for a radioactive substance in the body to lose one-half of its activity present through a combination of biological elimination and radioactive decay.

Exposure: means the amount of ionization in the air from X-rays and gamma rays.

Extremity: means hand, elbow, and arm below the elbow, foot, knee, or leg below the knee.

Gamma rays: are very penetrating electromagnetic radiations emitted form a nucleus and an atom during radioactive decay.

Half-life: is the length of time required for a radioactive substance to lose on-half of its activity by radioactive decay.

Limits (dose limits): means the permissible upper bounds of radiation doses.

Permitted Worker: is a laboratory worker who does not work with radioactive materials but works in a radiation laboratory.

Photon: means a type of radiation in the form of an electromagnetic wave.

Rad: is a unit of radiation absorbed dose. One rad is equal to 100 ergs per gram.

Radioactive decay: is the spontaneous process of unstable nuclei in an atom disintegrating into stable nuclei, releasing radiation in the process.

Radiation (ionizing radiation): means alpha particles, beta particles, gamma rays, X-rays, neutrons, high-speed electrons, high-speed protons, and other particles capable of producing ions.

Radiation Workers: are those personnel listed on the Authorized User Form of the supervisor to conduct work with radioactive materials.

Radioisotope: is a radioactive nuclide of a particular element.

Rem is a unit of dose equivalent. One rem is approximately equal to one rad of beta, gamma, or X-ray radiation, or 1/20 of alpha radiation.

Restricted area: means an area, access to which is limited by the licensee for the purpose of protecting individuals against undue risks from exposure to radiation and radioactive materials.

Roentgen: is a unit of radiation exposure. One roentgen is equal to 0.00025 Coulombs of electrical charge per kilogram of air.

Thermoluminescent Dosimeter (TLD): is a dosimeter worn by radiation workers to measure their radiation dose. The TLD contains crystalline material which stores a fraction of the absorbed ionizing radiation and releases this energy in the form of light photons when heated.

Total Effective Dose Equivalent (TEDE): means the sum of the deep-dose equivalent (for external exposures) and the committed effective dose equivalent (for internal exposures).

Unrestricted area: means an area, access to which is neither limited nor controlled by the licensee.

X-rays: is a penetrating type of photon radiation emitted from outside the nucleus of a target atom during bombardment of a metal with fast electrons.

IONIZING RADIATION

Ionization is the process by which atoms are made into ions by the removal or addition of one or more electrons; they produce this effect by the high kinetic energies of the quanta (discrete pulses) they emit. Simply, ionizing radiation is any radiation capable of producing ions by interaction with matter. Direct ionizing particles are charged particles (e.g., electrons, protons, alpha particles, etc.) having sufficient kinetic energy to produce ionization by collision. Indirect ionizing particles are uncharged particles (e.g., photons, neutrons, etc.) that can liberate direct ionizing particles. Ionizing radiation sources can be found in a wide range of occupational settings, including health care facilities, research institutions, nuclear reactors and their support facilities, nuclear weapon production facilities, and other various manufacturing settings, just to name a few.

These ionizing radiation sources can pose a considerable health risk to affected workers if not properly controlled. Ionization of cellular components can lead to functional changes in the tissues of the body. Alpha, beta, neutral particles, X-rays, gamma rays and cosmic rays are ionizing radiations.

Three mechanisms for external radiation protection include time, distance, and shielding. A shorter time in a radiation field means less dose. From a point source, dose rate is reduced by the square of the distance and expressed by the inverse square law:

Industrial Hygiene

$$I_1(d_1)^2 = I_2(d_2)^2$$

where:

I_1 = dose rate or radiation intensity at distance d_1
I_2 = dose rate or radiation intensity at distance d_2

Radiation is reduced exponentially by the thickness of shielding material.

Effective Half-Life

The half-life is the length of time required for one-half of a radioactive substance to disintegrate. The formula depicted below is used when the industrial hygienist is interested in determining how much radiation is left in a worker's stomach after a period of time. Effective half-life is a combination of radiological and biological half-lives and is expressed as:

$$T_{eff} = \frac{(T_b)(T_r)}{T_b + T_r}$$

where:

T_b = biological half-life
T_r = radiological half-life

It is important to point out that T_{eff} will always be shorter than either T_b or T_r. T_b may be modified by diet and physical activity.

ALPHA RADIATION

Alpha radiation is used for air ionization—elimination of static electricity (Po-210), clean room applications, and smoke detectors (Am-241). It is also used in air density measurement, moisture meters, non-destructive testing, and oil well logging. Naturally occurring alpha particles are also used for physical and chemical properties, including uranium (coloring of ceramic glaze, shielding) and thorium (high temperature materials).

The characteristics of Alpha radiation are listed below.

- Alpha (α) radiation is a particle composed of two protons and neutrons with source: Ra-226 → Rn 222 → Accelerators.
- Alpha radiation is not able to penetrate skin.
- Alpha-emitting materials can be harmful to humans if the materials are inhaled, swallowed, or absorbed through open wounds.
- A variety of instruments have been designed to measure alpha radiation. Special training in use of these instruments is essential for making accurate measurements.
- A civil defense instrument (CD V-700) cannot detect the presence of radioactive materials that produce alpha radiation unless the radioactive materials also produce beta and/or gamma radiation.
- Instruments cannot detect alpha radiation through even a thin layer of water, blood, dust, paper, or other material. Because alpha radiation is not penetrating.
- Alpha radiation travels a very short distance through the air.
- Alpha radiation is not able to penetrate turnout gear, clothing, or a cover on a probe. Turnout gear and dry clothing can keep alpha emitters off of the skin.

Alpha Radiation Detectors

The types of high-sensitivity portable equipment used to evaluate alpha radiation in the workplace include:

- Geiger-Mueller counter
- Scintillators
- Solid-state analysis
- Gas proportional devices

BETA RADIATION

Beta radiation is used for thickness measurements for coating operations; radio luminous signs; tracers for research; and for air ionization (gas chromatograph, nebulizers).

The characteristics of Beta radiation are listed below.

- Beta (β) is a high energy electron particle with source: Sr-90 → Y-90 → Electron beam machine.
- Beta radiation may travel meters in the air and is moderately penetrating.
- Beta radiation can penetrate human skin to the "germinal layer," where new skin cells are produced. If beta-emitting contaminants are allowed to remain on the skin for a prolonged period of time, they may cause skin injury.
- Beta-emitting contaminants may be harmful if deposited internally.
- Most beta emitters can be detected with a survey instrument (such as a DC V-700), provided the metal probe cover is open). Some beta emitters, however, produce very low energy, poorly penetrating radiation that may be difficult or impossible to detect. Examples of these are carbon-14, tritium, and sulfur-35
- Beta radiation cannot be detected with an ionization chamber such as CD V-715.
- Clothing and turnout gear provides some protection against most beta radiation. Turnout gear and dry clothing can keep beta emitters off of the skin.
- Beta radiation present two potential exposure methods, external and internal. External beta radiation hazards are primarily skin burns. Internal beta radiation hazards are similar to alpha emitters.

Beta Detection Instrumentation

The types of equipment used to evaluate beta radiation in the workplace include:

- Geiger-Mueller counter
- Gas proportional devices
- Scintillators
- Ion chambers
- Dosimeters

Shielding for Beta Radiation

Shielding for beta radiation is best accomplished by using materials with a low atomic number (low z materials) to reduce Bremsstrahlung radiation (i.e., secondary X-radiation produced when a beta particle is slowed down or stopped by a high-density surface). The thickness is critical to stop the maximum energy range and varies with the type of material used. Typical shielding material includes lead, water, wood, plastics, cement, Plexiglas, and wax.

Industrial Hygiene

GAMMA RADIATION AND X-RAYS

Gamma radiation and X-rays are used for sterilization of food and medical products; radiography of welds, castings, and assemblies; gauging of liquid levels and material density; and oil well logging, and material analysis.

The characteristics of gamma radiation and X-rays are listed below.

- Gamma (γ) is not a particle (electromagnetic wave) composed of high energy electron with source: Tc-99.
- X-Ray is composed of photons (generated by electrons leaving an orbit) with source: Most radioactive materials, X-ray machines, secondary to β.
- Gamma radiation and X-rays are electromagnetic radiation like visible light, radio waves, and ultraviolet light. These electromagnetic radiations differ only in the amount of energy they have. Gamma rays and X-rays are the most energetic of these.
- Gamma radiation is able to travel many meters in the air and many centimeters in human tissue. It readily penetrates most materials and is sometimes called "penetrating radiation."
- X-rays are like gamma rays. They, too, are penetrating radiation.
- Radioactive materials that emit gamma radiation and X-rays constitute both an external and internal hazard to humans.
- Dense materials are needed for shielding form gamma radiation. Clothing and turnout gear provide little shielding from penetrating radiation but will prevent contamination of the skin by radioactive materials.
- Gamma radiation is detected with survey instruments, including civil defense instruments. Low levels can be measured with a standard Geiger counter, such as the CD V-700. High levels can be measured with an ionization chamber, such as a CD V-715.
- Gamma radiation or X-rays frequently accompany the emission of alpha and beta radiation.
- Instruments designed solely for alpha detection (such as an alpha scintillation counter) will not detect gamma radiation.
- Pocket chamber (pencil) dosimeters, film badges, thermoluminescent, and other types of dosimeters can be used to measure accumulated exposure to gamma radiation.
- The principal health concern associated with gamma radiation is external exposure by penetrating radiation and physically strong source housing. Sensitive organs include the lens of the eye, the gonads, and damage to the bone marrow.

Gamma Detection Instrumentation

The types of equipment used to evaluate gamma radiation in the workplace include:

- Ion chamber
- Gas proportional
- Geiger Mueller

Shielding for Gamma and X-Rays

Shielding gamma and X-radiation depends on energy level. Protection follows an exponential function of shield thickness. At low energies, absorption can be achieved with millimeters of lead. At high energies, shielding can attenuate gamma radiation.

RADIOACTIVE DECAY EQUATIONS

Radioactive materials emit alpha particles, beta particles, and photon energy, and lose a proportion of their radioactivity with a characteristic half-life. This is known as radioactive decay. To calculate the amount of radioactivity remaining after a given period of time, use the following basic formulae for decay calculations:

$$\text{Later activity} = (\text{earlier activity})\, e^{-\lambda} (\text{elapsed time})$$

$$A = A_i e^{-\lambda} t$$

where:

$$\lambda = LN2 / T$$

and:

λ = lambda decay constant (probability of an atom decaying in a unit time)
t = time
LN2 = 0.693
T = radioactive half-life (time period in which half of a radioactive isotope Decays)
A = new or later radioactivity level
A_i = initial radioactivity level

In determining time required for a radioactive material to decay (A_0 to A) use:

$$t = (-LN\, A/A_i)(T/LN2)$$

where:

λ = lambda decay constant (probability of an atom decaying in a unit time)
t = time
LN2 = 0.693
T = radioactive half-life (time period in which half of a radioactive isotope decays)
A = new or later radioactivity level
A_i = initial radioactivity level

Basic Rule of Thumb: In seven half-lives, reduced to <1%; 10 half-lives <0.1%

In determining the rate of radioactive decay, keep in mind that radioactive disintegration is directly proportional to the number of nuclei present. Thus, the radioactive decay rate is expressed in nuclei disintegrated per unit time.

$$A_i = (0.693/T)(N_i)$$

where:

A_i = initial rate of decay
N_i = initial number of radionuclei
T = half life

As mentioned earlier, half-life is defined as the time it takes for a material to lose 50% of its radioactivity. The following equation can be used to determine half-life.

Industrial Hygiene

$$A = A_i(0.5)^{t/T}$$

where:

A = activity at time t
A_i = initial activity
t = time
T = half-life

RADIATION DOSE

In the U.S., radiation *absorbed dose, dose equivalent*, and *exposure* are often measured and stated in the traditional units called *rad, rem,* or *roentgen (R)*. For practical purposes with gamma and X-rays, these units of measure for exposure or dose are considered equal. This exposure can be from an external source irradiating the whole body, an extremity, or other organ or tissue resulting in an *external radiation dose*. Alternately, internally deposited radioactive material may cause an *internal radiation dose* to the whole body or other organ or tissue.

A prefix is often used for smaller measured fractional quantities, such as, milli (m) means 1/1,000. For example, 1 rad = 1,000 mrad. Micro (μ) means 1/1,000,000. So, 1,000,000 μrad = 1 rad, or 10 μR = 0.000010 R.

The SI system (System International) for radiation measurement is now the official system of measurement and uses the "gray" (Gy) and "sievert" (Sv) for absorbed dose and equivalent dose respectively. Conversions are as follows:

- 1 Gy = 100 rad
- 1 mGy = 100 mrad
- 1 Sv = 100 rem
- 1 mSv = 100 mrem

Radioactive transformation event (radiation counting systems) can be measured in units of "disintegrations per minute" (dpm) and, because instruments are not 100% efficient, "counts per minute" (cpm). Background radiation levels are typically less than 10 μR per hour, but due to differences in detector size and efficiency, the cpm reading on fixed monitors and various handheld survey meters will vary considerably.

NON-IONIZING RADIATION

Non-ionizing radiation is described as a series of energy waves composed of oscillating electric and magnetic fields traveling at the speed of light. Non-ionizing radiation includes those electromagnetic regions extending from ultraviolet to radio waves—and usually refers to the portion of the spectrum commonly known as the radio frequency range. Non-ionizing radiation does not cause ionization. In this text we are concerned with four types of nonionizing radiation that can cause injury: ultraviolet, light, infrared, laser, microwave, and radiofrequency radiation. Adverse effects on humans range from ultraviolet radiation causing problems that range from serious sun burns (sometimes ultimately causing skin cancers) to photochemical damage to the eyes; high intensity visible light damaging the eyes; infrared radiation leading to skin burns, dehydration, and eye damage; and microwave radiation causing thermal damage to body tissues and internal organs and leading to cataracts or other eye injury.

In comparison to ionizing radiation, non-ionizing radiation is incapable of dislodging orbital electrons, but may leave the atom in an "excited state." All lower energy (frequency) radiation is non-ionizing. Non-ionizing radiation is expressed as a relationship of frequency, wavelength, and the speed of light. The higher the frequency, the higher the energy.

Non-ionizing radiation is found in a wide range of occupational settings and can pose a considerable health risk to potentially exposed workers if not properly controlled. The various types of non-ionizing radiation sources are listed below.

- **Extremely Low Frequency Radiation (ELF)**: ELF radiation at 60 Hz is produced by power lines, electrical wiring, and electrical equipment. Common sources of intense exposure include ELF induction furnaces and high-voltage power line.

    Wavelength is in the 50–60 Hz range. ACFIH exposure standards are based on understood, verifiable health effects (e.g., magnetophosphenes, induced currents, and potential interference with electronic devices, like pacemakers). ELF electric extremely applied to surfaces of the body induce electric currents and fields inside the body and excite cells.
- **Radiofrequency (RF)/Microwave Radiation (MW)**: Microwave radiation is absorbed near the skin, while RF radiation may be absorbed throughout the body. At high enough intensities both will damage tissue through heating. Sources of RF and MW radiation include radio emitters and cell phones. Microwave and radio frequency radiation include frequencies ranging from 0.1 cm to 300 m or 1 to 300,000 MHz Microwaves create heat by causing water molecules to vibrate, get agitated, and heat up. Microwaves are reflected by metal but pass-through glass, paper, and plastic. Materials containing water absorb them.
- **Infrared Radiation (IR)**: All objects emit infrared radiation to other objects that have lower surface temperature. Infrared radiation has a wavelength of from 700 nm to 1 mm. The skin and eyes absorb infrared radiation as heat. Workers normally notice excessive exposure through heat sensation and pain. Sources of IR radiation include furnaces, glass blowing, heat lamps, and IR lasers. Infrared light is heat. Exposure standards can be found in the ACGIH TLV Booklet. To use this information, the wavelength, geometry of source, and length of exposure must be known.
- **Visible light radiation**: The different visible frequencies of the electromagnetic (EM) spectrum are "seen" by our eyes as different colors. Good lighting is conducive to increased production and can help prevent incidents related to poor lighting conditions. Excessive visible radiation can damage the eyes and skin. Visible light wavelength ranges from 400 to 700 nm. Lasers, compact arc lamps, quartz-iodide-tungsten lamps, gas and vapor discharge tubes, and flash lamps are all sources of visible light. Visible light exposure standards are outlined in the ACGIH TLV Booklet. They depend on wavelength and exposure duration.
- **Ultraviolet Radiation (UV)**: Ultraviolet radiation has a high photon energy range and is particularly hazardous because there are usually no immediate symptoms of excessive exposure. Sources of UV radiation include the Sun, black lights, fluorescent lamps, welding arcs, and UV lasers. The wavelength range of UV extends from 100 to 400 nm. The ozone layer only allows wavelengths greater than 290 nm to reach the Earth. Exposure standards for UV are wavelength dependent. UV-A: 1 mW/cm^2 for 10^3 seconds measuring UV-A at the source. UV-B & C: wavelength dependent on action spectrum, most active at 200 nm. Sunglasses, clothing, sunblock, and enclosing the source provides the best protection against UV.
- **Laser hazards**: LASER is an acronym for Light Amplification by Stimulated Emission of Radiation. The photon of one atom can cause an excited electron of a neighboring atom to drop to the same energy level, thus causing the emission of another identical photon. Lasers typically emit optical (UV, visible, IR) radiations and are primarily an eye and skin hazard. Common lasers include carbon dioxide IR laser; helium—neon, neodymium YAG, ruby visible lasers, and the Nitrogen UV laser. ANSI has classified lasers into specific categories. The categories range from I-IV. Class I is less hazardous than Class IV.

Industrial Hygiene

- Class I lasers are considered to be incapable of producing damaging radiation levels, such as laser printers, and are therefore exempt from most control measures or other forms of surveillance.
- Class II lasers emit radiation in the visible portion of the spectrum, and protection is normally afforded by the normal human aversion response (blink reflex) to bright radiant sources. They may be hazardous if viewed directly for extended periods of time. Example: laser printers.
- Class IIIa lasers are those that normally would not produce injury if viewed only momentarily with the unaided eye. They may present a hazard if viewed using collecting optics, e.g., telescopes, microscopes, or binoculars. Example: HeNe lasers above 1 mW but not exceeding 5 mW radiant power.
- Class IIIb lasers can cause severe eye injuries if beams are viewed directly or specular reflections are viewed. A Class 3 laser is not normally fire hazard. Example: visible HeNe lasers above 5 mW but not exceeding 500 mW radiant power. Class IIIa and IIIb lasers require "Caution" signs and well-lighted areas to decrease pupil size.
- Class IV lasers are a hazard to the eye from the direct beam and specular reflections and sometimes even from diffuse reflections. Class IV lasers can also start fires and can damage skin. Class IV lasers require "Danger" signs.

Optical Density (OD)

Optical density (OD) is a parameter for specifying the attenuation afforded by a given thickness of any transmitting medium. Since laser beam intensities may be a factor of a thousand or a million above safe exposure levels, percent transmission notation can be unwieldy and is not used. As a result, laser protective eyewear fitters are specified in terms of the logarithmic units of Optical Density.

Because of the logarithmic factor, a filter attenuating a beam by a factor of 1,000 (or 10(3)) has an optical density of 3 and attenuating a beam by 1,000,000 or (10(6)) has an optical density of 6. The required optical density is determined by the maximum laser beam intensity to which the individual could be exposed. The optical density of two highly absorbing filter when stacked together is essentially the linear sum of two individual optical densities. The optical density for welding goggles may be 14. A pair of specific protective goggles may have an OD of 7. The formula for calculating optical density is shown below.

$$OD = \mathrm{Log}(I_o/I)$$

where:

OD = optical density
I_o or I_1 = initial beam intensity
I or I_2 = final beam intensity

Finally, as with ionizing radiation (and all other workplace hazards), environmental practitioners must understand the principles of electromagnetic radiation, its uses in the workplace, its hazards and effective control measures. The environmental professional will usually find him or herself responsible for the radiation safety program if one is needed in the organization.

OSHA's Radiation Safety Requirements

OSHA has standards for both ionizing radiation (29 CFR 1910.96) and nonionizing radiation (29 CFR 1910.97). In order to understand the hazards associated with radiation, environmental engineers need to understand the basic terms and concepts summarized in the following paragraphs, adapted from 29 CFR 1910.96.

- **Radiation**: consists of energetic nuclear particles and includes alpha rays, beta rays, gamma rays, X-rays, neutrons, high-speed electrons, and high-speed protons.
- **Radioactive material**: is material that emits corpuscular or electromagnetic emanations as the result of spontaneous nuclear disintegration.
- **Restricted area**: is any are to which access is restricted in an attempt to protect employees from exposure to radiation or radioactive materials.
- **Unrestricted area**: is any area to which access is not controlled because there is no radio-activity hazard present.
- **Dose**: is the amount of ionizing radiation absorbed per unit of mass by part of the body or the whole body.
- **Rad**: is a measure of the dose of ionizing radiation absorbed by body tissues stated in terms of the amount of energy absorbed per unit of mass of tissue. One rad equals the absorption of 100 ergs per gram of tissue.
- **Rem**: is a measure of the dose of ionizing radiation to body tissue stated in terms of its estimated biological effect relative to a dose of one roentgen (r) to X-rays.
- **Air dose**: means that the dose is measured by an instrument in air at or near the surface of the body in the area that has received the highest dosage.
- **Personal monitoring devices**: are devices worn or carried by an individual to measure radiation doses received. Widely used devices include film badges, pocket chambers, pocket dosimeters, and film rings.
- **Radiation area**: is any accessible area in which radiation hazards exist that could deliver does as follows: (1) within 1 hour a major portion of the boy could receive more than 5 millirem; or (2) with five consecutive days a major portion of the body could receive more than 100 millirem.
- **High-radiation area**—is any accessible area in which radiation hazards exist that could deliver a dose in excess of 100 millirem within 1 hour.

OSHA's requirements for *ionizing* radiation (according to 29 CFR 1910.96) include the following:

- The employer must ensure that no individual in a restricted area receives higher levels of radiation than those summarized in Table 11.5.
- The employer is responsible for ensuring that no employee under 18 years of age receives, in one calendar year, a dose of ionizing radiation in excess of 10% of the values shown in Table 11.6.
- The employer is responsible for the provision and use of radiation, and the use of radiation monitoring devices such as film badges.
- Where a potential for exposure to radioactive materials exists, appropriate warning signs must be posted.

TABLE 11.5
Levels of Radiation

Part of Body	Dose, Rems/Quarter
Whole body; head and trunk, active, blood-forming organs, lens of eyes, or gonads	1.25
Hands and forearms; feet & ankles	8.75
Skin of whole body	0.5

Source: The Office of the Federal Register, *Code of Federal Regulations Title 29 Parts 1900-1910*, Office of Federal Register, Washington, D.C.: 1985.

TABLE 11.6
Controls for Ionizing Radiation

Types of Controls	Accomplished by
Limit radiation emissions at the source	Limiting the **quantity** of ionizing material
Limiting Time Exposure	Limit employees' time of exposure. Prevent access to locations where radiation sources exist. Written procedures to limit exposures.
Extending the distance from a source	Increased distance tends to dilute airborne particulates and gases. Radiation levels decrease with the square of the distance—the inverse square law.
Shielding	Reducing radiation levels with shielding made of concrete, lead, steel, or water.
Barriers	Walls or fences will keep people out who should not be near or around radiation sources.
Warnings	Radiation areas should be clearly marked.
Evacuation	If a significant release of radioactive material occurs, the site should have a well thought out evacuation plan that employees are familiar with.
Security	Physical monitoring & security procedures can be used.
Training	Employees who work with or around radiation must be trained on the hazards of ionizing radiation.

Source: 29 CFR 1910.96, OSHA.

For normal environmental conditions, OSHA requirements for non-ionizing radiation (according to 29 CFR 1910.97) include guidelines for electromagnetic energy of frequencies between 10 MHz and 100 GHz: Power density —10 mW/cm^2 for periods of 0.1 hour or more; energy density—1 mW-h/cm^2 (milliwatt hour per square centimeter) during any 0.1-hour period. Note that this guide applies whether the radiation is continuous or intermittent. Appropriate warning signs must also be posted.

RADIATION EXPOSURE CONTROLS

Controls, both engineering and administrative, are an important element in any Radiation Safety Program. The environmental practitioner can employ some controls (depending upon the situation) to protect employees and the public. Again, as we have stated throughout this text, engineering controls are the preferred methodology, when they are appropriate and possible. Tables 11.6 and 11.7 list the kinds of engineering and other control methods that can be employed to protect people from ionizing radiation, as well as the controls for nonionizing radiation. The information contained in these tables comes primarily from publications by the American National Standards Institute (ANSI), New York; readers should refer to a complete listing of ANSI standards.

RADIATION EXPOSURE TRAINING

Rationalizing the need for extensive employee training is not at all difficult when it comes to working with or around ionizing radiation sources and materials. However, to date, not enough emphasis has been placed on training employees on the hazards involved with nonionizing radiation sources. The safety engineer must ensure that training becomes a key component of the organizational Radiation Safety Program.

TABLE 11.7
Controls for Nonionizing Radiation Nonionizing Radiation

Source	Controls
Microwaves	Limiting the intensity of microwaves (frequency or wavelength one is exposed to) or limiting the duration of exposure. Increasing the distance from a source and shielding can also limit intensity of exposure. Signs to warn about radiation hazard or dangers. Employees should handle equipment near microwave sources with insulated gloves to minimize shock and burn hazards. Microwave equipment must be properly grounded to reduce hazards.
Ultraviolet Radiation	Limit exposure to most harmful wavelengths. Use absorbing materials to shield skin and eyes.
Infrared Radiation	Limit duration of exposure and the intensity of exposure. Looking into infrared sources must be avoided. Shielding (eyewear that absorbs and reflects the impact of infrared radiation on the eyes) reduces the intensity of exposure.
Lasers	Depends on the class of Laser (The Food and Drug Administration [FDA] has standards for the classification and safety design features of lasers). Controls may include enclosure of the laser source, control of potentially reflective surfaces, interlocks on doors to location where lasers are used, fail-safe pulsing controls to prevent accidental actuation, remote firing room and controls, use of baffles to limit location of beams and wearing suitable protective eyewear and clothing.

Source: 29 CFR 1910.97.

THERMAL STRESS

Appropriately controlling the temperature, humidity, and air distribution in work areas is an important part of providing a safe and healthy workplace. A work environment in which the temperature is not properly controlled can be uncomfortable. Outdoor work areas where extremes of either heat or cold are beyond human control can be more than uncomfortable—they can be dangerous. Heat stress and cold stress are major concerns of modern health and environmental practitioners responsible for worker safety and health in the workplace. This section provides information on how to recognize and overcome the hazards associated with extreme temperatures.

THERMAL COMFORT

Thermal comfort in the workplace is a function of a number of different factors. Temperature, humidity, air distribution, personal preference, and acclimatization are all determinants of comfort in the workplace. However, determining optimum conditions is not a simple process (Alpaugh, 1988). To fully understand the hazards posed by temperature extremes, engineers must be familiar with several basic concepts related to thermal energy. The most important of these include:

Industrial Hygiene

- **Conduction**: the transfer of heat between two bodies that are touching, or from one location to another within a body. For example, if an employee touches a work-piece that has just been welded and is still hot, heat will be conducted from the work-piece to the hand. Of course, the result of this heat transfer is a burn.
- **Convection**: the transfer of heat from one location to another by way of a moving medium (a gas or a liquid). Convection ovens use this principle to transfer heat from an electrode by way of gases in the air to whatever is being baked.
- **Metabolic heat**: produced within a body as a result of activity that burns energy. All humans produce metabolic heat. A room that is comfortable when occupied by just a few people may become uncomfortable when it is crowded, because of metabolic heat, unless the thermostat is lowered to compensate.
- **Environmental heat**: produced by external sources. Gas or electric heating systems produce environmental heat, as do sources of electricity and a number of industrial processes.
- **Radiant heat**: the result of electromagnetic nonionizing energy that is transmitted through space without the movement of matter within that space.

Some standard terminology is also important:

- Heat is a measure of energy in terms of quantity.
- A calorie is the amount of energy in terms of quantity.
- Evaporative cooling takes place when sweat evaporates from the skin. High humidity reduces the rate of evaporation and thus reduces the effectiveness of the body's primary cooling mechanism.
- Metabolic heat is a by-product of the body's activity.

THE HEAT INDEX

The Heat Index (see Table 11.8) combines temperature and humidity levels to determine how the combined conditions affect individuals. The formula for calculating the Heat Index is:

TABLE 11.8
Heat Index Chart

Heat Index Chart (Temperature and Relative Humidity)

RH (%)	Temperature (°F)															
	90	91	92	93	94	95	96	97	98	99	100	101	102	103	104	105
90	119	123	128	132	137	141	146	152	157	163	168	174	180	186	193	199
85	115	119	123	127	132	136	141	145	150	155	161	166	172	178	184	190
80	112	115	119	123	127	131	135	140	144	149	154	159	164	169	175	180
75	109	112	115	119	122	126	130	134	138	143	147	152	156	161	166	171
70	106	109	112	115	118	122	125	129	133	137	141	145	149	154	158	163
65	103	106	108	111	114	117	121	124	127	131	135	139	143	147	151	155
60	100	103	105	108	111	114	116	120	123	126	129	133	136	140	144	148
55	98	100	103	105	107	110	113	115	118	121	124	127	131	134	137	141
50	96	98	100	102	104	107	109	112	114	117	119	122	125	128	131	135
45	94	96	98	100	102	104	106	108	110	113	115	118	120	123	126	129
40	92	94	96	97	99	101	103	105	107	109	111	113	116	118	121	123
35	91	92	94	95	97	98	100	102	104	106	107	109	112	114	116	118
30	89	90	92	93	95	96	98	99	101	102	104	106	108	110	112	114

Source: Weather Images: http://www.weatherimages.org/data/heatindex.html.
Note: Exposure to full sunshine can increase HI values by up to 15°F.

$$\begin{aligned}
\text{HI} = &\ 16.923 + (1.85212 \times 10^1 * T) + (5.37941 * \text{RH}) - (1.00254 \times 10^1 * T * \text{RH}) \\
&+ (9.41695 \times 10^3 * T^2) + (7.28898 \times 10^3 * \text{RH}^2) + (3.45372 \times 10^4 * T^2 * \text{RH}) \\
&- (8.14971 \times 10^4 * T * \text{RH}^2) + (1.02102 \times 10^5 * T^2 * \text{RH}^2) - (3.8646 \times 10^5 * T^3) \\
&+ (2.91583 \times 10^5 * \text{RH}^3) + (1.42721 \times 10^6 * T^3 * \text{RH}) + (1.97483 \times 10^7 * T * \text{RH}^3) \\
&- (2.18429 \times 10^8 * T^3 * \text{RH}^2) + (8.43296 \times 10^{10} * T^2 * \text{RH}^3) - (4.81975 \times 10^{11} * T^3 * \text{RH}^3)
\end{aligned}$$

where:

HI = Heat Index
T = Temperature (°F)
RH = Relative Humidity (%)

THE BODY'S RESPONSE TO HEAT

Operations involving high air temperatures, radiant heat sources, high humidity, direct physical contact with hot objects, or strenuous physical activities have a high potential for inducing heat stress in employees engaged in such operations. Industries that involve processes that create such environments include Iron and steel foundries, nonferrous foundries, brick-firing and ceramic plants, glass products facilities, rubber products factories, electrical utilities (particularly boiler rooms), bakeries, confectioneries, commercial kitchens, laundries, food canneries, chemical plants, mining sites, smelters, and steam tunnels.

Outdoor operations conducted in hot weather, such as construction, refining, asbestos removal, and hazardous waste site activities, especially those that require workers to wear semipermeable or impermeable protective clothing, are also likely to cause heat stress among exposed workers (OSHA, 2003).

The human body is equipped to maintain an appropriate balance between the metabolic heat it produces and the environmental heat to which it is exposed. Sweating and the subsequent evaporation of the sweat are the body's way of trying to maintain an acceptable temperature balance.

According to Alpaugh (1988), this balance can be expressed as a function of the various factors in the following equation.

$$H = M \pm R \pm C E \tag{11.2}$$

where

H = body heat
M = internal heat gain (metabolic)
R = radiant heat gain
C = convection heat gain
E = evaporation (cooling)

The ideal balance when applying the equation is no new heat gain. As long as heat gained from radiation, convection, and metabolic processes do not exceed that lost through the evaporation induced by sweating, the body experiences no stress or hazard. However, when heat gain from any source of sources is more than the body can compensate for by sweating, the result is heat stress.

Heat stress involves several causal factors. These include:

- Age, weight, degree of physical fitness, degree of acclimatization, metabolism, use of alcohol or drugs, and a variety of medical conditions (such as hypertension) all affect a person's sensitivity to heat. However, even the type of clothing worn must be considered. Prior heat injury predisposes an individual to additional injury.
- Predicting just who will be affected and when is difficult because individual susceptibility varies widely. In addition, environmental factors include more than the ambient air temperature. Radiant heat, air movement, conduction, and relative humidity all affect an individual's response to heat (OSHA, 2003).

The American Conference of Governmental Industrial Hygienists (1992) states that workers should not be permitted to work when their deep body temperature exceeds 38°C (100.4°F).

Heat Disorders and Health Effects

According to OSHA (2003), heat stress can manifest itself in a number of ways, depending on the level of stress. The most common types of heat stress are heat stroke, heat exhaustion, heat cramps, heat rash, transient heat fatigue, chronic heat fatigue, and sunburn. These various types of heat stress can cause a number of undesirable bodily reactions, including prickly heat, inadequate venous return to the heart, inadequate blood flow to vital body parts, circulatory shock, cramps, thirst, and fatigue.

- *Heat stroke* occurs when the body's system of temperature regulation fails and body temperature rises to critical levels. This condition is caused by a combination of highly variable factors, and its occurrence is difficult to predict. Heat stroke is very dangerous and should be dealt with immediately because it can be fatal. The primary signs and symptoms of heat stroke are confusion; irrational behavior, loss of consciousness; convulsions; a lack of sweating (usually), hot, dry skin; and an abnormally high body temperature, e.g., a victim of heat stroke will have a rectal temperature of 104.5°F or higher that will typically continue to climb. If a worker shows signs of possible heat stroke, professional medical treatment should be obtained immediately. The worker should be placed in a shady area and the outer clothing should be removed. The worker's skin should be wetted and air movement around the worker should be increased to improve evaporative cooling until professional methods of cooling are initiated and the seriousness of the condition can be assessed. Fluids should be replaced as soon as possible. The medical outcome of an episode of heat stroke depends on the victim's physical fitness and the timing and effectiveness of first aid treatment.
- *Heat exhaustion* is a type of heat stress that occurs as a result of water and/or salt depletion. When people sweat in response to exertion and environmental heat, they lose water, salt and electrolytes, the minerals needed for the body to maintain the proper metabolism and for cells to produce energy. Loss of electrolytes causes these functions to break down. Electrolyte imbalance is a problem with heat exhaustion and heat cramps. For this reason, using commercially produced drinks that contain water, salt, sugar, potassium, or electrolytes to replace those lost through sweating is important. Employees working in the heat should have water and electrolyte replacement drinks readily available and drink them frequently.

  Symptoms of heat exhaustion are headache, nausea, vertigo, weakness, thirst, and giddiness. Fortunately, this condition responds readily to prompt treatment. Heat exhaustion should not be dismissed lightly. One principal reason should be apparent to the safety engineer: fainting associated with heat exhaustion can be dangerous because the victim may be operating machinery, controlling an operation that should not be left unattended, or in physical danger from falling. A victim of heat exhaustion should be moved to a cool

(but not cold) environment and should rest lying down. Give fluids slowly but steadily by mouth until the urine volume indicates that the body's fluid level is once again in balance.
- Performing hard physical labor in a hot environment can cause *heat cramps*, a type of heat stress occurs as a result of salt and potassium depletion. Primary observable symptoms are muscle spasms in the arms, legs, and abdomen. To prevent heat cramps, acclimatize workers to the hot environment gradually over a period of at least a week. Ensure that fluid replacement is accomplished with a commercially available carbohydrate-electrolyte replacement product that contains the appropriate amount of salt, potassium, and electrolytes.
- *Heat rashes* are the most common problem in hot work environments. This type of heat manifests itself as small, raised bumps or blisters that cover a portion of the body and give off a prickly sensation that can cause discomfort. Caused by prolonged exposure to hot, damp conditions in which the body is continuously covered with sweat that does not evaporate because of the high humidity, in most cases, heat rashes will disappear when the affected individual returns to a cool environment.
- *Heat Fatigue's* principle cause is the victim's lack of acclimatization. Well-conditioned, properly acclimatized employees will suffer this form of heat stress less frequently and less severely than poorly conditioned employees. Consequently, preventing heat fatigue involves physical conditioning and acclimatization, because removing the heat stress before a more serious heat-related condition develops is the only treatment for heat fatigue available.
- While getting sunburned isn't good for anyone, certain skin-types are particularly susceptible to damage from the Sun. High-level protection sunscreen (20 spf), lightweight long-sleeved protective clothing, and a hat are indicated for outdoor safety for these individuals. Workers should be reminded that the effects of over-exposure to the Sun take time to develop. Often the real problem doesn't show up until the workday is over. Though the short term is painful, safety measures should take long-term consequences of sun-exposure (skin problems that can include cancer) into consideration, while focusing on the Sun and heat-related problems that cause symptoms that must be addressed immediately.

COLD HAZARDS

Temperature hazards are generally thought of as relating to extremes of heat because most workplace temperature hazards do relate to heat. However, temperature extremes at the other end of the spectrum—cold—can also be hazardous. Employees who work outdoors in colder climates and employees who work indoors in such jobs as meatpacking are subjected to cold hazards.

Four factors contribute to cold stress: cold temperature, high or cold wind, dampness, and cold water. These factors, alone or in combination, draw heat away from the body (Greaney, 2000). OSHA (1998) expresses cold stress though its cold stress equation. That is,

Low Temperature + Wind Speed + Wetness = Injuries and Illness

The major injuries associated with extremes of cold can be classified as being either generalized or localized. A generalized injury from extremes of cold is hypothermia. Localized injuries include frostbite, frostnip, and trench foot.

- *Hypothermia* results when the body is unable to produce enough heat to replace the heat loss to the environment. It may occur at air temperatures up to 65°F. The body uses its defense mechanisms to help maintain its core temperature.

Industrial Hygiene

- *Frostbite* is an irreversible condition in which the skin freezes, causing ice crystals to form between cells. The toes, fingers, nose, ears, and cheeks are the most common sties of freezing cold injury.
- *Frostnip* is less severe than frostbite. It causes the skin to turn white and typically occurs on the face and other exposed parts of the body. No tissue damage occurs; however, if the exposed area is not either covered or removed from exposure to the cold, frostnip can become frostbite.
- *Trench foot* is caused by continuous exposure to cold water. It may occur in wet, cold environments or through actual immersion in water

Wind-Chill Factor

The wind-chill factor (see Table 11.9) increases the level of hazard posed by extremes of cold. Safety engineers need to understand this concept and how to make it part of their deliberations when developing strategies to prevent cold stress injuries.

Cold Stress Prevention

OSHA *Fact Sheet 98-55—Protecting Workers in Cold Environments* **suggests the following precautions, safe work practices, and engineering controls.**
Lower the Risk of Cold-Related Stress

- Inadequate or wet clothing increases the effects of cold on the body.
- Certain drugs or medications (including alcohol, nicotine, caffeine, and some kinds of over-the-counter and prescription medications) inhibit the body's response to cold and/or impairs judgment.
- A cold, or diseases (including diabetes, heart, vascular, and thyroid problems) may make someone more susceptible to winter weather.
- Men are at higher risk from cold-related stresses (in general) than women. Men experience far higher death rates from exposure to cold, for reasons that include performing more inherently risky activities, body-fat composition, and other physiological differences.

TABLE 11.9
Wind Chill Chart

Wind MPH	Temperature											
	30	25	20	15	10	5	0	−5	−10	−15	−20	−25
5	25	19	13	7	1	−5	−11	−16	−22	−28	−34	−40
10	21	15	9	3	−4	−10	−16	−22	−28	−35	−41	−47
15	19	13	3	0	−7	−13	−19	−26	−32	−39	−45	−51
20	17	11	4	−2	−9	−15	−22	−29	−35	−42	−48	−55
25	16	9	3	−4	−11	−17	−24	−31	−37	−44	−51	−58
30	15	8	1	−5	−12	−19	−26	−33	−39	−46	−53	−60
35	14	7	0	−7	−14	−21	−27	−34	−41	−48	−55	−62
40	13	6	−1	−8	−15	−22	−29	−36	−43	−50	−57	−64
45	12	5	−2	−9	−16	−23	−30	−37	−44	−51	−58	−65
50	12	4	−3	−19	−17	−24	−31	−38	−45	−52	−60	−67
55	11	4	−3	−11	−18	−25	−32	−39	−46	−54	−61	−68
60	10	3	−4	−11	−19	−26	−33	−40	−48	−55	−62	−69

Source: USA Today: http://www.usatoday.com/weather/resources/basics/windchill/wind-chill-chart.htm.

- Exhaustion or immobilization, especially due to injury or entrapment, can speed up cold weather's ill effects.
- In general, people are more vulnerable to the effects of harsh winter weather as they become older.

VENTILATION

Simply put, ventilation is "the" classic method, and the most powerful tool of control used in environmental engineering to control airborne environmental hazards. Experience has shown that the proper use of ventilation as a control mechanism can assure that workplace air remains free of potentially hazardous levels of airborne contaminants. In accomplishing this, ventilation works in two ways: (1) by physically removing the contaminated air from the workplace, or (2) by diluting the workplace atmospheric environment to a safe level by the addition of fresh air.

A ventilation system is all very well and good (virtually essential, actually), but an improperly designed ventilation system can make the hazard worse. This essential point cannot be over-emphasized. At the heart of a proper ventilation system are proper design, proper maintenance, and proper monitoring. The environmental engineer plays a critical role in ensuring that installed ventilation systems operate at their optimum level.

Because of the importance of ventilation in the workplace, the environmental engineer must be well versed in the general concepts of ventilation, the principles of air movement, and monitoring practices. The environmental engineer must be properly prepared (through training and experience) to evaluate existing systems and design new systems for control of the workplace environment. In the next few sections, we present the general principles and concepts of ventilation system design and evaluation. This material should provide the basic concepts and principles necessary for the proper application of industrial ventilation systems. This material also serves to refresh the knowledge of the practitioner in the field. Probably the best source of information on ventilation is the ACGIH's *Industrial Ventilation: A Manual of Recommended Practice* (current edition)—this text is a must-have reference for every safety engineer.

CONCEPTS OF VENTILATION

The purpose of industrial ventilation is essentially to (under control) recreate what occurs in natural ventilation. Natural ventilation results from differences in pressure. Air moves from high-pressure areas to low-pressure areas. This difference in pressure is the result of thermal conditions. We know that hot air rises, which (for example) allows smoke to escape from the smokestack in an industrial process, rather than disperse into areas where workers operate the process. Hot air rises because air expands as it is heated, becoming lighter. The same principle is in effect when air in the atmosphere becomes heated. The air rises and is replaced by air from a higher-pressure area. Thus, convection currents cause a natural ventilation effect through the resulting winds.

What does all of this have to do with industrial ventilation? Simply put, industrial ventilation is installed in a workplace to circulate the air within, to provide a supply of fresh air to replace air with undesirable characteristics. Could this be accomplished simply by natural workplace ventilation? That is, couldn't we just heat the air in the workplace so that it will rise and escape through natural ports—windows, doors, cracks in walls, or mechanical ventilators in the roof (installed wind-powered turbines, for example)? Yes, we could design a natural system like this, but in such a system, air does not circulate fast enough to remove contaminants before a hazardous level is reached, which defeats our purpose in providing a ventilation system in the first place. Thus, we use fans to provide an artificial, mechanical means of moving the air.

Along with controlling or removing toxic airborne contaminants from the air, installed ventilation systems perform several other functions within the workplace. These functions include:

Industrial Hygiene

- Maintaining an adequate oxygen supply in an area. In most workplaces, this is not a problem because natural ventilation usually provides an adequate volume of oxygen; however, some work environments (deep mining and thermal processes that use copious amounts of oxygen for combustion) the need for oxygen is the major reason for an installed ventilation system.
- Removing odors from a given area. This type of system (as you might guess) has applications in such places as athletic locker rooms, rest rooms, and kitchens. In performing this function, the noxious air may be replaced with fresh air, or odors may be masked with a chemical-masking agent.
- Providing heat, cooling, and humidity control.
- Removing undesirable contaminants at their source, before they enter the workplace air (e.g., from a chemical dipping or stripping tank). Obviously, this technique is an effective way to ensure that certain contaminants never enter the breathing zone of the worker—exactly the kind of function safety engineering is intended to accomplish.

Earlier we stated that installed ventilation is able to perform its designed function via the use of a mechanical fan. Actually, a mechanical fan is the heart of any ventilation system, but like the human heart, certain ancillaries are required to make it function as a system. Ventilation is no different. Four major components make up a ventilation system: (1) The fan forces the air to move; (2) An inlet or some type of opening allows air to enter the system; (3) An outlet must be provided for air to leave the system; and (4) a conduit or pathway (ducting) not only directs the air in the right direction but also limits the amount of flow to a predetermined level.

An important concept regarding ventilation systems is the difference between exhaust and supply ventilation. An exhaust ventilation system removes air and airborne contaminants from the workplace. Such a system may be designed to exhaust an entire work area, or it may be placed at the source to remove the contaminant prior to its release into the workplace air. The second type of ventilation system is the supply system, which (as the name implies) adds air to the work area, usually to dilute work area contaminants to lower the concentration of these contaminants. However, a supplied-air system does much more; it also provides movement to air within the space (especially when an area is equipped with both an exhaust and supply system—a usual practice, because it allows movement of air from inlet to outlet and is important in replenishing exhausted air with fresh air).

Air movement in a ventilation system is a result of differences in pressure. Note that pressures in a ventilation system are measured in relation to atmospheric pressure. In the workplace, the existing atmospheric pressure is assumed to be the zero point. In the supply system, the pressure created by the system is in addition to the atmospheric pressure that exists in the workplace (i.e., a positive pressure). In an exhaust system, the objective is to lower the pressure in the system below the atmospheric pressure (i.e., a negative pressure).

When we speak of increasing and decreasing pressure levels within a ventilation system, what we are really talking about is creating small differences in pressure—small when compared to the atmospheric pressure of the work area. For this reason, these differences are measured in terms of inches of water or water gauge, which results in the desired sensitivity of measurement. Air can be assumed to be incompressible, because of the small-scale differences in pressure.

Let's get back to the water gauge or inches of water. Since one pound per square inch of pressure is equal to 27 inches of water, one inch of water is equal to 0.036 pounds pressure, or 0.24% of standard atmospheric pressure. Remember the potential for error introduced by considering air to be incompressible is very small at the pressure that exists with a ventilation system.

The environmental engineer must be familiar with the three pressures important in ventilation: velocity pressure, static pressure, and total pressure. To understand these three pressures and their function in ventilation systems, you must first be familiar with pressure itself. In fluid mechanics, the energy of a fluid (air) that is flowing is termed head. Head is measured in terms of unit weight of the fluid or in foot-pounds/pound of fluid flowing.

NOTE: The usual convention is to describe the head in terms of feet of fluid that is flowing.

So, what is pressure? Pressure is the force per unit area exerted by the fluid. In the English system of measurement, this force is measured in lbs./ft^2. Since we have stated that the fluid in a ventilation system is incompressible, the pressure of the fluid is equal to the head.

Velocity pressure (VP) is created as air travels at a given velocity through a ventilation system. Velocity pressure is only exerted in the direction of airflow and is always positive (i.e., above atmospheric pressure). When you think about it, velocity pressure has to be positive, and obviously the force or pressure that causes it also must be positive.

The velocity of the air moving within a ventilation system is directly related to the velocity pressure of the system. This relationship can be derived into the standard equation for determining velocity (and clearly demonstrates the relationship between the velocity of moving air and the velocity pressure):

$$V = 4005 / \sqrt{VP} \tag{11.3}$$

Static pressure (SP) is the pressure that is exerted in all directions by the air within the system, which tends to burst or collapse the duct. It is expressed in inches of water gauge (wg). A simple example may help you grasp the concept of static pressure. Consider a balloon that is inflated at a given pressure. The pressure within the balloon is exerted equally on all sides of the balloon. No air velocity exits within the balloon itself. The pressure in the balloon is totally the result of static pressure. Note that static pressure can be both negative and positive with respect to the local atmospheric pressure.

Total pressure (TP) is defined as the algebraic sum of the static and velocity pressures or

$$TP = SP + VP \tag{11.4}$$

The total pressure of a ventilation system can be either positive or negative (i.e., above, or below atmospheric pressure). Generally, the total pressure is positive for a supply system, and negative for an exhaust system.

For the environmental engineer to evaluate the performance of any installed ventilation system, he or she must make measurements of pressures in the ventilation system. Measurements are normally made using instruments such as a manometer or a Pitot tube.

The manometer is often used to measure the static pressure in the ventilation system. The manometer is a simple, U-shaped tube, open at both ends, and usually constructed of clear glass or plastic so that the fluid level within can be observed. To facilitate measurement, a graduated scale is usually present on the surface of the manometer. The manometer is filled with a liquid (water, oil, or mercury). When pressure is exerted on the liquid within the manometer, the pressure causes the level of liquid to change as it relates to the atmospheric pressure external to the ventilation system. The pressure measured, therefore, is relative to atmospheric pressure as the zero point.

When manometer measurements are used to obtain positive pressure readings in a ventilation system, the leg of the manometer that opens to the atmosphere will contain the higher level of fluid. When a negative pressure is being read, the leg of the tube open to the atmosphere will be lower, thus indicating the difference between the atmospheric pressure and the pressure within the system.

The Pitot tube is another device used to measure static pressure in ventilation systems. The Pitot tube is constructed of two concentric tubes. The inner tube forms the impact portion, while the outer tube is closed at the end and has static pressure holes normal to the surface of the tube. When the inner and outer tubes are connected to opposite legs of a single manometer, the velocity pressure is obtained directly. If the engineer wishes to measure static pressure separately, two manometers can be used. Positive and negative pressure measurements are indicated on the manometer as above.

Local Exhaust Ventilation

Local exhaust ventilation (the most predominant method of controlling workplace air) is used to control air contaminants by trapping and removing them near the source. In contrast to dilution ventilation (which lets the contamination spread throughout the workplace, later to be diluted by exhausting quantities of air from the workspace), local exhaust ventilation surrounds the point of emission with an enclosure and attempts to capture and remove the emissions before they are released into the worker's breathing zone. The contaminated air is usually drawn through a system of ducting to a collector, where it is cleaned and delivered to the outside through the discharge end of the exhauster. A typical local exhaust system consists of a hood, ducting, an air-cleaning device, fan, and a stack. Hazard (1988) points out that a local exhaust system is usually the proper method of contaminant control if:

- the contaminant in the workplace atmosphere constitutes a health, fire, or explosion hazard.
- national or local codes require local exhaust ventilation at a particular process.
- maintenance of production machinery would otherwise be difficult.
- housekeeping or employee comfort will be improved.
- emission sources are large, few, fixed and/or widely dispersed.
- emission rates vary widely by time.
- emission sources are near the worker-breathing zone.

The environmental engineer must remember that determining beforehand precisely the effectiveness of a particular system is often difficult. Thus, measuring exposures and evaluating how much control has been achieved after a system is installed is essential. A good system may collect 80 to 90+ percent, but a poor system may capture only 50 percent or less.

A phenomenon that many practitioners in the environmental engineering field forget (or never knew in the first place) is that ventilation, when properly designed, installed, and maintained, can go a long way to ensure a healthy working environment. However, ventilation does have limitations. For example, the effects of blowing air from a supply system and removing air through an exhaust system are different. To better understand the difference and its significance, let's take an example of air supplied through a standard exhaust duct.

When air is exhausted through an opening, it is gathered equally from all directions around the opening. This includes the area behind the opening itself. Thus, the cross-sectional area of airflow approximates a spherical form, rather than the conical form that is typical when air is blown out of a supply system. To correct this problem, a flange is usually placed around the exhaust opening, which reduces the air contour, from the large spherical contour to that of a hemisphere. As a result, this increases the velocity of air at a given distance from the opening. This basic principle is used in designing exhaust hoods. Remember that the closer the exhaust hood is to the source, and the less uncontaminated air it gathers, the more efficient the hood's percentage of capture will be. Simply put, it is easier for a ventilation system to blow air than it is for one to exhaust it. Keep this in mind whenever you are dealing with ventilation systems and/or problems.

General and Dilution Ventilation

Along with local exhaust ventilation are two other major categories of ventilation systems: general and dilution ventilation. Each of these systems has a specific purpose and finding all three types of systems present in a given workplace location is not uncommon.

General ventilation systems (sometimes referred to as heat control ventilation systems) are used to control indoor atmospheric conditions associated with hot industrial environments (such as those found in foundries, laundries, bakeries, and other workplaces that generate excess heat) for the purpose of preventing acute discomfort or injury. General ventilation also functions to control the comfort level of the worker in just about any indoor working environment. Along with the removal of air that has become process-heated beyond a desired temperature level, a general ventilation system supplies air to the work area to condition (by heating or cooling) the air, or to make up for the air that has been exhausted by dilution ventilation in a local exhaust ventilation system.

A dilution ventilation system dilutes contaminated air with uncontaminated air, to reduce the concentration below a given level (usually the threshold limit value of the contaminant) to control potential airborne health hazards, fire and explosive conditions, odors, and nuisance-type contaminants. This is accomplished by removing or supplying air, to cause the air in the workplace to move, and as a result, mix the contaminated with incoming uncontaminated air.

This mixing operation is essential. To mix the air there must be, of course, air movement. Air movement can be accomplished by natural draft caused by prevailing winds moving through open doors and windows of the work area.

Thermal draft can also move air. Whether the thermal draft is the result of natural causes or is generated from process heat, the heated air rises, carrying any contaminant present upward with it. Vents in the roof allow this air to escape into the atmosphere. Makeup air is supplied to the work area through doors and windows.

A mechanical air-moving device provides the most reliable source for air movement in a dilution ventilation system. Such a system is rather simple. It requires a source of exhaust for contaminated air, a source of air supply to replace the air mixture that has been removed with uncontaminated air, and a duct system to supply or remove air throughout the workplace. Dilution ventilation systems often are equipped with filtering systems to clean and temper the incoming air.

PERSONAL PROTECTIVE EQUIPMENT (PPE)

While the goal of the environmental practitioner responsible for workplace safety and health is certainly to engineer out all workplace hazards, we realize that this goal is virtually impossible to achieve in every case. Even in this age of robotics, computers, and other automated equipment and processes, the man-machine-process interface still exists. When people are included in the work equation, the opportunity for their exposure to hazards is very real—as injury statistics make clear, injuries occur.

Experience shows us that when some workers put on their PPE, they also sometimes donn a "Superperson" mentality. What does this mean? Often, when workers use eye, hand, foot, head, hearing, protective clothing, or respiratory protection, they also adopt an "I can't be touched—I am safe from all hazards" attitude. They naturally feel safe, as if the PPE somehow magically protects them from the hazard, so they act as if they are protected, invincible, and beyond injury . . . They feel, however illogically, that they are well out of harm's way. Nothing could be further from the truth. Hazards are all workers' kryptonite.

OSHA's PPE Standard

In the past, many OSHA standards have included PPE requirements, ranging from very general to very specific requirements. It may surprise you to know, however, that not until recently (1993–1994) did OSHA incorporate a stand-alone primary PPE Standard into its 29 CFR 1910/1926 Guidelines. This relatively new *Personal Protective Equipment* standard is covered (General Industry) under 1910.132-. 138, but you can find PPE requirements elsewhere in the General Industry Standards. For example, 29 CFR 1910.156, OSHA's Fire Brigades Standard has requirements for firefighting gear. In addition, 29 CFR 1926.95-106 covers the construction industry. OSHA's general PPE requirements mandate that employers conduct a hazard assessment of their workplaces to determine what hazards are present that require the use of PPE, provide workers with appropriate PPE, and require them to use and maintain it in sanitary and reliable condition.

Industrial Hygiene

As currently written, the PPE standard focuses on the head, feet, eye, hand, body (clothing), respiratory, and hearing protection.

Common PPE classifications and examples include:

1. Head protection (hard hats, welding helmets)
2. Eye protection (safety glasses, goggles)
3. Face protection (face shields)
4. Respiratory protection (respirators)
5. Arm protection (protective sleeves)
6. Hearing protection (ear plugs, muffs)
7. Hand protection (gloves)
8. Finger protection (cots)
9. Torso protection (aprons)
10. Leg protection (chaps)
11. Knee protection (kneeling pads)
12. Ankle protection (boots)
13. Foot protection (boots, metatarsal shields)
14. Toe protection (safety shoes)
15. Body protection (coveralls, chemical suits)

Note that Respiratory and Hearing protection requirements have been covered extensively under their own standards for quite some time. Respiratory Protection is covered under 29 CFR 1910.134 and Hearing Protection is covered under 29 CFR 1910.95.

Using PPE is often essential, but, as mentioned, it generally is the last line of defense after engineering controls, work practices, and administrative controls. Recall that engineering controls involve physically changing a machine or work environment. Administrative controls involve changing how or when employees do their jobs, such as scheduling work and rotating employees to reduce exposure. Work practices involve training workers how to perform tasks in ways that reduce their exposure to workplace hazards.

OSHA's PPE Requirements

OSHA mandates several requirements for both the employer and the employee under its PPE Standard. OSHA's requirements include:

1. Employers are required to provide employees with personal protective equipment that is sanitary and in good working condition.
2. The employer is responsible for examining all PPE used on the job to ensure that it is of a safe (and approved) design and in proper condition.
3. The employer must ensure that employees use PPE.
4. The employer must provide a means for obtaining additional and replacement equipment; defective and damaged PPE is not to be used.
5. The employer must ensure that PPE is inspected on a regular basis.
6. The employee must ensure that he or she uses PPE when required.
7. Where employees provide their own PPE, the employer must ensure that it is adequate, including properly maintained and sanitized.

NOTE: While the employer must ensure the employee wears PPE when required, both the employer and employee should factor in three considerations: (1) The PPE used must not degrade performance unduly; (2) it must be dependable; and (3) it must be suitable for the hazard involved.

Hazard Assessment

How does an environmental professional determine when and where an employer should provide PPE, and when the employee should use it? This can be determined in three ways:

1. From the manufacturer's guidance (when it comes to equipment and processes produced by a manufacturer, the manufacturer is considered the "expert" on the equipment or process and is normally best-suited to determine the hazards associated with the equipment and/or processes they manufacture).
2. If the process or equipment the employee is working on/with involves chemicals, the Safety Data Sheets (SDS) for the chemicals involved list the required PPE to be used.
3. OSHA mandates that the employer perform a hazard assessment of the workplace.

The purpose of the *hazard assessment* is to determine if workplace hazards are present or likely to be present that necessitate the use of PPE. If a facility presents such hazards, the employer is required to (1) select, and have each affected employee use the types of PPE that will protect the affected employee from the hazards identified in the hazard assessment; (2) communicate selection decisions to each affected employee; and (3) select PPE that properly fits each affected employee.

The employer is required to verify that the workplace hazard assessment has been conducted through a written certification that identifies the workplace evaluated, the person certifying that the evaluation has been performed, the date of the hazard assessment, and that also identifies the document as a certification of hazard assessment.

PPE Training Requirement

OSHA requires the employer to provide training to each employee required to use PPE. This training must inform the employee on when the PPE is necessary, what PPE is necessary, how to properly don, doff, adjust, and wear PPE, the limitations of the PPE, and the proper care, maintenance, useful life, and disposal of the PPE.

OSHA's PPE employee training requirements are summarized below.

Employees must be trained to:

- Use PPE properly
- Be aware of when PPE is necessary
- Know what kind of PPE is necessary
- Understand the limitations of PPE in protecting employees from injury
- Don, adjust, wear, and doff PPE
- Maintain PPE Properly

During an OSHA audit, OSHA requires each employee to demonstrate his or her understanding of his/her training on PPE. This is usually best accomplished by demonstration (e.g., wearing and operating and SCBA) and through written or oral examination.

If the employer has reason to believe that any affected employee who has already been trained on PPE does not have the understanding and skill required, the employer must retrain such an employee. In this re-training requirement, remember that everything in life is dynamic (constantly changing), including the workplace and work assignments. OSHA understands this dynamic trend, and thus requires the employer to re-train employees who install new processes, equipment, or requirements—any new element in a job task that might render previous training obsolete. Changes also occur in PPE itself. Maybe a new type or model of PPE is introduced and used in the workplace. If this is the case, the employer must ensure that employees using such PPE are fully trained on the new PPE.

Industrial Hygiene

HEAD PROTECTION

OSHA requires employers to ensure that employees are protected from head injury whenever they work in areas where there is a possible danger of head injury from impact, or from falling or flying objects, or from electrical shock and burns.

According to 29 CFR 1926.100 (b), helmets for the protection of workers against impact and penetration of falling and flying objects must meet the specifications contained in American National Standards Institute, Z89.1-1969, *Safety Requirements for Industrial Head Protection*.

Helmets for the head protection of employees exposed to high voltage electrical shock and burns must meet the specification contained in American National Standards Institute, Z89.2-1971 (OSHA 29 CFR 1926.100 (c).

HAND PROTECTION

Under general requirements listed in 29 CFR 1910.138(a), *Hand Protection*, OSHA mandates that employers must select and require employees to use appropriate hand protection when employees' hands are exposed to hazards such as those from skin absorption of harmful substances; severe cuts or lacerations; severe abrasions; punctures; chemical burns, thermal burns; and harmful temperature extremes.

In the selection of protective gloves, 29 CFR 1910.100(b) mandates employers to base the selection of the appropriate hand protection on an evaluation of the performance characteristics of the hand protection relative to the task(s) to be performed, conditions present, duration of use, and the hazards and potential hazards identified.

EYE AND FACE PROTECTION

Eye and face protection requirements are outlined in OSHA 29 CFR 1910.133. The American National Standards Institute (ANSI), in its publication Practice for Occupational and Educational Eye and Face Protection for Occupational and Educational Eye and Face Protection (ANS Z87.1989), specifies the use and construction of protective eyewear.

Eye and face protection includes safety glasses, chemical goggles, and face shields. Appropriate selection is based on the type of hazard.

Face shields or chemical splash goggles are appropriate whenever a worker may be subject to splashing from chemicals. They are excellent for use whenever workers are dipping parts into open-surface tanks containing plating baths, cleaning solutions, organic chemicals, or corrosive chemicals. They are also appropriate whenever a worker may be subject to flying particles, such as from using a portable or pedestal grinder. Face shields are designed only to prevent direct splash exposures to the face and not to provide complete eye protection; they serve as "secondary" eye protection only.

Safety glasses are necessary when working with hazardous materials or when operating machinery, air guns, or when there is reasonable probability of injury that can be prevented by the use of such equipment. Safety glasses should be affixed with side-shields.

FOOT PROTECTION

OSHA, in its 29 CFR 1910.136, *Occupational Foot Protection*, states that the employer must ensure that each affected employee wears protective footwear when working in areas where there is a danger of foot injuries due to falling or rolling objects, or objects piercing the sole, and where such employee's feet are exposed to electrical hazards.

Protective footwear is required to be durable and comfortable and designed for expected exposures. Footwear deemed unacceptable in most work settings includes any footwear that permits direct contact between the foot and a foreign agent. Examples include open-toed shoes, open-heel shoes, and soft shoes (e.g., canvas athletic shoes). Note that canvas athletic safety shoes are not recommended for those who handle or work around corrosive chemical substances. Steel-toed safety shoes are required whenever there is the potential for items to be dropped onto the foot.

FULL BODY PROTECTION: CHEMICAL PROTECTIVE CLOTHING

According to OSHA (1998), the purpose of chemical protective clothing and equipment is to shield or isolate individuals form the chemical, physical, and biological hazards that may be encountered during hazardous materials operations. During chemical operations, it is not always apparent when exposure occurs. Many chemicals pose invisible hazards and off no warning properties.

The guidelines provided in this section describe the various types of clothing that are appropriate for use in various chemical operations and provide recommendations for their selection and use.

It is important that protective clothing users realize that no single combination of protective equipment and clothing is capable of protecting you against all hazards. Thus, protective clothing should be used in conjunction with other protective methods. For example, engineering or administrative controls to limit chemical contact with personnel should always be considered as an alternative measure for preventing chemical exposure. The use of protective clothing can itself create significant wearer hazards, such as heat stress, physical and psychological stress, in addition to impaired vision, mobility, and communication. In general, the greater the level of chemical protective clothing required the greater the associated risks. For any given situation, equipment and clothing should be selected that provide an adequate level of protection. Overprotection as well as under-protection can be hazardous and should be avoided.

DESCRIPTION OF PROTECTIVE CLOTHING

Protective Clothing Applications

Protective clothing must be worn whenever the wearer faces potential hazards arising from chemical exposure. Some examples include:

- Emergency response
- Chemical manufacturing and process industries
- Hazardous waste site cleanup and disposal
- Asbestos removal and other particulate operations
- Agricultural application of pesticides

Within each application, there are several operations which require chemical protective clothing. For example, in emergency response, the following activities dictate chemical protective clothing use:

- **Site survey**: The initial investigation of a hazardous materials incident; these situations are usually characterized by a large degree of uncertainty and mandate the highest levels of protection.
- **Rescue**: Entering a hazardous materials area for the purpose of removing an exposure victim; special considerations must be given to how the selected protective clothing may affect the ability of the wearer to carry out rescue and to the contamination of the victim.
- **Spill mitigation**: Entering a hazardous materials area to prevent a potential spill or to reduce the hazards form an existing spill (i.e., applying a chlorine repair kit on railroad tank car). Protective clothing must accommodate the required tasks without sacrificing adequate protection.

Industrial Hygiene

- **Emergency monitoring**: Outfitting personnel in protective clothing for the primary purpose of observing a hazardous materials incident without entry into the spill site. This may be applied to monitoring contract activity for spill cleanup.
- **Decontamination**: Applying decontamination procedures to personnel or equipment leaving the site; in general, a lower level of protective clothing is used by the personnel involved in decontamination.

The Clothing Ensemble

The approach in selecting personal protective clothing must encompass an "ensemble" of clothing and equipment items which are easily integrated to provide both an appropriate level of protection and still allow one to carry out activities involving chemicals. In many cases, simple protective clothing by itself may be sufficient to prevent chemical exposure, such as wearing gloves in combination with a splash apron and face shield (or safety goggles).

The following is a checklist of components that may form the chemical protective ensemble:

- Protective clothing (suit, coveralls, hoods, gloves, boots)
- Respiratory equipment (SCBA, combination SCBA/SAR, air-purifying respirators)
- Cooling system (ice vest, air circulation, water circulation)
- Communication device
- Head protection
- Eye protection
- Ear protection
- Inner garment
- Outer protection (over gloves, over boots, flash cover)

Factors that affect the selection of ensemble components include:

- How each item accommodates the integration of other ensemble components. Some ensemble components may be incompatible due to how they are worn (e.g., some SCBAs may not fit within a particular chemical protective suit or allow acceptable mobility when worn).
- The ease of interfacing ensemble components without sacrificing required performance (e.g., a poorly fitting over glove that greatly reduces wearer dexterity).
- Limiting the number of equipment items to reduce donning time and complexity (e.g., some communications devices are built into SCBAs which as a unit are not NIOSH certified).

Levels of Protection

Personal protective equipment (PPE) is categorized into levels A through D. Level A is the most complete and comprehensive level and level D is the lowest of protection.

The PPE levels listed below can be used as the starting point for ensemble creation; however, each ensemble must be tailored to the specific situation in order to provide the most appropriate level of protection. For example, if an emergency response activity involves a highly contaminated area or if the potential contamination is high, it may be advisable to wear a disposable covering such as Tyvek coveralls or PVC splash suits, over the protective ensemble.

- Level A Protection includes:
  - Vapor protective suit
  - Two-way radio
  - Pressure demand, Self-contained breathing apparatus (SCBA).
  - Hard hat.

- Two pair of gloves.
- Chemical-resistant steel toe shank and disposable booties.

Use Level A protection when:
- The chemical concentration is known to be above a safe level
- During a confined space entry.
- In the presence of extremely hazardous substances (e.g., cyanide).
- In the presence of skin destructive substances.

Limitations of Level A protection:
- Protective clothing must resist permeation by the chemical or mixture present. Ensemble items must allow integration without loss of performance.

- Level B Protection includes:
  - Two-piece liquid splash-protective suit with hood or disposable suit.
  - Pressure demand, full-facepiece SCBA.
  - Hard hat.
  - Two pairs of gloves.
  - Chemical-resistant steel toe shank and disposable booties

Use Level B protection when:
- Immediately dangerous to life and health (IDLH) conditions exist.
- Concentrations are above the protection factors provided by a full mask, and air-purifying respirator.
- Oxygen levels are less than 19.5%.
- Skin contact is unlikely to the head and neck.
- An unidentified vapor is suspected.

Limitations of level B protection:
- Protective clothing must resist penetration by the chemicals or mixtures present.
- Ensemble items must allow integration without loss of performance.

- Level C Protection includes:
  - Full-face air-purifying respirator.
  - Hard hat.
  - Two pairs of gloves.
  - chemical resistant steel toe with shank and disposable booties.
  - Two-piece suit or disposable suit.

Use Level C protection when:
- The air concentration is known.
- Assigned protective factors offer control with an air-purifying respirator.
- There is no threat of IDLH conditions.
- There is no skin hazard.
- There is no unidentified vapor present.

Limitation of Level C protection:
- Protective clothing items must resist penetration by the chemical or mixtures present.
- Chemical airborne concentration must be less than IDLH levels.
- The atmosphere must contain at least 19.5%% oxygen.

- Level D Protection includes:
  - Safety glasses
  - Hard hat.
  - One pair of gloves.
  - Safety shoes.
  - Coveralls.

Industrial Hygiene

  Use Level C Protection when:
  - There is no measurable concentration.
  - No exposure to splash or inhalation will occur.

  Limitation of Level D Protection:
  - This level should not be worn in the Hot Zone.
  - The atmosphere must contain at least 19.5% oxygen.
  - Not acceptable for Chemical Emergency Response.

The type of equipment used and the overall level of protection should be evaluated periodically as the amount of information about the chemical situation or process increases, and when workers are required to perform different tasks. Personnel should upgrade or downgrade their level of protection only with concurrence with site supervisor, safety officer, or plant industrial hygienist.

The recommendations listed above serve only as guidelines. It is important for you to realize that selecting items by how they are designed or configured alone is not sufficient to ensure adequate protection. In other words, just having the right components to form an ensemble is not enough. The USEPA levels of protection do not define what performance the selected clothing or equipment must offer.

Clothing Selection Factors

- Chemical hazards: Chemicals present a variety of hazards such as toxicity, corrosiveness, flammability, reactivity, and oxygen deficiency. Depending on the chemicals present, any combination of hazards may exist.
- **Physical environment**: Chemical exposure can happen anywhere: in industrial settings, on the highways, or in residential areas. It may occur either indoors or outdoors; the environment may be extremely hot, cold, or moderate; the exposure site may be relatively uncluttered or rugged, presenting a number of physical hazards; chemical handling activities may involve entering confined spaces, heavy lifting, climbing a ladder, or crawling on the ground. The choice of ensemble components must account for these conditions.
- **Duration of exposure**: The protective qualities of ensemble components may be limited to certain exposure levels (e.g., material chemical resistance, air supply). The decision for ensemble use time must be made assuming the worst-case exposure that safety margins can be applied to increase the protection available to the work.
- **Protective clothing or equipment available**: Hopefully, an array of different clothing or equipment is available to workers to meet all intended applications. Reliance on one particular clothing or equipment item may severely limit a facility's ability to handle a broad range of chemical exposures. In its acquisition of equipment and clothing, the safety department or other responsible authority should attempt to provide a high degree of flexibility while choosing protective clothing and equipment that is easily integrated and provides protection against each conceivable hazard.

Classification of Protective Clothing

Personal protective clothing includes the following:

- Fully encapsulating suits
- Nonencapsulating suits
- Gloves, boots, and hoods
- Firefighter's protective clothing
- Proximity, approach clothing
- Blast or fragmentation suits
- Radiation-protective suits

Firefighter turnout clothing, proximity gear, blast suits, and radiation suits by themselves are not acceptable for providing adequate protection from hazardous chemicals.

Material Chemical Resistance

Ideally, the chosen material(s) must resist permeation, degradation, and penetration by the respective chemicals. *Permeation* is the process by which a chemical dissolves in or moves through a material on a molecular basis. In most cases, there will be no visible evidence of chemicals permeating a material. Permeation breakthrough time is the most common result used to assess material-chemical compatibility. The rate of permeation is a function of several factors such as chemical concentration, material thickness, humidity, temperature, and pressure. Most material testing is done with 100% chemical over an estimated exposure period. The time it takes chemical to permeate through the material is the *breakthrough time*. An acceptable material is one where the breakthrough time exceeds the expected period of garment use. However, temperature and pressure effects may enhance permeation and reduce the magnitude of this safety factor. For example, small increases in ambient temperature can significantly reduce breakthrough time and the protective barrier properties of a protective clothing material.

Degradation involves physical changes in a material as the result of a chemical exposure, use, or ambient conditions (e.g., sunlight). The most common observations of material degradation are discoloration, swelling, loss of physical strength, or deterioration.

Penetration is the movement of chemicals through zippers, seams, or imperfections in a protective clothing material.

It is important to note that no material protects against all chemicals and combinations of chemicals, and that no currently available material is an effective barrier to any prolonged chemical exposure.

Decontamination Procedures

Decontamination is the process of removing or neutralizing contaminants that have accumulated on personnel and equipment. This process is critical to health and safety at hazardous material response sites. Decontamination protects end users from hazardous substances that may contaminate and eventually permeate the protective clothing, respiratory equipment, tools, vehicles, and other equipment used in the vicinity of the chemical hazard; it protects all plant or site personnel by minimizing the transfer of harmful material into clean areas; it helps prevent mixing of incompatible chemicals; and it protects the community by preventing uncontrolled transportation of contaminants from the site.

There are two types of decontamination:

- *Gross decontamination*: To allow end user to safely exist or doff the chemical protective clothing.
- *Decontamination* for reuse of chemical protective clothing.

The first step in decontamination is to establish Standard Operating Procedures that minimize contact with chemicals and thus the potential for contamination. For example:

- Stress work practices that minimize contact with hazardous substances (e.g., do not walk-through areas of obvious contamination; do not directly touch Potentially hazardous substances).
- Use remote sampling, handling, and container-opening techniques (e.g., drum grapples, pneumatic impact wrenches).
- Protect monitoring and sampling instruments by bagging. Make openings in the bags for sample ports and sensors that must contact site materials.
- Wear disposable outer garments and use disposable equipment where appropriate.

Industrial Hygiene

- Cover equipment and tools with s strippable coating that can be removed during decontamination.
- Encase the source of contaminants, e.g., with plastic sheeting or overpacks.
- Ensure all closures and ensemble component interfaces are completely secured; and that no open pockets that could serve to collect contaminant are present.

Types of Contamination

- *Surface contaminants* may be easy to detect and remove.
- *Permeated contaminants* are contaminants that are difficult or impossible to detect or remove. If contaminants that have permeated a material are not removed by decontamination, they may continue to permeate the material where they can cause an unexpected exposure.

Four major factors affect the extent of permeation:

- **Contact time**: The longer a contaminant is in contact with an object, the greater the probability and extent of permeation. For this reason, Minimizing contact time is one of the most important objectives of a decontamination program.
- **Concentration**: Molecules flow from areas of high concentration to areas of low concentration. As concentrations of chemicals increase, the potential for permeation of personal protective clothing increases.
- **Temperature**: An increase in temperature generally increases the permeation rate of contaminants.
- **Physical state of chemicals**: As a rule, gases, vapors, and low-viscosity liquids tend to permeate more readily than high-viscosity liquids or solids.

Decontamination Methods

Decontamination methods: (1) physically remove contaminants; (2) inactivate contaminants by chemical detoxification or disinfection/sterilization; or (3) remove contaminants by a combination of both physical and chemical means.

INSPECTION, STORAGE, AND MAINTENANCE OF PROTECTIVE CLOTHING

The end user in donning protective clothing and equipment must take all necessary steps to ensure that the protective ensemble will perform as expected. During emergencies is not the right time to discover discrepancies in the protective clothing. Teach end user care for his clothing and other protective equipment in the same manner as parachutists care for parachutes. Following a standard program for inspection, proper storage, and maintenance along with realizing protective clothing/equipment limitations is the best way to avoid chemical exposure during emergency response.

Inspection

An effective chemical protective clothing inspection program should feature five different inspections:

- Inspection and operational testing of equipment received as new form the factory or distributor.
- Inspection of equipment as it is selected for a particular chemical operation.
- Inspection of equipment after use or training and prior to maintenance.

- Periodic inspection of stored equipment.
- Periodic inspection when a question arises concerning the appropriateness of selected equipment, or when problems with similar equipment are discovered.

Storage

Clothing must be stored properly to prevent damage or malfunction from exposure to dust, moisture, sunlight, damaging chemicals, extreme temperatures, and impact. Procedures are needed for both the initial receipt of equipment and after use or exposure of that equipment. Many manufacturers specify recommended procedures for storing their products. These should be followed to avoid equipment failure resulting from improper storage.

Maintenance

Manufacturers frequently restrict the sale of certain protective suit parts to individuals or groups who are specially trained, equipped, or authorized by the manufacturer to purchase them. Explicit procedures should be adopted to ensure that the appropriate level of maintenance is performed only by those individuals who have this specialized training and equipment. In no case should you attempt to repair equipment without checking with the person in your facility that is responsible for chemical protective clothing maintenance.

RESPIRATORY PROTECTION

Wearing respiratory protective devices to reduce exposure to airborne contaminants is widespread in industry. An estimated 5.0 million workers wear respirators, either occasionally or routinely. Although it is preferred industrial hygiene practice to use engineering controls to reduce contaminant emissions at their source, there are operations where this type of control is not technologically or economically feasible or is otherwise inappropriate.

Respirators are devices that can allow workers to safely breathe without inhaling particles or toxic gases. Two basic types are (1) *air-purifying*, which filter dangerous substances from the air; and (2) *air-supplying*, which deliver a supply of safe breathing air from a tank (SCBA), or group of tanks (cascade system), or an uncontaminated area nearby via a hose or airline to your mask.

Because respirators are not as consistently dependable as engineering and work practice controls and may create additional problems, they are not the preferred method of reducing exposures below the occupational exposure levels. Accordingly, their use as a primary control is restricted to certain circumstances. In those circumstances where engineering and work practice controls cannot be used to reduce airborne contaminants below their occupational exposure levels (e.g., certain maintenance and repair operations, emergencies, or during periods when engineering controls are being installed), the use of respirators could be justified to reduce worker exposure. In other cases, where work practices and engineering controls alone cannot reduce exposure levels to below the occupational exposure level, the use of respirators would be essential for supplemental protection.

If the environmental professional responsible for safety and health in the workplace determines that respiratory protection is required, then it is incumbent upon him or her to implement a written respiratory protection program that follows OSHA's Respiratory Protection Standard (29 CFR 1910.134).

Respirators can only provide adequate protection if they are properly selected for the task; are fitted to the wearer and are consistently donned and worn properly; and are properly maintained so that they continue to provide the protection required for the work situation. These variations can only be controlled if a comprehensive respiratory protection program is developed and implemented in each workplace where respirators are used. When respirator use is augmented by an appropriate respiratory protection program, it can prevent fatalities and illnesses from both acute and chronic exposures to hazardous substances.

We have continuously stressed the vital need to attempt first to engineer-out any hazard. However, when engineering and other methods of control are not feasible, proper selection and use of respiratory protection can be used to protect against airborne hazards.

Unlike past practices, where respiratory protection entailed nothing more than providing respirators to workers who could be exposed to airborne hazards and expecting workers to use the respirator to protect themselves, today, supplying respirators without the proper training, paperwork, and testing is illegal. Employers are sometimes unaware that by supplying respirators to their employees without having a comprehensive respiratory protection program. This is a serious mistake—because by issuing respirators, they have implied that a hazard actually exists. In a lawsuit, they then become fodder for the lawyers.

OSHA mandates that an effective program must be put in place. This respiratory protection program must not only follow OSHA's guidelines but must also be well planned and properly managed. A well-planned, well-written respiratory protection program must include the elements listed below.

- Procedures for selecting respirators for use in the workplace
- Medical evaluations of employees required to use respirators
- Fit testing procedures for tight-fitting respirators
- Use of respirators in routine and reasonably foreseeable emergency situations
- Procedures and schedules for cleaning, disinfecting, storing, inspecting, repairing, and otherwise maintaining respirators
- Procedures to ensure adequate air quality, quantity, and flow of breathing air for atmosphere-supplying respirators
- Training of employees in the respiratory hazards to which they are potentially exposed
- Training of employees in the proper use of respirators, including putting on and removing them, any limitations on their use, and maintenance procedures
- Procedures for regularly evaluating the effectiveness of the program.

In this section, we discuss these elements and explain what they require by providing a sample written Respiratory Protection Program. Although each individual written program must be site-specific and germane to existing conditions, this sample information will aid the industrial hygienist to implement a respiratory protection program that complies with OSHA requirements.

NOTE: For permit-required confined space entry operations, respiratory protection is a key piece of safety equipment, one always required for entry into an Immediately Dangerous to Life or Health (IDLH) space, and one that must be readily available for emergency use and rescue if conditions change in a non-IDLH space. Remember, however, that *only air-supplying respirators should be used in confined spaces where there is not enough oxygen.*

Selecting the proper respirator for the job, the hazard, and the worker is very important, as is thorough training in the use and limitations of respirators. Compliance with OSHA's Respiratory Standard begins with developing written procedures covering all applicable aspects of respiratory protection.

ENGINEERING DESIGN AND CONTROLS FOR SAFETY

An engineer is charged with the responsibility for designing a new seatbelt that is comfortable, functional, inexpensive, and easy for factory workers to install. He designs a belt that meets all these requirements and it is installed in 10,000 new cars. As the cars are bought and accidents begin to occur, it becomes apparent that the new seatbelt fails in crashes involving speeds over 36 miles per hour. The engineer that designed the belt took all factors into consideration except one: *safety.*

D.L. Goetsch, *Occupational Safety and Health* (1996, pp. 147–48).

> The cost-effectiveness of safety is open to constant debate. What IS the most cost-effective way? "Dependent upon the individual situation" is often the answer. But in most cases, for the best long-term results, engineered controls put into place at the earliest possible stage is the most cost-effective, in terms of dollars and in terms of worker health and safety.
>
> **Frank R. Spellman (1998)**

The enduring question is "What is the best way employers can ensure the safety and health of their employees?" In our experience, the most common answer is "Provide them as much safety protective equipment as possible." When we ask specifically "What type of safety protective equipment do you mean?" We generally hear the same thing over and over again. "You know safety protective equipment like eye and face protection, head protection, hand protection, respiratory protection, fall protection and electrical protection." By now the user of this handbook should know that these respondents are referring to Personal Protective Equipment (PPE). You should also know that notwithstanding the efficacy of PPE in protecting employees on the job, PPE is the protection of the last resort. Engineering and/or Administrative Controls are always the preferred and recommended methods of protecting workers on the job—PPE should be used only when the other two are not feasible or impossible since PPE does not eliminate the hazardous condition. PPE is used to establish a barrier between the exposed employee and the hazard to reduce the probability and severity of an injury. The real work of safety engineering is in ensuring worker safety and health by identifying and implementing engineering and administrative controls and safe work practices.

The question remains "What is the best way employers can ensure the safety and health of their workers?" Following the guidelines provided in OSHA's standards is the best way. But let's take a broad overview of the problem(s) involved. Because most companies operate with one primary goal in mind (to make a profit), companies try to operate in the most cost-effective manner possible. What is the most cost-effective manner possible? Some would answer, quite simply, cut costs, cut costs, cut costs—because, obviously, costs are the steady state of concern of the business world. Typically, the cost of ensuring the safety of a workforce is seen as an add-on cost—a cost that does not contribute to the bottom line. Though this view is shortsighted, it is a view commonly held in the industry today. Why do managers feel that safety is a burden, analogous to taking money and dumping it down an endless drain?

This isn't a simple question to answer, and it becomes even more complex, complicated, and compounded when you factor in other issues. For example, law (OSHA) regulates the safety and health of workers—should the company manager think strictly in terms of OSHA or other regulations? Or should they take a broader view?

Because workplace environments have become more technologically complex, we feel the need for protecting workers from safety and health hazards becomes more pressing, more complex, and more necessary. This requires the manager to take the broader view and to make choices and decisions that require a broad background and greater level of knowledge.

Fortunately, most company executives eventually come to share this view (i.e., that being aware of and concerned with employee safety and health is important), either of their own volition or via the results of regulatory pressure. When this occurs (if it does), the focus shifts from "we must comply; we must ensure the safety and health of our workers" to the original question—"What is the best way to ensure the safety and health of our employees?" This question naturally leads to other questions. "Shall we adopt elaborate engineering controls or trust to the effectiveness of personal protective equipment?" "Should we undergo complete process or hardware redesign, or simple modification of existing systems?" The point is, in either case (in all cases), costs are still the main factor—the bottom line. The question shifts back to "What are the costs and benefits?" Others would take this question a step further and include "What are the limitations and risks of each possible approach?"

Industrial Hygiene

This section attempts to answer some of these questions for the manager (and the safety engineer). To provide the broadest possible grasp of the issue, we concentrate on plant design and layout for safety, using engineering controls instead of personal protective equipment to ensure the safety and health of workers.

Simply put, we feel that effective safety and health engineering begins long before the worker appears in the workplace—a critical fact that astute planners, managers, design engineers and safety engineers must remember. Why? Because it is less costly (remember, reducing costs is the bottom line) and more efficient to correct safety and health hazards (engineer them out—that is, eliminate them before they exist) before they become part of the workplace.

Is this an idealistic, impossible approach? Does it ever happen this way in the real world? Why not? The method for achieving it is relatively simple: proper attention to safety and health in the design phase is the answer. The environmental engineer's primary function in the workplace is to reduce or eliminate hazards. This can be accomplished through proper design and layout of the plant or facility. For an environmental engineer working for a company that is building or renovating a facility, this is a golden opportunity to do just that.

We first discuss codes and standards, then physical plant layout, illumination, high hazard areas, personal services and sanitation facilities, and finish with the concept of system engineering—all of which are important in the planning and design phase but are especially important in ensuring effective accident prevention.

CODES AND STANDARDS

Again, probably the first known written admonition regarding the need for accident prevention is contained in Hammurabi's Code, about 1750 B.C. It states: "If a builder constructs a house for a person and does not make it firm and the house collapses and causes the death of the owner, the builder shall be put to death. Today some of us would say that the justice rendered in such a case is rather severe. Though true that the penalties have become less severe, the need for care has increased exponentially with the growth of technology since Hammurabi's time.

Countless pages have been written relating laws, standards, and codes regarding safety, health, and the environment since Hammurabi's Code. The fact of the matter is codes and standards have become essential tools in any plan of operations—and in the design of any workplace. Standards and codes have as their primary intent to prescribe minimum requirements and controls to safeguard life, property, or public welfare from hazards.

To understand codes and standards used in accident control, you'll need to learn a few pertinent definitions (provided by Hammer (1989):

Criterion: any rule or set of rules that might be used for control, guidance, or judgment.

Standard: a set of criteria, requirements, or principles.

Code: a collection of laws, standards, or criteria relating to a particular subject, such as the National Electric Code (NEC), the Uniform Fire Code (UFC), or Building Officials and Code Administrators (BOCA) National Fire Prevention Code.

Regulation: a set of orders issued to control the conduct of persons within the authority of the regulatory authority.

Specification: a detailed description of requirements, usually technical.

Practice: a series of recommended methods, rules, or designs, generally on a single subject.

Design handbooks, guide, or manuals: contain non-mandatory practices, general concepts, and examples to assist a designer or operator (p. 74).

Let us point out that local or state laws also have many ordinances governing specific requirements that cover such items or systems as fire sprinklers, fire alarms, exhaust and ventilation systems, emergency lighting, and means of egress. City, county, state, and federal agencies may have specific standards for sanitation, building construction, and pollution control and prevention requirements. Criteria contained in such standards and other work rules and in building and operating permits can be extremely beneficial in accident prevention. Written standards aid in making designer's (and safety engineer's) jobs easier, by providing useful technical information and promoting consistency to provide a basic level of safety in similar operations, material, and equipment.

A large number of standards and voluntary safety codes (consensus standards) have been incorporated into law. The best-known example of this practice is the American National Standards Institute (ANSI)—many of the original OSHA standards originated from ANSI standards. ANSI has a wide range of standards for such items as ladders, stairs, sanitation, building load design, floor, and wall openings, marking hazards, accident prevention signs, and many others. The design and safety engineer must keep in mind that standards provided by ANSI, the NEC, National Fire Protection Association (NFPA), American Society of Mechanical Engineers, and others are only recommendations—a starting point for safe workplace design—but the design engineer and/or safety engineer/or environmental engineer who does not pay attention to various codes, standards, and local requirements is setting him or herself up for admonitions that may not be quite as severe as the ones recommended by Hammurabi, but a headache generator at the very least.

PLANT LAYOUT

During the design phase for a plant or facility, and especially for general working areas, several elements must be taken into consideration. With safety and efficient use of materials in various processes and methods as the primary goal, the location, size, shape, and layout of worksite buildings should be determined. Designers and safety/environmental engineers have learned from experience (generally from past mistakes) that when the worksite functions to produce a finished product, designing the worksite so that raw materials enter at one end of the worksite and the finished product is shipped at the other is more efficient, and sometimes safer. What we have basically described here is process flow—an important consideration that should not be overlooked. For example, consider the following; if a certain process calls for robotic welding to be conducted during a product's assembly phase, a process flow diagram should indicate this—to ensure that hazardous materials such as flammable cleaning solvents, gasoline, and/or explosives are not staged or stored in such an area. This is critical to ensuring safe operations. Process flow diagrams also aid in the proper positioning of equipment, electrical apparatus, heating, ventilation, and air conditioning (HVAC), storage spaces, and other appurtenances or add-ons.

ILLUMINATION

Care must be taken with lighting design—not only to ensure that enough lighting is provided for workers to perform their work tasks safely and efficiently but also to ensure that the lighting does not interfere with work or cause visual fatigue. To ensure that the proper quantity or amount of illumination (usually measured in foot candles) is installed in the worksite, you must determine exactly what kind of work is to be performed in the space. The amount of illumination will vary with the job function. Experience has shown that a lack of proper illumination (poor illumination) in various industrial areas (including office areas) is listed as a common cause of accidents. ANSI in its Practice for Office Lighting (ANSI/IER RP7, 1-1982/1983) lists the minimum levels of illumination for various industrial areas and tasks, in Table 11.10.

One aspect of lighting that is often overlooked in the design phase is emergency lighting. No one doubts that an emergency of just about any size is apt to involve the loss of electrical power, which, of course, would mean that shutdown of equipment and processes, evacuation of workers, and rescue must be performed in darkness—unless emergency lighting is provided. The design engineer should at least incorporate into the workplace design standby sources of light that come on automatically when the power fails, if only to allow for safe evacuation. Whatever type of emergency lighting system is chosen, remember that it must be designed to operate from an independent connection at the point where the main service line enters the workplace.

TABLE 11.10
Minimum Levels for Industrial Lighting

Area	Foot-Candles
Assembly-rough, easy seeing	0
Assembly-medium	100
Building construction-general	10
Corridors	20
Drafting rooms-detailed	200
Electrical equipment, testing	100
Elevators	20
Garages-repair areas	100
Garages-traffic areas	20
Inspection, ordinary	50
Inspection, highly difficult	200
Loading platforms	20
Machine shops-medium work	100
Materials-loading, trucking	20
Offices-general areas	100
Paint dipping, spraying	50
Service spaces-wash rooms	30
Sheet metal-presses, shears	50
Storage rooms-inactive	5
Storage rooms-active, medium	20
Welding-general	50
Woodworking-rough sawing	30

HIGH HAZARD WORK AREAS

Work areas involved with process operations typically include some areas or operations that have an inherent high hazard potential. Such areas may require special precautionary measures and planning, such as the need for sprinkler systems, containment dikes, alarms, electrical interlocks, and other precautionary measures. These areas include:

- Spray-painting areas;
- Explosives manufacturing, use, or storage;
- Manufacturing, use, or storage of flammable materials;
- Areas with process equipment of high-energy movement through a power source, such as steam, electrical, hydraulic, pneumatic, and mechanical;
- Radiation areas;
- Confined spaces;
- Chemical mixing areas.

PERSONAL AND SANITATION FACILITIES

Not only must the design engineer factor into any workplace design several sanitation and personal hygiene requirements (i.e., provisions for potable water for drinking and washing; sewage, solid waste, and garbage disposal; sanitary food services; and drinking fountains, washrooms, locker rooms, toilets, and showers), he or she must also design the facility for easy and correct housekeeping activities. Housekeeping and sanitation are closely related. Control of health hazards requires sanitation, and control is usually put into place through good housekeeping practices. While true that disease transmission and ingestion of toxic or hazardous materials are controlled through a variety of sanitation practices, it is also true that if the workplace is not properly designed with correct sanitary and storm sewers, availability of safe drinking water and sanitary dispensing equipment, then sound sanitary practices are made much more difficult to include within the workplace.

NOTE

1. Information in this section is primarily from OSHA's (2014) *Modification of the Hazard Communication Standard (HCS) to conform with the United Nations' (UN) Globally Harmonized System of Classification and Labeling of Chemicals (GHS)*. Accessed 10/31/2014 @ https://www.osha.gov/dsg/hazcome/hazcome.faq.html.

REFERENCES

Alpaugh, E.L. (1998). (Revised by T.J. Hogan) *Fundamentals of Industrial Hygiene*, 3rd ed. Chicago, IL: National Safety Council, pp. 259–260.
American Conference of Governmental Industrial Hygienists (ACGIH) (1992). *Threshold Limit Values for Chemical Substances and Physical Agents and Biological Exposure Indices*. Cincinnati, OH: American Conference of Governmental Industrial Hygienists.
American National Standards Institute (2014), 1430 Broadway, New York.
Manual on Uniform Traffic Control Devices for Streets and Highways, ANSI D6.1
Life Safety Code, A9.1 (NFPA 101)
Minimum Requirements for Sanitation in Places of Employment, Z4.1
Safety Requirements for Construction, A10 Series
Buildings and Facilities—Providing Accessibility and Usability for Physically Handicapped People, A117.1.
29 CFR 1910.96, OSHA.
29 CFR 1910.97, OSHA.
Goetsch, D.L. (1996). *Occupational Safety and Health*. Englewood Cliffs, NJ: Prentice-Hall.
Greaney, P.P. (2000). *Ensuring Employee Safety in Cold-Weather Working Environments*. Work Care. Accessed at www.workcare.com/Archive/News_Art_2000_Dec14.htm.
Hammer, W. (1989). *Occupational Safety Management and Engineering*, 4th ed. Englewood Cliffs, NJ: Prentice-Hall.
Hazard, W.G. (1988). Industrial ventilation. In: Plog, B. (ed.) *Fundamentals of Industrial Hygiene*, 3rd ed. Chicago, IL: National Safety Council.
McNeel, S. & Kreutzer, R. (1996). Fungi and indoor air quality. *Health and Environment Digest* 10(2), 9–12.
NIOSH (1973). *The Industrial Environment, Its Evaluation and Control*. Cincinnati, OH: NIOSH.
NIOSH (2004). *NIOSH Respirator Selection Logic*. The National Institute for Occupational Safety and Health. Cincinnati, OH: US Department of Health.
NIOSH (2014). *Manual of Analytical Methods*. Cincinnati, OH: US Department of Health.
Olishifski, J.B. & Plog, B.A. (1988). Overview of Industrial Hygiene. In: Plog, B.A. (ed.) *Fundamentals of Industrial Hygiene*, 3rd ed. Chicago, IL: National Safety Council.
OSHA Fact Sheet 98-55—*Protecting Workers in Cold Environments*. Washington, DC: U.S. Department of Labor. Occupational Safety and Health Administration.
OSHA (1998). *Informational Booklet on Industrial Hygiene, OSHA 3143*. Washington, DC: U.S. Department of Labor. Accessed June 2005 @ [www.osha.gov].
OSHA (2003). *OSHA Technical Manual 4: Heat stress*. Washington, DC: U.S. Department of Labor. Accessed at www.osha.gov.
Sanders, M.S. & McCormick, E.J. (1993). *Human Factors in Engineering and Design*. New York: McGraw-Hill.
Spellman, F.R. (1998). *Surviving an OSHA Audit*. Boca Raton, FL: CRC Press.

12 Green Engineering

If the hole we've dug ourselves into is consumerism, it doesn't particularly help that your shovel is a fashionable shade of green.

Earon Davis

Science is about understanding; engineering is about performing.

Recently, one of the major new buzzwords to emerge is Green Energy Jobs. Most supporters give thumbs up to the concept and many pundits profess in their authoritative manner the benefits to be derived from such practices. The standard verbiage they put forward touts the benefits of cleaning up the environment, controlling global warming and creating an entirely new sector of employment. In the current economy, it is difficult to argue against the benefits derived from Green Energy Jobs: decent wages, a decent career path with upward mobility, and reduction of pollution and waste to benefit the environment. On the surface all of these statements make sense. However, I feel there is a more important reason to create and sustain Green Energy Job growth and viability. That is, when Western economies begin to rebound from single- and double-dip recessions the bounce to be achieved will be limited by a heavy lid placed over such advances by our continuing appetite for liquid hydrocarbon fuels. The lion's share of these fuels, of course, are controlled by nations and entities unfriendly to us. Thus, it is my position that Green Jobs are not just important for the environment and for creating jobs, but also to wean us away from those who choose to be greedy, and hate everything western, who are controlling the oily flow from the tap. Again, as soon as the West begins to recover economically (if it ever does), the tap-controllers will simply raise the price of oil—and this is the bottom line as to why Green Energy Jobs are important to all of us in the West. We need to shut down the foreign tap.

Frank R. Spellman (2012)

INTRODUCTION

Green energy is a hot topic today. Shifting from fossil fuels to renewable green energy is mentioned in written and spoken news media sources almost on a daily basis. When people hear about or think about green energy, it is not uncommon for thoughts of wind turbines and solar power sources being forefront in their minds. This is understandable because wind and sun power sources are quite visible and often mentioned when it comes to green energy. While wind and sun sources of green energy are certainly major sources of renewable green energy, there are additional sources of green energy such as hydropower and others.

Other than mentioning the sources of green energy, many people can't state definitively and with accuracy what green energy is or is all about. Simply put, green energy is derived from green engineering, actually it is more correct to state that green energy/green engineering is a subset of sustainable engineering. Sustainable engineering develops new technologies focused on societal needs and wants, within the limitations imposed by natural resources and environmental systems, in one of the most pressing challenges of the twenty-first century. Environmental engineers will play a central role in addressing the challenges.

Speaking of challenges, there are several challenges that are the focus of environmental engineers. These focuses range from water supplies, food production, sanitation and waste management, housing and shelter, transportation, industrial processing, developing natural resources, cleaning up polluted waste sites, restoring natural environments such as forests, lakes, steams and wetlands, and other areas; however, the one major focus not mentioned to this point is energy development—that is safe, environmentally safe energy development.

Environmental engineers in energy development could be involved in the design, commercialization, and use of processes and products in a way that reduces pollution, promotes sustainability, and minimizes risk to human health and the environment without sacrificing economic viability and efficiency. Implementing anything to any project in the early planning stage is the key to ensuring that environmental factors are included and also have the greatest impact and cost-effectiveness—it is always less expensive to incorporate requirements in the planning, staging and construction phase than to go back later and add on to what should have been included in the first place.

WIND ENERGY[1]

Wind turbines generate electricity from wind and are being manufactured and installed all across the nation. *Wind energy* is the movement of wind to create power. Since early recorded history, people have been harnessing the energy of the wind for milling grain, pumping water, and other mechanical power applications. Wind energy propelled boats along the Nile River as early as 5000 B.C. By 200 B.C., simple windmills in china were pumping water, while vertical-axis windmills with woven reed sails were grinding grain in Persia and the Middle East.

The use of wind energy spread around the world and by the eleventh century, people in the Middle East were using windmills extensively for food production; returning merchants and crusaders carried this idea back to Europe. The Dutch refined the windmill and adapted it for draining lakes and marshes in the Rhine River Delta. When settlers took this technology to the New World in the later nineteenth century, they began using windmills to pump water for farms and ranches, and later, to generate electricity for homes and industry. Today, there are several hundred thousand windmills in operation around the world, and many of which are used for water pumping. But it is the use of wind energy as a pollution-free means of generating electricity on a significant scale that is attracting the most current interest in the subject. As a matter of fact, with the present and pending shortage and high cost of fossil fuels to generate electricity and the green movement toward the use of cleaner fuels, wind energy is the world's fastest growing energy source and will power industry, businesses, and home with clean, renewable electricity for many years to come. In the United States since 1970, wind-based electricity generating capacity has increased markedly although (at present) it remains a small fraction of total electric capacity. But this trend is beginning to change—with the advent of $4-7/gal of gasoline, high heating and cooling costs, and subsequent increases in the cost of electricity, worldwide political unrest or uncertainty in oil-supplying countries, one only needs to travel the "wind corridors" of the U.S. encompassing parts of Arizona, New Mexico, Texas, Missouri and north through the Great Plains to the Pembina Escarpment and Turtle Mountains of North Dakota and elsewhere witness the considerable activity the seemingly exponential increase in wind energy development and wind turbine installations; these machines are being installed to produce and provide electricity to the grid.

When you get right down to it, we can classify wind energy as a form of solar energy. Winds are caused by uneven heating of the atmosphere by the Sun, irregularities of the Earth's surface, and the rotation of the Earth. As a result, winds are strongly influenced and modified by local terrain, bodies of water, weather patterns, vegetative cover, and other factors. The wind flow, or motion of energy when harvested by wind turbines, can be used to generate electricity.

As with any other source of energy, non-renewable or renewable, there are advantages and disadvantages associated with their use. On the positive side, it should be noted that wind energy is a free, renewable resource, so no matter how much is used today, there will still be the same supply in the future. Wind energy is also a source of clean, non-polluting, electricity. One huge advantage of wind energy is that it is a domestic source of energy, produced in the U.S. or country where installed and where wind is abundant. In the U.S., the wind supply is abundant.

Wind turbines can be installed on farms or ranches, thus benefiting the economy in rural areas, where most of the best wind sites are found. Moreover, farmers and ranchers can continue to work the land because the wind turbines use only a fraction of the land.

On the other side of the coin, wind energy does have a few negatives and this is where engineering comes into play after the initial design and installation. Environmental, green, sustainable engineers need to look at and be involved with correcting or eliminating or reducing the negatives of wind power. Wind power must compete with conventional generation sources on a cost basis. Even though the cost of wind power has decreased dramatically in the past 10 years, the technology requires a higher initial investment than fossil-fueled generators. The challenge to using wind as a source of power is that the wind is intermittent, and it does not always blow when electricity is needed. Wind energy cannot be stored (unless batteries are being used), and not all winds can be harnessed to meet the timing of electricity demands. Another problem is that good sites are often located in remote locations, far from cities where the electricity is needed. Moreover, wind resource development may compete with other uses for the land and those alternative uses may be more highly valued than electricity generation. Finally, in regard to the environment, wind power plants have relatively little impact on the environment compared to other conventional power plants, there is some concern over the noise produced by the rotor blades, aesthetic (visual) impacts, and sometimes birds have been killed by flying into the rotors. Most of these problems have been resolved or greatly reduced through technological development or by properly siting wind plants. Again, environmental engineering plays a key role in technological development and proper siting of wind plants or farms.

In regard to wind energy and its future, one thing is certain; it continues to be one of the fastest growing energy technologies and it looks set to become a major generator of electricity throughout the world.

ENVIRONMENTAL IMPACT OF WIND TURBINES

The Good, Bad, and Ugly of Wind Energy

Good: As long as Earth exists, the wind will always exist. The energy in the winds that blow across the U.S. each year could produce more than 16 billion GJ of electricity—more than one and on-half times the electricity consumed in the U.S. in 2000.

Bad: Turbines are expensive. Wind doesn't blow all the time, so they have to be part of a larger plan. Turbines make noise. Turbine blades kill birds.

Ugly: Some look upon giant wind turbine blades cutting through the air as grotesque scars on the landscape, visible polluters.

The bottom line: Do not expect Don Quixote, mounted in armor on his old nag, Rocinante, with or without Sancho Panza, to charge those windmills. Instead, expect—you can count on it, bet on it, and rely on it—that the charge to build those windmills will be done by the rest of us, to satisfy our growing, inexorable need for renewable energy. What other choice do we have?

Frank R. Spellman (2012)

Air Quality (Including Global Climate Change and Carbon Footprint)

Emissions resulting from construction, installation and operation of wind turbines and associated activities include vehicle emissions, diesel emissions from large construction equipment and generators; volatile organic compound (VOC) releases from storage and transfer of vehicle/equipment fuels, small amounts of carbon monoxide, nitrogen oxides, and particulates from blasting activities, and fugitive dust. Fugitive dust would be caused by:

- Disturbing and moving soils (clearing, grading, excavation, trenching backfilling, dumping, and truck and equipment traffic)
- Mixing concrete and associated storage piles
- Drilling and pile driving.

A construction permit is needed from the state or local air agency to control or mitigate these emissions; therefore, these emissions would not likely cause an exceedance of air quality standards or have an impact on climate change.

CULTURAL RESOURCES

Direct impacts on cultural resources could occur from construction activities, and indirect impacts might be caused by soil erosion and increased accessibility to possible site locations. Potential impacts include:

- Complete destruction of the resource if present in areas undergoing surface disturbance or excavation
- Degradation or destruction of near-surface cultural resources on- and off-site resulting from topographic or hydrological pattern changes, or from soil movement (removal, erosion, sedimentation). (Note: the accumulation of sediment could protect some localities by increasing the amount of protective cover)
- Unauthorized removal of artifacts or vandalism at the site could occur as a result of increase in human access to previously inaccessible areas, if significant cultural resources are present
- Visual impacts resulting from vegetation clearing, increases in dust, and the presence of large-scale equipment, machinery, and vehicles (if the resources have an associated landscape component that contributes to their significance, such as a sacred landscape or historic trail).

ECOLOGICAL RESOURCES

Ecological resources that could be affected include vegetation, fish, and wildlife, as well as their habitats. Adverse ecological effects could occur during construction from:

- Erosion and runoff
- Fugitive dust
- Noise
- Introduction and spread of invasive vegetation
- Modification, fragmentation, and reduction of habitat
- Mortality of biota (i.e., death of plants and animals)
- Exposure to contaminants
- Interference with behavioral activities.

Site clearing and grading, along with the construction of access roads, towers, and support facilities, could reduce, fragment, or dramatically alter existing habitat in the disturbed portions of the project area. Ecological resources would be most affected during construction by the disturbance of habitat in areas near turbines, support facilities, and access roads. Wildlife in surrounding habitats might also be affected if the construction activity (and associated noise) disturbs normal behaviors, such as feeding and reproduction.

Green Engineering

WATER RESOURCES (SURFACE WATER AND GROUNDWATER)

Water Use

Water would be used for dust control when clearing vegetation and grading, and for road traffic; for making concrete for foundations of towers, substations, and other buildings; and for consumptive use by the construction crew. Water could be trucked in from off-site or obtained from local groundwater wells or nearby surface water bodies, depending on availability.

Water Quality

Water Quality could be affected by:

- Activities that cause soil erosion
- Weathering of newly exposed soils could cause leaching and oxidation, thereby releasing chemicals into the water
- Discharge of waste or sanitary water
- Pesticide applications.

Flow Alteration

Surface and groundwater flow systems could be affected by withdrawals made for water use, wastewater and stormwater discharges, and the diversions of surface water flow for access road construction of stormwater control systems. Excavation activities and the extraction of geological materials could affect surface and groundwater flow. The interaction between surface water and groundwater could also be affected if the surface water and groundwater were hydrologically connected, potentially resulting in unwanted dewatering or recharging of water resources.

LAND USE

Impacts on land use could occur during construction if there were conflicts with existing land use plans and community goals; conflicts with existing recreational, educational, religious, scientific, or other use areas: or conversion of the existing commercial and use for the area (e.g., mineral extraction).

During construction, impacts on most land uses would be temporary, such as the removal of livestock from grazing areas during blasting or heavy equipment operations, or temporary effects to the character of a recreation area because of construction noise, dust, and visual intrusions. Long-term land use impacts would occur if existing land uses are not compatible with wind energy development, such as remote recreational experiences; however, those uses could potentially be resumed if the land is reclaimed to pre-development conditions.

When wind farm construction spreads, local opposition to the mass towers (some over 400 ft tall) impacts landowners' opinions, land values and state regulations. However, this local opposition is relative in the sense that opinions are based on the receipt of or nonreceipt of monetary rewards for use of the land. That is, the opposition is based on whether the wind farms in a local area constructed on personal property accrue a monetary reward or usage fee for property owners for allowing the presence of turbines on their land. This view is typically different for residents who do not reap an economical benefit from the presence of wind turbines in their backyards.

Impacts on aviation could be possible if the project is located within 20,000 ft (6,100 m) or less of an existing public or military airport, or if the proposed construction involves objects greater than 200 ft (61 m) in height. The Federal Aviation Administration (FAA) must be notified if either of these two conditions occurs, and the FAA would be responsible for determining if the project would adversely affect commercial, military, or personal air navigation safety. Similarly, impacts on military operations could occur if a project were located near a military facility if that facility conducts low-altitude military testing and training activities.

Soils and Geologic Resources (Including Seismicity/Geo Hazards)

Sands, gravels, and quarry stone would be excavated for construction access roads; for concrete for buildings, substations, transformer pads, foundations, and other ancillary structures; and for improving ground surface for laydown areas and crane staging areas.

Possible geological hazards such as landslides could be activated by excavation and blasting for raw materials, increasing slopes during site grading and construction of access roads, altering natural drainage patterns, and toe-cutting bases of slopes. Altering drainage patterns could also accelerate erosion and create slope instability.

Surface disturbance, heavy equipment traffic, and changes to surface runoff patterns could cause soil erosion and impact special soils (e.g., cryptobiotic soils). Impacts of soil erosion could include soil nutrient loss and reduced water quality in nearby surface water bodies.

Paleontological Resources

Impacts on paleontological resources could occur directly from the construction activities or indirectly from soil erosion and increased accessibility to fossil locations. Potential impacts include:

- Complete destruction of the resource if present in areas undergoing surface disturbance or excavation
- Degradation or destruction of near-surface fossil resources on- and off-site due to changes in topography, changes in hydrological patterns, and soil movement (removal, erosion, sedimentation). (Note: the accumulation of sediment could serve to protect some locations by increasing the amount of protective cover.)
- Unauthorized removal of fossil resources or vandalism to the site could occur as a result of increased human access to previously inaccessible areas if significant paleontological resources are present.

Transportation

Short-term increases in the use of local roadways would occur during the construction period. Heavy equipment likely would remain at the site. Shipments of materials are unlikely to affect primary or secondary road networks significantly, but this would depend on the location of the project site relative to the material source. Oversized loads could cause temporary transportation disruptions and could require some modifications to roads or bridges (such as fortifying bridges to accommodate the size or weight). Shipment weight might also affect the design of access roads for grade determinations and turning clearance requirements.

Visual Resources

Although many of us consider wind turbines to be visually acceptable and in some cases even pleasing to look at, wind turbines disturb the visual area of other people by creating negative changes in the natural environment. The test on whether a wind turbine or wind farm is a visual pollutant is to ask the question? How many of us would seriously like one or more wind turbines a few hundred feet or meters from our homes? It is important to remember that wind turbines can be anywhere from a few meters to a hundred meters high. Having a wind turbine tower over one's home is the last thing many residents want. Having said this, the possible sources of visual impacts during construction include the following:

- Road development (e.g., new roads or expansion of existing roads) and parking areas could introduce strong visual contrasts in the landscape, depending on the route relative to surface contours, and the width, length, and surface treatment of the roads.

- Conspicuous and frequent small-vehicle traffic for worker access and frequent large-equipment (e.g., trucks, graders, excavators, and cranes) traffic for road construction, site preparation, and turbine installation could produce visible activity and dust in dry soils. Suspension and visibility of dust would be influenced by vehicle speeds and road surface materials.
- Site development could be intermittent, staged, or phased, giving the appearance that works starts and stops. Depending on the length of time required for development, the project site could appear to be "under construction" for an extended period. This could give rise to perceptions of lost benefit and productivity, like those alleged for the equipment. Timing and duration concerns may result.
- There would be a temporary presence of large cranes or other large machines to assemble towers, nacelles, and rotors. This equipment would also produce emissions while operational and could create visible exhaust plumes. Support facilities and fencing associated with the construction work world also be visible.
- Ground disturbance and vegetation removal could result in visual impacts that produce contrast of color, form, texture, and line. Excavation for turbine foundations and ancillary structures; trenching to bury electrical distribution system; grading and surfacing roads; cleaning and leveling staging areas; and stockpiling soil and spoils (if not removed) would (1) damage or remove vegetation (2) exposed bare soil, and (3) suspend dust. Soil scars and exposed slope faces would result from excavation, leveling, and equipment movement. Invasive species could colonize disturbed and stockpile soils and compacted areas.

Socioeconomics

Direct impacts would include the creation of new jobs for workers (approximately two workers per megawatt) at wind energy development projects, and the associated income and taxes paid. Indirect impacts would occur as a result of the new economic development and would include new jobs at businesses that support the expanded workforce or provide project materials, and associated income taxes. Wind energy development activities could also potentially affect property value, either positively from increased employment effects or the image of "clean energy," or negatively form proximity to the wind farm and any associated or perceived environmental effect (noise, visual, etc.).

Adverse impacts could occur it a large in-migrant workforce, culturally different form the local indigenous group, is brought in during construction. This influx of migrant workers could strain the existing community infrastructure and social services.

Environmental Justice

If significant impacts occurred in any resource areas, and the impact disproportionately affected minority or low-income populations, then there could be environmental justice concerns. Potential issues during construction are noise, dust, and visual impacts form the construction site and possible impacts associated with the construction of new access roads.

Hazardous Materials and Waste Management

Solid and industrial waste would be generated during construction activities. The solid waste would likely be nonhazardous and consist mostly of containers, packing material, and wastes from equipment assembly and construction crews. Industrial wastes would include minor amounts of paints, coatings, and spent solvents. Hazardous materials stored on-site for vehicle and equipment maintenance would include petroleum fluids (lubricating oils, hydraulic fluid, fuels), coolants, and battery electrolytes. Oils, transmission fluids, and dielectric fluids would be brought to the site to fill turbine components and other large electrical devices. Also, compressed gases would be used for welding, cutting, brazing, etc. These materials would be transported off-site for disposal, but impacts could result if the wastes were not properly handled and released into the environment.

WIND ENERGY OPERATIONS IMPACTS

Typical activities during the wind energy facility operations phase include turbine operation, power generation, and associate maintenance activities that would require vehicular access and heavy equipment operation when large components are being replaced. Potential impacts from these activities are presented below, by the type of affected resource.

AIR QUALITY (INCLUDING GLOBAL CLIMATE CHANGE AND CARBON FOOTPRINT)

There are no direct air emissions from operating a wind turbine. Minor volatile organic compound (VOC) emissions are possible during routine maintenance activities of applying lubricants, cooling fluids, and greases. Minor amounts of carbon monoxide and nitrogen oxides would be produced during the periodic operation of diesel emergency generators as part of preventative maintenance. Vehicular traffic would continue to produce small amounts of fugitive dust and tailpipe emissions during the operations phase. These emissions would not likely exceed air quality standards or have any impact on climate change.

CULTURAL RESOURCES

Impacts during the operations phase would be limited to unauthorized collection of artifacts and visual impacts. The threat of unauthorized collection would be present once access roads are constructed in the site evaluation or construction phase, making remote lands accessible to the public. Visual impacts resulting from the presence of large wind turbines and associated facilities and transmission lines could affect some cultural resources, such as sacred landscapes or historic trails.

ECOLOGICAL RESOURCES

During operation, adverse ecological effects could occur from (1) disturbance of wildlife by turbine noise and human activity; (2) site maintenance (e.g., mowing); (3) exposure of biota to contaminants; and (4) mortality of birds and bats from colliding with the turbines and meteorological towers.

During the operation of a wind facility, plant and animal habitats could still be affected by habitat fragmentation due to the presence of turbines, support facilities, and access roads. In addition, the presence of an energy development project and its associated access roads may increase human use of surrounding areas, which could in turn impact ecological resources in the surrounding areas through:

1. Introduction and spread of invasive vegetation
2. Fragmentation of habitat
3. Disturbance of biota
4. Increased potential for fire

As discussed in detail later, the presence of a wind energy project (and its associated infrastructure) could also interfere with migratory and other behaviors of some wildlife.

WATER RESOURCES (SURFACE WATER AND GROUNDWATER)

Impacts on water use and quality and flow systems during the operation phase would be limited to possible degradation of water quality resulting from vehicular traffic and pesticide application if conducted improperly.

Green Engineering

Land Use

Impacts on land use would be minimal, as many activities would be minimal, as many activities can continue to occur among the operating turbines, such as agriculture and grazing. It might be possible to collocate other forms of energy development, provided the necessary facilities could be installed without interfering with operation and maintenance of the wind farm. Collocation of other forms of energy development could include directionally drill oil and gas wells, underground mining, and geothermal or solar energy development. Recreation activities (e.g., off-highway vehicle [OHV] use and hunting) are also possible, but activities centered on solitude and scenic beauty could be affected. Military operations and aviation could be affected by radar interference associated with the operating turbines, and low-altitude activities could be affected by the presence of turbines over 200 ft high.

Soils and Geologic Resources (Including Seismicity/Geo Hazards)

Following construction, disturbed portions of the site would be revegetated, and the soil and geologic conditions would stabilize. Impacts during the operations phase would be limited largely to soil erosion impacts caused by vehicular traffic for operators' maintenance.

Paleontological Resources

Impacts during the operations phase would be limited to the unauthorized collection of fossils. This threat is present once the access roads are constituted in the site evaluation or construction phases, making remote land accessible to the public.

Transportation

No noticeable impacts on transportation are likely during operations. Low volumes of heavy- and medium-duty pickup trucks and personal vehicles are expected for routine maintenance and monitoring. Infrequent, but routine shipments of component replacements during maintenance procedures are likely over the period of operation.

Visual Resources

Wind energy development projects would be highly visible in rural or natural landscapes, many of which have few other comparable structures. The artificial appearance of wind turbines may have visually incongruous "industrial" associations for some, particularly in a predominantly natural landscape; however, other viewers may find wind turbines visually pleasing and consider them a positive visual impact. Visual evidence of wind turbines cannot easily be avoided, reduced, or concealed, owing to their size and exposed location; therefore, effective mitigation is often limited.

Additional issues of concern are shadow flicker (strobe-like effects from flickering shadows cast by the moving rotors); blade glint from the Sun reflecting off moving blades; visual contrasts from support facilities, and light pollution from the lighting on facilities and towers(which are required safety features).

Additional visual impacts from vehicular traffic would occur during maintenance, and as towers, nacelles, and rotors are upgraded or replaced. When replacing turbines and other facility components, the opportunity, and pressures to break the uniformity of spacing between turbines and uniformity of size, shape, and color among facility components could increase visual contrast and visual "clutter."

Infrequent outages, disassembly, and repair of equipment may occur, producing the appearance of idle or missing rotors, "headless" towers (when nacelles are removed), and lowered towers, negative visual perceptions of "lost benefits" (e.g., loss of wind power) and "bone yards" (for storage) may result.

Socioeconomics

Direct impacts would include the creation of approximately one new job for 3 MW of installed capacity for operations and maintenance workers at wind energy development projects, and the associated income and taxes paid. Indirect impacts would occur from new economic development and would include new jobs at businesses that support the expanded workforce or that provide project materials, and associated income and taxes. Wind energy development activities could also potentially affect property values, either positively from increased employment effects or image of "clean energy," or negatively from proximity to the wind farm and any associated or perceived environmental effects (noise, visual, etc.).

Environmental Justice

Possible environmental justice impacts during operation include the alteration of scenic quality in areas of traditional or cultural significance to minority or low-income populations. Noise impacts and health and safety impacts are also possible sources of disproportionate effects.

Hazardous Materials and Waste Management

Industrial and sanitary wastes are generated during routine operations (e.g., lubricating oils, hydraulic fluids, coolants, solvents, cleaning agent, and sanitary wastewaters). These wastes are typically put in containers, characterized, and tabled, possibly stored briefly, and transported by a licensed hauler to an appropriate permitted off-site disposal facility as a standard practice. Impacts could result if these wastes were not properly handled and released into the environment. Releases could also occur if individual turbine components or electrical equipment were to fail.

Impact on Wildlife

Human-made and installed wind turbines are responsible for bird deaths. This has been a less documented impact of wind turbines and has mainly been argued by wildlife groups. The United States Fish and Wildlife Service (USFWS, 2014) points out noise standards, for example, for wind turbines developed by countries such as Sweden and New Zealand and some specific site-level standards implemented in the U.S. focus primarily on sleep disturbance and annoyance to humans. However, noise standards do not generally exist for wildlife, except in a few instances, where federally listed species may be impacted. Findings from recent research clearly indicate the need to better address noise-wildlife issues. As such, noise impacts on wildlife should clearly be included as a factor in wind turbine siting, construction, and operation. Later in this section, a detailed description of eagle conservation guidance with regard to land-based wind energy is presented. For now, it is important to point out some of the key issues which include (1) how wind facilities affect background noise levels; (2) how and what fragmentation, including acoustical fragmentation, occurs especially to species sensitive to habitat fragmentation; (3) comparison of turbine noise levels at lower valley sites—where it may be quieter—to turbines placed on ridge lines above rolling terrain where significant topographic sound shadowing can occur having the potential to significantly elevate sound levels above ambient conditions; and (4) correction and accounting of a 15 decibel (dB) underestimate from daytime wind turbine noise readings used to estimate nighttime turbine noise levels (van den Berg, 2004; Barber et al., 2010). The sensitivities of various groups of wildlife can be summarized as:

- Birds (more uniform than mammals) 100 Hz to 8–10 kHz; sensitivity at 0.10 dB
- Mammals < 10 Hz to 150 kHz; sensitivity to −20 dB
- Reptiles (poorer than birds) 50 Hz to 2 kHz; sensitivity at 40–50 dB
- Amphibians 100 Hz to 2 kHz; sensitivity form 10–60 dB

As mentioned, turbine blades at normal operating speeds can generate significant levels of noise. How much noise? Based on a propagation model of an industrial-scale 1.5 MW wind turbine at 263 ft hub height, positioned approximately 1,000 ft apart from neighboring turbines, the flowing decibel levels were determined for peak sound production. At a distance 300 ft from the blades, 45–50 dBA were detected; at 2,000 ft, 40 dBA; and at 1 mile, 30–35 dBA (Kaliski, 2009). Declines in densities of woodland and grassland bird species have been shown to occur at noise thresholds between 45 and 48 dB, respectively; while the most sensitive woodland and grassland species showed declines between 35 and 43 dB, respectively. Songbirds specifically appear to be sensitive to very low sound levels equivalent to those in a library reading room (~30 dBA) (Foreman and Alexander, 1998). Given this knowledge, it is possible that effects on sensitive species may be occurring at ≥1 mile from the center of a wind facility at periods of peak sound production.

As pointed out earlier, noise does not have to be loud to have negative effects. Very low-frequency sounds including infrasound (sound lower in frequency than 20 Hz) are also being investigated for their possible effects on both humans and wildlife. Wind turbine noise results in a high infrasound component (Salt and Hullar, 2010). Infrasound is inaudible to the human ear, but this unheard sound can trigger human annoyance, sensitivity, disturbance, and disorientation. For birds, bats, and other wildlife, the effects may be more profound. Noise from traffic, wind and operating turbine blades produce low-frequency sounds (<1–2 kHz; Dooling, 2002). Bird vocalizations are generally within the 2–5 kHz frequency range (Dooling and Popper, 2007) and birds hear best between 1–5 kHz (Dooling, 2002). Although traffic noise generally falls below the frequency of bird communication and hearing, several studies have documented that traffic noise can have significant negative impacts on bird behavior, communication, and ultimately on avian health and survival (e.g., Lohr et al., 2003; Lengagne, 2008, Barber et al., 2010). Whether these effects are attributable to infrasound effects or to a combination of other noise factors is not yet fully understood. The fact is little is known about the combination effect of traffic noise and wind turbine noise. However, given that wind-generated noise including blade turbine noise produces a fairly persistent, low frequency sound similar to that generated by traffic noise (Dooling, 2002) it is plausible that wildlife effects from these two sounds could be similar. It is also plausible that wildlife effects from these two sounds combined could be detrimental to the wildlife of all kinds. Based on experience, this book supports this view.

Some may feel that the combination of road noise and wind turbine noise causing wildlife effects is plainly a stretch. Although the author has studied, observed, and measured-monitored this phenomenon, the truth is little is known about the effects of noise related to road noise and wind turbines combined; moreover, at present, little to nothing on the subject is reported in the peer-reviewed literature.

Let's get back to why some people feel the statement that a combination of road noise and wind turbine noise has a wildlife effect is plainly a stretch. This point of view seems to be prevalent for those who do not travel the east-west U.S. Interstates from East Coast areas to the Southwestern, Pacific Northwest or and the north-south route Indiana I-65 and others. On many roads (Interstates) it is not uncommon to drive in any direction and to view hundreds of tall wind turbines off in the far distance ahead and to the right and left of the highway. On the other hand, some of these wind turbines are very close to the shoulder of the roads. However, again, many wind turbine farms are viewed off in the vast, far distance as one travels the highways. It is this perception of distance on some outlying, remote hillside or ridge in some unoccupied wildness or high plains area that gives the unknowing viewer this wrong view of combined noise from road traffic and wind turbines. They ask if the wind turbines are off in some remote corner of nowhere, how can road noise contribute to and be added to normal wind turbine noise? The reality is very easy; so is the answer. Remember, service roads are built to perform maintenance and preventive maintenance [i.e., inspection of components, servicing items that require some types of activities on a regular basis—retorquing bolts—and replacing consumable items at or before a specified age (i.e., replacing filters or changing the oil in the gearbox)] are accessed by light and heavy trucks and other vehicles on a routine basis—no matter the location. The larger the wind farm, the more access—the more traffic; the more noise. All access vehicles, including helicopters used to transport parts and personnel, produce noise; in some cases, a lot of noise.

The counter-argument, of course, is that wind turbines run on their own and require little to no operation and maintenance (O&M). If this is the case (and it is not), then there is very little traffic on their associated service roads, and thus little noise added to wind turbine noise. In the first place, wind turbines do not operate by themselves; they are normally operated remotely from a plant operations room by a human operator, but they can also be operated within the turbine nacelle. Their operation is also monitored by Supervisory Control and Data Acquisition (SCADA) communication system designed to alert the appropriate service personnel through computer warning and automated telephone calls.

Based on experience, the major causes of downtime were O&M works and faults. The occurrence of O&M and faults was variable. Some sites have downtown of no more than 43 hours per month and others as much as 127 hours (NREL, 2000). O&M downtime includes all troubleshooting, inspections, adjustments, retrofits, and repairs performed on the turbines. Faults generally require no more than a reset, and most can be performed remotely. Increased downtime for O&M and fault reasons means more traffic to and from the turbine farm.

Fast forwarding to wildlife effects resulting from wind turbine noise alone, it is important to point out that a bird's inability to detect turbine noise at close range may also be problematic. Note that the threshold for hearing in birds is higher than for humans at all frequencies and the overlap in the discernible frequencies between species indicates that birds do not filter out other species by simply being unable to detect themes (i.e., birds can hear songs of other species). In their environment birds must be able to discriminate their own vocalizations and those of other species apart from any background noise (Dooling, 1982). Calls are important in the isolation of species, pair bond formation, pre-copulatory display, territorial defense, danger, advertisement of food sources and flock cohesion (Knight, 1974).

For the average bird in a signal frequency of 1–4 kHz, noise must be 24–30 dB above the ambient noise level in order for a bird to detect it. As noted above, turbine blade and wind noise frequencies generally fall below the optimal hearing frequency of birds. Additionally, by the inverse square law the sound pressure level decrease by 6 dB with every doubling of distance. Therefore, although the sound level of the blade may be significantly above the ambient wind noise level and detectable by birds at the source, as the distance from the source increases and the blade noise level decreases toward the ambient wind noise level, a bird may lose its ability to detect the blade and risk colliding with the moving blade.

Some researchers have attempted to explain avian collisions with turbine blades on birds' inability to divide their attention between surveying the ground for prey and monitoring the horizon and above for obstacles; i.e., they are so busy searching the ground that they do not notice the turbines. This hypothesis derives from substituting our knowledge of human vision for that of avian vision. Humans are foveate animals; we search the visual world with a small area of the retina known as the fovea, which is our area of sharpest vision, like someone searching a dark room with a narrow-beam searchlight. This results from our very low ratio (approximately 1:1) of photoreceptors to ganglion cells in the macular region of the retina. Outside the macular region, the ratio of receptors to ganglion cells increases progressively to 50:1–100:1, and our visual acuity drops sharply. Birds and many other animals, on the other hand, have universal masculinity, which means that they have a low ratio of receptors to ganglion cells (4:1–8:1) out to the periphery of the retina. Observation over time indicates that they maintain good acuity even in peripheral vision. In addition, raptors possess the specialization of two foveal regions: one for frontal vision and one for looking at the ground. Moreover, birds have various optical methods for keeping objects at different distances simultaneously in focus on the retina (Hodos et al., 1992). Because of these considerations, failure to divide attention seems like an unlikely hypothesis.

ENERGY FROM THE SUN

The Sun nourishes our planet. When we consider the Sun and solar energy, we quickly realize that there is nothing new about renewable energy. The Sun was the first energy source; it has been around for 4.5 billion years, as long as anything else we are familiar with. On the Earth without the Sun there is nothing—absolutely nothing. The Sun provided light and heat to the first humans. During daylight, the people searched for food. They hunted and gathered and probably stayed together for safety. When nightfall arrived and in the dark, we can only imagine that they huddled together for warmth and safety in light of the stars and moon, waiting for the Sun and its life-giving and sustaining light to return.

Solar energy (a term used interchangeably with solar power) uses the power of the Sun, using various technologies, "directly" to produce energy. Solar energy is one of the best renewable energy sources available because it is one of the cleanest sources of energy. Direct solar radiation absorbed in solar collectors can provide space heating and hot water. Passive solar can be used to enhance solar energy use in buildings for space heating and lighting requirements. Solar energy can also be used to produce electricity, and this is the green (renewable) energy area that is the focus of attention in this section.

ENVIRONMENTAL IMPACTS OF SOLAR ENERGY

All energy-generating technologies, including solar technologies, affect the environment in many ways. Solar energy has some obvious advantages in that the source is free; however, the initial investment in operating equipment is not. Solar energy is also environmentally friendly, requires almost no maintenance, and reduces our dependence on foreign energy supplies. Probably the greatest downside of solar energy use is that in areas without direct sunlight during certain times of the year, solar panels cannot capture enough energy to provide heat for home or office. Geographically speaking, the higher latitudes do not receive as much direct sunlight as tropical areas. Because of the position of the Sun in the sky, solar panels must be placed in sun-friendly locations such as the U.S. Desert Southwest and the Sahara region of northern Africa.

Land Use/Siting Impact

Most of the land used for larger utility-scale solar facilities, depending on their location, can raise concerns about land degradation and habitat loss. This is the case even if abandoned industrial, fallow agricultural, or former mining sites are used. Total land area requirements vary depending on the technology, the topography of the site, and the intensity of the solar resource. Estimates for utility-scale PV systems range from 3.5 to 10 acres per megawatt, while estimates for CSP facilities are between 4 and 16.5 acres per megawatt (UCS, 2013).

Unlike wind facilities, there is less opportunity for solar projects to share land with agricultural uses. However, land impacts for utility-scale solar systems can be minimized by siting them at lower-quality locations such as brownfields, existing transportation, and transmission corridors, or, as mentioned earlier, abandoned mining land (AML) (USEPA, 2011).

Water Use Impact (Water Footprint)

Like with the production of energy from all other sources, in one way or another, solar energy production has a water footprint. *Water footprint* is defined as the total volume of freshwater used to produce energy and the services consumed by the production process. Water consumption for solar generation varies by technology and location. For our purposes in this text, water consumption is defined as the amount of water that is "evaporated, transpired, incorporated into products or crops, consumed by humans or livestock, or otherwise removed from the immediate water environment" (Kenny et al., 2009). Water consumption is distinct from water withdrawal. Water withdrawal is the total amount of "water removed from the ground or diverted from a surface-water source for use"

(Kenny et al., 2009), but which may be returned to the sources. Both water withdrawal and consumption are important metrics, but consumption is a very useful metric for water-scarce regions, especially in the context of future resource development, because consumption effectively removes water from the system, so it is not available for other uses (e.g., agriculture or drinking).

Hazardous Waste Impact

Like all other technologies, solar technologies require proper waste management and recycling, PV is associated with a few particular waste management and recycling issues, whereas CSP shares issues with other technologies that use common materials such as concrete, glass, and steel. Waste management and hazardous cycling issues for each technology are discussed below, with a focus on the issues surrounding PV.

The PV cell manufacturing process includes a number of hazardous materials [e.g., compounds of cadmium (Cd), selenium (Se), and lead (Pb)], and there are concerns about potential emissions at the end of a module's useful life. Managing the disposal and/or recycling of these materials to avoid groundwater contamination (via landfills) and air pollution (via incinerators) is an important environmental consideration. Another important consideration is that in creating millions of solar panels each year millions of pounds of polluted sludge and contaminated water are produced. To dispose of the material (properly), the producers must transport it by truck or rail far from their own plants to waste facilities hundreds and, in some cases, thousands of miles away. The fossil fuels used to transport that waste are not typically considered in calculating Solar's carbon footprint, giving scientists and consumers who use the measurement to gauge a product's impact on global climate change the impression that solar is cleaner than it is (Toledo Blade, 2013).

In addition to materials contained within the completed module, a number of chemicals may be used during PV manufacturing. For crystalline silicon modules, feedstock materials are made through a purification process, the by-products of which typically include silicon tetrachloride ($SiCl_4$). To reduce costs and protect the environment, most of today's manufacturing plants use a closed-loop process that greatly minimizes waste products by converting, separating, and reusing trichlorosilane from the $SiCl_4$ by-product. Silicon nitride (SiN_4) is used as an antireflective-coating material and is generally deposited via chemical vapor deposition. This process requires the safe handling and management of pyrophoric silane gas—i.e., gas that can ignite spontaneously when exposed to air. Silane is also the major feedstock in thin-film amorphous silicon (a-Si) PV and is often used as a coupling agent to adhere fibers to polymer materials. The a-Si/thin-film tandem segment of the PV industry also uses nitrogen trifluoride (NF_3) for reactor cleaning, which has a global warming potential 17,000 times greater than CO_2. The controlled use and production of NF_3 have been proven for specific production and end-use systems (for example, in the liquid crystal display industry), and its use in the a-Si/microcrystalline silicon PV industry will not alter the environmental benefits of PV replacing fossil fuels if best practices are adopted globally (Fthenakis et al., 2010).

The greatest concern surrounding thin-film cadmium telluride (CdTe) and copper indium gallium selenide (CIGS) PV is potential exposure to Cd, which the EPA defines as a Class B1 carcinogen (i.e., probable human carcinogen). Typical CdTe PV material contains 5 g of Cd per m^2 of module, whereas typical CIGS material (which can contain cadmium sulfide) contains less than 1 g of Cd per m^2 of module (Fthenakis and Zweibel, 2003). Although Cd is not emitted during normal module operation, small emissions could occur during manufacturing or accidental fires. However, the life-cycle Cd emissions of CdTe and CIGS PV are orders of magnitude lower than Cd emissions from the operation of fossil-fuel power plants (Fthenakis, 2004; Fthenakis et al., 2005, 2008).

Not all the news about PV production and subsequent PV waste is negative. Recycling can help resolve end-of-life (cradle-to-grave) PV module issues, and the PV industry is proactively engaged in building recycling infrastructure. The technical and economic feasibility of recycling the semiconductor materials, metals, and glass from manufacturing scrap and spent PV modules has been established (Fthenakis, 2000). Furthermore, recycling can provide a significant secondary source of materials that may be used in the production of future PV technologies, such as tellurium, indium, and germanium (Fthenakis, 2009). First Solar, which manufactures thin-film CdTe PV, established the industry's first comprehensive pre-funded module collection and recycling program, which the company claims will result in recycling 90% of the weight of each recover First Solar PV module. In Europe, the PV industry has established PV Cycle, a voluntary program to recycle PV modules (PV Cycle, 2010). The United States could adopt this type of industry-wide approach to manage the large-scale recycling and management of PV materials.

The major constituents of CSP plants include glass, steel, and concrete. In addition, some CSP plants will contain a significant quantity of nitrate salt and organic heat transfer oil. All these materials are recyclable (DOE, 2012).

Ecological Impact

All development creates ecological and other land-use impacts. The primary impacts of solar development relate to land used for utility-scale PV and CSP. Even with the most careful land selection, the projected utility-scale solar development may have significant local land-use impacts, especially on portions of the southern United States. Solar development should be consistent with national and local land-use priorities.

With regard to direct ecological impacts of solar development, these include soil disturbance, habitat fragmentation, and noise. Indirect impacts include changes in surface water quality because of soil erosion at the construction site. The specific impacts of utility-scale solar development will depend on project location, solar technology employed, size of the development, and proximity to existing roads and transmission lines.

The potential ecological impacts in the southwestern United States are particularly important because of the large-scale solar development envisioned for this area. The Southwest supports a wide variety of plant communities and habitats, including arid and semiarid desert-scrub and shrub land, grasslands, woodlands, and savannas. The wildlife in these areas includes diverse species of amphibians, reptiles, birds, and small and large mammals. Government agencies and conservation groups have identified a significant list of species that may be affected by solar development (DOE and DOI, 2010).

Altering plant communities with development can strain wildlife living in or near these communities, making it more difficult to find shelter, hunt, forage, and reproduce. Fenced-in power plants can add further strain by affecting terrestrial and avian migration patterns. Aquatic species also can be affected—as can terrestrial and avian species that rely on aquatic habitats—if the water requirements of solar development result in a substantial diversion of local water sources. Large areas covered by solar collectors also may affect plants and animals by interfering with natural sunlight, rainfall, and drainage. Solar equipment may provide perches for birds of prey that could affect bird and prey populations.

The potential impacts of solar development are not limited to ecological impacts. Solar development could affect a variety of activities that take place on public and private land. For example, conflicts may arise if development impacts cultural sites, or interferes with U.S. Department of Defense (DOE) activities. In addition, loss of forage base could result in rescued grazing, which would disrupt the longstanding economic and cultural characteristics of ranching operations. Potential direct impacts include the conversion of land to provide support services and housing for people who move to the region to support the solar development, with associated increases in roads, traffic, and penetration into previously remote areas. The additional transmission infrastructure associated with solar development could create various impacts as well.

These are merely examples of the types of impacts that may be associated with solar development. For an exhaustive discussion, see DOE and DOI (2010) and other detailed environmental-impact studies. Less well-studied impacts are also important and must be evaluated as solar development progresses. For example, the local and global climate effects of changes in albedo due to widespread PV and CSP deployment are not well studied. *Albedo* is the ratio between the light

TABLE 12.1
The ALBEDO of Some Surface Types in % Reflected

Surface	Albedo
Water (low Sun)	10–100
Water (high Sun)	3–10
Grass	16–26
Glacier ice	20–40
Deciduous forest	15–20
Coniferous forest	5–15
Old snow	40–70
Fresh snow	75–95
Sea ice	30–40
Blacktopped tarmac	5–10
Desert	25–30
Crops	15–25

reflected from a surface and the total light falling on it. Albedo is a surface phenomenon—basically a radiation reflector. Albedo always has a value less than or equal to 1. An object with a high albedo, near 1, is very bright, while a body with a low albedo, near 0, is dark. For example, freshly fallen snow typically has an albedo that is between 75% and 90%; that is, 75% to 95% of the solar radiation that is incident on snow is reflected. At the other extreme, the albedo of a rough, dark surface, such as a green forest, may be as low as 5%. The albedo of some common surfaces is listed in Table 12.1. The portion of insolation not reflected is absorbed by the Earth's surface, warming it. This means Earth's albedo plays an important part in the Earth's radiation balance and influences the mean annual temperature and the climate on both local and global scales. One study evaluated the net balance between GHC emissions reduction resulting from PV replacing fossil-fuel-based power generation (with PV growing to meet 50% of world energy demand in 2100) and a decrease in desert albedo due to PV module covering, concluding that the PV albedo effect would have little impact on global warming (Nemet, 2009).

With regard to solar energy production and the possible impact on global warming (climate change), note that there are no global climate change emissions associated with generating electricity from solar energy. However, there are emissions associated with other stages of the solar life cycle, including manufacturing, materials transportation, installation, maintenance, and decommissioning and dismantlement. Most estimates of life-cycle emissions for photovoltaic systems are between 0.07 and 0.18 pounds of carbon dioxide equivalent per kilowatt-hour.

HYDROPOWER

When we look at rushing waterfalls and rivers, we may not immediately think of electricity. But hydroelectric(water-powered) power plants are responsible for lighting many of our homes and neighborhoods. *Hydropower* is the harnessing of water to perform work. The power of falling water has been used in industry for thousands of years. The Greeks used water wheels for grinding wheat into flour more than 2,000 years ago. Besides grinding flour, the power of the water was used to saw wood and power textile mills and manufacturing plants.

The technology for using falling water to create hydroelectricity has existed for more than a century. The evolution of the modern hydropower turbine began in the mid-1700s when a French hydraulic and military engineer, Bernard Forest de Belidor wrote a four-volume work describing using a vertical axis versus a horizontal-axis machine.

Water turbine development continued during the 1700s and 1800s. In 1880, a brush arc light dynamo driven by a water turbine was used to provide theatre and storefront lighting in Grand Rapids, Michigan; and in 1881, a brush dynamo connected to a turbine in a flour mill provided street lighting at Niagara Falls New York. These two projects used direct-current (DC) technology.

Alternating current (AC) is used today. That breakthrough came when the electric generator was coupled to the turbine, which resulted in the world, and the United States,' first hydroelectric plant located on the Fox River in Appleton, Wisconsin, in 1882. The U.S. Library of Congress (2009) lists the Appleton hydroelectric power plant as one of the major accomplishments of the Gilded Age (1878–1889). Soon, people across the U.S. were enjoying electricity in homes, schools, and offices, reading by electric lamp instead of candlelight or kerosene. Today, we take electricity for granted, not able to imagine life without it.

Some hydropower plants use dams, and some do not. Many dams were built for other purposes and hydropower was added later. In the United States, there are about 80,000 dams of which only 2,400 produce power. The other dams are for recreation, stock/farm ponds, flood control, water supply, and irrigation. The types of hydropower plants are described below.

Impoundment

The most common type of hydroelectric power plant is an impoundment facility. An impoundment facility, typically a large hydropower system, uses a dam to store river water in a reservoir. This type of facility works best in mountainous or hilly terrain where high dams can be built and deep reservoirs can be maintained. Potential energy available in a reservoir depends on the mass of water contained in it, as well as on the overall depth of the water. Water released from the reservoir flows through a turbine, spinning it, which in turn activates a generator to produce electricity. The water may be released either to meet changing electricity needs or to maintain a constant reservoir level.

Diversion

A diversion, sometimes called run-of-river, facility channels all or a portion of the flow of a river from its natural course through a canal or penstock, and the current through this medium is used to drive turbine. It may not require the use of a dam. This type of system is best suited for locations where a river drops considerably per unit of horizontal distance. The ideal location is near a natural waterfall or rapids. The chief advantage of a diversion system is the fact that, lacking a dam, it has far less impact on the environment than an impoundment has.

Pumped Storage

When the demand for electricity is low, a pumped storage facility stores energy by pumping water from a lower reservoir to an upper reservoir. During periods of high electrical demand, the water is released back to the lower reservoir to generate electricity.

Hydropower Generation: Dissolved Oxygen Concerns

In regard to the benefits derived from the use of hydropower: it is a clean fuel source; it is a fuel source that is domestically supplied; it relies on the water cycle and thus is a renewable power source; it is generally available as needed; it creates reservoirs that offer a variety of recreational opportunities, notably fishing, swimming, and boating; and they supply water where needed and assist in flood control—many of these are well known and often taken for granted.

Coins are two-sided, of course, and so are the facts about hydropower; that is, with the good side of anything there generally is an accompanying bad side. Many view this to be the case with hydropower. The bad side or disadvantages of hydropower include the impact on fish populations (e.g., salmon) if they can't migrate upstream past impoundment dams to spawning grounds or if they can't migrate downstream to the ocean. Hydropower can also be impacted by drought in that when water is not available, the plant can't produce electricity. Hydropower plants also compete with other uses for the land.

Other lesser-known negatives of hydropower plants concern their impact on water flow and quality; hydropower plants can cause low water levels that impact riparian habitats. Water quality is also affected by hydropower plants. The low dissolved oxygen levels in the water, a problem that is harmful to riparian (riverbank) habitats, can result when reservoirs stratify (develop layers of water of different temperatures). Stratification could affect the water temperature with resultant effects on dissolved oxygen levels, nutrient levels, productivity and the bio-availability of heavy metals. During the summer, stratification, a natural process, can divide the reservoir into distinct vertical strata, i.e., a warm, well-mixed upper layer (epilimnion) overlying a cooler, relativity stagnant lower layer (hypolimnion). Plant and animal respiration, bacterial decomposition of organic matter, and chemical oxidation can all act to progressively remove DO from hypolimnetic waters. This decrease in hypolimnetic DO is not generally offset by the renewal mechanisms of atmospheric diffusion, circulation, and photosynthesis that operate in the epilimnion/. In temperate regions, the decline in hypolimnetic DO concentrations begins at the onset of stratification (spring or summer) and continues until either anaerobic conditions predominate or reoxygenation occurs during the fall turnover of the water body.

There are numerous structural, operational, and regulatory techniques that a hydropower operator can use to resolve a low DO issue. Levels of DO can be increased through modifications in dam operations. These include such techniques as fluctuating the timing and duration of flow releases, spilling, or sluicing water, increasing minimum flows, flow mixing, turbine aeration, and, at some sites, injection of air or oxygen in weir aeration have proven effective. The most effective strategy for addressing the DO problem is dependent on the site-specific situation.

ECOLOGICAL IMPACT OF HYDROPOWER

During operation, adverse ecological effects could result from the disturbance of wildlife by equipment noise; site inspection and maintenance activities; exposure of biota to contaminants; and mortality of birds from colliding with the project facilities and/or electrocution by transmission lines. During operation, wildlife could still be affected by habitat fragmentation or the presence of barriers in fenced areas, canals or above-ground pipelines, utility rights-of-way (ROWs), and access roads. In addition, the presence of the hydropower project and its associated access roads and ROWs may increase human use of surrounding areas, which could, in turn, impact ecological resources in the surrounding areas through:

- Introduction and spread of invasive vegetation
- Disturbance
- Mortality of wildlife from vehicles
- Increase in hunting (including poaching)
- Increased potential for fire
- Physical barriers to fish migration
- Flow alteration and fluctuations
- Biological impacts of flow fluctuations.

In the discussion that follows, for this text, we concentrate on the last three impacts listed above related to the Pacific Northwest region of the United States.

Green Engineering

Physical Barrier to Fish Migration

The presence of a dam could impose a physical barrier to fish migration. This can be mitigated by the construction of a fish ladder.

Flow Alteration, Flow Fluctuation, Regulated and Unregulated Rivers and Salmonids[2]

Hydropower plants can, to varying capacities, change instream flow patterns in rivers below the dams and powerhouses. These changes can be classified into two categories, flow alterations and flow fluctuations.

Flow alterations are changes in flow over long periods of time (weeks, months, or seasons) resulting from the storage of water, irrigation diversions, municipal diversions, or the reactions of flow between dams and powerhouses. These changes in net flow usually change the availability of fish habitat, and thus change the fish production potential of a river. Flow alterations are evaluated by studying the fish habitat requirements and estimating the changes in habitat area at different flows using a hydraulic model. The Instream Flow Incremental Methodology (IFIM) has become a standard method for estimating habitat changes resulting from flow alterations. THE IFIM which was developed under the guidance of the USFWS is a process utilizing various technical methodologies to evaluate changes in the amount of estimated usable habitat for various species or groups of species as flow changes. The IFIM methodology is routinely used to facilitate the negation of instream flow requirements, usually minimum flow requirements that meet the habitat needs of economically important or threatened fish species.

Flow fluctuations are unnaturally rapid changes in the flow over periods of minutes, hours, and days. Flow fluctuations can be immediately lethal or have indirect and delayed biological effects. Flow fluctuations can be measured either by changes in *flow*, which is the volume of water passing a specific river transect, or by changes in *stage*, which is the water surface elevation or gage height. Both units are needed to understand the problem, and the terms are used interchangeably in this text. Hydrologists and engineers require flow measurements for many applications; however, the biological impact of flow fluctuations is best measured by stage. These two units do not have a simple functional relationship; thus, *rating tables* or *rating curves* are used to define the flow at each stage for a specific river transect.

Flows in *unregulated rivers* respond to changes in precipitation and snow melt. West of the Cascade Range, the peak flows occur from heavy rainstorms in November, December, and January. A lesser but more sustained peak occurs from a combination of rain and snow melt in the spring. The lowest flows coincide with the dry season that occurs in late summer and early fall. Glacial streams and streams on the east side of the Cascades have a somewhat different pattern. Here, the highest flows often occur in the spring and extend into the early summer. The lowest flows in some years occur during cold periods in the winter. In either case, periods of heavy rainfall or dry weather can create flows that are above or below seasonal averages. These natural flow variations indirectly affect fish production as a result of changes in the quantity and quality of instream habitat.

On a shorter time, scale, individual storms can rapidly increase river stage in less than a day. After the storm, the stage declines to a relatively stable level over a longer period of time, usually days or weeks. In addition to storm events, limited daily stage changes sometimes occur during sunny weather as a result of snowmelt run-off.

Flows in *regulated rivers* respond to measures taken to improve river channels so that they can be used more efficiently in the national economic interest. When properly injured and constructed, river regulation ensures the creation of favorable conditions for navigation and timber raft, the maintenance of the necessary water levels at water intake work, the protection of populated areas and agricultural land from flooding during spring floods and high water, the slow movement of river sediment, and the smooth flow of water toward the openings of hydraulic engineering structure such as dams.

BIOLOGICAL IMPACT OF FLOW FLUCTUATIONS

Increases in Flow

Evidence of biological impacts from rapid flow increase is scarce. Some impacts associated with rapid flow increases might be more appropriately associated with high flows. Rochester et al. (1984) noted that eggs and alevins (fry) can be killed when gravel scour occurs, and juvenile fish may be physically flushed down the river. Some species of aquatic insects that swim in pools can be physically flushed downstream from a sudden increase in flow (Trotzky and Gregory, 1974).

The biological effects of unnatural flow increases are usually irrelevant in regulating hydropower operations because public safety concerns justify more stringent regulations than biological concerns. Flow increase can strand and occasionally drown fishermen and other people located on bars, rocks, or in confined canyons. Boaters might also be at risk under some circumstances.

Stranding

Stranding is the separation of fish from flowing surface water as a result of the declining river stage. Stranding can occur during any drop-in stage. It is not exclusively associated with the complete or substantial dewatering of a river. Stranding can be classified into two categories: *Beaching* is when fish flounder out of water on the substrate. *Trapping* is the isolation of fish in pockets of water with no access to free-flowing surface water. Stranding cannot always be neatly classified as beaching or trapping. Thus, in this test, we use the term stranding unless a more specific term is appropriate.

Salmonid stranding associated with hydropower operations has been widely documented in Washington and Oregon (e.g., Thompson, 1970; Witty and Thompson, 1974; Phinney, 1974a,b; Bauersfeld, 1977, 1978a; Becker et al., 1981; Fiscus, 1977; Satterthwaite, 1987; Olson, 1990). Stranding can occur many miles downstream of the powerhouse (Phillips, 1969; Woodin, 1984). The estimated numbers of fish stranded in flow fluctuation events range from negligible to 120,000 fry (Phinney, 1974a). Stranding mortality is difficult or impossible to estimate. Estimates are usually very conservation and/or highly variable.

Stranding can also occur as a result of other events, including natural declines in flow (author's observation of Colorado River over time), ship wash (Bauersfeld, 1977), municipal water withdrawals and irrigation withdrawals. Many factors affect the incidence of stranding. A recurrent theme in much of the following discussion is the high vulnerability of small salmonid fry.

Juvenile salmonids are more vulnerable to stranding than adults. Salmonid fry that has just absorbed the yolk sac and has recently emerged from the gravel is by far the most vulnerable. They are poor swimmers and settle along shallow margins of rivers (Phinney, 1974a; Woodin, 1984), where they seek refuge from currents and larger fish. Once Chinook attains the size of 50–60 mm in length, vulnerability drops substantially. For steelhead, vulnerability drops significantly when the fry reaches 40 mm (Beck Associates, 1989). Larger juveniles are more included in inhabitant pools, glides, overhanging banks, and midchannel substrates, where they are less vulnerable to stranding. However, many juveniles still inhabit shoreline areas and remain vulnerable to stranding until they emigrate to saltwater (Hamilton and Buell, 1976). Adult stranding as a result of hydropower fluctuations has been documented (Hamilton and Buell, 1976).

The river channel configuration is a major factor in the incidence of stranding. A river channel with many side channels, potholes, and low gradient bars will have a much greater incidence of stranding than a river confined to a single channel with steep banks.

Large numbers of small fry die from beaching on gravel bars when unnatural flow fluctuations occur (Phillips, 1969; Phinney, 1974a; Woodin, 1984). Bauersfeld (1978a) observed beaching primarily on bars with slopes less than 4%. Beck Associates (1989) determined that beaching occurred primarily on bars with slopes less than 5%. Under laboratory conditions, Monk (1989) determined that Chinook fry stranded in significantly large numbers on 1.8% slopes than on 5.1% slopes, however, results were not significant for steelhead. Stranding on steep gravel bars (>5% slope) has not been thoroughly studied.

Long side channels with intermittent flows are notorious for trapping juvenile fish. Substantial trapping can occur even with unregulated flows (Hunter, 1992). Side channels are valuable rearing habitats, and juveniles of several species prefer side channels over the main channel. However, unnatural fluctuations will repeatedly trap fish, eventually killing some or all of them (Witty and Thompson, 1974; Hamilton and Buell, 1976; Woodin, 1984; Olson, 1990). Side channels can trap substantial numbers of fingerlings and smolts (up to 150 cm) as well as fry.

As water recedes from river margins, juvenile salmonids may become trapped in deep pools called *potholes* (Woodin, 1984). River potholes are formed at high flows from scouring (a process called corrosion); around boulders and rootwads and where opposing flows meet. Potholes may remain watered for hours or months depending on the depth of the pothole and the river stage. R.W. Beck Associates (1989) extensively studied pothole stranding in the Skagit River (Washington State). Among the conclusions were: (1) Only a small fraction of the potholes in a river channel posed a threat to fish if fluctuations are limited in ranger; (2) The incidence of stranding is independent of the rate of stage decrease; and (3) The incidence of stranding was inversely related to the depth of water over the top of each pothole at the start of the decline in flow.

Most documented observations of stranding have occurred on gravel; however, stranding has also occurred in mud (Becker et al., 1981) and vegetation (Phillips, 1969; Satterthwaite, 1987). Under laboratory conditions, Monk (1989) found significantly different rates of stranding on different types of gravel. In fact, substrate was statistically the most significant factor contributing to stranding of Chinook and steelhead fry. On cobble substrate, fry (especially steelhead fry) were inclined to maintain a stationary position over the streambed (i.e., rheotaxis); while over small gravel, fry swam around, often in schools. When the water surface dropped, fry maintaining their position became trapped in pockets of water between cobbles, whereas mobile fish were more included to retreat with the water margin. When beaching became imminent, fry over cobble substrate retreated into intergravel cavities, where they became trapped. The difference in stranding rate was facilitated by the flow of water along a receding margin of the stream. On cobble substrate, the water drained into the substrate, whereas on finer substrates, a significant portion of the water flowed off on the surface.

Fry of some species is more vulnerable to stranding than others. In Washington State, stranding of chinook and steelhead fry has been frequently observed. Although pink salmon fry and chum salmon fry occur in the same rivers, they strand in lower numbers than chinook fry and steelhead fry (Woodin, 1984). However, Beck Associates (1989) determined that the rate of chum and pink fry stranding per the available fry was substantially higher than for chinook. The low numbers of pink and chum salmon stranding is a result of the short freshwater residency; they emigrate to salt water shortly after emergence, whereas chinook and steelhead remain in the river for months or years.

Hamilton and Buell (1976) observed extensive coho stranding in the Campbell River (British Columbia) and coho stranding has been observed in incidental numbers in other studies (Olson, 1990). The overall incidence of coho stranding is rather low in the studies conducted to date. The likely reason for this is that coho prefer streams for spawning and reading, whereas the formal research and evaluation have taken place in large and medium rivers. Juvenile coho rear for a full year in freshwater, and thus, it is reasonable to assume that stranding would occur at rates similar to chinook and steelhead.

The total drop in stage from an episode of flow fluctuation (known as ramping range) affects the incidence of stranding by increasing the gravel bar area exposed. In addition, it increases the number of side channels and potholes that become isolated from surface flow (Beck Associates, 1989).

Stranding increases dramatically when flow drops below a certain water level, defined as the *critical flow* (Thompson, 1970; Phinney, 1974a; Bauersfeld, 1978a; Woodin, 1984). In hydropower mitigation settlements, the critical flow is defined as the minimum operating discharge, or as an upper end of a flow range where more restrictive operation criteria are applied. The factors that likely account for this response have been discussed above. The exposure of the lowest gradient gravel bars often occurs in a limited range of flows. The exposure of spawning gravel from which fry is emerging may also account for the higher incidence of stranding.

In rivers with seasonal side channels and off-channel sloughs (also, slew, and slues), even a natural flow reduction can trap fry and smolts. Under normal circumstances, the natural population can sustain a small loss several times a year. However, when a hydropower facility causes repeated flow fluctuations, these small losses can accumulate to a very significant cumulative loss (Bauersfeld, 1978a).

The *ramping rate* is the rate of change in the stage resulting from regulated discharges. Unless otherwise noted, it refers to the rate of state decline. The faster the ramping rate, the more like fish are to be stranded (Phinney, 1974a; Bauersfeld, 1978a). Ramping rates of less than one inch per hour were needed to protect steelhead fry on the Sultan River. Olson (1990) determined that ramping rate of 1 inch per hour was adequate to protect steelhead fry. However, the ramping rate was measured at a confined river transect, whereas the stranding was observed on lower gradient bars further downstream. Thus, the effective ramping rate at these bars was less than one inch per hour.

Although many hydropower mitigation settlements specify ramping rates, some research has indicated that ramping rates cannot always protect fish from stranding. Woodin (1984) determined that *any* daytime ramping stranded chinook fry. Beck Associates (1989) could not find any correlation between the ramping and the incidence of pothole trapping, nor was there any correlation between the ramping rate and steelhead fry stranding during the summer. In both cases, stranding occurred regardless of the ramping rate.

Small fry is highly vulnerable to stranding and is present in the steams only at certain times of the year. Chinook, coho, pink, and chum fry emerge during late winter and early spring while steelhead emerges in late spring through early fall (Olson, 1989). Fingerlings, smolts, and adults are vulnerable to stranding in other seasons; however, less restrictive ramping criteria are often sufficient to protect them.

For at least some species, the incidence of stranding is influenced by the time of day. Chinook fry is less dependent on the substrate for cover at night and thus is less vulnerable to stranding at night (Woodin, 1984). Two studies (Stober et al., 1982; Olson, 1990) concluded that steelhead fry is less vulnerable during the day, presumably because of this species fees during the day. However, two other studies (Beck Associates, 1989; Monk, 1989) found no difference in the rate of steelhead fry stranding relative to night and day.

Salmonids respire using their gills and do not survive out of water for more than ten minutes. Thus, beaching is always fatal. Juvenile salmonids trapped in side channels and potholes can survive for hours, days, or under favorable circumstances, months (Hunter, 1992). However, much trapped fish die from predation, temperature shock, and/or oxygen depletion. Survivors that are rescued by higher flows are probably in poorer condition than fish in the free-flowing channel.

Some observations suggest that a highly stable flow regime for a week or more prior to a flow fluctuation will increase the incidence of fry stranding (Phinney, 1974b). Two hypotheses might explain this observation. One hypothesis states that after long periods of stable flow, more fry are available for stranding. In other words, a major flow reduction after a week of stable flows strands seven daily cohorts of emerging fry at once, rather than one cohort when fluctuations occur daily. An alternative hypothesis is that juveniles become accustomed to residing and feeding along the margins of a stream either as a behavioral response to stable flows or in response to aquatic invertebrate populations that thrive along the water's edge under stable flows. These hypotheses should be thoroughly tested before they are applied to mitigation practices.

Increased Predation

Phillips (1969) suggested that juvenile fish forced from the river margins as a result of declining flows suffer from predation by larger fish. This effect appears not to have been documented anywhere; however, it is a credible hypothesis under some circumstances.

Aquatic Invertebrates

Like fish, aquatic invertebrates are not necessarily adapted to unnatural drops in flow. Cushman (1985) extensively reviewed the effects of flow fluctuations on aquatic life, especially aquatic invertebrates. Interested readers should read this review.

Research on the effects of flow fluctuations on aquatic invertebrates in the Pacific Northwest is limited, although more information is available elsewhere in North America. These studies suggest that aquatic invertebrates can be severely impacted by flow fluctuations. Fluctuations substantially reduce invertebrate diversity, total biomass and changes the species composition under most circumstances. One study from the Skagit River found that flow fluctuations had a greater adverse impact on the aquatic invertebrate community than a substantial reduction in average flow (Gislason, 1985). The reduction in aquatic invertebrate production can impact salmonid products as a result of reduced feeding (Cushman, 1985).

Additional research is needed on the effects of flow fluctuations on aquatic invertebrates in the Pacific Northwest. However, a thorough our study would be a formidable task. It would involve many species with different life cycles, behavioral response, lethal response, and contributions as prey to salmonids. Populations of some species may change rapidly under normal conditions, and thus may be difficult to associate cause and effect. Flow fluctuations can impact aquatic invertebrates in the following ways:

- **Stranding**: Flow fluctuations can strand many species of aquatic invertebrates, much in the same way fish can become stranded (Phillips, 1969; Gislason, 1985). Death may result from suffocation, desiccation, temperature shock, or predation.
- **Increased drift**: Many aquatic invertebrates are sensitive to reductions in flow and respond by leaving the substrate and floating downstream. This floating behavior is called *drift*. Nighttime drift is normal; however, drift becomes highly elevated under unnatural fluctuations in flow (McPhee and Brusven, 1975; Cushman, 1985). This elevated drift may be an emergency response to avoid stranding, or a response to overcrowding of the inter-gravel habitat, or it may be a response by aquatic species that are adapted to a narrow range of water velocity. This response may temporarily increase fish food supply (McPhee and Brusven, 1975), but when repeated fluctuations occur, many species are flushed out of river reach and the aquatic invertebrate biomass usually declines, often substantially (Cushman, 1985; Gislason, 1985). Elevated drift also occurs in response to sudden increases in flow, which captures terrestrial insects from the riverbanks and scours some aquatic invertebrates from the river substrate (Mundie and Mounce, 1976).
- **Detritus feeders**: Under stable flow conditions, floating detritus (leaves, woody debris) accumulates on the shores of the river as a result of current and wind action on sand or gravel substrate. This detritus remains close to the river margin and often remains damp for days or weeks at a time. Under fluctuating flows, this organic detritus becomes suspended (Mundie and Mounce, 1976) and is flushed out of the river of redeposited at the high waterline where it desiccates during low flow periods. As a result, the invertebrate detritus community is less capable of exploiting this resource.
- **Herbivorous invertebrates**: Impacts are similar to that of the detritus community. Algae grow on exposed rock surfaces on which herbivorous aquatic invertebrates graze. Fluctuations desiccate and disrupt the growth of the exposed algae (Gislason, 1985) and reduce access by herbivores.

Redd Dewatering

Research has extensively documented the lethal impact of redd dewatering on salmonid eggs and alevins (i.e., larval fish) (Frailey and Graham, 1982; Fraser, 1972; Satterthwaite et al., 1987; Fustish et al., 1988). Salmonid eggs can survive for weeks in dewatered gravel (Stober et al., 1982; Reiser and White, 1983; Becker and Neitzel, 1985; Neitzel et al., 1985), if they remain moist and are not subjected to freezing or high temperatures. The necessary moisture may originate from subsurface river water or from groundwater. If the subsurface water level drops too far, the inter-gravel spaces will dry out, and the eggs will desiccate and die. Thus, redd dewatering is not always lethal or even harmful to eggs. However, site conditions, weather and duration of exposure wall affect survival.

Because alevins rely on gills to respire, dewatering is lethal (Stober et al., 1982; Neitzel et al., 1985). Alevins can survive in subsurface, inter-gravel flow from a river or groundwater source. If inter-gravel spaces are not obstructed with pea gravel, sand, or fines, some alevins will survive by descending through inter-gravel spaces with a declining water surface (Strober et al., 1982). Both alevins and eggs may die from being submerged in stagnant water. Standing inter-gravel water may lose its oxygen to biotic decay, and metabolic wastes may build up to lethal levels.

A redd can be dewatered between spawning and hatching without harm to the eggs under some circumstances, and in one situation, a hydropower facility is operated to allow limited redd dewatering (Neitzel et al., 1985). However, in most Pacific Northwest Rivers, anadromous fish spawn over an extended period. Different species spawn in different seasons and individual species may spawn over a range of 2–6 months. As a result, when eggs are present, alevins and fry are also present, both of which are highly vulnerable to flow fluctuations.

Spawning Interference

Bauersfeld (1978b) found that repeated dewatering caused chinook salmon to abandon attempts to spawn and move elsewhere, often to less desirable or crowded locations. Hamilton and Buell (1976) performed a highly detailed study using observation towers situated over spawning beds to track the activity on the spawning bed and to observe individual tagged fish. They observed that spawning chinook was frequently interrupted by flow fluctuations. Females repeatedly initiated redd digging, and then abandoned and redd sites when flows changed. They concluded that flow fluctuations decrease viability due to the untimely release of eggs, failure to cover eggs once they were released, and a failure of males to properly fertilize eggs laid in incomplete redds. Other researchers had conflicting conclusions. Stober et al. (12982) noted that chinook salmon successfully spawned in an area that was dewatered several hours a day, and researchers found that 8 hours a day of dewatering still permitted successful spawning.

Hydraulic Response to Flow Fluctuations

The ramping rate attenuates as a function of the distance downstream from the source of a fluctuation event (Nestler et al., 1989). The characteristics of the river greatly influence this attenuation. A fluctuation in flow passing through a narrow bedrock river channel will experience little or no attenuation. Pools, side channels, and gravel bars attenuate the ramping rate by storing water from higher flows and releasing this water gradually. Tributary inflow will attenuate the ramping rate and the ramping range. Hydraulic equations (e.g., unsteady flows; Chow, 2009, pp. 700, described as one of the best texts written on open channel hydraulics) exist to describe these responses.

The time it takes for a fluctuation to pass from one place to another on a river is known as lag time. The river channel configuration, gradient, and flow all influence the speed at which the fluctuation travels downstream. Lag time can be determined by field observations as several flows. For projects with long penstocks, the term *bypass lag time* refers to the time flow fluctuations take to pass down the natural stream channel from the dam to the powerhouse tailrace.

Types of Hydropower Activity that Fluctuate Flows

Hydropower facilities cause flow fluctuations in a variety of ways. A brief overview of mechanical causes is provided in the following.

- **Load following**: a hydropower plant adjusts its power output as demand for electricity fluctuates throughout the day. This practice, following daily changes in power demand, is called load following.

- **Peaking**: a peaking hydropower plant operate only during times of peak demand. Peaking is the most widely documented source of fish stranding. Biologists and fishermen have observed major fish kills from peaking (Thompson, 1970; Graybill et al., 1979; Phinney, 1974a; Bauersfeld, 1977, 1978; Becker et al., 1981).
- **Low flow shutdowns**: when turbine flow is below the level for practical turbine operation, the plant must take various turbines offline. In addition, a minimum flow is usually required to maintain the aquatic habitat in the bypass reach. Dam facilities with seasonal storage can operate for years without a low-flow shutdown.
- **Low-flow startups**: run-of-river plants will cause a drop in flow in the bypass and downstream reaches during powerhouse start-ups. In these situations, operators must ramp flows at the start of power generation to reduce stranding.
- **Powerhouse failures**: powerhouse failures are rare disruptions of the penstock flow originating from the powerhouse. These disruptions usually result from mechanical problems but can also occur due to load rejection, which is the inability of the utility line to receive power generated from the turns. Powerhouse failures can occur at any facility. This type of failure can cause a sudden drop in flow level in the downstream reach.
- **Flow continuation**: is the mechanical capacity to maintain flow through the penstock during powerhouse failures.
- **Intake failures**: refer to all penstock flow disruptions that occur at the intake structure. Intake failures are less frequent than powerhouse failures; they usually occur as a result of the accumulation of debris, the failure of fish screen cleaning equipment, or failure of the dam and associated gates to deliver water into the intake.
- **Cycling**: is a way to generate power when the flow is not enough for continuous or efficient operation, and it is not an attempt to follow load demands. Cycling will normally occur at low steam flows when the salmonids would be most vulnerable to fluctuations.
- **Multiple turbine operations**: if a powerhouse has two or more turbines, operators can cause abrupt changes in flow when changing the number of turbines in operation.
- **Forebay surges**: occur when the powerhouse of some run-of-the-river plants starts generation and is probably caused by a drop in heat at the intake during start-up.
- **Reservoir stranding**: in large reservoirs, standing is routinely anticipated as one of the consequences of drawdowns, and it is sometimes employed as a method of eradicating undesirable fish.
- **Tailwater maintenance and repair activities**: all hydropower plants will eventually require inspection, maintenance, and repair. However, it is often impossible to inspect or repair the structure or equipment submerged in the tailwater without completely or substantially disrupting the flow of the river.

BIOMASS/BIOENERGY

A Nation that runs on oil can't afford to run short.

Old Oil Industry Slogan

Biofuels made from the leftovers of harvested corn plants are worse than gasoline for global warming in the short term, a study shows, challenging the Obama administration's conclusions that they are a much cleaner oil alternative and will help combat climate change.

Dina Cappielo (2014)
Associated Press

In a recent New York Times article, Thomas Friedman (2010) stated that "the fat lady has sung." Specifically, Friedman was speaking about America's transition from "The Greatest Generation" to what Kurt Anderson referred to as "The Grasshopper Generation." That is, according to Friedman, we are "going from the age of government handouts to the age of citizen givebacks, from the age of companions fly free to the age of paying for each bag." He goes on to say that we all accept that our parents were the greatest generation, but it is us that we are concerned about and that it is the 'we' that comprise the Grasshopper Generation; "we are eating through the prosperity that was bequeathed us like hungry locusts." What we are "eating through," among other things, is our readily available, relatively cheap source of energy. The point is we can, like the grasshopper, gobble it all up until it is all gone, or we can find alternatives—renewable alternatives of energy.

With regard to renewable alternatives of energy, the Nation is aggressively developing the capacity to meet some of our energy needs through biofuels and biopower. The Energy Independence and Security Act of 2007 (EISA) calls for 36 billion gallons per year (BGY) of renewable fuels by 2011 and establishes new categories of renewable fuels, each with specific volume requirements and life cycle greenhouse gas (GHF) performance thresholds).

One of the promising sources of energy is biomass (biofuel) or bioenergy (biopower). *Biomass* is the feedstock used to produce *bioenergy*, a general term for energy derived from materials such as straw, wood, or animal wastes (i.e., biomass), which, in contrast to fossil fuels, were living matter relatively recently. Such materials can be burned directly as solids (biomass) to produce heat or power but can also be converted into liquid biofuels. In the last few years, there has been much-increased interest in bioenergy fuels such as biofuels which can be used for transport; that is, biofuels, as mentioned, are a liquid fuel (biodiesel and bioethanol) used for transport. At the moment, transport has taken center stage in our search for renewable, alternate fuels to eventually replace hydrocarbon fuels. Unlike biofuels, solid biomass fuel is used primarily for electricity generation or heat supply.

Even though we have stated that bioenergy is a promising source of energy for the future, it is rather ironic whenever the experts (or anyone else for that matter) make this point without qualification. The qualification? The reality? Simply, keep in mind that it was only 100 years ago our economy was based primarily on bioenergy from biomass, or carbohydrates, rather than from hydrocarbons. In the late 1800s, the largest selling chemicals were alcohols made from wood and grain, the first plastics were produced from cotton, and about 65% of the nation's energy came from wood (USDOE, 2003a). By the 1920s, the economy started shifting toward the use of fossil resources, and after World War II this trend accelerated as technology breakthroughs were made. By the 1970s, fossil energy was established as the backbone of the U.S. economy, and all but a small portion of the carbohydrate economy remained. In the industrial sector, plants accounted for about 16% of input in 1989, compared with 35% in 1925.

Processing cost and the availability of inexpensive fossil energy resources continue to be driving factor in the dominance of hydrocarbon resources. In many cases, it is still more economical to produce goods from petroleum or natural gas than from plant matter. This trend is about to shift dramatically as we reach peak oil and as the world continues to demand unprecedented amounts of petroleum supplies from an ever-dwindling supply.

Assisting in this trend shift are the technological advances in the biological sciences and engineering, political change, and concern for the environment have begun to swing the economy back toward carbohydrates on a number of fronts. Consumption of biofuels in vehicles, for example, has risen from zero in 1977 to nearly 1.5 billion gallons in 1999. The use of inks produced from soybeans in the U.S. increased by four-fold between 1989 and 2000 and is now at more than 22% of total use (Morris, 2002).

Technological advances are also beginning to make an impact on reducing the cost of producing industrial products and fuels from biomass, making them more competitive with those produced from petroleum-based hydrocarbons. Developments in pyrolysis, ultra-centrifuges, membranes, and the use of enzymes and microbes as biological factories are enabling the extraction of valuable components to form plants at a much lower cost. As a result, industry is investing in the development of new bioproducts that are steadily gaining a share of current markets.

New technologies are helping the chemical and food processing industries to develop new processes that will enable more cost-effective production of all kinds of industrial products from biomass. One example is a plastic polymer derived from corn that is now being produced at a 300 million pound per year plant in Nebraska, a joint venture between Cargill, the largest grain merchant and Dow Chemical, the largest chemical producer (Fahey, 2001).

Other chemical companies are exploring the use of low-cost biomass processes to make chemicals and plastics that are now made from more expensive petrochemical processes (USDOE, 2003b). In this regard, new innovative processes such as biorefineries may become the foundation of the new bioindustry. The Biorefinery is similar in concept to the petroleum refinery, except that it is based on the conversion of biomass feedstocks rather than crude oil. Biorefineries in theory would use multiple forms of biomass to produce a flexible mix of products, including fuels, power, heat, chemicals, and materials. In a biorefinery, biomass would be converted into high-value chemical products and fuels (both gas and liquid). Byproducts and residues, as well as some portion of the fuels produced, would be used to fuel on-site power generation or cogeneration facilities producing heat and power. The biorefinery concept has already proven successful in the U.S. agricultural and forest products industries, where such facilities now produce food, feed fiber, or chemicals, as well as heat and electricity to run plant operations. Biorefineries offer the most potential for realizing the ultimate opportunities of the bioenergy industry. At present, biomass (biofuels, waste, wood, and wood-derived fuels) accounted for more than 4,000 quadrillions Btu.

It is important to note that biomass-produced bioenergy is not the panacea for solving our energy needs now and for the future. There are environmental impacts associated with biomass technology. There are tradeoffs. Are these tradeoffs significant? Are the tradeoffs worth it? Will biomass eventually replace fossil fuel usage? The jury is still out; however, with the proper technology and proper planning, proper usage and proper operation, biomass conversion to bioenergy can and will be effective, eventually. The point is there are probably more benefits than non-benefits to be derived from the use of biomass; however, this book is about the environmental impacts of renewable energy, including biomass-to-bioenergy conversion. Therefore, later, after a brief introduction to what biomass is, how it is presently used, plant basics, feedstocks, biodiesel, and biogas, the environmental impacts of biomass production, refining and usage are discussed.

Biomass (all Earth's living matter) consists of energy from plants and plant-derived organic-based materials; it is essentially stored energy from the Sun. Plants convert solar energy into chemical energy during the process of photosynthesis. Biomass can be biochemically processed to extract sugars, thermochemically processed to produce biofuels or biomaterial, or combusted to produce hear to electricity. Biomass is also an input into other end-use markets, such as forestry products (pulpwood) and other industrial applications. This complicates the economics of biomass feedstock and requires that we differentiate between what is technically possible from what is economically feasible, considering relative prices and intermarket competition.

Biomass has been used since people began burning wood to cook food and keep warm. Trees have been the principal fuel of almost every society for over 5,000 years, from the Bronze Age until the middle of the nineteenth century (Perlin, 2005). Wood is still the largest biomass energy resource today, but other sources of biomass can also be used. These include food crops grassy and woody plants, residues from agriculture or forestry, and the organic component of municipal and industrial wastes. Even the fumes from landfills (which are methane, a natural gas) can be used as a biomass energy source. Organic material that has been transformed by geological processes into substances such as coal or petroleum are excluded from this category.

Feedstock Types

A variety of biomass feedstocks can be used to produce transportation fuels, biobased products, and power. Feedstocks refer to crops or products, like waste vegetable oil, that can be used as or converted into biofuels and bioenergy. With regard to the advantages or disadvantages of one type of feedstock as compared to another, this is gauged in terms of how much usable material they yield, where they can grow and how energy and water-intensive they are. Feedstock types are listed as first-generation or second-generation feedstocks. First-generation feedstocks include those that are already widely grown and used for some form of bioenergy or biofuel production, which means that there are possible foods versus fuel conflicts. First-generation feedstocks include sugars (sugar beets, sugar cane, sugar palm, sweet sorghum, and *nipa* palm), starches (cassava, corn, milo, sorghum, sweet potato, and wheat), waste feedstocks such as whey and citrus peels, and oils and fats (coconut oil, oil palm, rapeseed, soybeans sunflower seed, castor beans jatropha, jojoba, karanj (native tree of India), waste vegetable oil and animal fat). Second-generation feedstocks refer broadly to crops that have high potential yields of biofuels, but that are not widely cultivated, or not cultivated as an energy crop. It refers to cellulosic feedstocks or conventional crops such as miscanthus, prairie grass and switch grass, will and hybrid poplar trees. Algae and halophytes (saltwater plants) are other second-generation feedstocks.

Currently, a majority of the ethanol produced in the U.S. is made from corn or other starch-based crops. The present focus, however, is on the development of cellulosic feedstocks—non-grain, non-food-based feedstocks such as switchgrass, corn stover, and wood material—and on technologies to convert cellulosic material into transportation fuels and other products. Using cellulosic feedstocks can not only alleviate the potential concern of diverting food crops to produce fuel but also has a variety of environmental benefits (EERE, 2008). Because such a wide variety of cellulosic feedstocks can be used for energy production, potential feedstocks are grouped into categories—or pathways.

Composition of Biomass
The ease with which biomass can be converted to useful products or intermediates is determined by the composition of the biomass feedstock. Biomass contains a variety of components, some of which are readily accessible and others that are much more difficult and costly to extract. The composition and subsequent conversion issues for current and potential biomass feedstock compounds are listed and described below.

- **Starch (Glucose)**: is readily recovered and converted from grain (corn, wheat, rice) into products. Starch from corn grain provides the primary feedstock for today's existing and emerging sugar-based bioproducts, such as polylactide as well as the entire fuel ethanol industry. Corn grain serves as the primary feedstock for starch used to manufacture today's biobased products. Core wet mills use a multi-step process to separate starch from the germ, gluten (protein), and fiber components of corn grain. The starch streams generated by wet milling are highly pure, and acid of enzymatic hydrolysis is used to break the glycosidic linkages of starch to yield glucose. Glucose is then converted into a multitude of useful products.
- **Lignocellulosic biomass**: the non-grain portion of biomass (e.g., cobs, stalks), often referred to as agricultural stover or resides, and energy crops such as switchgrass also contain valuable components, but they are not as readily accessible as starch. These lignocellulosic biomass resources (also called cellulosic) are comprised of cellulose, hemicellulose, and lignin. Generally, lignocellulosic material contains 30%–50% cellulose, 20%–30% hemicellulose, and 20%–30% lignin. Some exceptions to this are cotton (98% cellulose) and flax (80% cellulose). Lignocellulosic biomass is perceived as a valuable and largely untapped resource for the future bioindustry. However, recovering the components in a cost-effective way represents a significant technical challenge.
- **Cellulose**: is one of nature's polymers and is composed of glucose, a six-carbon sugar. The glucose molecules are joined by glycosidic linkages which allow the glucose chains to assume an extended ribbon conformation. Hydrogen bonding between chains leads to the formation of flat sheets that lay on top of one another in a staggered fashion, similar to

the way staggered bricks add strength and stability to a wall. As a result, cellulose is very chemically stable and insoluble and serves as a structural component in plant walls.
- **Hemicellulose**: is a polymer containing primarily 5-carbon sugars such as xylose and arabinose with some glucose and mannose dispersed throughout. It forms a short-chain polymer that interacts with cellulose and lignin to form a matrix in the plant wall, strengthening it. Hemicellulose is more easily hydrolyzed than cellulose. Much of the hemicellulose in lignocellulosic material is solubilized and hydrolyzed to pentose and hexose sugars
- **Lignin**: helps bind the cellulosic/hemicellulose matrix while adding flexibility to the mix. The molecular structure of lignin polymers is very random and disorganized and consists primarily of carbon ring structures (benzene rings with methoxyl, hydroxyl, and propyl groups) interconnected by polysaccharides (sugar polymers). The ring structures of lignin have great potential as valuable chemical intermediates. However, separation and recovery of the lignin are difficult.
- **Oils and protein**: the seeds of certain plants offer two families of compounds with great potential for bioproducts: oils and protein. Oils and protein are found in the seeds of certain plants (soybeans, castor beans), and can be extracted in a variety of ways. Plants raised for this purpose include soy, corn, sunflower, safflower, rapeseed, and others. A large portion of the oil and protein recovered from oilseeds and corn is processed for human or animal consumption, but they can also serve as raw materials for lubricants, hydraulic fluids, polymers, and a host of other products.
  - **Vegetable oils**: are composed primarily of triglycerides, also referred to as triacylglycerols. Triglycerides contain a glycerol molecule as the backbone with three fatty acids attached to glycerol's hydroxyl groups.
  - **Proteins**: are natural polymers with amino acids as the monomer unit. They are incredibly complex materials, and their functional properties depend on molecular structure. There are 20 amino acids each differentiated by their side chain or R-group, and they can be classified as nonpolar and hydrophobic, polar uncharged, and ionizable. The interactions among the side chains, the amide protons, and the carbonyl oxygen help create the protein's 3-D shape.

IMPACT OF BIOMASS CONSTRUCTION, PRODUCTION AND OPERATION[3]

The combination of constructing biomass facilities (and associated ancillaries), producing biomass feedstock, and operating biomass energy facilities may cause environmental impacts. For example, construction activities that may cause environmental impacts to include ground clearing, grading, excavation, blasting, trenching, drilling, facility construction, and vehicular and pedestrian traffic. Additionally, potential environmental impacts could result from biomass feedstock production activities such as the collection of waste materials and growth and harvesting of woody and agricultural crops or algae, preprocessing, and transportation activities. Finally, operations activities may cause environmental impacts including the operation of the biomass energy facility, power generation, biofuel production, and associated maintenance activities. Many of the environmental impacts resulting from these activities are discussed in the following.

As mentioned, during the biomass energy facility construction phase, typical construction activities include ground clearing (removal of vegetative cover), grading, excavation, blasting, trenching, drilling, vehicular and pedestrian traffic, and construction and installation of facilities. Biomass power plants and some biogas plants that produce more electricity than required to operate the facility need transformers and transmission lines to deliver electricity to the power grid. Landfill gas production would require the drilling of wells for extraction of the gas and might require pipeline construction to deliver the gas to the user. Activities conducted in locations other than the facility site might include excavation and blasting for construction materials (e.g., sand, gravel) and access road construction. Potential impacts from these activities are presented below, by the type of affected resources.

Air Quality

Emissions generated during the construction phase include vehicle emissions; diesel emissions from large construction equipment and generators; release of volatile organic compounds (VOCs) from storage and transfer of vehicle/equipment fuels; small amounts of carbon monoxide, nitrogen oxides, and particulates from blasting activities; and fugitive dust from any sources such as disturbing and removing soils (clearing, grading, excavating, trenching, backfilling, dumping, and truck and equipment traffic), mixing concrete, storage of unvegetated soil piles, and drilling and pile driving. Note that a permit is needed from the state or local air agency to control or mitigate these emissions; therefore, these emissions would not likely cause an exceedance of air quality standards nor have an impact on climate change. Moreover, a construction permit under the mandated prevention of significant deterioration (PSD) or air quality regulations might be required.

Cultural Resources

Direct impacts on cultural resources could occur from construction activities, and indirect impacts might be caused by soil erosion and increased accessibility to possible site locations. Potential impacts include:

- Complete destruction of the resource if present in areas undergoing surface disturbance or excavation.
- Degradation or destruction of near-surface cultural resources on- and off-site resulting from topographic or hydrological pattern changes, or from soil movement (removal, erosion, and sedimentation). (Note: the accumulation of sediment could protect some localities by increasing the amount of protective cover.
- Unauthorized removal of artifacts or vandalism to the site could occur as a result of increases in human access to previously inaccessible areas if significant cultural resources are present.
- Visual impacts resulting from vegetation clearing, increases in dust, and the presence of large-scale equipment, machinery, and vehicles (if the affected cultural resources have an associated landscape or other visual component that contributes to their significance, such as Native American sacred landscape or a historic trail).

Ecological Resources

Ecological resources that could be affected include vegetation, fish, and wildlife, and their habitats. Vegetation and topsoil would be removed for the construction of the biomass energy facility, associated access roads, transmission lines, pipelines, and other ancillary facilities. This would lead to a loss of wildlife habitat, reduction in plant diversity, potential for increased erosion, and potential for the introduction of invasive or noxious weeds. The recovery of vegetation flowing interim and final reclamation would vary by the type of plant community desired. Dust settling on vegetation may alter or limit plants' abilities to photosynthesize and/or reproduce. Although the potential for an increase in the spread of invasive and noxious weeds would occur during the construction phase due to increasing traffic and human activity, the potential impacts could be reduced by interim reclamation and implementation of mitigation measures. Adverse impacts on wildlife could occur during construction from:

- Erosion and runoff
- Fugitive dust
- Noise
- Introduction and spread of invasive vegetation

- Modification, fragmentation, and reduction of habitat
- Mortality of biota (i.e., death of plants and animals)
- Exposure to contaminants
- Interference with behavioral activities.

Wildlife would be most affected by habitat reduction within the project site, access roads, and gas and water pipeline rights-of-way. Wildlife within surrounding habitats might also be affected if the construction activity (and associated noise) disturbs normal behaviors, such as feeding and reproduction. Depletion of surface waters from perennial streams could result in a reduction of water flow, which could lead to habitat loss and/or degradation of aquatic species.

Water Resources

With regard to water resources (surface water and groundwater), water used would be used for dust control when clearing vegetation and grading and for road traffic; for making concrete for foundations and ancillary structures; and for consumptive use by the construction crew. Water is likely to be obtained from nearby surface water bodies or aquifers, depending on availability, but could be trucked in from off site.

The bottom line on water for potable use always comes down to the Q and Q factors: Quantity and Quality. The quantity of water used would be small relative to water availability. Water quality could be affected by:

- Activities that cause soil erosion.
- Weathering of newly exposed soils could cause leaching and oxidation, thereby releasing chemicals into the water.
- Discharges of waste or sanitary water.
- Untreated groundwater used to control dust could deposit dissolved salts on the surface, allowing the salts to enter surface water systems.
- Chemical spills.
- Pesticide applications.

Surface and groundwater flow systems could be affected by withdrawals made for water use, wastewater and stormwater discharges, and the diversion of surface water flow for access road construction or stormwater control systems. A stormwater discharge permit might be required. Excavation activities and the extraction of geological materials could affect surface and groundwater flow. The interaction between surface water and groundwater could also be affected if the surface water and groundwater were hydrologically connected, potentially resulting in unwanted dewatering, or recharging of water resources.

Land Resources

Impacts on land use could occur during construction if there were conflicts with existing land use plans and community goals; conflicts with existing recreational, educational, religious, scientific, or other use areas; or conversion of the existing commercial land use for the area (e.g., agriculture, grazing, mineral extraction).

Existing land use during construction would be affected by intrusive impacts such as ground clearing, increased traffic, noise, dust, and human activity, as well as by changes in the visual landscape. In particular, these impacts could affect recreationists seeking solitude or recreation opportunities in a relatively pristine landscape. Ranchers or farmers could be effected by loss of available grazing or crop lands, potential for the introduction of invasive plants that could affect livestock forage availability, and possible increases in livestock/vehicle collision. An expanded access road system could increase the numbers of off-highway vehicle users, hunters, and other recreationists in the surrounding area.

Impacts on aviation could be possible if the project is located within 20,000 ft (6,100 m) or less of an existing public or military airport, or if the proposed construction involves objects greater than 200 ft (61 m) in height. The Federal Aviation Administration (FAA) must be notified if either of these two conditions occurs and the FAA would be responsible for determining if the project would adversely affect commercial, military, or personal air navigation safety. Similarly, impacts on military operations could occur if a project were located near a military facility if that facility conducts low-altitude military testing and training activities.

Soils and Geologic Resources

Sands, gravels, and quarry stone for construction access roads, making concrete for foundations and ancillary structures, and improving ground surface for laydown areas and crane staging areas would either be brought in from off-site sources or would be excavated on-site.

Depending upon the extent of excavation and blasting required to install access roads and support facilities, there is a limited risk of triggering geological hazards (e.g., landslides). Altering drainage patterns could also accelerate erosion and create slope instability.

Disturbed soil surfaces (crusts) now cover vast areas in the western United States as a result of ever-increasing recreational and commercial uses of these semi-arid and arid areas. Based on the results of several studies (McKenna-Neumann et al., 1996; Williams et al., 1995; Belnap and Gillette, 1997), the tremendous land area currently affected by human activity may lead to a significant increase in regional global wind erosion rates. Surface disturbance, heavy equipment traffic, and changes to surface runoff patterns resulting from biomass energy construction activities could cause soil erosion and impacts on special soils (e.g., cryptobiotic soil crusts; discussed below). Impacts of soil erosion could include soil nutrient loss and reduced water quality in nearby surface water bodies.

Cryptobiotic Soils Crust

With regard to the disturbance of cryptobiotic soil crusts, this is an important but often overlooked and not fully appreciated and/or understood soil disturbance problem; this is especially the case within the western and southwestern United States. Whether the renewable energy source is solar, wind, hydro, or biomass, the western and southwestern states are key players in harnessing and processing these energy sources.

As stated, cryptobiotic soil crusts, consisting of soil cyanobacteria, lichens, and mosses, play an important ecological role in the arid Southwest. In the cold deserts of the Colorado Plateau region (parts of Utah, Arizona, Colorado, and New Mexico), these crusts are extraordinarily well-developed, often representing over 70% of the living ground cover. Cryptobiotic crusts increase the stability of otherwise easily eroded soils, increase water infiltration in regions that receive little precipitation, and increase fertility in soils often limited in essential nutrients such as nitrogen and carbon (Harper and Marble, 1988; Johansen, 1993; Metting, 1991; Belnap and Gardener, 1993; Belnap, 1994; Williams et al., 1995).

Cyanobacteria occur as single cells or as filaments. The most common type found in desert soils is the filamentous type. The cells or filaments are surrounded by sheaths that are extremely persistent in these soils. When moistened, the cyanobacterial filaments become active, moving through the soils, and leaving a trail of the sticky, mucilaginous sheath material behind. This sheath material sticks to surfaces such as rock or soil particles, forming an intricate webbing of fibers in the soil. In this way, loose soil particles are joined together, and otherwise unstable and highly erosion-prone surfaces become resistant to both wind and water erosion. The soil-binding action is not dependent on the presence of living filaments. Layers of abandoned sheaths, built up over long periods of time, can still be found clinging tenaciously to soil particles at depths greater than 15 cm in sandy soils. This provides cohesion and stability in these loose sandy soils even at depth.

Cyanobacteria and cyanolichen components of these soil crusts are important contributors of fixed nitrogen. These crusts appear to be the dominant source of nitrogen in cold-desert pinyon-juniper and grassland ecosystems over much of the Colorado Plateau (Evans and Ehleringer, 1993). Biological soil crusts are also important sources of fixed carbon in sparsely vegetated areas common throughout the arid West (Beymer and Klopatek, 1991). Plants growing on crusted soil often show higher concentrations and/or greater total accumulation of various essential nutrients when compared to plants growing in adjacent, uncrusted soils (Belnap and Harper, 1995; Harper and Pendleton, 1993).

Cryptobiotic soil crusts are highly susceptible to soil-surface disturbance such as trampling by hooves or feet, or driving of off-road vehicles, especially in soils with low aggregate stability such as areas of sand dunes and sheets in the southwest, in particular over much of the Colorado Plateau (Belnap and Gardener, 1993; Gillette et al., 1980; Webb and Wilshire, 1983). When crusts in sandy areas are broken in dry periods, previously stable areas can become moving sand dunes in a matter of only a few years.

Cyanobacterial filaments, lichens, and mosses are brittle when dry and crush easily when subjected to compressional or shear forces by activities such as trampling or vehicular traffic. Many soils in these areas are thin and are easily removed without crust protection. As most crustal biomass is concentrated in the top 3 mm of the soil, very little erosion can have profound consequences for ecosystem dynamics. Because crustal organisms are only metabolically active when wet, re-establishment time is slow in arid systems. While cyanobacteria are mobile and can often move up through disturbed sediments to reach needed light levels for photosynthesis, lichens and mosses are incapable of such movement, and often die as a result. On newly disturbed surfaces, mosses and lichens often have extremely slow colonization and growth rates. Assuming adjoining soils are stable and rainfall is average, recovery rates for lichen cover in southern Utah have been most recently estimated at a minimum of 45 years, while recovery of moss cover was estimated at 250 years (Belnap, 1993).

Because of such slow recolonization of soil surfaces by the different crustal components, underlying soils are left vulnerable to both wind and water erosion for at least 20 years after disturbance (Belnap and Gillette, 1997). Because soils take 5,000–10,000 years to form in arid areas such as in southern Utah (Webb and Wilshire, 1983), accelerated soil loss may be considered an irreversible loss. Loss of soil also means loss of site fertility through loss of organic matter, fine soil particles, nutrients, and microbial populations in soils (Harper and Marble, 1988; Schimel et al., 1985). Moving sediments further destabilize adjoining areas by burying adjacent crusts, leading to their death, or by providing material for "sandblasting" nearby surfaces, thus increasing wind erosion rates (Belnap, 1995; McKenna-Neumann et al., 1996).

Soil erosion in arid lands is a global problem. Beasley et al. (1984) estimated that in rangeland of the United States alone, 3.6 million ha has undergone some degree of accelerated wind erosion. Relatively undisturbed biological soil crusts can contribute a great deal of stability to otherwise high erodible soils. Unlike vascular plant cover, crustal cover is not reduced in drought, and unlike rain crusts, these organic crusts are present year-round. Consequently, they offer stability over time and under adverse conditions that are often lacking in other soil surface protectors.

Paleontological Resources

Impacts on paleontological resources could occur directly from the construction activities or indirectly from soil erosion and increased accessibility to fossil locations. Potential impacts include:

- Complete destruction of the resource if present in areas undergoing surface disturbance or excavation.
- Degradation or destruction of near-surface fossil resources on- and off-site caused by changes to topography, changes to hydrological patterns, and soil movement (removal, erosion, and sedimentation). (Note: the accumulation of sediment could serve to protect some locations by increasing the amount of protective cover.)

- Unauthorized removal of fossil resources or vandalism to the site could occur as a result of increased human access to previously inaccessible areas if significant paleontological resources are present.

Transportation

Short-term increases in the use of local roadways would occur during the construction period. Heavy equipment likely would remain at the site. Shipments of materials are unlikely to affect primary or secondary road networks significantly, but this would depend on the location of the project site relative to the material source. Oversized loads could cause temporary transportation disruptions and could require some modifications to roads or bridges (such as fortifying bridges to accommodate the size or weight). Shipment weight might also affect the design of access roads for grade determinations and turning clearance requirements.

Visual Resources

The magnitude of visual impacts of the construction of a biomass facility is dependent upon the distance of the construction activities from the viewer, the view duration, and the scenic quality of the landscape. Possible sources of visual impacts during construction include:

- Ground disturbance and vegetation removal could result in visual impacts that produce contrasts of color, form, texture, and line. Excavation for foundations and ancillary structures; trenching to bury pipelines; grading and surfacing roads; cleaning and level staging areas; and stockpiling soil and spoils might be visible to viewers in the vicinity of the site. Soil scars and exposed slope faces would result from excavation, leveling, and equipment movement.
- Road development (new roads or expansion of existing roads) and parking areas could introduce strong visual contrasts in the landscape, depending on the route relative to surface contours, and the width, length, and surface treatment of the roads.
- Conspicuous and frequent small-vehicle traffic for worker access and frequent large-equipment (trucks, graders, excavators, and cranes) traffic for road construction, site preparation, and biomass facility construction could produce visible activity and dust in dry soils. Suspension and visibility of dust would be influenced by vehicle speeds and road surface materials.
- There would be a temporary presence of large equipment, producing emissions while operation and creating visible exhaust plumes. Support facilities and fencing associated with the construction work would also be visible.
- Night lighting would change the nature of the visual environment in the vicinity.

Socioeconomics

Direct impacts would include the creation of new jobs for construction workers and the associated income and taxes generated by the biomass facility. Indirect impacts would occur as a result of the new economic development and would include new jobs at businesses that support the expanded workforce or provide project materials, and associated income and taxes. Proximity to biomass facilities could potentially affect property values, either positively from increased employment effects or negatively from proximity to residences or local businesses and any associated or perceived environmental effect (noise, visual, etc.). Adverse impacts could occur if a large in-migrant workforce, culturally different from the local indigenous group, is brought in during construction. This influx of migrant workers could strain the existing community infrastructure and social services.

Environmental Justice

If significant impacts occurred in any resource areas, and these impacts disproportionately affected minority or low-income populations, then there could be an environmental justice impact. Issues of potential concern during construction are noise, dust, and visual impacts from the construction site and possible impacts associated with the construction of new access roads. Addition impacts include limitations on access to the area for recreation, subsistence, and traditional activities.

BIOMASS FEEDSTOCK PRODUCTION IMPACT

The impacts of biomass production are essentially the same as those of farming and forestry. The biomass production phase can be broken down into feedstock production and feedstock coordination. Feedstock production is the cultivation of crops such as corn, soybeans, or grasses and the collection of crop residues and wood residues from forests. These can be further categorized as primary, secondary, and tertiary resources:

- Primary feedstock includes grain and oilseed crops, such as corn or soybeans that are grown specifically to make biofuels; crop residues such as corn stover and straw; perennial grasses and woody crops; algae; logging residue; and excess biomass from forests.
- Secondary feedstock consists of manure from farm animals, food residue, wood processing mill residue, and pulping liquors.
- Tertiary feedstock includes municipal solids waste, municipal sanitary waste sludge, landfill gases, urban wood waste, construction and demolition debris, and packaging waste.

Feedstock coordination consists of harvesting or collecting the feedstock from the production area, processing it for use in a biomass facility, storing it to provide for a steady supply, and delivering it to the plant. The following potential impacts may result from biomass energy production activities.

Air Quality

Emissions generated during the feedstock production phase include vehicle emissions, diesel emissions from large equipment, emissions from storage/dispensing of fuels, and fugitive dust from many sources. The level of emissions would vary with the scale of operations and may be greater for agriculture operations than forestry operations. For feedstocks that do not require annual replanting (e.g., switchgrass, hybrid poplars) and cultivation (e.g., mill residues), or for algae, which is grown in enclosed aquaculture facilities, potential air emissions would be greatly reduced. If all vehicles and equipment have emission control devices and dust control measures are implemented, air emissions are unlikely to cause an exceedance of air quality standards.

The removal of biomass from forests can reduce the potential for major forest fires and limit the need for prescribed burns, thereby eliminating some air pollution sources. However, from a climate change perspective, large reductions in forest mass (clear-cutting) can remove biomass that served to capture carbon dioxide. Carbon dioxide, a "greenhouse," is considered a major contributor to climate change. Mechanisms that can capture or contain carbon dioxide, such as forests, are considered to be a viable mitigation measure against climate change.

Cultural Resources

Any cultural material presents on the surface or buried below the surface of existing agricultural areas has already been disturbed, some for many decades. The conversion of uncultivated land to agricultural use to produce feedstock for biomass facilities would disturb previously undisturbed land and could affect cultural resources on or buried below the surface. Harvesting and collecting biomass from the forests could also affect cultural resources on or buried below the surface associated with the harvesting. If new access roads were required, this construction could also affect cultural resources. These agricultural and forestry activities could affect areas of interest to Native Americans depending on their physical placement and/or level of visual intrusion. Surveys conducted prior to the commencement of farming uncultivated land or harvesting in the forest to evaluate the presence and/or significance of cultural resources in the area would assist developers in properly managing cultural resources so they can plan their project to avoid or minimize impacts on these resources.

Ecological Resources

Vegetation and wildlife, including threatened, endangered, and sensitive species and their critical habitats, have been displaced from years of crop production. Converting uncultivated or fallow land to crop agriculture will result in additional displacement of native vegetation and wildlife. Forest stand thinning improves the growth of the remaining trees and reduces fire hazard; however, some native wildlife populations may decline as a result of habitat loss, fragmentation, and disturbance due to forest openings resulting from road construction and biomass collection. The presence of workers could increase human disturbance to wildlife. Limiting work activities in the vicinity of any known active nesting sites would help protect wildlife.

Habitat alteration, including canopy cover and soil compaction, can degrade the habitat for native plant populations and provide for the establishment of invasive plant species. After clearing a given area of biomass, additional seeing of highly disturbed soils with native grasses and taking steps to prevent the spread of noxious weeds could minimize impacts. However, a lower abundance of birds is sometimes found in reforested areas compared with natural forest or grassland.

Water Resources

Agricultural land use can degrade water quality which results in runoff or the migration of nutrients, pesticides, and other chemicals into surface water and groundwater. Conversion of idle land to agricultural use would add to the degradation. Converting annual crops to perennial crops reduces the requirement for pesticides and fertilizers. If the conversion of idle land requires irrigation, then water would need to be withdrawn from surface water or groundwater sources. Large withdrawals could affect the water availability for other uses. Sedimentation from road construction and other ground-disturbing activities in forested regions could increase sedimentation levels in streams.

Land Resources

Demand for increasing amounts of agricultural biomass feedstock would convert the land and cropland pasture to the cultivation of perennial crops such as grass and wood crops. No change in land use in forests would occur as a result of clearing and thinning to remove biomass.

Soils and Geologic Resources

Crop residue left in the field and tree residue left in the forest help to maintain soil moisture, soil organic matter content, and soil carbon levels, and to limit wind erosion. Removing too much residue would be detrimental to the soil. Soil compaction in agricultural operations would result from multiple passes of equipment for crop residue collection. These impacts can be partially mitigated by converting land from annual crop to perennial biomass crop production. This would increase the organic matter content of the soils and maximize the potential benefits listed above. The application of pesticides affects soil quality by adding toxic chemical to the soil. Converting annual crops to perennial crops reduces the requirement for pesticides and fertilizers. Proper management of the type and quantities of pesticides can reduce the impact. Soil compaction, erosion, and topsoil loss result from logging operations. Biomass removal would utilize the same footprint as commercial harvesting activities and would not add to the amount of compacted or disturbed soil.

Paleontological Resources

Any paleontological resources present on the surface or buried below the surface of existing agricultural areas have already been disturbed. The conversion of uncultivated land to agricultural use to produce feedstock for biomass facilities would disturb previously undisturbed land and double affect paleontological resources on or buried below the surface. Harvesting and collecting biomass from the forests could also affect paleontological resources on or buried below the surface associated with the harvesting. If new access roads were required, this construction could also affect these resources.

Surveys conducted prior to the commencement of farming uncultivated land or harvesting in the forest to evaluate the presence and/or significance of paleontological resources in the area would assist developers in properly managing paleontological resources so they can plan their projects to avoid or minimize impacts on these resources.

Transportation

Increased road congestion from agricultural vehicles, logging trucks, and workers would occur for the duration of the activities in a given area. Transportation of collected biomass from the point of generation to storage facilities or to biomass energy production facilities could also result in impacts on the transportation system.

Visual Resources

Converting idle land to agricultural use would change the visual aspect of an area and probably would be noticeable to any nearby residents or travelers that are familiar with the area. Major timber harvests change the lock and character of hillsides and the views that local residents and tourists have of the forest. Collection of the residue biomass would not add to the visual degradation. Changes in the character of the forest resulting from the collection of biomass in forest thinning projects would most likely not be observed from a distance.

Socioeconomics

Direct impacts would include the creation of new jobs for farmworkers and the associated income and taxes generated by increase production of crops and grasses and new markets for crop residue. Increase in biomass collection in forested regions would also create new jobs. Indirect impacts would include new jobs at business that support the expanded workforce or provide farm and logging equipment materials and associated income taxes.

Environmental Justice

If significant impacts occurred in any resource areas, and these impacts disproportionately affect minority or low-income populations, then there could be an environmental justice impact. Issues of potential concern are noise, dust, and visual impacts from biomass production and harvesting and the potential construction of new access roads. Additional impacts include limitations on access to the area for recreation, subsistence, and traditional activities.

BIOMASS ENERGY OPERATIONS IMPACT

Operations activities that may cause environmental impacts to include the operation of the biomass energy facility, power generation, biofuel production, and associated maintenance activities. Typical activities during biomass facility operation include power generation or production of biofuels, and associated maintenance activities that would require vehicular access and heavy equipment operation when components are being replaced. Biomass power plants require pollution control devices to reduce emissions from combustion and large cooling systems. Water requirements vary greatly among the various biomass facilities. Potential impacts for these activities are presented below, by the type of affected resource.

Air Quality

Operation of biomass facilities results in emissions of criteria air pollutants and hazardous air pollutants (HAPs). Criteria air pollutants include particulate matter, carbon monoxide, sulfur oxides, nitrogen oxides, lead, and volatile organic compounds (VOCs). HAPs are 189 toxic chemicals, known or suspected to be carcinogens, which are regulated by the U.S. Environmental Protection

Agency as directed by the 1990 Clean Air Act. If the facility is in an area designed as "attainment" for all state and national ambient air quality standards (NAAQS), then emissions from operation, when added to the natural background levels, must not cause, or contribute to ambient pollution levels that exceed the ambient air quality standards.

In particular, the combustion of municipal solid wastes could result in trace quantities of mercury, other heavy metals, and dioxins in the air emissions. The use of Best Available Control Technology (BACT) would minimize the potential for adverse air quality impacts from biomass facilities. A gas-fired regenerative thermal oxidizer would reduce VOCs by 95%. Baghouses, which are a type of dust collector using fabric filters, control particulate matter. Enclosing the processing equipment in a slight negative pressure envelope in addition to the use of a baghouse could minimize fugitive dust emissions from milling operations.

The use of cultivated biomass fuel (i.e., fuel specifically grown for energy production) in place of possible fuels like coal, oil, and natural gas could result in are reduction in the amount of carbon dioxide that accumulates in the atmosphere only if the carbon released by combustion of biomass fuels is effectively recaptured by the next generation of feedstock plants. If the biomass source is not replaced by growing more plants, the carbon released in biomass combustion is not recaptured; therefore, these forms of biomass energy can only be considered to be carbon-free if the energy production cycle includes replacing the feedstock. Using perennial or fast-growing biomass plants, such as switchgrass or poplar hybrids, can increase the rate of carbon recapture. While the combustion of biomass fuels under these conditions can be considered to be carbon-free, in practice, any gains in terms of reduced carbon dioxide emissions are offset by carbon dioxide emissions associated with the use of fossil fuels in the cultivation, harvesting, and transportation of biomass feedstock. Certain agricultural practices (e.g., no-till agriculture and use of perennial feedstock crops) produce fewer carbon dioxide emissions than conventional practices. Biomass energy derived from waste product fuels (e.g., residues from forestry operations, construction wastes, municipal wastes) is not considered to be carbon-free as the energy production cycle does not involve any cultivation of new biomass.

Cultural Resources

Impacts during the operations phase would be limited to unauthorized collection of artifacts and visual impacts. The threat of unauthorized collection would be present once the access roads are constructed in the construction phase, making remote lands accessible to the public. Visual impacts resulting from the presence of a biomass facility and transmission lines could affect some cultural resources, such as sacred landscapes or historic trails.

Ecological Resources

During operation, adverse ecological effects could occur from (1) disturbance of wildlife by equipment noise and human activity, (2) exposure of biota to chemical spills and other contaminants, and (3) mortality of wildfire from increased vehicular traffic and collisions with and/or electrocution by transmission lines. Disturbed wildlife would be expected to acclimatize to facility operations.

Deposition of water and salts from the operation of mechanical-draft cooling towers has the potential to impact vegetation.

Water intake structures for the withdrawal of water from lakes or rivers would result in the impingement and entrainment of aquatic species. Proper design of these structures can minimize these impacts. Discharge of heated cooling water into water bodies could be beneficial or adverse, depending upon the design of the discharge structure and the temperature of the effluent.

Water Resources

Withdrawals of surface water and/or groundwater are expected to continue during the operations phase of both biomass power plants and biofuel production and refinery facilities. The amount of water needed depends upon the type of facility.

In a typical biomass power plant, the primary consumptive use of water will be to support the cooling system used to condense spent steam for reuse. Once-through cooling systems require large quantities of water to be withdrawn from and returned to a surface body of water. Wet recirculating cooling systems recycle cooling water through cooling towers where some portion of water is allowed to evaporate and must be continuously replenished. Wet recirculating cooling systems also periodically discharge small volumes of water as blowdown and replace that amount with fresh water to control chemical and biological contaminants to acceptable levels. The third type of cooling system, the dry cooling system condenses, and cools steam using only ambient air and requires no water to operate. However, some dry cooling systems can also be hybridized into wet/dry systems that use minimal amounts of water that are allowed to evaporate to improve performance. Other consumptive uses of water at a biomass power plant include the initial filling and maintenance of the steam cycle, sanitary applications to support the workforce, and a wide variety of incidental maintenance-related industrial applications.

Most uses of water at a biomass power plant will ultimately result in the generation of some wastewater. Blowdown from both the steam cycle and the wet recirculating cooling system will represent the largest wastewater stream and, because water in both the steam cycle and the cooling system undergoes some chemical treatment, the discharge will contain chemical residuals. Its temperature will also be elevated. Water discharged from once-through systems does not undergo chemical treatment, but the temperature of the discharge will be elevated. All wastewater discharges from biomass power plants can be directed to a holding pond for evaporation, cooling, and/or further treatment, but are likely to be eventually discharged to surface waters. The Clean Water Act requires any facility that discharges from a point source into water of the United States to obtain a National Pollutant Discharge Elimination System (NPDES) permit. The NPDES permit assures that the state's water quality standards are being met.

Water is used in a wide variety of applications for biofuel production and refining facilities and can be consumed at rates as high as 400 gallons per minute (gpm). Some water used in production and refining activities can be recovered and recycled to reduce the demand for the water source. Algae production ponds can be large but are very shallow (about 12 inches). Only a small volume of water would need to be added to replace any evaporation. Bioreactors for algae production are closed systems and require very little additional water. As much as 100 gpm of wastewater can be discharged from a biofuel production and refining plant. The effluent discharge temperature would be at or slightly above ambient temperature and would often contain small amounts of chemicals. As with wastewaters from biomass power plants, such discharges can be directed to lined holding ponds for further treatment or discharged directly to a surface water body under the authority of an EPA-issued NPDES permit.

Land Use

Any land use impacts would occur during construction, and no further impacts would be expected to result from biomass facility operation.

Soils and Geologic Resources

During operation, the soil and geologic conditions would stabilize with time. Soil erosion and soil compaction are both likely to continue to occur along access roads. With the project footprint, soil erosion, surface runoff, and sedimentation of nearby water bodies will continue to occur during operation, but to a lesser degree than during the construction phase.

Paleontological Resources

Impacts during the operations phase would be limited to unauthorized collection of fossils. This threat is present once the access roads are constructed in the construction phase, making remote lands accessible to the public.

Transportation

Increases in the use of local roadways and rail lines would occur during operations. Biomass fuels for boilers and power plants would arrive daily by truck or rail. Feedstock for biofuel facilities, such as corn, soybeans, wood products, manure, and sludge, would also arrive by truck or rail, and ethanol and biodiesel produced would most likely be trucked to the end user that would blend or sell the product. Depending upon the size and function of the facility, truck traffic, could be about 250 trucks per day. Biogas facilities would either combust the gas at the production plant or send it by pipeline to the user. Landfill gas would either be used to produce electricity near the point of collection or sent by pipeline to the user.

Visual Resources

The magnitude of visual impacts from operation of a biomass facility is dependent upon the distance of the facility from the viewer, the view duration, and the scenic quality of the landscape. Facility lighting would adversely affect the view of the night sky in the immediate vicinity of the facility. Plumes from stacks of cooling towers might be visible, particularly on cold days. Additional visual impacts would occur from the increase in vehicular traffic.

Socioeconomics

Direct impacts would include the creation of new jobs for operation and maintenance workers and the associated income and taxes paid. Indirect impacts are those impacts that would occur as a result of the new economic development and would include new jobs at businesses that support the workforce or that provide project materials, and associated income and taxes. The number of project personnel required during the operation and maintenance phase would be fewer than during construction. Therefore, socioeconomic impacts related directly to jobs would be smaller than during construction.

Environmental Justice

Possible environmental justice impacts during operation include the alteration of scenic quality in areas of traditional or cultural significance to the minority of low-income populations and disruption of access to those areas. Noise impacts, health and safety impacts, and water consumption in areas are also possible sources of disproportionate effects.

GEOTHERMAL ENERGY[4]

> The U.S. Geological Survey has calculated the heat energy in the upper 10 kilometers of the earth's crust in the U.S. is equal to over 600,000 times the country's annual non-transportation energy consumption. Probably no more than a tiny fraction of this energy could ever be extracted economically. However, just one hundredth of 1% of the total is equal to half the country's current non-transportation energy needs for more than a century, with only a fraction of the pollution from fossil-fueled energy sources.
>
> —*McLarty et al. (2000)*

> If we utilize waste biomass, solar (passive and thermal), wind (on shore), photovoltaic, geothermal, and other renewable resources available to us in the United States, we will exceed the demand (what we need) by at least five times as much energy as we need, all from clean, renewable sources.
>
> —*Frank R. Spellman*

Approximately 4,000 miles below the Earth's surface is the core where temperatures can reach 9,000°F. This heat—geothermal energy (*geo*, meaning Earth, and thermos, meaning *heat*)—is continuously created and flows outward from the core, heating the surrounding area, which can form underground reservoirs of hot water and steam and when drilled for accessibility for use can release energy like a geyser if not closed off by valve or throttled. These reservoirs can be tapped for a variety of uses, such as to generate electricity or heat buildings.

The geothermal energy potential in the uppermost 6 miles of the Earth's crust amounts to 50,000 times the energy of all oil and gas resources in the world. In the U.S., most geothermal reservoirs are located in the western states, Alaska, and Hawaii. However, geothermal heat pumps (GHPs), which take advantage of the shallow ground's stable temperature for heating and cooling buildings can be used almost anywhere.

Again, it is important to point out that there is nothing new about renewable energy. From solar power to burning biomass (wood) in the cave and elsewhere, humans have taken advantage of renewable resources from time immemorial. For example, hot springs have been used for bathing since Paleolithic times or earlier. The early Romans used hot springs to feed public baths and under floor heating. The world's oldest geothermal district heating system, in France, has been operating since the fourteenth century (Lund, 2007). The history of geothermal energy use in the United States is interesting and lengthy.

As mentioned, geothermal energy processes evolve around the natural heat of the Earth that can be used for beneficial purposes when the heat is collected and transported to the surface. To gain proper understanding of geothermal energy as it is used at present, it is important to define enthalpy; that is, the heat content of a substance per unit mass. The term "enthalpy" is used because temperature alone is not sufficient to define the useful energy content of a steam/water mixture. A mass of steam at a given temperature and pressure can provide much more energy than the same mass of water under the same conditions. Enthalpy is a function of:

$$\text{Pressure} + \text{Volume} + \text{Temperature} = \text{Enthalpy}$$

Geothermal practitioners usually classify geothermal resources as "high enthalpy" (water and steam at temperatures above about 180°C–200°C), "medium enthalpy" (about 100°C–180°C) and "low enthalpy" (<100°C). For this text, it is sufficient to think of temperature and enthalpy as going hand in hand.

Environmental Impact of Geothermal Power Development[5]

There are several potential environmental impacts from any geothermal power development. These include:

- **Gaseous emissions**: result from the discharge of noncondensable gases (NCGs) that are carried in the source stream to the power plant. For hydrothermal installations, the most common NCGs are carbon dioxide (CO_2) and hydrogen sulfide (H_2S), although species such as methane, hydrogen, sulfur dioxide, and ammonia are often encountered in lower concentrations.
- **Water pollution**: liquid streams from well drilling, stimulation, and production may contain a variety of dissolved minerals, especially for high-temperature reservoirs (>230°C). The amount of dissolved solids increases significantly with temperature. Some of these dissolved minerals (e.g., boron and arsenic) could poison surface or ground waters and also harm local vegetation. Liquid streams may enter the environment through surface runoff or through breaks in the well casing.

- **Solids emissions**: there is practically no chance for contamination of surface facilities or the surrounding area by the discharge of solids *per se* from the geofluid. The only conceivable situation would be an accident associated with a fluid treatment or minerals recovery system that somehow failed in a catastrophic manner and spewed removed solids onto the area.
- **Noise pollution**: from geothermal operations is typical of many industrial activities (DiPippo, 1991). The highest noise levels are usually produced during the well drilling, stimulation, and testing phases when noise levels ranging from about 80 to 115 decibels A-weighted (dBA) may occur at the plant fence boundary. During normal operations of a geothermal power plant, noise levels are in the 71–83 dB range at a distance of 900 m (DiPippo, 2005).
- **Land use**: land footprints for hydrothermal power plants vary considerably by site because the properties of the geothermal reservoir fluid and the best options for waste stream discharge (usual reinjection) are highly site-specific. Typically, the power plant is built at or near the geothermal reservoir because long transmission lines degrade the pressure and temperature of the geofluid. Although well fields can cover a considerable area, typically 5–10 km^2 or more, the well pads themselves will only cover about 2% of the area.
- **Land subsidence**: if geothermal fluid production rates are much greater than recharge rates, the formation may experience consolidation, which will manifest itself as a lowering of the surface elevation, i.e., this may lead to surface subsidence. This was observed early in the history of geothermal power at the Wairakei field in New Zealand where reinjection was not used. Subsidence rates in one part of the field were as high as 0.45 m per year (Allis, 1990). Wairakei used shallow wells in a sedimentary basin. Subsidence in this case is very similar to mining activities at shallow depths where raw minerals are extracted, leaving a void that can manifest itself as subsidence on the surface.
- **Induced seismicity**: in normal hydrothermal settings has not been a problem because the injection of waste fluids does not require very high pressures. However, the situation in the case of many EGS reservoirs will be different and requires serious attention. Induced seismicity continues to be under active review and evaluation by researchers worldwide.
- **Induced landslides**: there have been instances of landslides at geothermal fields. The cause of the landslide is often unclear. Many geothermal fields are in rugged terrain that is prone to natural landsides, and some fields actually have been developed atop ancient landslides.
- **Water use**: geothermal projects, in general, require access to water during several stages of development and operation.
- **Disturbance of natural hydrothermal manifestations**: although numerous cases can be cited of the compromising or total destruction of natural hydrothermal manifestations such as geysers, hot springs, mud pots, etc. by geothermal developments (Jones, 2006; Keam et al., 2005), EGS projects will generally be sited in non-hydrothermal areas and will not have the opportunity to interfere with such manifestations.
- **Disturbance of wildlife habitat, vegetation, and scenic vistas**: it is undeniable that any power generation facility constructed where none previously existed will alter the view of the landscape.
- **Catastrophic events**: accidents can occur during various phases of geothermal activity including well blowouts, ruptured steam pipes, turbine failures, fires, etc.
- **Thermal pollution**: although thermal pollution is currently not a specially regulated quantity, it does represent an environmental impact for all power plants that rely on a heat source for their motive force.

Specific environmental impacts related to geothermal energy exploration, drilling, construction, and operation and maintenance activities are discussed in the following sections.

Geothermal Energy Exploration and Drilling Impact

Activities during the resource exploration and drilling phase are temporary and are conducted at a smaller scale than those during the construction, operations, and maintenance phases. The impacts described for each resource would occur from typical exploration and drilling activities, such as localized ground clearing, vehicular traffic, seismic testing, positing of equipment, and drilling. Most impacts during the resource exploration and drilling phase would be associated with the development (improving or constructing) of access roads and exploratory and flow testing wells. Many of these impacts would be reduced by implementing good industry practices and restoring disturbed areas once drilling activities have been completed.

Air Quality

Emissions generated during the exploration and drilling phase include exhaust from vehicular traffic, and drilling rigs, fugitive dust from traffic on paved and unpaved roads. And the release of geothermal fluid vapors (especially hydrogen sulfide, carbon dioxide, mercury, arsenic, and boron, if present in the reservoir). Initial exploration activities such as surveying, and sampling would have minimal air quality impacts. Activities such as site clearing and grading, road construction, well pad development, sump pit construction, and the drilling of production and injection wells would have more intense exhaust-related emissions for 1–5 years. Impacts would depend upon the amount, duration, location, and characteristics of the emissions and the meteorological conditions (e.g., wind speed and direction, precipitation, and relative humidity). Emissions during this phase would not have a measurable impact on climate change. State and local regulators may require permits and air monitoring programs.

Cultural Resources

Cultural resources could be impacted if additional roads or routes are developed across or within the historic landscape of a cultural resource. Additional roads could lead to increased surface and subsurface disturbance that could increase illegal collection and vandalism. The magnitude and extent of impacts would depend on the current state of the resource and their eligibility for the *National Register of Historic Places*. Drilling activities could result in long-term impacts on archeological artifacts and historic buildings or structures if present. Surveys condition during this phase to evaluate the presence and/or significance of cultural resources in the area would assist developers in locating sensitive resources and siting project facilities in order to avoid or minimize impacts on these resources.

Ecological Resources

Most impacts on ecological resources (vegetation, wildlife, aquatic biota, special status species, and their habitats) would be low to moderate and localized during exploration and drilling (although impacts due to noise could be high). Activities such as the clearing and grading, road construction, well drilling, ancillary facility construction, and vehicle traffic, have the potential to affect ecological resources by disturbing habitat, increasing erosion and runoff, and creating noise at the project site. Impacts on vegetation include loss of native species and species diversity; increased risk of invasive species; increased risk of topsoil erosion and seed bank depletion; increased risk of fire; and alteration of water and seed dispersal.

Exploration and drilling activities have the potential to destroy or injure wildlife (especially species with limited mobility); disrupt the breeding, migration, and foraging behavior of wildlife; reduce habitat quality and species diversity; disturb habitat (e.g., causing loss of cover or food source); reduce the reproductive success of some species (e.g., amphibians). Accidental spills could be toxic to fish and wildlife. The noise from seismic surveys and drilling has a high potential to disturb wildlife and affect breeding, foraging, and migrating behavior. If not fenced or covered in netting, sump pits containing high concentrations of minerals and chemical from drilling fluids could adversely impact animals (e.g., birds, wild horses and burrows, and grazing livestock). Surveys conducted during this phase to evaluate the presence and/or significance of ecological resources in the area would assist developers in locating sensitive resources and siting project facilities in order to avoid or minimize impact on these resources.

Water Resources

Impacts on water resources during the exploration and drilling phase would range from low to high. Survey activities would have little or no impact on surface water of groundwater. Exploration drilling would involve some ground-disturbing activities that could lead to increased erosions and surface runoff. Drilling into the reservoir can create pathways for geothermal fluids (which are under high pressure) to rise and mix with shallower groundwater. Impacts of these pathways may include the alteration of the natural circulation of geothermal fluids and the usefulness of the resource. Geothermal fluids may also degrade the quality of shallow aquifers. Best management practices based on stormwater pollution prevention requirements and other industry guidelines would ensure that soil erosion and surface runoff are controlled. Proper drilling practices and closure and capping of wells can reduce the potential for drilling-related impacts.

Temporary impacts on surface water may also occur as a result of the release of geothermal fluids during and testing if they are not contained. Geothermal fluids are hot and highly mineralized and, if released to surface water, could cause thermal changes and changes in water quality. Accidental spills of geothermal fluids could occur due to well blowouts during drilling, leaks in piping or well heads, or overflow form sump pits. Proper well casing and drilling techniques would minimize these risks.

Extracting geothermal fluids could also cause drawdowns in connected shallower aquifers, potentially affecting connected springs or streams. The potential for these types of adverse effects is moderate to high; but may be reduced through extensive aquifer testing and selection combined with compliance with the state and federal regulations that protect water quality and the limitations of water rights as issued.

During the exploration and drilling phase, water would be required for dust control, making concrete, consumptive use by the construction crew, and drilling of wells. Depending on availability, it may be trucked in from off-site or obtained from local groundwater wells of nearby municipal supplies.

Land Use

Temporary and localized impacts on land use would result from exploration and drilling activities. These activities could create a temporary disturbance in the immediate vicinity of a surveying or drilling site (e.g., to recreational activities or livestock grazing). The magnitude and extent of impacts from constructing additional roads would depend on the current land use in the area; however, long-term impacts on land use would be minimized by reclaiming all roads and routes that are not needed once exploration and drilling activities are completed. All other land uses on land under well pads would be precluded as long as they are in operation. Exploration activities are unlikely to affect mining and energy development activities, military operations, livestock grazing, or aviation on surrounding lands. Activities affecting resources and values identified for protection areas would likely be prohibited.

Soils and Geologic Resources

Impacts on soils and geologic resources would be proportional to the amount of disturbance. The amount of surface disturbance and use of geologic materials during exploration would be minimal. Surface effects from vehicular traffic could occur in areas that contain special soils. The loss of biological or desert crusts can substantially increase water and wind erosion. Also, soil compaction due to development activities at the exploratory well pads and along access roads would reduce aeration, permeability, and water-holding capacity of the soils and cause an increase in surface runoff, potentially causing increased sheet, rill, and gully erosion. The excavation and reapplication of surface soils could cause the mixing of shallow soil horizons, resulting in a blending of soil characteristics

Green Engineering

and types. The blending would modify the physical characteristics of the soils, including structure, texture, and rock content, which could lead to reduced permeability and increased runoff from these areas. Soil compaction and blinding could also impact the viability of future vegetation. Any geologic resources within the areas of disturbance would not be accessible during the life of the development. Possible geological hazards (earthquakes, landslides, and subsidence) could be activated by drilling and blasting. Altering drainage patterns could also accelerate erosion and create slope instability.

Paleontological Resources

Paleontological Resources are nonrenewable resources. Disturbance to such resources, whether through mechanical surface3 disturbance, erosion, or paleontological excavation, irrevocably alters or destroys them. The potential for impacts on paleontological resources is high where grading for access roads and drilling sites intercept geologic units with important fossil resources. Seismic surveys, ground clearing, and vehicular traffic have the potential to impact the fossil resources at the surface. The disturbance caused by all these activities could increase illegal collection and vandalism. Surveys conducted during this phase to evaluate the presence and/or significance of paleontological resources in the area would assist developers in locating significant resources so they can be studied and collected or so that project facilities can be sited in other areas.

Transportation

No impacts on transportation are anticipated during the exploration and drilling phase. Transportation activities would be temporary and intermittent and limited to the low volume of light utility trucks and personal vehicles.

Visual Resources

Impacts on visual resources would be considered adverse if the landscape were substantially degraded or modified. Exploration and drilling activities would have only temporary and minor visual effects, resulting from the presence of drill rigs, workers, vehicles, and other equipment (including lighting for safety); and from vegetation damage, scarring of the terrain, and altering landforms or contours. Reclamation following exploration and drilling to restore visual resources to pre-disturbance conditions would lessen these impacts.

Socioeconomics

As the activities conducted during the exploration and drilling phase are temporary and limited in scope, they would not result in significant socioeconomic impacts on equipment, local services, or property values.

Environmental Justice

Exploration activities are limited and would not result in significant long-term impacts in any resource area; therefore, environmental justice is not expected to be an issue during this phase.

GEOTHERMAL ENERGY CONSTRUCTION IMPACT

Activities that may cause environmental impacts during construction include site preparation (e.g., clearing and grading); facility construction (e.g., geothermal power plant, pipelines, and transmission lines); and vehicular and pedestrian traffic. The construction of the geothermal power plant would disturb about 15–25 acres of land. Transmission line construction would disturb about 1 acre of land per mile to line. Impacts would be similar to but more extensive than those addressed for the exploration and drilling phase; however, many of these impacts would be reduced by implementing good industry practices and restoring disturbed areas once construction activities have been completed.

Air Quality

Emissions generated during the construction phase include exhaust from vehicular traffic and construction equipment, fugitive dust from traffic on paved and unpaved roads, and the release of geothermal fluid vapors (especially hydrogen sulfide, carbon dioxide, mercury, arsenic, and boron, if present in the reservoir). Activities such as site clearing and grading, power plant and pipeline system construction, and transmission line construction would have more intense exhaust-related emissions for 2–10 years. Impacts would depend upon the amount, duration, location, and characteristics of the emissions and the meteorological conditions (e.g., wind speed and direction, precipitation, and relative humidity). Emissions during this phase would not have a measurable impact on climate change. State and local regulators may require permits and air monitoring programs.

Cultural Resources

Potential impacts on cultural resources during the construction phase could occur due to land disturbance related to the construction of the power plant and transmission lines. Impacts include the destruction of cultural resources in areas undergoing surface disturbance and unauthorized removal of artifacts or vandalism as a result of human access to previously inaccessible areas (resulting in lost opportunities to expand scientific study and education and interpretive uses of these resources). In addition, for cultural resources that have as associated landscape component that contributes to their significance (e.g., sacred landscapes or historic trails), visual impacts could result from large areas of the exposed surface, increases in dust, and the presence of large-scale equipment, machinery, and vehicles. While the potential for encountering buried sites is relatively low, the possibility that buried sites would be disturbed during construction does exist. Unless the buried site is detected early in the surface-disturbing activities, the impact on the site can be considerable. Disturbance that uncovers cultural resources of significant importance that would otherwise have remained buried and unavailable could be viewed as a beneficial impact, provided the discovery results in study, curation, or recordation of the resource. Vibration, resulting from increased traffic and drilling/development activities may also have effects on rock art and other associated sites (e.g., sites with standing architecture).

Ecological Resources

Most impacts on ecological resources (vegetation, wildlife, aquatic biota, special status species, and their habitats) would be low to moderate and localized during the construction phase (although impacts due to noise could be high). Activities such as site clearing and grading, road construction, power plant construction, ancillary facility construction, and vehicle traffic have the potential to affect ecological resources by disturbing habitat, increasing erosion and runoff, and creating noise at the project site. Impacts on vegetation include loss of native species and species diversity; increased risk of invasive species; increased risk of topsoil erosion and seed bank depletion; increased risk of fire; and alteration of water and seed dispersal.

Construction activities have the potential to destroy or injure wildlife (especially species with limited mobility); disrupt the breeding, migration, and foraging behavior of wildlife; reduce habitat quality and species diversity; disturb habitat (e.g., causing loss of cover or food source); reduce the reproductive success of some species (e.g., amphibians). Accidental spills could be toxic to fish and wildfire. The noise from construction and vehicle traffic has a high potential to disturb wildfire and affect breeding, foraging, and migrating behavior. Wild horses, burros, and grazing livestock could be adversely affected by the loss of forage and reduced forage palatability (due to dust settlement on vegetation) and restricted movement around the development area.

Water Resources

Impacts on water resources during the construction phase would be moderate because of ground-disturbing activities (related to road, well pad, and power plant construction) that could lead to an increase in soil erosion and surface runoff. Impacts on surface water would be moderate but temporary and could be reduced by implementing best management practices based on stormwater pollution-preventing requirements and other industry guidelines. During the construction phase, water would be required for dust control, making concrete, and consumptive use by the construction crew. Depending on availability, it may be trucked in from off-site or obtained from local groundwater wells or nearby municipal supplies.

Land Use

Temporary and localized impacts on land use would result from construction activities. These activities could create a temporary disturbance in the immediate vicinity of a construction site (e.g., to recreational activities or livestock grazing). The magnitude and extent of impacts from constructing power plants and pipeline systems would depend on the current land use in the area; however, long-term impacts on land use would be minimized by reclaiming all roads and routes that are not needed once construction is completed. All other land uses on land under well pads, buildings, and structures would be precluded as long as they are in operation. Construction activities are unlikely to affect mining and energy development activities, military operations, livestock grazing, or aviation on surrounding lands. Activities affecting resources and values identified for protection areas would likely be prohibited.

Soils and Geologic Resources

Impacts on soils and geologic resources would be greater during the construction phase than for other phases of development because of the increased footprint and would be particularly significant if biological or desert crusts are disturbed. Construction of additional roads, well pads, the geothermal power plant, and structures related to the power plant (e.g., the pipeline system and transmission lines) would occur during this phase. Construction of well pads, the geothermal power plant, and structures related to the power plant, the pipeline system, access roads, and other project facilities could cause topographic changes. These changes would be minor, but long term. Soil compaction due to construction activities would reduce aeration, permeability, and water-holding capacity of the soils and cause and increase in surface runoff, potentially causing increased sheet, rill, and gully erosion. The excavation and reapplication of surface soils could cause the mixing of shallow soil horizons, resulting in a blending of soil characteristics and types. This blending would modify the physical characteristics of the soils, including structure, texture, and rock content, which could lead to reduced permeability and increased runoff from these areas. Soil compaction and blending could also impact the viability of future vegetation. Any geologic resources within the areas of disturbance would not be accessible during the life of the development. It is unlikely that construction activities would activate geologic hazards. However, altering drainage patterns or building on steep slopes could accelerate erosion and create slope instability. It is unlikely that construction activities would activate geologic hazards. However, altering drainage patterns or building on steep slopes could accelerate erosion and create slope instability.

Paleontological Resources

The potential for impacts on paleontological resources is high where grading and excavation intercept geologic units with important fossil resources. Ground clearing and vehicular traffic have the potential to impact the fossil resources at the surface. The disturbance caused by all these activities could increase illegal collection and vandalism. Disturbance that uncovers paleontological resources of significant importance that would otherwise have remained buried and unavailable could be viewed as a beneficial impact, provided the discovery results in study, collection, or recordation of the resource.

Transportation

Geothermal development would result in the need to construct and/or improve access roads and would result in an increase in industrial traffic. Overweight and oversized loads could cause temporary disruptions and could require extensive modifications to roads or bridges (e.g., widening roads or fortifying bridges to accommodate the size or weight of truck loads). An overall increase in heavy truck traffic would accelerate the deterioration of pavement, and required local government agencies to schedule pavement repair or replacement more frequently than under the existing traffic conditions. Increased traffic would also result in a potential for increased accidents within the project area. The locations at which accidents are most likely to occur are intersections used by project-related vehicles to turn onto or off of highways from access roads. Conflicts between industrial traffic and other traffic are likely to occur, especially on weekends, holidays, and seasons of high use by recreationists. Increased recreational use of the area could contribute to a gradual increase in traffic on the access roads.

Visual Resources

Impacts on visual resources would be considered adverse if the landscape were substantially degraded or modified. Construction activities would have only temporary and minor visual effects, resulting from the presence of workers, vehicles, and construction equipment (including lighting for safety); and from vegetation damage, dust generation, scarring of the terrain, and altering landforms or contours. Reclamation following construction to restore visual resources to pre-disturbance conditions would lessen these impacts.

Socioeconomics

Construction phase activities would contribute to the local economy by providing employment opportunities, monies to local contractors, and recycled revenues through the local economy. The magnitude of these benefits would vary depending on the resource potential. Construction of a typical 50-megawatt (MW) power plant and related transmission lines would require an estimated 387 mobs and $22.5 million in income but would vary depending on the community. Job availability would vary with different stages of construction. Expenditures for equipment, materials fuel, lodging, food, and other needs would stimulate the local economy over the duration of construction.

Economic impacts may occur if other land use activities (e.g., recreation, grazing, or hunting) are altered by geothermal development. Constructing facilities alter the landscape and could affect the nonmarket values of the immediate area. Many of these land uses may be compatible; however, it's possible that some land uses will be displaced by geothermal development.

Environmental Justice

Environmental justice impacts occur only if significant impacts in other resource areas disproportionately affect minority or low-income populations. It is anticipated that the development of geothermal energy could benefit low-income, minority, and tribal populations by creating job opportunities and stimulating local economic growth via project revenues and increased tourism. However, noise, dust, visual impacts, and habitat destruction could have an adverse effect on traditional tribal life ways and religious and cultural sites. The development of wells and ancillary facilities could affect the natural character of previously undisturbed areas and transform the landscape into a more industrialized setting. Development activities could impact the use of cultural sites for traditional tribal activities (hunting and plant-gathering activities, and areas in which artifacts, rock art, or other significant cultural sites are located).

GEOTHERMAL ENERGY OPERATION AND MAINTENANCE IMPACT

Typical activities during the operations and maintenance phase include operation and maintenance of production and injection wells and pipeline systems, operation and maintenance of the power plant, waste management, and maintenance and replace of facility components.

Air Quality

Emissions generated during the operations and maintenance phase include exhaust from vehicular traffic and fugitive dust from traffic on paved and unpaved roads, most of which would be generally limited to worker and maintenance vehicle traffic. In addition, emission could include the release of geothermal fluid vapors (especially hydrogen sulfide, carbon dioxide, mercury, arsenic, and boron, if present in the reservoir). Impacts would depend upon the amount, duration, location, and characteristics of the emissions and the meteorological conditions (e.g., wind speed and direction, precipitation, and relative humidity). Carbon dioxide emissions would be considerably less than for comparable power plants using fossil fuel. State and local regulators may require permits and air monitoring programs.

Cultural Resources

During the operations and maintenance phase, impacts on cultural resources could occur primarily from unauthorized collection of artifacts and from visual impacts. In the latter case, the presence of the aboveground structures could impact cultural resources with an associated landscape component that contributes to their significance, such as a scared landscape or historic trail. The potential for indirect impacts (e.g., vandalism and unauthorized collection) would be greater duration of the operations and maintenance phase compared to prior phases, due to its longer duration.

Ecological Resources

Most impacts on ecological resources (vegetation, wildlife, aquatic biota, special status species, and their habitat) would be less during the operations and maintenance phase than for the exploration and drilling and construction phases because no new drilling or construction activities would take place. However, operations and maintenance activities have the potential to affect ecological resources mainly be reducing the acreage for foraging and migrating animals, fragmenting habitat, and creating noise at the project site during the life cycle of the project (which could last up to 50 years). Some of these impacts could be significant. Increased human activity also increases the risk of fire, especially in and or semiarid areas. Application of herbicides to control vegetation along access roads, buildings, and power plant structures, would increase the risk of wildfire exposure to contaminants.

Water Resources

Impacts on water resources during the operations and maintenance phase result mainly from the water demands associated with operating a geothermal power plant. Water resources during operations would be needed for replenishment of the geothermal Resour through reinjection. However, because some water would be consumed by evaporation, additional water would need to be added to the system from another source. Makeup water to replace the evaporative losses and blowdown in a water-cooled power plant system would also be needed, depending on the type of power plant used (e.g., flash steam facilities can lose up to 20% of its cooling water due to evaporation, but binary plants are nonconsumptive because they use a closed-loop system). Water can also be lost due to pipeline failures or surface discharge for monitoring and testing the geothermal reservoir. The availability of water resources could be a limiting factor in siting or expanding a geothermal development at a given location. Cooling water or water from geothermal wells that is discharged to the ground or to an evaporation pond could affect the quality of shallow groundwater if allowed to percolate through the ground. However, the potential for this type of impact is considered minor or negligible because the facility would have to comply with the terms of the discharge permit required by the state.

Land Use

Impacts on land uses during the operations and maintenance phase are an extension of those that occurred during the exploration and drilling and construction phases. While, to some extent, land use can revert to its original uses (e.g., livestock grazing), many other uses (e.g., mining, farming, or hunting) would be precluded during the life span of the geothermal development. Mineral resources would remain available for recovery and operation and maintenance activities are unlikely to affect mining and energy development activities, military operations, livestock grazing, or aviation on surrounding lands.

Soils and Geologic Resources

Impacts on soils and geologic resources would be minimal during the operations and maintenance phase. The initial areas disturbed during the construction phase would continue to be used during standard operation and maintenance activities, but no additional impacts would occur unless new construction projects or drill sites are needed. Impacts associated with new construction projects or drill sites would be similar to those described for the exploration and drilling construction phases.

Paleontological Resources

The potential for impacts on paleontological resources would be limited primarily to the unauthorized collection of fossils. This threat is present once the access roads are constructed, making remote areas more accessible to the public. Damage to locations caused by OHV use could also occur. The potential for indirect impacts (e.g., vandalism and unauthorized collection) would be greater during the production phase compared to the drilling/development phase, due to the longer duration of the production phase.

Transportation

Daily traffic levels, particularly heavy truck traffic, would be expected to be lower during the operations and maintenance phase compared to other phases of geothermal development. For the most part, heavy truck traffic would be limited to period monitoring and maintenance activities at the well pads and power plant.

Visual Resources

Adverse impacts on visual resources would occur during the 10–30-year life of the geothermal development. Impacts during the operations and maintenance phase would result from the presence of facility structures and roads (where undeveloped land once stood), increased vehicular traffic to the site, and releases of steam plumes from the geothermal power plant. Periodic construction projects occurring through the life of the development would have impacts similar to those described in the construction phase.

Socioeconomics

Activities during the operations and maintenance phase would contribute to the local economy by providing employment opportunities, monies to local contractors, and recycled revenues through the local economy. The magnitude of these benefits would vary depending on the resource potential. Operations of a typical 50-MW power plant and related transmission line would require an estimated 93 jobs and $8 million in income but would vary depending on the community. Job availability would vary with different stages of construction. Expenditures for equipment, materials, fuel, lodging, food, and other needs would simulate the local economy over the duration of the project, which could last up to 50 years.

Economic impacts may occur if other land use activities (e.g., recreation, grazing, or hunting) are altered by geothermal development. Constructing facilities will alter the landscape and could affect the nonmarket values of the immediate area during the life of the geothermal development.

Environmental Justice

Possible environmental justice impacts during the operations and maintenance phase include the alteration of scenic quality in areas of traditional or cultural significance to minority populations. Noise, water, and health and safety impacts are also potential sources of disproportionate effects on minority or low-income populations.

MARINE AND HYDROKINETIC ENERGY

> When you get right down to it; that is, right down to the barebones or to the bone marrow of it all, hydrokinetic generators are nothing more than underwater windmills.
>
> **Frank R. Spellman**

Broadly categorized as marine and hydrokinetic energy systems, a new generation of waterpower technologies offers the possibility of generating electricity from water with the need for dams and diversions. There are numerous plans, both in the United States and internationally, to develop these energy conversion technologies. However, because the concepts are new, few devices have been deployed and tested in rivers and oceans. Even fewer environmental studies of these technologies have been carried out, and thus potential environmental effects remain mostly speculative (Pelc and Fujita, 2002; Cada et al., 2007; Michel et al., 2007; Boehlert et al., 2008). The following account is based on what we presently know or what we think we know (speculation and/or conjecture and/or assumption—guesswork) about the potential environmental impacts of hydrokinetic technologies.

The ocean can produce two types of energy: *thermal energy* from the Sun's heat, and *mechanical energy* from the tides and waves. Generating technologies for deriving electrical power from the ocean include tidal power, wave power, ocean thermal energy conversion, ocean currents, ocean winds and salinity gradients. Of these, the three most well-developed technologies are tidal power, wave power and *ocean thermal energy conversion* (OTEC). Tidal power requires large tidal differences which, in the U.S., occur only in Main and Alaska. Ocean thermal energy conversion is limited to tropical regions, such as Hawaii, and to a portion of the Atlantic coast. Wave energy has a more general application, with potential along the California boast. The western coastline of the United States has the highest wave potential; in California, the greatest potential is along the northern coast.

It is important to distinguish tidal energy from hydro power. Recall that hydro power is derived from the hydrological climate cycle, powered by solar energy, which is usually harnessed via hydroelectric dams. In contrast, tidal energy is the result of the interaction of the gravitational pull of the moon and, to a lesser extent, the Sun, on the seas. Processes that use tidal energy rely on the twice-daily tides, and the resultant upstream flows and downstream ebbs in estuaries and the lower reaches of some rivers, as well, in some cases, tidal movement out at sea. A dam is typically used to convert tidal energy into electricity by forcing the water through turbines, activating a generator. Meanwhile, wave energy, a very large potential resource to be tapped, uses mechanical power to directly activate a generator, to transfer to a working fluid, water, or air, which then drives a turbine or generator. Before discussing the thermal and mechanical energy potential of the ocean, we provide a basic understanding of oceans and especially their margins, where most, if not all, ocean energy is harnessed using present technology. The following section provides a foundation for better understanding of the ocean energy concepts presented later in this chapter.

OCEAN TIDES, CURRENTS, AND WAVES

Water is the master sculptor of Earth's surfaces. The ceaseless, restless motion of the sea is an extremely effective geologic agent. Besides shaping inland surfaces, water sculpts the coast. Coasts include sea cliffs, shores, and beaches. Seawater set in motion erodes cliffs, transports eroded debris along shores, and dumps it on beaches. Therefore, most coats retreat or advance. In addition to the unceasing causes of motion—wind, density of sea water, and rotation of the Earth—the chief agents in this process are tides, currents, and waves.

Tides

The periodic rise and fall of the sea (once every 12 hours and 26 minutes) produce the tides. Tides are due to the gravitational attraction of the moon and to a lesser extent, the Sun on the Earth. The moon has a larger effect on tides and causes the Earth to bulge toward the moon. It is interesting to note that at the same time the moon causes a bulge on Earth, a bulge occurs on the opposite side of the Earth due to inertial forces (further explanation is beyond the scope of this text). The effect of the tides is not too noticeable in the open sea, the difference between high and low tide amounting to about 2 ft. The tidal range may be considerably greater near shore, however. It may range from less than 2 ft to as much as 50 ft. The tidal range will vary according to the phase of the moon and the distance of the moon from the Earth. The type of shoreline and the physical configuration of the ocean floor will also affect the tidal range.

Currents

The oceans have localized movements of masses of seawater called ocean currents. These are the result of drift of the upper 50–100 m of the ocean due to drag by wind. Thus, surface ocean currents generally follow the same patterns as atmospheric circulation with the exception that atmospheric currents continue over the land surface while ocean currents are deflected by the land. Along with wind action, current may also be caused by tides, variation in salinity of the water, rotation of the Earth, and concentrations of turbid or muddy water. Temperature changes in water affect water density which, in turn, causes currents—theses currents cause seawater to circulate vertically.

Waves

Waves, varying greatly in size, are produced by the friction of wind on open water. Wave height and power depend upon wind strength and fetch—the amount of unobstructed ocean over which the wind has blown. In a wave, water travels in loops. Essentially an up-and-done movement of the water, the diameter of the loops decreases with depth. The diameter of loops at the surface is equal to wave height (h). Breakers are formed when the wave comes into shallow water near the shore. The lower part of the wave is retarded by the ocean bottom, and the top, having greater momentum, is hurled forward causing the wave to break. These breaking waves may do great damage to coastal property as the race across coastal lowlands driven by winds or gale or hurricane velocities.

Wave Energy[6]

Waves are caused by the wind blowing over the surface of the ocean. In many areas of the world, the wind blows with enough consistency and force to provide continuous waves. Wave energy does not have the tremendous power of tidal fluctuations, but the regular pounding of the waves should not be underestimated because there is tremendous energy in the ocean waves. The total power of waves breaking on the world's coastlines is estimated at between 2 and 3 million megawatts. In optimal wave areas, more than 65 MW of electricity could be produced along a single mile of shoreline, according to the U.S. Office of Energy Efficiency and Renewable Energy (EERE). In essence, because the wind is originally derived from the Sun, we can actually consider the energy in ocean waves to be a stored, moderately high-density form of solar energy. According to certain estimates, wave technologies could feasibly fulfill 10% of the global electricity supply if fully developed (World Energy, 2004). The west coast of the United States and Europe and the coast of Japan and New Zealand are good sites for harnessing wave energy.

Wave Energy Conversion Technology

While the development of modern wave energy concerted dates back to 1799 (Ross, 1995), the technology did not receive worldwide attention until the 1970s when an oil crisis occurred, and Stephen Salter published a notable paper about the technology in *Nature* in 1974 (Salter, 1974). In the early 1980s, after a significant drop in oil prices, technical setbacks and a general lack of confidence, progress slowed in the development of wave energy devices as a commercial source of electrical power. In the late 1990s, awareness of the depletion of traditional energy resources and the environmental impacts of the large utilization of fossil fuels significantly increase, thereby facilitating the development of green energy resources. The development of wave energy technology grew rapidly, particularly in oceanic countries such as Ireland, Denmark, Portugal, the United Kingdom, and the United States. Quite a few pre-commercial ocean devices were deployed. For example, a United States company, Ocean Power Technology, deployed one of their 150 kW wave energy conversion (WEC) systems in Scotland in 2011 (OPT, 2011). An Irish company, Wavebob, tested a one-quarter model in Galway Bay, Ireland, in 2006 (Wavebob, 2011). In Denmark, the half scale 600 kW Wave Star energy system was deployed at Hanstholm in 2009 (Wavestar, 2014), and a quarter-and-a-half-size model Wave Dragon was tested at Nissum Bredning in 2003 (Wave Dragon, 2014). Furthermore, international organizations, such as the International Energy Agency and the International Electrotechnical Commission (IEC), are heavily involved in the development of wave energy devices. In 2001, the International Energy Agency established an Ocean Energy System Implementation Agreement to facilitate the coordination of ocean energy studies between countries (IEA-OES, 2011). In 2007, the IEC established an Ocean Energy Technical Committee to develop ocean energy standards (IEC, 2011).

In the early 1970s, the harnessing of wave power focused on using floating devices such as Cockerell Rafts (a wave power hydraulic device), Salter Duck (curved-cam-like device that can capture 90% of waves for energy conversion), Rectifier (concerts A-C to D-C electricity), and the Clam (a floating rigid toroid—i.e., doughnut-shaped—that converts wave energy to electrical energy). Wave energy converters can be classified in terms of their location: fixed to the seabed, generally in shallow water; floating offshore in deep water; or tethered in intermediate depths. At present, these floating devices are not cost-effective and have very difficult moving problems. So current practice is to move in shore, sacrificing some energy but fixed devices, according to Tovey (2005), have several advantages, including:

- easier maintenance
- easier to land on device
- no mooting problem
- easier power transmission
- enhanced productivity
- better design life

Wave energy devices can be classified by means of their reaction system, but it is often more instructive to discuss how they interact with the wave field. In this context, each moving body may have lasted as either displace or reactor:

- **Displacer**: this is the body moved by the waves. It might be a buoyant vessel or a mass of water. If buoyant, the displacer may pierce the surface of the waves or be submerged.
- **Reactor**: this is the body that provides reaction to the displacer. As suggested above, it could be a body fixed to the seabed, or the seabed itself. It could also be another structure or mass that is not fixed but moves in such a way that reaction forces are created (e.g., by moving by a different amount or at different times). A degree of control over the forces acting on each body and/or acting between the bodies (particularly stiffness and damping characteristics) is often required to optimize the amount of energy captured.

In some designs, the reactor is actually inside the displacer, while in others it is an external body. Internal reactors are not subject to wave forces, but external ones may experience loads that cause them to move in ways similar to a displacer. This can be extended to the view that some devices do not have dedicated reactors at all, but rather a system of displacers whose relative emotions create a reaction system. There are three types of well-known Wave Energy Conversion devices: Point absorbers, terminators, and attenuators.

Point Absorber

A point absorber is a floating structure that absorbs energy in all directions by virtue of its movements at or near the water surface. It may be designed so as to resonate—that it moves with larger amplitudes than the waves themselves. This feature is useful to maximize the amount of power that is available for capture. The power take-off system may take a number of forms, depending on the figuration of displacers/reactors.

Terminator

A terminator is also a floating structure that moves at or near the water surface, but it absorbs energy in only a single direction. The device extends in the direction normal to the predominant wave direction, sot that as waves arrive, the device restrains them. Again, resonance may be employed, and the power take-off system may take a variety of forms.

Attenuator

An attenuator device is a long floating structure like the terminator but is orientated parallel to the waves rather than normal to them. It rides the waves like a ship and movements of the device at its bow and along its length can be restrained so as to extract energy. A theoretical advantage of the attenuator over the terminator is that its area normal to the waves is small and therefore the forces it experiences are much lower.

TIDAL ENERGY

The tides rise and fall in eternal cycles. Tides are changes in the level of the oceans caused by the gravitational pull of the moon and Sun, and the rotation of the Earth. The relative motions of these cause several different tidal cycles, including a semidiurnal cycle (with period 12 hours 25 minutes); a semi-monthly cycle—i.e., Spring or Neap Tides corresponding with the position of the moon; a semi-annual cycle—period about 178 days which is associated with the inclination of the Moon's orbit. This causes the highest spring tides to occur in March and September; and other long-term cycles—e.g., a 19-year cycle of the moon. Nearshore water levels can vary up to 40 ft, depending on the season and local factors. Only about 20 locations have good inlets and a large enough tidal range—about 10 ft—to produce energy economically (USDOE, 2010). The tide ranges have been classified as follows:

- *Mesomareal*, when the tidal rang is between 2 and 4 m
- *Macromareal*, when the tidal range is higher than 4 m

Tidal Energy Technologies

Some of the oldest ocean energy technologies use tidal power. Tidal power is more predictable than solar power and wind energy. All coastal areas consistently experience two high and two low tides over a period of slightly greater than 24 hours. For those tidal differences to be harness into electricity, the difference between high and low tide must be at least 5 m, or more than 16 ft. There are only about 40 sites on the Earth with tidal ranges of this magnitude. Currently, there are no tidal power plants in the United States. However, conditions are good for tidal power generation in both the Pacific Northwest and the Atlantic Northeast regions of the country. Tidal energy technologies include the following:

- **Tidal barrages**: a barrage or dam is a simple generation system for tidal plants that involves a dam, known as a barrage, across an inlet. Sluice gates (gates commonly used to control water levels and flow rates) on the barrage allow the tidal basin to fill on the incoming high tides and to empty through the turbine system on the outgoing tide, also known as

the ebb tide. There are two-way systems that generate electricity on both the incoming and outgoing tides. A potential disadvantage of a barrage tidal power system is the effect a tidal station can have on plants and animals in estuaries. Tidal barrages can change the tidal level in the basin and increase the amount of matter in suspension in the water (turbidity). They can also affect navigation and recreation.
- **Tidal fences**: these look like giant turnstiles. A tidal fence has vertical axis turbines mounted in a fence. All the water that passes is forced through the turbines. Some of these currents run at 5–8 knots (5.6–9 mph) and generate as much energy as winds of much higher velocity. Tidal fences can be used in areas such as channels between two landmasses. Tidal fences are cheaper to install than tidal barrages and have less impact on the environment tidal barrages, although they can disrupt the movement of large marine animals.
- **Tidal turbine**: arewind turbines in the water that can be located anywhere there is a strong tidal flow (function best where coastal currents run at between 3.6 and 4.9 knots—4 to 5.5 mph). Because water is about 800 times denser than air, tidal turbines have to be much sturdier than wind turbines. Tidal turbines are heaver ad more expensive to build but capture more energy.

Ocean Thermal Energy Conversion

The most plentiful renewable energy source in our planet by far is solar radiation: 170,000 TW ($170,000 \times 10^{12}$ W) fall on Earth. Because of its dilute and erratic nature, however, it is difficult to harness. To do so, that is, to capture this energy, we must employ the use of large collecting areas and large storage capacities; these requirements are satisfied on Earth only by the tropical oceans. We are all taught at an early age that oceans (and water in general) cover about 71% (or 2/3rds) of Earth's surface. In a fitting reference to the vast oceans covering the majority of Earth, Ambrose Bierce (1842–1914) commented: "A body of water occupying about two-thirds of the world made for man who has no gills." So, true, we have no gills; thus, for those who look out upon those vast bodies of water that cover the surface they might ask: What is their purpose? And, of course, this is a good question with several possible answers. In regard to renewable energy, we can look out upon those vast seas and wonder: How can we use this massive storehouse of energy for our own needs? Because it is so vast and deep it absorbs much of the heat and light that comes from the Sun. One thing seems certain: Our origin, past, present, and future lies within those massive wet confines we call oceans.

Potential Environmental (General) Impacts[7]

This section summarizes the potential (generalized) environmental impacts of new ocean energy and hydrokinetic technologies. Environmental issues that apply to all technologies include:

- alteration of river or ocean current or waves
- alteration of bottom, substrates, and sediment transport/deposition
- alteration of bottom habitats
- impacts of noise
- effects of electromagnetic fields
- toxicity of chemicals
- interference with animal movements and migrations
- designs that incorporate moving rotors or blades also pose the potential for injury to aquatic organisms from strike or impingement
- ocean thermal energy conversion technologies have unique environmental impacts

Alteration of Current and Waves

The extraction of kinetic energy from river and ocean currents or tides will reduce water velocities in the vicinity (i.e., near field) of the project (Bryden et al., 2004). Large numbers of devices in a river will reduce water velocities, increase water surface elevations, and decrease flood conveyance capacity. These effects would be proportional to the number and size of structures installed in the water. Rotors, foils, mooring and electrical cables, and field structures will all act as impediments to water movement. The resulting reduction in water velocities could, in turn, affect the transport and deposition of sediment, organisms living on or in the bottom sediments, and plants and animals in the water column. Conversely, moving rotors and foils might increase mixing in systems where salinity or temperature gradients are well defined. Changes in water velocity and turbulence will vary greatly, depending on distance from the structure. For small numbers of units, the changes are expected to dissipate quickly with distance and are expected to be only localized; however, for large arrays, the cumulative effects may extend to a greater area. The alterations of circulation/mixing patterns caused by large numbers of structures might cause changes in nutrient inputs and water quality, which could in turn lead to eutrophication, hypoxia, and effects on the aquatic food web.

The presence of floating wave energy converters will alter wave heights and structures, both in the near field (within meters of the units or project) and, if installed in large numbers, potentially in the far field (extending meters to kilometers out form the project). The above-water structures of wave energy converters will act as a localize barrier to wind and, thus, reduce wind-wave interactions. Michel et al. (2007) noted that many of the changes would not directly relate to environmental impacts; for example, impacts on navigational conditions, wave loads on adjacent structures, and recreation on nearby beaches (e.g., surfing, swimming) might be expected. Reduced wave action could alter bottom erosion and sediment transport and deposition (Largier et al., 2008).

Wave measurements at operating wave energy conversion projects have not yet been made, and the data will be technology and project-size specific. The potential reductions in wave heights are probably smaller than those for wind turbines due to the low profiles of wave energy devices. For example, ASR Ltd. (2007) predicted that the operation of wave energy conversion devices at the proposed Wave Hub (a wave power research facility off the coast of Cornwall) would reduce wave height at shorelines 5–20 kilometers (km) away by 3%–6%. Operation of six-wave energy conversion buoys (WEC; a version of OPT's Power Buoys) in Hawaii was not predicted to impact oceanographic conditions (DON, 2003). This conclusion was based on modeling analyses of wave height reduction due to both wave scattering and energy absorption. The proposed large spacing of buoy cylinders (51.5 m apart, compared to a buoy diameter of 4.5 m) resulted in predicted wave height reductions of 0.5% for a wave period (i.e., time in seconds, between the passage of consecutive wave crests past a fixed point) of 9 seconds (s) and less than 0.3% for a wave period of 15 seconds. Boehlert et al. (2008) summarized the changes in wave heights that were predicted in various environmental assessments. Recognizing that impacts will be technology- and location-specific, estimated wave height reductions ranged from 3% to 15%, with maximum effects closet to the installation and near the shoreline. Millar et al. (2007) used a mathematical model to predict that the operation of the Wave Hub, with WECs covering a 1 km by 3 km area located 20 km from shore, could decrease average wave heights by about 1 to 2 centimeters (cm) at the coastline. This represents an average decrease in wave height of 1%; a maximum decrease in the wave height of 3% was predicted to occur with a 90% energy-transmitting wave farm (Smith et al., 2007). Other estimates in other environmental settings predict wave height reductions ranging from 3% to 13% (Nelson, 2008). Largier et al. (2008) concluded that height and incident angle are the most important wave parameters for determining the effects of reducing the energy supply to the coast.

The effects of reduced wave heights on coastal systems will vary from site to site. It is known that the richness and density of benthic organism are related to such factors as relative tidal range and sediment grain size (e.g., Rodil and Lastra, 2004), so changes in wave height can be expected to alter benthic sediments and habitat for the benthic organism. Coral reefs reduce wave heights and dissipate wave and tidal energy, thereby creating valuable ecosystems (Roberts et al., 1992; Lugo-Fernandez et al., 1998). In other cases, wave height reductions can have long-term adverse effects. Estuary and lagoon inlets may be particularly sensitive to change in wave heights. For example, the construction of a storm-surge barrier across an estuary in the Netherlands permanently reduce both the tidal range and mean high water level by about 12% from original values, and numerous changes to the affected salt marshes and wetlands soils were observed (de Jong et al., 1994).

Tidal energy converters can also modify wave heights and structure by extracting energy from the underlying current. The effects of structural drag on currents were not expected to be significant (MMS, 2007), but few measurements of the effects of tidal/current energy devices on water velocities have been reported. A few tidal velocity measurements were made near a single, 150-kilowatt (kW) Stingray demonstrator in Yell Sound in the Shetland Islands (The Engineering Business Ltd, 2005). Acoustic Doppler Current Profilers were installed near the oscillating hydroplane (which travels up and down the water column in response to lift and drag forces) as well as upstream and downstream of the device. Too few velocity measurements were taken for firm conclusions to be made, but the data suggest that 1.5–2.0 m/s tidal currents were slowed by about 0.5 m/s downstream from the Stingray. In practice, multiple units will be spaced far enough apart to prevent a drop in performance (turbine output) caused by extraction of kinetic energy and localized water velocity reductions.

Modeling of the Wave Hub project in the United Kingdom suggested a local reduction in marine current velocities of up to 0.8 m/s, with a simultaneous increase in velocities of 0.6 m/s elsewhere (Michel et al., 2007). Wave energy converters are expected to affect water velocities less than submerged rotors and other, similar designs because only cables and anchors will interfere with the movements of tides and currents.

Tidal energy conversion devices will increase turbulence, which in turn will alter mixing properties, sediment transport and, potentially, wave properties. In both the near field and far field, extraction of kinetic energy form tides will decrease tidal amplitude, current velocities, and water exchange in proportion to the number of units installed, potentially altering the hydrologic, sediment transport, and ecological relationships of rivers, estuaries, and oceans. For example, Polagye et al. (2008) used an idealized estuary to model the effects of kinetic power extraction on estuary-scale fluid mechanics. The predicted effects of kinetic power extraction included (1) reduction of the volume of water exchanged through the estuary over the tidal cycle, (2) reduction of the tidal range landward of the turbine array, and (3) reduction of the kinetic power density in the tidal channel. These impacts were strongly dependent on the magnitude of kinetic power extraction, estuary geometry, tidal regime, and non-linear turbine dynamics.

Karsten et al. (2008) estimated that extracting a maximum of 7 gigawatts (GW) of power from the Minas Passage (Bay of Fundy) with in-stream tidal turbines could result in large changes in the tides of the Minas Basin (greater than 30%) and significant far-field changes (greater than 15%). Extracting 4 GW of power was predicted to cause less than a 10% change in tidal amplitudes, and 2.5 GW could be extracted with less than a 5% change. The model of Blanchfield et al. (2007) predicted that extracting the maximum value of 54 megawatts (MW) from the tidal current of Masset Sound (British Columbia) would decrease the water surface elevation within a bay and the maximum flow rate through the channel by approximately 40%. On the other hand, the tidal regime could be kept within 90% of the undisturbed regime by limiting extracted power to approximately 12 MW.

In the extreme far field (i.e., thousands of km), there is an unknown potential for dozens or hundreds of tidal energy extraction devices to alter major ocean current such as the gulf stream (Michel et al., 2007). The significance of these potential impacts could be ascertained by predictive modeling and subsequent operational monitoring as projects are installed.

Alteration of Substrates and Sediment Transport and Deposition

Operation of hydrokinetic or ocean energy technologies will extract energy from the water, which will reduce the height of waves or velocity of currents in the local area. This loss of wave/current energy could, in turn, alter sediment transport and the wave climate of nearby shorelines. Moreover, installation of many of the technologies will entail attaching the devices to the bottom by means of pilings or anchors and cables. Transmission of electricity to the shore will be through cables that are either buried in or attaché to the seabed. Thus, project installation will temporarily disturb sediments, the significance of which will be proportional to the amount and type of bottom substrate disturbed. There have been few studies of the effects of burying cables from ocean energy technologies, but experience with other buried cables and trawl fishing indicate the possible severity of the impacts. For example, Kogan et al. (2006) surveyed the condition of an armored, 6.6-cm-diameter coaxial cable that was laid on the surface of the seafloor off Half Moon Bay, California. The cable was not anchored to the seabed. Whereas the impacts of laying the cable on the surface of the seabed were probably small, subsequent movements of the cable had continuing impacts on the bottom substrates. For example, cable strumming by wave action in shallower, nearshore areas created incisions in rocky siltstone outcrops ranging from superficial scrapes to vertical grooves and had minor effects on the habitats of aquatic organisms. At greater depths, there was little evidence of the effects of the cable on the seafloor, regardless of exposure. Limited self-burial of the unanchored cable occurred over 8 years, particularly in deeper waters of the continental shelf.

During operation, changes in current velocities or wave heights will alter sediment transport, erosion, and sedimentation. Due to the complexity of currents and their interaction with structures, the operation of the projects will likely increase scour and deposition of fine sediments on both localized and far-field scales. For example, turbulent vortices that are shed immediately downstream from a velocity-reducing structure (e.g., rotors, pilings, concrete anchor blocks) will cause scour, and this sediment is likely to be deposited further downstream. On average, extraction of kinetic energy from currents and waves is likely to increase sediment deposition in the shadow of the project (Michel et al., 2007), the depth and areal extent of which will depend on local topography, sediment types, and characteristics of the current and the project. Subsequent deposition of sediments is likely to cause shoaling and a shift to a finer sediment grain size on the lee side of wave energy arrays (Boehlert et al., 2008). Scour and deposition should be considered in project development, but many of the high energy (high velocity) river and nearshore marine sites that could be utilized for electrical energy production are likely to have substrates with few or no fine sediments. Changes in scour and deposition will alter the habitat for bottom-dwelling plants and animals.

Loss of wave energy may lead to changes in longshore currents, reductions in the width and energy of the surf zone, and changes in beach and erosion and deposition patterns. Millar et al. (2007) modeled the wave climate near the Wave Hub electrical grid connection point off the north coast of Cornwall. The installation would be located 20 km off the coast, in water depths of 50–60 m. Arrays of WECs connected to the Wave Hub would occupy a 1 km × 3 km site. The mathematical model predicted that an array of WECs would potentially affect the wave climate on the nearby coast, by about 1–2 cm. It is unknown whether such small reductions in the average wave height would measurably alter sediment dynamics along the shore, given the normal variations in waves due to wind and storms.

Water quality will be temporarily affected by increased suspended sediments (turbidity) during installation and initial operation. Suspension of anoxic sediments may result in a temporary and localized decline in the dissolved oxygen content of the water, but dilution by oxygenated water current would minimize the impacts. Water quality may also be compromised by the mobilization of buried contaminated sediments during both construction and operation of the projects. Excavation to install the turbines, anchoring structures, and cables could release contaminants adsorbed to sediments, posing a threat to water quality and aquatic organisms. Effects on aquatic biota may range from temporary degradation of water quality (e.g., a decline in dissolved oxygen content) to biotoxicity and bioaccumulation of previously buried contaminants such as metals.

Impact of Habitat Alternation on Benthic Organisms

Installation and operation of hydrokinetic and marine energy projects can directly displace benthic (i.e., bottom-dwelling) plants and animals or change their habitats by altering water flows, wave structures, or substrate composition. Many of the designs will include a large anchoring system made of concrete or metal, mooring cables, and electrical cables that lead from the offshore facility to the shoreline. Electrical cables might simply be laid on the bottom, or they more likely will be anchored or buried to prevent movement. Large bottom structures will alter water flow, which may result in localized scour and/or deposition. The new structures will affect bottom habitats, consequently, changes to the benthic community composition and species interactions in the area defined by the project may be expected.

Displacement of Benthic Organisms by Installation of the Project

Bottom disturbances will result from the temporary anchoring of construction vessels; digging and refilling the trenches for power cables; and installation of permanent anchors, pilings, or other mooring devices. Motile organisms will be displaced, and sessile organism destroyed in the limited areas affected by these activities. Displaced organisms may be able to relocate if similar habitats exist nearby and those habitats are not already at carrying capacity. That is, each population has an upper limit on size, called the carrying capacity. Carrying capacity can be defined as being the optimum number of species' individuals that can survive in a specific area over time. Stated differently, the carrying capacity is the maximum number of species that can be supported in a bioregion. A pond may be able to support only a dozen frogs depending on the food resources for the frogs in the pond. If there were thirty frogs in the same pond, at least half of them would probably die because the pond environment wouldn't have enough food for them to live. Carrying capacity, symbolized as K, is based on the quantity of food supplies, the physical space available, the degree of predation, and several other environmental factors.

Species with benthic-associated spawning or whose offspring settle into and inhabit benthic habitats are likely to be most vulnerable to disruption during project installation. Temporary increase in suspended sediments and sedimentation down current from the construction area can be expected. The potential effects of suspended sediments and sedimentation on aquatic organism are periodically reviews (e.g., Newcombe and Jensen, 1996) when construction is completed, disturbed areas are likely to be recolonized by these same organisms, assuming that the substrate and habitats are restored to a similar state. For example, Lewis et al. (2003) found that numbers of calms and burrowing polychaetes (worms) fully recovered within 1 year after the construction of an estuarine pipeline, although fewer wading birds returned to forage on these invertebrates during the same time period.

ALTERATION OF HABITATS FOR BENTHIC ORGANISMS DURING OPERATION

Installation of the project will alter benthic habitats over the longer term if the trenches containing electrical cables are backfilled with sediments of different sizes or compositions than the previous substrate. Permanent structures on the bottom (ranging in size from anchoring systems to seabed-mounted generators or turbine rotors) will supplant the existing habitats. These new structures would replace natural hard substrates or, in the case of previously sandy area, add to the amount of hard bottom habitat available to benthic algae, invertebrates, and fish.

This could attract a community of rocky reef fish and invertebrate species (including biofouling organisms) that would not normally exist at that site. Depending on the situation, the newly created habitat could increase biodiversity or have negative effects by enabling introduced (exotic) benthic species to spread. Marine fouling communities developed on monopiles for offshore wind power plants are significantly different from the benthic communities on adjacent hard substrates (Wilhelmsson et al., 2006; Wilhelmsson and Malm, 2008).

Changes in water velocities and sediment transport, erosion, and deposition caused by the presence of new structures will alter benthic habitats, at least on a local scale. This impact may be more extensive and long-lasting than the effects of anchor and cable installation. Deposition of sand may impact seagrass beds by increasing mortality and decreasing the growth rate of plant shoots (Craig et al., 2008). Conversely, the deposition of organic matter in the wakes of marine energy devices could encourage the growth of benthic invertebrate communities that are adapted to that substrate. Mussel shell mounts that slough off form oil and gas platforms may create surrounding artificial reefs that attract a large variety of invertebrates (e.g., crabs, sea stars, sea cucumbers, anemones) and fish (Love et al., 1999). Accumulation of shells and organic matter in the areas would depend on the wave and current energy, activities of biota, and numerous other factors (Widdows and Brinsley, 2002). While the new habitats created by energy conversion structures may enhance the abundance and diversity of invertebrates, predation by fish attracted to artificial structures can greatly reduce the numbers of benthic organism (Davis et al., 1982).

Movements of mooring or electrical transmission cables along the bottom (sweeping) could be a continual source of habitat disruption during the operation of the project. For example, Kogan et al. (2006) found that shallow water wave action shifted a 6.6-cm-diameter, armored coaxial cable that was laid on the surface of the seafloor. The strumming action caused incisions in rocky outcrops, but the effects on seafloor organisms were minor. Anemones colonized the cable itself, preferring the hard structure over the nearby sediment-dominated seafloor. Some flatfishes were more abundant near the cable than at control sites, probably because the cable created a more structurally heterogeneous habitat. Sensitive habitats that may be particularly vulnerable to the effects of cable movements include macroalgae and seagrass beds, coral habitats, and other biogenic habitats like worm reefs and mussel mounds.

Renewable energy projects may also have benefits to some aquatic habitats and populations. The presence of a marine energy conversion project will likely limit most fishing activities and other access in the immediate area. Bottom trawling can disrupt habitats, and benthic communities in areas that are heavily fished tend to be less complex and productive than in areas that are not fished in that way (Kaiser et al., 2000; Jennings et al., 2001). Blyth et al. (2004) found that cessation of towed-gear fishing resulted in significantly greater total species richness and biomass of benthic communities compared to sites that were still fished. The value of these areas in which fishing is precluded (or, at least limited to certain gear types) by the energy project would depend on the species of fish and their mobility. For relatively sedentary animals, reserves less than 1 km across have augmented local fisheries, and reserves in Florida of 16 and 24 km^2 have sustained more abundant and sizable fish than nearby exploited areas (Gell and Roberts, 2003). On the other hand, the protection of long-lived, late-maturing, or migratory marine fish species may require much larger marine protected areas (greater than 500 km^2) than those envisioned for most energy developments (Kaiser, 2005; Blyth-Skyrme et al., 2006; Nelson, 2008).

Impact of Noise

Freshwater and marine animals rely on sound for many aspects of their lives including reproduction, feeding, predator and hazard avoidance, communication, and navigation (Popper, 2003; Wielgart, 2007). Consequently, underwater noise generated during installation and operation of a hydrokinetic or ocean energy conversion device has the potential to impact these organisms. Noise may interfere with sounds animals make to communicate or may drive animals from the area. If severe enough, loud sounds could damage their hearing or cause mortalities. For example, it is known to form experience with other marine construction activities that the noise created by pile driving creates sound pressure levels high enough to impact the hearing of harbor porpoises and harbor seals (Thomsen et al., 2006). The effects are less certain for fish (Hastings and Popper, 2005), although fish mortalities have been reported for some pile-driving activities (Longmuir and Lively, 2001; Caltrans, 2001). Noise generated during normal operations is expected to be less powerful, but could still disrupt the behavior of marine mammals, sea turtles, and fish at great distances from the source. Changes in animal behavior or physiological stresses could lead to decreased foraging efficiency, abandonment of nearby habitats, decrease reproduction, and increased mortality (NRC, 2005)—all of which could have adverse effects on both individuals and populations.

Construction and operation noise may disturb seabirds using the offshore and intertidal environment. Shorebirds will be disturbed by onshore construction and operations, causing them to abandon breeding colonies (Thompson et al., 2008). Pinnipeds (seals, sea lions, and walruses) may abandon onshore sites used for reproduction (rookeries) because of noise and other disturbing activities during installation. On the other hand, some marine mammals and birds may be attracted to the area by underwater sounds, lights, or increase prey availability.

There are many sources of sound/noise in the aquatic environment (NRC, 2005; Simmonds et al., 2003). Natural sources include wind, waves, earthquakes, precipitation, cracking ice, and mammal and fish vocalizations. Human-generated ocean noise comes from such diverse sources as recreational, military, and commercial ship traffic; dredging; construction; oil drilling and production; geophysical surveys; sonar; explosions; and ocean research (Johnson et al., 2008). Many of these sounds will be present in an area of new energy developments. Noises generated by marine and hydrokinetic energy technologies should be considered in the context of these background sounds. The additional noises from these energy technologies could result from installation and maintenance of the units, movements of internal machinery, waves striking the buoys, water flow moving over mooring and transmission cables, synchronous and additive non-synchronous sound from multiple unit arrays, and environmental monitoring using hydroacoustic techniques.

There are many ways to express the intensity and frequency of underwater sound waves (Wahlberg and Westerberg, 2005; Thomsen et al., 2006). An underwater acoustic wave is generated by the displacement of water particles. Consequently, the passage of an acoustic wave creates local pressure oscillations that travel through water with a given sound velocity. These two parameters, pressure, and velocity are used to define the intensity of an acoustic field, and therefore are useful for considering the effects of noise on aquatic animals.

The intensity of the acoustic field is defined as the vector product of the local pressure fluctuations and the velocity of the particle displacement. A basic unit for measuring the intensity of underwater noise is the sound pressure level (SPL). The SPL of a sound, given in decibels (dB), is calculated by:

$$\text{SPL}(\text{dB}) = 20 \log_{10} \left(P/P_o \right)$$

where P is a pressure fluctuation caused by a sound source, and P_o is the reference pressure, defined in underwater acoustics as 1 µPa at 1 m from the sources (Thomsen et al., 2006). Using the above formula, doubling the pressure of a sound (P) results in a 6 dB increase in SPL.

The sound pressure of a continuous signal is often expressed by a root means square (rms) measure, which is the square root of the mean value of squared instantaneous sound pressures integrated over time (Madsen, 2005). Like SPL, the resulting integration of instantaneous sound pressure levels is also expressed in dB re 1 µPa (rms). A rms level of safe exposure to received noise has been established for marine mammals; the lower limits for concern about temporary or permanent hearing impairments in cetaceans and pinnipeds are currently 180 and 190 dB re 1 µPa (rms) respectively (NMFS, 2003). However, Madsen (2005) argues that rms safety measures are insufficient and should be supplement by other estimates of the magnitude of noise (e.g., maximum peak-to-peak SPL in concert with a maximum received energy flux level).

Sound intensity is greatest near the sound source and, in the far field, decreases smoothly with distance. As the acoustic wave propagates through the water, intensity is reduced by geometric spreading (dilution of the energy of the sound wave as it spreads out from the source over a larger and larger area) and, to a lesser extent, absorption, refraction, and reflection (Wahlberg and Westerberg, 2005). Attenuation of sound due to spherical spreading in deep water is estimated by $20 \log r$, where r is the distance in m from the source (NRC, 2000). Assuming simple spherical spreading (no reflection from the sea surface or bottom) and the consequent transmission loss of SPL, a 190 dB source level would be reduced to 150 dB at 100 m. close to the source, changes in sound intensity vary in a more complicated fashion, particularly in shallow water, as a result of acoustic interference from natural or man-made sounds or where there are reflective surfaces (seabed and water surface).

Sound exposure level (SEL) is a measure of the cumulative physical energy of the sound event which considers both intensity and duration. SELs are computed by summing the cumulative sound pressure squared (p^2) over time and normalizing the time to 1 second. Because the calculation of the SEL for a given underwater sound source is a way to normalize to 1 second the energy of noise that may be much briefer (such as the powerful, but short impulses caused by pile driving), SEL is typically used to compare noise events of varying durations and intensities.

In addition to intensity, underwater noise will have a range of frequencies (Hz or cycles per $). For convenience, measurements of the potentially wide range of individual frequencies associated with noise are integrated into "critical bands" or filters; the width of a band is often given in 1/3-octave levels (Thomsen et al., 2006). Thus, sounds can be expressed in terms of the intensities (dB) at particular frequency (Hz) bands.

The National Research Council (NRC, 2000) pointed out that there are four fundamental properties of sound transmission in water relevant to the consideration of the effects of noise on aquatic animals:

1. The transmission distance of sound in seawater is determined by a combination of geometric spreading loss and an absorptive loss that is proportional to the sound frequency. Thus, the attenuation (weakening) of sound increases as its frequency increases.
2. The speed of a sound wave in water is proportional to the temperature.
3. The sound intensity decreasees with distance from the sound source. Transmission loss of energy (intensity) due to spherical spreading in deep water is estimated by $20 \log_{10} r$, where r is the distance in m from the source.
4. The strength of sound is measured on a logarithmic scale.

From these properties, it can be seen that high-frequency sounds will dissipate faster than low-frequency sounds, and a sound level may decrease by as much as 60 dB at 1 km from the source. Acoustic wave intensity of 180 dB is 10 times less intense than 190 dB, and 170 dB is 100 times less than 190 dB (NRC, 2000).

There is very little information available on sound levels produced by the construction and operation of ocean energy conversion structures (Michel et al., 2007). However, reviews of the construction and operation of European offshore wind farms provide useful information on the sensitivity of aquatic organism to underwater noise. For example, Thomsen et al. (2006) reported that pile-driving activities generate brief, but very high sound pressure levels over a broad band of frequencies (20–20,000 Hz). Single pulses are about 50–100 ms in duration and occur approximately 30–60 times per minute. The SEL at 400 m from the driving of a 1.5-m-diameter pile exceeded 140 dB re 1 µPa over a frequency range of 40–3,000 Hz (Betke et al., 2004). It usually takes 1–2 hours to drive one pile to the bottom. Sounds produced by the pile-driving impacts above the water's surface enter the water from the air and from the submerged portion of the pile, propagate through the water column, and into the sediments, from which they pass successively back into the water column. Larger-diameter, longer piles require relatively more energy to drive into the sediments, which results in higher noise levels. For example, the SPL associated with driving 3.5-m-diameter piles is expected to be roughly 10 dB greater than for a 1.5-m-diameter pile (Thomsen et al., 2006). Pile driving sound, while intense and potential damaging, would occur only during the installation of some marine and hydrokinetic energy devices.

Some ocean energy technologies will be secured to the bottom by means of moorings and anchors drilled into rock. Like pile-driving hydraulic drilling will occur during a limited time period, and noise generation will be intermittent. DON (2003) summarized underwater SPL measurements of three hydraulic rock drills; frequencies ranged from about 15 Hz to over 39 Hz and SPLs ranged from about 120 to 170 dB re 1 µPa. SPLs were relatively consistent across the entire frequency range.

During operation, the vibration of the device's gearbox, generator, and other moving components are radiated as sound into the surrounding water. Noise during the operation of wind farms is of much lower intensity than noise during construction (Thomsen et al., 2006; Betke et al., 2004), and the same may be true for hydrokinetic and ocean energy farms. However, this source of noise will be continuous. Measurements of sound levels associated with the operation of hydrokinetic and ocean energy farms have not yet been published. One example of a wave energy technology, the WEC buoy (a version of OPT's Power Buoy) that has been tested in Hawaii, has many of the mechanical parts contained within an equipment canister or mounted to a structure through mounting pads. Thus, the acoustic energy produced by the equipment is not well coupled to the seawater, which is expected to reduce the amount of radiated noise (EERE, 2003). Although no measurements had been made, it was predicted that the acoustic output from the WEC buoy system would probably be in the range of 75–80 dB re 1 µPa. This SPL is equivalent to light to normal density shipping nose, although the frequency spectrum of the WEC buoy is expected to be shifted to higher frequencies than typical shipping noise. By comparison, Thomsen et al., (2006) reported the ambient noise measured at fire at different locations in the North Sea. Depending on Frequency, SPL ranged from 85 to 115 dB, with most energy occurring at frequencies less than 100 Hz.

The Environmental Statement for the proposed installation of the Wave Dragon wave energy demonstrator off the coast of Pembrokeshire, UK predicted noise levels associated with the installation of concrete caisson (gravity) block and steel cable mooring arrangement, installation of subsea cable, and support activity (Wave Dragon Wales Ltd., 2007). The installation of gravity blocks is not expected to generate additional noise over and above that of the vessel conducting the operation. Vessel noise will depend on the size and design of the ship, but is expected to be up to 180 dB re 1 µPa at 1 m. Other predicted installation noise sources and levels stem from the operation of the ship's echosounder (220 dB re 1 µPa at 1 m peak-to-peak), cable laying and fixing (159–181 dB re 1 µPa at 1 m), and directional drilling (129 dB re 1 µPa rms at 40 m above the drill). There are no measurements available for the noise associated with the operation of an overtopping device such as the Wave Dragon. Wave Dragon Wales Ltd. (2007) predicted that operational noise would result from the Kaplan-style hydro turbines (an estimated 143 dB re 1 µPa at 1 m), as well as unknown levels and frequencies of sound from wave interactions with the body of the device, hydraulic pumps, and the mooing system.

In April 2008, the Ocean Renewable Power Company (ORPC) made limited measurements of underwater noise associated with the operation of their 1/3-scale working prototype instream tidal energy conversion device, its Turbine Generation Unit (TFU). The TGU is a single horizontal axis device with two advanced design cross-flow turbines that drive a permanent magnet generator. An omnidirectional hydrophone, calibrated for a frequency range of 20–250 kHz, was used to make near field measurement adjacent to the barge from which the turbine was suspended and at approximately 15 m from the turbine. Multiple far field measurements were also made at distance out to 2.0 km from the barge. Noise measurements were made over one full tidal cycle, with supplemental measurement taken later (EERE, 2009). Sound pressure levels at 1/3-octave frequency bands were used to calculate rms levels and SELs. During times when the turbine generator unit was not operating, background noise ranged from 112 to 138 dB re 1 µPa rms and SELs ranged from 120 to 140 dB re 1 µPa. A single measurement made when the turbine blades were rotating (at 52 rpm) resulted in an estimate of 132 dB re 1 µPa (rms) and an SEL of 126 dB re 1 µPa at a horizontal distance of 15 m and a water depth of 10 m. These very limited readings suggest that the single 1/3-scale turbine generator unit did not increase noise above ambient levels.

In addition to the sound intensity and frequency spectrum produced by the operation of individual machines, impacts of noise will depend on the geographic location of the project (water depth, type of substrate), the number of units, and the arrangement of multiple-unit arrays. For example, due to noise from surf and surface waves, noise levels in shallow, nearshore areas (≤ 100 m deep and within 5 km of the shore) are typically somewhat higher for low frequencies (≤ 1 kHz) and much higher frequencies above 1 kHz.

Because of the complexity of describing underwater sounds, investigators have often used different units to express the effects of sound on aquatic animals and have not always precisely reported the experimental conditions. For example, acoustic signal characteristics that might be relevant to biological effects include frequency content, rise time, pressure and particle velocity time series, zero-to-peak and peak-to-peak amplitude, means squared amplitude, duration, integral of mean squared amplitude over duration, sound exposure level, and repetition rate (NRC, 2005; Thomsen et al., 2006). Each of these sound characteristics may differentially impact different species of aquatic animals, but the relationships are not sufficiently understood to specify which are the most important. Many studies of the effects of noise report the frequency spectrum and some measure of sound intensity (SPL, rms, and/or SEL).

Underwater noise can be detected by fish and marine mammals if the frequency and intensity fall within the range of hearing for the particular species. An organism's hearing ability can be displayed as an audiogram, which plots sound pressure level (dB) against frequency (Hz). Nedwell et al. (2004) compiled audiograms for a number of aquatic organisms. If the pressure level of a generated sound is transmitted at these frequencies and exceeds the sound pressure level (i.e., above the line) on a given species' audiogram, the organism will be able to detect the sound. There is a wide range of sensitivity to sound among marine fish. The herrings (Clupeoidea) are highly sensitive to sound due to the structure of their swim bladder and auditory apparatus, whereas flatfish such as plaice and dab (Pleuronectidae) that have no swim bladder are relatively insensitive to sound (Nedwell et al., 2004). Possible responses to the received sound may include altered behavior (i.e., attraction, avoidance, interference with normal activities) Nelson, 2008) or, if the intensity is great enough, hearing damage or mortality. For example, fish kills have been reported in the vicinity of pile-driving activities (Longmuir and Lively, 2001; Caltrans, 2001).

The National Research Council (2000) reviewed studies that demonstrated a wide range of susceptibilities to exposure-induced hearing damage among different marine species. The implications are that critical sound levels will not be able to be extrapolated from studies of a few species (although a set of representative species might be identified), and it will not be possible to identify a single sound level value at which damage to the auditory system will begin at all, or even most, marine mammals. Participants in a recent NOAA workshop (Boehlert et al., 2008) suggested that sounds that are within the range of hearing and "sweep" in frequency are more likely to disturb marine mammals than constant-frequency sounds. Thus, devices that emit a constant frequency may be preferable to ones that vary. They believed that the same may be true, although perhaps to a lesser extent, for sounds that change in amplitude.

Moore and Clarke (2002) complied information on the reactions of gray whales (*Eschrichtius robustus*) to noise associated with offshore oil and gas development and vessel traffic. Gray whale response included changes in swim speed and direction to avoid sound sources, abrupt but temporary cessation of feeding, changes in calling rates and call structure, and changes in surface behavior. They reported a 0.5 probability of avoidance when continuous noise levels exceeded about 120 dB re 1 µPa and when intermittent noise levels exceeded about 170 dB re 1 µPa. They found little evidence that gray whales travel far or remain disturbed for long as a result of noises of this nature

Weilgart (2007) reviewed the literature on the effects of ocean noise on cetaceans (whales, dolphins, porpoises), focusing on underwater explosions, shipping, seismic exploration by the oil and gas industries, and naval sonar operations. She noted that strandings and mortalities of cetaceans have been observed even when estimated received sound levels were not high enough to cause hearing damage. This suggests that a change in diving patterns may have resulted in injuries due to gas and fat emboli (a fat droplet that enters the blood stream). That is, aversive noise may prompt cetaceans to rise to the surface too rapidly, and the rapid decompression causes nitrogen gas supersaturation and the subsequent formation of bubbles (emboli) in their tissues (Fernandez et al., 2005).

Other adverse (but not directly lethal) impacts could include increase stress elves, abandonment of important habitats, masking of important sounds, and changes in vocal behavior that may lead to reduced foraging efficiency or mating opportunities. Weilgart (2007) pointed out that responses of cetaceans to ocean noise are highly variable between species, age classes, and behavioral states, and many examples of apparent tolerance of noise have been documented.

Nowacek et al. (2007) reviewed the literature on the behavioral, acoustic, and physiological effects of anthropogenic noise on cetaceans, and concluded that the noise source of primary concern are ships, seismic exploration, sonars, and some acoustic harassment devices (AHDs) that are employed to reduce the by-catch of small cetaceans and seals by commercial fishing gear.

Two marine mammals whose hearing and susceptibility to noise have been studied are the harbor porpoise (*Phocoena phocoena*) and the harbor seal (*Phoca vitulina*). Both species inhabit shallow coastal waters in the north Atlantic and North Pacific. Harbor porpoises are found as far south as Central California on the West Coast. The hearing of the harbor porpoise ranges from below 1 kHz to around 140 kHz. In the United States, harbor seals range from Alaska to Southern California on the West Coast, and as far south as South Carolina on the East Coast. Harbor seal hearing ranges from less than 0.1 kHz to around 100 kHz (Thomsen et al., 2006). Sounds produced by marine energy devices that are outside of these frequency ranges would not be detected by these species.

Thomsen et al. (2006) compared the underwater noise associated with pile driving to the audiograms of harbor porpoises and harbor seals and concluded that pile-driving noise would likely be detectable at least 80 km away from the source. The zone of masking (the area within which the noise is strong enough to interfere with the detection of other sounds) may differ between the two species. Because the echolocation (sonar) used by harbor porpoises is in a frequency range (120–150 kHz) where pile-driving noises have little or no energy, they considered masking of echolocation to be unlikely. On the other hand, harbor seals communicate at frequencies ranging from 0.2 to 3.5 kHz, which is within the range of highest pile-driving sound pressure levels; thus, harbor seals may have their communications masked at considerable distances by pile-driving activities.

The responses of green turtles (*Chelonia mydas*) and loggerhead turtles (*Caretta caretta*) to the sounds of air guns used for marine seismic surveys were studied by McCauley et al. (2000a,b). They found that above a noise level of 166 dB re 1 µPa rms the turtles noticeably increased their swimming activity, and above 175 dB re 1 µPa rms their behavior became more erratic, possibly indicating that the turtles were in an agitated state. On the other hand, they were not able to detect an impact on turtles of the sounds produced by air guns in geophysical seismic surveys. Caged squid (*Sepioteuthis australis*) showed a strong startle response to an air gun at a received level of 174 dB re 1 µPa rms. When sound levels were ramped up (rather than a sudden nearby startup), the squid showed behavioral response (e.g., rapid swimming) at sound levels as low as approximately 156 dB re 1 µPa rms but did not display the startle response seen in the other tests.

Hastings and Popper (2005) reviewed the literature on the effects of underwater sounds on fish, particularly noises associated with pile driving. The limited number of quantitative studies found evidence of changes in the hearing capabilities of some fish, damage to the sensory structure of the inner ear, or, of fish close to the sources, mortality. They concluded that the body of scientific and commercial data is inadequate to develop more than the most preliminary criteria to protect fish from pile driving sounds and suggested the types of studies that could be conducted to address the information gaps. Similarly, Viada et al. (2008) found very little information on the potential impacts on sea turtles of underwater explosives. Although explosives produce greater sound pressures than pile driving and are unlikely to be used in most ocean energy installations, studies of their effects provide general information about the peak pressures and distances that have been used to establish safety zones for turtles.

Wahlberg and Westerberg (2005) compared source level and underwater measurements of sounds from offshore windmills to information about the heaving capabilities of three species of fish: goldfish, Atlantic salmon, and cod. They predicted that these fish could detect offshore windmills at a maximum distance of about 0.4–25 km, depending on wind speed, type and number of windmills, water depth, and substrate. They could find no evidence that the underwater sounds emitted by windmill operation would cause temporary or permanent hearing loss in these species, even at a distance of a few meters, although sound intensities might cause permanent avoidance within ranges of about 4 m. They noted that shipping causes considerably higher sound intensities than operating windmills (although the noise from shipping is transient), and noises from installation may have much more significant impacts on fish than those from the operation.

In the Environmental Assessment of the proposed Wave Energy Technology (WET) Project, DON (2003) considered the sounds made by hydraulic rock drilling to be detectable by humpback whales, bottlenose dolphins, Hawaiian spinner dolphins, and green sea turtles. Assuming a transmission loss due to spherical spreading, drilling sound pressure levels of 160 dB re 1 µPa would decrease by about 40 dB at 100 m from the source. They regarded a SPL of 120 dB re 1 µPa to be below the level that would affect these four species. They reported that other construction activities involving similar drilling attracted marine life, fish, and sea turtles in particular, perhaps because bottom organisms were stirred up by the drilling.

There are considerable information gaps regarding the effects of noise generated by marine and hydrokinetic energy technologies on cetaceans, pinnipeds, turtles, and fish. Sound levels from these devices have not been measured, but it is likely that installation will create more noise than operation, at least for those technologies that require pile driving. Operational noise form generators, rotating equipment, and other moving parts may have comparable frequencies and magnitudes to those measured at offshore wind farms; however, the underwater noise created by a wind turbine is transmitted down through the pilings, whereas noises from marine and hydrokinetic devices are likely to be greater because they are at least partially submerged. It is probable that noise from marine energy projects may be less than the intermittent noises associated with shipping and many other anthropogenic sound sources (e.g., seismic exploration, explosions, commercial, and naval sonar).

The resolution of noise impacts will require information about the device's acoustic signature (e.g., sound pressure levels across the full range of frequencies) for both individual units and multiple-unit arrays, similar characterization of ambient (background) noise in the vicinity of the project, the hearing sensitivity (e.g., audiograms) of fish and marine mammals that inhabit the area, and information about the behavioral response to anthropogenic noise (e.g., avoidance, attraction, changes in schooling behavior or migration routes). Simmonds et al. (2003) describe the types of *in situ* monitoring that could be carried out to develop information on the effects of underwater noise arising from a variety of activities. The studies include monitoring marine mammal activity in parallel with sound level monitoring during construction and operation. Baseline sound surveys would be needed against which to measure the added effects of energy generation. It will be important to measure the acoustic characteristics produced by both single units and multiple units in an array, due to the possibility of synchronous or asynchronous, additive noise produced by the array (Boehlert et al., 2008). Minimally, the operational monitoring would quantify the sound pressure levels across the entire range of sound frequencies for a variety of ocean/river conditions in order to assess how meteorological, current strength, and/or wave height conditions affect the sound generation and sound masking. The monitoring effort should consider the effects of marine fouling on noise production, particularly as it relates to mooring cables.

Impact of Electromagnetic Fields (EMF)

Underwater cables will be used to transmit electricity between turbines in an array (inter-turbine cables), between the array and a submerged step-up transformer (if part of the design), and from the transformer or array to the shore (CMAC, 2003). Ohman et al. (2007) categorize submarine electric cables into the following types: telecommunications cables; high voltage, direct current (HVDC) cables; alternating current three-phase power cables, and low-voltage cables. All types of cable will emit EMF in the surrounding water. The electric current traveling through the cables will induce magnetic fields in the immediate vicinity, which can in turn induce a secondary electrical field when animals move through the magnetic fields (CMAC, 2003).

In 1819, Hans Christian Oersted, a Danish scientist, discovered that a field of magnetic force exists around a single wire conductor carrying an electric current. The electromagnetic field (EMF) created by electric current passing through a cable is composed of both an electric field (E field) and an induced magnetic field (B field). Although E can be contained within undamaged insulation surrounding the cable, B fields are unavoidable and will in turn induce a secondary electric field (iE field). Thus, it is important to distinguish between the two constituents of the EMF (E and B) and the induced field, iE. Because the electric field is a measure of how the voltage changes when a measurement point is moved in a given direction, E and iE are expressed volts/m (V/m).

The intensity of a magnetic field can be expressed as magnetic field strength or magnetic flux density (CMAC, 2003). The magnetic field can be visualized as field lines, and the field strength (measured in amperes/m [A/m]) corresponds to the density of the field lines. Magnetic flux density is a measure of the density of magnetic lines or force, or magnetic flux lines, passing through an area. Magnetic flux density (measured in teslas[T]) diminishes with increasing distance form a straight current-carrying wire. At a given location in the vicinity of a current-carrying wire, the magnetic flux density is directly proportional to the current in amperes. Thus, the magnetic field B is directly linked to the magnetic flux density that is flowing in a given direction.

When electricity flows (electron flow) through the wire in a cable, every section of the wire has this field of force around it in a plane perpendicular to the wire. The strength of the magnetic field around a wire (cable) carrying a current depends on the current because it is the current that produces the field. The greater the current flow in a wire, the greater the strength of the magnetic field. A large current will produce many lines of force extending far from the wire, while a small current will produce only a few lines close to the wire.

The EMF associated with new marine and hydrokinetic energy designs has not been quantified. However, there is considerable experience with submarine electrical transmission cables, with some predictions and measurements of their associated electrical and magnetic fields. For example, the Wave Energy Technology (WET) generator will be housed in a canister buoy and connected to shore by a 1,190-m-long, 6.5-cm-diameter electrical cable. The cable is designed for three-phase AC transmission, can carry up to 250 kW, and has multiple layers of insulation and armoring to contain the electrical current. Depending on current flow (amperage), at 1 m from the cable, the magnetic field strength was predicted to range from 0.1 to 0.8 A/m and the magnetic flux density would range from 0.16 to 1.0 µT. The estimated strength of the electric field at the surface of the cable (apparently the iE) would range from 1.5 to 10.5 mV/m. The electric field strength, magnetic field strength, and magnetic flux density would all decrease exponentially with distance from the cable.

The Centre for Marine and Coastal Studies (CMAC, 2003) surveyed cable manufacturers and independent investigators to compile estimates of the magnitudes of E, B, and iE fields. Most agreed that the E field can be completely contained within the cable by insulation. Estimates of the B field strength ranged from zero (by one manufacturer) to 1.7 and 0.61 µT at distances of 0 and 2.5 m from the cable respectively. By comparison, the Earth's geomagnetic field strength ranges from approximately 20 to 75 µT (Bochert and Zettler, 2006). In another study cited by CMAC (2003), a 150 kV cable carrying a current of 600 A generated an induced electric field (iE) of more than 1 mV/m at a distance of 4 m from the cable; the field extended for approximately 100 m before dissipating. Lower voltage/amperage cables generated similarly large iE fields near the cable, but the fields dissipated much more rapidly with distance.

For short-distance undersea transmission of electricity, three-phase AC power cables are most common; HVDC is used for longer-distance, high-power applications (Ohman et al., 2007). In AC cables the voltage and current alternate sinusoidally at a given frequency (50 or 60 Hz), and therefore the E and B fields are also time-varying. That is, like AC current, the magnetic field induced by a three-phase AC current has a cycling polarity, which is not like the natural geomagnetic fields. On the other hand, the E and B fields produced by a direct current (DC) cable (e.g., HVDC) are static. Because the magnetic fields induced by DC and AC cables are different, they are likely to be perceived differently by aquatic organisms.

Because neither sand nor seawater has magnetic properties, burying a cable will not affect the magnitude of the magnetic (B) field; that is, the B fields at the same distance from the cable are identical, whether in water or sediment (CMAC, 2003). On the other hand, due to the higher conductivity of seawater compared to sand, the iE field associated with a buried cable is discontinuous across the sand/water boundary; the iE field strength is greater in water than in sand at a given distance from the cable. For example, for the three-phase AC cable modeled by CMAC (2003), the estimated iE field strengths at 8 m from the cable were 10 μV/m and 1 to 2 μV/m in water and sand, respectively.

The EMF generated by a multi-unit array of marine or hydrokinetic devices will differ from EMF associated with a single unit or from the single cable sources that have been surveyed. Depending on the power generation device, a project may have an electrical cable running vertically through the water column in addition to multiple cables running along the seabed or converging on a subsea pod. The EMF created by a matrix of cables has not been predicted or quantified.

Effect of Electromagnetic Fields on Aquatic Organisms
Electrical Fields

Natural electric fields can occur in the aquatic environment as a result of biochemical, physiological, and neurological processes within an organism or as a result of an organism swimming through a magnetic field (Gill et al., 2005). Some of the elasmobranchs (e.g., sharks, skates, rays) have specialized tissues that enable them to detect electric fields (i.e., electroreception), an ability which allows them to detect prey and potential predators and competitors. Two species of Asian sturgeon have been reported to alter their behavior in changing electric fields (Basov, 1999, 2007). Other fish species (e.g., eels, cod, Atlantic salmon, catfish, paddlefish) will respond to induced voltage gradients associated with water movement and geomagnetic emissions (Colin and Whitehead, 2004; Wilkens and Hoffman, 2005), but their electrosensitivity does not appear to be based on the same mechanism as sharks (Gill et al., 2005).

Balayev and Fursa (1980) observed the reaction of 23 species of marine fish to electric currents in the laboratory. Visible reactions occurred following exposure to electric fields ranging from 0.6 to 7.2 V/m and varied depending on the species and orientation to the field. They noted that changes in the fishes' electrocardiograms occurred at field strength 20 times lower than those that elicited observable behavioral response. Enger et al. (1976) found that European eels (*Anguilla anguilla*) exhibited a decelerated heart rate when exposed to a direct current electrical field with a voltage gradient of about 400–600 μV/cm. In contrast, Rommel and McCleave (1972) observed much lower voltage thresholds of response (0.07–0.67 μV/cm) in American eels (*Anguilla rostrata*). The eels' electrosensitivity measured by Rommel and McCleave is well within the range of naturally occurring oceanic electric fields of at least 0.10 μV/cm in many currents in the Atlantic Ocean and up to 0.46 μV/cm in the Gulf Stream.

Kalmijn (1982) described the extreme sensitivity of some elasmobranchs to electric fields. For example, the skate (*Raja clavata*) exhibited cardiac response to uniform square-wave fields of 5 Hz at voltage gradients as low as 0.01 μV/cm. Dogfish (*Mustelus canis*) initiated attacks on electrodes form distances in excess of 38 cm and voltage gradients as small as 0.005 μV/cm.

Marra (1989) described the interactions of elasmobranchs with submarine optical communications cables. The cable created an iE field (1 μV/m at 0.1 m) when sharks crossed the magnetic field induced by the cable. The sharks respond by attacking and biting the cable. Marra (1989) was unable to identify the specific stimuli that elicited the attacks, but he suggested that at close range the shark interpreted the electrical stimulus of the iE field as prey, which it then attacked.

The weak electric fields produced by swimming movements of zooplankton can be deterred by juvenile freshwater paddlefish (*Polyodon spathula*). Woytenek et al. (2001) used dipole electrodes to create electric fields that simulated those created by water flea (*Daphnia sp.*) swimming. They tested the effects of alteration current oscillations at frequencies ranging from 0.1 to 50 Hz and stimulus intensities ranging from 0.125 to 1.25 µA peak-to-peak amplitude. Paddlefish made significantly more feeding strike sat the electrodes at sinusoidal frequencies of 5–15 Hz compared to lower and higher frequencies. Similarly, the highest strike rate occurred at the intermediate electric field strength (stimulus intensity of 0.25 µA peak-to-peak amplitude). Strike rate was reduced at higher water conductivity, and their fish habituated (ceased to react) to repetitive dipole stimuli that were not reinforced by prey capture.

Gill and Taylor (2002; cited in CMAC, 2003) carried out a pilot study of the effects on dogfish of electric fields generated by a DC electrode in a laboratory tank. They reported that the dogfish avoided constant electric fields as small as 1,000 µV/m, which would be produced by 150 kV cables with a current of 600 A. Conversely, the dogfish were attracted to a field of 10 µV/m at 0.1 m from the source, which is similar to the bioelectric fields emitted by dogfish prey. The electrical field created by the three-phase, AC cable modeled by CMAC (2003) would likely be detectable by a dogfish (or other similarly sensitive elasmobranchs) at a radial distance of 20 m. It is possible that the ability of fish to discriminate an electrical field is a function of not only the size/intensity but also the frequency (Hz) of the emitted field.

Like elasmobranches, sturgeon (closely related to paddlefish) can utilize electroreceptor senses to locate prey and may exhibit varying behavior at different electric field frequencies (Bullock, 2005). For this reason, electrical fields are a concern as they may impact migration or ability to find prey. The National Marine Fisheries Service (NMFS) proposed critical habitat for the Southern distinct population segment of the threatened North American green sturgeon (*Acipenser medirostris*) along the coastline out to the 110 m isobath line (70 FR 52084-52110; September 8, 2008). One of the principal constituent elements in the proposal is safe passage along the migratory corridor. Green sturgeons migrate extensively along the nearshore coast from California to Alaska, and there is concern that these fish may be deterred from migration by either low-frequency sounds or electromagnetic fields created during the operation of marine energy facilities.

Magnetic Fields

Many terrestrial and aquatic animals can sense the Earth's magnetic field and appear to use this magneto sensitivity for long-distance migrations. Aquatic species whose long-distance migrations or spatial orientation appear to involve magnetoreception include eels (cited in CMAC, 2003), spiny lobsters (Boles and Lohmann, 2003), elasmobranchs (Kalmijn, 2000), sea turtles (Lohmann and Lohmann, 1996), rainbow trout (Walker et al., 1988), tuna, and cetaceans (Wiltschko and Wiltschko, 1995; Lohmann et al., 2008a). Four species of Pacific salmon were found to have crystals of magnetite within them, and it is believed that these crystals serve as a compass that orients to the Earth's magnetic field (Mann et al., 1988; Walker et al., 1988). Because some aquatic species use the Earth's magnetic field to navigate or orient themselves in space, there is a potential for the magnetic fields created by the numerous electrical cables associated with offshore power projects to disrupt these movements.

Gill et al. (2005) placed magneto-sensitive organisms into two categories: (1) those able to detect the iE field caused by movement through a natural or anthropogenic magnetic field, and (2) those with detection systems based on ferromagnetic minerals (i.e., magnetite or greigite). Johnsen and Lohmann (2005, 2008) add a third possible mechanism for magnetosensitivity—chemical reactions involving proteins known as crytochromes (i.e., a class of flavoproteins that are sensitive to blue light; involved in circadian rhythm entrainment in plants, insects, and mammals). Those species using the iE mode may either do it passively (i.e., the animal estimates its drift from the electric fields produced by the interaction between tidal/wind-driven currents and the vertical component of the Earth's magnetic field) or actively (i.e., the animal derives its magnetic compass heading form its own interaction with the horizontal component of the Earth's magnetic field). For example, Kalmijn (1982) suggested that the electric fields that elasmobranchs induce by swimming through the Earth's magnetic field may allow them to detect their magnetic compass headings; the resulting voltage gradients may range from 0.05 to 0.5 µV/cm. Detection of a magnetic field based on internal deposits of magnetite occurs in a wide range of animals, including birds, insects, fish, sea turtles, and cetaceans (Gould, 1984; Bochert and Zettler, 2006). There is no evidence to suggest that seals are sensitive to magnetic fields (Gill et al., 2005).

Westerberg and Begout-Aranas (1999; cited in CMAC, 2003) studied the effects of a B field generated by a HVDC power cable on eels (*Anguilla anguilla*). The B field was on the same order of magnitude as the Earth's geomagnetic field and, coming from a DC cable, was also a static field. Approximately 60% of the 25 eels tracked crossed the cable, and the authors concluded that the cable did not appear to act as a barrier to the eel migration. In another behavioral study, Meyer et al. (2004) showed that conditioned sandbar and scalloped hammerhead sharks readily respond to localized magnetic fields of 25–100 µT, a range of values that encompasses the strength of the Earth's magnetic field.

Some sea turtles (see Sidebar 12.1) undergo transoceanic migrations before returning to nest on or near the same beaches where they were hatched. Lohmann and Lohmann (1996) showed that sea turtles have the sensory abilities necessary to approximate their global position of a magnetic map. This would allow them to exploit unique combinations of magnetic field intensity and field line inclination in the ocean environmental to determine the direction and/or position during their long-distance migrations. Irwin and Lohmann (1996) found that magnetic orientation in loggerhead sea turtles (*Caretta caretta*) can be disrupted at least temporarily by strong magnetic pulses (i.e., five brief pulses of 40,000 µT with a 4 ms rise time). The impact of a changed magnetic environment would depend upon the role of magnetic information in the hierarchy of cues used to orient/navigated (Wiltschko and Wiltschko, 1995). Juvenile loggerheads deprived of either magnetic or visual information were still able to maintain a direction of orientation, but when both cues were removed, the turtles were disoriented (Avens and Lohmann, 2003). The magnetic map sense exhibited by hatchlings is also thought to allow female sea turtles to imprint upon the location of their natal beaches so that later in life they can return there to nest. This phenomenon is termed "natal homing" (Lohmann et al., 2008b), and it serves to drive genetic division among subpopulations of the same species. As a result, altering magnetic fields near nesting beaches could potentially result in altered nesting patterns. Given the important role of magnetic information in the movements of sea turtles, impacts of magnetic field disruption could range from minimal (i.e., temporary disorientation near a cable or structure) to significant (i.e., altered nesting patterns and corresponding demographic shifts resulting from large-scale magnetic field changes) and should be carefully considered with siting projects.

SIDEBAR 12.1 WITH REGARD TO TURTLES

Marine turtles have outlived almost all of the prehistoric animals with which they once shared the planet. Five species of marine turtles frequent the beaches and offshore waters of the southeastern United States (Escambia, 2007):

- **Loggerhead**: is the most common turtle to nest in Florida. Over 50,000 loggerhead nests are recorded annually in Florida. This turtle is named for its disproportionately large head and feeds on crabs, mollusks, and jellyfish.
- **Green**: is the second most common turtle in Florida waters. Green sea turtles are the only herbivorous sea turtles. They feed on seagrasses in shallow areas through the Gulf. The lower jaw is serrated to help cut the seagrasses it eats.
- **Kemp's Ridley**: are the rarest sea turtle in the world. They primarily nest on one beach on the gulf coast of Mexico and are the smallest species of sea turtle. Scientists have been trying to transplant Kemp's Ridley eggs to Texas to establish a new nesting colony. They are the only species of sea turtle known to lay their eggs during the day.

- **Leatherback**: is the largest sea turtle in the world and can be over 6 ft long and weigh 1,400 pounds. It does not have a hard shell, but rather a leather-like carapace with bony ridges underneath the skin. The leatherback makes long migrations to and from its nesting beaches in the tropics as far north as Canada. Jelly fish are the favored prey to these turtles.
- **Hawksbill**: is usually found feeding primarily on sponges in the southern Gulf of Mexico and the Caribbean. The hawksbill sea turtle was hunted to near extinction for its beautiful shell which features overlapping scales.

All five are reported to nest, but only the loggerhead and green turtle do so in substantial numbers. Most nesting occurs from southern North Carolina to the middle west coast of Florida, but scattered nesting occurs from Virginia through southern Texas. The beaches of Florida, particularly in Brevard and Indian River counties, host wheat may be the world's largest population of loggerheads (Dodd, 1995).

Marine turtles, especially juveniles and subadults, use lagoons, estuaries, and bays as feeding grounds. Areas of particular importance include Chesapeake Bay, Virginia (for loggerheads and Kemp's Ridleys); Pamlico sound, North Carolina (for loggerheads); and Mosquito Lagoon, Florida, and Laguna Madre, Texas (for greens). Offshore waters also support important feeding grounds such as Florida Bay and the Cedar Keys, Florida (for green turtles), and the mouth of the Mississippi River and the northeast Gulf of Mexico (for Kemp's Ridleys).

Offshore reefs provide feeding and resting habitat (for loggerheads, greens, and hawksbills), and offshore currents, especially the Gulf Stream, are important migratory corridors (for all species, but especially leatherbacks).

- **Interesting Point**: Raccoons destroy thousands of sea turtle eggs each year and are the single greatest cause of sea turtle mortality in Florida.

Most marine turtles spend only part of their lives in U.S. waters. For example, hatchling loggerheads ride oceanic currents and gyres (giant circular oceanic surface currents) for many years before returning to feed as subadults in southeastern lagoons. They travel as far as Europe and the Azores and even enter the Mediterranean Sea, where they are susceptible to longline fishing mortality. Adult loggerheads may leave U.S. waters after nesting and spend years in feeding grounds in the Bahamas and Cuba before returning. Nearly the encircle world population of Kemp's Ridleys uses a single Mexican beach for nesting, although juveniles and subadults, in particular, spend much time in U.S. offshore waters (Dodd, 1995).

The biological characteristics that make sea turtles difficult to conserve and manage include a long-life span, delayed sexual maturity, differential use of habitats both among species and life stages, adult migratory travel, high egg and juvenile mortality, concentrated nesting, and vast areal dispersal of young and subadults. Genetic analyses have confirmed that females of most species return to their natal beaches to nest (Bowen et al., 1992; Bowen et al., 1993). Nesting assemblages contain unique genetic markers showing a tendency toward isolation from other assemblages (Bowen et al., 1993); thus, Florida green turtles are genetically different from green turtles nesting in Costa Rica and Brazil (Bowen et al., 1992). Nesting on warm sandy beaches puts the turtles in direct conflict with human beach use and their use of rich off-shore waters subjects them to mortality from commercial fisheries (National Research Council, 1990).

Marine turtles have suffered catastrophic declines since European discovery of the New World (National Research Council, 1990). In a relatively short time, the huge nesting assemblages in the Cayman Islands, Jamaica, and Bermuda were decimated. In the United States, commercial turtle fisheries once operated in south Texas (Doughty, 1984), Cedar Keys, Florida Keys, and Mosquito Lagoon; these fisheries collapsed from overexploitation of the mostly juvenile green turtle populations. Today, marine turtle populations are threatened world-wide and are under intense pressure in the Caribbean basin and Gulf of Mexico, including Cuba, Mexico, Hispaniola, and Bahamas, and Nicaragua. Marine turtles can be conserved only though international efforts and cooperation (Dodd, 1995).

A number of interesting questions related to turtle migration remain unsolved. For example, how do turtles find their way precisely back to their natal beach over their vast travel distance? Do turtles imprint, as salmon do, on olfactory features in the water or is the location pinpointed using geomagnetic information?

Goff et al. (1998) point out that sea turtles have migration pattern somewhat similar to that of salmon. After hatching and entering the sea and facing and surviving the tribulations presented by the elements and predators, and after spending time in their sea-feeding grounds, the females return to their natal grounds. Adult females lay eggs in the sand. Goff et al. (1998) point out that turtles may use the geomagnetic field to tell them their location and to lead them to their natal grounds.

The emphasis of most of these studies is on the value of magnetoreception for navigation; marine and hydrokinetic energy technologies are unlikely to create magnetic fields strong enough to cause physical damage. For example, Bochert and Zettler (2006) summarized several studies of the potential injurious effects of magnetic fields on marine organisms. They subjected several marine benthic species (i.e.., flounder, blue mussel, prawn, isopods, and crabs) to static (DC-induced) magnetic fields of $3,700\,\mu T$ for several weeks and detected no differences in survival compared to controls. In addition, they exposed shrimp, isopods, echinoderms, polychaetes, and young flounder to a static, $2,700\,\mu T$ magnetic field in laboratory aquaria where the animals could move away from or toward the source of the field. At the end of the 24-hours test period, most of the test species showed a uniform distribution relative to the source, not significantly different from controls. Only one of the species, the benthic isopod *Saduria entomon*, showed a tendency to leave the area of the magnetic field. The oxygen consumption of two North Sea prawn species exposed to both static (DC) and cycling (AC) magnetic fields were not significantly different from controls. Based on these limited studies, Bochert and Zettler (2006) could not detect changes in marine benthic organisms' survival, behavior, or a physiological response parameter (e.g., oxygen consumption) resulting from magnetic flux densities that might be encountered near an undersea electrical cable.

The current state of knowledge about the EMF emitted by submarine power cables is too variable and inconclusive to make an informed assessment of the effects on aquatic organisms (CMAC, 2003). Following a thorough review of the literature related to EMF and extensive contacts with the electrical cable and offshore wind industries, Gill et al. (2005) concluded that there are significant gaps in knowledge regarding sources and effects of electrical and magnetic fields in the marine environment. They recommended developing information about likely electrical and magnetic field strengths associated with existing sources (e.g., telecommunications cables, power cables, electrical heating cables for oil and gas pipelines), as well as the generating units, offshore sub-stations and transformers, and submarine cables that are a part of offshore renewable energy projects. They cautioned that networks of cables in close proximity to each other (as would be substations) are likely to have overlapping, and potentially additive, EMF fields. These combined EMF fields would be more difficult to evaluate than those emitted from a single, electrical cable. The small, time-varying B field emitted by a submarine three-phase AC cable may be perceived differently by sensitive marine organisms than the persistent, static, geomagnetic field generated by the Earth (CMAC, 2003).

Toxic Effect of Chemicals

Chemicals that are accidentally or chronically released from hydrokinetic and ocean energy installations could have toxic effects on aquatic organisms. Accidental releases include leaks of hydraulic fluids from a damaged unit or fuel from a vessel due to a collision with the unit; such events are unlikely but could potentially have a high impact (Boehlert et al., 2008). On the other hand, chronic releases of dissolved metals or organic compounds used to control biofouling in marine applications would result in low, predictable concentrations of contaminants over time. Even at low concentrations that are not directly lethal, some contaminants can cause sublethal effects on sensory systems, growth, and behavior of animals; they may also be bioaccumulated (i.e., bioaccumulation is the biological concentration mechanism whereby filter feeders such as limpets, oysters and other shellfish concentrate heavy metals or other stable compounds present in dilute concentrations in the sea or freshwater).

Toxicity of Paints, Anti-Fouling Coatings, and Other Chemicals

Biofouling (growth on external surfaces by algae, barnacles, mussels, and other marine organisms) will occur rapidly in ocean applications (Wilhelmsson and Malm, 2008). Sundberg and Langhamer (2005) observed that a 3-m-diameter buoy may accumulate as much as 300 kg of biomass on the buoy and mooring cables, whereas siting devices in deeper water with even slight currents will exhibit reduced biofouling. The encrustation of biofouling organism could cause undesirable mechanical wear or changes in the weight, shape, and performance of energy conversion devices that would require increased maintenance or the application of antifouling measures. Encrustation by barnacles and other organisms could increase corrosion and fatigue and decrease electrical generating efficiency.

Michel et al. (2007) noted that there are three-options for removing marine biofouling: (1) use of antifouling coatings, (2) in situ cleaning using a high-pressure jet spray, and (3) removal of the device from the water for cleaning on a floating platform or onshore. Antifouling coating hinders the development of marine encrustations by slowly releasing a biocide such as tribuyltin (TBT), copper, or arsenic. As the coatings wear away, they must be reapplied periodically. There are concerns about the immediate toxicity of these biocides to other, non-targeted organisms, and numerous countries and organizations have called for the ban of TBT as an anti-fouling coating (Antizar-Ladislao, 2008). As a result, alternative coatings are being explored. The release of toxic contaminants from a single unit may be relatively minor, but the cumulative impacts of persistent toxic compounds from dozens or hundreds of units may be considerable (Boehlert et al., 2008). Accumulations of biofouling organisms (e.g., barnacles) removed from the project structures may alter nearby bottom substrates and habitats.

Accidental releases of hydraulic fluids and lubricating oils from inside the energy conversion device or from vessels used to install and service the equipment could have toxic effects. At the least, leaks of inert (non-toxic) oils could cause physical/mechanical effects by coating organisms and blanketing the sediments.

Interference with Animal Movement and Migration

Energy developments will add new structures to rivers and oceans that may affect the movements and migrations of aquatic organisms. Hydrokinetic devices, and their associated anchors and cables in a river, could attract or repel animals or interfere with their movements. In addition to seabed structures (e.g., anchors, turbines), many of the ocean energy devices would use mooring lines to attach a floating generator to the ocean bottom and electrical transmission lines to connect multiple devices to each other and to the shoreline. For example, MMS (2007) estimated that wave energy facilities may have as many as 200–300 mooring lines securing the wave energy devices to the ocean floor (based on 2–3 mooring lines per device and a 100-device facility). Mooring and transmission lines that extend from a floating structure to the ocean floor will create new fish attraction devices in the pelagic zone (i.e., the entire water column of the water body), pose a threat of collision for entanglement to some organisms, and potentially alter both local movements and long-distance migrations of marine animals (Nelson, 2008; Thompson et al., 2008). Because the transport of planktonic (drifting) life stages is affected by water velocity (Epifanio, 1988; DiBacco et al., 2001), localized reduction of water velocities by large, multi-unit projects could influence the recruitment of some species. A variety of aquatic organisms use magnetic, chemical, and hydrodynamic cues for navigation (Cain et al., 2005; Lohmann et al., 2008a). Thus, in addition to mechanical obstructions, the electrical and magnetic fields and current and wave alterations produced by energy technologies could interfere with local movement or long-distance migrations.

As mentioned, anchors and other permanent structures on the bottom will create new habitats, and thus may act as artificial reefs (Wilhelmsson et al., 2006). Artificial reefs are often constructed in order to increase fish production, but some studies suggest that they may be less effective than natural refs (Carr and Hixon, 1997) and that they may even have deleterious effects on reef fish populations by stimulating overfishing and overexploitation (Grossman et al., 1997).

Similarly, new structures in the pelagic zone (e.g., pilings or mooring cables for floating devices) will create habitat that may act as fish aggregation/attraction devices (FADs). These devices are extremely effective in concentrating fish and making them susceptible to harvest (Dempster and Taquet, 2004; Michel et al., 2007; Myers et al., 1986). Sea turtles are also known to be attracted to floating objects (Arenas and Hall, 1992). Fish are attracted to the devices as physical structure/shelter, and they may feed on organisms attached to the structures (Boehlert et al., 2008). Artificial lighting used to distinguish structures at night may also attract aquatic organisms.

The aggregation of predators near FADs may adversely affect juvenile salmonids or Dungeness crabs moving through the project area. Wilhelmsson et al. (2006) found that fish abundance in the vicinity of monopolies that supported wind turbines was greater than in surrounding areas, although species richness and diversity were similar. Most of the fish they observed near the structure were small (juvenile gobies), which may in turn attract commercially important fish looking for prey. Dempster (2005) observed considerable temporal variability in the abundance and diversity of fish associated with FADs moored between 3 and 10km offshore. The variability was often related to the seasonal appearance of large schools of juvenile fish. Fish assemblages differed between times when predators were present or absent; few small fishes were observed near the FADs when predators were present, regardless of the season. Using FADs as an experimental tool, Nelson (2003) found that fish formed larger, more species-rich assemblages around large FADs compared to small ones, and they formed larger assemblages around FADs with fouling biota. Devices enriched with fish accumulated additional recruits more quickly than those in which fish were removed.

It is likely that floating wave energy devices will act as FADs, but the effect on fish populations may be difficult to determine. FADs are attractive to fish because they provide food and shelter (Castro et al., 2002); subsequently, they also attract predators (Dempster, 2005) that can in turn attract commercial and sport fisheries. Without well-designed monitoring, it will be difficult to determine whether an energy park will enhance populations of aquatic organisms (by providing more habitat to support more fish), will have no overall effect (because it simply draws fish from other, nearby areas), or will decrease fish populations (by facilitating harvest by predators and fishermen). Kingsford (1999) pointed out that the determination of the effects of FADs at a particular location is complicated by the influence of non-independent factors: proximity of other FADs (e.g., other wave energy units), interconnection of multiple FADs to provide routes for the movement of associated fishes, and temporal dependence (the number of fish present at one sampling date influencing the number at the next sampling date due to fish becoming residents). Statistical approaches that could be applied to experiments on the effects of FADs on fish populations and solutions to the independent factor problems were also described.

Since anchoring systems and mooring lines will likely exclude fishing activities, energy parks could serve as marine protected areas. The PFMC (Pacific Fisheries Management Council) (2008) expressed concerns related to the prohibition of commercial fishing at wave energy test areas and suggested that there may be either a reduction in total fishing effort and lost productivity or a displacement of fishing effort to areas outside the areas closed to fishing. Displaced fishermen would likely concentrate their efforts in areas immediately outside the wave park boundaries, resulting in increased pressures on fish and habitats in those nearby areas.

Floating offshore wave energy facilities could create artificial haul-out sites for marine mammals (pinnipeds). Devices with a low profile above the waterline (desirable for aesthetic reasons) may enable seals and sea lions to use them as a haul-out site, particularly if the installations attract the marine mammals by acting as fish-concentrating devices. NOAA considers the creation of such artificial haul-outs as undesirable and recommends the use of deterrents to discourage use by marine mammals.

Floating devices could potentially impede the movements of floating marine habitat communities, such as *Sargassum* communities. Masses of floating Sargassum algae form unique communities of organism that serve as important habitat for hatchling sea turtles and juvenile fish (Coston-Clements et al., 1991). Strong current from the Sargasso Sea in the middle of the Atlantic Ocean carry these *Sargassum* communities around the world.

Floating devices with above-water structures may attract seabirds by creating artificial roosting sites or encouraging predation on fish near the FAD (Michel et al., 2007). There is particular concern about collision injuries to marine birds that are attracted to lighted structures at night or in inclement weather (Boehlert et al., 2008; Thompson et al., 2008). Peterson et al. (2006) monitored the interactions of birds and above-water structures at a Danish offshore wind farm from 1999 to 2005 and found that birds generally avoided the wind farms by flying around them, although there were considerable differences among species. The monitoring data suggested that avoidance was reduced at night. The authors obtained few data under conditions of poor visibility because bird migrations slowed or ceased during such times. Birds typically showed avoidance responses to the rotating wind turbine blades. A stochastic model predicted very low of Eider collisions with the offshore wind turbines, and the predictions were confirmed by subsequent monitoring (Petersen et al., 2006). Desholm (2003) provided a series of papers that describe techniques for predicting and monitoring interactions of birds and wind turbine structures at sea.

The numerous floating and submerged structures, mooing lines, and transmission cables associated with large ocean energy facilities could interfere with the long-distance migrations of marine animals (e.g., juvenile, and adult salmonids, Dungeness crabs, Green sturgeon, elasmobranchs, sea turtles, marine mammals, birds) if they are sited along migration corridors. On the U.S. Pacific Coast, effects on gray whales (*Eschrichtius robustus*) may be of particular concern because they migrate within 3.0 km of the shoreline. Boehlert et al. (2008) noted that buoys attached to commercial crab pots already comprise a major exiting risk to gray whales off the coast of Oregon. Lines associated with lobster pots and other fishing gears are a source of injury and mortality to endangered North Atlantic right whales (*Eubalaena glacialis*) on the East Coast of the U.S. (Caswell et al., 1999; Kraus et al., 2005). Many marine fish species drift or actively migrate long distance in the sea and may interact with ocean energy developments. Anadromous fish (e.g., green sturgeon, salmon, steelhead) and catadromous fish (e.g., eels) migrate through both rivers and oceans and therefore may encounter buoy hydrokinetic devices in the rivers and ocean energy projects (Dadswell et al., 1987).

Entanglement of large, planktonic jellyfish with long tentacles (as well as actively swimming sea turtles and marine mammals) is a potential issue for energy technologies with mooring lines in the pelagic zone. Thin mooring cables are expected to be more dangerous than thick ones because they are more likely to cause lacerations and entanglements, and slack cables are more likely to cause entanglements than taut ones (Boehlert et al., 2008).

Michel et al. (2007) expect that smaller dolphins and pinnipeds could easily move around mooring cables, but larger whales may have difficulty passing through an energy facility with numerous, closely spaced lines. Marine species with proportionately large pectoral fins or flippers may be relatively more vulnerable to mooring lines, based on information from humpback whale entanglements with pot and gill net lines (Johnson et al., 2005). Boehlert et al. (2008) suggested that whales probably do not sense the presence of mooring cables, and as a result could strike them or become entangled. In addition, they believed that if the cable density is sufficiently great and spacing is close, cables could have a "wall effect" that could force whales around them, potentially changing their migration routes. Whales and dolphins traveling or feeding together may be at a greater risk than solitary individuals because "ground response" many lead some individuals to follow others into danger (Faber Maunsell and Metoc, 2007).

Wave energy converters deployed near sea turtle nesting beaches have the potential to interfere with the offshore migration of hatchlings. Interference with migration could occur if the energy project acts as a physical barrier or alters wave action, which has been demonstrated to guide hatchlings away from the beach toward the open ocean (Lohmann et al., 1995; Goff et al., 1998; Wang et al., 1998).

Some marine fish species form spawning aggregations at specific sites or times (Cushing, 1969; Sinclair and Tremblay, 1984; Crawford and Carey, 1985; Coleman et al., 1996; Domeier and Colin, 1997). Smith (1972) reported a spawning aggregation consisting of 30,000–100,000 Nassau groupers (*Epinephelus striatus*) in the Bahamas. Since spawning success is important to the viability of populations, the siting and operation of ocean energy facilities would need to avoid interfering with these activities.

Collision and Strike

Submerged structures present a collision risk to aquatic organisms and diving birds, and the above-water components of floating structures may be a risk to flying animals. Wilson et al. (2007) defined collision as physical contact between a device or its pressure field and an organism that may result in an injury to the organism. They noted that collisions can occur between animals and fixed submerged structures, mooring equipment, surface structures, horizontal and vertical axis turbines, and structures that, by their individual design or in combination, may form traps. Harmful effects on animal populations could occur directly (e.g., from strike mortality) or indirectly (e.g., if the loss of prey species to strike reduces food for predators). Attraction of marine mammals and other predators to fish congregation near structures may also expose them to increase risk of collision or blade strike.

In an attempt to define the risk of collisions from marine renewable energy devices, Wilson et al. (2007) reviewed information from other industrial and natural activities: power plant cooling intakes, shipping, fishing gear, fish aggregation devices, and wind turbines. They concluded that although animals may strike any of the physical structures associated with marine renewable energy devices (i.e., vertical, or horizontal support piles, duct, nacelles, anchor locks, chains, cables, and floating structures), turbine rotors are the most intuitive sources of significant collision risks with marine vertebrates.

Effect of Rotor Blade Strike on Aquatic Animals

Many of the hydrokinetic and ocean current technologies extract kinetic energy by means of moving/rotating blades. A wide variety of swimming and drifting organisms (e.g., fish, sea turtles, driving birds, cetaceans, seals, and otters) may be struck by the blades, and suffer injury or mortality (Wilson et al., 2007). Mortality is a function of the probability of strike and the force of the strike. The seriousness of strike is related to the animal's swimming ability (i.e., ability to avoid the blade), water velocity, number of blades, blade design (i.e., leading edge shape), blade length and thickness, blade spacing, blade movement (rotation) rate, and the part of the rotor that the animals strikes. A vertical-axis turbine will have the same leading-edge velocity along the entire length of the blade. On the other hand, blade velocity on a horizontal axis turbine will increase from the hub out to the tip. The rotor blade tip has a much higher velocity than the hub because of the greater distance that is covered in each revolution. For example, on a rotor spinning at 20 rpm, the leading edge of the blade 1 m from the center point will be traveling at about 2 m/s—a speed that is likely to be avoidable or undamaging to most organisms. However, a 20-m-diameter rotor spinning at 20 rpm would have a tip velocity of nearly 21 m/s. Fraenkel (2006, 2007b) described a horizontal axis turbine with a maximum rotation speed of 12–15 rpm, which results in a maximum blade tip velocity of 12 m/s. Wilson et al. (2007) suggested that rotor blade tips will likely move at or below 12 m/s because greater speeds will incur efficiency losses through cavitation.

The force of the strike is expected to be proportional to the strike velocity. Consequently, the potential for injury from a strike would be greatest at the outer periphery of the rotor. Unfortunately, little is known about the magnitude of impact forces that cause injuries to most marine and freshwater organism (Cada et al., 2005, 2006) or the swimming behavior (e.g., burst speeds) that organisms may use to avoid strike. Although the blade tip will be moving at the highest velocity and exhibit the greatest strike force, animals may be able to avoid the tip of an unducted rotor. Relatively safe areas of passage through the rotor would be nearest the hub (because of low velocities) and potentially nearest the tip (because of the opportunity for the animal to move outward to avoid strike). The central zone of relatively high blade velocity and relatively less opportunity to avoid strike may be the most dangerous area (Coutant and Cada, 2005). For rotors contained in housings, there would be no opportunity for an organism entrained in the intake flow to escape strike by moving outward from the periphery; safe passage would depend on sensing and evading the intake flow or passing through the rotor between the blades. This suggestion of relatively high and low-risk passage zones has not been tested and remains speculative until the phenomenon is investigated in field applications.

There have been several studies to estimate the potential of fish strike by rotating blades (e.g., Cada et al., 2007; Deng et al., 2005), but all involve conventional hydroelectric turbines that are enclosed in turbine housings and afford little opportunity for flow-entrained organisms to avoid strike. It is likely that both the probability and consequences of organisms striking the rotor blade are greater for a conventional turbine than for an unducted current energy turbine, due to the greater opportunities for organisms to avoid approaching the turbine rotor or moving outward from the periphery. However, passage through a conventional turbine poses only a single exposure to the rotor, whereas passage through a project consisting of large numbers of hydrokinetic energy turbines represents a larger risk of strike that has not been investigated.

Wilson et al. (2007) described a simple model to estimate the probability of aquatic animals entering the path of a marine turbine. The mode is based on the density of the animals and the water volume swept by the rotor. The volume swept by the turbine can be estimated from the radius of the rotor and the velocity of the animals and the turbine blades. They emphasized that their model predicts the probability of an animal entering the region swept by a rotor, not collisions. Entry into the path toward the animal does not take evasive action or has not already sensed the presence of the turbine and avoided the encounter. Applying this simplified model (no avoidance or evasive action) to a hypothetical field of 100 turbines, each with a 2-bladed rotor 16 m in diameter, they predicted that 2% of the herring population and 3.6%–10.7% of the porpoise population near the Scottish coast would encounter a rotating blade. At this time, there is no information about the degree to which marine animals may sense the presence of turbines, take appropriate evasive maneuvers, or suffer injury in response to a collision. Wilson et al. (2007) suggested that marine vertebrates may see or hear the device at some distance and avoid the area, or they may evade the structure by dodging or swerving when in closer range.

The potential injurious effects of turbine rotors have been compared to those of ship propellers, which are common in the aquatic environment. Fraenkel (2007a) pointed out that in contrast to ship propellers, the rotors of hydrokinetic and current energy devices are much less energetic. He estimated that a tidal turbine rotor at a good site will absorb about 4 kW/m^2 of swept area from the current, whereas typical ship propellers release over 100 kW/m^2 of swept area into the water column. In addition to the greater power density, a ship propeller and ship hull generate suction that can pull objects toward it, increasing the area of influence for strike (Fraenkel, 2006).

In addition to direct strike, there is a potential for adverse effects due to sudden water pressure changes associated with the movement of the blade. For example, if the local water pressures immediately behind the turbine blades drop below the vapor pressure of water, calibration will occur. Cavitation is the process of forming water vapor bubble sin areas of extremely low pressure within liquids. As a turbine blade rotates, cavitation can occur in areas of low pressure (i.e., downstream surface of blades) causing increased local velocities, abrupt changes in the direction of flow, and roughness or surface irregularities (USACE, 1995). Once formed, cavitation bubbles stream from the area of formation and flow to regions of higher pressure where they collapse. The violent collapse of cavitation bubbles creates shock waves, the intensity of which depends on bubble size, water pressure in the region of collapse, and dissolved gas content. Within enclosed, conventional hydroelectric turbines, forces generated by cavitation bubble collapse may reach tens of thousands of kilopascals at the instant and point of collapse (Hamilton, 1983; Rodrigue, 1986). Cavitation is an undesirable condition that will reduce the efficiency of the turbine and damage blades as well as nearby organisms (Cada et al., 2007). Properly operating turns would not cavitate, and the zone of low pressure that might be injurious to organisms would be relatively small.

The pressure drops associated with the blades of hydrokinetic turbine shave not been measured in field applications, but experimental evidence suggests that tidal turbines may experience strong and unstable sheet and cloud cavitation, as well as tip vortices at a shallow depth of submergence (Wang et al., 2007). If this occurs, aquatic organisms passing near the cavitation zones in the immediate blade area may be injured. The likelihood of cavitation-related injuries would depend on the extent of cavitation and the ability of aquatic organisms to avoid the area—the collapse of cavitation vapor bubbles creates noise which may act as a deterrent.

Impact of Ocean Thermal Energy Conversion (OTEC)

An OTEC technology operates a low-temperature heat engine based on the temperature differences between warm, surface water and cold, deep water (Holdren et al., 1980). This type of project consists of pumps and ducts for transferring large volumes of water (several times more flow than is needed for a once-through cooling system of a comparably-sized steam electric power plant), large heat exchangers, and a working fluid that can be vaporized and recondensed (i.e., ammonia, propane, Freon®, or water). Electrical energy could be transported from offshore systems via subsea cables, or alternatively could be converted to chemical energy *in situ* (e.g., hydrogen, ammonia, methanol) and transported to shore in tankers (Pelc and Fujita, 2002).

Impacts of the construction of an OTEC facility will depend on whether the project is located onshore or offshore. An offshore facility would require the installation of large, long water conduits on the seabed to access deep water. Alternatively, OTEC projects located on offshore platforms would depend on subsea cables to transfer electricity to shore. The installation and maintenance of pipelines and electrical cables would disturb bottom habitats and generate EMF. Structures could become colonized with marine organisms and attract fish. Depending on the location of the warm water intake and discharges, these fish might be more susceptible to entrainment, impingement, or contact with the discharge plume.

The potential environmental effects of OTEC operation have been considered by a number of authors (Holdren et al., 1980; Myers et al., 1986; Harrison, 1987; Abbasi and Abbasi, 2000; Pelc and Fujita, 2002). Myers et al. (1986) provided the most comprehensive assessment of the possible effects on the marine environment resulting from the operation of the types of OTEC facilities that were contemplated in the early 1980s. Most of the likely effects were expected to be physical and chemical changes in the ocean surface waters arising from the transfer of large volumes of cool, deep water. Abbasi and Abbasi (2000) suggested that OTEC plants will displace about $4\,m^3/s$ of water per MW of electrify output from both the surface layer and the deep ocean layer, and then discharge the water at some intermediate depth. The warm water intake would be located at about 10 to 20 m depth, and the cold-water intake might extend to a depth of 750–1,000 m (Myers et al., 1986). The large transfer of water may disturb the thermal structure of the ocean near the plant, change salinity gradients, and change the amounts of dissolved gases, dissolved minerals, and turbidity. The transfer will result in an artificial upwelling of nutrient-rich deep water, which may increase marine productivity in the area. The stimulation of marine productivity may be especially strong in tropical waters, where nutrient levels are often low, and could have detrimental effects on nearby sensitive habitats like coral refs. Moreover, carbon dioxide will also be released when the deep water is warmed and subjected to lower pressures at the surface. The possible amounts of carbon dioxide released have not been rigorously quantified; some estimate that the quantities will be minute (Pelc and Fujita, 2002) and others suggest that the contribution will be relatively large (Holdren et al., 1980). The relatively high carbon dioxide and low dissolved oxygen content of the deep water may alter pH and dissolved oxygen concentrations in a surface mixing zone.

The large heat exchangers will need to be treated with biocides (e.g., chlorine or hypochlorite) in order to prevent the growth of bacterial slimes and other biofouling organisms; volumes of biocides would be proportional to the large volume of heating and cooling water. Degradation of the heat exchanger materials will result in chronic releases of metals (e.g., copper, nickel, and aluminum). Accidental release of the working fluid that is evaporated and condensed to drive the turbine could have toxic effects. The potential for acute and chronic toxicity and bioaccumulation of metals form deep ocean water will need to be considered (Fast et al., 1990).

Ocean thermal energy conversion projects would be sources of waterborne noise, arising from operation of ammonia turbines, seawater pumps, support system associated with the energy-producing cycle, and in some cases propulsion machinery for dynamic positioning of the OTEC platform. Janota and Thompson (1983) measured noise form OTEC-1, a 1-MWe test facility that was moored near Keahole Point, Hawaii. The most significant sources of noise from the small project resulted from the interaction of inflow turbulence with the sweater pumps and form thrusters used for dynamic positioning. Based on their measurements, Janota and Thompson (1983) predicted that a 160-MWe OTEC plant would radiate less than 0.05 acoustic W of broadband sound in the frequency range of 10–1,000 Hz, which is a least an order of magnitude less than that which is produced by a typical ocean-going freighter. Similarly, Rucker and Friedl (1985) predicted that pump noise (at 10 Hz) from a 40 MWe OTEC plant would be reduced from 136 to 78 dB at about 0.8 km; this is less than ambient noise at a sea state of 1 (very gentle sea with waves less than 0.3 m in height).

Large marine organisms may be impinged on the screens that protect the OTEC intakes, and smaller organisms (e.g., zooplankton, fish eggs, and larvae) will pass through the screens and be entrained in the heat exchanger system (Abbasi and Abbasi, 2000). The number of organisms entrained in the water will depend on their concentrations in the intake areas; more aquatic organisms are likely to be impinged and entrained at the surface water intake than from the deep-water intake. Due to the large flow rates of water at the warm water intake, impingement and entrainment will especially need to be monitored there. As with steam electric power plants, the heat exchanger-entrained organisms will be susceptible to mechanical damage in the piping and to rapid changes in temperature, pressure, salinity, and dissolved gases that my cause mortality. For example, the temperature of cold, deep water is expected to increase by about 2°C–3°C after passing through the heat exchangers; likewise, the temperature of shallow, warm water is expected to decrease by the same amount. Myers et al. (1986) noted that there is insufficient information to judge the impacts of a 2°C–3°C temperature shock but assumed that most organisms will probably not be directly impacted by this amount of temperature change. However, secondary entrainment into the discharge plume will also expose marine organisms to chemical, physical, and temperature stresses. A mixed discharge of warm and cold water could subject organisms entrained from the warm surface waters to a drop of 10°C, which would likely cause lethal cold shock for some species. Few organisms are expected to be entrained in the deep, cold-water flow, but those that do will be subjected to potentially lethal pressure decreases of 70–100 atmospheres (7,100–10,100 kilopascals) (Myers et al., 1986).

Potential Environmental (Specific) Impacts

The previous section provided a general discussion of potential environmental impacts of hydrokinetic energy systems. This section provides a discussion of many of the specific impacts related to site evaluation, construction, and operations and maintenance (O&M) activities.

Hydrokinetic Energy Site Evaluation Impact

Site evaluation phase activities, such as monitoring and site characterization, are temporary and are conducted at a smaller scale than those at the construction and operation phases. Potential impacts from these activities are presented below, by the type of affected resource. The impacts described are for typical site evaluation activities, such as drilling to characterize the seabed or riverbed. Onshore site characterization activities would be limited to a topographic survey to establish onshore site design and placement for an operations and maintenance facility, substation, and electric transmission lines. If road construction were necessary during this phase, potential impacts would be similar in character to those for the construction phase, but generally of smaller magnitude.

Air Quality (Including Carbon Footprint and Global Climate Change)

Impacts on air quality during site evaluation activities would be limited to barges conducting surveys, and vehicular traffic to proposed sites for all hydrokinetic energy land-based facilities. These air pollutant emissions would be minor, of short duration, and intermittent.

Cultural Resources

Cultural material present within the project area could be impacted by any seafloor, riverbed, or ground disturbance. Such disturbance could result from drilling and sampling activities and, for land-based activities, vehicular and pedestrian traffic. These activities would be relatively limited in scope during this phase. Surveys conducted during this phase to evaluate the presence and/or significance of cultural resources in the area would assist developers in designing the project to avoid or minimize impacts to these resources.

Ecological Resources

Impacts on ecological resources would be minimal during site evaluation because of the limited nature of the activities. For offshore projects (e.g., wave barrage, and tidal turbine projects) and river projects, the potential effects of low-energy geological and geophysical surveys on marine mammals, sea turtles, and fish could include behavior response such as avoidance and deflections in travel direction. A few individuals could be injured or killed by collisions with the survey vessels. Those individuals displaced because of avoidance behaviors during survey are likely to return within relatively short periods following the cessation of survey activities. Marine mammals, sea turtles, and fish could be exposed to discharges or accidental fuel releases from survey vessels and to accidentally released solid debris. Such spills would be small and would not be expected to measurably affect marine or river wildlife. Land-based activities could give rise to the introduction and spread of invasive vegetation as a result of vehicular traffic. Soil borings would destroy vegetation and disturb wildlife. Overall, site evaluations are not expected to cause significant impacts on terrestrial or aquatic biota. Surveys conducted during this phase to evaluate the presence and/or significance of ecological resources in the area would assist developers in properly locating the facility and its components.

Water Resources (Surface Water and Groundwater)

Survey ships could contribute small amounts of fuel or oil to the ocean or river through bilge discharges or leaks. Anchoring of the ships can cause sediment from the seabed or riverbed to enter the water column. Negligible to minor impact on water quality would be expected. Relatively limited amounts of water would be used if drilling were required; this water could be obtained locally, or it could be trucked in with the drilling equipment. Land-based site evaluation activities are anticipated to have minimal to no impact on water resources, local water quality, water flows, and surface water/groundwater interactions.

Land Use

Very few offshore and onshore site evaluation activities are expected; consequently, no impacts on existing land uses are anticipated.

Soils and Geologic Resources

Seabed, riverbed, and onshore ground disturbances would be minimal during the site evaluation phase and, as a result, impacts on seabed and riverbed sediments or soils are unlikely to occur. Site characterization activities would also be unlikely to activate geological hazards.

Paleontological Resources

Paleontological resources present within the project area could be impacted by any seafloor, riverbed, or ground disturbance. Such disturbance could result from drilling and sampling activities and, for land-based activities, vehicular and pedestrian traffic. These activities would be very limited in scope during this phase and would not be likely to affect paleontological resources. Surveys conducted during this phase to evaluate the presence and/or significance of paleontological resources in the area would assist developers in designing the project to avoid or minimize impacts on these resources.

Transportation

Impacts on transportation are anticipated to be insignificant during the site evaluation phase from the one or two survey vessels that might be deployed at any one time. Vehicular traffic would be temporary and intermittent and limited to very low volumes of heavy- and medium-duty equipment and personal vehicles.

Visual Resources

Site evaluation activities would have temporary and minor visual effects caused by the presence of survey vessels, workers, vehicles, and equipment.

Socioeconomics

Site evaluation activities are temporary and limited and would not result in socioeconomic impacts on employment, local serves, or property values.

Environmental Justice

Site evaluation activities are limited and would not result in significant adverse impacts in any resource area; therefore, environmental justice impacts are not expected at this phase.

Acoustics (Noise)

Onshore and offshore drilling activities for all hydrokinetic energy facilities, if required, would generate the most noise during this phase, but impacts would be much lower than those that could occur during construction. Surveys using air-gun arrays may generate low-frequency noise that may be detected by marine mammals, sea turtles, and fish within the survey area. Other sea and river geophysical surveys and the installation of wave measuring devices equipped with recording equipment would generate some ship and boat noise.

Hazardous Materials and Waste Management

The only hazardous material associated with site evaluation activities would be the fuel for boats, barges, and vehicles. Impacts from operational discharges, accidental fuel releases, and accidentally released solid debris are expected to be small or nonexistent if appropriate management practices are followed.

HYDROKINETIC ENERGY FACILITY CONSTRUCTION IMPACT

Typical activities during the wave or tidal turbine energy farm construction phase include assembling hydrokinetic units on shore; transporting each device to its designated location offshore; anchoring it to the seabed; connecting each device electrically to a central junction box; laying or burying submarine transmission and signal cable; and construction of onshore substation and electrical transmission lines to connect to the grid. Activities required for river in-stream facilities are essentially the same but are conducted in a river rather than offshore. For a barrage facility, a dam would be constructed across the inlet or estuary to contain the incoming tidal flow and a powerhouse would be constructed to produce hydroelectric energy. Onshore activities include ground clearing, grading, excavation, vehicular traffic, and construction of facilities.

Air Quality (Including Global Climate Change and Carbon Footprint)

Offshore activities that generate emissions include ship, boat, and barge traffic to and from the hydrokinetic energy facility site, installation of the hydrokinetic energy devices and their associated anchoring devices, and the laying of the underwater cables. Air emissions result from the operation of ship engines and on-ship equipment such as cranes, generators, and air compressors. In most cases, an air quality permit would not be required for offshore facility construction. However, in areas of non-attainment for any criteria pollutants, the states have authority to regulate near-shore activities.

Emissions generated during the construction phase of land-based facilities, (including docks, equipment storage, and assembly area) include:

- Vehicle emissions
- Diesel emissions from large construction equipment and generators
- Volatile organic compound (VOC) releases from storage and transfer of vehicle/equipment fuels
- Small amounts of carbon monoxide, nitrogen oxides, and particulates form blasting activities
- Fugitive dust from many sources, such as disturbing and moving soils (clearing, grading, excavation, trenching, backfilling, dumping, and truck and equipment traffic), mixing concrete, use of un-vegetated soil piles, drilling, and pile driving.

These emissions would also be expected during the construction of a barrage facility. A permit may be required from the state or local air agency to control or mitigate these emissions, especially in non-attainment areas.

Cultural Resources

For offshore projects, trenching, dredging, and placement of hydrokinetic energy devices and associated components could impact shipwrecks or buried archeological artifacts. For onshore projects, impacts on cultural resources could occur from site preparation (e.g., clearing, excavation, and grading) and construction of transmission-related facilities. For either offshore or onshore projects, visual impacts could also result from the disruption of a historical setting that is important to the integrity of a historic structure, such as a lighthouse. Potential cultural resource impacts include:

- Complete destruction of the resource if present in areas undergoing surface disturbance or excavation
- Unauthorized removal of artifacts or vandalism to cultural resource sites resulting from increases in human access to previously inaccessible areas
- Visual impacts resulting from vegetation cleaning, increase industry, and the presence of large-scale equipment, machinery, and vehicles (if the affected cultural resources have an associated landscape or other visual components that contributes to their significance, such as a scared landscape or historic trail)

Ecological Resources

Wave and Tidal Turbine Energy Farms

The potential effects of construction activities associated with the placement of wave and tidal turbine energy devices on marine mammals, sea turtles, and fish may include behavioral responses such as avoidance and deflections in travel direction. Noise and vibrations generated during the various activities could disturb the normal behaviors and mask sounds from other members of the same species or from predators. Coastal birds could be displaced from offshore feeding habitats; however, most birds would be likely to return within relatively short periods following cessation of construction activities. A few could be injured or killed by collisions with the survey vessels.

The movement and deposition of sediment during construction activities on the seafloor could kill benthic organisms, a source of food for fish. Effects on fish could potentially occur if spawning or nursery grounds are disturbed during construction or if re-suspended sediments cause smothering of habitat. The area of seafloor disturbance from anchoring systems relative to the surface area occupied would be small.

Marine mammals, sea turtles, fish, and seabirds could be exposed to discharges or accidental fuel releases from construction vessels and to accidentally released solid debris. Such spills would be small and quickly diluted and would not be expected to measurably affect marine mammal or fish populations.

Onshore impacts from construction could affect terrestrial vegetation and wildlife, but the overall impact is anticipated to be minimal because permanent on-shore facilities are expected to be small. Wildlife would be most affected by habitat reduction within the project site, access roads, and transmission line right-of-way. Wildlife within surrounding habitats might also be affected if the construction activity (and associated noise) disturbs normal behaviors, such as feeding and reproduction. Impacts on wildlife are expected to be minor.

Turtles nest along the south Atlantic and Gulf coastlines. Those nests containing eggs and emerging hatchlings could be affected by construction activities onshore. Lighting from the construction areas could disorient the hatchlings and increase their exposure to predators. The minimal amount of onsite construction would limit the impact to no more than a few nests.

River In-stream Facilities

The potential effects of the placement of river in-stream energy devices and associated construction on fish may include behavioral response such as avoidance and deflections in travel direction. Noise and vibrations generated during the various construction activities, especially placement of supporting structures and installation of submarine transmission lines, could disturb the normal behavior. Those displaced because of avoidance behaviors during construction are like to return within relatively short periods following the cessation of construction activities.

The movement and deposition of sediment during construction activities on the riverbed could kill benthic organism, a source of food for fish. Effects to fish could potentially occur if spawning or nursery grounds are disturbed during construction or if re-suspended sediments cause smothering of habitat. The area of riverbed disturbance would be very small relative to the availability of similar habitats in surrounding areas.

Terrestrial wildlife would be most affected by habitat reduction within the project site, access roads, and transmission lien rights-of-way. Wildlife within surrounding habitats might also be affected if the construction activity (and associated noise) disturbs normal behaviors, such as feeding and reproduction. Impacts on wildlife are expected to be minor.

Barrage Facilities

Dam construction at a barrage facility would not increase the amount of wetted area inundated within the embayment, but it would alter the period of time that water is held in the embayment and could alter the aquatic environment of the embayment. These alterations could lead to habitat loss for terrestrial wildlife and bird species and/or degradation for aquatic species. Underwater habitat would be altered, and marine species could be injured or killed during the construction of the intake and dam. The ability of fish and marine mammals to enter and leave the embayment would be substantially altered. The significance of construction impacts on fish, marine mammals, and saltwater wetland-dependent birds and terrestrial species is likely to be site-specific.

Terrestrial wildlife also would be affected by habitat reduction caused by the construction of land-based facilities within the project site, including access roads and transmission lien right-of-way. Wildlife within surrounding habitats might also be affected if the construction activity (and associated noise) disturbs normal behaviors, such as feeding and reproduction.

Water Resources (Surface Water and Groundwater)

Water Use

Water would be used onshore for dust control when clearing vegetation and grading and for road traffic; for making concrete; and for domestic use by the construction crew. Water would likely be traced in from off-site. The quantity of water required would be small.

Water Quality

Vessels used for the transport and installation of hydrokinetic energy devices and components could contribute small amounts of fuel or oil to the ocean through bilge discharges or leaks. Anchoring of construction ships, and installation of anchoring devices and electrics cables, can cause sediment from the seabed to enter the water column. Onshore activities that cause soil erosion or discharges of waste or sanitary water could affect water quality. Negligible to minor impact on water quality would be expected.

Land Use

The wave and tidal turbine energy farm could occupy from 17 to 250 acres of the ocean surface. The facility would exclude commercial shipping and, possibly, fishing activities. A river in-stream facility could occupy about 5 acres and could affect commercial shipping, recreation, and fishing activities. A barrage facility would impact the area behind the dam and would exclude commercial and recreational ships and boats from entering the previously accessible estuary unless a ship lock was constructed.

Existing onshore land use during construction would be affected by intrusive impacts such as ground clearing, increased traffic, noise, dust, and human activity, as well as by changes in the visual landscape. In particular, these impacts could affect those seeking recreational opportunities on the shore or in the water. Generally, offshore, and onshore impacts associated with these types of facilities are expected to be minor.

Soils and Geologic Resources

Seabed and riverbed disturbance would result from drilling or pile driving required for anchoring the hydrokinetic energy devices and from excavation required to bury electrical cable. The area disturbed is a small portion of the area occupied by the hydrokinetic energy facility. Construction activities would also be unlike to activate geological hazards. Surface disturbance, heavy equipment traffic, and changes to surface runoff patterns during the construction of onshore facilities could cause soil erosion. Impacts of soil erosion could include soil nutrient loss and reduced water quality in nearby surface water bodies.

Visual Resources

Viewers on onshore and offshore would observe an increase in vessel traffic transporting hydrokinetic energy devices, components, and works to the site. The activity during installation would also be noticed. Wave, tidal, and in-river facilities are generally low-profile structures, and although visible, may not be as objectionable as larger, more visible structures.

The overall effect on visual resources is also related to existing uses (especially land-based residential uses) that would have a view of the hydrokinetic energy farm areas and will require site-specific assessment. Construction of the dam at a barrage site would change the character of the water basin and could create visual concerns for nearby residents or recreation users.

Possible sources of visual impacts during the construction of onshore facilities include ground disturbance, construction of highly visible facilities, vegetation removal, road construction, and increased traffic. Increased truck and vessel traffic and human activity at the port facility supporting project construction would also be visible, although it is anticipated this would only be a short-term impact.

Paleontological Resources

For offshore projects, trenching dredging, and placement of hydrokinetic energy devices and associated components could impact paleontological resources. For onshore projects, impacts on paleontological resources could occur directly from the construction activities and increased accessibility to fossil locations. Potential impacts include:

- Complete destruction of the resource if present in areas undergoing surface disturbance or excavation
- Unauthorized removal of paleontological resources or vandalism to the site as a result of increased human access to previously inaccessible areas if significant paleontological resources are present.

Transportation

Traffic at the port would increase as wave or tidal energy devices and components are delivered prior to assembly transport to the project site. Vessel traffic will increase during the construction phase.

The same effect would occur with river in-stream projects although, because they are generally much smaller installations, the impacts on transportation would be less significant. Short-term increases in the use of local roadways would occur during the onshore construction period.

Barrage projects would be relatively large projects and would likely require more labor for construction. This would cause an increase in traffic on local roads and potentially could disrupt local traffic use.

Socioeconomics

Direct impacts would include the creation of new jobs for construction workers and the associated income and taxes generated by the hydrokinetic energy facilities. Indirect impacts would occur as a result of the new economic development and would include new jobs at businesses that support the expanded workforce to provide project materials, and associated income and taxes. An influx of new workers could strain the existing community infrastructure and social services; however, since most hydrokinetic projects are relatively small, the workforce required is also expected to be relatively small and impacts, therefore, are expected to be minor.

Environmental Justice

If significant impacts occurred in any resource areas, and these impacts disproportionately affected minority or low-income populations, then there could be an environmental justice impact. Issues of potential concern during construction are noise, dust, and visual impacts from the construction site and possible impacts associated with the construction of new access roads.

Additional impacts could include limitations on access to the area for tribal recreation, subsistence, and traditional activities. Environmental justice impacts are dependent upon vulnerable populations being located within the area of influence of a project and are, therefore, site-specific.

Acoustics (Noise)

Underwater and above-water noise sources include boat, ship and barge activity associated with transporting workers, materials, and hydrokinetic energy devices to the offshore site, installing hydrokinetic facilities, and laying of electrical and signal cables. Human receptors on the ocean shore likely would be far enough away for any impacts to be minor. Human receptors on the river shore could be close to the activities required for locating and anchoring river in-stream turbines.

If pile driving is required for anchoring hydrokinetic energy devices or for the construction of offshore power-gathering stations, the noise could be audible at the shoreline and might be annoying to populations. This impact would be intermittent. Techniques for laying cable could require use of air guns, rock cutters, or shaped explosive charges. The noise could be intense but would occur over a very short time period.

Onshore noise would result from pre-assembly of the hydrokinetic energy devices and the construction of onshore facilities. The primary source of noise during the construction of onshore facilities, transmission lines, and a barrage facility would be from equipment operation (e.g., rollers, bulldozers, diesel engines). Other sources of noise include vehicular traffic, tree felling, and blasting.

Whether the noise levels from these activities exceed U.S. Environmental Protection Agency (EPA) guidelines or local ordinances would depend on the distance to the nearest residence and the effectiveness of any mitigating measures to reduce noise levels. If near a residential area, noise levels from blasting and some equipment operation could exceed the EPA guideline but would be intermittent and extend for only a limited time.

Adverse impacts due to noise could occur if the site is located near a sensitive area, such as a park, wilderness, or other protected areas. The primary impacts from noise would be localized disturbances to wildlife, recreationists, and residents.

Hazardous Materials and Waste Management

Hazardous material associated with installation of hydrokinetic energy devices and construction of associated support components would include fuels, lubricants, and hydraulic fluids contained in the hydrokinetic energy devices or used in ships and construction equipment. Impacts from accidental spills, accidental fuel releases, and release of solid debris are expected to be minor if appropriate management practices are followed. Garbage and sanitary waste generated onboard the vessels and barges would be returned to shore for disposal.

Solid and industrial waste would be generated during onshore construction activities. The solid wastes would likely be nonhazardous and consist mostly of containers, packaging materials, and waste from equipment assembly and construction crews. Industrial wastes would include minor amounts of fuels, spent vehicle and equipment fluids (lubricating oils, hydraulic fluids, battery electrolytes, glycol coolants), and spent solvents.

These materials would be transported off-site for disposal, but impacts could result in the wastes were not properly handled and were released to the environment. No impacts are expected from the proper handling of all wastes.

HYDROKINETIC ENERGY FACILITY OPERATIONS & MAINTENANCE IMPACT

Typical activities during the hydrokinetic energy facility operational phase include the operation of the hydrokinetic energy devices, power generation, and associated maintenance activities that would require operations from a vessel or barge when components are being maintained, repaired, or replaced.

Air Quality (Including Global Climate Change and Carbon Footprint)

There are no direct air emissions from the operation of hydrokinetic energy facilities. Air emissions result from the operation of maintenance ship engines and on-ship equipment such as cranes, generators, and air compressors. Onshore vehicular traffic will continue to produce small amounts of fugitive dust and tailpipe emission during maintenance activities. These emissions would not likely exceed air quality standards and impacts on air quality would be minor.

Cultural Resources

The operation and maintenance of offshore facilities would have no direct impact on cultural resources unless previously undisturbed areas are disturbed. Potential indirect impacts associated with the operation and maintenance of onshore facilities would be limited to the unauthorized collection of artifacts made possible by access roads if they make remote lands accessible to the public. Visual impacts resulting from the presence of a large wave, tidal turbine, river in-stream energy facility, or barrage facility and transmission lines could affect some cultural resources, such as sacred landscapes or historic trails.

Ecological Resources

Wave and Tidal Energy Farms

The presence of a wave or tidal turbine energy farm could cause some marine mammals to avoid the area. Collisions with maintenance vessels and underground structures are anticipated to be rated. Overtopping wave energy devices could trap hatching sea turtles and fish, resulting in injury or death. Impacts on the marine populations are expected to be minor to moderate depending on the specific site.

Some whale species migrate along the pacific coast from 1.5 to 2 miles (2.5 to 3 km) offshore. Any wave or tidal turbine facility located in this zone could impact whale migration.

Noise levels from wave energy devices would be similar to those from ship traffic but would be continuous for the life of the wave farm. This could result in long-term avoidance by wildlife, which could lead to the abandonment of feeding or mating grounds. Noise from submerged tidal turbines would be low due to the low rotational speed of the turbine blades.

Marine mammals, sea turtles, fish, and seabirds could be exposed to discharges or accidental fuel releases from maintenance vessels and to accidentally released solid debris. Such spills would be small and quickly diluted and would not be expected to measurably affect these wildlife populations.

Depending on the design of the facilities, wave energy devices or any above-water portions of tidal energy devices could become the host for seabird colonies or may be used as haul out areas for seals lions. Underwater structures may also create an artificial habitat for benthic species. This could complicate the maintenance and repair of wave energy devices.

Electromagnetic fields from the transmission cable can be detected by some fish and might result in attraction or avoidance. Such impacts would be negligible to minor.

Onshore impacts form operation could affect vegetation and wildlife by habitat reduction within the project site, access roads, and transmission line rights-of-way. Turtles nest along the south Atlantic and Gulf coastline and nests and emerging hatchlings could be affected by maintenance activities onshore. Outdoor lighting from the onshore facilities could disorient the hatchlings and increase their exposure to predators.

River In-stream Facilities

Fish behavior is influenced primarily by the natural current in the river and only secondarily by the rotating mechanisms in the turbines. Proper location of the turbines would have minimal impact on fish movement or abundance. Allowance of sufficient turbine spacing in a turbine farm or in the river may minimize the impact to fish. Noise generated by the turbines would have minimal effects on aquatic biota.

The turbines are not expected to affect terrestrial wildlife since they are mostly, or completely, submerged. Wildlife would be most affected by habitat reduction within the onshore project site, access roads, and transmission line rights-of-way. Impacts on wildlife are expected to be minor.

Barrage Facilities

Dam construction at a barrage factory would not increase the amount of wetted area inundated within embayment, but it would alter the period of time that water is held in the embayment and could alter the aquatic environment of the embayment. These alterations could lead to habitat loss for terrestrial wildlife and bird species and/or degradation for aquatic species. Fish could be injured or killed during the operation of the intake or during power generation. Because of a reduction in natural fishing of sediment, increases in sedimentation within the embayment may also adversely affect embayment ecosystems. The ability of fish and marine mammals to enter and leave the embayment would likely be substantially altered. The significance of operational impacts on fish, marine mammals, and saltwater wetland depends on birds and terrestrial species would likely be site-specific.

Water Resources (Surface Water and Groundwater)

Water Use

Water use at the port would be required for normal operation, including fire protection, cleaning and maintenance of equipment, and consumptive use of personnel. Water would also be required for consumptive use on the vessels.

Water Quality

Vessels used for the maintenance of hydrokinetic energy devices and components could contribute small amounts of fuel or oil to the ocean or river through bilge discharges or leaks. Damage to a hydrokinetic energy device, which may contain petroleum-based materials, could result in water contamination. Anchoring of the ships can cause sediment from the seabed to enter the water column. Onshore activities that could affect water quality are those that cause soil erosion, or discharges of waste or sanitary water. Negligible to minor impact on water quality would be expected.

Land Use

All types of hydrokinetic facilities would likely exclude traditional uses of the areas where they are constructed. Commercial shipping, fishing, and recreation uses may be the most likely uses that could be affected. Depending on the nature and visibility of the facilities, the visual impacts of hydrokinetic development might also create conflicts with existing shore-based uses.

Soils and Geologic Resources

Seabed or riverbed disturbance would be minimal from maintenance activities and sediments are unlikely to be affected. Maintenance activities would also be unlikely to activate geological hazards. Maintenance activities would also be unlikely to activate geological hazards. A wave energy farm could cause a reduction in wave height of 10%–15%, and this reduction could result in an interruption of the natural sediment transport along the shore, increasing erosion and drift. The impact is greater the closer the wave farm is to the shore and is greater for floating devices oriented parallel to the shore. Tidal and river in-stream turbines will cause turbulence downstream that might cause scour of the seabed or riverbed if the units are located near the bottom.

During operation, the soil and geologic conditions would stabilize at onshore facilities. Soil erosion and soil compaction are both likely to continue to occur along access roads.

Paleontological Resources

The operation and maintenance of offshore facilities would have no direct impacts on paleontological resources unless additional undisturbed areas are developed. Potential indict impacts associated with the operation and maintenance of onshore facilities would be limited to unauthorized collection of fossils made possible by access roads if they make remote lands accessible to the public.

Green Engineering

Transportation
No noticeable impacts on transportation are likely during operations. Maintenance vessels would service the hydrokinetic energy devices at regular intervals, but the additional levels of activity resulting from this are anticipated to be small. Infrequent, but routine, truck shipments of components replacements to the dock during maintenance procedures are likely over the period of operation.

Visual Resources
Visual impacts of the operation of hydrokinetic energy devices would be the same as identified for construction activities, with the exception of the increase in vessel and vehicle traffic associated with construction.

Socioeconomics
Direct impacts would include the creation of new jobs for operation and maintenance workers and the associated income and taxes paid. Indirect impacts are those impacts that would occur as a result of the new economic development and would include things such as new jobs at businesses that support the workforce or that provide project materials, and associated income and taxes. However, the total number of operations and maintenance jobs likely would be small and therefore the associated socioeconomic impacts are anticipated to be minimal.

Environmental Justice
If significant impacts occurred in any resource areas as a result of the operations of hydrokinetic facilities, and these impacts disproportionately affect minority or low-income populations, then there could be an environmental justice impact. Issues of potential concern during operations are noise, ecological, and visual impacts. Additional impacts include limitations on access to the area for tribal activities.

Acoustics (Noise)
Underwater and above-water noise sources include ship and barge noise associated with transporting workers for maintenance activities, which would require frequent (and possibly daily) trips to the wave or tidal turbine energy farm or to the river in-stream turbines. Wave energy device noise would result from the flexing action of attenuators and point absorbers, from compressed air released from oscillating water column turbines, and from the impact of waves on terminator and overtopping devices. Underwater noise from the operation of tidal or rive in-stream turbines is expected to be low because the rotational speed of the turbine blades is low. Overall, noise from the operation of the hydrokinetic energy devices is expected to be low.

Onshore transformers would produce a humming noise and cooling fan noise. Sources of noise during the operation of a barrage facility would be the turbines, generators, and transformers

Hazardous Materials and Waste Management
Hazardous material associated with the operation and maintenance of hydrokinetic energy devices and associated support compounds would include the fuel for boats, vessels, and barges, and lubricants and hydraulic fluids contained in the wave or tidal energy devices. Impacts from accidental spills, accidental fuel releases, and releases of solid debris are expected to be minor if appropriate management practices are followed. Garbage and sanitary waste generated onboard the vessels and barges would be returned to shore for disposal.

Industrial wastes are generated during routine maintenance (used fluids, cleaning agents, and solvents). These wastes typically would be put in containers, characterized, and labeled, possibly stored briefly, and transported by a licensed hauler to an appropriate permitted off-site disposal facility as a standard practice.

Adverse impacts could result if these wastes were not properly handled were released into the environment. Given current standards, the impact of this is expected to be minor.

FUEL CELLS

> I believe that water will one day be employed as a fuel, that hydrogen and oxygen which constitute it, used singly or together, will furnish an inexhaustible source of heat and light.
>
> **Jules Verne,** *Mysterious Island* **(1874)**

Hydrogen Fuel Cell[8]

Containing only one electron and one proton, Hydrogen, chemical symbol H, is the simplest element on Earth. Hydrogen is a diatomic molecule—each molecule has two atoms of hydrogen (which is why pure hydrogen is commonly expressed as H_2). Although abundant on Earth as an element, hydrogen combines readily with other elements and is almost always found as part of another substance, such as water hydrocarbons, or alcohols. Hydrogen is also found in biomass, which includes all plants and animals.

- Hydrogen is an energy carrier, not an energy source. Hydrogen can store and deliver usable energy, but it doesn't typically exist by itself in nature; it must be produced from compounds that contain it. Its production is not inexpensive.
- Hydrogen can be produced using diverse, domestic resources including nuclear; natural gas and coal; and biomass and other renewables including solar, wind, hydro-electric or geothermal energy. This diversity of domestic energy sources makes hydrogen a promising energy carrier and important to our nation's energy security. It is expected and desirable for hydrogen to be produced using a variety or resources and process technologies (or pathways).
- DOE focuses on hydrogen-production technologies that result in near-zero, net greenhouse gas emissions and use renewable energy sources, nuclear energy, and coal (when combined with carbon sequestration). To ensure sufficient clean energy for our overall energy needs, energy efficiency is also important.
- Hydrogen can be produced via various process technologies, including thermal (natural gas reforming, renewable liquid and bio-oil processing, and biomass and coal gasification), electrolytic (water splitting using a variety of energy resources), and photolytic (splitting water using sunlight via biological and electrochemical materials).
- Hydrogen can be produced in large, central facilities (50–300 miles from point of use), smaller semi-central (located within 25–100 miles of use) and distributed (near or at point of use). Learn more about distributed vs. centralized production.
- In order for hydrogen to be successful in the marketplace, it must be cost-competitive with the available alternatives. In the light-duty vehicle transportation market, this competitive requirement means that hydrogen needs to be available untaxed at $2-$3/gge (gasoline gallon equivalent). This price would result in hydrogen fuel cell vehicles having the same cost to the consumer on a cost-per-mile-driven basis as a comparable conventional internal-combustion engine or hybrid vehicle.
- DOE is engaged in research and development of a variety of hydrogen production technologies. Some are further along in development than others—some can be cost-competitive for the transition period (beginning in 2015), and others are considered long-term technologies (cost-competitive after 2030).

Infrastructure is required to move hydrogen from the location where it's produced to the dispenser at a refueling station or stationary power site. Infrastructure includes the pipelines, trucks, railcars, ships, and barges that deliver fuel, as well as the facilities and equipment needed to load and unload them.

Delivery technology for hydrogen infrastructure is currently available commercially, and several U.S. companies deliver bulk hydrogen today. Some of the infrastructure is already in place because hydrogen has long been used in industrial applications, but it's not sufficient to support widespread consumer use of hydrogen as an energy carrier. Because hydrogen has a relatively low volumetric energy density, its transportation, storage, and final delivery to the point of use comprise a significant cost and result in some of the energy inefficiencies associated with using it as an energy carrier.

Options and trade-offs for hydrogen delivery from central, semi-central, and distributed production facilities to the point of use are complex. The choice of a hydrogen production strategy greatly affects the cost and method of delivery.

For example, larger, centralized facilities can produce hydrogen at relatively low costs due to economies of scale, but the delivery costs for centrally produced hydrogen are higher than the delivery costs for semi-central or distributed production options (because the point of use is farther away). In comparison, distributed production facilities have relatively low delivery costs, but the hydrogen production costs are likely to be higher—lower volume production means higher equipment costs on a per-unit-of-hydrogen basis.

Key challenges to hydrogen delivery include reducing delivery cost, increasing energy efficiency, maintaining hydrogen purity, and minimizing hydrogen leakage. Further research is needed to analyze the trade-offs between the hydrogen production options and the hydrogen delivery options taken together as a system. Building a national hydrogen delivery infrastructure is a big challenge. It will take time to develop and will likely include combinations of various technologies. Delivery infrastructure needs and resources will vary by region and type of market (e.g., urban, interstate, or rural). Infrastructure options will also evolve as the demand for hydrogen grows and as delivery technologies develop and improve.

Hydrogen Storage

Storing enough hydrogen on-board a vehicle to achieve a driving range of greater than 300 miles is a significant challenge. On a weight basis, hydrogen has nearly three times the energy content of gasoline (120 MJ/kg for hydrogen versus 44 MJ/kg for gasoline). However, on a volume basis the situation is reversed (8 MJ/L for liquid hydrogen versus 32 MJ/L for gasoline). On-board hydrogen storage in the range of 5–13 kg H_2 is required to encompass the full platform of light-duty vehicles. Hydrogen can be stored in a variety of ways, but for hydrogen to be a competitive fuel for vehicles, the hydrogen vehicle must be able to travel a comparable distance to conventional hydrocarbon-fueled vehicles. Hydrogen can be physically stored as either a gas or a liquid. Storage as a gas typically requires high-pressure tanks (5,000–10,000 psi tank pressure). Storage of hydrogen as a liquid requires cryogenic temperatures because the boiling point of hydrogen at one-atmosphere pressure −252.8°C. Hydrogen can also be stored on the surfaces of solids (by adsorption) or within solids (by absorption). In adsorption, hydrogen is attached to the surface of material either as hydrogen molecules or as hydrogen atoms. In absorption, hydrogen is dissociated into H-atoms, and then the hydrogen atoms are incorporated into the solid lattice framework.

Hydrogen storage in solids may make it possible to store large quantities of hydrogen in smaller volumes at low pressures and at temperatures close to room temperature. It is also possible to achieve volumetric storage densities greater than liquid hydrogen because the hydrogen molecule is dissociated into atomic hydrogen within the metal hydride lattice structure. Finally, hydrogen can be stored through the reaction of hydrogen-containing materials with water (or other compounds such as alcohols). In this case, the hydrogen is effectively stored in both the material and in the water. The term "chemical hydrogen storage" or chemical hydrides is used to describe this form of hydrogen storage. It is also possible to store hydrogen in the chemical structures of liquids and solids.

How a Hydrogen Fuel Cell Works

The fuel cell uses the chemical energy of hydrogen to cleanly and efficiently produce electricity with water and heat as byproducts. Fuel cells are unique in terms of variety of their potential applications; they can provide energy for systems as large as a utility power station and as small as a laptop computer. Fuel cells have several benefits over conventional combustion-based technologies currently used in many power plants and passenger vehicles. They produce much smaller quantities of greenhouse gases and none of the air pollutants that create smog and cause health problems. If pure hydrogen is used as a fuel, fuel cells emit only heat and water as byproducts.

A hydrogen fuel cell is a device that uses hydrogen (or hydrogen-rich fuel) and oxygen to create electricity by an electrochemical process. A single fuel cell consists of an electrolyte and two catalyst-coated electrodes (a porous anode and cathode). While there are different fuel cell types, all fuel cells work similarly:

- Hydrogen, or a Hydrogen-rich fuel, is fed to the anode where a catalyst separates hydrogen's negatively charged electrons from positively charged ions (protons).
- At the cathode, oxygen combines with electrons and, in some cases, with species such as protons or water, resulting in water or hydroxide ions, respectively.
- For polymer electrolyte membrane and phosphoric acid fuel cells, protons move through the electrolyte to the cathode to combine with oxygen and electrons, producing water and heat.
- For alkaline, molten carbonate, and solid oxide fuel cells, negative ions travel through the electrolyte to the anode where they combine with hydrogen to generate water and electrons.
- The electrons from the anode cannot pass through the electrolyte to the positively charged cathode; they must travel around it via an electrical circuit to reach the other side of the cell. This movement of electrons is an electrical current.

ENVIRONMENTAL IMPACT OF FUEL CELLS

Beyond the expectation that hydrogen leakage from its use in fuel cells could greatly impact the hydrogen cycle and could, when oxidized in the stratosphere, cool the stratosphere, and create more clouds, delaying the breakup of the polar vortex at the poles, making the holes in the ozone layer larger and longer lasting, little is understood about how hydrogen leakage would affect the environment. For example, much uncertainty exists over the extent of hydrogen emissions impact on soil absorption of hydrogen from the atmosphere. This concept or principle is important because if we use extensive quantities of hydrogen for fuel cells, absorption of hydrogen by soils could have a compensatory effect on any possible anthropogenic emissions. Again, little is understood about how hydrogen leakage would affect the environment.

CARBON CAPTURE AND SEQUESTRATION

> You cannot get through a single day without having an impact on the world around you—for better or for worse. What you do makes a difference.
>
> **Jane Goodall**

Human activities, especially the burning of fossil fuels such as coal, oil, and gas, have caused a substantial increase in the concentration of carbon dioxide (CO_2) in the atmosphere. This increase in atmospheric CO_2—from about 280 to more than 380 parts per million (ppm) over the last 250 years—is causing measurable global warming. Potential adverse impacts include sea-level rise; increased frequency and intensity of wildfires, floods, droughts, and tropical storms; changes in the amount, timing, and distribution of rain, snow, and runoff; and disturbance of coastal marine and other ecosystems. Rising atmospheric CO_2 is also increasing the absorption of CO_2 by seawater, causing the ocean to become more acidic, with potentially disruptive effects on marine plankton and coral reefs. Technically and economically feasible strategies are needed to mitigate the consequences of increased atmospheric CO_2.

USGS (2008)

Above all we should, in the century since Darwin, have come to know that man, while captain of the adventuring ship, is hardly the sole object of its quest, and that his prior assumptions to this effect arose from the simple necessity of whistling in the dark. These things, I say, should have come to us. I fear they have not come to many.

Aldo Leopold, *A Sand County Almanac* (1948)

THE 411 ON CARBON CAPTURE AND SEQUESTRATION

The reader might wonder what does carbon capture and sequestration (CCS) have to do with the environmental impacts of renewable energy? Renewable energy has two pluses: (1) it is a possible source of energy now and in the future (it is renewable and sustainable) and will be called on to replace nonrenewable hydrocarbon energy sources as they are depleted; (2) renewable energy produces little or no waste products such as carbon dioxide or other chemical pollutants, so has minimal impact on the environment. It is the later of the two that is related to carbon capture and sequestration. That is, at the present time, and in the near future, we have and will continue to have an ongoing increase of atmospheric carbon dioxide. Many scientists agree that global climate change is occurring and that to prevent its most serious effects we must begin immediately to significantly reduce our greenhouse gas (GHG) emissions. One major contributor to climate change is the release of the greenhouse gas carbon dioxide (CO_2). This is the essence of the carbon capture and sequestration process: capture and sequester carbon dioxide. Further, to control atmospheric carbon dioxide will require deliberate mitigation with an approach the combines reducing emissions by utilizing renewable sources and by increasing capture and storage.

The term *carbon sequestration* is used to describe both natural and deliberate processes by which CO_2 is either removed from the atmosphere or diverted from emission sources and stored in the ocean, terrestrial environments (vegetation, soils, and sediments), and geologic formations. Before human-caused CO_2 emissions began, the natural processes that make up the global carbon cycle-maintained a near balance between the uptake of CO_2 and its release back into the atmosphere. However, existing CO_2 uptake mechanisms (sometimes called CO_2 or carbon *sinks*) are insufficient to offset the accelerating pace of emissions related to human activities. Annual carbon emissions from burning fossil fuels in the United States are about 1.6 gigatons (billion metric tons), whereas annual uptake amounts are only about 0.5 gigatons, resulting in a net release of about 1.1 gigatons per year.

Scientists at the U.S. Geological Survey (USGS) and elsewhere are working to assess both the potential capacities and the potential limitations of the various forms of carbon sequestration and to evaluate their geologic, hydrologic, and ecological consequences. USGS is providing information needed by decision-makers and resource managers to maximize carbon storage while minimizing undesirable impacts on humans and their physical and biological environment.

Terrestrial Carbon Sequestration

Terrestrial sequestration (sometimes termed "biological sequestration") is the removal of gaseous cab on dioxide from the atmosphere and binding it in living tissue by plants. Terrestrial sequestration is typically accomplished through forest and soil conservation practices that enhance the storage of carbon (such as restoring and establishing new forests, wetlands, and grasslands) or reduce CO_2 emissions (such as reducing agricultural tillage and suppressing wildfires). In the United States, these practices are implemented to meet a variety of land-management objectives. Although the net terrestrial uptake fluxes shown in Figure 9.2 offset about 30% of U.S. fossil-fuel CO_2 emissions, only a small fraction of this uptake results from activities undertaken specifically to sequester carbon. The largest net uptake is due primarily to the ongoing natural regrowth of forests that were harvested during the nineteenth and early twentieth centuries.

Existing terrestrial carbon storage is susceptible to disturbances such as fire, disease, and change in climate and land use. Boreal forests, also known as taiga, and northern peatlands, which store nearly half the total terrestrial carbon in North American, are already experiencing substantial warming, resulting in large-scale thawing of permafrost and dramatic changes in aquatic and forest ecosystems. USGS scientists have estimated that at least 10 gigatons of soil carbon in Alaska are stored in organic soils that are extremely vulnerable to fire and decomposition under warming conditions.

The capacity of terrestrial ecosystems to sequester additional carbon is uncertain. An upper estimate of potential terrestrial sequestration in the U.S might be the amount of carbon that would be accumulated if U.S. forests and soils were restored to their historic levels before they were depleted by logging and cultivation. These amounts (about 32 and 7 gigatons for forests and soils, respectively) are probably not attainable by deliberate sequestration because restoration on this scale would displace a large percentage of U.S. agriculture and disrupt many other present-day activities. Decisions about terrestrial carbon sequestration require careful consideration of priorities and tradeoffs among multiple resources. For example, converting farmlands to forests or wetlands may increase carbon sequestration, enhance wildlife habitat, and water quality, and increase flood storage and recreational potential—but the loss of farmlands will decrease crop production. Converting existing conservation lands to intensive cultivation, while perhaps producing valuable corps (for example, for biofuels), may diminish wildlife habitat, reduce water quality, and supply, and increase CO_2 emissions. Scientists are working to determine the effects of climate and land-use change on potential carbon sequestration and ecosystem benefits, and to provide information about these effects for use in resource planning.

SIDEBAR 12.2 URBAN FORESTS AND CARBON SEQUESTRATION[9]

The urban environment presents important considerations for terrestrial carbon sequestration and global climate change. Over half of the world's population lives in urban areas (Population Reference Bureau, 2012). Because cities are denser and walkable, urban per capita emissions of greenhouse gases (GHGs) are almost always substantially lower than average per capita emissions for the counties in which they are located (The Cities Alliance, 2007). Urban areas are more likely than non-urban areas to have adequate emergency services, and so may be better equipped to provide critical assistance to residents in the case of climate-related stress and events such as heat waves, floods, storms, and disease outbreaks (Myers et al., 2013). However, cities are still major sources of GHG emissions (Dodman, 2009). Studies suggest that cities account for 40%–70% of all GHG emissions worldwide due to resource consumption and energy, infrastructure, and transportation demands (USEPA, 2009). Highly concentrated urban areas, especially in coastal regions and in developing countries, are disproportionately vulnerable to extreme weather and infectious disease.

The term "urban forest" refers to all trees within a densely populated area, including trees in parks, on streetways, and on private property. Though the composition, health, age, extent, and costs of urban forests vary considerably among different cities, all urban forests offer some common environmental, economic, and social benefits. Urban forests play an important role in climate change mitigation and adaptation. Active stewardship of a community's forestry assets can strengthen local resilience to climate change while creating more sustainable and desirable places to live. Trees in a community help to reduce air and water pollution, alter heating cooling costs, and increase real estate values. Trees can improve physical and mental health, strengthen social connections, and are associated with reduced crime rates. Trees, community gardens, and other green spaces get people outside, helping foster active living and neighborhood pride.

Like any forest, urban forests help mitigate climate change by capturing and storing atmospheric carbon dioxide during photosynthesis, and by influencing energy needs for heating cooling buildings; trees typically reduce cooling costs but can increase or decrease winter heating use depending on their location around a building and whether they are evergreen or deciduous. In the contiguous United States alone, urban trees store over 708 million tons of carbon (approximately 12.6% of annual carbon dioxide emissions in the United States) and capture an additional 28.2 million tons of carbon (approximately 0.05% of annual emissions) per year (Nowak et al., 2013; USEPA, 2013). The value of urban carbon sequestration is substantial: approximately $2 billion per year, with a total current carbon storage value of over $50 billion (Nowak et al., 2010). Shading and reduction of wind speed by trees can help to reduce carbon emissions by reducing summer air conditioning and winter heating demand and, in turn, the level of emissions from supplying power plants (Nowak et al., 2010). Shading can also extend the useful life of street pavement by as much as 10 years, thereby reducing emissions associated with the petroleum-intensive materials and operation of heavy equipment required to repave roads and haul away waste (McPherson and Muchnick, 2005). Establishing 100 million mature trees around residences in the United States would save an estimated $2 billion annually in reduced energy costs (Akbari et al., 1992); Population Reference Bureau, 2012). However, this level of tree planting would only offset less than 1% of United States emissions over a 50-year period (Nowak and Crane, 2002).

The sustainable use of wood, food, and other goods provided by the local urban forest may also help mitigate climate change by displacing imports associated with higher levels of carbon dioxide emitted during production and transport. Urban wood is a valuable and underutilized resource. At current utilization rates, forest products manufactured from felled urban trees are estimated to save several hundred million tons of CO_2 over a 30-year period. Furthermore, wood chips made from low-quality urban wood may be combusted for heat and/or power to displace an additional 2.1 million tons of fossil fuel emissions per year (Sherril and Bratkovitch, 2011).

Urban forests enable cities to better adapt to the effect of climate change on temperature patterns and weather events. Cities are generally warmer than their surroundings (typically by about 1°C–2°C, though this difference can be as high as 10°C under certain climatic conditions), meaning that average temperature increases caused by global warming are frequently amplified in urban areas (Bristow et al., 2012; Kovats and Akhtar, 2008). Urban forests help control this "heat island" effect by providing shade and by reducing urban albedo (the fraction of solar radiation reflected back into the environment), and through cooling evapotranspiration (Romero-Lankao and Gratz, 2008; Bristow et al., 2012). Cities are also particularly susceptible to climate-related threats such as storms and flooding. Urban trees can help control runoff from these by catching rain in their canopies and increasing the infiltration rate of deposited precipitation. Reducing stormwater flow reduces stress on urban sewer systems by limiting the risk of hazardous combined sewer overflows (Fazio, 2010). Furthermore, well-maintained urban forests help buffer high winds, control erosion, and reduce drought (Nowak et al., 2010; Fazio, 2010; Cullington and Gye, 2010).

Urban forests provide critical social and cultural benefits that may strengthen community resilience to climate change. Street trees can hold spiritual value, promote social interaction, and contribute to a sense of place and family for local residents (Dandy, 2010). Overall, forested urban areas appear to have potentially stronger and more stable communities (Dandy,

2010). Community stability is essential to the development of effective long-term sustainable strategies for addressing climate change (Williamson et al., 2010). For example, neighborhoods with stronger social networks are more likely to check on the elderly and other vulnerable residents during heat waves and other emergencies (Klinenberg, 2002).

Urban forests help control the causes and consequences of climate-related threats. However, forests may also be negatively impacted by climate change. While it is true that carbon dioxide levels and water temperature may initially promote urban tree growth by accelerating photosynthesis, it is also true that too warming in the absence of adequate water and nutrients stresses trees and retards future development (Tubby and Webber, 2010). Warmer winter temperatures increase the likelihood of winter kill, in which trees, responding to their altered environment, prematurely begin to circulate water and nutrients in their vascular tissue. If rapid cooling follows these unnatural warm periods, tissues will freeze, and tress will sustain injury or death.

Warmer winter temperatures favor many populations of tress pest and pathogen species normally kept at low levels by cold winter temperatures (Tubby and Webber, 2010). Although climate change may reduce the populations of some species, many others are better able than their arboreal host to adapt to changing environments due to their short lifecycles and rapid evolutionary capacity (Cullington and Gye, 2010; Tubby and Webber, 2010). The consequences of these population changes are compounded by the fact that hot, dry environments enrich carbohydrate concentrations in tree foliage, making urban trees more attractive to pests and pathogens (Tubby and Webber, 2010).

Climate change alters water cycles in ways that impact urban forests. Increased winter precipitation puts urban forests at greater risk from physical damage due to increased snow and ice loading (Johnston, 2004). Increased summer evaporation and transpiration create water shortages often exacerbated by urban soil compaction and impermeable surfaces. More frequent and intense extreme weather events increase the likelihood of severe flooding, which may uproot trees and cause injury or death to tree root systems if waterlogged soils persist for prolonged periods (Johnston, 2004).

Especially cold regions may benefit from increased tourism, agricultural productively, and ease of transport as a result of climate change. However, the potential positive implications of climate change are far eclipsed by the negative (Parry et al., 2007). Rising temperatures, increased pest and pathogen activity, and water cycle changes impose physiological stresses on urban forests that compromise forest ability to deliver ecosystem services that protect against climate change. Climate change will also continue to alter species ranges and regeneration rates, further affecting the health and composition of urban forests (Nowak, 2010; Ordonez et al., 2010). Proactive management is necessary to protect urban forests against climate-related threats, and to sustain desired urban forest structures for future generations.

Geologic Carbon Sequestration

Geologic sequestration begins with capturing carbon dioxide from the exhaust of fossil-fuel power plants and other major sources. The captured carbon dioxide is piped 1–4 km below the land surface and injected into porous rock formations. Compared to the rates of terrestrial carbon uptake shown in Figures 9.1 and 9.2, geologic sequestration is currently used to store only small amounts of carbon per year. Much larger rates of sequestration are envisioned to take advantage of the potential permanence and capacity of geologic storage.

The permanence of geologic sequestration depends on the effectiveness of several carbon dioxide trapping mechanisms. After carbon dioxide is injected underground, it will rise buoyantly until it is trapped beneath an impermeable barrier, or seal. In principle, this physical trapping mechanism, which is identical to the natural geologic trapping of oil and gas, can retain carbon dioxide for thousands to millions of years. Some of the injected carbon dioxide will eventually dissolve in groundwater, and some may be trapped in the form of carbonate minerals formed by chemical reactions with the surrounding rock. All of these processes are susceptible to change over time following carbon dioxide injection. Scientists are studying the permanence of these trapping mechanisms and developing methods to determine the potential for geologically sequestered carbon dioxide to leak back into the atmosphere.

The capacity for geologic carbon sequestration is constrained by the volume and distribution of potential storage sites. According to the U.S. Department of energy, the total storage capacity of physical traps associated with depleted oil and gas reservoirs in the United States is limited to about 38 gigatons of carbon and is geographically distributed in locations that are distant from most U.S. fossil-fuel power plants. The potential U.S. storage capacity of deep porous rock formations that contain saline groundwater is much larger (estimated by the U.S. Department of Energy to be about 900–3,400 gigatons of carbon) and more widely distributed, but less is known about the effectiveness of trapping mechanisms at these sites. Unmineable coal beds have also been proposed for potential carbon dioxide storage, but more information is needed about the storage characteristics and impacts of carbon dioxide injection in these formations. Scientists are developing methods to refine estimates of the national capacity for geologic carbon sequestration.

To fully assess the potential for geologic carbon sequestration, economic costs and environmental risks must be considered. Infrastructure costs will depend on the locations of suitable storage sites. Environmental risks may include seismic disturbances, deformation of the land surface, contamination of portable water supplies, and adverse effects on ecosystems and human health. Many of these environmental risks and potential environmental impacts are discussed in the sections below.

POTENTIAL IMPACT OF TERRESTRIAL SEQUESTRATION

Potential environmental impacts associated with terrestrial sequestration include ground disturbance and the loss of soil resources due to erosion; equipment-related noise, visual impact and air emissions; the disturbance of ecological, cultural and paleontological resources, and conflicts with current or proposed land use.

Establishing and managing a terrestrial sequestration plot could involve ground clearing (removal of vegetative cover) to prepare the ground for planting, grading, vehicular traffic, and pedestrian traffic. Management could require the use of water for dust control and in some cases, water could be required to establish and maintain seeds, seedlings, or crops. The addition of soil additives like fertilizer and pesticides could have an impact on water quality. Equipment used to maintain a terrestrial sequestration plot could be a source of noise and air emissions and create a visual impact if frequent and conspicuous use were required.

Ecological, cultural, and paleontological resources could be impacted, especially if a terrestrial sequestration plot were going to replace an established ecological habitat or otherwise impact undisturbed land that hosts important cultural or paleontological resources. Impacts on land use could occur if there were conflicts with existing land use plans; for example, if land zoned for future commercial or housing development is used to establish a forest sequestration plot.

Soil resources can also be impacted by terrestrial sequestration. The careful management of a sequestration plot should result in an improvement of soil resources, but poor management practices could adversely impact soils and the viability of the sequestration project. Practices like no-till cultivation and plant, crop rotation, and the use of cover crops, should result in the maintenance of soil organic material and nutrients and an increase in the relative health of soil resources. Some management practices, however, could involve the use of hazardous materials like herbicides to kill a cover crop before planting the terrestrial sequestration crop.

POTENTIAL IMPACT OF GEOLOGIC SEQUESTRATION

The potential impacts of geologic sequestration, including the transportation of carbon, are discussed in this section. For this discussion, we have assumed that carbon capture would likely occur at a single power-generating station. Because captured carbon may have to be transported for some distance away from the power station, transport, in general, has been evaluated. The significance of the impacts depends upon factors such as the number and size of transport pipelines and injection wells, the amount of land disturbed by drilling and transport activities, the amount of land occupied by facilities over the life of the sequestration project, the project's location with respect to other resources (e.g., wildlife use, distant to surface water bodies), and so forth.

GEOLOGIC SEQUESTRATION EXPLORATION IMPACT

Activities during the exploration phase (including seismic surveys, testing, and exploratory drilling) are temporary and are conducted at a smaller scale than those at the drilling/construction, sequestration, and decommissioning/reclamation phases. The impacts described for each resource would result from typical exploration activities, such as localized ground clearing, vehicular traffic, seismic testing, positioning of equipment, and exploratory drilling. Most impacts during the exploration phase would be associated with the development of access roads and exploratory wells. Impacts on resources would be similar in character, but lesser in magnitude, to those for the drilling phase. Potential impacts from these activities are presented below, by the type of affected resource.

Air Quality

Impacts on air quality during exploration activities would include emissions and dust from earth-moving equipment, vehicles, seismic surveys, well completion and testing, and drill rig exhaust. Pollutants would include particulates, oxides of nitrogen, carbon monoxide, sulfur dioxide, and volatile organic compounds (VOCs). Nitrogen oxides and VOCs may combine to form ground-level ozone. Impacts would depend upon the amount, duration, location, and characteristics of the emission and the meteorological conditions (e.g., wind speed and direction, precipitation, and relative humidity). Emissions during this phase would not have a measurable impact on climate change.

Cultural Resources

During the exploration phase, soil surface and subsurface disturbance are minimal. Cultural resources buried below the surface are unlikely to be affected; while material present on the surface could be disturbed by vehicular traffic, ground clearing, and pedestrian activity (including the collection of artifacts). Exploration activities could affect areas of interest to Native Americans, depending on the placement of equipment and/or level of visual intrusion. Surveys conducted during this phase to evaluate the presence and/or significance of cultural resources in the area would assist developers in siting project facilities, in order to avoid or minimize impacts on these resources.

Ecological Resources

Impacts on ecological resources (vegetation, wildlife, aquatic biota, special status species, and their habitats) would be minimal and localized during exploration because of the limited nature of the activities. The introduction or spread of some nonnative invasive vegetation could occur as a result of vehicular traffic, but this would be relatively limited in extent. Seismic surveys could disturb wildlife. Exploratory well establishment would destroy vegetation and impact wildlife. Surveys conducted during this phase to evaluate the presence and/or significance of ecological resources in the area would assist developers in siting project facilities in order to avoid or minimize impacts to these resources.

Water Resources (Surface Water and Groundwater)

Minimal impact on water resources (water quality, water flows, and surface water/groundwater interactions) would be anticipated from exploration activities. Exploratory wellbores may provide a path for surface contaminants to come into contact with groundwater or for waters from subsurface formations to commingle. They may also decrease pressure in water wells and affect their quality. Very little produced water would likely be generated during the exploration phase. Most water needed to support drilling operations could be trucked in from off-site.

Land Use

Temporary and localized impacts on land use would result from exploration activities. These activities could create a temporary disturbance in the immediate vicinity of a surveying or monitoring site or an exploratory well (e.g., disturbing recreational activities or livestock grazing). Exploration activities are unlikely to affect mining activities, military operations, or aviation.

Soils and Geologic Resources

Surface effects from vehicular traffic could occur in areas that contain special (e.g., cryptobiotic) soils. The loss of biological crusts can substantially increase water and wind erosion. Also, soil compaction due to development activities at the exploratory well pads and along access roads would reduce aeration, permeability, and water-holding capacity of the soils and cause an increase in surface runoff, potentially causing increased sheet, rill, and gully erosion. The excavation and reapplication of surface soils could cause the mixing of shallow soil horizons, resulting in a blending of soil characteristics and types. This blending would modify the physical characteristics of the soils, including structure, texture, and rock content, which could lead to reduced permeability and increased runoff from these areas. Potential impacts on geologic and mineral resources would include the depletion of hydrocarbons and sand and gravel resources. It is unlikely that exploration activities would activate geological hazards. Impacts on soils and geologic resources would be proportional to the amount of disturbance. The amount of surface disturbance and use of geologic materials using exploration would be minimal.

Paleontological Resources

Paleontological resources are nonrenewable resources. Disturbance of such resources, whether it is through mechanical surface disturbance, erosion, or paleontological excavation, irrevocably alters or destroys them. Direct impacts on paleontological resources would include surface disturbance during seismic surveys and the drilling of exploratory wells and the construction of access roads and other ancillary facilities. The amount of subsurface disturbance is minimal during the exploration phase, and paleontological resources buried below the surface are unlikely to be affected. Fossil material present on the surface could be disturbed by vehicular traffic, ground clearing, and pedestrian activities (including the collection of fossils).

Surveys conducted during this phase to evaluate the presence and/or significance of paleontological resources in the area would assist developers in siting project facilities in order to avoid or minimize impacts on these resources.

Transportation

No impacts on transportation are anticipated during the exploration phase. Transportation activities would be temporary and intermittent and limited to low volumes of light utility trucks and personal vehicles.

Visual Resources

Impacts on visual resources would be considered adverse if the landscape were substantially degraded or modified. Exploration activities would have only temporary and minor visual effects, resulting from the presence of drill rigs, workers, vehicles, and other equipment.

Socioeconomics

As the activities conducted during the exploration phase are temporary and limited in scope, they would not result in significant socioeconomic impacts on employment, local services, or property values.

Environmental Justice

Exploration activities are limited and would not result in significant adverse impacts in any resource area; therefore, environmental justice is not expected to be an issue during this phase.

Acoustics (Noise)

Primary sources of noise associated with exploration include earth-moving equipment, vehicle traffic, seismic survey, blasting, and drill rig operations.

Hazardous Materials and Waste Management

Seismic and exploratory well crews may generate waste (plastic, paper, containers, fuel leaks/spills, food, and human waste). Wastes produced by exploratory drilling would be similar but occur to a lesser extent than those produced during drilling and operation of injection wells. The would include drilling fluid and muds, used oil and filters, spilled fuel, drill cuttings, spent and unused solvents, scrap metal, solid waste, and garbage.

GEOLOGIC SEQUESTRATION DRILLING/CONSTRUCTION IMPACT

Typical activities during the drilling/construction phase of a sequestration project include ground clearing and removal of vegetative cover, grading, drilling, waste management, vehicular and pedestrian traffic, and construction and installation of facilities. Activities conducted in locations other than at the injection well pad site may include excavation/blasting for construction materials (sands, gravels), access road and storage area construction, and construction of pipelines, compressor stations or pumping stations, and other facilities (e.g., office buildings). Potential impacts from these activities are presented below, by the type of affected resource.

Air Quality

Emissions generated during the drilling/construction phase include vehicle emissions; diesel emissions from large construction equipment and generators, storage/dispensing of fuel, and, if installed at this stage, flare stacks; small amounts of carbon monoxide, nitrogen oxides, and particulates from blasting activities; and dust from many source, such as disturbing and moving soils (clearing, grading, excavation, trenching, backfilling, dumping, and truck and equipment traffic), mixing concrete, and drilling. During windless conditions (especially in areas of thermal inversion), project-related odors may be detectable at more than a mile from the source. Excess increases in dust could decrease forage palatability for wildlife and livestock and increase the potential for dust pneumonia.

Cultural Resources

Potential impacts on cultural resources during the drilling/construction phase could include:

- Destruction of cultural resources in areas undergoing surface disturbance
- Unauthorized removal of artifacts or vandalism as a result of human access to previously inaccessible areas (resulting in lost opportunities to expand scientific study and educational and interpretive uses of these resources)
- Visual impacts resulting from large areas of the exposed surface, increases in dust, and the presence of large-scale equipment, machinery, and vehicles for cultural resources that have an associated landscape component that contributes to their significant (e.g., sacred landscapes or historic trails).

Green Engineering

While the potential for encountering buried sites is relatively low, the possibility that buried sites would be disturbed during pipeline, access road, or well pad construction does exist. Unless the buried site is detected early in the surface-disturbing activities, the impact on the site can be considerable. Disturbance that uncovers cultural resources of significant important that would otherwise have remained buried and unavailable could be viewed as a beneficial impact. Vibration, resulting from increased traffic and drilling/development activities, may also have effects on rock art and other associated sites (e.g., sites with standing architecture).

Ecological Resources

Impacts on ecological resources would be proportional to the amount of surface disturbance and habitat fragmentation. Vegetation and topsoil would be removed for the development of well pads, access roads, pipelines, and other ancillary facilities. This would lead to a loss of wildlife habitat, reduction in plant diversity, potential for increased erosion, and potential for the introduction of invasive or noxious weeds. The recovery of vegetation following interim and final reclamation would vary by community (e.g., grasslands would recover before sagebrush or forest habitats). Indirect impacts on vegetation would include increased deposition of dust, spread of invasive and noxious weeds, and the increased potential for wildfires. Dust settling on vegetation may alter or limit plants' abilities to photosynthesize and/or reproduce. Over time, a composition of native and/or invasive vegetation would become established in areas disturbed by wildfire. Although injection field development would likely increase the spread of invasive and noxious weeds by increasing traffic and human activity, the potential impacts could be partially reduced by interim reclamation and implementation of mitigation measures. Adverse impacts on fish and wildlife could occur during the drilling/construction phase from:

- Erosion and runoff
- Dust
- Noise
- Introduction and spread on invasive nonnative vegetation
- Modification, fragmentation, and reduction of habitat
- Mortality of biota
- Exposure to contaminants
- Interference with behavioral activities
- Increased harassment and/or poaching

Depletion of surface waters from perennial streams could result in a reduction of water flow, which could lead to habitat loss and/or degradation of aquatic species.

Water Resources (Surface Water and Groundwater)

Impacts on water resources could occur due to water quality degradation form increases in turbidity, sedimentation, and salinity; spills; cross-aquifer missing; and water quantity depletion. During the drilling/construction phase, water would be required for dust control, making concrete, consumptive use by the construction crew, and in drilling of wells. Depending on availability, it may be trucked in from off-site or obtained from local groundwater wells or nearby surface water bodies. Where surface waters are used to meet drilling and construction needs, depletion of stream flows could occur. Drilling and well completion can require the use of drilling fluids that could, if not managed properly, contaminate soils and surface water features. Drilling activities may affect surface and groundwater flows. If a well is completed improperly, such that subsurface formations are not sealed off by the well casing and cement, aquifers can be impacted by other nonpotable formation waters. The interaction between surface water and groundwater may also be affected if the two are hydrologically connected, potentially resulting in unwanted dewatering or recharging. Soils compacted on existing roads, new access roads, and well pads generated more runoff than undisturbed sites. The increased runoff could lead to slightly higher peak storm flows into streams, potentially increasing erosion of the channel banks. The increased runoff could also lead to more efficient sediment delivery and increased turbidity during storm events. During drilling and construction, water quality can be affected by:

- Activities that cause soil erosion or dust that can be washed into water bodies
- Weathering of newly exposed soils, causing leaching and oxidation that can release chemicals into the water
- Increase salinity levels resulting from increased sediment loading
- Discharges of waste or sanitary water
- Use of herbicide and dust suppressants (e.g., magnesium chloride)
- Contaminant spills

Land Use

Land use impacts would occur during the drilling/construction phase if there are conflicts with existing land use plants and community goals; existing recreational, educational, religious, scientific, or other use areas; or existing commercial and use (e.g., agriculture, grazing, or mineral extraction). In general, the development of large-scale geologic sequestration facilities and transport pipelines is expected to change the character of the landscape from a rural to a more industrialized setting. Existing land use would be affected by intrusive impacts such as increased traffic, noise, dust, and human activity, as well as by changes in the visual landscape. In particular, these impacts could affect recreationists seeking solitude or recreational opportunities in a relatively pristine landscape. Ranchers or farmers could be affected by loss of available grazing or crop lands, potential for the introduction of invasive and noxious plants that could affect livestock forage availability, and possible increase in livestock/vehicle collisions. In forested areas, drilling could result in the long-term loss of timber resources. The expanded access road system could increase the number of off-highway vehicle (OHV) users, hunters, and other recreationists in the area. While the change in landscape character could discourage hunters who prefer a more remote backcountry setting; the potential for illegal hinting activities could increase due to the expected access road system. Construction and drilling noise could potentially be heard 20 miles (32 km) or more from the project area. While it would be barely audible at this distance, it could affect residents' and recreationists' perception of solitude.

Most land use impacts that occur during the drilling/construction phase would continue throughout the life of the sequestration project. Overall, land use impacts could range from minimal to significant depending upon the areal extent of the project, the density of injection wells and other ancillary facilities, and the compatibility of the project with the existing land uses.

Soils and Geologic Resources

Potential impacts on soils during the drilling/construction phase would occur due to the removal of vegetation, mixing of soil horizons, soil compaction, increased susceptibility of the soils to wind and water erosion, contamination of soils form spills of hazardous materials (e.g., drilling mud, fluids used to hydraulically fracture subsurface formations), loss of topsoil productivity, and disturbance of biological soil crusts. Impacts on soils would be proportionate to the amount of disturbance. Sands, gravels, and quarry stone could be excavated for use in the construction of access roads; foundations and ancillary structures; and for well pad and storage areas. Construction of well pads, pipelines, compressors or pumping stations, access roads, and other project facilities could cause topographic changes. These changes would be minor, but long term. Well pads located on canyon rims of the side slopes of canyons could result in bedrock disturbances. Additional bedrock disturbance could occur due to the construction of access roads, pipelines, rock borrow pits, and other ancillary facilities. Possible geological hazards (earthquakes, landslides, and subsidence) could be activated by drilling and blasting. Altering drainage patterns could also accelerate erosion and crate slope instability.

Paleontological Resources

Impacts on paleontological resources can occur directly from construction and drilling activities or indirectly as a result of soil erosion and increased accessibility to fossil localities (e.g., unauthorized removal of fossil resources or vandalism to the resource). This would result in lost opportunities to expand the scientific study and educational interpretive uses of these resources. Disturbance that uncovers paleontological resources of significant importance that would otherwise have remained buried and unavailable could be viewed as a beneficial impact. Direct impacts on unknown paleontological resources can be anticipated to be proportional to the total area impacted by construction and drilling activities.

Transportation

Development of a geologic sequestration project would result in the need to construct and/or improve access roads and would result in an increase in industrial traffic (e.g., hundreds of truck loads or more per well site). Overweight and oversized loads could cause temporary disruptions and could require extensive modifications to roads or bridges (e.g., widening roads or fortifying bridges to accommodate the size or weigh of truck loads). An overall increase in heavy truck traffic would accelerate the deterioration of pavement, requiring local government agencies to schedule pavement repair or replacement more frequently than under the existing traffic conditions. Increased traffic would also result in a potential for increased accidents within the project area. The locations at which accidents are most likely to occur are intersections used by project-related vehicles to turn onto or off of highways from access roads. Conflicts between industrial traffic and other traffic are likely to occur, especially on weekends, holidays, and seasons of high use by recreationists. Increased recreational use of the area could contribute to a gradual increase in traffic on the access roads. Over 1,000 truckloads per well could be expected during the drilling/construction phase.

Visual Resources

During the drilling/construction phase, impacts on visual resources would occur as a result of the addition of well pads, pipelines, access roads, and other facilities which would result in an industrial landscape throughout the project area. Additional components that would adversely affect the visual character of the landscape are pumping units, compressor stations, aggregate borrow areas, equipment storage areas, and, if needed, worker housing units and airstrips. Project facilities would introduce new elements of form, line, color, and texture into the landscape, which would dominate foreground views. In some instances, the facilities would also be visible form greater distances and could, occasionally, dominate the view. Vehicles and the dust they generate would also contribute to visual impacts. Because drilling activities typically take place 24 hours per day, visual impacts would include lighting of drill rigs during nighttime hours. Nighttime lighting on drill rigs would be visible from long distances.

Socioeconomics

Drilling/construction phase activities would contribute to the local economy by providing employment opportunities, monies to local contractors, and recycled revenues through the local economy. Additional revenues could be generated in the form of carbon avoidance-type emission credits sold by the sequestration facility operator in a commodity market. Taxes collected by federal, state, and local governments could also be involved for transportation, injection, and regulation of the sequestration project. Indirect impacts could occur as a result of the new economic development (e.g., new jobs at businesses that support the expanded workforce or that provide project materials). Depending on the source of the workforce, local increases in population could occur. Development of an injection well field also could potentially affect property values, either positively from increased employment effects or negatively from proximity to the field and any associated or perceived adverse environmental effects (noise of compressor stations, visual effects, air quality, etc.). Some economic losses could occur if recreationists (including hunters and fishermen) avoid the area. Increased growth of the transient population could contribute to increased criminal activities in the project area (e.g., robberies, drugs).

Environmental Justice

If significant impacts were to occur in any of the resource areas and these were to disproportionately affect minority or low-income populations, there could be an environmental justice impact. It is anticipated that the drilling/construction phase could benefit low-income, minority, and tribal populations by creating job opportunities and simulating local economic growth via project revenues and increased tourism. However, noise, dust, visual impacts, and habitat destruction could have an adverse effect on traditional tribal life ways and religious and cultural sites. The development of wells and ancillary facilities could affect the natural character of previously undisturbed areas and transform the landscape into a more industrialized setting. Drilling and construction activities could impact the use of cultural sites for traditional tribal activities (hunting and plant-gathering activities, and areas in which artifacts, rock art, or other significant cultural sites are located).

ACOUSTICS (NOISE)

Primary sources of noise during the drilling/construction phase would be equipment (bulldozers, drill rigs, and diesel engines). Other sources of noise include vehicular traffic and blasting. Blasting activities typically would be very limited, the possible exception being in areas where the terrain is hilly and bedrock shallow. With the exception of blasting, noise would be restricted to the immediate vicinity of the work in progress. Noise form blasting would be sporadic and of short duration but would carry for long distances. If noise-producing activities occur near a residential area, noise levels form blasting, drilling, and other activities could exceed the U.S. Environmental Protection Agency (EPA) guidelines. The movement of heavy vehicles and drilling could result in frequent-to-continuous noise. Drilling noise would occur continuously for 24 hours per day for 1–2 months or more depending on the depth of the formation. Exploratory wells that end up becoming injection wells would continue to generate noise during the sequestration phase.

HAZARDOUS MATERIALS AND WASTE MANAGEMENT

Solid and industrial waste would be generated during the drilling/construction phase. Much of the solid wastes would be expected to be nonhazardous, considering of containers and packaging materials, miscellaneous wastes from equipment assembly and the presence of construction crews (food wrappers and scraps), and woody vegetation. Industrial wastes would include minor amounts of paints, coatings and spent solvents. Most of these materials would likely be transported off-site for disposal. In forested areas, commercial-grade timber could be sold, while slash may be spread or burned near the well site.

Drilling wastes include hydraulic fluids, pipe dope, used oils and oil filters, rigwash, spilled fuel, drill cuttings, drums, and containers, spent and unused solvents, pain and paint washes, sandblast media, scrap metal, solid waste, and garbage. Wastes associated with drilling fluids include oil derivatives (e.g., polycyclic aromatic hydrocarbons [PAHs], spilled chemicals, suspended and dissolved solids, phenols, cadmium, chromium, copper, lead, mercury, nickel, and drilling mud additives [including potentially harmful contaminants such as chromate and barite]). Adverse impacts could result if hazardous wastes are not properly handled and are released into the environment.

GEOLOGIC SEQUESTRATION OPERATIONS IMPACT

Typical activities during the operations phase include the operation of wells and compressor stations or pump stations, waste management, and maintenance and replacement of facility components.

Green Engineering

Impacts could also result from the fact that a geologic sequestration project could be linked to an enhanced oil recovery or enhanced coalbed methane recovery project.

Air Quality

The primary emission sources during the operations phase would include compressor and pumping station operations, vehicle traffic, and operating wells. Venting of carbon dioxide may occur during injection and pipeline maintenance operations.

Cultural Resources

During the operations phase, impacts on cultural resources could occur primarily from the unauthorized collection of artifacts and from visual impacts. In the latter case, the presence of the aboveground structures could impact cultural resources with an associated landscape component that contributes to their significance, such as a sacred landscape or historic trail. Damage to localities caused by off-highway vehicle (OHV) use could also occur. The potential for indirect impacts (e.g., vandalism and unauthorized collecting) would be greater during the operations phase compared to the drilling/construction phase, due to the longer duration of the operations phase.

Ecological Resources

During the operations phase, adverse impacts on ecological resources could result from:

- Disturbance of wildlife from noise and human activity
- Exposure of biota to contaminants
- Mortality of biota from colliding with aboveground facilities or vehicles

Ecological resources may continue to be affected by the reduction in habitat quality associated with habitat fragmentation due to the presence of operating wells, pipelines, ancillary facilities, and access roads. In addition, the presence of access roads any increase human use of surrounding areas, which, in turn, could impact ecological resources in the surrounding areas through:

- Introduction and spread of invasive nonnative vegetation
- Fragmentation of habitat
- Disturbance of biota
- Increase in hunting (including poaching)
- Increased potential for fire

The presence of an injection well field could also interfere with migratory and other behaviors of some wildlife.

In some coal bed methane production areas, methane gas or carbon dioxide gas could seep up into fields and create dead zones. High levels of carbon dioxide could asphyxiate wildfire in their burrows.

Water Resources (Surface Water and Groundwater)

During the life of an injection well, the integrity of the well casing and cement will determine the potential for adverse impacts on groundwater. If subsurface formations are not sealed off by the well casing and cement, aquifers can be impacted by other nonpotable formation waters, hydraulic fracturing fluids, or the injected carbon dioxide.

Other potential impacts on water availability and quality during the operations phase would include possible minor degradation of water quality resulting from vehicular traffic and machinery operations during maintenance (e.g., erosion and sedimentation) or herbicide contamination resulting from improper application. A spill or blowout could potentially cause extensive contamination of surface waters or a shallow aquifer. Contaminated groundwater could potentially be discharged into springs or as base flow into stream channels, leading to surface water contamination.

Recovered waters used for hydraulic fracturing could cause altered surface water quality or an increase in flows in normally dry water bodies such as ephemeral drainages if they are disposed of by discharge to the surface.

With regard to hydraulic fracturing wastewater discharge into the environment, a point of interest is that the Clean Water Act (CWA) made it unlawful to discharge any pollutant from a point source into the navigable waters of the U.S., unless done in accordance with a specific approved permit. The NPDES permit program controls discharges from point sources that are discrete conveyances, such as pipes or man-made ditches. Industrial, municipal, and other facilities such as shale gas production sites or commercial facilities that handle the disposal or treatment of shale gas produced water must obtain permits if they intend to discharge directly into surface water. Large facilities usually have individual NPDES permits. Discharge from some smaller facilities may be eligible for inclusion under general permits that authorize a category of discharge under the CWA within a geographic area. A general permit is not specifically tailored to an individual discharger. Most oil and gas production facilities with related discharges are authorized under general permits because there are typically numerous sites with common discharges in a geographic area.

Land Use

Land use impacts during the operations phase would be an extension of those that occurred during the drilling/construction phase. However, to some extent, land can revert to its original uses after the major drilling/construction phase is over. For example, farmers can graze livestock or grow crops around the well sites. Other industrial projects would likely be excluded within the sequestration project area. Recreation activities (e.g., OHV use and hunting) are possible, although gun and archery restrictions would probably exist. Operations may conflict with livestock and farming operations.

Soils and Geologic Resources

Following construction and drilling, disturbed portions of well and ancillary facility sites not required for operations would be revegetated. This would help to stabilize soil and geologic conditions. Routine impacts on soils during the operations phase would be limited largely to soil erosion impacts caused by vehicular traffic. Any excavations required for maintenance would cause impacts similar to those from the drilling/construction phase, but at a lesser spatial and temporal extent. The accidental spill of product or other wastes would likely cause soil contamination. Except in the case of a large spill, soil contamination would be localized and limited in extent and magnitude. In areas where interim reclamation is implemented (e.g., reclamation of an individual well that is no longer needed), ground cover by herbaceous species could re-establish within 1–5 years following seeding of native plant species and diligent weed control efforts, consequently reducing soil erosion. Operations might preclude or interfere with mineral development activities in the project area, including oil and gas development and mining activities. Possible geological hazards (earthquakes, landslide, and subsidence) could be activated by injection activities.

Paleontological Resources

Impacts on paleontological resources during the operations phase would be limited primarily to the unauthorized collection of fossils. This threat is present once the access roads are constructed, making remote areas more accessible to the public. Damage to localities caused by OHV use could also occur. The potential for indirect impacts (e.g., vandalism and unauthorized collecting) would be greater during the operations phase compared to the drilling/construction phase, due to its longer duration.

Transportation

Impacts on transportation during the operations phase would be similar to those for the drilling/construction phase. However, unless carbon dioxide is transported to the site by truck or rail, daily traffic levels, particularly heavy truck traffic, would be expected to be lower during the operations phase compared to the drilling/construction phase. For the most part, heavy truck traffic would be limited to periodic visits to a well site for workovers, and formation treatment. The use of pipelines to convey carbon dioxide to the operating site would reduce the volume of traffic during the operations phase. If a pipeline is not used for the injection well field, multiple truckloads per day would be needed.

Visual Resources

Once operating facilities are installed, portions of well pads, access roads, and pipeline rights-of-way (ROWs) that are not needed for operations would be reclaimed; however, much of the disturbed area would continue to contrast with the natural form, line, color, and texture of the surrounding landscape. This would impact undisturbed vistas and areas of solitude. The aboveground portions of an injection well would be highly visible in rural or natural landscapes, many of which may have a few other comparable structures. The artificial appearance of an injection well may have visually incongruous "industrial" associations for some, particularly in a predominately natural landscape. Any nighttime lighting would be visible form long distances. During the operations phase, indirect impacts on visual resources would occur as a result of sequestration activities (e.g., industrial traffic, heavy equipment use, and dust). However, human activity would be substantially lower than during the drilling/construction phase.

Socioeconomics

Direct socioeconomic impact would include the creation of new jobs and the associated royalties and taxes paid for carbon emission avoidance created by the sequestration project. Indirect impacts are those impacts that would occur as a result of the new economic development and would include new jobs at businesses that support the expanded workforce or that provide project materials, and associated taxes. Potential impacts on the value of residential properties located adjacent to an oil or gas field would continue during this phase.

Environmental Justice

Possible environmental justice impacts during the operations phase include the alteration of scenic quality in areas of traditional or cultural significance to minority populations. Noise and health and safety impacts are also potential sources of disproportionate effects on minority or low-income populations.

Acoustics (Noise)

The main sources of noise during the operations phase would include compressor and pumping stations, producing wells (including occasional flaring), and vehicle traffic. Compressor stations produce noise levels between 64 and 86 dBA at the station to between 58 and 75 dBA at about 1 mile (1.6 km) from the station. The use of remote telemetry equipment would reduce daily traffic and associated noise levels within the project area. The primary impacts from noise would be localized disturbance to wildlife, recreationists, and residents. Noise associated with cavitation is a major concern for landowners, livestock, and wildlife.

Hazardous Materials and Waste Management

Industrial wastes are generated during routine operations (lubricating oils, hydraulic fluids, coolants, solvents, and cleaning agents). These wastes are typically placed in containers, characterized, and labeled, possibly stored briefly, and transported by a licensed hauler to an appropriate permitted off-site disposal facility as a standard practice. Impacts could result if these wastes were not properly handled and were released to the environment. Environmental contamination could result from accidental spills of herbicides or other chemicals. Chemicals in open pits used to store wastes may pose a threat to wildlife and livestock.

Should geologic sequestration become common, a wide diversity of geologic formations is likely to be encountered. Depending upon the nature of the formation targeted for injection, it may be necessary to increase the injectivity of carbon dioxide by using hydraulic fracturing. Hydraulic fracturing fluids can contain innocuous constituents, like sand and water, and potentially toxic substances, such as diesel fuel (which contains benzene, ethylbenzene, toluene, xylenes, naphthalene, and other chemicals), polycyclic aromatic hydrocarbons (PAHs), methanol, formaldehyde, ethylene glycol, glycol ethers, hydrochloric acid, and sodium hydroxide. Since some aspects of a hydraulic fracturing operation are considered proprietary, information about the specific constituents used in a given hydrofracturing operation may not be available, thus causing some concern over the risks presented by this practice. Some of the hydrofracture fluids used to increase the injectivity of a carbon dioxide injection well would probably be pumped out of the well and then be managed at the surface in tanks of impoundments. However, some of the fluids would remain in the underground formation. For example, in the process of producing hydrocarbons and produced water, about 20%–40% of the fluids used for hydrofracturing may remain underground. Thus, should hydraulic fracturing be used in a carbon sequestration injection well, these fluids could have an impact on underground water sources that are close (horizontally or vertically) to the injection well.

During the operations phase, scale and sludge wastes can accumulate inside pipelines and storage vessels. They must be removed periodically from the equipment for disposal. These wastes may be transported to off-site disposal facilities. In some instances, they may be disposed of via land spreading, a practice that entails spreading the wastes over the surface of the disposal area and mixing it with the top few inches of soil.

THE BOTTOM LINE

Current and future environmental engineers and other environmental professionals need to add Green Energy and all that it entails to their professional toolboxes. Remember, there is nothing, absolutely nothing static about the environment and those who work to maintain it.

NOTES

1. Much of the information in this section is from USDOE-EERE 2005. *History of Wind Energy.* Accessed 06/14/22 @ http://www1/eere/emergu/gpv/womdamdjudrp/printable_versions/wind_hisotry.htm.
2. Much of the information in section is from State of Washington Department of Fisheries (1992). Mark A. Hunter. *Hydropower Flow Fluctuations and Salmonids: A Review of the Biological Effects, mechanical Causes, and Options for Mitigation.* Olympia, WA. Technical Report 119, 46 p.
3. Information contained herein derived and adapted from *Tribal Energy Information Clearinghouse* accessed 03/13/14 @ http://teeic.anl.gov/er/biomass/impact/construct/index.cfm.
4. Based on information from DOE (2001). *Renewable Energy*: An Overview. Washington, DC: U.S. Department of Energy.
5. From *Environmental Impacts, Attributes, and Feasibility Criteria.* Accessed 03/19/14 @ www1.eere.energy.gov/geothermalpdfs/egs_chapter_8.pdf. Modified from Tribal Energy and Environmental Information Clearinghouse: *Geothermal Energy.* Accessed 03/19/14 @ http://teeic.anl.gov/er/geothermal/impact/siteeval/index.cfm.
6. From DOI (2010) *Ocean Energy.* Accessed 03/31/10 @ www.mms.gov.
7. Adapted from USEPA (2009) Report to Congress on the Potential Environmental Effects of Marine and Hydrokinetic Energy Technologies. Washington, DC: U.S. Environmental Protection Agency.
8. Information in this section from USDOE 2008, *Hydrogen, Fuel Cells & Infrastructure Technologies Program.* Accessed @ http://www1.eere.energy.gov/hydrogenandfuelcells/production/basics.html; F.R. Spellman & R. Bieber. *The Science of Renewable Energy.* (2011). CRC Press, Boca Raton, FL.
9. From Safford et al. (2013). *Urban Forests and Climate Change.* USDA.

REFERENCES

Abbasi, S.A., & Abbasi, N. (2000). The likely adverse environmental impacts of renewable energy sources. *Applied Energy* **65**:121–144.

Akbari, H., Davis, S., Dorsano, S., Huang, J., & Winnett, S. (1992). *Cooling our Communities: A Guidebook on Tree Planting and Light-Colored Surfacing*. Washington, DC: U.S. Environmental Protection Agency.

Allis, R.G. (1990). Subsidence at Wairakei field, New Zealand. *Transactions Geothermal Resources Council* **14**:1081–1087.

Antizar-Ladislao, B. (2008). Environmental levels, toxicity, and human exposure to tributyltin (TBT)-contaminated marine environment: A review. *Environmental International* **34**:292–308.

Arenas, P., & Hall, M. (1992). The association of sea turtles and other pelagic fauna with floating objects in the eastern tropical Pacific Ocean. *Proceedings of the Eleventh Annual Workshop on Sea Turtle Biology and Conservation*. U.S. Department of Commerce. NOAS Technical Memorandum NMGS-SEFSC-302. http://www.nfms.noaa.gov/pr/pdfs/speices/turtlesymposium1991.pdf (accessed 04/08/22).

ASR, Ltd. (2007). Review of wave hub technical studies: Impacts on inshore surfing beaches. Version 3. Final Report to Southwest of England Regional Development Agency, Sutton Harbor, Plymouth, United Kingdom. http://www.sas.org.uk/pr/2007/docs07/Review-of-Wave-Hub-Technical-Stiudies-Apr-071.pdf (accessed January 9, 2022).

Avens, L., & Lohmann, K.J. (2003). Use of multiple orientation cues by juvenile loggerhead sea turtles, Caretta caretta. *The Journal of Experimental Biology* **206**:4317–4325.

Balayev, L.A., & Fursa, N.N. (1980). The behavior of ecologically different fish in electric fields I. Threshold of first reaction in fish. *Journal of Ichthyology* **20**(4):147–152.

Barber, J.R., Cooks, K.R., & Fristrup, K. (2010). The costs of chronic noise exposure for terrestrial organisms. *Trends Ecology and Evolution* **25**(3): 180–189. Available at: http://www.sciencedirect.com/.

Basov, B.M. (1999). Behavior of sterlet Acipenser ruthenus and Russian Sturgeon A. gueldenstaedii in low-frequency electric fields. *Journal of Ichthyology* **39**(9):782–787.

Basov, B.M. (2007). On electric fields of power lines and on their perception by freshwater fish. *Journal of Ichthyology* **47**(8):656–661.

Bauersfeld, K. (1977). Effects of peaking (stranding) of Columbia River Dams o Juvenile Anadromous Fish below the Dalles Dam, 1974 and 1975. WDF, Olympia, WA. Tech. Rep. 31:117 pp.

Bauersfeld, K. (1978a). Stranding of Juvenile Salmon by flow reductions at mayfield dam on the Cowlitz River. WDF, Olympia, WA, Tech. Rep. 36:32 pp.

Bauersfeld, K. (1978b). The effect of daily flow fluctuations on spawning fall chinook in the Columbia River. WDF, Olympia, WA, Tech. Rep. 38:32 pp.

Beasley, R.P., Gregory, M., & McCarty, T.R. (1984). *Erosion and Sediment Pollution Control*, 2nd ed. Ames, IA: Iowa State University Press.

Beck Associates, R.W. (1989). Skagit River Salmon and steelhead fry stranding studies. Prepared by R.W. Beck Associates for the Seattle City Light Environmental Affairs Division, March 1989, Seattle, WA, 300 pp.

Becker, C.D., & Neitzel, D.A. (1985). Assessment of intergravel conditions influencing egg and alvein survival during Salmonid Redd Dewatering. *Environmental Biology of Fishes* **12**:33–46.

Becker, C.D., Fickeison, D.H., & Montgomery, J.C. (1981). *Assessments of Impacts form Water Level Fluctuations on Fish in the Hanford Reach, Columbia River*. Richland, Washington : Pacific Northwest Laboratory, Batelle Memorial Institute.

Belnap, J. (1993). Recovery rates of cryptiobiotic crusts: Inoculant use and assessment methods. *Great Basin Naturalist* **53**:80–95.

Belnap, J. (1994). Potential value of cyanobacterial inoculation in revegetation efforts. In: *Proceedings-Ecology and Management of Annual Rangelands: U.S. Department of Agriculture, Forest Service*. Monsen, S.B. & Kitchen, S.G. (Eds.),Technical report INT-GRR-313, Ogden, Utah, pp. 179–185.

Belnap, J. (1995). Surface disturbances: Their role in accelerating desertification. *Environmental Monitoring and Assessment* **37**:39–57.

Belnap, J., & Gardener, J.S. (1993). Soils microstructure in soils of the Colorado Plateau: The role of the cyanobacterium Microcoleus vaginatus. *Great Basin Naturalist* **53**:40–47.

Belnap, J., & Gillette, D.A. (1997). Disturbance of biological soil crusts: Impacts on potential wind erodibility of sandy desert soils in SE Utah, USA: Land Degradation and Development.

Belnap, J., & Harper, K.T. (1995). The influence of cryptobiotic soil crusts on elemental content of tissue of two desert seed plants. *Arid Soil Research and Rehabilitation* **9**:107–115.

Betke, K., Schultz-von Glahn, M., & Matuschek, R. (2004). Underwater noise emission from offshore wind turbines. *Proceedings of the Joint Congress of CFA/DAGA'04*, Strasbourg, France. http://www.itap.de/dago04owea.pdf (accessed 04/09/14).

Beymer, R.J., & Klopatek, J.M. (1991). Potential contribution of carbon by macrophytic crusts in pinyon-juniper woodlands. *Arid Soil Research and Rehabilitation* **5**:187–198.

Blanchfield, J., Rowe, A., Wild, P., & Garrett, C. (2007). The power potential of tidal streams including a case study for Masset Sound. *Proceedings of the 7th European Wave and Tidal energy Conference*, Porto, Portugal, 10 p.

Blyth, R.E., Kaiser, M.J., Edwards-Jones, G., & Hard, P.J.B. (2004). Implications of a zoned fishery management system for marine benthic communities. *Journal of Applied Ecology* **41**:951–961.

Blyth-Skyrme, R.E., Kaiser, M.J., Hiddink, J.G., Edwards-Jones, G., & Hard, P.J.B. (2006). Conservation benefits of temperate marine protected areas: Variation among fish species. *Conservation Biology* **20**(3):811–820.

Bochert, R., & Zettler, M.I. (2006). Effect of electromagnetic fields on marine organisms, Chapter 14. In: Offshore Wind Energy, pp. 343–351. Koller, J., Koppel, J. & Peters, W. (Eds.). Berlin: Springer-Verlag.

Boehlert, G.W., McMurray, G.R., & Tortorici, C.E. (Eds.) (2008). Ecological Effects of Wave Energy Development in the Pacific Northwest. U.S. Department of Commerce, NOAA Technical Memorandum NMFS-F/SPO-92, 174 p.

Boles, L.C., & Lohmann, K.J. (2003). True navigation and magnetic maps in spiny lobster. *Nature* **421**:60–63.

Bowen, B.W., Meylan, A.B., Rose, J.P., Limpus, C.J., Balazs, G.H., & Avise, J.C. (1992). Global population structure and natural history of the green turtle in terms of matriarchal phylogeny. *Evolution* **46**:865–881.

Bowen, F., Avise, J.C., Richardson, J.I., Meylan, A.B., Margaritoulis, D., & Hopkins-Humpy, S.R. (1993). Population structure of loggerhead turtles in the northwestern Atlantic Ocean and Mediterranean Seas. *Conservation Biology* **7**:834–844.

Bristow, R.S., Blackie, R., & Brown, N. (2012). Parks and the urban heat Island: A longitudinal study in Westfield, Massachusetts. In: *Proceedings of the Northeastern Recreation Research Symposium*. Fisher, C.I. & Watts, C.E., Jr. (Eds.) Gen. Tech. Rep. NRS-P-94. Newtown Square, PA: U.S. Department of Agriculture, Forest Service, Northern Research station, pp. 224–230.

Bryden, I.G., Grinstead, T., & Melville, G.T. (2004). Assessing the potential of a simple tidal channel to deliver useful energy. *Applied Ocean Research* **26**:198–204.

Bullock, T.H. (2005). *Electroreception*. New York: Springer.

Cada, G.F., Smith, J., & Busey, J. (2005). Use of pressure sensitive film to quantify sources of injury to fish. *North American Journal of Fisheries Management* **25**(2):57–66.

Cada, G.F., Loar, J.M., Garrison, L., Fisher, R.K., & Neitzel, D. (2006). Efforts to reduce mortality to hydroelectric turbine-passed fish: Locating and quantifying damaging shear stresses. *Environmental Management* **37**(6):898–906.

Cada, G.F., Ahlgrimm, J., Bahleda, M., Bigford, T., Damiani Stavrakas, S., Hall, D., Moursund, R., & Sale, M. (2007). Potential impacts of hydrokinetic and wave energy conversion technologies on aquatic environments. *Fisheries* **32**(4):174–181.

Cain, S.D., Boles, L.C., Wang, J.H., & Lohmann, K.J. (2005). Magnetic orientation and navigation in marine turtles, lobsters, and mollusks: Concepts and conundrums. *Integrative and Comparative Biology* **45**:539–546.

Caltrans. (2001). Fisheries Impact Assessment. *Pile Installation Demonstration Project*. San Francisco-Oakland Bay Bridge East Span Seismic Safety Project. PIDP EA 012081, 59 p. http://www.biomitigation.org/reports/files/PIDP_Fisheris_Impact_Assessment_0_1240.pdf (accessed 04/09/22).

Carr, M.H., & Hixon, M.A. (1997). Artificial reefs: The importance of comparisons with natural reefs. *Fisheries* **22**(4):28–33.

Castro, J.J., Santiago, J.A., & Santana-Ortega, A.T. (2002). A general theory on fish aggregation to floating objects: An alternative to the meeting pint hypothesis. *Reviews in Fish and Biology and Fisheries* **11**:255–277.

Caswell, H., Fujiwara, M., & Brault, S. (1999). Declining survival probability threatens the North American right whale. *Proceedings of the National Academy of Sciences USA* **96**:3308–3313.

Chow, V.T. (2009). *Open Channel Hydraulics*. Caldwell, NJ: The Blackburn Press.

CMAC (Centre for Marine and Coastal Studies) (2003). A baseline assessment of electromagnetic fields generated by offshore Windfarm Cables. COWRIE Report EMG-01-2002 66. Liverpool, UK. http://www.offshorewidn.co.uk. (accessed April 9, 2022).

Coleman, F.C., Koenig, C.C., & Collins, L.A. (1996). Reproductive styles of shallow-water groupers (Pisces: Serranidae) in the eastern gulf of Mexico and the consequences of fishing spawning aggregations. *Environmental Biology of Fishes* **47**:129–141.

Colin, S.P., & Whitehead, D. (2004). The functional roles of passive electroreception in non-electric fishes. *Animal Biology* **54**(1):1–25.

Coston-Clements, L., Settle, L.R., Hoss, D.E., & Cross, F.A. (1991). Utilization of the sargassum habitat by marine invertebrates and vertebrates: A review. NOAS Technical Memorandum NMFS-SEFSC-296. National Marine Fisheries Service, Southeast Fisheries Science Center, Beaufort, SC, 32 p. http://www.aoml.noaa.gov/general/lib/seagrss.html (accessed April 9, 2022).

Coutant, C.C., & Cada, G.F. (2005). What's the future of instream hydro? *Hydro Review* **XXIV**(6):42–49.

Craig, C., Wyllie-Escheverria, S., Carrington, E., & Shafer, D. (2008). Short-term sediment burial effects on the seagrass Phyllospadix scouleri. EMRPP Technical Notes Collection (ERDC TN-EMRRP-EI-03). Vicksburg, MS: U.S. Army Engineer Research and Development Center, 10 p.

Crawford, R.E., & Carey, C.G. (1985). Retention of winter flounder larvae with a Rhode Island salt pond. *Estuaries* **8**(2B):217–227.

Cullington, J., & Gye, J. (2010). Urban Forests: A Climate Adaptation Guide. British Columbia, Canada: Part of the BC Regional Adaption Collaborative (RAC), Ministry of Community, Sport, and Cultural Development.

Cushing, D.H. (1969). The regularity of spawning season of some fishes. *Journal du Conseil: Conseil International pour l'Exploriation de la Mer* **33**(1):81–92.

Cushman, R.M. (1985). Review of ecological effects of rapidly varying flows downstream from hydroelectric facilities. *North American Journal of Fisheries Management* **5**:330–339.

Dadswell, M.J.U., Klauda, R.J., Moffitt, C.M., Saunders, R.L., Rulifson, A., & Cooper, J.E. (1987). Common strategies of anadromous and catadromous fishes. *American Fisheries Society Symposium 1*, Bethesda, MD.

Dandy, N. (2010). The social and cultural values, and governance, of street trees. Climate Change & Street Trees Project: Social Research Report. The Research Agency of the Forestry Commission.

Davis, N., VanBlaricom, G.R., & Dayton, P.K. (1982). Man-made structures on marine sediments: Effects on adjacent benthic communities. *Marine Biology* **70**:295–3030.

de Jong, D.J., de Jong, Z., & Mulder, J.P.M. (1994). Changes in area, geomorphology, and sediment nature of salt marshes in the Oosterschelde estuary (SW Netherlands) due to tidal changes. *Hydrobiologia* 281/283:303–316.

Dempster, T. (2005). Temporal variability of pelagic fish assemblages around fish aggregation devices: Biological and physical influences. *Journal of Fish Biology* **66**:1237–1260.

Dempster, T., & Taquet, M. (2004). Fish aggregation device (FAD) research: Gaps in current knowledge and future directions for ecological studies. *Review in Fish Biology and Fisheries* **14**:21–42.

Deng, D.L., Carlson, T.J., Ploskey, G.R., & Richmond, M.C. (2005). Evaluation of blade-strike models for estimating the biological performance of large Kaplan turbines. PNNL-15370, Pacific Northwest National Laboratory, Richland, WA.

Desholm, M. (2003). Thermal Animal Detection System (TADS). Development of a method for estimating collisions frequency of migrating birds at offshore wind turbines. NETI Technical Report No. 440. National Environmental Research Institute, Denmark, 27 p. http://www2.dmu.dk/1_Viend/2_Publickationer/3_fagrpoporter/rapporter?FR440.pdf.

DiBacco, C., Sutton, D., & McConnico, L. (2001). Vertical migration behavior and horizontal distribution of brachyuran larvae in a low-inflow estuary: Implication for bay-ocean exchange. *Marine Ecology Progress Series* **217**:191–206.

DiPippo, R. (1991). Geothermal energy: Electricity production and environmental impact. A Worldwide Perspective, *Energy and Environment in the 21st Century*, pp. 741–754, Cambridge: MIT Press.

DiPippo, R. (2005). *Geothermal Power Plants: Principles, Applications and Case Studies*. Oxford, U.K.: Elsevier.

Dodd, C.K. (1995). Marine turtles in the Southeast. In: *Our Living Resources*. La Roe, E.T. (Ed.) Washington, DC: USGS, pp. 124–140.

Dodman, D. (2009). Blaming cities for climate change? An analysis of urban greenhouse gas emissions inventories. *Environment and Urbanization* 21(1):185–201.

DOE (2012). *SunShot Vision Study*. Washington, DC: U.S. Department of Energy.

DOE and US Department of the Interior (DOI) (2010). Draft Programmatic Environmental impact Statement on Solar Energy Development on BLM-Administered Lands in the Southwestern United States. Available at http://solarieis.anl.gov. Accessed March 2011.

Domeier, M.L., & Colin, P.L. (1997). Tropical reef fish spawning aggregations: Defined and reviewed. *Bulletin of Marine Science* **60**(3):698–726.

DON (U.S. Department of Navy). (2003). *Environmental Assessment: Proposed Wave Energy Technology Project*. Marine Corps Base Hawaii, Kaneohe Bay, Hawaii: Office of Naval Research.

Dooling, R.J. (1982). Auditory perception in birds. In: *Acoustic Communications in Birds*. Kroodsma, D., Miller, E.H., & Ouellet, H. (Eds). New York: Academic Press, vol. 1, pp. 95–129.

Dooling, R.J. (2002). Avian hearing and the avoidance of wind turbines. National Renewable Energy Laboratory, NREL/TP-500-30844, **83** p. Available at: http://www.nrel.gov/wind/pdfs/30844.pdf.

Dooling, R.J., & Popper, A.N. (2007). The effects of highway noise on birds. Report to the California Department of Transportation, contract 43A0139. California Department of Transportation, Division of Environmental Analysis, Sacramento, California, USA.

Doughty, R.W. (1984). Sea turtles in Texas: A forgotten commerce. *Southwestern Historical Quarterly* **88**:43–70.

EERE (2003). Acoustic effect hydrokinetic tidal turbines. Accessed 12/22/21 @ https://stage.energy.gov/eere/water/articles/acousitic_effect_hydrokinetic_tidal_turbine.

EERE (2008). Biomass program. Accessed 03/04/22 @ http://www1.eere.energy.gov/biomass/feedstocks_types.html.

EERE (2009). Geothermal heat pumps. Accessed 03/27/22 @ http://www1.eere.energy.gov/geothermal/heat-pumps.html?print.

Enger, P.S., Kristensen, L., & Sand, O. (1976). The perception of weak electric d.c. currents by the European eel (Anguilla anguilla). *Comparative Biochemistry and Physiology* **54A**:101–103.

Epifanio, C.E. (1988). Transport of invertebrate larvae between estuaries and the continental shelf. Chapter 10 in Larval Fish and Shellfish Transport Through Inlets. *American Fisheries Society Symposium* **3**:104–114.

Escambia (2007). Turtle types. Accessed 03/18/22 @ http://escambia.ifas.ufl.edu/marine/types_ofsea_turtles.htm.

Evans, R.D., & Ehleringer, J.R. (1993). Broken nitrogen cycles in and lands: Evidence for 15N of soils. *Oecologia* **94**:314–317.

Faber Maunsell and Metoc. (2007). Scottish marine SEA: Environmental report section C. Chapter C9: Marine Mammals. Scottish Executive, 42 p. + figures.

Fahey, J. (2001). Shucking petroleum. *Forbes Magazine*, November 26th, p. 206.

Fast, A.W., D'Itri, F.M., Barclay, D.K., Katase, S.A., & Madenjian, C. (1990). Heavy metal content of coho Onchorhynchus kisutch and Chinook salmon O. tscharwytscha reared in deep upwelled ocean waters in Hawaii. *Journal of the World Aquaculture Society* **21**(4):271–276.

Fazio, J.R. (Ed). (2010). *How Trees Can Retain Stormwater Runoff*. Tree City USA Bulletin No. 55. Arbor Day Foundation: Nebraska City, NE.

Fernandez, A., Edwards, J.F., Rodriguez, F., Espniosa de los Monteros, A., Harraez, P., Castro, P., Jaber, J.R., Martin, V., & Arbelo, M. (2005). Gas and fat embolic syndrome involving a mass stranding of beaked whales (Family Ziphiidae) expose to anthropogenic sonar signals. *Veterinary Pathology* **42**:446–457.

Fiscus, G. (1977). This citation includes four short reports: Cedar River Fish Damage Observations and Reports on February 18, 1977; (2) Report of Cedar River Fish Kills on March 1, 1977; (3) Report of Cedar River Fish Kill on March 8, 1977 (No. 3); and (4) investigation of Cedar River Flow Fluctuation on May 21, 1977. WDF. Internal memos, Olympia, WA.

Foreman, R.T.T., & Alexander, L.E. (1998). Roads and their major ecological effects. *Annual Review of Ecological Systems* **29**:207–231.

Fraenkel, P.L. (2006). Tidal current energy technologies. *Ibis* **148**:145–151.

Fraenkel, P.L. (2007a). Marine current turbines: Pioneering the development of marine kinetic energy converters. *Proceedings of the Institution of Mechanical Engineers, Part A: Journal of Power and Energy* **221**(2):159–169.

Fraenkel, P.L. (2007b). Marine current turbines: Marine from experimental test rigs to a commercial technology. *26th International Conference on Offshore Mechanics & Arctic Engineering*. ASME-OMAE07, 10 p.

Frailey, J.J., & Graham, P.J. (1982). The impact of hungry horse dam on the fishery in the Fathead River. Final Report. US Bureau of Reclamation, Boise, Idaho.

Fraser, J.C. (1972). Regulated discharge and the stream environment. In: Oglesby, R.T., Carlson, C.A., & McCann, J.A. (Eds). *River Ecology and Man*, pp. 111–117. New York: Academic press.

Friedman, T.L. (2010). The fat lady has sung. *New York Times*, 02/20/22.

Fthenakis, V.M. (2000). End-of-life management and recycling of PV modules. *Energy Policy* **28**:1050–1058.

Fthenakis, V.M. (2004). Life cycle impact analysis of cadmium in CdTe PV production. *Renewable and Sustainable Energy Reviews* **8**:303–334.

Fthenakis, V.M. (2009). Sustainability of photovoltaics: The case for thin-film solar cells. *Renewable and Sustainable Energy Reviews* **13**:2746–2750.

Fthenakis, V., & Zweibel, K. (2003). CdTe PV: Real and perceived EHS risks. Prepared for the NCPV and Solar Program Review Meeting 2003. Upton, NY: Brookhaven National Laboratory (BNL). http://www.nrel.gov/docs/fy03osti/33561.pdf, Accessed March 2022.

Fthenakis, V., Fuhrmann, M., Heiser, J., Lanzirotti, A., Fitts, J., & Wang, W. (2005). Emissions and encapsulation of cadmium in CdTe PV modules during fires. *Progress in Photovoltaics: Research and Application* 13:713–723.

Fthenakis, V., Kim, H.C., & Alsema, E. (2008). Emissions from photovoltaic life cycles. *Environmental Science and Technology* **42**(6):2168–2174.

Fthenakis, V., Clark, C., Moalem, M., Chandler, P., Ridgeway, R., Hulbert, F., Cooper, D., & Maroulls, P. (2010). Life-cycle nitrogen trifluoride emissions form photovoltaics. *Environmental Science and Technology* **44**(22):8750–8757.

Fustish, C.A., Jacobs, S.E., McPherson, B.P., & Frazier, P.A. (1988). Effects of the applegate dam on the biology of Anadromous Salmonids in the Applegate river. Prepared by Research and Development Section, Oregon Department of Fish and Wildlife for US Army Corps of Engineers, DAWCW5-77-C-0033, 105 pp.

Gell, F.R., & Roberts, C.M. (2003). Benefits beyond boundaries: The fishery effects of marine reserves. *Trends in Ecology and Evolution* **18**(9):448–455.

Gill, A.B., & Taylor, H. (2002). The potential effects of electromagnetic fields generated by enabling between offshore wind turbines upon elasmobranch fishes. Report to the Countryside Council for Wales (CCW Contract Science Report No. 488), 60 p.

Gill, A.B., Gloyne-Phillips, I., Neal, K.J., & Kimber, J.A. (2005). The potential effects of electromagnetic fields generated by sub-sea power cables associated with offshore wind farm developments on electrically and magnetically sensitive marine organisms: A review. COWRIE Report EM Field 2-06-2004. http://www.offshorewind.co.uk (accessed April 9, 2014).

Gillette, D.A., Adams, J., Endo, A., Smith, D., & Kihl, R. (1980). Threshold velocities for input of soil particles into the air by desert soils. *Journal of Geophysical Research* **85**:5621–5630.

Gislason, J.C. (1985). Aquatic insert abundance in a regulated stream under fluctuating and stable diel flow patterns. *North American Journal of Fisheries Management* **5**:39–46.

Goff, M., Salmon, M., & Lohmann, K.J. (1998). Hatchling sea turtles use surface waves to establish a magnetic compass direction. *Animal Behavior* **55**:69–77.

Gould, J.L. (1984). Magnetic field sensitivity in animals. *Annual Reviews in Physiology* **46**:585–598.

Graybill, J.P., Burgner, R.L., Gislason, J.C., Huffman, P.E., Wyman, K.H., Gibbons, R.G., Kirko, K.W., Stober, Q.J., Fagnan, T.W., Sayman, A.P., & Eggers, D.M. (1979). Assessment of reservoir related effects of the Skagit River project on downstream fishery resources of the Skagit River, Washington. Final Report to Seattle City Light, FRI-UW-7905, 602 pp.

Grossman, G.D., Jones, G.P., & Seaman, W.J., Jr. (1997). Do artificial reefs increase regional fish production? A review of existing data. *Fisheries* **22**(4):17–23.

Hamilton, W.W. (1983). Preventing cavitation damage to hydraulic structures Part one. *Waterpower and Dam Construction.* **1983**:48–53.

Hamilton, R., & Buell, J.W. (1976). Effects of modified hydrology on Campbell River Salmonids. Technical Report Series No. Pac/T-76-20. Canada Department of the Environment, Fisheries and Marine Service, Vancouver, B.C., 156 pp.

Harper, K.T., & Marble, J.R. (1988). A role for nonvascular plants in management of arid and semiarid rangelands. In: *Vegetation Science Applications for Rangeland Analysis and Management.* Tueller, P.T. (Ed.) Dordrecht, Germany, Kluwer Academic Publish, pp. 135–169.

Harper, K.T., & Pendleton, R.L. (1993). Cyanobacteria and cyanolichens: Can they enhance availability of essential minerals for higher plants? *Great Basin Naturalist* **53**:89–95.

Harrison, J.T. (1987). The 40 MWe OTEC plant at Kahe point, Oahu, Hawaii: A case study of potential biological impacts. NOAA Technical Memorandum NMFS NOAA-TM-NMFS-SWFC-68. Southwest Fisheries Center, Honolulu, HI, 105 p.

Hastings, M.C., & Popper, A.N. (2005). Effects of sound on fish. Report to California Department of Transportation. January 28, 2005, 82 p. http://www.dot.ca.gov/hq/env/bio/files/Effects_of_Sound_on_Fish23Aug05.pdf (accessed April 9, 2022).

Hodos, W., et al. (1992). *Life Span Changes in the Visual Acuity and Retina in Birds.* New York: Springer.

Holdren, J.P., Morris, G., & Mintzer, I. (1980). Environmental aspects of renewable energy sources. *Annual Review of Energy* **5**:241–291.

Hunter, M.A. (1992). Hydropower flow fluctuations and Salmonids: A review of the biological effects, mechanical causes, and options for mitigation. Olympia, WA: State of Washington Department of Fisheries.

IEA-OES (2011). International Energy Agency ocean energy system. Accessed 03/25/22 @ http://www.iea-oceans.org/index.asp.
IEC (2011). International Electrotechnical Commission Marine Energy Technical Committee, 2014. Accessed 03/25/14 @ http://www.bsigroup.com/enStandards-andPublications/committee-Members/Copmmittee-member-news/Summer-2007/New-Committee-IECTC-114-Marine-Energy/.
Irwin, W.P., & Lohmann, K.J. (1996). Disruption of magnetic orientation in hatchling loggerhead sea turtles by pulsed magnetic fields. *Journal of Comparative Physiology A* **191**:475–480.
Janota, C.P., & Thompson, D.E. (1983). Waterborne noise due to ocean thermal energy conversion plants. *Journal of the Acoustic Society of America* **74**(1):256–266.
Jennings, S., Pinnegar, J.K., Polunin, N.V.C. & Warr, K.J. (2001). Impacts of trawling disturbance on the trophic structure of benthic invertebrate communities. *Marine Ecology Progress Series* 213:127–142.
Johansen, J.R. (1993). Cryptogamic crusts of semiarid and arid lands of North America. *Journal of Phycology* **29**:140–147.
Johnsen, S. & Lohmann, K.J. (2005). The physics and neurobiology of magnetoreception. *Neuroscience* **6**:703–712.
Johnsen, S., & Lohmann, K.J. (2008). Magnetoreception in animals. *Physics Today* **61**(3):29–35.
Johnson, A., Salvador, G., Kenney, J., Robbins, J., Kraus, S., Landry, S., & Clapham, P. (2005). ishing gear involved in entanglements of right and humpback whales. *Marine Mammal Science* **21**(4):635–645.
Johnson, M.R., Boelke, C., Chiarella, L.A., Colosi, P.D., Green, K., Lellis-Dibble, K., Ludemann, H., Ludwig, M., Mc Dermott, S., Ortiz, J., Rusanowsky, D., Scott, M., & Smith, J. (2008). Impacts to marine fisheries habitat form nonfishing activities in the northeastern United States. NOAA Technical Memorandum NMGS-NE-209. U.S. Department of Commerce, National Marine Fisheries Service, Gloucester, MA, 322 p. http://www.nefsc.noaa.gov/nefsc/publications/tm/tm209.pdf (accessed 04/10/22).
Johnston, M. (2004). Impacts and adaptation for climate change in urban forests. *Proceedings of 6th Canadian Urban Forest Conference*, Kelowna, B.C.
Jones, G.L. (2006). Geysers/hot springs damaged or destroyed by man, http://www.wyohones.com/desitroyer.htm.
Kaiser, M.J. (2005). Are marine protected areas a red herring or fisheries panacea. *Canadian Journal of Fisheries and Aquatic Sciences* **62**:1194–1199.
Kaiser, M.J., Spence, F.E., & Hart, P.J.B. (2000). Fishing-gear restrictions and conservation of benthic habitat complexity. *Conservation Biology* **14**(5):1512–1525.
Kaliski, K. (2009). Calibrated sound propagation models for wind power projects. *State of the Art in Wind Siting Seminar*, Washington, DC: National Wind Coordinating Collaborative.
Kalmijn, A.T. (1982). Electric and magnetic field detection in elasmobranch fishes. *Science* **218**(4575):916–918.
Kalmijn, A.T. (2000). Detection and processing of electromagnetic and near-field acoustic signals in elasmobranch fishes. *Philosophical Transactions of the Royal Society of London B.* **355**:1135–1141.
Karsten, R.H., McMillan, J.M., Lickley, M.J., & Haynes, R.D. (2008). Assessment of tidal current energy in the Minas Passage, bay of Fundy. Proceedings of the Institution of Mechanical Engineers. *Part A: Journal of Power and Energy* **222**(5):493–507.
Keam, R.R., Luketina, K.M., & Pipe, L.Z. (2005). Definition and listing of significant geothermal feature types in the Waikato region, New Zealand. *Proceedings of World Geothermal Congress 2005*, Antalya, Turkey, 24–29 April 2005.
Kenny, J.G., Barber, N.L, Hutson, S.S., Linsey, K.S., Lovelace, J.K., & Maupin, M.A., (2009). Estimated use of water in the United States in 2005. USGS Ciruclar 1344. Reston, VA: U.S. Geological Survey.
Kingsford, M.J. (1999). Fish attraction devices (FADs) and experimental designs. *Scientia Marina* 63(3–4):181–190.
Klinenberg, E. (2002). *Heat Wave: A Social Autopsy of Disaster in Chicago*. Chicago, IL: University of Chicago Press.
Knight, T.A. (1974). A review of hearing and song in birds with comments on the significance of song in display. *Emu* **74**:5–8.
Kogan, I., Paull, C.K., Kuhnz, L.A., Burton, E.J., Von Thun, S., Green, H.G., & Barry, J.P. (2006). ATOC/Pioneer Seamount cable after 8 years on the seafloor: Observations, environmental impact. *Continental Shelf Research* **26**(2006):771–787.
Kovats, S., & Akhtar, R. (2008). Climate, climate change and human health in Asian Cities. *Environment and Urbanization* 20:165–75.
Kraus, S.D., Brown, M.W., Caswell, H., Clark, C.W., Fujiwara, M., Hamilton, P.K., Keeney, R.D., Knowlton, A.R., Landry, S., Mayo, C.A., McLellan, W.A., Moore, M.J., Pl Nowacek, D., Pabst, D.A., Read, A.J. & Rolland, R.M. (2005). North Atlantic right whales in crisis. *Science* **309**:561–562.

Largier, J., Behrens, D., & Robart, M. (2008). The potential impact of WEC development on Nearshore and shoreline environments through a reduction in Nearshore wave energy, Chapter 3. In: *Developing Wave Energy in Coastal California: Potential Socio-Economic and Environmental Effects*. Nelson, P.A., et al. (Eds). California Energy Commission, PIER Energy-Related Environmental Research Program & California Ocean Protection Council CEC-500-2008-083. http://www.resources.ca.gov/copc/docs/ca_wec_effects.pdf (accessed April 10, 2014).

Lengagne, T. (2008). Traffic noise affects communication behavior in a breeding anuran, Hyla arborea. *Biological Conservation* 141:2023–2031.

Lewis, L.J., Davenport, J., & Kelly, T.C. (2003). A study of the impact of a pipeline construction on estuarine benthic invertebrate communities. Part 2. Recolonization by benthic invertebrates after 1 year and response of estuarine birds. *Estuarine, Coastal and Shelf Sciences* 57(2003):201–208.

Lohmann, K.J., & Lohmann, C.M.F. (1996). Detection of magnetic field intensity by sea turtles. *Nature* **380**:59–61.

Lohmann, K.J., Lohmann, C.M.F., & Endres, C.S. (2008a). The sensory ecology of ocean navigation. *The Journal of Experimental Biology* **211**:1719–1728.

Lohmann, K.J., Putman, N.F., & Lohmann, C.M.F. (2008b). Geomagnetic imprinting: A unifying hypothesis of long-distance natal homing in salmon and sea turtles. *The Proceedings of the National Academy of sciences USA* **105**(49):19096–19101.

Lohmann, K.J., Swartz, A.W., & Lohmann, C.M.F. (1995). Perception of ocean wave direction by sea turtles. *The Journal of Experimental Biology* **198**:1079–1085.

Lohr, B., Wright, T.F., & Dooling, R.J. (2003). Detection and discrimination of natural calls in masking noise by birds: Estimating the active space of a signal. *Animal Behavior* **65**:763–777.

Longmuir, C., & Lively, T. (2001). Bubble curtain systems help protect the marine environment. *Pile Driver Magazine*. Summer 2001, pp. 11–16.

Love, M.S., Caselle, J., & Snook, L. (1999). Fish assemblages on mussel mounds surround seven oil platforms in the Santa Barbara Channel and Santa Maria Basin. *Bulletin of Marine Science* **65**(2):497–513.

Lugo-Fernandez, A., Roberts, H.H., & Wiseman, W.J., Jr. (1998). Tide effects on wave attenuation and wave set-up on a Caribbean coral reef. *Estuarine, Coastal and Shelf Science* **47**:385–393.

Lund, J.W. (2007). Characteristics, development and utilization of geothermal sources. Klamath falls, ore: Oregon Institute of Technology. *Geo-Heat Centre Quarterly Bulletin* **28**(2):1–9.

Madsen, P.T. (2005). Marine mammals and noise: Problems with root mean square sound pressure levels for transients. *Journal of the Acoustic Society of America* **117**(6):3952–3957.

Mann, S., Sparks, N.H.C., Walker, M.M., & Kirschvink, J.L. (1988). Ultrastructure, morphology, and organization of biogenic magnetite from sockeye salmon, Oncorhynchus nerka: implications for magnetoreception. *Journal of Experimental Biology* **140**:35–49.

Marra, L.J. (1989). Sharkbite on the SL submarine lightwave cable system: History, causes, and resolution IEEE. *Journal of Oceanic Engineering* **14**(3):230–237.

McCauley, R.D., Fewtrell, J., Duncan, A.J., Jenner, C., Jenner, M.-N., Penrose, J.D., Prince, R.I.T., Adhitya, A., Murdoch, J., & McCabe, K. (2000a). Marine seismic surveys: Analysis and propagation of air-gun signals; and effects of air-gun exposure on humpback whales, sea turtles, fishes, and squid. Report R99-15. Centre for Marine Science and Technology, Curtin University of Technology, Western Australia.

McCauley, R.D., Fewtrell, J., Duncan, A.J., Jenner, C., Jenner, M-N., Penrose, J.D., Prince, R.I.T., Adhitya, A., Murdoch, & McCabe, K. (2000b). Marine seismic surveys: A study of environmental implications. *Australian Petroleum Production Exploration Association Journal* **2000**:692–708.

McKenna-Neumann, C.M., Maxwell, C.D., & Bolton, J.W. (1996). Wind transport of sand surface with photo-autotrophic microorganism. *Catena* **27**:229–247.

McLarty, L., Grabowski, P., Entingh, D., & Robertson-Tait, A. (2000). Enhanced geothermal systems R&D in the United States. *WGC* **2000**:3793–96.

McPhee, C., & Brusven, M.A. (1975). The effect of river fluctuations resulting from hydroelectric peaking on selected aquatic invertebrates and fish. September, 1976. Submitted to the Office of Water Research and Technology, US Department of Interior, Idaho Water Resources Research Institute, University of Idaho, Moscow, ID, 46 p.

McPherson, G., & Muchnick, J. (2005). Effects of street tree shade on asphalt concrete pavement performance. *Journal of Arboriculture* 31(6):303–310.

Metting, B. (1991). Biological surface features of semiarid lands and deserts. In: *Semiarid Lands and Deserts: Soil Resource and Reclamation*. Skujins, J. (Ed.), New York, Marcel Dekker, Inc., pp. 257–293.

Meyer, C.G., Holland, K.N., & Papastamatiou, Y.P. (2004). Sharks can detect changes in the geomagnetic field. *Journal of the Royal Society Interface* **2**(2):129–130.

Michel, J., Dunagan, H., Boring, C., Healy, E., Evans, W., Dean, J., McGillis, A., & Hain, J. (2007). Worldwide synthesis and analysis of existing information regarding environmental effects of alternative energy uses on the outer continental shelf. OCS Report MMS 2007-038. Mineral Management Service, U.S. Department of the Interior, Washington, DC. http://www.mms.gov/offshroe/AlternativeEnergy/Studies.htm (access April 10, 2022).

Millar, D.L., Smith, H.C.M., & Reeve, D.E. (2007). Modeling analysis of the sensitivity of shoreline change to a wave farm. *Ocean Engineering* **34**(2007):884–901.

MMS (Minerals Management Service). (2007). Programmatic environmental impact statement for alternative energy development and production and alternate uses of facilities on the outer continental shelf. Final EIS. MMS 2007-046. October 2007. http://ocsenergy.anl.gov (accessed April 10, 2012).

Monk, C.L. (1989). Factors that influence stranding of Juvenile Chinook Salmon and steelhead trout. Master's Thesis. University of Washington, Seattle, WA, 81 pp.

Moore, S.E., & Clarke, J.T. (2002). Potential impact of offshore human activities on gray whale (Eschrichtius robustus). *Journal of Cetacean Research and Management* **4**(1):19–25.

Morris, D. (2002). *Accelerating the Shift to a Carbohydrate Economy: The Federal Role*. Washington, DC: Institute for Local Self-Reliance.

Mundie, J.H., & Mounce, D.E. (1976). Effects of changes in discharge in the lower Campbell River on the Transport of Food Organisms of Juvenile Salmon. Appendix Report In: Hamilton, R., & J.W. Buell. 1976. Effects of Modified Hydrology on Campbell River Salmonids. Technical Report Series No. Pac/T-76-20. Canada Department of the Environment, Fisheries and Marine Service, Vancouver, C.C, 156 pp.

Myers, E.P., Hoss, D.E., Peters, D.S., Matsumoto, W.M., Seki, M.P., Uchida, R.N., Ditmars, J.D., & Paddock, R.A. (1986). The potential impact of ocean thermal energy conversion (OTEC) on fisheries. NOAA Technical Report NMFS 40. U.S. Department of Commerce, Seattle, WA, 33 pp.

Myers, S.R., Branas, C.C., French, B.C., Kallan, M.L., Wiebe, D.J., & Carr, H.G. (2013). Safety in numbers: Are major cities the safety places in the United States? Annals of Emergency Management. Accessed from http://www.annemergmed.com/webfiles/images/jounrnals/ymem/FA-5548.pdf (April16, 2014).

National Research Council. (1990). *Decline of the Sea Turtles: Causes and Prevention*. Washington, DC: National Academy Press.

Nedwell, J.R., Edwards, B., Turnpenny, A.W.H., & Gordon, J. (2004). Fish and marine mammals audiograms: A summary of available information. Subacoustech Report ref: 534R0214 to Chevron Texaco Ltd., TotalfFrianElf Exploration UK Plc, DSTL, DT1 and Shell U.K. Exploration and Production Ltd, 281 p. http://www.suacoustech.com/information/publications.shtml (accessed April 10, 2014).

Neitzel, D.A., Becker, C.D., & Abernathy, C.S. (1985). *Proceedings of the Symposium on Small Hydropower and Fisheries*. Aurora, CO: American Fisheries Society.

Nelson, P.A. (2003). Marine fish assemblages associated with fish aggregating devices: Effects of fish removal, FAD size, fouling communities, and prior recruits. *Fishery Bulletin* **101**(40):835–850.

Nelson, P.A. (2008). Ecological effects of wave energy conversion technology on California's Marine and Anadromous Fishes, Chapter 5. In: *Developing Wave Energy in Coastal California: Potential Socio-Economic and Environmental Effects*. Nelson, P.A., et al. (Eds) California Energy Commission, PIER Energy-Related Environmental Research Program & California Ocean Protection Council CEC-500-2008-083. http://www.resources.ca.gov/copc/docs/ca_wec_effects.pdf (accessed April 10, 2022).

Nemet, G.F. (2009). Net radiative forcing from widespread deployment of photovoltaics. *Environmental Science and Technology* **43**(6):2173–2178.

Nestler, M.N., Milhous, R.T., & Layzer, J.B. (1989). Instream habitat modeling techniques. In: *Alternatives in Relative River Management*, pp. 201–211. Gore, J.A., & Petts, G.E. (Eds.) Boca Raton, FL: CRC Press.

Newcombe, C.P., & Jensen, J.O.T. (1996). Channel suspended sediment and fisheries: A synthesis for quantitative assessment of risk and impact. *North American Journal of Fisheries Management* **16**(4):693–727.

NMFS (National Marine Fisheries Service). (2003). Taking marine mammals incidental to conducting oil and gas exploration. Activities in the Gulf of Mexico. March 3, 2003. *Federal Register* **68**(41):9991–9996.

Nowacek, D.P., Thorne, L.H., Johnston, D.W., & Tyack, P.I. (2007). Responses of cetaceans to anthropogenic noise. *Mammal Review* **37**(2):81–115.

Nowak, D.J. (2010). Urban biodiversity and climate change. In: *Urban Biodiversity and Design*. Miller, N., Werner, P., & Kelcey, J.G. (Eds). Hoboken, NJ: Wiley-Blackwell Publishing, pp. 101–117.

Nowak, D.J., & Crane, D.E. (2002). Carbon storage and sequestration by urban trees in the USA. *Environmental Pollution* 116:381–389.

Nowak, D.J. et al. (2010). Sustaining America's Urban trees and forests: A forests on the edge report. Gen. Tech. Rep. NRS-62. Newtown Square, PA: U.S. Department of Agriculture, Forest Service, Northern Research stations.

Nowak, D.J., Greenfield, E.J., Hoehn, R., & LaPoint, E. (2013). Carbon storage and sequestration by trees in urban and community areas of the United States. *Environmental Pollution* 178:229–236.

NRC (National Research Council). (2000). *Marine Mammals and Low-Frequency Sound. Progress Since 1994*. Washington, DC: National Academy Press.

NRC National Research Council). (2005). Marine Mammal Populations and Oceans Noise: Determining when Noise Causes Biologically Significant Effects. Washington, DC: National Academy Press.

NREL (2000). *Review of Operation and Maintenance Experience in the DOE-EPR Wind Turbine Verification Program*. Golden, CO: National Renewable Energy Laboratory.

Ohman, M.C., Sigray, P., & Westerberg, H. (2007). Offshore windmills and the effects of electromagnetic fields on fish. *Ambio* **36**(8):630–633.

Olson, A. (1989). Vertical and Horizontal movements of Chinook salmon. Accessed 12/12/21 @ https://fishery-bulletin.nmfs.NOAA.gov/sites/default/files/pdf.

Olson, F.W. (1990). Down ramping regime for power operations to minimize stranding of salmon fry in the Sultan River. Contract report by CH2M Hill (Bellevue, WA.) for Snohomish County PUD 1, 70 pp.

OPT (Ocean Power Technologies). (2011). Accessed 03/25/22 @ http://www.oceanpowertechnologies.com.

Ordonez, C., Dunker, P.N., & Steenberg, J. (2010). Climate change mitigation and adaptation in urban forests: A framework for sustainable urban forest management. *Paper Prepared for Presentation at the 18th Commonwealth Forestry Conference*, Edinburgh.

Parry, M.L., et al. (Eds.) (2007). Impacts, adaptation, and vulnerability. Contribution of Working Group II to the Fourth Assessment, Report of the Intergovernmental Panel on Climate Change. Cambridge, MA: Cambridge University Press.

Pelc, R., & Fujita, R.M. (2002). Renewable energy form the ocean. *Marine Policy* **26**:471–479.

Perlin, J. (2005). *A Forest Journey: The Story of Wood and Civilization*. Woodstock, VT: The Countryman Press.

Petersen, I.K., Christensen, T.K., Kahlert, J., Desholm, M., & Fox, A.D. (2006). *Final Results of Bird Studies at the Offshore Wind Farms at Nysted and Horns Rev*. Denmark: National Environmental Research Institute.

Peterson, A.T. et al. (2006). Distribution and conservation of birds of northern Central America. Accessed 12/10/22 @ https://www.jstor.org/stable/4164001

PFMC (Pacific Fisheries Management Council). (2008). Letter from D.O. McIsaac, Executive Director of PFMC to Director Randall Luthi, Minerals Management Service, regarding docket ID MMS-2008-) MM-0020.

Phillips, R.W. (1969). Effects of an unusually low discharge from Pelton regulating reservoir, Deschutes River, on Fish and Other Aquatic Organisms. Special Report 1. Basin Investigation Section, Oregon State Game Commission.

Phinney, L.A. (1974a). Further observations on juvenile salmon stranding in the Skagit River, March 1973. WDF, Olympia, WA, Progress Report, 26:34 pp.

Phinney, L.A. (1974b). Report on the 1972 study of the effect of river flow fluctuations below Merwin Dam on Downstream Migrant Salmon, WDF, 23 pp.

Polagye, B., Malte, P., Kawase, M., & Durran, D. (2008). Effect of large-scale kinetic power extraction on time-dependent estuaries. Proceedings of the Institution of Mechanical Engineers. *Part A: Journal of Power and Energy* **222**(5):471–484.

Popper, A.N. (2003). Effects of anthropogenic sound on fishes. Adaptive management of large rivers with special reference to the Missouri River. *Journal of the American Water Resources Association (JAWRA)* **39**(4):935–946.

Population Reference Bureau. (2012). 2012 World population data sheet. Accessed from http://www.prb.org/pdf/2012-population-data-sheet_eng.pdf (April 16, **2022**).

PV Cycle. (2010). PV cycle website. www.pvcyle.org. (Accessed March 2022).

Reiser, D.W., & White, R.G. (1983). Effects of complete redd dewatering on salmonid egg-hatching and development of juveniles. *Transactions of the American Fisheries Society* **112**:532–540.

Roberts, H.H., Wilson, P.A., & Lugo-Fernandez, A. (1992). Biologic and geologic responses to physical processes: Examples form modern reef systems of the Caribbean-Atlantic region. *Continental Shelf Research* 12(7/8):809–834.

Rochester, H. Jr., Lloyd, T., & Farr, M. (1984). Physical impacts of small-scale hydroelectric facilities and their effects on fish and wildlife. Division of Biological Services, US Fish and Wildlife Service, Department of Interior. Washington, DC (FWS/OBS-84/19) 192 p.

Rodil, I.F., & Lastra, M. (2004). Environmental factors affecting benthic macrofauna along a gradient of intermediate sandy beaches in northern Spain. *Estuarine, Coastal and Shelf Science* **61**:37–44.

Rodrigue, P.R. (1986). Cavitation pitting mitigation in hydraulic turbines. Volume 2: cavitation review and assessment. EPRI AP-4719, Electric Power Research Institute, Palo Alto, CA.

Romero-Lankao, P., & Gratz, D.M. (2008). Urban areas and climate change: Review of current issues and trends. Issues paper for the 2011 Global Report on Human Settlements.

Rommel, S.A., Jr., & McCleave, J.D. (1972). Oceanic electric fields: Perception by American eels? *Science* **176**:1233–1235.

Ross, D., (1995). *Power from Sea Waves*, 1st ed. United Kingdom: Oxford University Press.

Rucker, J.B., & Friedl, W.A. (1985). Potential impacts for OTEC-generated underwater sounds. *Oceans* **17**:1279–1283.

Safford, H, Larry, E., McPherson, E.G., Nowak, D.J., & Westphal, L.M (2013). Urban forests and climate change. U.S. Department of Agriculture, Forest Service, Climate Change Resource Center.

Salt, A.N., & Hullar, T.E. (2010). Responses of the ear to low frequency sounds, infrasound, and wind turbines. *Hearing Research* 268:12–21.

Salter, S. (1974). Wave power. *Nature* **249**:720–724.

Satterthwaite, T.D. (1987). Effects of lost creek dam on spring chinook in the Rouge River, Oregon. An Update. Prepared by Research and Development Section, Oregon Department of Fish and Wildlife for US Army Corps of Engineers, DACW57-77-C-0027. 72 pp.

Schimel, D.S., Kelly, E.F., Yonker, C., Aquilar, R., & Heil, R.D. (1985). Effects of erosional processes on nutrient cycling in semiarid landscapes. In: *Planetary Ecology*. Caldwell, D.E., Brierley, J.A., & Brierley, C.L. (Eds.) New York: Van Nostrand Reinhold, pp. 571–580.

Sherril, S., & Bratkovitch, S. (2011). *Carbon and Carbon Dioxide Equivalent Sequestration in Urban Forest Products*. Minneapolis, MN: Dovetail Partners, Inc.

Simmonds, M.P., Dolman, S.J., & Weilgart, L. (Eds.). (2003). Oceans of noise: A whale and dolphin conservation society science report, 164 p. http://www.wdes.org (accessed April 11, 2022).

Sinclair, M., & Tremblay, M.J. (1984). Timing of spawning of Atlantic herring (Clupea harengus harengus) populations and the match-mismatch theory. *Canadian Journal of Fisheries and Aquatic Sciences* **41**:1055–1065.

Smith, C.L. (1972). A spawning aggregation of Nassau grouper, Epinephelus striatus (Bloch). *Transactions of the American Fisheries Society* **101**(2):257–261.

Smith, H.C.M., Millar, D.L., & Reeve, D.E. (2007). Generalization of wave farm impact assessment on inshore wave climate. *Proceedings of the 7th European Wave and Tidal* Energy Conference, September 11–13, 2007, Porto, Portugal, 7 p.

Spellman, F.R. (2012). *Safe Work Practices for Green Energy Personnel*. Lancaster, PA: Technomic Publishing Company.

Stober, Q.J., Crumley, S.C., Fast, D.E., Killebrew, E.S., Woodin, R.M., Engman, G.E., & Tutmark, G. (1982). Effects of hydroelectric discharge fluctuation on salmon and steelhead in the Skagit River, Washington. Final Report for period December 1979 to December 1982. Seattle City Light Contract. University of Washington, Fish. Res. Inst. FRI-UW-8218, 302 pp.

Strober, M. et al. (1982). Behavioral responses of juvenile salmonoids. Accessed 12/14/21 @ https://digital.lib. Washington,edu/researchworks.

Sundberg, J., & Langhamer, O. (2005). Environmental questions related to point-absorbing linear wave-generators: Impact, effects, and fouling. In: *Proceedings of the 6th European Wave and tidal energy Conference*, 30th August-2nd September 2005, Glasgow, Scotland.

The Cities Alliance. (2007). Liveable Cities: The Benefits of Urban Environmental Planning. Accessed from http://www.citiesalliance.org/sites/citiesalliance.org/files/CA_Docs/resources/cds/liveable/liveablecities_web_7dec07.pdf (April 15, **2022**).

The Engineering Business Ltd. (2005). Stingray tidal steam energy device: Phase 3. T/06/00230/00/REPURN05/864, 110 p. http://www.engb.com/downloads/Stingray%Phase%201r.pdf (assessed April 11, 2022).

Thompson, J.S. (1970). *The Effect of Water Regulation at Gorge Dam on Standing of Salmon Fry in the Skagit River, 1969–1970*. Olympia, WA: WDF, 46 p.

Thompson, S.A., Castle, J., Mills, K.L., & Sydeman, W.J. (2008). Wave energy conversion technology development in Coastal California: Potential impacts on marine birds and mammals, Chapter 6. In: *Developing Wave Energy in Coastal California: Potential Socio-Economic and Environmental Effects*. Nelson, P.A., et al. (Eds.) California Energy Commission, PIER Energy-Related Environmental Research Program & California Ocean Protection Council CEF-500-2008-083. http://www.resources.ca.gov/copc/docs/ca_wec_effects.pdf (accessed April 11, 2022).

Thomsen, F., Ludemann, K., Kafemann, R., & Piper, W. (2006). Effects of offshore wind farm noise on marine mammals and fish, biola, Hamburg, Germany on behalf of COWRES, Ltd. http://www.offshroewind.co.uk/Assets/BIOLAReport06072006FINAL.pdf (accessed April 11, 2022).

Toledo Blade (2013). Solar power boom fuels increase in hazardous waste sent to dumps. The Toledo Times, February 11, 2013.

Tovey, N.K. (2005). ENV-2E02 Energy resources 2005 Lecture. Accessed 03/01/10 @ www2.env.ac.UK.gmmc/energy. Trotzky, H.M., & Gregory, R.W. (1974). The effects of water flow manipulation below a hydroelectric power dam on the bottom fauna of the upper Kennebec River, Maine. *Transactions of the American Fisheries Society* **103**:318–324.

Tubby, K.V., & Webber, J.F. (2010). Pests and diseases threatening urban trees under a changing climate. *Forestry* 83(4):451–59.

UCS (2013). Environmental impacts of solar power. Union of Concerned Scientists. Accessed 02/22/14 @ http://www.ucsusa.org/clean_energy/our-energy-choices/renewable-energy/enivornmenta.

United States Fish and Wildlife Service (USFWS) (2014). The effects of noise on wildlife. Accessed 02/18/22 @www.fws.gov/windenergy/docs/noise.pdf.

USACE (U.S. Army Corps of Engineers). (1995). Proceedings: 1995 turbine passage survival workshop. U.S. Army Corps of Engineers, Portland District, Portland, OR, 212 p.

USDOE (2010). Ocean energy. Washington, DC: Department of Interior. Accessed @ www.mms.gov.

USDOE (2003a). *Industrial Bioproducts: Today and Tomorrow*. Washington, DC: U.S. Department of Energy.

USDOE (2003b). *The Bioproducts Industry: Today and Tomorrow*. Washington, DC: U.S. Department of Energy.

USEPA (2011). *Shining Light on a Bright Opportunity*. Washington, DC: U.S. Environmental Protection Agency.

USEPA. (2009). Buildings and their impact on the environment: A statistical summary. Accessed from http://www.epa.gov/greenbuilding/pubs/gbstats.pdf (April 16, 2022).

USEPA. (2013). DRAFT inventory of U.S. greenhouse gas emissions and sinks: 1990–2011. Accessed from http://www.epa.gov/climatechange/Downloads/ghgemissions/US-GHG-Inventory-2011pd (April 16, 2014).

Van den Berg, G.P. (2004). Effects of the wind profile at night on wind turbine sound. *Journal of Sound and Vibration* **277**:955–970.

Viada, S.T., Hammer, R.M., Racca, R., Hannay, D., Thompson, M.J., Balcom, B.J., & Phillips, N.W. (2008). Review of potential impacts to sea turtles form underwater explosive removal of offshore structures. *Environmental Impact Assessment Review* 28(2008):267–285.

Wahlberg, M., & Westerberg, H. (2005). Hearing in fish and their reactions to sounds from offshore wind farms. *Marine Ecology Progress Series* **288**:295–309.

Walker, M.M., Quinn, T.P., Kirschvink, J.L., & Groot, C. (1988). Production of single-domain magnetite throughout life by sockeye salmon, Onchorhynchus nerka. *Journal of Experimental Biology* **140**:51–63.

Wang, J.H., Jackson, J.K., & Lohmann, K.J. (1998). Perception of wave surge motion by hatchling sea turtles. *Journal of Experimental Marine Biology and Ecology* **229**:177186.

Wang, D., Altar, M., & Sampson, R. (2007). An experimental investigation on cavitation, noise, and slipstream characteristics of ocean stream turbines. *Proceedings of the Institution of Mechanical Engineers, Part A: Journal of Power and Energy* **221**(2):219–231.

Wave Dragon (2014). Accessed 03/25/14 @ http://www.wavedragon.net/.

Wave Dragon Wales Ltd. (2007). Wave dragon pre-commercial wave energy device. Volume 2, Environmental Statement, April, 2007. Http://www.wavedragon.co.uk/ (accessed April 11, 2022).

Wavebob. (2011). Accessed 03/25/22 @ http://wavbob.com/home/.

Wavestar (2014). Accessed 03/25/22 @ http://wavestarenergy.com/.

Webb, R.H., & Wilshire, H.G. (1983). *Environmental Effects of Off-Road Vehicles: Impacts and Management in Arid Regions*. New York, Springer-Verlag.

Weilgart, L.S. (2007). The impacts of anthropogenic ocean noise on cetaceans and implications for management. *Canadian Journal of Zoology* **85**:1091–1116. Accessed 12/20/21 @ https://www.semanticsscholar.org/paper/the-impacts-of-anthrogpdenci-ocean-noise-on-and-for-weilgart.

Widdows, J., & Brinsley, M. (2002). Impact of biotic and abiotic processes on sediment dynamics and the consequences to the structure and functioning of the intertidal zone. *Journal of Sea Research* **48**:143–156.

Wilhelmsson, D., & Malm, T. (2008). Fouling assemblages on offshore wind power plants and adjacent substrata. *Estuarine, Coastal and Shelf Science* **79**(3):459–466.

Wilhelmsson, D., Malm, T., & Ohman, M.C. (2006). The influence of offshore windpower on demersal fish. *ICES Journal of Marine Science* **63**:775–784.

Wilkens, L.A., & Hoffman, M.H. (2005). Behavior of animals with passive, low-frequency electrosensory systems. In: *Handbook of Auditory Research*. Bullock, T.H., Hopkins, C.D., Popper, A.N., & Fay, R.R. (Eds). New York: Springer, vol., 21, pp. 229–263.

Williams, J.D., Dobrowolski, J.P., West, N.E., & Gillette, D.A. (1995). Microphytic crust influences on wind erosion. *Transactions of the American Society of Agricultural Engineers* 38:131–137.

Williamson, T., Dubb, S., & Alperovitz, G. (2010). *Climate Change, Community Stability, and the Next 150 Million Americans*. College Park, MD: The Democracy Collaboration.

Wilson, R., Batty, R.S., Daunt, F., & Carter, C. (2007). Collision risks between marine renewable energy devices and mammals, fish and diving birds. Report to the Scottish Executive. Scottish Association for marine Science, Oban, Scotland, PA25 IQA, 110 p.

Wiltschko, R., & Wiltschko, W. (1995). *Magnetic Orientation in Animals*. Berlin: Springer Verlag.

Witty, K., & Thompson, K. (1974). Fish stranding surveys, Chapter 10. In: *Anatomy of a River: An Evaluation of Water Requirements of the Hells Canyon Reach of the Snake River Conducted March*. Bayha, K., & Koski, C. (Eds). Vancouver, WA: Pacific Northwest River Basins Commission, pp. 113–120.

Woodin, R.M. (1984). Evaluation of salmon fry stranding induced by fluctuating hydroelectric discharge in the Skagit River, 1980–1983. WDF Tech. Rep. 83-38 pp.

World Energy (2004). Survey of energy sources: Wave energy. Accessed 03/31/22 @ http://www.worldenergy.org/wec-gies/publications.

Woytenek, W., Pei, X., & Wilkens, L.A. (2001). Paddlefish strike at artificial dipoles simulating the weak electric fields of planktonic prey. *The Journal of Experimental Biology* **204**:1391–1399.

Index

Note: **Bold** page numbers refer to tables and *italic* page numbers refer to figures.

abandoned mining land (AML) 723
abatement period 613
abiotic organisms 5
abrasive blasting 613
absolute pressure 122, 235, 254
absolute temperature 122, 250, 251
absolute viscosity 256
absorbed dose, of radiation 673
absorption 488, 613, 656
    of gaseous pollutants 300–301
    skin 636, 647
acceleration 222
accidental spills 573
accident analysis/prevention 614
accommodation, eye 614
accuracy, sampling 649
acid
    defined 235
    deposition 275–276
    mine drainage 576
    precipitation 235
    rain 11, 267, 272, 273, 275, 276, 282, 290, 291
    surge 235
acidification, ocean 12
acid mine drainage 576
Acoustic Doppler Current Profilers 767
acoustics 614
acoustic trauma 661
actinides 11, 12
action level 614, 647
    asbestos 319
    formaldehyde 320–321
    lead 321, 384
    noise 616, 661, 666–667
activated alumina 434
activated carbon 71, 301, 375, 376, 407, 412, 413, 515, 580
activated charcoal 614
activated sludge 438, *438,* 439, 463, 487–501, 521, **521**
    formation 493–494
    modifications 494–500, **496**
    process 491–492
        control parameters 500–501
    return 441, 490, 497, 501, 517–518
    waste 441, 488, 491, 495, 497, 501
acute exposure 653
acute health effects 614
acute toxicity 614, 636, 647
adiabatic lapse rate 255–256, 283, 284, 285
adjusted means 206–207
administrative controls 614, 638–639, 655, 695
Administrative Procedure Act (APA) 23
adsorption 375, 391, 434, 488, 614
    arsenic removal, and 431
    of gaseous pollutants 301

site density 580
advanced waste treatment 439, 444
advanced wastewater treatment 511–518
advection 71
advective transport 124
aerated ponds 464–465, 467
aerated static pile (ASP) 537–538
aeration 375–376, 381, 382, 386, 455, 467, 488, 489, 491–495, 517–518, 530
    extended 495–497
    tank 488, 491
aerobic
    bacteria 468
    defined 439, 489
    digestion **530,** 530–533
    ponds 465–467
    processes 606
aerodynamic diameter 259–260, 265, 653
aerosol photometry 657
aerosols 247, 614, 635
afterburners 302, 303
aggregates 568
Aggressive Index (A.I.) 385
air
    binding 392
    change efficiency 314
    composition of 5, 233, 237–247
    diffusion performance index 315
    -line respirator 615
    monitoring 648–655
        analytical methods 649–650
        confined space entry 659–660
        vs. sampling 651–652
    parcels 282, 283
    pollutants 571
    pollution 21, 233–325, 615
        control technology 294–303
        mechanics 281–287
    -purifying respirator 615, 699, 700, 704
    quality 306
    sample volume 648
    sampling 651–652
        collection process 656
        direct-reading instruments 657–659
        gases and vapors 655–660
        methods 656
        types of 655–656
    stripping 607
    -supplying respirator 615, 704, 705
    -to-solids ratio 529
airborne particulate matter 647
    air sampling for 652
airborne particulate soil contamination 571
airborne pathogens 522

airborne toxins 235, 631
air contaminants 635
airflow pathways 315, **315**
air quality
    biomass energy operations impact 747–748
    biomass feedstock production impact 745
    biomass, impact of 740
    geothermal energy construction impact 756
    geothermal energy exploration and drilling impact 753
    geothermal energy O&M impact 759
    marine energy impact 790
    wind energy impacts 718
    wind turbine impacts 713–714
Air Quality Criteria 234
Air quality index 11
albedo 235, 270
algae 407, 416, 468–471, 473
algal blooms 469, 474
algorithms 125–129
    vs. computations 125
alkalinity 297, 333, 385–386, 389, 404, 427, 430, 434, 446, 548; *see also* volatile acids to alkalinity ratio
    activated sludge, and 493, 500
    aerobic digestion, and 530, 532
    anaerobic digestion, and 533–535
    nitrification, and 514
    source water **413**
    wastewater pond 471–472
allergens 324, 615
allocation, sample 151
allowable stress 218
alpha particle 615, 673, 675
alternating current (AC) 727
alum 388, 412
alveoli 615, 626
ambient, defined 615
ammonia 233, 238, 419, 463, 502, 512
ammonification 470
anaerobic
    bacteria 468
    defined 439
    digestion **533,** 533–536
    ponds 465, 467
    processes 606
analysis of variance 162–186, 189, 196
    vs. t test 165
animal feedlots 572; *see also* concentrated animal feeding operation
anoxic, defined 439
anthropogenic greenhouse effect 269
antibiotics, in the environment 344, 345, 357–358
antidepressants 359–361
antiepileptics 361
antineoplastics 361–362
antiseptics, in the environment 366
aquatic invertebrates 733
aquatic organisms
    electromagnetic fields 778–782
    rotor blade strike 786–788
aquifers 354, 370–372, *371,* 522, 571, 572
Archimedes number 315
area air sampling 649, 656
area measurements 91–94
area of a circle 92

argon **242,** 242–243
arithmetic mean 76–78, 235
arithmetic scale 235
arsenate 431
arsenic 263, 334, 572, 574
    removal 430–437
        adsorptive processes 431, 434
        ion exchange processes 431, 434
        prescriptive processes 431–434
        separation processes 431, 434–435
artificial media drying beds 541
asbestos 309, 316–319, **318,** 335–336, 624
    airborne fibers 653
asbestosis 615
Asiatic clams **404,** 404–405
asphyxiation 615
assimilable organic carbon (AOC) 409
atmosphere 5
atmosphere-supplying respirator 615
atmospheric chemistry 266–268
atmospheric dispersion 281–287
    models 11, 287–289
atom 235
atomic number 235
atomic weight 235, 616
attenuation 616, 661
attenuator 764
audible range 661
audiogram 616, 661, 667
audiometric testing 616, 667–668
authorized person 616
authorized user 673
    alternate 673
auto emissions, soil pollution and 571
auto-ignition temperature 616
available chlorine 69, 417, 418, 420, 421, 424, 425, 427–428, 505
average 76–78, 134
    monthly discharge limitation 439
    weekly discharge limitation 439
Avogadro's number 616

background noise 616, 661
backwashing 392, 393
bacteria 401, 468, 470, 472, 489, 513–514, 522
    chlorine, and 427
    nitrification 238
baghouse filters 298–299, **299**
ballasts 51–52, 508, 509
bare rock succession 569
barminution 453
barrage facilities 793–794, 798
bar screen 368, 450, 451
Bartlett's test of homogeneity of variance 157–158, 201
base 235
beaching 730
bearing force 211
behavior-based management models 616
belt filter 544–546
belt thickeners 527–530
benching, excavation 230
benchmarking 616
bending forces 211
bending strength 219

beneficial reuse, biosolids 519
benthic organisms
  displacement 769
  habitats alteration 769–770
    anti-fouling coatings 783
    aquatic organisms 778–782
    collision and strike 786
    electromagnetic fields impact 776–778
    interference with animal movement and migration 783–786
    noise impact 770–776
    ocean thermal energy conversion 788–789
    rotor blade strike 786–788
    toxic effect of chemicals 782–783
best available control measures (BACMs) 274
Best Available Control Technology (BACT) 748
best management practices (BMPs) 24
beta-blockers 359
beta particle 616, 673, 676
Bhopal 15, 640
bias, sampling 649
binomial distribution 152
bioassay 673
biochemical oxygen demand (BOD) 77, 332–333, 389, 439, 446, 457, 483, 486, 489, 493, 494, 507, 511
  carbon, and 474
  land application, and 515–516
  microbioreactors, and 513–514
  ozone, and 509
  removal 463, 465, 466, 480
biodegradable, defined 489
biodegradable dissolved organic carbon (BDOC) 409
biodegradable pollutant 4
biodegradation 353, 364, 406, 581–582
bio-floc 493
biofouling 783
biofuels 735
biogeochemical cycles 5
biological contaminants 308
biological denitrification 514
biological exposure limits (BEl) 648
biological half-life 673, 675
biological nitrification 514
biological nutrient removal (BNR) 515–518, 516–517
biological stability 406
biological stressors 631, 632
biological towers 477
biological treatment 463, 605–606
biomass/bioenergy
  composition of 738–739
  feedstock production impact 745–747
  feedstock types 737–738
  impact of construction, production and operation 739–745
  operations impact 747–750
biorefinery 737
bioremediation 11
biosolids 518–526
  alkaline stabilized 538
  defined 439
  disposal cost 548
  freezing 542
biosphere 5
biotic organisms 5

bleach 420, 423–424
blood lipid regulators 358–359
BNR *see* biological nutrient removal (BNR)
BOD *see* biochemical oxygen demand (BOD)
Boyle's law 249–250, 617
breakpoint chlorination 419, 502
breakthrough, filter 392
breakthrough time, protective clothing 702
breast cancer 338, 339
Bremsstrahlung 673
brick manufacturing process, contaminated soil and 586
Briggs' plume rise formula 286–287
brine, geothermal 575
brittleness 221
bromide ion 408–411, 412, 429
bromine chloride 501, 511
Brownian diffusion 265
buffer 439
bulking 489
buoyancy flux 315
buoyancy, of air parcels 283–284
Buoyant Line and Point Source Model 129
bypass lag time 734
Byzantine Generals algorithm 127

cadmium 384
cadmium telluride 724
calcium carbonate 382, 386
calcium hypochlorite 69, 417, 420–421, 424, 505
calcium removal 433
calibration 115, 649, 656–657, 659
CALINE3 129
CALPUFF 129
CAL3QHC/CAL3QHCR 130
cancer, endocrine disruptors and 337–339
capture velocity 617
carbon 6, 248, 474–475, 490, 493, 571
  activated 71, 301, 375, 376, 407, 412, 413, 515, 580
  adsorption 515, 581
  capture and sequestration 802–818
  cycle 6, 571
  organic 57, 58, 59, 355, **408**, 409, 410, 430, 447, 474, 510, 548
  oxides 299
  sorption, activated 607
  total organic 410, 413, 430, 510, 548
carbonaceous biochemical oxygen demand (CBOD) 439
carbon capture and sequestration 802–806
  acoustics 814
  drilling/construction impacts 810–814
  exploration impacts 808–810
  geologic 806–807
  hazardous materials and waste management 814
  operations impact 814–818
  potential impacts 807–808
  terrestrial 803–804
carbon dioxide 12, 233, 234, 236, 241–242, **242**, 248, 252, 263, 266, 268, 269, 275, 281, 289, 293, 382, 385, 386, 433, 444, 464, 466, 468, 533, 547, 581, 680, 745, 751
carbon footprint 713–714
carbonic acid 382
carbon monoxide 273, 274, 275, 278, 281, 286, 289, 291, 308, 310, 617, 629, 635, 718

carbon sequestration 803
carcinogens 275, 321, 335, 337, 338, 361, 364, 377, 408, **409**, 575, 617, 640, 641
Carson, Rachel 335
cartridge filter systems 394
catalytic combustion 303
catastrophe, defined 617
catastrophic events 752
cathodic protection 386, 578
cation exchange 384
cellulose 738
Celsius 39
central nervous system (CNS) 340
central tendency 76, 133
centrifugal force 223
centrifugation 546–547
centripetal force 223
channel length required 455–456
Charles's law 250–251, 617
chemical addition 375–377, 386, 457–458
chemical bond 236
chemical feeder calibration 379–380
chemical oxygen demand (COD) 439, 446, 489
chemical process audit 602
chemical protective clothing 698–704
chemical reaction 236
chemical sludges 520
chemical solutions 377–379
chemical spills 573, 617
    protective clothing, and 698
    risk assessment for 573
chemical stressors 631
chemical transport systems 124–125
chemical treatment 512
chemical weathering 569
chemiluminescence 657
chi-square test 155–157, 158
chloramine 399, 406, 409, 411–415, **414**, 417, 419, 503
chlorination 405, 413, 416–430, 501–506
    breakpoint 502
    gas 419–420
    process calculations 504–506
chlorine 381, 399, 406, 409–411, 414, 416–430, 458
    activated sludge, and 494
    attributes of 416
    available 69, 417, 419, 420–424, 427–428, 505
    chemistry 418–419
    contact time 418
    demand 418, 502, 504
    disinfection 501–506
    dose 417, 421–425, 504, 505
    efficiency, pH and 428
    feed rate 504
    forms of 417
    gas 425
    generation 424–425
    mussels, and 405
    ordering supply of 506
    oxidation 538
    pathogen inactivation 428–429
    points of application 425
    reaction 417
    residual 406, 407, 415, 417–419, 422, 427, 428, 501, 503, 504
    systems 503–504
    usage of 425
chlorofluorocarbons (CFCs) 11, 277, 281
chloropicrin 411
chronic exposure 653
circumference 66, 87, 88–91, 93
clams, Asiatic **404**, 404–405
clarification 390, 458–463
clarifier 439, 488, 497, 517–519
clay particles 567
Clean Air Act 13, 23, 240, 272–280, 748
clean energy 46–63
cleaner products 11
Clean Water Act (CWA) 13, 23, 332, 334, 464
climate, defined 236
clofibric acid 358
cloud seeding 247
cluster sampling 154–155
c-multipliers 198
coagulation 387–390, 407
    -assisted microfiltration 432–433
    enhanced 433
    test 389
coagulation/flocculation (C/F) 431, 443
coal, carbon dioxide emissions and 61–62
COD *see* chemical oxygen demand (COD)
code, defined 707
Code of Federal Regulations (CFR) 24
coefficient of determination 190, 197
coefficient of friction 618
coefficient of variation 141
cohesive soil 227–228
coil filter 544
cold-mix asphalt (CMA) 585
cold stress 688–690
coliform 439, 465
collection mechanisms, particle 264–266
colloidal particles 388
colloidal solids 447, 490, 508, 519–520
color
    comparator 427
    fluorescent lamp 51–52
    as nontoxic pollutant 334
    retention index 52
    slime 485
    temperature 52
    wastewater 445, 463
    water 334, 373, 385, 388, 404, 408, 425
colorimetry 618, 651, 657–659
combined chlorine residual 417, 419, 503
combined sewer 439
combustible liquid 618
combustion 247–248, 302–303, **303**, 618, 657
    byproducts 308–309
    particle formation, and 262
commercial chemical products 599
commercial sources, of MSW 594
comminution 439, 453
common divisor 65
common factor 65
compact fluorescent light (CFL) 50–53
compaction, soil 229
competent person 317, 618
COMPLEX 1 130
Complex Terrain Screening model 130
composite number 65

Index

composite sample 439
composting 537–539, 573
Comprehensive Environmental Response, Compensation, and Liability Act (CERCLA) 576, 592, 601, 617
comprehensive performance evaluation (CPE) 396
compressed air 248
compressibility 249
    air 249
    fluid 233
    soil 227, 228
compression forces 211
compression stress 223
compressive strength 219
computations vs. algorithms 125
concentrated animal feeding operation (CAFO) 332, 344, 352, 356
concentration 115–117, 420, 618, 703
    calculations 31–33, **33**, *33*
    chemical 377–378
    factor 529
condensable particulate matter 261
condensation, of gaseous pollutants 301–302
conduction, energy 269, 271
confidence limits 147, 151, 152, 153, 155, 190–191
confined aquifer 370
confined spaces 449, 534, 618, 648
    air monitoring 659–660
coning 489
conservation of matter 249
consistency, soil 227
construction and demolition sources, of MSW 594
construction, soil for 224–225
contact condenser 302
contact tank
    chlorine 502–504
    hypochlorite 503
    ozonation 509
    ultraviolet disinfection 507
contact time 414–415, 418, 502, 703
contaminants, personal protective equipment and 703
contamination 14, 355, 373, 543
    air 131, 614, 639, 693, 699
    background 287
    defined 673
    drinking water 14, 355, 399, 400, 543, 572
    groundwater 345, 370, 372, 398
    indoor air 308
    microorganism 333, 398
    mold 323
    organochlorine 342
    radioactive 6, 673, 677
    soil 6, 12, 16, 22, 563, 570–576
    surface water 332, 333
    types of 703
    underground storage tank 576–577
content
    atmosphere 246
    soil 227, 228, 229, 580, 581
contrast media 362–363
convection 268, 269, 618, 685
conventional pollutant 346
conversion factors 30–41, **32**
copper indium gallium selenide (CIGS) 724
coral reefs 767
correction term 163, 169, 174

correlation coefficient 143–144
corrosion
    control 376, 384–387, 433
    underground storage tank 577
corrosive-resistant pipe materials 387
corrosive substances 597
coulometry 657
coupons, corrosion detection 385
covalent bond 236
covariance 142–143
covariance of the means 145
Cradle-to-Grave Act 576
critical flow 731
cross-sectional area 91
cryptobiotic soils crust 742–743
Cryptosporidium 391, 394, 402–403, 418, 426, 427, 429
CTDMPLUS 129
CT factor 413–414, **414**, 418
CTSCREEN 130
cubic feet per second to million gallons per day 37
cubic feet to gallons 34
cubic units 66, 87
cultural resources
    biomass energy operations impact 748
    biomass feedstock production impact 745–746
    biomass, impact of 740
    geothermal energy construction impact 756
    geothermal energy exploration and drilling impact 753
    geothermal energy O&M impact 759
    marine energy impact 790
    wind energy impacts 718
    wind turbine impacts 714
Cunningham slip correction factor 265
currents 762, 766–767
curvilinear regressions 198–207
Cuyahoga River 15
CWA *see* Clean Water Act (CWA)
cyanobacteria 468, 470, 742
cycling 735
cyclone collectors 297–298
cyclone degritter 456

daily discharge 439
Dalton's law 122–123, 253, 618
data, qualitative/quantitative 136
DBPs *see* disinfection byproducts (DBPs)
DDT *see* dichlorodiphenyltrichloroethane (DDT)
decibel (dB) 618, 720
decomposition 489, 521
decontamination 319, 619, 639, 699
    personal protective equipment 702–703
deep-well injection 608
deferrization 381
degradation 353, 582, 702; *see also* biodegradation
    bacterial 470, 473, 496
    environmental 18, 20, 21
    formaldehyde resins 320
    hydrocarbon 582
    microbial 353–354, 361, 366
    musks, and 364
    pharmaceutical and personal care products, and 355, 358, 361
    products 352, 354
degrees of freedom 147, 148, 156, 157, 160, 161, 163, 164, 166, 167, 169, 170, 176, 177, 178, 184, 191, 196

demanganization 381
denitrification 467, 470, 471, 493, 497, 498, 514, 517, 518
denominator 82
density 31, 222, 378, 619
    defined 236
    gas 254, 256
Department of Defense (DOE) 725
deposition 281–287
desertification 20
detention time 439, 443–444, 461, 471
    facultative pond 467
detritus feeders 733
dewatering 439, 518, 519, 520, 522, 526, 533, 539–547
dew point 236
diagnostic contrast media 362–363
diameter 66, 87, 90
    aerodynamic 653
diatomaceous earth 391, 393
diazepam 362
1,1-dichloro-2,2-bis(p-chlorophenyl)ethylene (DDE) 338, 339
dichlorodiphenyltrichloroethane (DDT) 334–337, 339, 341, 342, 364
diethylstilbestrol (DES) 338, 342
differential heating 271
diffuser 489, 503
diffusion 124, 297
diffusiophoresis 266
diffusivity 265
digester performance 442
dike 619
dilution ventilation 693–694
dimensional analysis 81–86
dimming, global 11
direct filtration 391, 394
direct-flame combustion 302
direct-reading air sampling instruments 657–659
Discharge Monitoring Report (DMR) 439
discovery, invention, innovation 16
discrete variables 152–153
disinfectant residuals 408, **408**
disinfectants, in the environment 366
disinfection 398–430, 444; *see also* disinfection byproducts
    methods 416
    need for 400–402
    regulatory requirements 414–415
    vs. sterilization 399
    wastewater 501–506
disinfection byproducts (DBPs) 399, 404, **408–409**, 408–413, **411**, 415
    chlorine, and 429–430
    control strategies 411–413
    inorganic 411
    precursors of 404, 408, 410–413
dispersion models 129
    air quality 287–289
dispersive transport 124
displacer 763
disposal cost, biosolids 548
dissolved gases 446
dissolved organic carbon (DOC) 410
dissolved oxygen (DO) 9, 11, 135, 140, 381, 385, 386, 439, 457, 460, 466, 471, 480, 485, 489, 492, 493, 511

activated sludge, and 500
aeration, and 498, 530
aerobic digestion, and 530–531
biological nitrification, and 514
biological nutrient removal, and 516–518
ozone, and 509
distributions, statistical *137*, 137–139, **138**, *138*
diversion 727
dividend 66
dividing by a fraction 82
divisor 65, 126
domestic animals, pharmaceuticals and 356
dose equivalent 673
dose-response 619, 634
dosimeter 662, 665, 673, 676
DPD color comparator 427
drainage basins 331, 372
drift 733, 762
drilling wastes 814
drinking water
    arsenic removal from 430–437
    chemicals used in 377
    contamination 14, 355, 399, 400, 543, 572
    corrosion, and 384
    CT values 418
    disinfection, and 399, 403, 408, 416, 420, 425, 516
    lead in 384
    membrane filtration 394
    pathogen inactivation 428
    pharmaceuticals, and 355, 358, 359
    road salt spreading, and 572
    standards 397
    surface water 333
    taste and odors 407
    wells 370, 571, 572
droplet evaporation 262
drug classes 356–363
drug disposal terminology 346
drugs in the environment 346–352, **348–351**
dry chemical feeder 379, 503, 504
dry deposition 280, 282
drying beds 539–542
dry well pumping station 449
ductility 220
dumps, soil contamination and 572
dusts 619, 635, 652–653
    airborne concentration of 654
    classification of 652–653
    nuisance 625, 647, 652
dynamics 224

Earth's natural capital 19
ecological resources
    biomass energy operations impact 748
    biomass feedstock production impact 746
    biomass, impact of 740–741
    geothermal energy construction impact 756
    geothermal energy exploration and drilling impact 753–754
    geothermal energy O&M impact 759
    marine energy impact 790
    wind energy impacts 718
    wind turbine impacts 714
ecology 6

Index

ecosystem 7
EDCs *see* endocrine-disrupting chemicals (EDCs)
effective half-life 673, 675
effluent
    activated sludge 488, 500
    defined 440
    disinfection byproducts in 415
    drugs in 354–355, 356, 357, 359, 361, 366
    gas 300, 301, 303
    oxidation ditch 498
    pond 465, 467, 469, 474
    rotating biological contactor 485, *485*
    settling tank 463
    sewage 344, 356, 359, 365–366, 425
    turbidity 393, 394, 407
    wastewater 501–504, 507–509, 511–520
    weir 458, 460, 494, 497
EIS step-by-step process 25, *26*
elastic limit 219, *219*, 223
electrical fields 778–779
electrical resistance heating remediation 11
electricity reduction 46
electrodialysis reversal 435
electrolytic recovery technique 607
electromagnetic fields (EMF) 776–778
    on aquatic organisms 778–782
electrostatic attraction 266
electrostatic precipitators 253, 264, 266, 297
emerging contaminants 11, 334–369
emission factor 47
emissions 45, 46–63, 113
    compact fluorescent lights, and 50–53
    forests, and 56–58
    home energy use, and 54–55
    per barrel of oil consumed 50
    per coal-fired power plant 63
    per gallon of gasoline combusted 47
    per garbage truck of waste recycled instead of landfilled 62–63
    per mile driven by passenger vehicle 48
    per passenger vehicle per year 47–48
    per pound of coal burned 61–62
    per propane cylinder burned 60–61
    per railcar of coal burned 61
    per tanker truck 50
    per therm of natural gas 49
    per ton of waste recycled instead of landfilled 62
    per wind turbine 63
    sequestration 57
    tree seedlings, and 56
Emissions & Generation Resource Integrated Database (eGRID) 46
emission standards 236
emissivity 236
empirical rule, normal distribution 137
endocrine-disrupting chemicals (EDCs) 334, 336–343, 356
    biological effects 337–339
    immunological effects 342–343
    neurological effects 340–341
    reproductive effects 340
energy 42–63, **46**
    atmospheric 255–256
    clean 46–63
    defined 236
    defining 43
    environmental engineering, perspective on 42
    home electricity use of 54
    home use of 54–55
    measuring 45–46
    non-benefits 44, 45
    nonrenewable benefits 44, 45
    from Sun 723–726
    types of 43–44
    units and math operations 42
    wind 712–713
Energy Independence and Security Act of 2007 (EISA) 736
engineering controls 620, 637, 655, 661, 666, 704–710
English system 29
enhanced biological nutrient removal (EBNR) 518
enhanced coagulation 433
enthalpy 254
environment 16–18
    defined 2
environmental controls, occupational hazard 637
environmental degradation 20
environmental effect, of pollutants 281
environmental endocrine disruptors 336
environmental engineering components 7
environmental equilibrium 2–4
environmental health *17*, 17–18
environmental heat 685
Environmental Impact Statement (EIS) 24–25
environmental justice
    biomass energy operations impact 750
    biomass feedstock production impact 747
    biomass, impact of 744–745
    geothermal energy construction impact 758
    geothermal energy exploration and drilling impact 755
    geothermal energy O&M impact 760
    wind energy impacts 720
    wind turbine impacts 717
environmental laws and regulations 22–23
environmental limits, occupational 646–648
environmentally preferable products 11
environmentally safe products 11
environmental modeling 113–130
    algorithms, and 125–129, *128*
    chemical transport systems 124–125
    development, steps for 114
    dispersion models, and 129
    media material content, and 115–119
    phase equilibrium and steady state 120
    screening tools 130
    uses for 114–115
environmental pollution, technology and 18–22, *19*
environmental radioactivity 11
environmental science 16–18, *17*
environmental stressors 630–633
environmental terrorism 574
equal-energy rule 662
equalization, flow 458
equilibrium 120–124
    laws of 122–124
    problems 120–121
ergonomics 620
ergonomic stressors 631–633
erosion 565

Escherichia coli 402, 427
ESPs *see* electrostatic precipitators (ESPs)
estimation, environmental modeling 114
estuaries 331
euphotic zone 472
eutrophication 11, 333, 512, 516
evaporation rate 620
evaporative cooling 685–687
even number 65
excavation 229–230
    of contaminated soil 587
excretion, of pharmaceuticals and personal care products 344, 352
exhaust fumes, soil pollution and 571
exhaust ventilation 691–692
    local 693
exponent 74–76, **76**, 82, 83
exposure limits 646–647, 651
exposure pathways, chemical spill 579
extended aeration 495–497, *496*
extensive property 118
extraction 597
extrapolations 114
extremely hazardous substance 596
extremely low frequency radiation (ELF) 680
eye protection 325, 507, 695, 697, 699

fabric filters 299
face protection 695, 697, 699
factorial experiments 173–179
factor, math 65
facultative
    bacteria 489
    defined 440
    ponds 464–465, 467, 468, 471, 473
Fahrenheit 39
fall arresting system 620
fatigue strength 220
faulty construction 577
faulty installation 577
fecal coliform 440, 465, 466, 504, 518
Federal Aviation Administration (FAA) 715, 742
Federal Coal Mine Safety and Health Act 629
feedlots 572
feedstock
    first-generation 738
    second-generation 738
    types 737–738
fertilizers 573
fibers 620, 635
filamentous bacteria 489
filtration 377, 390–398, 407, 412, 414, 418, 513
    direct 391, 394
    rotary vacuum 543–544
    technologies 391–394
filtration hazardous waste 608
finite population correction 142
first law of thermodynamics 236
fish migration, physical barrier to 729
fish, tumors in 337–338
fission product 11
fixed bed adsorber 301
fixed-film systems 464, 476, 484
flame ionization detectors (FIDs) 651, 658

flammable liquid 620
flash point 621
flexible-membrane liner (FML) 610
float cake 527
floc 389–390, 440
flocculation 390, 431, 443, 493, 513
flotation thickening 527
flow
    alterations 715, 729
    calculations 31–33, **33**, *33*
    characteristics 256–257
    continuation 735
    critical 731
    equalization 458
    fluctuations 729
    rate, gas 252
    in regulated rivers 729
    in unregulated rivers 729
    wastewater 446
fluidized bed incinerator 547
fluidized bed reactor 436
fluidized sand bed 514
flume 440
fluorescent bulbs 50–53
    operating costs 52
fluoxetine 360–361
Flux Richardson number 315
food-to-microorganism ratio (F/M) 81, 440, 489, 493
foot protection 695, 697–698
force 98–104, *99*
    defined 211
    magnitude of 211
    pressure, and 98
    resolution of 210–216, *211, 213, 214, 215–217*
force main collection system 448
forebay surges 735
forests, carbon dioxide emissions and 56–58
formaldehyde 267, 291, 303, 304, 310, 316, 320–321
fossil fuels 276, 281, 290
fraction, dividing by 82
fragrances 363–366
free available chlorine 419
free chlorine residual 404, 407, 414, 418, 419, 428, 503
freeze-assisted dewatering 542
frequency distribution 137
friction 216, 221–222, 251
    coefficient of 618
    head 99
frostbite 688
frostnip 688, 689
Froude number 315
F test 158, 164, 165–166, 167, 168, 183, 184, 189, 192, 206, 207
fuel cells
    environmental impact 802
    hydrogen 800–802
fumes 621, 635
fumigation, pollutant 286

gage pressure 250
gallons per day to million gallons per day 37
gallons per minute to million gallons per day 36
gallons to cubic feet 34
gallons to pounds 34

galvanic corrosion 385, 386
gamma radiation 621, 626, 658, 674, 677
gas
    air sampling 655–660
    chlorination 419–420
    chromatography 621, 650, 657–658
    conversions 252–253
    density 254, 256
    effluent 300, 301, 303
    flow rate 252
    laws 249–252
    physics 249–257
    velocity 253–254
gaseous cycle 6
gaseous emissions 751
gaseous pollutants, removal of 299, **300**
Gaussian models 288
Geiger-Müller counter 621, 658, 676
general ventilation 693–694
geologic sequestration exploration
    acoustics 810, 814, 817
    air quality 808, 810, 815
    cultural resources 808, 810–811, 815
    drilling/construction phase 810–814
    ecological resources 808, 811, 815
    environmental justice 810, 814, 817
    hazardous materials and waste management 810, 817–818
    land use 809, 812, 816
    operations phase 814–818
    paleontological resources 809, 813, 816
    socioeconomics 810, 813–814, 817
    soils and geologic resources 809, 812, 816
    transportation 809, 813, 817
    visual resources 809, 813, 817
    water resources 809, 811–812, 815–816
geometrical measurements 87–97, *89, 90, 92, 93, 94, 97*
geosphere 5
geothermal energy
    construction impact 755–758
    environmental impact of geothermal power development 751–752
    exploration and drilling impact 753–755
    operation and maintenance impact 758–760
geothermal fluids 754
geothermal sites, soil contamination and 575
Giardia 388, 391, 394, 402, 414, 418, 429
global dimming 11
global distillation 11
Globally Harmonized System (GHS) 641–644
global warming 11, 46, 236
good housekeeping practices 638–639, 710
Gould sludge age 489, 500
grab sample 440, 621
granular activated carbon (GAC) 407, 412, 415, 430
granular ferric hydroxide (GFH) 436
Grashof number 315
grasshopper effect 11
gravel 567
gravimetric chemical feeder 379
gravitational settling 264, 265, 280, 520
gravity collection system 448
gravity settlers 297
gravity thickening 526–527

gravity/velocity-controlled grit removal 454–455
greatest common divisor 65
green energy 711
green engineering 711
greenhouse effect 236, 241, 267
greenhouse gas emissions 47, 48, 57, 58, 113
green products 11
greensand 382, 383, 432
grit 440, 445, 447, 450, 451, 463
    removal 453–458, 498
ground-fault circuit interrupter (GFCI) 621
grounding, electrical 619
groundwater 330, 369–373, 398, 415, 445, 582, 583
    contamination 345, 370, 372, 398
    flow 371–372
    infiltration 445
    wind turbine impacts 715
groundwater under the direct influence (GWUDI) 394
group regressions 200–201

habitat 7
habitats alteration, in benthic organisms 769–770
    anti-fouling coatings 783
    aquatic organisms 778–782
    collision and strike 786
    electromagnetic fields impact 776–778
    interference with animal movement and migration 783–786
    noise impact 770–776
    ocean thermal energy conversion 788–789
    rotor blade strike 786–788
    toxic effect of chemicals 782–783
half-life 674, 675, 677–679
haloacetic acid (HAA) 410, 415, 429
halogenated organic byproducts 408
Hammurabi's Code 707
hand protection 695, 697, 706
hardness, of steel 221
hardness, water 382–383, 433
Hawthorne effect 52
hazard
    assessment 696
    classification 642
        labels 644
    defined 621
    and operability analysis 621
    vs. toxicity 634
Hazard Communication Standard (HazCom) 621, 631, 640–645, *645*
hazardous air pollutants (HAPs) 747
Hazardous and Solid Waste Act 600
hazardous materials and waste management
    wind energy operation impacts 720
    wind turbine impacts 717–718
hazardous materials/substances 595, 596, 622, 636
hazardous task inventory 662
hazardous waste 22, 340, 574, 587, 589–590, 595–611, 622, 698
    categories 598–599
    CERCLA 601
    control technology 601–605
    deep-well injection 608–609
    disposal 608–611
    EPA-list 598–599

hazardous waste (cont.)
  incinerators 298, 606
  landfilling *610*, 610–611
  legislation 599, 600–601
  management 814
  minimization 602–603
  piles 609–610
  recycling 603–605
  surface impoundments 609
  treatment technologies 605–608
    activated carbon sorption 607
    air stripping 607
    biological process 605–606
    electrolytic recovery technique 607
    filtration and separation 608
    stabilization and solidification 607
    thermal process 606
head 98–104, *99*
  pressure, and 98–100, *99*
  protection 697
health hazards
  engineering controls, and 710
  human and animal waste 445
  industrial 636–639
  lead 384
  permit-required confined space 659
  Safety Data Sheets, and 641–642
  silica 320
  vibration 671
hearing conservation programs 660–661
hearing conservation record 662
hearing threshold level (HTL) 662
heat 14, 42, 45, 236, 416, 684
  atmospheric 255–256
  balance, of Earth 247, 268–269
  capacity 254
  content, of fuel 47, 49
  control ventilation 693
  cramps 622, 688
  distribution 270
    in composting mixture 537–538
  environmental 685
  exchanger 303
  exhaustion 622, 687–688
  fatigue 688
  index **685**, 685–686
  metabolic 624, 685, 686
  as nontoxic pollutant 334
  radiant 626, 632, 685
  rash 622, 688
  stress 686–687
  stroke 622, 687
  transfer 271, 284, 618, 685
heavy metals 494, 521, 548
helium 243, **244**
helminths 522
hemicellulose 739
Henry's law 124, **125**
herbal remedies, in the environment 367
herbicide 11
herbivorous invertebrates 733
heterogeneous nucleation 263
high-hazard work areas 709
high-rate aerobic ponds 466

high-rate trickling filters 478
home electricity use 54
home energy use 54–55
home-generated pharmaceuticals 347
homogeneous nucleation 263
homoscedasticity 192
Hooke's law 223
horizons, soil 568, 569
hormones 334, 336, 338–341, 356–357, 359
hot-mix asphalt (HMA) 585
hot work 622
household products, indoor air quality and 309
housekeeping 639
human-generated ocean noise 771
hydraulic detention time (HDT) 443–444, 461, 473, 475
hydraulic fracturing fluids 818
hydraulic loading 477, 478, 480
  activated sludge 494
  rate 480, 482
    belt filter press 546
hydraulic response to flow fluctuations 734
hydraulics 224
hydrocarbon analyzer 658
hydrocarbons 574
hydrochloric acid (HCl) 418
hydrogen 7, 44, 120, 233, 238, 243, 245, **245**, 248, 256, 291, 292, 418, 441, 447, 468, 471, 472, 490
  fuel cells 800–802
  storage 801
hydrogen chloride 300
hydrogen peroxide 292, 379, 405, 407, 507, 581
hydrogen sulfide 373, 375, 407, 417, 427, 441, 446, 456, 457, 466, 467, 474, 533, 571, 635, 641, 751
hydrokinetic energy
  acoustics 796, 799
  air quality 792, 796–797
  alteration of habitats, for benthic organisms 769–789
  construction impact 791–796
  cultural resources 792, 797
  currents 762
  ecological resources 793–794, 797–798
  environmental impacts 765–769, 789–791
  environmental justice 795, 799
  hazardous materials and waste management 796, 799
  land use 794, 798
  ocean thermal energy conversion 765
  operations & maintenance impact 796–799
  paleontological resources 795, 798
  socioeconomics 795, 799
  soils and geologic resources 794, 798
  tidal energy 764–765
  tides 761–762
  transportation 795, 799
  visual resources 794–795, 799
  water resources 794, 798
  wave energy 762–764
  waves 762
hydropower
  biological impact of flow fluctuations
    aquatic invertebrates 733
    hydraulic response 734
    increased predation 732
    increases in flow 730
    redd dewatering 733–734

Index   841

spawning interference 734
stranding 730–732
types of hydropower activity 734–735
diversion 727
ecological impact 728
flow alterations 729
flow fluctuations 729
generation 727–728
hydropower generation 727–728
impoundment 727–728
physical barrier to fish migration 729
pumped storage 727
unregulated rivers and salmonids 729
hydrosphere 5
hypochlorination 420–421
hypochlorite 418, 428, 503, 505
hypochlorous acid (HOCl) 418, 425, 428, 503
hypothermia 688
hypothesis testing 132, 133
hypothesized count, test of 156–157
hypoxia 11

ibuprofen 359
ideal gas law 122, 251–252
ignitability 597
ignition temperature 547, 622
illicit drugs, in wastewater 368–369
illumination, as engineering control 708
Imhoff tank 459–460
immediately dangerous to life or health (IDLH) 622, 700, 705
immunological effects of endocrine disruptors 342–343
impaction 264–265, 297
impact strength 220
impingement 623
impotence drugs 362
impoundment 727
impulsive noise 663
incandescent bulbs 50–53
incineration 299, 302–303, **303**
inclined plane 215–216, *215–217*
incohesive soil 227, **227**
incomplete combustion 240, 248, 291, 617
indoor air quality 11, 303–325, 623
building factors affecting 309–310
contaminants 316–325
contaminant transport 311, **311–312**, 312, **313–314**
dispersion parameters 314–315
Legionnaires' disease, and 304–305
pollutants found in homes 307–309
sick building syndrome 305–306
industrial health hazards 636–639
industrial hygiene 613–710
areas of concern 633–636
defined 623
history of 628–629
industrial practices 574
Industrial Source Complex Model (ISC3) 129
industrial toxicology 633–635
industrial waste 814
inertial impaction 264–265, 297
infiltration 440
air 248, 304, 306, 310
contaminated surface water 570, 571

groundwater 14, 445, 494, 581
infiltration-percolation 515, 516
inflow 440
influent
defined 440
flow 355, 443, 454, 476, 480, 482, 495, **496,** 517, 523, 528
raw wastewater 447, 452, 453
infrared radiation (IR) 680
ingestion 521, 522, 578, 623, 631, 635
of arsenic 430–431
of lead 321
of toxic substance 636, 710
inhalable particulate mass (IPM) 647
inhalation 623, 636, 646, 652, 655
injection, of toxic substance 636
Innovative Cleanup Technologies 576
inorganic
byproducts 408
defined 440
dusts 652
solids 333, 440, 447, 453, 456, 490, 521
in situ soil remediation technologies 580–583
biodegradation 581–582
isolation/containment 583
leaching and chemical reaction 582–583
passive remediation 583
vitrification 583
volatilization 580–581
insolation 236
inspirable particles 653
institutional sources, of MSW 594
Instream Flow Incremental Methodology (IFIM) 729
intake failures 735
integer 65
intensive property 118
interception 297
particle 265
Interim Enhanced Surface Water Treatment Rule (IESWTR) 394, 396–397, 418, 426
International Electrotechnical Commission (IEC) 763
International System of Units (SI) 29, 41
interpolations 114
invasive species 12
inversions, temperature 285
invertebrates 469, 470
ion, defined 236
ion exchange 382, 383, 432, 434
arsenic removal, and 431
ionization 120, 620, 651, 658, 674–677, 679
ionizing radiation 616, 623, 631, 672–675, 681
OSHA requirements for 682–683
iron bacteria 387
iron filings 436–437
iron-oxide-coated sand 435
iron removal 375, 380–381, 391, 393, 405–406, 417, 431–432
irrigation 20, 22, 344, 514, 515, 573

jar test 389–390
Javelle water 425
job hazard analysis 623

K factor 486

kinematic viscosity 256
kinetic energy 43, 122, 224, 265, 266, 620, 623, 674, 766
krypton 243–244, **244**

labels, hazard classification 644
Laboratory Safety Standard 623
lagoons 463, 521, 541, 574
lakes 331
laminar flow 256
land application 539
    biosolids 548–550
    secondary effluent 515–516
    soil contamination, and 571
landfills 355, 362, 574, *610,* 610–611
land resources
    biomass feedstock production impact 746
    biomass, impact of 741–742
landslides 752
land subsidence 752
land use
    biomass energy operations impact 749
    geothermal energy construction impact 757
    geothermal energy exploration and drilling impact 754
    geothermal energy O&M impact 759–760
    wind energy operations impacts 719
    wind turbine impacts 715
Langelier Saturation Index (LSI) 385
lapse rate 236, 255
lasers 623, 680–681
Latin square 171–173, 186
law of conservation of energy 43
laws of thermodynamics 21
leachate 539
    drugs in 355
leach liquors 576
lead 289, 629, 653
    corrosion, and 384
    exposure 321–323
    poisoning 628
Lead and Copper Rule 384
Lead and Copper Rule Compliance 411
lead-mine scale 575
least squares 188, 190, 193, 196
Le Chatelier, principle of 120
leftover medications 347
Legionnaires' disease 304–305, 308, 403
lentic 331–332
life-cycle analysis 11
lighting, as engineering control 708
light pollution 12
light, wastewater ponds and 472
lignin 739
lignocellulosic biomass 738
lime, for water treatment 381, 382, 384, 386, 433
limit of detection (LOD) 648
limit of quantification (LOQ) 648–649
limnology 331
linear regression 186–189, 194
linear systems 105
linear units 66, 87
liquid phases, material content of 117–118
lithosphere 5
load following 734
lockout/tagout procedure 624

LONGZ 130
lotic 331–332
lower explosive limit (LEL) 624, 657
lowest common multiple 65
low flow shutdowns 735
low-flow startups 735
Luvox 359, 360

magnesium removal 433
magnetic fields 779–780
magnitude of force 211
malleability 221
manganese 653
    greensand 382
    removal 375, 380–381, 391, 393, 405–406, 417, 431–432
manually cleaned screens 451
manufactured gas plants 575
marine debris 12
marine energy
    acoustics 796, 799
    air quality 792, 796–797
    alteration of habitats, for benthic organisms 769–789
    construction impact 791–796
    cultural resources 792, 797
    currents 762
    ecological resources 793–794, 797–798
    environmental impacts 765–769
    environmental justice 795, 799
    hazardous materials and waste management 796, 799
    land use 794, 798
    marine energy 789–791
    ocean thermal energy conversion 765
    operations & maintenance impact 796–799
    paleontological resources 795, 798
    socioeconomics 795, 799
    soils and geologic resources 794, 798
    tidal energy 764–765
    tides 761–762
    transportation 795, 799
    visual resources 794–795, 799
    water resources 794, 798
    wave energy 762–764
    waves 762
marine pollution 12
mass per unit volume 115
mass spectrometer (MS) 650
material balance 248–249
material chemical resistance 702
material content 115–119
Material Safety Data Sheet (MSDS) 624, 631, 644
materials handling procedures 639
materials, properties of 209, 217–223, **218,** *218–221*
mathematical operations 65–112
    algebraic 104–106
    averages 76–78
    dimensional analysis 81–86
    force, pressure, and head calculations 98–104, *99*
    geometrical measurements 87–97, *89, 90, 92, 93, 94, 97*
    laws of equilibrium, and 122–124
    percent 68–72
    powers and exponents 74–76, **76,** 82, 83
    quadratic equations 107–110

Index 843

ratio 78–81
sequence of 66–68
significant digits 72–74
threshold odor number 86–87
trigonometric 110–112, **111,** *111, 112*
matrices 104–106
maximum achievable control technology (MACT) 275
maximum contaminant level (MCL) 373, 413
   arsenic 436
   odor 86
   total trihalomethanes 411, 429
maximum contaminant level goal (MCLG)
   Cryptosporidium 394
   lead and copper 384
maximum residual disinfectant levels (MRDLs) 413
maximum sustainable yield 20
mean 134–135, 136
   arithmetic 76–78, 235
   comparing 168
   normal distribution 137
   sample estimates of 146, 148
   squares 164–165, 166, 167, 168, 170, 171, 177, 184, 186, 192, 197, 201, 202, 205
   standard error of 141–142, 146–147, 150, 151
mean cell residence time (MCRT) 440, 489, 514
measurement, units of 30–41, **32**
mechanically cleaned screens 451–452
mechanics, principles of 223–231, **227, 231**
median 76, 134, 135, 136
membrane bioreactors 513–514
membrane filtration systems 394, 412, 434–435, **435**
mesosphere 236
mesothelioma 624, 653
metabolic heat 624, 685, 686
Metal and Nonmetallic Mines Safety Act 629
metals loading 549
methane 47–48, 236, 266, 267, 291, 467, 468, 533, 574, 641
microbial degradation 353–354, 361, 366
microfiltration (MF) 432–433, 434, 513–514
microorganisms 333, 418, 427, 463, 480, 481, 492, 516
micropollutants 344
microscreening 512–513
microwave radiation 680
milligrams per liter to kilograms per day 35
milligrams per liter to pounds 35
milligrams per liter to pounds per day 35
million gallons per day to cubic feet per second 37
million gallons per day to gallons per day 36
million gallons per day to gallons per minute 36
mimics 356–357
mining site soil contamination 575–576
missing plots 185–186
mists 624, 635
mixed liquor 490, 491, 497
mixed liquor suspended solids (MLSS) 440, 489, 490, 500
mixed liquor total suspended solids (MLTSS) 500
mixed liquor volatile suspended solids (MLVSS) 440, 489, 500
mixing, turbulent in atmosphere 285
mixis 473
MKS system 33, 41, **41**
mode 134, 135, 136
modeling, environmental 113–130
modulus of elasticity 219

molality *(m)* 116
molarity *(M)* 116
molar volume of a gas 122
mold 304, 308, 323–324
   control 323–324
mole fraction 253
morphogenesis 569
mosquitoes 469
multiple hearth furnace 547
multiple, of a number 65
multiple regression 192–198
multiple turbine operations 735
municipal services sources, of MSW 594
municipal solid wastes (MSW) 593–595
   commercial sources of 594
   construction and demolition sources of 594
   institutional sources of 594
   municipal services sources 594
   residential sources of 593
   treatment plant site sources 594
musks 363–366
mycotoxins 324, 625

nanofiltration (NF) 412, 434
National Ambient Air Quality Standards (NAAQS) 272, 273–274, **274,** 277, 278, 279, 289, 748
National Environmental Policy Act (NEPA) 24
National Institute for Occupational Safety and Health (NIOSH) 630
National Pollutant Discharge Elimination System (NPDES) 276, 441, 749
National Research Council 772, 774
natural hydrothermal manifestation 752
natural organic matter (NOM) 387, 409, 411, 412, 429, 431, 434
natural resources 18
negative lapse rate 284
neon 243, **243**
neuroendocrine disruptors 340–341
neurotoxicity 340–341
New Source Review (NSR) 278–279
Newton's second law of motion 211, 222
niche 7
nitric oxide 290–291, **291**
nitrification 238, 470, 471, 484, 485, 493, 497–499, 500, 512, 517, 518
   bacteria 238
   biological 514
nitrite bacteria 238
nitrogen 238, 409, 419, 429, 446, 468, 474, 511–512, 515–516
   breakpoint chlorination, and 502
   crop requirement 549
   cycle 238, 470–471, *471,* 571
   fertilizers 573
   organic 409, 429, 447, 468, 470, 474, 485, 493, 495, 548–549
   oxides 238–240, **239,** *239,* **240,** 275, 276, 281, 288, 289, 290–291
nitrogen dioxide 290–291, **291**
nitrogen fixation 571
nitrogenous oxygen demand (NOD) 440, 512
nitrogen trifluoride 724
nitro musks 364–366

noise 660–672
    calculations 668–671
    control of 632, 660
    dose 663
    exposure, occupational 665
    -induced permanent threshold shift 664
    levels 670
    pollution 12, 752
    reduction rating 663
    workplace levels of 665–666
nonbiodegradable pollutant 4
noncondensable gases (NCGs) 751
non-ionizing radiation 625, 632, 633, 672, 679–681
nonopioid analgesics, in the environment 359
nonpoint source 4, 9, 332, 333
nonrenewable resources 20–21, 44, 45
nonspecific source wastes 599
nonsteroidal anti-inflammatory drugs (NSAIDs) 355, 359
nontoxic pollutants 332, 334
normal distribution *137*, 137–139, **138,** *138*
nuclear fallout 12
nucleation 263
nuisance dusts 625, 647, 652
numerator 82
numerical models 288
nutraceuticals, in the environment 367
nutrients 238, 332, 333, 353, 490, 511, 538, 548, 575, 581, 655
    activated sludge, and 493
    defined 441
    gaseous cycle 6
    removal of 516–518

obsessive-compulsive regulators 359–361
occupational environmental limits (OELS) 646–648
occupational noise exposure 665
Occupational Safety and Health Act (OSH Act) 625, 629–630
ocean acidification 12
oceanic currents 271
Ocean Renewable Power Company (ORPC) 773
ocean thermal energy conversion 765
Ocean Thermal Energy Conversion (OTEC) 788–789
octave-band noise analyzers 665–666
octave bands, frequency of 669
odd number 65
odor 14, 86; *see also* threshold odor number
    formaldehyde 320
    maximum contaminant level 373, 413
    rotten egg (*see* hydrogen sulfide)
    sick building syndrome, and 306
    sulfur dioxide 289
    ventilation, and 691
    wastewater 446, 448, 451, 457, 467, 468, 473, 480, 502, 509, 540
    water 375, 385, 404, 407, 411, 415, 416, 419
Offshore and Coastal Dispersion Model (OCD) 129
oil field site contamination 574
oils and protein 739
oil spill 12, 333
oligotrophic lakes 353
operation and maintenance (O&M) 722
optical density (OD) 681
optimization, environmental modeling 115

oral contraceptives 356
organic
    arsenic 431
    carbon 57, 58, 59, 355, **408,** 409, 410, 430, 447, 474, 510, 548
    compounds 261, 262, 263, 267, 268, 281, 288, 289, 300, 301, 309, 387, 409, 410, 416, 464, 468, 469, 470, 515, 574, 625
        biodegradable 406, 409, 498
        nitrogenous 240
        synthetic 431
        volatile 281, 288, 289, 291, 310, 324, 574, 580, 651
    defined 441
    dusts 652
    lead compounds 321
    loading 458, 467, 471, 476–478, 480, 482, 497, 534
        activated sludge 494, 495
        rate 482, 483, 485, 487
    matter 224, 231, 332, 353, 368, 381, 413, 425, 427, 439, 440, 444, 446, 466, 480, 481, 491, 492, 533, 537, 538, 547, 565, 567, 568
        natural 387, 409, 411, 412, 429, 431, 434
    nitrogen 409, 429, 447, 468, 470, 474, 485, 493, 495, 548–549
    oxidation byproducts 408, 411
    phosphorus 474
    pollutants
        persistent 11, 358
        polychlorinated 365
        toxic 334
    solids, wastewater 447, 453, 455–456, 486, 490–492, 521
    vapors 263
    waste 332, 463, 469, 470, 489, 493, 495, 574
ORPC *see* Ocean Renewable Power Company (ORPC)
OSHA Form 300 625
OSHA-200 Log 624
OTEC *see* Ocean Thermal Energy Conversion (OTEC)
over-cultivation 20
overgrazing 20
overland flow 515–516
over-the-counter (OTC) drugs 347
oxidation 375, 381, 404–411, **406,** 415, 431–432, 538–539, 625
    ditch 495, *498,* 498–500, 517
    pond 463, 465, 466
    towers 477
    vs. combustion 247
oxygen 240–241
ozonation 501, 509–510
ozone 236, 237, 240–241, 267, 268, 273, 274, 275, 276–277, 278, 279, 281, 282, 288, 289, 291, 292–293, **293,** 404, 410, 416–417, 509–511
    depletion 12, 277, 292

package plant 495–497
packed-bed reactors 436
packed tower scrubbing system 300, 301
paints, toxicity 783
paleontological resources
    biomass energy operations impact 749
    biomass feedstock production impact 746–747
    biomass, impact of 743–744
    geologic sequestration exploration 809, 813, 816

Index   845

geothermal energy construction impact 757
geothermal energy exploration and drilling impact 755
geothermal energy O&M impact 760
hydrokinetic energy 795, 798
marine energy impact 791, 795, 798
wind energy operations impacts 719
wind turbine impacts 716
parallelogram law 212
parent material, soil 567
partial pressure 253
particle physics 257–263, **258, 260, 262**
particle size 653–654
particulate matter 12, 129, 237, 247, 249, 253, 257, 260, 261, 262, 268, 273, 274, 275, 278, 281, 287, 289, 293, 296–297, 298, 387–388, 625, 653–654
    aerodynamic diameter 259–260, 265
    airborne 647
    analysis of 654
    atmospheric 247
    collection mechanisms 264–266
    control of 655
    defined 293
    dry, removal of 296–299, **299**
    formation 262
    gas streams containing 253
    haze, and 655
    health/environmental impacts of 654–655
    respirable size 626
    sample collection of 654
    size categories 260–261
    size distribution 261–262
    total suspended 260
parts per billion (ppb) 29, 116
parts per million (ppm) 29, 116, 378, 441
parts per trillion (ppt) 29, 116–117
Pascal's law 249
Pasquill–Gifford dispersion model 279, 289
pathogenic organisms 333, 521–523; *see also* waterborne pathogens
paved drying beds 540–541
peaking 735
pedologists 566
peds 568
penetrating radiation 677
penetration, chemical 702
percent 68–72
percent by mass 115
percent by volume 116
percent moisture reduction 536
percent removal, clarifier 460
percent total solids 462
percent volatile matter reduction 531, 536
perimeter 66, 87, 88–91
permanent threshold shift 664
permeated contaminants 703
permissible exposure limit (PEL) 625, 636, 638, 646–648
    asbestos 316
    lead 321
    radiation 672
    silica 320
permit-required confined space 659, 705
perpetual resources 19
persistent bioaccumulative toxicants 358
persistent organic pollutants 11, 358

personal air sampling 649, 656
personal care products 363–367; *see also* pharmaceuticals and personal care products
personal hygiene 632
    as administrative control 639
    engineering controls, and 709
personal protective equipment (PPE) 325, 625, 637, 639, 655, 694–704, 706
    classifications 695
    decontamination 702–703
    hazard assessment 696
    levels of protection 699–701
    OSHA's requirements 695
    training requirement 696
pesticides 12, 234, 309, 311, 334, 335, 339, 341, 345–347, 352, 356, 357, 366, 372, 548, 573, 650, 698
petroleum products, soil contamination and 570, 578
pets, pharmaceuticals and 356
pH 236, 446, 625
    acid rain 275
    activated sludge, and 492, 493, 500
    adjustment 416, 433, 532, 538, 584
    aerobic digestion, and 530, 532
    alum, and 412
    anaerobic digestion, and 533, 534
    arsenic removal, and 432–434, 436
    calcium removal, and 433
    carbon dioxide, and 242
    chlorine inactivation efficiency, and 428–429
    coagulation, and 389
    color removal, and 408
    corrosion, and 386, 433
    Cryptosporidium, and 429
    disinfection byproducts, and 411, 412, 429, 433
    enhanced coagulation, and 433
    extended aeration, and 496
    Giardia, and 402, 414, 418, 429
    ocean 12
    ozone, and 404
    as pollutant 332, 333
    rainfall 275
    soil 581, 582
    stabilization pond 466
    surface water, and 333
    wastewater pond 466, 471–472
    water treatment, and 375, 381, 383, 389, 538
pharmaceutical manufacturing facilities (PMFs) 354
pharmaceuticals and personal care products (PPCPs) 11, 334, 343–369, **348–351**
    drinking water, and 355
    drug classes 356–363
    environment, and 346–352, 363–367
    wastewater treatment, and 353–355, 357–363, 364, 366
phase equilibrium 120–124
phosphates 372, 386, 390, 436, 447, 474, 575
phosphorus 389, 446, 463, 468, 474, 511, 515–516, 518
photochemical reaction 268
photochemical smog 236, 267–268, 292–293, **293**
photoionization detectors (PIDs) 651
photometry 658
photooxidation 437
photo reactivation 508
photosynthesis 234, 236, 469–470, 472
physical attrition particle formation 262

physical models 288
physical stressors 631–632
physical weathering 569
pi 66, 88
pipe materials, corrosive-resistant 387
piping failure, underground storage tank 577–578
pitting 384, 385
plant available nitrogen (PAN) 548–549
plant layout, as engineering control 708
plate-and-frame filter 545
plume behavior 286
plutonium 12
pneumatic pumping station 449
pneumatics 224
pneumoconiosis 653
podzol-type soils 569
point absorber 764
point source 3, 9, 129, 130, 286, 287, 332, 333, 440, 441, 516, 674
Poisson distribution 133
Poisson's ratio 219
polarography 658
polishing pond 465, 466
pollutant 12–13, 18, 281
    air 11, 21, 22, 23, 129, 233–235, 615
    biodegradable 4
    defined 4, 12, 237
    dispersion 280
    emerging 11
    gaseous 12, 249, 296, 298, 299
    National Ambient Air Quality Standards, and 273–274, **274**
    New Source Review, and 278–279
    nonbiodegradable 4
    persistent organic 11
    primary 4, 237, 275, 289, 291
    secondary 4, 237, 282, 283, 289, 291, 292
    soil 14, 570–576
    stratification 257
    toxic air 275
    water 22, 329–550
pollution
    air 11, 21, 22, 23, 129, 233–235, 615
    categories of 6, **6**
    defined 4–18
    effects 15
    environmental science, and 16–18, *17*
    environmental, technology and 18–21, *19*
    light 12
    manifestations 5
    marine 12
    noise 12
    prevention 12
    radio spectrum 12
    soil 14, 570–576
    thermal 9, 13
    unexpected 16
    visual 13
    water 22, 329–550
polyaromatic hydrocarbons (PAHs) 337
polychlorinated biphenyls (PCBs) 336, 337, 339, 341, 342, 574
polycyclic aromatic hydrocarbons (PAHs) 365
polycyclic musks 364–365

polymerization 625
ponds 464–487, 574
    biochemistry in 469–472
    facultative 464–465, 467, 468, 471, 473
    nutritional requirements of 473–474
    organisms in 467–469
    physical factors of 472–473
    process control calculations for 475–476
positive lapse rate 284
postconsumer content 604
potassium permanganate 375, 381, 382, 399, 404, 406, 407, 413, 415, 416, 432
potential energy 43, 619–620
potholes 731
pounds per day to milligrams per liter 35
pounds to flow in million gallons per day 36
pounds to gallons 34
pounds to milligrams per liter 35
powerhouse failures 735
powers, mathematical 74–76, **76**
PPCPs *see* pharmaceuticals and personal care products (PPCPs)
PPE *see* personal protective equipment (PPE)
prairie chickens 20
preaeration 457
prechlorination 376–377, 407, 412
precipitation 437
    acid 11, 14, 235, 569
    atmospheric 237, 238, 280, 282, 331, 541, 571
    chemical 381, 437
        arsenic 431
        arsenopyrite 436
        calcium carbonate 386, 433
        iron 431
        manganese 432
        phosphorus 474
    preliminary treatment
        wastewater 444, 450–458, 463–464, 498
        water 375–382
    scavenging 280
    from solution 120
precision, sampling 649
prediction, environmental modeling 114–115, 732
preliminary assessment 625–626
presbycusis 664
preservatives, in the environment 366
pressure 98–104, *99*
    defined 222, 692
    filtration 390–391, 393, 545
    force, and 98
    head, and 98–100, *99*
    volume, and 249
pretreatment 375–382, 413
Prevention of Significant Deterioration (PSD) 278, 288, 740
primary effluent 463
primary pollutant 4, 237, 275, 289, 291
primary sludge 519, 526
primary treatment 444, 447, 458–463, 494, 495, 519, 526
prime number 65
probability 132–133
product, multiplication 65
propane cylinders, carbon dioxide emissions and 60–61
properties of materials 209
    mechanics, and 217–223, **218,** *218–221*

Index

proportions 79
proposed maximum contaminant level (PMCL), cadmium 384
prostate cancer 339
protective clothing 319, 632, 639, 655, 686, 688, 697–704; *see also* personal protective equipment (PPE)
    classification of 701–702
    inspection 703–704
    maintenance 704
    material chemical resistance 702
    selection factors 701
    storage 704
proteins 739
proto-air 233
proto-atmosphere 233
protozoa 391, 394, 401, 402, 428, 429, 447, 468, 470, 477, 480, 490, 492, 522, 568
Prozac 359, 360
publicly owned treatment works (POTW) 356, 359, 361–363, 437
    diazepam in 362
pumped storage 727
pumping stations 449–450
purple sulfur bacteria 468
PV cell manufacturing process 724

quadratic equations 107–110
qualitative/quantitative data 136
quality 268–271
    dispersion models 287–290
    impact analysis 278, 279
    indoor 303–325
    management 272
    modeling/monitoring 279
    water 741, 794, 798
quotient 66

Rachel River, salmon and 8–10
radiant heat 626, 632, 685, 687
radiation 269–270, 626, 631–632, 672–684, **684**
    beta 616
    electromagnetic 267, 268, 631
    exposure controls 683
    exposure training 683
    extremely low frequency 680
    infrared 246, 268, 269, 680
    inversions 285
    ionizing 616, 623, 631, 672–675, 681
    non-ionizing 625, 632, 633, 672, 679–681
    nuclear fallout 12
    poisoning 12
    radiofrequency 680
    safety requirements, OSHA 681–682
    solar 236, 267, 268, 269, 270, 282, 292, 472
    suits 701, 702
    ultraviolet 5, 241, 267, 268, 276, 281, 292, 501, 507–509, 540, 680
radioactive decay 674
    equations 677–679
radioactivity 658
    environmental 11
radiofrequency radiation 680
radioisotope 674
radiological half-life 675

radio spectrum pollution 12
radiotelemetry 20
radium 12
radius 66, 88, 92
radon 12, 237, 307–308
rain shadow effect 237
Raleigh scattering 237
ramping rate 732
randomization, statistical 162–165, 185
randomized block design 168–171, 174
random sampling 141, 146–152, 153
    simple 146–149
    stratified 149–152
range 134, 135, 136
Raoult's law 124
rapid sand filtration 391–393
ratio calculations 78–81
raw sewage stabilization ponds 465
raw sludge 519, 532, 533, **533**
Rayleigh number 315
Raynaud's syndrome 626, 671
reactive substances 597
reactivity hazard 626
reactor 763
real ear attenuation at threshold (REAT) 661
real-time air monitoring 651
reasonably available control measures (RACMs) 274
recirculation ratio 482
rectangle 92
rectifier 763
recyclable product 604
recycled-content product 604
recycle flow 529
recycling 21
    collection and process 603
    manufacture 604
    purchasing new products made from 604
redd dewatering 733–734
reeds 539, 541, 543
regeneration
    adsorption 434
    continuous 382
    granular activated carbon 412
    granular ferric hydroxide 436
    greensand 432
    ion exchange 434
    iron oxide coated sand 435
    packed bed 434
regression 186–207, *188, 191*
    curvilinear 198–207
regulation
    defined 707
    environmental 22–23
relative humidity 237, 306, 309, 320, 685–687
remediation
    arsenic 436
    electrical resistance heating 11
    mold 304, 308, 323–324
    soil 11, 229, 563, 567, 569, 570, 576, 579–587
    subsurface 576
renewable energy projects 770
renewable resources 19, 20, 44, 45
replications 186
    unequal 167–168

reportable quantity 626
reproductive effects of endocrine disruptors 340
reservoir stranding 735
residence time 254, 429
    arsenic removal 435, 436
    brick manufacturing 586
    mean cell 440, 489, 514
residential sources, of MSW 593
resistivity 266
Resource Conservation and Recovery Act (RCRA) 13, 23, 576, 591, 600–601, 626
resource, defined 18–19, *19*
resource protection 12
respirable particles 653
respirable particulate mass (RPM) 647
respiration 470
respirators 317, 319, 320, 322, 699, 700, 704–705
resultant, force 211
retention time 459
    aeration 530
    hydraulic 473, 500, 513
    sludge 440, 489, 514, 534, 535
    solids 498, 527
retinoids 362
return activated sludge (RAS) 490, 499, 501, 517
return activated sludge solids (RASS) 441
return sludge system 491
reuse 21
reverse osmosis (RO) 434
Reynolds number 257, 315
Richardson number 315
rights-of-way (ROWs) 728
Right to Know Law 640
risk assessment 626
    chemical release 578
river in-stream facilities 793, 797
Rivers and Harbors Act 591
rivers and streams 331
rotary vacuum filtration 543–544
rotating biological contactor (RBC) 463, 484–487, 517
    process control calculations 485–487
rotifers 490
roughing filters 478, 480
roughing towers 477
routes of entry, chemical 635–636
runoff 11, 331, 332, 344, 352, 355, 445
    acid 333
    domestic animals, and 356
    land application site 522
    livestock operations, and 333, 345, 572
    slope, and 567
    soil 333
    stormwater 14, 333, 344, 440, 441, 445
    surface 13, 331, 332

Safe Drinking Water Act (SDWA) 394
Safety Data Sheets (SDS) 641–645
safety factor 211, 626
safety glasses 697, 700
salmon, Rachel River and 8–10
salt spreading 572
saltwater 330, 372, 412
sample allocation 151
sample size 148–149, 151, 153, 160

sampling, statistical 146–155
sand drying beds 539–540
sand particles 567
sanitary landfills 610
scenic vistas 752
Scheffe's test 166–167
science, defined 17
scientific method 5
SCREEN3 130
screening
    of chemicals for endocrine disruption potential 356–357
    illicit drugs in wastewater 368–369
    wastewater treatment 368, 450–453, 465, 497
    water treatment 374, **374,** 376
screening tools, analytical 130, 279
scrubbers 300
scum 383, 390, 441, 458–460, 466, 490, 519
Secondary Maximum Contaminant Level (SMCL) 86
secondary pollutants 4, 237, 282, 283, 289, 291, 292
secondary sludge 519
secondary treatment 444, 447, 463–487, 490, 511
second law of thermodynamics 237
secure landfills *610*, 610–611
sedimentary cycle 6
sedimentation 377, 390, 391, 437, 443, 458–463
    basin 439
    calculations 460–463
    tanks 460
    test 230
sediment transport 768
segmental vibration 671
seismicity 716, 752
selective serotonin reuptake inhibitors (SSRIs) 359–360
self-contained breathing apparatus (SCBA) 627, 696, 699–700, 704
self-purification 26, 444
semen quality 337
sensitivity, soil 227
sensorineural hearing loss 663
separation hazardous waste 608
separation processes, arsenic removal and 432
September 11, 2001 16
septic, defined 441
septic tanks 344, 372, 459, 571
sequestration 352, 354, 382, 386
    carbon 57
serotonin 359–361
settleable solids 390, 441, 447, 458, 459, 461, 480, 492, 520
settled sludge volume (SSV) 441, 490
settling 381, 388–391, 431, 441, 447, 457–460, 466, 478, 490, 495, 498, 512, 514, 654, 655
    gravitational 264, 265, 280, 520
    particle 129, 260, 310, 388, 469, 653
    process testing 460
    tank/basin 34, 69, 96, 439, 441, 443, 458–462, 466, 476, 477, 479, 480, 484, 485, 488, 490, 491, 495, 501, 523, 526, 528
        effluent 463
    time 455, 459
    velocity 260, 265, 653
sewage sludge 355, 365, 520, 521, 522, 523
shear forces 211
shear strength 219

Index

shear stress 223
shorebirds 771
shoring 230
short-circuiting 473
short-term exposure limit (STEL) 627, 646–647, 651
    formaldehyde 321
SHORTZ 130
shredding 453
SI *see* International System of Units (SI)
sick building syndrome 305–306
significance, tests of 196–197
significant digits 72–74
Silent Spring 335
silica 301, 433, 434, 627, 652, 654, 656
silicon nitride 724
silicon tetrachloride 724
silt particles 567
simple correlation coefficient 143–144
simple payback 53
simple random sampling 146–149
sinks, nutrient 6
SKI calculus 126–127
skin absorption 636
slime 441, 476–477, 480, 484
slings 212–215, *213, 214*
slope, soil layer 567
slope winds 285
sloughings 477, 478, 484
slow sand filtration 391–392
sludge 441, 518–526
    age 493
    blanket 493, 501
        depth 501
    characteristics of 520–523
    dewatering 518, 539–547
    incineration 547
    process control calculations 523–526
    pumping rate 462
    retention time 440, 489, 514, 534, 535
    rising 490
    sources of 519–520
    thickening 526–530
    treatment 444, 526–550
    volume index 441, 490
    wasting 493
smog 12, 236, 267–268, 273, 274, 290, 292–293, **293, 294**
socioeconomics
    biomass energy operations impact 750
    biomass feedstock production impact 747
    biomass, impact of 744
    geologic sequestration exploration 810, 813–814, 817
    geothermal energy construction impact 758
    geothermal energy exploration and drilling impact 755
    geothermal energy O&M impact 760
    marine energy impact 791
    wind energy operations impacts 720
    wind turbine impacts 717
sodium hypochlorite 405, 417, 420–421, 420–422, 425
soil 5, 563–587
    basics 567
    characteristics 225–226
    classification 230
    compaction 229
        and blending 757
        and blinding 755
    compressibility 228
    contamination
        asphalt incorporation, and 584–586
        chemical extraction, and 586
        excavation, and 587
        in situ treatment 580–583
        land treatment 584
        solidification/stabilization, and 586
        thermal treatment 584
    failure 229–230
    fertility 569–570
    formation *568,* 568–569
    and geologic resources 809, 812, 816
        biomass energy operations impact 749
        biomass feedstock production impact 746
        biomass, impact of 742–743
        geothermal energy construction impact 757
        geothermal energy exploration and drilling impact 754–755
        geothermal energy O&M impact 760
        wind energy operations impacts 719
        wind turbine impacts 716
    horizons 568, 569
    in situ treatment 580–583
    marine energy impact 791
    mechanics 223–231, **227, 231**
    organisms 567
    particle characteristics 227–228
    physics 231
    pollutant 14
    pollution 570–576
    profile 569
    properties 567–568
    remediation 11, 229, 563, 567, 569, 570, 576, 577, 579–587
    resistivity 387
    strain 228
    stress 228
    structure *vs.* texture 567–568
    texture 567
    water content 227, 228, 229
    weight, volume and 226–227
soil engineers 566
soil erosion in arid lands 743
Soil Guideline Values (SGVs) 12
solar constant 269
solar energy, environmental impacts of
    ecological impact 725–726
    hazardous waste impact 724–725
    land use/siting impact 723
    water use impact 723–724
solar radiation 236, 267, 268, 269, 270, 282, 292, 472
solidification 607
solids 446–447, 490–491
    colloidal 447, 490, 508, 519
    concentrators 527–530
    dissolved 385, 447, 490, 520
    inorganic 333, 440, 447, 453, 456, 490, 521
    loading rate 528
    mixed liquor suspended 440, 489, 490, 500
    mixed liquor total suspended 500
    mixed liquor volatile suspended 440, 489, 500
    organic 447, 453, 455–456, 486, 490–492, 521

percent total 462
retention times 498, 527
return activated sludge 441
settleable 390, 441, 447, 458, 459, 461, 480, 492, 520
suspended 119, 332, 333, 353, 372, 390–392, 407, 425, 427, 431, 440, 441, 443, 446, 447, 455, 456, 458, 460, 463–466, 476, 480, 485, 486, 489–491, 493, 500–501, 507, 508, 510–516, 518, 519
total dissolved 385, 432, 447
total suspended 389, 447, 458, 460, 464, 465, 467, 469, 480, 486, 500–501, 507, 508, 511–516, 518, 523, 528
volatile suspended 81, 447, 489
waste activated sludge 441
solids emissions 752
solid waste 814
   characteristics 592, 592–593, *593*
   classification **592**
   defined 592
   regulatory history 591–592
   stream 593
Solid Waste Disposal Act 591
solubility 115, 124, 377, 386, 581, 585, 627
   chlorine 427
   of a gas 124, 280
   oxygen 9, 471
   ozone 510
   parent drug 354
solute 115, 237, 377
solution feeders 379
solutions 117, 118, 119, 124, 378–380, 388, 420–425
solvents 115, 309, 377, 627
sorbent 627
sorption 353
sound exposure level (SEL) 772
sound intensity 772
sound level meter (SLM) 664, 665
sound power 664, 668–670
sound pressure level (SPL) 664, 669–671
source reduction 13
source water control strategies 412
spawning interference 734
specification, defined 707
specific gravity 222, 237, 378, 627
specific source wastes 599
specific weight 31
speed of sound 669
split plot design 179–185, 186
spoils, excavation 572
square units 66, 88
stabilization 607
   contact 495
   iron/manganese removal, and 381
   ponds 464–465, *465*
      process control calculations for 475–476
   sludge 519, 530–547
Stachybotrys chartarum 324
stagnation, water 14
standard conditions (SC) 249, 251
standard, defined 707
standard deviation 137, 139–141, *140*, **141**, 144
standard error 141–142, 146–147, 150, 151, 152, 154
standard gas flow rate 252

standard temperature and pressure (STP) 122, 249
standard threshold shift (STS) 664, 667–668
starch 738
State Implementation Plans (SIPs) 277–278
static head 99
static pressure 692
statics 223
statistical models 288
statistics 131–207
   analysis of variance 162–186
   Bartlett's test of homogeneity of variance 157–158
   chi-square test 155–157, 158
   cluster sampling 154–155
   coefficient of variation 141
   correlation coefficient 143–144
   covariance 142–143
   discrete variables 152–153
   distributions 137–139
   objectives of 131–132
   probability, and 132–133
   regression 186–207
   sampling measurement variables 146–152
   standard deviation 139–141
   standard error of the mean 141–142
   t test 158–162
   variance of a linear function 144–145
steady state 120
   vs. equilibrium 120
sterilization 399, 502
stockpiles, soil pollution and 571
Stokes particle diameter 260, 264, 265
Stokes's law 653–654
stormwater runoff 14, 333, 344, 440, 441, 445
strain 218, 223, 228
straining 391
stranding 730–732, 733
stratification, pollutant 257
stratified random sampling 149–152
stratospheric 292
   ozone depletion 237
stress 218, 223, 228
   allowable 218
   intensity of 218
   ultimate 218
stressors, workplace 630–633
structural failure, soil 229
structurally related substances (SRSs) 352
subadiabatic lapse rate 284
sub-chronic exposure 653
subplot error 181
subscripts, statistics and 134
subsidence inversion 286
subsoil 567
substrates, alteration of 768
subsurface remediation 576
sulfur 19, 234, 237, 241, 275, 276, 279, 281, 289, 435, 441, 468, 474, 485, 490, 571, 628, 635, 658
   cycle 6, 571
   -modified iron (SMI) 435–436
sulfur dioxide 11, 12, 23, 233, 275, 276, 279, 281, 289, 290, 291, 295, 300, 571
sulfuric acid 235, 275, 276, 290, 388, 434, 435, 474, 572, 576
sulfurous smog 290

summation, statistical 134, 135
sum of squares 143, 157, 159, 163–164, 165, 166, 167, 169–170, 172, 175–176, 177, 178, 179, 180, 181, 182, 187–188, 189, 190, 194, 196–198, 201, 203, 204, 205, 206
sunscreen 688
    presence of in the environment 366–367
super adiabatic lapse rate 284
Superfund Amendments and Reauthorization Act (SARA) 275, 627, 640
supernatant 389, 390, 441, 488, 526, 527, 530, 533, 534, 539, 541, 545
super-rate trickling filters 477
Supervisory Control and Data Acquisition (SCADA) 722
supplied-air respirator 615, 704, 705
supply ventilation 691
surface area 91
    particle 257–258
surface condenser 302
surface contaminants 703
surface impoundments 609
surface loading rate 460, 461, 528–529
surface runoff 13, 331, 332
surface water 331–369, 415
    contaminated 572
    defined 331
    pollutants 332–334, 369
    wind turbine impacts 715
Surface Water Treatment Rule (SWTR) 391, 394, 398, 407, 414, 418, 426
suspended growth systems 464
suspended solids 119, 332, 333, 353, 372, 390–392, 407, 425, 427, 431, 440, 441, 443, 446, 447, 455, 456, 458, 460, 463–466, 476, 480, 485, 486, 489–491, 493, 500–501, 507, 508, 510–516, 518, 519
symbols, statistical 133–134, **136**
synthetic organic compounds (SOCs) 431

tailings 572
tailwater maintenance 735
target organs 636, **637**
taste, water 5, 86, 373, 375, 380, 385, 388–400, 404, 405, 407, 411, 415, 416, 419, 425
technological development 21–22
temperature; *see also* cold stress; thermal stress
    absolute pressure, and 249
    activated sludge, and 493, 501
    altitude, and 236, 256, 269, 271, 283–285
    anaerobic digestion, and 534
    biodegradation, and 582
    chlorine solubility, and 427
    coagulation, and 389
    color 52
    contaminant permeation, and 703
    conversions 39
    corrosion, and 385
    disinfection byproducts, and 410
    dissolved oxygen, and 9, 471
    effectiveness 315
    equilibrium, and 120–124
    friction, and 222
    gas flow rate, and 252
    hydrocarbon compounds, and 581
    ignition 547, 622
    as intensive property 118
    inversions 285
    kinetic energy, and 266
    molality, and 116
    molarity, and 116
    permeation rate of contaminants, and 703
    relative humidity, and 237
    slime, and 485
    solution stability, and 377
    viscosity, and 256
    volume, and 251
    wastewater 446, 459
    wastewater pond 472–473
temporary threshold shift (TTS) 627, 664
tensile force 211
tensile strength 219
tensile stress 223
terminator 764
terrorism, environmental 574
test of independence 155–156
tests of significance 196–197
thermal incinerator 303, **303**
thermal pollution 9, 13, 752
thermal stress 684–690
thermal treatment 538, 584, 606
thermoluminescent dosimeter (TLD) 674
thermophoresis 266
thermosphere 237
thoracic particulate mass 647
thread test 230
threshold levels 634
threshold limit value (TLV) 617, 627, 646–648, 694
threshold odor number (TON) 86–87
threshold shift, hearing sensitivity 664, 667
    temporary 627, 662
thyroid hormones 356
tidal barrages 764–765
tidal energy 764
    technologies 764–765
tidal fences 765
tidal turbine 765
tides 761–762
tilth 567
time-integrated air sampling 651
time-weighted average (TWA) 627, 646, 647, 649, 667, 671
tobacco smoke 308
topography, air motion and 285
topsoil 567
torsional forces 211
torsional strength 219
total chlorine residual 417, 419
total dissolved solids (TDS) 385, 432, 447
total dynamic head 100
total effective dose equivalent (TEDE) 674
total Kjeldahl nitrogen (TKN) 447
total organic carbon (TOC) 410, 413, 430, 510, 548
total pressure (TP) 692
total quality management (TQM) 627
total suspended particulate matter 260
total suspended solids (TSS) 389, 447, 458, 460, 464, 465, 467, 469, 480, 486, 500–501, 507, 508, 511–516, 518, 523, 528

total trihalomethanes (TTHMs) 376, 377, 381, 410–413, 429, 430
toughness, material 221
TOX 409, 411, 412
toxic air pollutants 275, 690–691
toxic chemicals 619, 629, 631, 647
toxic chemical use substitution 13
toxicity 597
    acute 614, 636, 647
    defined 627
    vs. hazard 634
Toxicity Characteristics Leaching Procedure (TCLP) 597, **598**
toxicological evaluation 578
toxicology, industrial 633–635
Toxic Substances Control Act (TSCA) 23, 335
toxic waste 372, 494, 501, 521
toxic water pollutants 334, 361–362, 364–366, 370, 372
trace gases 571
Tragedy of the Commons 21
training, as administrative control 639
tranquilizers 362
transformations, statistical 155
transparency, of atmosphere 270
transportation
    biomass energy operations impact 750
    biomass feedstock production impact 747
    biomass, impact of 744
    geologic sequestration exploration 809, 813, 817
    geothermal energy construction impact 757–758
    geothermal energy exploration and drilling impact 755
    geothermal energy O&M impact 760
    marine energy impact 791
    wind energy operations impacts 719
    wind turbine impacts 716
transport, pollutant 281–287
trapping 730
Traveling Salesman algorithm 127
treatment plant site sources, of MSW 594
treatment ponds 464–487
treatment time 254
tree seedlings, carbon dioxide emissions and 56
triangle law 212
trickling filter 463, 476–484, *477*, 521
    classification 480–482
    equipment 479
    process control calculations 482–484
triclosan 366
trigonometric ratios 110–112, **111**, *111, 112*
trihalomethane (THM) **374,** 375, 376, **408,** 410–413, 415, 429; *see also* total trihalomethanes (TTHMs)
troposphere 237, 293
t test 158–162, 171, 184–185
    vs. analysis of variance 165
TTHMs *see* total trihalomethanes (TTHMs)
tuberculation 380, 384–387
tumors, in fish 337–338
turbidity 333, 388, 389, 392–396, 397, 403, 407, 416, 417, 425–428, 432, 469, 479, 507–508, 516, 518
    ozone, and 509
    ultraviolet irradiation, and 507
Turbine Generation Unit (TFU) 773
turbulence, atmospheric 283
turbulent flow 256

two-story tank 459–460
Type I/II error 133

ultimate stress 218
ultrafiltration 394, 412, 434
ultra-high-rate filter 392
ultraviolet radiation 5, 241, 267, 268, 276, 281, 292, 363, 399, 410, 414, 416, 467, 497, 501, 507–509, 540, 680
unconfined aquifer 370
underground storage tanks (USTs) 574, 576–578
underwater noise 774
unequal replication 167–168
United States Code 23
United States Fish and Wildlife Service (USFWS) 720
units of measurement 30–41, **32**
upper explosive limit (UEL) 628
uranium 13
U.S. Customary System (USCS) 29
U.S. Environmental Protection Agency (USEPA) 4, 747–748
U.S. Office of Energy Efficiency and Renewable Energy (EERE) 762

vacuum-assisted drying beds 542
vacuum collection system 448
VALLEY 130
valley winds 285
vandalism 760
vapor 120, 261, 263, 266, 268, 270, 280–281, 284, 285, 289, 290, 301, 302, 614, 615, 617, 619–621, 625, 627, 628, 631, 635, 636, 646, 647, 649–651, 700, 703
    air sampling for 655–660
    analytical methods for 649–650
    as chemical stressor 631
    density, chlorine gas 502
    diffusion 580
    plume 280
    pressure 124, 581, 616, 628, 650, 651, 655
    vs. gases 655
    water 5, 233, 237, 241, 246, 247, 252, 266, 268, 270, 280, 281, 284, 285, 289, 290, 547, 614
variance 144–145, 146, 148, 149
    analysis of 162–207
variation, statistics and 131
vector attraction 521–523
vectors, mathematical 105
vegetable oils 739
vegetation 752
vehicle emissions 275, 292
velocity
    acceleration, and 222, 223
    air 306, 619, 620, 623, 692, 693
    capture 617
    channel flow 454
    gas 253–254
    head 100
    particle 264, 265, 266
    pressure 692
    settling 260, 265, 454, 653
    sound 664
    terminal 653
    wastewater 448, 454–455, 458, 461

Index

ventilation 690–694
    effectiveness factor 314
Venturi scrubbers 298
vertisol-type soils 569
Viagra 362
vibration 671–672
viruses 402, 522
    chlorine, and 428–429
viscosity 256, 628
VISCREEN 130
visibility impairment, USEPA and 280
visible light radiation 680
visual pollution 13
visual resources
    biomass energy operations impact 750
    biomass feedstock production impact 747
    biomass, impact of 744
    geologic sequestration exploration 809, 813, 817
    geothermal energy construction impact 758
    geothermal energy exploration and drilling impact 755
    geothermal energy O&M impact 760
    marine energy impact 791
    wind energy operations impacts 719
    wind turbine impacts 716–717
void space, soil 226–227
volatile acids to alkalinity ratio 533–535
volatile organic compounds (VOCs) 281, 288, 289, 291, 310, 324, 574, 580, 651, 713–714, 718, 740, 747, 748
volatile solids loading 531
volatile suspended solids 81, 447, 489
volume 66, 74, 87, 115–116, 118, 421
    air sample 648
    bed 434
    calculations 31–33, **33**, *33*, 88, 94–97, *98*
    concentration, and 420
    density, and 222, 236, 378, 619
    molar 122
    particle 257–258
    percent 116, 252, 366
    pond 475
    pressure, and 122, 123, 249, 617
    seed 534
    settled sludge 441, 490
    soil, and 226–227
    temperature, and 251
volumetric chemical feeder 379
vulnerability assessment 628
vulnerability zones 288

washout 282
waste activated sludge (WAS) 441, 488, 491, 495, 497, 501
waste activated sludge solids (WASS) 441
waste, defined 13
waste minimization 13, 602
waste piles 609–610
waste reduction 13
wastewater 441
    characteristics of 445–448, 508
    classification of 445
    collection systems 448–450
    defined 14
    disinfection 501–506
    genotoxicity of 361
    industrial 440, 445
    odor 446, 467, 473
    organic solids 447, 453, 455–456, 486, 490–492, 521
    sanitary 441, 445
    solids handling 518–526
    sources of 444–445
wastewater treatment 437–550
    advanced 511–518
    illicit drugs, and 368–369
    model 438
    pharmaceuticals and personal care products, and 353–355, 362–364, 366
    plant 353
        measuring performance of 441–444
        purpose of 444
    preliminary 450–458
    primary 458–463, 494, 495
    process residuals 518–527, 533
    secondary 463–487, 490, 511–512
water 5
    arsenic removal from 430–437
    atmospheric 246, *246*, 268, 270, 289
    color of 334, 408
    content, sludge 519, 520
    density of 378
    odor 375, 385, 404, 407, 411, 415, 416, 419
    pollution 21, 329–550
    quality 14, 741, 794, 798
    quantity of on Earth 330
    as renewable resource 18
    specific gravity of 378
    stagnation 14
    table 370
    taste 5, 86, 373, 375, 380, 385, 388–400, 404, 405, 407, 411, 415, 416, 419, 425
    treatment 373–437
        defined 373
        pretreatment 375–382
        purpose of 373–374
    vapor 5, 233, 237, 241, 246, 247, 252, 266, 268, 270, 280, 281, 284, 285, 289, 290, 547, 614
    weight of 101
waterborne diseases 14, 402–403
waterborne pathogens 394, 399, 401–402
    inactivation of 403
water footprint 723–724
water pollution 751
water resources
    biomass energy operations impact 748–749
    biomass feedstock production impact 746
    biomass, impact of 741
    geologic sequestration exploration 809, 811–812, 815–816
    geothermal energy construction impact 756–757
    geothermal energy exploration and drilling impact 754
    geothermal energy O&M impact 759
    marine energy impact 790
    wind energy operations impacts 718
    wind turbine impacts 715
watersheds 331
water use 715, 752
wave and tidal energy farms 797
wave and tidal turbine energy farms 793
wave energy 762

conversion technology 762–764
wave energy conversion (WEC) 762–764
Wave Energy Technology (WET) Project 776
Wave Hub project 767
waves 762
    alteration of 766–767
weather, air pollution, and 282–283
weathering 569
weight calculations 31–33, **33,** *33*
weir 441
    overflow rate 460–462, 499
welding 223–224
West Nile Virus 594, 595
wet adiabatic lapse rate 285
wet deposition 280
wetlands 372–373
wet scrubbers 298
wet well pumping station 449
whole-body vibration 671
wildlife habitat 752
wildlife, wind energy operations impacts 720–722
wind 15, 19, 44, 45, 237, 255, 279, 280, 281, 282, 283, 285, 287, 288, **314,** 471, 564, 565, 569, 580, 711
    chill 689
    cold stress, and 688–689
    energy 712–713
    turbines 15, 63, 712, 713
    types of 285
    wastewater ponds, and 473
wind energy, operations impacts
    air quality 718
    cultural resources 718
    ecological resources 718
    environmental justice 720
    hazardous materials and waste management 720
    land use 719
    paleontological resources 719
    socioeconomics 720
    soils and geologic resources 719
    transportation 719
    visual resources 719
    water resources 718
    on wildlife 720–722
wind turbines, environmental impact of
    air quality 713–714
    cultural resources 714
    ecological resources 714
    environmental justice 717
    flow alteration 715
    hazardous materials and waste management 717–718
    land use 715
    paleontological resources 716
    socioeconomics 717
    soils and geologic resources 716
    transportation 716
    visual resources 716–717
    water quality 715
    water resources 715
    water use 715
    wind energy 713
workers' compensation 628, 629
working stress 218
workplace air pollutants 310–311
workplace stressors 630–633
work practice controls 637, 638, 704

xenon 244, **245**
x-rays 362–363, 621, 673, 674, 677

yield point 219

zebra mussel 404–405
zoogleal slime 441, 478, *478*, 484, *484*
zoonoses 628